Regions

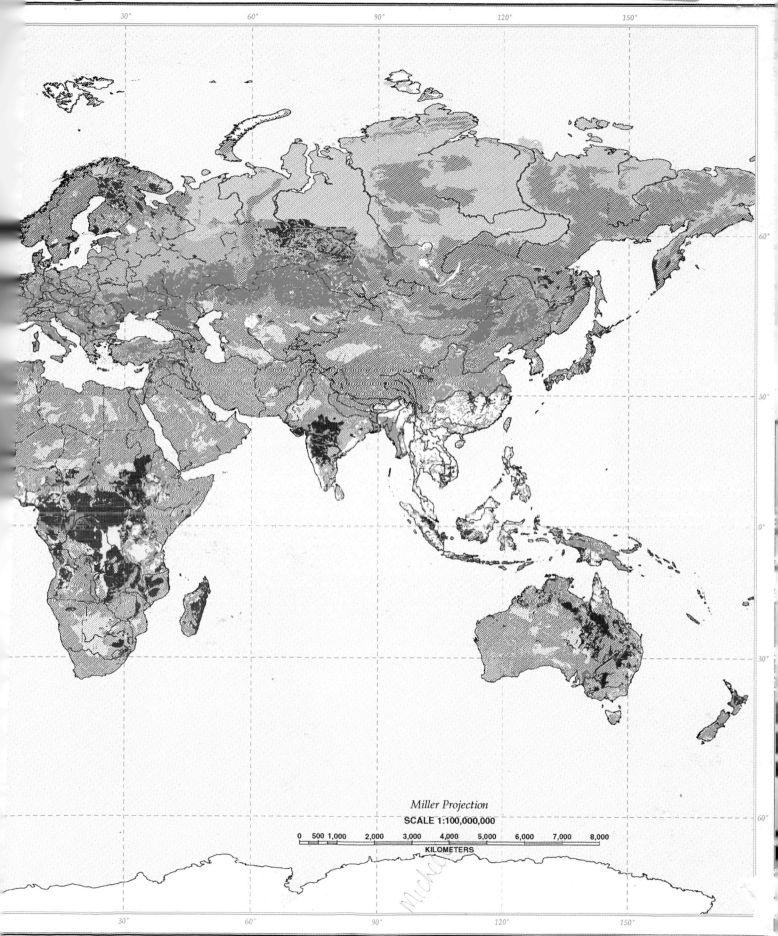

Miller Projection

SCALE 1:100,000,000

| 0 | 500 | 1,000 | 2,000 | 3,000 | 4,000 | 5,000 | 6,000 | 7,000 | 8,000 |

KILOMETERS

May 1997

The Nature and Properties of Soils

TWELFTH EDITION

NYLE C. BRADY
EMERITUS PROFESSOR OF SOIL SCIENCE
CORNELL UNIVERSITY

RAY R. WEIL
PROFESSOR OF SOIL SCIENCE
UNIVERSITY OF MARYLAND AT COLLEGE PARK

PRENTICE HALL
UPPER SADDLE RIVER, NEW JERSEY 07458

Library of Congress Cataloging-in-Publication Data

Brady, Nyle C.
    The nature and properties of soils / Nyle C. Brady, Ray R. Weil. —
12th ed.
            p.   cm.
    Includes bibliographical references and index.
    ISBN 0-13-852444-0 (hardcover)
    1. Soil science.   2. Soils.   I. Weil, Ray R.   II. Title.
S591.B79   1999
631.4—dc21   98-13008
                                                   CIP

Acquisitions Editor: *Charles Stewart*
Production Coordinator: *Adele Kupchik*
Managing Editor: *Mary Carnis*
Manufacturing Buyer: *Marc Bove*
Director of Manufacturing & Production: *Bruce Johnson*
Marketing Manager: *Melissa Brunner*
Formatting/page make-up, production management: *North Market Street Graphics*
Interior illustrations: *Mark Ammerman*
Printer/Binder: *Courier Westford*
Creative Director: *Marianne Frasco*
Cover Designer: *Marianne Frasco*
Cover Illustration: *Connie Hayes*

 © 1999, 1996 by Prentice-Hall, Inc.
Simon & Schuster A Viacon Company
Upper Saddle River, New Jersey 07458

Earlier editions by T. Lyttleton Lyon and Harry O. Buckman copyright 1922, 1929,
1937, and 1943 by Macmillan Publishing Co., Inc. Earlier edition by T. Lyttleton,
Harry O. Buckman, and Nyle C. Brady copyright 1952 by Macmillan Publishing
Co., Inc. Earlier editions by Harry O. Buckman and Nyle C. Brady copyright ©
1960 and 1969 by Macmillan Publishing Co., Inc. Copyright renewed 1950 by
Bertha C. Lyon and Harry O. Buckman, 1957 and 1965 by Harry O. Buckman,
1961 by Rita S. Buckman. Earlier editions by Nyle C. Brady copyright © 1974,
1984 and 1990 by Macmillan Publishing Company.

Printed in the United States of America

10 9 8 7 6 5 4 3

ISBN 0-13-852444-0

Prentice-Hall International (UK) Limited, *London*
Prentice-Hall of Australia Pty. Limited, *Sydney*
Prentice-Hall Canada Inc., *Toronto*
Prentice-Hall Hispanoamericana, S.A., *Mexico*
Prentice-Hall of India Private Limited, *New Delhi*
Prentice-Hall of Japan, Inc., *Tokyo*
Simon & Schuster Asia Pte. Ltd., *Singapore*
Editora Prentice-Hall do Brasil, Ltda., *Río de Janeiro*

*To all the students and colleagues in soil science who have
shared their inspirations, camaraderie, and deep love of the Earth.*

CONTENTS

12 SOIL ORGANIC MATTER   446

13 NITROGEN AND SULFUR ECONOMY OF SOILS   491

A fundamental knowledge of soil science is a prerequisite to meeting the many natural-resource challenges that will face humanity in the 21st century. We also believe that the study of soils can be both fascinating and intellectually satisfying. The soil provides an ideal system in which to observe practical applications for basic principles of biology, chemistry, and physics. In turn, these principles can be used to minimize the degradation and destruction of one of our most important natural resources.

In this new edition of the classic book in the field, our priority is to explain the fundamental principles of soil science in a manner that is relevant to students in many fields of study. Throughout, the text emphasizes the soil as a natural resource and highlights the many interactions between the soil and other components of forest, range, agricultural, wetland, and constructed ecosystems. We recognize that for some readers this book will be their only formal exposure to soil science, while for others it represents the initial step in a comprehensive soil science education. We have, therefore, sought the advice of professors and students to help us make this book serve both types of readers as an exciting introduction to the fascinating world of soil science, and as a reliable reference in the years to come.

While we have had to reduce the detail in a few areas to make room for new topics and information, we have been careful to maintain the level of rigor and thoroughness that has made previous editions so valuable. New to this edition are sections on soil quality, soil degradation, soil suitability for septic tank drainfields, wetlands and their hydric soils, engineering properties of soils, Gelisols and other changes in *Soil Taxonomy*, distribution coefficients for organic compounds in soils, genetically engineered microorganisms, factors affecting litter quality, precision agriculture technology, and many other topics of current interest in soil science. In response to their popularity in the 11th edition, we have added many new "boxes" that present either fascinating examples and applications or technical details and calculations.

As with the last edition, the artwork in this edition utilizes two colors to make the figures clear, informative and attractive to the reader. We have found that problems in photocopying figures to generate overhead transparencies can be largely overcome by using the photocopier on its lightest setting. To further facilitate using the book figures in the classroom, a CD-ROM is available that contains electronic versions of more than 180 of the line drawings and all 35 of the color plates from the book. Some figures could not be included on the CD-ROM at this time because of copyright limitations.

We are thankful for the many constructive comments we constantly receive by letter, e-mail, and in person from soil scientists, instructors, and students. The devotion and camaraderie we have shared with students and practitioners of soil science is a valued source of inspiration. The book has greatly benefited from suggestions contributed by many colleagues, especially those in universities around the country who responded to our questionnaire on the 11th edition. The able research and editorial assistance provided by Missy Stine, Ashley Gaede, Joel Gruver, and Rafiq Islam is much appreciated.

We would like to give special thanks to the following colleagues who reviewed portions of the text in detail and made valuable suggestions for improvement: Bob Ahrens, Susan Davis, Hari Eswaran, Paul Reich, and Sharon Waltman (USDA/Natural Resources Conservation Service); Charles Tarnocai (Agriculture and Agri-Food Canada); Kudjo Dzantor, Delvin Fanning, Robert Hill, Bruce James, Margaret Mayers Norton, and Martin Rabenhorst (University of Maryland); Duane Wolf (University of Arkansas); J. Kenneth Torrence (Carleton University); Dan Richter (Duke University); Daniel Hillel (University of Massachusetts); Jimmie Richardson (North Dakota State University); Murray Milford (Texas A & M University); Lyle Nelson (Mississippi State University); Rattan Lal (Ohio State University); Fred Magdoff and Wendy Sue Harper (University of Vermont); Pedro Sanchez (International Center for Research on Agroforestry); Darrell G. Schultze (Purdue University); Joyce Torio (American Chemical Society); and Mike Swift (UNESCO Tropical Soil Biology and Fertility Program).

Last, but not least, we wish to express our deep appreciation to our wives, Martha and Trish, for their constant encouragement and patience as we utilized almost every free moment during the past 18 months to concentrate on the revision of this textbook.

N. C. B. and R. R. W.

1

THE SOILS AROUND US

For in the end we will conserve only what we love.
We will love only what we understand.
And we will understand only what we are taught.
—BABA DIOUM, AFRICAN CONSERVATIONIST

The Earth, our unique home in the vastness of the universe, is in crisis. Depletion of the ozone layer in the upper atmosphere is threatening us with an overload of ultraviolet radiation. Tropical rain forests, and the incredible array of plant and animal species they contain, are disappearing at an unprecedented rate. Groundwater supplies are being contaminated in many areas and depleted in others. In parts of the world, the capacity of soils to produce food is being degraded, even as the number of people needing food is increasing. It will be a great challenge for the current generation to bring the global environment back into balance (Figure 1.1).

FIGURE 1.1 Our planet is unique in that it is covered with life-sustaining water and soil. Great care will be required in preserving the quality of both if our species is to continue to thrive. (Photo courtesy of NASA)

1

Soils are crucial to life on earth. From ozone depletion to rain forest destruction to water pollution, the global ecosystem is impacted in far-reaching ways by the processes carried out in the soil. To a great degree, soil quality determines the nature of plant ecosystems and the capacity of land to support animal life and society. As human societies become increasingly urbanized, fewer people have intimate contact with the soil, and individuals tend to lose sight of the many ways in which they depend upon soils for their prosperity and survival. The degree to which we are dependent on soils is likely to increase, not decrease, in the future. Of course, soils will continue to supply us with nearly all of our food and much of our fiber. On a hot day, would you rather wear a cotton shirt or one made of polyester? A large percentage of our medicines, including anti-cancer drugs, will continue to be derived from native and cultivated plants and soil organisms. In addition, biomass grown on soils is likely to become an increasingly important source of energy and industrial feedstocks, as the world's finite supplies of petroleum are depleted over the coming century. The early signs of this trend can be seen in the soybean oil–based inks, the cornstarch plastics, and the wood alcohol fuels that are becoming increasingly important on the market (Figure 1.2).

The art of soil management is as old as civilization. As we move into the 21st century, new understandings and new technologies will be needed to protect the environment and at the same time produce food and biomass to support society. The study of soil science has never been more important to foresters, farmers, engineers, natural resource managers, and ecologists alike.

1.1 FUNCTIONS OF SOILS IN OUR ECOSYSTEM

In any ecosystem, whether your backyard, a farm, a forest, or a regional watershed, soils have five key roles to play (Figure 1.3). *First*, soil supports the growth of higher plants,

FIGURE 1.2 In the future, we will be increasingly dependent on soils to grow renewable resources that can substitute for dwindling supplies of crude oil. Plastics and inks, for example, can be manufactured from soybean oil instead of from petroleum. Soybean oil is edible and far less toxic to the environment. Plastics made from cornstarch, such as packaging foam "peanuts," are readily biodegradable. (Photos courtesy of R. Weil)

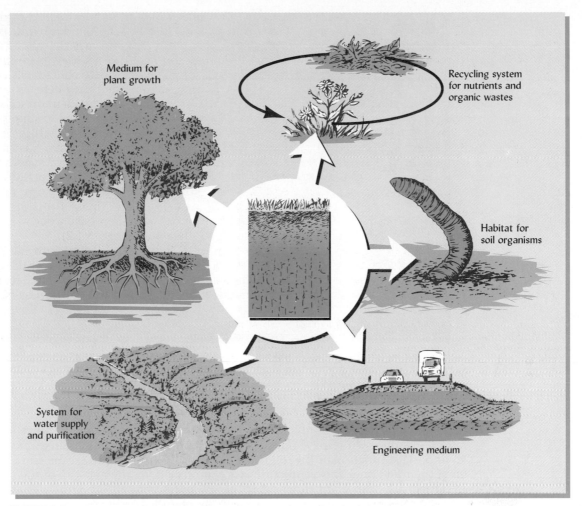

Medium for
plant growth

Recycling system
for nutrients and
organic wastes

Habitat for
soil organisms

System for
water supply
and purification

Engineering medium

FIGURE 1.3   The many functions of soil can be grouped into five crucial ecological roles.

mainly by providing a medium for plant roots and supplying nutrient elements that are essential to the entire plant. Properties of the soil often determine the nature of the vegetation present and, indirectly, the number and types of animals (including people) that the vegetation can support. *Second*, soil properties are the principal factor controlling the fate of water in the hydrologic system. Water loss, utilization, contamination, and purification are all affected by the soil. *Third*, the soil functions as nature's recycling system. Within the soil, waste products and dead bodies of plants, animals, and people are assimilated, and their basic elements are made available for reuse by the next generation of life. *Fourth*, soils provide habitats for a myriad of living organisms, from small mammals and reptiles to tiny insects to microscopic cells of unimaginable numbers and diversity. *Finally*, in human-built ecosystems, soil plays an important role as an engineering medium. Soil is not only an important building material in the form of earth fill and bricks (baked soil material), but provides the foundation for virtually every road, airport, and house we build.

1.2   MEDIUM FOR PLANT GROWTH

Imagine a growing tree or a corn plant. The aboveground portion of this plant may be most familiar, but the portion of the plant growing below the soil surface, its root system, may be nearly as large as the portion we see above ground. What things do these plants obtain from the soils in which their roots proliferate? It is clear that the soil mass provides physical support, anchoring the root system so that the plant does not fall over. Occasionally, as in Figure 1.4, strong wind or heavy snow does topple a plant whose root system has been restricted by shallow or inhospitable soil conditions.

FIGURE 1.4 This wet, shallow soil failed to allow sufficiently deep roots to develop to prevent this tree from blowing over when snow-laden branches made it top-heavy during a winter storm. (Photo courtesy of R. Weil)

Plant roots depend on the process of respiration to obtain energy. Since root respiration, like our own respiration, produces carbon dioxide (CO_2) and uses oxygen (O_2), an important function of the soil is *ventilation*—allowing CO_2 to escape and fresh O_2 to enter the root zone. This ventilation is accomplished via the network of soil pores.

An equally important function of the soil pores is to absorb rainwater and hold it where it can be used by plant roots. As long as plant leaves are exposed to sunlight, the plant requires a continuous stream of water to use in cooling, nutrient transport, turgor maintenance, and photosynthesis. Since plants use water continuously, but in most places it rains only occasionally, the water-holding capacity of soils is essential for plant survival. A deep soil may store enough water to allow plants to survive long periods without rain (see Figure 1.5).

As well as moderating moisture changes in the root environment, the soil also moderates temperature fluctuations. Perhaps you can recall digging in garden soil on a summer afternoon and feeling how hot the soil was at the surface and how much cooler just a few centimeters below. The insulating properties of soil protect the deeper portion of the root system from the extremes of hot and cold that often occur at the soil surface. For example, it is not unusual for the temperature at the surface of bare soil to exceed 35 or 40°C in midafternoon, a condition that would be lethal to most plant roots. Just a few centimeters deeper, however, the temperature may be 10°C cooler, allowing roots to function normally.

There are many potential sources of **phytotoxic substances** in soils. These toxins may result from human activity, or they may be produced by plant roots, by microorganisms, or by natural chemical reactions. Many soil managers consider a function of a good soil to be protection of plants from toxic concentrations of such substances by ventilating gases, by decomposing or adsorbing organic toxins, or by suppressing toxin-producing organisms. At the same time, it is true that some microorganisms in soil produce organic, growth-stimulating compounds. These substances, when taken up by plants in small amounts, may improve plant vigor.

Soils supply plants with inorganic, mineral nutrients in the form of dissolved ions. These mineral nutrients include such metallic elements as potassium, calcium, iron, and copper, as well as such nonmetallic elements as nitrogen, sulfur, phosphorus, and boron. By eating plants, humans and other animals usually obtain the minerals they need (including several elements that plants take up but do not appear to use themselves) indirectly from the soil (see periodic table, Appendix C). Under some circumstances, animals

FIGURE 1.5   A family of African elephants finds welcome shade under the leafy canopy of a huge acacia tree in this East African savanna. The photo was taken in the middle of a long dry season and no rain had fallen for almost five months. The tree roots are still using water from the previous rainy season stored several meters deep in the soil. The light-colored grasses are more shallow-rooted and have either set seed and died or gone into a dried-up dormant condition. (Photo courtesy of R. Weil)

satisfy their craving for minerals by ingesting or licking soil directly (Figure 1.6). The plant takes these elements out of the soil solution and incorporates most of them into the thousands of different organic compounds that constitute plant tissue. A fundamental role of soil in supporting plant growth is to provide a continuing supply of dissolved mineral nutrients in amounts and relative proportions appropriate for plant growth.

FIGURE 1.6   A mountain goat visits a natural *salt lick* where it ingests needed minerals directly from the soil. Animals normally obtain their dietary minerals indirectly from soils by eating plants. (Photo courtesy of R. Weil)

TABLE 1.1   Elements Essential for Plant Growth and Their Sources[a]

The chemical forms most commonly taken in by plants are shown in parentheses, with the chemical symbol for the element in bold type.

Macronutrients: Used in relatively large amounts (>0.1% of dry plant tissue)		Micronutrients: Used in relatively small amounts (<0.1% of dry plant tissue)
Mostly from air and water	*Mostly from soil solids*	*From soil solids*
Carbon (CO_2)	Nitrogen (NO_3^-, NH_4^+)	Iron (Fe^{2+})
Hydrogen (H_2O)	Phosphorus ($H_2PO_4^-$, HPO_4^{2-})	Manganesse (Mn^{2+})
Oxygen (O_2, H_2O)	Potassium (K^+)	Boron (HBO_3)
	Calcium (Ca_2^+)	Zinc (Zn^{2+})
	Magnesium (Mg_2^+)	Copper (Cu^{2+})
	Sulfur (SO_4^{2-})	Chlorine (Cl^-)
		Cobalt (Co^{2+})
		Molybdenum (MoO_4^{2-})
		Nickel (Ni^{2+})

[a] Many other elements are taken up from soils by plants, but are not essential for plant growth. Some of these (such as sodium, silicon, iodine, fluorine, barium, and strontium) do enhance the growth of certain plants, but do not appear to be as universally required for normal growth as are the 18 listed in this table.

Of the 92 naturally occurring chemical elements, only the 18 listed in Table 1.1 have been shown to be **essential elements** without which plants cannot grow and complete their life cycles. A message that may have been found on a sign hanging on a cafe door may help you remember the 18 elements essential for plant growth. Most of the chemical symbols are obvious, but finding those for copper (Cu) and zinc (Zn) may require some imagination.

C.B. HOPKiNS CaFe, Co.
Closed Monday Morning and Night
See You Zoon, the Mg.

Essential elements used by plants in relatively large amounts are called **macronutrients**; those used in smaller amounts are known as **micronutrients.**

In addition to the mineral nutrients just listed, plants may also use minute quantities of organic compounds from soils. However, uptake of these substances is certainly not necessary for normal plant growth. The organic metabolites, enzymes, and structural compounds making up a plant's dry matter consist mainly of carbon, hydrogen, and oxygen, which the plant obtains by photosynthesis from air and water, not from the soil.

It is true that plants can be grown in nutrient solutions without soil (**hydroponics**), but then the plant-support functions of soils must be engineered into the system and maintained at a high cost of time, effort, and management. Although hydroponic production can be feasible on a small scale for a few high-value plants, the production of the world's food and fiber and the maintenance of natural ecosystems will always depend on the use of millions of square kilometers of productive soils.

1.3   REGULATOR OF WATER SUPPLIES

There is much concern about the quality and quantity of the water in our rivers, lakes, and underground aquifers. Governments and citizens everywhere are working to stem the pollution that threatens the value of our waters for fishing, swimming, and drinking. For progress to be made in improving water quality, we must recognize that nearly every drop of water in our rivers, lakes, estuaries, and aquifers has either traveled through the soil or flowed over its surface.[1] Imagine, for example, a heavy rain falling on the hills surrounding the river in Figure 1.7. If the soil allows the rain to soak in,

---

[1] This excludes the relatively minor quantity of precipitation that falls directly into bodies of fresh surface water.

FIGURE 1.7 The condition of the soils covering these foothills of the Blue Ridge in West Virginia and Maryland will greatly influence the quality and quantity of water flowing down the Potomac River past Washington, D.C., over 100 kilometers downstream. (Photo courtesy of R. Weil)

some of the water may be stored in the soil and used by the trees and other plants, while some may seep slowly down through the soil layers to the groundwater, eventually entering the river over a period of months or years as base flow. If the water is contaminated, as it soaks through the upper layers of soil it is purified and cleansed by soil processes that remove many impurities and kill potential disease organisms.

Contrast the preceding scenario with what would occur if the soil were so shallow or impermeable that most of the rain could not penetrate the soil, but ran off the hillsides on the soil surface, scouring surface soil and debris as it picked up speed, and entering the river rapidly and nearly all at once. The result would be a destructive flash flood of muddy water. Clearly, the nature and management of soils in a watershed will have a major influence on the purity and amount of water finding its way to aquatic systems.

1.4 RECYCLER OF RAW MATERIALS

What would a world be like without the recycling functions performed by soils? Without reuse of nutrients, plants and animals would have run out of nourishment long ago. The world would be covered with a layer, possibly hundreds of meters high, of plant and animal wastes and corpses. Obviously, recycling must be a vital process in ecosystems, whether forests, farms, or cities. The soil system plays a pivotal role in the major geochemical cycles. Soils have the capacity to assimilate great quantities of organic waste, turning it into beneficial **humus**, converting the mineral nutrients in the wastes to forms that can be utilized by plants and animals, and returning the carbon to the atmosphere as carbon dioxide, where it again will become a part of living organisms through plant photosynthesis. Some soils can accumulate large amounts of carbon as soil organic matter, thus having a major impact on such global changes as the much-discussed *greenhouse effect* (see Sections 1.12 and 12.2).

1.5 HABITAT FOR SOIL ORGANISMS

When we speak of protecting ecosystems, most people envision a stand of old-growth forest with its abundant wildlife, or perhaps an estuary such as the Chesapeake Bay with its oyster beds and fisheries. (Perhaps, once you have read this book, you will envision a handful of soil when someone speaks of an ecosystem.) Soil is not a mere pile of broken rock and dead debris. A handful of soil may be home to *billions* of organisms, belonging to thousands of species. In even this small quantity of soil, there are likely to exist predators, prey, producers, consumers, and parasites (Figure 1.8).

How is it possible for such a diversity of organisms to live and interact in such a small space? One explanation is the tremendous range of niches and habitats in even a uniform-appearing soil. Some pores of the soil will be filled with water in which swim organisms such as roundworms, diatoms, and rotifers. Tiny insects and mites may be crawling about in other, larger pores filled with moist air. Micro-zones of good aeration may be only millimeters from areas of **anoxic** conditions. Different areas may be enriched with decaying organic materials; some places may be highly acidic, some more basic. Temperature, too, may vary widely.

Hidden from view in the world's soils are communities of living organisms every bit as complex and intrinsically valuable as their counterparts that roam the savannas, forests, and oceans of the earth. Soils harbor much of the earth's genetic diversity. Soils, like air and water, are important components of the larger ecosystem. Yet only now is soil quality taking its place, with air quality and water quality, in discussions of environmental protection.

1.6 ENGINEERING MEDIUM

"*Terra firma,* solid ground." We usually think of the soil as being firm and solid, a good base on which to build roads and all kinds of structures. Indeed, most structures rest on the soil, and many construction projects require excavation into the soil. Unfortunately, as can be seen in Figure 1.9, some soils are not as stable as others. Reliable construction on soils, and with soil materials, requires knowledge of the diversity of soil properties, as discussed later in this chapter. Designs for roadbeds or building foundations that work well in one location on one type of soil may be inadequate for another location with different soils.

Working with natural soils or excavated soil materials is not like working with concrete or steel. Properties such as bearing strength, compressibility, shear strength, and stability are much more variable and difficult to predict for soils than for manufactured

FIGURE 1.8 The soil is home to a wide variety of organisms, both relatively large and very small. Here, a predatory centipede is about to encounter a detritus-eating sowbug. (Photo courtesy of R. Weil)

FIGURE 1.9 Better knowledge of the soils on which this road was built may have allowed its engineers to develop a more stable design, thus avoiding this costly and dangerous situation. (Photo courtesy of R. Weil)

building materials. Chapter 4 provides an introduction to some engineering properties of soils. Many other physical properties discussed will have direct application to engineering uses of soil. For example, Chapter 8 discusses the swelling properties of certain types of clays in soils. The engineer should be aware that when soils with swelling clays are wetted they expand with sufficient force to crack foundations and buckle pavements. Much of the information on soil properties and soil classification discussed in later chapters will be of great value to people planning land uses that involve construction or excavation.

1.7   SOIL AS A NATURAL BODY

You may have noticed that this book sometimes refers to "the soil," sometimes to "a soil," and sometimes to "soils." *The soil* is often said to cover the land as the peel covers an orange. However, while the peel is relatively uniform around the orange, the soil is highly variable from place to place on Earth. In fact, the soil is a collection of individually different soil bodies. One of these individual bodies, *a soil,* is to *the soil* as an individual tree is to the earth's vegetation. Just as one may find sugar maples, oaks, hemlocks, and many other species of trees in a particular forest, so, too, might one find Christiana clay loams, Sunnyside sandy loams, Elkton silt loams, and other kinds of soils in a particular landscape.

A soil is a three-dimensional natural body in the same sense that a mountain, lake, or valley is. By dipping a bucket into a lake you may sample some of its water. In the same way, by digging or augering a hole into a soil, you may retrieve some soil material. Thus you can take a sample of soil material or water into a laboratory and analyze its contents, but you must go out into the field to study a soil or a lake.

In most places, the rock exposed at the earth's surface has crumbled and decayed to produce a layer of unconsolidated debris overlying the hard, unweathered rock. This unconsolidated layer is called the *regolith,* and varies in thickness from virtually nonexistent in some places (i.e., exposed bare rock) to tens of meters in other places. The regolith material, in many instances, has been transported many kilometers from the site of its initial formation and then deposited over the bedrock which it now covers. Thus, all or part of the regolith may or may not be related to the rock now found below it. Where the underlying rock has weathered in place to the degree that it is loose enough to be dug with a spade, the term **saprolite** is used (see Plate 11).

Through their biochemical and physical effects, living organisms such as bacteria, fungi, and plant roots have altered the upper part—and, in many cases, the entire depth—of the regolith. Here, at the interface between the worlds of rock (the lithosphere), air (the atmosphere), water (the hydrosphere), and living things (the biosphere), soil is born. The transformation of inorganic rock and debris into a living soil is one of nature's most fascinating displays. Although generally hidden from everyday view, the soil and regolith can often be seen in road cuts and other excavations (Figure 1.10).

A soil is the product of both destructive and creative (synthetic) processes. Weathering of rock and microbial decay of organic residues are examples of destructive processes, whereas the formation of new minerals, such as certain clays, and of new stable organic compounds are examples of synthesis. Perhaps the most striking result of synthetic processes is the formation of horizontal layers called **soil horizons.** The develop-

FIGURE 1.10 Relative positions of the regolith, its soil, and the underlying bedrock. Note that the soil is a part of the regolith, and that the A and B horizons are part of the **solum** (from the Latin word *solum*, which means soil or land). The C horizon is the part of the regolith that underlies the solum, but may be slowly changing into soil in its upper parts. Sometimes the regolith is so thin that it has been changed entirely to soil; in such a case, soil rests directly on the bedrock. (Photo courtesy of R. Weil)

ment of these horizons in the upper regolith is a unique characteristic of soil that sets it apart from the deeper regolith materials.

1.8 THE SOIL PROFILE AND ITS LAYERS (HORIZONS)

Soil scientists often dig a large hole, called a *soil pit,* usually several meters deep and about a meter wide, to expose soil horizons for study. The vertical section exposing a set of horizons in the wall of such a pit is termed a **soil profile.** Road cuts and other ready-made excavations can expose soil profiles and serve as windows to the soil. In an excavation open for some time, horizons are often obscured by soil material that has been washed by rain from upper horizons to cover the exposed face of lower horizons. For this reason, horizons may be more clearly seen if a fresh face is exposed by scraping off a layer of material several centimeters thick from the pit wall. Observing how soils exposed in road cuts vary from place to place can add a fascinating new dimension to travel. Once you have learned to interpret the different horizons (see Chapter 2), soil profiles can warn you about potential problems in using the land, as well as tell you much about the environment and history of a region. For example, soils developed in a dry region will have very different horizons from those developed in a humid region.

Horizons within a soil may vary in thickness and have somewhat irregular boundaries, but generally they parallel the land surface (Figure 1.11). This horizontal alignment is expected since the differentiation of the regolith into distinct horizons is largely the result of influences, such as air, water, solar radiation, and plant material, originating at the soil—atmosphere interface. Since the weathering of the regolith occurs first at the surface and works its way down, the uppermost layers have been changed the most, while the deepest layers are most similar to the original regolith, which is referred to as the soil's *parent material.* In places where the regolith was originally rather uniform in composition, the material below the soil may have a similar composition to the parent

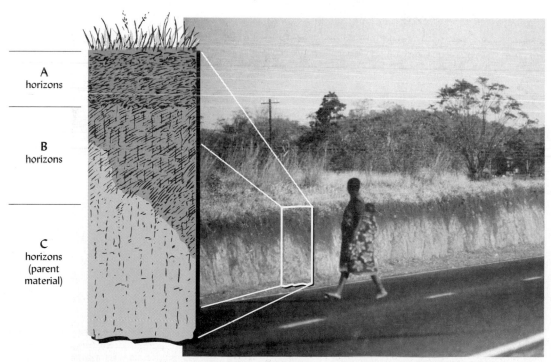

A horizons

B horizons

C horizons (parent material)

FIGURE 1.11   This road cut in central Africa reveals soil layers or horizons which parallel the land surface. Taken together, these horizons comprise the profile of this soil, as shown in the enlarged diagram. The upper horizons are designated *A horizons.* They are usually higher in organic matter and darker in color than the lower horizons. Some constituents, such as iron oxides and clays, have been moved downward from the A horizons by percolating rainwater. The lower horizon, called a *B horizon,* is sometimes a zone in which clays and iron oxides have accumulated, and in which distinctive structure has formed. The presence and characteristics of the horizons in this profile distinguish this soil from the thousands of other soils in the world. (Photo courtesy of R. Weil)

material from which the soil formed. In other cases, the regolith material has been transported long distances by wind, water, or glaciers and deposited on top of dissimilar material. In such a case, the regolith material found below a soil may be quite different from the upper layer of regolith in which the soil formed.

Organic matter from decomposed plant leaves and roots tends to accumulate in the uppermost horizons of a soil profile, giving these horizons a darker color than the lower horizons. Also, as weathering tends to be most intense in the upper layers, in many soils these layers have lost some clay or other weathering products by leaching to the horizons below. The organic-matter-enriched horizons nearest the soil surface are called the *A horizons*. In some soils, intensely weathered and leached horizons that have not accumulated organic matter occur in the upper part of the profile, usually just below the A horizon. These horizons may be designated *E horizons* (Figure 1.12).

The layers underlying the A and O horizons contain comparatively less organic matter than the horizons nearer the surface. Varying amounts of silicate clays, iron and aluminum oxides, gypsum, or calcium carbonate may accumulate in the underlying horizons. The accumulated materials may have been washed down from the horizons above, or they may have been formed in place through the weathering process. These underlying layers are referred to as *B horizons* (Figure 1.12).

Plant roots and microorganisms often extend below the B horizon, especially in humid regions, causing chemical changes in the soil water, some biochemical weathering of the regolith, and the formation of *C horizons*. The C horizons are the least weathered part of the soil profile.

In some soil profiles, the component horizons are very distinct in color, with sharp boundaries that can be seen easily by even novice observers. In other soils, the color changes between horizons may be very gradual, and the boundaries more difficult to locate. However, color is only one of many properties by which one horizon may be distinguished from the horizon above or below it (see Figure 1.13). The study of soils in the

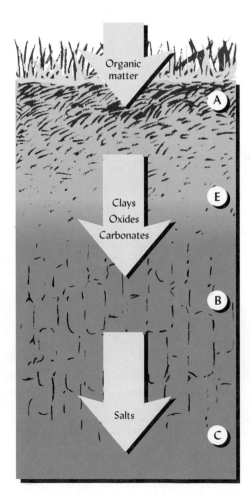

FIGURE 1.12 Horizons begin to differentiate as materials are added to the upper part of the profile and other materials are translocated to deeper zones. Under certain conditions, usually associated with forest vegetation and high rainfall, a leached E horizon forms between organic-matter-rich A and the B horizons. If sufficient rainfall occurs, soluble salts will be carried below the soil profile, perhaps all the way to the groundwater.

FIGURE 1.13 (Left) This soil profile was exposed by digging a pit about 2 meters deep in a well-developed soil (a Hapludalf) in southern Michigan. The top horizon can be easily distinguished because it has a darker color than those below it. However, some of the horizons in this profile are difficult to discern on the basis of color differences, especially in this black-and-white photo. The white string was attached to the profile to clearly demarcate some of the horizon boundaries. Then a trowel full of soil material was removed from each horizon and placed on a board, at right. It is clear from the way the soil either crumbled or held together that soil material from horizons with very similar colors may have very different physical properties. (Photos courtesy of R. Well)

field is a sensual as well as an intellectual activity. Delineation of the horizons present in a soil profile often requires a careful examination, using all the senses. In addition to seeing the colors in a profile, a soil scientist may feel, smell, and listen[2] to the soil, as well as conduct chemical tests, in order to distinguish the horizons present.

1.9 TOPSOIL AND SUBSOIL

The organically enriched A horizon at the soil surface is sometimes referred to as *topsoil.* When a soil is plowed and cultivated, the natural state of the upper 12 to 25 centimeters (5 to 10 inches) is modified. In this case, the topsoil may also be called the *plow layer* (or the *furrow slice* in situations where a moldboard plow has turned or "sliced" the upper part of the soil). Even where a plow is no longer used, the plow layer will remain evident for many years. For example, if you stroll through a typical New England forest you will see towering trees some 100 years old, but should you dig a shallow pit in the forest floor, you might still be able to see the smooth boundary between the century-old plowed layer and the lighter-colored, undisturbed soil below.

In cultivated soils, the majority of plant roots can be found in the topsoil (Figure 1.14). The topsoil contains a large part of the nutrient and water supplies needed by plants. The chemical properties and nutrient supply of the topsoil may be easily altered by mixing in organic and inorganic amendments, thereby making it possible to improve or maintain the soil's fertility and, to a lesser degree, its productivity. The physical structure of the topsoil, especially the part nearest the surface, is also readily affected by management operations such as tillage and application of organic materials. Maintaining an open structure at the soil surface is especially critical for providing balanced air and water supplies to plant roots and for avoiding excessive losses of soil and water by runoff. Sometimes the plow layer is removed from a soil and sold as topsoil for use at another site. This use of topsoil is especially common to provide a rooting medium suitable for lawns and shrubs around newly constructed buildings, where the original topsoil was removed or buried, and the underlying soil layers were exposed during grading operations (Figure 1.15).

---

[2] For example, the grinding sound emitted by wet soil rubbed between one's fingers indicates the sandy nature of the soil.

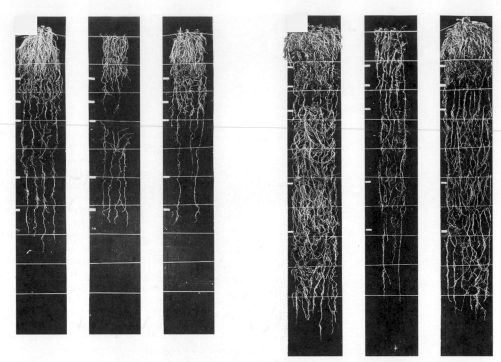

FIGURE 1.14 Plant roots respond to the varying conditions they find in the horizons of a soil profile. Roots tend to proliferate in the better-aerated, more fertile, and looser A horizon than in the horizons below. Improvement of soil fertility in the A horizon not only enhances root growth there, but may also increase the vigor and extent of the root system deeper in the profile as well. The roots shown are from a corn crop grown on an Illinois soil (Cisne series) that received no fertilizers or crop residues (left) or that received both fertilizers and crop residues (right). (Photos courtesy of J. B. Fehrenbacher, University of Illinois)

The soil layers that underlie the topsoil are referred to as *subsoil*. The subsoil is not normally seen from the surface, even on plowed cropland, for it lies below the usual depth of plowing. However, the characteristics of the subsoil horizons can greatly influence most land uses. Much of the water needed by plants is stored in the subsoil. Many subsoils also supply important quantities of certain plant nutrients. In some soils there is an abrupt change in properties between the topsoil and the subsoil. In other soils, the change is gradual and the upper part of the subsoil may be quite similar to the topsoil. In most soils the properties of the topsoil are far more conducive to plant growth than those of the subsoil. That is why there is often a good correlation between the productivity of a soil and the thickness of the topsoil in a profile.

Impermeable subsoil layers can impede root penetration, as can very acid subsoils. Poor drainage in the subsoil can result in waterlogged conditions in the topsoil. Because of its relative inaccessibility, it is usually much more difficult and expensive to physically or chemically modify the subsoil. On the other hand, good fertilization of the topsoil can produce vigorous plants whose roots are capable of greater exploration in subsoil layers (Figure 1.14; Box 1.1).

Many of the chemical, biological, and physical processes that characterize topsoils also take place to some degree in the C horizons of soils, which may extend deep into the underlying saprolite or other regolith material. Traditionally, the lower boundary of the soil has been considered to occur at the greatest rooting depth of the natural vegetation, but soil scientists are increasingly studying layers below this in order to understand ecological processes such as groundwater pollution, parent material weathering, and geochemical cycles.

1.10 SOIL: THE INTERFACE OF AIR, MINERALS, WATER, AND LIFE

We stated that where the regolith meets the atmosphere, the worlds of air, rock, water, and living things are intermingled. In fact, the four major components of soil are air,

FIGURE 1.15   The large mound of material at this construction site consists of topsoil (A horizon material) carefully separated from the lower horizons and pushed aside during initial grading operations. This stockpile was then seeded with grasses to give it a protective cover. After the construction activities are complete, the stockpiled topsoil will be used in landscaping the grounds around the new building. (Photo courtesy of R. Weil)

water, mineral matter, and organic matter. The relative proportions of these four components greatly influence the behavior and productivity of soils. In a soil, the four components are mixed in complex patterns; however, the proportion of soil volume occupied by each component can be represented in a simple pie chart. Figure 1.17 shows the approximate proportions (by volume) of the components found in a loam surface soil in good condition for plant growth. Although a handful of soil may at first seem to be a solid thing, it should be noted that only about half the soil volume consists of solid material (mineral and organic); the other half consists of pore spaces filled with air or water. Of the solid material, typically most is mineral matter derived from the rocks of the earth's crust. Only about 5% of the *volume* in this ideal soil consists of organic matter. However, the influence of the organic component on soil properties is often far greater than its small proportion would suggest. Since it is far less

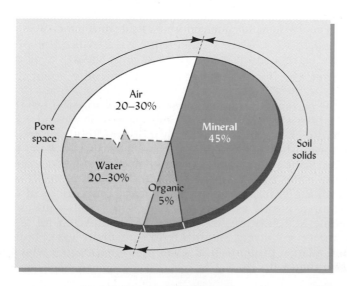

FIGURE 1.17   Volume composition of a loam surface soil when conditions are good for plant growth. The broken line between water and air indicates that the proportions of these two components fluctuate as the soil becomes wetter or drier. Nonetheless, a nearly equal proportion of air and water is generally ideal for plant growth.

BOX 1.1   USING INFORMATION FROM THE ENTIRE SOIL PROFILE

Soils are three-dimensional bodies that carry out important ecosystem processes at all depths in their profiles. Depending on the particular application, the information needed to make proper land management decisions may come from soil layers as shallow as the upper 1 or 2 centimeters, or as deep as the lowest layers of saprolite (Figure 1.16).

For example, the upper few centimeters of soil often hold the keys to plant growth and biological diversity, as well as to certain hydrologic processes. Here, at the interface between the soil and the atmosphere, living things are most numerous and diverse. Forest trees largely depend for nutrient uptake on a dense mat of fine roots growing in this zone. The physical condition of this thin surface layer may also determine whether rain will soak in or run downhill on the land surface. Certain pollutants, such as lead from highway exhaust, are also concentrated in this zone. For many types of soil investigations it will be necessary to sample the upper few centimeters separately so that important conditions are not overlooked.

On the other hand, it is equally important not to confine one's attention to the easily accessible "topsoil," for many soil properties are to be discovered only in the deeper layers. Plant-growth problems are often related to inhospitable conditions in the B or C horizons that restrict the penetration of roots. Similarly, the great volume of these deeper layers may control the amount of plant-available water held by a soil. For the purposes of recognizing or mapping different types of soils, the properties of the B horizons are often paramount. Not only is this the zone of major accumulations of minerals and clays, but the layers nearer the soil surface are too quickly altered by management and soil erosion to be a reliable source of information for the classification of soils.

In deeply weathered regoliths, the lower C horizons and saprolite play important roles. These layers, generally at depths below 1 or 2 meters, and often as deep as 5 to 10 meters, greatly affect the suitability of soils for most urban uses that involve construction or excavation. The proper functioning of on-site sewage disposal systems and the stability of building foundations are often determined by regolith properties at these depths. Likewise, processes that control the movement of pollutants to groundwater or the weathering of geologic materials may occur at depths of many meters. These deep layers also have major ecological influences because, although the intensity of biological activity and plant rooting may be quite low, the total impact can be great as a result of the enormous volume of soil that may be involved. This is especially true of forest systems in warm climates.

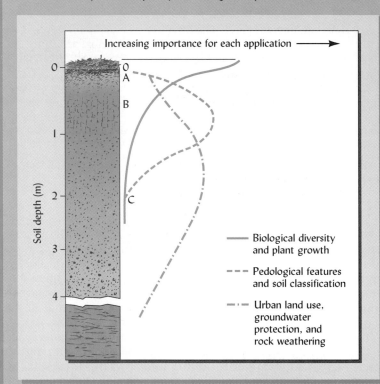

FIGURE 1.16   *Information important to different soil functions and applications is most likely to be obtained by studying different layers of the soil profile.*

dense than mineral matter, the organic matter accounts for only about 2% of the *weight* of this soil.

The spaces between the particles of solid material are just as important to the nature of a soil as are the solids themselves. It is in these pore spaces that air and water circulate, roots grow, and microscopic creatures live. Plant roots need both air and water. In an optimum condition for most plants, the pore space will be divided roughly equally among the two, with 25% of the soil volume consisting of water and 25% consisting of air. If there is much more water than this, the soil will be waterlogged. If much less water is present, plants will suffer from drought. The relative proportions of water and air in a soil typically fluctuate greatly as water is added or lost. Soils with much more than 50% of their volume in solids are likely to be too compacted for good plant growth. Compared to surface soil layers, subsoils tend to contain less organic matter, less total pore space, and a larger proportion of small pores (*micropores*) which tend to be filled with water rather than with air.

1.11 MINERAL (INORGANIC) CONSTITUENTS OF SOILS

Except in the case of organic soils, most of a soil's solid framework consists of **mineral**[3] particles. The larger soil particles, which include stones, gravel, and coarse sands, are generally rock fragments of various kinds. That is, these larger particles are often aggregates of several different minerals. Most smaller particles tend to be made of a single mineral. Thus, any particular soil is made up of particles that vary greatly in both size and composition.

The mineral particles present in soils are extremely variable in size. Excluding, for the moment, the larger rock fragments such as stones and gravel, soil particles range in size over four orders of magnitude: from 2.0 millimeters (mm) to smaller than 0.0002 mm in diameter (Table 1.2). **Sand** particles are probably most familiar to us. Individual sand particles are large enough (2.0 to 0.05 mm) to be seen by the naked eye and feel gritty when rubbed between the fingers. Sand particles do not adhere to one another; therefore, sands do not feel sticky. **Silt** particles are somewhat smaller (0.05 to 0.002 mm). Silt particles are too small to see without a microscope or to feel individually, so silt feels smooth but not sticky, even when wet. The smallest class of mineral particles are the **clays** (<0.002 mm), which adhere together to form a sticky mass when wet and hard clods when dry. The smaller particles (<0.001 mm) of clay (and similar-sized organic particles) have **colloidal**[4] properties and can be seen only with the aid of an electron microscope. Because of their extremely small size, colloidal particles possess a tremendous amount of surface area per unit of mass. Since the surfaces of soil colloids (both mineral and organic) exhibit electromagnetic charges that attract positive and negative ions as well as water, this fraction of the soil is the seat of most of the soil's chemical and physical activity.

The proportions of particles in these different size ranges is called **soil texture.** Terms such as *sandy loam, silty clay,* and *clay loam* are used to identify the soil texture. Texture has a profound influence on many soil properties, and it affects the suitability of a soil for most uses. To understand the degree to which soil properties can be influenced by texture, imagine dressing in a bathing suit and lying first on a sandy beach, and then in a clayey mud puddle. The difference in these two experiences would be due largely to the properties described in lines 5 and 7 of Table 1.2. Other properties related to particle size are also listed in Table 1.2. Note that clay-sized particles play a dominant role in holding certain inorganic chemicals and supplying nutrients to plants.

To anticipate the effect of clay on the way a soil will behave, it is not enough to know only the *amount* of clay in a soil. It is also necessary to know the *kinds of clays*

---

[3] The word *mineral* is used in soil science in three ways: (1) as a general adjective to describe inorganic materials derived from rocks; (2) as a specific noun to refer to distinct minerals found in nature, such as quartz and feldspars (see Chapter 2 for detailed discussions of soil-forming minerals and the rocks in which they are found); and (3) as an adjective to describe chemical elements, such as nitrogen and phosphorus, in their inorganic state in contrast to their occurrence as part of organic compounds.

[4] Colloidal systems are two-phase systems in which very small particles of one substance are dispersed in a medium of a different substance. Clay and organic soil particles smaller than about 0.001 mm (1 micrometer, μm) in diameter are generally considered to be colloidal in size. Milk and blood are other examples of colloidal systems in which very small solid particles are dispersed in a liquid medium.

TABLE 1.2   Some General Properties of the Three Major Size Classes of Inorganic Soil Particles

	Property	*Sand*	*Silt*	*Clay*
1.	Range of particle diameters in mm	2.0–0.05	0.05–0.002	Smaller than 0.002
2.	Means of observation	Naked eye	Microscope	Electron microscope
3.	Dominant minerals	Primary	Primary and secondary	Secondary
4.	Attraction of particles for each other	Low	Medium	High
5.	Attraction of particles for water	Low	Medium	High
6.	Ability to hold chemicals and nutrients in plant-available form	Very low	Low	High
7.	Consistency when wet	Loose, gritty	Smooth	Sticky, malleable
8.	Consistency when dry	Very loose, gritty	Powdery, some clods	Hard clods

present. As home builders and highway engineers know all too well, certain clayey soils, such as those high in smectite clays, make very unstable material on which to build because the clays swell when the soil is wet and shrink when the soil dries. This shrink-and-swell action can easily crack foundations and cause even heavy retaining walls to collapse. These clays also become extremely sticky and difficult to work when they are wet. Other types of clays, formed under different conditions, can be very stable and easy to work with. Learning about the different types of clay minerals will help us understand many of the physical and chemical differences among soils in various parts of the world (see Box 1.2).

Primary and Secondary Minerals

Minerals that have persisted with little change in composition since they were extruded in molten lava (e.g., quartz, micas, and feldspars) are known as *primary minerals.* They

---

BOX 1.2 OBSERVING SOILS IN DAILY LIFE

Your study of soils can be enriched if you make an effort to become aware of the many daily encounters with soils and their influences that go unnoticed by most people. When you dig a hole to plant a tree or set a fence post, note the different layers encountered, and note how the soil from each layer looks and feels different. If you pass a construction site, take a moment to observe the horizons exposed by the excavations. An airplane trip is a great opportunity to observe how soils vary across landscapes and climatic zones. If you are flying during daylight hours, ask for a window seat. Look for the shapes of individual soils in plowed fields if you are flying in spring or fall (Figure 1.18).

Soils can give you clues to understanding the natural processes going on around you. Down by the stream, use a magnifying glass to examine the sand deposited on the banks or bottom. It may contain minerals not found in local rocks and soils, but originating many kilometers upstream. When you wash your car, see if the mud clinging to the tires and fenders is of a different color or consistency than the soils near your home. Does the "dirt" on your car tell you where you have been driving? Forensic investigators have been known to consult with soil scientists to locate crime victims or establish guilt by matching soil clinging to shoes, tires, or tools with the soils at a crime scene.

Other examples of soil hints can be found even closer to home. The next time you bring home celery or leaf lettuce from the supermarket, look carefully for bits of soil clinging to the bottom of the stalk or leaves (Figure 1.19). Rub the soil between your thumb and fingers. Smooth, very black soil may indicate that the lettuce was grown in mucky soils, such as those in New York state or southern Florida. Light brown, smooth-feeling soil with only a very fine grittiness is more typical of California-grown produce, while light-colored, gritty soil is common on produce from the southern Georgia–northern Florida vegetable-growing region. In a bag of dry pinto beans, you may come across a few lumps of soil that escaped removal in the cleaning process because of being the same size as the beans. Often this soil is dark-colored and very sticky, coming from the "thumb" area of Michigan, where a large portion of the U.S. dry bean crop is grown.

Opportunities to observe soils in daily life range from the remote and large-scale to the close-up and intimate. As you learn more about soils, you will undoubtedly be able to see more examples of their influence in your surroundings.

FIGURE 1.18 *The light and dark colored soil bodies, as seen from an airliner flying over central Texas, reflect differences in drainage and topography in the landscape. (Photo courtesy of R. Weil)*

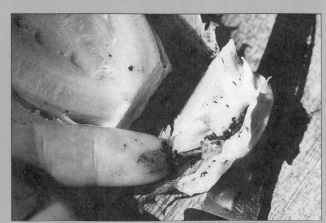

FIGURE 1.19 *Although this Romaine lettuce was purchased in a Virginia grocery store in early Fall, the black, mucky soil clinging to the base of the leaves indicates that it was grown on organic soils, probably in New York state. (Photo courtesy of R. Weil)*

are prominent in the sand and silt fractions of soils. Other mineral⹂
clays and iron oxides, were formed by the breakdown and weatherin⹂
minerals as soil formation progressed. These minerals are called *sec⹂*
and tend to dominate the clay and, in some cases, silt fractions.

The inorganic minerals in the soil are the original source of most of ⹂
ments essential for plant growth. Although the bulk of these nutrients ⹂
components of the basic crystalline structure of the minerals, a smal⹂ ⹂ut important
portion is in the form of charged ions on the surface of fine colloidal particles (clays and
organic matter). Mechanisms of critical importance to growing plants allow plant roots
to have access to these surface-held nutrient ions (see Section 1.16).

Soil Structure

Sand, silt, and clay particles can be thought of as the building blocks from which soil is
constructed. The manner in which these building blocks are arranged together is called
soil structure. The particles may remain relatively independent of each other, but more
commonly they are associated together in aggregates of different-size particles. These
aggregates may take the form of roundish granules, cubelike blocks, flat plates, or other
shapes. Soil structure (the way particles are arranged together) is just as important as soil
texture (the relative amounts of different sizes of particles) in governing how water and
air move in soils. Both structure and texture fundamentally influence the suitability of
soils for the growth of plant roots.

1.12 SOIL ORGANIC MATTER

Soil organic matter consists of a wide range of organic (carbonaceous) substances,
including living organisms (the soil *biomass*), carbonaceous remains of organisms
which once occupied the soil, and organic compounds produced by current and past
metabolism in the soil. The remains of plants, animals, and microorganisms are continuously broken down in the soil and new substances are synthesized by other
microorganisms. Over periods of time ranging from hours to centuries, organic matter
is lost from the soil as carbon dioxide produced by microbial respiration. Because of
such loss, repeated additions of new plant and/or animal residues are necessary to maintain soil organic matter.

Under conditions that favor plant production more than microbial decay, large
quantities of atmospheric carbon dioxide used by plants in photosynthesis are
sequestered in the abundant plant tissues which eventually become part of the soil
organic matter. Since carbon dioxide is a major cause of the greenhouse effect which is
believed to be warming the earth's climate, the balance between accumulation of soil
organic matter and its loss through microbial respiration has global implications. In
fact, more carbon is stored in the world's soils than in the world's plant biomass and
atmosphere combined.

Even so, organic matter comprises only a small fraction of the mass of a typical soil.
By weight, typical well-drained mineral surface soils contain from 1 to 6% organic matter. The organic matter content of subsoils is even smaller. However, the influence of
organic matter on soil properties, and consequently on plant growth, is far greater than
the low percentage would indicate.

Organic matter binds mineral particles into a granular soil structure that is largely
responsible for the loose, easily managed condition of productive soils. Part of the soil
organic matter that is especially effective in stabilizing these granules consists of certain gluelike substances produced by various soil organisms, including plant roots (Figure 1.20).

Organic matter also increases the amount of water a soil can hold and the proportion of water available for plant growth (Figure 1.21). In addition, it is a major soil
source of the plant nutrients phosphorus and sulfur, and the primary source of nitrogen
for most plants. As soil organic matter decays, these nutrient elements, which are
present in organic combinations, are released as soluble ions that can be taken up by
plant roots. Finally, organic matter, including plant and animal residues, is the main
food that supplies carbon and energy to soil organisms. Without it, biochemical activity so essential for ecosystem functioning would come to a near standstill.

FIGURE 1.20 Abundant organic matter, including plant roots, helps create physical conditions favorable for the growth of higher plants as well as microbes (left). In contrast, soils low in organic matter, especially if they are high in silt and clay, are often cloddy (right) and not suitable for optimum plant growth.

Humus, usually black or brown in color, is a collection of very complex organic compounds which accumulate in soil because they are relatively resistant to decay. Just as clay is the colloidal fraction of soil mineral matter, so humus is the colloidal fraction of soil organic matter. Because of their charged surfaces, both humus and clay act as contact bridges between larger soil particles; thus, both play an important role in the for-

FIGURE 1.21 Soils higher in organic matter are darker in color and have greater water-holding capacities than soils low in organic matter. The soil in each container has the same texture, but the one on the right has been depleted of much of its organic matter. The same amount of water was applied to each container. As the lower photo shows, the depth of water penetration was less in the high organic matter soil (left) because of its greater water-holding capacity. It required a greater volume of the low organic matter soil (right) to hold the same amount of water.

mation of soil structure. The surface charges of humus, like those of clay, attract and hold both nutrient ions and water molecules. However, gram for gram, the capacity of humus to hold nutrients and water is far greater than that of clay. Unlike clay, humus contains certain components that can have a hormone-like stimulatory effect on plants. All in all, small amounts of humus may remarkably increase the soil's capacity to promote plant growth.

1.13 SOIL WATER: A DYNAMIC SOLUTION

Water is of vital importance in the ecological functioning of soils. The presence of water in soils is essential for the survival and growth of plants and other soil organisms. The soil moisture regime, often reflective of climatic factors, is a major determinant of the productivity of terrestrial ecosystems, including agricultural systems. Movement of water, and substances dissolved in it, through the soil profile is of great consequence to the quality and quantity of local and regional water resources. Water moving through the regolith is also a major driving force in soil formation.

Two main factors help explain why **soil water** is different from our everyday concept of, say, drinking water in a glass.

1. Water is held within soil pores with varying degrees of tenacity depending on the amount of water present and the size of the pores. The attraction between water and the surfaces of soil particles greatly restricts the ability of water to flow as it would flow in a drinking glass.

2. Because soil water is never pure water, but contains hundreds of dissolved organic and inorganic substances, it may be more accurately called the *soil solution.* An important function of the soil solution is to serve as a constantly replenished, dilute nutrient solution bringing dissolved nutrient elements (e.g., calcium, potassium, nitrogen, and phosphorus) to plant roots.

When the soil moisture content is optimum for plant growth (Figure 1.17), the water in the large and intermediate-sized pores can move in the soil and can be used by plants. As some of the moisture is removed by the growing plants, however, that which remains is in the tiny pores and in thin films around soil particles. The soil solids strongly attract this soil water and consequently compete with plant roots for it. Thus, not all soil water is *available* to plants. Depending on the soil, one-fourth to two-thirds of the moisture remains in the soil after the plants have wilted or died for lack of water.

Soil Solution

The soil solution contains small but significant quantities of soluble inorganic compounds, some of which supply elements that are essential for plant growth. Refer to Table 1.1 for a listing of the 18 **essential elements**, along with their sources. The soil solids, particularly the fine organic and inorganic colloidal particles, release these elements to the soil solution, from which they are taken up by growing plants. Such exchanges, which are critical for higher plants, are dependent on both soil water and the fine soil solids.

One other critical property of the soil solution is its *acidity* or *alkalinity.* Many chemical and biological reactions are dependent on the levels of the H^+ and OH^- ions in the soil. The levels of these ions also influence the solubility, and in turn the availability, of several essential nutrient elements (including iron and manganese) to plants.

The concentrations of H^+ and OH^- ions in the soil solution are commonly ascertained by measuring the *pH* of the soil solution. Technically, the pH is the negative logarithm of the activity of H^+ ions in the soil solution. Figure 1.22 shows very simply the relationship between pH and the concentration of H^+ and OH^- ions. It should be studied carefully along with Figure 1.23, which shows ranges in pH commonly encountered in soils from different climatic regions. Sometimes referred to as a *master variable,* the pH controls the nature of many chemical and microbial reactions in the soil. It is of great significance in essentially all aspects of soil science.

FIGURE 1.22 Diagrammatic representation of acidity, neutrality, and alkalinity. At neutrality the H$^+$ and OH$^-$ ions of a solution are balanced, their respective numbers being the same (pH 7). At pH 6, the H$^+$ ions are dominant, being 10 times greater, whereas the OH$^-$ ions have decreased proportionately, being only one-tenth as numerous. The solution therefore is acid at pH 6, there being 100 times more H$^+$ ions than OH$^-$ ions present. At pH 8, the exact reverse is true; the OH$^-$ ions are 100 times more numerous than the H$^+$ ions. Hence, the pH 8 solution is alkaline. This mutually inverse relationship must always be kept in mind when pH data are used.

1.14 SOIL AIR: A CHANGING MIXTURE OF GASES

Approximately half of the volume of the soil consists of pore spaces of varying sizes (refer to Figure 1.17), which are filled with either water or air. When water enters the soil, it displaces air from some of the pores; the air content of a soil is therefore inversely related to its water content. If we think of the network of soil pores as the ventilation system of the soil connecting airspaces to the atmosphere, we can understand that when the smaller pores are filled with water the ventilation system becomes clogged. Think how stuffy the air would become if the ventilation ducts of a classroom became clogged. Because oxygen could not enter the room, nor carbon dioxide leave it, the air in the room would soon become depleted of oxygen and enriched in carbon dioxide and water vapor by the respiration (breathing) of the people in it. In an air-filled soil

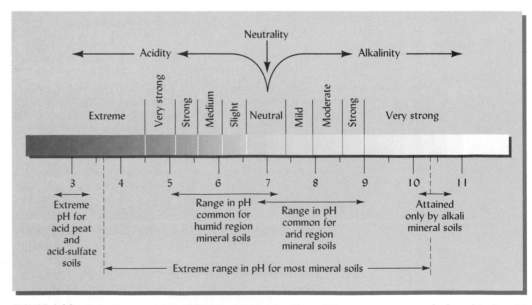

FIGURE 1.23 Extreme range in pH for most mineral soils and the ranges commonly found in humid region and arid region soils. Also indicated are the maximum alkalinity for alkali soils and the minimum pH likely to be encountered in very acid peat soils.

pore surrounded by water-filled smaller pores, the metabolic activities of plant roots and microorganisms have a similar effect.

Therefore, soil air differs from atmospheric air in several respects. First, the composition of soil air varies greatly from place to place in the soil. In local pockets, some gases are consumed by plant roots and by microbial reactions, and others are released, thereby greatly modifying the composition of the soil air. Second, soil air generally has a higher moisture content than the atmosphere; the relative humidity of soil air approaches 100% unless the soil is very dry. Third, the content of carbon dioxide (CO_2) is usually much higher, and that of oxygen (O_2) lower, than contents of these gases found in the atmosphere. Carbon dioxide in soil air is often several hundred times more concentrated than the 0.035% commonly found in the atmosphere. Oxygen decreases accordingly and, in extreme cases, may be 5 to 10%, or even less, compared to about 20% for atmospheric air.

The amount and composition of air in a soil are determined to a large degree by the water content of the soil. The air occupies those soil pores not filled with water. As the soil drains from a heavy rain or irrigation, large pores are the first to be filled with air, followed by medium-sized pores, and finally the small pores, as water is removed by evaporation and plant use. This explains the tendency for soils with a high proportion of tiny pores to be poorly aerated. In such soils, water dominates, and the soil air content is low, as is the rate of diffusion of the air into and out of the soil from the atmosphere. The result is high levels of CO_2 and low levels of O_2, unsatisfactory conditions for the growth of most plants. In extreme cases, lack of oxygen both in the soil air and dissolved in the soil water may fundamentally alter the chemical reactions that take place in the soil solution. This is of particular importance to understanding the functions of wetland soils.

1.15   INTERACTION OF FOUR COMPONENTS TO SUPPLY PLANT NUTRIENTS

As you read our discussion of each of the four major soil components, you may have noticed that the impact of one component on soil properties is seldom expressed independently from that of the others. Rather, the four components interact with each other to determine the nature of a soil. Thus, soil moisture, which directly meets the needs of plants for water, simultaneously controls much of the air and nutrient supply to the plant roots. The mineral particles, especially the finest ones, attract soil water, thus determining its movement and availability to plants. Likewise, organic matter, because of its physical binding power, influences the arrangement of the mineral particles into clusters and, in so doing, increases the number of large soil pores, thereby influencing the water and air relationships.

Essential Element Availability

Perhaps the most important interactive process involving the four soil components is the provision of essential nutrient elements to plants. Plants absorb essential nutrients, along with water, directly from one of these components: the soil solution. However, the amount of essential nutrients in the soil solution at any one time is far less than is needed to produce a mature plant. Consequently, the soil solution nutrient levels must be constantly replenished from the inorganic or organic parts of the soil and from fertilizers or manures added to agricultural soils.

Fortunately, relatively large quantities of these nutrients are associated with both inorganic and organic soil solids. By a series of chemical and biochemical processes, nutrients are released from these solid forms to replenish those in the soil solution. For example, through *ion exchange,* essential elements such as Ca^{2+} and K^+ are released from the colloidal surfaces of clay and humus to the soil solution. The intimate contact between soil solution ions and adsorbed[5] ions makes this exchange possible. In the example below, an H^+ ion in the soil solution is shown to exchange readily with an

---

[5] *Adsorption* refers to the attraction of ions to the surface of particles, in contrast to *absorption,* the process by which ions are taken *into* plant roots. The adsorbed ions are exchangeable with ions in the soil solution.

adsorbed K$^+$ ion on the colloidal surface. The K$^+$ ion is then available in the soil solution for uptake by crop plants.

$$\boxed{\text{colloid}}\,K^+ + H^+ \text{ ion} \longrightarrow \boxed{\text{colloid}}\,H^+ + K^+ \text{ ion}$$

Adsorbed   Soil   Adsorbed   Soil
solution   solution

The K$^+$ ion thus released can be readily taken up (absorbed) by plants. Some scientists consider that this ion exchange process is among the most important of chemical reactions in nature.

Nutrient ions are also released to the soil solution as soil microorganisms decompose organic tissues. Plant roots can readily absorb all of these nutrients from the soil solution, provided there is enough O$_2$ in the soil air to support root metabolism.

Most soils contain large amounts of plant nutrients relative to the annual needs of growing vegetation. However, the bulk of most nutrient elements is held in the structural framework of primary and secondary minerals and organic matter. Only a small fraction of the nutrient content of a soil is present in forms which are readily available to plants. Table 1.3 will give you some idea of the quantities of various essential elements present in different forms in typical soils of humid and arid regions.

Figure 1.24 illustrates how the two solid soil components interact with the liquid component (soil solution) to provide essential elements to plants. Plant roots do not ingest soil particles, no matter how fine, but are able to absorb only nutrients that are dissolved in the soil solution. Because elements in the coarser soil framework of the soil are only slowly released into the soil solution over long periods of time, the bulk of most nutrients in a soil is not readily available for plant use. Nutrient elements in the framework of colloid particles are somewhat more readily available to plants, as these particles break down much faster because of their greater surface area. Thus, the structural framework is the major storehouse and, to some extent, a significant source of essential elements in many soils.

The distribution of nutrients among the various components of a fertile soil, as illustrated in Figure 1.24, may be likened to the distribution of financial assets in the portfolio of a wealthy individual. In such an analogy, nutrients readily available for plant use would be analogous to cash in the individual's pocket. A millionaire would likely keep most of his or her assets in long-term investments such as real estate or bonds (the coarse fraction solid framework), while investing a smaller amount in short-term stocks and bonds (colloidal framework). For more immediate use, an even smaller amount might be kept in a checking account (exchangeable nutrients), while a tiny fraction of the overall wealth might be carried to spend as currency and coins (nutrients in the soil solution). As the cash is used up, the supply is replenished by making a withdrawal from the checking account. The checking account, in turn, is replenished occasionally by the sale of long-term investments. It is possible for a wealthy person to run short of cash even though he or she may own a great deal of valuable real estate. In an analogous way, plants may use up the readily available supply of a nutrient even though the total supply of that nutrient in the soil is very large. Luckily, in a fertile soil, the process

TABLE 1.3   **Quantities of Six Essential Elements Found in Upper 15 cm of Representative Soils in Temperate Regions**

Essential element	Humid region soil			Arid region soil		
	In solid framework, kg/ha	Exchangeable, kg/ha	In soil solution, kg/ha	In solid framework, kg/ha	Exchangeable, kg/ha	In soil solution, kg/ha
Ca	8,000	2,250	60–120	20,000	5,625	140–280
Mg	6,000	450	10–20	14,000	900	25–40
K	38,000	190	10–30	45,000	250	15–40
P	900	—	0.05–0.15	1,600	—	0.1–0.2
S	700	—	2–10	1,800	—	6–30
N	3,500	—	7–25	2,500	—	5–20

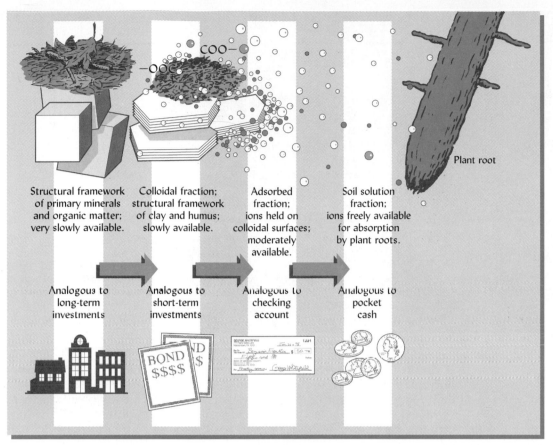

FIGURE 1.24 Nutrient elements exist in soils in various forms characterized by different accessibility to plant roots. The bulk of the nutrients is locked up in the structural framework of primary minerals, organic matter, clay, and humus. A smaller proportion of each nutrient is adsorbed in a swarm of ions near the surfaces of soil colloids (clay and organic matter). From the swarm of adsorbed ions, a still smaller amount is released into the bulk soil solution, where uptake by plant roots can take place.

described in Figure 1.24 can help replenish the soil solution as quickly as plant roots remove essential elements.

1.16 NUTRIENT UPTAKE BY PLANT ROOTS

To be taken up by a plant, a nutrient element must be in a soluble form and must be located *at the root surface*. Often, parts of a root are in such intimate contact with soil particles (see Figure 1.25) that a direct exchange may take place between nutrient ions adsorbed on the surface of soil colloids and H^+ ions from the surface of root cell walls. In any case, the supply of nutrients in contact with the root will soon be depleted. This fact raises the question of how a root can obtain additional supplies once the nutrient ions at the root surface have all been taken up into the root. There are three basic mechanisms by which the concentration of nutrient ions at the root surface is maintained (Figure 1.26).

First, **root interception** comes into play as roots continually grow into new, undepleted soil. For the most part, however, nutrient ions must travel some distance in the soil solution to reach the root surface. This movement can take place by **mass flow**, as when dissolved nutrients are carried along with the flow soil water toward a root that is actively drawing water from the soil. In this type of movement, the nutrient ions are somewhat analogous to leaves floating down a stream. On the other hand, plants can continue to take up nutrients even at night, when water is only slowly absorbed into the roots. Nutrient ions continually move by **diffusion** from areas of greater concentration toward the nutrient-depleted areas of lower concentration around the root surface.

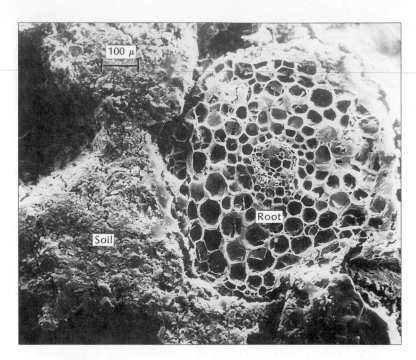

FIGURE 1.25 Scanning electron micrograph of a cross section of a peanut root surrounded by soil. Note the intimacy of contact. [Courtesy Tan and Nopamornbodi (1981)]

In the diffusion process, the random movements of ions in all directions causes a *net* movement from areas of high concentration to areas of lower concentrations, independent of any mass flow of the water in which the ions are dissolved. Factors such as soil compaction, cold temperatures, and low soil moisture content, which reduce root interception, mass flow, or diffusion, can result in poor nutrient uptake by plants even in soils with adequate supplies of soluble nutrients. Furthermore, the availability of nutrients for uptake can also be negatively or positively influenced by the activities of microorganisms that thrive in the immediate vicinity of roots. Maintaining the supply of available nutrients at the plant root surface is thus a process that involves complex interactions among different soil components.

It should be noted that the plant membrane separating the inside of the root cell from the soil solution is permeable to dissolved ions only under special circum-

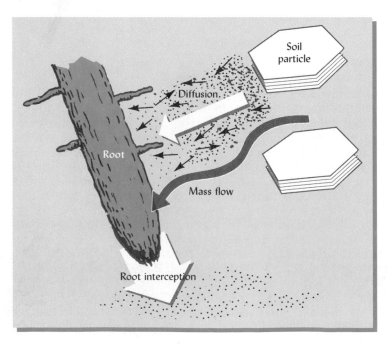

FIGURE 1.26 Three principal mechanisms by which nutrient ions dissolved in the soil solution come into contact with plant roots. All three mechanisms may operate simultaneously, but one mechanism or another may be most important for a particular nutrient. For example, in the case of calcium, which is generally plentiful in the soil solution, mass flow alone can usually bring sufficient amounts to the root surface. However, in the case of phosphorus, diffusion is needed to supplement mass flow because the soil solution is very low in this element in comparison to the amounts needed by plants.

stances. Plants do not merely take up, by mass flow, those nutrients that happen to be in the water that roots are removing from the soil. Nor do dissolved nutrient ions brought to the root's outer surface by mass flow or diffusion cross the root cell membrane and enter the root passively by diffusion. On the contrary, a nutrient is taken up when a chemical carrier molecule in the root cell membrane forms an activated complex with the nutrient and then travels across the membrane to the interior of the root cell before releasing the nutrient. The carrier mechanism, activated by root metabolic energy, allows the plant to accumulate concentrations of nutrients inside the root cell that far exceed the nutrient concentrations in the soil solution. Because different nutrients are taken up by specific types of carrier molecules, the plant is able to exert some control over how much and in what relative proportions essential elements are taken up.

Since nutrient uptake is an active metabolic process, conditions that inhibit root metabolism may also inhibit nutrient uptake. Examples of such conditions include excessive soil water content or soil compaction resulting in poor soil aeration, excessively hot or cold soil temperatures, and aboveground conditions which result in low translocation of sugars to plant roots. We can see that plant nutrition involves biological, physical, and chemical processes and interactions among many different components of soils and the environment.

1.17   SOIL QUALITY, DEGRADATION, AND RESILIENCE

Soil is a basic resource underpinning all terrestrial ecosystems. Managed carefully, soils are a reusable resource, but in the scale of human lifetimes they cannot be considered a renewable resource. As we shall see in the next chapter, most soil profiles are thousands of years in the making. In all regions of the world, human activities are destroying some soils far faster than nature can rebuild them. Soils completely washed away by erosion or excavated and paved over by urban sprawl are lost, for all practical purposes. More often, soils are degraded in quality rather than totally destroyed.

Soil quality is a measure of the ability of a soil to carry out particular ecological functions, such as those described in Sections 1.2 to 1.6. Soil quality reflects a combination of *chemical*, *physical*, and *biological* properties. Some of these properties are relatively unchangeable, inherent properties that help define a particular type of soil. Soil texture and mineral makeup (Section 1.11) are examples. Other soil properties, such as structure (Section 1.11) and organic matter content (Section 1.12) can be significantly changed by management. These more changeable soil properties can indicate the status of a soil's quality relative to its potential, much the way water turbidity or oxygen content indicate the water quality status of a river.

Mismanagement of forests, farms, and rangeland causes widespread degradation of soil quality by erosion that removes the topsoil, little by little (see Chapter 17). Another widespread cause of soil degradation is the accumulation of salts in improperly irrigated soils in arid regions. When people cultivate soils and harvest the crops without returning organic residues and mineral nutrients, the soil supply of organic matter and nutrients becomes depleted (see Chapter 12). Such depletion is particularly widespread in sub-Saharan Africa, where degrading soil quality is reflected in diminished capacity to produce food (see Chapter 20). Contamination of a soil with toxic substances from industrial processes or chemical spills can degrade its capacity to provide habitat for soil organisms, to grow plants that are safe to eat, or to safely recharge ground and surface waters (see Chapter 18). Degradation of soil quality by pollution is usually localized, but the environmental impacts and costs involved are very large.

While protecting soil quality must be the first priority, it is often necessary to attempt to restore the quality of soils that have already been degraded. Some soils have sufficient **resilience** to recover from minor degradation if left to revegetate on their own. In other cases, more effort is required to restore degraded soils. Organic and inorganic amendments may have to be applied, vegetation may have to be planted, physical alterations by tillage or grading may have to be made, or contaminants may have to be removed. As societies around the world assess the damage already done to their soil resources, the job of **soil restoration** is becoming serious business.

1.18 CONCLUSION

The earth's soil is comprised of numerous soil individuals, each of which is a three-dimensional natural body in the landscape. Each individual soil is characterized by a unique set of properties and soil horizons as expressed in its profile. The nature of the soil layers seen in a particular profile is closely related to the nature of the environmental conditions at a site.

Soils perform five broad ecological functions: they act as the principal medium for plant growth, regulate water supplies, recycle raw materials and waste products, and serve as a major engineering medium for human-built structures. They are also home to many kinds of living organisms. Soil is thus a major ecosystem in its own right. The soils of the world are extremely diverse, each type of soil being characterized by a unique set of soil horizons. A typical surface soil in good condition for plant growth consists of about half solid material (mostly mineral, but with a crucial organic component, too) and half pore spaces filled with varying proportions of water and air. These components interact to influence a myriad of complex soil functions, a good understanding of which is essential for wise management of our terrestrial resources.

STUDY QUESTIONS

1. As a society, is our reliance on soils likely to increase or decrease in the decades ahead? Explain.

2. Discuss how *a soil,* a natural body, differs from *soil,* a material that is used in building a roadbed?

3. What are the five main roles of soil in an ecosystem? For each of these ecological roles, suggest one way in which interactions occur with another of the five roles.

4. Think back over your activities during the past week. List as many incidents as you can in which you came into direct or indirect contact with soil.

5. Figure 1.17 shows the volume composition of a loam surface soil in ideal condition for plant growth. To help you understand the relationships among the four components redraw this pie chart to represent what the situation might be after the soil has been compacted by heavy traffic. Redraw the original pie chart again, but show how the four components would be related on a mass (weight) basis rather than on a volume basis.

6. Explain in your own words how the soil's nutrient supply is held in different forms, much the way that a person's financial assets might be held in different forms.

7. List the essential nutrient elements that plants derive mainly from the soil.

8. Are all elements contained in plants essential nutrients? Explain.

9. Define these terms: *soil texture, soil structure, soil pH, humus, soil profile, B horizon, soil quality, solum,* and *saprolite.*

10. Describe four processes that commonly lead to degradation of soil quality.

REFERENCE

Tan, K. H., and O. Nopamornbodi. 1981. "Electron microbeam analysis and scanning electron microscopy of soil–root interfaces," *Soil Sci.,* **131**:100–106.

2

FORMATION OF SOILS FROM PARENT MATERIALS

It is a poem of existence . . . not a lyric but a slow-moving epic whose beat has been set by eons of the world's experience. . . .
—JAMES MICHENER, CENTENNIAL

The first astronauts to explore the moon labored in their clumsy pressurized suits to collect samples of rocks and dust from the lunar surface. These they carried back to Earth for analysis. It turned out that moon rocks are similar in composition to those found deep in the Earth—so similar that scientists concluded that the moon itself began as a large chunk of molten Earth that broke away eons ago, when the young planet nearly melted in a stupendous collision with a Mars-sized object, leaving the Pacific Ocean as a scar. On the moon, this rock remained unchanged or crumbled into dust with the impact of meteors. On Earth, the rock at the surface, coming in contact with water, air, and living things, was transformed into something new, into many different kinds of living soils. This chapter reveals the story of how rock and dust become "the ecstatic skin of the Earth."[1]

We will study the processes of soil formation that develop, from relatively uniform regolith, the variegated layers of the soil profile. We will also learn about the environmental factors that influence these processes to produce soils in Belgium so different from those in Brazil, soils on limestone so different from those on sandstone, and soils in the valley bottoms so different from those on the hills.

Every landscape is comprised of a suite of different soils, each influencing ecological processes in its own way. Whether we intend to modify, exploit, preserve, or simply understand the landscape, our success will depend on our knowing how soil profile properties relate to the environment on each site and to the landscape as a whole.

2.1 WEATHERING OF ROCKS AND MINERALS

The influence of **weathering**, the physical and chemical breakdown of particles, is evident everywhere. Nothing escapes it. It breaks up rocks and minerals, modifies or destroys their physical and chemical characteristics, and carries away the soluble products. Likewise, it synthesizes new minerals of great significance in soils. The nature of the rocks and minerals being weathered determines the rates and results of the breakdown and synthesis (Figure 2.1).

---

[1] From Logan (1995).

29

FIGURE 2.1 Two stone markers, photographed on the same day in the same cemetery, illustrate the effect of rock type on weathering rates. The date and initials carved in the slate marker in 1798 are still sharp and clear, while the date and figure of a lamb carved in the marble marker in 1875 have weathered almost beyond recognition. The slate rock consists largely of resistant silicate clay minerals, while the marble consists mainly of calcite, which is much more easily attacked by acids in rainwater. (Photo courtesy of R. Weil)

Characteristics of Rocks and Minerals

The rocks in the earth's outer surface are commonly classified as **igneous, sedimentary,** and **metamorphic.** Those of igneous origin are formed from molten magma and include such common rocks as granite and diorite (Figure 2.2).

Igneous rock is composed of primary minerals[2] such as light-colored quartz, muscovite, and feldspars and dark-colored biotite, augite, and hornblende. In general, dark-

---

[2] *Primary minerals* have not been altered chemically since they formed as molten lava solidified. *Secondary minerals* are recrystallized products of the chemical breakdown and/or alteration of primary minerals.

Rock texture	Quartz	Light-colored minerals (e.g., feldspars, muscovite)		Dark-colored minerals (e.g., hornblende, augite, biotite)
Coarse	Granite	Diorite	Gabbro	Peridotite / Hornblendite
Intermediate	Rhyolite	Andesite	Basalt	
Fine	Felsite / Obsidian		Basalt glass	

FIGURE 2.2 Classification of some igneous rocks in relation to mineralogical composition and the size of mineral grains in the rock (rock texture). Worldwide, light-colored minerals and quartz are generally more prominent than are the dark-colored minerals.

FIGURE 2.3 Primary minerals are randomly interlocked in igneous rocks, as in the syenite on the left. High heat and pressure may partially remelt the rock, causing the lighter minerals to separate from the heavier ones, thus forming the light- and dark-colored bands typical of gneiss, a metamorphic rock (right). In this case, the primary mineral content of both rocks is similar, the light-colored minerals being mainly orthoclase and the darker ones hornblende. (Photo courtesy of R. Weil)

colored minerals contain iron and magnesium and are more easily weathered. Therefore, dark-colored igneous rocks such as gabbro and basalt are more easily broken down than are granites and other lighter-colored igneous rocks. The mineral grains in igneous rocks are randomly dispersed and interlocked, giving a salt-and-pepper appearance if they are coarse enough to see with the unaided eye (Figure 2.3, left).

When weathering products from old rocks are compacted or cemented, they form new sedimentary rocks. For example, quartz sand weathered from a granite and deposited near the shore of a prehistoric sea may become cemented into a solid mass called *sandstone*. Similarly, clays may be compacted into *shale*. Other important sedimentary rocks are listed in Table 2.1, along with their dominant minerals. Because most of what is presently dry land was at some time in the past covered by water, sedimentary rocks are the most common type of rock encountered, covering about 75% of the Earth's land surface. The resistance of a given sedimentary rock to weathering is determined by its particular dominant minerals and by the cementing agent.

Metamorphic rocks are those that have formed by the metamorphism or change in form of other rocks. As the Earth's continental plates shift, and sometimes collide, forces are generated that can uplift great mountain ranges. As a result, igneous and sedimentary masses are subjected to tremendous heat and pressure that compress, distort, and/or partially remelt the original rocks. Igneous rocks are commonly modified to form schist or gneiss in which light and dark minerals have been reoriented into bands (Figure 2.3, right). Sedimentary rocks, such as limestone and shale, may be metamorphosed to marble and slate, respectively (Table 2.1.) As is the case for igneous and sedimentary rock, the particular minerals that dominate a given metamorphic rock influence its resistance to chemical weathering (see Table 2.2 and Figure 2.1). A high degree of metamorphism may also physically weaken the rock mass, hastening its breakdown into smaller fragments.

Weathering: A General Case

Weathering combines the processes of destruction and synthesis (Figure 2.4). Moving from left to right in the weathering diagram, the original rocks and minerals are

TABLE 2.1 Some of the More Important Sedimentary and Metamorphic Rocks and the Minerals Commonly Dominant in Them

	Type of rock	
Dominant mineral	Sedimentary	Metamorphic
Calcite ($CaCO_3$)	Limestone	Marble
Dolomite ($CaCO_3 \cdot MgCO_3$)	Dolomite	Marble
Quartz (SiO_2)	Sandstone	Quartzite
Clays	Shale	Slate
Variable	Conglomerate[a]	Gneiss[b]
Variable		Schist[b]

[a] Small stones of various mineralogical makeup are cemented into conglomerate.
[b] The minerals present are determined by the original rock, which has been changed by metamorphism. Primary minerals present in the igneous rocks commonly dominate these rocks, although some secondary minerals are also present.

destroyed by both *physical disintegration* and *chemical decomposition*. Without appreciably affecting their composition, physical disintegration breaks down rock into smaller rocks and eventually into sand and silt particles that are commonly made up of individual minerals. Simultaneously, the minerals decompose chemically, releasing soluble materials and synthesizing new minerals, some of which are resistant end products. New minerals form either by minor chemical alterations or by complete chemical breakdown of the original mineral and resynthesis of new minerals. During the chemical changes, particle size continues to decrease, and constituents continue to dissolve in the aqueous weathering solution. The dissolved substances may be lost in drainage waters or they may recombine into new (secondary) minerals.

Three groups of minerals that remain in well-weathered soils are shown on the right side of Figure 2.4: (1) silicate clays, (2) very resistant end products, including iron and aluminum oxide clays, and (3) very resistant primary minerals, such as quartz. In highly weathered soils of humid tropical and subtropical regions, the oxides of iron and aluminum and certain silicate clays with low Si/Al ratios predominate because most other constituents have been broken down and removed (see Table 2.3).

TABLE 2.2 The More Important Primary and Secondary Minerals Found in Soils Listed in Order of Decreasing Resistance to Weathering Under Conditions Common in Humid Temperate Regions

The primary minerals are also found abundantly in igneous and metamorphic rocks. Secondary minerals are commonly found in sedimentary rocks.

Primary minerals		Secondary minerals		
		Goethite	FeOOH	Most resistant
		Hematite	Fe_2O_3	
		Gibbsite	$Al_2O_3 \cdot 3H_2O$	
Quartz	SiO_2			
		Clay minerals	Al silicates	
Muscovite	$KAl_3Si_3O_{10}(OH)_2$			
Microcline	$KAlSi_3O_8$			
Orthoclase	$KAlSi_3O_8$			
Biotite	$KAl(Mg,Fe)_3Si_3O_{10}(OH)_2$			
Albite	$NaAlSi_3O_8$			
Hornblende[a]	$Ca_2Al_2Mg_2Fe_3Si_6O_{22}(OH)_2$			
Augite[a]	$Ca_2(Al,Fe)_4(Mg,Fe)_4Si_6O_{24}$			
Anorthite	$CaAl_2Si_2O_8$			
Olivine	$(Mg,Fe)_2SiO_4$			
		Dolomite[b]	$CaCO_3 \cdot MgCO_3$	
		Calcite[b]	$CaCO_3$	
		Gypsum	$CaSO_4 \cdot 2H_2O$	Least resistant

[a] The given formula is only approximate since the mineral is so variable in composition.
[b] In semiarid grasslands dolomite and calcite are more resistant to weathering than suggested because of low rates of carbonation weathering (see Section 2.3).

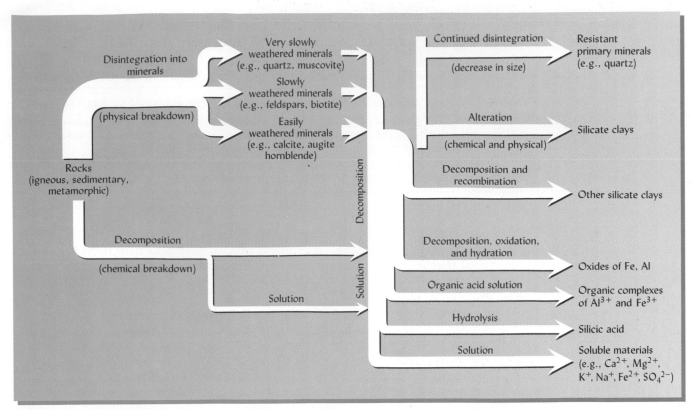

FIGURE 2.4 Pathways of weathering that occur under moderately acid conditions common in humid temperate regions. The disintegration of rocks into small individual mineral grains is a physical process, whereas decomposition, recombination, and solution are chemical processes. Alteration of minerals involves both physical and chemical processes. Note that resistant primary minerals, newly synthesized secondary minerals, and soluble materials are products of weathering. In arid regions the physical processes predominate, but in humid topical areas decomposition and recombination are most prominent.

2.2 PHYSICAL WEATHERING (DISINTEGRATION)

TEMPERATURE.   Rocks heat up during the day and cool down at night, causing alternate expansion and contraction of their constituent minerals. As some minerals expand more than others, temperature changes set up differential stresses that eventually cause the rock to crack apart.

TABLE 2.3 Partial Elemental Analysis of an Igneous Rock (a Granite) and the Mature Soil Developed from It in a Warm Humid Region (Percent of Total Composition)

Note that during soil formation there was a relative loss of phosphorus, calcium, potassium and silicon, but a relative increase in iron and aluminum. A declining ratio of silicon to aluminum is considered to be an indicator of more complete weathering.

Element[a]	In rock	In soil	Percent change
P	0.057	0.035	−38
K	3.11	0.44	−86
Ca	1.07	0.23	−79
Si	32.9	21.8	−34
Al	7.68	14.36	+87
Fe	2.50	8.54	+242
Si/Al	4.28	1.52	−64

[a] Elemental analysis is given here, though many geologists report these values in terms of the oxides of the element: for example, Al_2O_3, rather than simply Al. [Data adapted from Jackson (1964).]

Because the outer surface of a rock is often warmer or colder than the inner, more protected portions, some rocks may weather by *exfoliation*—the peeling away of outer layers (Figure 2.5). This process may be sharply accelerated if ice forms in the surface cracks. When water freezes, it expands with a force of about 1465 Mg/m², disintegrating huge rock masses (Figure 2.6, bottom) and dislodging mineral grains from smaller fragments.

ABRASION BY WATER, ICE, AND WIND.   When loaded with sediment, water has tremendous cutting power (Figure 2.6, top), as is amply demonstrated by the gorges, ravines, and valleys around the world. The rounding of riverbed rocks and beach sand grains is further evidence of the abrasion that accompanies water movement.

Windblown dust and sand also can wear down rocks by abrasion, as can be seen in the many picturesque rounded rock formations in certain arid regions. In glacial areas, huge moving ice masses embedded with soil and rock fragments grind down rocks in their path and carry away large volumes of material (see Section 2.10).

PLANTS AND ANIMALS.   Plant roots sometimes enter cracks in rocks and pry them apart, resulting in some disintegration. Burrowing animals may also help disintegrate rock somewhat. However, such influences are of little importance in producing parent material when compared to the drastic physical effects of water, ice, wind, and temperature change.

2.3   CHEMICAL WEATHERING (DECOMPOSITION)

While physical weathering is accentuated in dry, cool regions, chemical weathering is most intense where the climate is wet and hot. However, both types of weathering occur together, and each tends to accelerate the other.

Chemical weathering is enhanced by water (with its omnipresent solutes), oxygen, and the organic and inorganic acids that result from biochemical activity. These agents act in concert to convert primary minerals (e.g., feldspars and micas) to secondary minerals (e.g., silicate clays) and release plant nutrient elements in soluble forms (see Figure 2.7). Note the importance of water in each of the six types of chemical weathering reactions discussed in the following.

HYDRATION.   Intact water molecules may bind to a mineral by the process called *hydration.*

$$5Fe_2O_3 + 9H_2O \xrightarrow{\text{Hydration}} Fe_{10}O_{15} \cdot 9H_2O$$

Hematite   Water   Ferrihydrite

FIGURE 2.5   Two illustrations of rock weathering. (Left) An illustration of concentric weathering called *exfoliation*. A combination of physical and chemical processes stimulate the mechanical breakdown, which produces layers that appear much like the leaves of a cabbage. (Right) Concentric bands of light and dark colors indicate that chemical weathering (oxidation and hydration) has occurred from the outside inward, producing iron compounds that differ in color. (Right photo courtesy of R. Weil)

FIGURE 2.6 Effects of water on weathering and the breakdown of rock. (Upper) The V-like notch carved by water in this sandstone cliff in Montana is evidence of the cutting power of water laden with sediment. The cave below the notch was carved by eddies under the ancient waterfall. (Lower) The expansion of water as it freezes has broken up these Appalachian sedimentary rocks into ever smaller fragments. (Photos courtesy of R. Weil)

Hydrated oxides of iron and aluminum (e.g., $Al_2O_3 \cdot 3H_2O$) exemplify common products of hydration reactions.

HYDROLYSIS. In hydrolysis reactions, water molecules split into their hydrogen and hydroxyl components and the hydrogen often replaces a cation from the mineral structure. A simple example is the action of water on microcline, a potassium-containing feldspar.

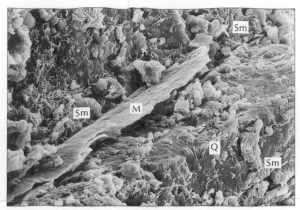

FIGURE 2.7 Scanning electron micrographs illustrating silicate clay formation from weathering of a granite rock in southern California. (*Left*) a potassium feldspar (K-spar) is surrounded by the silicate clays, smectite (Sm) and vermiculite (Vm). (*Right*) Mica (M) and quartz (Q) associated with smectite (Sm). (Courtesy of J. R. Glasmann)

$$KAlSi_3O_8 + H_2O \underset{}{\overset{Hydrolysis}{\rightleftharpoons}} HAlSi_3O_8 + K^+ + OH^-$$
(solid)   Water   (solid)   (solution)

$$2HAlSi_3O_8 + 11H_2O \underset{}{\overset{Hydrolysis}{\rightleftharpoons}} Al_2O_3 + 6H_4SiO_4$$
(solid)   Water   (solid)   (solution)

The potassium released is soluble and is subject to adsorption by soil colloids, uptake by plants, and removal in the drainage water. Likewise, the silicic acid (H_4SiO_4) is soluble. It can be removed slowly in drainage water, or it can recombine with other compounds to form secondary minerals such as the silicate clays.

DISSOLUTION.   Water is capable of dissolving many minerals by hydrating the cations and anions until they become dissociated from each other and surrounded by water molecules. Gypsum dissolving in water provides an example.

$$CaSO_4 \cdot 2H_2O + 2H_2O \underset{}{\overset{Dissolution}{\rightleftharpoons}} Ca^{2+} + SO_4^{2-} + 4H_2O$$
(solid)   Water   (solution)   Water

CARBONATION AND OTHER ACID REACTIONS.   Weathering is accelerated by the presence of acids, which increase the activity of hydrogen ions in water. For example, when carbon dioxide dissolves in water (a process enhanced by microbial and root respiration) the carbonic acid (H_2CO_3) produced hastens the chemical dissolution of calcite in limestone or marble, as illustrated when the following reactions go to the right:

$$CO_2 + H_2O \rightleftharpoons H_2CO_3$$

$$H_2CO_3 + CaCO_3 \underset{}{\overset{Carbonation}{\rightleftharpoons}} Ca^{2+} + 2HCO_3^-$$
Carbonic acid   Calcite   (solution)   (solution)
            (solid)

Soils also contain other, much stronger acids, such as nitric acid (HNO_3), sulfuric acid (H_2SO_4), and many organic acids. Hydrogen ions are also associated with soil clays. Each of these sources of acidity is available for reaction with soil minerals.

OXIDATION-REDUCTION.   Minerals that contain iron, manganese, or sulfur are especially susceptible to oxidation-reduction reactions. Iron is usually laid down in primary minerals in the divalent Fe(II) (ferrous) form. When rocks containing such minerals are exposed to air and water during soil formation, the iron is easily oxidized (loses an electron) and becomes trivalent Fe(III) (ferric). If iron is oxidized from Fe(II) to Fe(III), the change in valence and ionic radius causes destabilizing adjustments in the crystal structure of the mineral.

In other cases, Fe(II) may be released from the mineral and almost simultaneously oxidized to Fe(III). For example, the hydration of olivine releases ferrous oxide, which may be oxidized immediately to ferric oxyhydroxide (goethite).

$$3MgFeSiO_4 + 2H_2O \overset{\text{Hydrolysis}}{\rightleftharpoons} H_4Mg_3Si_2O_9 + 2SiO_2 + 3FeO$$

Olivine (solid)   Water   Serpentine (solid)   (solution)   Fe(II) oxide (solid)

$$4FeO + O_2 + 2H_2O \overset{\text{Oxidation}}{\rightleftharpoons} 4FeOOH$$

Fe(II) oxide   Goethite [Fe(III) oxide]

The oxidation and/or removal of iron during weathering is often made visible by changes in the colors of the resulting altered minerals (see Figure 2.5, right).

COMPLEXATION. Soil biological processes produce organic acids such as oxalic, citric, and tartaric acids, as well as the much larger fulvic and humic acid molecules (see Section 12.4). In addition to providing H^+ ions that help solubilize aluminum and silicon, they also form organic complexes (chelates) with the Al^{3+} ions held within the structure of silicate minerals. By so doing they remove the Al^{3+} from the mineral, which then is subject to further disintegration. In the following example, oxalic acid forms a soluble complex with aluminum from the mineral, muscovite. As this reaction proceeds to the right, it destroys the muscovite structure and releases dissolved ions of the plant nutrient, potassium.

$$K_2[Si_6Al_2]Al_4O_{20}(OH)_4 + 6C_2O_4H_2 + 8H_2O \overset{\text{Complexation}}{\rightleftharpoons} 2K^+ + 8OH^- + 6C_2O_4Al^+ + 6Si(OH)_4^0$$

Muscovite (solid)   Oxalic acid   Water   Potassium hydroxide (solution)   Complex (solution)   (solution)

Had there been no living organisms on Earth, the chemical weathering processes we have just outlined would probably have proceeded 1000 times more slowly, with the result that little, if any, soil would have developed on our planet.

INTEGRATED WEATHERING PROCESSES. The various chemical weathering processes occur simultaneously and are interdependent. For example, hydrolysis of a given primary mineral may release ferrous iron [Fe(II)] that is quickly oxidized to the ferric [Fe(III)] form, which, in turn, is hydrated to give a hydrous oxide of iron. Hydrolysis or complexation also may release soluble cations, silicic acid, and aluminum or iron compounds. In humid environments, some of the soluble cations and silicic acid are likely to be lost from the weathering mass in drainage waters. The released substances can also be recombined to form silicate clays and other secondary silicate minerals. In this manner, the biochemical processes of weathering transform primary geologic materials into the compounds of which soils are made (Figure 2.8).

2.4 FACTORS INFLUENCING SOIL FORMATION[3]

We learned in Chapter 1 that *the soil* is a collection of *individual soils*, each with distinctive profile characteristics. This concept of soils as organized natural bodies derived initially from late-19th-century field studies by a brilliant Russian team of soil scientists led by V. V. Dukochaev. They noted similar profile layering in soils hundreds of kilometers apart, provided that the climate and vegetation were similar at the two locations. Such observations and much careful subsequent field and laboratory research led to the recognition of five major factors that control the formation of soils.

---

[3] Many of our modern concepts concerning the factors of soil formation are derived from the work of Hans Jenny, an American soil scientist whose books published in 1941 and 1980 are considered classics in the field.

FIGURE 2.8 First stages of soil development: biochemical weathering of a boulder under the influence of lichens and mosses in a humid climate. These pioneering plants produce organic acids which break down minerals in the rock by hydrolysis and complexation reactions. Note the etched surface of the rock (arrow) where the mat of moss was pulled away to expose the weathering underneath. The resulting loose mineral material and soluble nutrients, in conjunction with trapped dust and the organic debris left by the mosses and lichens, will eventually provide a medium for the growth of higher plants; these in turn will further accelerate soil formation. Scale is in centimeters. (Photo courtesy of R. Weil)

1. *Parent materials* (geological or organic precursors to the soil)
2. *Climate* (primarily precipitation and temperature)
3. *Biota* (living organisms, especially native vegetation, microbes, soil animals, and human beings)
4. *Topography* (slope, aspect, and landscape position)
5. *Time* (the period of time since the parent materials became exposed to soil formation)

Soils are often defined in terms of these factors as "dynamic natural bodies having properties derived from the combined effects of climate and biotic activities, as modified by topography, acting on parent materials over periods of time."

We will now examine how each of these five factors affects the outcome of soil formation. However, as we do, we must keep in mind that these factors do not exert their influences independently. Indeed, interdependence is the rule. For example, contrasting climatic regimes are likely to be associated with contrasting types of vegetation, and perhaps differing topography and parent material, as well. Nonetheless, in certain situations one of the factors has had the dominant influence in determining differences among a set of soils. Soil scientists refer to such a set of soils as a **lithosequence, climosequence, biosequence, toposequence,** or **chronosequence.**

2.5 PARENT MATERIALS

Geological processes have brought to the earth's surface numerous parent materials in which soils form (Figure 2.9). The nature of the parent material profoundly influences soil characteristics. For example, a soil might inherit a sandy texture (see Section 4.2) from a coarse-grained, quartz-rich parent material such as granite or sandstone. Soil texture, in turn, helps control the percolation of water through the soil profile, thereby affecting the translocation of fine soil particles and plant nutrients.

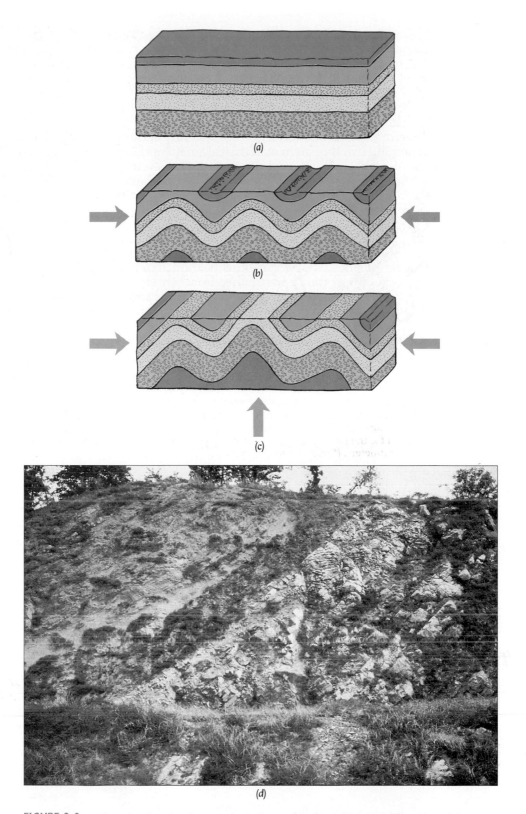

(a)

(b)

(c)

(d)

FIGURE 2.9 Diagrams showing how geological processes have brought different rock layers to the surface in a given area. (*a*) Unaltered layers of sedimentary rock with only the uppermost layer exposed. (*b*) Lateral geological pressures deform the rock layers through a process called *crustal warping*. At the same time, erosion removes much of the top layer, exposing part of the first underlying layer. (*c*) Localized upward pressures further reform the layers, thereby exposing two more underlying layers. As these four rock layers are weathered, they give rise to the parent materials on which different kinds of soils can form. (*d*) Crustal warping that lifted up the Appalachian Mountains tilted these sedimentary rock formations that were originally laid down horizontally. This deep roadcut in Virginia illustrates the abrupt change in soil parent material (lithosequence) as one walks along the ground surface at the top of this photograph. (Photo courtesy of R. Weil)

The chemical and mineralogical composition of parent material also influences both chemical weathering and the natural vegetation. For example, the presence of limestone in parent material will delay the development of acidity that typically occurs in humid climates. In addition, trees growing in limestone materials produce leaf litter that is relatively high in calcium. Incorporation of the calcium-rich litter into the soil further delays the process of acidification and, in humid temperate areas, the progress of soil development.

Parent material also influences the quantity and type of clay minerals present in the soil profile. First, the parent material itself may contain varying amounts and types of clay minerals, perhaps from a previous weathering cycle. Second, the nature of the parent material greatly influences the kinds of clays that can develop as the soil evolves (see Section 8.6). In turn, the nature of the clay minerals present markedly affects the kind of soil that develops.

Classification of Parent Materials

Inorganic parent materials can either be formed in place as residual material weathered from rock, or they can be transported from one location and deposited at another (Figure 2.10). In wet environments (such as swamps and marshes), incomplete decomposition may allow organic parent materials to accumulate from the residues of many generations of vegetation. Although it is their chemical and physical properties that most influence soil development, parent materials are often classified with regard to the mode of placement in their current location:

1. Formed in place from rock   **Residual**

2. Transported
   - By gravity   **Colluvial**
   - By water
     - Rivers   **Alluvial**
     - Oceans   **Marine**
     - Lakes   **Lacustrine**
   - By ice   **Glacial**
   - By wind   **Eolian**

3. Accumulated plant debris   **Organic**

Although these terms properly relate only to the placement of the parent materials, people sometimes refer to the soils that form from these deposits as *organic soils, glacial soils, alluvial soils,* and so forth. These terms are quite nonspecific because parent material properties vary widely within each group, and because the effect of parent material is modified by the influence of climate, organisms, topography, and time.

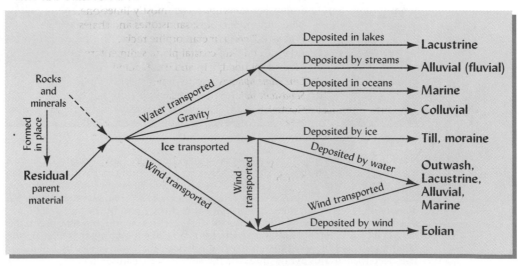

FIGURE 2.10 How various kinds of parent material are formed, transported, and deposited.

Residual parent material develops in place from weathering of the underlying rock. In stable landscapes it may have experienced long and possibly intense weathering. Where the climate is warm and humid, residual parent materials are typically thoroughly leached and oxidized, and show the red and yellow colors of various oxidized iron compounds (see Plates 9, 11, and 15). In cooler and especially drier climates, the color and chemical composition of residual parent material tends to resemble more closely the rock from which it formed.

Residual materials are widely distributed on all continents. The physiographic map of the United States (Figure 2.11) shows nine great provinces where residual materials are prominent (italicized in the map key).

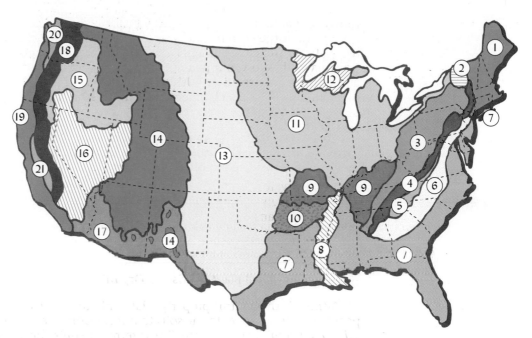

FIGURE 2.11 Generalized physiographic and regolith map of the United States. The regions are as follows (major residual areas italicized).

1. New England: mostly glaciated metamorphic rocks.
2. Adirondacks: glaciated metamorphic and sedimentary rocks.
3. *Appalachian Mountains and plateaus:* shales and sandstones.
4. *Limestone valleys and ridges:* mostly limestone.
5. Blue Ridge mountains: sandstones and shales.
6. *Piedmont Plateau:* metamorphic rocks.
7. Atlantic and Gulf coastal plain: sedimentary rocks with sands, clays, and limestones.
8. Mississippi floodplain and delta: alluvium.
9. *Limestone uplands:* mostly limestone and shale.
10. *Sandstone uplands:* mostly sandstone and shale.
11. Central lowlands: mostly glaciated sedimentary rocks with till and loess, a wind deposit of great agricultural importance (see Figure 2.17).
12. Superior uplands: glaciated metamorphic and sedimentary rocks.
13. *Great Plains region:* sedimentary rocks.
14. *Rocky Mountain region:* sedimentary, metamorphic, and igneous rocks.
15. Northwest intermountain: mostly igneous rocks; loess in Columbia and Snake river basins (see Figure 2.21).
16. Great Basin: gravels, sands, alluvial fans from various rocks; igneous and sedimentary rocks.
17. Southwest arid region: gravel, sand, and other debris of desert and mountain.
18. *Sierra Nevada and Cascade mountains:* igneous and volcanic rocks.
19. *Pacific Coast province:* mostly sedimentary rocks.
20. Puget Sound lowlands: glaciated sedimentary.
21. California central valley: alluvium and outwash.

A great variety of soils occupy the regions covered by residual debris because of the marked differences in the nature of the rocks from which these materials evolved. The varied soils are also a reflection of wide differences in other soil-forming factors, such as climate and vegetation (Sections 2.13 and 2.14).

2.7 COLLUVIAL DEBRIS

Colluvial debris, or colluvium, is made up of poorly sorted rock fragments detached from the heights above and carried downslope, mostly by gravity, assisted in some cases by frost action. Rock fragment (talus) slopes, cliff rock debris (detritus), and similar heterogeneous materials are good examples. Avalanches are made up largely of such accumulations.

Colluvial parent materials are frequently coarse and stony because physical rather than chemical weathering has been dominant. Stones, gravel, and fine materials are interspersed (not layered), and the coarse fragments are rather angular (Figure 2.12). Packing voids, spaces created when tumbling rocks come to rest against each other (sometimes at precarious angles), help account for the easy drainage of many colluvial deposits and also for their tendency to be unstable and prone to slumping and landslides, especially if disturbed by excavations.

2.8 ALLUVIAL STREAM DEPOSITS

There are three general classes of alluvial deposits: *floodplains, alluvial fans,* and *deltas.* They will be considered in order.

Floodplains

That part of a river valley that is inundated during floods is a floodplain. Sediment carried by the swollen stream is deposited during the flood, with the coarser materials being laid down near the river channel where the water is deeper and flowing with more turbulence and energy. Finer materials settle out in the calmer flood waters farther from the channel. Each major flooding episode lays down a distinctive layer of sediment, creating the stratification that characterizes alluvial soils (Figure 2.13).

FIGURE 2.12 A productive soil formed in colluvial parent material in the Appalachian mountains of the eastern United States. The ridge in the background was at one time much taller, but material from the crest tumbled downslope and came to rest in the configuration seen in the soil profile. Note the unstratified mix of particle sizes and the rather angular nature of the coarse fragments. (Photo courtesy of R. Weil)

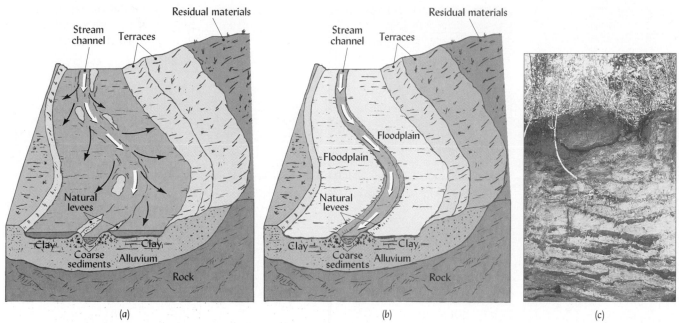

FIGURE 2.13 Illustration of floodplain development. (*a*) A stream is at flood stage, has overflowed its banks, and is depositing sediment in the floodplain. The coarser particles are being deposited near the stream channel where water is moving most rapidly, while the finer clay particles are being deposited where water movement is slower. (*b*) After the flood the sediments are in place and vegetation is growing. (*c*) Contrasting layers of sand, silt, and clay characterize the alluvial floodplain. Each layer resulted from separate flooding episodes. (Redrawn from *Physical Geology*, 2d ed., by F. R. Flint and B. J. Skinner, copyright © 1977 John Wiley & Sons, Inc. Reprinted by permission of John Wiley & Sons, Inc.) (Photo courtesy of R. Weil)

If, over a period of time, there is a change in grade, a stream may cut down through its already well-formed alluvial deposits. This cutting action leaves **terraces** above the floodplain on one or both sides. Some river valleys feature two or more terraces at different elevations, each reflecting a past period of alluvial deposition and stream cutting.

Major areas of alluvial parent materials are found along the Nile River in Egypt and the Sudan; the Euphrates, Ganges, Indus, Brahmaputra, and Hwang Ho river valleys of Asia; and the Amazon River of Brazil. The floodplain along the Mississippi River is the largest in the United States (Figure 2.14), varying from 30 to 125 km in width. Floodplains of smaller streams also provide parent materials for locally important soil areas.

To some degree, nutrient-rich materials lost by upland soils are deposited on the river floodplain and delta (see following). Soils derived from alluvial sediments generally have characteristics seen as desirable for human settlement and agriculture. These characteristics include nearly level topography, proximity to water, high fertility, and high productivity. Although many alluvial soils are well drained, in some cases artificial drainage may be necessary for upland crops as well as for stable building foundations.

While alluvial soils are often uniquely suited to forestry and crop production, their use for home sites and urban development should generally be avoided. Unfortunately, the desirable properties of many alluvial soils have already led civilizations to found cities and towns on floodplains. As the great Mississippi floods of 1993 and the Ohio River and Red River floods of 1997 have illustrated, building on a floodplain, no matter how great the investment in flood-control measures, all too often leads to tragic loss of life and property during serious flooding.

In many areas, installation of systems for drainage and flood protection has proven costly and ineffective. Farmers and the general public pay high costs to keep such areas in agricultural or urban uses. Steps are therefore being taken to reestablish the wetland conditions of certain flood-prone agricultural areas that originally were natural wetlands. These and other alluvial soils can provide natural habitats, such as bottomland forests, which are very productive of timber and support a high diversity of birds and other wildlife.

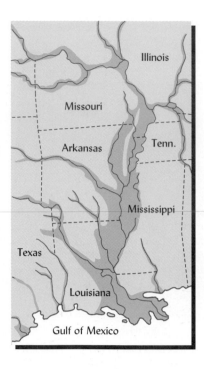

FIGURE 2.14 Floodplain and delta of the Mississippi River. This is the largest continuous area of alluvial soil in the United States.

Alluvial Fans

Streams that leave a narrow valley in an upland area and suddenly descend to a much broader valley below deposit sediment in the shape of a fan, as the water spreads out and slows down (see Figures 2.15 and 19.13). The rushing water tends to sort the sediment particles by size, first dropping the gravel and coarse sand, then depositing the finer materials toward the bottom of the alluvial fan.

Alluvial fan debris is found in widely scattered areas in mountainous and hilly regions. The soils derived from this debris often prove very productive, although they may be quite coarse-textured. The Sacramento Valley in California and the Willamette Valley in Oregon are examples of large, agriculturally important areas with alluvial fan materials.

Delta Deposits

Much of the finer sediment carried by streams is not deposited in the floodplain but is discharged into the lake, reservoir, or ocean into which the streams flow. Some of the suspended material settles near the mouth of the river, forming a delta. Such delta deposits are by no means universal, being found at the mouths of only a few rivers of

FIGURE 2.15 Characteristically shaped alluvial fans alongside a river valley in Alaska. Although the areas are small and sloping, they can develop into well-drained soils. (Courtesy U.S. Geological Survey)

the world. A delta often is a continuation of a floodplain (its front, so to speak). It is clayey in nature and is likely to be poorly drained as well.

Delta marshes are among the most extensive and biologically important of wetland habitats. Many of these habitats are today being protected or restored, but civilizations both ancient and modern have also developed important agricultural areas (often specializing in the production of rice) by creating drainage and flood-control systems on the deltas of such rivers as the Amazon, Euphrates, Ganges, Hwang Ho, Mississippi, Nile, Po, and Tigris.

2.9 MARINE SEDIMENTS

Streams eventually deposit much of their sediment loads in oceans, estuaries, and gulfs. The coarser fragments settle out near the shore and the finer particles at a distance (Figure 2.16). Over long periods of time, these underwater sediments build up, in some

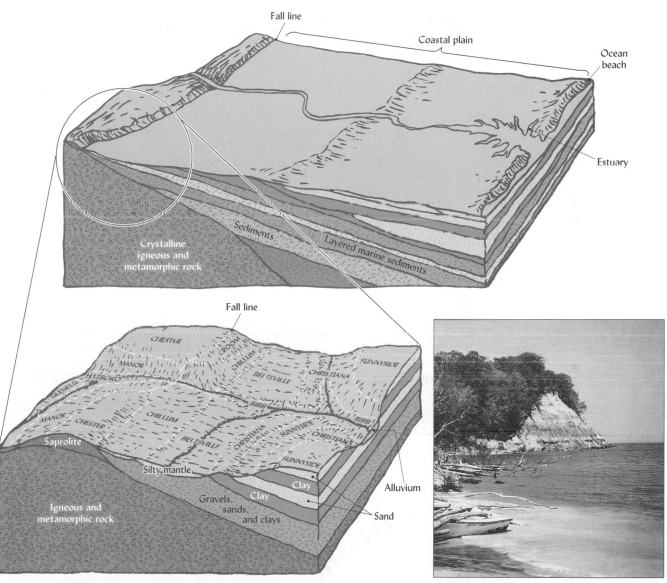

FIGURE 2.16 Diagrams showing sediments laid down by marine waters adjacent to coastal residual igneous and metamorphic rocks. Note that the marine sediments are alternate layers of fine clay and coarse-textured sands and gravels. The photo in the lower right shows such layering on coastal marine sediment. The soils developing from the marine and residual parent materials are also shown (names in capital letters). The river had cut through the marine sediments and has deposited some alluvial material in its basin. This diagram and photo illustrate the relationship between marine sediments and residual materials in the Coastal Plain of southeastern United States. [Photo courtesy of R. Weil; soils diagram from Miller (1976)]

cases becoming hundreds of meters thick. Changes in the relative elevations of sea and land may later raise these marine deposits above sea level, creating a coastal plain. The deposits are then subject to a new cycle of weathering and soil formation.

A coastal plain usually has only moderate slopes, being more level in the low-lying parts nearer the coastline and more hilly farther inland, where streams and rivers flowing down the steeper grades have more deeply dissected the landscape. The land surface in the lower coastal portion may be only slightly above the water table during at least part of the year, so wetland forests and marshes often characterize such parent materials.

Marine deposits are quite variable in texture. Some are sandy, as is the case in much of the Atlantic seaboard coastal plain. Others are high in clay, as are deposits found in the Atlantic and Gulf coastal flatwoods and in the interior pinelands of Alabama and Mississippi. Where streams have cut down through layers of marine sediments (as in the detailed block diagram in Figure 2.16), clays, silts, and sand may be encountered side by side. All of these sediments came from the erosion of upland areas, some of which were highly weathered before the transport took place. However, the marine sediments generally have been subjected to soil-forming processes for a shorter period of time than their upland counterparts. As a consequence, the properties of the soils that form are heavily influenced by those of the marine parent materials. Many marine sediments are high in sulfur and go through a period of acid-forming sulfur oxidation at some stage of soil formation (see Section 13.21 and Plate 24, before page 499).

2.10 PARENT MATERIALS TRANSPORTED BY GLACIAL ICE AND MELTWATERS

During the Pleistocene epoch (about 10^4 to 10^7 years ago), up to 20% of the world's land surface—northern North America, northern and central Europe, and parts of northern Asia—was invaded by a succession of great ice sheets, some more than 1 km thick. Present-day glaciers in polar regions and high mountains cover about a third as much area, but are not nearly so thick as the glaciers of the Great Pleistocene Ice Age. Even so, if all present-day glaciers were to melt, the world sea level would rise by about 65 m. Some scientists predict that if the current global warming trend continues, these present-day glaciers could partially melt, causing an increase in sea level and flooding many coastal areas around the world.

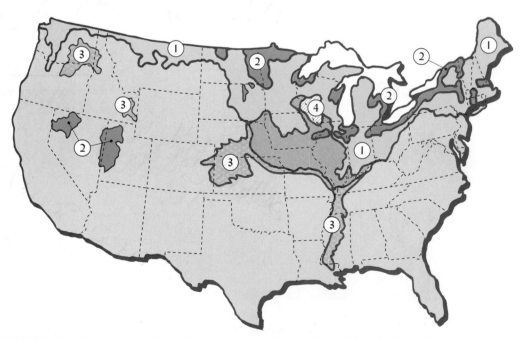

FIGURE 2.17 Areas in the United States covered by the continental ice sheet and the deposits either directly from, or associated with, the glacial ice. (1) Till deposits of various kinds; (2) glacial-lacustrine deposits; (3) the loessial blanket (note that the loess overlies great areas of till in the Midwest); (4) an area, mostly in Wisconsin, that escaped glaciation and is partially loess covered.

In North America, Pleistocene-epoch glaciers covered most of what is now Canada, southern Alaska, and the northern part of the contiguous United States. The southernmost extension went down the Mississippi Valley, where the least resistance was met because of the lower and smoother topography (Figure 2.17).

Europe and central North America apparently sustained several distinct ice invasions over a period of 1 to 1.5 million years. These invasions were separated by long, interglacial, ice-free intervals of warm or even semitropical climates. We now may be enjoying the mildness of another interglacial period.

As the glacial ice pushed forward, the existing regolith with much of its mantle of soil was swept away, hills were rounded, valleys were filled, and, in some cases, the underlying rocks were severely ground and gouged. Thus, the glacier became filled with rock and all kinds of unconsolidated materials, carrying great masses of these materials as it pushed ahead (Figure 2.18). Finally, as the ice melted and the glacier retreated, a

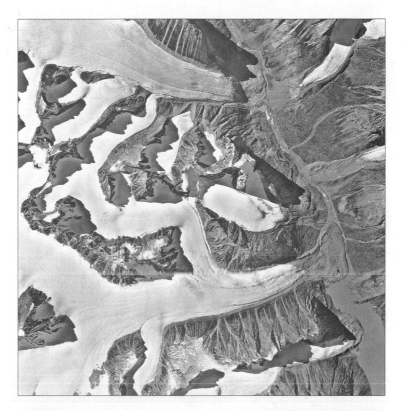

FIGURE 2.18 (Upper) Tongues of a modern-day glacier in Canada. Note the evidence of transport of materials by the ice and the "glowing" appearance of the two major ice lobes. (Lower) This U-shaped valley in the Rocky Mountains illustrates the work of glaciers in carving out land forms. The glacier left the valley floor covered with glacial till. Some of the material gouged out by the glacier was deposited many miles down the valley. (Upper photo A-16817-102 courtesy National Air Photo Library, Surveys and Mapping Branch, Canadian Department of Energy, Mines, and Resources; Lower photo courtesy of R. Weil)

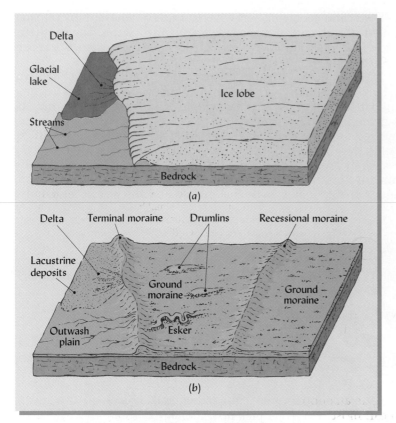

FIGURE 2.19 Illustration of how several glacial materials were deposited. (*a*) A glacier ice lobe moving to the left, feeding water and sediments into a glacial lake and streams, and building up glacial till near its front. (*b*) After the ice retreats, terminal, ground, and recessional moraines are uncovered along with cigar-shaped hills (drumlins), the beds of rivers that flowed under the glacier (eskers), and lacustrine, delta, outwash deposits.

mantle of glacial debris or drift remained. This provided a new regolith and fresh parent material for soil formation.

Glacial Till and Associated Deposits

The name **drift** is applied to all material of glacial origin, whether deposited by the ice or by associated waters. The materials deposited directly by the ice, called **glacial till**, are heterogeneous (unstratified) mixtures of debris, which vary in size from boulders to clay. Glacial till may therefore be somewhat similar in appearance to colluvial materials, except that the coarse fragments are more rounded from their grinding journey in the ice, and the deposits are often much more densely compacted because of the great weight of the overlying ice sheets. Much glacial till is deposited in irregular ridges called **moraines.** Figure 2.19 shows how glacial sheets deposited several types of soil parent materials.

Glacial Outwash and Lacustrine Sediments

The torrents of water gushing forth from melting glaciers carried vast loads of sediment. In valleys and on plains, where the glacial waters were able to flow away freely, the sediment formed an **outwash plain** (Figure 2.19). Such sediments, with sands and gravels sorted by flowing water, are common **valley fills** in parts of the United States. Figure 2.20 shows the sorted layering of coarse and fine materials in glacial outwash overlaid by mixed materials of glacial till.

When the ice front came to a standstill where there was no ready escape for the water, ponding began; ultimately, very large lakes were formed (Figure 2.19). Particularly prominent in North America were those south of the Great Lakes and in the Red River Valley of Minnesota and Manitoba (see Figure 2.17). The latter lake, called Glacial Lake Agassiz, was about 1200 km long and 400 km wide at its maximum extension.

The **lacustrine deposits** formed in these glacial lakes range from coarse delta materials and beach deposits near the shore to larger areas of fine silts and clay deposited from

FIGURE 2.20 The stratification of different-size particles can be seen in the glacial outwash in the lower part of this soil profile in North Dakota. The outwash is overlain by a layer of glacial till containing a random assortment of particles, ranging in size from small boulders to clays. Note the rounded edges of the rocks, evidence of the churning action within the glacier. (Photo courtesy of R. Weil)

the deeper, more still waters at the center of the lake. Areas of inherently fertile (though not always well-drained) soils developed from these materials as the lakes dried.

2.11 PARENT MATERIALS TRANSPORTED BY WIND

Wind is capable of picking up an enormous quantity of material at one site and depositing it at another. Wind can most effectively pick up material from soil or regolith that is loose, dry, and unprotected by vegetation. Dry, barren landscapes have served, and continue to serve, as sources of parent material for soils forming as far away as the opposite side of the globe. The smaller the particles, the higher and farther the wind will carry them. Wind-transported (**eolian**) materials important as parent material for soil formation include, from largest to smallest particle size: **dune sand, loess** (pronounced "luss"), and **aerosolic dust.** Windblown **volcanic ash** from erupting volcanoes is a special case that is also worthy of mention.

Dune Sand

Along the beaches of the world's oceans and large lakes, and over vast barren deserts, strong winds pick up medium and fine sand grains and pile them into hills of sand called *dunes*. The dunes, ranging up to 100 m in height, may continue to slowly shift their locations in response to the prevailing winds. Because most other minerals have been broken down and carried away by the waves, beach sand usually consists mainly of quartz, which is devoid of plant nutrients and highly resistant to weathering action. Nonetheless, over time dune grasses and other pioneering vegetation may take root and soil formation may begin. The sandy soils that extend for many kilometers east of Lake Michigan provide an example of this process. Some of the very deep sandy soils on the Atlantic coastal plain are thought to have formed on dunes marking the location of an ancient beach.

Desert sands, too, are usually dominated by quartz, but they may also include substantial amounts of other minerals that could contribute more to the establishment of vegetation and the formation of soils, should sufficient rainfall occur. The pure-white dunes of sand-sized gypsum at White Sands, New Mexico are a dramatic example of weatherable minerals in desert sands.

Loess

The windblown materials called *loess* are composed primarily of silt with some very fine sand and clay. They cover wide areas in the central United States, eastern Europe, Argentina, and central Asia (Figure 2.21*b*). Loess may be blown for hundreds of kilometers. The deposits farthest from the source are thinnest and consist of the finest particles.

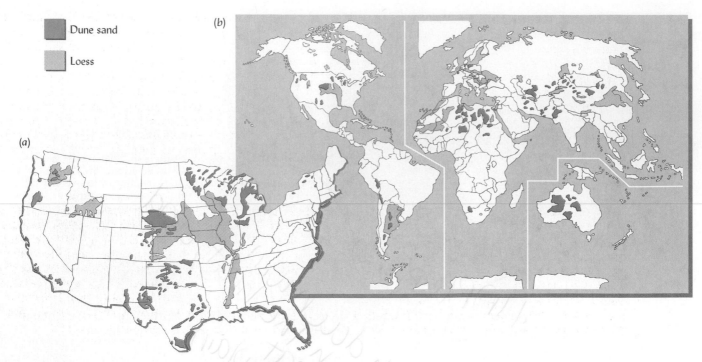

Dune sand

Loess

(a)

(b)

FIGURE 2.21 (*a*) Approximate distribution of loess and dune sand in the United States. The soils that have developed from loess are generally silt loams, often quite high in fine sands. Note especially the extension of the central loess deposit down the eastern side of the Mississippi River and the smaller areas of loess in Washington, Oregon, and Idaho. The most prominent areas of dune sands are the Sand Hills of Nebraska and the dunes along the eastern shore of Lake Michigan. (*b*) Other major eolian deposits of the world include the loess deposits in Argentina, eastern Europe, and northern China, and the large areas of dune sands in north Africa and Australia.

In the United States (Figure 2.21*a*), the main sources of loess were the great barren expanses of till and outwash left in the Missouri and Mississippi River valleys by the retreating glaciers of the last Ice Age. During the winter months, winds picked up fine materials and moved them southward, covering the existing soils and parent materials with a blanket of loess that accumulated to as much as 8 m thick.

In central and western China, loess deposits reaching 30 to 100 m in depth cover some 800,000 km² (Figure 2.22). These materials have been windblown from the deserts of central Asia and are generally not associated directly with glaciers. These and other loess deposits tend to form silty soils of rather high fertility and potential productivity.

Aerosolic Dust

Very fine particles (about 1 to 10 μm) carried high into the air may travel for thousands of kilometers before being deposited, usually with rainfall. These fine particles are called *aerosolic dust* because they can remain suspended in air, due to their very small size. While this dust has not blanketed the receiving landscapes as thickly as is typical for loess, it does accumulate at rates that make significant contributions to soil formation. Much of the calcium carbonate in soils of the western United States probably originated as windblown dust. Recent studies have shown that dust, originating in the Sahara Desert of northern Africa and transported over the Atlantic Ocean in the upper atmosphere, is the source of much of the calcium and other nutrients found in the highly leached soils of the Amazon basin in South America.

Volcanic Ash

During volcanic eruptions cinders fall in the immediate vicinity of the volcano, while fine, often glassy, ash particles may blanket extensive areas downwind. Soils developed from volcanic ash are most prominent within a few hundred kilometers of the volcanoes that ring the Pacific Ocean. Important areas of volcanic ash parent materials occur in

FIGURE 2.22 Villagers carve houses out of thick loess deposits in Xian, China. The loess consists mainly of silty materials bound together with a small amount of clay. The clay binder helps stabilize the loess when excavated, but only if the material is protected from rain, as in these vertical walls. Sloping excavations of this material would quickly slump and wash away when saturated by rain. Vertical road cuts are therefore a common feature of loessal landscapes around the world. (Photos courtesy of Raymond Miller, University of Maryland)

Japan, Indonesia, New Zealand, western United States (in Hawaii, Montana, Oregon, Washington, and Idaho), Mexico, and Chile. The soils formed are uniquely light and porous, and tend to accumulate organic matter more rapidly than other soils in the area.

2.12 ORGANIC DEPOSITS

Organic material accumulates in wet places where plant growth exceeds the rate of residue decomposition. In such areas residues accumulate over the centuries from wetland plants such as pondweeds, cattails, sedges, reeds, mosses, shrubs, and certain trees. These residues sink into the water, where their decomposition is limited by lack of oxygen. As a result, organic deposits often accumulate up to several meters in depth (Figure 2.23). Collectively, these organic deposits are called **peat.**

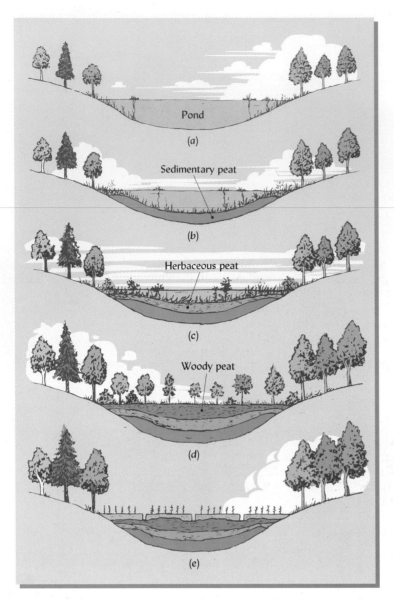

FIGURE 2.23 Stages in the development of a typical woody peat bog and the area after clearing and draining. (*a*) Nutrient runoff from the surrounding uplands encourages aquatic plant growth, especially around the pond edges. (*b, c*) Organic debris fills in the bottom of the pond. (*d*) Eventually trees cover the entire area. (*e*) If the land is cleared and a drainage system installed, the area becomes a most productive muck soil.

Distribution and Accumulation of Peats

Peat deposits are found all over the world, but most extensively in the cool climates and in areas that have been glaciated. About 75% of the 340 million hectares of peat lands in the world are found in Canada and northern Russia. The United States is a distant third, with about 20 million hectares.

The rate of peat accumulation varies from one area to another. In the warm climate of Florida, peat deposits of the everglades accumulated for 3000 to 4000 years at the rate of about 0.8 mm/yr. In Wisconsin, accumulation rates have been estimated to be 0.2 to 0.4 mm/yr.

Types of Peat Materials

Based on the nature of the parent materials, four kinds of peat are recognized:

1. Moss peat, the remains of mosses such as sphagnum
2. Herbaceous peat, residues of herbaceous plants such as sedges, reeds, and cattails
3. Woody peat, from the remains of woody plants, including trees and shrubs
4. Sedimentary peat, remains of aquatic plants (e.g., algae) and of fecal material of aquatic animals

---

Organic deposits generally contain two or more of these kinds of peats. Alternating layers of different peats are common, as are mixtures of the peats. Because the succession of plants, as the residues accumulate, tends to favor trees (see Figure 2.23), woody peats often dominate the surface layers of organic materials.

In cases where a wetland area has been drained, woody peats tend to make very productive agricultural soils that are especially well suited for vegetable production. While moss peats have high water-holding capacities, they tend to be quite acid. Sedimentary peat is generally undesirable as an agricultural soil. This material is highly colloidal and compact and is rubbery when wet. Upon drying, it resists rewetting and remains in a hard, lumpy condition. Fortunately, it occurs mostly deep in the profile and is unnoticed unless it interferes with drainage of the bog area.

The organic material is called **peat** or **fibric** if the residues are sufficiently intact to permit the plant fibers to be identified. If most of the material has decomposed sufficiently so that little fiber remains, the term **muck** or **sapric** is used. In mucky peats (**hemic** materials) some of the plant fibers can be recognized and some cannot.

WETLAND PRESERVATION. Wetland areas are important natural habitats for wildlife, and their drainage reduces the area of such habitats. While organic material is the foundation for some very productive agricultural soils, environmentalists argue that such use is unsustainable because, once drained, the organic deposits will decompose and disappear after a century or so; therefore, these areas might be better left in (or returned to) their natural state (see Section 7.8).

Recognizing that the effects of **parent materials** on soil properties are modified by the combined influences of **climate**, **biotic activities**, **topography**, and **time**, we will now turn to these other four factors of soil formation, starting with climate.

2.13   CLIMATE

Climate is perhaps the most influential of the four factors acting on parent material because it determines the nature and intensity of the weathering that occurs over large geographic areas. The principal climatic variables influencing soil formation are *effective precipitation* (see Box 2.1) and *temperature,* both of which affect the rates of chemical, physical, and biological processes.

Effective Precipitation

We have already seen (Section 2.3) that water is essential for all the major chemical weathering reactions. To be effective in soil formation, water must penetrate into the regolith. The greater the depth of water penetration, the greater the depth of weathering soil and development. Surplus water percolating through the soil profile transports soluble and suspended materials from the upper to the lower layers. It may also carry away soluble materials in the drainage waters. Thus, percolating water stimulates weathering reactions and helps differentiate soil horizons.

Likewise, a deficiency of water is a major factor in determining the characteristics of soils of dry regions. Soluble salts are not leached from these soils, and in some cases they build up to levels that curtail plant growth. Soil profiles in arid and semiarid regions are also apt to accumulate carbonates and certain types of cracking clays.

Temperature

For every 10°C rise in temperature, the rates of biochemical reactions more than double. Temperature and moisture both influence the organic matter content of soil through their effects on the balance between plant growth and microbial decomposition (see Figure 12.15). If warm temperatures and abundant water are present in the profile at the same time, the processes of weathering, leaching, and plant growth will be maximized. The very modest profile development characteristic of cold areas contrasts sharply with the deeply weathered profiles of the humid tropics (Figure 2.24).

Climate also influences the natural vegetation. Humid climates favor the growth of trees (Figure 2.25). In contrast, grasses are the dominant native vegetation in subhumid and semiarid regions, while shrubs and brush of various kinds dominate in arid areas. Thus, climate also exerts its influence through a second soil-forming factor, the living organisms.

BOX 2.1 EFFECTIVE PRECIPITATION FOR SOIL FORMATION

Water from rain and melting snow is a primary requisite for parent material weathering and soil development. To fully promote soil development, water must not only enter the profile and participate in weathering reactions; it must also percolate through the profile and translocate soluble weathering products.

Let's consider a site that receives an average of 600 mm of rainfall per year. The amount of water leaching through a soil is determined not only by the total annual precipitation but by at least four other factors, as well.

a. **Seasonal distribution** of precipitation. The 600 mm of rainfall distributed evenly throughout the year, with about 50 mm each month, is likely to cause less soil leaching or erosion than the same annual amount of rain falling at the rate of 100 mm per month during a six-month rainy season.

b. **Temperature and evaporation.** In a hot climate, evaporation from soils and vegetation is much higher than in a cool climate. Therefore, in the hot climate, much less of the 600 mm will be available for percolation and leaching. Most or all will evaporate soon after it falls on the land. Thus, 600 mm of rain may cause more leaching and profile development in a cool climate than in a warmer one. Similar reasoning would suggest that rainfall concentrated during a mild winter (as in California) may be more effective in leaching the soil than the same amount of rain concentrated in a hot summer (as in the Great Plains).

c. **Topography.** Water falling on a steep slope will run downhill so rapidly that only a small portion will enter the soil where it falls. Therefore, even though they receive the same rainfall, a level or concave site will experience more percolation and leaching than a steeply sloping site. The effective rainfall can be said to be greater on the level site than on the sloping one. The concave site will receive the greatest effective rainfall because, in addition to direct rainfall, it will collect the runoff from the adjacent sloping site, as well.

d. **Permeability.** Even if all three of the above conditions are the same, more rain water will infiltrate and leach through a coarse, sandy profile than a tight, clayey one. Therefore the sandy profile can be said to experience a greater effective precipitation, and more rapid soil development may be expected.

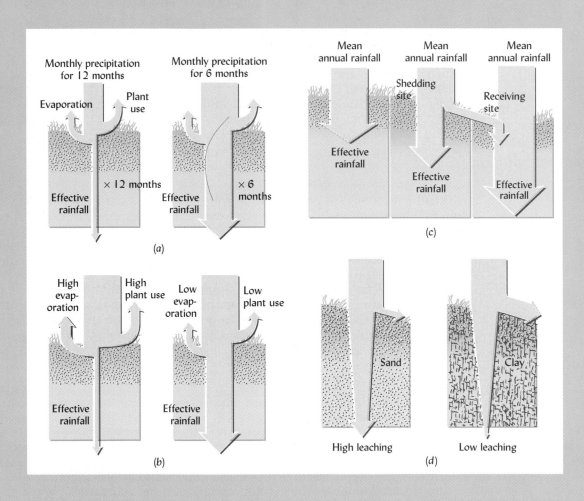

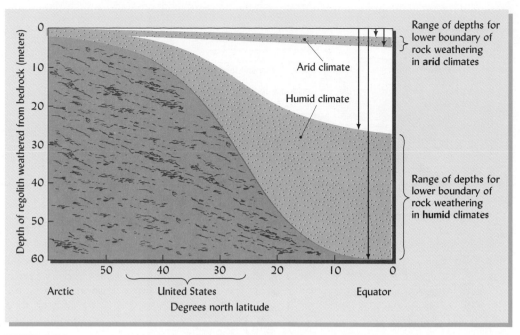

FIGURE 2.24   A generalized illustration of the effects of two climatic variables, temperature and precipitation, on the depths of regolith weathered in bedrock. The stippled areas represent the range of depths to which the regolith typically extends. In cold climates (arctic regions), the regolith is shallow under both humid and arid conditions. In the warmer climates of the lower latitudes, the depth of the residual regolith increases sharply in humid regions, but is little affected in arid regions. Under humid tropical conditions, the regolith may be 50 or more meters in depth. The vertical arrows represent depths of weathering near the Equator. Remember that soil depth may not be as great as regolith depth.

Considering soils with similar temperature regime, parent material, topography, and age, increasing effective annual precipitation generally leads to increasing clay and organic matter contents, higher acidity, and lower ratio of Si/Al (an indication of more highly weathered minerals; see Table 2.3). However, many places have experienced climates in past geologic epochs that were not at all similar to the climate evident today. This fact is illustrated in certain old landscapes in arid regions, where highly leached and weathered soils stand as relics of the humid tropical climate that prevailed there many thousands of years ago.

2.14   BIOTA: LIVING ORGANISMS

Organic matter accumulation, biochemical weathering, profile mixing, nutrient cycling, and aggregate stability are all enhanced by the activities of organisms in the soil. Vegetative cover reduces natural soil erosion rates, thereby slowing down the rate of mineral surface soil removal. Organic acids produced from certain types of plant leaf litter bring iron and aluminum into solution by complexation and accelerate the downward movement of these metals and their accumulation in the B horizon.

Role of Natural Vegetation

A-HORIZON DEVELOPMENT.   The effect of vegetation on soil formation can be seen by comparing properties of soils formed under grassland and forest vegetation near the boundary between these two ecosystems (Figure 2.26). In the grassland, much of the organic matter added to the soil is from the deep fibrous grass root systems. By contrast, tree leaves falling on the forest floor are the principal source of soil organic matter in the forest. As a result, the soils under grassland generally have a thicker A horizon with a deeper distribution of organic matter than in comparable soils under forests, which

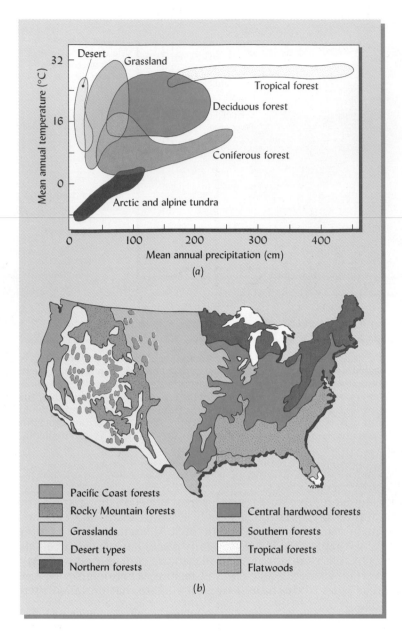

FIGURE 2.25 The effect of climate on vegetation. (*a*) General relationship among mean annual temperatures and precipitation and types of vegetation. (*b*) General types of vegetation in the United States. [(*a*) From NSF (1975); (*b*) redrawn from a more detailed map of U.S. Geological Survey]

characteristically store most of their organic matter in the forest floor and a thin A horizon. The light-colored, leached E horizon typically found under the O or A horizon of a forested soil is generally not found in a grassland soil. Structural stability of the mineral soil aggregates tends to be greatest in the grassland soils.

CATION CYCLING BY TREES. The ability of natural vegetation to take up mineral elements strongly influences the characteristics of the soils that develop, especially their acidity. Litter falling from coniferous trees (e.g., pines and firs) will recycle only small quantities of calcium, magnesium, and potassium compared to those recycled by litter from some deciduous trees (e.g., beech, oaks, and maples) that take up and store much larger amounts of such metallic cations (Figure 2.27). Because conifer tree roots allow more of these base-forming cations to be lost by leaching, soil acidity is more strongly developed under most coniferous vegetation than under most deciduous trees. The acidic, resinous needles from coniferous trees tend to accumulate in a thick O horizon with distinct layers of fibric (undecomposed) and sapric (highly decomposed) material. The more readily broken down leaves of deciduous trees generally form a thinner forest floor with less distinct layers.

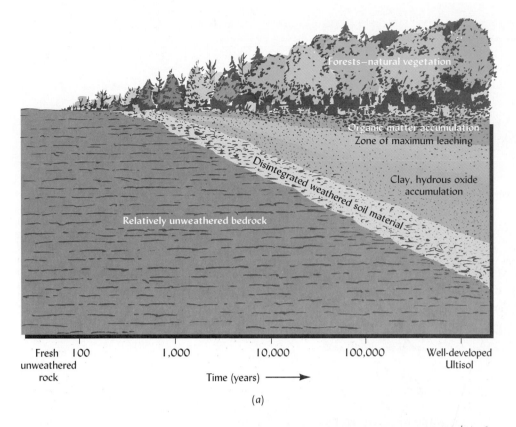

Forests—natural vegetation

Organic matter accumulation

Zone of maximum leaching

Disintegrated weathered soil material

Clay, hydrous oxide accumulation

Relatively unweathered bedrock

Fresh 100   1,000   10,000   100,000   Well-developed
unweathered   Ultisol
rock

Time (years) ⟶

(a)

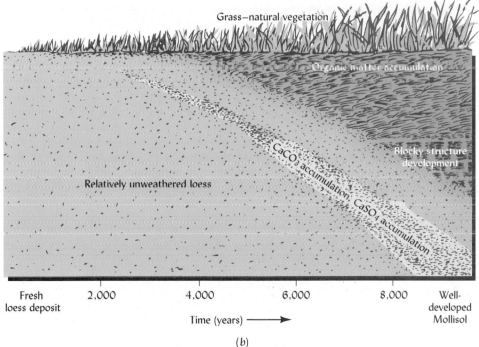

Grass—natural vegetation

Organic matter accumulation

CaCO₃ accumulation, CaSO₄ accumulation

Blocky structure development

Relatively unweathered loess

Fresh   2,000   4,000   6,000   8,000   Well-
loess deposit   developed
                                                                         Mollisol

Time (years) ⟶

(b)

FIGURE 2.26 How two soil profiles may have developed in climatic areas that encouraged as natural vegetation either (a) forests or (b) grasslands. Organic matter accumulation in the upper horizons occurs in time, the amount and distribution depending on the type of natural vegetation present. Clay and iron oxide accumulate and characteristic structures develop in the lower horizons. The end products differ markedly from the soil materials from which they formed. Note that the time scales for soil development differ markedly for the two parent materials.

HETEROGENEITY IN RANGELANDS. In arid and semiarid rangelands, competition for limited soil water does not permit vegetation dense enough to completely cover the soil surface. Scattered shrubs or bunch grasses are interspersed with openings in the plant canopy where the soil is bare or partially covered with plant litter. The widely scattered vegetation alters soil properties in several ways. Plant canopies trap windblown dust that is often relatively rich in silt and clay. Roots scavenge nutrients such as nitrogen, phosphorus, potassium and sulfur from the interplant areas. These nutrients are then deposited with the leaf litter under the plant canopies. The decaying litter adds organic

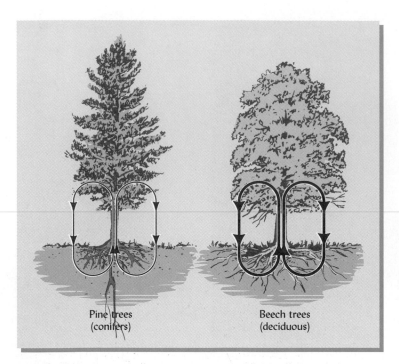

FIGURE 2.27 Nutrient cycling is an important factor in determining the relationship between vegetation and the soil that develops. If residues from the vegetation are low in bases, as is the case for most conifers such as pine species (left), acid weathering conditions are favored. High base-containing plant residues, such as many deciduous trees including the European beach (right), tend to neutralize acids, thereby favoring only slightly acid to neutral weathering conditions.

Pine trees (conifers)

Beech trees (deciduous)

acids, which lower the soil pH and stimulate mineral weathering. As time goes on, the relatively bare soil areas between plants decline in fertility and may increase in size as they become impoverished and even less inviting for the establishment of plants. Simultaneously, the vegetation creates "islands" of enhanced fertility, thicker A horizons, and often more deeply leached calcium carbonate (see Figure 2.28).

Role of Animals

The role of animals in soil-formation processes must not be overlooked. For example, large animals such as gophers, moles, and prairie dogs bore into the lower soil horizons, bringing materials to the surface. Their tunnels are often open to the surface, encouraging movement of water and air into the subsurface layers. In localized areas, they enhance mixing of the lower and upper horizons by creating, and later refilling, underground tunnels. For example, dense populations of prairie dogs may completely turn over the upper meter of soil in the course of several thousand years. Old animal burrows in the lower horizons often become filled with soil material from the overlying A horizon, creating profile features known as *crotovinas* (Figure 2.29).

EARTHWORMS. Earthworms and other small animals bring about considerable soil mixing as they burrow through the soil, significantly affecting soil formation. Earthworms ingest soil particles and organic residues, enhancing the availability of plant nutrients in the material that passes through their bodies. They aerate and stir the soil and increase the stability of soil aggregates, thereby assuring ready infiltration of water. Ants and termites, as they build mounds, also transport soil materials from one horizon to another. In general, the mixing activities of animals, sometimes called **pedoturbation**, tends to undo or counteract the tendency of other soil-forming processes to accentuate the differences among soil horizons.

HUMAN INFLUENCES. Human activities also influence soil formation. For example, it is believed that Native Americans regularly set fires to maintain several large areas of prairie grasslands in Indiana and Michigan. In more recent times, human destruction of natural vegetation (trees and grass) and subsequent tillage of the soil for crop production has abruptly modified soil formation. Likewise, irrigating an arid region soil drastically influences the soil-forming factors, as does adding fertilizer and lime to soils of low fertility. In surface mining and urbanizing areas today, bulldozers may have an effect on soils almost akin to that of the ancient glaciers; they level and mix soil horizons and set the clock of soil formation back to zero.

FIGURE 2.28 The scattered bunch grasses of this semiarid rangeland in the Patagonia region of Argentina have created "islands" of soil with enhanced fertility and thicker A horizons. A lens cap placed at the edge of one of these islands provides scale and highlights the increased soil thickness under the plant canopies. Such small-scale, plant-associated soil heterogeneity is common where soil water limitations prevent complete plant ground cover. (Photo courtesy of Ingrid C. Burke, Short-Grass Steppe Long-Term Ecological Research Program, Colorado State University)

2.15 TOPOGRAPHY

Topography relates to the configuration of the land surface and is described in terms of differences in elevation, slope, and landscape position—in other words, the lay of the land. Topography can hasten or delay the work of climatic forces (see Box 2.1). Steep slopes generally encourage erosion of the surface layers and allow less rainfall to enter the soil before running off, thus preventing soil formation from getting very far ahead of soil destruction. In semiarid regions, less effective moisture on the steeper slopes also results in a more sparse, less diverse plant cover. Therefore, soils on steep terrain have relatively shallow, poorly differentiated soil profiles (Figure 2.30) compared to nearby soils on more level terrain.

INTERACTION WITH VEGETATION. Topography often interacts with vegetation to influence soil formation. In grassland–forest transition zones, trees are commonly confined to the slight depressions where soil is generally wetter than in upland positions. As would be expected, the nature of the soil in the depressions is quite different from that in the uplands. If water stands for part or all of the year in a depression in the landscape, cli-

FIGURE 2.29 Abandoned animal burrows in one horizon filled with soil material from another horizon are called *crotovinas*. In this Illinois prairie soil, dark, organic-matter-rich material from the A horizon has filled in old prairie dog burrows that extend into the B horizon. The dark circular shapes in the subsoil mark where the pit excavation cut through these burrows. Scale marked every 10 cm. (Photo courtesy of R. Weil)

mate is less influential in regulating soil development. Low-lying areas may give rise to peat bogs and, in turn, to organic soils (see Figure 2.23).

SLOPE ASPECT. Topography affects the absorbance of solar energy in a given landscape. In the northern hemisphere, south-facing slopes are more perpendicular to the sun's rays and are generally warmer and thereby commonly lower in moisture than their north-facing counterparts (see also Figure 7.22). Consequently, soils on the south slopes tend to be lower in organic matter and are not so deeply weathered.

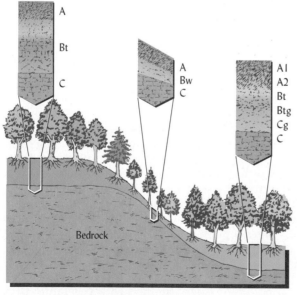

FIGURE 2.30 Topography influences soil properties, including soil depth. The diagram on the left shows the effect of slope on the profile characteristics and the depth of a soil on which forest trees are the natural vegetation. The photo on the right illustrates the same principle under grassland vegetation. Often a relatively small change in slope can have a great effect on soil development. See Section 2.18 for explanation of horizon symbols. (Photo courtesy of R. Weil)

SALT BUILDUP. In arid and semiarid areas, topography influences the buildup of soluble salts. Dissolved salts from surrounding upland soils move on the surface and through the underground water table to the lower-lying areas. There they rise to the surface as the water evaporates, often accumulating to plant-toxic levels.

PARENT MATERIAL INTERACTIONS. Topography can also interact with parent material. For example, in areas of tilted beds of sedimentary rock, the ridges often consist of resistant sandstone, while the valleys are underlaid by more weatherable limestone. In many landscapes, topography reflects the distribution of residual, colluvial, and alluvial parent materials, with residual materials on the upper slopes, colluvium covering the lower slopes, and alluvium filling the valley bottom (Figure 2.31).

2.16   TIME

Soil-forming processes take time to show their effects. The clock of soil formation starts ticking when a landslide exposes new rock to the weathering environment at the surface, when a flooding river deposits a new layer of sediment on its floodplain, when a glacier melts and dumps its load of mineral debris, or when a bulldozer cuts and fills a landscape to level a construction or mine-reclamation site.

RATES OF WEATHERING. Where the other factors of soil formation are favorable, organic matter may accumulate to form a darkened A horizon in a mere decade or two. In some cases, an incipient B horizon has become discernible on humid-region mine spoils in as few as 40 years. However, the formation of a simple B horizon with altered colors and structure is generally likely to require centuries. The accumulation of silicate clays usually becomes noticeable only after thousands of years. A mature, deeply weathered soil may be hundreds of thousands of years in the making.

When we speak of a "young" or a "mature" soil, we are not so much referring to the age of the soil in years, as to the degree of weathering and profile development. Time interacts with the other factors of soil formation. On a site in a warm climate, with much rain falling on a permeable parent material that is rich in weatherable minerals, weathering and soil profile differentiation will proceed far more rapidly than on a site with steep slopes and resistant parent material in a cold, dry climate.

There are a few instances in which soil formation is so rapid that the process can be observed, and the effect of time measured, in a human life span. Examples include the

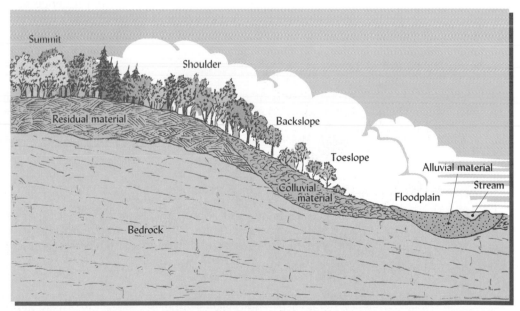

FIGURE 2.31   An interaction of topography and parent material as factors of soil formation. The soils on the summit, toe-slope, and floodplain in this idealized landscape have formed from residual, colluvial, and alluvial parent materials, respectively.

rapid formation of an A horizon on freshly deposited, fertile alluvium and the mineralogical, color, and structural changes that occur within a few years when certain sulfide-containing materials are first exposed to air by mining, land drainage, or dredging operations (see Sections 9.6 and 13.21).

CHRONOSEQUENCE. Most soil-profile features develop so slowly that it is not possible to directly measure time-related changes in their formation. Indirect methods, such as carbon dating or the presence of fossils and human artifacts, must be turned to for evidence about the time required for different aspects of soil development to occur. In a different approach to studying the effects of time on soil development, soil scientists look for a **chronosequence**—a set of soils that share a common community of organisms, climate, parent material, and slope, but differ with regard to the length of time that the materials have been subjected to weathering and soil formation. A chronosequence can sometimes be found among the soils forming on alluvial terraces of differing age. The highest terraces have been exposed for the longest time, those in lower positions have been more recently exposed by the cutting action of the stream, and those on the current flood plain are the youngest, being still subject to periodic additions of new material.

INTERACTION WITH PARENT MATERIALS. Residual parent materials have generally been subject to soil-forming processes for longer periods of time than have materials transported from one site to another. Soils of the uplands in the southeast part of the United States, for example, have been developing for a much longer period of time than nearby soils in low-lying areas that are developed on marine or alluvial materials. Soils forming on glacial materials in the northern parts of North America, Europe, and Asia have generally had far less time to develop than those soils farther south that escaped disturbance by the glaciers. Often the mineralogy and other properties of the "younger" soils in glaciated regions are more similar to those of their parent materials than is the case in the "older" soils of unglaciated regions. But comparisons are complicated because climate, vegetation, and parent material mineralogy also often differ between soils in glaciated and unglaciated areas.

The last example again emphasizes that the five factors influencing soil formation usually act simultaneously and interdependently; thus, vegetation varies with climate and parent material may be related to topographic position, which may also influence vegetation, and so on. The interdependence of these factors presents a challenge to those who would like to understand just how a given soil was formed or predict what soil properties are likely to be encountered in a given environment.

With this in mind, we will now focus on the processes that change parent material into a soil, processes that operate under the influence of the five major factors of soil formation we have just considered.

2.17 SOIL FORMATION IN ACTION[4]

Accumulation of regolith by the breakdown of bedrock or the deposition (by wind, water, ice, etc.) of unconsolidated geologic materials may precede or, more commonly, occur simultaneously with the development of the distinctive horizons of a soil profile. During the formation (**genesis**) of a soil from parent material, the regolith undergoes many profound changes.

Four Broad Processes of Soil Genesis

Soil genesis is brought about by a series of specific changes that can be grouped into four broad processes (Figure 2.32): (1) **transformations**, such as mineral weathering and organic matter breakdown, by which some soil constituents are modified or

---

[4] For the original discussion of many of the concepts in this section, see Simonson (1958). In addition, detailed discussions of soil genesis can be found in Fanning and Fanning (1989) and Boul, Hole, McCraken, and Southard (1997).

FIGURE 2.32 A schematic illustration of additions, losses, translocations, and transformations as the fundamental processes driving soil-profile development.

destroyed and others are synthesized; (2) **translocations** or movements of inorganic and organic materials from one horizon up or down to another, the materials being moved mostly by water but also by soil organisms; (3) **additions** of materials to the developing soil profile from outside sources, such as organic matter from leaves, dust from the atmosphere, or soluble salts from the groundwater; and (4) **losses** of materials from the soil profile by leaching to groundwater, erosion of surface materials, or other forms of removal. Often, transformation and translocation result in the accumulation of materials in a particular horizon.

These processes of soil genesis, operating under the influence of the environmental factors discussed previously, give us a logical framework for understanding the relationships between particular soils and the landscapes and ecosystems in which they function. In analyzing these relationships for a given site, ask yourself: What are the materials being added to this soil? What transformations and translocations are taking place in this profile? What materials are being removed? And how have the climate, organisms, topography, and parent material at this site affected these processes over time?

Consider the changes that might take place as a soil develops from a thick layer of relatively uniform parent material. Although some physical weathering and leaching may be necessary to allow plants to grow in certain parent materials, soil formation truly begins when plants become established and begin to deposit their litter and root residues on and in the surface layers of the parent material. The plant residues are disintegrated and partly decomposed by soil organisms that also synthesize humus and other new organic substances. The accumulation of humus enhances the availability of water and nutrients, providing a positive feedback for accelerated plant growth and further humus buildup. Earthworms, ants, termites, and a host of smaller animals come to live in the soil and feed on the newly accumulating organic resources; in so doing they burrow into the soil, mixing the plant residues deeper into the loose mineral material.

A-HORIZON DEVELOPMENT.   The resulting organic-mineral mixture near the soil surface, which comes into being rather quickly, is commonly the first soil horizon developed, the A horizon. It is darker in color and its chemical and physical properties differ from those of the original parent material. Individual soil particles in this horizon commonly clump together under the influence of organic substances to form granules, differentiating this layer from the deeper layers and from the original parent material (see Figure 2.33). On sloping land, erosion may remove some materials from the newly forming upper horizon, retarding, somewhat, the progress of horizon development.

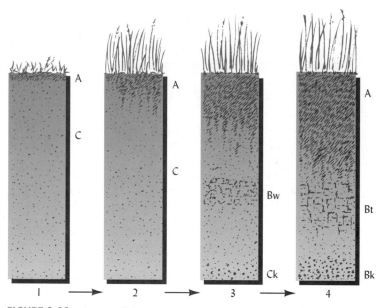

FIGURE 2.33   Stages of development of a hypothetical soil forming from relatively uniform parent material such as loess in a humid temperate climate (see Section 2.18 and Table 2.4 for explanation of the horizon symbols). At stage 1, no layering exists. Organic matter has just begun to accumulate from plant residues near the surface. By stage 2, the activity of earthworms and the decay of plant roots have incorporated organic matter into the upper few centimeters of soil, and soluble ions (e.g. Ca^{2+}) have moved downward. Clays are beginning to form from minerals in the loess and to disperse in the percolating waters. By stage 3, more organic matter has accumulated and moved deeply into the soil. Further leaching of soluble ions has occurred, and additional clay is forming and beginning to translocate into lower horizons. A Bw horizon is now distinguished by a change in color and the development of some blocklike structure. Deeper in the regolith, ions leached from the upper layers have precipitated, forming a Ck horizon enriched in calcium carbonates. By stage 4, even higher levels of organic matter are found deeper in the profile. Clays have accumulated sufficiently to form a Bt horizon, and have stimulated the formation of distinct blocklike structures. The pedogenic processes have extended the B horizon down to the upper part of the carbonate accumulation, forming a Bk horizon.

LEACHING. Organic acids that form from decaying plant residues are carried by percolating waters into the soil, where they stimulate weathering reactions. The acid-charged percolating water dissolves various chemicals and translocates (leaches) them from upper to lower horizons where they may precipitate, creating respective zones of depletion and accumulation. The dissolved substances include both positively charged ions (cations; e.g., Ca^{2+}) and negatively charged ions (anions; e.g., CO_3^- and SO_4^{2-}) released from the breakdown of minerals and organic matter. In semiarid and arid regions, precipitation of these ions at the maximum depth of water penetration commonly produces horizons enriched in calcite ($CaCO_3$) or gypsum ($CaSO_4 \cdot 2H_2O$).

Where rainfall is great enough, some of the dissolved materials may be completely removed from the developing soil profile as drainage occurs. Removal of weathering products encourages more weathering to proceed, and the profile becomes increasingly acidic and depleted of Ca^{2+}, Mg^{2+}, and K^+. On the other hand, deep-growing plant roots may intercept some of these soluble weathering products and return them, through leaf- and litter-fall, to the soil surface, thus retarding somewhat the processes of acid weathering and horizon differentiation.

MINERAL WEATHERING. Biochemical processes accelerate the weathering of primary minerals, disintegrating and altering some to form different kinds of silicate clays. As other primary minerals decompose, the decomposition products recombine into new minerals that include additional types of silicate clays and hydrous oxides of iron and aluminum (see Figure 2.4).

These newly formed clay minerals may accumulate where they are formed, or they may move downward and accumulate deeper in the profile. As materials are removed from one layer and accumulate in another, adjacent layers become more distinct from each other and soil horizons are formed. In the zones of clay accumulation, blocklike or prismatic units of soil structure commonly develop when the clayey mass periodically dries out and cracks (see Section 4.7). As the soil matures, the various horizons within the profile generally become more numerous and more distinctly different from each other.

Soil Genesis in Nature

Not all of the contrasting layers of material found in soil profiles are **genetic horizons** resulting from processes of soil genesis, such as those just described. The parent materials from which many soils develop contain contrasting layers *before* soil genesis starts. For example, glacial outwash or marine deposits may consist of various layers of fine and coarse particles laid down by separate episodes of sedimentation. Consequently, in characterizing soils, we must recognize not only the genetic horizons and properties that come into being during soil genesis, but also those layers or properties that may have been inherited from the parent material.

Furthermore, we must recognize the dynamic nature of soil bodies whose genetic horizons continue to developed and change. Consequently, in some soils the process of horizon differentiation has only begun while in others it is well advanced.

2.18   THE SOIL PROFILE

At each location on the land, the earth's surface has experienced a particular combination of influences from the five soil-forming factors, causing a different set of layers (horizons) to form in each part of the landscape, thus slowly giving rise to the natural bodies we call **soils.** Each soil is characterized by a given sequence of these horizons. A vertical exposure of this sequence is termed a **soil profile.** We will now consider the major horizons making up soil profiles and the terminology used to describe them.

The Master Horizons and Layers

Five **master** soil horizons are recognized and are designated using the capital letters O, A, E, B, and C (Figure 2.34). Subordinate horizons may occur within some master horizons and these are designated by lowercase letters following the capital master horizon letter (e.g., Bt, Ap, or Oi).

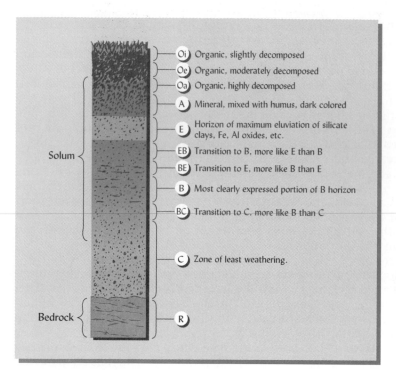

FIGURE 2.34 Hypothetical mineral soil profile showing the major horizons that may be present in a well-drained soil in the temperate humid region. Any particular profile may exhibit only some of these horizons, and the relative depths vary. In addition, however, a soil profile may exhibit more detailed subhorizons than indicated here. The solum includes the A, E, and B horizons plus some cemented layers of the C horizon.

O HORIZONS.   The O group is comprised of organic horizons that generally form above the mineral soil or occur in an organic soil profile. They derive from dead plant and animal residues. Generally absent in grassland regions, O horizons usually occur in forested areas and are commonly referred to as the **forest floor.** Often three subordinate O horizons can be distinguished.

> The **Oi horizon** is an organic horizon of **fibric** materials—recognizable plant and animal parts (leaves, twigs, and needles), only slightly decomposed. (The Oi horizon is referred to as the *litter* or *L layer* by some foresters.)

> The **Oe horizon** consists of **hemic** materials—finely fragmented residues intermediately decomposed, but still with much fiber evident when rubbed between the fingers. (This layer corresponds to the *fermentation* or *F layer* described by some foresters.)

> The **Oa horizon** contains **sapric** materials—highly decomposed, smooth, amorphous residues that do not retain much fiber or recognizable tissue structures. (This is the *humidified* or *H layer* designated by some foresters.)

A HORIZONS.   The topmost mineral horizons, designated A horizons, generally contain enough partially decomposed (humified) organic matter to give the soil a color darker than that of the lower horizons. The A horizons are often coarser in texture, having lost some of the finer materials by translocation to lower horizons and by erosion.

E HORIZONS.   These are zones of maximum leaching or ***eluviation*** (from Latin *ex* or *e,* out, and *lavere,* to wash) of clay, iron, and aluminum oxides, which leaves a concentration of resistant minerals, such as quartz, in the sand and silt sizes. An E horizon is usually found underneath the A horizon and is generally lighter in color than either the A horizon above it or the horizon below. Such E horizons are quite common in soils developed under forests, but they rarely occur in soils developed under grassland. Distinct E horizons can be seen in the soils in Plates 10 and 11 (after page 82).

B HORIZONS.   B horizons form below an O, A, or E horizon and have undergone sufficient changes during soil genesis so that the original parent material structure is no longer discernable. In many B horizons materials have accumulated, typically by illuviation (from the Latin *il,* in, and *lavere,* to wash) from the horizons above. In humid regions, B horizons are the layers of maximum accumulation of materials such as iron and aluminum oxides (Bo or Bs horizons) and silicate clays (Bt horizons), some of which may have illuviated from upper horizons and some of which may have formed in place. Such B horizons can be clearly seen in the middle depths of the profiles shown in Plates 1, 10, and 11 (after page 82). In arid and semiarid regions, calcium carbonate or calcium sulfate may accumulate in the B horizon (giving Bk and By horizons, respectively).[5]

The B horizons are sometimes referred to incorrectly as the subsoil, a term that lacks precision. In soils with shallow A horizons, part of the B horizon may become incorporated into the plow layer and thus become part of the topsoil. In other soils with deep A horizons, the plow layer or topsoil may include only the upper part of the A horizons, and the subsoil would include the lower part of the A horizon along with the B horizon. This emphasizes the need to differentiate between colloquial terms (*topsoil* and *subsoil*) and technical terms used by soil scientists to describe the soil profile.

C HORIZON.   The C horizon is the unconsolidated material underlying the solum (A and B horizons). It may or may not be the same as the parent material from which the solum formed. The C horizon is below the zones of greatest biological activity and has not been sufficiently altered by soil genesis to qualify as a B horizon. While loose enough to be dug with a shovel, C horizon material often retains some of the structural features of the parent rock or geologic deposits from which it formed (see, for example, the lower third of the profiles shown in Plates 7 and 11 after page 82). Its upper layers may in time become a part of the solum as weathering and erosion continue.

R LAYERS.   These are consolidated rock, with little evidence of weathering.

Subdivisions within Master Horizons

Often distinctive layers exist *within* a given master horizon, and these are indicated by a numeral *following* the letter designation. For example, if three different combinations of structure and colors can be seen in the B horizon, then the profile may include a B1–B2–B3 sequence.

If two different geologic parent materials (e.g., loess over glacial till) are present within the soil profile, the numeral 2 is placed in front of the master horizon symbols for horizons developed in the second layer of parent material. For example, a soil would have a sequence of horizons designated O–A–B–2C if the C horizon developed in glacial till while the upper horizons developed in loess.

Transition Horizons

Transitional layers between the master horizons (O, A, E, B, and C) may be dominated by properties of one horizon but also have prominent characteristics of another. The two applicable capital letters are used to designate the transition horizons (e.g., AE, EB, BE, and BC), the dominant horizon being listed before the subordinate one. Letter combinations such as E/B are used to designate transition horizons where distinct parts of the horizon have properties of E while other parts have properties of B.

Subordinate Distinctions

Since the capital letter designates the nature of a master horizon in only a very general way, specific horizon characteristics may be indicated by a lowercase letter following the master horizon designation. For example, three types of O horizons (Oi, Oe, and Oa) are indicated in the profile shown in Figure 2.34, which presents a commonly encountered sequence of horizons. Other subordinate distinctions include special physical properties and the accumulation of particular materials, such as clays and salts. A list of the recognized subordinate letter designations and their meanings is given in Table 2.4. We suggest that you mark this table for future reference, and study it now to get an idea of the distinctive soil properties that can be indicated by horizon designations. By way of illustration, a Bt horizon is a B horizon characterized by clay accumulation (t from the German *ton*, meaning clay); likewise, in a Bk horizon, carbonates (k) have accumulated. We have already used the suffixes i, a, and e to distinguish different types of O horizons. The significance of several other subordinate horizon designations will be discussed in the next chapter.

---

[5] In some soils of arid and semiarid regions, these compounds are also concentrated in the C horizon.

TABLE 2.4 Lowercase Letter Symbols to Designate Subordinate Distinctions Within Master Horizons

Letter	Distinction	Letter	Distinction
a	Organic matter, highly decomposed	n	Accumulation of sodium
b	Buried soil horizon	o	Accumulation of Fe and Al oxides
c	Concretions or nodules	p	Plowing or other disturbance
d	Dense unconsolidated materials	q	Accumulation of silica
e	Organic matter, intermediate decomposition	r	Weathered or soft bedrock
f	Frozen soil	s	Illuvial accumulation of O.M.[a] and Fe and Al oxides
g	Strong gleying (mottling)	ss	Slickensides
h	Illuvial accumulation of organic matter	t	Accumulation of silicate clays
i	Organic matter, slightly decomposed	v	Plinthite (high iron, red material)
j	Jarosite	w	Distinctive color or structure
jj	Cryoturbation (frost churning)	x	Fragipan (high bulk density, brittle)
k	Accumulation of carbonates	y	Accumulation of gypsum
m	Cementation or induration	z	Accumulation of soluble salts

[a] O.M. = organic matter.

Horizons in a Given Profile

It is not likely that the profile of any one soil will show all of the horizons that collectively are shown in Figure 2.34. The ones most commonly found in well-drained soils are Oi and Oe (or Oa) if the land is forested; A or E (or both, depending on circumstances); Bt or Bw; and C. Conditions of soil genesis will determine which others are present and their clarity of definition.

When a virgin (never-cultivated) soil is plowed and cultivated for the first time, the upper 15 to 20 cm becomes the plow layer or Ap horizon. Cultivation, of course, obliterates the original layered condition of the upper portion of the profile, and the Ap horizon becomes more or less homogeneous. An example of an Ap horizon can be seen in the upper 20 cm of the soil shown in Plate 1 (following page 82). In some soils, the A and E horizons are deeper than the plow layer (Figure 2.35). In other cases, where the upper horizons are quite thin, the plow line is just at the top of, or even down in, the B horizon.

In some cultivated land, serious erosion produces a **truncated profile.** As the surface soil is swept away over the years, the plow reaches deeper and deeper into the profile. Hence, the plowed zone in many cases consists almost entirely of former B horizon material, and the C horizon is correspondingly near the surface. Comparison to a nearby noneroded site can show how much erosion has occurred. Another, sometimes perplexing profile feature, is the presence of a buried soil resulting from natural or human action. In profile study and description, such a situation requires careful analysis.

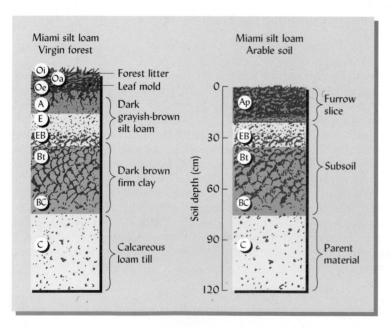

FIGURE 2.35 Generalized profile of the Miami silt loam, one of the Alfisols of the eastern United States, before and after land is plowed and cultivated. The surface layers (O, A, and E) are mixed by tillage and are termed the *Ap* (plowed) *horizon*. If erosion occurs, they may disappear, at least in part, and some of the B horizon will be included in the furrow slice.

2.19 CONCLUSION

The parent materials from which soils develop vary widely around the world and from one location to another only a few meters apart. Knowledge of these materials, their sources or origins, mechanisms for their weathering, and means of transport and deposition are essential to understanding soil genesis.

Soil formation is stimulated by *climate* and living *organisms* acting on *parent materials* over periods of *time* and under the modifying influence of *topography*. These five major factors of soil formation determine the kinds of soil that will develop at a given site. When all of these factors are the same at two locations, the kind of soil at these locations should be the same.

Soil genesis starts when layers or horizons not present in the parent material begin to appear in the soil profile. Organic matter accumulation in the upper horizons, the downward movement of soluble ions, the synthesis and downward movement of clays, and the development of specific soil particle groupings (structure) in both the upper and lower horizons are signs that the process of soil formation is under way.

The four general processes of soil formation (gains, losses, transformations, and translocations) and the five major factors influencing these processes provide us with an invaluable logical framework in site selection and in predicting the nature of soil bodies likely to be found on a particular site. Conversely, analysis of the horizon properties of a soil profile can tell us much about the nature of the climatic, biological, and geological conditions (past and present) at the site.

Characterization of the horizons in the profile leads to the identity of a soil individual, which is then subject to classification—the topic of our next chapter.

STUDY QUESTIONS

1. What is meant by the statement, *weathering combines the processes of destruction and synthesis?* Give an example of these two processes in the weathering of a primary mineral.

2. How is water involved in the main types of chemical weathering reactions?

3. Explain the significance of the ratio of silicon to aluminum in soil minerals.

4. Give an example of how parent material may vary across large geographic regions on one hand, but may also vary within a small parcel of land, on the other.

5. Name the five factors affecting soil formation. With regard to each of these factors of soil formation, compare a forested Rocky Mountain slope to the semiarid grassland plains far below.

6. How do *colluvium, glacial till,* and *alluvium* differ in appearance and agency of transport?

7. What is *loess,* and what are some of its properties as a parent material?

8. Give two specific examples for each of the four broad processes of soil formation.

9. Assuming a level area of granite rock was the parent material in both cases, describe in general terms how you would expect two soil profiles to differ, one in a warm, semiarid grassland and the other in a cool, humid pine forest.

10. For the two soils described in question 5, make a profile sketch using master horizon symbols and subordinate suffixes to show the approximate depths, sequence, and nature of the horizons you would expect to find in each soil.

REFERENCES

Boul, S. W., F. D. Hole, R. J. McCraken, and R. J. Southard. 1997. *Soil Genesis and Classification,* 4th ed. (Ames, Iowa: Iowa State University Press).

Fanning, D. S., and C. B. Fanning. 1989. *Soil: Morphology, Genesis, and Classification.* (New York: John Wiley and Sons).

Haering, K. C., W. L. Daniels, and J. A. Roberts. 1993. "Changes in mine soil properties resulting from overburden weathering," *J. Environ. Quality* **22**:194–200.

Jackson, M. L. 1964. "Chemical composition of soils," in F. E. Bear (ed.), *Chemistry of the Soil,* 2nd ed. (New York: Reinhold).

Jenny, Hans. 1941. *Factors of Soil Formation: A System of Quantitative Pedology.* (originally published by McGraw-Hill; Mineola, N.Y.: Dover).

Jenny, Hans. 1980. *The Soil Resource—Origins and Behavior.* Ecological Studies, vol. 37. (New York: Springer-Verlag).

Logan, William Bryant. 1995. *Dirt: The Ecstatic Skin of the Earth.* (New York: Riverhead Books).

Miller, F. P. 1976. *Maryland Soils Bulletin 212.* (College Park, Md.: University of Maryland Cooperative Extension Service).

National Science Foundation. 1975. "All that unplowed land," *Mosaic,* 6:17–21. (Washington, D.C.: National Science Foundation).

Petersen, G. W., R. L. Cummingham, and R. P. Matelski. 1971. "Moisture characteristics of Pennsylvania soils: III. Parent material and drainage relationships," *Soil Sci. Soc. Amer. Proc.,* **35**:115–119.

Simonson, R. W. 1958. "Outline of a generalized theory of soil genesis," *Soil Sci. Soc. Amer. Proc.,* **22**:152–156.

Soil Survey Division Staff. 1993. *Soil Survey Manual.* Agricultural Handbook 18. (Washington, D.C.: U.S. Government Printing Office).

Soil Survey Staff. 1996. *Keys to Soil Taxonomy,* 7th ed. (Washington, D.C.: USDA Natural Resources Conservation Service).

3

SOIL CLASSIFICATION

It is embarrassing not to be able to agree on
what soil is. In this the pedologists are not alone.
Biologists cannot agree on a definition
of life and philosophers on philosophy.
—HANS JENNY, THE SOIL RESOURCE:
ORIGIN AND BEHAVIOR

People classify things in order to make sense of their world. We do it all the time, whenever we call things by group names, giving them labels that inform us on important properties. Imagine a world without classifications. Imagine surviving in the woods knowing only that each plant was a plant, not which are edible by people, which attract wildlife, or which are poisonous to touch. Imagine surviving in a city knowing only that each person was a person, not a child or an adult, a male or a female, a police officer, a hoodlum, a friend, a teacher, a potential date, or any of the other categories into which we classify people. So, too, our understanding and management of soils and terrestrial systems would be hobbled if we knew only that a soil was a soil. How could we organize our information about soils? How could we learn from others' experience or communicate our knowledge to clients, colleagues, or workers?

Throughout history people have used various systems to name and classify soils. From the time crops were first cultivated, humans noticed differences in soils and classified them, grouping them according to their suitability for different uses by giving them descriptive names such as *black cotton soils, rice soils,* or *olive soils.* Other soil names still in common use today have geological connotations, suggesting the parent materials from which the soils formed: *limestone soils, piedmont soils,* and *alluvial soils.* Such terms, while they may convey some meaning to local users, do little to help us organize our knowledge of soils or to see the relationships among soils in the world.

In this chapter we will learn how soils are classified as natural bodies, on the basis of their profile characteristics, not merely on the basis of their suitability for a particular use. Such a soil classification system is essential to foster global communications about soils among soil scientists and all people concerned with the management of land and the conservation of the soil resource. Through such systems we can take advantage of research and experience at one location to predict the behavior of similarly classified soils at another location. Soil names such as *Mollisols* or *Oxisols* conjure up similar mental images in the minds of soil scientists everywhere, whether they live in the United States, Europe, Japan, developing countries, or elsewhere. Through the classification system, we can create a universal language of soils that enhances communication among users of soils around the world.

3.1 CONCEPT OF INDIVIDUAL SOILS

In the 1880s, when the Russian soil scientist V. V. Dukochaev and his associates conceived the idea that soils exist as natural bodies, they opened up an entirely new vision for the study of soils. Russian soil scientists soon developed a system for classifying natural soil bodies, but poor international communications and the reluctance of some scientists to acknowledge such radical ideas delayed the acceptance of the natural bodies concept by scientists in Europe and the United States. It was not until the late 1920s that C. F. Marbut of the U.S. Department of Agriculture, one of the few scientists who grasped the concept of soils as natural bodies, developed a soil classification scheme based on these principles.

The natural body concept of soils recognizes the existence of individual entities, each of which we call *a soil*. Just as human individuals differ from one another, soil individuals have characteristics distinguishing each from the others. Likewise, just as human individuals may be grouped according to characteristics such as height or weight, soil individuals having one or more characteristics in common may be grouped together. In turn, we may aggregate these groups into higher-level categories of soils, each having some characteristic that sets them apart from the others. Increasingly broad soil groups are defined as one moves up the classification pyramid to *the soil*.

Pedon and Polypedon

There are no sharp demarcations between one soil individual and another. Rather, there is a gradation in properties as one moves from one soil individual to an adjacent one. The gradation in soil properties can be compared to the gradation in the wavelengths of light as you move from one color to another in a rainbow. The change is gradual, and yet we identify a boundary differentiating between what we call green and what we call blue. Soils in the field are heterogenous; that is, the profile characteristics are not exactly the same in any two points within the soil individual you may choose to examine. Consequently, it is necessary to characterize a soil individual in terms of an imaginary three-dimensional unit called a *pedon* (rhymes with "head on," from the Greek *pedon*, ground; see Figure 3.1). It is the smallest sampling unit that displays the full range of properties characteristic of a particular soil.

Pedons occupy from about 1 to 10 m² of land area. Because it is what is actually examined during field investigation of soils, the pedon serves as the fundamental unit of soil classification. However, a soil unit in a landscape usually consists of a group of very similar pedons, closely associated together in the field. Such a group of similar pedons, or *polypedon,* is of sufficient size to be recognized as a landscape component termed a *soil individual.*

All the soil individuals in the world that have in common a suite of soil profile properties and horizons that fall within a particular range are said to belong to the same *soil series.* A soil series, then, is class of soils, not a soil individual, in the same way that *Pinus sylvestrus* is a species of tree, not a particular individual tree. The nearly 19,000 soil series characterized in the United States are the basic units used to classify the nation's soils. As we shall see in Section 19.7, units delineated on detailed soil maps are not purely one soil, but are usually named for the soil series to which *most* of the pedons within the unit belong.

Groupings of Soil Individuals

We have now identified the most specific and most general extremes in the concept of soils. One extreme is that of a natural body called *a soil,* characterized by a three-dimensional sampling unit (pedon), related groups of which (polypedons) are included in a soil individual. At the other extreme is *the soil,* a collection of all these natural bodies that is distinct from water, solid rock, and other natural parts of the earth's crust. Hierarchical soil classification schemes generally group soils into classes at increasing levels of generality between these two extremes.

There are a number of soil classification systems in use in different parts of the world.[1] A refined Russian system with heavy emphasis on factors of soil formation con-

---

[1] See Appendix B for summaries of the Canadian and United Nations Food and Agriculture Organization systems.

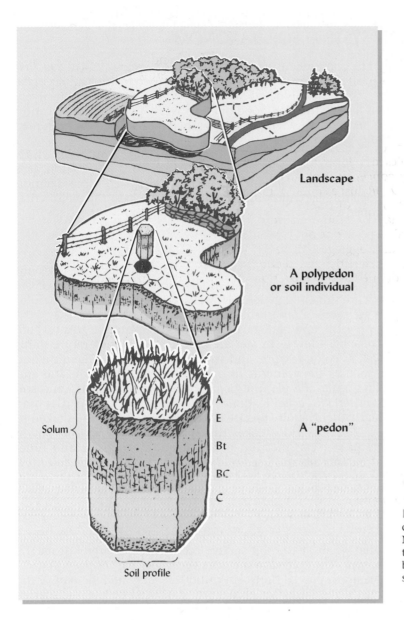

Landscape

A polypedon
or soil individual

A "pedon"

Solum {

A
E
Bt
BC
C

Soil profile

FIGURE 3.1 A schematic diagram to illustrate the concept of pedon and of the soil profile that characterizes it. Note that several contiguous pedons with similar characteristics are grouped together in a larger area (outlined by broken lines) called a *polypedon* or soil *individual*. Several soil individuals are present in this landscape.

tinues to be used, not only in countries of the former Soviet Union, but in some other European countries as well. Also, classification systems have been developed and are used in France and its former colonies, Belgium, the United Kingdom, Australia, Canada, South Africa, and Brazil. In each case the system was designed to meet the needs of a particular country or region. The Food and Agriculture Organization (FAO) and the United Nations Education, Science, and Cultural Organization (UNESCO), both associated with the United Nations, have produced a world soil resource map involving a partial classification system used to inventory and describe the world's soil resources.

In the United States, Marbut's 1927 classification scheme was improved in 1935, 1938, and 1949, the latter revision being widely used for about 25 years. But in 1951 the Soil Survey Staff of the U.S. Department of Agriculture began a cooperative effort with soil scientists in the United States and in other countries to develop a new comprehensive system of soil classification. This system, with modifications, has been in use in the United States since 1965, and it is used, at least to some degree, by scientists in some 55 other countries. This system will be used in this text.

Like some earlier systems, the comprehensive soil classification system, called *Soil Taxonomy,*[3] provides a hierarchical grouping of natural soil bodies. Two major features make *Soil Taxonomy* unique. First, the system is based on *soil properties* that can be objectively observed or measured. This lessens the likelihood of controversy over the classification of a given soil, which can occur when scientists deal with systems based on presumed mechanisms of soil formation.

The second unique feature of *Soil Taxonomy* is the *nomenclature* employed, which gives a definite connotation of the major characteristics of the soils in question. It is truly international since it is not based on any one national language. Consideration will be given to the nomenclature used after brief reference is made to the major criteria for the system—soil properties.

Bases of Soil Classification

Soil Taxonomy is based on the properties of soils as they are found today. This does not mean that the processes of soil genesis are ignored. In fact, one of the objectives of the system is to group soils that are similar in genesis. However, the specific criteria used to place soils in these groups are those of observable soil properties.

Most of the chemical, physical, and biological properties presented in this text are used as criteria for *Soil Taxonomy*. A few examples are moisture and temperature status throughout the year, as well as color, texture, and structure of the soil. Chemical and mineralogical properties, such as the contents of organic matter, clay, iron and aluminum oxides, silicate clays, salts, the pH, the percentage base saturation,[4] and soil depth are other important criteria for classification. While many of the properties used may be observed in the field, others require precise measurements on samples taken to a sophisticated laboratory. This precision makes the system more objective, but in some cases may make the proper classification of a soil quite expensive and time-consuming. Precise measurements are also used to define certain **diagnostic soil horizons,** the presence or absence of which help determine the place of a soil in the classification system.

Diagnostic Surface Horizons

The diagnostic horizons that occur at the soil surface are called **epipedons** (from the Greek *epi*, over, and *pedon*, soil). The epipedon includes the upper part of the soil darkened by organic matter, the upper eluvial horizons, or both. It may include part of the B horizon (see Section 2.18) if the latter is significantly darkened by organic matter. Seven epipedons are recognized (Table 3.1), but only five occur naturally over wide areas (Figure 3.2). The other two, anthropic and plaggen, are the result of intensive human use. They are common in parts of Europe and Asia where soils have been utilized for many centuries.

The **mollic epipedon** (Latin *mollis*, soft) is a mineral surface horizon noted for its dark color (see Plate 8) associated with its accumulated organic matter (>0.6% organic C throughout the thickness), for its thickness (generally >25 cm), and for its softness even when dry. It has a high base saturation (greater than 50% base-forming cations). Mollic epipedons are moist at least three months a year when the soil temperature is usually 5°C or higher to a depth of 50 cm. These epipedons are characteristic of soils developed under native prairies (Figure 3.3 and Plate 8, after page 82).

The **umbric epipedon** (Latin *umbra*, shade; hence, dark) has the same general characteristics as the mollic epipedon except the percentage base saturation is less than 50%. In comparison with the mollic epipedon, this mineral horizon commonly devel-

---

[2] For a complete description of *Soil Taxonomy* see Soil Survey Staff (1998). The first edition of *Soil Taxonomy* was published as Soil Survey Staff (1975).

[3] Taxonomy is the science of the principles of classification. For a review of the achievements and challenges of *Soil Taxonomy,* see SSSA (1984).

[4] The percentage base saturation is the percentage of the soil's negatively charged sites (cation exchange capacity) that are satisfied by attracting metal cations (such as Ca^{2+}, Mg^{2+} and K^+) that form strong bases (e.g., $Ca(OH)_2$) upon hydrolysis (see Section 8.12).

TABLE 3.1   Major Features of Diagnostic Horizons in Mineral Soils Used to Differentiate at the Higher Levels of *Soil Taxonomy*

Diagnostic horizon (and typical genetic horizon designation)	Major features
Surface horizons = epipedons	
Mollic (A)	Thick, dark-colored, high base saturation, strong structure
Umbric (A)	Same as mollic except low base saturation
Ochric (A)	Light-colored, low organic content, may be hard and massive when dry
Melanic (A)	Thick, black, high in organic matter (>6% organic C), common in volcanic ash soils
Histic (O)	Very high in organic content, wet during some part of year
Anthropic (A)	Human-modified molliclike horizon, high in available P
Plaggen (A)	Human-made sodlike horizon created by years of manuring
Subsurface horizons	
Argillic (Bt)	Silicate clay accumulation
Natric (Btn)	Argillic, high in sodium, columnar or prismatic structure
Spodic (Bh, Bs)	Organic matter, Fe and Al oxides accumulation
Cambic (Bw, Bg)	Changed or altered by physical movement or by chemical reactions, generally nonilluvial
Agric (A or B)	Organic and clay accumulation just below plow layer resulting from cultivation
Oxic (Bo)	Highly weathered, primarily mixture of Fe, Al oxides and non-sticky-type silicate clays
Duripan (qm)	Hardpan, strongly cemented by silica
Fragipan (x)	Brittle pan, usually loamy textured, dense
Albic (E)	Light-colored, clay and Fe and Al oxides mostly removed
Calcic (k)	Accumulation of $CaCO_3$ or $CaCO_3 \cdot MgCO_3$
Gypsic (y)	Accumulation of gypsum
Salic (z)	Accumulation of salts
Kandic	Accumulation of low-activity clays
Petrocalcic (km)	Cemented calcic horizon
Petrogypsic (ym)	Cemented gypsic horizon
Placic (sm)	Thin pan cemented with iron alone or with manganese or organic matter
Sombric (Bh)	Organic matter accumulation
Sulfuric	Highly acid with Jarosite mottles

ops in areas with somewhat higher rainfall and where the parent material has lower content of calcium and magnesium.

The **ochric epipedon** (Greek *ochros,* pale) is a mineral horizon that is either too thin, too light in color, or too low in organic matter to be either a mollic or umbric horizon. It is usually not as deep as the mollic or umbric epipedons. As a consequence of its low organic matter content, it may be hard and massive when dry, (see Plates 1, 7, and 11).

The **melanic epipedon** (Greek *melas,* melan, black) is a mineral horizon that is very black in color due to its high organic matter content (organic carbon >6%). It is characteristic of soils high in such minerals as allophane, developed from volcanic ash. It is more than 30 cm thick, and is extremely light in weight and fluffy for a mineral soil (see Plate 2, after page 82).

The **histic epipedon** (Greek *histos,* tissue) is a relatively thin layer of *organic soil materials*[5] overlying a mineral soil. Formed in wet areas, the Histic epipedon is commonly a layer of peat or muck about 20 to 30 cm thick, with a black to dark brown color, and a very low density.

---

[5] *Organic soil material,* such as peat and muck, may consist almost entirely of organic matter, but it is technically defined as containing more than a certain *minimum amount of organic matter* as follows: If the material is not water-saturated in its natural state, its organic matter content must be at least 35% (about 200 g/kg organic C). If it is water-saturated during part of the year in its natural state, then the minimum organic matter content varies with the amount of clay in the material, ranging from 20% (120 g/kg organic C) if no clay is present to 30% (180 g/kg organic C) if the clay content exceeds 600 g/kg. For a discussion of the relationship between organic matter and organic carbon, see Section 12.3.

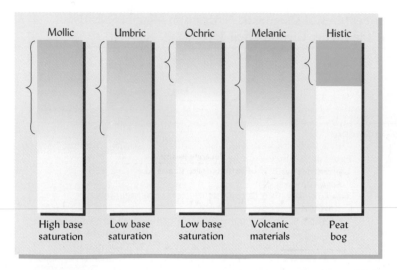

FIGURE 3.2 Representative profile characteristics of five surface diagnostic horizons (epipedons). The comparative organic matter levels and distribution are indicated by the darkness of colors. The mollic and umbric epipedons have similar organic matter distribution but the percentage base saturation is higher (greater than 50%) in the mollic epipedon and lower (less than 50%) in the umbric epipedon. The ochric epipedon is lower in organic matter content; consequently, it is light in color and sometimes hard when dry. Two other epipedons have very high organic matter contents and are very dark in color. The melanic epipedon is formed on recently deposited volcanic materials, usually in cool wet areas. The histic epipedon is formed from organic deposits laid down over mineral soils, usually in wet, boggy conditions. The relative depth of each epipedon is shown by the brackets.

Diagnostic Subsurface Horizons

Many subsurface horizons are used to characterize different soils in *Soil Taxonomy*. The 18 that are considered diagnostic horizons are listed along with their major features in Table 3.1 and Figure 3.4. Each diagnostic horizon provides a characteristic that helps place a soil in its proper class in the system. We will briefly discuss a few of the more commonly encountered subsurface diagnostic horizons.

The **argillic horizon** is a subsurface accumulation of high-activity silicate clays that have moved downward from the upper horizons or have formed in place. An example

FIGURE 3.3 The mollic epipedon (a diagnostic horizon) in this soil includes genetic horizons designated Ap, an A2, and a Bt, all darkened by the accumulation of organic matter. Scale marked every 10 cm. (Photo courtesy of R. Weil).

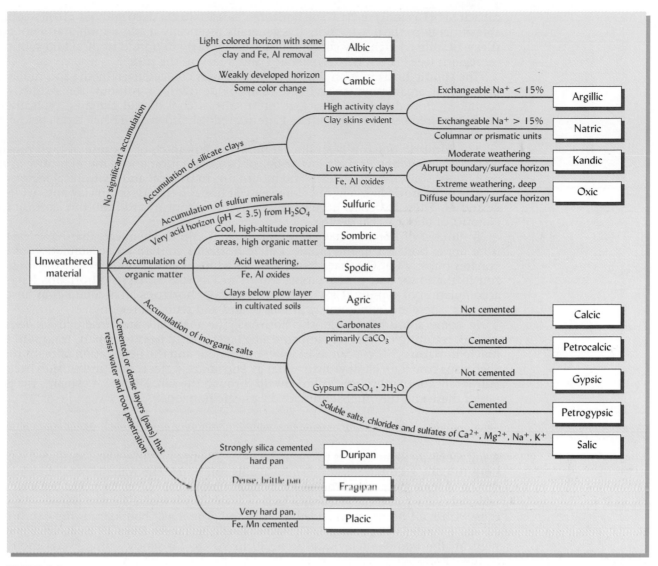

FIGURE 3.4 Names and major distinguishing characteristics of subsurface diagnostic horizons. Note that the focus is primarily on soil properties, not on how the properties have presumably evolved. Among the characteristics emphasized is the accumulation of silicate clays, organic matter, Fe and Al oxides, calcium compounds, and soluble salts, as well as materials that become cemented or highly acidified, thereby constraining root growth. The presence or absence of these horizons plays a major role in determining in which class a soil falls in *Soil Taxonomy*. See Chapter 8 for discussion of low- and high-activity clays.

is shown in Plate 1 between 25 and 70 cm on the scale in the photograph. The clays often are found as coatings on pore walls (as shown in Figure 4.3) and surfaces of the structural groupings. The coatings usually appear as shiny surfaces or as clay bridges between sand grains. Termed *argillans* or *clay skins,* they are concentrations of clay translocated from upper horizons (see Plate 23, before page 499).

The **natric horizon** likewise has silicate clay accumulation (with clay skins), but the clays are accompanied by more than 15% exchangeable sodium on the colloidal complex and by columnar or prismatic soil structural units. The natric horizon is found mostly in arid and semiarid areas. An example is shown in Figure 4.11*e*.

The **kandic horizon** has an accumulation of Fe and Al oxides as well as low-activity silicate clays (e.g., kaolinite), but clay skins need not be evident. The clays are low in activity as shown by their low cation-holding capacities (<16 $cmol_c$/kg clay). The epipedon that overlies a kandic horizon has commonly lost much of its clay content.

The **oxic horizon** is a highly weathered subsurface horizon that is very high in Fe and Al, oxides, and in low-activity silicate clays (e.g., kaolinite). The cation-holding capacity is <16 cmol$_c$/kg clay. The horizon is at least 30 cm deep and has <10% weatherable minerals in the fine fraction. It is generally physically stable, crumbly, and not very sticky, despite its high clay content. It is found mostly in humid tropical and subtropical regions (see Plate 9, between about 1 and 3 feet on the scale).

The **spodic horizon** is an illuvial horizon that is characterized by the accumulation of colloidal organic matter and aluminum oxide (with or without iron oxide). It is commonly found in highly leached forest soils of cool humid climates, typically on sandy-textured parent materials (see Plate 10, reddish brown and black layers below the whitish layer).

The **sombric horizon** is an illuvial horizon, dark in color because of high organic matter accumulation. It has a low degree of base saturation and is found mostly in the cool, moist soils of high plateaus and mountains in tropical and subtropical regions.

The **albic horizon** is a light-colored eluvial horizon that is low in clay and oxides of Fe and Al. These materials have largely been moved downward from this horizon (see Plate 10, at about 10 cm depth).

A number of horizons have accumulations of saltlike chemicals that have leached from upper horizons in the profile. **Calcic horizons** contain an accumulation of carbonates (mostly CaCO$_3$) that often appear as white chalklike nodules (see the Bk horizon in the lower part of the profile shown in Figure 3.3). **Gypsic horizons** have an accumulation of gypsum (CaSO$_4 \cdot$ 2H$_2$O), and **salic horizons** an accumulation of soluble salts. These are found mostly in soils of arid and semiarid regions.

In some subsurface diagnostic horizons, the materials are cemented or densely packed, resulting in relatively impermeable layers called *pans* (**duripan, fragipan,** and **placic horizons**).[6] These can resist water movement and the penetration of plant roots. Such pans constrain plant growth and may encourage water runoff and erosion because rainwater cannot move readily downward through the soil. Figure 3.4 explains the genesis of these and the other subsurface diagnostic horizons.

Soil Moisture Regimes (SMR)

A soil moisture regime refers to the presence or absence of either water-saturated conditions (usually groundwater) or plant-available soil water. The water is present or absent during specified periods in the year in what is termed the *control section* of the soil. The upper boundary of the SMR control section is the depth that 2.5 cm of water will penetrate within 24 hours when added to a dry soil. The lower boundary is the depth that 7.5 cm of water will penetrate. The control section ranges from 10 to 30 cm for soils high in fine particles (clay) and from 30 to 90 cm for sandy soils. Several moisture regime classes are used to characterize soils. The percentage distribution of areas with different soil moisture regimes is shown in Table 3.2.

Aquic. Soil is saturated with water and virtually free of gaseous oxygen for sufficient periods of time for evidence of poor aeration (gleying and mottling) to occur.

Udic. Soil moisture is sufficiently high year-round in most years to meet plant needs. This regime is common for soils in humid climatic regions and characterizes about one-third of the worldwide land area. An extremely wet moisture regime with excess moisture for leaching throughout the year is termed **perudic.**

Ustic. Soil moisture is intermediate between Udic and Aridic regimes—generally there is some plant-available moisture during the growing season, although significant periods of drought may occur.

Aridic. The soil is dry for at least half of the growing season and moist for less than 90 consecutive days. This regime is characteristic of arid regions. The term *torric* is used to indicate the same moisture condition in certain soils that are both hot and dry in summer, though they may not be hot in winter.

Xeric. This soil moisture regime is found in typical Mediterranean-type climates, with cool, moist winters and warm, dry summers. Like the Ustic regime, it is characterized by having long periods of drought in the summer.

---

[6] Well-developed argillic horizons may present such a great and abrupt increase in clay content that water and root movement are severely restricted; such an horizon is commonly referred to by the nontaxonomic term, *claypan.*

TABLE 3.2 Percent of Global Area Occupied by Soils with Different Soil Moisture and Temperature Regimes

Soil temperature regimes	Soil moisture regimes						
	Aridic	Xeric	Ustic	Udic	Perudic	Aquic	Total
Pergelic	3.7			5.3		1.9	10.9
Cryic	4.3			8.5	0.0	0.5	13.5
Frigid	0.5		0.2	0.4			1.2
Mesic	5.4	0.8	0.6	5.3	0.1	0.2	12.5
Thermic	3.3	2.6	1.5	3.0	0.2	0.6	11.4
Hyperthermic	15.7		1.5	1.0	0.0	0.3	18.5
Isofrigid	0.0			0.0	0.0	0.0	0.1
Isomesic				0.2	0.0	0.0	0.3
Isothermic	0.6		1.0	0.7	0.0	0.0	2.4
Isohyperthermic	2.1		13.0	8.4	0.4	1.9	26.0
Water						1.2	1.2
Ice						1.4	1.4
Total	35.9	3.5	18.0	33.1	1.0	8.3	100.0

From Eswaran (1993).

These terms are used to diagnose the soil moisture regime and are helpful not only in classifying soils but in suggesting the most sustainable long-term use of soils.

Soil Temperature Regimes

Soil temperature regimes, such as frigid, mesic, and thermic, are used to classify soils at some of the lower levels in *Soil Taxonomy*. The cryic (Greek *kryos,* very cold) temperature regime distinguishes some higher-level groups. These regimes are based on mean annual soil temperature, mean summer temperature, and the difference between mean summer and winter temperatures, all at 50 cm depth. The specific temperature regimes will be described in the discussion of soil families (Section 3.17).

3.3 CATEGORIES AND NOMENCLATURE OF SOIL TAXONOMY

There are six categories of classification in *Soil Taxonomy:* (1) *order,* the highest (broadest) category, (2) *suborder,* (3) *great group,* (4) *subgroup,* (5) *family,* and (6) *series* (the most specific category). These categories are hierarchical because the lower categories fit within the higher categories (Figure 3.5). Thus, each order has several suborders, each suborder has several subgroups, and so forth. This system may be compared with those used for the classification of plants or animals, as shown in Table 3.3. Just as *Trifolium repens* identifies a specific kind of plant, the Miami series identifies a specific kind of soil. The similarity continues up the classification scale to the highest categories—phylum for plants and order for soils.

Nomenclature of Soil Taxonomy

A unique feature of *Soil Taxonomy* is the unusual-sounding nomenclature used to identify different soil classes. Although unfamiliar at first sight, the nomenclature system has a very logical construction and conveys a great deal of information about the nature of the soils named. The system is easy to learn after a bit of study. The nomenclature is used throughout this book, especially to identify the kinds of soils shown in illustrations. When reading, if you make a conscious effort to identify the parts of each soil class mentioned in the text and figure captions, and recognize the level of category indicated, the system will become second nature.

The names of the classification units are combinations of syllables, most of which are derived from Latin or Greek and are root words in several modern languages. Since each part of a soil name conveys a concept of soil character or genesis, the name automatically describes the general kind of soil being classified. For example, soils of the order **Aridisols** (from the Latin *aridus,* dry, and *solum,* soil) are characteristically dry

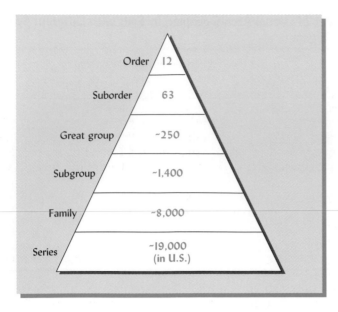

FIGURE 3.5 The categories of *Soil Taxonomy* and approximate number of units in each category. (Data from U.S. Department of Agriculture, personal correspondence)

soils in arid regions. Those of the order **Inceptisols** (from the Latin *inceptum,* beginning, and *solum,* soil) are soils with only the beginnings or inception of profile development. Thus, the names of orders are combinations of (1) formative elements, which generally define the characteristics of the soils, and (2) the ending *sols.*

The names of suborders automatically identify the order of which they are a part. For example, soils of the suborder **Aquolls** are the wetter soils (from the Latin *aqua,* water) of the Mollisols order. Likewise, the name of the great group identifies the suborder and order of which it is a part. **Argiaquolls** are Aquolls with clay or argillic (Latin *argilla,* white clay) horizons. In the following illustration, note that the three letters *oll* identify each of the lower categories as being in the Mo**ll**isols order.

Moll**isols**	Order
Aqu**olls**	Suborder
Argiaqu**olls**	Great group
Typic Argiaqu**olls**	Subgroup

Likewise, the suborder name Aquolls is included as part of the great group and subgroup name. If one is given only the subgroup name, the great group, suborder, and order to which the soil belongs are automatically known.

Family names in general identify subsets of the subgroup that are similar in texture, mineral composition, and mean soil temperature at a depth of 50 cm. Thus the name **Typic Argiaquolls, fine, mixed, mesic, active** identifies a family in the Typic

TABLE 3.3 **Comparison of the Classification of a Common Cultivated Plant, White Clover (*Trifolium repens*), and a Soil, Miami Series**

Plant classification			*Soil classification*	
Phylum	Pterophyta		Order	Alfisols
Class	Angiospermae		Suborder	Udalfs
Subclass	Dicotyledoneae	Increase specificity	Great Group	Hapludalfs
Order	Rosales		Subgroup	Oxyaquic Hapludalfs
Family	Leguminosae		Family	Fine loamy, mixed, mesic, active
Genus	*Trifolium*		Series	Miami
Species	*repens*		Phase[a]	Miami silt loam

[a] Technically not a category in *Soil Taxonomy* but used in field surveying. *Silt loam* refers to the texture of the A horizon.

Argiaquolls subgroup with a fine texture, mixed clay mineral content, mesic (8 to 15°C) soil temperature, and clays active in cation exchange.

Soil series are named after a geographic feature (town, river, etc.) near where they were first recognized. Thus, names such as *Fort Collins, Cecil, Miami, Norfolk,* and *Ontario* identify soil series first described near the town or geographic feature named. Approximately 19,000 soil series have been classified in the United States alone.

In detailed field soil surveying, soil series are sometimes further differentiated on the basis of surface soil texture, degree of erosion, slope, or other characteristics. These practical subunits are called soil **phases.** Names such as *Fort Collins loam, Cecil clay,* or *Cecil clay loam, eroded phase* are used to identify such phases. Note, however, that soil phases, practical as they may be in local situations, are *not* a category in the *Soil Taxonomy* system.

With this brief explanation of the nomenclature of the new system, we will now consider the general nature of soils in each of the soil orders.

3.4 SOIL ORDERS

Each of the world's soils is assigned to one of 12 **orders**, largely on the basis of soil properties that reflect a major course of development, with considerable emphasis placed on the presence or absence of major diagnostic horizons (Table 3.4). As an example, many soils that developed under grassland vegetation have the same general sequence of horizons and are characterized by a mollic epipedon—a thick, dark, surface horizon that is high in base-forming cations. Soils with these properties are thought to have been formed by the same general genetic processes, but it is because of the properties they have in common that they are included in the same order: Mollisols. The names and major characteristics of each soil order are shown in Table 3.4. Note that all order names have a common ending, *sols* (from the Latin *solum,* soil)

The general conditions that enhance the formation of soils in the different orders are shown in Figure 3.6. From soil profile characteristics, soil scientists can ascertain the relative degree of soil development in the different orders, as shown in this figure. Note that soils with essentially no profile layering (Entisols) have the least development, while the deeply weathered soils of the humid tropics (Oxisols and Ultisols) show the greatest soil development. The effect of climate (temperature and moisture) and of vegetation (forests or grasslands) on the kinds of soils that develop is also indicated in Figure 3.6. Study Table 3.4 and Figure 3.6 to better understand the relationship between soil properties and the terminology used in *Soil Taxonomy.*

To some degree, most of the soil orders occur in climatic regions that can be described by moisture and temperature regimes. Figure 3.7 illustrates some of the rela-

TABLE 3.4 **Names of Soil Orders in *Soil Taxonomy* with Their Derivation and Major Characteristics**

The bold letters in the order names indicate the formative element used as the ending for suborders and lower taxa within that order.

Name	Formative element	Derivation	Pronunciation	Major characteristics
Alfisols	alf	Nonsense symbol	Ped*alf*er	Argillic, natric, or kandic horizon; high to medium base saturation
Andisols	and	Jap. *ando,* blacksoil	*And*esite	From volcanic ejecta, dominated by allophane or Al-humic complexes
Ar**id**isols	id	L. *aridus,* dry	A*rid*	Dry soil, ochric epipedon, sometimes argillic or natric horizon
Entisols	ent	Nonsense symbol	Rec*ent*	Little profile development, ochric epipedon common
Gel**el**isols	el	Gk. *gelid,* very cold	J*ell*y	Permafrost, often with cryoturbation (frost churning)
H**ist**osols	ist	Gk. *histos,* tissue	*Hist*ology	Peat or bog; >20% organic matter
Inc**ept**isols	ept	L. *inceptum,* beginning	Inc*ept*ion	Embryonic soils with few diagnostic features, ochric or umbric epipedon, cambic horizon
M**oll**isols	oll	L. *mollis,* soft	M*oll*ify	Mollic epipedon, high base saturation, dark soils, some with argillic or natric horizons
Oxisols	ox	Fr. *oxide,* oxide	O*x*ide	Oxic horizon, no argillic horizon, highly weathered
Sp**od**osols	od	Gk. *spodos,* wood ash	P*od*zol; odd	Spodic horizon commonly with Fe, Al oxides and humus accumulation
Ultisols	ult	L. *ultimus,* last	*Ult*imate	Argillic or kandic horizon, low base saturation
V**ert**isols	ert	L. *verto,* turn	In*vert*	High in swelling clays; deep cracks when soil dry

tionships among the soil orders with regard to these climatic factors. While only the Gelisols and Aridisols orders are defined directly in relation to climate, Figure 3.7 indicates that orders with the most highly weathered soils tend to be associated with the warmer and wetter climates. Figure 3.8 is a simplified general soil map showing the major areas of the United States dominated by each order. Profiles for each soil order are shown in color on Plates 1 to 12 (after page 82). A more detailed soil map (along with the map legend) for the United States can be found in Appendix A. A general world map of the 12 soil orders is printed in color on the front end papers.

Although a detailed description of all the lower levels of soil categories is far beyond the scope of this (or any other) book, a general knowledge of the 12 soil orders is essential for understanding the nature and function of soils in different environments. The simplified key given in Figure 3.9 helps illustrate how *Soil Taxonomy* can be used to key out the order of any soil based on observable and measurable properties of the soil profile. Because certain diagnostic properties take precedence over others, the key must always be used starting at the top, and working down. It will be useful to review this key after reading about the general characteristics, nature, and occurrence of each soil order.

We will now consider each of the soil orders, beginning with those characterized by little profile development and progressing to those with the most highly weathered profiles (as represented from left to right in Figure 3.6).

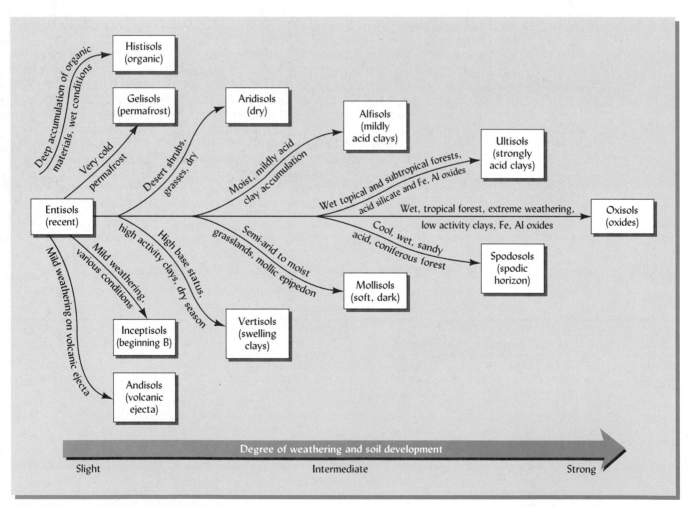

FIGURE 3.6 Diagram showing general degree of weathering and soil development in the different orders of mineral soils classified in *Soil Taxonomy*. Also shown are the general climatic and vegetative conditions under which soils in each order are formed.

PLATE 1 Alfisols—an Aeric Epiaqualf from western New York. Argillic horizon between 20–80 cm. Scale in centimeters.

PLATE 2 Andisols—a Typic Melanudand from western Tanzania. Scale in 10 cm.

PLATE 3 Aridisols—a Typic Haplocambid from western Nevada. Scale in feet.

PLATE 4 Entisols—a Typic Quartzipsamment from eastern Texas. Scale in feet.

PLATE 5 Gelisols—a Typic Aquaturbel from Alaska. Permafrost below 32 cm on scale.

PLATE 6 Histosols—a Limnic Haplosaprist from southern Michigan. Buried mineral soil at bottom of scale. Scale in feet.

PLATE 7 Inceptisols—a Typic Dystrudept from West Virginia. Knife is in cambic horizon.

PLATE 8 Mollisols—a Typic Hapludoll from central Iowa. Mollic epipedon to 1.8 ft. Scale in feet.

PLATE 9 Oxisols—a Udeptic Hapludox from central Puerto Rico. Scale in meters.

PLATE 10 Spodosols—a Humic Cryorthod from southern Quebec. Albic horizon at about 10 cm. Bar − 10 cm.

PLATE 11 Ultisols—a Typic Hapludult from central Virginia showing metamorphic rock structure in the saprolite below the 60-cm-long shovel.

PLATE 12 Vertisols—a Typic Haplustert from Queensland, Australia, during wet season. Scale in meters.

PLATE 13 Typic Argiustolls in eastern Montana with a chalky white calcic horizon (Bk and Ck) overlain by a Mollic epipedon (Ap, A2, and Bt).

PLATE 14 Ice wedge and permafrost underlying Gelisols in the Seward Peninsula of Alaska. Shovel is 1 m long.

PLATE 15 A Typic Plinthudult in central Sri Lanka. Mottled zone is plinthite, in which ferric iron concentrations will harden irreversibly if allowed to dry.

PLATE 16 A soil catena or toposequence in central Zimbabwe. Redder colors indicate better internal drainage. Inset: B-horizon clods from each soil in the catena.

Photo credits follow plate 35, preceding page 499

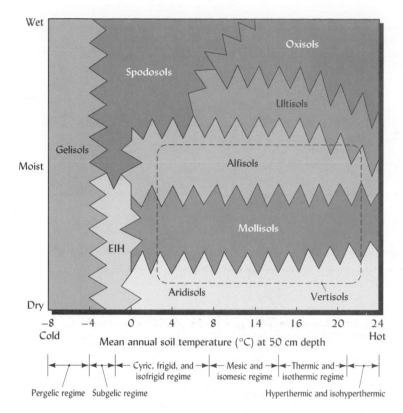

FIGURE 3.7 Diagram showing the general soil moisture and soil temperature regimes that characterize the most extensive soils in each of eight soil orders. Soils of the other four orders (Andisols, Entisols, Inceptisols, and Histosols) may be found under any of the soil moisture and temperature conditions (including the area marked EIH). Major areas of Vertisols are found only where clayey materials are in abundance and are most extensive where the soil moisture and temperature conditions approximate those shown inside the box with broken lines. Note that these relationships are only approximate and that less extensive areas of soils in each order may be found outside the indicated ranges. For example, some Ultisols (Ustults) and Oxisols (Ustox) have soil moisture levels for at least part of the year that are much lower than this graph would indicate. (The terms used at the bottom to describe the soil temperature regimes are those used in helping to identify soil families.)

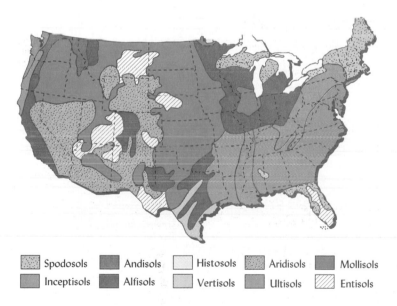

Spodosols Andisols Histosols Aridisols Mollisols
Inceptisols Alfisols Vertisols Ultisols Entisols

FIGURE 3.8 Simplified general soil map of the United States showing patterns of soil orders based on *Soil Taxonomy*. For more detailed soil maps showing the distribution of both soil orders and suborders for the United States, see Appendix A.

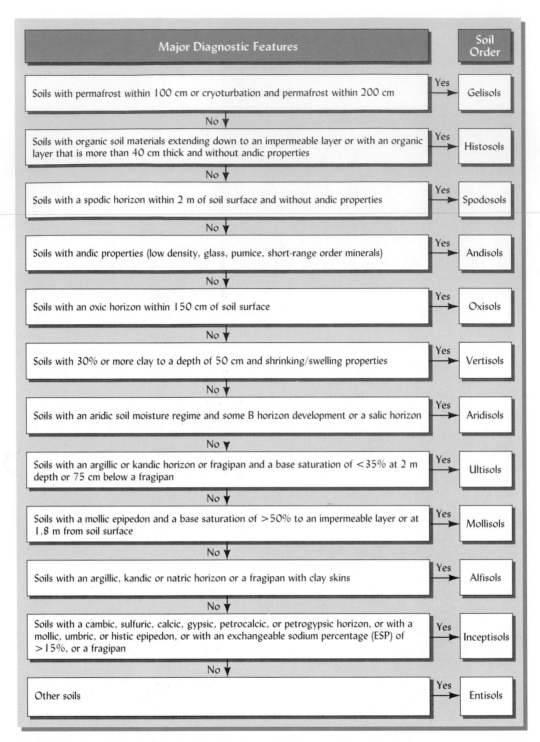

Major Diagnostic Features		Soil Order
Soils with permafrost within 100 cm or cryoturbation and permafrost within 200 cm	Yes →	Gelisols
Soils with organic soil materials extending down to an impermeable layer or with an organic layer that is more than 40 cm thick and without andic properties	Yes →	Histosols
Soils with a spodic horizon within 2 m of soil surface and without andic properties	Yes →	Spodosols
Soils with andic properties (low density, glass, pumice, short-range order minerals)	Yes →	Andisols
Soils with an oxic horizon within 150 cm of soil surface	Yes →	Oxisols
Soils with 30% or more clay to a depth of 50 cm and shrinking/swelling properties	Yes →	Vertisols
Soils with an aridic soil moisture regime and some B horizon development or a salic horizon	Yes →	Aridisols
Soils with an argillic or kandic horizon or fragipan and a base saturation of <35% at 2 m depth or 75 cm below a fragipan	Yes →	Ultisols
Soils with a mollic epipedon and a base saturation of >50% to an impermeable layer or at 1.8 m from soil surface	Yes →	Mollisols
Soils with an argillic, kandic or natric horizon or a fragipan with clay skins	Yes →	Alfisols
Soils with a cambic, sulfuric, calcic, gypsic, petrocalcic, or petrogypsic horizon, or with a mollic, umbric, or histic epipedon, or with an exchangeable sodium percentage (ESP) of >15%, or a fragipan	Yes →	Inceptisols
Other soils	Yes →	Entisols

FIGURE 3.9 A simplified key to the 12 soil orders in *Soil Taxonomy*. In using the key, always begin at the top. Note how diagnostic horizons and other profile features are used to distinguish each soil order from the remaining orders. Entisols, having no such special diagnostic features, key out last. Also note that the sequence of soil orders in this key bears no relationship to the degree of profile development and adjacent soil orders may not be more similar than nonadjacent ones. See Section 3.2 for explanations of the diagnostic horizons.

3.5 ENTISOLS (RECENT: LITTLE IF ANY PROFILE DEVELOPMENT)

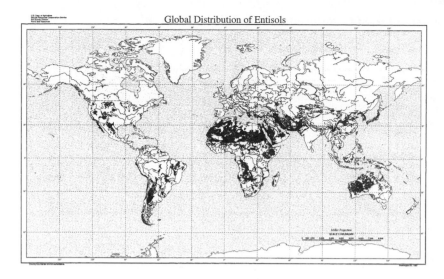

Global Distribution of Entisols

Suborders

Aquents (wet)

Arents (mixed horizons)

Fluvents (alluvial deposits)

Orthents (typical)

Psamments (sandy)

Weakly developed mineral soils without natural genetic (subsurface) horizons, or with only the beginnings of such horizons (see Plate 4, after page 82) belong to the Entisols order. Most have an ochric epipedon and a few have human-made anthropic or agric epipedons. Some have albic subsurface horizons. Soil productivity ranges from very high for certain Entisols formed in recent alluvium to very low for those forming in shifting sand or on steep rocky slopes.

This is an extremely diverse group of soils with little in common, other than the lack of evidence for all but the earliest stages of soil formation. Entisols are either young in years or their parent materials have not reacted to soil-forming factors. On parent materials such as fresh lava flows or recent alluvium (Fluvents), there has been too little time for much soil formation. In extremely dry areas, scarcity of water and vegetation may inhibit soil formation. Likewise, frequent saturation with water (Aquents) may delay soil formation. Some Entisols occur on steep slopes, where the rates of erosion may exceed the rates of soil formation, preventing horizon development. Others occur on construction sites where bulldozers destroy or mix together the soil horizons, causing the existing soils to become Entisols (some have suggested that these be called *urbents,* or urban Entisols).

Distribution and Use

Globally, Entisols occupy about 16% of the total ice-free land area. In the United States, they comprise 12% of the soil area (Table 3.5). Entisols are found under a wide variety of environmental conditions in the United States (Figure 3.8 and end papers). For example, in the Rocky Mountain region and in southwest Texas, shallow, medium-textured Entisols (Orthents) over hard rock are common. These are used mostly as rangeland. Sandy Entisols (Psamments; Figure 3.10) are found in Florida, Alabama, and Georgia, and typify the sand hill section of Nebraska. Psamments are used for cropland in humid areas. Some of the citrus-, vegetable-, and peanut-producing areas of the South are typified by Psamments. Poorly drained and seasonally flooded Entisols (Aquents) occur in major river valleys.

Entisols are probably found under even more widely varied environmental conditions outside the United States (see front papers). Psamments are typical of the Sahara desert and Saudi Arabia and dominate parts of southern Africa and central and north central Australia. Fluvents on recent alluviums in Asia have produced rice crops for generations. Entisols having medium to fine textures (Orthents) are found in northern Quebec and parts of Alaska, Siberia, and Tibet. Orthents are typical of some mountain

TABLE 3.5 Approximate Land Areas of Different Soil Orders and Suborders as Percentage of the Ice-Free Land in the World and in the United States

The major land use and natural fertility status of these soils are also given.

Soil order	Suborder[a]	Percent of ice-free land		Major land uses	Natural fertility
		Global[b]	United States[c]		
Alfisols		**9.65**	**14.51**		
	Aqualfs	0.64	2.47	Crops, forests	High
	Cryalfs	1.94	0.68	Forest	High
	Udalfs	2.09	7.21	Crops, forest	High
	Ustalfs	4.36	3.23	Crops	High
	Xeralfs	0.69	0.92	Range	High
Andisols		**0.70**	**1.74**		
	Cryands	0.20	0.98	Tundra, forest	Moderate
	Xerands	<0.01	0.34	Forest, range	Moderate
Aridisols		**12.10**	**8.78**		
	Argids	4.17	3.77	Crops, range	Low to moderate
	Calcids	3.75	2.54	Range	Low
	Cambids	2.23	1.18	Crops, range	Low
	Cryids	0.73	0.01	Range	Low
	Durids	<0.01	0.85	Range	Low
	Gypsids	0.53	0.16	Range	Low
	Salids	0.69	0.27	Range	Low
Entisols		**16.29**	**12.16**		
	Aquents	<0.01	1.40	Wetland, crops	Moderate
	Fluvents	2.20	1.78	Crops	Moderate
	Psamments	3.41	2.81	Crops, range	Low
	Orthents	10.58	5.93	Forest, range, crops	Low to moderate
Gelisols		**8.61**	**7.50**		
	Histels	0.77	0.02	Bogs	Moderate
	Orthels	3.02	7.47	Tundra	Moderate
	Turbels	4.88	<0.01	Tundra	Moderate
Histosols		**1.18**	**1.28**		
	Hemists	0.76	0.21	Wetland, crops	Moderate to high
	Saprists	0.26	0.94	Wetland, crops	High
Inceptisols		**9.91**	**9.11**		
	Aquepts	2.42	1.18	Crops	Low to high
	Cryepts	~1.5	1.47	Tundra, forest	Moderate
	Udepts	~4.0	3.90	Forest, crops	Low to moderate
	Ustepts	~1.0	1.67	Forest, range	Low to high
	Xerepts	~1.0	0.89	Range, forest	Moderate to high
Mollisols		**6.94**	**22.40**		
	Aquolls	<0.01	2.64	Crops, wetland	High
	Cryolls	0.90	1.51	Crops, range	High
	Udolls	0.97	4.09	Crops	High
	Ustolls	4.04	9.76	Crops, range	High
	Xerolls	0.71	4.20	Crops, range	High
Oxisols		**7.56**	**<0.01**		
	Perox	0.90	0.0	Forest	Low
	Udox	4.01	<0.01	Forest, crops	Low
	Ustox	2.39	<0.01	Forest, crops	Low
Spodosols		**2.58**	**3.27**		
	Aquods	<0.01	0.58	Forest	Low
	Cryods	1.90	0.64	Forest	Low
	Orthods	0.51	2.00	Forest	Low
Ultisols		**8.52**	**9.61**		
	Aquults	0.99	0.86	Forest	Low to moderate
	Udults	4.27	8.49	Forest, crops	Low
	Ustults	2.98	0.01	Forest, crops	Low
Vertisols		**2.44**	**1.72**		
	Aquerts	<0.05	0.64	Wetland, crops	High
	Torrerts	0.69	0.11	Range	High
	Uderts	0.31	<0.01	Crops	High
	Usterts	1.36	0.84	Crops, range	High
Shifting sands or rock		**14.07**	**7.81**		

[a] To save space, suborders representing less than 0.25% of the land area in both the world and in the United States are omitted from this table. Total global ice-free land area = 129,788,231 km^2. Total United States land area estimated from STATGO as 8,739,275 km^2.
[b] Global areas calculated from FAO world database by USDA/NRCS Soil Survey Division, World Soils Resources, Washington, D.C.
[c] U.S. areas calculated from State Soil Geographic Data Base (STATSGO) taxonomically amended in 1997 by USDA/NRCS Soil Survey Division, National Soil Survey Center, Lincoln, Nebr.

A

C

FIGURE 3.10 Profile of a Psamment formed on sandy alluvium in Virginia. Note the accumulation of organic matter in the A horizon but no other evidence of profile development. The knife is 24 cm long. (Photo courtesy of R. Weil)

areas, such as the Andes in South America, and some of the uplands of an area extending from Turkey eastward to Pakistan.

The agricultural productivity of the Entisols varies greatly depending on their location and properties. With adequate fertilization and a controlled water supply, some Entisols are quite productive. Entisols developed on alluvial floodplains are among the world's most productive soils. Such soils, with their level topography, proximity to water for irrigation, and periodic nutrient replenishment by floodwater sediments, have supported the development of many major civilizations. However, the productivity of most Entisols is restricted by inadequate soil depth, clay content, or water availability.

3.6 INCEPTISOLS (FEW DIAGNOSTIC FEATURES: INCEPTION OF B HORIZON)

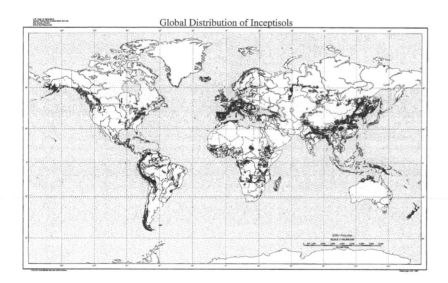

Global Distribution of Inceptisols

Suborders

Anthrepts (human-made, high phosphorus, dark surface)

Aquepts (wet)

Cryepts (very cold)

Udepts (humid climate)

Ustepts (semiarid)

Xerepts (dry summers, wet winters)

In Inceptisols the beginning or *inception* of profile development is evident, and some diagnostic features are present. However, the well-defined profile characteristics of soils thought to be more mature have not yet developed. For example, a cambic horizon showing some color or structural change is common in Inceptisols, but a more mature illuvial B horizon such as an argillic cannot be present. Other subsurface diagnostic horizons that may be present in Inceptisols include duripans, fragipans, and calcic, gypsic, and sulfidic horizons. The epipedon in most Inceptisols is an ochric, although a plaggen or weakly expressed mollic or umbric epipedon may be present. Inceptisols show more significant profile development than Entisols, but are defined to exclude soils with diagnostic horizons or properties that characterize certain other soil orders. Thus, soils with only slight profile development occurring in arid regions, or containing permafrost or andic properties, are excluded from the Inceptisols. They fall, instead, in the soil orders Aridisols, Gelisols, or Andisols, as discussed in later sections.

Distribution and Use

Inceptisols are widely distributed throughout the world and constitute more than 9% of the world's land area (Table 3.5). As with Entisols, Inceptisols are found in most climatic and physiographic conditions. They are often prominent in mountainous areas, especially in the tropics. They are also probably the most important soil order in the lowland rice-growing areas of Asia.

Inceptisols are found in each of the continents (see front papers). Inceptisols of humid regions, called *Udepts*, often have only thin, light-colored surface horizons (ochric epipedons). Such Udepts extend from southern New York through central and western Pennsylvania, West Virginia, and eastern Ohio. Udepts, along with Xerepts (Inceptisols in xeric climates), also dominate an area extending from southern Spain through central France to Germany, and are present as well in Chile, North Africa, eastern China, and western Siberia. Wet Inceptisols or Aquepts are found in areas along the Amazon and Ganges rivers.

The natural productivity of Inceptisols varies considerably. For instance, those found in the Pacific Northwest of the United States are quite fertile and provide some of the world's best land for producing wheat. In contrast, some of the low-organic-matter Udepts in southern New York and northern Pennsylvania are not naturally productive. They have been allowed to reforest following earlier periods of crop production.

3.7 ANDISOLS

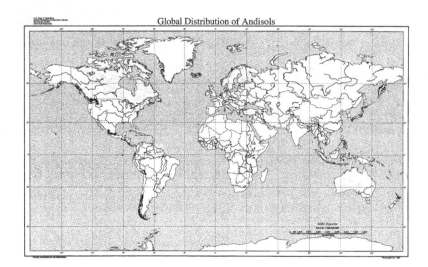

Global Distribution of Andisols

Suborders

Aquands (wet)	Ustands (moist/dry)
Cryands (cold)	Vitrands (volcanic glass)
Torrands (hot, dry)	Xerands (dry summers, moist winters)
Udands (humid)	

Andisols are usually formed on volcanic ash and cinders and are commonly found near the volcano source or in areas downwind from the volcano, where a sufficiently thick layer of ash has been deposited during eruptions. They are generally found in volcanic materials deposited in recent geological time. Andisols have not had time to become highly weathered. The principal soil-forming process has been the rapid weathering (transformation) of volcanic ash to produce amorphous or poorly crystallized silicate minerals such as *allophane* and *imogolite* and the iron oxy-hydroxide, *ferrihydrite*. The accumulation of organic matter is also quite pronounced (but not enough to form a histic epipedon), due largely to its protection in aluminum–humus complexes. Little downward translocation of these colloids, or other profile development, has yet to take place. Like the Entisols and Inceptisols, Andisols are young soils, usually having developed for only 5000 to 10,000 years. Unlike the previous two orders of immature soils, Andisols have a unique set of *andic properties* due to a common type of parent materials.

Andisols have andic soil properties in at least 35 cm of the upper 60 cm of soil. Materials with andic properties are characterized by a high content of volcanic glass and/or a high content of amorphous or poorly crystalline iron and aluminum minerals. The combination of these minerals and the high organic matter results in light, fluffy soils that are easily tilled, but have a high water-holding capacity and resist erosion by water. They are mostly found in regions where rainfall keeps them from being susceptible to erosion by wind. Andisols are usually of high natural fertility, except that phosphorus availability is severely limited by the extremely high phosphorus retention capacity of the andic materials (see Section 14.8). Fortunately, proper management of plant residues and fertilizers can usually overcome this difficulty. Some Andisols have a melanic epipedon, a surface diagnostic horizon that has a high organic matter content and dark color (see Plate 2, after page 82).

Distribution and Use

Andisols are found in areas where significant depths of volcanic ash and other ejecta have accumulated (Figure 3.11). Globally, they make up less than 1% of the soil area (Table 3.5). However, in the Pacific rim area they are important and productive soils that support intensive agriculture, especially in cool, high-elevation areas.

Andisols having a Udic (moist) soil moisture regime (Udands) are widely cultivated in Asia, producing enough food to support very high population densities. The somewhat drier Ustands are also used intensively for agriculture. Significant areas of Udands and Ustands occur along the Rift Valley of eastern Africa. Andisols are found to a minor extent in cold climates (Cryands) in Canada and Russia, and in hot, dry climates (Torrands) in Mexico and Syria. Very recent eruptions, such as that of Mount Saint Helens in the northwestern part of the United States and Mount Pinetabo in the Phillipines, are giving rise to Vitrands that still have much volcanic glass and lower water-holding capacities.

In the United States, the area of Andisols is not extensive since recent volcanic action is not widespread. However, Andisols do occur in some very productive wheat- and timber-producing areas of Washington, Idaho, Montana, and Oregon. Likewise, this soil order represents some of the best farmland found in Chile, Ecuador, Colombia, and much of Central America.

3.8 GELISOLS (PERMAFROST AND FROST CHURNING)

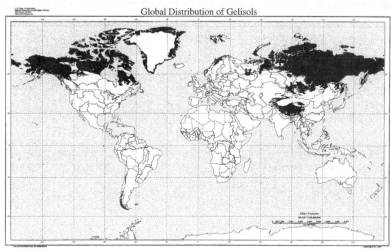

Global Distribution of Gelisols

Gelisols are young soils with little profile development. Cold temperatures and frozen conditions for much of the year slow the process of soil formation. The principal defining feature of these soils is the presence of a ***permafrost*** layer (see Plates 5 and 14 after page 82). Permafrost is a layer of material that remains at temperatures below 0°C for more than two consecutive years. It may be a hard, ice-cemented layer of soil material (e.g., designated Cfm in profile descriptions), or, if dry, it may be uncemented (e.g., designated Cf). In Gelisols, the permafrost layer lies within 100 cm of the soil surface, unless ***cryoturbation*** is evident within the upper 100 cm, in which case the permafrost may begin as deep as 200 cm from the soil surface.

Cryoturbation is the physical disturbance of soil materials caused by the formation of ice wedges and by the expansion and contraction of water as it freezes and thaws. This *frost churning* action moves the soil material so as to orient rock fragments along the lines of force and to form broken, convoluted horizons (e.g., designated Cjj), and/or accumulations of organic matter at the top of the permafrost. An example of "soil art" formed by cryoturbation is shown in Figure 3.12. The frost churning also may form patterns on the ground surface, such as hummocks and ice-rich polygons that may be several meters across. In some cases rocks forced to the surface form rings or netlike patterns.

Gelisols showing evidence of cryoturbation are called *Turbels*. Other Gelisols, often found in wet environments, have developed in accumulations of mainly organic materials, making them ***Histels*** (Greek *histos,* tissue; Figure 3.13). Most of the soil-forming processes that occur take place above the permafrost in the ***active layer*** that thaws every year or two. Various types of diagnostic horizons may have developed in different Gelisols, including mollic, histic, umbric, calcic, and, occasionally, argillic horizons.

Melanic
Epipedon

Pumice layer

Weathered
layers of
volcanic
ash and
pumice

Buried
A horizon

Oldest
layers of
volcanic
pumice

Underlying
layer of
expanding
clay

FIGURE 3.11   An Andisol developed on layers of volcanic ash and pumice in central Africa. (Photo courtesy of R. Weil)

FIGURE 3.12   The broken and involuted patterns formed by cryoturbation. This example is actually a relic of cryoturbation during the Pleistocene ice age in Europe, when this soil would most likely have been in the Turbels suborder. However, it is located in Hungary where permafrost no longer occurs, and today is found buried beneath a modern Mollisol formed in an overlying layer of windblown silt (loess). The dark round spots are filled-in animal burrows called *crotovinas* (see Section 2.14). Photograph is about 60 cm across. (Photo courtesy of Ericka Michéli, University of Agricultural Sciences, Gödöllö, Hungary.)

Active layer

Permafrost

FIGURE 3.13   A Gelisol with a histic epipedon and permafrost. This Histel was photographed in Alaska in July. Scale in centimeters. (Photo courtesy of Bockheim, University of Wisconsin)

Gelisols cover over 11 million km² or 8.6% of the Earth's land area—about 8 million km² in Northern Russia, and another 4 million in Canada and Alaska. Blanketed under snow and ice for much of the year, most Gelisols support tundra vegetation of lichens, grasses, and low shrubs that grow during the brief summers. Large areas of Gelisols consist of bogs, some literally floating on layers of frozen or unfrozen water. Millions of caribou, reindeer, and muskox survive on this vegetation during the summer, then migrate to the boreal forests during the coldest seasons. The many bogs and pools serve as nesting sites for migratory birds, which feed on the thick clouds of biting flies and mosquitoes. Human populations are very sparse in these inhospitable environments.

Gelisols present unique problems for construction projects. Many of the Histels are very wet and have little bearing strength to hold up roads or foundations. The permafrost in some Gelisols is unstable once disturbed. As a result, the trans-Alaska pipeline had to be built on stilts, rather than buried. If the pipe were in contact with the soils, heat from the flowing oil in the pipe would melt the permafrost, causing soil collapse that could break the pipes and spill the oil. Very few areas of Gelisols are used for agriculture. Plant productivity is low because of the extremely short potential growing season in the far Northern latitudes, the low levels of solar radiation (except during the fleeting summer), and the waterlogged condition of many Gelisols in which permafrost inhibits internal drainage during the summer thaw (Figure 3.13).

3.9 HISTOSOLS (ORGANIC SOILS WITHOUT PERMAFROST)

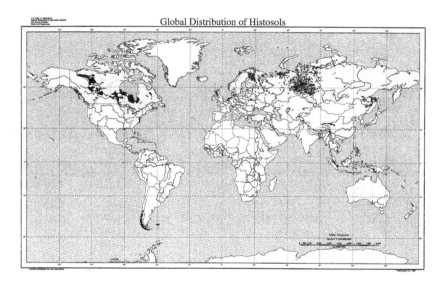

Global Distribution of Histosols

Suborders

Fibrists (fibers of plants obvious)   Hemists (fibers partly decomposed)

Folists (leaf mat accumulations)   Saprists (fibers not recognizable)

Histosols are soils that have undergone little profile development because of the anaerobic environment in which they form. The main process of soil formation evident in Histosols is the accumulation of partially decomposed organic parent material without permafrost (which would cause the soil to be classified in the Histels suborder of Gelisols). Histosols consist of one or more thick layers of *organic soil material* (see footnote 5, page 75). Generally, Histosols have organic soil materials in more than half of the upper 80 cm of soil (Plate 6 after page 82), or in two-thirds of the soil overlying shallow rock.

While not all wetlands contain Histosols, all Histosols (except Folists) occur in wetland environments. They can form in almost any moist climate in which plants can grow, from equatorial to arctic regions, but they are most prevalent in cold climates, up to the limit of permafrost. Horizons are differentiated by the type of vegetation contributing the residues, rather than by translocations and accumulations within the profile.

Organic deposits accumulate in marshes, bogs, and swamps, which are habitats for water-loving plants such as pond-weeds, cattails, sedges, reeds, mosses, shrubs, and even some trees. Generation after generation, the residues of these plants sink into the water, which inhibits their oxidation by reducing oxygen availability and, consequently, acts as a partial preservative (see Figure 2.23).

Because of their high content of organic matter, Histosols are generally black to dark brown in color. They are extremely lightweight when dry, being only about 10 to 20% as dense as most mineral soils. Histosols also have high water-holding capacities on a mass basis. While a mineral soil will absorb and hold from 20 to 40% of its weight of water, a cultivated Histosol may hold a mass of water equal to 200 to 400% of its dry weight. However, because of the low density of the organic material, most Histosols will hold only about the same amount of plant-available water as will a good mineral soil on a per-hectare or per-unit-volume basis.

Distribution and Use

Even though they cover only about 1% of the world's land area, Histosols, or *peat lands,* comprise significant areas in cold, wet regions of Alaska, Canada, Finland, and Russia, as well as in swampy areas such as Iceland. Of approximately 2 million km² of Histosols worldwide, about 75,000 are found in the contiguous United States. Three-quarters of this peat land is in glaciated areas such as Wisconsin, Minnesota, New York, and Michigan (Figure 3.14). Other important areas of Histosols are found in the tule-reed beds of California and in low-lying parts of the Atlantic and Gulf coastal plains, especially in the Everglades of Florida, the bayous of Louisiana, and the tidal marshes of the mid-Atlantic states.

Because the ecological roles of natural wetland environments have not always been appreciated (or protected by law), more than 50% of the original wetland area in the United States has been drained for agricultural or other uses, especially for vegetable and flower production. Some Histosols make very productive farmlands, but the

FIGURE 3.14 Profile of a drained Histosol on which onions are being produced in New York State. The organic soil rests on lake-laid (lacustrine) mineral material. (Photo courtesy of R. Weil)

organic nature of the materials requires liming, fertilization, tillage, and drainage practices quite different from those applied to soils in the other 11 orders.

If other than wetland plants are to be grown, the water table is usually lowered to provide an aerated zone for root growth. This practice, of course, alters the soil environment and causes the organic material to oxidize, resulting in the disappearance of as much as 5 cm of soil per year in warm climates (see Figure 12.27). To slow the loss of valuable soil resources and avoid unnecessarily aggravating the global greenhouse effect (see Section 12.11), the water table in forested or agricultural Histosols should be kept no lower than is needed to assure adequate root aeration. A more sustainable approach would be to allow some Histosol areas to revert to their native wetland condition (see Section 7.8).

In some places, Histosols are also mined for their peat, which is sold for use in potting media, as a mulch, and to make peat-fiber pots. Peat deposits are also used for fuel in some countries, especially in Russia, where several power stations are fueled by this material.

3.10 ARIDISOLS (DRY SOILS)

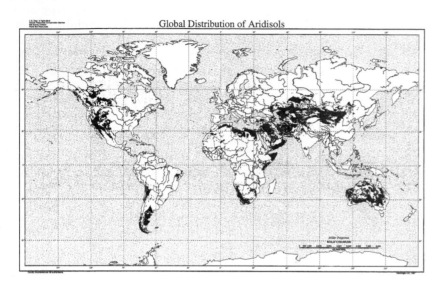

Global Distribution of Aridisols

Suborders

Argids (clay)	Durids (duripan)
Calcids (carbonate)	Gypsids (gypsum)
Cambids (typical)	Salids (salty)
Cryids (cold)	

Aridisols occupy a larger area globally than any other soil order (more than 12%) except Entisols. Water deficiency is a major characteristic of these soils. The soil moisture level is sufficiently high to support plant growth for no longer than 90 consecutive days. The natural vegetation consists mainly of scattered desert shrubs and short bunchgrasses. Soil properties, especially in the surface horizons, may differ substantially between interspersed bare and vegetated areas (see Section 2.14).

Aridisols are characterized by an ochric epipedon that is generally light in color and low in organic matter (see Plate 3 after page 82). The processes of soil formation have brought about a redistribution of soluble materials, but there is generally not enough water to leach these materials completely out of the profile. Therefore, they often accumulate at a lower level in the profile. These soils may have a horizon of accumulation of calcium carbonate (calcic), gypsum (gypsic), soluble salts (salic), or exchangeable sodium (natric). Under certain circumstances, carbonates may cement together the soil particles and coarse fragments in the layer of accumulation, producing hard layers known as ***petrocalcic*** horizons (Figure 3.15). These hard layers act as impediments to plant root growth and also greatly increase the cost of excavations for buildings.

FIGURE 3.15 Two features characteristic of some Aridisols. (Left) Wind-rounded pebbles have given rise to a desert pavement. (Right) A petrocalcic horizon of cemented calcium carbonate. (Photo courtesy of R. Weil)

Some Aridisols (the *Argids*) have an argillic horizon, most probably formed under a wetter climate that long ago prevailed in many areas that are deserts today. With time, and the addition of carbonates from calcareous dust and other sources, many argillic horizons become engulfed by carbonates (*Calcids*). On steeper land surfaces subject to erosion, argillic horizons do not get a chance to form, and the dominant soils are often *Cambids* (Aridisols with only weakly differentiated cambic subsurface B horizons).

In rocky soils, erosion may remove all the fine particles from the surface layers, leaving behind a layer of wind-rounded pebbles that is called **desert pavement** (see Figure 3.15). The surfaces of the pebbles in desert pavement often have a shiny coating called **desert varnish.** This coating is thought to be produced by algae that extract iron and manganese from the minerals and leave an oxide coating on the pebbles.

Except where there is groundwater or irrigation, the soil layers are only moist for short periods during the year. These short moist periods may be sufficient for drought adapted desert shrubs and annual plants, but not for conventional crop production. If groundwater is present near the soil surface, soluble salts may accumulate in the upper horizons to levels that most crop plants cannot tolerate.

Distribution and Use

Aridisols occur on nearly 12% of the land area worldwide and 9% in the United States, mostly in the western region. A large area dominated by Argids (Aridisols with an argillic horizon) occupies much of the southern parts of California, Nevada, Arizona, and central New Mexico (Appendix A). The Argids also extend down into northern Mexico. Smaller areas of Cambids (simple Aridisols without horizons of clay or salt accumulation) are found in several western states (see Plate 3, after page 82).

Vast areas of Aridisols are present in the Sahara desert in Africa, the Gobi and Taklamakan deserts in China, and the Turkestan desert of the former Soviet Union. Most of the soils of southern and central Australia are Aridisols, as are those of southern Argentina, southwestern Africa, Pakistan, and the Middle East countries.

Without irrigation, Aridisols are not suitable for growing cultivated crops. Some areas are used for low-intensity grazing, especially with sheep or goats, but the production per unit area is low. The overgrazing of Aridisols leads to increased heterogeneity of both soils and vegetation. The animals graze the relatively even cover of palatable grasses, giving a competitive advantage to various shrubs not eaten by the grazing animals. The scattered shrubs compete against the struggling grasses for water and nutrients. The once-grassy areas become increasingly bare, and the soils between the scattered shrubs succumb to erosion by the desert winds and occasional thunderstorms. The desertification of areas of sub-Saharan Africa and the western United States is evidence of such degradation (Figure 3.16).

Some xerophytic plants, such as a jojoba, have been cultivated on Aridisols to produce various industrial feedstocks, such as oil and rubber. Where irrigation water and fertilizers are available, some Aridisols can be made highly productive. Irrigated valleys

FIGURE 3.16 It is easy to overgraze the vegetation on an Aridisol, and an overgrazed Aridisol is subject to ready wind and water erosion. This gully was formed during infrequent but heavy rainstorms in an area in Arizona that at one time had been well covered with desert plants such as alkali sacaton and vine mesquite, plants that were destroyed by the overgrazing. (Photo courtesy U.S. Natural Resource Conservation Service)

in arid areas are among the most productive in the world. However, they must be carefully managed to prevent the accumulation of soluble salts (see Section 10.3).

3.11 VERTISOLS (DARK, SWELLING AND CRACKING CLAYS)[7]

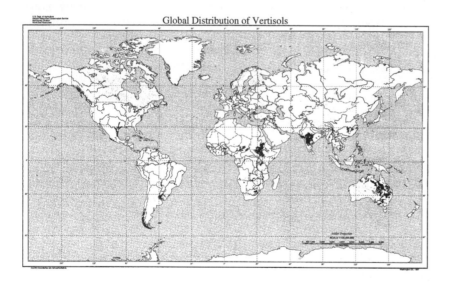

Suborders

Aquerts (wet)	Uderts (humid)
Cryerts (cold)	Usterts (moist/dry)
Torrerts (hot summer, very dry)	Xererts (dry summers, moist winters)

The main soil-forming process affecting Vertisols is the shrinking and swelling of clay as these soils go through periods of drying and wetting. Vertisols have a high content (>30%) of sticky, swelling and shrinking-type clays to a depth of 1 m or more. Most Vertisols are dark, even blackish in color, to a depth of 1 m or more (Plate 12 after page 82).

---

[7] Knowledge of the properties and mode of formation of Vertisols has increased greatly in recent years. See Coulombe, et al. (1996) for a detailed review.

However, unlike for most other soils, the dark color of Vertisols is not necessarily indicative of a high organic matter content. The organic matter content of dark Vertisols typically ranges from as much as 5 or 6% to as little as 1%.

Vertisols typically develop from limestone, basalt or other calcium and magnesium rich parent materials. In east Africa, they typically form in landscape depressions which collect the calcium and magnesium leached out of the surrounding upland soils. The presence of these cations encourages the formation of swelling-type clays (see Section 8.6).

Vertisols are found mostly in subhumid to semiarid environments in warm regions, but a few (Cryerts) occur where the average soil temperatures are as low as 0°C (see Figure 3.7). The native vegetation is usually grassland. Vertisols generally occur where the climate features dry periods of several months. In dry seasons the clay shrinks, causing the soils to develop deep, wide cracks that are diagnostic for this order (Figure 3.17). The surface soil generally forms granules, of which a significant number may slough off into the cracks, giving rise to a partial inversion of the soil (Figure 3.18). This accounts for the association with the term *invert,* from which this order derives its name.

When the rains come, water entering the large cracks moistens the clay in the subsoils, causing it to swell. The repeated shrinking and swelling of the subsoil clay results in a kind of imperceptively slow "rocking" movement of great masses of soil. As the subsoil swells, blocks of soil shear off from the mass and rub past each other under pressure, giving rise in the subsoil to shiny, grooved, tilted surfaces called *slickensides* (Figure 3.19). Eventually, this back-and-forth motion may form bowl-shaped depressions with relatively deep profiles surrounded by slightly raised areas in which little soil development has occurred and in which the parent material remains close to the surface (see Figure 3.18). The resulting pattern of microhighs and microlows on the land surface, called *Gilgai,* is usually discernable only where the soil is untilled.

FIGURE 3.17 Wide cracks formed during the dry season in the surface layers of this Vertisol in India. Surface debris can slough off into these cracks and move to subsoil. When the rains come, water can move quickly to the lower horizons, but the cracks are soon sealed, making the soils relatively impervious to the water.

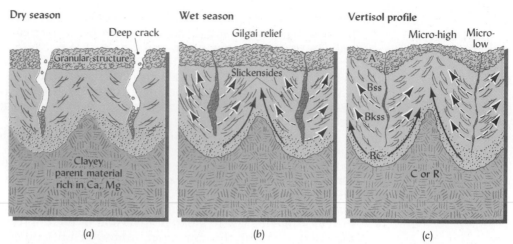

FIGURE 3.18 Vertisols typically are high in swelling-type clay and have developed wedgelike structures in the subsoil horizons. (*a*) During the dry season, large cracks appear as the clay shrinks upon drying. Some of the surface soil granules fall into cracks under the influence of wind and animals. This action causes a partial mixing, or *inversion,* of the horizons. (*b*) During the wet season, rainwater pours down the cracks, wetting the soil near the bottom of the cracks first, and then the entire profile. As the clay absorbs water, it swells the cracks shut, entrapping the collected granular soil. The increased soil volume causes lateral and upward movement of the soil mass. The soil is pushed up between the cracked areas. As the subsoil mass shears from the strain, smooth surfaces or *slickensides* form. (*c*) These processes result in a Vertisol profile that typically exhibits gilgai, cracks more than 1 m deep, and slickensides in a Bss horizon. Calcium carbonate concretions often accumulate in a Bkss horizon.

Distribution and Use

Globally, Vertisols comprise about 2.5% of the total land area. Large areas of Vertisols are found in India, Ethiopia, the Sudan, and northern and eastern Australia (see front papers). Smaller areas occur in sub-Saharan Africa and in Mexico, Venezuela, Bolivia, and Paraguay. These latter soils probably are of the Usterts or Xererts suborders, since dry conditions persist long enough for the wide cracks to stay open for periods of three months or longer.

There are several small but significant areas of Vertisols in the United States (see Figure 3.8). They make up less than 2% of the land area in this country. Two areas are located in humid areas, one in eastern Mississippi and western Alabama (the so-called

FIGURE 3.19 An example of a slickenside in a Vertisol. Massive rubbing of soil blocks against each other in response to the swelling clays have brought about this shiny appearance. The white concretions of carbonate minerals are evidence of the high base status of these soils. (Photo courtesy of R. Weil)

black belt) and the other along the southeast coast of Texas. These soils are of the Uderts suborder, because their moist condition prevents cracks from persisting for more than three months of the year.

Two other Vertisol areas are found in east central and southern Texas, where the soils are drier. Since cracks persist for more than three months of the year in these areas, the soils belong to the Usterts suborder, characteristic of areas with hot, dry summers. An area of Cryerts is located in the Dakotas and Saskatchewan. Xererts (Greek *xeros,* dry) are also located in California.

The high shrink-swell potential of Vertisols makes them extremely problematic for any kind of highway or building construction (Figure 8.20). This property also makes agricultural management very difficult. Because they are very sticky and plastic when wet and become very hard when dry, the timing of tillage operations is critical. Some farmers refer to Vertisols as *24-hour soils,* because they are said to be too wet to plow one day and too dry the next.

Even when the soil moisture is near optimal, the energy requirement for tillage is high. Therefore, except where heavy equipment is used for tillage, the operations are slow, and the amount of land a farmer can cultivate is much smaller than for other soil orders. In areas such as those in India and the Sudan, where slow-moving animals or human power are commonly used to till the soil, farmers cannot perform tillage operations on time and are limited to the use of very small tillage implements that their animals can pull through the heavy soil.

Recent research shows that the large areas of Vertisols in the tropics can produce greatly increased yields of food crops with improved soil management practices. Soils in this order are, however, very susceptible to physical degradation and erosion (despite their mainly gentle slopes), and conservation practices or reversion to rangeland are important management options to consider.

3.12 MOLLISOLS (DARK, SOFT SOILS OF GRASSLANDS)

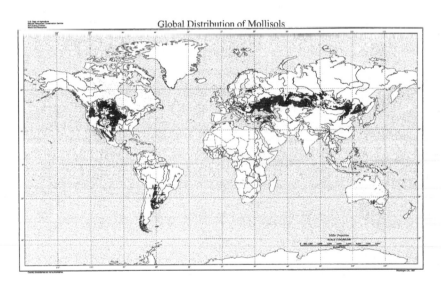

Global Distribution of Mollisols

Suborders

Albolls (albic horizon)   Udolls (humid)
Aquolls (wet)   Ustolls (moist/dry)
Cryolls (very cold)   Xerolls (dry summers, moist winters)
Rendolls (calcareous)

The principal process in the formation of Mollisols is the accumulation of calcium-rich organic matter, largely from the dense root systems of prairie grasses, to form the thick, soft Mollic epipedon that characterizes soils in this order (Plate 8 after page 82). This humus-rich surface horizon is often 60 to 80 cm in depth and high in calcium and magnesium. Its cation exchange capacity is more than 50% saturated with base-forming cations (Ca^{2+}, Mg^{2+}, etc.). Mollisols in humid regions generally have higher organic matter and darker, thicker mollic epipedons than their lower-moisture-regime counterparts (see Section 12.10).

The surface horizon generally has granular or crumb structures, largely resulting from an abundance of organic matter and swelling-type clays. The aggregates are not hard when the soils are dry, hence the name *Mollisol*, implying softness (Table 3.4). In addition to the mollic epipedon, Mollisols may have an argillic (clay), natric, albic, or cambic subsurface horizon, but not an oxic or spodic horizon.

Most Mollisols have developed under grass vegetation (Figure 3.20). Grassland soils of the central part of the United States, lying between Aridisols on the west and the Alfisols on the east, make up the central concept of this order. However, a few soils developed under forest vegetation (primarily in depressions) have a mollic epipedon and are included among the Mollisols.

Distribution and Use

Mollisols cover a larger land area in the United States than any other soil order (about 22%) (see Table 3.5). Mollisols are dominant in the Great Plains of the United States, as well as in Illinois (see Figure 3.8). Where soil moisture is not limiting, Udolls are found. They are associated with nearby wet Mollisols termed Aquolls. A region characterized by Ustolls (intermittently dry during the summer) extends from Manitoba and Saskatchewan in Canada to southern Texas. Farther west are found sizable areas of Xerolls (with a Xeric moisture regime, which is very dry in summer but moist in winter). Landscapes common for Udolls and Ustolls can be seen in Figure 3.21. Conservation of soil water is a major consideration in the management of Ustolls, in particular. Two Mollisols profiles are included in Figure 3.22.

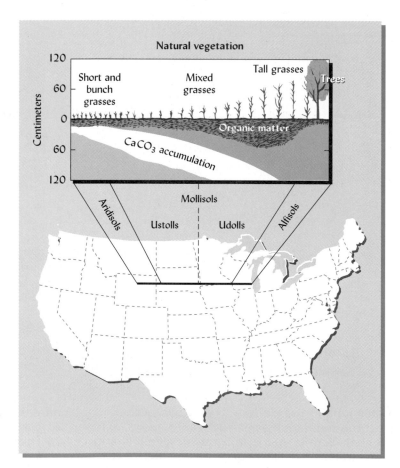

FIGURE 3.20 Correlation between natural grassland vegetation and certain soil orders is graphically shown for a transect across north central United States. The control, of course, is climate. Note the deeper organic matter and deeper zone of calcium accumulation, sometimes underlain by gypsum, as one proceeds from the drier areas in the west toward the more humid region where prairie soils are found. Alfisols may develop under grassland vegetation, but more commonly occur under forests and have lighter-colored surface horizons.

FIGURE 3.21 Typical landscapes dominated by Ustolls (top) and Udolls (bottom). These productive soils produce much of the food and feed in the United States. (Photos courtesy of R. Weil)

The largest area of Mollisols in the world stretches from east to west across the heartlands of Kazakhstan, Ukraine, and Russia. Other sizable areas are found in Mongolia and northern China and in northern Argentina, Paraguay, and Uruguay. Mollisols occupy only about 7% of the world's total soil area, but because of their generally high fertility, they account for a much higher percentage of total crop production.

In the United States, efforts are underway to preserve the few remnants of the once vast and diverse prairie ecosystem. Because the high native fertility of Mollisols makes them among the world's most productive soils, few Mollisols have been left uncultivated in regions with sufficient rainfall for crop production. When they were first cleared and plowed, much of their native organic matter was oxidized, releasing nitrogen and other nutrients in sufficient quantities to produce high crop yields even without the use of fertilizers. Even after more than a century of cultivation, Mollisols are among the most productive soils, although some fertilization is generally required. However, continuous cultivation with row crops has led to serious deterioration of soil structure and to soil erosion where the land is sloping.

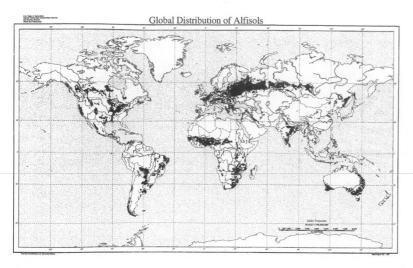

Global Distribution of Alfisols

Suborders

Aqualfs (wet)   Ustalfs (moist/dry)

Cryalf (cold)   Xeralfs (dry summers, moist winters)

Udalfs (humid)

The Alfisols appear to be more strongly weathered than soils in the orders just discussed, but less so than Spodosols (see following). They are formed in cool to hot humid areas (see Figure 3.7) but also are found in the semiarid tropics and Mediterranean climates. Most often, Alfisols develop under native deciduous forests, although in some cases, as in California and parts of Africa, grass savanna is the native vegetation.

Alfisols are characterized by a subsurface diagnostic horizon in which silicate clay has accumulated by illuviation (see Plate 1, after page 82). Clay skins or other signs of clay movement are present in such a B horizon (see Plate 23). In Alfisols, this clay-rich

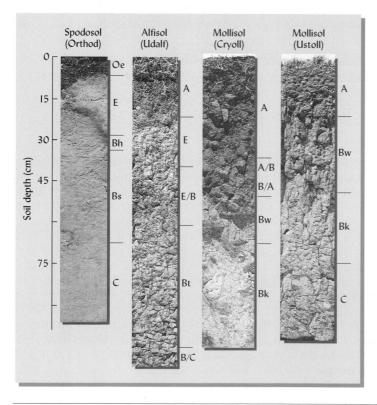

FIGURE 3.22   Monoliths of profiles representing three soil orders. The suborder names are in parentheses. Genetic (not diagnostic) horizon designations are also shown. Note the spodic horizons in the Spodosol characterized by humus (Bh) and iron (Bs) accumulation. In the Alfisol is found the illuvial clay horizon (Bt), and the structural B horizon (Bw) is indicated in the Mollisols. The thick dark surface horizon (mollic epipedon) characterizes both Mollisols. Note that the zone of calcium carbonate accumulation (Bk) is near the surface in the Ustoll, which has developed in a dry climate. The E/B horizon in the Alfisol has characteristics of both E and B horizons.

horizon is only moderately leached, and its cation exchange capacity is more than 35% saturated with base-forming cations (Ca^{2+}, Mg^{2+}, etc.). In most Alfisols this horizon is termed *argillic* because of its accumulation of silicate clays. The horizon is termed *natric* if, in addition to having an accumulation of clay, it is more than 15% saturated with sodium and has prismatic or columnar structure (see Figure 4.11). In some Alfisols in subhumid tropical regions, the accumulation is termed a *kandic* (from the mineral kandite) horizon because the clays have a low cation exchange capacity.

Alfisols very rarely have a mollic epipedon, for such soils would be classified in the Argiudolls or other suborder of Mollisols with an argillic horizon. Instead, Alfisols typically have a relatively thin, gray to brown ochric epipedon (Plate 1, after page 82, shows an example) or an umbric epipedon. Those formed under deciduous temperate forests commonly have a light-colored, leached *albic* E horizon immediately under the A horizon (see Figures 1.3 and 3.22).

Distribution and Use

Alfisols occupy about 14% of the land area in the United States and 10% globally. Udalfs (humid region Alfisols) dominate large areas in Ohio, Indiana, Michigan, Wisconsin, Minnesota, Pennsylvania, and New York in the United States (see Figure 3.8), as well as in central China, England, France, central Europe, and southeastern Australia. There are sizable areas of Xeralfs (Alfisols in regions of dry summers and moist winters) in central California, southwestern Australia, Italy, and central Spain and Portugal (Figure 3.23). Cryalfs (very cold) can be found in the Rocky Mountains; in south-central Canada; in Minnesota; in northern Europe, extending from the Baltic States through western Russia; and in Siberia. Where summers are hot and dry, including areas in Texas, New Mexico, sub-Saharan Africa, eastern Brazil, eastern India, and southeastern Asia, Ustalfs are prominent. Many Alfisols landscapes include wet depressions characterized by Aqualfs.

In general, Alfisols are productive soils. Good hardwood forest growth and crop yields are favored by their medium- to high-base status, generally favorable texture, and location (except for some Xeralfs) in regions with sufficient rainfall for plants for at least part of the year. In the United States these soils rank favorably with the Mollisols and Ultisols in their productive capacity. Many, especially the sandier ones, are quite susceptible to erosion by heavy rains if deprived of their natural surface litter. Alfisols in udic moisture regimes are sufficiently acidic to require amendment with limestone for many kind of plants (see Chapter 9).

FIGURE 3.23 A landscape in southern Portugal dominated by Xeralfs. Most of the original forest has been cleared and the soils plowed for production of winter cereals and grapes. (Photo courtesy of R. Weil)

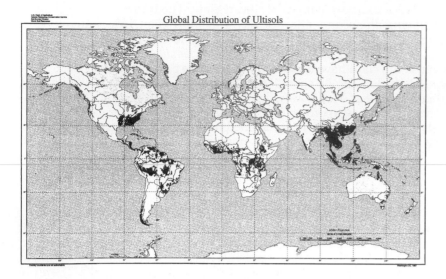

Suborders

Aquults (wet)   Ustults (moist/dry)

Humults (high humus)   Xerults (dry summers, moist winters)

Udults (humid)

The principal processes involved in forming Ultisols are clay mineral weathering, translocation of clays to accumulate in an argillic or kandic horizon, and leaching of base-forming cations from the profile. Most Ultisols have developed under moist conditions in warm to tropical climates. Ultisols are formed on old land surfaces, usually under forest vegetation, although savanna or even swamp vegetation is also common. They often have an ochric or umbric epipedon, but are characterized by a relatively acidic B horizon that has less than 35% of the exchange capacity satisfied with base-forming cations. The clay accumulation may be either an argillic horizon or, if the clay is of low activity, a kandic horizon. Ultisols commonly have both an epipedon and a subsoil that is quite acid and low in plant nutrients.

Ultisols are more highly weathered and acidic than Alfisols, but less acid than Spodosols and less highly weathered than the Oxisols. Except for the wetter members of the order, their subsurface horizons are commonly red or yellow in color, evidence of accumulations of oxides of iron (see Plate 11). Certain Ultisols that formed under fluctuating wetness conditions have horizons of iron-rich mottled material called *plinthite* (see Plate 15). This material is soft and can be easily dug from the profile so long as it remains moist. When dried in the air, however, plinthite hardens irreversibly into a kind of ironstone that is virtually useless for cultivation, but can be used to make durable bricks for building.

Distribution and Use

About 9% of the soil area in the United States and in the world is classified in the Ultisols order. Most of the soils of the southeastern part of the United States fall in the suborder Udults (see Figure 3.8 and Appendix A). Large areas of Udults are also located in southeastern Asia and in southern China. Extensive areas of Ultisols are found in the humid tropics in close association with some Oxisols. Important agricultural areas are found in southern Brazil and Paraguay.

Humults (high in organic matter) are found in the United States in Hawaii and in western California, Oregon, and Washington. Humults are also present in the highlands of some tropical countries. Xerults (Ultisols in Mediterranean-type climates) occur locally in southern Oregon and northern and eastern California. Ustults are found in semiarid areas with a marked dry season. Together with the Ustalfs, the Ustults occupy large areas in Africa and India. Ultisols are prominent on the east and northeast coasts of Australia (see front papers).

FIGURE 3.24 The soils in this high-elevation, tropical area of south Asia are Ultisols in the suborder Humults. These soils are being intensively used both for house construction and for market gardens. The combination of a favorable climate and soils that are high in organic matter (Humults have at least 9% down to the upper part of the B horizon) and respond well to fertilizer has encouraged local residents to use every bit of the land in producing vegetables to supplement their incomes. (Photo courtesy of R. Weil)

Although Ultisols are not naturally as fertile as Alfisols or Mollisols, they respond well to good management. They are located mostly in regions of long growing seasons and of ample moisture for good crop production (Figure 3.24). The silicate clays of Ultisols are usually of the nonsticky type, which, along with the presence of iron oxides and aluminum, assures ready workability. Where adequate levels of fertilizers and lime are applied, Ultisols are quite productive. In the United States, well-managed Ultisols compete well with Mollisols and the Alfisols as first-class agricultural soils. They also support the most productive commercial softwood and hardwood forests in the country.

3.15 SPODOSOLS (ACID, SANDY, FOREST SOILS, LOW BASES)

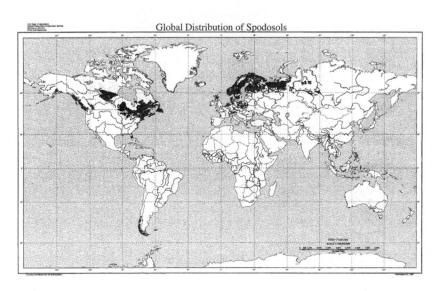

Global Distribution of Spodosols

Suborders

    Aquods (wet)   Humods (humus)

    Cryods (icy cold)   Orthods (typical)

Spodosols occur mostly on coarse-textured, acid parent materials subject to ready leaching. They occur only in moist to wet areas, commonly where it is cold or temperate (see Figure 3.7), but also in some tropical and subtropical areas. Intensive acid leaching is the principal soil-forming process. They are mineral soils with a *spodic* horizon, a subsurface accumulation of illuviated organic matter, and an accumulation of aluminum oxides with or without iron oxides (see Plate 10 and Figure 3.22). This usually thin, dark, illuvial horizon typically underlies a light, ash-colored, eluvial *albic* horizon.

Spodosols form under forest vegetation, especially under coniferous species whose needles are low in base-forming cations like calcium and high in acid resins. As this acid litter decomposes, strongly acid organic compounds are released and carried down into the permeable profile by percolating waters. These acids bind with iron and aluminum and carry them downward until they precipitate in the spodic horizon. Similarly, soluble organic compounds move downward and are likewise precipitated above or intermixed with the iron and aluminum compounds. Most minerals except quartz are removed by this acid leaching, generally removing the coloring agents from the E horizon. Deeper in the profile, precipitated oxides of aluminum and iron form a Bs spodic horizon that is often reddish brown in color. The leaching organic compounds may precipitate and form a black-colored Bh spodic horizon. The precipitation often occurs along wavy wetting fronts, thus yielding striking Spodosol profiles (Figure 3.25). The depth of these weathering and leaching processes, hence the spodic horizons, can vary from less than 20 cm to several meters.

Distribution and Use

Large areas of spodosols are found in northern Europe and Russia and central and eastern Canada. Many of the soils in the northeastern United States, as well as those of

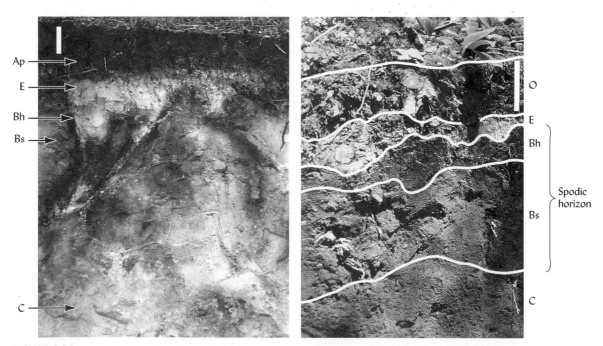

FIGURE 3.25 Two Spodosol profiles. The Spodosol in the Upper Penninsula of Michigan (left) shows a relatively deep, very wavy spodic horizon under a discontinuous albic horizon. It has a uniform ochric Ap horizon, indicating that it has been cleared and plowed. The Spodosol in Quebec (right) has a very shallow spodic horizon and an undisturbed O horizon consisting of conifer needless in various states of decay. Typical of Spodosols, both soils are sandy, very acidic, and best suited to supporting coniferous forests such as influenced their formation. The bar in each photo is 10 cm long. (Photos courtesy of R. Weil)

northern Michigan and Wisconsin and southern Alaska, belong to this order (see Figures 3.8 and 3.25). Spodosols are found on about 3% of the land area both globally and in the United States. Small but important areas occur in the southern part of South America and in the cool mountainous areas of temperate regions.

Most Spodosols are Orthods, soils that typify the central concept of Spodosols described previously. Some, however, are Aquods because they are seasonally saturated with water and possess characteristics associated with this wetness. Important areas of Aquods occur in Florida and other areas with warm climates.

Spodosols are not naturally fertile. When properly fertilized, however, these soils can become quite productive. For example, most potato-producing soils of northern Maine are Spodosols, as are some of the vegetable- and fruit-producing soils of Florida, Michigan, and Wisconsin. Because of their sandy nature and occurrence in regions of high rainfall, groundwater contamination by leaching of soluble fertilizers and pesticides has proved to be a serious problem where these soils are used in crop production. They are now covered mostly with forests, the vegetation under which they originally developed. Most Spodosols should remain as forest habitats. Because they are already quite acid and poorly buffered, many Spodosols and the lakes in watersheds dominated by soils of this order are susceptible to damage from acid rain (see Section 9.6).

3.16 OXISOLS (OXIC HORIZON, HIGHLY WEATHERED)

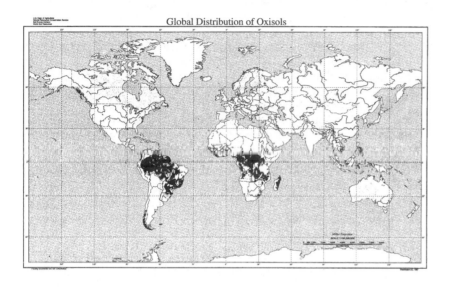

Global Distribution of Oxisols

Suborders

Aquox (wet)	Udox (humid)
Perox (very humid)	Ustox (moist/dry)
Torrox (hot, dry)	

The Oxisols are the most highly weathered soils in the classification system (see Figure 3.6). They form in hot climates with nearly year-round moist conditions; hence, the native vegetation is generally thought to be tropical rain forest. However, some Oxisols (Ustox) are found in areas which are today much drier than was the case when the soils were forming their oxic characteristics. Their most important diagnostic feature is a deep oxic subsurface horizon. This horizon is generally very high in clay-size particles dominated by hydrous oxides of iron and aluminum. Weathering and intense leaching have removed a large part of the silica from the silicate materials in this horizon. Some quartz and 1:1-type silicate clay minerals remain, but the hydrous oxides are dominant (see Chapter 8 for information on the various clay minerals). The epipedon in most Oxisols is either ochric or umbric. Usually the boundaries between subsurface horizons are indistinct, giving the subsoil a relatively uniform appearance with depth.

The clay content of Oxisols is generally high, but the clays are of the low-activity, nonsticky type. Consequently, when the clay dries out it is not hard and cloddy, but is easily worked. Also, Oxisols are resistant to compaction, so water moves freely through the profile. The depth of weathering in Oxisols is typically much greater than for most of the other soils, 20 m or more having been observed. The low-activity clays have a very limited capacity to hold nutrient cations such as Ca^{2+}, Mg^{2+}, and K^+, so they are typically of low natural fertility and moderately acid. The high concentration of iron and aluminum oxides also gives these soils a capacity to bind so tightly with what little phosphorus is present, that phosphorus deficiency often limits plant growth once the natural vegetation is disturbed.

Road and building construction is relatively easily accomplished on most Oxisols because these soils are easily excavated, do not shrink and swell, and are physically very stable on slopes. The very stable aggregation of the clays, stimulated largely by iron compounds, makes these soils quite resistant to erosion.

Distribution and Use

Oxisols are found on about 8% of the world's land (see Table 3.5). They occupy old land surfaces and occur mostly in the tropics. Although nearly all Oxisols occur in the tropics, most tropical soils are *not* Oxisols. Large areas of Oxisols occur in South America and Africa (see front papers). New data from Brazilian soil scientists suggests that some of the areas of the Amazon basin currently mapped as Oxisols are in reality dominated by Ultisols and other soils. Udox (Oxisols having a short dry season or none) occur in northern Brazil and neighboring countries as well as in the Caribbean area (see Plate 9, after page 82). Important areas of Ustox (hot, dry summers) occur in Brazil to the south of the Udox. In the humid areas of central Africa, Oxisols are prominent and in some cases dominant.

Relatively less is known about Oxisols than about most other soil orders. They occur in large geographic areas, often associated with Ultisols. Millions of people in the tropics depend on them for food and fiber production. However, because of their low natural fertility, most Oxisols have been left under forest vegetation or are farmed by shifting cultivation methods. Nutrient cycling by deep-rooted trees is especially important to the productivity of these soils. Probably the best use of Oxisols, other than supporting rain forests, is the culture of mixed-canopy perennial crops, especially tree crops. Such cultures can restore the nutrient cycling system that characterized the soil–plant relationships before the rain forest was removed.

3.17 LOWER-LEVEL CATEGORIES IN SOIL TAXONOMY

Suborders

Within each soil order just described, soils are grouped into suborders (Table 3.6) on the basis of soil properties that reflect major environmental controls on current soil-forming processes. Many suborders are indicative of the moisture regime or, less frequently, the temperature regime under which the soils are found. Thus, soils formed under wet conditions generally are identified under separate suborders (e.g., Aquents, Aquerts, and Aquepts), as being wet soils.

To determine the relationship between suborder names and soil characteristics, refer to Table 3.7. Here the formative elements for suborder names are identified and their connotations given. Thus, the Ustolls are dry Mollisols. Likewise, soils in the Udults suborder (from the Latin *udus,* humid) are moist Ultisols. Identification of the primary characteristics of each of the other suborders can be made by cross-reference to Tables 3.6 and 3.7.

Great Groups

The great groups are subdivisions of suborders. More than 240 great groups are recognized. They are defined largely by the presence or absence of diagnostic horizons and the arrangements of those horizons. These horizon designations are included in the list of formative elements for the names of great groups shown in Table 3.8. Note that these formative elements refer to epipedons such as umbric and ochric (see Table 3.1 and Figure 3.2), to subsurface horizons such as argillic and natric, and to certain diagnostic impervious layers such as duripans and fragipans (see Figure 3.26).

TABLE 3.6 Soil Orders and Suborders in *Soil Taxonomy*

Note that the ending of the suborder name identifies the order in which the soils are found.

Order	Suborder	Order	Suborder	Order	Suborder
Alfisols	Aqualfs	Andisols	Aquands	Aridisols	Argids
	Cryalfs		Cryands		Calcids
	Udalfs		Torrands		Cambids
	Ustalfs		Udands		Cryids
	Xeralfs		Ustands		Durids
			Vitrands		Gypsids
			Xerands		Salids
Entisols	Aquents	Gelisols	Histels	Histosols	Fibrists
	Arents		Orthels		Folists
	Fluvents		Turbels		Hemists
	Orthents				Saprists
	Psamments				
Inceptisols	Anthrepts	Mollisols	Albolls	Oxisols	Aquox
	Aquepts		Aquolls		Perox
	Cryepts		Cryolls		Torrox
	Udepts		Rendolls		Udox
	Ustepts		Udolls		Ustox
	Xerepts		Ustolls		
			Xerolls		
Spodosols	Aquods	Ultisols	Aquults	Vertisols	Aquerts
	Cryods		Humults		Cryerts
	Humods		Udults		Uderts
	Orthods		Ustults		Usterts
			Xerults		Xererts

TABLE 3.7 Formative Elements in Names of Suborders in *Soil Taxonomy*

Formative element	Derivation	Connotation of formative element
alb	L. *albus*, white	Presence of albic horizon (a bleached eluvial horizon)
anthr	Gk. *anthropos*, human	Presence of anthropic or plaggen epipedon
aqu	L. *aqua*, water	Characteristics associated with wetness
ar	L. *arare*, to plow	Mixed horizons
arg	L. *argilla*, white clay	Presence of argillic horizon (a horizon with illuvial clay)
calc	L. *calcis*, lime	Presence of calcic horizon
camb	L. *cambriare*, to change	Presence of cambric horizon
cry	Gk. *kryos*, icy cold	Cold
dur	L. *durus*, hard	Presence of a duripan
fibr	L. *fibra*, fiber	Least decomposed stage
fluv	L. *fluvius*, river	Floodplains
fol	L. *folia*, leaf	Mass of leaves
gyps	L. *gypsum*, gypsum	Presence of gypsic horizon
hem	Gk. *hemi*, half	Intermediate stage of decomposition
hist	Gk. *histos*, tissue	Presence of histic epipedon
hum	L. *humus*, earth	Presence of organic matter
orth	Gk. *orthos*, true	The common ones
per	L. *per*, throughout time	Of year-round humid climates, perudic moisture regime
psamm	Gk. *psammos*, sand	Sand textures
rend	Modified from Rendzina	Rendzinalike—high in carbonates
sal	L. *sal*, salt	Presence of salic (saline) horizon
sapr	Gk. *sapros*, rotten	Most decomposed stage
torr	L. *torridus*, hot and dry	Usually dry
turb	L. *turbidus*, disturbed	Cryoturbation
ud	L. *udus*, humid	Of humid climates
ust	L. *ustus*, burnt	Of dry climates, usually hot in summer
vitr	L. *vitreus*, glass	Resembling glass
xer	Gk. *xeros*, dry	Dry summers, moist winters

TABLE 3.8   Formative Elements for Names of Great Groups and Their Connotation

These formative elements combined with the appropriate suborder names give the great group names.

Formative element	Connotation	Formative element	Connotation	Formative element	Connotation
acr	Extreme weathering	fol	Mass of leaves	petr	Cemented horizon
agr	Agric horizon	fragi	Fragipan	plac	Thin pan
al	High aluminum, low iron	fragloss	See *frag* and *gloss*	plagg	Plaggen horizon
alb	Albic horizon	fulv	light-colored melanic horizon	plinth	Plinthite
and	Ando-like	gyps	Gypsic horizon	psamm	Sand texture
anhy	Anhydrous	gloss	Tongued	quartz	High quartz
aqu	Water saturated	hal	Salty	rhod	Dark red colors
argi	Argillic horizon	hapl	Minimum horizon	sal	Salic horizon
calc, calci	Calcic horizon	hem	Intermediate decomposition	sapr	Most decomposed
camb	Cambic horizon	hist	Presence of organic materials	somb	Dark horizon
chrom	High chroma	hum	Humus	sphagn	Sphagnum moss
cry	Cold	hydr	Water	sulf	Sulfur
dur	Duripan	kand	Low-activity 1:1 silicate clay	torr	Usually dry and hot
dystr, dys	Low base saturation	lithic	Near stone	ud	Humid climates
endo	Fully water saturated	luv, lu	Illuvial	umbr	Umbric epipedon
epi	Perched water table	melan	Melanic epipedon	ust	Dry climate, usually hot in summer
eutr	High base saturation	molli	With a mollic epipedon	verm	Wormy, or mixed by animals
ferr	Iron	natr	Presence of a natric horizon	vitr	Glass
fibr	Least decomposed	pale	Old development	xer	Dry summers, moist winters
fluv	Floodplain				

Remember that the great group names are made up of these formative elements attached as prefixes to the names of suborders in which the great groups occur. Thus, Ustolls with a natric horizon (high in sodium) belong to the Natrustolls great group. As can be seen in the example discussed in Box 3.1, soil descriptions at the great group level can provide important information not indicated at the higher, more general levels of classification.

The names of selected great groups from three orders are given in Table 3.9. This list illustrates again the usefulness of *Soil Taxonomy*, especially the nomenclature it employs. The names identify the suborder and order in which the great groups are

FIGURE 3.26 A forested Fragiudalf in Missouri containing a typical well-developed fragipan with coarse prismatic structure (outlined by gray, iron-depleted coatings). Fragipans (usually Bx or Cx horizons) are extremely dense and brittle. They consist mainly of silt, often with considerable sand, but not very much clay. One sign of encountering a fragipan in the field is the ringing noise that your shovel will make when you attempt to excavate it. Digging through a fragipan is almost like digging concrete. Plant roots cannot penetrate this layer. Yet, once a piece of a fragipan is broken loose, it fairly easily crushes with hand pressure. But it does not squash or act in a plastic manner as a claypan would; instead, it bursts in a *brittle* manner. (Photo courtesy of Fred Rhoton, Agricultural Research Service, U.S. Department of Agriculture)

BOX 3.1   GREAT GROUPS, FRAGIPANS, AND ARCHAEOLOGIC DIGS

Soil Taxonomy is a communications tool that helps scientists and land managers share information. In this box we will see how *misclassification*, even at a lower level in *Soil Taxonomy* such as the great group, can have costly ramifications.

In order to preserve our national heritage, both federal and state laws require that an archaeological impact statement be prepared prior to starting major construction works on the land. The archaeological impact is usually assessed in three phases. Selected sites are then studied by archaeologists, with the hope that at least some of the artifacts can be preserved and interpreted before construction activities obliterate them forever. Only a few relatively small sites can be subjected to actual archaeological digs because of the expensive skilled hand labor involved (Figure 3.27).

Such an archaeological impact study was ordered as a precursor to construction of a new highway in a mid-Atlantic state. In the first phase, a consulting company gathered soils and other information from maps, aerial photographs, and field investigations to determine where neolithic people may have occupied sites. Then the consultants identified about 12 ha of land where artifacts indicated significant neolithic activities. The soils in one area were mapped mainly as Typic Dystrudepts. These soils formed in old colluvial and alluvial materials that, many thousands of years ago, had been along a river bank. Several representative soil profiles were examined by digging pits with a backhoe. The different horizons were described, and it was determined in which horizons artifacts were most likely to be found. What was not noted was the presence in these soils of a **fragipan**, a dense, brittle layer that is extremely difficult to excavate using hand tools.

A fragipan is a subsurface diagnostic horizon used to classify soils, usually at the great group or subgroup level (see Figure 3.26). Its presence would distinguish Fragiudepts from Dystrudepts.

When it came time for the actual hand excavation of sites to recover artifacts, a second consulting company was awarded the contract. Unfortunately, their bid on the contract was based on soil descriptions that did not specifically classify the soils as Fragiudepts—soils with very dense, brittle, hard fragipans in the layer that would need to be excavated by hand. So difficult was this layer to excavate and sift through by hand, that it nearly doubled the cost of the excavation—an additional expense of about $1 million. Needless to say, there ensued a controversy as to whether this cost would be borne by the consulting firm that bid with faulty soils data, the original consulting firm that failed to adequately describe the presence of the fragipan, or the highway construction company that was paying for the survey.

This episode gives us an example of the practical importance of soil classification. The formative element *Fragi* in a soil great group name warns of the presence of a dense, impermeable layer that will be very difficult to excavate, will restrict root growth (often causing trees to topple in the wind or become severely stunted), may cause a perched water table (**epiaquic** conditions), and will interfere with proper percolation in a septic drain field.

FIGURE 3.27   *An archaeological dig. (Photo courtesy of Antonio Segovia, University of Maryland)*

found. Thus, Argiudolls are Mollisols of the Udolls suborder characterized by an argillic horizon. Cross-reference to Table 3.8 identifies the specific characteristics separating the great group classes from each other.

Note from Table 3.9 that not all possible combinations of great group prefixes and suborders are used. In some cases a particular combination does not exist. For example, Aquolls occur in lowland areas but not on very old landscapes. Hence, there are no "Paleaquolls." Also, since *all* Alfisols and Ultisols contain an argillic horizon, the use of terms such as "Argiudults" would be redundant.

Subgroups

Subgroups are subdivisions of the great groups. More than 1300 subgroups are recognized. The central concept of a great group makes up one subgroup, termed *Typic*. Thus, the Typic Hapludolls subgroup typifies the Hapludolls great group. Other subgroups may have characteristics that intergrade between those of the central concept and soils of other orders, suborders, or great groups. A Hapludoll with restricted drainage would be classified as an Aquic Hapludoll. One with evidence of intense earthworm activity would fall in the Vermic Hapludolls subgroup. Some intergrades may have properties in common with other orders or with other great groups. Thus, soils in the Entic Hapludolls subgroup are very weakly developed Mollisols, close to being in the Entisols order. The subgroup concept illustrates very well the flexibility of this classification system.

Families

Within a subgroup, soils fall into a particular family if, at a specified depth, they have similar physical and chemical properties affecting the growth of plant roots. About 8000 families have been identified. The criteria used include broad classes of particle size, mineralogy, cation exchange activity of the clay, temperature, and depth of the soil penetrable by roots. Table 3.10 gives examples of the classes used. Terms such as *loamy*, *sandy*, and *clayey* are used to identify the broad particle size classes. Terms used to describe the mineralogical classes include *smectitic*, *kaolinitic*, *siliceous*, *carbonatic*, and *mixed*. The clays are described as *superactive*, *active*, *semiactive*, or *subactive* with regard to their capacity to hold cations. For temperature classes, terms such as *cryic*, *mesic*, and *thermic* are used. The terms *shallow* and *micro* are sometimes used at the family level to indicate unusual soil depths.

TABLE 3.9 **Examples of Names of Great Groups of Selected Suborders of the Mollisol, Alfisol, and Ultisol Orders**

The suborder name is identified as the italicized portion of the great group name.

	Dominant feature of great group		
	Argillic horizon	Archetypical with no distinguishing features	Old land surfaces
Mollisols			
1. Aquolls (wet)	Argi*aquolls*	Hapl*aquolls*	—
2. Udolls (moist)	Argi*udolls*	Hapl*udolls*	Pale*udolls*
3. Ustolls (dry)	Argi*ustolls*	Hapl*ustolls*	Pale*ustolls*
4. Xerolls (Med.)[a]	Argi*xerolls*	Haplo*xerolls*	Pale*xerolls*
Alfisols			
1. Aqualfs (wet)	—	—	—
2. Udalfs (moist)	—	Hapl*udalfs*	Pale*udalfs*
3. Ustalfs (dry)	—	Hapl*ustalfs*	Pale*ustalfs*
4. Xeralfs (Med.)[a]	—	Haplo*xeralfs*	Pale*xeralfs*
Ultisols			
1. Aquults (wet)	—	—	Pale*aqults*
2. Udults (moist)	—	Hapl*udults*	Pale*udults*
3. Ustults (dry)	—	Hapl*ustults*	Pale*ustults*
4. Xerults (Med.)[a]	—	Haplo*xerults*	Pale*xerults*

[a] Med. = Mediterranean climate; distinct dry period in summer.

TABLE 3.10 Some Commonly Used Particle-Size, Mineralogy, Cation Exchange Activity, and Temperature Classes Used to Differentiate Soil Families.

The characteristics generally apply to the subsoil or 50 cm depth. Other criteria used to differentiate soil families (but not shown here) include the presence of calcareous or highly aluminum toxic (allic) properties, extremely shallow depth (shallow or micro), degree of cementation, coatings on sand grains, and the presence of permanent cracks.

Particle-size class	Mineralogy class	Cation exchange activity class[b] Term	CEC / % clay	Mean annual temperature, °C	Soil temperature regime class >6 °C difference between summer and winter	<6 °C difference between summer and winter
Ashy	Mixed	Superactive	0.60	<−10	Hypergelic[c]	—
Fragmental	Micaceous	Active	0.4 to 0.6	−4 to −10	Pergelic[c]	—
Sandy-skeletal[a]	Siliceous	Semiactive	0.24 to 0.4	+1 to −4	Subgelic[c]	—
Sandy	Kaolinitic	Subactive	<0.24	<+8	Cryic	—
Loamy	Smectitic			<+8	Frigid[d]	Isofrigid
Clayey	Gibbsitic			+8 to +15	Mesic	Isomesic
Fine-silty	Gypsic			+15 to +22	Thermic	Isothermic
Fine-loamy	Carbonic			>+22	Hyperthermic	Isohyperthermic
Etc.	Etc.					

[a] Skeletal refers to presence of up to 35% rock fragments by volume.
[b] Cation exchange activity class is not used for taxa already defined by low CEC (e.g., kandic or oxic groups).
[c] Permafrost present.
[d] Frigid is warmer in summer than Cryic.

Thus, a Typic Argiudoll from Iowa, loamy in texture, having a mixture of moderately active clay minerals and with annual soil temperatures (at 50 cm depth) between 8 and 15°C, is classed in the *Typic Argiudolls loamy, mixed, active, mesic* family. In contrast, a sandy-textured Typic Haplorthod, high in quartz and located in a cold area in eastern Canada, is classed in the *Typic Haplorthods sandy, siliceous, frigid* family (note that clay activity classes are not used for soils in sandy textural classes).

Series

The series category is the most specific unit of the classification system. It is a subdivision of the family, and each series is defined by a specific range of soil properties involving primarily the kind, thickness, and arrangement of horizons. Features such as a hardpan within a certain distance below the surface, a distinct zone of calcium carbonate accumulation at a certain depth, or striking color characteristics greatly aid in series identification.

In the United States, each series is given a name, usually from some town, river, or county, such as Fargo, Muscatine, Cecil, Mohave, or Ontario. There are about 19,000 soil series in the United States.

The complete classification of a Mollisol, the Kokomo series, is given in Figure 3.28. This figure illustrates how *Soil Taxonomy* can be used to show the relationship between *the soil*, a comprehensive term covering all soils, and a specific soil series. The figure deserves study because it reveals much about the structure and use of *Soil Taxonomy*. If a soil series name is known, the complete *Soil Taxonomy* classification of the soil may be found on the World Wide Web at http://www.statlab.iastate.edu/soils-info/osd/.

Field Configurations

In reviewing different units in *Soil Taxonomy* we have made little reference to the fact that soils exist alongside each other, often in complex patterns. A given tract of land, even as small as a few hectares, will likely contain several different kinds of soils. On detailed soil survey maps, map units are named for the soil series dominant in the area delineated. Adjacent soils may be of different series, families, subgroups, and so forth. In some cases they may even be classified in different soil orders. An example of such a field situation is shown in Figure 3.29 where three map units dominated by soil series from the Mollisols order are found in association with two map units dominated by soil

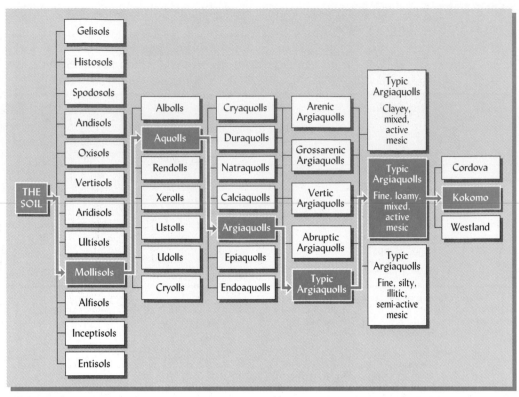

FIGURE 3.28 Diagram illustrating how one soil (Kokomo) keys out in the overall classification scheme. The shaded boxes show that this soil is in the Mollisols order, Aquolls suborder, Argiaquolls great group, and so on. In each category, other classification units are shown in the order in which they key out in *Soil Taxonomy.*

series from the Alfisols order. Recognition of such field associations is important in making soil surveys and interpreting geographic soils information (see Chapter 19).

3.18 CONCLUSION

The soil which covers the earth is actually comprised of many individual soils, each with distinctive properties. Among the most important of these properties are those associated with the horizontal layers, or *horizons,* found in a soil profile. These horizons reflect the physical, chemical, and biological processes soils have undergone during their development. Horizon properties greatly influence how soils can and should be used.

Knowledge of the kinds and properties of soils around the world is critical to humanity's struggle for survival and well-being. A soil classification system based on these properties is equally critical if we expect to use knowledge gained at one location to solve problems at other locations where similarly classed soils are found. *Soil Taxonomy,* a classification system based on measurable soil properties, helps fill this need in some 55 countries. Scientists constantly update the system as they learn more about the nature and properties of the world's soils and the relationships among them. In the remaining chapters of this book we will use taxonomic names whenever appropriate to indicate the kinds of soils to which a concept or illustration may apply.

STUDY QUESTIONS

1. Diagnostic horizons are used to classify soils in *Soil Taxonomy.* Explain the difference between a diagnostic horizon (such as an argillic horizon) and a genetic horizon designation (such as a Bt1 horizon). Give an field example of a diagnostic horizon that contains several genetic horizon designations.

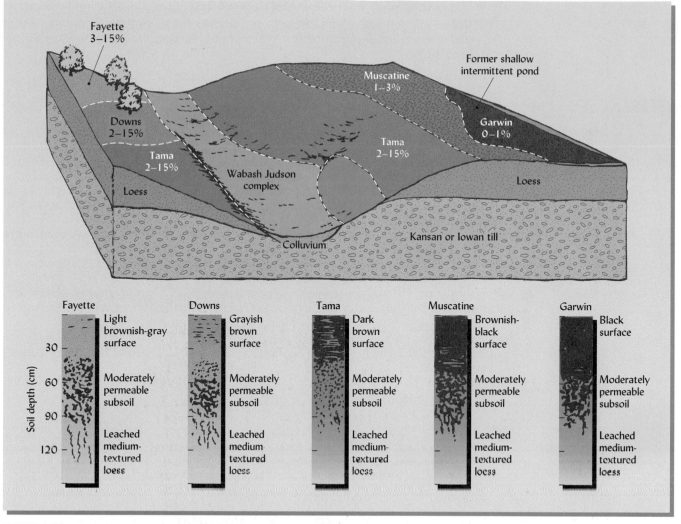

FIGURE 3.29 Association of soils in Iowa. Note the relationship of soil type to (1) parent material, (2) vegetation, (3) topography, and (4) drainage. Two Alfisols (Fayette and Downs) and three Mollisols (Tama, Muscatine, and Garwin) are shown. [From Riecken and Smith (1949)]

2. Explain the relationships among a *soil individual,* a *polypedon,* a *pedon,* and a *landscape.*

3. Rearrange the following soil orders from the *least* to the *most* highly weathered: Oxisols, Alfisols, Mollisolls, Entisols, and Inceptisols.

4. What is the principal soil property by which Ultisols differ from Alfisols? Inceptisols from Entisols?

5. Use the key given in Figure 3.9 to determine the soil order of a soil with the following characteristics: a spodic horizon at 30 cm depth, permafrost at 80 cm depth. Explain your choice of soil order.

6. Of the five soil forming factors discussed in Chapter 2 (parent material, climate, organisms, topography, and time), choose *two* that have had the dominant influence on developing soil properties characterizing each of the following soil orders: Vertisols, Mollisols, Spodosols, and Oxisols.

7. To which soil order does each of the following belong: Psamments, Udolls, Argids, Udepts, Fragiudalfs, Haplustox, and Calciusterts.

8. What's in a name? Write a hypothetical soil profile description and land-use suitability interpretation for a hypothetical soil that is classified in the Aquic Argixerolls subgroup.

9. Explain why *Soil Taxonomy* is said to be a hierarchical classification system.

10. Name the soil taxonomy category and discuss the engineering implications of these soil taxonomy classes: Aquic Paleudults, Fragiudults, Haplusterts, Saprists, and Turbels.

REFERENCES

Coulombe, C. E., L. P. Wilding, and J. B. Dixon. 1996. "Overview of Vertisols: Characteristics and impacts on society," *Advances in Agronomy,* **17:**289–375.

Eswaran, H. 1993. "Assessment of global resources: Current status and future needs," *Pedologie,* **43**(1):19–39.

McCracken, R. J., et al. 1985. "An appraisal of soil resources in the USA," in R. F. Follet and B. A. Stewart (eds.), *Soil Erosion and Crop Productivity.* (Madison, Wis.: Amer. Soc. Agron.).

Riecken, F. F., and G. D. Smith. 1949. "Principal upland soils of Iowa, their occurrence and important properties," *Agron.,* **49** (revised). Iowa Agr. Exp. Sta.

SSSA. 1984. *Soil Taxonomy, Achievements and Challenges.* SSSA Special Publication no. 14. (Madison, Wis.: Soil Sci. Soc. Amer.).

Soil Survey Staff. 1975. *Soil Taxonomy: A Basic System of Soil Classification for Making and Interpreting Soil Surveys.* (Washington, D.C.: USDA Natural Resources Conservation Service).

Soil Survey Staff. 1998. *Soil Taxonomy: A Basic System of Soil Classification for Making and Interpreting Soil Surveys,* 2nd ed. (Washington, D.C.: USDA Natural Resources Conservation Service).

Soil Survey Staff. 1996. *Keys to Soil Taxonomy.* (Washington, D.C.: USDA Natural Resources Conservation Service).

4

SOIL ARCHITECTURE AND PHYSICAL PROPERTIES

And when that crop grew, and was harvested, no man had crumbled a hot clod in his fingers and let the earth sift past his fingertips.
—JOHN STEINBECK, THE GRAPES OF WRATH

Soil physical properties profoundly influence how soils function in an ecosystem and how they can best be managed. Success or failure of both agricultural and engineering projects often hinges on the physical properties of the soil used. The occurrence and growth of many plant species and the movement of water and solutes over and through the soil are closely related to soil physical properties.

Soil scientists use the color, texture, and other physical properties of soil horizons in classifying soil profiles and in making field determinations about soil suitability for agricultural and environmental projects. Knowledge of the basic soil physical properties is not only of great practical value in itself, but will also help in understanding many aspects of soils considered in later chapters.

The physical properties discussed in this chapter relate to the solid particles of the soil and the manner in which they are aggregated together. If we think of the soil as a house, the solid particles in soil are the building blocks from which the house is constructed. **Soil texture** describes the size of the soil particles. The larger mineral fragments usually are embedded in, and coated with, clay and other colloidal materials. Where the larger mineral particles predominate, the soil is gravelly, or sandy; where the mineral colloids are dominant, the soil is claylike. All gradations between these extremes are found in nature.

In building a house, the manner in which the building blocks are put together determines the nature of the walls, rooms, and passageways. Organic matter and other substances act as cement between individual particles, encouraging the formation of clumps or aggregates of soil. **Soil structure** describes the manner in which soil particles are aggregated together. This property, therefore, defines the nature of the system of pores and channels in a soil.

The physical properties considered in this chapter directly describe the nature of soil solids and their impacts on the soil water and air which reside in the pore spaces between the solid particles. Together, soil texture and structure help determine the nutrient-supplying ability of soil solids, as well as the ability of the soil to hold and conduct the water and air necessary for plant root activity. These factors also determine how soils behave when used for highways, building construction and foundations, or when manipulated by tillage. In fact, through their influence on water movement through and off soils, soil physical properties also exert considerable control over the destruction of the soil itself by erosion.

4.1 SOIL COLOR[1]

The first thing we are likely to notice about a soil is its color. In and of themselves, soil colors have little effect on the behavior and use of soils. An important exception to this statement is the fact that dark-colored surface soils absorb more solar energy than lighter-colored soils, and therefore may warm up faster.

The main reason for studying soil colors is that they provide valuable clues to the nature of other soil properties and conditions. Because of the importance of color in soil classification and interpretation, a standard system for accurate color description has been developed using Munsell color charts (Figure 4.1). In this system, a small piece of soil is compared to standard color chips in a soil color book. Each color chip is described by the three components of color: the **hue** (in soils, usually redness or yellowness), the **chroma** (intensity or brightness, a chroma of 0 being neutral gray), and the **value** (lightness or darkness, a value of 0 being black).

Soils display a wide range of reds, browns, yellows, and even greens (see Plates 16 and 19, following pages 82 and 498). Some soils are nearly black, others nearly white. Some soil colors are very bright, others are dull grays. Soil colors may vary from place to

---

[1] For an excellent collection of papers on the causes and measurement of soil color, see Bigham and Ciolkosz (eds.; 1993).

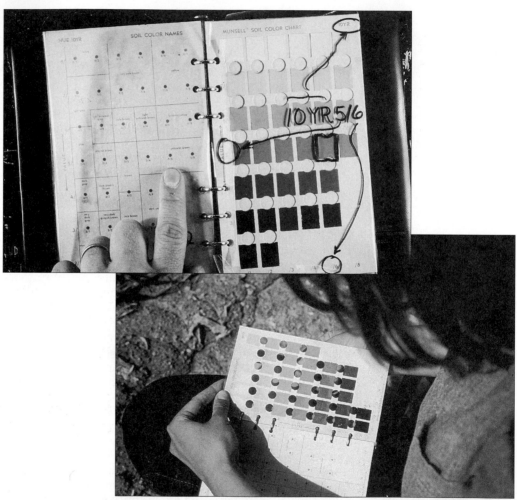

FIGURE 4.1 Determining soil color by comparison to color chips in a Munsell Color Book. Each page shows a different hue, ranging from 5R (red) to 5Y (yellow). On a single page, higher-value (lighter) colors are nearer the top and lower-value (darker) colors are near the bottom. Higher-chroma (brighter) colors are nearer the right-hand side and duller, grayish (low-chroma) colors are nearer the left-hand side. The complete Munsell color description of the soil pictured is 10YR 5/6 (yellowish brown), moist. (Photos by R. Weil)

place in the landscape, as when adjacent soils have differing surface horizon colors (e.g., Plate 17). Colors also typically change with depth through the various layers (horizons) within a soil profile. In many soils, the horizons in a given profile have colors that are similar in hue, but vary with respect to chroma and value. Even within a single horizon or clod of soil, colors may vary from spot to spot (see Plate 15).

Causes of Soil Colors

Most soil colors are derived from the colors of iron oxides and organic matter that coat the surfaces of soil particles. Organic coatings tend to darken and mask the colors derived from iron oxides (see Plate 8, after page 82). Subsoil horizons with little organic matter, therefore, often most clearly display the iron oxide colors, such as the yellow of goethite, the red of hematite, and the brown of maghematite. Other minerals that sometimes give soils distinctive colors are manganese oxide (black) and glauconite (green). Carbonates, such as calcite, that typically accumulate in soils of semiarid regions may impart a whitish color (see Plate 29).

Interpreting Soil Colors

Colors can tell us a great deal about a soil, and are therefore used extensively in classifying soils (see Chapter 3). Colors often help us distinguish the different horizons of a soil profile. Typically an A horizon is darker and a B horizon is a brighter color than adjacent horizons. In some cases, color is a diagnostic criterion for classification. For example, a mollic epipedon (see Chapter 3 for definition) is defined as having a color so dark that both its value and chroma are 3 or less (i.e., in the lower left or blackish corner of a Munsell color book page, Figure 4.1). Another example is that of the Rhodic subgroups of certain soil orders, which have B horizons characterized by very red colors having hues between 2.5YR (the most red of the yellowish red pages in the Munsell book) and 10R (the most red of all the pages in the Munsell color book). Color can also provide qualitative information about the current moisture status of a soil, dry soils generally having lighter (higher-value) colors than moist soils.

Because of the color changes that take place when various iron-containing minerals undergo oxidation and reduction, soil colors can provide extremely valuable insights into the hydrologic regime or drainage status of a soil. Bright (high-chroma) colors throughout the profile are typical of well-drained soils through which water easily passes and in which oxygen is generally plentiful. Prolonged anaerobic conditions can cause iron oxide coatings to become chemically reduced, changing high-chroma (red or brown) colors to low-chroma (gray, bluish or gray-green) colors, a condition referred to as **gley** (as in Plate 20, after page 498). The presence in upper horizons of gley (low-chroma colors), either alone or mixed in a mottled pattern with brighter colors (see Plates 18 and 19), is used in delineating **wetlands**, for it is indicative of waterlogged conditions during at least a major part of the plant growing season (see Section 7.8. The depth in the profile at which gley colors are found helps to define the **drainage class** of the soil (see Figure 19.3).

One common misconception about soil color is the idea that there is a consistent relationship between soil color and soil texture (see Section 4.2). For example, people in warm climates often speak of "red clay," but reddish colors do not necessarily indicate clayey soils. Very sandy materials also commonly have reddish colors because iron minerals coat the sand particles (e.g., Plate 10).

Finally, it is worth mentioning that soils, with their distinctive colors, are important aesthetic components of the landscape. For example, warm, reddish colors are characteristic of many tropical and subtropical landscapes, while dark grays and browns typify cooler, more temperate regions. Thus, a native of Georgia would probably have a very different mental image of soil than a person from New York.

4.2   SOIL TEXTURE (SIZE DISTRIBUTION OF SOIL PARTICLES)

The size of mineral particles in soil may seem too mundane a subject to warrant much attention, yet knowledge of the proportions of different-sized particles in a soil (i.e., the **soil texture**) is critical for understanding soil behavior and management. When investigating soils on a site, the texture of various soil horizons is often the first and most important property to determine, for a soil scientist can draw many conclusions from this information. Furthermore, the texture of a soil in the field is not readily subject to change, so it is considered a basic property of a soil.

Nature of Soil Separates

Diameters of individual soil particles range over six orders of magnitude, from boulders (1 m) to submicroscopic clays (<10⁻⁶ m). Scientists group these particles into soil **separates** according to several classification systems, as shown in Figure 4.2. The classification established by the U.S. Department of Agriculture is used in this text. The size ranges for these separates are not purely arbitrary, but reflect major changes in how the particles behave and in the physical properties they impart to soils.

Gravels, cobbles, boulders and other **coarse fragments** >2 mm in diameter may affect the behavior of a soil, but they are not considered to be part of the **fine earth fraction** to which the term *soil texture* properly applies. Coarse fragments reduce the volume of the soil available for water retention and root growth; however, in dense soils the spaces between them may create paths for water drainage and root penetration. Larger coarse fragments, especially if consisting of hard minerals such as quartz, interfere with tillage or excavation.

SAND.   Sand particles are those smaller than 2 mm but larger than 0.05 mm. They may be rounded or angular (Figure 4.3), depending on the extent to which they have been worn down by abrasive processes during soil formation. The coarsest sand particles may be rock fragments containing several minerals, but most sand grains consist of a single mineral, usually quartz (SiO_2) or other primary silicate (Figure 4.4). Sand grains may appear brown, yellow, or red as a result of iron or aluminum oxide coatings. In any case, the dominance of quartz means that the sand separate generally has a far smaller total content of plant nutrients than do the finer separates.

Sand feels gritty between the fingers and the particles are generally visible to the naked eye (Figure 4.5). As sand particles are relatively large, so, too, the voids between them are relatively large and promote free drainage of water and entry of air into the soil. The relationship between particle size and **specific surface area** (the surface area for a given volume or mass of particles) is illustrated in Figure 4.6. Because of their large size, particles of sand have relatively low specific surface area. Therefore, sand particles can hold little water, and soils dominated by sand are prone to drought. Sand particles are considered noncohesive; that is, they do not tend stick together in a mass (see Section 4.9).

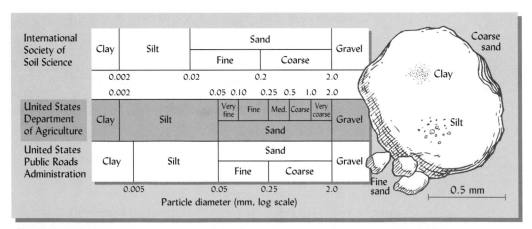

FIGURE 4.2   Classification of soil particles according to their size. The shaded scale in the center and the names on the drawings of particles follow the United States Department of Agriculture system, which is widely used throughout the world. The USDA system is also used in this book. The other two systems shown are also widely used by soil scientists and by highway construction engineers. The drawing illustrates the size of soil separates (note scale).

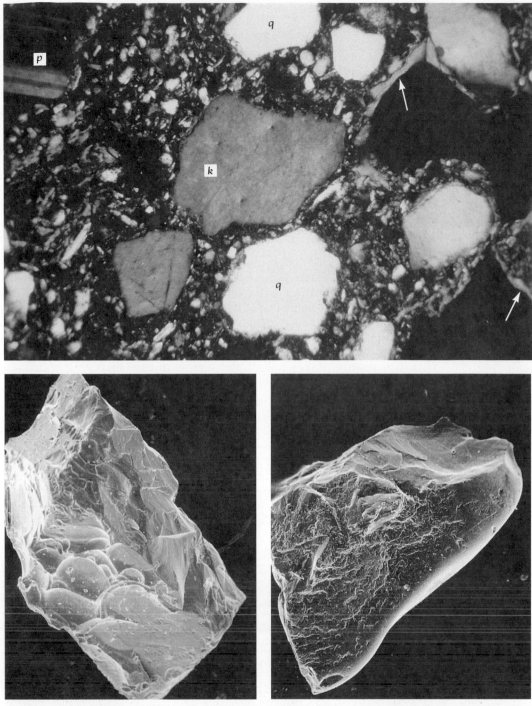

FIGURE 4.3 (top) A thin section of a loamy soil as seen through a microscope using polarized light (empty pores appear black). Both the sand and silt particles shown are irregular in size and shape, the silt being only smaller. Although quartz (*q*) dominates the sand and silt fractions in this soil, several other silicate minerals can be seen (*p* = plagioclase, *k* = feldspar). Clay films can be seen to coat the walls of the large pores (arrow). Scanning electron micrographs of sand grains show quartz sand (bottom left) and a feldspar grain (bottom right) magnified about 40 times. (Upper photo courtesy of Martin Rabenhorst, University of Maryland; lower photos courtesy of J. Reed Glasmann, Union Oil Research)

SILT. Particles smaller than 0.05 mm but larger than 0.002 mm in diameter are classified as silt. Individual silt particles are not visible to the unaided eye (see Figure 4.5), nor do they feel gritty when rubbed between the fingers. These are essentially microsand particles, with quartz generally the dominant mineral. Where silt is composed of weatherable minerals, the smaller size of the particles allows weathering to proceed rapidly enough to release significant amounts of plant nutrients.

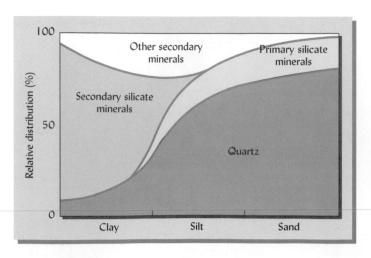

Although silt is composed of particles similar in shape to sand, it feels smooth or silky, like flour. The pores between silt particles are much smaller (and much more numerous) than those in sand, so silt retains more water and lets less drain through. However, even when wet, silt itself does not exhibit much **stickiness** or **plasticity** (malleability). What little plasticity, cohesion (stickiness), and adsorptive capacity some silt fractions exhibit is largely due to a film of adhering clay. Because of their low stickiness and plasticity, silty soils generally are easily washed away by flowing water in a process called **piping**. Box 4.1 illustrates a consequence of piping and highlights the importance of distinguishing between silt and clay.

(a)

(b)

FIGURE 4.5 Separation of soil particles by size during sedimentation. The soil in the cylinder was agitated into a suspension, then allowed to settle out. After about one minute, the sand particles have settled out and the amount of silt and clay remaining in suspension can be determined by floating a hydrometer (a). After about a day the clay remains in suspension, but the sand and silt have formed layers on the bottom (b). The boundary between silt and sand can be seen as a line between the zone in which the individual particles are visible and the zone in which they are not (see arrow). (Photos by R. Weil)

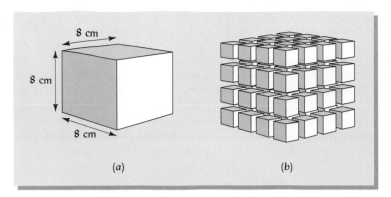

(a)   (b)

FIGURE 4.6   The relationship between the surface area of a given mass of material and the size of its particles. In the single large cube (a) each face has 64 cm² of surface area. The cube has six faces, so the cube has a total of 384 cm² surface area (6 faces × 64 cm² per face). If the same cube of material was cut into smaller cubes (b) so that each cube was only 2 cm on each side, then the same mass of material would now be present as 64 smaller cubes (4 × 4 × 4). Each face of each small cube would have 4 cm² (2 × 2 cm) of surface area, giving 24 cm² of surface area for each cube (6 faces × 4 cm² per face). The total mass would therefore have 1536 cm² (24 cm² per cube × 64 cubes) of surface area. This is four times as much surface area as the single large cube. Since clay particles are very very small and usually platelike in shape, their surface area is thousands of times greater than that of the same mass of sand particles.

BOX 4.1   SILT AND THE FAILURE OF THE TETON DAM[a]

One of the most tragic and costly engineering failures in American history occurred in southern Idaho on 5 June 1977, less than a year after construction was completed on a large earth-fill dam across the Teton River. Eleven people were killed and 25,000 made homeless in the five hours it took to empty the 28-km-long lake that had been held in place by the dam. Some $400 million (1977 dollars) worth of damages were caused as the massive wall of water surged through the collapsed dam and the valley below. The dam failed with little warning as small seepage leaks quickly turned into raging torrents that swept away a team of bulldozers sent to make repairs.

The Teton dam was built according to a standard, time-tested design for zoned earth-fill embankments. Essentially, after preparing a base in the rhyolite rock below the soil, a core (zone 1) of tightly compacted soil material was constructed and covered with a layer (zone 2) of coarser alluvial soil material to protect it from water and wind erosion.

The core is meant to be the watertight seal that prevents water from seeping through the dam. Normally, clayey material is chosen for the core, since the sticky, plastic qualities of moist clay allow it to be compacted into a malleable, watertight mass that holds together and does not crack so long as it is kept moist. Silt, on the other hand, though it may appear similar to clay in the field, has little or no stickiness or plasticity and therefore cannot be compacted into

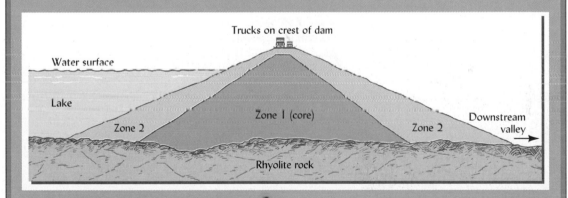

a coherent mass as clay can. In fact, a moist mass of compacted silt will crack as it settles because it lacks plasticity. Also, if water seeps into these cracks, the silty material will rapidly wash away with the flowing water, enlarging the crack and inviting more water to flow through, which will wash away still more of the silt. This process of rapidly enlarging seepage channels is termed *piping*. Such piping was almost certainly a major cause of the Teton Dam failure, for the engineers built the zone 1 core of the dam using windblown silt deposits (termed *loess*—see Section 2.11) rather than clay.

This was a tragic but useful lesson about the importance of texture in determining soil behavior.

[a] Based on the report by the U.S. Department of Interior Teton Dam Failure Group (1977).

CLAY.   Particles smaller than 0.002 mm are classified as clay and have a very large specific surface area, giving them a tremendous capacity to adsorb water and other substances. A spoonful of clay may have a surface area the size of a football field (see Section 8.1). This large adsorptive surface causes clay particles to cohere together in a hard mass after drying. When wet, clay is sticky and can be easily molded.

Clay size particles are so small that they behave as *colloids*—if suspended in water they do not readily settle out. Unlike most sand and silt particles, clay particles tend to be shaped like tiny flakes or flat platelets. The pores between clay particles are very small and convoluted, so movement of both water and air is very slow. Each unique clay mineral (see Chapter 8) imparts very different properties to the soils in which they are prominent. Therefore, soil properties such as shrink-swell behavior, plasticity, water-holding capacity, soil strength, and chemical adsorption depend on the *kind* of clay present as well as the *amount*.

Influence of Surface Area on Other Soil Properties

When particle size decreases, specific surface area and related properties increase greatly, as shown graphically in Figure 4.7. Fine colloidal clay has about 10,000 times as much surface area as the same weight of medium-sized sand. Soil texture influences many other soil properties in far-reaching ways (see Table 4.1) as a result of five fundamental surface phenomena:

1. Water is retained in soils as thin films on the surfaces of soil particles. The greater the surface area, the greater the soil's capacity for holding water.

2. Both gases and dissolved chemicals are attracted to and adsorbed by mineral particle surfaces. The greater the surface area, the greater the soil's capacity to retain nutrients and other chemicals.

3. Weathering takes place at the surface of mineral particles, releasing constituent elements into the soil solution. The greater the surface area, the greater the rate of release of plant nutrients from weatherable minerals.

4. The surfaces of mineral particles often carry both negative and some positive electromagnetic charges so that particle surfaces and the water films between them tend to attract each other (see Section 4.7). The greater the surface area, the greater the propensity for soil particles to stick together in a coherent mass, or as discrete aggregates.

5. Microorganisms tend to grow on and colonize particle surfaces. For this and other reasons, microbial reactions in soils are greatly affected by the specific surface area.

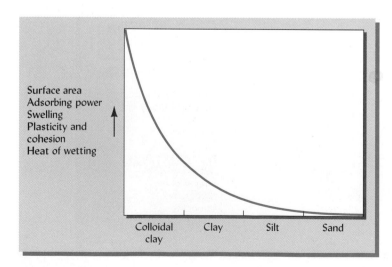

Surface area
Adsorbing power
Swelling
Plasticity and
cohesion
Heat of wetting

Colloidal clay   Clay   Silt   Sand

FIGURE 4.7   The finer the texture of a soil, the greater is the effective surface exposed by its particles. Note that adsorption, swelling, and the other physical properties cited follow the same general trend and that their intensities go up rapidly as the colloidal size is approached.

TABLE 4.1   Generalized Influence of Soil Separates on Some Properties and Behavior of Soils.[a]

Property/behavior	Rating associated with soil separates		
	Sand	*Silt*	*Clay*
Water-holding capacity	Low	Medium to High	High
Aeration	Good	Medium	Poor
Drainage rate	High	Slow to Medium	Very slow
Soil organic matter level	Low	Medium to High	High to Medium
Decomposition of organic matter	Rapid	Medium	Slow
Warm-up in spring	Rapid	Moderate	Slow
Compactability	Low	Medium	High
Susceptibility to wind erosion	Moderate (high if fine sand)	High	Low
Susceptibility to water erosion	Low (unless fine sand)	High	Low if aggregated, high if not
Shrink-swell potential	Very Low	Low	Moderate to very high
Sealing of ponds, dams, and landfills	Poor	Poor	Good
Suitability for tillage after rain	Good	Medium	Poor
Pollutant leaching potential	High	Medium	Low (unless cracked)
Ability to store plant nutrients	Poor	Medium to high	High
Resistance to pH change	Low	Medium	High

[a] Exceptions to these generalizations do occur, especially as a result of soil structure and clay minerology.

4.3   SOIL TEXTURAL CLASSES

Three broad groups of textural classes are recognized: *sandy soils, clayey soils,* and *loamy soils.* Within each group, specific **textural class** names convey an idea of the size distribution of particles and indicate the general nature of soil physical properties. The 12 textural classes named in Table 4.2 form a graduated sequence from the sands, which are coarse in texture and easy to handle, to the clays, which are very fine and difficult to manage physically.

Sands and loamy sands are dominated by the properties of sand, for the sand separate comprises at least 70% of the material by weight (less than 15% of the material is clay). Characteristics of the clay separate are distinctly dominant in clays, sandy clays, and silty clays.

TABLE 4.2   General Terms Used to Describe Soil Texture in Relation to the Basic Soil Textural Class Names

U.S. Department of Agriculture Classification System

General terms		Basic soil textural class names
Common names	Texture	
Sandy soils	Coarse	Sands
		Loamy sands
	Moderately coarse	Sandy loam
		Fine sandy loam[a]
Loamy soils	Medium	Very fine sandy loam[a]
		Loam
		Silt loam
		Silt
	Moderately fine	Sandy clay loam
		Silty clay loam
		Clay loam
Clayey soils	Fine	Sandy clay
		Silty clay
		Clay

[a] Although not included as class names in Figure 4.8, these soils are usually treated separately because of their fine sand content.

LOAMS. The loam group contains many subdivisions. An ideal **loam** may be defined as a mixture of sand, silt, and clay particles that exhibits the *properties* of those separates in about equal proportions. This definition does not mean that the three separates are present in equal *amounts* (as will be revealed by careful study of Figure 4.8). This anomaly exists because a relatively small percentage of clay is required to engender clayey properties in a soil, whereas small amounts of sand and silt have a lesser influence on how a soil behaves. Thus, the clay modifier is used in the class name of soils with as little as 20% clay; but to qualify for the modifiers sandy or silt, a soil must have at least 40 or 45% of those separates, respectively.

Most soils are some type of loam. They may possess the ideal makeup of equal proportions previously described and be classed simply as loam. However, a loam in which

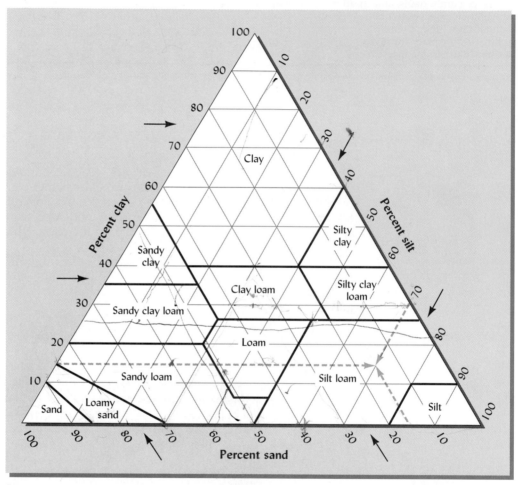

FIGURE 4.8 The major soil textural classes are defined by the percentages of sand, silt, and clay according to the heavy boundary lines shown on the textural triangle. If these percentages have been determined for a soil sample by particle size analysis, then the triangle can be used to determine the soil textural class name that applies to that soil sample. To use the graph, first find the appropriate clay percentage along the left side of the triangle, then draw a line from that location across the graph going parallel to the base of the triangle. Next find the sand percentage along the base of the triangle, then draw a line inward going parallel to the triangle side labeled "Percent silt." The small arrows indicate the proper direction in which to draw the lines. The name of the compartment in which these two lines intersect indicates the textural class of the soil sample. Percentages for any two of the three soil separates is all that is required. Because the percentages for sand, silt, and clay add up to 100%, the third percentage can easily be calculated if the other two are known. If all three percentages are used, the three lines will all intersect at the same point. Consider, as an example, a soil which has been determined to contain 15% sand, 15% clay, and 70% silt. This example is indicated by the light dashed lines that intersect in the compartment labeled "Silt loam." What is the textural class of another soil sample which has 33% clay, 33% silt, and 33% clay? The lines (not shown) for this second example would intersect in the center of the "Clay loam" compartment.

sand is dominant is classified as a *sandy loam.* In the same way, there may occur silt loams, silty clay loams, sandy clay loams, and clay loams.

COARSE FRAGMENT MODIFIERS.   For some soils, qualifying factors such as stone, gravel, and the various grades of sand become part of the textural class name. Fragments that range from 2 to 75 mm along their greatest diameter are termed *gravel* or *pebbles;* those ranging from 75 to 250 mm are called *cobbles* (if round) or *flags* (if flat); and those more than 250 mm across are called *stones* or *boulders.* A *cobbly, fine sandy loam* is an example of such a modified textural class.

Alteration of Soil Textural Class

Over very long periods of time, pedologic processes (see Chapter 2) such as erosion, deposition, illuviation, and weathering can alter the textures of various soil horizons. However, management practices generally do not alter the textural class of a soil on a field scale. The texture of a given soil can be changed only by mixing it with another soil of a different textural class. For example, the incorporation of large quantities of sand to change the physical properties of a clayey soil for use in greenhouse pots or for turf would bring about such a change. However, where specifications (as for a landscape design) call for soil materials of a certain textural class, it is advisable to find a naturally occurring soil that meets the specification, rather than attempt to alter the textural class by mixing in sand or clay.[2]

Also, it should be noted that adding peat or compost to a soil while mixing a potting medium does not constitute a change in texture, since this property refers only to the mineral particles. The term *soil texture* is not relevant to artificial media that contain mainly perlite, peat, styrofoam, or other nonsoil materials.

Determination of Textural Class by the "Feel" Method

Textural class determination is one of the first field skills a soil scientist should develop. Determining the textural class of a soil by its feel is of great practical value in soil survey, land classification, and any investigation in which soil texture may play a role. Accuracy depends largely on experience, so practice whenever you can, beginning with soils of known texture to calibrate your fingers.

The textural triangle (see Figure 4.8) should be kept in mind when determining the textural class by the feel method as explained in Box 4.2.

Laboratory Particle-Size Analyses

The first and sometimes most difficult step in a particle-size analysis is the complete dispersion of a soil sample in water, so even the tiniest clumps are broken down into individual, primary particles. Dispersion is usually accomplished using chemical treatments along with a high speed blender or sonicator.

While a set of sieves can be used to separate out the sand, a sedimentation procedure is usually used to determine the amounts of silt and clay. The principle involved is simple. Because soil particles are more dense than water, they tend to sink, settling at a velocity that is proportional to their size. In other words: "The bigger they are, the faster they fall." The equation that describes this relationship is referred to as *Stokes' law.* The complete equation is given in Box 4.3, but in its simplest form it tells us that the velocity of settling V is proportional to the square of a particle's diameter d

$$V = kd^2$$

---

[2] Great care must be exercised in attempting to ameliorate physical properties of fine textured soils by adding sand. If the sand is not of the proper size and not added in sufficient amounts, it may make matters worse, rather than better. While adjacent coarse sand grains form large pores between them, sand grains embedded in a silty or clayey matrix do not. Therefore, moderate amounts of fine sand or sand ranging widely in size may yield a product more akin to concrete than to a sandy soil.

where k is a constant related to the acceleration due to gravity and the density and viscosity of water. By measuring the amount of soil still in suspension after various amounts of settling time (using a pipette or a hydrometer, as shown in Figure 4.5), the percentages of each size fraction can be determined so as to identify the soil textural class and generate particle-size distribution curves such as those shown in Figure 4.10.

Figure 4.10 presents particle-size distribution curves for soils representative of three textural classes. The fact that these curves are smooth emphasizes that there is no sharp line of demarcation in the distribution of sand, silt, and clay fractions, and suggests a gradual change of properties with change in particle size.

It is important to note that soils are assigned to textural classes *solely* on the basis of the mineral particles of sand size and smaller; therefore, the percentages of sand, silt, and clay always add up to 100%. The amounts of stone and gravel are rated separately. Organic matter is usually removed from a soil sample by oxidation before the mechanical separation.

The relationship between such analyses and textural class names is commonly shown diagrammatically as a triangular graph (see Figure 4.8). This **textural triangle**

BOX 4.2 A METHOD FOR DETERMINING TEXTURE BY FEEL

The first, and most critical, step in the texture-by-feel method is to knead a walnut-sized sample of moist soil into a uniform puttylike consistency, slowly adding water if necessary. This step may take a few minutes, but a premature determination is likely to be in error as hard clumps of clay and silt may feel like sand grains. The soil should be moist, but not quite glistening. Try to do this with only one hand so as to keep your other hand clean for writing in a field notebook (and shaking hands with your client).

While squeezing and kneading the sample, note its maleability, stickiness, and stiffness, all properties associated with the clay content. A high silt content makes a sample feel smooth and silky, with little stickiness or resistance to deformation. A soil with a significant content of sand feels rough and gritty, and makes a grinding noise when rubbed near one's ear.

Get a feel for the amount of clay by attempting to squeeze a ball of properly moistened soil between your thumb and the side of your forefinger, making a ribbon of soil. Make the ribbon as long as possible until it breaks from its own weight (see Figure 4.9).

Interpret your observations as follows:

1. Soil will not cohere into a ball, falls apart: **sand**
2. Soil forms a ball, but will not form a ribbon: **loamy sand**
3. Soil ribbon is dull and breaks off when less than 2.5 cm long and
   a. Grinding noise is audible; grittiness is prominent feel: **sandy loam**
   b. Smooth, floury feel prominent; no grinding audible: **silt loam**
   c. Only slight grittiness and smoothness; grinding not clearly audible: **loam**
4. Soil exhibits moderate stickiness and firmness, forms ribbons 2.5 to 5 cm long, and
   a. Grinding noise is audible; grittiness is prominent feel: **sandy clay loam**
   b. Smooth, floury feel prominent; no grinding audible: **silty clay loam**
   c. Only slight grittiness and smoothness; grinding not clearly audible: **clay loam**
5. Soil exhibits dominant stickiness and firmness, forms shiny ribbons longer than 5 cm, and
   a. Grinding noise is audible; grittiness is dominant feel: **sandy clay**
   b. Smooth, floury feel prominent; no grinding audible: **silty clay**
   c. Only slight grittiness and smoothness; grinding not clearly audible: **clay**

A more precise estimate of sand content (and hence more accurate placement in the horizontal dimension of the textural class triangle) can be made by wetting a pea-sized clump of soil in the palm of your hand and smearing it around with your finger until it your palm becomes coated with a souplike suspension of soil. The sand grains will stand out visibly and their volume as compared to the original "pea" can be estimated, as can their relative size (fine, medium, coarse etc.).

It is best to learn the method using samples of known textural class. With practice, accurate textural class determinations can be made on the spot.

(continued)

BOX 4.2 (Cont.) A METHOD FOR DETERMINING TEXTURE BY FEEL

FIGURE 4.9   The "feel" method for determining soil textural class. A moist soil sample is rubbed between the thumb and forefingers (insets) and squeezed out to make a "ribbon." (*a*) The gritty, noncohesive appearance and short ribbon of a sandy loam with about 15% clay. (*b*) The smooth, dull appearance and crumbly ribbon characteristic of a silt loam. (*c*) The smooth, shiny appearance and long, flexible ribbon of a clay. See Box 4.2 for details. (Large photos courtesy of R. Weil)

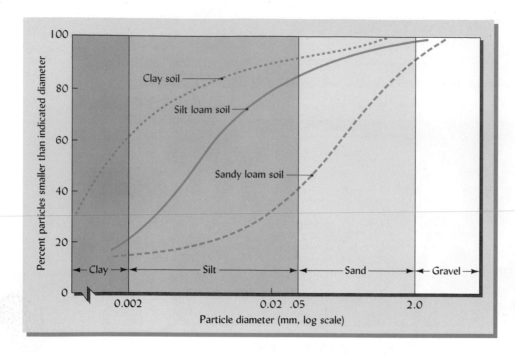

FIGURE 4.10 Particle-size distribution in three soils varying widely in their textures. Note that there is a gradual transition in the particle-size distribution in each of these soils.

also enables us to use laboratory particle-size analysis data to check the accuracy of field textural determinations by feel.

4.4 STRUCTURE OF MINERAL SOILS

The term **structure** relates to the arrangement of primary soil particles into groupings called **aggregates** or **peds.** The pattern of pores and peds defined by soil structure greatly influences water movement, heat transfer, aeration, and porosity in soils. Activities such as timber harvesting, grazing, tillage, trafficking, drainage, liming, and manuring impact soils largely through their effect on soil structure, especially in the surface horizons.

The processes involved in the formation, stabilization, and management of soil structure will be discussed in Sections 4.7 and 4.8. Here we will examine the nature of the various types of structural peds.

Types of Soil Structure

Many types of structural peds occur in soils, often within different horizons of a particular soil profile. Soil structure is characterized in terms of the shape (or *type*), size, and distinctness (or *grade*) of the peds. The four principal shapes of soil structure are *spheroidal, platy, prismlike,* and *blocklike.* These structural types (and some subtypes) are illustrated in Figure 4.11 and are described following.

SPHEROIDAL. **Granular** structure consists of spheroidal peds or **granules** that are usually separated from each other in a loosely packed arrangement (see Figure 4.11*a*). When the spheroidal peds are especially porous, the term **crumb** is sometimes applied.[3] They typically range from <1 to >10 mm in diameter.

Granular and crumb structures are characteristic of many surface soils (usually A horizons), particularly those high in organic matter. Consequently, they are the principal types of soil structure affected by management. They are especially prominent in grassland soils and soils that have been worked by earthworms.

---

[3] The term *crumb* is no longer an official USDA designation.

BOX 4.3 STOKE'S LAW IN CALCULATING PARTICLE SIZE BY SEDIMENTATION METHODS

The complete expression of Stoke's Law tells us the velocity V of a particle falling through a fluid is directly proportional to the gravitational force g, the difference between the density of the particle and the density of the fluid $(D_s - D_f)$ and the square of the effective[a] particle diameter (d^2). The settling velocity is *inversely* proportional to the viscosity or "thickness" of the fluid η. Since velocity equals distance h divided by time t we can write Stoke's Law as:

$$V = \frac{h}{t} = \frac{d^2 g(D_s - D_f)}{18\eta}$$

Where: g = gravitational force = 9.81 Newtons per kilogram (9.81 N/kg)

η = viscosity of water at 20°C = 1/1000 Newton-seconds per m² (10^{-3} Ns/m²)

D_s = density of the solid particles, for most soils = 2.65×10^3 kg/m³

D_f = density of the fluid (i.e., water) = 1.0×10^3 kg/m³

Substituting these values into the equation, we can write:

$$V = \frac{h}{t} = \frac{d^2 * 9.81 \text{ N/kg} * (2.65 * 10^3 \text{ kg/m}^3 - 1.0 * 10^3 \text{ kg/m}^3)}{18 * 10^{-3} \text{ Ns/m}^2}$$

$$= \frac{9.81 \text{ N/kg} * 1.65 * 10^3 \text{ kg/m}^3}{18 * 10^{-3} \text{ Ns/m}^2} * d^2$$

$$= \frac{16.19 * 10^3 \text{ N/m}^3}{0.018 \text{ Ns/m}^2} * d^2$$

$$= \frac{9 \times 10^5}{sm} * d^2$$

$$= kd^2$$

where $k = \dfrac{9 \times 10^5}{sm}$

Note that $V = kd^2$ is the same as the simplified formula given in the text.

Let's choose to sample a soil suspension at 0.1 m (10 cm) depth. We can calculate the seconds of settling time we must allow if we want the smallest silt particle to have just passed our sampling depth so our sample will contain only clay.

Chosen: $h = 0.1$ m
$d - 2 * 10^{-6}$ m (0.002 mm, smallest silt)

Solving for t we can write:

$$\frac{h}{t} = d^2 k \quad \Rightarrow \quad \frac{t}{h} = \frac{1}{d^2 k} \quad \Rightarrow \quad t = \frac{h}{d^2 k}$$

Therefore:

$$t = \frac{0.1 \text{ m}}{(2 * 10^{-6} \text{ m})^2 * 9 * 10^5 \text{ s }^1/\text{m}^{-1}}$$

$$t = 27{,}777 \text{ seconds} = 463 \text{ minutes} = 7.72 \text{ hours}$$

By comparison, the smallest sand particle ($d = 0.05$ mm) would make the same journey in only 44 seconds.

---

[a] Stoke's Law applies to smooth round particles. Since most soil particles are neither smooth nor round, sedimentation techniques determine the *effective* diameters, not necessarily the actual diameter of the soil particle.

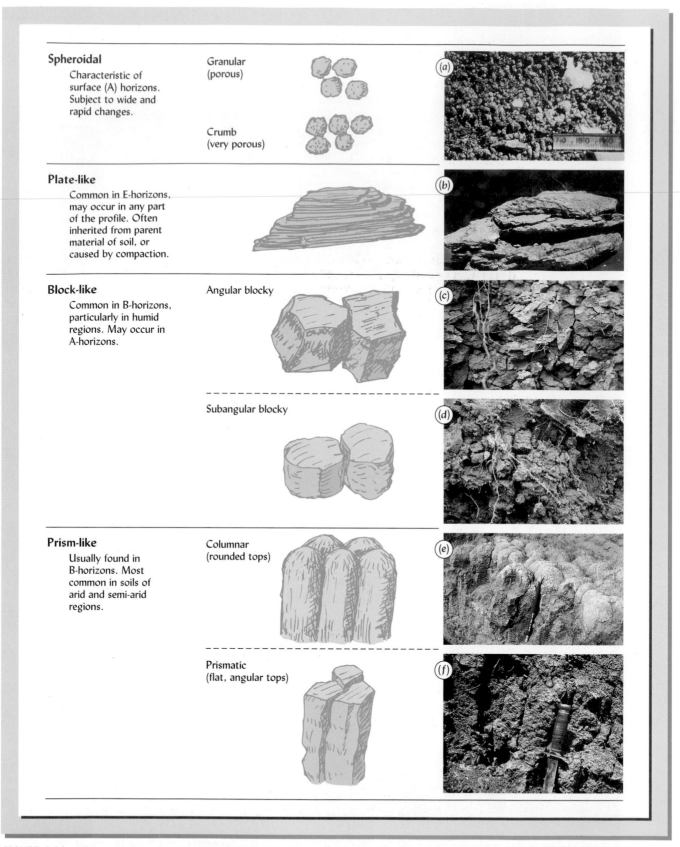

Spheroidal

Characteristic of surface (A) horizons. Subject to wide and rapid changes.

Granular (porous)

Crumb (very porous)

(a)

Plate-like

Common in E-horizons, may occur in any part of the profile. Often inherited from parent material of soil, or caused by compaction.

(b)

Block-like

Common in B-horizons, particularly in humid regions. May occur in A-horizons.

Angular blocky

(c)

Subangular blocky

(d)

Prism-like

Usually found in B-horizons. Most common in soils of arid and semi-arid regions.

Columnar (rounded tops)

(e)

Prismatic (flat, angular tops)

(f)

FIGURE 4.11 The various structure types (shapes) found in mineral soils. Their typical location is suggested. The drawings illustrate their essential features and the photos indicate how they look in situ. For scale, note the 15-cm-long pencil in (*e*) and the 3-cm-wide knife blade in (*d*) and (*f*). (Photo (*e*) courtesy of J. L. Arndt, now with Petersen Environmental Consulting; North Dakota State University. Others courtesy of R. Weil)

PLATE-LIKE. **Platy** structure, characterized by relatively thin horizontal peds or plates, may be found in both surface and subsurface horizons. In most instances, the plates have developed as a result of soil-forming processes. However, unlike other structure types, platy structure may also be inherited from soil parent materials, especially those laid down by water or ice. In some cases compaction of clayey soils by heavy machinery can create platy structure (see Figure 4.11*b*).

BLOCKLIKE. **Blocky** peds are irregular, roughly cubelike, (Figure 4.12), and range from about 5 to 50 mm across. The individual blocks are not shaped independently, but are molded by the shapes of the surrounding blocks. When the edges of the blocks are sharp and the rectangular faces distinct, the subtype is designated **angular blocky** (see Figure 4.11*c*). When some rounding has occurred, the aggregates are referred to as **subangular blocky** (see Figure 4.11*d*). These types are usually found in B horizons where they promote good drainage, aeration, and root penetration.

PRISMLIKE. **Columnar** and **prismatic** structure are characterized by vertically oriented prisms or pillarlike peds that vary in height among different soils and may have a diameter of 150 mm or more. Columnar structure (see Figure 4.11*e*), which has pillars with distinct, rounded tops, is especially common in subsoils high in sodium (e.g., natric horizons; see Section 3.2). When the tops of the prisms are relatively angular and flat horizontally, the structure is designated as prismatic (see Figure 4.11*f*). Both prismlike structures are often associated with swelling types of clay. They commonly occur in subsurface horizons in arid and semiarid regions and, when well developed, are a very striking feature of the profile. In humid regions, prismatic structure sometimes occurs in poorly drained soils and in fragipans (see Figure 3.26), the latter typically having prisms 200 to 300 mm across.

Description of Soil Structure in the Field

In describing soil structure (see Table 19.1), soil scientists note not only the *type* (shape) of the structural peds present, but also the relative *size* (fine, medium, coarse) and degree of development or distinctness of the peds (*grades* such as strong, moderate, or

FIGURE 4.12 Strong, medium angular blocky peds in the B horizon of a Alfisol (Ustalf) in a semiarid region. The knifepoint is prying loose an individual blocky ped. Note the lighter-colored surface coatings of illuvial clay that help define and bind together the ped. (Photo courtesy of R. Weil)

weak). For example, the soil horizon shown in Figure 4.11*d* might be described as having "weak, fine, subangular blocky structure." Generally, the structure of a soil is easier to observe when the soil is relatively dry. When wet, structural peds may swell and press closer together, making the individual peds less well defined. In any case, the structural arrangement of soil particles and the pore spaces between structural peds greatly influence soil density, an aspect of soil architecture that we will now examine in detail.

4.5 SOIL DENSITY

Particle Density

Soil **particle density** D_p is defined as the mass per unit volume of soil *solids* (in contrast to the volume of the *soil,* which would also include spaces between particles). Thus, if 1 cubic meter (m³) of soil solids weighs 2.6 megagrams (Mg), the particle density is 2.6 Mg/m³ (which can also be expressed as 2.6 grams per cubic centimeter).[4]

Particle density is essentially the same as the **specific gravity** of a solid substance. The chemical composition and crystal structure of a mineral determines its particle density. Particle density is *not* affected by pore space, and therefore is not related to particle size or to the arrangement of particles (soil structure).

Particle densities for most mineral soils vary between the narrow limits of 2.60 to 2.75 Mg/m³ because quartz, feldspar, micas, and the colloidal silicates that usually make up the major portion of mineral soils all have densities within this range. For general calculations concerning arable mineral surface soils (1 to 5% organic matter), a particle density of about 2.65 Mg/m³ may be assumed if the actual particle density is not known.

This estimated particle density of 2.65 Mg/m³ should be adjusted for certain soils. For instance, when large amounts of high-density minerals (such as magnetite, garnet, epidote, zircon, tourmaline, and hornblende) are present, the particle density of the soil may exceed 3.0 Mg/m³. Organic matter, having a particle density of 0.9 to 1.3 Mg/m³, is much less dense than soil minerals. Consequently, mineral surface soils with very high organic matter (15 to 20%) may have particle densities lower than 2.4 Mg/m³. Typically, organic soils (Histosols) have particle densities between 1.1 and 2.0 Mg/m³.

Bulk Density

A second important mass measurement of soils is **bulk density** D_b, which is defined as the mass of a unit volume of dry soil. This volume includes both solids and pores. A careful study of Figure 4.13 should make clear the distinction between particle and bulk density. Both expressions of density use only the mass of the solids in a soil; therefore, any water present is excluded from consideration.

There are several methods of determining soil bulk density by obtaining a known volume of soil, drying it to remove the water, and weighing the dry mass.[5] A special coring instrument (Figure 4.14) can obtain a sample of known volume without disturbing the natural soil structure. For surface soils, perhaps the simplest method is to dig a small hole, collect all the excavated soil, and then line the hole with plastic film and fill it completely with a measured volume of water. Still another method involves coating a clod of soil with a waterproof film. The volume of the odd-shaped clod is determined by its buoyancy when suspended in water.

Factors Affecting Bulk Density

Soils with a high proportion of pore space to solids have lower bulk densities than those that are more compact and have less pore space. Consequently, any factor that influences soil pore space will affect bulk density. Typical ranges of bulk density for various soil materials and conditions are illustrated in Figure 4.15. It would be worthwhile to study this figure until you have a good feel for these ranges of bulk density.

---

[4] Since 1 Mg = 1 million grams and 1 m³ = 1 million cubic centimeters, 1 Mg/m³ = 1 g/cm³.

[5] An instrumental method that measures the soil's resistance to the passage of gamma rays is used in soil research, but is not discussed here.

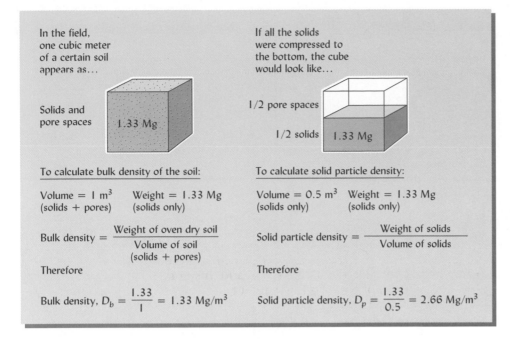

In the field, one cubic meter of a certain soil appears as...

Solids and pore spaces

1.33 Mg

If all the solids were compressed to the bottom, the cube would look like...

1/2 pore spaces

1/2 solids   1.33 Mg

To calculate bulk density of the soil:

Volume = 1 m³   Weight = 1.33 Mg
(solids + pores)   (solids only)

$$\text{Bulk density} = \frac{\text{Weight of oven dry soil}}{\text{Volume of soil (solids + pores)}}$$

Therefore

$$\text{Bulk density, } D_b = \frac{1.33}{1} = 1.33 \text{ Mg/m}^3$$

To calculate solid particle density:

Volume = 0.5 m³   Weight = 1.33 Mg
(solids only)   (solids only)

$$\text{Solid particle density} = \frac{\text{Weight of solids}}{\text{Volume of solids}}$$

Therefore

$$\text{Solid particle density, } D_p = \frac{1.33}{0.5} = 2.66 \text{ Mg/m}^3$$

FIGURE 4.13 Bulk density D_b and particle density D_p of soil. Bulk density is the weight of the solid particles in a standard volume of field soil (solids plus pore space occupied by air and water). Particle density is the weight of solid particles in a standard volume of those solid particles. Follow the calculations through carefully and the terminology should be clear. In this particular case the bulk density is one-half the particle density, and the percent pore space is 50.

(a)

(b)

FIGURE 4.14 A special sampler designed to remove a cylindrical core of soil without causing disturbance or compaction (a). The sampler head contains an inner cylinder and is driven into the soil with blows from a drop hammer. The inner cylinder (b) containing an undisturbed soil core is then removed and trimmed on the end with a knife to yield a core whose volume can easily be calculated from its length and diameter. The weight of this soil core is then determined after drying in an oven. (Photo courtesy of R. Weil)

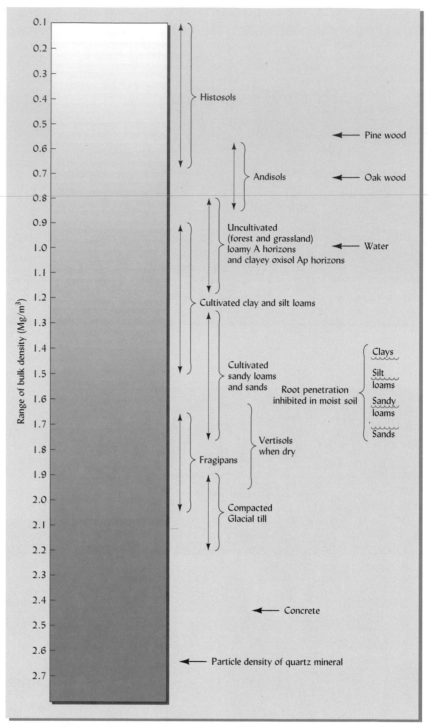

FIGURE 4.15   Bulk densities typical of a variety of soils and soil materials.

EFFECT OF SOIL TEXTURE.   As illustrated in Figure 4.15, fine-textured soils such as silt loams, clays, and clay loams generally have lower bulk densities than do sandy soils.[6] This is true because the solid particles of the fine-textured soils tend to be organized in porous granules, especially if adequate organic matter is present. In these aggregated soils,

---

[6] This fact may seem counterintuitive at first because sandy soils are commonly referred to as "light" soils, while clays and clay loams are referred to as "heavy" soils. The terms *heavy* and *light* in this context refer not to the mass per unit volume of the soils, but to the amount of effort that must be exerted to manipulate these soils with tillage implements—the sticky clays being much more difficult to till.

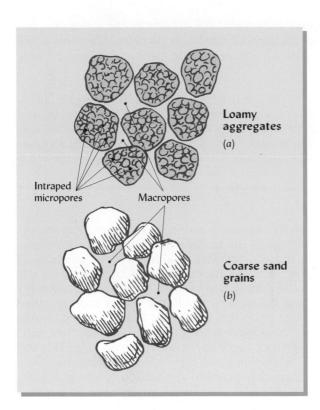

Loamy
aggregates
(a)

Intraped
micropores

Macropores

Coarse sand
grains
(b)

FIGURE 4.16 A schematic comparison of sandy and clayey soils showing the relative amounts of large (macro-) pores and small (micro-) pores in each. There is less total pore space in the sandy soils than in the clayey one because the clayey soil contains a large number of fine pores within each aggregate (a), but the sand particles (b), while similar in size to the clayey aggregates, are solid and contain no pore spaces within them. This is the reason why, among surface soils, those with coarse texture are usually more dense than those with finer textures.

pores exist both between *and* within the granules. This condition assures high total pore space and a low bulk density. In sandy soil, however, organic matter contents generally are low, the solid particles are less likely to be aggregated together, and the bulk densities are commonly higher than in the finer-textured soils. Figure 4.16 illustrates the concept that similar amounts of large pores are present in both sandy and well-aggregated fine-textured soils, but that sandy soils have few of the fine, within-ped pores, and so have less total porosity.

While sandy soils generally have high bulk densities, the packing arrangement of the sand grains also affects their bulk density (see Figure 4.17). Loosely packed grains may fill as little as 52% of the bulk volume, while tightly packed grains may fill as much as 75% of the volume.[7] The bulk density is generally lower if the sand particles are mostly of one size class (i.e., *well-sorted* sand), while a mixture of different sizes (i.e.,

---

[7] If we assume that grains consist of quartz with a particle density of 2.65 Mg/m, then the corresponding range of bulk densities for loose to tightly packed sand would be 1.38 to 1.99 Mg/m³ (0.52 * 2.65 = 1.38 and 0.75 * 2.65 = 1.99), not too different from the range actually encountered in very sandy soils.

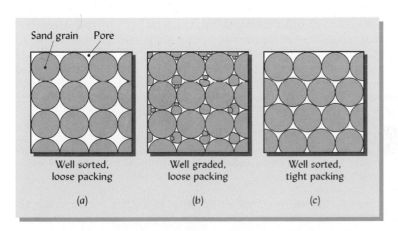

Sand grain   Pore

Well sorted,   Well graded,   Well sorted,
loose packing   loose packing   tight packing

(a)   (b)   (c)

FIGURE 4.17 The uniformity of grain size and the type of packing arrangement significantly affect the bulk density of sandy materials. Materials consisting of all similar-sized grains are termed *well sorted* (or poorly graded). Those with a variety of grain sizes are *well graded* (or poorly sorted). In either case, compaction of the particles into a tight packing arrangement markedly increases the bulk density of the material and decreases its porosity. Note that the size distribution of sand and gravel particles can be described as either graded or sorted, the two terms having essentially opposite meanings. Geologists usually speak of rivers having sorted the sand grains by size as deposits were laid down. Engineers usually are concerned as to whether or not sand consists of a gradation of sizes (i.e., is well graded or not).

well-graded sand) is likely to have an especially high bulk density. In the latter case, the smaller particles partially fill in the spaces between the larger particles. The most dense materials are those characterized by both a mixture of sand sizes and a tight packing arrangement.

DEPTH IN SOIL PROFILE. Deeper in the soil profile, bulk densities are generally higher, probably as a result of lower organic matter contents, less aggregation, fewer roots, and compaction caused by the weight of the overlying layers. Very compact subsoils may have bulk densities of 2.0 Mg/m³ or even greater. Many soils formed from glacial till (see Section 2.10) have extremely dense subsoils as a result of past compaction by the enormous mass of ice.

Useful Density Figures

For engineers involved with moving soil during construction, or for landscapers bringing in topsoil by the truckload, a knowledge of the bulk density of various soils is useful in estimating the weight of soil to be moved. A typical medium-textured mineral soil might have a bulk density of 1.25 Mg/m³, or 1250 kilograms in a cubic meter.[8] People are often surprised by how heavy soil is. Imagine driving your pickup truck to a nursery where natural topsoil is sold in bulk and filling your truck bed with a nice, rounded load. Of course, you would not really want to do this as the load might break your rear axle; you certainly would not be able to drive away with the load. A typical "half-ton" (1000 lb or 454 kg) load capacity pickup truck could carry less than 0.4 m³ of this soil even though the truck bed has room for about six times this volume of material.

The design of rooftop gardens offers another practical application for these density figures. In this instance, the mass of soil involved must be known in order to design a structure of sufficient strength to carry the load. One might choose to grow only turf and other shallow-rooted plants so that a relatively thin layer of soil (say, 30 cm) could be used to keep the total mass of soil from being too great. It might also be possible to reduce the cost of construction by selecting a soil having a low bulk density. However, if anchoring trees and other plants is an important function of the soil in such an installation, a very low density soil such as a peat or a perlite/soil mix would not be suitable.

The mass of soil in 1 ha to a depth of normal plowing (15 cm) can be calculated from soil bulk density. If we assume a bulk density of 1.3 Mg/m³ for a typical arable surface soil, such a hectare–furrow slice 15 cm deep weighs about 2 million kg.[9] This estimate of the mass of surface soil in a hectare of land is very useful in calculating lime and fertilizer application rates and organic matter mineralization rates (see Boxes 8.1, 9.3, 10.3, and 13.1 for detailed examples). However, this estimated mass must be adjusted if the bulk density is other than 1.3 Mg/cm³ or the depth of the layer under consideration is more or less than 15 cm.

Management Practices Affecting Bulk Density

Changes in bulk density for a given soil are easily measured and can alert soil managers to changes in soil quality and ecosystem function. Increases in bulk density usually indicate a poorer environment for root growth, reduced aeration, and undesirable changes in hydrologic function, such as reduced water infiltration.

FOREST LANDS. The surface horizons of most forested soils have rather low bulk densities (see Figure 4.15). Tree growth and forest ecosystem function are particularly sensitive to increases in bulk density. Conventional timber harvest generally disturbs and compacts 20 to 40% of the forest floor (Figure 4.18) and is especially damaging along the skid trails where logs are dragged and at the landing decks—areas where logs are piled and loaded onto trucks (Table 4.3). An expensive, but effective, means of moving logs while

---

[8] Most commercial landscapers and engineers in the United States still use English units. To convert values of bulk density given in units of Mg/m³ into values of lb/yd³, multiply by 1686. Therefore, 1 yd³ of a typical medium-textured mineral soil with a bulk density of 1.25 Mg/m³ would weigh over a ton (1686 * 1.25 = 2108 lb/yd³).

[9] 10,000 m²/ha × 1.3 Mg/m³ × 0.15 m = 1950 Mg/ha ≈ 2 million kg per ha 15 cm deep. A comparable figure in the English system is 2 million lb per acre–furrow slice 6 to 7 in. deep.

FIGURE 4.18   (a) Timber harvest with a conventional rubber-tired skidder in a boreal forest in western Alberta, Canada, and (b) result-ing disturbance to the surface soil horizons (forest floor). Such practices cause significant soil compaction that can impair soil ecosystem functions for many years. Timber harvest practices that can reduce such damage to forest soils include selective cutting, use of flexible-track vehicles and overhead cable transport of logs, and abstaining from harvest during wet conditions. (Photo courtesy of Andrei Start-sev, Alberta Environmental Center).

TABLE 4.3   Effects of Timber Harvest on Bulk Density at Different Depths in Two Forested Ultisols in Georgia

Rubber-wheeled skidders were used to harvest the logs.
Note the generally higher bulk densities of the sandy loam soil compared to the clay loam, and the greater effect of timber harvest on the skidder trails.

Soil Depth, cm	Bulk density, Mg/m^3		
	Preharvest	Postharvest, off trails	Postharvest, skidder trails
Upper coastal plain, sandy loam			
0–8	1.25	1.50	1.47
8–15	1.40	1.55	1.71
15–23	1.54	1.61	1.81
23–30	1.58	1.62	1.77
Piedmont, clay loam			
0–8	1.16	1.36	1.52
8–15	1.39	1.49	1.67
15–23	1.51	1.51	1.66
23–30	1.49	1.46	1.61

Data from Gent, et al. (1984, 1986).

minimizing compactive degradation of forest lands is the use of cables strung between towers or hung from large balloons.

Intensive recreational use of forested soils can also lead to increased bulk densities in areas with heavy foot traffic, such as campsites and trails (Figure 4.19). An important consequence of increased bulk density is a diminished capacity of the soil to take in water, hence increased losses by surface runoff. Damage from hikers can be minimized by restricting foot traffic to well-designed, established trails that may include a thick layer of wood chips, or even a raised boardwalk in the case of heavily traveled paths over very fragile soils, such as in wetlands.

Trees planted into compacted soil may fail to grow normally and may even die. However, several practices can assist normal root development and growth. In preparation for replanting after clear cutting, some forest soils (especially on relatively level sites) may be loosened by tillage and bedding operations similar to those applied to agriculture land (see following). However, such preparation may also cause subsoil compaction and loss of soil organic matter.

In urban areas, trees planted for landscaping purposes must often contend with severely compacted soils. While it is usually not practical to modify the entire root zone of a tree, several practices can help (see also Section 7.7). First, making the planting hole as large as possible will provide a zone of loose soil for early root growth. Second, a thick

FIGURE 4.19 Impact of campers on the bulk density of forest soils, and the consequent effects on rainwater infiltration rates and runoff losses (see white arrows). At most campsites the high-impact area extends for about 10 m from the fire circle or tent pad. Managers of recreational land must carefully consider how to protect sensitive soils from compaction that may lead to death of vegetation and increased erosion. [Data from Vimmerstadt, et al. (1982)].

layer of mulch spread out to the drip line (but not too near the trunk) will enhance root growth, at least near the surface. Third, the tree roots may be given paths for expansion by digging a series of narrow trenches radiating out from the planting hole and back-filled with loose, enriched soil.

AGRICULTURAL LAND. Although tillage may temporarily loosen the surface soil, in the long term intense tillage increases soil bulk density because it depletes soil organic matter and weakens soil structure (see Section 4.8). The data in Table 4.4 illustrate this trend. These data are from long-term studies in different locations where relatively undisturbed soils were compared to adjacent areas that had been cultivated for 20 to 90 years. In all cases, cropping increased the bulk density of the topsoils. The effect of cultivation can be minimized by adding crop residues or farm manure in large amounts and rotating cultivated crops with a grass sod.

In modern agriculture, heavy machines used to pull implements, apply amendments, or harvest crops can create yield-limiting soil compaction. Certain tillage implements, such as the moldboard plow and the disk harrow, compact the soil below their working depth even as they lift and loosen the soil above. Use of these implements or repeated trips over the field by heavy machinery can form **plow pans** or **traffic pans**, dense zones immediately below the plowed layer (Figure 4.20). Other tillage implements, such as the chisel plow and the spring-tooth harrow, do not press down upon the soil beneath them, and so are useful in breaking up plow pans and stirring the soil with a minimum of compaction. Large chisel-type plows can be used in **subsoiling** to break up dense subsoil layers (Figure 4.21). However, in many soils the effects of subsoiling are quite temporary (Figure 4.22). Any tillage tends to reduce soil strength, thus making the soil less resistant to subsequent compaction.

Traffic is particularly damaging on wet soil. Generally, with heavier loads and on wetter soils, compactive effects are more pronounced and penetrate more deeply into the profile. To prevent compaction, which can result in yield reductions and loss of profitability, the number of tillage operations and heavy equipment trips over the field should be minimized and timed to avoid periods when the soil is wet. Unfortunately, traffic on wet agricultural soils is sometimes unavoidable in humid temperate regions in spring and fall.

Another approach for minimizing compaction is to carefully restrict all wheel traffic to specific lanes, leaving the rest of the field (usually 90% or more of the area) free from compaction. Such **controlled traffic** systems are widely used in Europe, especially on clayey soils. Gardeners can practice controlled traffic by establishing permanent foot paths between planting beds. The paths may be enhanced by covering with a thick mulch, planting to sod grass, or paving with flat stones.

Some managers attempt to reduce compaction using an opposite strategy in which special wide tires are fitted to heavy equipment so as to spread the weight over more soil surface, thus reducing the force applied per unit area (Figure 4.23a). Wider tires do lessen the compactive effect, but they also increase the percentage of the soil surface that is impacted. In a practice analogous to using wide wheels, home gardeners can avoid concentrating their body weight on just the few square centimeters of their foot-

TABLE 4.4 Bulk Density and Pore Space of Some Surface Soils from Cultivated and Nearby Uncultivated Areas (One Subsoil Included)

The bulk density was increased, and the pore space proportionately decreased, in every case.

Soil	Texture	Years cropped	Bulk density, Mg/m³		Pore space, %	
			Cultivated soil	Uncultivated soil	Cultivated soil	Uncultivated soil
Udalf (Pennsylvania)	Loam	58	1.25	1.07	50	57.2
Udoll (Iowa)	Silt loam	50+	1.13	0.93	56.2	62.7
Aqualf (Ohio)	Silt loam	40	1.31	1.05	50.5	60.3
Ustoll (Canada)	Silt loam	90	1.30	1.04	50.9	60.8
Cambid (Canada)	Clay	70	1.28	0.98	51.7	63.0
Cambid, subsoil (Canada)	Clay	70	1.38	1.21	47.9	54.3
Mean of 3 Ustalfs (Zimbabwe)	Clay	20–50	1.44	1.20	54.1	62.6
Mean of 3 Ustalfs (Zimbabwe)	Sandy loam	20–50	1.54	1.43	42.9	47.2

Data for Canadian soils from Tiessen, et al. (1982), for Zimbabwe soils from Weil (unpublished), and for other soils from Lyon, et al. (1952).

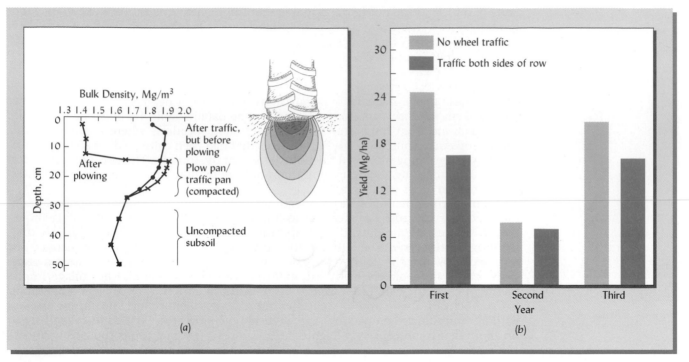

FIGURE 4.20 Tractors and other heavy equipment compact the soil to considerable depths, increasing bulk density and reducing plant growth and crop yields. The effects are especially damaging if the soil is wet when trafficked. (*a*) The tires of a heavy vehicle compact a sandy loam soil to about 30 cm, creating a traffic pan. Plowing temporarily loosens the compacted surface soil (plow layer), but increases compaction just below the plowed layer, creating a combined traffic pan and plow pan. Bulk densities in excess of 1.8 Mg/m³ prevented the penetration of cotton roots in this case. (*b*) The yield of potatoes was reduced in two out of three years in this test on a clay loam in Minnesota. Yield reductions are often most pronounced in relatively dry years when plants have the greatest need for subsoil moisture. [Based on data from Camp and Lund (1964) and Voorhees (1984)]

prints by standing on wooden boards when preparing seedbeds in relatively wet soil (Figure 4.23*b*).

Influence of Bulk Density on Soil Strength and Root Growth

High bulk density may occur as a natural soil profile feature (for example, a fragipan), or it may be an indication of human-induced soil compaction. In any case, root growth is inhibited by excessively dense soils for a number of reasons, including the soil's resistance to penetration, poor aeration, slow movement of nutrients and water, and the buildup of toxic gases and root exudates.

Roots penetrate the soil by pushing their way into pores. If a pore is too small to accommodate the root cap, the root must push the soil particles aside and enlarge the pore. To some degree, the density per se restricts root growth, as the roots encounter fewer and smaller pores. However, root penetration is also limited by **soil strength**, the property of the soil that causes it to resist deformation. One way to quantify soil strength is to measure the force needed to push a standard cone-tipped rod (a **penetrometer**) into the soil (see also Section 4.9). Compaction generally increases both bulk density and soil strength. At least two factors (both related to soil strength) must be considered to determine the effect of bulk density on the ability of roots to penetrate soil.

EFFECT OF SOIL WATER CONTENT.   Soil water content and bulk density both affect *soil strength* (see Section 4.9). Soil strength is increased when a soil is compacted to a higher bulk density, and also when a soil dries out and hardens. Therefore, the effect of bulk density

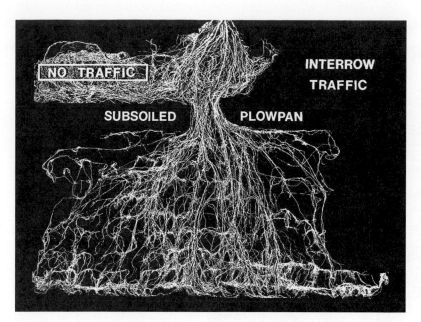

FIGURE 4.21 Root distribution of a cotton plant. On the right, interrow tractor traffic and plowing have caused a plowpan that restricts root growth. Roots are more prolific on the left where there had been no recent tractor traffic. The roots are seen to enter the subsoil through a loosened zone created by a subsoiling chisel-type implement. (Courtesy USDA National Tillage Machinery Laboratory)

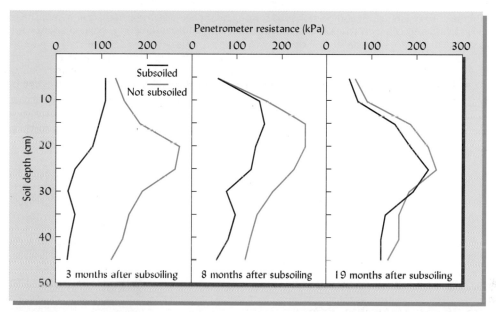

FIGURE 4.22 Deep tillage with a large chisel plow, sometimes called *subsoiling*, can break up a plowpan in the subsoil, reducing the soil strength and density as measured by resistance to penetration. Such subsoiling is most effective when a clayey subsoil is dry enough that pulling the implement through it shatters the mass of soil, creating a network of cracks. Even then, the effects are generally short-lived. In the Florida coastal plain Ultisol represented here, the subsoiling operation resulted in much-reduced penetration resistance for a few months. However, the difference between soil that had been subsoiled and soil that had not was less noticeable after 8 months and virtually unapparent after 27 months. Accordingly, in this experiment, subsoiling resulted in higher corn yield in the first year, but not in the second year after the deep tillage was performed. [From Wright, et al. (1980)]

(a)   (b)

FIGURE 4.23   One approach to reducing soil compaction is to spread the applied weight over a larger area of the soil surface. Examples are extra-wide wheels on heavy vehicles used to apply soil amendments (left) and standing on a wooden board while preparing a garden seedbed in early spring (right). (Photos courtesy of R. Weil)

on root growth is most pronounced in a relatively dry soil, a higher bulk density being necessary to prevent root penetration when the soil is moist. For example, a traffic pan having a bulk density of 1.6 Mg/m^3 may completely prevent the penetration of roots when the soil is rather dry, yet roots may readily penetrate this same layer when it is in a moist condition.

EFFECT OF SOIL TEXTURE.   The more clay present in a soil, the smaller the average pore size, and the greater the resistance to penetration at at a given bulk density. Therefore, if the bulk density is the same, roots more easily penetrate a moist sandy soil than a moist clayey one. The growth of roots into moist soil is generally limited by bulk densities ranging from 1.45 Mg/m^3 in clays to 1.85 Mg/m^3 in loamy sands (see Figure 4.15). Viewed in this context, root growth was probably inhibited by the bulk density of the skid trails in both soils illustrated in Table 4.3.

4.6   PORE SPACE OF MINERAL SOILS

One of the main reasons for measuring soil bulk density is that this value can be used to calculate pore space. For soils with the same particle density, the lower the bulk density, the higher the percent pore space (**total porosity**). See Box 4.4 (page 145) for derivation of the formula expressing this relationship.

Factors Influencing Total Pore Space

In Chapter 1 (Figure 1.17) we noted that for an "ideal" medium-textured, well-granulated surface soil in good condition for plant growth, approximately 50% of the soil volume would consist of pore space, and that the pore space would be about half-filled with air and half-filled with water. Actually, total porosity varies widely among soils for the same reasons that bulk density varies. Values range from as low as 25% in compacted subsoils to more than 60% in well-aggregated, high-organic-matter surface soils. As is the case for bulk density, agricultural management can exert a decided influ-

BOX 4.4 CALCULATION OF PERCENT PORE SPACE IN SOILS

The bulk density of a soil can be easily measured and particle density can usually be assumed to be 2.65 Mg/m³ for most silicate-dominated mineral soils. Direct measurement of the pore space in soil requires the use of much more tedious and expensive techniques. Therefore, when information on the percent pore space is needed, it is often desirable to calculate the pore space from data on bulk and particle densities.

The derivation of the formula used to calculate the percentage of total pore space in soil follows:

Let D_b = bulk density, Mg/m³ V_s = volume of solids, m³

       D_p = particle density, Mg/m³ V_p = volume of pores, m³

       W_s = Weight of soil (solids), Mg $V_s + V_p$ = total soil volume V_t, m³

By definition,

$$\frac{W_s}{V_s} = D_p \qquad \text{and} \qquad \frac{W_s}{V_s + V_p} = D_b$$

Solving for W_s gives

$$W_s = D_p \times V_s \qquad \text{and} \qquad W_s = D_b(V_s + V_p)$$

Therefore

$$D_p \times V_s = D_b(V_s + V_p) \qquad \text{and} \qquad \frac{V_s}{V_s + V_p} = \frac{D_b}{D_p}$$

Since

$$\frac{V_s}{V_s + V_p} \times 100 = \%\ \text{solid space} \qquad \text{then} \qquad \%\ \text{solid space} = \frac{D_b}{D_p} \times 100$$

Since % pore space + % solid space = 100, and % pore space = 100 − % solid space, then

$$\%\ \text{pore space} = 100 - \left(\frac{D_b}{D_p} \times 100\right)$$

EXAMPLE

Consider the cultivated clay soil from Canada in Table 4.4 (the Cambid). The bulk density was determined to be 1.28 Mg/m³. Since we have no information on the particle density, we assume that the particle density is approximately that of the common silicate minerals (i.e., 2.65 Mg/m³). We calculate the percent pore space using the formula derived above:

$$\%\ \text{pore space} = 100 - \left(\frac{1.28\ \text{Mg/m}^3}{2.65\ \text{Mg/m}^3} \times 100\right) = 100 - 48.3 = 51.7$$

This value of pore space, 51.7%, is quite close to the typical percentage of air and water space described in Figure 1.17 for a well-granulated, medium- to fine-textured soil in good condition for plant growth. This simple calculation tells us nothing about the relative amounts of large and small pores, however, and so must be interpreted with caution.

For certain soils it is inaccurate to assume that the soil particle density is 2.65 Mg/m³. For instance, a soil with a high organic matter content can be expected to have a particle density somewhat lower than 2.65. Similarly, a soil rich in iron oxide minerals will have a particle density greater than 2.65 because these minerals have particle densities as high as 3.5. As an example of the latter type of soil, let us consider the uncultivated clay soils from Zimbabwe (Ustalfs) described in Table 4.4. These are red-colored clays, high in iron oxides. The particle density for these soils was determined to be 3.21 Mg/m³ (not shown in Table 4.4). Using this value and the bulk density value from Table 4.4, we calculate the pore space as follows:

$$\%\ \text{pore space} = 100 - \left(\frac{1.20}{3.21} \times 100\right) = 100 - 37.4 = 62.6$$

Such a high percentage pore space is an indication that this soil is in an uncompacted, very well granulated condition typical for soils found under undisturbed natural vegetation.

ence on the pore space of soils (see Table 4.4). Data from a wide range of soils show that cultivation tends to lower the total pore space compared to that of uncultivated soils. This reduction usually is associated with a decrease in organic matter content and a consequent lowering of **granulation**.

Size of Pores

Bulk density values help us predict only *total* porosity. However, soil pores occur in a wide variety of sizes and shapes. The size of a pore largely determines what role the pore can play in the soil. Pores can be grouped by size into macropores, mesopores, micropores, and so on (Table 4.5 illustrates one such grouping). We will simplify our discussion at this point by referring only to **macropores** (larger than about 0.08 mm) and **micropores** (smaller than about 0.08 mm).

MACROPORES.   The macropores characteristically allow the ready movement of air and the drainage of water. They also are large enough to accommodate plant roots and the wide range of tiny animals that inhabit the soil (see Chapter 11). Several types of macropores are illustrated in Figure 4.24.

Macropores can occur as the spaces between individual sand grains in coarse textured soils. Thus, even though a sandy soil has relatively low total porosity, the movement of air and water through such a soil is surprisingly rapid because of the dominance of the macropores.

In well-structured soils, the macropores are generally found between peds. These **interped pores** may occur as spaces between loosely packed granules or as the planar cracks between tight-fitting blocky and prismatic peds.

Macropores created by roots, earthworms and other organisms constitute a very important type of pores termed **biopores**. These are usually tubular in shape, and may be continuous for lengths of a meter or more. In some clayey soils, biopores are the principal form of macropores, greatly facilitating the growth of plant roots (Table 4.6). Perennial vegetation, such as forest trees and certain grasses, are particularly effective at creating channels that serve as conduits for roots, long after the death and decay of the roots that originally created them. Two such old root channels, each about 8 mm in diameter, can be seen perforating the clay slickenside shown in Figure 3.19.

It is clear that both soil structure and texture influence the balance between macropores and micropores in a soil. Figure 4.25 shows that the decrease in organic matter and increase in clay that occur with depth in many profiles are associated with a shift from macropores to micropores.

MICROPORES.   In contrast to macropores, micropores are usually filled with water in field soils. Even when not water-filled, they are too small to permit much air movement. Water movement in micropores is slow, and much of the water retained in these pores is not available to plants (see Chapter 5). Fine-textured soils, especially those without a stable granular structure, may have a preponderance of micropores, thus allowing relatively slow gas and water movement, despite the relatively large volume of total pore

TABLE 4.5   A Size Classification of Soil Pores and Some Functions of Each Size Class

Pore sizes are actually a continuum and the boundaries between classes given here are inexact and somewhat arbitrary. The term micropore *is often broadened to refer to all the pores smaller than macropores.*

Simplified class	Class[a]	Effective diameter range (mm)	Characteristics and functions
Macropores	Macropores	0.08–5+	Generally found between soil peds (interped); water drains by gravity; effectively transmit air; large enough to accommodate plant roots, habitat for certain soil animals.
Micropores	Mesopores	0.03–0.08	Retain water after drainage; transmit water by capillary action, accommodate fungi and root hairs.
	Micropores	0.005–0.03	Generally found within peds (intraped); retain water that plants can use; accommodate most bacteria.
	Ultramicropores	0.0001–0.005	Found largely with clay groupings; retain water that plants cannot use; exclude most microorganisms.
	Cryptopores	<0.0001	Exclude all microorganisms, too small for large molecules to enter.

[a] The pore size classes and boundary diameters are those of Brewer, 1964 as cited in Soil Sci. Soc. Amer. (1996).

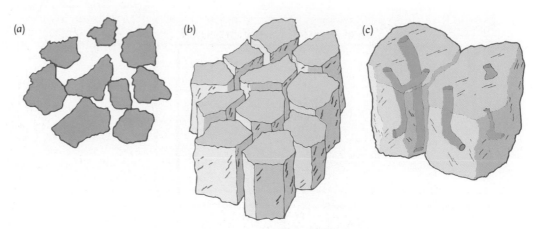

FIGURE 4.24   Various types of soil pores. (*a*) Many soil pores occur as *packing pores,* spaces left between primary soil particles. The size and shape of these spaces is largely dependent on the size and shape of the primary sand, silt, and clay particles and their packing arrangement. (*b*) In soils with structural peds, the spaces between the peds form *interped pores.* These may be rather planar in shape, as with the cracks between prismatic peds, or they may be more irregular, like those between loosely packed granular aggregates. (*c*) *Biopores* are formed by organisms such as earthworms, insects, and plant roots. Most of these are long, sometimes branched channels, but some are round cavities left by insect nests and the like.

space. Aeration, especially in the subsoil, may be inadequate for satisfactory root development and desirable microbial activity. While the larger micropores accommodate plant root hairs and microorganisms, the smaller micropores (sometimes termed *ultramicropores* and *cryptopores*) are too small to allow the entry of even the smallest bacteria. They therefore can serve as protected sites in which organic compounds may remain untouched for centuries (see Chapter 12).

Clearly, the size of individual pore spaces rather than their combined volume is the important factor in determining soil drainage, aeration and other processes. The loosening and granulating of fine textured soils promotes aeration, not so much by increasing the total pore space as by raising the proportion of macropores.

Cultivation and Pore Size

Continuous cropping, particularly of soils originally high in organic matter, often results in a reduction of macropore spaces. Data from a fine-textured soil in Texas (Table 4.7) clearly illustrate this effect. Cropping with plow tillage significantly reduced soil organic matter content and total pore space. But most striking is the effect of cropping on the size of the soil pores: The amount of macropore space, needed for ready air movement, was reduced by about one-half. Samples taken from 1 m deep showed that the reduction in pore size extended far into the soil profile (data not shown).

TABLE 4.6   **Distribution of Different-Sized Loblolly Pine Roots in the Soil Matrix and in Old Root Channels in the Uppermost Meter of an Ultisol in South Carolina**

The root channels were generally from 1 to 5 cm in diameter and filled with loose surface soil and decaying organic matter. They are easy for roots to penetrate and have better fertility and aeration than the surrounding soil matrix.

	Numbers of roots counted per 1 m² of the soil profile		
Root size, diameter	*Soil matrix*	*Old root channels*	*Comparative increase in root density in the old channels, %*
Fine roots, <4 mm	211	3617	94
Medium roots, 4–20 mm	20	361	95
Coarse roots, >20 mm	3	155	98

Calculated from Parker and Van Lear (1996).

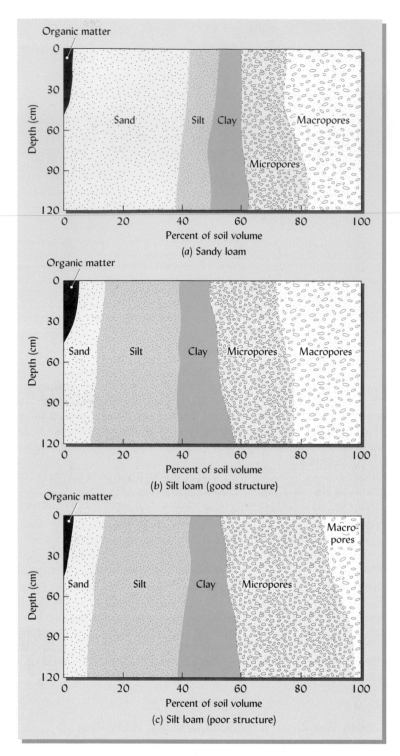

FIGURE 4.25 Volume distribution of organic matter, sand, silt, clay, and pores of macro- and microsizes in a representative sandy loam (a) and in two representative silt loams, one with good soil structure (b) and the other with poor soil structure (c). Both silt loam soils have more total pore space than the sandy loam, but the silt loam with poor structure has a smaller volume of larger (macro) pores than either of the other two soils. Note that at the lowest depths in both silt loams, about one-third of the mineral matter is clay, giving the lower horizons enough clay to be classified as silty clay loams.

In recent years, conservation tillage practices, which minimize plowing and associated soil manipulations, have been widely adopted in the United States (see Sections 4.8 and 17.6). Because of increased accumulation of organic matter near the soil surface and the development of a long-lived network of macropores (especially biopores), some conservation tillage systems lead to greater macroporosity of the surface layers. These benefits are particularly likely to accrue in soils with extensive production of earthworm burrows, which may remain undisturbed in the absence of tillage. Unfortunately, such improvements in porosity do not always occur. Under some circumstances, less pore space has been found with conservation tillage than with conventional tillage, a matter of some concern in soils with poor internal drainage.

TABLE 4.7 Effect of About 50 Years of Continuous Cropping on the Macropore and Micropore Spaces of a Fine-Textured Vertisol (Houston Black Clay) in Texas

Compared to the undisturbed prairie (virgin soil), the cultivated soil has far less macropore space, but has gained some micropore space as aggregates were destroyed, changing large interped pores into much smaller micropores. The loss of macropore space probably resulted from the loss of organic matter that is evident.

Soil history	Organic matter, %	Total pore space, %	Macropore space, %	Micropore space, %	Bulk density, Mg/m^3
			0–15 cm depth		
Virgin prairie	5.6	58.3	32.7	25.6	1.11
Tilled 50 years	2.9	50.2	16.0	34.2	1.33
			15–30 cm depth		
Virgin prairie	4.2	56.1	27.0	29.1	1.16
Tilled 50 years	2.8	50.7	14.7	36.0	1.31

Data from Laws and Evans (1949).

4.7 FORMATION AND STABILIZATION OF SOIL AGGREGATES

The formation and maintenance of a high degree of aggregation is one of the most difficult tasks of soil management, yet it is also one of the most important, since it is a potent means of influencing ecosystem function. The organization of surface soils into relatively large structural aggregates provides for the low bulk density and a high proportion of macropores so desirable for most soil uses.

Some aggregates readily succumb to the beating of rain and the rough and tumble of plowing and tilling the land. Others resist disintegration, thus making the maintenance of a suitable soil structure comparatively easy (see Figures 4.30 and 4.31, pages 153 and 154). Generally, the smaller aggregates are more stable than the larger ones, so maintaining the much-prized larger aggregates requires much care.

We will discuss practical means of managing soil structure after we consider those factors responsible for aggregate formation and those that give the aggregates stability once they are formed. Since both sets of factors are operating simultaneously, it is sometimes difficult to distinguish their relative effects on the development of stable aggregates in soils.

Hierarchical Organization of Soil Aggregates[10]

The large aggregates (>1 mm) so desirable for most soil uses are typically composed of smaller aggregates, which in turn are composed of still smaller units, down to clusters of clay and humus less than 0.001 mm in size. You may easily demonstrate the existence of this *hierarchy of aggregation* by selecting a few of the largest aggregates in a soil and gently crushing or picking them apart to separate them into many smaller-sized aggregates. Then try rubbing the tiniest of these granules between your thumb and forefinger. You will find that most of these break down into a smear of still smaller aggregates composed of silt, clay, and humus. The hierarchical organization of aggregates (Figure 4.26) seems to be characteristic of most soils, with the exception of certain Oxisols and some very young Entisols. At each level in the hierarchy of aggregates, different factors are responsible for binding together the subunits.

Factors Influencing Aggregate Formation and Stability in Soils

Both biological and physical-chemical (abiotic) processes are involved in the formation of soil aggregates. The physical-chemical processes tend to be most important at the smaller end of the scale, biological processes at the larger end. Also, the physical-

---

[10] The role of organic matter and biological processes in the hierarchical organization of soil aggregates was originally put forward by Tisdall and Oades (1982) and elaborated on by Oades (1993) and Tisdall (1994).

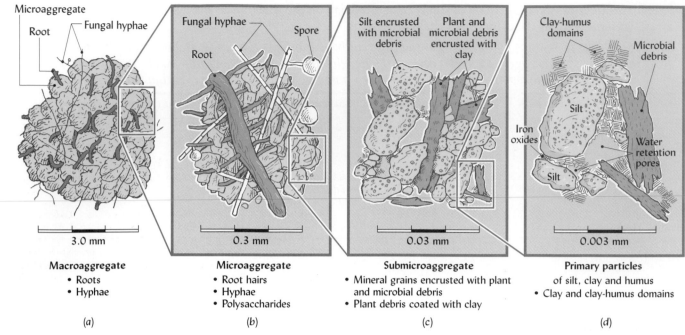

Microaggregate
Root ┐ ┌ Fungal hyphae

Fungal hyphae ─┐   Spore
Root ┐

Silt encrusted   Plant and
with microbial   microbial debris
debris   encrusted with
                 clay

Clay-humus
domains   Microbial
                     debris

Iron
oxides
        Silt

Silt   Water
                        retention
                        pores

| 3.0 mm | 0.3 mm | 0.03 mm | 0.003 mm |

Macroaggregate
- Roots
- Hyphae

Microaggregate
- Root hairs
- Hyphae
- Polysaccharides

Submicroaggregate
- Mineral grains encrusted with plant and microbial debris
- Plant debris coated with clay

Primary particles
of silt, clay and humus
- Clay and clay-humus domains

(a)   (b)   (c)   (d)

FIGURE 4.26 Larger aggregates are often composed of an agglomeration of smaller aggregates. This illustration shows four levels in this hierarchy of soil aggregates. The different factors important for aggregation at each level are indicated. (a) A *macroaggregate* composed of many microaggregates bound together mainly by a kind of sticky network formed from fungal hyphae and fine roots. (b) A *microaggregate* consisting mainly of fine sand grains and smaller clumps of silt grains, clay, and organic debris bound together by root hairs, fungal hyphae, and microbial gums. (c) A very small submicroaggregate consisting of fine silt particles encrusted with organic debris and tiny bits of plant and microbial debris (called *particulate organic matter*) encrusted with even smaller packets of clay, humus, and Fe or Al oxides. (d) Clusters of parallel and random clay platelets interacting with Fe or Al oxides and organic polymers at the smallest scale. These organoclay clusters or *domains* bind to the surfaces of humus particles and the smallest of mineral grains.

chemical processes of aggregate formation are mainly associated with clays and, hence, tend to be of greater importance in finer-textured soils. In sandy soils that have little clay, aggregation is almost entirely dependent on biological processes.

Physical-Chemical Processes

Most important among the physical-chemical processes are (1) the mutual attraction among clay particles and (2) the swelling and shrinking of clay masses.

FLOCCULATION OF CLAYS AND THE ROLE OF ADSORBED CATIONS. Except in very sandy soils that are almost devoid of clay, aggregation begins with the **flocculation** of clay particles into microscopic clumps or *floccules* (Figure 4.27). Flocculation may be explained by the fact that most types of clay particles possess electronegatively charged surfaces that normally attract a swarm of positively charged cations from the soil solution (see Section 8.1). If two clay platelets come close enough to each other, the cations compressed in a layer between them will attract the negative charges on both platelets, thus serving as bridges to hold the platelets together. This process is repeated until a small "stack" of parallel clay platelets, termed a *clay domain,* is formed. Other types of clay domains are more random in orientation, resembling a house of cards. These form when the positive charges on the edges of the clay platelets attract the negative charges on the planar surfaces (Figure 4.27). Clay floccules or domains, along with charged organic colloids (humus), form bridges that bind to each other and to fine silt particles (mainly quartz), creating the smallest size groupings in the hierarchy of soil aggregates (Figure 4.26d). These domains, aided by the flocculating influence of polyvalent cations (e.g., Ca^{2+}, Fe^{2+}, and Al^{3+}) and humus, provide much of the long-term stability for the smaller (< 0.03 mm) **microaggregates.** The cementing action of inorganic compounds, such as iron oxides, produces very stable small aggregates sometimes called **pseudosand** in certain clayey soils (Ultisols and Oxisols) of hot, humid regions.

When Na^+ (rather than polyvalent cations such as Ca^{2+} or Al^{3+}) is a prominent adsorbed ion, as in some soils of arid and semiarid areas, the attractive forces are not

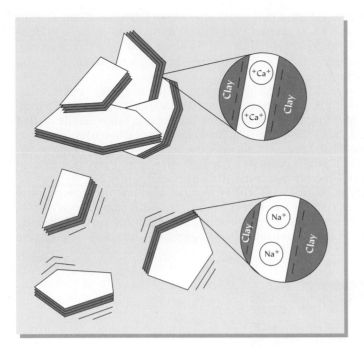

FIGURE 4.27 The role of cations in the flocculation of soil clays. The di- and trivalent cations, such as Ca^{2+} and Al^{3-}, are tightly adsorbed and can effectively neutralize the negative surface charge on clay particles. These cations can also form bridges that bring clay particles close together. Monovalent ions, especially Na^+, with relatively large hydrated radii, can cause clay particles to repel each other and create a dispersed condition. Two things contribute to the dispersion: (1) the large hydrated sodium ion does not get close enough to the clay to effectively neutralize the negative charges, and (2) the single charge on sodium is not effective in forming a bridge between clay particles.

able to overcome the natural repulsion of one negatively charged clay by another. The clay platelets cannot approach closely enough to flocculate, so remain dispersed as far apart from one another as possible. Clay in this dispersed, gel-like condition causes the soil to become almost structureless, impervious to water and air, and very undesirable from the standpoint of plant growth (See Section 10.10).

VOLUME CHANGES IN CLAYEY MATERIALS. As a soil dries out and water is withdrawn, the platelets in clay domains move closer together, causing the domains and, hence, the soil mass to shrink in volume. As a soil mass shrinks, cracks will open up along planes of weakness. Over the course of many cycles (as occur between rain or irrigation events in the field) cracks form and reform along the same planes of weakness. The network of cracks becomes more extensive and the aggregates between the cracks better defined. Plant roots also have a distinct drying effect as they take up soil moisture in their immediate vicinity. Water uptake, especially by fibrous-rooted perennial grasses, accentuates the physical aggregation processes associated with wetting and drying. This effect is but one of many examples of ways in which physical and biological soil processes interact.

Freezing and thawing cycles have a similar effect, since the formation of ice crystals is a drying process that also draws water out of clay domains. The swelling and shrinking actions that accompany freeze-thaw and wet-dry cycles in soils create fissures and pressures that alternately break apart large soil masses and compress soil particles into defined structural peds. The aggregating effects of these water and temperature cycles are most pronounced in soils with a high content of swelling-type clays (see Chapter 8), especially Vertisols, Mollisols, and some Alfisols.

Farmers have long recognized that large, hard clods of soil will break down into a "mellow" seedbed if left to overwinter under conditions in which they will alternately freeze and thaw or become dry and wet (under gentle rains). For clayey soils, the cold winters of temperate regions and the long dry seasons of certain warmer regions can do much to undo the ill effects of compaction and poor structural management (see Section 4.8).

Biological Processes

ACTIVITIES OF SOIL ORGANISMS. Among the biological processes of aggregation, the most prominent are: (1) the burrowing and molding activities of earthworms, (2) the enmeshment of particles by sticky networks of roots and fungal hyphae, and (3) the production of organic glues by microorganisms, especially bacteria and fungi. In both cultivated and uncultivated soils, earthworms (and termites) move soil particles around, often ingesting them and forming them into pellets or casts (see Chapter 11). In some forested soils, the surface horizon consists primarily of aggregates formed as earthworm castings (see, for example, Figure 4.11*a*). Plant roots also move particles about as they push their way through the soil. This movement forces soil particles to come into close contact with each other, encouraging aggregation. At the same time, the channels created by plant roots and soil animals serve as macropores, breaking up large clods and helping to define larger soil structural units.

Plant roots (particularly root hairs) and fungal hyphae exude sugarlike polysaccharides and other organic compounds, forming sticky networks that bind together individual soil particles and tiny microaggregates into larger agglomerations called **macroaggregates** (see Figure 4.26*a*). The threadlike fungi that associate with plant roots (called *mycorrhizae;* see Section 11.9) are especially effective in providing this type of relatively short-term stabilization of large aggregates, because they secrete a gooey protein called **glomulin**, which is very effective as a cementing agent (Figure 4.28).

Bacteria also produce polysaccharides and other organic glues as they decompose plant residues. Bacterial polysaccharides are shown intermixed with clay at a very small scale in Figure 4.29. Many of these root and microbial organic glues resist dissolution by water and so not only enhance the formation of soil aggregates but also help ensure

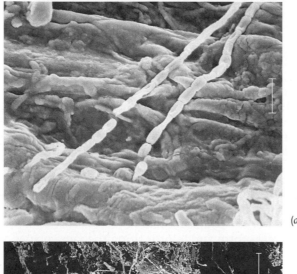

(a)

(b)

(c)

FIGURE 4.28 Fungal hyphae binding soil particles into aggregates. (*a*) Close-up of a hyphae growing over the surface of a mineral grain encrusted with microbial cells and debris. Bar = 10 μm. (*b*) An advanced stage of aggregation during the formation of soil from dune sands. Note the net of fungal hyphae and the encrustation of the mineral grains with organic debris. Bar = 50 μm. (*c*) Hyphae of the root-associated fungus from the genus *Gigaspora* interconnecting particles in a sandy loam from Oregon. Note also the fungal spore and the plant root. Bar = 320 μm. [Photos (*a*) and (*b*) courtesy of Sharon L. Rose, Willamette University; photo (*c*) courtesy of R. P. Schreiner, USDA-ARS, Corvallis, Ore.]

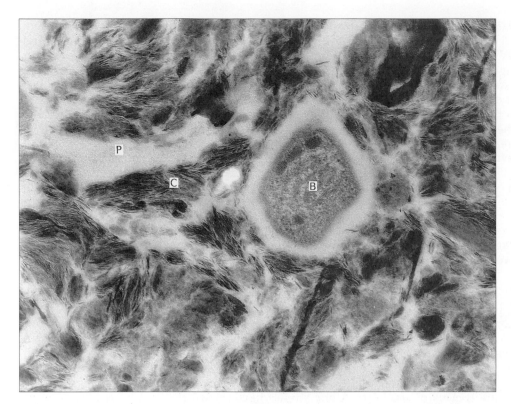

FIGURE 4.29 An ultrathin section illustrating the interaction among organic materials and silicate clays in a water-stable aggregate. The dark-colored materials (C) are groups of clay particles that are interacting with organic polysaccharides (P). A bacterial cell (B) is also surrounded by polysaccharides. Note the generally horizontal orientation of the clay particles, an orientation encouraged by the organic materials. [From Emerson, et al. (1986); photograph provided by R. C. Foster, CSIRO, Glen Osmond, Australia])

their stability over a period of months to a few years. These processes are most notable in surface soils where root and animal activities and organic matter accumulation are greatest.

INFLUENCE OF ORGANIC MATTER. In most soils, organic matter is *the* major agent stimulating the formation and stabilization of granular and crumb-type aggregates (see Figure 4.30). In the first place, organic matter provides the energy substrate that makes possible the previously mentioned activities of the fungi, bacteria, and soil animals. Second, as organic residues decompose, gels and other viscous microbial products, along with associated bacteria and fungi, encourage crumb formation. Organic exudates from plant roots also participate in this aggregating action.

Organic products of decay, such as complex polymers, chemically interact with particles of silicate clays and iron and aluminum oxides. These compounds orient the clays into packets (domains) which form bridges between individual soil particles, thereby binding them together in water-stable aggregates (see Figure 4.26d).

During the aggregation process, soil mineral particles (silts and fine sands) become coated and encrusted with bits of decomposed plant residue and other organic materials. At an even smaller scale, microscopic bits of decomposed residues and humus particles become encrusted with clay packets. In either case, the resulting organomineral complexes promote the formation of aggregates. Figure 4.29 shows direct evidence of the organomineral domains that bind soil particles.

FIGURE 4.30 The aggregates of soils high in organic matter are much more stable than are those low in this constituent. The low-organic-matter soil aggregates fall apart when they are wetted; those high in organic matter maintain their stability.

Modern advances in herbicides and planting equipment allow good weed control and seedling establishment without tillage. Nonetheless, many soil managers still consider tillage to be normal and necessary for the agricultural use of soils. Tillage can have both favorable and unfavorable effects on aggregation. If the soil is not too wet or too dry when the tillage is performed, the short-term effect of tillage is generally favorable. Tillage implements break up large clods, incorporate organic matter into the soil, kill weeds, and generally create a more favorable seedbed (See Box 4.5). Immediately after plowing, the surface soil is loosened (its cohesive strength is decreased) and total porosity is increased.

Over longer periods, tillage operations have detrimental effects on surface soil structure. In the first place, by mixing and stirring the soil, tillage greatly hastens the oxidation of soil organic matter, thus reducing the aggregating effects of this soil component. Second, tillage operations, especially if carried out when the soil is wet, tend to crush or smear stable soil aggregates, resulting in loss of macroporosity and the creation of a *puddled* condition. This action also exposes organic matter that had been protected inside the aggregates, hastening its loss by decomposition. The dramatic structural difference between an overly tilled and an untilled (grassland) soil is shown in Figure 4.31.

4.8 TILLAGE AND STRUCTURAL MANAGEMENT OF SOILS

When protected under dense vegetation and undisturbed by tillage, most soils (except perhaps some sparsely vegetated soils in arid regions) possess a surface structure sufficiently stable to allow rapid infiltration of water and to prevent crusting. However, for the manager of cultivated soils, the development and maintenance of stable surface soil structure is a major challenge. Many studies have shown that aggregation and associated desirable soil properties such as water infiltration rate decline under long periods of tilled row-crop cultivation (Table 4.8).

Tillage and Soil Tilth

Simply defined, **tilth** refers to the physical condition of the soil in relation to plant growth. Tilth depends not only on aggregate formation and stability, but also on such factors as bulk density, soil moisture content, degree of aeration, rate of water infiltration, drainage, and capillary water capacity. As might be expected, tilth often changes rapidly and markedly. For instance, the workability of fine-textured soils may be altered abruptly by a slight change in moisture.

Clayey soils are especially prone to puddling and compaction because of their high plasticity and cohesion. When puddled clayey soils dry, they usually become dense and hard. Proper timing of trafficking is more difficult for clayey than for sandy soils, because the former take much longer to dry to a suitable moisture content and may also become too dry to work easily.

FIGURE 4.31 Puddled soil (left) and well-granulated soil (right). Plant roots and especially humus play the major role in soil granulation. Thus a sod tends to encourage development of a granular structure in the surface horizon of cultivated land. (Courtesy USDA Natural Resources Conservation Service)

BOX 4.5   PREPARING A GOOD SEEDBED

Early in the growing season, one of the main activities of a farmer or gardener is the preparation of a good seedbed to assure that the sowing operation goes smoothly, and that the plants come up quickly, evenly, and well spaced.

A good seedbed consists of soil loose enough to allow easy root elongation and seedling emergence. [see photo (a)]. At the same time, a seedbed should be packed firmly enough to ensure good contact between the seed and moist soil so that the seed can easily imbibe water to begin the germination process. The seedbed should also be relatively free of large clods. Seeds could fall between such clods and become lodged too deeply for proper emergence and without sufficient soil contact.

Tillage may be needed to loosen compacted soil, help control weeds, and, in cool climates, help the soil dry out so it will warm more rapidly. On the other hand, the objectives of seedbed preparation may be achieved with little or no tillage if soil and climatic conditions are favorable and a mulch or herbicide is used to control early weeds.

Mechanical planters can assist in maintaining a good seedbed. Most planters are equipped with coulters (sharp steel disks) designed to cut a path through plant residues on the soil surface. No-till planters usually follow the coulter with a pair of sharp cutting wheels called a **double disk opener** that opens a groove in the soil into which the seeds can be dropped [see photos (b) and (c)].

(a) A bean seed emerges from a seedbed. (Photo courtesy R. Weil)

Most planters also have a press wheel that follows behind the seed dropper and packs the loosened soil just enough to ensure that the groove is closed and the seed is pressed into contact with moist soil.

Ideally, only a narrow strip in the seed row is packed down to create a seed *germination* zone, while the soil between the crop rows is left as loose as possible to provide a good *rooting* zone. The surface of the interrow rooting zone may be left in a rough condition to encourage water infiltration and discourage erosion. The above principles also apply to the home gardener who may be sowing seeds by hand.

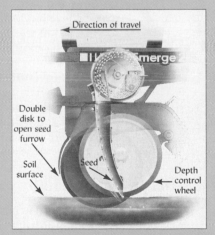

(b) A no-till planter in action and (c) a diagram showing how it works. [Photo (b) and diagram (c) courtesy of Deere & Company, Moline, Ill.]

TABLE 4.8 **Effect of Period of Corn Cultivation on Soil Organic Matter, Aggregate Stability, and Water Infiltration[a] in Five Silt Loam Inceptisols from Southwest France**

In soils with depleted organic-matter levels, aggregates easily broke down under the influence of water, forming smaller aggregates and dispersed materials that sealed the soil surface and inhibited infiltration. A level of 3% soil organic matter seems to be sufficient for good structural stability in these temperate region silt loam soils.

Period of cultivation, years	Organic matter, %	Aggregate stability, mm MWD[b]	Infiltration	
			Of total rain, %	Prior to ponding, mm
100	0.7	.35	25	6
47	1.6	.61	34	9
32	2.6	.76	38	15
27	3.1	1.38	47	25
15	4.2	1.52	44	23

[a] Infiltration is the amount of water that entered the soil when 64 mm of "rain" was applied during a 2-hour period.
[b] MWD = mean weighted diameter or average size of the aggregates that remained intact after sieving under water. [Data from Le Bissonnais and Arrouays (1997)]

Some clayey soils of humid tropical regions are much more easily managed than those just described. The clay fraction of these soils is dominated by hydrous oxides of iron and aluminum, which are not as sticky, plastic, and difficult to work. These soils may have very favorable physical properties, since they hold large amounts of water but have such stable aggregates that they respond to tillage after rainfall much like sandy soils.

In tropical and subtropical regions with a long dry season, soil often must be tilled in a very dry state in order to prepare the land for planting with the onset of the first rains. Tillage under such dry conditions can be very difficult and can result in hard clods if the soils contain much sticky-type silicate clay. Thus farmers in temperate regions typically find their soils too wet for tillage just prior to planting time (early spring), while farmers in tropical regions may face the opposite problem of soils too dry for easy tillage just prior to planting (end of dry season).

Conventional Tillage and Crop Production

Since the Middle Ages, the moldboard plow has been the primary tillage implement most used in the Western world.[11] Its purpose is to lift, twist, and invert the soil while incorporating crop residues and animal wastes into the plow layer (Figure 4.32). The moldboard plow is often supplemented by the disk plow, which is used to cut up residues and partially incorporate them into the soil. In conventional practice, such primary tillage has been followed by a number of secondary tillage operations, such as harrowing to kill weeds and to break up clods, thereby preparing a suitable seedbed.

After the crop is planted, the soil may receive further secondary tillage to control weeds and to break up crusting of the immediate soil surface. In mechanized agriculture, all conventional tillage operations are performed with tractors and other heavy equipment that may pass over the land several times before the crop is finally harvested. In many parts of the world, farmers use hand hoes or animal-drawn implements to stir the soil. Although humans and draft animals are not as heavy as tractors, their weight is applied to the soil in a relatively small area (foot- or hoofprint), and so can also cause considerable compaction.

Conservation Tillage and Soil Tilth

In recent years, land-management systems have been developed that minimize the need for soil tillage. Since these systems also leave considerable plant residues on or near the soil surface, they protect the soil from erosion (see Section 17.6 for a detailed discussion). For this reason, the tillage practices followed in these systems are called *conservation tillage*.[12] Figure 4.33 illustrates a no-till operation, where one crop is planted in the residue of another with virtually no tillage. Other minimum-tillage systems permit

---

[11] For an early but still valuable critique of the moldboard plow, see Faulkner (1943).

[12] The U.S. Department of Agriculture defines *conservation tillage* as any system that leaves at least 30% of the soil surface covered by residues.

FIGURE 4.32 While the action of the moldboard plow lifts, turns, and loosens the upper 15 to 20 cm of soil (the furrow slice), the counterbalancing downward force compacts the next lower layer of soil. This compacted zone can develop into a *plowpan*. Compactive action can be understood by imagining that you are lifting a heavy weight—as you lift the weight your feet press down on the floor below. (Photo courtesy of R. Weil)

FIGURE 4.33 An illustration of one conservation tillage system. Wheat is being harvested (background) and soybeans are planted (foreground) with no intervening tillage operation. This no-tillage system permits double-cropping, saves fuel costs and time, and helps conserve the soil. (Courtesy Allis-Chalmers Corporation)

some stirring of the soil, but still leave a high proportion of the crop residues on the surface. These organic residues protect the soil from the beating action of raindrops and the abrasive action of the wind, thereby reducing water and wind erosion and maintaining soil structure.

Soil Crusting

Falling drops of water during heavy rains or sprinkler irrigation beat apart the aggregates exposed at the soil surface. In some soils the dilution of salts by this water stimulates the dispersion of clays. Once the aggregates become dispersed, small particles and dispersed clay tend to wash into and clog the soil pores. Soon the soil surface is covered with a thin layer of fine, structureless material called a **surface seal.** The surface seal inhibits water infiltration and increases erosion losses.

As the surface seal dries, it forms a hard **crust.** Seedlings, if they emerge at all, can do so only through cracks in the crust. A crust-forming soil is compared to one with stable aggregates in Figure 4.34. Formation of a crust soon after a crop is sown may allow so few seeds to emerge that the crop has to be replanted. In arid and semiarid regions, soil sealing and crusting can have disastrous consequences because high runoff losses leave little water available to support plant growth.

Crusting can be minimized by keeping some vegetative or mulch cover on the land to reduce the impact of raindrops. Once a crust has formed, it may be necessary to rescue a newly planted crop by breaking up the crust with light tillage (as with a rotary hoe), preferably while the soil is still moist. Improved management of soil organic matter and use of certain soil amendments can "condition" the soil and help prevent clay dispersion and crust formation (see also Section 10.10).

Soil Conditioners

GYPSUM. Gypsum (calcium sulfate) is widely available in its relatively pure mined form, or as a major component of various industrial by-products. Gypsum has been shown effective in improving the physical condition of many types of soils, from some highly weathered acid soils to some low-salinity, high-sodium soils of semiarid regions (see Chapter 10). The more soluble gypsum products provide enough electroytes (cations and anions) to promote flocculation and inhibit the dispersion of aggregates, thus preventing surface crusting. Field trials have shown that gypsum-treated soils permit

(c)

FIGURE 4.34 Scanning electron micrographs of the upper 1 mm of a soil with stable aggregation (*a*) compared to one with unstable aggregates (*b*). Note that the aggregates in the immediate surface have been destroyed and a surface crust has formed. The bean seedling (*c*) must break the soil crust as it emerges from the seedbed. [Photos (*a*) and (*b*) from O'Nofiok and Singer (1984), used with permission of Soil Science Society of America; photo (*c*) courtesy of R. Weil]

greater water infiltration and are less subject to erosion than untreated soils. Similarly, gypsum can reduce the strength of hard subsurface layers, thereby allowing greater root penetration and subsequent plant uptake of water from the subsoil.

SYNTHETIC ORGANIC POLYMERS. Certain synthetic organic polymers can stabilize soil structure in much the same way as do natural organic polymers such as polysaccharides. While large applications of these polymers would be uneconomical, it has been shown that even very small amounts can effectively inhibit crust formation if applied properly. Since crusting is a surface phenomenon, its prevention requires application of only enough polymer to treat the surface few millimeters of soil. Adding the polymer to irrigation water has proved to be a convenient and effective way to apply the small amounts needed. For example, polyacrylamide (PAM) is effective in stabilizing surface aggregates when applied at rates as low as 1 to 15 mg/L of irrigation water or sprayed on at rates as low as 1 to 10 kg/ha. Figure 4.35 shows the dramatic stabilizing effect of synthetic polyacrylamides used in irrigation water. A number of research reports indicate that the best results can be obtained by combining the use of PAM and gypsum products.

OTHER SOIL CONDITIONERS. Several species of algae that live near the soil surface are known to produce quite effective aggregate-stabilizing compounds. Application of small quantities of commercial preparations containing such algae may bring about a significant improvement in surface soil structure. The amount of amendment required is very small because the algae, once established in the soil, can multiply.

Various humic materials are marketed for their soil conditioning effects when incorporated at low rates (< 500 kg/ha). However, carefully conducted research at many universities has failed to show that these materials have significantly affected aggregate stability or crop yield, as claimed.

General Guidelines for Managing Soil Tilth

Although each soil presents unique problems and opportunities, the following principles are generally relevant to managing soil tilth:

1. Minimizing tillage, especially moldboard plowing, disk harrowing or rototilling, reduces the loss of aggregate-stabilizing organic matter.
2. Timing traffic activities to occur when the soil is as dry as possible and restricting tillage to periods of optimum soil moisture conditions will minimize destruction of soil structure.

FIGURE 4.35 The remarkable stabilizing effect of a synthetic polyacrylamide is seen in the furrow on the right compared to the left, untreated, furrow. Irrigation water broke down much of the structure of the untreated soil but had no effect on the treated row. [From Mitchell (1986)]

(a)　　　　　　　　　　　　　　　　　(b)

FIGURE 4.36 The incorporation of animal manures and other types of readily decomposable organic material can have dramatic effects on soil aggregation, especially the formation of large, stable aggregates. The soil pictured is a Davidson clay loam (Paleudults) from Virginia which had been fertilized with mineral fertilizer (a) or poultry manure (b) for five years. While both soils have a high proportion of water-stable aggregates, the aggregates and associated interped pores are much larger in the manure-treated soil. Each photo is 10 cm across. [From Weil and Kroontje (1979)]

3. Mulching the soil surface with crop residues or plant litter adds organic matter, encourages earthworm activity, and protects aggregates from beating rain.

4. Adding crop residues, composts, and animal manures to the soil is effective in stimulating microbial supply of the decomposition products that help stabilize soil aggregates. Figure 4.36 shows enhanced aggregation resulting from addition of poultry manure to a clay loam soil.

5. Including sod crops in the rotation favors stable aggregation by helping to maintain soil organic matter, providing maximal aggregating influence of fine plant roots, and assuring a period without tillage. The data in Table 4.9 illustrate the detrimental effect of continuously growing tilled corn and the importance of growing other than row crops if water-stable aggregates are to be maintained.

6. Using cover crops and green manure crops, where practical, provides another good source of root action and organic matter for structural management.

7. Applying gypsum (or calcareous limestone if the soil is acidic) by itself or in combination with synthetic polymers can be very useful in stabilizing surface aggregates, especially in irrigated soils.

TABLE 4.9   Water-Stable Aggregation of a Udoll (Marshall Silt Loam) near Clarinda, Iowa, under Different Cropping Systems

| | Water-stable aggregates, % | |
Crop	Large (1 mm and above)	Small (less than 1 mm)
Corn continuously	8.8	91.2
Corn in rotation	23.3	76.7
Meadow in rotation	42.2	57.8
Bluegrass continuously	57.0	43.0

From Wilson, et al. (1947).

Field Rating of Soil Consistence and Consistency

CONSISTENCE. Soil **consistence** is a term used by soil scientists to describe the resistance of a soil to mechanical stresses or manipulations at various moisture contents. Soils are rated for consistence as part of describing a soil profile and for estimating suitability for traffic and tillage. This property is a composite expression of those forces of mutual attraction among soil particles, and between particles and pore water, that determine the ease with which a soil can be reshaped or ruptured. Consistence is commonly determined in the field by feeling and manipulating the soil by hand. As a clod of soil is squeezed between the thumb and forefinger (or crushed underfoot, if necessary), observations are made on the amount of force needed to crush the clod and on the manner in which the soil responds to the force. The degree of cementation of the soil by such materials as silica, calcite, or iron is also considered in identifying soil consistence.

Moisture content greatly influences how a soil responds to stress; hence, moist and dry soils are given separate consistence ratings (Table 4.10). A dry clayey soil that cannot be crushed between the thumb and forefinger but can be crushed easily underfoot would be designated as *hard.* It would exhibit much greater resistance to deformation than the same soil in a wet, plastic state. Anyone who has gotten a vehicle stuck up to its axles in sticky mud has first-hand experience of these principles. While most of the terms in Table 4.10 are self-explanatory, the term **friable** requires some comment. If a clod of moist soil crumbles into aggregates when crushed with only light pressure, it is said to be friable. Friable soils are easily tilled or excavated.

The degrees of *stickiness* and *plasticity* (malleability) of soil in the wet condition are often included in describing soil consistence (although not shown in Table 4.10).

CONSISTENCY. The term **consistency** is used in a similar manner by soil engineers to describe the degree to which a soil resists deformation when a force is applied. However, consistency is determined by the soil's resistance to *penetration* by an object, while the soil scientist's consistence describes resistence to *rupture*. Instead of crushing a clod of soil, the engineer attempts to penetrate it with either the blunt end of a pencil (some use their thumb) or a thumbnail. For example, if the blunt end of a pencil makes only a slight indentation, but the thumbnail penetrates easily, the soil is rated as *very firm* (Table 4.10).

TABLE 4.10 Some Field Tests and Terms Used to Describe the Consistence and Consistency of Soils

The consistency of cohesive materials is closely related to, but not exactly the same as, their consistence. Conditions of least coherence are represented by terms at the top of each column, those of greater coherence near the bottom.

	Soil consistence[a]			Soil consistency[b]	
Dry soil	Moist to wet soil	Soil dried then submerged in water	Field rupture (crushing) test	Soil at in situ moisture	Field penetration test
Loose	Loose	Not applicable	Specimen not obtainable	Soft	Blunt end of pencil penetrates deeply with ease
Soft	Very friable	Noncemented	Crumbles under very slight force between thumb and forefinger	Medium firm	Blunt end of pencil can penetrate about 1.25 cm with moderate effort
Slightly hard	Friable	Extremely weakly cemented	Crumbles under slight force between thumb and forefinger	Firm	Blunt end of pencil can penetrate about 0.5 cm
Hard	Firm	Weakly cemented	Crushes with difficulty between thumb and forefinger	Very firm	Blunt end of pencil makes slight indentation; thumbnail easily penetrates
Very hard	Extremely firm	Moderately cemented	Cannot be crushed between thumb and forefinger, but can be crushed slowly underfoot	Hard	Blunt end of pencil makes no indentation; thumbnail barely penetrates
Extremely hard	Slightly rigid	Strongly cemented	Cannot be crushed by full body weight underfoot		

[a] Abstracted from Soil Survey Division Staff (1993).
[b] Modified from McCarthy (1993).

Field observations of both consistence and consistency provide valuable information to guide decisions about loading and manipulating soils. For construction purposes, however, soil engineers usually must make more precise measurements of a number of related soil properties that help predict how a soil will respond to applied stress.

Soil Strength and Sudden Failure

Perhaps the most important property of a soil for engineering uses is its **strength.** This is a measure of the capacity of a soil mass to withstand stresses without giving way to those stresses by rupturing or becoming deformed. The failure of a soil to withstand stress can be seen where a structure topples as its weight exceeds the soil's bearing strength, where earthen dams give way under the pressure of impounded water, or where pavements and structures slide down unstable hillsides (Figure 1.9).

COHESIVE SOILS.   Two components of strength apply to **cohesive soils** (essentially soils with a clay content of more than about 15%): (1) the inherent electrostatic attractive forces between clay platelets and between clay surfaces and the water in very fine pores (see the discussion of clay flocculation in Section 4.7), and (2) the frictional resistance to movement between soil particles of all sizes. While many different laboratory tests are used to estimate soil strength, perhaps the simplest to understand is the direct **unconfined compression test** using the apparatus illustrated in Figure 4.37a. A cylindrical specimen of cohesive soil is placed vertically between two flat porous stones (which allow water to escape from the compressed soil pores) and a slowly increasing downward force is applied. The soil column will first bulge out a bit and then fail—that is, give way suddenly and collapse—when the force exceeds the soil strength.

The strength of cohesive soils declines dramatically if the material is very wet and the pores are nearly filled with water. Then the particles are forced apart so that neither the cohesive nor the frictional component is very strong, making the soil prone to failure, often with catastrophic results such as the mudslide shown in Figure 4.38. On the other hand, if cohesive soils become dry or more compacted, their strength increases as particles are forced into closer contact with one another—a result that has implications for plant root growth as well as for engineering (see Section 4.5).

NONCOHESIVE SOILS.   The strength of dry, noncohesive soil materials such as sand depends entirely on frictional forces, including the interlocking of rough particle surfaces. One reflection of the strength of a noncohesive material is its **angle of repose**, the steepest angle to which it can be piled without slumping. Smooth rounded sands grains cannot be piled as steeply as can rough, interlocking sands. If a small amount of water bridges the gaps between particles, electrostatic attraction of the water for the mineral surfaces will increase the soil strength. Interparticle water bridges explain why cars can drive along the

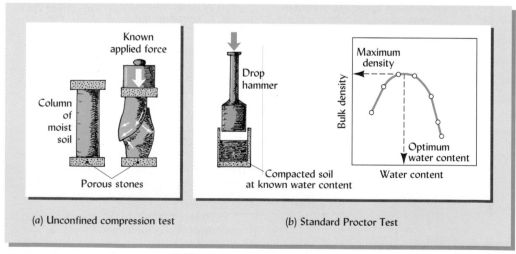

(a) Unconfined compression test   (b) Standard Proctor Test

FIGURE 4.37   Two important tests to determine engineering properties of soil materials. (a) An unconfined compression test for soil strength. (b) The Proctor test for maximum density and optimum water content for compaction control.

FIGURE 4.38 Houses damaged by a mudslide that occurred when the soils of a steep hillside in Oregon became saturated with water after a period of heavy rains. The weight of the wet soil exceeded its shear strength, causing the slope to fail. Excavations for roads and houses near the foot of a slope can contribute to the lack of slope stability, as can removal of tree roots by large-scale clear-cutting on the slope itself. (Photo courtesy of John Griffith, Coas Bay, Ore.)

edge of the beach where the sand is moist, but their tires sink in and lose traction on loose dry sand or in saturated quicksand (see Figure 5.28 for an example of the latter).

COLLAPSIBLE SOILS.   Certain soils that exhibit considerable strength at low in situ water contents lose their strength suddenly if they become wet. Such soils may collapse without warning under a roadway or building foundation. A special case of soil collapse is **thixotropy,** the sudden liquification of a wet soil mass subjected to vibrations, such as those accompanying earthquakes and blasting.

Most **collapsible soils** are noncohesive materials in which loosely packed sand grains are cemented at their contact points by small amounts of gypsum, clay, or water under tension. These soils usually occur in arid and semiarid regions where such cementing agents are relatively stable. Many collapsible soils have derived their open particle arrangement from the process of sedimentation beneath past or present bodies of water. When these soils are wetted, excess water may dissolve cements such as gypsum or disperse clays that form bridges between particles, causing a sudden loss of strength. In some cases, similar behavior is exhibited by highly weathered Oxisols in humid tropical regions; however, since dispersible clay or water-soluble cements are not present, the mechanism for collapse in these Oxisols is not yet clear. In any case, a soil engineer would certainly want to know when this type of soil is present.

Settlement—Gradual Compression

While embankments and hill slopes commonly fail due to stresses that exceed the soil's strength, most buildings and roads are unlikely to provide loads that cause the soil to rupture. Instead, most foundation problems result from slow, often uneven, vertical subsidence or **settlement** of the soil.

COMPACTION CONTROL.   For the purposes of growing plants, soil compaction is to be avoided; however, soils are purposely compacted prior to being used as a roadbed or for a building foundation. Compaction occurring later under the heavy load would result in uneven settlement and a cracked pavement or foundation. Compaction to an optimum density is usually achieved on clayey soils by heavy sheepsfoot rollers that knead the soil like a baker kneads bread dough (Figure 4.39). For sandy soils, vibrating rollers or impact hammers do a better job because they can jar the particles into a tight packing arrangement (see Figure 4.17).

Some soil particles, such as those of certain colloidal silicate clays and of micas of all sizes, can be compressed when a load is placed upon them. When that load is removed, however, these particles tend to regain their original shape, in effect reversing their

FIGURE 4.39 Compaction of soils used as foundations and roadbeds is accomplished by heavy equipment such as this sheepsfoot roller. The knobs ("sheepsfeet") concentrate the mass on a small impact area, punching and kneading the loose, freshly graded soil to optimum density. (Photo courtesy of R. Weil)

compression. As a result, soils rich in these particles are not easily compacted into a stable base for roads and foundations.

The **Proctor test** is the most common method used to obtain data that can guide efforts at compacting soil materials before construction. A specimen of soil is mixed to a given water content and placed in a holder where it is compacted by a drop hammer (weighing 2.5 kg for the standard Proctor test). The bulk density (usually referred to as the *dry density* by engineers) is then measured. The process is repeated several times with increasing water contents until enough data is collected to produce a *Proctor curve* such as that shown in Figure 4.37b. The curve indicates the maximum bulk density to which the soil may be compacted by a given force. Just as important, the test also indicates the water content of the soil that is optimum for maximum compaction. When the soil is either drier or wetter than this, compaction will be more difficult. On the construction site, water tank trucks will be used, if needed, to bring the water content of the soil to the determined optimum level, and compaction equipment such as that shown in Figure 4.39 will be used to achieve the desired density.

COMPRESSIBILITY. A **consolidation test** may be conducted on a soil specimen to determine its **compressibility**—how much its volume will be reduced by a given applied force. Because of the relatively low porosity and equidimensional shape of the individual mineral grains, very sandy soils resist compression once the particles have settled into a tight packing arrangement. They make excellent soils to bear foundations. The high porosity of clay floccules and the flakelike shape of clay particles gives clayey soils much greater compressibility. Soils consisting mainly of organic matter (peats) have the highest compressibilities and generally are unsuitable for foundations.

In the field, compression of wet clayey soils may occur very slowly after a load (e.g., a building) is applied because compression can occur only as fast as water can escape from the soil pores—which for the fine pores in clayey materials is not very fast. Perhaps the most famous example of uneven settlement due to slow compression is the Leaning Tower of Pisa in Italy. Unfortunately, most cases of uneven settlement result in headaches, not tourism.

Expansive Soils

Damages caused by expansive soils in the United States rarely makes the evening news programs, although the total cost annually exceeds that caused by tornados, floods, earthquakes, or any other type of natural disaster. Expansive clays occur on about 20% of the land area in the United States and cause upwards of $4 billion in damages annually to pavements, foundations, and utility lines. The damages can be severe in certain sites in all parts of the country, but are most extensive in regions that have long dry periods alternating with periods of rain (e.g., California, Texas, Wyoming, and Colorado).

Some clays, particularly the smectites, swell when wet and shrink when dry (see Section 8.5). Expansive soils are rich in these types of clay. The electrostatic charges on the clay surfaces attract water molecules from the larger pores into the spaces between clay platelets. Also, the adsorbed cations associated with the clay surfaces tend to hydrate, drawing in additional water. The water pushes the platelets apart, causing the mass of soil to swell in volume. The reverse of these processes occurs when the soil dries and water is withdrawn from packets of clay platelets, causing shrinkage and cracking. After a prolonged dry spell, soils high in smectites can be recognized in the field by the criss-crossed pattern of wide, deep cracks (Figure 4.40). The swelling and shrinkage cause sufficient movement of the soil to crack building foundations, burst pipelines, and buckle pavements.

Atterberg Limits

As a dry, clayey soil takes on increasing amounts of water, it undergoes dramatic and distinct changes in behavior and consistency. A hard, rigid solid in the dry state, it becomes a crumbly (friable) semisolid when a certain moisture content (termed the **shrinkage limit**) is reached. If it contains expansive clays, the soil also begins to swell in volume as this moisture content is exceeded. Increasing the water content beyond the **plastic limit** will transform the soil into a malleable, plastic mass and cause additional swelling. The soil will remain in this plastic state until its **liquid limit** is exceeded, causing it to transform into a viscous liquid that will flow when jarred. These critical water contents (measured in units of percent) are termed the **Atterberg limits**, after the German engineer who developed the system. Soil engineers determine the Atterberg limits as part of their investigation of soil consistency, to help predict the behavior of particular soils and the suitability of these materials for different construction purposes.

PLASTICITY INDEX. The plasticity index (PI) is the difference between the plastic limit (PL) and liquid limit (LL) and indicates the water-content range over which the soil has plastic properties:

$$PI = LL - PL$$

Soils with a high plasticity index (greater than about 25) are usually expansive clays that make poor roadbeds or foundations. Figure 4.41 shows the relationship among the

FIGURE 4.40 Certain types of clays, especially the smectites, undergo significant volume changes in conjunction with changes in water content. Here, an expansive soil rich in smectitic clay has shrunk during a dry period, causing a network of large cracks to open up in the soil surface. (Courtesy of USDA Natural Resources Conservation Service)

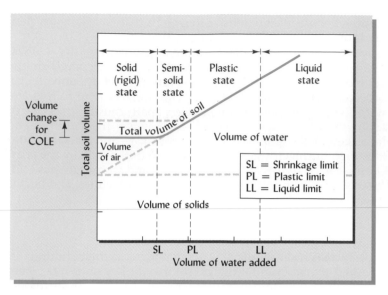

FIGURE 4.41 A common depiction of The Atterberg limits, which mark major shifts in the behavior of a cohesive soil as its water content changes (from left to right). As water is added to a certain volume of dry solids, first air is displaced; then, as more water is added, the total volume of the soil increases (if the soil has some expansive properties). When the shrinkage limit (SL) is reached, the once rigid, hard solid becomes a crumbly, semisolid. With more water, the plastic limit (PL) is reached, after which the soil becomes plastic and can be molded. It remains in a plastic stage over a range of water contents until the liquid limit (LL) is exceeded, at which point the soil begins to behave as a viscous liquid that will flow when jarred. The volume change for calculating the coefficient of linear extensibility (COLE) is shown at the left.

Atterberg limits and the changes in soil volume associated with increasing water contents for a hypothetical soil.

Smectite clays (see Section 8.5) generally have high liquid limits and plasticity indices, especially if saturated with sodium. Kaolinite and other nonexpansive clays have low liquid limit values. The plastic and liquid limits of several soils and three clay samples are shown in Table 4.11.

COEFFICIENT OF LINEAR EXTENSIBILITY. The expansiveness of a soil (and therefore the hazard of its destroying foundations and pavements) can be quantified as the *coefficient of linear extensibility* (COLE). Suppose a sample of soil is moistened to its plastic limit and molded into the shape of a bar with length LM. If the bar of soil is allowed to air dry, it will shrink to length LD. The COLE is the percent reduction in length of the soil bar upon shrinking:

$$COLE = \frac{LM - LD}{LM} \times 100$$

Figure 4.41 indicates how the volume change used to calculate the COLE relates to the Atterberg limits.

TABLE 4.11 **Plastic and Liquid Limits of Several Soils and of Na- and Ca-Saturated Smectite**

Clay soils with large amounts of smectites (Susquehanna and Bashaw) have high liquid limits, as do Na-saturated clays.

Soil	Location	Plastic limit	Liquid limit	Plasticity index
Davidson (Udults)	Georgia	19	27	8
Cecil (Udults)	Georgia	29	49	20
Putnam (Aqualfs)	Georgia	24	37	13
Susquehanna (Udalfs)	Georgia	29	57	28
Sliprock (Udepts)	Oregon	46	59	13
Jory (Udults)	Oregon	30	45	15
Bashaw (Xererts)	Oregon	18	71	53
Na-saturated smectite	—	—	950	—
Ca-saturated smectite	—	—	360	—
Na-saturated kaolinite	—	—	36	—

Data for Georgia soils from Hammel, et al. (1983); Oregon soils from McNabb (1979); clays from Warkentin (1961).

Unified Classification System for Soil Materials

The U.S. Army Corps of Engineers and the U.S. Bureau of Reclamation have established a widely used system of classifying soil materials in order to aid in predicting the engineering behavior of different soils. Each type of soil is given a two-letter designation based primarily on its particle-size distribution (texture), Atterberg limits, and organic-matter content (Figure 4.42).

The system first groups soils into coarse-grained soils (more than 50% retained on a 0.075 mm sieve) and fine-grained soils (at least half smaller than 0.075 mm). The coarse materials are further divided on the basis of grain size (gravels and sands), amount of fines present, and uniformity of grain size (well or poorly graded). The fine-grained soils are divided into silts, clays, and organic materials. These classes are further subdivided on the basis of their liquid limit (above or below 50) and their plasticity index.

Major division			Group symbols	Typical names
Coarse-grained soils More than 50% retained on No. 200 sieve	Gravels 50% or more of coarse fraction retained on No. 4 sieve	Clean gravels	GW	Well-graded gravels and gravel-sand mixtures, little or no fines
			GP	Poorly graded gravels and gravel-sand mixtures, little or no fines
		Gravels with fines	GM	Silty gravels, gravel-sand-silt mixture
			GC	Clayey gravels, gravel-sand-clay mixtures
	Sands More than 50% of coarse fraction passes No. 4 sieve	Clean sands	SW	Well-graded sands and gravelly sands, little or no fines
			SP	Poorly graded gravels and gravel sand mixtures, little or no fines
		Sands with fines	SM	Silty sands, sand-silt mixtures
			SC	Clayey sands, sand-clay mixtures
Fine-grained soils 50% or more passes No. 200 sieve	Silts and clays Liquid limit 50% or less		ML	Inorganic silts, very fine sands, rock flour, silty or clayey fine sands
			CL	Inorganic clays of low to medium plasticity, gravelly clays, sandy clays, silty clays, lean clays
			OL	Organic silts and organic silty clays of low plasticity
	Silts and clays Liquid limit greater than 50%		MH	Inorganic silts, micaceous, or diatomaceous fine sand or silts, elastic silts
			CH	Inorganic clays of high plasticity, fat clays
			OH	Organic clays of medium to high plasticity
Highly organic soils			Pt	Peat, muck, and other highly organic soils

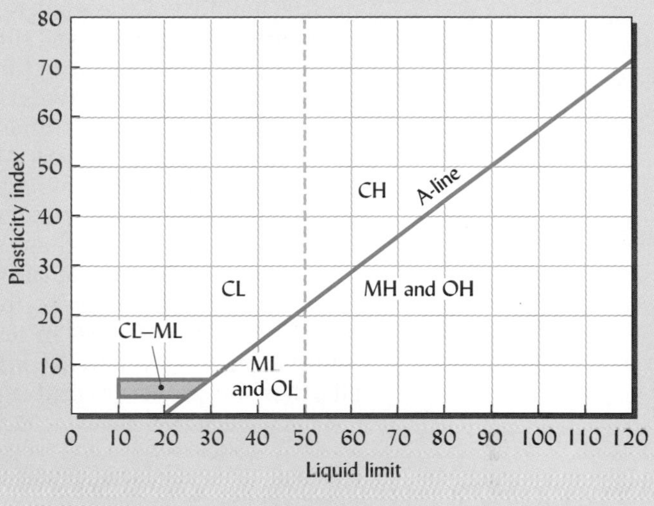

FIGURE 4.42 The Unified System of Classification. Note that this is a system to classify soil materials, not natural soil bodies. The two-letter designations (SW, MH, etc.) help engineers predict the behavior of the soil material when used for construction purposes. The first letter is one of the following: G = gravel, S = sand, M = silts, C = clays, and O = organic-rich materials. The second letter indicates whether the sand or gravels are well graded (W) or poorly graded (P), and whether the silts, clays, and organic-rich materials have a high plasticity index (H) or a low plasticity index (I). Among the fine-grained materials, those closer to the top of the table are better suited for foundations and roadbeds.

This classification of soil materials helps engineers predict the soil strength, expansiveness, compressibility, and other properties so that appropriate engineering designs can be made for the soil at hand.

4.10 CONCLUSION

Physical properties exert a marked influence on the behavior of soils with regard to plant growth, hydrology, environmental management, and engineering uses. The nature and properties of the individual particles, their size distribution, and their arrangement in soils determine the total volume of nonsolid pore space, as well as the pore sizes, thereby impacting on water and air relationships.

The properties of individual particles and their proportionate distribution (soil texture) are subject to little human control in field soils. However, it is possible to exert some control over the arrangement of these particles into aggregates (soil structure) and on the stability of these aggregates. Tillage and traffic must be carefully controlled to avoid undue damage to soil tilth, especially when soils are rather wet. Generally, nature takes good care of soil structure, and humans can learn much about soil management by studying natural systems. Vigorous and diverse plant growth, generous return of organic residues, and minimal physical disturbance are attributes of natural systems worthy of emulation. Proper plant species selection, crop rotation, and management of chemical, physical, and biological factors can help assure maintenance of soil physical quality. In recent years, these management goals have been made more practical by the advent of conservation tillage systems that minimize soil manipulations while decreasing soil erosion and water runoff.

Particle size, moisture content, and plasticity of the colloidal fraction all help determine the stability of soil in response to loading forces from traffic, tillage, or building foundations. The physical properties presented in this chapter greatly influence nearly all other soil properties and uses, as discussed throughout this book.

STUDY QUESTIONS

1. If you were investigating a site for a proposed housing development, how could you use soil colors to help predict where problems might be encountered?

2. You are considering the purchase of some farmland in a region with variable soil textures. The soils on one farm are mostly of sandy loams and loamy sands, while those on a second farm are mostly clay loams and clays. List the potential advantages and disadvantages of each farm as suggested by the texture of its soils.

3. Revisit your answer to question 2. Explain how soil structure in both the surface and subsurface horizons might modify your opinion of the merits of each farm.

4. Two different timber-harvest methods are being tested on adjacent forest plots with clay loam surface soils. Initially, the bulk density of the surface soil in both plots was 1.1 Mg/m^3. One year after the harvest operations, plot A soil had a bulk density of 1.48 Mg/m^3, while that in plot B was 1.29 Mg/m^3. Interpret these values with regard to the relative merits of systems A and B, and the likely effects on the soil's function in the forest ecosystem.

5. What are the textural classes of two soils, the first with 15% clay and 45% silt, and the second with 80% sand and 10% clay? (Hint: Use Figure 4.8)

6. For the forest plot B in question 4, what was the change in percent pore space of the surface soil caused by timber harvest? Would you expect that most of this change was in the micropores or in the macropores? Explain.

7. Discuss the positive and negative impacts of tillage on soil structure. What is another physical consideration would you have to take into account in deciding whether or not to change from a conventional to a conservation tillage system?

8. What would you, as a home gardener, consider to be the three best and three worst things that you could do with regard to managing the soil structure in your home garden?

9. What does the Proctor test tell an engineer about a soil, and why would this information be important?

10. In a humid region characterized by expansive soils, a homeowner experienced burst water pipes, doors that no longer closed properly, and large vertical cracks in the brick walls. The house had had no problems for over 20 years, and a consulting soil scientist blamed the problems on a large tree that was planted near the house some 10 years before the problems began to occur. Explain.

REFERENCES

Bigham, J. M., and E. J. Ciolkosz (eds.). 1993. *Soil Color.* SSSA Special Publication no. 31. (Madison, Wis.: Soil Sci. Soc. of Amer.) 172 pp.

Brewer, R. 1964. *Fabric and Mineral Analysis of Soils.* (New York: John Wiley and Sons).

Camp, C. R., and J. F. Lund. 1964. "Effects of soil compaction on cotton roots," *Crops and Soils,* **17**:13–14.

Emerson, W. W., R. C. Foster, and J. M. Oades. 1986. "Organomineral complexes in relation to soil aggregation and structure," in P. M. Huang and M. Schnitzer (eds.), *Interaction of Soil Minerals with Natural Organics and Microbes.* SSSA Special Publication no. 17. (Madison, Wis.: Soil Sci. Soc. Amer.).

Faulkner, E. H. 1943. *Plowman's Folly* (Norman, Okla.: University of Oklahoma Press).

Gent, J. A., Jr., and L.A. Morris. 1986. "Soil compaction from harvesting and site preparation in the Upper Gulf Coastal Plain," *Soil Sci. Soc. Amer. J.,* **50**:443–446.

Gent, J. A., Jr., R. Ballard, A. E. Hassan, and D. K. Cassel. 1984. "Impact of harvésting and site preparation on physical properties of Piedmont forest soils," *Soil Sci. Soc. Amer. J.,* **48**:173–177.

Hammel, J. E., M. E. Summer, and J. Burema. 1983. "Atterberg limits as indices of external areas of soils," *Soil Sci. Soc. Amer. J.,* **47**:1054–1056.

Laws, W. D., and D. D. Evans. 1949. "The effects of long-time cultivation on some physical and chemical properties of two rendzina soil," *Soil Sci. Soc. Amer. Proc.,* **14**:15–19.

Le Bissonnais, Y., and D. Arrouays. 1997. "Aggregate stability and assessment of soil crustability and erodibility: II. Application to humic loamy soils with various organic carbon contents," *European J. Soil Sci.,* **48**:39–48.

Lyon, T. L., H. O. Buckman, and N. C. Brady. 1952. *The Nature and Properties of Soils,* 5th ed. (New York: Macmillan), p. 60.

McCarthy, D. F. 1993. *Essentials of Soil Mechanics and Foundations,* 4th ed. (Englewood Cliffs, NJ: Prentice Hall).

McNabb, D. H. 1979. "Correlation of soil plasticity with amorphous clay constituents," *Soil Sci. Soc. Amer. J.,* **46**:450–456.

Mitchell, A. R. 1986. "Polyacrylamide application in irrigation water to increase infiltration," *Soil Sci.,* **141**:353–358.

Oades, J. M. 1993. "The role of biology in the formation, stabilization, and degradation of soil structure," *Geoderma,* **56**:377–400.

O'Nofiok, O., and M. J. Singer. 1984. "Scanning electron microscope studies of surface crusts formed by simulated rainfall," *Soil Sci. Soc. Amer. J.,* **48**:1137–1143.

Parker, M. M., and D. H. Van Lear. 1996. "Soil heterogeneity and root distribution of mature loblolly pine stands in Piedmont soils," *Soil Sci. Soc. Amer. J.,* **60**:1920–1925.

Soil Survey Division Staff. 1993. *Soil Survey Manual.* USDA-NRSC Agricultural Handbook 18. (Washington, D.C.: U.S. Government Printing Office), pp. 174–175.

Tiessen, H., J. W. B. Stewart, and J. R. Bettany. 1982. "Cultivation effects on the amounts and concentration of carbon, nitrogen, and phosphorus in grassland soils," *Agron. J.,* **74**:831–835.

Tisdall, J. M. 1994. "Possible role of soil microorganisms in aggregation in soils," *Plant and Soil,* **159**:115–121.

Tisdall, J. M., and J. M. Oades. 1982. "Organic matter and water-stable aggregates in soils," *J. Soil Sci.,* **33**:141–163.

Unger, P. W., and T. C. Kaspar. 1994. "Soil compaction and root growth: A review," *Agron. J.,* **86**:759–766.

U.S. Department of Interior Teton Dam Failure Review Group. 1977. *Failure of Teton Dam.* Stock no. 024-003-00112-1. (Washington, D.C.: U.S. Government Printing Office).

Vimmerstadt, J., F. Scoles, J. Brown, and M. Schmittgen. 1982. "Effects of use pattern, cover, soil drainage class, and overwinter changes on rain infiltration on campsites," *J. Environ. Qual.*, **11**:25–28.

Voorhees, W. B. 1984. "Soil compaction, a curse or a cure?" *Solutions*, **28**:42–47 (Peoria, Ill.: Solutions Magazine Inc.).

Warkentin, B. P. 1961. "Interpretation of the upper plastic limits of clays," *Nature*, **190**:287–288.

Weil, R. R., and W. Kroontje. 1979. "Physical condition of a Davidson clay loam after 5 years of heavy poultry manure applications," *J. Environ. Qual.*, **8**:389–392.

Wilson, H. A., R. Gish, and G. M. Browning. 1947. "Cropping systems and season as factors affecting aggregate stability," *Soil Sci. Soc. Amer. Proc.*, **12**:36–43.

Wright, D. L., F. M. Rhoads, and R. L. Stanley, Jr. 1980. "High level of management needed on irrigated corn," *Solutions*, May/June: 24–36 (Peoria, Ill.: Solutions Magazine Inc.).

5

SOIL WATER: CHARACTERISTICS AND BEHAVIOR

*When the earth will . . . drink up
the rain as fast as it falls.*
—*H. D. THOREAU, THE JOURNAL*

Water is a vital component of every living thing. Although it is one of nature's simplest chemicals, water has unique properties that promote a wide variety of physical, chemical, and biological processes. These processes greatly influence almost every aspect of soil development and behavior, from the weathering of minerals to the decomposition of organic matter, from the growth of plants to the pollution of groundwater.

We are all familiar with water. We drink it, wash with it, swim in it, and irrigate our crops with it. But water in the soil is something quite different from water in a drinking glass. In the soil, water is intimately associated with solid particles, particularly those that are colloidal in size. The interaction between water and soil solids changes the behavior of both.

Water causes soil particles to swell and shrink, to adhere to each other, and to form structural aggregates. Water participates in innumerable chemical reactions that release or tie up nutrients, create acidity, and wear down minerals so that their constituent elements eventually contribute to the saltiness of the oceans.

Attraction to solid surfaces restricts some of the free movement of water molecules, making it less liquid and more solidlike in its behavior. In the soil, water can flow up as well as down. Plants may wilt and die in a soil whose profile contains a million kilograms of water in a hectare. A layer of sand and gravel in a soil profile may actually inhibit drainage, causing the upper horizons to become muddy and saturated with water during much of the year. These and other soil water phenomena seem to contradict our intuition about how water ought to behave.

Soil–water interactions influence many of the ecological functions of soils and practices of soil management. These interactions determine how much rainwater runs into and through the soil and how much runs off the surface. Control of these processes in turn determines the movement of chemicals to the groundwater, and of both chemicals and eroded soil particles to streams and lakes. The interactions affect the rate of water loss through leaching and evapotranspiration, the balance between air and water in soil pores, the rate of change in soil temperature, the rate and kind of metabolism of soil organisms, and the capacity of soil to store and provide water for plant growth.

The characteristics and behavior of water in the soil is a common thread that interrelates nearly every chapter in this book. The principles contained in this chapter will help us understand why mudslides occur in water-saturated soils (Chapter 4), why earthworms improve soil quality (Chapter 11), why rice paddies contribute to global ozone depletion (Chapter 13), and why famine stalks humanity in certain regions of the world (Chapter 20). Mastery of the principles presented in this chapter are fundamental to a working knowledge of the soil system.

5.1 STRUCTURE AND RELATED PROPERTIES OF WATER

The ability of water to influence so many soil processes is determined primarily by the structure of the water molecule. This structure also is responsible for the fact that water is a liquid, not a gas, at temperatures found on earth. Water is, with the exception of mercury, the *only* inorganic (not carbon-based) liquid found on Earth. Water is a simple compound, its individual molecules containing one oxygen atom and two much smaller hydrogen atoms. The elements are bonded together covalently, each hydrogen atom sharing its single electron with the oxygen.

Polarity

The arrangement of the three atoms in a water molecule is not symmetrical, as one might expect. Instead of the atoms being arranged linearly (H-O-H), the hydrogen atoms are attached to the oxygen in a V-shaped arrangement at an angle of only 105°. As shown in Figure 5.1, this results in an asymmetrical molecule with the shared electrons spending most of the time nearer to the oxygen than to the hydrogen. Consequently, the water molecule exhibits *polarity*; that is, the charges are not evenly distributed. Rather, the side on which the hydrogen atoms are located tends to be electropositive and the opposite side electronegative. The fact that water consists of polar molecules accounts for many of the properties that allow water to play its unique roles in the soil environment.

The property of polarity helps explain how water molecules interact with each other. Each water molecule does not act independently but rather is coupled with other neighboring molecules. The hydrogen (positive) end of one molecule attracts the oxygen (negative) end of another, resulting in a chainlike (polymer) grouping. Because its molecules cluster together, water has a much higher boiling point than other liquids with comparably low molecular weights (e.g., methyl alcohol).

Polarity also explains why water molecules are attracted to electrostatically charged ions and to colloidal surfaces. Cations such as H^+, Na^+, K^+, and Ca^{2+} become hydrated through their attraction to the oxygen (negative) end of water molecules. Likewise, negatively charged clay surfaces attract water, this time through the hydrogen (positive) end of the molecule. Polarity of water molecules also encourages the dissolution of salts in water since the ionic components have greater attraction for water molecules than for each other.

When water molecules become attracted to electrostatically charged ions or clay surfaces, they are more closely packed than in pure water. In this packed state their free-

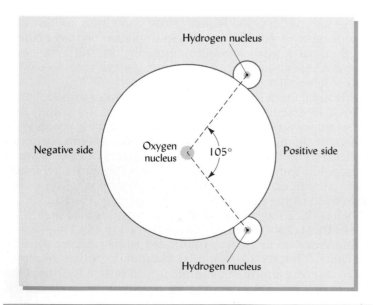

FIGURE 5.1 Two-dimensional representation of a water molecule showing a large oxygen atom and two much smaller hydrogen atoms. The HOH angle of 105° results in an asymmetrical arrangement. One side of the water molecule (that with the two hydrogens) is electropositive; the other is electronegative. This accounts for the polarity of water.

dom of movement is restricted and their energy status is lower than in pure water. Thus, when ions or clay particles become hydrated, energy must be released. That released energy is evidenced as *heat of solution* when ions hydrate or as *heat of wetting* when clay particles become wet. The latter phenomenon can be demonstrated by placing some dry, fine clay in the palm of the hand and then adding a few drops of water. A slight rise in temperature can be felt.

Hydrogen Bonding

Through a phenomenon called **hydrogen bonding**, a hydrogen atom may be shared between two electronegative atoms such as O and N, forming a relatively low-energy link. Because of its high electronegativity, an O atom in one water molecule exerts some attraction for the H atom in a neighboring water molecule. This type of bonding accounts for the polymerization of water. Hydrogen bonding also accounts for the relatively high boiling point, specific heat, and viscosity of water compared to the same properties of other hydrogen-containing compounds, such as H_2S, which has a higher molecular weight but no hydrogen bonding. It is also responsible for the structural rigidity of some clay crystals and for the structure of some organic compounds, such as proteins.

Cohesion versus Adhesion

Hydrogen bonding accounts for two basic forces responsible for water retention and movement in soils: the attraction of water molecules for each other (**cohesion**) and the attraction of water molecules for solid surfaces (**adhesion**). By adhesion (also called *adsorption*), some water molecules are held rigidly at the surfaces of soil-solids. In turn, these tightly bound water molecules hold by cohesion other water molecules farther removed from the solid surfaces (Figure 5.2). Together, the forces of adhesion and cohesion make it possible for the soil solids to retain water and control its movement and use. Adhesion and cohesion also make possible the property of plasticity possessed by clays (see Section 4.9).

Surface Tension

Another important property of water that markedly influences its behavior in soils is that of **surface tension**. This property is commonly evidenced at liquid–air interfaces and results from the greater attraction of water molecules for each other (cohesion) than for the air above (Figure 5.3). The net effect is an inward force at the surface that

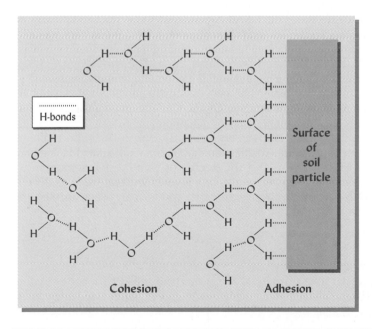

Cohesion   Adhesion

FIGURE 5.2   The forces of adhesion (between water molecules) and cohesion (between water and solid surface) in a soil–water system. The forces are largely a result of H-bonding shown as broken lines. The adhesive or adsorptive force diminishes rapidly with distance from the solid surface. The cohesion of one water molecule to another results in water molecules forming temporary clusters that are constantly changing in size and shape as individual water molecules break free or join up with others. The cohesion between water molecules also allows the solid to indirectly restrict the freedom of water for some distance beyond the solid–liquid interface.

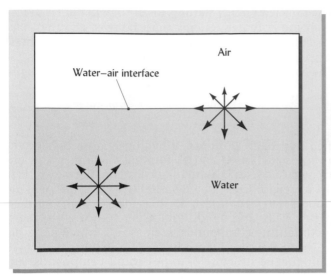

FIGURE 5.3 Comparative forces acting on water molecules at the surface and beneath the surface. Forces acting below the surface are equal in all directions since each water molecule is attracted equally by neighboring water molecules. At the surface, however, the attraction of the air for the water molecules is much less than that of water molecules for each other. Consequently, there is a net downward force on the surface molecules, and the result is something like a compressed film or membrane at the surface. This phenomenon is called *surface tension.*

causes water to behave as if its surface were covered with a stretched elastic membrane, an observation familiar to those who have seen insects walking on water in a pond (Figure 5.4). Because of the relatively high attraction of water molecules for each other, water has a high surface tension (72.8 newtons/mm at 20°C) compared to that of most other liquids (e.g., ethyl alcohol, 22.4 N/mm). As we shall see, surface tension is an important factor in the phenomenon of capillarity, which determines how water moves and is retained in soil.

5.2   CAPILLARY FUNDAMENTALS AND SOIL WATER

The movement of water up a wick typifies the phenomenon of capillarity. Two forces cause capillarity: (1) the attraction of water for the solid (adhesion or adsorption), and (2) the surface tension of water, which is due largely to the attraction of water molecules for each other (cohesion).

Capillary Mechanism

Capillarity can be demonstrated by placing one end of a fine, clean glass tube in water. The water rises in the tube; the smaller the tube bore, the higher the water rises. The water molecules are attracted to the sides of the tube (adhesion) and start to spread out along the glass in response to this attraction. At the same time, the cohesive forces hold the water molecules together and create surface tension, causing a curved surface (called a *meniscus*) to form at the interface between water and air in the tube (Figure 5.5c). Lower pressure under the meniscus in the glass tube (P2) allows the higher pressure (P1) on the free water to push water up the tube. The process continues until the water in the tube has risen high enough that its weight just balances the pressure differential across the meniscus (see Box 5.1 for details).

The height of rise in a capillary tube is inversely proportional to the tube radius r. Capillary rise is also inversely proportional to the density of the liquid, and is directly proportional to the liquid's surface tension and the degree of its adhesive attraction to the soil surface. If we limit our consideration to water at a given temperature (e.g., 20°C), then these factors can be combined into a single constant, and we can use a simple capillary equation to calculate the height of rise h:

$$h = \frac{0.15}{r}$$

where both h and r are expressed in centimeters. This equation tells us that the smaller the tube bore, the greater the capillary force and the higher the water rise in the tube (Figure 5.6a).

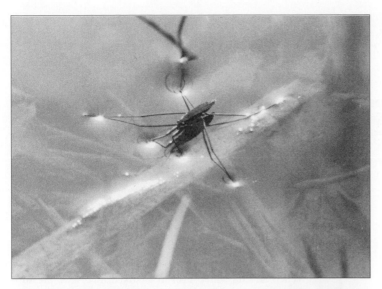

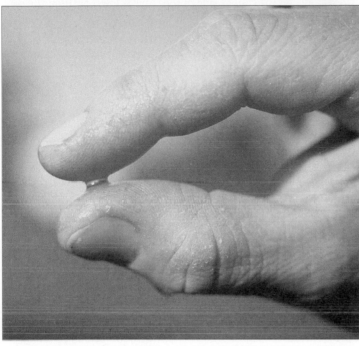

FIGURE 5.4 Everyday evidences of water's surface tension (above) as insects land on water and do not sink, and of forces of cohesion and adhesion (below) as a drop of water is held between the fingers. (Photos courtesy of R. Weil)

Height of Rise in Soils

Capillary forces are at work in all moist soils. However, the rate of movement and the rise in height are less than one would expect on the basis of soil pore size alone. One reason is that soil pores are not straight, uniform openings like glass tubes. Furthermore, some soil pores are filled with air, which may be entrapped, slowing down or preventing the movement of water by capillarity (see Figure 5.6b).

The upward movement due to capillarity in soils is illustrated in Figure 5.6c. Usually the eventual height of rise resulting by capillarity is greater with fine-textured soils, but the rate of flow may be very slow because of frictional forces in the tiny pores. The large pores in sandy soils present little frictional resistance to rapid capillary water movement. However, as expected from our discussion of the capillary equation, the large radii of the pores between sand grains result in relatively little height of capillary rise.[1]

---

[1] For example, if water rises by capillarity to a height of 37 cm above a free-water surface in a sand (as shown in the example in Figure 5.6c), then it can be estimated (by rearranging the capillary equation to $r = 0.15/h$) that the smallest continuous pores must have a radius of about 0.004 cm (0.15/37 = 0.004). This calculation gives an approximation of the minimum effective pore radius in a soil.

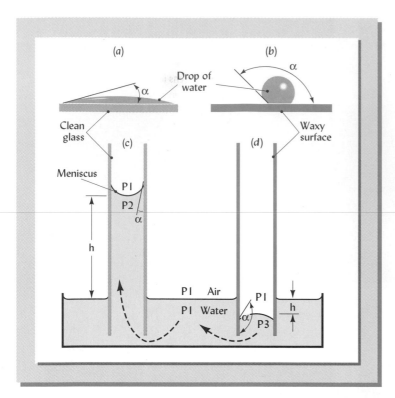

FIGURE 5.5 The interaction of water with a hydrophillic (*a, c*) or a hydrophobic (*b, d*) surface results in a characteristic *contact angle* (α). If the solid surface surrounds the water as in a tube, a curved water–air interface termed the *meniscus* forms because of adhesive and cohesive forces. When air and water meet in a curved meniscus, pressure on the convex side of the curve is lower than on the concave side. (*c*) Capillary rise occurs in a fine hydrophillic (e.g., glass) tube because pressure under the meniscus (P2) is less than pressure on the free water. (*d*) Capillary depression occurs if the tube is hydrophobic, and the meniscus is inverted.

Capillarity is traditionally illustrated as an upward adjustment. But movement in any direction takes place, since the attractions between soil pores and water are as effective in forming a water meniscus in horizontal pores as in vertical ones (Figure 5.7, page 179). The significance of capillarity in controlling water movement in small pores will become evident as we turn to soil water energy concepts.

5.3 SOIL WATER ENERGY CONCEPTS

The retention and movement of water in soils, its uptake and translocation in plants, and its loss to the atmosphere are all energy-related phenomena. Different kinds of energy are involved, including *potential energy* and *kinetic energy*. Kinetic energy is certainly an important factor in the rapid, turbulent flow of water in a river, but the movement of water in soil is so slow that the kinetic energy component is usually negligible. Potential energy is most important in determining the status and movement of soil water. For the sake of simplicity, in this text we will use the term *energy* to refer to potential energy.

As we consider energy, we should keep in mind that all substances, including water, tend to move or change from a higher to a lower energy state. Therefore, if we know the pertinent energy levels at various points in a soil, we can predict the direction of water movement. It is the *differences* in energy levels from one contiguous site to another that influence this water movement.

Forces Affecting Potential Energy

The discussion of the structure and properties of water in the previous section suggests three important forces affecting the energy level of soil water. First, adhesion, or the attraction of water to the soil solids (matrix), provides a **matric** force (responsible for adsorption and capillarity) that markedly reduces the energy state of water near particle surfaces. Second, the attraction of water to ions and other solutes, resulting in **osmotic** forces, tends to reduce the energy state of water in the soil solution. Osmotic movement of pure water across a semipermeable membrane into a solution (osmosis) is evidence of the lower energy state of water in the solution. The third major force acting on soil water is **gravity**, which always pulls the water downward. The energy level of soil water

BOX 5.1   THE MECHANISM OF CAPILLARITY

Capillary action is due to the combined forces of adhesion and cohesion, as seen when a drop of water is placed on a solid surface. Solid substances that have an electronegative surface (for example, due to the oxygens in the silica tetrahedra of quartz or glass) strongly attract the electropositive H-end of the water molecule. These substances are said to be *hydrophillic* (water-loving) because attraction of the water molecules for the solid surface (adhesion) is much greater than the attraction of the water molecules for each other (cohesion). Adhesion will cause a drop of water placed on a hydrophillic solid, such as clean glass, to spread out along the surface, thus forming an acute (<90°) angle between the water–air interface and the solid surface (see Figure 5.5a). This *contact angle* is characteristic for a particular liquid–solid pair (e.g., water on glass). The more strongly the water molecules are attracted to the solid, the closer to zero the contact angle.

In contrast, water molecules placed on a hydrophobic (water-hating) surface will pull themselves into a spherical mass. The resulting contact angle is obtuse (>90°), indicating that the adhesion is not as strong as the cohesion (see Figure 5.5b). This relationship explains why water beads up on a freshly waxed automobile.

Now instead of a flat surface and a drop of water, consider a small-diameter tube of clean glass dipped into a pool of water. Adhesion will again cause the water to spread out on the glass surface, forming the same contact angle α with the glass as was the case for the water drop. At the same time, cohesion among water molecules creates a surface tension that causes a curved surface (called a *meniscus*) to form at the interface between the water and the air in the tube (see Figure 5.5c). If the contact angle is nearly zero, the curvature of the meniscus will approximate a hemisphere.

The curved (rather than a flat) interface between water and air causes the pressure to be lower on the convex side (labeled P2 in Figure 5.5c) than on the concave side of the meniscus. Normal atmospheric pressure P1 is exerted both above the meniscus and on the pool of free water. Because the pressure under the meniscus P2 is less than the pressure on the free-water pool, water is pushed up the capillary tube. The water will rise in the tube until the meniscus reaches the height h at which the weight of the water in the tube just balances the pressure difference P2 – P1. In this condition, the forces pushing water up the tube will be just balanced by the forces pulling it down.

The upward forces are determined by the product of surface tension T, the *length* of the contact between the tube and the meniscus (tube *circumference* – $2\pi r$), and the upward component of this force (cos α).

The downward forces are determined by the product of the water density d, the water volume above the free-water surface $h\pi r^2$, and the acceleration of gravity g.

Thus, when capillary rise ceases we can equate:

$$\text{Upward-acting force} = \text{Downward-acting force}$$
$$T * 2\pi r * \cos \alpha = d * h * \pi r^2 * g$$

Note that if the tube radius were made half as large (0.5r), the force acting upward would be cut in half, but the downward force would be ¼ as great $[(0.5r)^2 = 0.5r*0.5r = 0.25r]$—hence, the height of rise would be twice as great when the forces come into balance again. Herein lies the reason why capillarity rise is greater in finer tubes. The equation balancing the upward- and downward-acting forces can be algebraically rearranged to give an equation describing the height of capillary rise:

$$h = \frac{2T \cos \alpha}{rdg}$$

Most water–solid interactions in soils are of the hydrophillic type shown in Figure 5.5a and c. The attraction between water and soil particle surfaces is usually so strong that the angle of contact is very close to zero, making its cosine ≈ 1. The cos α can therefore be ignored under these circumstances. Three of the other factors affecting capillary rise (T, d, and g) are constants at a given temperature and can therefore be combined into a single constant. Thus, we can write the simplified capillary rise equation given on page 174:

$$h \text{ (cm)} = \frac{0.15 \text{ (cm}^2)}{r \text{ (cm)}}$$

As one would expect, capillary rise will only occur if the tube is made of hydrophillic material. If a *hydrophobic* tube (such as one with a waxed surface) is dipped into a pool of water, the meniscus will be convex rather than concave to the air, so that the situation is reversed and capillary *depression* rather rise will occur (see Figure 5.5d). This is the case in certain water-repellent soil layers (see Figure 7.20).

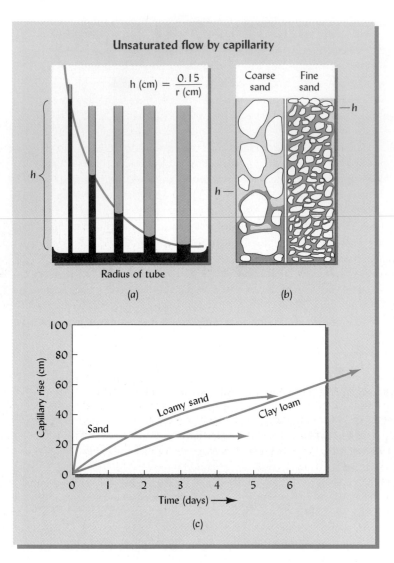

Unsaturated flow by capillarity

$$h \text{ (cm)} = \frac{0.15}{r \text{ (cm)}}$$

Radius of tube

(a)

Coarse sand   Fine sand

(b)

(c)

FIGURE 5.6 Upward capillary movement of water through tubes of different bore and soils with different pore sizes. (a) The capillary equation can be graphed to show that the height of rise h doubles when the tube inside radius is halved. The same relationship can be demonstrated using glass tubes of different bore size. (b) The same principle also relates pores sizes in a soil and height of capillary rise, but the rise of water in a soil is rather jerky and irregular because of the tortuous shape and variability in size of the soil pores (as well as because of pockets of trapped air). (c) The finer the soil texture, the smaller the average pore diameter and, hence, the higher the ultimate rise of water above a free water table. However, because of the much greater frictional forces in the smaller pores, the capillary rise is much slower in the finer-textured soil than in the sand.

at a given elevation in the profile is thus higher than that of water at some lower elevation. This difference in energy level causes water to flow downward.

Soil Water Potential

The *difference* in energy level of water from one site or one condition (e.g., in wet soil) to another (e.g., in dry soil) determines the direction and rate of water movement in soils and in plants. In a wet soil, most of the water is retained in large pores or thick water films around particles. Therefore, most of the water molecules in a wet soil are not very close to a particle surface and so are not held very tightly by the soil solids (the matrix). In this condition, the water molecules have considerable freedom of movement, so their energy level is near that of water molecules in a pool of pure water outside the soil. In a drier soil, however, the water that remains is located in small pores and thin water films, and is therefore held tightly by the soil solids. Thus the water molecules in a drier soil have little freedom of movement, and the energy level of the water is much lower than that of the water in wet soil. If wet and dry soil samples are brought in touch with each other, water will move from the wet soil (higher energy state) to the drier soil (lower energy).

Determining the absolute energy level of soil water is a difficult and sometimes impossible task. Fortunately, it is not necessary to know the absolute energy level of water to be able to predict how it will move in soils and in the environment. Relative

FIGURE 5.7 As this field irrigation scene in Arizona shows (left), water has moved up by capillarity from the irrigation furrow toward the top of the ridge. The photo on the right illustrates some horizontal movement to both sides and away from the irrigation water.

values of soil water energy are all that is needed. Usually the energy status of soil water in a particular location in the profile is compared to that of pure water at standard pressure and temperature, unaffected by the soil and located at some reference elevation. The *difference* in energy levels between this pure water in the reference state and that of the soil water is termed **soil water potential** (Figure 5.8), the term *potential*, like the term *pressure*, implying a difference in energy status.

If all water potential values under consideration have a common reference point (the energy state of pure water), differences in the water potential of two soil samples in fact reflect differences in their absolute energy levels. This means that water will move from a soil zone having a high soil water potential to one having a lower soil water potential. This fact should be always kept in mind in considering the behavior of water in soils.

The soil water potential is due to several forces, each of which is component of the **total soil water potential** ψ_t. These components are due to differences in energy levels resulting from gravitational, matric, submerged hydrostatic and osmotic forces and are termed **gravitational potential** ψ_g, **matric potential** ψ_m, **submergence potential**, and **osmotic potential** ψ_o, respectively. All of these components act simultaneously to influence water behavior in soils. The general relationship of soil water potential to potential energy levels is shown in Figure 5.8 and can be expressed as:

$$\psi_t = \psi_g + \psi_m + \psi_o + \ldots$$

where the ellipses (. . .) indicate the possible contribution of additional potentials not yet mentioned.

Gravitational Potential

The force of gravity acts on soil water the same as it does on any other body (Figure 5.9), the attraction being toward the earth's center. The gravitational potential ψ_g of soil water may be expressed mathematically as:

$$\psi_g = gh$$

where g is the acceleration due to gravity and h is the height of the soil water above a reference elevation. The reference elevation is usually chosen within the soil profile or at its lower boundary to ensure that the gravitational potential of soil water above the reference point will always be positive.

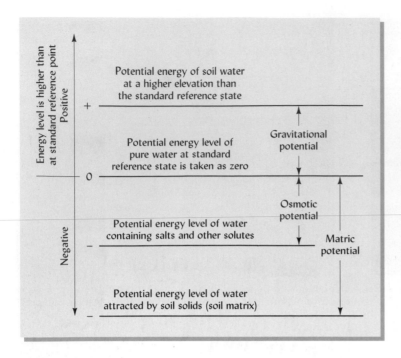

FIGURE 5.8 Relationship between the potential energy of pure water at a standard reference state (pressure, temperature, and elevation) and that of soil water. If the soil water contains salts and other solutes, the mutual attraction between water molecules and these chemicals reduces the potential energy of the water, the degree of the reduction being termed *osmotic potential*. Similarly, the mutual attraction between soil solids (soil matrix) and soil water molecules also reduces the water's potential energy. In this case the reduction is called *matric potential*. Since both of these interactions reduce the water's potential energy level compared to that of pure water, the changes in energy level (osmotic potential and matric potential) are both considered to be negative. In contrast, differences in energy due to gravity (gravitational potential) are always positive. This is because the reference elevation of the pure water is purposely designated at a site in the soil profile below that of the soil water. A plant root attempting to remove water from a moist soil would have to overcome all three forces simultaneously.

Following heavy precipitation or irrigation, gravity plays an important role in removing excess water from the upper horizons and in recharging groundwater below the soil profile. It will be given further attention when the movement of soil water is discussed (see Section 5.5).

Pressure Potential (Including Submergence and Matric Potentials)

This component accounts for the effects on soil water potential of all factors other than gravity and solute levels. It most commonly includes (1) the positive hydrostatic pressure due to the weight of water in saturated soils and aquifers, and (2) the negative pressure due to the attractive forces between the water and the soil solids or the soil matrix.[2]

The hydrostatic pressures give rise to what is often termed the **submergence potential** ψ_s, a component that is operational only for water in saturated zones. Anyone who has dived to the bottom of a swimming pool has felt hydrostatic pressure on the eardrums.

The attraction of water to solid surfaces gives rise to the **matric potential** ψ_m, which is always negative because the water attracted by the soil matrix has an energy state lower than that of pure water. (These negative pressures are sometimes referred to as *suction* or *tension*.) The matric potential is operational in unsaturated soil above the water table, while the submergence potential applies to water in saturated soil or below the water table (Figure 5.10).

While each of these pressures is significant in specific field situations, the matric potential is important in all unsaturated soils because there are omnipresent interactions between soil solids and water. The movement of soil water, the availability of water to plants, and the solutions to many civil engineering problems are determined to a considerable extent by matric potential. Consequently, matric potential will receive primary attention in this text, along with gravitational and osmotic potential.

Matric potential ψ_m, which results from the phenomena of adhesion (or adsorption) and of capillarity, influences soil moisture retention as well as soil water movement. Differences between the ψ_m of two adjoining zones of a soil encourage the movement of water from moist zones (high energy state) to dry zones (low energy state) or from large pores to small pores. Although this movement may be slow, it is extremely important, especially in supplying water to plant roots.

---

[2] In addition to matric and hydrostatic forces, in some situations the weight of the overburden soil and the pressure of air in the soil also make a contribution to the total soil water potential.

FIGURE 5.9 Whether concerning matric potential, osmotic potential, or gravitational potential (as shown here), water always moves to where its energy state will be lower. In this case the energy lost by the water is used to turn the historic Mabry Mills waterwheel and grind flour. (Photo courtesy of R. Weil)

Osmotic Potential

The osmotic potential ψ_o is attributable to the presence of solutes in the soil solution. The solutes may be inorganic salts or organic compounds. Their presence reduces the potential energy of water, primarily because of the reduced freedom of movement of the water molecules that cluster around each solute ion or molecule. The greater the concentration of solutes, the more the osmotic potential is lowered. As always, water will tend to move to where its energy level will be lower, in this case to the zone of higher solute concentration. However, liquid water will move in response to differences in osmotic potential (the process termed **osmosis**) only if a *semipermeable membrane* exists between the zones of high and low osmotic potential, allowing water through but *preventing the movement of the solute*. If no membrane is present, the solute, rather than the water, generally moves to equalize concentrations.

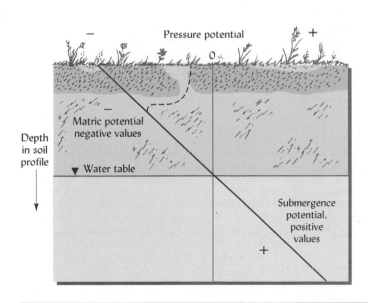

FIGURE 5.10 The matric potential and submergence potential are both pressure potentials that may contribute to total water potential. The matric potential is always negative and the submergence potential is positive. When water is in unsaturated soil above the water table (top of the saturated zone) it is subject to the influence of matric potentials. Water below the water table in saturated soil is subject to submergence potentials. In the example shown here, the matric potential decreases linearly with elevation above the water table, signifying that water rising by capillary attraction up from the water table is the only source of water in this profile. Rainfall or irrigation (see dotted line) would alter or curve the straight line, but would not change the fundamental relationship described.

Because soil zones are *not* generally separated by membranes, the osmotic potential ψ_o has little effect on the mass movement of water in soils. Its major effect is on the uptake of water by plant root cells that *are* isolated from the soil solution by their semipermeable cell membranes. In soils high in soluble salts, ψ_o may be lower (have a greater negative value) in the soil solution than in the plant root cells. This leads to constraints in the uptake of water by the plants. In very salty soil, the soil water osmotic potential may be low enough to cause cells in young seedlings to collapse (plasmolyze) as water moves from the cells to the lower osmotic potential zone in the soil.

The random movement of water molecules causes a few of them to escape a body of liquid water, enter the atmosphere, and become water vapor. Since the presence of solutes restricts the movement of water molecules, fewer water molecules escape into the air as the solute concentration of liquid water is increased. Therefore, water vapor pressure is lower in the air over salty water than in the air over pure water. By affecting water vapor pressure, ψ_o affects the movement of water vapor in soils (see Section 5.7). The process of osmosis and the relationship between the matric and osmotic components of total soil water potential is shown in Figure 5.11.

Methods of Expressing Energy Levels

Several units can be used to express differences in energy levels of soil water. One is the *height of a water column* (usually in centimeters) whose weight just equals the potential

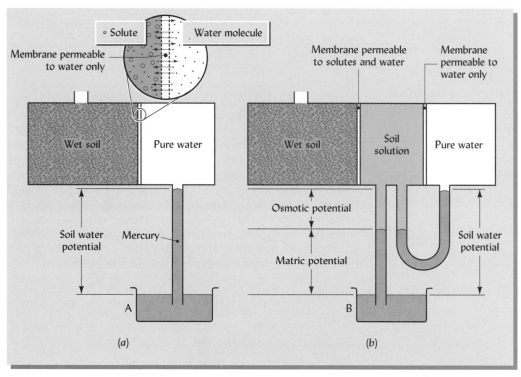

FIGURE 5.11   Relationships among osmotic, matric, and combined soil water potentials. (Left) Assume a container of soil separated from pure water by a membrane permeable only to water (see inset showing osmosis across the membrane). The pure water is connected to a vessel of mercury through a tube. Water will move into the soil in response to the matric forces attracting water to soil solids and the osmotic forces attracting water to solutes. At equilibrium the height of the mercury column above vessel A is a measure of this combined soil water potential (matric plus osmotic). (Right). Assume a second container is placed between the pure water and the soil, and this container is separated from the soil by a fine screen permeable to both solutes and water. Ions will move from the soil into this second container until the concentration of solutes in this water and in the soil water have equalized. Then the difference between the potential energies of this solution and of the pure water gives a measure of the *osmotic potential.* The *matric potential,* as measured by the column of mercury above vessel B, would then be the difference between the combined soil water potential and the osmotic component. The gravitational potential (not shown) is the same for all compartments and does not affect the outcome since the water movement is horizontal. [Modified from Richards (1965)]

under consideration. We have already encountered this means of expression since the *h* in the capillary equation (page 174) tells us the matric potential of the water in a capillary pore. A second unit is the standard *atmosphere* pressure at sea level, which is 760 mm Hg or 1020 cm of water. The unit termed *bar* approximates the pressure of a standard atmosphere. Energy may be expressed per unit of mass (**joules/kg**) or per unit of volume (**newtons/m²**). In the International System of Units (SI), 1 pascal (Pa) equals 1 newton (N) acting over an area of 1 m². In this text we will use Pa or kilopascals (kPa) to express soil water potential. Since other publications may use other units, Table 5.1 is provided to show the equivalency among common means of expressing soil water potential.

5.4 SOIL MOISTURE CONTENT AND SOIL WATER POTENTIAL

The previous discussions suggest an inverse relationship between the water content of soils and the tenacity with which the water is held in soils. Water is more likely to flow out of a wet soil than from one low in moisture. Many factors affect the relationship between soil water potential ψ and moisture content θ. A few examples will illustrate this point.

Soil Moisture versus Energy Curves

The relationship between soil water potential ψ and moisture content θ of three soils of different textures is shown in Figure 5.12. Such curves are sometimes termed *water release characteristic curves*, or simply *water characteristic curves*. The absence of sharp breaks in the curves indicates a gradual change in the water potential with increased soil water content and vice versa. The clay soil holds much more water at a given potential than does the loam or sand. Likewise, at a given moisture content, the water is held much more tenaciously in the clay than in the other two soils (note that soil water potential is plotted on a log scale). The amount of clay in a soil largely determines the proportion of very small micropores in that soil. As we shall see, about half of the water held by clay soils is held so tightly in these micropores that it cannot be removed by growing plants. Soil texture clearly exerts a major influence on soil moisture retention.

Soil structure also influences soil water content–energy relationships. A well-granulated soil has more total pore space and greater overall water-holding capacity than one with poor granulation or one that has been compacted. The greater total pore space indicates a greater overall water-holding capacity. Furthermore, as was explained in Section 4.6, the increased porosity of well-structured soils results mainly from greater amounts of large pores in which water is held with little tenacity. The compacted soil will hold less total water but is likely to have a higher proportion of small- and medium-sized pores, which hold water with greater tenacity than do larger pores. Therefore, soil structure predominantly influences the shape of the water characteristic curve in the portion where the potentials are between 0 and about 100 kPa. The shape of the remainder of the curve generally reflects the influence of soil texture.

TABLE 5.1 **Approximate Equivalents Among Expressions of Soil Water Potential**

Height of unit column of water, cm	Soil water potential, bars	Soil water potential, kPa[a]
0	0	0
10.2	−0.01	−1
102	−0.1	−10
306	−0.3	−30
1,020	−1.0	−100
15,300	−15	−1,500
31,700	−31	−3,100
102,000	−100	−10,000

[a] The SI unit kilopascal (kPa) is equivalent to 0.01 bars.

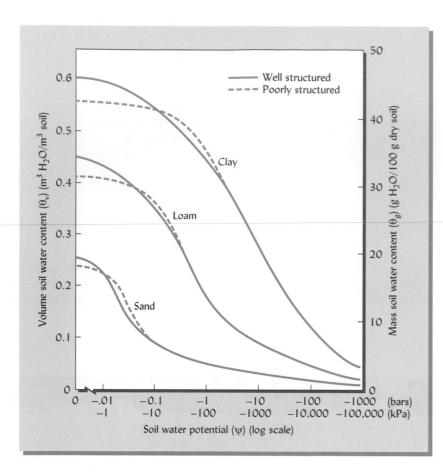

FIGURE 5.12 Soil water potential curves for three representative mineral soils. The curves show the relationship obtained by slowly drying completely saturated soils. The dashed lines show the effect of compaction or poor structure. The soil water potential ψ (which is negative) is expressed in terms of bars (upper scale) and kilopascals (kPa) (lower scale). Note that the soil water potential is plotted on a log scale.

The soil water characteristic curves in Figure 5.12 have marked practical significance for various field measurements and processes. These curves will be useful to refer back to as we consider the applied aspects of soil water behavior in the following sections.

Hysteresis

The relationship between soil water content and potential, determined as a soil dries out, will differ somewhat from the relationship measured as the same soil is rewetted. This phenomena, known as **hysteresis**, is illustrated in Figure 5.13. Hysteresis is caused by a number of factors, including the nonuniformity of soil pores. As soils are wetted some of the smaller pores are bypassed, leaving entrapped air that prevents water penetration. Some of the macropores in a soil may be surrounded only by micropores, creating a bottleneck effect. In this case, the macropore will not lose its water until the matric potential is low enough to empty the water from the surrounding smaller pores (see Figure 5.13). Also, the swelling and shrinking of clays as the soil is dried and rewetted brings about changes in soil structure that affect the soil–water relationships. Because of hysteresis, it is important to know whether soils are being wetted or dried when properties of one soil are compared with those of another.

Measurement of Soil Water Status

The soil water characteristic curves just discussed highlight the importance of making two general kinds of soil water measurements: the *amount* of water present (water content) and the *energy status* of the water (soil water potential). In order to understand or manage water supply and movement in soils it is essential to have information (directly measured or inferred) on *both* types of measurements. For example, a soil water potential measurement might tell us whether water will move toward the groundwater, but

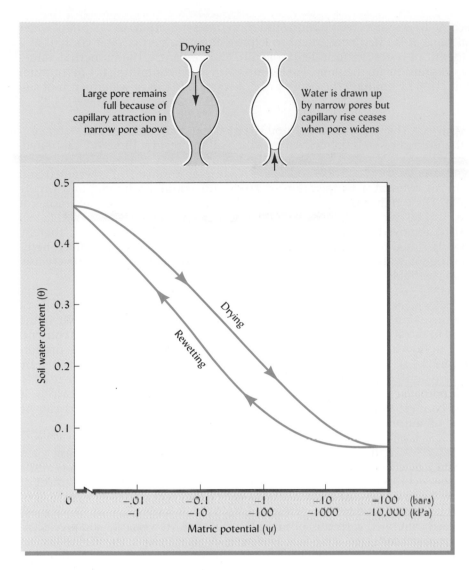

FIGURE 5.13 The relationship between soil water content and matric potential of a soil upon being dried and then rewetted. The phenomenon, known as *hysteresis*, is apparently due to factors such as the nonuniformity of individual soil pores, entrapped air, and the swelling and shrinking that might affect soil structure. The drawings show the effect of nonuniformity of pores.

without a corresponding measurement of the soil water content, we would not know the potential significance of the contribution to groundwater.

Generally, the behavior of soil water is most closely related to the energy status of the water, not to the amount of water in a soil. Thus, a clay soil and a loamy sand will both feel moist and will easily supply water to plants when the ψ_m is, say, −10 kPa. However, the amount of water held by the clay loam, and thus the length of time it could supply water to plants, would be far greater at this potential than would be the case for the loamy sand.

We will consider several methods for making each of these two types of soil water measurements. Researchers, land managers, and engineers may use a combination of several of these methods to study the storage and movement of water in soil, manage irrigation systems, and predict the physical behavior of soils.

Measuring Water Content

The **volumetric water content** θ is defined as the volume of water associated with a given volume (usually 1 m³) of dry soil (see Figure 5.12). A comparable expression is the **mass water content** θ_m, or the mass of water associated with a given mass (usually 1 kg) of dry soil. Both of these expressions have advantages for different uses. In most cases we shall use the volumetric water content θ in this text.

Because in the field we think of plant root systems as exploring a certain depth of soil, and because we express precipitation (and sometimes irrigation) as a depth of water (e.g., mm of rain), it is often convenient to express the volumetric water content as a *depth ratio* (depth of water per unit depth of soil). Conveniently, the numerical values for these two expression are the same. For example, for a soil containing 0.1 m^3 of water per m^3 of soil (10% by volume) the depth ratio of water is 0.1 m of water per m of soil depth.[3]

GRAVIMETRIC METHOD.   The gravimetric method is a direct measurement of soil water content and is therefore the standard method by which all indirect methods are calibrated. The water associated with a given mass (and, if the bulk density of the soil is known, a given volume) of dry soil solids is determined. A sample of moist soil is weighed and then dried in an oven at a temperature of 105°C for about 24 hours,[4] and finally weighed again. The weight loss represents the soil water. Box 5.2 provides examples of how θ and θ_m can be calculated. The gravimetric method is a *destructive* method (a soil sample must be removed for each measurement) and cannot be automated, thereby making it poorly suited to monitoring changes in soil moisture. Several indirect methods of measuring soil water content are nondestructive, easily automated, and very useful in the field (see Table 5.2).

ELECTRICAL RESISTANCE BLOCKS.   The **electrical resistance block** method uses small blocks of porous gypsum, nylon, or fiberglass, suitably embedded with electrodes. When the block is placed in moist soil, it absorbs water in proportion to the soil moisture content. The resistance to flow of electricity between the embedded electrodes decreases proportionately (Figure 5.14). The accuracy and range of soil moisture contents measured by these devices are limited (Table 5.2). However, they are inexpensive and can be used to measure approximate changes in soil moisture during one or more growing seasons. It is possible to connect them to electronic switches so that irrigation systems can be turned on and off automatically at set soil moisture levels.

NEUTRON SCATTERING.   A **neutron scattering** probe, which is lowered into the soil via a previously installed access tube (Figure 5.15), contains a source of fast neutrons and a detector for slow neutrons. When fast neutrons collide with hydrogen atoms (most of which are part of water molecules), the neutrons slow down and scatter. The number of slow neutrons counted by a detector corresponds to the soil water content. Once these meters have been calibrated with the soils in question, they are versatile and give accurate results in mineral soils (see Table 5.2). However, in organic soils, the method is less precise because the neutrons collide with many hydrogen atoms that are combined in organic substances rather than in water.

TIME-DOMAIN REFLECTOMETRY.   A relatively recent technique known as **time-domain reflectometry** (TDR) measures two parameters: (1) the time it takes for a electromagnetic impulse to travel down two parallel metal transmission rods (wave guides) buried in the soil, and (2) the degree of dissipation of the impulse as it impacts with the soil at the end of the lines. The transit time is related to the soil's apparent dielectric constant, which in turn is proportional to the amount of water in the soil. The dissipation of the signal is related to the level of salts in the soil solution. Thus, both soil *moisture content* and *salinity* can be measured using TDR.

The TDR wave guides may be portable (inserted into the soil for each reading) or may be installed in the soil at various depths and connected by wire to a junction box where the meter or a computerized data logger can be attached for monitoring. The TDR instrument incorporates sophisticated electronics and computer software capable of measuring and interpreting minute voltage changes over precise picosecond time

---

[3] When measuring amounts of water added to soil by irrigation, it is customary to use units of volume such as m^3 and hectare-meter (the volume of water that would cover a hectare of land to a depth of 1 m). Generally, farmers and ranchers in the irrigated regions of the United States use the English units ft^3 or acre-foot (the volume of water needed to cover an acre of land to a depth of 1 ft).

[4] Enough drying time must be allowed so that the soil has stopped losing water and has reached a constant weight. To save time, a microwave oven may be used. About a dozen small samples of soil (about 20 g each) in glass beakers may be dried on a turntable in a 1000-W microwave oven using three or more consecutive 3-minute periods with the oven power set at high.

BOX 5.2 GRAVIMETRIC DETERMINATION OF SOIL WATER CONTENT

The gravimetric procedures for determining soil water content are relatively simple. Assume that you want to determine the water content of a 100-g sample of moist soil. You dry the sample in an oven kept at 105°C and then weigh it again. Assume that the dried soil now weighs 70 g, which indicates that 30 g of water has been removed from the moist soil. Expressed in kilograms, this is 30 kg water associated with 70 kg dry soil.

Since the mass soil water content θ_m is commonly expressed in terms of kg water associated with 1 kg dry soil (not 1 kg of wet soil), it can be calculated as follows:

$$\frac{30 \text{ kg water}}{70 \text{ kg dry soil}} = \frac{X \text{ kg water}}{1 \text{ kg dry soil}}$$

$$X = \frac{30}{70} = 0.428 \text{ kg water/kg dry soil} = \theta_m$$

To calculate the volume soil water content θ, we need to know the bulk density of the dried soil, which in this case we shall assume to be 1.3 Mg/m^3. In other words, a cubic meter of this soil has a mass of 1300 kg. From the above calculations we know that the mass of water associated with this 1300 kg is 0.428 × 1300 or 556 kg.

Since 1 m^3 of water has a mass of 1000 kg, the 556 kg of water will occupy 556/1000 or 0.556 m^3.

Thus, the volume water content is 0.556 m^3/m^3 of dry soil:

$$\frac{1300 \text{ kg soil}}{\text{m}^3 \text{ soil}} * \frac{\text{m}^3 \text{ water}}{1000 \text{ kg water}} * \frac{0.428 \text{ kg water}}{\text{kg soil}} = \frac{0.556 \text{ m}^3 \text{ water}}{\text{m}^3 \text{ soil}}$$

The relationship between the mass and volume water contents for a soil can be summarized as:

$$\theta = D_b \times \theta_m$$

TABLE 5.2 Some Methods of Measuring Soil Water

Note that more than one method may be needed to cover the entire range of soil moisture conditions.

Method	Measures soil water Content	Measures soil water Potential	Useful range, kPa	Used mainly in Field	Used mainly in Lab	Comments
1. Gravimetric	x		0 to < 10,000		x	Destructive sampling; slow (1 to 2 days) unless microwave used. The standard for calibration.
2. Resistance blocks	x		−100 to <−1,500	x		Can be automated; not sensitive near optimum plant water contents.
3. Neutron scattering	x		0 to <−1,500	x		Radiation permit needed; expensive equipment; not good in high-organic-matter soils; requires access tube.
4. Time domain reflectometry (TDR)	x		0 to <−10,000	x	x	May be automated; accurate to 1 kPa; requires wave guides; expensive instrument.
5. Tensiometer		x	0 to −85	x		Accurate to 0.1 to 1 kPa; limited range; inexpensive; can be automated; needs periodic servicing.
6. Thermocouple psychrometer		x	50 to <−10,000	x	x	Moderately expensive; wide range; accurate only to ±50 kPa.
7. Pressure membrane apparatus		x	50 to <−10,000		x	Used in conjunction with gravimetric method to construct water characteristic curve.

FIGURE 5.14 A cutaway view of a commercial gypsum electrical resistance block placed about 45 cm below the soil surface. Thin wires lead from the block to the surface, where they can be connected to a resistance meter. For most applications several blocks should be buried at different depths throughout the root zone. (Photo courtesy of R. Weil)

intervals (Figure 5.16). While quite expensive, the TDR instrument can be used (without repeated calibration) in most types of soils to obtain accurate readings for the entire range of soil water contents.

Measuring Soil Water Potentials

TENSIOMETERS. The tenacity with which water is held in soils is an expression of soil water potential ψ. Field **tensiometers** (Figure 5.17) measure this attraction or *tension*. The tensiometer is basically a water-filled tube closed at the bottom with a porous ceramic cup and at the top with an airtight seal. Once placed in the soil, water in the tensiometer moves through the porous cup into the adjacent soil until the water potential in the tensiometer is the same as the matric water potential in the soil. As the water is drawn out, a vacuum develops under the top seal, which can be measured by a vacuum gauge or an electronic transducer. If rain or irrigation rewets the soil, water will enter the tensiometer through the ceramic tip, reducing the vacuum or tension recorded by the gauge.

Tensiometers are useful between 0 and −85 kPa potential, a range that includes half or more of the water stored in most soils. Laboratory tensiometers called **tension plates** operate over a similar range of potentials. As the soil dries beyond −80 to −85 kPa, tensiometers fail because air is drawn in though the pores of the ceramic, relieving the vacuum. A solenoid switch can be fitted to a field tensiometer in order to automatically turn an irrigation system on and off.

THERMOCOUPLE PSYCHROMETER. Since plant roots must overcome both matric and osmotic forces when they draw water from the soil, there is sometimes need for an instrument that measures both. The relative humidity of soil air is affected by both matric and osmotic forces, for both constrain the escape of water molecules from liquid water.

In a **thermocouple psychrometer**, a thermocouple junction housed in a tiny (about 5-mm) porous ceramic chamber is cooled sufficiently to cause a drop of water to con-

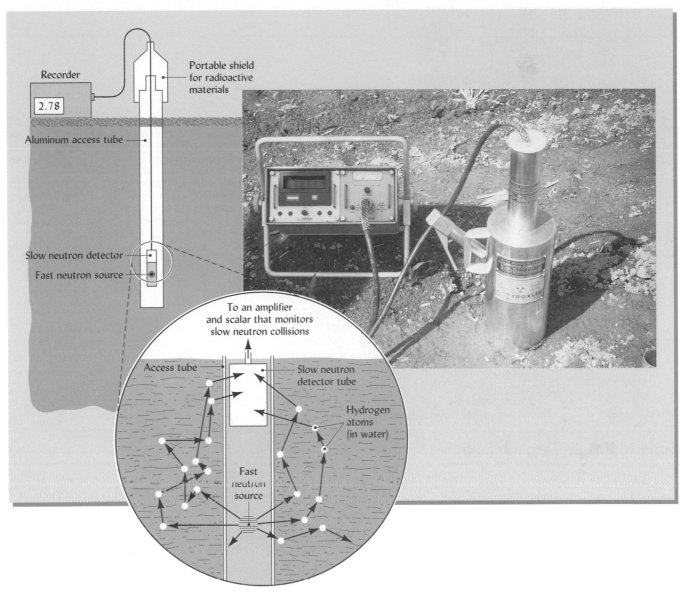

FIGURE 5.15 How a neutron moisture meter operates. The probe, containing a source of fast neutrons and a slow neutron detector, is lowered into the soil through an access tube. Neutrons are emitted by the source (e.g., radium or americium-beryllium) at a very high speed (fast neutrons). When these neutrons collide with a small atom, such as the hydrogen contained in soil water, their direction of movement is changed and they lose part of their energy. These slowed neutrons are measured by a detector tube and a scalar. The reading is related to the soil moisture content. The photograph shows a neutron probe in the field. The heavy metal cylinder is a shield to protect the operator from irradiation. It will be placed over the aluminum-lined hole (extreme lower right) and the neutron source will then be lowered down into the hole for measurement. (Photo courtesy of R. Weil)

dense on it. When the current is switched off, the water drop evaporates at a rate inversely related to the relative humidity of the surrounding air, which in turn is related to the soil moisture potential. A voltage generated by the evaporation of the water drop is converted into a readout of soil water potential ($\psi_o + \psi_o$). The thermocouple psychrometer is most useful in relatively dry soils in which imprecisions of ±50 kPa involve negligible quantities of water.

PRESSURE MEMBRANE APPARATUS.   A **pressure membrane apparatus** (Figure 5.18) is used to subject soils to matric potentials as low as −10,000 kPa. After application of a specific matric potential to a set of soil samples, their soil water contents are determined gravimetrically. This important laboratory tool makes possible accurate measurement of water content over a wide range of matric potentials in a relatively short time. It is used, along with the tension plate, to obtain data to construct soil water characteristic curves such as those shown in Figure 5.12.

FIGURE 5.16 Instrumental measurement of soil water content using time domain reflectometry (TDR). The electronic instrument sends a pulse of electromagnetic energy down the two parallel metal rods of a waveguide that the hydrologist is pushing into the soil. The TDR instrument makes precise picosecond measurements of the speed at which the pulse travels down the rods, a speed influenced by the nature of the surrounding soil. Microprocessors in the instrument analyze the wave patterns generated and calculate the apparent dielectric constant of the soil. Since the dielectric constant of a soil is mainly influenced by its water content, the instrument can accurately convert its measurements into volumetric water content of the soil. (Photo courtesy of Soilmoisture Equipment Corporation, Goleta, Calif., and Dr. Stephen J. Cullen, hydrologist)

5.5 THE FLOW OF LIQUID WATER IN SOIL

Three types of water movement within the soil are recognized: (1) saturated flow, (2) unsaturated flow, and (3) vapor movement. In all cases water flows in response to energy gradients, with water moving from a zone of higher to one of lower water potential. *Saturated flow* takes place when the soil pores are completely filled (or saturated) with water. *Unsaturated flow* occurs when the larger pores in the soil are filled with air, leaving only the smaller pores to hold and transmit water. *Vapor movement* occurs as vapor pressure differences develop in relatively dry soils.

Saturated Flow Through Soils

Under some conditions, at least part of a soil profile may be completely saturated; that is, all pores, large and small, are filled with water. The lower horizons of poorly drained

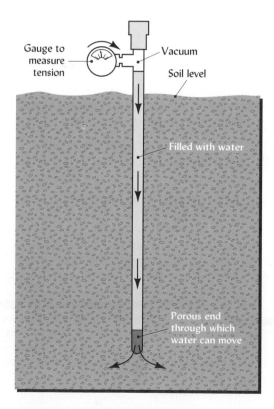

FIGURE 5.17 Tensiometer method of determining water potential in the field. Cross section showing the essential components of a tensiometer. Water moves through the porous end of the instrument in response to the pull (matric potential) of the soil. The vacuum so created is measured by a gauge that reads in kPa of tension (–kPa water potential).

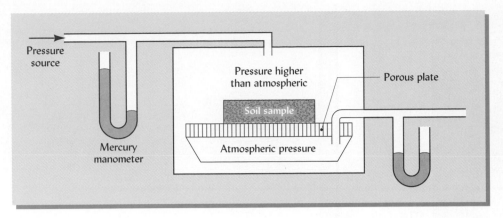

FIGURE 5.18 Pressure membrane apparatus used to determine water content–matric potential relations in soils. An outside source of gas creates a pressure inside the cell. Water is forced out of the soil through a porous plate into a cell at atmospheric pressure. The applied pressure when the downward flow ceases gives a measure of the water potential in the soil. This apparatus will measure much lower soil water potential values (drier soils) than will tensiometers or tension plates.

soils are often saturated, as are portions of well-drained soils above stratified layers of clay. During and immediately following a heavy rain or irrigation, pores in the upper soil zones are often filled entirely with water.

SATURATED HYDRAULIC CONDUCTIVITY. The quantity of water per unit of time Q that flows through a column of saturated soil can be expressed by Darcy's law, as follows:

$$Q = \frac{K_{sat}A\Delta P}{L}$$

where K_{sat} is the saturated hydraulic conductivity (a property of the particular soil), A is the cross-sectional area of the column through which the water flows, ΔP is the hydrostatic pressure difference from the top to the bottom of the column, and L is the length of the column. Since area A and length L of a given column are fixed, the rate of flow is determined by the hydraulic force ΔP driving the water through the soil (commonly gravity) and the saturated hydraulic conductivity K_{sat} or ease with which the soil pores permit water movement. By analogy, one might think of pumping water through a garden hose, with K_{sat} representing the diameter of the hose (water flows more readily though a large hose) and ΔP representing the size of the pump that drives the water through the hose.

The saturated hydraulic conductivity K_{sat} of a uniform soil remains fairly constant over time (assuming the soil is not compacted or otherwise disturbed during this time). The value of K_{sat} depends on the size and configuration of the soil pores, all of which are filled with water. This is in contrast to the hydraulic conductivity K in an unsaturated soil, which decreases as the water content decreases.

HYDRAULIC GRADIENT. In the case of vertical saturated flow (Figure 5.19), the driving force, known as the hydraulic gradient ΔP is equal to h cm of water, the difference in height of the water above and below the soil column. The volume of water moving down the column is determined by this force as well as by the cross-sectional area A through which it is flowing and the K_{sat} of the soil.

It should not be inferred from Figure 5.19 that saturated flow occurs only down the profile. The hydraulic force can also cause horizontal and even upward flow, as occurs when groundwater wells up under a stream (see Section 6.8). The rate of such flow is usually not quite as rapid, however, since the force of gravity does not assist horizontal flow and hinders upward flow. Downward and horizontal flow is illustrated in Figure 5.20, which records the flow of water from an irrigation furrow into two soils, a sandy loam and a clay loam. The water moved down much more rapidly in the sandy loam than in the clay loam. On the other hand, horizontal movement (which would have been largely by unsaturated flow) was much more evident in the clay loam.

Factors Influencing the Hydraulic Conductivity of Saturated Soils

Any factor affecting the size and configuration of soil pores will influence hydraulic conductivity. The total flow rate in soil pores is proportional to the fourth power of the radius. Thus, flow through a pore 1 mm in radius (say, a small earthworm channel) is equivalent to that in 10,000 pores with a radius of 0.1 mm even though it takes only

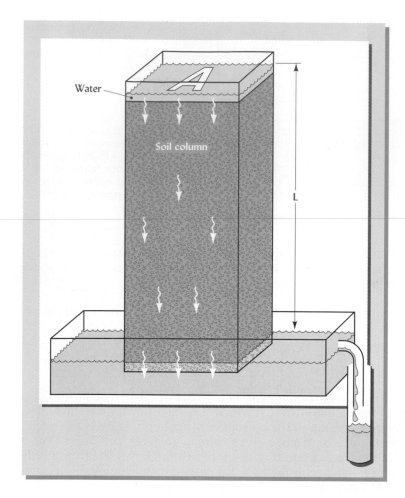

FIGURE 5.19 Saturated (percolation) flow in a column of soil with cross-sectional area *A*. All soil pores are filled with water. The force drawing the water through the soil is *L*, the difference in the heights of water above and below the soil layer. This same force could be applied horizontally. The water is shown running off into a side container to illustrate that water is actually moving down the profile.

100 pores of radius 0.1 mm to give the same cross-sectional area as a 1-mm pore. As a result, macropores (radius > 0.04 mm) account for most water movement in saturated soils. The presence of biopores, such as root channels and earthworm burrows (typically >1 mm in radius), may have a marked influence on the saturated hydraulic conductivity of different soil horizons (Table 5.3).

The texture and structure of a soil horizon most directly determine its saturated hydraulic conductivity. Because they usually have more macropore space, sandy soils generally have higher saturated conductivities than finer-textured soils. Likewise, soils with stable granular structure conduct water much more rapidly than do those with unstable structural units, which break down upon being wetted. Fine clay and silt can

TABLE 5.3   The Saturated Hydraulic Conductivity K_{sat} and Related Properties of Various Horizons in a Typic Hapludult Profile

The upper horizons had many biopores (mainly earthworm burrows), which resulted in high values of K_{sat} as well as extreme variability from sample to sample. The presence of a clay-enriched argillic horizon resulted in reduced K_{sat} values. Apparently most large biopores in this soil did not extend below 30 cm.

Horizon	Depth, cm	Clay, %	Bulk density, Mg/m³	Mean K_{sat}, cm/h	Range of K_{sat} values[a], cm/h
Ap	0–15	12.6	1.42	22.4	0.80–70
E	15–30	11.1	1.44	7.9	0.50–24
E/B	30–45	14.5	1.47	0.93	0.53–1.33
Bt	45–60	22.2	1.40	0.49	0.19–0.79
Bt	60–75	27.2	1.38	0.17	0.07–0.27
Bt	75–90	24.1	1.28	0.04	0.01–0.07

[a] For each soil layer K_{sat} was determined on 5 soil cores of 7.5 cm diameter.
Data from Waddell and Weil (1996).

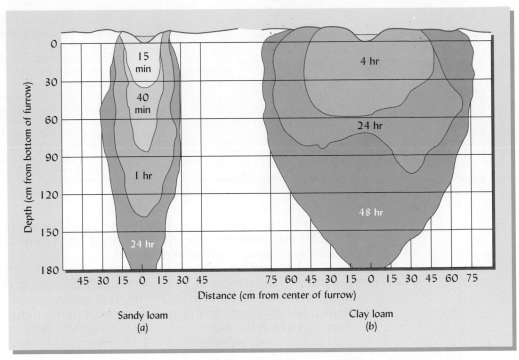

FIGURE 5.20 Comparative rates of irrigation water movement into a sandy loam and a clay loam. Note the much more rapid rate of movement in the sandy loam, especially in a downward direction. [Redrawn from Cooney and Peterson (1955)]

clog the small connecting channels between the larger pores. Entrapped air, which is common in recently wetted soils, can also slow down the movement of water and thereby reduce hydraulic conductivity.

PREFERENTIAL FLOW. Concern over the movement of pesticides and other toxic chemicals through the soil and into the groundwater has called attention to the structural hetero- geneity of some soils. Measurements of hydraulic conductivity made on small cores of soil or on sieved soil packed into laboratory columns may dramatically underestimate the actual movement of water and dissolved chemicals in the field. In the natural field soil, water may flow rapidly through certain pathways in preference to slower, more uniform movement through the bulk of the soil. These **preferential flow** pathways may allow for rapid movement of chemical-laden water deep into the soil profile, increasing the likelihood of groundwater contamination.

In certain sandy soils this flow occurs as "fingers" of rapidly wetted soil, much the way rain drops falling on a window glass will coalesce and flow down the window in tiny streams rather than as a uniform sheet of water. Evidence of this type of flow dur- ing soil formation can be seen in the fingerlike horizon boundaries in the sandy Spo- dosol shown in Figure 3.25. In finer-textured soils, the shrinkage of clays during dry spells may produce a deep network of cracks, often between blocky or prismatic struc- tural peds (see Figure 4.12). These cracks may serve as pathways for preferential water flow (Figure 5.21) that permit ready downward movement of rainwater and associated chemicals before the surrounding soil can become wetted and the cracks closed. Bio- pores, such as deep earthworm burrows and old root channels, may also serve as pref- erential flow paths.

Unsaturated Flow in Soils

Most of the time, water movement takes place when upland soils are *un*saturated. Such movement occurs in a more complicated environment than that which characterizes saturated water flow. In saturated soils, essentially all the pores are filled with water, although the most rapid water movement is through the large and continuous pores. But in unsaturated soils, these macropores are filled with air, leaving only the finer pores to accommodate water movement. Also, in unsaturated soils the water content

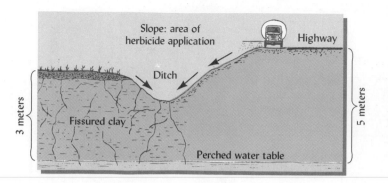

FIGURE 5.21 An illustration of the effects of natural cracks in soils on the conductivity of water and pesticides downward to the water table. An herbicide (weed killer) was applied alongside a highway (right) with the expectation that downward movement into the water table would not be a serious problem since the surrounding soils were fine-textured and would not be expected to readily permit infiltration of the chemical. Because of the wide cracks in this swelling-type clay, however, the first heavy rain carried the chemicals into the groundwater before the soil could swell and shut the cracks. Through the groundwater the herbicide could move into nearby streams. [From DeMartinis and Cooper (1994) with permission of Lewis Publishers]

and, in turn, the tightness with which water is held (water potential) can be highly variable. This influences the rate and direction of water movement and also makes it more difficult to measure the flow of soil water.

As was the case for saturated water movement, the driving force for unsaturated water flow is differences in water potential. This time, however, the difference in the matric potential, not gravity, is the primary driving force. This **matric potential gradient** is the difference in the matric potential of the moist soil areas and nearby drier areas into which the water is moving. Movement will be from a zone of thick moisture films (high matric potential, e.g., −1 kPa) to one of thin films (lower matric potential, e.g., −100 kPa).

INFLUENCE OF TEXTURE. Figure 5.22 shows the general relationship between matric potential ψ_m (and, in turn, water content) and hydraulic conductivity of a sandy loam and clay soil. Note that at or near zero potential (which characterizes the saturated flow region), the hydraulic conductivity is thousands of times greater than at potentials that characterize typical unsaturated flow (−10 kPa and below).

At high potential levels (high moisture contents), hydraulic conductivity is higher in the sand than in the clay. The opposite is true at low potential values (low moisture contents). This relationship is to be expected because the sandy soil contains many large pores which are water-filled when the soil water potential is high (and the soil is quite wet), but most of these have been emptied by the time the soil water potential

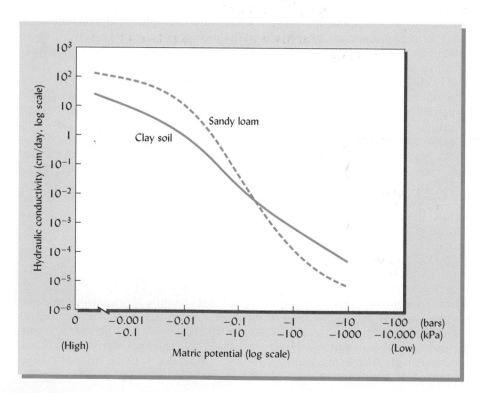

FIGURE 5.22 Generalized relationship between matric potential and hydraulic conductivity for a sandy soil and a clay soil (note log scales). Saturation flow takes place at or near zero potential, while much of the unsaturated flow occurs at a potential of −0.1 bar (−10 kPa) or below.

becomes lower than about 10 kPa. The clay soil has many more micropores which are still water-filled at lower soil water potentials (drier soil conditions) and can participate in unsaturated flow.

The influence of the magnitude of the potential gradient on water movement is illustrated by Figure 5.23. Laboratory measurements made on three moist soil samples adjacent to a dry soil showed that the higher the water content in the moist soil, the greater the matric potential gradient between the moist and dry soil and, in turn, the more rapid the flow. Note that the curves tend to level off as the water moves into the dry soil, because the distance L between the wettest soil and the as yet unwetted soil increases, reducing the potential gradient $\Delta\psi/L$.

5.6   INFILTRATION AND PERCOLATION

A special case of water movement is the entry of free water into the soil at the soil–atmosphere interface. As we shall explain in Chapter 6, this is a pivotal process in landscape hydrology that greatly influences the moisture regime for plants and the potential for soil degradation, chemical runoff, and down-valley flooding. The source of free water at the soil surface may be rainfall, snowmelt, or irrigation.

Infiltration

The process by which water enters the soil pore spaces and becomes soil water is termed **infiltration**, and the rate at which water can enter the soil is termed the **infiltration capacity** I:

$$I = \frac{Q}{A * t}$$

where Q is the volume quantity of water (m³) infiltrating, A is the area of the soil surface (m²) exposed to infiltration, and t is time (s). Since m³ appears in the numerator and m² in the denominator, the units of infiltration can be simplified to m/s or, more commonly, cm/h. The infiltration capacity is not constant over time, but generally decreases during an irrigation or rainfall episode. If the soil is quite dry when infiltration begins, all the macropores open to the surface will be available to conduct water into the soil. In soils with expanding types of clays, the initial infiltration rate may be particularly high as water pours into the network of shrinkage cracks. However, as infiltration proceeds, many macropores fill with water and shrinkage cracks close up. The infiltration capacity declines sharply at first and then tends to level off, remaining fairly constant thereafter (Figure 5.24).

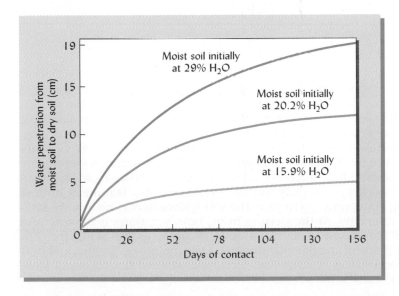

FIGURE 5.23   Rate of water movement from a moist soil at three moisture levels to a drier one. The higher the water content of the moist soil, the greater the potential gradient and the more rapid the delivery. Water adjustment between two slightly moist soils at about the same water content will be exceedingly slow. [After Gardner and Widtsoe (1921)]

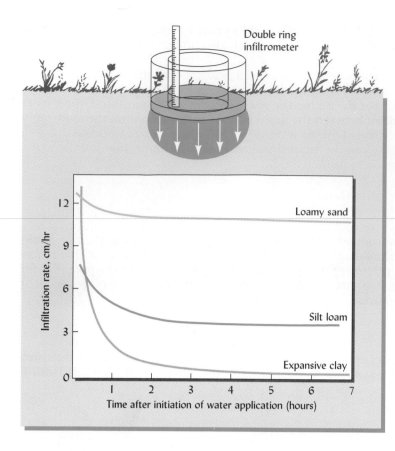

FIGURE 5.24 The potential rate of water entry into the soil, or infiltration capacity, can be measured by recording the drop in water level in a double ring infiltrometer (top). Changes in the infiltration rate of several soils during a period of water application by rainfall or irrigation are shown (bottom). Generally, water enters a dry soil rapidly at first, but its infiltration rate slows as the soil becomes saturated. The decline is least for very sandy soils with macropores that do not depend on stable structure or clay shrinkage. In contrast, a soil high in expansive clays may have a very high initial infiltration rate when large cracks are open, but a very low infiltration rate once the clays swell with water and close the cracks. Most soils fall between these extremes, exhibiting a pattern similar to that shown for the silt loam soil.

MEASUREMENT. Like hydraulic conductivity, infiltration capacity is a characteristic of each particular soil and depends mainly upon the texture and structure at the soil surface, but also upon the presence of any layers in the soil profile that might restrict the downward movement of water. The infiltration capacity of a soil may be easily measured using a simple device known as a **double ring infiltrometer.** Two heavy metal cylinders, one smaller in diameter than the other, are pressed partially into the soil so that the smaller is inside the larger (see Figure 5.24). A layer of cheesecloth is placed inside the rings to protect the soil surface from disturbance, and water is poured into both cylinders. The depth of water in the central cylinder is then recorded periodically as the water infiltrates the soil. The water infiltrating in the outer cylinder is not measured, but it ensures that the surrounding soil will be equally moist and that the movement of water form the central cylinder will be principally downward, not horizontal.

Percolation

Infiltration is a transitional phenomenon that takes place at the soil surface. Once the water has infiltrated the soil, the water moves downward into the profile by the process termed **percolation.** Both saturated and unsaturated flow are involved in percolation of water down the profile, and rate of percolation is related to the soil's hydraulic conductivity. In the case of water that has infiltrated a relatively dry soil, the progress of water movement can be observed by the darkened color of the soil as it becomes wet (Figure 5.25). There usually appears to be a sharp boundary, termed a **wetting front,** between the dry underlying soil and the soil already wetted (see Figure 5.26). During an intense rain or heavy irrigation, water movement near the soil surface occurs mainly by saturated flow in response to gravity. At the wetting front, however, water is moving into the underlying drier soil in response to matric potential gradients as well as gravity. During a light rain, both infiltration and percolation may occur mainly by unsaturated flow as water is drawn by matric forces into the fine entrapped pores without accumulating at the soil surface or in the macropores.

Soil surface

Ap horizon

←Wetting front

Subsoil dried
by plant roots

Moist
substrata

FIGURE 5.25 The wetting front 24 hours after a 5 cm rainfall. Water removal by plant roots had dried the upper 70 to 80 cm of this humid-region (Alabama) profile during a previous three-week dry spell. The clearly visible boundary results from the rather abrupt change in soil water content at the wetting front between the dry, lighter-colored soil and the soil darkened by the percolating water. The wavy nature of the wetting front in this natural field soil is evidence of the heterogeneity of pore sizes. Scale in 10-cm intervals. (Photo courtesy of R. Weil)

Water Movement in Stratified Soils

The fact that, at the wetting front, water is moving by unsaturated flow has important ramifications for how percolating water behaves when it encounters an abrupt change in pore sizes. In the field, many soil profiles contain subsurface layers with pore sizes that contrast markedly with adjacent layers in the profile. Common examples include relatively impervious horizons, such as fragipans or claypans, and coarse layers, such as

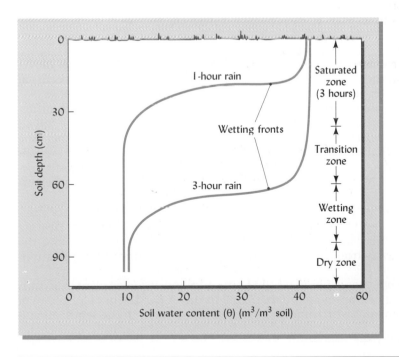

FIGURE 5.26 Water infiltration into a relatively dry soil after one and three hours of a steady rain. The wetting fronts indicate the depth of water penetration. After three hours, the upper 30 to 40 cm of soil is saturated with water. There is a transition zone of near saturation with water above the wetting zone which, in turn, is above the dry zone.

sand and gravel lenses. In some cases, such pore-size stratification may be created by soil managers, as when coarse plant residues are plowed under in a layer, or a layer of gravel is placed under finer soil in a planting container. In all cases, the effect on water percolation is similar—that is, the downward movement is impeded—even though the causal mechanism may vary. It is not surprising that percolating water should slow down markedly when it reaches a layer with finer pores, which therefore has a lower hydraulic conductivity. However, the fact that a layer of *coarser* pores will temporarily stop the movement of water may not be obvious.

In Figure 5.27, a layer of coarse sand impedes downward movement of water in an otherwise fine-textured soil. Intuitively, one might expect the sand layer to speed, rather than impede, percolation. However, it has the opposite effect because the macropores of the sand offer less attraction for the water than do the finer pores of the overlying material. Therefore, when the unsaturated wetting front reaches the sand layer the matric potential is lower in the overlying material than in the sand. Since water always moves from higher to lower potential (to where it will be held more tightly), it cannot move readily into the sand. Eventually the downward-moving water will accumulate above the sand layer and nearly saturate the pores at the soil–sand interface (i.e., the matric potential of the water at the wetting front will fall to nearly zero). Once this occurs, the water will be so loosely held by the fine-textured soil that gravitational forces will be able to pull water into the sand layer.

Interestingly, a coarse sand layer in an otherwise fine-textured soil profile would also inhibit the *rise* of water from moist subsoil layers up to the surface soil, a situation that could be illustrated by turning Figure 5.27b upside down. The large pores in the coarse layer will not be able to support capillary movement up from the smaller pores in a finer layer. Consequently, water rises by capillarity up to the coarse-textured layer but cannot cross it to supply moisture to overlying layers. Thus, plants growing on some soils with buried gravel lenses are subject to drought since they are unable to exploit the lower soil layers. This principle also allows a layer of gravel to act as a capillary barrier under a concrete slab foundation to prevent water from soaking up from the soil and through the concrete floor of a home basement.

The fact that coarse-textured layers (e.g., gravel, sand, coarse organic materials, or geotextile fabrics) can hinder both downward and upward unsaturated flow of water must be considered when using such materials in planting containers or landscape drainage schemes (see Section 6.9). For example, because it retards downward movement, stratification markedly influences the amount of water the upper part of the soil holds in the field. The contrasting layer acts as a moisture barrier until a relatively high water content is built up. This may result in a much higher field-moisture level than that normally encountered in freely drained soils (Figure 5.28).

5.7 WATER VAPOR MOVEMENT IN SOILS

Two types of water vapor movement occur in soils, *internal* and *external*. Internal movement takes place within the soil, that is, in the soil pores. External movement occurs at the land surface, and water vapor is lost by surface evaporation (see Section 6.6).

Water vapor moves from one point to another within the soil in response to differences in vapor pressure. Thus, water vapor will move from a moist soil where the soil air is nearly 100% saturated with water vapor (high vapor pressure) to a drier soil where the vapor pressure is somewhat lower. Also, water vapor will move from a zone of low salt content to one with a higher salt content (e.g., around a fertilizer granule). The salt lowers the vapor pressure of the water and encourages water movement from the surrounding soil.

If the temperature of one part of a uniformly moist soil is lowered, the vapor pressure will decrease and water vapor will tend to move toward this cooler part. Heating will have the opposite effect in that heating will increase the vapor pressure and the water vapor will move away from the heated area. Figure 5.29 illustrates these relationships.

The actual amount of water vapor in a soil at optimum moisture for plant growth is surprisingly small, being perhaps no more than 10 L in the upper 15 cm of a hectare of a silt loam soil. This compares with some 600,000 L of liquid water in the same soil vol-

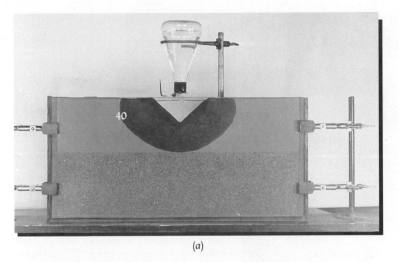

(a)

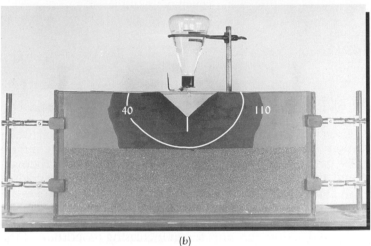

(b)

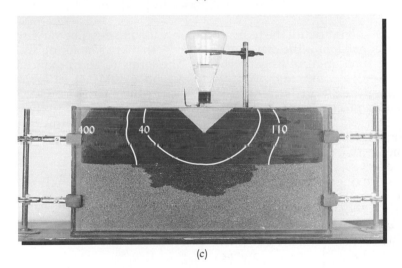

(c)

FIGURE 5.27 Downward water movement in soils with a stratified layer of coarse material. (*a*) Water is applied to the surface of a medium-textured topsoil. Note that after 40 min, downward movement is no greater than movement to the sides, indicating that in this case the gravitational force is insignificant compared to the matric potential gradient between dry and wet soil. (*b*) The *downward* movement stops when a coarse-textured layer is encountered. After 110 min, no movement into the sandy layer has occurred. The macropores of the sand provide less attraction for water than the finer-textured soil above. Only when the water content (and in turn the matric potential gradient) is raised sufficiently will the water move into the sand. (*c*) After 400 min, the water content of the overlying layer becomes sufficiently high to give a water potential of about –1 kPa or more, and downward movement into the coarse material takes place. (Courtesy W. H. Gardner, Washington State University)

ume. Because the amount of water vapor is small, its movement in soils is of limited practical importance if the soil moisture is kept near optimum for plant growth. In dry soils, however, water vapor movement may be of considerable significance to drought-resistant desert plants (*xerophytes*), many of which can exist at extremely low soil water contents. For instance, at night the surface horizon of a desert soil may cool sufficiently to cause vapor movement up from deeper layers. If cooled enough, the vapor may then condense as dewdrops in the soil pores, supplying certain shallow-rooted xerophytes with water for survival.

FIGURE 5.28 One result of soil layers with contrasting texture. This North Carolina soil has about 50 cm of loamy sand coastal plain material atop deeper layers of silty clay-loam-textured material derived from the piedmont. Rainwater rapidly infiltrates the sandy surface horizons, but its downward movement is arrested at the finer-textured layer, resulting in saturated conditions near the surface and a quicksandlike behavior. (Photo courtesy of R. Weil)

5.8 QUALITATIVE DESCRIPTION OF SOIL WETNESS

The measurement of soil water potential and the observable behavior of soil water always depend on that portion of the soil water that is farthest from a particle surface and therefore has the highest potential. As an initially water-saturated soil dries down, both the soil as a whole and the soil water it contains undergo a series of gradual changes in physical behavior and in their relationships with plants. These changes are due mainly to the fact that the water remaining in the drying soil is found in smaller pores and thinner films where the water potential is lowered principally by the action of matric forces. Matric potential therefore accounts for an increasing proportion of the total soil water potential, while the proportion attributable to gravitational potential decreases.

To study these changes and introduce the terms commonly used to describe varying degrees of soil wetness, we shall follow the moisture and energy status of soil during and after a heavy rain or the application of irrigation water. The terms to be introduced describe various stages along a continuum of soil wetness, and should not be interpreted to imply that soil water exists in different "forms." Because these terms are essentially qualitative and lack a precise scientific basis, some soil physicists object to their use. However, it would be a serious disservice to the reader to omit them from this text as they are widely used in practical soil management and help communicate important facts about soil water behavior.

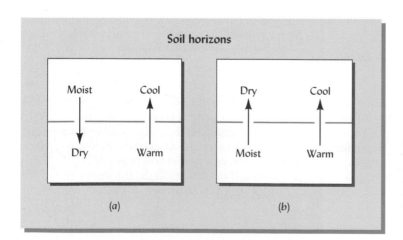

FIGURE 5.29 Vapor movement tendencies that may be expected between soil horizons differing in temperature and moisture. In (a) the tendencies more or less negate each other, but in (b) they are coordinated and considerable vapor transfer might be possible if the liquid water in the soil capillaries does not interfere.

Maximum Retentive Capacity

When all soil pores are filled with water from rainfall or irrigation, the soil is said to be **saturated** with respect to water (Figure 5.30) and at its **maximum retentive capacity**. The matric potential is close to zero, nearly the same as that of pure water. The volumetric water content is essentially the same as the total porosity. The soil will remain at maximum retentive capacity only so long as water continues to infiltrate, for the water in the largest pores (sometimes termed **gravitational water**) will percolate downward, mainly under the influence of gravitational forces (hydrostatic and gravitational potentials). It is slowed in its percolation through macropores mainly by frictional forces associated with the viscosity of water. Since water viscosity decreases (the water appears to become thinner) as temperature increases, drainage will be more rapid in warmer soils. Data on maximum retentive capacities and the average depth of soils in a watershed are useful in predicting how much rainwater can be stored in the soil temporarily, thus possibly avoiding downstream floods.

Field Capacity

Once the rain or irrigation has ceased, water in the largest soil pores will drain downward quite rapidly in response to the hydraulic gradient (mostly gravity). After one to three days, this rapid downward movement will become negligible as matric forces play a greater role in the movement of the remaining water (Figure 5.31). The soil then is said to be at its **field capacity**. In this condition, water has moved out of the macropores and air has moved in to take its place. The micropores or capillary pores are still filled with water and can supply plants with needed water. The matric potential will vary slightly from soil to soil but is generally in the range of –10 to –30 kPa, assuming

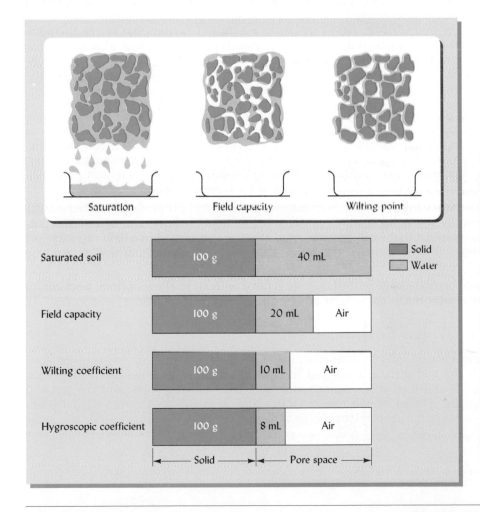

FIGURE 5.30 Volumes of water and air associated with 100 g of a well-granulated silt loam at different moisture levels. The top bar shows the situation when a representative soil is completely saturated with water. This situation will usually occur for short periods of time during a rain or when the soil is being irrigated. Water will soon drain out of the larger pores (*macropores*). The soil is then said to be at the *field capacity*. Plants will remove water from the soil quite rapidly until they begin to wilt. When permanent wilting of the plants occurs, the soil water content is said to be at the *wilting coefficient*. There is still considerable water in the soil, but it is held too tightly to permit its absorption by plant roots. A further reduction in water content to the *hygroscopic coefficient* is illustrated in the bottom bar. At this point the water is held very tightly, mostly by the soil colloids. (Top drawings modified from *Irrigation on Western Farms,* published by the U.S. Departments of Agriculture and Interior)

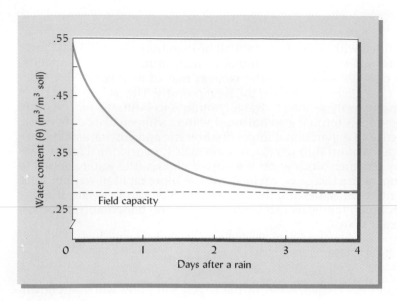

FIGURE 5.31 The water content of a soil drops quite rapidly by drainage following a period of saturation by rain or irrigation. After two or three days the rate of water movement out of the soil is quite slow and the soil is said to be at the *field capacity*.

drainage into a less-moist zone of similar porosity.[5] Water movement will continue to take place by unsaturated flow, but the rate of movement is very slow since it now is due primarily to capillary forces, which are effective only in micropores (Figure 5.30). The water found in pores small enough to retain it against rapid gravitational drainage, but large enough to allow capillary flow in response to matric potential gradients, is sometimes termed **capillary water.**

While all soil water is affected by gravity, the term *gravitational water* refers to the portion of soil water that readily drains away between the states of maximum retentive capacity and field capacity. Most soil leaching occurs as gravitational water that drains from the larger pores before field capacity is reached. Gravitational water therefore includes much of the water that transports chemicals such as nutrient ions, pesticides, and organic contaminants into the groundwater and, ultimately, into streams and rivers.

Field capacity is a very useful term because it refers to an approximate degree of soil wetness at which several important soil properties are in transition:

1. At field capacity, a soil is holding the maximal amount of water useful to plants. Additional water, while held with low energy of retention, would be of limited use to plants because it would remain in the soil for only a short time before draining, and, while in the soil, it would occupy the larger pores, thereby reducing soil aeration. Drainage of gravitational water from the soil is generally a requisite for optimum plant growth (hydrophilic plants, such as rice or cattails, excepted).

2. At field capacity, the soil is near its lower plastic limit—that is, the soil behaves as a crumbly semisolid at water contents below field capacity, and as a plastic puttylike material that easily turns to mud at water contents above field capacity (see Section 4.9). Therefore, field capacity approximates the optimal wetness for ease of tillage or excavation.

3. At field capacity, sufficient pore space is filled with air to allow optimal aeration for most aerobic microbial activity and for the growth of most plants (see Section 7.7).

Permanent Wilting Percentage or Wilting Coefficient

Once an unvegetated soil has drained to its field capacity, further drying is quite slow, especially if the soil surface is covered to reduce evaporation. However, if plants are growing in the soil they will remove water from their rooting zone, and the soil will continue to dry. The roots will remove water first from the largest water-filled pores where the water potential is relatively high. As these pores are emptied, roots will draw

---

[5] Note that because of the relationships pertaining to water movement in stratified soils (see Section 5.6), soil in a flower pot will cease drainage while much wetter than field capacity.

their water from the progressively smaller pores and thinner water films in which the matric water potential is lower and the forces attracting water to the solid surfaces are greater. Hence, it will become progressively more difficult for plants to remove water from the soil at a rate sufficient to meet their needs.

As the soil dries, the rate of plant water removal may fail to keep up with plant needs, and plants may begin to wilt during the daytime to conserve moisture. At first the plants will regain their turgor at night when water is not being lost through the leaves and the roots can catch up with the plant's demand. Ultimately, however, the plant will remain wilted night and day when the roots cannot generate water potentials low enough to coax the remaining water from the soil. Although not yet dead, the plants are now in a permanently wilted condition and will die if water is not provided. For most plants this condition develops when the soil water potential ψ has a value of about −1500 kPa (−15 bars). A few plants, especially xerophytes (desert-type plants) can continue to remove water at even −1800 or −2000 kPa, but the amount of water available between −1500 kPa and −2000 kPa is very small (Figure 5.32).

The water content of the soil at this stage is called the **wilting coefficient** or **permanent wilting percentage** and by convention is taken to be that amount of water retained by the soil when the water potential is −1500 kPa. The soil will appear to be dusty dry, although some water remains in the smallest of the micropores and in very thin films (perhaps only 10 molecules thick) around individual soil particles (see Figure 5.30). As illustrated in Figure 5.32, **plant available water** is considered to be that water retained in soils between the states of field capacity and wilting coefficient (between −10 to −30 kPa and −1500 kPa). The amount of capillary water remaining in the soil that is unavailable to higher plants can be substantial, especially in fine-textured soils and those high in organic matter.

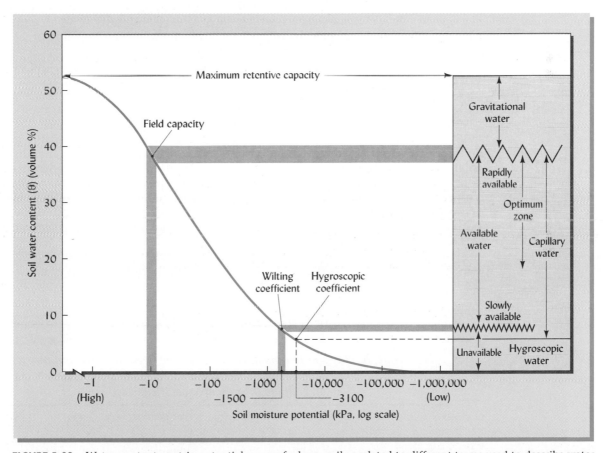

FIGURE 5.32 Water content–matric potential curve of a loam soil as related to different terms used to describe water in soils. The wavy lines in the diagram to the right suggest that measurements such as field capacity are only approximations. The gradual change in potential with soil moisture change discourages the concept of different "forms" of water in soils. At the same time, such terms as *gravitational* and *available* assist in the qualitative description of moisture utilization in soils.

Hygroscopic Coefficient

Although plant roots do not generally dry the soil beyond the permanent wilting percentage, if the soil is exposed to the air, water will continue to be lost by evaporation. When soil moisture is lowered below the wilting point, the water molecules that remain are very tightly held, mostly being adsorbed by colloidal soil surfaces. This state is approximated when the atmosphere above a soil sample is essentially saturated with water vapor (98% relative humidity) and equilibrium is established at a water potential of −3100 kPa. The water is thought to be in films only 4 or 5 molecules thick and is held so rigidly that much of it is considered nonliquid and can move only in the vapor phase. The moisture content of the soil at this point is termed the **hygroscopic coefficient.** Soils high in colloidal materials (clay and humus) will hold more water under these conditions than will sandy soils that are low in clay and humus (Table 5.4). Soil water considered to be **unavailable** to plants includes the hygroscopic water as well as that portion of capillary water retained at potentials below −1500 kPa (see Figure 5.32).

5.9 FACTORS AFFECTING AMOUNT OF PLANT-AVAILABLE SOIL WATER

The amount of soil water available for plant uptake is determined by a number of factors, including water content–potential relationship for each soil horizon, soil strength and density effects on root growth, soil depth, rooting depth, and soil stratification or layering. Each will be discussed briefly.

Matric Potential

Matric potential ψ_m influences the amount of soil water plants can take up because it affects the amounts of water at the field capacity and at the permanent wilting percentage. These two characteristics, which determine the quantity of water a given soil can supply to growing plants, are influenced by the texture, structure, and organic matter content of the soil.

The general influence of texture on field capacity, wilting coefficient, and **available water holding capacity** is shown in Figure 5.33. Note that as fineness of texture increases, there is a general increase in available moisture storage from sands to loams and silt loams. However, clay soils frequently provide less available water than do well-granulated silt loams since the clays tend to have a high wilting coefficient.

The influence of organic matter deserves special attention. The available moisture content of a well-drained mineral soil containing 5% organic matter is generally higher than that of a comparable soil with 3% organic matter. There has been considerable controversy over the degree to which this favorable effect is due directly to the water-supplying ability of organic matter and how much comes from the indirect effects of soil organic matter on soil structure and total pore space. Evidence now suggests that both the direct and indirect factors contribute to the favorable effects of organic matter on soil water availability.

The direct effects are due to the very high water-holding capacity of organic matter which, when the soil is at the field capacity, is much higher than that of an equal vol-

TABLE 5.4 **Volumetric Water Content θ at Field Capacity and Hygroscopic Coefficient for Three Representative Soils and the Calculated Capillary Water**

Note that the clay soil retains most water at the field capacity, but much of that water is held tightly in the soil at −31 bars potential by soil colloids (hygroscopic coefficient).

	Volume % (θ)		
Soil	Field capacity 10–30 kPa	Hygroscopic coefficient, −3100 kPa	Capillary water, col. 1 − col. 2
Sandy loam	12	3	9
Silt loam	30	10	20
Clay	35	18	17

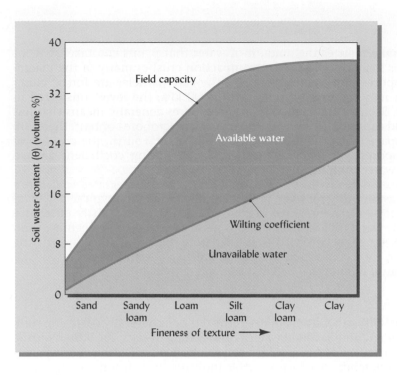

FIGURE 5.33 General relationship between soil water characteristics and soil texture. Note that the wilting coefficient increases as the texture becomes finer. The field capacity increases until we reach the silt loams, then levels off. Remember these are representative curves; individual soils would probably have values different from those shown.

ume of mineral matter. Even though the water held by organic matter at the wilting point is also somewhat higher than that held by mineral matter, the amount of water available for plant uptake is still greater from the organic fraction. Figure 5.34 provides data from a series of experiments to justify this conclusion.

Organic matter indirectly affects the amount of water available to plants through its influence on soil structure and total pore space. We learned in Section 4.7 that organic matter helps stabilize soil structure and that it increases the total volume as well as the size of the pores. This results in an increase in water infiltration and water-holding capacity with a simultaneous increase in the amount of water held at the wilting coefficient. Recognizing the beneficial effects of organic matter on plant-available water is essential for wise soil management.

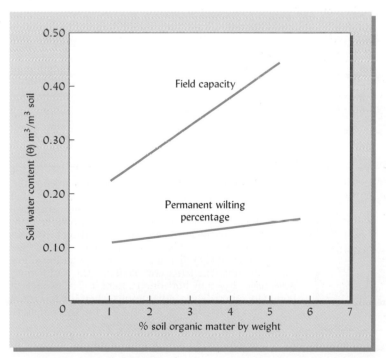

FIGURE 5.34 The effects of organic matter content on the field capacity and permanent wilting percentage of a number of silt loam soils. The differences between the two lines shown is the available soil moisture content, which was obviously greater in the soils with higher organic matter levels. [Redrawn from Hudson (1994); used with permission of the Soil & Water Conservation Society]

Compaction Effects on Matric Potential, Aeration and Root Growth

Soil compaction generally reduces the amount of water that plants can take up. Four factors account for this negative effect. First, compaction crushes many of the macropores and large micropores into smaller pores. As the clay particles are forced closer together, soil strength may increase beyond about 2000 kPa, the level considered to limit root penetration. Second, the reduction in macropores generally means that less water is retained at field capacity. Third, with the reduced macropore content, there will be less aeration pore space when the soil is near field capacity. Fourth, the creation of more very fine micropores will increase the permanent wilting coefficient and so decrease the available water content.

These four factors associated with soil compaction have been integrated to define a water content range least limiting for plant growth, as illustrated for a typical clay loam soil in Figure 5.35. For a relatively uncompacted soil, there will be little difference between plant-available water (field capacity minus permanent wilting percentage) and the least-limiting water range (between 10% air-filled pore space and at root-restricting soil strength). However, for a compacted soil, significantly less water would be considered available by the latter criteria (see hatched area in Figure 5.35).

Osmotic Potential

The presence of soluble salts, either from applied fertilizers or as naturally occurring compounds, can influence plant uptake of soil water. For soils high in salts, the total moisture stress will include the osmotic potential ψ_o of the soil solution as well as the matric potential. The osmotic potential tends to reduce available moisture in such soils because more water is retained in the soil at the permanent wilting coefficient than would be the case due to matric potential alone. In most humid region soils these osmotic potential effects are insignificant, but they become of considerable importance for certain soils in dry regions which may accumulate soluble salts through irrigation or natural processes.

Soil Depth and Layering

Our discussion thus far has referred to available water holding capacity as the percentage of the soil volume consisting of pores that can retain water at potentials between field capacity and wilting percentage. The total volume of available water will depend on the total volume of soil explored by plant roots. This volume may be governed by the total depth of soil above root-restricting layers, by the greatest rooting depth characteristic of a particular plant species, or even by the size of a flower pot chosen for con-

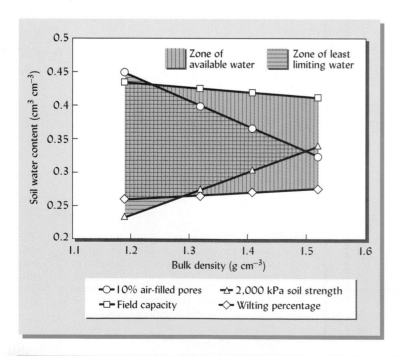

FIGURE 5.35 Influence of increased bulk density on the range of soil water contents available for plant uptake. Traditionally, plant-available water is defined as that retained between field capacity and wilting percentage (vertical hatching). If soils are compacted, however, plant use of water may be restricted by poor aeration (<10% air-filled pore space) at high water contents and by soil strength (>2000 kPa) that restricts root penetration at low water contents. The latter criteria define the *least limiting water range* shown by horizontal hatching. The two sets of boundary criteria give similar results when the soil is not compacted (bulk density about 1.25 for the soil illustrated). [Adapted from Da Silva and Kay (1997)]

tainerized plants. The depth of soil available for root exploration is of particular significance for deep-rooted plants, especially in subhumid to arid regions where perennial vegetation depends on water stored in the soils for survival during long periods without precipitation (see Figure 1.5).

Soil stratification or layering can markedly influence the available water and its movement in the soil. Impervious layers drastically slow down the rate of water movement and also restrict the penetration of plant roots, thereby reducing the soil depth from which moisture is drawn. Sandy layers also act as barriers to soil moisture movement from the finer-textured layers above, as explained in Section 5.7 and Figure 5.27.

The capacity of soils to store available water determines to a great extent their usefulness for plant growth. The productivity of forest sites is often related to soil water-holding capacity. This capacity provides a buffer between an adverse climate and plant production. In irrigated soils, it helps determine the frequency with which water must be applied. Soil water-holding capacity becomes more significant as the use of water for all purposes—industrial and domestic, as well as irrigation—begins to tax the supply of this all-important natural resource. To estimate the water-holding capacity of a soil, each soil horizon to which roots have access may be considered separately and then summed to give a total water-holding capacity for the profile (see Box 5.3).

BOX 5.3 CALCULATION OF THE TOTAL AVAILABLE WATER HOLDING CAPACITY OF A SOIL PROFILE

The total amount of water available to a plant growing in a field soil is a function of the rooting depth of the plant and the sum of the water held between field capacity and wilting percentage in each of the soil horizons explored by the roots. For each soil horizon, then, the mass available water holding capacity is estimated as the difference between the mass water contents θ_m (Mg water per 100 Mg soil) at field capacity and permanent wilting percentage. This value can be converted into a volume water content θ by multiplying by the ratio of the bulk density of the soil to the density of water. Finally, this volume ratio is multiplied by the thickness of the horizon to give the total centimeters of available water capacity (AWC) in that horizon. For the first horizon described in Table 5.5 the calculation (with units) is:

$$\left(\frac{22 \text{ g}}{100 \text{ g}} - \frac{8 \text{ g}}{100 \text{ g}}\right) * \frac{1.2 \text{ Mg}}{\text{m}^3} * \frac{1 \text{ m}^3}{1 \text{ Mg}} * 20 \text{ cm} = 3.36 \text{ cm AWC}$$

Note that all units cancel out except cm, resulting in the depth of available water (cm) held by the horizon. In Table 5.5, the available water holding capacity of all horizons within the rooting zone are summed to give a total AWC for the soil–plant system. Since no roots penetrated to the last horizon (1.0 to 1.25 m), this horizon was not included in the calculation. We can conclude that for the soil–plant system illustrated, 14.13 cm of water could be stored for plant use. At a typical summertime water-use rate of 0.5 cm of water per day, this soil could hold about a four-week supply.

TABLE 5.5 Calculation of Estimated Soil Profile Available Water Holding Capacity

Soil depth, cm	Relative root length	Soil bulk density, Mg/m³	Field capacity (FC), g/100g	Wilting percentage (WP), g/100g	Available water holding capacity (AWC), cm
0–20	xxxxxxxxx	1.2	22	8	$20 * 1.2 \left(\frac{22}{100} - \frac{8}{100}\right) = 3.36$ cm
20–40	xxxx	1.4	16	7	$20 * 1.4 \left(\frac{16}{100} - \frac{7}{100}\right) = 2.52$ cm
40–75	xx	1.5	20	10	$35 * 1.5 \left(\frac{20}{100} - \frac{10}{100}\right) = 5.25$ cm
75–100	xx	1.5	18	10	$25 * 1.5 \left(\frac{18}{100} - \frac{10}{100}\right) = 3.00$ cm
100–125	—	1.6	15	11	No roots
Total					$3.36 + 2.52 + 5.25 + 3.00 = 14.13$ cm

At any one time, only a small proportion of the soil water is adjacent to the absorptive plant-root surfaces. How then do the roots get access to the immense amount of water (see Section 6.3) used by vigorously growing plants? Two phenomena seem to account for this access: the capillary movement of the soil water to plant roots and the growth of the roots into moist soil.

Rate of Capillary Movement

When plant rootlets absorb water, they reduce the soil moisture content, thus reducing the water potential in the soil immediately surrounding them (see Figure 5.36). In response to this lower potential, water tends to move toward the plant roots. The rate of movement depends on the magnitude of the potential gradients developed and the conductivity of the soil pores. With some sandy soils, the adjustment may be comparatively rapid and the flow appreciable if the soil is near field capacity. In fine-textured and poorly granulated clays, the movement will be sluggish and only a meager amount of water will be delivered. However, as indicated by the relative changes in hydraulic conductivity (see Figure 5.22), in drier conditions with water held at lower potentials, the clay soils will be able to deliver more water by capillarity than the sand because the later will have very few pores still filled with water.

The total distance that water flows by capillarity on a day-to-day basis may be only a few centimeters (Figure 5.37). This might suggest that capillary movement is not a significant means of enhancing moisture uptake by plants. However, if the roots have penetrated much of the soil volume so that individual roots are rarely more than a few centimeters apart, movement over greater distances may not be necessary. Even during periods of hot, dry weather when evaporative demand is high, capillary movement can be an important means of providing water to plants. It is of special significance during periods of low moisture content when plant root extension is low.

Rate of Root Extension

Capillary movement of water is complemented by rapid rates of root extension, which ensure that new root–soil contacts are constantly being established. Such root penetration may be rapid enough to take care of most of the water needs of a plant growing in a soil at optimum moisture. The mats of roots, rootlets, and root hairs in a forest floor or a meadow sod exemplify successful adaptations of terrestrial plants for exploitation of soil water storage. Table 5.6 provides data on the length of roots of soybeans in one experiment. These figures do not include the length of thousands of root hairs which are known to permeate the soil several millimeters beyond the main root surfaces.

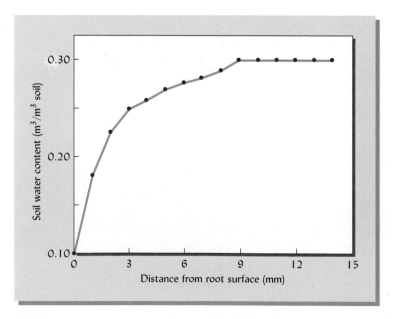

FIGURE 5.36 The drawdown of soil water levels surrounding a radish root after only two hours of transpiration. Water has moved by capillarity from a distance of at least 9 mm from the surface of the root. [Modified from Hamza and Aylmore (1992); used with permission of Kluwer Academic Publishers, The Netherlands]

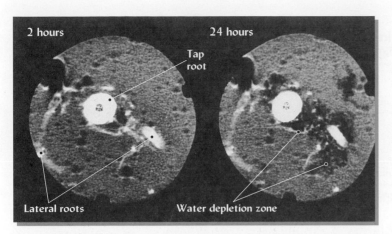

FIGURE 5.37 The intimate soil–root relationship and the rapid depletion of water near roots is illustrated in these two magnetic resonance images (MRIs) of a 2-mm slice of a 25-mm cross-sectional view of moist sand surrounding the roots of a loblolly pine seedling. (Left) Image of a tap-root (large white circle) and two woody lateral roots (elongated bright area to the right and L-shaped somewhat bright area to the left) only two hours after water was supplied. Note some water depletion immediately around the taproot (dark area to the right of the taproot). (Right) After 24 hours, the zone of water depletion (dark area around the roots) has expanded around the taproot and the lateral root to the right. Ready movement of water to the roots is obvious. (Photos courtesy Dr. Janet S. MacFall, Duke University Medical Center)

The primary limitation of root extension is the small proportion of the soil with which roots are in contact at any one time. Even though the root surface is considerable, as shown in Table 5.6, root–soil contacts commonly account for less than 1% of the total soil surface area. This suggests that most of the water must move from the soil to the root even though the distance of movement may be no more than a few millimeters. It also suggests the complimentarity of capillarity and root extension as means of providing soil water for plants.

Root Distribution

The distribution of roots in the soil profile determines to a considerable degree the plant's ability to absorb soil water. Most plants, both annuals and perennials, have the bulk of their roots in the upper 25 to 30 cm of the profile (Table 5.7). Perennial plants such as alfalfa and trees have some roots that grow very deeply (>3 m) and are able to absorb a considerable proportion of their moisture from subsoil layers. Even in these cases, however, it is likely that much of the root absorption is from the upper layers of the soil, provided these layers are well supplied with water. On the other hand, if the upper soil layers are moisture-deficient, even annual plants such as corn and soybeans will absorb much of their water from the lower horizons, provided that adverse physical or chemical conditions do not inhibit their exploration of these horizons.

Root–Soil Contact

As roots grow into the soil, they move into pores of sufficient size to accommodate them. Contact between the outer cells of the root and the soil permits ready movement of water from the soil into the plant in response to differences in energy levels (Figure 5.38). When the plant is under moisture stress, however, the roots tend to shrink in size as their cortical cells lose water in response to this stress. Such conditions exist during a hot, dry spell and are most severe during the daytime, when water loss through plant leaves is at a maximum. The diameter of roots under these conditions may shrink by 30%. This reduces considerably the direct root–soil contact, as well as the movement of liquid

TABLE 5.6 **Length of Soybean Roots at Different Soil Depths in a Captina Silt Loam (Typic Fragiudult) in Arkansas**

	Root length, km/m³	
Soil depth, cm	Nonirrigated	Irrigated
0–16	76	89
16–32	30	37
32–48	21	27
48–64	14	16

Calculated from Brown, et al. (1985).

TABLE 5.7 Percentage of Root Mass of Three Crops and Two Trees Found in the Upper 30 cm Compared with Deeper Depths (30–180 cm)

Plant species	Percentage of roots	
	Upper 30 cm	30–180 cm
Soybeans	71	29
Corn	64	36
Sorghum	86	14
Pinus radiata	82	18
Eucalyptus marginata	86	14

Data for crops from Mayaki, et al. (1976); for trees, estimated from Bowen (1985).

water and nutrients into the plants. While water vapor can still be absorbed by the plant, its rate of absorption is too low to keep any but the most drought-tolerant plants alive.

5.11 CONCLUSION

Water impacts all life. The interactions and movement of this simple compound in soils help determine whether these impacts are positive or negative. An understanding of the principles that govern the attraction of water for soil solids and for dissolved ions can help maximize the positive impacts while minimizing the less-desirable ones.

The water molecule has a polar structure that results in electrostatic attraction of water to both soluble cations and soil solids. These attractive forces tend to reduce the potential energy level of soil water below that of pure water. The extent of this reduction, called soil water potential ψ, has a profound influence on a number of soil properties, but especially on the movement of soil water and its uptake by plants.

The water potential due to the attraction between soil solids and water (the matric potential ψ_m) combines with the force of gravity ψ_g to largely control water movement. This movement is relatively rapid in soils high in moisture and with an abundance of macropores. In drier soils, however, the adsorption of water on the soil solids is so strong that its movement in the soil and its uptake by plants are greatly reduced. As a

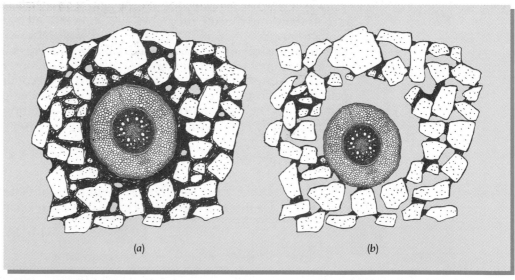

(a)   (b)

FIGURE 5.38   Cross-section of a root surrounded by soil. (*a*) During periods of adequate moisture and low plant moisture stress the root completely fills the soil pores and is in close contact with the soil water films. (*b*) When the plant is under severe moisture stress, such as during hot, dry weather, the root shrinks (mainly in the cortical cells), significantly reducing root–soil contact. Such root shrinkage can occur on a hot day even if soil water content is high.

consequence, plants die for lack of water—even though there are still significant quantities of water in the soil—because that water is unavailable to plants.

Water is supplied to plants by capillary movement toward the root surfaces and by growth of the roots into moist soil areas. In addition, vapor movement may be of significance in supplying water for drought-resistant desert species (xerophytes). The osmotic potential ψ_o becomes important in soils with high soluble salt levels that can impede plant uptake of water from the soil. Such conditions occur most often in soils with restricted drainage in areas of low rainfall and in potted indoor plants.

The characteristics and behavior of soil water are very complex. As we have gained more knowledge, however, it has become apparent that soil water is governed by relatively simple, basic physical principles. Furthermore, researchers are discovering the similarity between these principles and those governing the movement of groundwater and the uptake and use of soil moisture by plants—the subject of the next chapter.

STUDY QUESTIONS

1. What is the role of the *reference state of water* in defining soil water potential? Describe the properties of this reference state of water.

2. Imagine a root of a cotton plant growing in the upper horizon of an irrigated soil in California's Imperial Valley. As the root attempts to draw water molecules from this soil, what forces (potentials) must it overcome? If this soil were compacted by a heavy vehicle, which of these forces would be most affected? Explain.

3. Using the terms *adhesion, cohesion, meniscus, surface tension, atmospheric pressure,* and *hydrophilic surface,* write a brief essay to explain why water rises up from the water table in a mineral soil.

4. Suppose you were hired to design an automatic irrigating system for a wealthy homeowner's garden. You determine that the flower beds should be kept at a water potential above –60 kPa, but not wetter than –10 kPa as the annual flowers here are sensitive to both drought and lack of good aeration. The rough turf areas, however, can do well if the soil dries to as low as –300 kPa. Your budget allows either tensiometers or electrical resistance blocks to be hooked up to electronic switching valves. Which instruments would you use and where? Explain.

5. Suppose the homeowner referred to in question 4 increased your budget and asked to use the TDR method to measure soil water contents. What additional information about the soils, not necessary for using the tensiometer, would you have to obtain to use either the resistance blocks or the TDR instrument? Explain.

6. A greenhouse operator was growing ornamental woody plants in 15-cm-tall plastic containers filled with a loamy sand. He watered the containers daily with a sprinkler system. His first batch of 1000 plants yellowed and died from too much water and not enough air. As an employee of the greenhouse, you suggest that he use 30-cm-tall pots for the next batch of plants. Explain your reasoning.

7. Suppose you measured the following data for a soil. A horizon from 0 to 30 cm with bulk density = 1.2 Mg/m^3: θ_m = 28% at –10 kPa, 20% at –100 kPa, and 8% at –1500 kPa. Bt horizon 30 to 70 cm with bulk density = 1.4 Mg/m^3: θ_m = 30% at –10 kPa, 25% at –100 kPa, and 15% at –1500 kPa. Bx horizon from 70 to 120 cm with bulk density = 1.95 Mg/m^3: θ_m = 20% at –10 kPa, 15% at –100 kPa, and 5% at –1500 kPa. Estimate the total available water holding capacity (AWC) of the soil in question.

8. A forester obtained a cylindrical core (L = 15 cm, r = 3.25 cm) of soil from a field site. She placed all the soil in a metal can with a tight-fitting lid. The empty metal can weighed 300 g and when filled with the field-moist soil weighed 972 g. Back in the lab, she placed the can of soil, with lid removed, in an oven for several days until it ceased to lose weight. The weight of the dried can with soil (including the lid) was 870 g. Calculate both θ_m and θ.

9. Give four reasons why compacting a soil is likely to reduce the amount of water available to growing plants.

10. Since even rapidly growing, finely branched root systems rarely contact more than 1 or 2% of the soil particle surfaces, how is it that the roots can utilize much more than 1 or 2% of the water held on these surfaces?

REFERENCES

Bowen, G. D. 1985. "Roots as components of tree productivity," in M. G. R. Cannell and J. E. Jackson (eds.), *Attributes of Trees as Crop Plants*. (Midlothian, Scotland: Institute of Terrestrial Ecology).

Brown, E. A., C. E. Caviness, and D. A. Brown. 1985. "Response of selected soybean cultivars to soil moisture deficit," *Agron. J.*, **77**:274–278.

Cooney, J. J., and J. E. Peterson. 1955. *Avocado Irrigation*. Leaflet 50. California Agricultural Extension Service.

Da Silva, A. P., and B. D. Kay. 1997. "Estimating the least limiting water range of soil from properties and management," *Soil Sci. Soc. Amer. J.*, **61**:877–883.

DeMartinis, J. M., and S. C. Cooper. 1994. "Natural and man-made modes of entry in agronomic areas," in R. Honeycutt and D. Schabacker (eds.), *Mechanisms of Pesticide Movement in Ground Water*. (Boca Raton, Fla.: Lewis Publishers), pp. 165–175.

Gardner, W., and J. A. Widtsoe. 1921. "The movement of soil moisture," *Soil Sci.*, **11**:230.

Hamza, M., and L. A. G. Aylmore. 1992. "Soil solute concentrations and water uptake by single lupin and radish plant roots: 1. Water extraction and solute accumulation," *Plant Soil,* **145**:187–196.

Huck, M. G., B. Klepper, and H. M. Taylor. 1970. "Diurnal variations in root diameter," *Plant Physiol.*, **45**:529.

Hudson, B. D. 1994. "Soil organic matter and available water capacity," *J. Soil and Water Cons.*, **49**:189–194.

MacFall, J. S., and G. A. Johnson. 1994. "Use of magnetic resonance imaging of plants and soils," in S. H. Anderson and J. W. Hopmans (eds.), *Tomography of Soil–Water–Root Processes*. SSSA Special Publication no. 36. (Madison, Wis.: Soil Sci. Soc. Amer.), pp. 99–113.

Mayaki, W. C., L. R. Stone, and I. D. Teare. 1976. "Irrigated and nonirrigated soybean, corn, and grain sorghum root systems," *Agron. J.*, **68**:532–534.

Richards, L. A. 1965. "Physical condition of water in soil," in Agronomy 9: *Methods of Soil Analysis, Part 1*. (Madison, Wis.: American Society of Agronomy).

Waddell, J. T. and Weil, R. R. 1996. "Water distribution in soil under ridge-till and no-till corn," *Soil Sci. Soc. Amer. J.*, **60**:230–237.

6

SOIL AND THE HYDROLOGIC CYCLE

Both soil and water belong to the biosphere, to the order of nature, and—as one species among many, as one generation among many yet to come—we have no right to destroy them.
—DANIEL HILLEL, OUT OF EARTH

One of the most striking—and troubling—features of human society is the yawning gap in wealth between the world's rich and poor. The life experience of the one group is quite incomprehensible to the other. So it is with the distribution of the world's water resources. The rain forests of the Amazon and Congo basins are drenched by more than 2000 mm of rain each year, while the deserts of North Africa and central Asia get by with less than 200. Nor is the supply of water distributed evenly throughout the year. Rather, periods of high rainfall and flooding alternate with dry spells or periods of drought.

Yet, one could say that everywhere the supply of water is adequate to meet the needs of the plants and animals native to the natural communities of the area. Of course, this is so only because the plants and animals have adapted to the local availability of water. Early human populations, too, adapted to the local water supplies by settling where water was plentiful from rain or rivers, by learning which underground gourds and plant stems could quench one's thirst, by developing techniques to harvest water for agriculture and store it in underground cisterns, by adopting nomadic lifestyles that allowed them and their herds to follow the rains and the grass supply, and, sometimes, by picking up and moving when water supplies gave out.

But "civilized" humans have seemingly lost the ability and desire to adapt their needs and cultures to their environment. Rather, it is a mark of civilization that humans join in organized efforts bent on adapting their environment to their desires. Hence, the ancients tamed the flows of the Tigris and Euphrates. We moderns dig wells in the Sahel, bottle up the mighty Nile at Aswan, channel the waters of the Colorado to the chaparral region of southern California, pump out the aquifers under farms and suburbs, and create sprawling cities (with swimming pools and bluegrass lawns!) in the deserts of the American Southwest or on the sands of Arabia. Truly, cities like Las Vegas are gambling in more ways than one.

If the 750 mm of water that, on average, falls annually on the earth's land areas were evenly distributed and properly managed, there would be enough water to support a world human population several times the 6 billion living today. However, as we have just seen, water resources are *not* evenly distributed in space and time, and rarely do we humans manage them with either efficiency or wisdom. South America and the Caribbean receive nearly one-third of the annual global precipitation, Australia only 1%. Africa's total precipitation exceeds that of Europe, but most of it falls in the Congo River basin in west central Africa, leaving the countries of the north, east, and south with water supplies grossly inadequate to support their burgeoning populations. In the Middle East, water, as much as religion, is at the root of much political strife.

Yet, most farmers in arid regions still lose 7 out of 10 L of water allocated for irrigation, farmers in semiarid regions still allow nearly half their precious rain to slip away as surface runoff, and most city dwellers still flush away 20 L of water to transport every few grams of human waste.

There is plenty of room for improvement in managing water resources, and many of the improvements are likely to come as a result of better management of soils. The soil plays a central role in the cycling and use of water. For instance, by serving as a massive reservoir, soil helps moderate the adverse effects of excesses and deficiencies of water. It takes in water during times of surplus and then releases it in due time, either to satisfy the transpiration requirements of plants or to replenish groundwater. Stable structure at the surface of the soil ensures that a large fraction of the precipitation received will move slowly into the groundwater and from there to nearby streams or to deeper reservoirs under the earth. The soil can help us treat and reuse waste waters from animal, domestic, and industrial sources. The flow of water through the soil in these and other circumstances connects the chemical pollution of soils to the possible contamination of groundwater.

In Chapter 5 we considered principles governing the nature and movement of water in soils. In this chapter we will see how those principles apply to the cycling of water between the soil, the atmosphere, and vegetation. We will then examine the unique role of the soil in water-resource management and will determine how well-managed soils can help make this cycle most useful to every living creature.

6.1 THE GLOBAL HYDROLOGIC CYCLE

Global Stocks of Water

There are nearly 1400 million km^3 of water on the earth—enough (if it were all aboveground at a uniform depth) to cover the earth's surface to a depth of some 3 km. Most of this water, however, is relatively inaccessible and is not active in the annual cycling of water that supplies rivers, lakes, and living things. More than 97% is found in oceans (Figure 6.1) where the water is not only salty, but has an average residence time of several thousand years. Only the near-surface ocean layers take part in annual water cycling. An additional 2% of the water is in glaciers and ice caps of mountains with similarly long residence times (about 10,000 years). Some 0.7% is found in groundwater, most of which is more than 750 m underground. Except where it is tapped for pumping by humans, it, too, has a long average residence time of several hundred years or more.

The more actively cycling water is in the surface layer of the oceans, in shallow groundwater, in lakes and rivers, in the atmosphere, and in the soil (see Figure 6.1). Although the combined volume is a tiny fraction of the water on Earth, these pools of water are accessible for movement in and out of the atmosphere and from one place on the Earth's surface area to another. The average residence time for water in the atmosphere is about 10 days, that for the longest rivers is 20 days or less, and for soil moisture is about one month. While water in large lakes (Lake Baikal in Siberia, Lake Tanganyika in Africa, and the five Great Lakes in North America) that accounts for

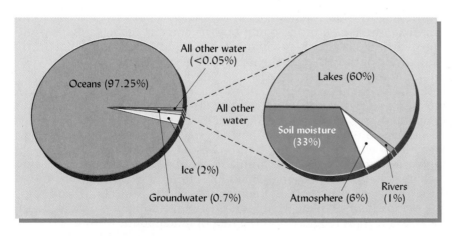

FIGURE 6.1 The sources of the earth's water. The preponderance of water is found in the oceans, glaciers and ice caps, and deep groundwater (left) but most of these waters are inaccessible for rapid exchange with the atmosphere and the land. The sources on the right, though much smaller in quantity, are actively involved in water movement through the hydrologic cycle. (Data from several sources)

about half of all lake water has longer residence times (90 to 200 years), the smaller lakes and reservoirs have much shorter periods of turnover. The accessibility of water in these latter sources makes them primary participants in the global water cycle, which will now receive our attention.

The Hydrologic Cycle

Solar energy drives the cycling of water from the earth's surface to the atmosphere and back again in what is termed the **hydrologic cycle** (Figure 6.2). About one-third of the solar energy that reaches the earth is absorbed by water on or near the earth's surface, stimulating *evaporation*—the conversion of liquid water into water vapor. The water vapor moves up into the atmosphere, eventually forming clouds that can move from one region of the globe to another. Within an average of about 10 days, pressure and temperature differences in the atmosphere cause the water vapor to condense into liquid droplets or solid particles, which return to the earth as rain or snow.

As Figure 6.2 illustrates, about 500,000 km^3 of water are evaporated from the earth surfaces and vegetation each year. More than 85% (430,000 km^3) of this water comes from the oceans, the remainder from the lakes, rivers, land, and vegetation on the continents. There is also a net migration of about 40,000 km^3 of water in the clouds above the oceans to the land areas.

The bulk of the precipitation also occurs over oceans, some 390,000 km^3 or 78% of the total falling there each year. About 110,000 km^3 of water falls as rain or snowfall on the continents. Some of the water falling on land moves over the surface of the soil (surface runoff) and some infiltrates into the soil and drains into the groundwater. Both the surface runoff and groundwater seepage enter streams and rivers that, in turn, flow downstream into the oceans. The volume of water returned in this way is about 40,000 km^3, which balances the same quantity of water that is transferred annually in clouds from the oceans to the continents.

Water Balance Equation

It is often useful to consider the components of the hydrologic cycle as they apply to a given **watershed,** an area of land drained by a single system of streams and bounded by

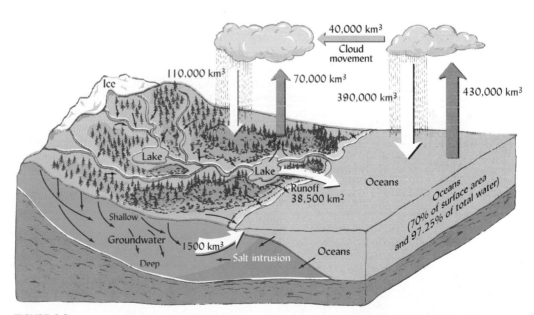

FIGURE 6.2   The hydrologic cycle upon which all life depends is very simple in principle. Water evaporates from the earth's surface, both the oceans and continents, and returns in the form of rain or snowfall. The net movement of clouds brings some 40,000 km^3 of water to the continents and an equal amount of water is returned through runoff and groundwater seepage that is channeled through rivers to the ocean. About 86% of the evaporation and 78% of the precipitation occurs in the ocean areas. However, the processes occurring on land areas where the soils are influential have impacts not only on humans but on all other forms of life, including those residing in the sea.

ridges that separate it from adjacent watersheds. All the precipitation falling on a watershed is either stored in the soil, returned to the atmosphere (see Section 6.2), or discharged from the watershed as surface or subsurface flow (runoff). Water is returned to the atmosphere either by **evaporation** from the land surface (vaporization of soil water) or, after plant uptake and use, by vaporization from the stomata on the surfaces of leaves (a process termed **transpiration**). Together, these two pathways of evaporative loss to the atmosphere are called **evapotranspiration.**

The disposition of water in a watershed is often expressed by the **water-balance equation**, which in its simplest form is:

$$P = ET + SS + D$$

where P = precipitation, ET = evapotranspiration, SS = soil storage, and D = discharge.

For a forested watershed (sometimes termed a *catchment*), management may aim to maximize D so as to provide more water to downstream users. The equation makes it clear that discharge can be increased only if ET and/or SS are decreased, changes that may or may not be desirable. In the case of an irrigated field, water applied in irrigation would be included on the left side of the equation. Irrigation managers may want to save water applied by minimizing unnecessary losses in D and allowing negative values for soil storage (withdrawals of soil water) during parts of the year.

6.2 FATE OF PRECIPITATION AND IRRIGATION WATER

Water supplied to soils by rain, snowfall, and irrigation moves by a number of pathways. In areas covered with vegetation, some of the precipitation is intercepted by plant foliage (Table 6.1) and returned to the atmosphere by evaporation without ever reaching the soil. In some forested areas, **interception** may prevent 30 to 50% of the precipitation from reaching the soil. You may have personally experienced the interception capacity of tree foliage by seeking shelter under a tree during a light rain (with no chance of lightning!). The interception and subsequent sublimation (vaporization directly from the solid state) of snow is especially important in coniferous forests. Even in agricultural areas, the quantity of plant interception is significant. Although intercepted water may temporarily reduce transpiration from leaves, it does not contribute to the supply of water that supports plant growth.

Most of the water that does reach the soil penetrates downward by the process of **infiltration,** especially if the soil surface structure is loose and open. If the rate of rainfall or snowmelt exceeds the infiltration capacity of the soil, the excess water unable to penetrate will begin to pond on the soil surface. In relatively level areas, large quantities of water may remain ponded in depressions, forming temporary wetland conditions that may benefit wildfowl or foul agriculture, as the case may be (Figure 6.3).

In sloping areas, especially if the soils are not loose and open, considerable runoff and erosion may take place, thereby reducing the proportion of the water that moves

TABLE 6.1   **Percentage Interception of Precipitation by Several Crop and Tree Species at Different Locations in the United States**

Note the high interception for the forest species and close-growing crops such as alfalfa.

Species	Location	Percent of precipitation intercepted by plant
Alfalfa	Mo.	22
Corn	Mo.	7
Soybeans	N.J.	15
Ponderosa pine	Idaho	22
Douglas fir	Wash.	34
Maple, beech	N.Y.	43

Crop data from Haynes (1954); forest data from Kittridge (1948).

FIGURE 6.3 A small depressional wetland in North Dakota called a *prairie pothole*. When rain falls faster than the soil infiltration capacity, water will begin to pond on the soil surface and then run off into the depressions. In some landscapes with relatively low infiltration capacities, water may stand in depressions for several months during the year. Farmers who cultivate these landscapes with large machinery may view these wetlands as nuisances in need of drainage. However, small and temporary though they are, wetlands like these scattered across the northern Great Plains provide nesting sites for about half of the ducks in North America. (Photo courtesy of R. Weil)

into the soil and contributes to plant growth. In extreme cases, more than 50% of the precipitation may be lost in this manner. While some of this loss may be desirable to prevent excessive water saturation of the soil, most of it is undesirable, especially if this **surface runoff** carries with it considerable amounts of dissolved chemicals and detached soil particles (**sediment**; see Chapter 17).

Once the water penetrates the soil, some of it is subject to downward percolation and eventual loss from the root zone by **drainage.** In humid areas and in some irrigated areas of arid and semiarid regions, up to 50% of the precipitation may be lost as drainage below the root zone. However, during subsequent periods of low rainfall, some of this water may move back up into the plant-root zone by **capillary rise.** Such movement is important to plants in areas with deep soils, especially in dry climates.

The water retained by the soil is referred to as **soil storage** water, some of which eventually moves upward by capillarity and is lost by evaporation from the soil surface. Much of the remainder is absorbed by plants and then moves through the roots and stems to the leaves, where it is lost by transpiration. The water thus lost to the atmosphere by evapotranspiration may later return to the soil as precipitation or irrigation water, and the cycle starts again.

Effects of Precipitation

The amount of water moving through each of the channels just discussed is influenced greatly by the timing, rate, and form of the precipitation. Heavy rainfall, even if of short duration, can supply water faster than most soils can absorb it. This accounts for the fact that in some arid regions with only 200 mm of annual rainfall, a cloudburst that brings 20 to 50 mm of water in a few minutes can result in flash flooding and gully erosion. A larger amount of precipitation spread over several days of gentle rain could move slowly into the soil, thereby increasing the stored water available for plant absorption, as well as replenishing the underlying groundwater.

As illustrated in Figure 6.4, the timing of snowfall in early winter can affect the partitioning of spring snow meltwater between surface runoff and infiltration. A blanket of snow is good insulation. Therefore, if the soil becomes frozen *before* the first heavy

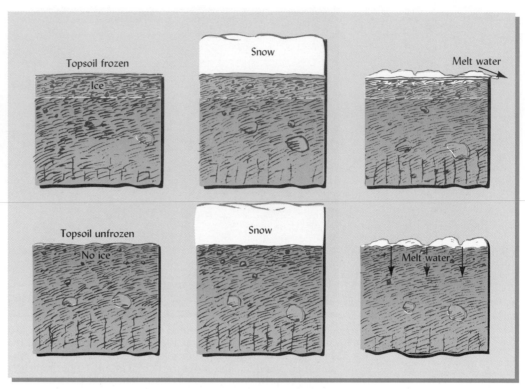

FIGURE 6.4   The relative timing of freezing temperatures and snowfall in the fall in some temperate regions drastically influences water runoff and infiltration into soils in the spring. The upper three diagrams illustrate what happens when the surface soil freezes before the first heavy snowfall. The snow insulates the soil so that it is still frozen and impermeable as the snow melts in the spring. The lower sequence of diagrams illustrates the situation when the soil is unfrozen in the fall when it is covered by the first deep snowfall.

snow, the subsequent snow will keep it frozen, even when warm spring air temperatures begin to melt the snow. In this situation, the snow meltwater is unable to penetrate the underlying frozen soil and runs off. On the other hand, if a soil becomes blanketed by snow while still unfrozen, the insulating snow will keep the soil from freezing during the winter. In this case, when the snow melts in spring the meltwater can easily infiltrate the unfrozen soil, and little surface runoff occurs.

Effects of Vegetation and Soil Properties on Infiltration

TYPE OF VEGETATION.   In addition to intercepting rain and snowfall before they reach the soil, plants help determine the proportion of water that runs off and that which penetrates the soil. The vegetation and surface residues of perennial grasslands and dense forests protect the porous soil structure from the beating action of raindrops. Therefore, they encourage water infiltration and reduce the likelihood that soil will be carried off by any runoff that does occur. In general, very little runoff occurs from land under undisturbed forests or well-managed turf grass. However, as Figure 6.5 indicates, differences in plant species, even among grasses, can influence runoff.

STEM FLOW.   Many plant canopies direct rainfall toward the plant stem, thus altering the spatial distribution of rain reaching the soil. Under a forest canopy, more than half of the rainfall may trickle down the leaves, twigs, and branches to the tree trunk, there to progress downward as **stem flow**. Likewise, certain crop canopies, such as that of corn, funnel a large proportion of the rainfall to the soil in the crop row (Figure 6.6). The concentration of water in limited zones around plant stems increases the opportunity for saturated flow to occur. If fertilizers are placed near the plant row, this nonuniform infiltration may increase the chances of leaching soluble nutrients below the root zone. Stem flow must be considered in studying the hydrology and nutrient cycling in many plant ecosystems.

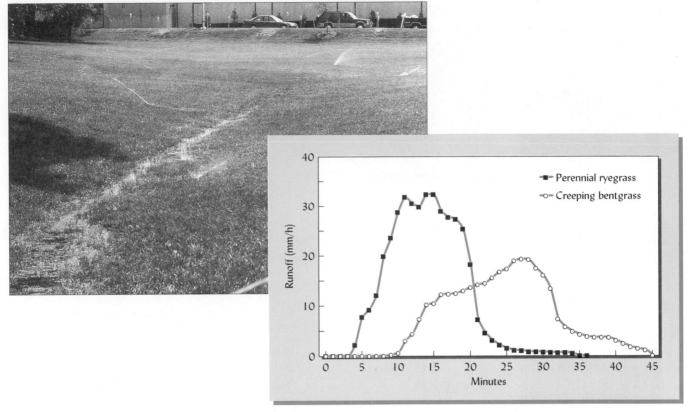

FIGURE 6.5 Little water generally runs off turf grass, except during very intense rainfall, where the soils have been compacted, or where irrigation water is applied unevenly or at too high a rate (as shown in the photo). The graph shows runoff from two golf fairway grasses following irrigation at the rate of 150 mm per hour. Note that the runoff peak was much lower on the creeping bent grass, which was characterized by a dense thatch of plant stems and many biopores near the soil surface. A lower rate of irrigation on either type of turf grass could have eliminated the runoff, which represents a waste of water and a potential for soluble lawn chemicals be carried to streams and rivers. [Data from Linde, et al. (1995), used with permission of the American Society of Agronomy; photo courtesy of R. Weil]

SOIL MANAGEMENT. A major objective of soil- and water-management systems, especially in semiarid and subhumid regions, is to encourage water infiltration rather than runoff. This may be achieved by enhancing soil surface storage to allow more time for infiltration to take place (Figure 6.7a). Another approach is to establish dense vegetation during periods of high rainfall. This may include the use of **cover crops**, plants established between the principle crop-growing seasons (Figure 6.7b). Such cover crops encourage the activities of earthworms and protect the soil structure, leading to greatly enhanced water infiltration. However, remember that cover crops also transpire water. If the following crop will be dependent on soil storage water, care may be needed to kill the cover crop before it can dry out the soil profile.

Bulldozers and other heavy equipment used to clear forests can sharply reduce water infiltration rates. Figure 6.8 shows the results of one comparison between bulldozing and traditional hand methods of clearing tropical forest lands. The traditional methods did little to disturb the soil, so the infiltration rate remained essentially the same as before clearing. Bulldozing, however, destroyed the forest floor and compacted the soil, dramatically reducing infiltration. While other research suggests that the ill effects of the compaction can be partially overcome by deep tillage, these results caution against indiscriminate use of heavy equipment to clear land, especially in the humid tropics.

SOIL POROSITY. Soil properties will also affect the fate of precipitation. If the soil is loose and open (e.g., sands and well-granulated soils), a high proportion of the incoming water will infiltrate the soil, and relatively little will run off. In contrast, heavy clay soils with unstable soil structures resist infiltration and encourage runoff. These differences, attributable to soil properties as well as vegetation, are illustrated in Figure 6.9.

Having considered the overall hydrologic cycle, we will now focus our attention on the component of the cycle for which soils and plants play the most prominent roles.

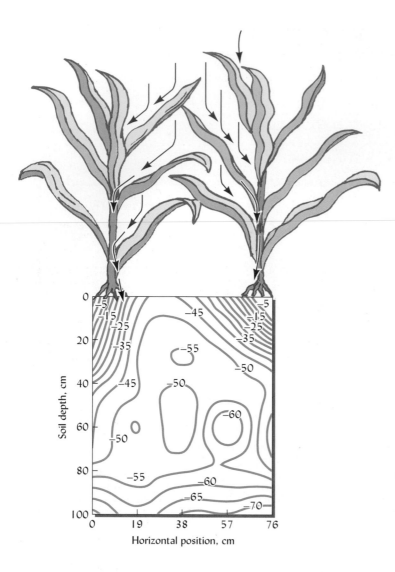

FIGURE 6.6 Vertical and horizontal distribution of soil water resulting from stem flow. The contours indicate the soil water potential in kPa between two corn rows in a sandy loam soil. During the previous two days, 26 mm of rain fell on this field. Many plant canopies, including that of the corn crop shown, direct a large proportion of rainfall toward the plant stem. Stem flow results in uneven spatial distribution of water. In cropland this may have ramifications for the leaching of chemicals such as fertilizers, depending on whether they are applied in or away from the zone of highest wetting near the plant stems. The concentration of water by stem flow may also increase the likelihood of macropore flow in soils after only moderate rainfall. [Data from Waddell and Weil (1996); used by permission of the Soil Science Society of America]

6.3   THE SOIL–PLANT–ATMOSPHERE CONTINUUM

The flow of water through the soil–plant–atmosphere continuum (SPAC) is a major component of the overall hydrologic cycle. Figure 6.10 ties together many of the processes we have just discussed: *interception, surface runoff, percolation, drainage, evaporation, plant water uptake, water movement to plant leaves,* and *transpiration* of water from the leaves back into the atmosphere.

In studying the SPAC, scientists have discovered that the same basic principles govern the retention and movement of water whether it is in soil, in plants, or in the atmosphere. In Chapter 5, the potential energy level of water was seen to be a major factor determining water behavior in soils. The same can be said for water movement between the soil and the plant root and between the plant and the atmosphere (see Figure 6.10).

If a plant is to absorb water from the soil, the water potential must be lower in the plant root than in the soil adjacent to the root. Likewise, movement up the stem to the leaf cells is in response to differences in water potential, as is the movement from leaf surfaces to the atmosphere. To illustrate the movement of water to sites of lower and lower water potential, Figure 6.10 shows that the water potential drops from –50 kPa in the soil, to –70 in the root, to –500 kPa at the leaf surfaces, and, finally, to –20,000 kPa in the atmosphere.

Two Points of Resistance

Changes in water potential illustrated in Figure 6.11 suggest major resistance at two points as the water moves through the SPAC: the root–soil water interface and the leaf

(a)

(b)

FIGURE 6.7   Managing soils to increase infiltration of rainwater. (a) Small furrow dikes on the right side of this field in Texas retain rainwater long enough for it to infiltrate rather than runoff. (b) This saturated soil under a winter cover crop of hairy vetch is riddled with earthworm burrows that greatly increased the infiltration of water from a recent heavy rain. Scale in centimeters. [Photo (a) courtesy of O. R. Jones, USDA Agricultural Research Service, Bushland, Texas; photo (b) courtesy of R. Weil]

cell–atmosphere interface. This means that two primary factors determine whether plants are well supplied with water: (1) the rate at which water is supplied by the soil to the absorbing roots, and (2) the rate at which water is evaporated from the plant leaves. Since factors affecting the soil's ability to supply water were discussed in Section 5.9, we will now address the loss of water by evaporation from soil–plant systems.

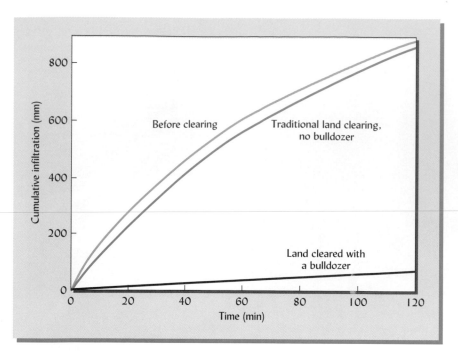

FIGURE 6.8 The effect of traditional and mechanical means of land clearing on the cumulative infiltration rate into an Ultisol located in the Amazon region of Peru. The measurements were made 14 weeks after land clearing. Apparently, the surface soil disturbance and compaction by the bull-dozer reduced the quantity and size of pores, thereby drastically reducing the infil-tration rate. The traditional land clearing was by hand, leaving the soil uncompacted. [Redrawn from Alegre, et al. (1986); used with permission of the Soil Science Society of America]

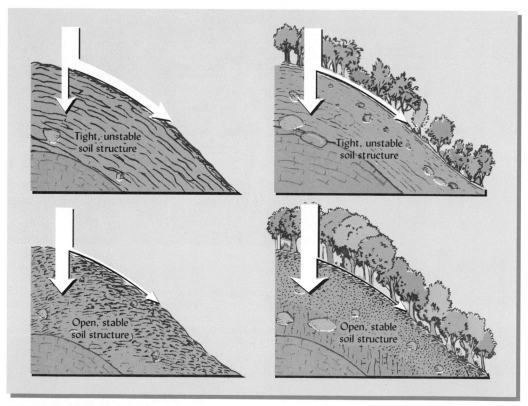

FIGURE 6.9 Influence of soil structure and vegetation on the partitioning of rainfall into infiltration and runoff. The upper two diagrams show soils with tight, unstable structure that resists infiltration and percolation. The bare soil is especially prone to surface sealing and resulting high losses by runoff. Even with forest cover, the low-permeability soils cannot accept all the rain in an intense storm. The two lower diagrams show much greater infiltration into soils that have open, stable structures with signifi-cant macropore space. The more open soil structure combined with the protective effects of the forest floor and canopy nearly eliminate surface runoff.

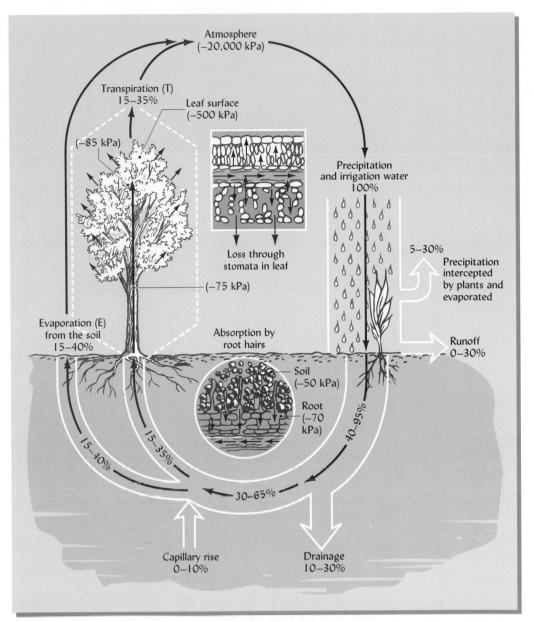

FIGURE 6.10   Soil–plant–atmosphere continuum (SPAC) showing water movement from soil to plants to the atmosphere and back to the soil in a humid to subhumid region. Water behavior through the continuum is subject to the same energy relations covering soil water that were discussed in Chapter 5. Note that the moisture potential in the soil is –50 kPa, dropping to –70 kPa in the root, declining still further as it moves upward in the stem and into the leaf, and is very low (–500 kPa) at the leaf–atmosphere interface, from whence it moves into the atmosphere where the moisture potential is –20,000 kPa. Moisture moves from a higher to lower moisture potential. Note the suggested ranges for partitioning of the precipitation and irrigation water as it moves through the continuum.

Evapotranspiration

While it is relatively easy to measure the total change in soil water content due to vapor losses, it is quite difficult to determine just how much of that loss occurred directly from the soil (by **evaporation**, E) and how much occurred from the leaf surfaces after plant uptake (by **transpiration**, T). Therefore, information is most commonly available on **evapotranspiration** (ET), the combined loss resulting from these two processes. The evaporation component of ET may be viewed as a "waste" of water from the standpoint of plant productivity. However, at least some of the transpiration component is essential for plant growth, providing the water that plants need for cooling, nutrient transport, photosynthesis and turgor maintenance.

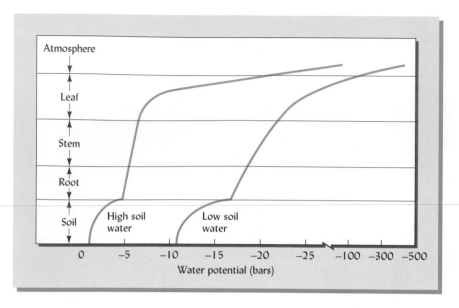

FIGURE 6.11 Change in water potential as water moves from the soil through the root, stem, and leaf to the atmosphere. Note that water potential decreases as the water moves through the system. [Adapted from Hillel (1980)]

The **potential evapotranspiration** rate (PET) tells us how fast water vapor *would* be lost from a densely vegetated plant–soil system *if* soil water content were continuously maintained at an optimal level. The PET is largely determined by climatic variables, such as temperature, relative humidity, cloud cover, and wind speed, that influence the *vapor pressure gradient* between a wet soil, leaf, or body of water and the atmosphere.

A number of mathematical models have been devised to estimate PET from climatological data, but in practice, PET can be most easily estimated by applying a correction factor to the amount of water evaporated from an open pan of water of standard design (a class-A evaporation pan—see Figure 6.12). Dense, well-watered vegetation typically transpires water about 65% as rapidly as water evaporates from an open pan; hence, the correction factor for dense vegetation such as a lawn is typically 0.65 (it is lower for less-dense vegetation):

$$PET = 0.65 * pan\ evaporation$$

Values of PET range from more than 1500 mm per year in hot, arid areas, to less than 40 mm in very cold regions. During the winter in temperate regions, PET may be less than 1 mm per day. By contrast, hot, dry wind will continually sweep away water vapor from a wet surface, creating a particularly steep vapor pressure gradient and PET levels as high as 10 to 12 mm per day. For this reason, farmers in both the North American Great Plains and the North African Sahel dread the hot winds characteristic of those regions.

Effect of Soil Moisture Supply on ET

Evaporation from the soil surface E at a given temperature is determined to a large extent by soil surface wetness and by the ability of the soil to replenish this surface water as it evaporates. In most cases, the upper 15 to 25 cm of soil provides most of the water for surface evaporation E. Unless a shallow water table exists, the upward capillary movement of water is very limited and the surface soil soon dries out, greatly reducing further evaporation loss.

However, because plant roots penetrate deep into the profile, a significant portion of the water lost by evapotranspiration ET comes from the subsoil layers (see Figure 6.13). Water stored deep in the profile is especially important to vegetation in regions having alternating moist and dry seasons (such as Ustic or Xeric moisture regimes). Water stored in the subsoil during rainy periods is available for evapotranspiration during dry periods.

FIGURE 6.12 A class-A evaporation pan used to help estimate potential evapotranspiration (PET). Once a day, the water level is determined in the stilling well (small cylinder) and a measured amount of water is added to bring the level back up to the original mark. Evaporation from the pan integrates the effects of relative humidity, temperature, wind speed, and other climatic variables related to the vapor pressure gradient. Also shown are an anemometer to measure wind speed and a shelter containing instruments to measure temperature and relative humidity. (Photo courtesy of R. Weil)

The photo labels read: Instrument shelter, Anemometer, Stilling well, Class A pan.

Water Deficit and Plant Water Stress

For dense vegetation growing in soil well supplied with water, ET will nearly equal PET. When soil water content is less than optimal, the plant will not be able to withdraw water from the soil fast enough to satisfy the PET. If water evaporates from the leaves faster than it enters the roots, the plant will lose turgor pressure and wilt. Under these conditions, actual evapotranspiration is less than potential evapotranspiration and the plant experiences **water stress**. The difference between PET and actual ET is termed the **water deficit**. A large deficit is indicative of high water stress and aridity.

Under water-stress conditions, plants first close the **stomata** (openings) on their leaf surfaces to reduce the vapor loss of water and prevent wilting. However, the closure of the stomata has two detrimental side effects: (1) *plant growth* is arrested because insufficient CO_2 for photosynthesis can get in through the closed stomata and (2) the reduction in evaporative cooling results in detrimental *heating* of leaves as they continue to absorb solar radiation. The latter effect allows infrared-detecting instruments to estimate water stress in plants by sensing the increase in leaf temperature over air temperature.

Influence of Radiant Energy PET

Solar radiant energy provides the 2260 joules (540 calories) needed to evaporate each gram of water, whether from the soil (E) or from leaf surfaces (T). Direct sunlight stimulates the greatest evaporation. On cloudy days, less solar radiation strikes soil and

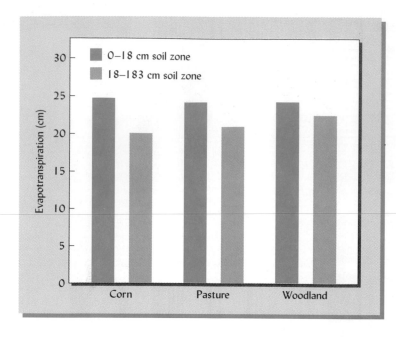

FIGURE 6.13 Evapotranspiration loss from the surface layer (0 to 18 cm) compared with loss from the subsoil (18 to 183 cm). Note that more than half the water loss came from the upper 18 cm and only half from the *next 165 cm* of depth. Periods of measurement: corn, May 23 to September 25; pasture, April 15 to August 23; woodland, May 25 to September 28. [Calculated from Dreibelbis and Amerman (1965)]

plant surfaces; hence, the evaporative potential is not as great. Sunlight striking the land surface at a low angle spreads its energy over a larger area, and therefore stimulates less evaporation per unit land area than more perpendicular solar radiation. Hence, evaporation is relatively low in winter and on slopes that face away from the sun (north-facing slopes in the Northern Hemisphere—see Figure 7.22).

Influence of Plant Canopy Development on ET

In vegetated areas, incoming solar radiation is either absorbed by plant leaves on its way through the plant canopy, or it reaches the ground and is absorbed by the soil. Therefore, as the leaf area per unit land area (a ratio termed the **leaf area index**) increases, more radiation will be absorbed by the foliage to stimulate transpiration (T), and less will reach the soil to promote evaporation (E).

For a monoculture of annual plants, the leaf area index typically varies from 0 at planting to a peak of perhaps 3 to 5 at flowering, then declines as the plant senesces, and finally drops back to 0 when the plant is removed at harvest (assuming no weeds are allowed to grow). Transpiration takes place during a short (3- to 5-month) growing period, whereas evaporation from the soil surface continues to occur as long as the soil is unfrozen (Figure 6.14). In contrast, perennial vegetation, such as pastures and forests, have very high leaf area indices both early and late in the growing season. Where the forest floor (leaf litter) is undisturbed, even less direct sunlight strikes the soil and E is very low throughout the year.

In summary, water losses from the soil surface and from transpiration are determined by: (1) climatic conditions, (2) plant cover in relation to soil surface (leaf area index), (3) efficiency of water use by different plants, and (4) length and season of the plant growing period.

Influence of Plant Characteristics on ET

Plant characteristics, including rooting depth, length of life cycle, and leaf morphology, can influence the amount of water lost by ET over a growing season. In a natural plant community, plants with different characteristics are likely to fill different ecological niches. Some plant habits that reduce ET as a percentage of PET are dry season dormancy (e.g., bluegrass during hot summer weather), dry season loss of leaves (e.g., most savanna trees), low leaf area index as a result of wide plant spacing (e.g., creosote bush in semiarid rangeland), and the ability to utilize carbon dioxide at night when open stomata will not result in much water loss (e.g., succulent desert plants).

For agricultural crops, the amount of water used is closely related to the growing season of a particular species. Table 6.2 illustrates this point for several crops studied in Cal-

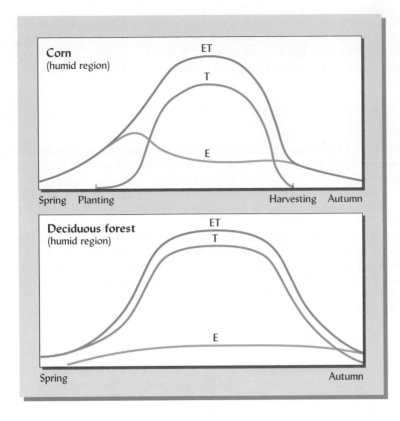

FIGURE 6.14 Relative rates of evaporation from the soil surface E, of transpiration from the plant leaves T, and the combined vapor loss ET for two field situations. (Upper) a field of corn in a humid region. Until the plants are well established, most of the vapor loss is from the soil surface E, but as the plants grow transpiration T soon dominates. The soil surface is shaded and E may actually decrease somewhat since most of the moisture is moving through the plants. As the plants reach maturity, T declines as does the combined ET. For a nearby deciduous forested area (lower) the same general trend is illustrated, except there is relatively less evaporation from the soil surface and a higher proportion of the vapor loss is by transpiration. The soil is shaded by the leaf canopy most of the growing season in the forested area. It should be noted that these figures relate to a representative situation and that the actual field losses would be influenced by rainfall distribution, temperate fluctuations, and soil properties.

ifornia. For example, barley grown for seven months mainly during the moist, cool winter used less water than did a bean crop grown for only three months during the hot, dry summer. Most of the differences in ET were due to differences in the PET of the period during which the crops were grown. However, some crops used only 70% of the PET while others used close to 90%. Such results suggest that soil scientists and plant breeders can work together to determine plant characteristics that might enhance drought resistance and efficiency of water use.

6.4 EFFICIENCY OF WATER USE

The dry matter produced by a given plant while using a given amount of water is an important measure of efficiency, especially in areas where moisture is scarce. This efficiency may be expressed in terms of dry matter yield per unit of water transpired (*T* efficiency) or the dry matter yield per unit of water lost by evapotranspiration (*ET* efficiency).

TABLE 6.2 Evapotranspiration Water Use by Various Crops Grown in Well-Watered Soils

Generally, ET was controlled by the seasonal climate, but differences in the ratio
ET/PET did occur among crops during their growing seasons.
The PET for the entire year at the California site studied was 1316 mm.

Crop	Growing period		ET, mm	PET for period, mm	ET/PET for period
	Planting date	Harvest date			
Barley	Nov. 1	May 31	384	504	0.76
Tomatoes	April 30	Sept. 24	681	866	0.79
Corn	May 15	Sept. 20	640	775	0.83
Beans	June 21	Sept. 24	403	568	0.71
Sugar beet	June 15	March 15	790	896	0.88

Data calculated from Pruitt, et al. (1972).

Another expression of water use efficiency is the **transpiration ratio**, which is the inverse of T efficiency and is expressed as kilograms of water transpired to produce 1 kg of dry matter. The transpiration ratio for a given plant is markedly affected by climatic conditions, ranging from 200 to 500 in humid regions and almost twice as much in arid regions (Table 6.3).

Transpiration Ratio

At a given location there are considerable differences in transpiration ratio among different plant species. For example, the data in Table 6.3 show that in semiarid Akron, Colorado, millet and maize (corn) have relatively low transpiration ratios; that is, they require relatively small quantities of water to produce 1 kg of dry matter. In contrast, some legume forages such as clover use twice as much water per kilogram of dry matter produced.

We can see the significance of these figures by considering a wheat crop growing in a semiarid region and having a transpiration ratio of 500. For every kilogram of aboveground dry matter produced, about 500 kg (or liters) of water were transpired. If we assume that only about 40% of the dry matter produced by the time of harvest is in the grain (the rest is in the straw), we see that it takes about 1250 L of water to produce 1 kg of wheat grain. When we take into account the water used to grow fruits, vegetables, and feed for cattle, as well as grains, no wonder it is estimated that almost 7000 L (1700 gal) of water are required to grow a single day's food supply for one adult in the United States.

Indeed, the amount of water necessary to bring a crop to maturity is very large. For example, a representative crop of wheat containing 5000 kg/ha of dry matter and having a transpiration ratio of 500 will withdraw water from the soil during the growing season equivalent to about 250 mm of rain. This amount of water, *in addition* to that evaporated from the surface, must be supplied during the growing season. It is not surprising that water—and the ability of the soil to store it—is often the most critical factor in plant growth.

ET Efficiency

Since evapotranspiration (ET) includes both transpiration (T) from plants and evaporation (E) from the soil surface, ET efficiency is more subject to management than T efficiency. Highest ET efficiency is attained where plant density and other growth factors are optimum for plant growth. Increases in leaf area index (LAI) lead to increases in plant production, up to the point at which virtually all available solar radiation is absorbed by leaf surfaces. Furthermore, as shown in Figure 6.15, when the LAI is increased over the range of 0 to about 4, plant production increases by a greater amount than does ET. The plant production per unit of water vaporized (ET efficiency) is therefore increased.

TABLE 6.3 **Transpiration Ratios of Various Crops as Determined at Different Locations**

Kilograms of water used in production of 1 kg of dry matter.

Crop	Harpenden, England	Dahme, Germany	Madison, Wisconsin	Pusa, India	Akron, Colorado
Beans	209	282	—	—	736
Clover	269	310	576	—	797
Maize (corn)	—	—	271	337	368
Millet	—	—	—	—	310
Oats	—	376	503	469	597
Peas	259	273	477	563	788
Potatoes	—	—	385	—	636
Wheat	247	338	—	544	513

Data compiled by Lyon, et al. (1952).

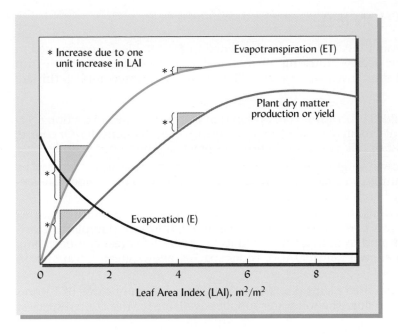

FIGURE 6.15 Generalized effect of vegetative leaf area index (LAI) on plant production, evapotranspiration (ET), and evaporation (E). The stippled areas show the effects of increasing LAI by one unit. Note that ET increases with increasing LAI up to a point and then levels off when the foliage intercepts nearly all the incoming solar radiation. Simultaneously, E decreases and levels off at a very low value. Except at either very low or very high LAI levels, practices (such as fertilization or closer plant spacing) that increase LAI are likely to increase plant productivity by a greater percentage than the corresponding increase in ET, resulting in improved water-use efficiency.

Figure 6.16 illustrates the effect on ET efficiency of adding phosphorus and nitrogen fertilizers to a nutrient-deficient soil. The increased production per unit of water vaporized is the result of the fertilized, more vigorous plants (1) sending their roots deeper into the soil profile, and (2) producing a greater leaf area, which intercepts a larger portion of the solar radiation, leaving less radiation to evaporate water from the soil (see also Figure 6.15). Therefore, as plant production is increased by improved nutrition, T is increased substantially, but E is decreased, so that ET is increased only a little. We can conclude that so long as the supply of water is not too limited, maintaining optimum conditions for plant growth (by closer plant spacing, fertilization, or selection of more vigorous varieties) increases the efficiency of water use by plants.

In harsh semiarid environments, one should apply this principle with caution, however, if irrigation water is not available and the period of rainfall is very short. The modest increase in total water use (ET) by the more vigorously growing plants may deplete the stored soil water before the life cycle of the plant is complete, resulting in serious water stress, or even plant death before any harvestable yield has been produced.

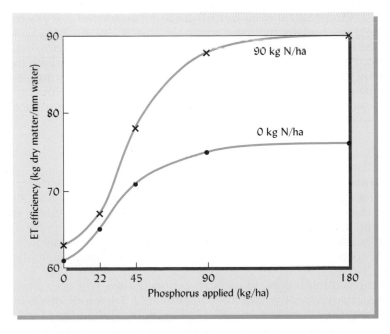

FIGURE 6.16 Water-use efficiency of wheat as affected by application of phosphorus fertilizer with and without nitrogen fertilizer (12-year average). ET efficiency increased with increased fertilizer rates. [Data from Black (1982)]

6.5 CONTROL OF EVAPOTRANSPIRATION (ET)

As we have just seen, water loss by ET is closely related to the total leaf area exposed to solar radiation. To some degree, adjusting the total leaf area exposed can bring ET into balance with PET (and thus lessen water stress). Five approaches to accomplish this are the following:

1. If water can be added by irrigation (an option that will be discussed in Section 6.11) both ET and plant production are likely to increase, often dramatically. Of course, for reasons of cost or lack of a water source, this practice is often not feasible.

2. Where rapid growth might prematurely deplete the available water, it may be wise to limit plant-growth factors such as nutrient supply to only moderate levels, thus keeping LAI in check.

3. The LAI of desired plants can be limited by sowing fewer seeds per unit land area or by spacing plants farther apart. It should be noted, however, that plants growing farther apart tend to individually produce a greater leaf area, compensating somewhat for the lower density. Also, wider spacing allows greater evaporation losses from the soil.

4. Elimination of undesired plants (i.e., weeds) can remove a major cause of T losses that could greatly reduce the soil water supplies available for the desired species.

5. In semiarid regions, elimination of all vegetation for a period of time may allow rainwater to accumulate in the soil profile for use by subsequent plantings.

We will now discuss the last two approaches listed.

Control of Unwanted Vegetation

Transpiration by weeds may seriously deplete soil water supplies needed by more desirable plants. It is largely through their heavy use of soil water that weeds interfere with the establishment and growth of desireable forest, range and crop plants. Weeds have traditionally been controlled by cultivation of the soil, but in recent decades weeds have been more widely controlled by spraying herbicides (see Section 18.2). Weed control with herbicides has several advantages over cultivation, among which are that it requires less labor and energy, and it allows the soil to be left undisturbed with plant residues covering the soil surface (see Section 6.6). Chemical weed control also has serious disadvantages in some situations, including high material costs, eventual evolution of weed resistance to specific compounds, damage to desired plants, and environmental toxicity. With regard to the latter, some herbicides are toxic to soil organisms, fish, or land animals, and have accumulated to undesirable levels in downstream waterways and underground drinking-water sources (see Section 18.3).

Consequently, alternatives to chemical weed control are being developed. For example, some weeds can be held in check by biological controls; that is, encouraging specific insects or diseases that attack only the weeds. Well-timed mowing or grazing can also reduce weed problems. Furthermore, cultivation techniques to help control weeds are being improved, thereby minimizing the need for chemical herbicides. New implements capable of cultivating through high levels of plant residue hold promise for controlling weeds without herbicides while still maintaining some of the desirable surface residue cover.

Fallow Periods in Dryland Cropping Systems

Farming systems that alternate summer **fallow** (unvegetated period) one year with traditional cropping the next are sometimes used to conserve soil moisture in low-rainfall environments. To protect the soil and help capture snow, the previous year's stubble is commonly allowed to stand at least until the spring of the fallow year. To prevent T water loss, weeds are eliminated in the fallow years by light tillage and/or herbicides. The top part of Figure 3.21 (page 101) shows typical alternating strips of summer fallow and wheat in a region of Ustolls.

During the year that the soil lies fallow, some evaporation (E) loss will occur, but since no plants are growing, transpiration losses will be eliminated. Therefore, much of the rain that infiltrates during the fallow year is stored in the profile. When the crop is planted the following year, it will have access to significantly more stored water in the profile than if a crop had been grown the previous year. Yields are therefore generally higher after a fallow, but not always sufficiently so to make up for the harvest foregone during the fallow year (Table 6.4). Summer fallow certainly reduces the risk of crop failure where rainfall is unreliable or barely adequate for crop production.

6.6 CONTROL OF SURFACE EVAPORATION (E)

More than half the precipitation in semiarid and subhumid areas is usually returned to the atmosphere by evaporation (E) directly from the soil surface. In natural rangeland systems, E is a large part of ET because plant communities tend to self-regulate toward low plant densities and leaf area configurations that minimize the deficit between PET and ET. In addition, plant residues on the soil surface are sparse. Evaporation losses are also high in arid-region irrigated soil, especially if inefficient practices are used (see Section 6.11). Even in humid-region rain-fed areas, E losses are significant during hot rainless periods. Such moisture losses rob the plant community of much of its growth potential and reduce the water available for discharge to streams. Careful study of Figure 6.17 will clarify these relationships and the principles discussed in Sections 6.4 and 6.5.

For arable soils, the most effective practices aimed at controlling E are those that provide some cover for the soil. This cover can best be provided by mulches and by selected conservation tillage practices that leave plant residues on the soil surface, mimicking the soil cover of natural ecosystems.

Vegetative Mulches

A *mulch* is any material placed on the soil surface primarily for the purpose of reducing evaporation or controlling weeds. Examples include sawdust, manure, straw, leaves, and crop residues. Mulches can be highly effective in checking evaporation, but they may be expensive and labor-intensive to produce or purchase, transport to the field, and apply to the soil. Mulching is therefore most practical for small areas (gardens and landscaping beds) and for high-value crops such as cut flowers, berries, fruit trees, and certain vegetables. In addition to reducing evaporation, vegetative mulches may provide these benefits: (1) reducing the spread of soilborne diseases; (2) providing a clean path for foot traffic; (3) reducing weed growth (if applied thickly); (4) moderating soil temperatures, especially preventing overheating in summer months (see Section 7.12); (5) increasing water infiltration; (6) providing organic matter and, possibly, plant nutrients to the soil; (7) encouraging earthworm populations; and (8) reducing soil erosion. Most of these ancillary benefits do not accrue from the use of plastic or paper mulches, discussed next.

Paper and Plastic Mulches

Specially prepared paper and plastics are also used as mulches to control evaporative water losses. In the case of black plastic and dark paper, they also effectively control weeds. The mulch is often applied by machine and plants grow through slits or holes in

TABLE 6.4 **Wheat Yields at Seven Locations in the Great Plains Area of the United States Where Continuous Cropping and Fallow Cropping Were Compared**

Location	Years of record	Yield, kh/ha	
		After fallow	After wheat
Havre, Mont.	35	2100	540
Dickinson, N. Dak.	44	1400	780
Newell, S. Dak.	40	1420	910
Akron, Colo.	60	1420	500
North Platte, Nebr.	56	2140	830
Colby, Kans.	49	1320	620
Bushland, Tex.	29	1010	640

From USDA (1974).

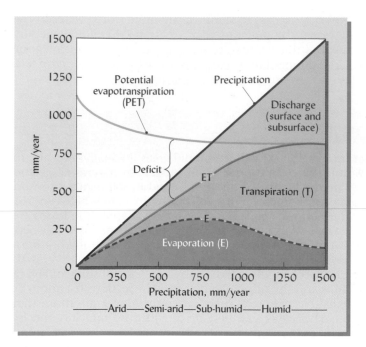

FIGURE 6.17 Partitioning of liquid water losses (discharge) and vapor losses (evaporation and transpiration) in regions varying from low (arid) to high (humid) levels of annual precipitation. The example shown assumes that temperatures are constant across the regions of differing rainfall. Potential evapotranspiration (PET) is somewhat higher in the low-rainfall zones because the lower relative humidity there increases the vapor pressure gradient at a given temperature. Evaporation (E) represents a much greater proportion of total vapor losses ET in the drier regions due to sparse plant cover caused by interplant competition for water. The greater the gap between PET and ET, the greater the deficit and the more serious the water stress to which plants are subject.

the sheeting (Figure 6.18). These mulches are widely used for vegetable and small fruit crops and in landscaping beds, where they are often covered with a layer of tree-bark mulch or gravel for a more pleasing appearance. As long as the ground is covered, evaporation is checked, and in some cases remarkable increases in plant growth have been reported. Unless rainfall is very heavy, the mulch does not seriously interfere with the infiltration of rainwater into the soil.

The temperature of the soil under plastic mulch is commonly 8 to 10°C higher than under a straw mulch. The higher temperature may be helpful for crops established early in spring, but it can reduce the yields of heat-sensitive plants in summer (see Section 7.12). A common problem with plastic mulches is that they can rarely be completely removed at the end of the growing season, so after a number of years scraps of plastic accumulate in the soil, causing an unsightly mess and interfering with water movement and cultivation.

FIGURE 6.18 For crops with high cash value, plastic mulches are commonly used. The plastic is installed by machine (left) and at the same time the plants are transplanted (right). Plastic mulches help control weeds, conserve moisture, encourage rapid early growth, and eliminate need for cultivation. The high cost of plastic makes it practical only with the highest-value crops. (Courtesy K. Q. Stephenson, Pennsylvania State University)

Crop Residue and Conservation Tillage

Plant residues left on the soil surface are effective in reducing E and, in turn, in conserving soil moisture. *Conservation tillage* practices leave a high percentage of the residues from the previous crop on or near the surface (Figure 6.19). A conservation tillage practice widely used in subhumid and semiarid regions is **stubble mulch** tillage. With this method, residues such as wheat stubble or cornstalks from the previous crop are uniformly spread on the soil surface. The land is then tilled with special implements that permit much of the plant residue to remain on or near the surface. Conservation tillage planters are capable of planting through the stubble and allow much of it to remain on the surface during the establishment of the next crop (see Figure 6.19). Combining summer fallow with stubble mulching saves soil moisture in dry areas, as shown in Table 6.5. Unfortunately, plant growth in dry regions is usually insufficient to produce residue mulches at the levels needed to reduce E.

Other conservation tillage systems that leave residues on the soil surface include no-tillage (see Figure 6.19), where the new crop is planted directly into the sod or residues of the previous crop, with no plowing or disking. This and other systems will receive further attention in Section 17.6.

6.7 LIQUID LOSSES OF WATER FROM THE SOIL

In our discussion of the hydrologic cycle we noted two types of liquid losses of water from soils: (1) percolation or drainage water, and (2) runoff water (see Figure 6.2). *Percolation water* recharges the groundwater and moves chemicals out of the soil. *Runoff water* often carries appreciable amounts of soil (erosion) as well as dissoved chemicals.

Percolation and Leaching: Methods of Study

Two general methods are used to study percolation and leaching losses: (1) underground perforated pipes, often termed **tile drains**, specially installed for the purpose, and (2) specially constructed **lysimeters**.[1] For the first method, an area should be cho-

---

[1] For reviews of lysimeter work, see Aboukhaled, et al. (1982) and Allen, et al. (1991).

FIGURE 6.19 Conservation tillage leaves plant residues on the soil surface, reducing both evaporation losses and erosion. (Left) In a semiarid region (South Dakota) the straw from the previous year's wheat crop was only partially buried to anchor it against the wind while still allowing it to cover much of the soil surface. In the next year the left half of the field, now shown growing wheat, will be thusly *stubble mulched* and the right half sown to wheat. (Right) *No-till* planted corn in a more humid region grows up through the straw left on the surface by a previous wheat crop. Note that with no-till, almost no soil is directly exposed to solar radiation, rain or wind. (Photos courtesy of R. Weil, left, and USDA Natural Resources Conservation Service, right)

Location	Av. annual precipitation, mm	Soil water gain, mm, at each mulch rate			
		0 Mg/ha	2.2 Mg/ha	4.4 Mg/ha	6.6 Mg/ha
Bushland, Tex.	508	71	99	99	107
Akron, Colo.	476	134	150	165	185
North Platte, Nebr.	462	165	193	216	234
Sidney, Mont.	379	53	69	94	102
Average		107	127	145	157
Average gain by mulching			20	38	50

From Greb (1983).

sen in which the tile drain receives only the water from the land under study. The advantage of the tile method is that water and nutrient losses can be determined from relatively large areas of soil under normal field conditions.

In the lysimeter method, a small volume of undisturbed field soil is isolated from surrounding areas by concrete or metal dividers (Figure 6.20). In either case, water percolating through the soil is collected and measured. The advantages of lysimeters over tile-drain systems are that the variations in a large field are avoided, and numerous lysimeters can be used in a controlled experiment in which many different soil–plant management systems are compared.

Percolation Losses of Water

When the amount of rainfall entering a soil exceeds the water-holding capacity of the soil, losses by percolation will occur. Percolation losses are influenced by (1) the amount of rainfall and its distribution, (2) runoff from the soil, (3) evaporation, (4) the character of the soil, and (5) the nature of the vegetation.

Percolation-Evaporation Balance

Figure 6.21 illustrates the relationships among precipitation, runoff, soil storage, soil water depletion, and percolation for representative humid and semiarid regions and for

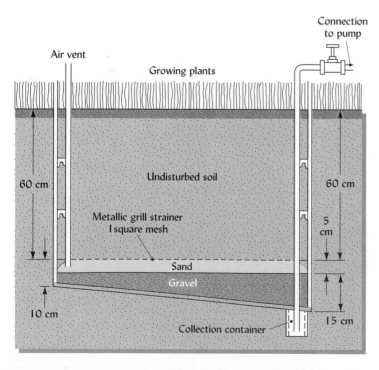

FIGURE 6.20 Field lysimeter used to collect percolation water from an undisturbed soil. The water moves down through the soil to the sand and gravel, then down the inclined slope to the container (lower right) from which it can be pumped and collected. Some more sophisticated systems are equipped with devices to weigh the entire lysimeter, thereby permitting the additional measurement of water intake and evapotranspiration. [Redrawn from UNDP/WMO (1974)]

an irrigated arid region. In the humid temperate region, the rate of water infiltration into the soil (precipitation minus runoff) is greater, at least during certain seasons, than the rate of evapotranspiration. As soon as the soil field capacity is reached, percolation into the substrata occurs.

In the example shown in Figure 6.21*a*, maximum percolation occurs during the winter and early spring, when evaporation is lowest. During the summer little percolation occurs. In fact, evapotranspiration exceeds precipitation, resulting in a depletion of soil water. Normal plant growth is possible only because of water stored in the soil from the previous winter and early spring.

In the semiarid region, as in the humid region, water is stored in the soil during the winter months and is used to meet the moisture deficit in the summer. But because of the low rainfall, little runoff and essentially no percolation out of the profile occurs. Water may move to the lower horizons, but it is absorbed by plant roots and ultimately is lost by transpiration.

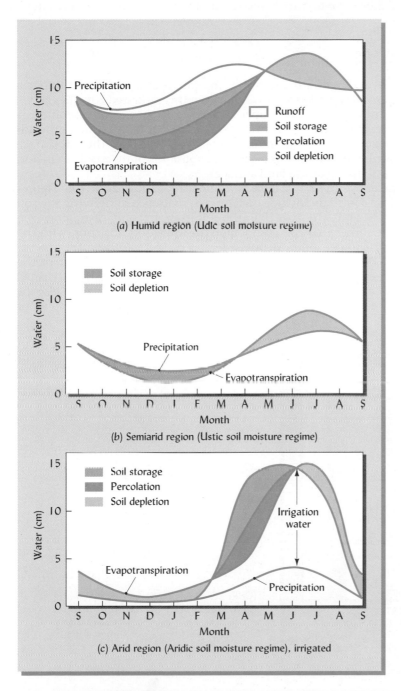

(a) Humid region (Udic soil moisture regime)

(b) Semiarid region (Ustic soil moisture regime)

(c) Arid region (Aridic soil moisture regime), irrigated

FIGURE 6.21 Generalized curves for precipitation and evapotranspiration for three temperate zone regions: (*a*) a humid region, (*b*) a semiarid region, and (*c*) an irrigated arid region. Note the absence of percolation through the soil in the semiarid region. In each case water is stored in the soil. This water is released later when evapotranspiration demands exceed the precipitation. In the semiarid region evapotranspiration would likely be much higher if ample soil moisture were available. In the irrigated arid region soil, the very high evapotranspiration needs are supplied by irrigation. Soil moisture stored in the spring is utilized by later summer growth and lost through evaporation during the late fall and winter.

The irrigated soil in the arid region shows a unique pattern. Irrigation in the early spring, along with a little rainfall, provides more water than is being lost by evapotranspiration. The soil is charged with water, and some percolation may occur. As we shall see in Section 10.9, irrigation systems must provide enough water for some percolation, in order to remove excess soluble salts. During the summer, fall, and winter months, the stored water is depleted because the amount being added is less than the very high evapotranspiration that takes place in response to large vapor pressure gradients.

The situations depicted in Figure 6.21 are typical of temperate zones where potential evapotranspiration (PET) varies seasonally with temperature. In the tropics, where temperatures are somewhat higher and less variable, PET is somewhat more uniform throughout the year, although it does vary with seasonal changes in humidity. In very high rainfall tropical areas (perudic moisture regimes) much more runoff and somewhat more percolation occurs than shown in Figure 6.21. In irrigated areas of the arid tropics, the relationships are similar to those shown for arid temperate regions.

The comparative losses of water by evapotranspiration and percolation through soils found in different climatic regions are shown in Figure 6.22. These differences should be kept in mind while reading the following section on percolation and groundwaters.

6.8 PERCOLATION AND GROUNDWATERS

When drainage water moves downward and out of the soil, it eventually encounters a zone in which the pores are all saturated with water. Often this saturated zone lies above a layer of impermeable rock or clay. The upper surface of this zone of saturation is known as the **water table**, and the water within the saturated zone is termed **groundwater**. The water table (Figure 6.23) is commonly only 1 to 10 m below the soil surface in humid regions but may be several hundred or even thousands of meters deep in arid regions. In swamps it is essentially at the land surface.

The unsaturated zone above the water table is termed the **vadose zone** (see Figures 6.23 and 6.24). The vadose zone may include unsaturated materials underlying the soil profile, and so may be considerably deeper than the soil itself. In some cases, however, the saturated zone may be sufficiently high to include the lower soil horizons, with the vadose zone confined to the upper soil horizons.

Shallow groundwater receives downward-percolating drainage water. Most of the groundwater, in turn, seeps laterally through porous geological materials (termed **aquifers**) until it is discharged into springs and streams. Groundwater may also be removed by pumping for domestic and irrigation uses. The water table will move up or

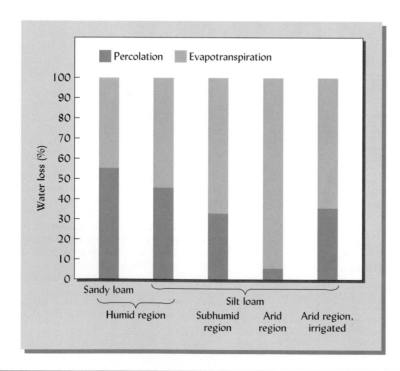

FIGURE 6.22 Percentage of the water entering the soil that is lost by downward percolation and by evapotranspiration. Representative figures are shown for different climatic regions.

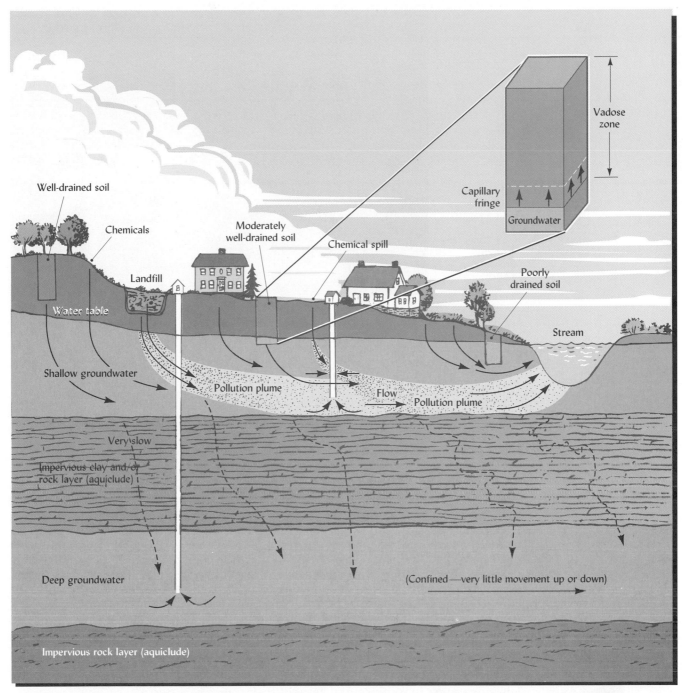

FIGURE 6.23 Relationship of the water table and groundwater to water movement into and out of the soil. Precipitation and irrigation water move down the soil profile under the influence of gravity (gravitational water), ultimately reaching the water table and underlying shallow groundwater. The unsaturated zone above the water table is known as the *vadose zone*. As water is removed from the soil by evapotranspiration, groundwater moves up from the water table by capillarity in what is termed the *capillary fringe*. Groundwater also moves horizontally down the slope toward a nearby stream, carrying with it chemicals that have leached through the soil, including essential plant nutrients (N, P, Ca, etc.) as well as pesticides and other pollutants from domestic, industrial, and agricultural wastes. Groundwaters are major sources of water for wells, the shallow ones removing water from the groundwater near the surface, while deep wells exploit deep and usually large groundwater reserves. Two plumes of pollution are shown, one originating from landfill leachate, the other from a chemical spill. The former appears to be contaminating the shallow well.

down in response to the balance between the amount of drainage water coming in through the soil and the amount lost through pumped wells and natural seepage to springs and streams. In humid temperate regions, the water table is usually highest in early spring following winter rains and snow melt, but before evapotranspiration begins to withdraw the water stored in the soil above.

Vadose
zone

Capillary
fringe

Water table

Ground-
water

FIGURE 6.24 The water table, capillary fringe, zone of unsaturated material above the water table (vadose zone), and groundwater are illustrated in this photograph. The groundwater can provide significant quantities of water for plant uptake. (Photo courtesy of R. Weil)

Groundwater Resources

Groundwater is a significant source of water for domestic, industrial, and agricultural use. For example, some 20% of all water used in the United States comes from groundwater sources, and about 50% of the people use groundwater to meet at least some of their needs. Shallow aquifers, which are replenished annually, commonly provide water for farm and rural dwellings (see Figure 6.23). The larger groundwater stores in deeper aquifers, which may take decades or centuries to recharge, are commonly pumped to meet municipal, industrial, and irrigation needs. In many cases deep wells are used to tap water from aquifers confined between impermeable strata (termed **aquicludes**). Water is replenished by slow, mainly horizontal seepage from **recharge areas** where the aquifer is exposed to the soil.

The regional importance of groundwater aquifers is illustrated by the enormous Ogallala aquifer that underlies much of the Great Plains of the United States. It has provided sufficient water to (temporarily?) transform the regional economy from one based on range cattle and dryland farming to one based on much more productive irrigated agriculture. Similarly, the major cities of San Antonio and Austin depend on the Edwards aquifer in south central Texas as their source of water.

Water draining from soils is the main source of replenishment for most underground water resources. So long as water is not withdrawn faster than percolation allows it to be replenished, groundwater may be considered to be a renewable resource. However, in some areas people are pumping water out faster than it can be replenished, an activity that is lowering water tables and depleting the resource in a manner akin to mining an ore. The rate of pumping from these poorly managed (or unmanaged) aquifers ranges from 2 to 200 times the rate of recharge. In deep aquifers such as the Ogallala, water that filtered downward thousands of years ago is being removed far more rapidly than it can be replaced. Consequently, these water tables are falling and the aquifers may eventually cease to provide water at all.

Another problem occurs when aquifers are overpumped near ocean coastlines. As the freshwater is pumped out faster than new freshwater seeps in from the land surface, seawater under pressure pushes its way into the aquifer, a process termed **saltwater intrusion** (as illustrated in Figure 6.2). Soon, deep municipal wells begin to pull in salty instead of fresh water.

Shallow Groundwater

Groundwater that is near the surface can serve as a reciprocal water reservoir for the soil. As plants remove water from the soil, it may be replaced by upward capillary movement from a shallow water table. The zone of wetting by capillary movement is known as the **capillary fringe** (see Figure 6.24). Such movement can provide a steady and significant supply of water that enables plants to survive during periods of low rainfall.[2]

Movement of Chemicals in the Drainage Water[3]

Percolation of water through the soil to the water table not only replenishes the groundwater, it also dissolves and carries downward a variety of inorganic and organic chemicals found in the soil or on the land surface. Chemicals *leached* from the soil to the groundwater (and eventually to streams and rivers) in this manner include elements weathered from minerals, natural organic compounds resulting from the decay of plant residues, plant nutrients derived from natural and human sources, and various synthetic chemicals applied intentionally or inadvertently to soils.

In the case of plant nutrients, especially nitrogen, downward movement through the soil and into underlying groundwaters has two serious implications. First, the leaching of these chemicals represents a depletion of plant nutrients from the root zone (see Section 16.2). Second, accumulation of these chemical nutrients in ponds, lakes, reservoirs, and groundwater downstream may stimulate a process called *eutrophication,* which ultimately depletes the oxygen content of the water, with disastrous effects on fish and other aquatic life (see also Section 14.2). Also, in some areas underground sources of drinking water may become contaminated with excess nitrates to levels unsafe for human consumption (see Section 13.8).

Of even more concern is the contamination of groundwater with various highly toxic synthetic compounds, such as pesticides and their breakdown products or chemicals leached out of waste disposal sites (see Chapter 18 for a detailed discussion of these pollution hazards). Figure 6.23 illustrates how the groundwater is charged with these chemicals, and how a plume of contamination spreads to downstream wells and bodies of water.

Chemical Movement Through Macropores

Studies of the movement of chemicals in soils and from soils into the groundwater and downstream bodies of water have called attention to the critical role played by large macropores in determining field hydraulic conductivity (see Section 5.5). The pore configuration in most soils is nonuniform. Old root channels, earthworm burrows, and clay shrinkage cracks commonly contribute large macropores that may provide channels for rapid water flow from the soil surface to depths of 1 m or more.

Once chemicals are carried below the zone of greatest root and microbial activity, they are less likely to be removed or degraded before being carried further down to the groundwater. This means that chemicals that are normally broken down in the soil by microorganisms within a few weeks may move down through large macropores to the groundwater before their degradation can occur.

BYPASS FLOW.   In some cases, leaching of chemicals is most serious if the chemicals are merely applied on the soil surface. As shown in Figure 6.25, chemicals may be washed from the soil surface into large pores, through which they can quickly move downward. Research suggests that most of the water flowing through large macropores does not come into contact with the bulk of the soil. Such flow is sometimes termed **bypass flow** as it tends to move rapidly around, rather than through, the soil matrix. As a result, if chemicals have been incorporated into the upper few centimeters of soil, their movement into the larger pores is reduced, and downward leaching is greatly curtailed.

---

[2] Capillary rise from shallow groundwater may also bring a steady supply of salts to the surface if the groundwater is brackish (see Section 10.3 for details on this soil-degrading process).

[3] For a review of chemical transport through field soils see Jury and Fluhler (1992).

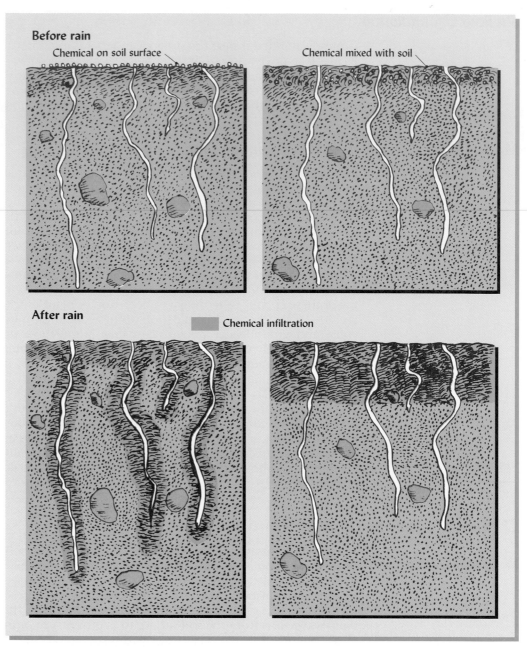

Before rain

Chemical on soil surface

Chemical mixed with soil

After rain

Chemical infiltration

FIGURE 6.25 Preferential or bypass flow in macropores transports soluble chemicals downward through a soil profile. Where the chemical is on the soil surface (left), and can dissolve in surface-ponded water when it rains, it may rapidly be transported down cracks, earthworm channels, and other macropores. Where the chemical is dispersed within the soil matrix in the upper horizon (right), most of the water moving down through the macropores will bypass the chemical, and thus little of the chemical will be carried downward.

INTENSITY OF RAIN OR IRRIGATION. Flow of water and chemicals through macropores, such as the vertical cracks and channels just mentioned, is much more pronounced when water is applied by high-intensity rainfall or irrigation. Apparently, when a large volume of water is rapidly applied, some localized saturation occurs that encourages flow through macropores. Table 6.6 illustrates the effects of rainfall *intensity* on the leaching of pesticides applied to turf grass.

Manipulations of soil macropores and irrigation intensity provide us with two examples of how environmental quality can be impacted by human activities that influence the hydrologic cycle. Likewise, we have considered steps that can be taken to influence the infiltration of water into soils and to maximize plant biomass production from the water stored in soils. The picture would not be complete if we did not consider

TABLE 6.6 Influence of Water Application Intensity on the Leaching of Pesticides Through the Upper 50 cm of a Grass Sod Covered Mollisol

The tests were conducted on 20-cm-diameter soil columns with the natural structure, including worm channels, undisturbed. All columns received 2.54 cm of simulated rain per week for four weeks, but the heavy rain was applied in four doses of 2.54 cm each, while the light rain was applied as 16 evenly spaced doses of 0.64 cm each. Metalaxyl is much more highly water soluble than Isazofos, but in each case the heavy rain stimulated much more pesticide leaching through the macropores.

	Pesticide leached as percentage of that applied to surface	
Pesticide	Heavy rains	Light rains
Isazofos	8.8	3.4
Metalaxyl	23.8	13.9

Data from Starrett, et al. (1996).

briefly three other anthropogenic modifications to the hydrologic cycle: (1) the use of artificial drainage to enhance the downward percolation of excess water from some soils; conversely, (2) the application of additional water to soils as a means of waste-water disposal; and (3) the use of irrigation to supplement water available for plant growth. We shall consider artificial drainage first.

6.9   ENHANCING SOIL DRAINAGE[4]

Some soils tend to be water-saturated in the upper part of their profile for extended periods during part or all of the year. The prolonged saturation may be due to the low-lying landscape position in which the soil is found, such that the regional water table is at or near the soil surface for extended periods. In other soils, water may accumulate above an impermeable layer in the soil profile, creating a *perched water table* (see Figure 6.26). Soils with either type of saturation may be components of wetlands (see Section 7.8), transitional ecosystems between land and water.

Reasons for Enhancing Soil Drainage

Water-saturated, poorly aerated soil conditions are essential to the normal functioning of wetland ecosystems and to the survival of many wetland plant species. However, for most other land uses, these conditions are a distinct detriment.

ENGINEERING PROBLEMS.   During construction, the muddy, low-bearing-strength conditions of saturated soils make it very difficult to operate machinery (see Section 4.9). Houses built on poorly drained soils may suffer from uneven settlement and flooded basements during wet periods. Similarly, a high water table will result in capillary rise of water into roadbeds and around foundations, lowering the soil strength and leading to damage from frost-heaving (see Section 7.9) if the water freezes in winter. Heavy trucks traveling over a paved road underlaid by a high water table create potholes and eventually destroy the pavement.

PLANT PRODUCTION.   Water-saturated soils make the production of most upland crops and forest species difficult, if not impossible. Prolonged saturation makes it difficult to carry out operations in a timely fashion as tractors and other equipment used for planting, tillage, and harvest operations tend to form deep ruts and bog down frequently. Even if a crop or forest stand can be established, the poorly aerated conditions may stunt its growth because the roots of all but a few specially adapted plants (bald cypress trees, rice, cattails, etc.) require aerated soil for respiration (see Section 7.2). Furthermore, a high water table early in the growing season will confine the plant roots to a shallow layer of

---

[4] For a classic and practical treatment of artifical drainage systems, see Beauchamp (1955).

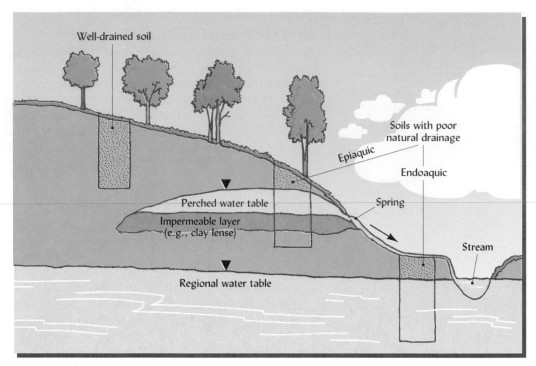

FIGURE 6.26 Cross-section of a landscape showing the regional and perched water tables in relation to three soils, one well-drained and two with poor internal drainage. By convention, a triangle (▼) is used to identify the level of the water table. The soil containing the perched water table is wet in the upper part, but unsaturated below the impermeable layer, and therefore is said to be *epiaquic* (Greek *epi,* upper), while the soil saturated by the regional water table is said to be *endoaquic* (*Greek* endo, under). Artificial drainage can help to lower both types of water tables.

partially aerated soil; the resulting restricted root system can lead to water stress later in the year, when the weather turns dry and the water table drops rapidly (Figure 6.27).[5]

For these and other reasons, artificial drainage systems have been widely used to remove excess (gravitational) water and lower the water table in poorly drained soils. Land drainage is practiced in select areas in almost every climatic region, but is most widely used to enhance the agricultural productivity of clayey alluvial and lacustrine soils. Drainage systems are also a vital, if sometimes neglected, component of arid region irrigation systems, where they are needed to remove excess salts and prevent waterlogging.

Artificial drainage is a major alteration of the soil system, and the following list of potential beneficial and detrimental effects should be carefully considered. Also take note that in many instances laws designed to protect wetlands require that a special permit be obtained for the installation of a new artificial drainage system.

Benefits of Artificial Drainage

1. Increased bearing strength and improved soil workability, which allow more timely field operations and greater access to vehicular or foot traffic (e.g., in recreational facilities).

2. Reduced frost heaving of foundations, pavements, and crop plants (e.g., see Figure 7.19).

3. Enhanced rooting depth, growth, and productivity of most upland plants due to improved oxygen supply and, in acid soils, lessened toxicity of manganese and iron (see Sections 7.4 and 7.6).

4. Reduced levels of fungal disease infestation in seeds and on young plants.

5. More rapid soil warming in spring, resulting in earlier maturing crops (see Section 7.12).

---

[5] On the other hand, if the groundwater is not too near the surface, it may be of great benefit in providing water by capillarity to plant roots just above the water table.

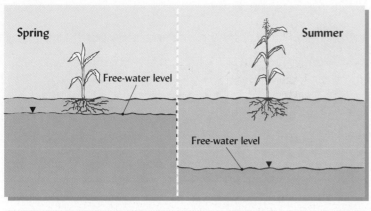

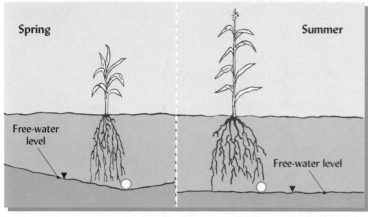

FIGURE 6.27 Illustration of water levels of undrained and tile-drained land in the spring and summer. Benefits of the drainage include more vigorous early growth, as well as a more extensive root system capable of exploiting a large volume of soil when the profile dries out in summer. [Redrawn from Hughes (1980); used with permission of Deere & Company, Moline, Ill.]

6. Less production of methane and nitrogen gases that cause global environmental damages (see Sections 12.11 and 13.9).

7. Removal of excess salts from irrigated soils, and prevention of salt accumulation by capillary rise in areas of salty groundwater (see Section 10.3).

Detrimental Effects of Artificial Drainage

1. Loss of wildlife habitat, especially waterfowl breeding and overwintering sites.

2. Reduction in nutrient assimilation and other biochemical functions of wetlands (see Section 7.8).

3. Increased leaching of nitrates and other contaminants to groundwater.

4. Accelerated loss of soil organic matter, leading to subsidence of certain soils (see Sections 12.10 and 12.12).

5. Increased frequency and severity of flooding due to loss of runoff water retention capacity.

6. Greater cost of damages when flooding occurs on alluvial lands developed after drainage.

Artificial drainage systems are designed to promote two general types of drainage: (1) *surface drainage,* and (2) internal or *subsurface drainage.* Each will be discussed briefly.

Surface Drainage Systems

This type of drainage is extensively used, especially where the landscape is nearly level and soils are fine-textured with slow internal drainage (percolation). Its purpose is to remove water from the land before it infiltrates the soil.

SURFACE DRAINAGE DITCHES. Most surface drainage systems hasten the surface runoff of water by constructing shallow ditches with gentle side slopes that do not interfere with equipment traffic. If there is some slope on the land, the shallow ditches are usually oriented across the slope and across the direction of planting and cultivating, thereby per-

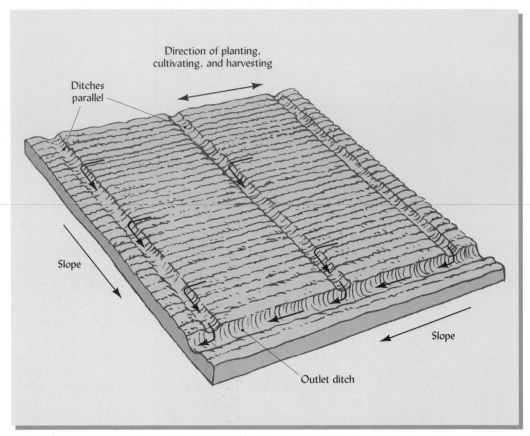

FIGURE 6.28 Example of an open ditch drainage system to promote surface drainage on a field with gentle slope. [From Hughes (1980); used with permission of Deere & Company, Moline, Ill.]

mitting the interception of water as it runs off down the slope (Figure 6.28). These ditches can be made at low cost with simple equipment. For removing surface water from landscaped lawns, this system of drainage can be modified by constructing gently sloping swales rather than ditches.

LAND FORMING. Often, surface drainage ditches are combined with *land forming* or smoothing to eliminate the ponding of water and facilitate its removal from the land (Figure 6.29). Small ridges are cut down and depressions are filled in using precision, laser-guided field-leveling equipment. The resulting land configuration permits excess

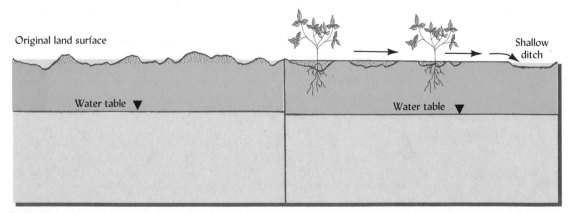

FIGURE 6.29 Land surface before (left) and after (right) land forming or smoothing. The original soil surface (left) is uneven and water ponds in depressions. After land smoothing (right), the land has a uniform slope and water runs off the surface at a controlled rate, to be carried away by a shallow ditch. Note that the soil from the ridges (stippled) is used to fill in the depressions.

water to move at a controlled rate over the soil surface to the outlet ditch and then on to a natural drainage channel. Land smoothing is also commonly used to prepare a field for flood irrigation (see Section 6.11).

Subsurface (Internal) Drainage

The water table rises when the rate of water input (usually rain or irrigation water percolating down from the soil surface) exceeds the rate of groundwater movement away from the area. Therefore, the purpose of a subsurface drainage system is to provide a series of "giant pores" or pathways that vastly increase the rate of groundwater movement and subsequently lower the level of the water table.

The design of a drainage system must be compatible with the basic principle that water moves to where its energy level will be lower. That is, water will enter a pipe or ditch where its potential will be essentially zero only if it is under a positive pressure ($\psi \geq 0$) (due to the sum of the positive gravitational and submergence potentials). Therefore, internal drainage will occur only when the pathway for drainage is located *below* the level of the water table (Figure 6.30). No water will be removed if the pipe or ditch bottom is in wet but not saturated soil, because the matric component in such soil results in a negative water potential ($\psi < 0$). The flow of water from a saturated soil into a drainage outlet is illustrated in Figure 6.31. Box 6.1 provides an example of how knowledge of basic soil properties and water-movement principles can be appplied in designing a system to alleviate drainage problems.

The network of drainage channels may be laid out in several types of patterns, depending on the nature of the area to be drained (Figure 6.32). Where the landscape is too level or at too low an elevation to provide for sufficient fall, expensive pumping operations are sometimes used to remove the drainage water (e.g., in parts of Florida and the Netherlands).

OPEN DITCH DRAINAGE. If a ditch is excavated to a depth below the water table (Figures 6.30 and 6.33), water will seep from the saturated soil, where it is under a positive pressure, into the ditch, where its potential will be essentially zero. Once in the ditch, the water no longer must overcome the frictional forces that delay its twisting journey through tiny soil pores but can flow rapidly off the field. The ditches, being 1 m or more deep, present barriers to equipment travel and so generally are practical only for coarse-textured soils in which they may be spaced quite far apart. The spacing is based on the principle that in a soil with a high saturated hydraulic conductivity, movement of water to the ditch will be quite rapid and so the effect on lowering the water table will extend out for a considerable distance from the ditch.[6] Open ditches need regular maintenance to control vegetation and sediment buildup.

BURIED PERFORATED PIPES (DRAIN TILES).[7] A network of perforated plastic pipe can be laid underground using specialized equipment (Figure 6.34, page 250). Water moves into the pipe through the perforations. The pipe should be laid with the slotted or perforated side down. This allows water to flow up into the pipe (as shown in Figure 6.31) but protects against soil falling into and clogging the pipe (see Figure 6.35, page 251). Sediment buildup will also be avoided if the pipe has the proper slope (usually a 0.5 to 1% drop) so water flows rapidly to the outlet ditch or stream. The pipe outlet should be covered by a gate or wire mesh, so as to prevent the entrance of rodents in dry weather, but allow the free flow of water. A properly installed subsurface drainage system may continue to operate for many decades.

BUILDING FOUNDATION DRAINS. Surplus water around foundations and underneath basement floors of houses and other buildings can cause serious damage. The removal of this excess water is commonly accomplished using buried perforated pipe, placed alongside and slightly below the foundation or underneath the floor (Figure 6.30c). The

---

[6] The required spacing of ditches and buried drain pipes ranges from about 10 m apart for very low permeability soils to more than 100 m apart for high-permeability soils such as peats and sands.

[7] Before the adoption of plastic perforated pipe in the 1960s, short sections of ceramic pipe known as *tiles* were laid down end to end with small gaps between them through which water could enter. The terms *drain tile* or *tile line* are still often used in reference to the newer plastic pipe.

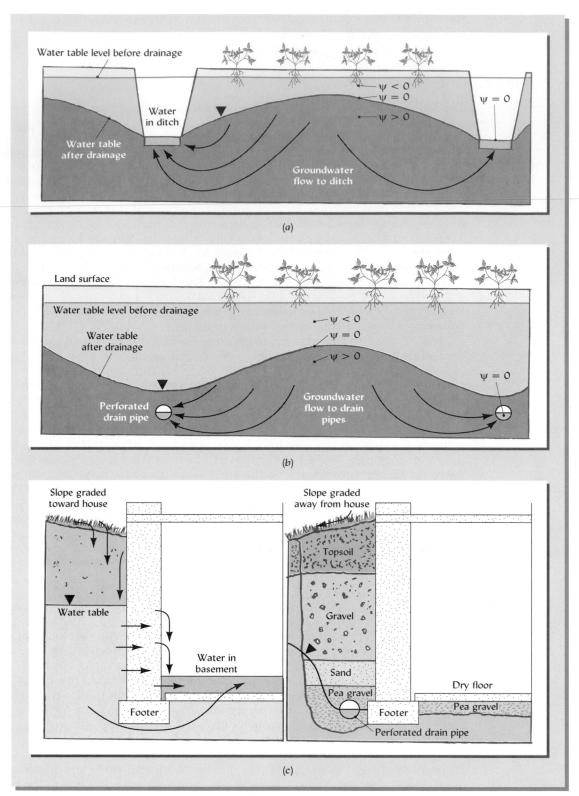

FIGURE 6.30 Three types of subsurface drainage systems. (*a*) Open ditches are used to lower the water table in a poorly drained soil. The wet season levels of the water table before and after ditch installation are shown. The water table is deepest next to the ditch, and the drainage effect diminishes with distance from the ditch. (*b*) Buried "tile lines" made of perforated plastic pipe act very much as the ditches in *a*, but have two advantages: They are not visible after installation and they do not present any obstacle for surface equipment. Note the flow lines indicating the paths taken by water moving to the drainage ditches or pipes in response to the water potential gradients between the submerged water ($\psi > 0$) and free water in the drainage ditch and pipes ($\psi = 0$). (*c*) The water table around a building foundation before (left) and after (right) installation of a *footer drain* and correction of surface grading. The principles of water movement in soils are applied to keep the basement dry.

FIGURE 6.31 Demonstration of the saturated flow patterns of water toward a drainage tile. Water containing a colored dye was added to the surface of the saturated soil, and drainage was allowed through the simulation drainage tile shown on the extreme right. (Courtesy G. S. Taylor, The Ohio State University)

perforated pipe must be sloped to allow water to move rapidly to an outlet ditch or sewer. If the drain successfully prevents the water table from rising above the floor level, water will not seep into the basement for the same reason that it will not seep into a drainpipe placed above the water table.

MOLE DRAINAGE. A **mole drain** system can be created by pulling a pointed shank followed by an attached bullet-shaped steel plug about 7 to 10 cm in diameter through the soil at the desired depth. The compressed wall channel thus formed provides a pathway for the removal of excess water, similar to a buried pipe. Mole drainage is quite inexpensive to install, but is efficient only in fine-textured soils in which the channel is likely to remain open for a number of years.

6.10   SEPTIC TANK DRAIN FIELDS

Thousands of ordinary suburbanites get their first exposure to the importance of water movement through soils when they choose a beautiful home site "in the country," make plans to build a dream home, and apply for the required local government building permit. If the home site is beyond the reach of the urban sewage system, treatment of household wastewater to prevent environmental pollution and health hazards will be the responsibility of the homeowner. The local authorities will usually not allow a home to be built until arrangements are made for wastewater treatment. Typically, a soil scientist will come out to inspect the soils at the homesite and judge their suitability for use as a septic tank **drain field**. If the soils fail to perc, or fall short in some other attribute described later in this section, the landowner may be denied the permit to build.

Thus, in many regions, it is through their wastewater treatment function that soil properties influence the value of land and the spread of residential development. For consulting soil scientists in industrialized nations, determination of soil suitability for septic tank drain fields is probably the single most commonly rendered service. Furthermore, in rapidly suburbanizing areas improperly sited septic tank drain fields may contribute significantly to pollution of groundwater and streams. For all these reasons, our discussion of practical water management in soil systems would be incomplete without consideration of the role of soil and, in particular, water movement through soil in treating wastewater in areas not served by a centralized sewage treatment system.

Operation of a Septic System

The most common type of on-site wastewater treatment for homes not connected to municipal sewage systems is the *septic tank* and associated *drain field* (sometimes called *filter field* or *absorption field*). Over 10 million homes in the United States use septic tank drain fields to treat sewage and wastewater. This method of sewage and wastewater treatment depends on several soil processes, of which the most fundamental is the movement of water through the soil.

BOX 6.1   SUCCESS OR FAILURE
IN LANDSCAPE DRAINAGE DESIGN

A hedge of hemlock trees, all carefully pruned into ornamental shapes as part of an intricate landscape design, were dying again because of poor drainage. A very expensive effort to improve the drainage under the hemlocks had proved to be a failure and the replanted hemlocks were again causing an unsightly blemish in an otherwise picture-perfect garden.

Finally, the landscape architect for the world-famous ornamental garden called in a soil scientist to assist her in finding a solution to the dying hemlock problem. Records showed that in the previous, failed attempt to correct the drainage problem, contractors had removed all the hemlock trees in the hedge and had dug a trench under the hedge some 3 m deep (a). They had then backfilled the trench with gravel up to about 1 m from the soil surface, completing the backfill with a high-organic-matter, silt loam topsoil. It was into this silt loam that the new hemlock trees had been planted. Finally a bark mulch had been applied to the surface.

The landscaper who had designed and installed the drainage system apparently had little understanding of the various soil horizons and their relation to the local hydrology. When the soil scientist examined the problem site, he found an impermeable claypan that was causing water from upslope areas to move laterally into the hemlock root zone.

Basic principles of soil water movement also told him that water would not drain from the fine pores in the silt loam topsoil into the large pores of the gravel, and therefore the gravel in the trench would do no good in draining the silt loam topsoil (compare the situation to that in figure 5.27). In fact, the water moving laterally over the impermeable layer created a perched water table that poured water into the gravel-filled trench, soon saturating both the gravel and the silt loam topsoil.

To cure the problem, the previous "solution" had to be undone (b). The dead hemlocks were removed, the ditches were reexcavated, and the gravel was removed from the trenches. Then the trench, except for the upper ½ m, was filled with a sandy loam subsoil to provide a suitable rooting medium for the replacement evergreen trees (a different species was chosen for reasons unrelated to drainage). The upper ½ m of the trench was filled with a sandy loam topsoil, which was also acid but contained a higher level of organic matter. The interface between the subsoil and surface soil was mixed so that there would be no abrupt change in pore configuration. This would allow an unsaturated wetting front to move down from the upper to the lower layers, drawing down any excess water.

About 1 m uphill from the trench an *interceptor drain* was installed (b). This involved digging a small trench through the impermeable clay layer that was guiding water to the area. A perforated drainage pipe surrounded by a layer of gravel was laid in the bottom of this trench with about a 1% slope to allow water to flow away from the area to a suitable outlet. The interceptor drain prevented the water moving laterally over the impermeable soil layer from reaching the evergreen hedge root zone.

Even though the replanting of the hedge was followed by an exceptionally rainy year, the new drainage system kept the soil well aerated, and the trees thrived. The principles of water movement explained in Sections 5.6 and 6.9 of this text were applied successfully in the field.

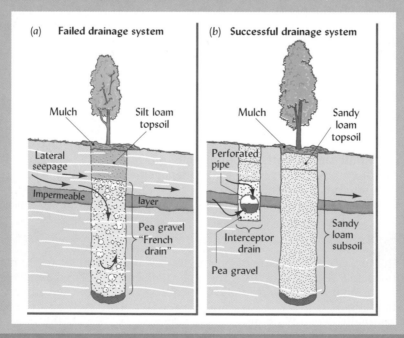

(a) **Failed drainage system**   (b) **Successful drainage system**

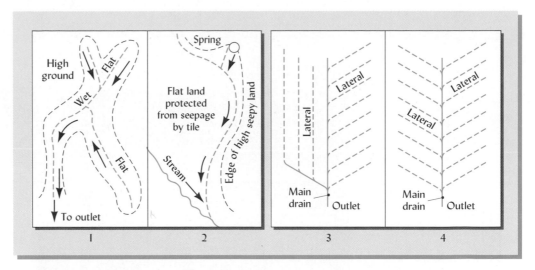

FIGURE 6.32 Tile drainage systems. (Top) Four typical systems for laying tile drain: (1) *natural,* which merely follows the natural drainage pattern; (2) *interception,* which cuts off water seeping into lower ground from higher lands above; and (3) *gridiron* and (4) *fishbone* systems, which uniformly drain the entirety of an area. (Bottom) A freshly installed drainage system. Tile lines feed into main drains at the bottom, the direction of flow being shown by the arrows. (Photo courtesy of U.S. Department of Agriculture)

In essence, a septic drain field operates like artificial soil drainage in reverse. A network of perforated underground pipes is laid in trenches, very much like the network of drainage pipes used to lower the water table in a poorly drained soil. But instead of draining water away from the soil, the pipes in a septic drain field carry wastewater *to* the soil, the water entering the soil via slits or perforations in the pipes. In a properly functioning septic drain field the wastewater will enter the soil and percolate downward, undergoing several purifying processes before it reaches the groundwater. One of the advantages of this method of sewage treatment is that it has the potential to replenish groundwater supplies for other uses.

THE SEPTIC TANK.   The basic design of a standard septic tank and drain field is shown in Figure 6.36*a*. Water carrying wastes from toilets, sinks, and bathtubs flows by gravity

FIGURE 6.33 An open-ditch drainage system designed to lower the water table during the wet season. The water flowing in the ditch has seeped in from the saturated soil. Such ditches also speed the removal of surface runoff water. The capillary fringe above the water table can be seen as a dark band of soil. In order to protect water quality from chemicals and sediments that might reach the ditch via surface runoff, buffer strips of grass or forest vegetation should be established on both sides. Note in the background the piles of *spoil,* subsoil excavated to make the ditches. This material is usually spread thinly and mixed by tillage with the surface soil in the field, but it should first be tested for extreme acidity or other detrimental properties that could impair the quality of the surface soil in which it is to be mixed. (Photo courtesy of R. Weil)

FIGURE 6.34 Specialized, laser-guided equipment laying a drain line made of corrugated plastic pipe. The pipe has perforations on the underside that allow water to seep in from a saturated soil. The trench will be backfilled with the soil shown piled on the left. The drain line must be laid deeper than the seasonal high water table if it is to remove any water. (Photo courtesy of USDA Natural Resources Conservation Service).

FIGURE 6.35   A perforated plastic drainage pipe or "tile line" clogged with sediment. The sediment entered the pipe because of improper installation with the perforations facing upward instead of downward.

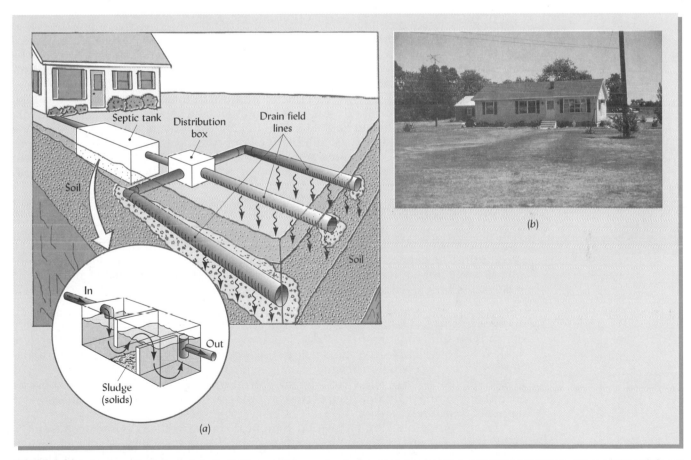

(b)

(a)

FIGURE 6.36   (a) A septic tank and drain field comprising a standard system for on-site wastewater treatment. Most of the solids suspended in the household wastewater settle out in the concrete septic tank. Effluent from the tank flows to the drain field, where it seeps out of the perforated pipes and into the soil. In the soil, the effluent is purified by microbial, chemical, and physical processes as it percolates toward the groundwater. (b) Photo shows the tell-tale dark strips of lawn where poorly functioning septic tank drain lines are stimulating the grass with wastewater and nitrogen. (Photo courtesy of R. Weil)

through sealed pipes to a large concrete box buried downslope from the dwelling. This large box, called the *septic tank,* has one or two removable inspection covers and one or more interior baffles or dividers. The baffles cause the inflowing wastewater to slow down and drop most (70%) of its load of suspended solid materials, which subsequently settle to the bottom of the septic tank. As these organic solids partially decompose by microbial action in the septic tank, their volume is reduced so that many years usually pass before the septic tank becomes too full and the accumulated **sludge** (also called *septage*) has to be pumped out.

THE DRAIN FIELD.   The water exiting the septic tank via a pipe near the top is termed the septic tank **effluent.** Although its load of suspended solids has been much reduced, it still carries organic particles, dissolved chemicals (including nitrogen), and microorganisms (including pathogens). The septic tank effluent flows by gravity to another, much smaller, concrete box called a *distribution box,* from which the flow is directed to one or more buried pipes that comprise the **drain field.** Blanketed in gravel and buried in trenches about 0.6 to 1 m under the soil surface, these pipes are perforated on the bottom to allow the wastewater to seep out and enter the soil.

It is at this point that soil properties play a crucial role. Septic systems depend on the soil in the drain field to (1) keep the effluent out of sight and out of contact with people, (2) treat or purify the effluent, and (3) conduct the purified effluent to the groundwater.

Soil Properties Influencing Suitability for a Septic Drain Field

The soil should have a *saturated hydraulic conductivity* (see Section 5.5) that will allow the wastewater to enter and pass through the soil profile rapidly enough to avoid backups that might saturate the surface soil with effluent, but slowly enough to allow the soil to purify the effluent before it reaches the groundwater. The soil should be sufficiently *well aerated* to encourage *microbial breakdown* of the wastes and *destruction of pathogens.* The soil should have some fine pores and clay or organic matter to adsorb and filter contaminants from the wastewater.

Properties that can be observed in the soil profile that indicate the presence of a high water table at some time of the year, such as mottled and gleyed colors, may disqualify a site for use as a septic drain field. Likewise, impermeable layers such as a fragipan or a heavy claypan could eliminate a building site from consideration, although restrictions caused by impermeable layers can sometimes be countered by placing the drain-field pipe below the restricted layer. This is an option, of course, only where there is considerable distance to the regional groundwater table.

Septic tank drain fields installed where soil properties are not appropriate may result in extensive pollution of groundwater and in health hazards caused by seepage of untreated wastewater. The photograph in Figure 6.36*b* shows one result of installing a septic drain field on a marginally suitable soil. The dark lines in the lawn show where the grass has responded to the water and nitrogen in the waste stream. Such signs indicate that the soil has too slow a percolation rate or too high a water table and that the wastewater has moved upward rather than downward.

SUITABILITY RATING.   The suitability of a site for septic drain-field installation largely depends on soil properties that affect water movement and the ease of installation (see Table 6.7). For example, too steep a slope may interfere both with the ease of installation and with the operation of a septic drain field. A septic drain field laid out on a slope greater than 15% may allow considerable lateral movement of the percolating water such that at some point downslope, the wastewater will seep to the surface and present a potential health hazard.

The soil properties ideal for a septic drain field are nearly the opposite of those associated with the need for tile drainage. For example, instead of a high water table which requires lowering by drainage, septic drain-field sites should have a low water table so that there is plenty of well-aerated soil to purify the wastewater before it reaches the groundwater. Application of large quantities of wastewater through septic drain fields will actually raise the water table somewhat under the drain field.

TABLE 6.7 Soil Properties Influencing Suitability for a Septic Tank Drain Field

Note that most of these soil properties pertain to the movement of water through the soil profile.

Soil property[a]	Limitations		
	Slight	Moderate	Severe
Flooding	—	—	Floods frequent to occasional
Depth to bedrock or impermeable pan, cm	>183	102–183	<102
Ponding of water	No	No	Yes
Depth to seasonal high water table, cm	>183	122–183	<122
Permeability (perc test) at 60 to 152 cm soil depth, mm/h	50–150	15–50	<15 or >150[b]
Slope of land, %	<8	8–15	>15
Stones >7.6 cm, % of dry soil by weight	<25	25–50	>50

[a] Assumes soil does not contain permafrost and has not subsided more than 60 cm.
[b] Soil permeability (as determined by a perc test) greater than 150 mm/h is considered too fast to allow for sufficient filtering and treatment of wastes.
Adapted from Soil Survey Staff (1993), Table 620-17.

PERC TEST. One of the most important tests conducted to determine the suitability of a soil for the installation of a septic tank is called the **perc test**. This is a test that determines the *percolation rate* (which is related to the saturated hydraulic conductivity described in Section 5.5). The percolation rate is expressed in millimeters (or other unit of depth) of water entering the soil per hour and indicates whether or not the soil can accept wastewater rapidly enough to provide a practical disposal medium (Table 6.7). The test is simple to conduct (Figure 6.37) and should be carried out during the wettest part of the year.

To some degree, a low percolation rate can be compensated for by increasing the total length of drain-field pipes, and hence increasing the area of land devoted to the drain field. By putting in more and longer trenches, a given amount of wastewater will be spread out over a larger soil volume. Generally, the size of the septic drain field is determined by the amount of wastewater that is likely to be generated. In many jurisdictions the size of the drain field is therefore determined by the number of bedrooms in the dwelling that it serves.

Alternative Systems

In certain low-lying regions, it is virtually impossible to find sites where the water table is deep enough to be suitable for installation of a standard septic tank drain field. When this is the case, alternatives to the traditional underground septic drain field must be found.

One of these alternatives is the *mound drain-field system,* which involves the construction of a septic drain field above ground (Figure 6.38). The perforated drain-field pipes are laid above ground on a bed of sand and are covered by a mound of sandy soil material. A final covering of loamy soil is used to support grass vegetation. This type of system must rely on pumps to deliver the wastewater up to the perforated pipes in the mound. Vegetation established on the mound will use some of the water for evapotranspiration, but most of the wastewater will seep through the porous mound material into the underlying soil.

A few communities are experimenting with artificial wetlands constructed to treat septic tank effluent where soils are too impermeable or low-lying to be suitable for a standard drain field. The effluent is made to slowly flow through a series of shallow, vegetated ponds that purify the wastewater by physically filtering some of the solids and by removing or destroying nutrients and organic compounds by plant uptake and biochemical reactions.

Such alternatives to septic drain fields require a special permit, and are not allowed in all localities. Still more uncommon, but more environmentally friendly, are composting toilets that convert human wastes directly into a humuslike soil amendment rather than using water to flush away wastes.

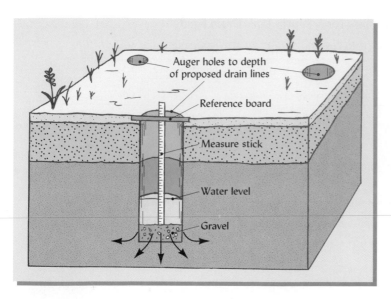

FIGURE 6.37 The perc test used to help determine soil suitability for a septic tank drain field. On the site proposed for the drain field, a number of holes are drilled to the depth where the perforated pipes are to be laid. The bottom of each hole is lined with a few centimeters of gravel, and the holes are filled with water as a pretreatment to assure that the soil is wet when the test is conducted. After the water has drained, the hole is refilled with water and a measuring rod is used to determine how rapidly the water level drops during a period of several hours or a day.

In the diagram: Auger holes to depth of proposed drain lines; Reference board; Measure stick; Water level; Gravel.

6.11 IRRIGATION PRINCIPLES AND PRACTICES[8]

In most regions of the world, insufficient water is the prime limitation to agricultural productivity. In semiarid and arid regions intensive crop production is all but impossible without supplementing the meager rainfall provided by nature. However, if given supplemental water through irrigation, the sunny skies and fertile soils of some arid regions stimulate extremely high crop yields. It is no wonder, then, that many of the earliest civilizations and city-states depended on irrigated agriculture (and vice versa).

[8] For a fascinating account of water resources and irrigation management in the Middle East, see Hillel (1995). For a practical manual on small-scale irrigation with simple but efficient microtechnology, see Hillel (1997). CAST (1988 and 1996) provides overviews of methods and prospects for commercial irrigated agriculture in the Western United States.

FIGURE 6.38 A mound-type septic tank effluent treatment system installed for a newly built house in a flat, poorly drained region. Note the air vent on the top of the mound and the ponded rainwater in the foreground. (Photo courtesy of R. Weil)

The history of irrigation is nearly as old as the history of agriculture itself. Rice producers in Asia, wheat and barley producers in the Middle East, and corn producers in Central and South America were irrigating their crops well over 2000 years ago.

The means of conveying water to crops and the style of irrigation have varied with time and place. The ancient Mesopotamian civilizations of the Tigris and Euphrates river valleys created complex networks of canals and ditches to divert water from the great rivers to extensive areas of cultivated land. Some desert inhabitants used a system of water harvesting—they channeled the storm runoff water from large areas of barren slopes to small valley bottoms of deep, level soils, where enough water could be stored in the soil profile to grow a crop. In other places, farmers filled simple buckets with water from a stream or open well and carried it to the thirsty plants in their gardens. Today these methods of water conveyance can still be seen, although in many agricultural and landscaping systems they have been replaced by modern pumps and pipes.

Importance of Irrigation Today

LANDSCAPING.   Irrigation is an integral part of landscaping in many residential and institutional areas, such as golf courses, home lawns, and flower beds (see photo in Figure 6.5). Irrigation serves to keep the grass green during summer dry spells in humid regions, but is necessary almost year-round to keep certain species thriving in drier regions. In arid regions, irrigation can alter a glaring, brown landscape into one of cooling shade trees and colorful flowers. In most cases, the use of irrigation for landscaping is predicated on the desire to maintain vegetation that conforms to an ideal notion of perpetual green lushness. For example, turf grass species such as Kentucky bluegrass, which is naturally verdant in Kentucky only during a few cool, wet months, are made to stay green year-round in some desert gardens.

In contrast, lawns, landscapes, and golf courses in many parts of the world utilize only adapted vegetation that is capable of surviving the periods of dry weather and other adverse conditions characteristic of the local climate without irrigation (though not necessarily in a lush green state). In the United States, landscaping irrigation has increased even faster than the growth in population. However, increasing environmental awareness has engendered a small but growing trend toward more xerophytic landscaping (utilizing desert plants and rocks) in arid regions and generally toward more use of locally native vegetation that requires little or no irrigation.

FOOD PRODUCTION.   During the 20th century, the area of irrigated cropland expanded greatly in many parts of the world, including the 17 semiarid states of the western United States. The total area under irrigation seems to be leveling off at about 250 million hectares worldwide, about 16% of the world's agricultural land. The high productivity induced by irrigation is evident in the fact that this irrigated land is responsible for some 40% of global crop production.

Expanded and improved irrigation, especially in Asia, has been a major factor in helping global food supplies keep up with, and even surpass, the global growth in population. As a result, irrigated agriculture remains the largest *consumptive* user[9] of water resources, accounting for about 80% of all water consumed, both worldwide and in the United States (Figure 6.39).

FUTURE PROSPECTS.   The water for expanded irrigation has come largely from reservoirs formed by the construction of dams and from the pumping of groundwater out of deep aquifers. Among the problems facing irrigated agriculture in the future is the slowly dwindling availability of irrigation water from both of these sources. The reasons for reduction include: (1) increased competition for water from a growing population of urban water users, (2) the overpumping of aquifers that has led to falling water tables,

---

[9] A distinction is made between *withdrawals* and *consumptive use*. The latter term refers to water withdrawn from surface or groundwater sources and used in a way that does not return the water directly to those sources for reuse. Thus, water withdrawn for irrigation and lost to the atmosphere by evapotranspiration is a consumptive use, but water withdrawn to cool an electric generation turbine and then returned to the river from which it came is not a consumptive use. On average, about 30 to 35% of the water withdrawals for irrigation are returned to water sources for reuse and are not part of the consumptive use, while 90 to 95% of the withdrawals for industrial use are returned and so are not considered consumptive use.

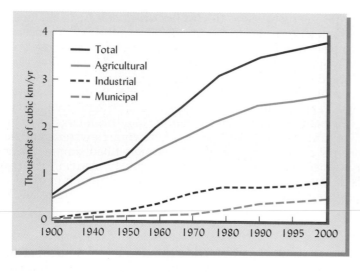

FIGURE 6.39 Trends in global consumptive water use by major categories of users. Note that agriculture (mainly irrigation) uses by far the greatest amount of water, although its share has declined in recent years. [Redrawn from Kuylenstierna, et al., (1997)]

(3) reduction of storage capacity of existing reservoirs by siltation with eroded sediments (see Section 17.2), and (4) increased recognition of the need to allow a portion of river flows to go unused by irrigation in order to maintain fish habitats downstream.

Water resources for irrigation are most rapidly becoming scarce and/or expensive in arid regions where they are most needed. Reducing waste and achieving greater efficiency of water use in irrigation are increasingly important aspects that will be emphasized in this section. One of the other major problems associated with irrigation, the salinization of soils and drainage waters, is considered in Chapter 10.

Water-Use Efficiency

Various measures of water-use efficiency are used to compare the relative benefits of different irrigation practices and systems. The most meaningful overall measure of efficiency would compare the output of a system (crop biomass or value of marketable product) to the amount of water allocated as an input into the system. There are many factors to consider (such as type of plants grown, reuse of "wasted" water by others downstream, etc.), so such comparisons must be made with caution.

APPLICATION EFFICIENCY.   A simpler measure of water-use efficiency, sometimes termed the **water application efficiency**, compares the amount of water allocated to irrigate a field to the amount of water actually used by the irrigated plants. In this regard, most irrigation systems are very inefficient, with as little as 30 to 50% of the water that is taken from the source ever reaching the plant roots.

Much of the water loss occurs in the canal and ditch system used to deliver the water to the irrigated field. If conveyance is by an open unlined ditch, water moves rapidly downward under the ditch and into the groundwater (however, some of this water may be used by irrigators farther downstream) and horizontally into the soil alongside the ditches. Substantial volumes of water are also lost by evaporation from the surface of water in canals and ditches. To reduce these losses during water distribution, ditches can be lined with concrete or plastic (see Figure 6.40). In some cases evaporative losses during distribution are eliminated by the use of closed conduits (pipelines), a means of distribution that is very much more expensive than open canals.

FIELD WATER EFFICIENCY.   Water-use efficiency *in the field* may be expressed as:

$$\text{Field water efficiency} = \frac{\text{Water transpired by the crop}}{\text{Water applied to the field}} \times 100$$

Values are usually quite low, especially if traditional systems are used. Commonly, more than half of the water actually delivered to the field is not transpired by the crop, but is lost as surface runoff, deep percolation below the root zone, and/or evaporation from the soil surface. Since some deep percolation is necessary to remove excess salts (see Section 10.9), the best **field water efficiency** possible is probably not much greater than

FIGURE 6.40 Concrete-lined irrigation ditches (center) and standard-sized siphon pipes (right) can increase the efficiency of water delivery to the field. Unlined ditches (left) lose much of the water to adjacent soil areas or to the groundwater. Note the evidence of capillary movement above the water level in the unlined ditch.

80 to 90%. Achieving a high level of field water efficiency (or a low level of water wastage) is very dependent on the skill of the irrigation manager and on the methods of irrigation used. We will therefore now turn our attention to the principal methods of irrigation in use today (Table 6.8).

Surface Irrigation

In these systems water is applied to the upper end of a field and allowed to distribute itself by gravity flow. Usually the land must be leveled and shaped so that the water will flow uniformly across the field. The water may be distributed in **furrows** graded to a slight slope so that water applied to the upper end of the field will flow down the furrows at a controlled rate (Figure 6.41). In **border irrigation** systems, the land is shaped into broad strips 10 to 30 m wide, bordered by low dikes.

WATER CONTROL. Water is usually brought to surface-irrigated fields in supply ditches or gated pipes (such as are shown in Figures 6.40 and 6.41). The amount of water that enters the soil is determined by the permeability of the soil and by the length of time a given spot in the field is inundated with water. Achieving a uniform infiltration of exactly the required amount of water is very difficult, and depends on controlling the slope and length of the irrigation runs across the field. If the soil is highly permeable (e.g., some sandy loams), too much water may infiltrate near the upper end of the field and too little may reach the lower end (Figure 6.42). On the other hand, for a fine-textured soil, infiltration may be so slow that water flows across the field and ponds up or runs off the lower end without a sufficient amount soaking into the soil. Therefore, in the very sandy soils, leaching loss of water and chemicals is a problem at the upper end of the field. In clayey soils, erosion, runoff, and waterlogging may be problems at the lower end.

The **level basin** technique of surface irrigation, as is used for paddy rice and certain tree crops, alleviates these problems because each basin has no slope and is completely surrounded by dikes that allow water to stand on the area until infiltration is complete.

TABLE 6.8 Some Characteristics of the Three Principal Methods of Irrigation

Methods and specific examples	Direct costs of installation, 1996 dollars[a]	Labor requirements	Field water efficiency, %[b]	Suitable soils
Surface: basin, flood, furrow	400–700	High to low, depending on system	40–50	Nearly level land; not too sandy or rocky
Sprinkler: center pivot, movable pipe, solid set	600–1200	Medium to low	60–70	Level to moderately sloping; not too clayey
Microirrigation: drip, porous pipe, spitter, bubbler	700–1500	Low	80–90	Steep to level slopes; any texture, including rocky or gravelly soils

[a] Average ranges from many sources. Costs of required drainage systems not included.
[b] Values from Hillel (1997). Field water efficiency = 100 × (water transpired by crop/water applied to field).

FIGURE 6.41 Typical irrigation scene. The use of easily installed siphons or gated pipes reduces the labor of irrigation and makes it easier to control the rate of application of water. Note the upward capillary movement of water along the sides of the rows. (Courtesy of USDA Natural Resources Conservation Service)

This method is not practical for highly permeable soils into which water would infiltrate so fast that the basin would never fill. On sloping land, terraces can be built in a modification of the level-basin method (Figure 6.43).

Variants of the surface systems have been in use for 5000 years. They require little equipment and are relatively inexpensive to operate. The principal capital cost is usually the initial land shaping, which may be quite expensive if the land is not nearly level to begin with. However, control over leaching and runoff losses is difficult, and the entire soil surface is wetted so that much water is lost by evaporation from the soil and by weed transpiration.

Sprinkler Systems

In sprinkler irrigation, water is sprayed through the air onto a field, simulating rainfall. Thus the entire soil surface, as well as plant foliage (if present), is wetted. This leads to evaporative losses similar to those described for surface systems. Furthermore, an additional 5 to 20% of the applied water may be lost by evaporation or windblown mist as the drops fly through the air. One advantage is that plants often respond positively to the cooler, better-aerated sprinkler water. A disadvantage is that wet leaves may increase the incidence of fungal diseases in some plants, such as grapes, fruit trees, and roses, so sprinkler systems are not often used for these plants.

WATER CONTROL.   A sprinkler system should be designed to deliver water at a rate that is less than the infiltration capacity of the soil, so that runoff or excessive percolation will not occur. In practice, runoff and erosion may be problems if the soil infiltration capac-

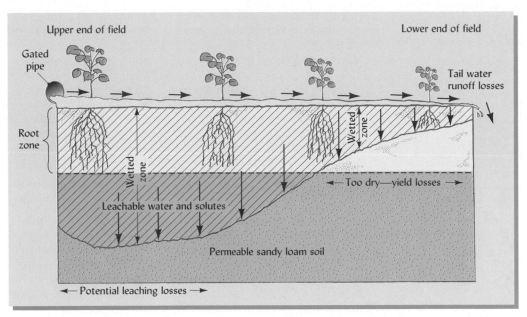

FIGURE 6.42 Penetration of water into a coarse-textured soil under surface irrigation. The high infiltration rate causes most of the water to soak in near the gated pipe at the upper end of the field. The uneven penetration of water results in the potential for leaching losses of water and dissolved chemicals at the upper end of the field, while plants at the lower end may receive insufficient water to moisten the entire potential root zone. On a less-permeable soil, or on a field with a steep slope, the tail water runoff losses would likely be greater at the lower end and leaching potential less at the upper end.

ity is low, the land is relatively steep, or too much water is applied in one place. Water is sometimes lost by deep percolation because more water falls near the sprinkler than farther from it. Overlapping of spray circles can help achieve a more even distribution of water. Because of better control over application rates, the field water-use efficiency is generally higher for sprinkler systems than for surface systems, especially on coarse-textured soils.

FIGURE 6.43 The level-basin type of surface irrigation modified for sloping land in South Asia by construction of terraces. Paddy rice is growing in the flooded terrace basins. This type of irrigation is practical only on soils of low permeability. (Photo courtesy of R. Weil)

SUITABLE SOILS. Sprinkler irrigation is practical on a wider range of soil conditions than is the case for the surface systems. Various types of sprinkler systems are adapted to moderately sloping as well as level land. They can be used on soils with a wide range of textures, even those too sandy for surface irrigation systems.

EQUIPMENT. The equipment costs for sprinkler systems are higher than those for surface-flow systems. Large pressure pumps and specialized pipes and nozzles are required. Some types of sprinkler systems are set in place, others are moved by hand, and still others are self-propelled, either moving in large circles around a central pivot (Figure 6.44) or rolling slowly across a rectangular field. Most systems can be automated and adapted to deliver doses of pesticides or soluble fertilizers to plants.

Microirrigation

The "micro" in microirrigation suggests that only a small portion of the soil is wetted by these systems (in contrast to the complete wetting accomplished by most surface and sprinkler systems). The name may also refer to the tiny amounts of water applied at any one time and the miniature size of the equipment involved.

Perhaps the best-established microirrigation system is *drip* (or *trickle*) *irrigation*, in which tiny emitters attached to plastic tubing apply water to the soil surface alongside individual plants. In some cases the tubing and emitters are buried 20 to 50 cm deep so the water soaks directly into the root zone. In either case, water is applied at a low rate (sometimes drop by drop) but at a high frequency, with the objective of maintaining optimal soil water availability in the immediate root zone while leaving most of the soil volume dry (Figure 6.45a).

Other forms of microirrigation that are especially well adapted for irrigating individual trees include *spitters* (microsprayers) and *bubblers* (small vertical standpipes); (Figure 6.45b). The bubblers (and usually the spitters) require that a small level basin be formed in the soil under each tree.

WATER CONTROL. Water is normally carried to the field in pipes, run through special filters to remove any grit or chemicals that might clog the tiny holes in the emitters (filtering is not necessary for bubblers), and then distributed throughout the field by means of a network of plastic pipes. Soluble fertilizers may be added to the water as needed.

If properly maintained and managed, microirrigation allows much more control over water application rates and spatial distribution than do either surface or sprinkler sys-

FIGURE 6.44 A center-pivot irrigation system in eastern Colorado. The system is rotating toward the left (note the dry soil at the extreme left), and the center pivot of the system is located approximately 0.5 km in the distance. Each self-propelled tower moves in coordination with all the others. The shed at the right houses a diesel-powered water pump that taps a deep aquifer. The tank next to the shed holds liquid fertilizer that can be metered into the irrigation water. (Photo courtesy of R. Weil)

FIGURE 6.45 Two examples of microirrigation. (Left) *Drip* or *trickle* irrigation with a single emitter for each seedling in a cabbage field in Africa. (Right) A *microsprayer* or *spitter* irrigating an individual tree in a home garden in Arizona. In both cases, irrigation wets only the small portion of the soil in the immediate root zone. Small quantities of water applied at high frequency (such as once or twice a day) assures that the root zone is kept almost continuously at an optimal moisture content. (Left photo courtesy of R. Weil)

tems. Losses by supply-ditch seepage, sprinkler-drop evaporation, runoff, drainage (in excess of that needed to remove salts), soil evaporation, and weed transpiration can be greatly reduced or eliminated. Once in place, the labor required for operation is modest.

Microirrigation often produces healthier plants and higher crop yields because the plant is never stressed by low water potentials or low aeration conditions that are associated with the feast or famine regime of infrequent, heavy water applications made by all surface irrigaton systems and most sprinkler systems. A disadvantage or risk is that there is very little water stored in the soil at any time, so even a brief breakdown of the system could be disastrous in hot, dry weather.

EQUIPMENT. The capital costs for microirrigation tend to be higher than for other systems (see Table 6.8), but the differences are not so great if the cost of drainage systems for control of salinity and waterlogging in surface systems, the costs of high-pressure pumping in sprinkler systems, and the real value of wasted water is taken into account. Because of its high water-use efficiency, microirrigation is most profitable where water supplies are scarce and expensive, and where high-valued plants such as fruit trees are being grown.

Irrigation Water Management

The two most serious irrigation management problems relate to the quality of the water being applied and to the efficiency of the irrigation water in stimulating plant production.

SALINITY BUILDUP. Most irrigation systems are located in semiarid and arid regions where the levels of soluble salts in the drainage water and, in turn, in the streams and rivers are relatively high. When this water is added to the soil and percolation takes place, still more salts are dissolved from the soil itself, making the drainage water even more saline than the originally added water. As the drainage water is repeatedly reused downstream, the salt buildup in the water can become very damaging to both the physical and chemical properties of the soils to which the water is applied (see Box 10.1).

EFFICIENT WATER USE. Practices to enhance water-use efficiency must begin with the collection of water in watersheds and the careful pumping of water from aquifers. Next, distribution canals must be lined or, better yet, replaced with pipelines, to avoid the major losses that occur before the water even gets to the farm field. Once the water arrives at the farm field, individual farmers must select the most efficient irrigation system feasible and keep it in top operating condition. The microirrigation systems are clearly the most efficient in water use, but whether or not they are most economical to use for a given situation depends on many factors.

In the field, the use of crop residues or mulches to reduce evaporation from the soil surface while simultaneously reducing the soil temperature (see Section 7.12) can also enhance overall productivity and water-use efficiency. It is also wise to concentrate irrigation on those crops that produce high-value products with relatively low levels of water use. Much water is currently squandered in producing low-value, high-transpiration-ratio crops (such as forages, cereals, and cotton) that would be more economically imported from rain-fed areas if all the costs of obtaining irrigation water were actually paid for by the irrigators (as opposed to being subsidized by governments).

Likewise, arid-region homeowners' attempts to sustain humid-region landscaping (e.g., green lawns and high-water-use shrubs) around their homes rather than using xerophytic (desert) plants, mulches and stones is equally wasteful.

Finally, to maintain or increase plant production in the face of dwindling water supplies, the application of water must be scheduled according to plant needs, which change with the weather, the LAI, and other factors. The basic principles of water retention and movement in soils and water use by plants, as outlined in this and the previous chapter, must be applied in developing efficient irrigation schedules.

6.12   CONCLUSION

The hydrologic cycle encompasses all movements of water on or near the earth's surface. It is driven by solar energy, which evaporates water from the ocean, the soil and vegetation. The water cycles into the atmosphere, returning elsewhere to the soil and the oceans in rain and snow.

The soil is an essential component of the hydrologic cycle. It receives precipitation from the atmosphere, rejecting some of it, which is then forced to run off into streams and rivers, and absorbing the remainder, which then moves downward, to be either transmitted to the groundwater, taken up and later transpired by plants, or evaporated directly from soil surfaces and returned to the atmosphere.

The behavior and movement of water in soils and plants are governed by the same set of principles: Water moves in response to differences in energy levels, moving from higher to lower water potential. These principles can be used to manage water more effectively and to increase the efficiency of its use.

Management practices should encourage movement of water into well-drained soils while minimizing evaporative (E) losses from the soil surface. These two objectives will provide as much water as possible for plant uptake and groundwater recharge. Water from the soil must satisfy the transpiration (T) requirements of healthy leaf surfaces; otherwise, plant growth will be limited by water stress. Practices that leave plant residues on the soil surface and that maximize plant shading of this surface will help achieve high efficiency of water use.

Extreme soil wetness, characterized by surface ponding and saturated conditions, is a natural and necessary condition for wetland ecosystems. However, for most other land uses extreme wetness is detrimental. Drainage systems have therefore been developed to hasten the removal of excess water from soil and lower the water table so that upland plants can grow without aeration stress, and so the soil can better bear the weight of vehicular and foot traffic.

A septic tank drain field operates as a drainage system in reverse. Septic waste waters can be disposed of and treated by soils if the soils are freely draining. Soils with low permeability or high water tables may indicate good conditions for wetland creation or appropriate sites for installation of artificial drainage for agricultural use, but they are not generally suited for septic tank drain fields.

Irrigation waters from streams or wells greatly enhance plant growth, especially in regions with scarce precipitation. With increasing competition for limited water resources, it is essential that irrigators manage water with maximal efficiency so that the greatest production can be achieved with the least waste of water resources. Such efficiency is encouraged by practices that favor transpiration over evaporation, such as mulching and the use of microirrigation.

As the operation of the hydrologic cycle causes constant changes in soil water content, other soil properties are also affected, most notably soil aeration and temperature, the subjects of the next chapter.

STUDY QUESTIONS

1. You know that the forest vegetation that covers a 120 km² wildland watershed uses an average of 4 mm of water per day during the summer. You also know that the soil averages 150 cm in depth and at field capacity can store 0.2 mm of water per mm of soil depth. However, at the beginning of the season the soil was quite dry, holding a average of only 0.1 mm/mm. As the watershed manager, you are asked to predict how much water will be carried by the streams draining the watershed during the 90-day summer period when 450 mm of precipitation falls on the area. Use the water balance equation to make a rough prediction of the stream discharge as a percentage of the precipitation and in cubic meters of water.

2. Draw a simple diagram of the hydrologic cycle using a separate arrow to represent these processes: *evaporation, transpiration, infiltration, interception, percolation, surface runoff,* and *soil storage.*

3. Describe and give an example of the *indirect* effects of plants on the hydrologic balance though their effects on the soil.

4. State the basic principle that governs how water moves through the SPAC. Give two examples, one at the soil–root interface and one at the leaf–atmosphere interface.

5. Define *potential evapotranspiration* and explain its significance to water management.

6. What is the role of evaporation from the soil (E) in determining water-use efficiency, and how does it affect ET? List three practices that can be used to control losses by E.

7. Weed control should reduce water losses by what process?

8. Comment on the relative advantages and disadvantages of organic versus plastic mulches.

9. What does conservation tillage conserve? How does it do it?

10. What is a lysimeter? For what is it used?

11. Explain under what circumstances earthworm channels might increase downward saturated water flow, but not have much effect on the leaching of soluble chemicals applied to the soil.

12. What will be the effect of placing a perforated drainage pipe in the capillary fringe zone just above the water table in a wet soil? Explain in terms of water potentials.

13. What soil features may limit the use of a site for a septic tank drain field?

14. Which irrigation systems are likely to be used where: (1) Water is expensive and the market value of crops produced per hectare is high, and (2) The cost of irrigation water is subsidized and the value of crop products that can be produced per hectare is low. Explain.

REFERENCES

Aboukhaled, A., A. Alfaro, and M. Smith. 1982. *Lysimeters.* FAO Irrigation and Drainage Paper 39. (Rome: U.N. Food and Agriculture Organization).

Alegre, J. C., D. K. Cassel, and D. E. Bandy. 1986. "Effects of land clearing and subsequent management on soil physical properties," *Soil Sci. Soc. Amer. J.,* 50:1379–1384.

Allen, R. G., T. A. Howell, W. O. Pruitt, W. I. Walter, and M. E. Jensen (eds.). 1991. *Lysimeters for Evapotranspiration and Environmental Measurements.* (New York: Am. Soc. Civ. Eng.).

Beauchamp, K. H. 1955. "Tile drainage—its installation and upkeep," *The Yearbook of Agriculture (Water).* (Washington, D.C.: U.S. Department of Agriculture), p. 513.

Black, A. L. 1982. "Long-term N-P fertilizer and climate influences on morphology and yield components of spring wheat," *Agron. J.,* 74:651–57.

CAST. 1988. *Effective Use of Water in Irrigated Agriculture.* Task Force Report no. 113. (Ames, Iowa: Council for Agricultural Science and Technology).

CAST. 1996. *Future of Irrigated Agriculture*. Task Force Report no. 127. (Ames, Iowa: Council for Agricultural Science and Technology).

Dreibelbis, F. R., and C. R. Amerman. 1965. "How much topsoil moisture is available to your crops?," *Crops and Soils*, **17**:8–9.

Fouss, J. L. 1974. "Drain tube materials and installation," in J. Van Schilfgaarde (ed.), *Drainage for Agriculture*. Agronomy Series no. 17. (Madison, Wis.: Amer. Soc. Agron.), pp. 147–177.

Greb, B. W. 1983. "Water conservation: Central Great Plains," in H. E. Dregue and W. O. Willis (eds.), *Dryland Agriculture*. Agronomy Series no. 23. (Madison, Wis.: Amer. Soc. of Agron.).

Haynes, J. L. 1954. "Ground rainfall under vegetative canopy of crops," *J. Amer. Soc. Agron.*, **46**:67–94.

Hillel, D. 1980. *Applications of Soil Physics*. (New York: Academic Press).

Hillel, D. 1995. *The Rivers of Eden* (New York: Oxford University Press).

Hillel, D. 1997. *Small-Scale Irrigation for Arid Zones*. FAO Development Series 2. (Rome: U.N. Food and Agriculture Organization).

Hughes, H. S. 1980. *Conservation Farming*. (Moline, Ill.: John Deere and Company).

Jury, W. A., and H. Fluhler. 1992. "Transport of chemicals through soil: Mechanisms, models, and field applications," *Advances in Agronomy*, **47**:141–201.

Kittridge, J. 1948. *Forest Influences: The Effects of Woody Vegetation on Climate, Water and Soil*. (New York: McGraw-Hill).

Kuylenstierna, J. L., G. Björklund, and P. Najlis. 1997. "Future sustainable water use: Challenges and constraints," *J. Soil Water Conserv.*, **52**:151–156.

Linde, D. T., T. L. Watschke, A. R. Jarrett, and J. A. Borger. 1995. "Surface runoff assessment from creeping bent grass and perennial ryegrass turf," *Agron. J.*, **87**:176–182.

Lyon, T. L., H. O. Buckman, and N. C. Brady. 1952. *The Nature and Properties of Soils*, 5th ed. (New York: Macmillan).

Pruitt, W. O., F. J. Lourence, and S. Von Oettingen. 1972. "Water use by crops as affected by climate and plant factors," *California Agriculture*, **26**:10–14.

Ritchie, J. T. 1983. "Efficient water use in crop production: Discussion on the generality of relations between biomass production and evapotranspiration," in H. M. Taylor, et al. (eds.), *Limitations to Efficient Water Use in Crop Production*. (Madison, Wis.: Amer. Soc. Agron., Crop Sci. Soc. Amer., Soil Sci. Soc. Amer.).

Starrett, S. K., N. E. Christians, and T. Al Austin. 1996. "Movement of pesticides under two irrigation regimes applied to turfgrass," *J. Environ. Qual.*, **25**:566–571.

Soil Survey Staff. 1993. *National Soil Survey Handbook*. Title 430-VI. (Washington, D.C.: USDA Natural Resources Conservation Service).

UNDP/WMO. 1974. *Hydrometeorological Survey of the Catchments of Lakes Victoria, Kyoja, and Albert*. Project RAF 66/025 (4 vols.), pp. 498–509.

USDA 1974. *Summer Fallow in the Western United States*. Cons. Res. Rep. no. 17. (Washington, D.C.: USDA Agricultural Research Service).

Waddell, J., and R. Weil. 1996. "Water distribution in soil under ridge-till and no-till corn," *Soil Sci. Soc. Amer. J.*, **60**:230–237.

7

SOIL AERATION AND TEMPERATURE

The naked earth is warm with Spring. . . .
—JULIAN GRENFELL, INTO BATTLE

It is a central maxim of ecology that "everything is connected to everything else." This interconnectedness is one of the reasons that soils are such fascinating (and challenging) objects of study. In this chapter we shall explore two aspects of the soil environment, aeration and temperature, that are not only closely connected to each other, but are both also intimately influenced by many of the soil properties discussed in other chapters.

Since air and water share the pore space of soils, it is not surprising that much of what we learned about the texture, structure, and porosity of soils (Chapter 4) and the retention and movement of water in soils (Chapters 5 and 6) will have direct bearing on soil aeration. These are some of the physical parameters affecting aeration status, but chemical and biological processes also affect, and are affected by, soil aeration.

For the growth of plants and the activity of microorganisms, soil aeration status can be just as important as soil moisture status, and can sometimes be even more difficult to manage. In most forest, range, agricultural, and ornamental applications, a major management objective is to maintain a high level of oxygen in the soil for root respiration. Yet it is also vital that we understand the chemical and biological changes that take place when the oxygen supply in the soil is depleted.

Soil temperatures affect plant and microorganism growth and also influence soil drying by evaporation. The movement and retention of heat energy in soils are often ignored but hold the key to understanding many important soil phenomena, from frost-damaged pipelines and pavements to the spring awakening of biological activity in soils. The unusually high soil temperatures that result from fires on forest-, range-, or croplands can markedly change critical physical and chemical soil properties.

We will see that increasing soil temperatures influence soil aeration largely through their stimulating effects on the growth of plants and soil organisms and on the rates of biochemical reactions. Nowhere are these interrelationships more critical than in the water-saturated soils of wetlands, ecosystems that will, therefore, receive special attention in this chapter.

7.1 THE NATURE OF SOIL AERATION

Aeration involves the ventilation of the soil, with gases moving both into and out of the soil. Aeration determines the rate of gas exchange with the atmosphere, the proportion of pore spaces filled with air, the composition of that soil air, and the resulting chemical oxidation or reduction potential in the soil environment.

The gases oxygen (O_2) and carbon dioxide (CO_2), along with water, are primary ingredients for two vital biological processes: (1) the **respiration** of all plant and animal cells, and (2) **photosynthesis** that creates sugars, a fundamental building block for all food. Respiration, which consumes O_2 and produces CO_2, oxidizes organic compounds as follows (using sugar as an example organic compound):

$$C_6H_{12}O_6 + 6O_2 \longrightarrow 6CO_2 + 6H_2O$$
$$\text{Sugar}$$

Through photosynthesis this reaction is reversed. Carbon dioxide and water are combined by green plants to form sugars, and oxygen is released, to the benefit of all respiring organisms—including people.

Soil aeration is a critical component of this overall system. For respiration to be carried out by soil organisms, oxygen must be supplied and carbon dioxide removed. In a well-aerated soil, the exchange of these two gases between the soil and the atmosphere is sufficiently rapid to prevent the deficiency of oxygen or the toxicity of excess carbon dioxide. For most upland plants, the supply of oxygen in the soil air must be kept above 0.1 L/L (as compared to 0.2 L/L in the atmosphere). In turn, the concentrations of CO_2 and other potentially toxic gases, such as methane or ethylene, must not be allowed to build up excessively.

7.2 SOIL AERATION IN THE FIELD

Oxygen availability in field soils is regulated by three principal factors: (1) *soil macroporosity* (as affected by texture and structure), (2) *soil water content* (as it affects the proportion of porosity that is filled with air), and (3) *O_2 consumption* by respiring organisms (including plant roots and microorganisms). The term *poor soil aeration* refers to a condition in which the availability of O_2 in the root zone is insufficient to support optimal growth of most plants. Typically, poor aeration becomes a serious impediment to plant growth when more than 80 to 90% of the soil pore space is filled with water (leaving less than 10 to 20% of the pore space filled with air). The high soil water content not only leaves little pore space for air storage, but, more important, the water blocks the pathways by which gases could exchange with the atmosphere. Compaction can also cut off gas exchange, even if the soil is not very wet and has a large percentage of air-filled pores.

Excess Moisture

The extreme case of excess water occurs when all or nearly all of the soil pores are filled with water. The soil is then said to be **water saturated** or **waterlogged.** Waterlogged soil conditions are typical of wetlands, and may also occur for short periods of time in depressions and flat areas on upland sites. In well-drained soils, saturated conditions may occur temporarily during a heavy rainstorm, when excess irrigation water is applied, or if wet soil has been compacted by plowing or by heavy machinery.

Such complete saturation of the soil with water is not a problem for some plant species whose roots have unique means of obtaining oxygen even when they are surrounded by water. Plants so adapted to life in waterlogged soils are termed **hydrophytes.** For example, a number of grass species, including rice, eastern gama grass, and spartina marsh grasses transport oxygen for respiration down to their roots via hollow structures in their stems and roots known as **aerenchyma** tissues. Mangroves (Figure 7.1), and other hydrophytic trees produce aerial roots and other structures that allow their roots to obtain O_2 while growing in water-saturated soils.

Most plants, however, are dependent on a supply of oxygen from the soil, and so suffer dramatically if good soil aeration is not maintained by drainage or other means (Figure 7.2). Some plants succumb to O_2 deficiency or toxicity of other gases within hours after the soil is saturated.

Gaseous Interchange

The more rapidly roots and microbes use up oxygen and release carbon dioxide, the greater is the need for the exchange of gases between the soil and the atmosphere. This

FIGURE 7.1 Relatively few plants, such as these mangrove trees along the coast of the Indian Ocean, are able to grow in saturated soils that are virtually devoid of oxygen. The mangroves are partially submerged during high tide, but have developed aerial "knees" on their roots that enable them to access oxygen directly from the atmosphere. To survive in this environment, the mangroves have also evolved means of excluding the salts in the ocean water. (Photos courtesy of R. Weil)

exchange is facilitated by two mechanisms, **mass flow** and **diffusion.** Mass flow of air is much less important than diffusion in determining the total exchange that occurs. It is enhanced, however, by fluctuations in soil moisture content that force air in or out of the soil or by wind and changes in barometric pressure.

The great bulk of the gaseous interchange in soils occurs by *diffusion.* Through this process, each gas moves in a direction determined by its own partial pressure. The *partial pressure* of a gas in a mixture is simply the pressure this gas would exert if it alone were present in the volume occupied by the mixture. Thus, if the pressure of air is 1 atmosphere (~100 kPa), the partial pressure of oxygen, which makes up about 21% (.21 L/L) of the air by volume, is approximately 21 kPa.

Diffusion allows extensive gas movement from one area to another even though there is no overall pressure gradient for the total mixture of gases. There is, however, a concentration gradient for each individual gas, which may be expressed as a *partial pressure gradient.* As a consequence, the higher concentration of oxygen in the atmosphere will result in a net movement of this particular gas into the soil. Carbon dioxide and water vapor normally move in the opposite direction, since the partial pressures of these two gases are generally higher in the soil air than in the atmosphere. A representation of the principles involved in diffusion is given in Figure 7.3.

FIGURE 7.2   Most plants depend on the soil to supply oxygen for root respiration, and therefore are disastrously affected by even relatively brief periods of soil saturation during which oxygen becomes depleted. (Left) Sugar beets on a clay loam soil dying where the soil has become water-saturated in a compacted area. (Right) Pine trees dying in a sandy soil area that has become saturated as a result of flooding by beavers. A new community of plants better adapted to poorly aerated soil conditions is taking over the site. (Photos courtesy of R. Weil)

7.3   MEANS OF CHARACTERIZING SOIL AERATION

The aeration status of a soil can be characterized in several ways, including (1) the content of oxygen and other gases in the soil atmosphere, (2) the air-filled soil porosity, and (3) the chemical oxidation-reduction (redox) potential.

Gaseous Composition of the Soil Air

OXYGEN.   The atmosphere above the soil contains nearly 21% O_2, 0.035% CO_2, and more than 78% N_2. In comparison, soil air has about the same level of N_2 but is consistently lower in O_2 and higher in CO_2. The O_2 content may be only slightly below 20% in the upper layers of a soil with a stable structure and an abundance of macropores. It may drop to less than 5% or even to near zero in the lower horizons of a poorly drained

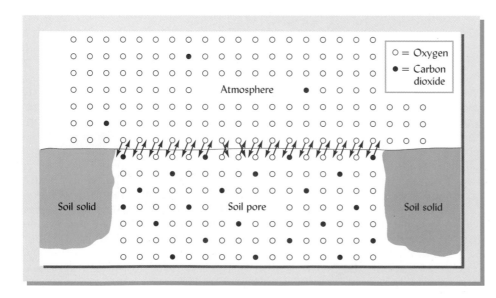

FIGURE 7.3   The process of diffusion between gases in a soil pore and in the atmosphere. The total gas pressure is the same on both sides of the boundary. The partial pressure of oxygen is greater, however, in the atmosphere. Therefore, oxygen tends to diffuse into the soil pore where fewer oxygen molecules per unit volume are found. The carbon dioxide molecules, on the other hand, move in the opposite direction owing to the higher partial pressure of this gas in the soil pore. This diffusion of O_2 into the soil pore and of CO_2 into the atmosphere will continue as long as the respiration of root cells and microorganisms consumes O_2 and releases CO_2.

soil with few macropores. Once the supply of O_2 is virtually exhausted, the soil environment is said to be **anaerobic.**

Low O_2 contents are typical of wet soils. Even in well-drained soils, marked reductions in the O_2 content of soil air may follow a heavy rain, especially if oxygen is being rapidly consumed by actively growing plant roots or by microbes decomposing readily available supplies of organic materials (Figure 7.4). Oxygen depletion in this manner occurs most rapidly when the soil is warm.

It is fortunate that the water in many soils contains small but significant quantities of dissolved O_2. When all the soil pores are filled with water, soil microorganisms can extract most of the O_2 dissolved in the water for metabolic purposes. This small amount of dissolved O_2 soon is used up, however, and if the excess water is not removed, aerobic microbial activities as well as plant growth are jeopardized.

CARBON DIOXIDE. Since the N_2 content of soil air is relatively constant, there is a general inverse relationship between the contents of the other two major components of soil air—O_2 and CO_2—with O_2 decreasing as CO_2 increases. Although the actual differences in CO_2 amounts may not be impressive, they are significant, comparatively speaking. Thus, when the soil air contains only 0.35% CO_2, this gas is about 10 times as concentrated as it is in the atmosphere. In cases where the CO_2 content becomes as high as 10%, it may be toxic to some plant processes.

OTHER GASES. Soil air usually is much higher in water vapor than is the atmosphere, being essentially saturated except at or very near the surface of the soil (see Section 5.7). Also, under waterlogged conditions, the concentrations of gases such as methane (CH_4) and hydrogen sulfide (H_2S), which are formed as organic matter decomposes, are notably higher in soil air. Another gas produced by anaerobic microbial metabolism is ethylene (C_2H_4). This gas is particularly toxic to plant roots, even in concentrations lower than 1 μ L/L (0.0001%). Root growth of a number of plants has been shown to be inhibited by ethylene that accumulates when gas exchange rates between the atmosphere and the soil are too slow.

Air-Filled Porosity

In Chapter 1 (Figure 1.17) we noted that the ideal soil composition for plant growth would include close to a 50:50 mix of air and water in the soil pore space, or about 25% air in the soil, by volume (assuming a total porosity of 50%). Many researchers believe

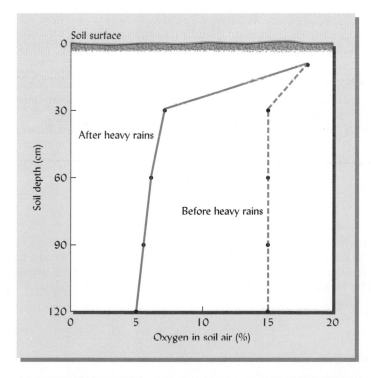

FIGURE 7.4 Oxygen content of soil air before and after heavy rains in a soil on which cotton was being grown. The rainwater replaced most of the soil air. The small amount of oxygen remaining was consumed in the respiration of roots and soil organisms. The carbon dioxide content (not reported) probably increased accordingly. [Redrawn from Patrick (1977); used with permission of the Soil Science Society of America]

that, in most soils, microbiological activity and plant growth become severely inhibited when air-filled porosity falls below 20% of the pore space or 10% of the total soil volume (with correspondingly high water contents).

One of the principal reasons that high water contents cause oxygen deficiencies for roots is that water-filled pores block the diffusion of oxygen into the soil to replace that used by respiration. In fact, oxygen diffuses 10,000 times faster through a pore filled with air than through a similar pore filled with water.

7.4 OXIDATION-REDUCTION (REDOX) POTENTIAL[1]

One important chemical characteristic of soils that is related to soil aeration is the reduction and oxidation states of the chemical elements in these soils. If a soil is well aerated, oxidized states such as those of Fe(III) in FeOOH and N(V) in NO_3^- (nitrate) are dominant. In poorly aerated soils, the reduced forms of such elements are found; for example, Fe(II) in FeO and N(III) in NH_4^+ (ammonium). The presence of these reduced forms is an indication of restricted drainage and poor aeration.

Redox Reactions

The reaction that takes place as the reduced state of an element is changed to the oxidized state may be illustrated by the oxidation of two-valent iron [Fe^{2+} or Fe(II)] in FeO to the trivalent form [Fe^{3+} or Fe(III)] in FeOOH.

$$\underset{Fe(II)}{\overset{(2+)}{2FeO}} + 2H_2O \rightleftharpoons \underset{Fe(III)}{\overset{(3+)}{2FeOOH}} + 2H^+ + 2e^-$$

Note that Fe(II) loses an electron e^- as it changes to Fe(III), and that H^+ ions are formed in the process. The loss of an electron suggests that there are potentials for the transfer of electrons from one substance to another. This **redox potential** can be measured using a platinum electrode.

The redox potential E_h provides a measure of the tendency of a substance to accept or donate electrons. It is usually measured in volts or millivolts. As is the case for water potential (Section 5.3), redox potential is related to a reference state, in this case the hydrogen couple $\frac{1}{2} H_2 \rightleftharpoons H^+ + e^-$, whose redox potential is arbitrarily taken as zero. If a substance will accept electrons easily, it is known as an *oxidizing agent*; if a substance supplies electrons easily, it is a *reducing agent*.

Role of Oxygen (O_2)

Oxygen gas (O_2) is an important example of a strong oxidizing agent, since it rapidly accepts electrons from many other elements. All aerobic respiration requires O_2 to serve as the electron acceptor as living organisms oxidize organic carbon to release energy for life.

Oxygen can oxidize both organic and inorganic substances. Keep in mind, however, that as it oxidizes another substance, O_2 is in turn reduced. This reduction process can be seen in the following reaction.

$$\overset{(0)}{\frac{1}{2}O_2} + 2H^+ + 2e^- \rightleftharpoons \overset{(2-)}{H_2O}$$

Note that the oxygen atom having zero charge in O_2 accepts two electrons, taking on a charge of −2 when it becomes part of the water molecule. These electrons could have been donated by two molecules of FeO undergoing oxidation, as shown in the previous reaction. If we combine the two equations we can see the overall effect of oxidation and reduction.

---

[1] For a review of redox reactions in soils, see Bartlett and James (1993).

$$2FeO + 2H_2O \rightleftharpoons 2FeOOH + 2H^+ + 2e^-$$
$$\frac{1}{2}O_2 + 2H^+ + 2e^- \rightleftharpoons H_2O$$

$$2FeO + \frac{1}{2}O_2 + H_2O \rightleftharpoons 2FeOOH$$

The donation and acceptance of electrons (e$^-$) and H$^+$ ions on each side of the equation have balanced each other and therefore do not appear in the combined reaction, but for the specific reduction and oxidation reactions they are both very important.

The redox potential E_h of a soil is dependent on both the presence of electron acceptors (oxygen or other oxidizing agents) and pH. The positive correlation in one soil between O_2 content of soil air and E_h (redox potential) is shown in Figure 7.5. In a well-drained soil, the E_h is in the 0.4 to 0.7 volt (V) range. As aeration is reduced, the E_h declines to a level of about 0.3 to 0.35 V when gaseous oxygen is depleted. Under flooded conditions, in warm, organic-matter-rich soils, E_h values as low as –0.3 V can be found.

Other Electron Acceptors

Other elements in addition to oxygen can act as terminal electron acceptors (oxidizers). For example, N(V) in nitrate accepts two electrons when it is reduced to N(III) in nitrite:

$$\underset{N(V)}{\overset{(5+)}{NO_3}} + 2e^- + 2H^+ \rightleftharpoons \underset{N(III)}{\overset{(3+)}{NO_2^-}} + H_2O$$

Similar reactions involve the reduction or oxidation of Fe, Mn, and S (see reactions in Figure 7.6).

The effect of pH on redox potentials relating to several important reactions taking place in soils is shown in Figure 7.6. Note that in all cases the E_h decreases as the pH rises from 2 to 8. Since both pH and E_h are easily measured, it is not too difficult to ascertain in a given soil whether a specified reaction would be likely to occur. For example, at pH 6 the E_h would need to be somewhat lower than +0.5 V to encourage the reduction of nitrates to nitrites, and about +0.2 volts to stimulate the reduction of FeOOH to the Fe^{2+} ion. For methane to form in a waterlogged soil at the same pH (6), an E_h of about –0.2 V would be required.

The E_h value at which oxidation-reduction reactions occur varies with the specific chemical to be oxidized or reduced. In Table 7.1 are listed oxidized and reduced forms of several elements important in soils, along with the approximate redox potentials at which the oxidation-reduction reactions occur. The E_h values explain the sequence of

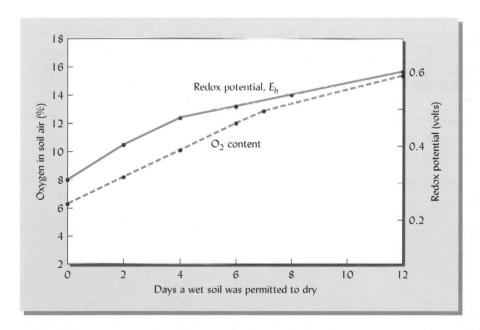

FIGURE 7.5 Relationship between oxygen content of soil air and the redox potential E_h. Measurements were taken at a 28-cm depth in a soil that had been irrigated continuously for 14 days prior to the drying. Note the general relationship between these two parameters. [Data selected from Meek and Grass (1975); used with permission of Soil Science Society of America]

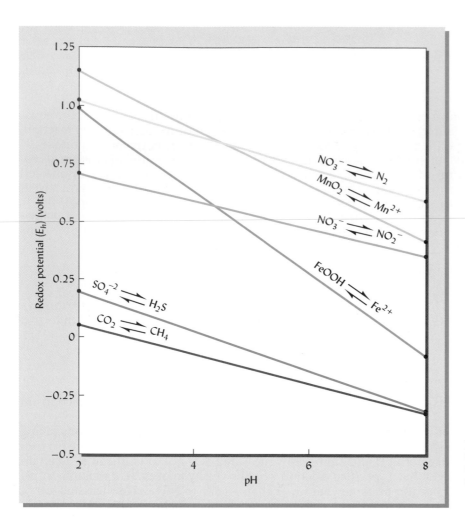

FIGURE 7.6 The effect of pH on the redox potential E_h at which several important reduction-oxidation reactions take place in soil. [From McBride (1994); used with permission of Oxford University Press]

reactions that are known to occur when a well-aerated soil becomes saturated with water.

At first, respiration will reduce the concentrations of O_2 in the soil air and dissolved in the soil water. As the concentration of this electron acceptor is lowered, so, too, the redox potential of the soil is lowered. Since O_2 is reduced to water at E_h levels of 0.38 to 0.32 V, once the soil redox potential falls below this level the soil is essentially devoid

TABLE 7.1 Oxidized and Reduced Forms of Certain Elements in Soils and the Redox Potentials E_h at Which Changes in Form Occur in a Soil at pH 6.5

Note that gaseous oxygen is depleted at E_h levels of 0.38 to 0.32 V. At lower E_h levels microorganisms utilize elements other than oxygen as the electron acceptor in their metabolism. By donating electrons they transform these elements into their reduced valence state.

Oxidized form	Reduced form	E_h at which change of form occurs, V
O_2	H_2O	0.38 to 0.32
NO_3^-	N_2	0.28 to 0.22
Mn^{4+}	Mn^{2+}	0.22 to 0.18
Fe^{3+}	Fe^{2+}	0.11 to 0.08
SO_4^{2-}	S^{2-}	−0.14 to −0.17
CO_2	CH_4	−0.20 to −0.28

From Patrick and Jugsujinda (1992).

of O_2. At lower E_h values, the only microorganisms that can function are those able to use elements other than oxygen as their metabolic electron acceptors.

The next most easily reduced element present is usually N(V) (in nitrate, NO_3^-). If the soil contains much nitrate, the E_h will remain near 0.28 to 0.22 V as the nitrate is reduced. Once nearly all the nitrate has disappeared [N(V) been transformed into N(III) and other N species], the E_h will drop further. At this point, organisms capable of reducing Mn will become active, and so on. Thus as E_h values fall, the elements N, Mn, Fe, and S (in SO_4^{-2}) and C (in CO_2) accept electrons and become reduced, predominantly in the order listed.

In other words, the soil E_h must be lowered to zero or less before methane is produced, but an E_h value of 0.28 to 0.22 is low enough to result in the reduction of nitrate-N. Thus, soil aeration helps determine the specific chemical species present and, in turn, the availability, mobility, and possible toxicity of various elements in soils.

7.5 FACTORS AFFECTING SOIL AERATION

Drainage of Excess Water

Drainage of gravitational water out of the profile and concomitant diffusion of air into the soil takes place most readily in macropores. The most important factors influencing the aeration of well-drained soils are therefore those that determine the volume of the soil macropores. Macropore content has a major influence on the total air space as well as on gaseous exchange and biochemical reactions. Soil texture, bulk density, aggregate stability, organic matter content, and biopore formation are among the soil properties that help determine macropore content and, in turn, soil aeration (see Section 4.6).

Rates of Respiration in the Soil

The concentrations of both oxygen and carbon dioxide are largely dependent on microbial activity, which in turn depends on the availability of organic carbon compounds as food. Incorporation of large quantities of manure, crop residues, or sewage sludge may alter the soil air composition appreciably. Likewise, the cycling of plant residues by leaf fall, root mass decay, and root excretion in natural ecosystems provides the substrate for microbial activity. Respiration by plant roots and enhanced respiration by soil organisms near the roots are also significant processes (Figure 7.7). All these processes are very much enhanced as soil temperature increases (see Section 7.9).

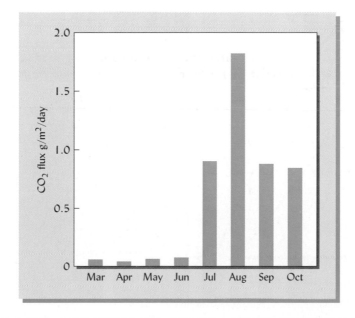

FIGURE 7.7 The rate of CO_2 movement (flux) from the surface of a soil within a northern hardwood forest ecosystem in New York State. Note the high rates from July to October when forest growth and microbical action were highest. [From Yavitt, et al. (1995); used with permission of Soil Science Society of America]

Subsoils are usually more deficient in oxygen than are topsoils. Not only is the water content usually higher (in humid climates), but the total pore space, as well as the macropore space, is generally much lower in the deeper horizons. In addition, the pathway for diffusion of gases into and out of the soil is longer for deeper horizons. However, if organic substrates are in low supply in the subsoil, it may still be aerobic. For this reason, certain recently flooded soils are anaerobic in the upper 50 to 100 cm and are aerobic below.

In some forested areas and in fruit orchards the tree roots extend into layers that are very low in oxygen and high in carbon dioxide, especially if the subsoil is high in clay. Carbon dioxide levels of nearly 15% (150 mL/L) have been observed in some such subsoils. Studies in deep, highly weathered soils under tropical rain forests indicate that respiration is carried out, and the concentration of CO_2 continues to rise, down into the deepest subsoil layers (Figure 7.8).

Soil Heterogeneity

PROFILE. As seen in Figures 7.3 and 7.8, the aeration status varies greatly in different locations in a soil profile. In well-drained, relatively uniform soils the trend is a general reduction in O_2 and increase in CO_2 as one moves down the profile. However, poorly aerated zones or pockets may be found in any horizon of an otherwise well-drained and well-aerated soil.

TILLAGE. One cause of soil heterogeneity is tillage, which has both short-term and long-term effects on soil aeration. In the short term, stirring the soil often allows it to dry out faster and mixes in large quantities of air. These effects are especially evident on somewhat compacted, fine-textured soils, on which plant growth often responds immediately after a cultivation to control weeds or "knife in" fertilizer. In the long term, however, tillage may reduce macroporosity (see Section 4.6).

LARGE MACROPORES. Poorly aerated zones may result from a heavy-textured or compacted soil layer, or it may be merely the inside of a soil ped (structural unit) where the smallness of pores may limit ready air exchange (Figure 7.9). In well-drained soils, the large pores (cracks) between peds, and old root channels in the subsoil, may periodically fill

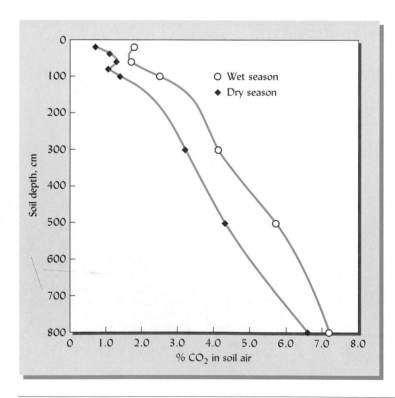

FIGURE 7.8 Changes in the concentration of soil air with depth into the profile of a Haplustox soil under a tropical rainforest in the Brazilian Amazon region. The source of the CO_2 was likely a combination of root and microbial respiration. Although by far the highest rates of CO_2 production were in the upper soil layers, gas produced there had little distance to travel to reach the atmosphere and many large pores to travel through. The concentration of CO_2 increased with depth because the increasing travel distance to the surface and the much lower macroporosity at depth greatly slowed the movement of gases, causing CO_2 to accumulate. [Data from Davidson and Trumbore (1995)]

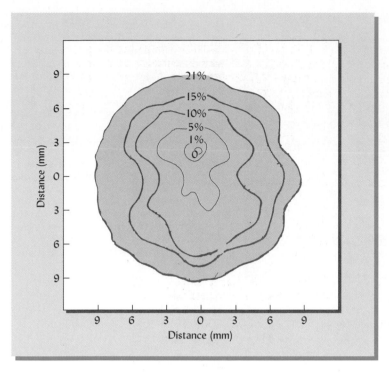

FIGURE 7.9   A map showing the oxygen content of soil air in a wet aggregate from an Aquic Hapludoll (Muscatine silty clay loam) from Iowa. The measurements were made with a unique microelectrode. Note that the oxygen content near the aggregate center was zero, while that near the edge of the aggregate was 21%. Thus pockets of oxygen deficiency can be found in a soil whose overall oxygen content may not be low. [From Sexstone, et al. (1985)]

with water, causing localized zones of poor aeration. In saturated soil, such large pores may cause the opposite effect, as they facilitate O_2 diffusion into the soil during periods of drying.

PLANT ROOTS.   Likewise, the roots of growing plants may either reduce or increase the O_2 concentration in their immediate vicinity. In somewhat poorly drained soils, respiration by roots of upland plants may deplete the O_2 in nearby soil (as can be seen in Plate 19, after page 498). In contrast, hydrophytic plants with aerenchyma tissues may transport surplus O_2 into their roots, allowing some to diffuse into the soil and produce an oxidized zone in an otherwise anaerobic soil (see for example, Plate 20).

For these reasons, oxidation reactions may be occurring within a few centimeters or millimeters from another location where reducing conditions exist. This heterogeneity of soil aeration should be kept in mind when attempting to understand the role that soil plays in elemental cycling and ecosystem function.

Seasonal Differences

There is marked seasonal variation in the composition of soil air. In the springtime in temperate humid regions, the soils are often wet and opportunities for ready gas exchange are poor. But due to low soil temperatures, the respiration of plant roots and soil microorganisms is restricted, so the utilization of oxygen and release of CO_2 are also restrained. In the summer months, the soils are commonly lower in moisture content, and the opportunity for gaseous exchange is increased. However, more favorable temperatures stimulate vigorous respiration by plant roots and microorganisms, releasing copious quantities of carbon dioxide. This is illustrated in Figure 7.10, which shows carbon dioxide levels in an Alfisol in Missouri during the corn-growing season and later into early winter.

Effects of Vegetation

In addition to the root respiration effects mentioned previously, vegetation may affect soil aeration by removing large quantities of water via transpiration, enough to lower the water table in some poorly drained soils. The data in Table 7.2 illustrate this effect for a pine forest. The effects of soil depth and season are also evident.

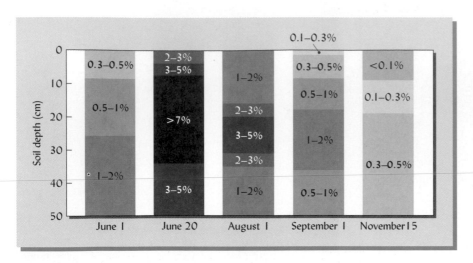

FIGURE 7.10 Seasonal changes in carbon dioxide content in the upper 50 centimeters of an Alfisol in Missouri on which corn was grown from early May until about September 1. By June 20, the corn was growing vigorously, the soil moisture level was still quite high, and the CO_2 level increased to more than 7% in the 10 to 30 cm zone. By August 1 gaseous exchange had probably increased and the CO_2 level had declined to 3 to 5% in the 20 to 30 cm layer, but even this is 100 times the level in the atmosphere. By November 15 activities of plants and microorganisms had declined because of lower temperatures and the CO_2 below 20 cm was only 10 times that in the atmosphere. [Data from Buyanovsky and Wagner (1983); used with permission of the Soil Science Society of America]

7.6 ECOLOGICAL EFFECTS OF SOIL AERATION

Effects on Organic Residue Degradation

Soil aeration influences many soil reactions and, in turn, many soil properties. The most obvious of these reactions are associated with microbial activity, especially the breakdown of organic residues and other microbial reactions. Poor aeration slows down the rate of decay, as evidenced by the relatively high levels of organic matter that accumulate in poorly drained soils.

The nature as well as the rate of microbial activity is determined by the O_2 content of the soil. Where O_2 is present, aerobic organisms are active, and oxidation reactions such as shown on page 266 occurs.

In the absence of gaseous oxygen, anaerobic organisms take over. Much slower breakdown occurs through reactions such as the following (see Section 12.2 for further details):

TABLE 7.2 Effect of Timber Harvest with Minimal Compaction on Soil Aeration and Temperature Regimes in a Subtropical Pine Forest

Once the 55-year-old loblolly pine trees were cut, evapotranspiration and shading were reduced, resulting in a higher water table, lower redox potentials, and warmer spring temperatures. The latter stimulated microbial use of oxygen and further lowered redox potentials in this Vertic Ochraqualf. Note that the redox potentials were lower with the warmer temperatures in spring, even though the soil was not as wet in spring as in winter.

Site treatment	Time soil is saturated, %	Soil temperature, °C		Soil redox potentials E_h, V	
		Winter	Spring	Winter	Spring
Measured at 50-cm depth					
Undisturbed pine stand	31	11.8	18.3	0.83	0.65
Trees cut, not compacted	64	11.7	20.5	0.51	0.11
Measured at 100-cm depth					
Undisturbed pine stand	54	13.3	17.3	0.83	0.49
Trees cut, not compacted	46	13.2	18.7	0.54	0.22

Data from Tiarks, et al. (1996).

$$C_6H_{12}O_6 \longrightarrow 2CO_2 + 2CH_3CH_2OH$$

$$\text{Sugar} \qquad\qquad\qquad\qquad \text{Ethanol}$$

Poorly aerated soils therefore tend to contain a wide variety of only partially oxidized products such as ethylene gas (C_2H_4), alcohols, and organic acids, many of which can be toxic to higher plants and to many decomposer organisms. The latter effect helps account for the formation of Histosols in wet areas where inhibition of decomposition allows thick layers of organic matter to accumulate. In summary, the presence or absence of oxygen gas completely modifies the nature of the decay process and its effect on plant growth.

Oxidation-Reduction of Elements

NUTRIENTS. Through its effects on the redox potential, the level of soil oxygen largely determines the forms of several inorganic elements, as shown in Table 7.3. The oxidized states of the nitrogen and sulfur are readily utilizable by higher plants. In general, the oxidized conditions are desirable for iron and manganese nutrition of most plants in acid soils of humid regions because, in these soils, the reduced forms of these elements are so soluble that toxicities may occur. However, some reduction of iron may be beneficial as it will release phosphorus from insoluble iron-phosphate compounds. Such phosphorus release has implications for eutrophication (see Section 14.2) when it occurs in saturated soils or in underwater sediments.

In drier areas, the opposite is generally true, and reduced forms of elements such as iron and manganese are preferred. In the neutral to alkaline soils of these drier areas, oxidized forms of iron and manganese are tied up in highly insoluble compounds, resulting in deficiencies of these elements. Such differences illustrate the interaction of aeration and soil pH in supplying available nutrients to plants (see Chapter 15).

OTHER ELEMENTS. Redox potential determines the species of such toxic elements as chromium, arsenic, and selenium, markedly affecting their impact on the environment and food chain (see Section 18.7).

SOIL COLORS. As was discussed in Section 4.1, soil color is influenced markedly by the oxidation status of iron and manganese. Colors such as red, yellow, and reddish brown are characteristic of well-oxidized conditions. More subdued shades such as grays and blues predominate if insufficient oxygen is present. Soil color can be used in field methods for determining the status of soil drainage. Imperfectly drained soils are characterized by contrasting streaks of oxidized and reduced materials (see Plates 15 and 18, after pages 82 and 498). Such a mottled condition indicates a zone of alternating good and poor aeration, a condition not conducive to the optimum growth of most plants.

METHANE PRODUCTION. The production of the organic compound methane in submerged soils is of universal significance, since this gas is one of those contributing to the greenhouse effect and global warming (see Section 12.11), and has been increasing in concentration by about 1% each year since 1980. Methane gas is produced by the reduction of CO_2. Its formation occurs when the E_h is reduced to about −0.2 V, a condition common in natural wetlands and in rice paddies. It is estimated that wetlands in the United States emit about 100 million metric tons of methane annually. Because of the biological productivity and diversity of these environments (see Section 7.8), soil scientists are seeking means of managing methane release without resorting to drainage of the wetlands.

TABLE 7.3 **Oxidized and Reduced Forms of Several Important Elements**

Element	Normal form in well-oxidized soils	Reduced form found in waterlogged soils
Carbon	CO_2, $C_6H_{12}O_6$	CH_4, C_2H_4, CH_3CH_2OH
Nitrogen	NO_3^-	N_2, NH_4^+
Sulfur	SO_4^{2-}	H_2S, S^{2-}
Iron	Fe^{3+} [Fe(III) oxides]	Fe^{2+} [Fe(II) oxides]
Manganese	Mn^{4+} [Mn(IV) oxides]	Mn^{2+} [Mn(II) oxides]

Effects on Activities of Higher Plants

Plants are adversely affected in at least three ways by conditions of poor aeration: (1) the growth of the plant, particularly the roots, is curtailed; (2) the absorption of nutrients and water is decreased; and (3), as discussed in the previous section, the formation of certain inorganic compounds toxic to plant growth is favored.

PLANT GROWTH.   Different plant species vary in their ability to tolerate poor aeration (Table 7.4). Sugar beets and barley are examples of crop species that require high air porosities for best growth (see Figure 7.2a). In contrast, ladino clover and reed canary grass can grow with very low air porosity. Rice and cranberries are crops that can grow with their roots submerged in water.

Furthermore, the tolerance of a given plant to low porosity may be different for seedlings than for rapidly growing plants. A case in point is the tolerance of red pine to restricted drainage during its early development and its poor growth or even death on the same site at later stages (see Figure 7.2b).

Knowledge of plant tolerance to poor aeration is useful in choosing appropriate species to revegetate wet sites. The occurrence of plants specially adapted to anaerobic conditions is useful in identifying wetland sites (see Section 7.8).

NUTRIENTS AND WATER UPTAKE.   Low O_2 levels constrain root respiration, a process that provides the energy needed for nutrient and water absorption. As a result, excess water ponding in low spots is sometimes seen to cause plants to wilt and reduce their uptake of water. Likewise, plants may exhibit nutrient deficiency symptoms on poorly drained soils even though the nutrients may be in good supply.

Soil Compaction and Aeration

Soil compaction does decrease the exchange of gases; however, the negative effects of soil compaction are not all owing to poor aeration. Soil layers can become so dense as to impede the growth of roots even if an adequate oxygen supply is available (see Section 4.5).

TABLE 7.4   **Examples of Plants with Varying Degrees of Tolerance to a High Water Table and Accompanying Restricted Aeration**

The plants in the leftmost column commonly thrive in wetlands. Those in the rightmost column are very sensitive to poor aeration.

Plants adapted to grow well with a water table at the stated depth				
<10 cm	*15 to 30 cm*	*40 to 60 cm*	*75 to 90 cm*	*>100 cm*
Bald cypress	Alsike clover	Birdsfoot trefoil	Beech	Arborvitae
Black spruce	Black Willow	Black locust	Birch	Barley
Common cattail	Cottonwood	Bluegrass	Cabbage	Beans
Cranberries	Deer tongue	Linden	Corn	Cherry
Duckgrass	Eastern gama grass	Mulberry	Hairy vetch	Hemlock
Fragmites grass	Ladino clover	Mustard	Millet	Oats
Maiden cane	Loblolly pine	Red maple	Peas	Peach
Mangrove	Orchard grass	Sorghum	Red oak	Sand lovegrass
Pitcher plant	Redtop grass	Sycamore		Sugar beets
Reed canarygrass	Tall fescue	Weeping lovegrass		Walnut
Rice		Willow oak		Wheat
Skunk cabbage				White pine
Spartina grass				
Swamp white oak				
Swamp rosemalow				
Water tupelo				

7.7 AERATION IN RELATION TO SOIL AND PLANT MANAGEMENT

Both surface and subsurface drainage are essential if an aerobic soil environment is to be maintained (see Section 6.9). Under irrigation, and for containerized plants, overwatering, which leads to poor aeration, is one of the most common problems encountered.

Soil Structure and Cultivation

The maintenance of a stable soil structure is an important means of augmenting good aeration. Pores of macro size, which are encouraged by large, stable aggregates, are soon drained of water following a rain, thereby allowing gases to move into the soil from the atmosphere. On cropland, the maintenance of organic matter by the addition of farm manure and crop residues and by the growth of close-growing grasses and legumes is perhaps the most practical means of encouraging aggregate stability, which in turn encourages good drainage and better aeration. In poorly drained, heavy-textured soils, however, it is often impossible to maintain optimum aeration without resorting to some cultivation of the soil. No-tillage systems, once established for 5 to 10 years, often provide enough continuous earthworm burrows and root channels to greatly assist in drainage. However, the mulch of surface residues may slow the evaporation of water in spring, resulting in undesirably cool, wet soils at planting time (see also Section 7.12).

Container-Grown Plants

Potted plants frequently suffer from waterlogging because it is difficult to supply the exact amount of water the plant needs. To prevent waterlogging, most containers have holes at the bottom to drain excess water. As was the case with stratified soils in the field (see Section 5.6), water drains out of the holes at the bottom of the pot *only* when the soil at the bottom is saturated with water. If the potting medium is mostly mineral soil, the fine soil pores remain filled with water, leaving no room for air, and anaerobic conditions soon prevail. Use of taller pots will allow for better aeration in the upper part of the medium.

To manage these problems, potting mixes are engineered to meet the requirements of the containerized plants. Mineral soil generally makes up no more than one-third of the volume of most potting mixes, the remainder being composed of inert lightweight but coarse-grained materials such as perlite (expanded volcanic glass), vermiculite (expanded mica—not soil vermiculite), or pumice (porous volcanic rock). Most modern mixes also contain some stable organic material, such as peat, shredded bark, wood chips, or compost, that holds water as well as adding macroporosity.

Tree and Lawn Management

In transplanting a young tree seedling or any woody species, special caution must be taken to prevent poor aeration or waterlogging immediately around the young roots. Figure 7.11 illustrates the right and the wrong way to manage the transplanting of trees in compacted soils.

The aeration of well-established, mature trees must also be safeguarded. If operators push surplus excavated soil around the base of a tree during landscape grading, serious consequences are soon noticed (Figure 7.12). The tree's feeder roots near the original soil surface are soon deficient in oxygen even if the overburden is no more than 5 to 10 cm in depth. Building a protective wall (a *dry well*) around the base of a valuable tree before grading operations begin will preserve enough of the original surface to allow the tree roots access to the O_2 they need, thereby saving the tree.

Management systems for heavily trafficked lawns commonly have components relating to soil aeration. For example, one means of increasing the aeration in compacted lawn areas is to use *core cultivation* that actually removes small cores of soil from the surface horizon, thereby permitting gas exchange to take place more easily (Figure 7.13). Care must be taken not to merely punch holes in the soil, however, since such a practice would thereby increase soil compaction in the surrounding soil.

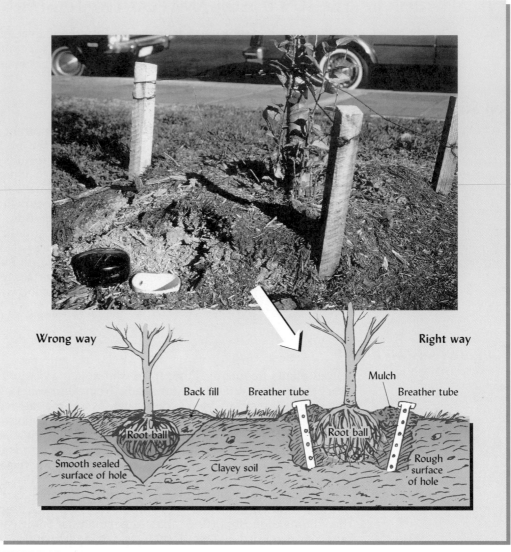

Wrong way
Right way
Back fill
Breather tube
Mulch
Breather tube
Root ball
Root ball
Smooth sealed surface of hole
Clayey soil
Rough surface of hole

FIGURE 7.11 Providing a good supply of air to tree roots can be a problem, especially when trees are planted in fine-textured, compacted soils of urban areas. A machine-dug hole with smooth sides will act as a "tea cup" and fill with water, suffocating tree roots. Breather tubes, a larger rough-surfaced hole, and a layer of surface mulch in which some fine tree roots can grow are all measures that can improve the aeration status of the root zone. (Photo courtesy of R. Weil)

7.8 WETLANDS AND THEIR POORLY AERATED SOILS[2]

Poorly aerated areas called **wetlands** cover approximately 14% of the world's ice-free land, with the greatest areas occurring in the cold regions of Canada, Alaska, and Russia (Table 7.5). In the continental United States, about 400,000 km² exist today, less than half of the area that is estimated to have existed when European settlement of the nation began. Most wetland losses occurred as farmers used artificial drainage (see Section 6.9) to convert them into cropland. This conversion process was assisted by some of the same government agencies (e.g., U.S. Army Corps of Engineers and USDA) that are now working to protect wetlands from further damage. In recent decades, filling and drainage for urban development has also taken its toll of wetland areas. Since environmental consciousness has become a force in modern societies, wetland preservation has become a major issue, and the loss of wetlands has been slowed.

---

[2] Two well-illustrated, nontechnical, yet informative publications on wetlands are Welsh, et al. (1995), and CAST (1994). For a compilation of technical papers on hydric soils and wetlands, see Rabenhorst, et al., (1998).

FIGURE 7.12   Protection of valuable trees during landscape grading operations. Even a thin layer of soil spread over a large tree's root system can suffocate the roots and kill the tree. (Inset) In order to preserve the original ground surface so that tree feeder roots can obtain sufficient oxygen, a dry well may be constructed of brick or any decorative material. The dry well may be incorporated into the final landscape design, or filled in at a rate of a few centimeters per year. (Photos courtesy of R. Weil)

FIGURE 7.13   One means of increasing the aeration of compacted soil is by core cultivation. The machine removes small cores of soil, leaving holes about 2 cm in diameter and 5 to 8 cm deep. This method is commonly used on high-traffic turf areas. Note that the machine *removes* the cores and does not simply punch holes in the soil, a process that would increase compaction around the hole and impede air diffusion into the soil. (Photos courtesy of R. Weil)

TABLE 7.5 Major Types of Wetlands and Their Global Areas

All together, wetlands constitute perhaps 14% of the world's land area.

Wetland type	Global area, 1000⁶ km²	Percent of ice-free land area	Percent of all wetland areas
Inland (swamps, bogs, etc.)	5415	3.9	28.8
Riparian or ephemeral	3102	2.3	16.5
Organic (Histosols)	1366	1.0	7.3
Salt-affected, including coastal	2230	1.6	11.9
Permafrost-affected (Histels)	6697	4.9	35.6

Data from Eswaren, et al. (1996).

Defining a Wetland

Swamp, bog, marsh, fen . . . these terms for particular types of wet lands have been in common use for centuries, and most people have a pretty good idea what these areas are like. It may be puzzling, therefore, that despite decades of effort by scientists and environmental regulators, the precise definition of a wetland is still controversial.[3] *Wetland* is a scientific term for ecosystems that are transitional between land and water. These systems are neither strictly terrestrial (land-based) nor aquatic (water-based). While there are many different types of wetlands, they all share a key feature, namely *soils that are water-saturated near the surface for prolonged periods when the soil temperature is high enough to result in anaerobic conditions*. It is largely the prevalence of anaerobic conditions that determines the kinds of plants, animals, and soils found in these areas. There is widespread agreement that the wetter end of a wetland occurs where the water is too deep for rooted, emergent vegetation to take hold. The difficulty is in precisely defining the so-called *drier end* of the wetland, the boundary beyond which exist nonwetland, upland systems in which the plant–soil–animal community is no longer predominantly influenced by the presence of water-saturated soils.

The controversy is probably more political than scientific, however. Since uses and management of wetlands are regulated by governments in the United States and in many other countries, billions of dollars are at stake in determining what is and what is not protected as a wetland. (Consider, for example, a developer who wants to buy a 100-ha tract of land on which to build a shopping center. If 20 versus 60 of those hectares are declared off-limits to development, how will that affect the developer's willingness to pay?)

Because so much money is at stake, thousands of environmental professionals are employed in the process of **wetland delineation**—finding the exact drier-end boundaries of wetlands on the ground. Wetland delineation is *not* done in front of a computer screen, but is a sweaty, muddy, tick- and mosquito-ridden business that those trained in soil science are uniquely qualified to carry out.

What are the characteristics these scientists look for to indicate the existence of a wetland system? Most authorities agree that three characteristics can be found in any wetland:

1. A wetland hydrology or water regime
2. Hydric soils
3. Hydrophytic plants

We shall briefly examine each.

Wetland Hydrology

WATER BALANCE.   Water is obviously a major component of wetlands. However, different types of wetlands differ with regard to how much water they contain and how that water is supplied. Water flows into wetlands from surface runoff (e.g., bogs and

---

[3] In 1987, the United States Army Corps of Engineers and the Environmental Protection Agency agreed on the following definition to be used in enforcing the Clean Water Act: "The term wetlands means those areas that are inundated or saturated by surface or ground water at a frequency and duration sufficient to support, and that under normal circumstances do support, a prevalence of vegetation typically adapted for life in saturated soil conditions."

marshes), from groundwater seepage, and from direct precipitation. It flows out by surface and subsurface flows, as well by evaporation and transpiration (see Section 6.3 and Table 7.2). The balance between inflows and outflows, as well as the water storage capacity of the wetland itself, determines how wet it will be and for how long.

HYDROPERIOD. The degree of saturation in wetlands is controlled by the level of the water table (above or below the soil surface), and this level fluctuates with time. The temporal pattern of these water table changes is termed the **hydroperiod.** For a coastal marsh, the hydroperiod may be daily, as the tides rise and fall (Figure 7.14). For inland swamps, bogs, or marshes, the hydroperiod is more likely to be seasonal, with the highest water table levels occurring during the seasons of high rainfall and/or snow melt and low evapotranspiration. Some wetlands may be flooded for only a month or so each year, while some may never be flooded, although they are saturated within the upper soil horizons.

Also, if the period of saturation occurs when the soil is too cold for microbial or plant-root activity to take place, anaerobic conditions may not develop, even in flooded soils. This temperature requirement is sometimes referred to as the *growing season* in reference to the time of year during which plants are actively growing. As we shall see in Section 7.9, microbial activity continues down to about 5°C, so O_2 may be depleted when the soil is warmer than this (even if plants are dormant). Remember, it is the anaerobic condition, not just saturation, that makes a wetland a wetland.

RESIDENCE TIME. The more slowly water moves through and out of the wetland, the longer the period that a given portion of water is subjected to the wetland environment. A long *residence time* makes it more likely that wetland functions and reactions will be carried out. For this reason, actions that speed water flow, such as creating ditches or straightening stream meanders, are degrading to wetlands and should be avoided in wetland management.

INDICATORS. All wetlands are water-saturated some of the time, but many are not saturated all of the time. During wet periods, inundation with shallow water may be obvious. However, systematic field observations, assisted by instruments to monitor the changing level of the water table, are required to ascertain the frequency and duration of flooding or saturated conditions.

FIGURE 7.14 An example of a wetland with a daily hydroperiod that follows the rise and fall of the slightly brackish estuary tides. This tidal marsh with a beaver house is seen at low tide when some saturated soils and emergent plants are exposed. The drier-end boundary of the wetland is probably just beyond the treeline in the background. (Photo courtesy of R. Weil).

In the field, even during dry periods, there are many signs one can look for to indicate where saturated conditions frequently occur. Past periods of flooding will leave water stains on trees and rocks, and a coating of sediment on the plant leaves and litter. Drift lines of once-floating branches, twigs, and other debris also suggest previous flooding. Trees with extensive root masses above ground indicate an adaptation to saturated conditions. But perhaps the best indicator of saturated conditions is the presence of **hydric soils.**

Hydric Soils[4]

In order to assist in delineating wetlands, soil scientists developed the concept of hydric soils. In *Soil Taxonomy* (see Chapter 3) these soils are mostly (but not exclusively) classified in the order Histosols, in Aquic suborders such as Aquents, Aquepts, and Aqualfs, or in Aquic subgroups. These soils generally have an aquic or peraquic moisture regime (see Section 3.2).

DEFINED. Three properties help define hydric soils. First, they are subject to *periods of saturation* that inhibit the diffusion of O_2 into the soil. Second, for substantial periods of time they undergo *reduced conditions* (see Section 7.4); that is, electron acceptors other than O_2 are reduced. Third, they exhibit certain features termed *hydric soil indicators*.

INDICATORS. **Hydric soil indicators** are features associated (sometimes only in specific geographic regions) with the occurrence of saturation and reduction. Most of the indicators can be observed in the field by digging a small pit to a depth of about 50 cm. They principally involve the loss or accumulation of various forms of Fe, Mn, S, or C. The carbon (organic matter) accumulations are most evident in Histosols, but thick, *dark surface layers* in other soils can also be indicators of hydric conditions in which organic matter decomposition has been inhibited (see, for example, Plates 6 and 21, after pages 82 and 498).

Iron, when reduced to Fe(II), becomes sufficiently soluble that it migrates away from reduced zones and may precipitate as Fe(III) compounds in more aerobic zones. Zones where reduction has removed or depleted the iron coatings from mineral grains are termed *redox depletions.* They commonly exhibit the gray, low-chroma colors of the bare, underlying minerals (see Section 4.1 for an explanation of chroma). Also, iron itself turns gray to blue-green when reduced. The contrasting colors of redox depletions or reduced iron and zones of reddish oxidized iron result in unique mottled *redoximorphic features* (see, for example, Plates 15 and 18). Other redoximorphic features involve reduced Mn. These include the presence of hard black *nodules* that sometimes resemble shotgun pellets. Under severely reduced conditions the entire soil matrix may exhibit *low-chroma colors,* termed *gley.* Colors with a chroma of 1 or less quite reliably indicate reduced conditions (Figure 7.15).

Always keep in mind that redoximorphic features are indicative of hydric soils only when they occur in the upper horizons. Many soils of upland areas exhibit redoximorphic features only in their deeper horizons, due to the presence of a fluctuating water table at depth. Upland soils that are saturated or even flooded for short periods, especially if during cold weather, are *not* wetland (hydric) soils.

A unique redoximorphic feature associated with certain wetland plants is the presence, in an otherwise gray matrix, of reddish oxidized iron around root channels where O_2 diffused out from the aerenchyma-fed roots of a hydrophyte (see Plate 20). These *oxidized root zones* exemplify the close relationship between hydric soils and *hydrophytic vegetation.*

Hydrophytic Vegetation

Although the vast majority of plant species cannot survive the conditions characteristic of wetlands, there do exist varied and diverse communities of plants that have evolved

---

[4] The U.S. Department of Agriculture Natural Resources Conservation Service defines a hydric soil as one "that formed under conditions of saturation, flooding or ponding long enough during the growing season to develop anaerobic conditions in the upper part." For an illustrated field guide to features that indicate hydric soils, see Hurt, et al. (1996). For a current list of soil series considered to be Hydric soils, see the Website http://www.statlab.iastate.edu/soils/hydric/.

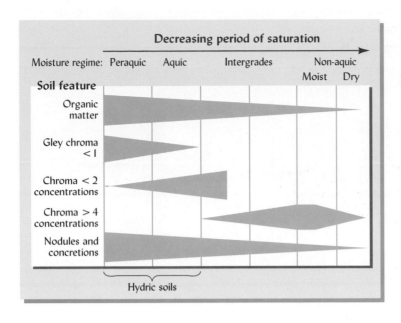

FIGURE 7.15 The relationship between the occurrence of some soil features and the annual duration of water-saturated conditions. The *absence* of iron concentrations (mottles) with colors of chroma >4, and the *presence* of strong expressions of the other features are indications that a soil may be hydric. [Adapted from Veneman (1998)]

special mechanisms to adapt to life in saturated, anaerobic soils. These plants comprise the **hydrophytic vegetation** that distinguishes wetlands from other systems.

Typical adaptive features include hollow aerenchyma tissues that allow plants like spartina grass to transport O_2 down to their roots. Certain trees (such as bald cypress) produce adventitious roots, buttress roots, or knees (see Figure 7.1). Other species spread their roots in a shallow mass on or just under the soil surface, where some O_2 can diffuse even under a layer of ponded water. The leftmost column in Table 7.4 lists a few common hydrophytes. Not all the plants in a wetland are likely to be hydrophytes, but the majority usually are.

Wetland Chemistry

Hydric soils and hydrophytic vegetation are both partly the result and partly the cause of the unique chemistry of wetland systems. The central characteristic of wetland chemistry is the low redox potential (see Section 7.4) that pertains. Furthermore, many wetland functions depend on *variations* in the redox potential; that is, in certain zones or for certain periods of time oxidizing conditions alternate with reducing ones.

LOW OXYGEN. For example, even in a flooded wetland, O_2 will be able to diffuse from the atmosphere or from oxygenated water into the upper 1 or 2 cm of soil, creating a thin *oxidizing zone* (see Figure 7.16). The diffusion of O_2 within the saturated soil is extremely limited, so that a few centimeters deeper into the profile, O_2 is eliminated and the redox potential becomes low enough for reactions such as nitrate reduction to take place. The close proximity of the oxidized and anaerobic zones allows water passing through wetlands to be stripped of N by the sequential oxidation of ammonium N to nitrate N, and then the reduction of the nitrate to various nitrogen gases that escape into the atmosphere (see Section 13.9).

REDOX. Redox potentials may become low enough for iron reduction to produce redoximorphic features, and for sulfate reduction to produce rotten-egg-smelling hydrogen sulfide (H_2S) gas. The anaerobic zone may extend downward, or in some cases may be limited to the upper horizons where microbial activity is high. The anaerobic carbon reactions discussed in Section 7.6 are characteristic of this zone, including methane (swamp gas) production. These and other chemical reactions involving the cycling of C, N, and S are explained in Sections 12.2, 13.9, and 13.21. Toxic elements such as chromium and selenium undergo redox reactions that may help remove them from the water before it leaves the wetland. Acids from industry or mine drainage may also be neutralized by reactions in hydric soils.

This array of unique chemical reactions contributes greatly to the benefits that society and the environment gain from wetlands (see Box 7.1)

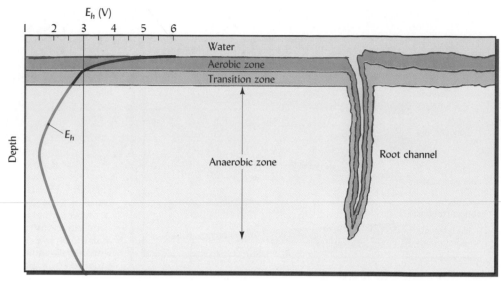

FIGURE 7.16 Representative redox potentials within the profile of an inundated hydric soil. Many of the biological and chemical functions of wetlands depend on the close proximity of reduced and oxidized zones in the soil. The changes in redox potential at the lower depths depend largely on the vertical distribution of organic matter. In some cases, low subsoil organic matter results in a second oxidized zone beneath the reduced zone.

Constructed Wetlands

Realizing all the beneficial functions of wetlands, scientists and engineers have begun not only to find ways to preserve natural wetlands, but to construct artificial ones for specific purposes, such as wastewater treatment (see, for example, Box 14.2).

Another reason for attempting to construct wetlands is the provision in several regulations that allows for the destruction of certain natural wetland areas, provided that an equal or larger area of new wetlands is constructed or that previously degraded wetlands are restored. As would be expected, this process, termed **wetland mitigation,** has been only partially successful, as scientists cannot be expected to create what they do not fully understand.

We have seen the influence of soil water on soil aeration. We now turn to another soil physical property, soil temperature, that is also greatly influenced by the content of water in a soil.

7.9 PROCESSES AFFECTED BY SOIL TEMPERATURE

The temperature of a soil greatly affects the physical, biological, and chemical processes occurring in that soil, and in plants growing on it. In cold soils, rates of chemical and biological reactions are slow. Biological decomposition is at a near standstill, thereby limiting the rate at which nutrients such as nitrogen, phosphorus, sulfur, and calcium are made available. Also, absorption and transport of water and nutrient ions by higher plants are inhibited by low temperatures. Temperatures that are too high can also inhibit plant and microbial processes. The temperature ranges for many soil processes, along with several benchmark soil temperatures, are given in Figure 7.17.

Plant Processes

The growth rates of most plants are actually more sensitive to *soil* temperature than to aboveground *air* temperature, but this is not often appreciated since air temperature is more commonly measured. Most plants have a rather narrow range of soil temperatures for optimal growth. For example, two species that evolved in warm regions, corn and loblolly pine, grow best when the soil temperature is about 25 to 30°C. In contrast, the optimal soil temperature for cereal rye and red maple, two species that evolved in cool regions, is in the range of 12 to 18°C.

In cool temperate regions, soil temperature often limits the productivity of crops and natural vegetation. The yields of some vegetables and small fruits can be markedly increased by warming the soil (Figure 7.18). The life cycles of plants are also influenced greatly by soil temperature. For example, tulip bulbs require chilling in early winter, to develop flower buds although flower development is suppressed until the soil warms up the following spring.

In warm regions, and in the summer in temperate regions, soil temperatures may be too high for optimal plant growth, especially in the upper few centimeters of soil. Even plants of tropical origin, such as corn, are adversely affected by soil temperatures higher than 35°C. Seed germination may also be reduced by high soil temperatures. Root growth near the surface may be encouraged by shading the soil with either live vegetation, plant residues, or the use of a mulch.

SEED GERMINATION. Different plant processes have different optimal temperatures. One of the processes most sensitive to soil temperature is seed germination. For example, farmers know that if they plant corn seed into soils cooler than 7 to 10°C, germination will not occur and the seed will likely rot. For the same species, optimum root growth occurs at a soil temperature of about 23 to 25°C, somewhat cooler than the optimum for shoot growth. For potatoes, tubers develop best when the soil temperature is 16 to 21°C, although the foliage grows quite well at warmer soil temperatures.

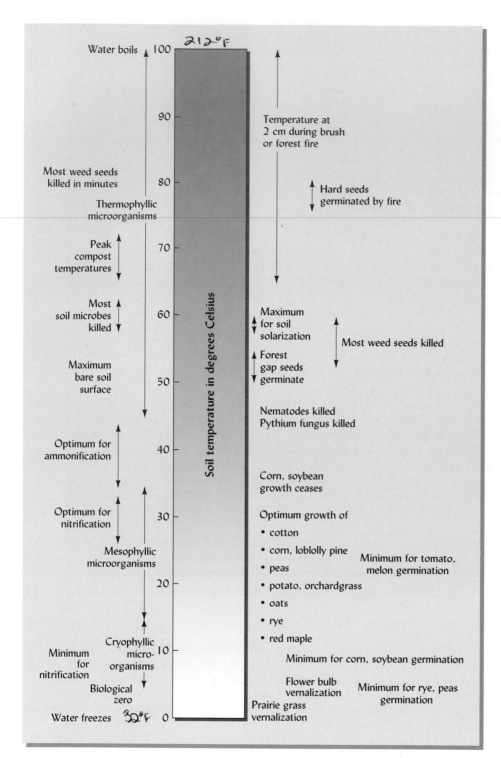

Water boils ↑ 100 212°F

90

Temperature at
2 cm during brush
or forest fire

Most weed seeds
killed in minutes — 80

Hard seeds
germinated by fire

Thermophyllic
microorganisms

Peak
compost — 70
temperatures

Most ↑
soil microbes — 60
killed ↓

Maximum
for soil
solarization

Most weed seeds killed

Forest
gap seeds
germinate

Maximum
bare soil — 50
surface

Nematodes killed
Pythium fungus killed

Optimum for — 40
ammonification

Corn, soybean
growth ceases

Optimum for — 30
nitrification

Optimum growth of

• cotton

Mesophyllic
microorganisms

• corn, loblolly pine   Minimum for tomato,
                             melon germination

• peas

— 20

• potato, orchardgrass

• oats

• rye

Cryophyllic
micro- — 10
organisms

• red maple

Minimum for corn, soybean germination

Minimum
for
nitrification

Biological
zero

Flower bulb
vernalization

Minimum for rye, peas
germination

Prairie grass
vernalization

Water freezes 32°F 0

Soil temperature in degrees Celsius

FIGURE 7.17 Soil temperature
ranges associated with a variety
of soil processes.

Many herbaceous annual plants require specific soil temperatures to trigger seed germination, accounting for much of the difference in species between early and late season weeds in cultivated land. Likewise, the seeds of certain plants adapted to open gaps in a forest stand are stimulated to germinate by the higher daily maximum soil temperatures and greater fluctuations in soil temperature that occur when the forest canopy is disturbed by timber harvest or wind-thrown trees. The seeds of certain prairie grasses require a period of cold soil temperatures (2 to 4°C) to enable them to germinate the following spring, a process termed *vernalization*.

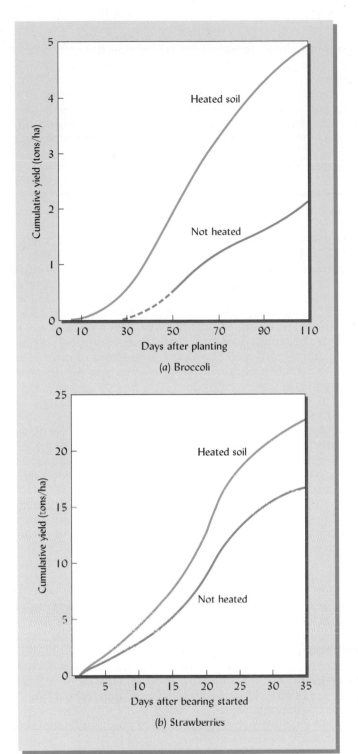

FIGURE 7.18 The influence of heating the soil in an experiment in Oregon on the yield of broccoli and strawberries. Heating cables were buried about 92 cm deep. They increased the average temperature of the 0 to 100 cm layer by about 10°C, although the upper 10 cm of soil was warmed by only about 3°C. [From Rykbost, et al. (1975); used with permission of American Society of Agronomy]

ROOT FUNCTIONS. Root functions such as nutrient uptake and water uptake are sluggish in cool soils with temperatures below the optimum for the particular species. One result is that nutrient deficiencies, especially of phosphorus, often occur in young plants in early spring, only to disappear when the soil warms later in the season. The phenomenon of *winter burn* of plant foliage is another consequence of low soil temperature that particularly affects evergreen shrubs. On bright sunny days in winter and early spring when the soil is still cold, evergreen plants may become desiccated and even die because the slow water uptake by roots in the cold soils cannot keep up with the high evaporative demand of bright sun on the foliage. The problem can be prevented by covering the shrubs with a shade cloth.

Microbiological processes are influenced markedly by soil temperature changes. General soil microbial activity and organic matter decomposition virtually ceases below about 5°C, a benchmark temperature sometimes referred to as *biological zero.* The rates of microbial processes such as respiration typically more than double for every 10°C rise in temperature up to an optimum of about 35 to 40°C (considerably higher than the optimum for plant growth). The dependence of microbial respiration on warm soil temperatures has important implications for soil aeration (see Section 7.8) and soil organic matter accumulation.

Not only is respiration reduced at low temperatures, but so is the decomposition of plant residues and, hence, the cycling of the nutrients they contain. The productivity of northern (boreal) forests is probably limited by the inhibiting effect of low soil temperatures on microbial recycling and release of nitrogen from the tree litter and soil organic matter.

The microbial oxidation of ammonium ions to nitrate ions, which occurs most readily at temperatures near 30°C, is also negligible when the soil temperature is low, below about 8 to 10°C. Farmers in cool regions can take advantage of this fact by injecting ammonia fertilizers into cold soils in the spring, knowing that ammonium ions will not be readily oxidized to nitrate ions until the soil temperature rises. Since nitrate ions are readily leached from the soil, while ammonium ions are not (see Section 13.8), the nitrogen is conserved until the soil warms up later in the year. Then nitrates will begin to appear, but plants should also be rapidly growing and using the nitrates (see Chapter 13). Unfortunately, in some years a warm spell allows the production of nitrates earlier than expected, with the result that much nitrate is lost by leaching, to the detriment of both the farmer and the downstream water quality.

High soil temperatures can be used to control certain plant diseases. In environments with hot, sunny summers—maximum daily air temperatures >35°C—covering the ground with transparent plastic sheeting can raise the temperature of the upper few centimeters of soil to as high as 50 to 60°C. Such temperatures markedly reduce certain wilt-causing fungal diseases of vegetables and fruits and adversely affect some weed seeds and insects. This heating process, called *soil solarization,* is used to control pests and diseases in some high-value crops.

As we shall see in Chapter 18, warm soil temperatures are critical for the microbial destruction of toxic organic pesticides and pollutants in soil. Temperature control is critical for some new technologies that take advantage of the ability of certain microorganisms to degrade petroleum products, pesticides, and other compounds.

Freezing and Thawing

When soil temperatures fluctuate above and below 0°C, the water in the soil undergoes cycles of freezing and thawing. Alternate freezing and thawing subject the soil aggregates to pressures as zones of pure ice, called *ice lenses,* form within the soil and as ice crystals form and expand.[5] These pressures alter the physical structure of the soil. In a saturated soil with a puddled structure, the frost action breaks up the large masses and greatly improves granulation. In contrast, for soils with good aggregation to begin with, freeze-thaw action when the soil is very wet can lead to structural deterioration.

Alternate freezing and thawing can force objects upward in the soil, a process termed **frost heaving.** Objects subject to heaving include stones, fence posts, and perennial taprooted plants. This action, which is most severe where the soil is silty in texture, wet, and lacking a covering of snow or dense vegetation, can drastically reduce the stand of alfalfa (Figure 7.19), some clovers, and trefoil.

Freezing can also heave shallow foundations, roads, and runways which have fine material as a base. Gravels and pure sands are normally resistant to frost damage, but silts and sandy soils with modest amounts of finer particles are particularly susceptible.

---

[5] The pressure is due mainly to the growth of the ice lenses rather than to the 9% increase in volume that water undergoes when it freezes. The ice lenses grow as water is drawn to the freezing zone from adjacent unfrozen areas. The flow to the growing ice lenses is encouraged by the fact that fine soil particles remain coated with a film of liquid (unfrozen) water at temperatures below the normal freezing point. The lowered freezing point occurs for two reasons: (1) the influence of water–solid interactions near the particle surface, and (2) the presence of dissolved and exchangeable ions in this film of water.

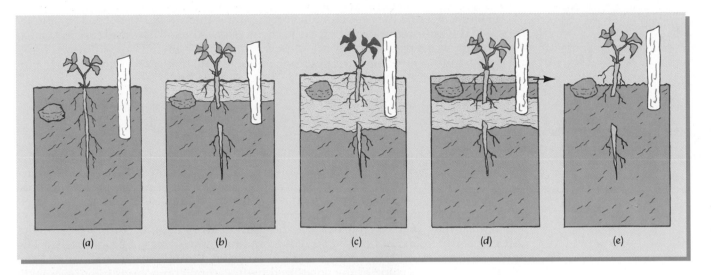

FIGURE 7.19 How frost heaving moves objects upward. (a) Position of the object (stone, plant, or fence post) before the soil freezes. (b) As lenses of pure ice form in the freezing soil by attraction of water from the unfrozen soil below, the frozen soil tightens around the upper part of the object, lifting it somewhat—enough to break the root in the case of the plant. (c) the objects are lifted upwards as ice-lens formation continues with deeper penetration of the freezing front. (d) As for freezing, thawing commences from the surface downward. Water from thawing ice lenses escapes to the surface because it cannot drain downward through the frozen soil. The soil surface subsides while the heaved objects are held in the "jacked-up" position by the still-frozen soil around their lower parts. (e) After complete thaw, the stone is closer to the surface than previously (although rarely at the surface unless erosion of the thawed soil has occurred), and the upper part of the broken plant's root is exposed, so that is likely to die. (f) Alfalfa plants lifted out of the ground by frost action. (g) Fence posts encased in concrete that have been progressively "jacked out" of the ground by frost action over several years. [Photo (f) courtesy of R. Weil; photo (g) courtesy of R. L. Berg, Corps of Engineers, Cold Regions Research and Engineering Laboratory, Hanover, N.H.]

Very clay-rich soils do not usually exhibit much frost heave, but ice-lens segregation can still occur and can lead to severe loss of strength when thawing occurs. To avoid damage by freezing soil temperatures, foundation footings (as well as water pipelines) should be set into the soil below the maximum depth to which the soil freezes—a depth that ranges from less than 10 cm in subtropical zones, such as South Texas and Florida, to more than 200 cm in very cold climates.

Soil with ice lenses may contain much more water than would be needed to saturate the soil in the unfrozen state. When the ice lenses thaw, the soil becomes supersaturated because the excess water cannot drain away through the underlying still-frozen soil. Soil in this condition readily turns into noncohesive mud that is very susceptible to erosion and movement by mudslides.

Soil Heating by Fire

Fire is one of the most far-reaching ecosystem disturbances in nature. In addition to the obvious aboveground effects of forest, range, or crop-residue fires, the brief, but sometimes dramatic, changes in soil temperature also may have lasting impacts. Unless the fire is artificially stoked with added fuel, the temperature rise itself is usually very brief and is limited to the upper few centimeters of soil.

The heat may affect the breakdown and movement of organic compounds. Figure 7.20 shows the results of one such wildfire on a lodgepole pine stand in Oregon. The high temperatures (>125°C being common) essentially distill various fractions of the organic matter, with some of the volatilized hydrocarbon compounds moving quickly through the soil pores to deeper, cooler areas. As these compounds reach cooler soil particles deeper in the soil, they condense (solidify) on the surface of the soil particles and fill some of the surrounding pore spaces. Some of the condensed compounds are water-repellent (hydrophobic) hydrocarbons. Consequently, when rain comes, water infiltration in even a sandy soil is greatly reduced over that of unburned areas. This effect of soil temperature is quite common on chaparral lands in semiarid regions and may be responsible for the disastrous mudslides that occur when the layer of soil above the hydrophobic zone becomes saturated with rainwater.

Fires also affect the germination of certain seeds, which have hard coatings that prevent them from germinating until they are heated above 70 to 80°C. On the other hand, burning of straw in wheat fields generates similar soil temperatures, but with the effect of killing most of the weed seeds near the surface and thus greatly reducing subsequent weed infestation. The heat and ash may also hasten the cycling of nutrient plant nutrients (see Chapter 16). Fires set to clear land of timber slash may burn long and hot enough to seriously deplete soil organic matter and kill so many soil organisms that forest regrowth is inhibited.

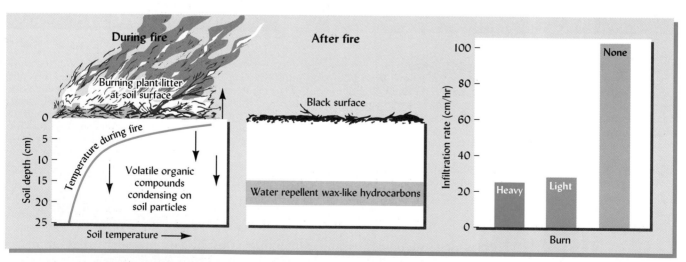

FIGURE 7.20   (Left) Wildfires of a lodgepole pine stand heat up the surface layers of this sandy soil (an Inceptisol) in Oregon. (Center) Note that the soil temperature is increased sufficiently near the surface to volatilize organic compounds, some of which then move down into the soil and condense (solidify) on the surface of cooler soil particles. These condensed compounds are waxlike hydrocarbons that are water repellent. As a consequence (right) the infiltration of water into the soil is drastically reduced and remains so for a period of at least six years. [From Dryness (1976)]

7.10 ABSORPTION AND LOSS OF SOLAR ENERGY[6]

The temperature of soils in the field is directly or indirectly dependent on at least three factors: (1) the net amount of heat energy the soil absorbs; (2) the heat energy required to bring about a given change in the temperature of a soil; and (3) the energy required for processes such as evaporation, which are constantly occurring at or near the surface of soils.

Solar radiation is the primary source of energy to heat soils. But clouds and dust particles intercept the sun's rays and absorb, scatter, or reflect most of the energy (Figure 7.21). Only about 35 to 40% of the solar radiation actually reaches the earth in cloudy humid regions, and 75% in cloud-free arid areas. The global average is about 50%.

Little of the solar energy reaching the earth actually results in soil warming. The energy is used primarily to evaporate water from the soil or leaf surfaces, or is radiated or reflected back to the sky. Only about 10% is absorbed by the soil and can be used to warm it. Even so, this energy is of critical importance to soil processes and to plants growing on the soils.

ALBEDO.   The fraction of incident radiation that is reflected by the land surface is termed the **albedo**, and ranges from as low as 0.1 to 0.2 for dark-colored, rough soil surfaces to as high as 0.5 or more for smooth, light-colored surfaces. Vegetation may affect the surface albedo either way, depending on whether it is dark green and growing or yellow and dormant.

The fact that dark-colored soils absorb more energy than lighter-colored ones does not necessarily imply, however, that dark soils are always warmer. In fact, the opposite

---

[6] For a review of models that describe these processes on a global scale, see Sellers, et al. (1997).

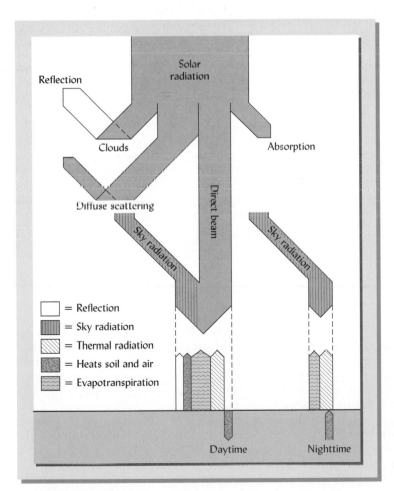

FIGURE 7.21   Schematic representation of the radiation balance in daytime and nighttime in the spring or early summer in a temperate region. About half the solar radiation reaches the earth, either directly or indirectly, from sky radiation. Most radiation that strikes the earth in the daytime is used as energy for evapotranspiration or is radiated back to the atmosphere. Only a small portion, perhaps 10%, actually heats the soil. At night the soil loses some heat, and some evaporation and thermal radiation occur.

is often true. In most landscapes, the darkest soils are those found in the low spots where excessive wetness has caused organic matter to accumulate. Therefore, the darkest soils are also usually the wettest. The water in these soils requires much energy to be warmed and it also cools the soil when it evaporates.

ASPECT. The angle at which the sun's rays strike the soil also influences soil temperature. If the sun is directly overhead, the incoming path of the rays is perpendicular to the soil surface and energy absorption (and soil temperature increase) is greatest (Figure 7.22). As an example, three soils, one on a southerly slope of 20°, one on a nearby level site, and one on a northerly slope of 20°, will receive energy from the sun's rays on June 21 (at the 42nd parallel north) in the proportion of 106:100:81. The effect of the direction of slope, or **aspect**, on forest species is illustrated in the photo in Figure 7.22.

Planting crops on soil ridges is one method of controlling the soil aspect on a microscale. This is most effectively done at high latitudes by planting crops on the south- or southwest-facing sides of ridges. The ridges need be only about 25 cm tall to have a major effect. In Fairbanks, Alaska, midafternoon soil temperatures (at 1 cm depth) in early May can be about 15°C warmer on the south side of such a ridge than on the north side, and about 8°C warmer than on level ground.

RAIN. Mention should also be made of the effect of rain or irrigation water on soil temperature. For example, in temperate zones, spring rains definitely warm the surface soil as the water moves into it. Conversely, in the summer, rainfall cools the soil, since it is often cooler than the soil it penetrates. In practice, however, the spring rains, by increasing the amount of solar energy used in evaporating water from the soil, often accentuate low-temperature problems.

SOIL COVER. Whether the soil is bare or is covered with vegetation, mulch or snow is another factor markedly influencing the amount of solar radiation reaching the soil. Bare soils warm up more quickly and cool off more rapidly than those covered with vegetation, with snow, or with plastic mulches. Frost penetration during the winter is considerably greater in bare, noninsulated land.

Even low-growing vegetation such as turf grass has a very noticeable influence on soil temperature and on the temperature of the surroundings (Table 7.6). Much of the cooling effect is due to heat dissipated by transpiration of water. To experience this effect, on a blistering hot day, try having a picnic on an asphalt parking lot instead of on a growing green lawn!

The effect of a dense forest is universally recognized. Timber-harvest practices that leave sufficient canopy to provide about 50% shade will likely prevent undue soil warming that could hasten the loss of soil organic matter or the onset of anaerobic conditions in wet soils. The effect of timber harvest on soil temperature as deep as 50 cm is seen in Table 7.2. In the case presented in Table 7.2, tree removal warmed the soil in spring, even though it also raised the soil water content. However, as we shall see in the next section, a higher water content normally slows the warming of soils in spring.

7.11 THERMAL PROPERTIES OF SOILS

Specific Heat of Soils

A dry soil is more easily heated than a wet one. This is because the amount of energy required to raise the temperature of water by 1°C (its heat capacity) is much higher than that required to warm soil solids by 1°C. When heat capacity is expressed per unit mass—for example, in calories per gram (cal/g)—it is called **specific heat** or heat capacity c. The specific heat of pure water is about 1.00 cal/g (or 4.18 joules per gram, J/g); that of dry soil is about 0.2 cal/g (0.8 J/g).

The specific heat largely controls the degree to which soils warm up in the spring, wetter soils warming more slowly than drier ones (see Box 7.2). Furthermore, if the water does not drain freely from the wet soil, it must be evaporated, a process that is very energy consuming, as the next section will show.

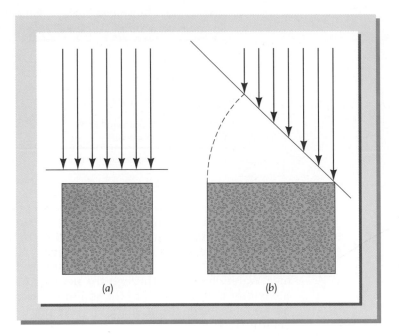

FIGURE 7.22 (Upper) Effect of the angle at which the sun's rays strike the soil on the area of soil that is warmed. (*a*) If a given amount of radiation from the sun strikes the soil at right angles, the radiation is concentrated in a relatively small area, and the soil warms quite rapidly. (*b*) If the same amount of radiation strikes the soil at a 45° angle, the area affected is larger by about 40%, the radiation is not so concentrated, and the soil warms up more slowly. This is one of the reasons why north slopes tend to have cooler soils than south slopes. It also accounts for the colder soils in winter than in summer. (Lower) A view looking eastward of a forested area in Virginia illustrates the temperature effect. The main ridge (left to right) is running north and south and the smaller side ridges east and west (up and down). The dark patches are pine trees in this predominantly hardwood deciduous forest. The pines dominate the southern (warmer and drier) slopes on each east-west ridge. (Photo courtesy of R. Weil)

TABLE 7.6 Maximum Surface Temperatures for Four Types of Surfaces on a Sunny August Day in College Station, Texas

	Maximum temperature, °C	
Type of surface	*Day*	*Night*
Green, growing turf grass	31	24
Dry, bare soil	39	26
Brown, summer-dormant grass	52	27
Dry synthetic sports turf	70	29

Data from Beard and Green (1994).

Temperature control systems referred to as *heat pumps,* which are designed to both warm and cool buildings, take advantage of the high specific heat of soils. A network of pipes is laid underground near the building to be heated and cooled to maximize heat-exchange contact with the soil. Advantage is taken of the fact that subsoils are generally warmer than the atmosphere in the winter and cooler than the atmosphere in the summer. Water circulating through the network of pipes absorbs heat from the soil during the winter and releases it to the soil in the summer. The high specific heat of soils permits a large exchange of energy to take place without greatly modifying the soil temperature (Figure 7.23).

Heat of Vaporization

The evaporation of water from soil surfaces requires a large amount of energy, 540 kilocalories (kcal) or 2.257 J for every kilogram of water vaporized. This energy must be provided by solar radiation or it must come from the surrounding soil. In either case, evaporation has the potential of cooling the soil, much the way it chills a person who comes out from swimming on a windy day.

For example, if the amount of water associated with 100 g of dry soil was reduced by evaporation from 25 g to 24 g (only about a 1% decrease) and if all the thermal energy needed to evaporate the water came from the moist soil, the soil would be cooled by about 12°C. Such a figure is hypothetical because only a part of the heat of vaporization comes from the soil itself. Nevertheless, it indicates the tremendous cooling influence of evaporation.

The low temperature of a wet soil is due partially to evaporation and partially to high specific heat. The temperature of the upper few centimeters of wet soil is commonly 3 to 6°C lower than that of a moist or dry soil. This is a significant factor in the spring in a temperate zone, when a few degrees will make the difference between the germination or lack of germination of seeds, or the microbial release or lack of release of nutrients from organic matter.

Thermal Conductivity of Soils

As shown in Section 7.10, some of the solar radiation that reaches the earth slowly penetrates the profile largely by conduction; this is the same process by which heat moves along an iron pipe when one end is placed in a fire. The movement of heat in soil is analogous to the movement of water (see Section 5.5), the rate of flow being determined by a driving force and by the ease with which heat flows through the soil. This can be expressed as Fourier's Law:

$$Q_h = K * \frac{\Delta T}{x}$$

where Q_h is the *thermal flux,* the quantity of heat transferred across a unit cross-sectional area in a unit time; K is the **thermal conductivity** of the soil; and $\Delta T/x$ is the temperature gradient over distance x that serves as the driving force for the conduction of heat.

The thermal conductivity K of soil is influenced by a number of factors, the most important being the moisture content of the soil and the degree of compaction. Heat passes through water many times faster than through air. As the water content increases in a soil, the air content decreases, and the transfer resistance is decidedly lowered.

BOX 7.2   CALCULATING THE SPECIFIC HEAT OR HEAT CAPACITY OF MOIST SOILS

Soil water content markedly impacts soil temperature changes through its effect on the specific heat or heat capacity c of a soil. For example, consider two soils with comparable characteristics, *soil A*, a wet soil with 30 g water/100 g soil solids, and *soil B*, a drier soil with only 10 g water/100 g soil solids.

We can assume the following values for specific heat:

$$\text{Water} = 1.0 \text{ cal/g}$$
$$\text{Dry mineral soil} = 0.2 \text{ cal/g}$$

For soil A with 10 g water/100 g dry soil, or 0.1 g water/g dry soil, the number of calories required to raise the temperature of 0.1 g of water by 1°C is

$$0.1 \text{ g} \times 1.0 \text{ cal/g} = 0.1 \text{ cal}$$

The corresponding figure for the 1.0 g of soil solids is

$$1 \text{ g} \times 0.2 \text{ cal/g} = 0.2 \text{ cal}$$

Thus, a total of 0.3 cal (0.1 + 0.2) is required to raise the temperature of 1.1 g (1.0 + 0.1) of the moist soil by 1°C. Since the specific heat is the number of calories required to raise the temperature of 1 g of moist soil by 1°C, we can calculate the specific heat of soil A as follows:

$$c_{soil\ A} = \frac{0.3}{1.1} = 0.273 \text{ cal/g}$$

These calculations can be expressed as a simple equation to calculate the weighted average specific heat of a mixture of substances:

$$c_{moist\ soil} = \frac{c_1 m_1 + c_2 m_2}{m_1 + m_2}$$

where c_1 and m_1 are the specific heat and mass of substance 1 (the dry mineral soil, in this case), and c_2 and m_2 are the specific heat of substance 2 (the water, in this case).

Applying this equation to soil A, we again calculate that $c_{soil\ B}$ is 0.273 cal/g, as follows:

$$c_{soil\ A} = \frac{0.2 \text{ cal/g} * 1.0 \text{ g} + 1.0 \text{ cal/g} * 0.10 \text{ g}}{1.0 \text{ g} + 0.10 \text{ g}}$$

$$= \frac{0.30 \text{ cal/g}}{1.1 \text{ g}}$$

$$= 0.273 \text{ cal/g}$$

In the same manner, we calculate the specific heat of the wetter soil B:

$$c_{soil\ B} = \frac{0.2 \text{ cal/g} * 1.0 \text{ g} + 1.0 \text{ cal/g} * 0.30 \text{ g}}{1.0 \text{ g} + 0.30 \text{ g}}$$

$$- \frac{0.50 \text{ cal/g}}{1.3 \text{ g}}$$

$$= 0.385 \text{ cal/g}$$

The wet soil B has a specific heat c_B of 0.385 cal/g, whereas the drier soil A has a specific heat c_A of 0.273 cal/g. Because it must absorb an additional 0.112 cal (0.385 − 0.273) of solar radiation for every degree of temperature rise, the wet soil will warm up much more slowly than the drier soil.

FIGURE 7.23 This high school in Oklahoma has been built into the ground with only one side exposed. This design takes advantage of the high specific heat and low thermal conductivity of the overlying soils, keeping the school warm in winter and cool in summer with a minimum of energy used for heating or air-conditioning. (Photo courtesy of R. Weil)

When sufficient water is present to form a bridge between most of the soil particles, further additions will have little effect on heat conduction. Heat moves through mineral particles even faster than through water, so when particle-to-particle contact is increased by soil compaction, heat-transfer rates are also increased. Therefore, a wet, compacted soil would be the poorest insulator or the best conductor of heat. Here again the interconnectedness of soil properties is demonstrated.

Relatively dry soil makes a good insulating material. Buildings built mostly underground can take advantage of both the low thermal conductivity and relatively high heat capacity of large volumes of soil (see Figure 7.23).

The significance of heat conduction with respect to field soil temperature is not difficult to comprehend. It provides a means of temperature adjustment, but, because it is slow, changes in subsoil temperature lag behind those of the surface layers. Moreover, temperature changes are always less in the subsoil. In temperate regions, surface soils in general are expected to be warmer in summer and cooler in winter than the subsoil, especially the lower horizons of the subsoil. Soil thermal conductivity can also affect air temperature above the soil, as shown in Figure 7.24.

Soil Temperature Fluctuations

The temperature of the soil at any time depends on the ratio of the energy absorbed to that being lost. The constant change in this relationship is reflected in the seasonal, monthly, and daily temperatures. The accompanying data (Figures 7.25 and 7.26) from College Station, Texas, and Lincoln, Nebraska, are representative of average seasonal and monthly temperatures in relation to soil depth in subhumid temperate regions.

Vertical and Seasonal Temperature Changes

It is apparent from Figures 7.25 and 7.26 that considerable seasonal and monthly variations of soil temperature occur, even at the lower depths. The surface layer temperatures vary more or less according to the temperature of the air, although these layers are generally warmer than the air throughout the year.

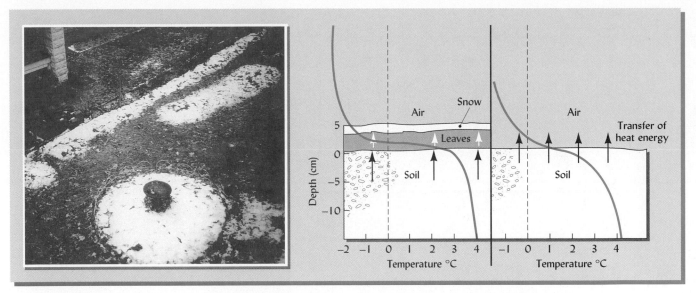

FIGURE 7.24   Transfer of heat energy from soil to air. The scene, looking down on a garden after an early fall snow storm, shows snow on the leaf-mulched flower beds, but not on areas where the soil is bare or covered with thin turf. The reason for this uneven accumulation of snow can be seen in the temperature profiles. Having stored heat from the sun, the soil layers are often warmer than the air as temperatures drop in fall (this is also true at night during other seasons). On bare soil, heat energy is transferred rapidly from the deeper layers to the surface, the rate of transfer being enhanced by high moisture content or compaction, which increase the *thermal conductivity* of the soil. As a result, the soil surface and the air above it are warmed to above freezing, so snow melts and does not accumulate. The leaf mulch, which has a low thermal conductivity, acts as an insulating blanket that slows the transfer of stored heat energy from the soil to the air. The upper surface of the mulch is therefore hardly warmed by the soil, and the snow remains frozen and accumulates. A heavy covering of snow can itself act as an insulating blanket. (Photo courtesy of R. Weil)

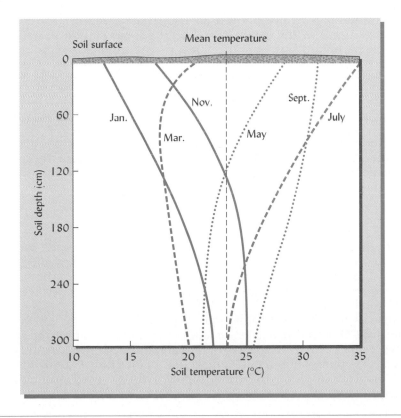

FIGURE 7.25   Average monthly soil temperatures for 6 of the 12 months of the year at different soil depths at College Station, Tex. (1951–1955). Note the lag in soil temperature change at the lower depths. [From Fluker (1958)]

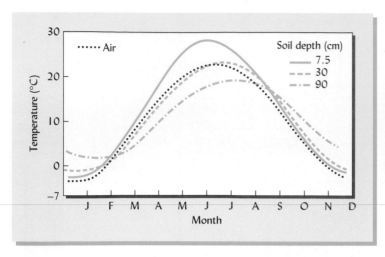

FIGURE 7.26 Average monthly air and soil temperatures at Lincoln, Nebr. (12 years). Note that the 7.5-cm soil layer is consistently warmer than the air above and that the 90-cm soil horizon is cooler in spring and summer, but warmer in the fall and winter, than surface soil.

In the subsoil, the seasonal temperature increases and decreases lag behind changes registered in the surface soil and in the air. Accordingly, the temperature data for March at College Station suggest that the surface soil temperatures have already begun to respond to the warming of the spring, while temperatures of the deep subsoil seem to still be responding to the cold winter weather.

The subsoil temperatures are less variable than the air and surface soil temperatures, although there is some temperature variation even at the 300-cm depth. The subsoils are generally warmer in the late fall and winter and cooler in the spring and summer than the surface soil layers and the air. This is to be expected since the subsoils are not subject to direct solar radiation.

Daily Variations

With a clear sky, the air temperature in temperate regions rises from lowest in the morning to a maximum at about 2 P.M. The surface soil, however, does not reach its maximum until later in the afternoon because of the usual lag. This retardation is greater and the temperature change is less as the depth increases. The lower subsoil shows little daily or weekly fluctuation; the variation there, as already emphasized, is a slow monthly or seasonal change.

7.12 SOIL TEMPERATURE CONTROL

The temperature of field soils is not subject to radical human regulation. However, two kinds of management practice have significant effects on soil temperature: those that affect the cover or mulch on the soil, and those that reduce excess soil moisture. These effects have meaningful biological implications.

Organic Mulches and Plant-Residue Management

Soil temperatures are influenced by soil cover and especially by organic residues or other types of mulch on the soil surface. Figure 7.27 shows that mulches effectively buffer extremes in soil temperatures. In periods of hot weather, they keep the surface soil cooler than where no cover is used; in contrast, during cold weather they keep the soil warmer than it would be if bare.

The forest floor is a prime example of a natural temperature-modifying mulch. It is not surprising, therefore, that timber harvest practices can markedly affect forest soil temperature regimes (Figure 7.28). Disturbance of the leaf mulch, changes in water content due to reduced evapotranspiration, and compaction by machinery are all factors that influence soil temperatures through thermal conductivity. Reduced shading after tree removal also lets in more solar radiation.

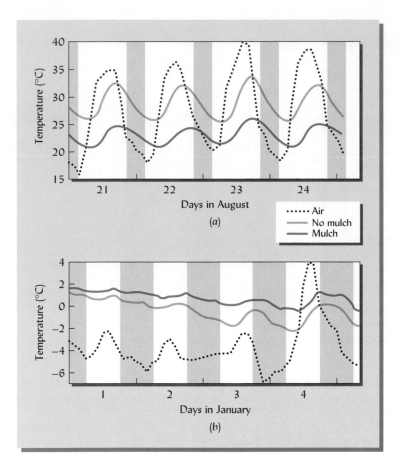

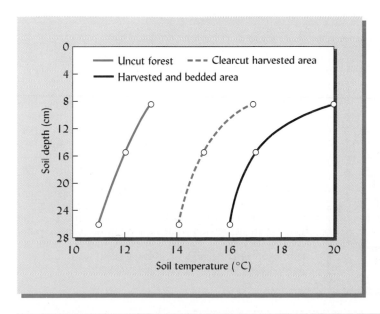

FIGURE 7.27 (*a*) Influence of straw mulch (8 tons/ha) on air temperature at a depth of 10 cm during an August hot spell in Bushland, Tex. Note that the soil temperatures in the mulched area are consistently lower than where no mulch was applied. (*b*) During a cold period in January the soil temperature was higher in the mulched than in the unmulched area. The shaded bars represent nighttime. [Redrawn from Unger (1978); used with permission of American Society of Agronomy]

MULCH FROM CONSERVATION TILLAGE. Until fairly recently, the labor and expense of carrying and spreading mulch materials limited their use in modifying soil temperature extremes mostly to home gardens and flower beds. Although these uses are still important, the use of mulches has been extended to field-crop culture in areas that have adopted conservation tillage practices. Conservation tillage leaves most or all of the crop residues at or near the soil surface, thereby growing the mulch in place, rather than transporting it to the field. The influence of surface residues on soil temperature is illustrated by data

FIGURE 7.28 Influence of timber harvest and site preparation on soil temperature profiles in June on a forested wetland in Michigan. Clear-cut timber harvest greatly disturbed the site and raised soil temperatures by 4 to 5°C. Tillage to form beds for tree replanting eliminated the remaining forest floor mulch and dried the soil, thus causing temperatures to rise another 2 to 3°C. The higher temperatures increased the rate of soil organic matter decomposition. [Redrawn from Trettin, et al. (1996)]

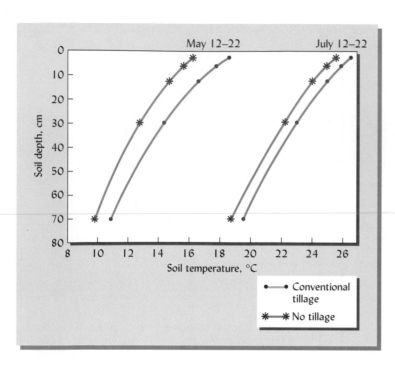

FIGURE 7.29 The influence of tillage system on soil temperature profile in a North Dakota wheat field. Compared to the unmulched soil of the conventionally plowed system, the soil with a surface residue mulch created by the no-tillage system was cooler in spring and summer to a depth of at least 70 cm. The difference was more than 3°C near the surface in early May. [Data from Merrill, et al. (1996)]

shown in Figure 7.29. Soil temperatures to depths as great as 70 cm were consistently lower during May and July in the no-tillage system, which left all crop residues on the surface as a mulch.

CONCERNS IN COOL CLIMATES. While the mulch provides great control over erosion (see Section 17.6), the soil-temperature-depressing effects of some mulch practices have a serious negative impact on the production of crops like corn in cold climates, such as in Canada and the northern United States. The lower temperatures in May and early June resulting from these practices inhibit seed germination, seedling performance, and, often, the yields of corn. The effect of the residue mulch is most pronounced in lowering the midday maximum temperature, and has much less effect on the minimum temperature reached at night. This effect is well illustrated by the data in Figure 7.30, which also features an innovative way to alleviate this problem by pushing aside the residues in just a narrow band over the seed row in the no-tillage system. Another approach to solving this problem is to ridge the soil, permit water to drain out of the ridge, and then plant on the drier, warmer ridgetop (or on the south side of the ridge—see Section 7.10).

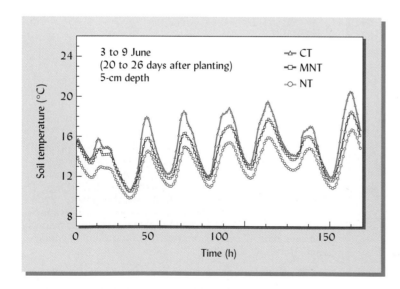

FIGURE 7.30 Tillage effects on hourly temperature changes near the surface of a cold Alfisol in northern British Columbia. The soil had been managed to grow barley under no-tillage (NT) and conventional (clean surface without residues) tillage (CT) systems for the previous 14 years. In the clean-tilled soil, midafternoon temperatures peaked at 4°C higher than those in the residue mulch covered no-tillage soil. A modification of the no-tillage system (MNT) that pushed aside the residues in a narrow (7.5-cm-wide) band over the seeding row eliminated much of the temperature depression while keeping most of surface covered by the soil- and water-conserving mulch. Note the daily temperature changes and the general warming trend during the six days shown. [From Arshad and Azooz (1996)]

ADVANTAGES IN WARM CLIMATES.   In warm regions, delayed planting is not a problem. In fact, the cooler near-surface soil temperatures under a mulch may reduce heat stress on roots during summer. Plant-residue mulches also conserve soil moisture by decreasing evaporation. The resulting cooler, moist surface layer of soil is an important part of no-tillage systems because it allows roots to proliferate in this zone, where nutrient and aeration conditions are optimal.

Plastic Mulches

One of the reasons for the popularity of plastic mulches for gardens and high-value specialty crops is their effect on soil temperature. In contrast to the organic mulches, plastic mulches generally increase soil temperature, clear plastic having a greater heating effect than black plastic. In temperate regions, this effect can be used to extend the growing season or to hasten production to take advantage of the higher prices offered by early-season markets. Figure 7.31 shows the use of a plastic mulch for winter-grown strawberries in southern California.

In warmer climates, and during the summer months, the soil-heating effect of plastic mulches may be quite detrimental, inhibiting root growth in the upper soil layers and sometimes seriously decreasing crop yields (see Table 7.7, for example). Obviously, the weed-control and moisture-conservation effects of plastic mulches may be outweighed in some circumstances by their excessive heating effects.

Moisture Control

Another means of exercising some control over soil temperature is by controlling soil moisture. Poorly drained soils in temperate regions that are wet in the spring have temperatures 3 to 6°C lower than comparable well-drained soils. Only by removing this water can temperature depression be alleviated. Water can be removed by installing drainage systems using ditches and underground pipes (see Section 6.9). Where this is not feasible, the ridging systems of tillage just referred to can be used.

As is the case with soil air, the controlling influence of soil water on soil temperature is apparent everywhere. Whether a problem concerns capturing solar energy, loss of energy to the atmosphere, or the movement of heat within the soil, the amount of water present is always important. Water regulation seems to be a key to what little practical temperature control is possible for field soils.

FIGURE 7.31   These winter-grown southern California strawberries will come to market when prices are still high because of the effect of the clear plastic mulch on soil temperature. (Photo courtesy of R. Weil)

TABLE 7.7 Soil Temperature and Tomato Yield with Straw or Black Plastic Mulch

The data are averages for two years of tomato production on a sandy loam Ultisol near Griffin, Georgia. The straw kept the surface soil from rising to detrimentally high temperatures, while it also increased infiltration of rainwater and reduced soil compaction. Daily drip irrigation supplied plenty of water, but could not overcome the temperature effects of the black plastic mulch.

	Not irrigated		Irrigated daily	
	Straw mulch	Plastic mulch	Straw mulch	Plastic mulch
Average soil temperature, °C[a]	24	37	24	35
Tomato yield, Mg/ha	68	30	70	24

[a] Soil temperature measured at 5 cm below the soil surface, average of weeks 2–10 of the growing season.
Data calculated from Tindall, et al. (1991).

7.13 CONCLUSION

Soil aeration and soil temperature critically affect the quality of soils as habitats for plants and other organisms. Most plants have definite requirements for soil oxygen along with limited tolerance for carbon dioxide, methane, and other such gases found in poorly aerated soils. Some microbes, such as the nitrifiers and general-purpose decay organisms, are also constrained by low levels of soil oxygen.

Soils with extremely wet moisture regimes are unique with respect to their morphology and chemistry and to the plant communities they support. Such hydric soils are characteristic of wetlands and help these ecosystems perform a myriad of valuable functions.

Plants as well as microbes are also quite sensitive to differences in soil temperature, particularly in temperate climates where low temperatures can limit essential biological processes. Soil temperature also impacts on the use of soils for engineering purposes, again primarily in the cooler climates. Frost action, which can move perennial plants such as alfalfa out of the ground, can do likewise to building foundations, fence posts, sidewalks, and highways.

Soil water exerts a major influence over both soil aeration and soil temperature. It competes with soil air for the occupancy of soil pores and interferes with the diffusion of gases into and out of the soil. Soil water also resists changes in soil temperature by virtue of its high specific heat and its high energy requirement for evaporation.

STUDY QUESTIONS

1. What are the two principle gases involved with soil aeration, and how do their relative amounts change as one samples deeper into a soil profile?

2. What is aerenchyma tissue and how does it affect plant–soil relationships?

3. If the redox potential for a soil at pH 6 is near zero, write two reactions that you would expect to take place. How would the presence of a great deal of nitrate compounds affect the occurrence of these reactions?

4. It is sometimes said that organisms in anaerobic environments will use the combined oxygen in nitrate or sulfate instead of the oxygen in O_2. Why is this statement incorrect? What actually happens when organisms reduce sulfate or nitrate?

5. If an alluvial forest soil were flooded for 10 days and you sampled the gases evolving from the wet soil, what gases would you expect to find (other than oxygen and carbon dioxide)? In what order of appearance? Explain.

6. Explain why warm weather during periods of saturation is required in order to form a hydric soil.

7. If you were in the field trying to delineate the so-called drier end of a wetland area, what are three soil properties and three other indicators that you might look for?

8. For each of these gases, write a sentence to explain its relationship to wetland conditions: *ethylene, methane, nitrous oxide, oxygen,* and *hydrogen sulfide.*

9. What are the three major components that define a wetland?

10. Discuss four plant processes that are influenced by soil temperature.

11. Explain how a brush fire might lead to subsequent mudslides, as often occurs in California.

12. If you were to build a house below ground in order to save heating and cooling costs, would you firmly compact the soil around the house? Explain your answer.

13. If you measured a daily maximum air temperature of 28°C at 1 P.M., what might you expect the daily maximum temperature to be at a 15-cm depth in the soil? At about what time of day would the maximum temperature occur at this depth? Explain.

14. In relation to soil temperature, explain why conservation tillage has been more popular in Missouri than in Minnesota.

REFERENCES

Arshad, A., and R. H. Azooz. 1996. "Tillage effects on soil thermal properties in a semi-arid cold region," *Soil Sci. Soc. Am. J.,* **60**:561–567.

Bartlett, R. J., and B. R. James. 1993. "Redox chemistry of soils," *Advances in Agronomy,* **50**:151–208.

Beard, J. B. and R. L. Green. 1994. "The role of turfgrasses in environmental protection and their benefits to humans," *J. Environ. Qual.,* **23**:452–460.

Buyanovsky, G. A., and G. H. Wagner. 1983. "Annual cycles of carbon dioxide level in soil air," *Soil Sci. Soc. Amer. J.,* **47**:1139–1145.

CAST. 1994. *Wetland Policy Issues.* Publication no. CC1994-1. (Ames, Iowa: Council for Agricultural Science and Technology).

Davidson, E. A., and S. E. Trumbore. 1995. "Gas diffusivity and production of CO_2 in deep soils of the eastern Amazon," *Tellus,* **47B**:550–565.

Dryness, C. T. 1976. "Effects of wildfire on soil wetability in the high cascades of Oregon," USDA Forest Service Research Paper PNW-202. (Washington, D.C.: USDA).

Eswaren, H., P. Reich, P. Zdruli, and T. Levermann. 1996. "Global distribution of wetlands," *Amer. Soc. Agron. Abstracts,* 328.

Fluker, B. J. 1958. "Soil temperature," *Soil Sci.,* **86**:35–46.

Hurt, G. W., P. M. Whited, and R. F. Pringle (eds.). 1996. *Field Indicators of Hydric Soils in the United States.* (Fort Worth, Tex.: USDA Natural Resources Conservation Service).

McBride, M. B. 1994. *Environmental Chemistry of Soils.* (New York: Oxford University Press).

Meek, B. D., and L. B. Grass. 1975. "Redox potential in irrigated desert soils as an indicator of aeration status," *Soil Sci. Soc. Amer. J.,* **39**:870–875.

Meek, B. D., E. C. Owen-Bartlett, L. H. Stolzy, and C. K. Labanauskas. 1980. "Cotton yield and nutrient uptake in relation to water table depth," *Soil Sci. Soc. Amer. J.,* **44**:301–305.

Merrill, S. D., A. L. Black, and A. Bauer. 1996. "Conservation tillage affects root growth of dryland spring wheat under drought," *Soil Sci. Soc. Am. J.,* **60**:575–583.

Patrick, W. H., Jr. 1977. "Oxygen content of soil air by a field method," *Soil Sci. Soc. Amer. J.,* **41**:651–652.

Patrick, W. H., Jr., and A. Jugsujinda. 1992. "Sequential reduction and oxidation of inorganic nitrogen, manganese, and iron in flooded soil," *Soil Sci. Soc. Amer. J.,* **56**:1071–1073.

Rabenhorst, M. C., J. Bell, and P. McDaniel (eds.). 1998. *Quantifying Soil Hydromorphology.* Special Publication (in press). (Madison, Wis.: Soil Sci. Soc. Amer.).

Rykbost, K. A., L. Boersma, J. J. Mack, and W. E. Schmisseur. 1975. "Yield response to soil warming: Vegetable crops," *Agron. J.,* **67**:738–743.

Sellers, P., R. Dickinson, D. Randall, K. Betts, F. Hall, J. Berry, G. Collatz, A. Denning, H. Mooney, C. Nobre, N. Sato, C. Field, and A. Henderson-Sellers. 1997. "Modeling the exchange of energy, water, and carbon between continents and the atmosphere," *Science,* **275**:502–509.

Sexstone, A. J., et al. 1985. "Direct measurement of oxygen profiles and denitrification rates in soil aggregates," *Soil Sci. Soc. Amer. J.*, **49**:645–651.

Tiarks, A. E., W. H. Hudnall, J. F. Ragus, and W. B. Patterson. 1996. "Effect of pine plantation harvesting and soil compaction on soil water and temperature regimes in a semi-tropical environment," in A. Schulte and D. Ruhiyat (eds.), *Proceedings of International Congress on Soils of Tropical Forest Ecosystems 3rd Conference on Forest Soils: Vol. 3, Soil and Water Relationships.* (Samarinda, Indonesia: Mulawarmon University Press).

Tindall, J. A., R. B. Beverly, and D. E. Radcliff. 1991. "Mulch effect on soil properties and tomato growth using micro-irrigation," *Agron. J.*, **83**:1028–1034.

Trettin, C. C., M. Davidian, M. F. Jurgensen, and R. Lea. 1996. "Organic matter decomposition following harvesting and site preparation of a forested wetland," *Soil Sci. Soc. Am. J.*, **60**:1994–2003.

Unger, P. W. 1978. "Straw mulch effects on soil temperatures and sorghum germination and growth," *Agron. J.*, **70**:858–864.

Veneman, P. L. M., D. L. Lindbo, and L. A. Spoks. 1998. "Soil moisture and redoximorphic features: A historical perspective," in M. J. Rabenhorst (ed.), TITLE. Spec. Publ. XXX. (Madison, Wis.: Soil Sci. Soc. Amer.).

Welsh, D., D. Smart, J. Boyer, P. Minkin, H. Smith, and T. McCandless (eds.). 1995. *Forested Wetlands: Functions, Benefits, and Use of Best Management Practices.* (Radnor, Pa.: USDA Forest Service).

Yavitt, J. B., T. J. Fahley, and J. A. Simmons. 1995. "Methane and carbon dioxide dynamics in a northern hardwood ecosystem," *Soil. Sci. Soc. Am. J.*, **59**:796–804.

8

SOIL COLLOIDS: THEIR NATURE AND PRACTICAL SIGNIFICANCE

The landscape of the clays is like—the intricate folds of the womb—whose activity is to receive, contain, enfold, and give birth.
—WILLIAM BRYANT LOGAN

Next to photosynthesis and respiration, no process in nature is more vital to plant and animal life than the exchange of ions between soil particles and plant roots. These **cation** and **anion exchanges** occur mostly on the surfaces of the finer or **colloidal**[1] **fractions** of both the inorganic and organic matter—clays and humus.

Such colloidal particles serve very much as does a modern bank. They are the sites within the soil where ions of essential mineral elements such as calcium, potassium, and sulfur are held and protected from excessive loss by percolating rain or irrigation water. Subsequently, the essential ions can be withdrawn from the colloidal bank sites and taken up by plant roots. In turn, these elements can be deposited or returned to the colloids through plant residues and through the addition of commercial fertilizers, lime, and manures.

Cation and anion exchange mechanisms are also significant in controlling the movement of some organic chemicals through the soil and into nearby streams or well water. Any chemical that is attracted by positive or negative charges will be held to some extent by the soil colloids, later to be released slowly into the surrounding environment.

But cation and anion exchange capabilities are not the only soil properties dependent upon soil colloids. These fine soil particles markedly influence soil structure formation and stability, and the retention and movement of water, with their implications for the control of soil air and soil temperature. Likewise, the use of soils for building sites and for highway construction is dependent to a considerable degree on the amount and nature of the soil colloids.

No understanding of soils is complete without at least a general knowledge of the nature and constitution of colloidal particles (clays and humus). Their common properties will be covered first and then their individual characteristics will receive attention.

---

[1] Some of the clay and organic particles are somewhat larger then the upper limit of the colloidal state (1 μm). However, they do have some colloidlike properties, such as positive and negative surface charges, the chemical attraction of ions to their surfaces, and the ability to disperse. Accordingly, the fine fractions of soils (clays and humus) will be referred to as *soil colloids,* recognizing that technically some of these particles are not small enough to be classed as colloids. For a review of the composition and properties of clays and humus, see Dixon and Weed (1989).

8.1 GENERAL PROPERTIES OF SOIL COLLOIDS

Size

The most important common property of colloids is their extremely small size. They are too small to be seen with an ordinary light microscope. Only with an electron microscope can they be photographed. Most are smaller than 2 millionths of a meter [micrometers (μm)] in diameter.

Surface Area

Because of their small size, all soil colloids expose a large **external** surface per unit mass. The external surface area of 1 g of colloidal clay is at least 1000 times that of 1 g of coarse sand. Some colloids, especially certain silicate clays, have extensive **internal** surfaces as well. These internal surfaces occur between platelike crystal units that make up each particle and commonly greatly exceed the external surface area. The total surface area of soil colloids ranges from 10 m^2/g for clays with only external surfaces to more than 800 m^2/g for clays with extensive internal surfaces. The colloidal surface area in the upper 15 cm of 1 ha of a clay soil could be as high as 700,000 km^2 (270,000 mi^2).

Surface Charges

Soil colloidal surfaces, both external and internal, carry negative and/or positive charges. For most soil colloids, electronegative charges predominate, although some mineral colloids in very acid soils have a net electropositive charge. The presence and intensity of the particle charges influence the attraction and repulsion of the particles toward each other, thereby influencing both physical and chemical properties. The source of these charges will be considered later.

Adsorption of Cations and Water

The charges on soil colloids attract ions of an opposite charge to the colloidal surfaces. Such attraction is of particular significance for negatively charged colloids. The colloidal particles, referred to as *micelles* (microcells), attract hundreds of thousands of positively charged ions or *cations,*[2] such as H^+, Al^{3+}, Ca^{2+} and Mg^{2+}. This gives rise to an **ionic double layer** (Figure 8.1). The colloidal particle constitutes the *inner* ionic layer, being essentially a huge anion, the external and internal surfaces of which are negative in charge. The *outer* ionic layer is made up of a swarm of rather loosely held (adsorbed) cations that are attracted to the negatively charged surfaces. Thus, a colloidal particle is accompanied by a swarm of cations that are **adsorbed** or held on the particle surfaces.

In addition to the adsorbed cations, a large number of water molecules are associated with soil colloidal particles. Some are attracted to the adsorbed cations, each of which is hydrated. Others are held in the internal surfaces of the colloidal particles. These water molecules play a critical role in determining both the physical and chemical properties of soil.

Anions such as NO_3^-, Cl^-, and SO_4^{2-} may also be adsorbed on some soil colloids, this time the attraction being positive charges on the colloidal surfaces. While the adsorption of anions is not as extensive as that of cations, it is an important mechanism for holding negatively charged constituents.

8.2 TYPES OF SOIL COLLOIDS

There are four major types of colloids present in soils: (1) layer silicate clays, (2) iron and aluminum oxide clays, (3) allophane and associated clays, and (4) humus. Each group possesses the general colloidal characteristics described in the preceding section, but each has specific characteristics that make it distinctive and useful.

---

[2] In water solutions the cations are always hydrated, giving rise, in the case of hydrogen, to the **hydronium** ion H_3O^+ [sometimes written H^+ (aq)]. For simplicity, however, the H^+ ion designation will be used in this text.

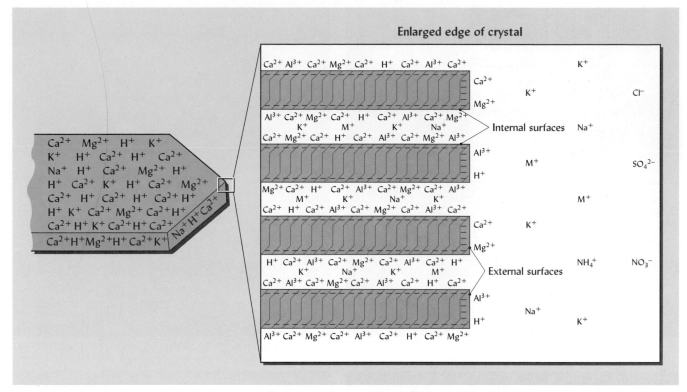

FIGURE 8.1 Diagrammatic representation of a silicate clay crystal with its sheetlike structure, negative surface charges, and associated adsorbed cations. The enlarged schematic view of the edge of the crystal illustrates the negatively charged internal as well as external surfaces to which cations and water molecules are attracted. Note that the orderly structured particle provides an inner negatively charged ionic layer that attracts a swarm of positively charged cations. Together they constitute what is known as an *ionic double layer.* Some clays exhibit positive charges (not shown) that can attract negative ions, such as NO_3^- and SO_4^{2-}.

Layer Silicate Clays

Silicate clay minerals are the dominant inorganic colloids in most all soils, especially those of temperate areas. Their most important properties are their layerlike, crystalline structures and their negative charges. Each particle is comprised of a series of layers, much like the pages of a book (Figure 8.2). The layers are comprised of planes of closely packed oxygen atoms held together by silicon, aluminum, magnesium, hydrogen, and/or iron atoms that occupy spaces between the oxygens. The formula of one of these clays, kaolinite $[Si_2Al_2O_5(OH)_4]$ illustrates their general chemical makeup.

The exact chemical composition and the internal arrangement of the atoms in each layer account for the particle's surface charge and its ability to hold and exchange ions, as well as its physical properties, including its stickiness and plasticity. Each of the types of silicate clays will be given more detailed consideration in Section 8.4.

Allophane and Imogolite

In many soils there are significant quantities of colloidal silicates whose crystalline structure is not sufficiently ordered to be detected easily by X-ray diffraction. Because they lack ordered three-dimensional crystalline structure over significant distances along the particle framework, they are sometimes referred to as ***short-range order minerals.***[3] They are more difficult to study than well-crystallized minerals, and, consequently, less is known about them.

The most significant of such poorly crystallized silicate colloids are **allophane** and its more weathered companion **imogolite.** These minerals are somewhat poorly defined

---

[3] Minerals that have an orderly crystalline structure over long distances [10 to 100 nanometers (nm) or 10^{-9} m] within the particle are said to exhibit **long-range order.** Others, such as allophane and imogolite, that have only short distances of well-ordered structure interspersed with materials that are noncrystalline, exhibit **short-range order.**

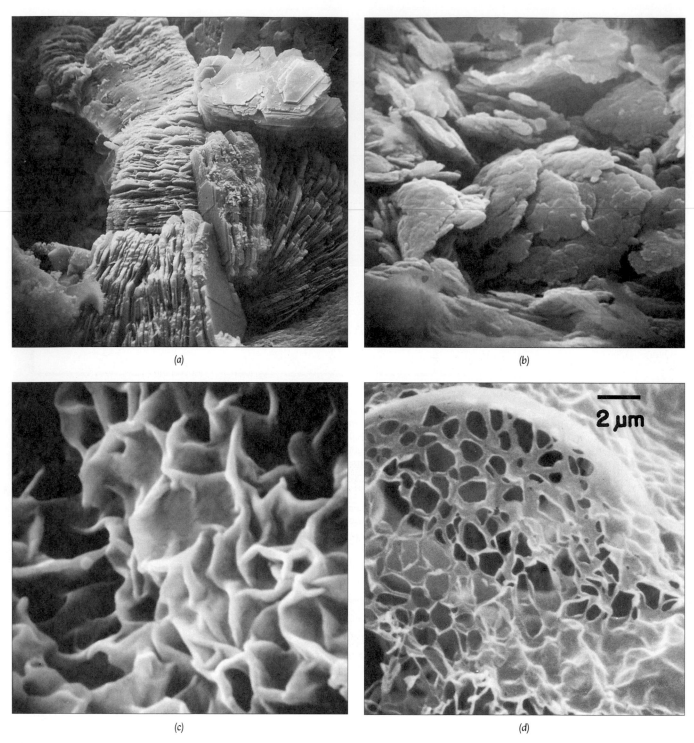

FIGURE 8.2 Crystals of three silicate clay minerals and a photomicrograph of humic acid found in soils. (*a*) Kaolinite from Illinois magnified about 1900 times (note hexagonal crystal at upper right). (*b*) A fine-grained mica from Wisconsin magnified about 17,600 times. (*c*) Montmorillonite (a smectite group mineral) from Wyoming magnified about 21,000 times. (*d*) Fulvic acid (a humic acid) from Georgia magnified about 23,000 times. [(*a*)–(*c*) Courtesy Dr. Bruce F. Bohor, Illinois State Geological Survey; (*d*) from Dr. Kim H. Tan, University of Georgia; used with permission of Soil Science Society of America]

aluminum silicates with a general composition approximating $Al_2O_3 \cdot 2SiO_2H_2O$. They are most prevalent in soils developed from volcanic ash (Andisols), such as those found in the Northwestern part of the United States. Their capacity to adsorb ions varies with pH, cations being adsorbed mostly at high pH, anions mostly at low pH. Allophane and imogolite are known for their high phosphate adsorbing capacities, especially in acid soils.

Iron and Aluminum Oxide Clays

These clays occur in greater quantities in the highly weathered Ultisols and Oxisols of the tropics and semitropics but are also present in significant quantities in the Alfisols and Inceptisols of some temperate regions. Properties of the red and yellow soils (Ultisols) so common in the southeastern United States are strongly influenced by these clays.

Examples of iron and aluminum oxides common in soils are goethite (FeOOH), hematite (Fe_2O_3), and gibbsite [$Al(OH)_3$]. For simplicity, they will be referred to as the **Fe, Al oxide** clays. Some have definite crystalline structures, but others are amorphous. They are not as sticky and plastic as the layer silicate clays on which they are commonly found as coatings. The surface charge of Fe, Al oxide minerals varies with pH. At high pH values, the particles carry a small negative charge that is balanced by adsorbed cations. In very acid soils some Fe, Al oxides carry a net positive charge and attract negatively charged anions instead of cations (anion exchange is covered in Section 8.15).

Organic Soil Colloids: Humus

The colloidal organization of humus has some similarities to that of clay. A highly charged micelle is surrounded by a swarm of cations. The humus colloids are not crystalline, however. They are convoluted chains of carbon bonded to hydrogen, oxygen, and nitrogen. The organic colloidal particles vary in size, but they may be at least as small as the silicate clay particles.

The negative charges of humus are associated with partially dissociated enolic (—OH), carboxyl (—COOH), and phenolic (◯—OH) groups; these groups, in turn, are associated with central units of varying size and complexity. The relationship is illustrated in Figure 8.3.

As is the case for the Fe, Al oxides, the negative charge associated with humus is dependent on the soil pH. Under very acid conditions, the negative charge is not very high—lower than that of some of the silicate clays. With a rise in pH to neutral and alkaline conditions, however, the electronegativity of humus per unit weight greatly exceeds that of the silicate clays. In these high-pH soils, the adsorbed hydrogen has been replaced by calcium, magnesium, and other such cations (see Figure 8.3).

8.3 ADSORBED CATIONS

In humid regions, the cations of calcium, aluminum, and, to a lesser extent, hydrogen are the most numerous; whereas in an arid-region soil, calcium, magnesium, potassium,

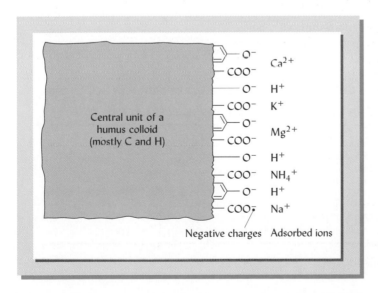

FIGURE 8.3 Adsorption of cations by humus colloids. The phenolic hydroxy groups are attached to aromatic rings,

Other —OH and the carboxyl (—COOH) groups are bonded to carbon atoms in the central unit. Note the general similarity to the adsorption situation in silicate clays.

and sodium predominate (Table 8.1). A colloidal complex may be represented in the following simple and convenient way for each region:

$$
\begin{array}{l}
a \ Ca^{2+} \\
b \ Al^{3+} \\
c \ H^{+} \\
d \ M^{+}
\end{array} \boxed{\text{Micelle}} \qquad
\begin{array}{l}
e \ Ca^{2+} \\
f \ Mg^{2+} \\
g \ K^{+} \\
h \ M^{+}
\end{array} \boxed{\text{Micelle}}
$$

Humid region   Arid region

The M^{+} stands for the small amounts of other base-forming cations (e.g., Na^{+} and NH_4^{+}) adsorbed by the colloids. The a through h indicate that the numbers of cations are variable.

These examples illustrate that soil colloids and their associated exchangeable ions can be considered, although perhaps in an oversimplified way, as *complex salts* with the large anion (micelle) surrounded by numerous cations.

Cation Prominence

Two major factors will determine the relative proportion of the different cations adsorbed by clays. First, these ions are not all held with equal tightness by the soil colloids. The order of strength of adsorption[4] when they are present in equivalent quantities is $Al^{3+} > Ca^{2+} > Mg^{2+} > K^{+} = NH_4^{+} > Na^{+}$.

Second, the relative concentration of the cations in the soil solution will help determine the degree to which adsorption occurs. Thus, in the soil solution of very acid soils, the concentration of Al^{3+} is high and some H^{+} ion is present; consequently, these ions dominate the adsorbed cations. At neutral pH and above, however, the Al^{3+} and H^{+} ion concentrations in the soil solution are very low; consequently, the adsorption of these ions is minimal. In these soils, Ca^{2+} and Mg^{2+} dominate. In some poorly drained soils of arid regions, salts high in sodium accumulate, and the adsorption of Na^{+} becomes much more prominent. The typical proportions of cations in different soil orders are shown in Table 8.1.

Cation Exchange

As pointed out in Chapter 1, the surface-held adsorbed cations are subject to exchange with other cations held in the soil solution. For example, a calcium ion held on the surface of a colloidal particle (micelle) is subject to exchange with two H^{+} ions in the soil solution.

$$
\boxed{\text{Micelle}} \ Ca^{2+} + 2H^{+} \rightleftharpoons \boxed{\text{Micelle}} \begin{array}{l} H^{+} \\ H^{+} \end{array} + Ca^{2+}
$$

(colloid) (soil solution) (colloid) (soil solution)

---

[4] The strength of adsorption of the H^{+} ion is difficult to determine since hydrogen-dominated mineral colloids break down to form aluminum-saturated colloids.

TABLE 8.1 **Typical Proportions of Major Adsorbed Cations in the Surface Layers of Different Soil Orders**

The percentage figures in each case are based on the sum of the cation equivalents taken as 100.

Soil order[a]	Typical location	H^{+} and Al^{3+}, %[b]	Ca^{2+}, %	Mg^{2+}, %	K^{+}, %	Na^{+}, %
Oxisols	Hawaii	85	10	3	2	tr[c]
Spodosols	New England	80	15	3	2	tr
Ultisols	Southeast United States	65	25	6	3	1
Alfisols	Pennsylvania to Wisconsin	45	35	13	5	2
Vertisols	Alabama to Texas	40	38	15	5	2
Mollisols	Midwest United States	30	43	18	6	3
Aridisols	Southwest United States	—	65	20	10	5

[a] See Chapter 3 for soil descriptions.
[b] Al^{3+} adsorption includes that of complex aluminum hydroxy ions.
[c] tr = trace.

Thus, soil colloids are focal points for cation exchange reactions, which have profound effects on soil–plant relations. These will be discussed in greater detail (see Section 8.11) after we give more detailed consideration to silicate clays.

8.4 FUNDAMENTALS OF LAYER SILICATE CLAY STRUCTURE[5]

The use of X-ray diffraction, electron microscopy, and other techniques has demonstrated that silicate clay particles are crystalline. The mineralogical organization of these particles varies from one type of clay to another, but there are some common features that we will first examine.

Silica Tetrahedral and Aluminium-Magnesium Octahedral Sheets

The most important silicate clays are known as **phyllosilicates** (Greek *phyllon,* leaf), because of their leaflike or planar structure. As shown in Figure 8.4, they are comprised of two kinds of **sheets**, one consisting of two planes of oxygen with mainly silicon in the spaces between the oxygens, and the other having two planes of oxygen and hydroxyls with aluminum or magnesium in the spaces between the oxygens and hydroxyls.

The basic building block for the silica-dominated sheet is a unit composed of one silicon atom surrounded by four oxygen atoms. It is called the silica **tetrahedron** because the centers of the four oxygens define the apices of a four-sided geometric solid (see Figure 8.4). An array of tetrahedra that share oxygens with one another give a **tetrahedral sheet** as shown in Figure 8.4.

Aluminum and/or magnesium ions are the key cations in the second type of sheet. In the basic units, an aluminum or magnesium ion is surrounded by six oxygen atoms or hydroxy groups, the centers of which define the apices of an eight-sided solid or **octahedron** (see Figure 8.4). Numerous octahedra linked together comprise the **octahedral sheet.** An aluminum-dominated sheet is known as a *dioctahedral* sheet, whereas one dominated by magnesium is called a *trioctahedral* sheet. The distinction is due to the fact that *two* aluminum ions in a *di*octahedral sheet satisfy the same negative charge from surrounding oxygens and hydroxys as *three* magnesium ions in a *tri*octahedral sheet.

The tetrahedral and octahedral sheets are the fundamental structural units of silicate clays. They, in turn, are bound together within the crystals by shared oxygen atoms into different **layers.** The specific nature and combination of sheets in these layers vary from one type of clay to another and largely control the physical and chemical properties of each clay. The relationship between sheets and layers shown in Figure 8.4 is important and should be well understood.

Isomorphous Substitution

The structural arrangements just described suggest a very simple relationship among the elements making up silicate clays. In nature, however, more complex formulas result. The weathering of a wide variety of rocks and minerals permits cations of comparable size to substitute for silicon, aluminum, and magnesium ions in the respective tetrahedral and octahedral sheets.

The ionic radii of a number of ions common in clays are listed in Table 8.2 to illustrate this point. Note that aluminum is only slightly larger than silicon. Consequently, aluminum can fit into the center of the tetrahedron in the place of the silicon without changing the basic structure of the crystal. The process, called **isomorphous substitution,** is common and accounts for the wide variability in the nature of silicate clays.

Isomorphous substitution also occurs in the octahedral sheet. Note from Table 8.2 that ions such as iron and zinc are not too different in size from aluminum and magnesium ions. As a result, these ions can fit into the position of either the aluminum or magnesium as the central ion in the octahedral sheet. It should be emphasized that some layer silicates are characterized by isomorphous substitution in either or both of the tetrahedral or octahedral sheets.

---

[5] The authors are indebted to Dr. Darrell G. Schultze of Purdue University for kindly providing the structural models for the silicate clay minerals.

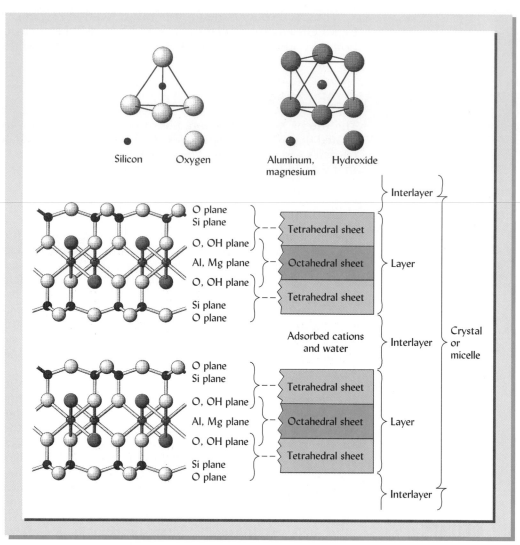

FIGURE 8.4 The basic molecular and structural components of silicate clays. (*a*) A single *tetrahedron,* a four-sided building block comprised of a silicon ion surrounded by four oxygen atoms; and a single eight-sided *octahedron,* in which an aluminum (or magnesium) ion is surrounded by six hydroxy groups or oxygen atoms. (*b*) In clay crystals thousands of these tetrahedral and octahedral building blocks are connected to give planes of silicon and aluminum (or magnesium) ions. These planes alternate with planes of oxygen atoms and hydroxy groups. Note that apical oxygen atoms are common to adjoining tetrahedral and octahedral sheets. The silicon plane and associated oxygen-hydroxy planes make up a *tetrahedral sheet.* Similarly, the aluminum-magnesium plane and associated oxygen-hydroxy planes comprise the *octahedral sheet.* Different combinations of tetrahedral and octahedral sheets are termed *layers.* In some silicate clays these layers are separated by *interlayers* in which water and adsorbed cations are found. Many layers are found in each *crystal* or *micelle* (microcell).

Source of Charges

Isomorphous substitution is of vital importance because it is the primary source of both negative and positive charges of silicate clays. For example, the substitution of one Al^{3+} for a Si^{4+} in the tetrahedral sheet leaves one unsatisfied negative charge. Likewise, the substitution of one Al^{3+} for one Mg^{2+} in a trioctahedral sheet results in one excess positive charge. The **net charge** associated with a clay micelle is the balance between the positive and negative charges. In layer silicate clays, the net charge is almost always negative. This subject will receive more attention later (see Section 8.8).

TABLE 8.2   Ionic Radii of Elements Found in Silicate Clays and an Indication of Which Are Found in the Tetrahedral and Octahedral Sheets

Note that Al, Fe, O, and OH can fit in either.

Ion	Radius, nm[a]	Found in
Si^{4+}	0.042	
Al^{3+}	0.051	Tetrahedral sheet
Fe^{3+}	0.064	
Mg^{2+}	0.066	Octahedral sheet
Zn^{2+}	0.074	Exchange sites
Fe^{2+}	0.070	
Na^+	0.097	
Ca^{2+}	0.099	
K^+	0.133	
O^{2-}	0.140	Both sheets
OH^-	0.155	

[a] 1 nm = 10^{-9}m.

8.5   MINERALOGICAL ORGANIZATION OF SILICATE CLAYS

On the basis of the number and arrangement of tetrahedral (silica) and octahedral (aluminium-magnesium) sheets contained in the crystal units or layers, silicate clays are classified into two different groups, 1:1-type minerals and 2:1-type minerals. For illustrative purposes, each of these is discussed briefly in terms of one member of each group.

1:1-Type Minerals

The layers of the 1:1-type minerals are made up of one tetrahedral (silica) sheet and one octahedral (alumina) sheet—hence, the terminology *1:1-type crystal* (Figure 8.5). In soils, **kaolinite** is the most prominent member of this group, which includes *halloysite, nacrite,* and *dickite.*

The tetrahedral and octahedral sheets in a layer of a kaolinite crystal are held together tightly by oxygen atoms, which are mutually shared by the silicon and aluminum cations in their respective sheets. These layers, in turn, are bound to other adjacent lay-

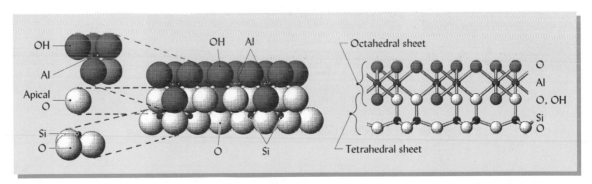

FIGURE 8.5   Models of ions that constitute a layer of the 1:1-type clay kaolinite. The primary elements of the octahedral (upper left) and tetrahedral (lower left) sheets are depicted as they might appear separately. In the crystal structure, however, these sheets are held together by common oxygen ions that are known as *apical* oxygen ions. Note that each layer consists of alternating octahedral (alumina) and tetrahedral (silica) sheets—hence, the designation 1:1. Aluminum ions surrounded by six hydroxy groups and/or oxygen atoms (counting the apical oxygen) make up the octahedral sheet (upper left). Smaller silicon ions associated with four oxygen atoms (counting the apical oxygen) constitute the tetrahedral sheet (lower left). The octahedral and tetrahedral sheets are bound together (center) by mutually shared (apical) oxygen atoms. The result is a layer with hydroxys on one surface and oxygens on the other. A schematic drawing of the ionic arrangement (right) shows a cross-sectional view of a crystal layer. Note the common apical oxygen ions that hold the sheets together. The kaolinite mineral is comprised of a series of these flat layers tightly held together with no interlayer spaces. [Note: To permit us to view the front silicon atoms in the silicon plane (center drawing), we have not shown the bottom oxygen atoms that are normally present].

ers by **hydrogen bonding** (see Section 5.1). Consequently, the structure is *fixed,* and no expansion ordinarily occurs between layers when the clay is wetted. Cations and water do not enter between the structural layers of a 1:1-type mineral particle. The effective surface of kaolinite is thus restricted to its outer faces or to its **external surface** area. Also, there is little isomorphous substitution in this 1:1-type mineral. Along with the relatively low surface area of kaolinite, this accounts for its low capacity to adsorb cations.[6]

Kaolinite crystals usually are hexagonal in shape (see Figure 8.2). In comparison to other clay particles, they are large in size, ranging from 0.1 to 5 μm across, with the majority falling within the 0.2- to 2-μm range. Because of the strong binding forces between their structural layers, kaolinite particles are not readily broken down into extremely thin plates.

In contrast with the other silicate groups, kaolinite exhibits less plasticity (capacity of being molded), stickiness, cohesion, shrinkage, or swelling. Its restricted surface and limited adsorptive capacity for cations and water molecules suggest that kaolinite does not exhibit colloidal properties of a high order of intensity (Table 8.3). At the same time, kaolinite-containing soils make good bases for roadbeds and building foundations, and are commonly used in making bricks. They are easy to cultivate for agriculture and, with some nutrient supplementation from manure and fertilizer, can be very productive.

2:1-Type Minerals

The crystal units (layers) of these minerals are characterized by an octahedral sheet sandwiched between two tetrahedral sheets. Four general groups have this basic crystal structure. Two of them, **smectite** and **vermiculite,** include expanding-type minerals; the other two, **fine-grained micas** (**illite**) and **chlorite,** are relatively nonexpanding.

EXPANDING MINERALS. The **smectite** group is noted for **interlayer expansion,** which occurs by swelling when the minerals are wetted, the water entering the interlayer space and forcing the layers apart. **Montmorillonite** is the most prominent member of this group in soils, although *beidellite, nontronite,* and *saponite* also are found.

The flakelike crystals of smectites (see Figure 8.2) are composed of 2:1-type layers, as shown in Figure 8.6. In turn, these layers are loosely held together by very weak oxygen-to-oxygen and cation-to-oxygen linkages. Exchangeable cations and associated water molecules are attracted between layers (the interlayer space), causing **expansion** of the crystal lattice. The **internal surface** thus exposed by far exceeds the external surface area of these minerals. For example, the **specific surface** or total surface area per unit mass (external and internal) of one smectite mineral (montmorillonite) is 650 to 800 m^2/g. A comparable figure for kaolinite is only 10 to 30 m^2/g (see Table 8.3). Commonly, these smectite crystals range in size from 0.01 to 1 μm (10^{-6} m), much smaller than the average kaolinite particle.

---

[6] Since the adsorbed cations may be freely exchanged with other cations, the capacity to adsorb cations is usually referred to as **cation exchange capacity** (see Section 8.12).

TABLE 8.3 **Major Properties of Selected Silicate Clay Minerals[a] and of Humus**

Property	2:1-type clays				1:1-type clay	
	Smectite	*Vermiculite*	*Fine mica*	*Chlorite*	*Kaolinite*	*Humus*
Size, μm	0.01–1.0	0.1–5.0	0.2–2.0	0.1–2.0	0.1–5.0	0.1–1.0
Shape	Flakes	Plates; flakes	Flakes	Variable	Hexagonal crystals	Variable
External surface, m^2/g	80–140	70–120	70–100	70–100	10–30	Variable[d]
Internal surface, m^2/g	570–660	600–700	—	—	—	
Intralayer spacing[b], nm	1.0–2.0	1.0–1.5	1.0	1.4	0.7	—
Net negative charge, $cmol_c/kg$[c]	80–120	100–180	15–40	15–40	2–5	100–550

[a] Both dioctahedral and trioctahedral forms of the 2:1-type clays are found in soils, but trioctahedral sheets are most prominent in chlorites while dioctahedral sheets are generally more common in the others.
[b] From the top of one layer to the top of the next similar layer; 1 nm = 10^{-9} m.
[c] Centimoles charge per kilogram (1 $cmol_c$ = 0.01 mol_c), a measure of the cation exchange capacity (see Section 8.12).
[d] It is very difficult to accurately determine the surface area of organic matter. Different procedures give values ranging from 20 to 800 m^2/g.

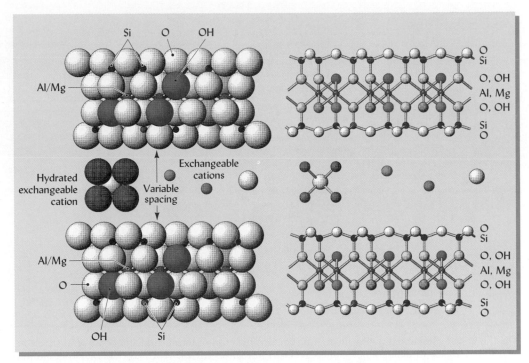

FIGURE 8.6 Model of two crystal layers and an interlayer characteristic of montmorillonite, a smectite expanding-lattice 2:1-type clay mineral. Each layer is made up of an octahedral (alumina) sheet sandwiched between two tetrahedral (silica) sheets with shared apical oxygen ions that hold the sheets together. There is little attraction between oxygen atoms in the bottom tetrahedral sheet of one unit and those in the top tetrahedral sheet of another. This permits a variable space between layers, which is occupied by water and exchangeable cations. The internal surface area thus exposed far exceeds the surface around the outside of the crystal. Note that magnesium has replaced aluminum in some sites of the octahedral sheet. Likewise, some silicon atoms in the tetrahedral sheet may be replaced by aluminum (not shown). These substitutions give rise to a negative charge, which accounts for the high cation exchange capacity of this clay mineral. A schematic drawing of the ionic arrangement is shown at the right.

Isomorphous substitution of Mg^{2+} for some of the Al^{3+} ions in the dioctahedral sheet accounts for most of the negative charge for smectites, although some substitution of Al^{3+} for Si^{4+} has occurred in the tetrahedral sheet. The smectites commonly show a high cation exchange capacity, perhaps 20 to 40 times that of kaolinite (see Table 8.3).

Smectites also are noted for their high plasticity and cohesion and their marked swelling when wet and shrinkage on drying. Wide cracks commonly form as smectite-dominated soils (e.g., Vertisols) are dried (see Figure 4.40). The dry aggregates or clods are very hard, making such soils difficult to till. Also, soils high in smectite make very poor bases for building foundations or roads.

Vermiculites are also 2:1-type minerals, an octahedral sheet being found between two tetrahedral sheets. In most soil vermiculites, the octahedral sheet is aluminum-dominated (**dioctahedral**), although magnesium-dominated (**trioctahedral**) vermiculites are also found. In the tetrahedral sheet of most vermiculites, considerable substitution of aluminum for silicon has taken place. This accounts for most of the very high net negative charge associated with these minerals.

Water molecules, along with magnesium and other ions, including Al-hydroxy ions, are strongly adsorbed in the interlayer space of vermiculites (Figure 8.7). However, these interlayer constituents act primarily as bridges holding the units together rather than as wedges driving them apart. The degree of swelling is, therefore, considerably less for vermiculites than for smectites. For this reason, vermiculites are considered to be **limited-expansion** clay minerals, expanding more than kaolinite but much less than the smectites.

The cation-adsorbing capacity of vermiculites usually exceeds that of all other silicate clays, including montmorillonite and other smectites (see Table 8.3), because of the very high negative charge in the tetrahedral sheet. Vermiculite crystals are larger than those of the smectites but much smaller than those of kaolinite.

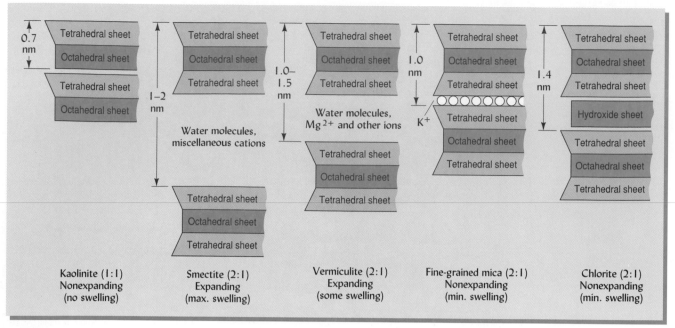

FIGURE 8.7 Schematic drawing illustrating the organization of tetrahedral and octahedral sheets in one 1:1-type mineral (kaolinite) and four 2:1-type minerals. The octahedral sheets in each of the 2:1 type clays can be either aluminum dominated (dioctahedral) or magnesium dominated (trioctahedral). However, in most chlorites the trioctahedral sheets are dominant while the dioctahedral sheets are generally most prominent in the other three 2:1-types. Note that kaolinite is nonexpanding, the layers being held together by hydrogen bonds. Maximum interlayer expansion is found in smectite, with somewhat less expansion in vermiculite because of the moderate binding power of numerous Mg^{2+} ions. Fine-grained mica and chlorite do not expand because K^+ ions (fine-grained mica) or an octahedral-like sheet of hydroxides of Al, Mg, Fe, and so forth (chlorite) tightly bind the 2:1 layers together. The interlayer spacings are shown in nanometers (1 nm = 10^{-9} m).

NONEXPANDING MINERALS. **Micas** and **chlorites** are the types of minerals in this group. Muscovite and biotite are examples of unweathered micas often found in the sand and silt separates. Weathered minerals similar in structure to these micas are found in the clay fraction of soils. They are called ***fine-grained micas.***[7]

Like smectites, fine-grained micas have a 2:1-type crystal. However, the particles are much larger than those of the smectites. Also, the major source of charge is in the tetrahedral sheet, where about 20% of the silicon sites are occupied by aluminum atoms. This results in a high net negative charge in the tetrahedral sheet, even higher than that found in vermiculites. To satisfy this charge, potassium ions are strongly attracted in the interlayer space. They are just the right size to fit snugly into certain spaces in the adjoining tetrahedral sheets (Figures 8.7 and 8.8). The potassium, being strongly attracted to the adjacent sheets, acts as a binding agent, preventing expansion of the crystal. Hence, fine-grained micas are quite **nonexpansive.**

Such properties as hydration, cation adsorption, swelling, shrinkage, and plasticity are much less intense in fine-grained micas than in smectites (see Table 8.3). The fined-grained micas exceed kaolinite with respect to these characteristics, but this may be due in part to the presence of interstratified layers of smectite or vermiculite. In size, too, fine-grained mica crystals are intermediate between the smectites and kaolinites (see Table 8.3). Their specific surface area varies from 70 to 100 m^2/g, about one-eighth that for the smectites.

Soil **chlorites** are basically iron-magnesium silicates with some aluminum present. In a typical chlorite clay crystal, 2:1 layers, such as are found in vermiculites, alternate with a hydroxide sheet (see Figure 8.7) that can be either dioctahedral or trioctahedral. However, magnesium commonly dominates both the interlayer hydroxide (trioctahedral) sheet as well as the trioctahedral sheet of the 2:1 layer of chlorites.

---

[7] In the past, the term **illite** has been used to identify these clay minerals. Current nomenclature identifies illite, along with others such as **glauconite**, as more specific clay minerals in the fine-grained mica group.

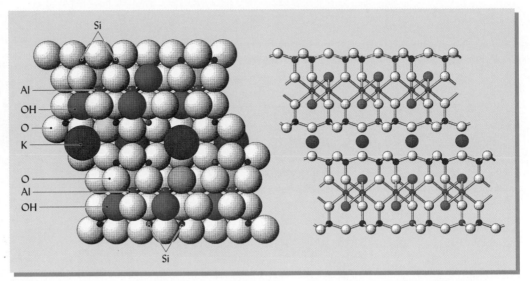

FIGURE 8.8 Model of a 2:1-type nonexpanding lattice mineral of the fine-grained mica group. The general constitution of the layers is similar to that in the smectites, one octahedral (alumina) sheet between two tetrahedral (silica) sheets. However, potassium ions are tightly held between layers, giving the mineral a more or less rigid type of structure that prevents the movement of water and cations into the space between layers. The internal surface and exchange capacity of fine-grained micas are thus far below those of the smectites.

The negative charge of chlorites is about the same as that of fine-grained micas and considerably less than that of the smectites or vermiculites. Like fine micas, chlorites may be interstratified with vermiculites or smectites in a single crystal. Particle size and surface area for chlorites are also about the same as for fine-grained micas. There is no water adsorption between the chlorite crystal units, which is in keeping with the non-expansive nature of this mineral.

Mixed and Interstratified Layers

Specific groups of clay minerals do not occur independently of one another. In a given soil, it is common to find several clay minerals in an intimate mixture. Furthermore, some mineral colloids have properties and compositions intermediate between those of any two of the well-defined minerals just described. Such minerals are termed *mixed layer* or *interstratified* because the individual layers within a given crystal may be of more than one type. Terms such as *chlorite-vermiculite* and *fine-grained mica-smectite* are used to describe mixed-layer minerals (Figure 8.9). In some soils, they are more common than single-structured minerals such as montmorillonite.

8.6 GENESIS OF SOIL COLLOIDS

Silicate Clays

The silicate clays are developed from the weathering of a wide variety of minerals by at least two distinct processes: (1) a slight physical and chemical **alteration** of certain primary minerals, and (2) a **decomposition** of primary minerals with the subsequent **recrystallization** of certain of their products into the silicate clays. These processes will each be given brief consideration.

ALTERATION.   The changes that occur as muscovite mica is altered to fine-grained mica represent a good example of alteration. Muscovite is a 2:1-type primary mineral with a nonexpanding crystal structure and a formula of $KAl_2(Si_3Al)O_{10}(OH)_2$. As weathering occurs, the mineral is broken down in size to the colloidal range, part of the potassium is lost, and some silicon along with some base-forming cations, represented as M^+, are added from weathering solutions. The net result is a less rigid crystal structure and the availability of free electronegative charges at sites formerly occupied by the fixed potas-

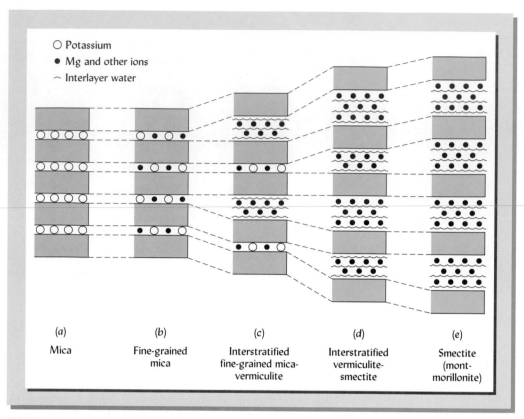

○ Potassium
● Mg and other ions
~ Interlayer water

(a)	(b)	(c)	(d)	(e)
Mica	Fine-grained mica	Interstratified fine-grained mica-vermiculite	Interstratified vermiculite-smectite	Smectite (mont-morillonite)

FIGURE 8.9   Structural differences among silicate minerals and their mixtures. Potassium-containing micas (*a*) with rigid crystals lose part of their potassium and weather to fine-grained mica, which is less rigid and attracts exchangeable cations to the interlayer space (*b*). At a more advanced stage of weathering (*c*), the potassium is leached from between some of the 2:1 layers; water and magnesium ions bind the layers together and an interstratified fine-grained mica-vermiculite is present. Further weathering removes more potassium (*d*), water and exchangeable ions push in between the 2:1 layers, and an interstratified vermiculite—smectite is formed. More weathering produces smectite (*e*), the highly expanded mineral. Smectite in turn is subject to weathering to kaolinite and iron and aluminum oxides. Although most smectites actually are formed by other processes, the sequence here illustrates the structural relationships of smectites to the other minerals.

sium. The fine mica colloid that emerges still has a 2:1-type structure, only having been *altered* in the process. Some of these changes, perhaps oversimplified, can be shown as

$$KAl_2(AlSi_3)O_{10}(OH)_2 + 0.2Si^{4+} + 0.1M^+ \xrightarrow{H_2O}$$

Muscovite (soil solution)
(rigid crystal)

$$M_{0.1}^+(K_{0.7})Al_2(Al_{0.8}Si_{3.2})O_{10}(OH)_2 + 0.3K^+ + 0.2Al^{3+}$$

Fine mica (soil solution)
(semirigid crystal)

RECRYSTALLIZATION.   This process involves the complete breakdown of the crystal structure and recrystallization of clay minerals from products of this breakdown. It is the result of much more intense weathering than that required for the alteration process just described.

An example of recrystallization is the formation of kaolinite (a 1:1-type clay mineral) from solutions containing soluble aluminum and silicon that came from the breakdown of primary minerals having a 2:1-type structure. Such recrystallization makes possible the formation of more than one kind of clay from a given primary mineral. The specific clay mineral that forms depends on weathering conditions and the specific ions present in the weathering solution as crystallization occurs.

RELATIVE STAGES OF WEATHERING. The more specific conditions conducive to the formation of important clay types are shown in Figure 8.10. Note that fine-grained micas and magnesium-rich chlorites represent earlier weathering stages of the silicates, and kaolinite and (ultimately) iron and aluminum oxides the most advanced stages. The smectites (e.g., montmorillonite) represent intermediate stages. As noted in Section 2.1, silicon tends to be lost as weathering progresses, leaving a lower Si:Al ratio in more highly weathered soil horizons.

GENESIS OF INDIVIDUAL SILICATE CLAYS. There are a variety of processes by which individual clays are formed. For example, fine-grained micas and chlorite are commonly formed from the alteration of muscovite and biotite micas, respectively. Vermiculites also can be formed through this process, although they as well as the smectites can be products of weathering of the fine-grained micas and chlorites. Smectites may also result from the process of recrystallization under neutral to alkaline weathering conditions. Kaolinite is formed by recrystallization under reasonably intense acid weathering conditions, which remove most of the metallic cations. Under hot, humid conditions of the tropics, the intense weathering commonly produces the oxides of iron and aluminum (see Figure 8.10).

Iron and Aluminum Oxides

Goethite ($FeOOH$) and hematite (Fe_2O_3) are the two most common iron oxides in soils. They are both produced by the weathering of iron-containing primary silicate minerals. The iron is commonly in the low-valent state, which is released but quickly oxidized to the very insoluble three-valent form upon the breakdown of the mineral framework as previously described. Goethite tends to be dominant in the temperate and more moist

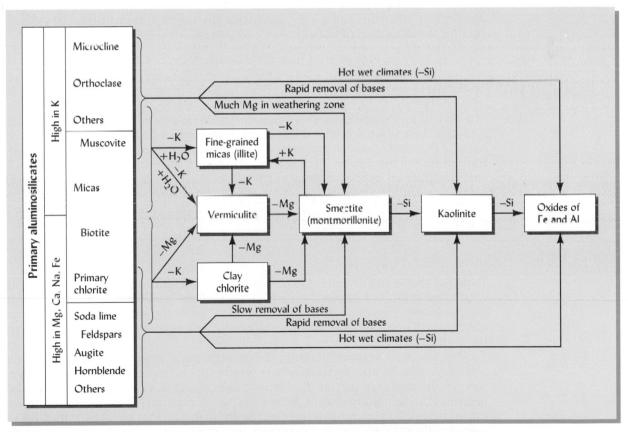

FIGURE 8.10   General conditions for the formation of the various layer silicate clays and oxides of iron and aluminum. Fine-grained micas, chlorite, and vermiculite are formed through rather mild weathering of primary aluminosilicate minerals, whereas kaolinite and oxides of iron and aluminum are products of much more intense weathering. Conditions of intermediate weathering intensity encourage the formation of smectite. In each case silicate clay genesis is accompanied by the removal in solution of such elements as K, Na, Ca, and Mg.

zones, while hematite with its deep red color is more prominent (even dominant) in tropical or drier conditions.

Gibbsite [(AlOH)$_3$], the most common oxide of aluminum found in soils, is a product of weathering of a variety of primary aluminosilicates. Hydrogen ions replace the base-forming cations and the mineral framework breaks down, releasing the aluminum and silicon. The aluminum released by the breakdown of some basic rocks, such as gabbro and basalt, commonly precipitates as gibbsite directly. The weathering of acid rocks, such as granite and gneiss, may first produce kaolinite or halloysite that, upon further weathering, yields gibbsite. Gibbsite represents the most advanced stage of weathering in soils.

It is interesting to note that isomorphic substitution of Al^{3+} for Fe^{3+} and vice versa may take place to some degree among oxides of these elements. However, since Fe^{3+} and Al^{3+} have the same valence, this substitution has no effect on the charge on the micelle.

Allophane and Imogolite

Relatively little is known of factors influencing the formation of allophane and imogolite. While they are commonly associated with materials of volcanic origin, they are also formed from igneous rocks and are found in some Spodosols. Apparently, volcanic ashes release significant quantities of $Si(OH)_x$ and $Al(OH)_x$ materials that precipitate as gels in a relatively short period of time. These minerals are generally poorly crystalline in nature, imogolite being the product of a more advanced state of weathering than that which produces allophane.

Humus

The breakdown and alteration of plant residues by microorganisms and the concurrent synthesis of new, more stable, organic compounds results in the formation of the dark-colored colloidal organic material called *humus* (see Section 12.4 for details). The various organic structural units associated with the decay and synthesis provide charged sites for the attraction of both cations and anions.

8.7 GEOGRAPHIC DISTRIBUTION OF CLAYS

The clay of any particular soil is generally made up of a mixture of different colloidal minerals. In a given soil, the mixture may vary from horizon to horizon. This occurs because the kind of clay that develops depends not only on climatic influences and profile conditions but also on the nature of the parent material. The situation may be further complicated by the presence in the parent material itself of clays that were formed under a preceding and perhaps an entirely different type of climatic regime. Nevertheless, some general deductions seem possible, based on the relationships shown in Figure 8.10.

Soil Order Differences

Table 8.4 shows the dominant clay minerals in different soil orders, the description of which was given in Chapter 3. The well-drained and well-weathered Oxisols of humid and subhumid tropics tend to be dominated by kaolinite, along with oxides of iron and aluminum. These clays also dominate Ultisols, such as those in the southeastern part of the United States.

The smectite, vermiculite, and fine-grained mica groups are more prominent in Alfisols, Mollisols, and Vertisols, where weathering is less intense. These clays are common in the northern part of the United States, Canada, and regions with similar temperatures throughout the world. Where the parent material is high in micas, fine-grained micas such as illite are apt to be formed. Parent materials that are high in metallic cations (particularly magnesium) or are subject to restricted drainage, which discourages the leaching of these cations, encourage smectite formation. Thus, fine-grained micas and smectites are more likely to be prominent in Aridisols than in the more humid areas.

The strong influence of parent material on the geographic distribution of clays can be seen in the black-belt Vertisols of Alabama, Mississippi, and Texas. These soils, which are

TABLE 8.4 Prominent Occurrence of Clay Minerals in Different Soil Orders in the United States and Typical Locations for These Soils

Soil order[a]	General weathering intensity	Typical location in U.S.	Fe, Al oxides	Kaolinite	Smectite	Fine-grained mica	Vermiculite	Chlorite	Intergrades
Aridisols	Low	Dry areas			XX	XX		X	X
Vertisols[b]	↑	Alabama, Texas			XXX				X
Mollisols		Central		X	XX	X	X	X	X
Alfisols		Ohio, Pennsylvania, New York		X	X	X	X	X	X
Spodosols	↓	New England	X	X					
Ultisols		Southeast	XX	XXX			X	X	X
Oxisols	High	Hawaii, Puerto Rico	XX	XXX					

[a] See Chapter 3 for soil descriptions.
[b] By definition these soils have swelling-type clays, which account for the dominance of smectites.

dark in color, have developed from base-rich marine parent materials and are dominated by smectite clays. The surrounding soils, which have developed from lower-base parent materials, are high in kaolinite and hydrous oxides, clays that are more representative of this warm, humid region. Similar situations exist in central India and Sudan.

Although a few broad generalizations relating to the geographic distribution of clays are possible, these examples suggest that local parent materials and weathering conditions tend to dictate the kinds of clay minerals found in soils.

8.8 SOURCES OF CHARGES ON SOIL COLLOIDS

There are two major sources of charges on soil colloids: (1) hydroxyls and other such functional groups on the surfaces of the colloidal particles that by releasing or accepting H^+ ions can provide either negative or positive charges, and (2) the charge imbalance brought about by the isomorphous substitution in some clay crystal structures of one cation by another of similar size but differing in charge.

All colloids, organic or inorganic, exhibit the surface charges associated with OH groups, charges that are largely **pH dependent.** Most of the charges associated with humus, 1:1-type clays, the oxides of iron and aluminum, and allophane are of this type. In the case of the 2:1-type clays, however, these surface charges are complemented by a much larger number of charges emanating from the isomorphous substitution of one cation for another. Since they are not dependent on the pH, they are termed **permanent** or **constant charges.** We will consider these constant charges first.

8.9 CONSTANT CHARGES ON SILICATE CLAYS

We noted in Section 8.4 that isomorphous substitution could be the source of both negative and positive charges. Examples of specific substitutions will now be considered.

Negative Charges

A net negative charge is found in minerals where there has been an isomorphous substitution of a lower-charged ion (e.g., Mg^{2+}) for a higher-charged ion (e.g., Al^{3+}). Such substitution commonly occurs in some aluminum-dominated dioctahedral sheets. As shown in Figure 8.11, this leaves an unsatisfied negative charge. The substitution of Mg^{2+} for Al^{3+} is an important source of the negative charge on the smectite, vermiculite, and chlorite clay micelles.

A second example is the substitution of an Al^{3+} for an Si^{4+} in the tetrahedral sheet, which also leaves one negative charge unsatisfied. This can be illustrated as follows, assuming that on an average two silicon atoms are associated with four oxygens in a charge-neutral unit.

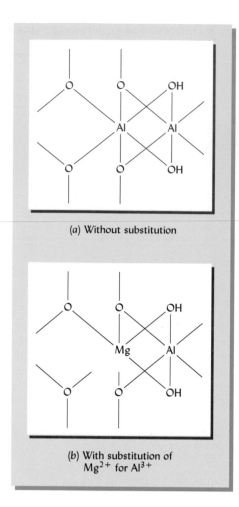

(a) Without substitution

(b) With substitution of
Mg^{2+} for Al^{3+}

FIGURE 8.11 Atomic configuration in the octahedral sheet of silicate clays (a) without substitution and (b) with a magnesium ion substituted for one aluminum. Where no substitution has occurred, the three positive charges on aluminum are balanced by an equivalent of three negative charges from six oxygens or hydroxy groups. With magnesium in the place of aluminum, only two of those negative charges are balanced, leaving one negative charge unsatisfied. This negative charge can be satisfied by an adsorbed cation.

Tetrahedral sheet
(no substitution)

Tetrahedral sheet
(Al^{3+} substituted for Si^{4+})

| Si$_2$O$_4$ | SiAlO$_4^-$ |

No charge

One excess negative charge

Such a substitution is common in several of the important soil silicate clay minerals, such as the fine-grained micas, vermiculites, and even some smectites.

Positive Charges

Isomorphous substitution can also be a source of positive charges if the substituting cation has a higher charge than the ion for which it substitutes. In a trioctahedral sheet, there are three magnesium ions surrounded by oxygen and hydroxy groups, and the sheet has no charge. However, if an Al^{3+} ion substitutes for one of the Mg^{2+} ions, a positive charge results.

Trioctahedral sheet
(no substitution)

Trioctahedral sheet
(Al^{3+} substituted for Mg^{2+})

| Mg$_3$O$_2$(OH)$_2$ | Mg$_2$AlO$_2$(OH)$_2^+$ |

No charge

One excess positive charge

Such positive charges are characteristic of the trioctahedral hydroxide sheet in the interlayer of clay minerals such as chlorites, a charge that is overbalanced by negative charges in the tetrahedral sheet. Indeed, in several 2:1-type silicate clays, including chlorites and smectites, substitutions in both the tetrahedral and octahedral sheets can occur. The net charge in these clays is the balance between the negative and positive charges. In all 2:1-type silicate clays, however, the **net charge** is negative since those substitutions leading to negative charges far outweigh those producing positive charges.

Chemical Composition and Charge

Because of the numerous ionic substitutions just discussed, simple chemical formulas cannot be used to identify specifically the clay in a given soil. However, type formulas for the major silicate clays shown in Table 8.5 can be used to illustrate the sources of both positive and negative charges in the tetrahedral and octahedral sheets. These formulas are commonly referred to as *structural formulae* or *unit layer formulae.*

Note that Table 8.5 shows no ionic substitution in kaolinite. Although a minor amount of such substitution may occur, the bulk of the negative charge on this colloid is provided by the surface OH groups. All other clay minerals shown in Table 8.5 owe most of their charge (negative and/or positive) to the isomorphous substitutions. This table should be studied to ascertain the source of these charges as well as the net charge on the micelles of each of the minerals listed.

8.10 pH-DEPENDENT CHARGES

The second source of charges noted on some layer silicate clays (e.g., kaolinite) and on humus, allophane, and Fe, Al oxides, is dependent on the soil pH and consequently is termed **variable** or **pH-dependent**. Both negative and positive charges come from this source.

Negative Charges

The pH-dependent charges are associated primarily with hydroxyl (OH) groups on the edges and surfaces of the inorganic and organic colloids (Figure 8.12). The OH groups

TABLE 8.5 Typical Unit Layer Formulas of Several Clay and Other Silicate Minerals Showing Octahedral and Tetrahedral Cations as Well as Coordinating Anions, Charge per Unit Formula, and Fixed and Exchangeable Interlayer Components

Note that the charge per unit formula is the sum of the charges on the octahedral and tetrahedral sheets and that this negative charge is counterbalanced by equivalent positive charges in interlayer areas.

Mineral	Octahedral sheet	Tetrahedral sheet	Coordinating anions	Charge per unit formula	Interlayer components Fixed	Interlayer components Exchangeable[a]
1:1-Type						
Kaolinite (dioctahedral)	Al_2	Si_2	$O_5(OH)_4$	0	None	None
Serpentine (trioctahedral)	Mg_3	Si_2	$O_5(OH)_4$	0	None	None
2:1-Type Dioctahedral Minerals						
Pyrophyllite	Al_2	Si_4	$O_{10}(OH)_2$	0	None	None
Montmorillonite	$Al_{1.7}Mg_{0.3}$ ⫶−0.3	$Si_{3.9}Al_{0.1}$ ⫶−0.1	$O_{10}(OH)_2$	−0.4	None	$M_{0.4}^+$
Beidellite	Al_2	$Si_{3.6}Al_{0.4}$ ⫶−0.4	$O_{10}(OH)_2$	−0.4	None	$M_{0.4}^+$
Nontronite	Fe_2	$Si_{3.6}Al_{0.4}$ ⫶−0.4	$O_{10}(OH)_2$	−0.4	None	$M_{0.4}^+$
Vermiculite	$Al_{1.7}Mg_{0.3}$ ⫶−0.3	$Si_{3.6}Al_{0.4}$ ⫶−0.4	$O_{10}(OH)_2$	−0.7	xH_2O	$M_{0.7}^+$
Fine mica (illite)	Al_2	$Si_{3.2}Al_{0.8}$ ⫶−0.8	$O_{10}(OH)_2$	−0.8	$K_{0.7}^+$	$M_{0.1}^+$
Muscovite	Al_2	Si_3Al ⫶−1.0	$O_{10}(OH)_2$	−1.0	K^+	None
2:1-Type Trioctahedral Minerals						
Talc	Mg_3	Si_4	$O_{10}(OH)_2$	0	None	None
Vermiculite	$Mg_{2.7}Fe_{0.3}^{3+}$ ⫶+0.3	Si_3Al ⫶−1.0	$O_{10}(OH)_2$	−0.7	xH_2O	$M_{0.7}^+$
Chlorite	$Mg_{2.6}Fe_{0.4}^{3+}$ ⫶+0.4	$Si_{2.5}(Al,Fe)_{1.5}$ ⫶−1.5	$O_{10}(OH)_2$	−1.1	$Mg_2Al(OH)_6^+$	$M_{0.1}^+$

[a] Exchangeable cations such as Ca^{2+}, Mg^{2+}, and H^+ are indicated by the singly charged cation M^+.

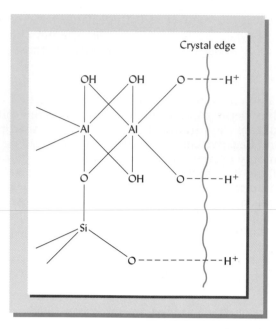

FIGURE 8.12 Diagram of a broken edge of a kaolinite crystal showing oxygens as the source of negative charge. At high pH values the hydrogen ions tend to be held loosely and can be exchanged for other cations.

are attached to iron and/or aluminum in the inorganic colloids (e.g., $\diagdown$Al—OH) and to the carbon in CO groups in humus (e.g., —CO—OH). Under moderately acid conditions, there is little or no charge on these particles, but as the pH increases, the hydrogen dissociates from the colloid OH group, and negative charges result.

$$\diagdown \text{Al—OH} + \text{OH}^- \rightleftharpoons \diagdown \text{Al—O}^- + \text{H}_2\text{O}$$

$$\text{—CO—OH} + \text{OH}^- \rightleftharpoons \text{—CO—O}^- + \text{H}_2\text{O}$$

No charge   (soil solution)   Negative charge   (soil solution)
(soil solids)   (soil solids)

As indicated by the $\rightleftharpoons$ arrows, the reactions are reversible. If the pH increases, more OH^- ions are available to force the reactions to the right, and the negative charge on the particle surfaces increases. If the pH is lowered, OH^- ion concentrations are reduced, the reaction goes back to the left, and the negativity is reduced.

Another source of increased negative charges as the pH is increased is the removal of positively charged complex aluminum hydroxy ions [e.g., Al(OH)_2^+]. At low pH levels, these ions block negative sites on the silicate clays (e.g., vermiculite) and make them unavailable for cation exchange. As the pH is raised, the Al(OH)_2^+ ions react with the OH^- ion in the soil solution to form insoluble Al(OH)_3, thereby freeing the negatively charged sites.

$$\diagdown \text{Al—(OH)}_2^- \text{Al(OH)}_2^+ + \text{OH}^- \longrightarrow \diagdown \text{Al—(OH)}_2^- + \text{Al(OH)}_3$$

Negative charged   Negative charge   No charge
site is blocked   site is freed

This mechanism of increasing negative charges is of special importance in soils high in Fe, Al oxides, such as are found in the southeastern United States and in the tropics.

Positive Charges

Under moderate to extreme acid soil conditions, some silicate clays and Fe, Al oxides may exhibit net positive charges. Once again, exposed OH groups are involved. In this case, however, as the soils become more acid, **protonation**—the attachment of H^+ ions to the surface OH groups—takes place. The reaction for silicate clays may be shown simply as:

$$\begin{array}{ccc} >\!\!Al\!-\!OH & +\ H^+ \rightleftharpoons & >\!\!Al\!-\!OH_2{}^+ \\ \text{No charge} & \text{(soil solution)} & \text{Positive charge} \\ \text{(soil solids)} & & \text{(soil solids)} \end{array}$$

Thus, in some cases, the same site on the inorganic soil colloid may be responsible for negative charge (high pH), no charge (intermediate pH), or positive charge (very low pH). The reaction as the pH of a highly alkaline soil is reduced may be illustrated as follows:

$$\begin{array}{ccccc} >\!\!Al\!-\!O^- & +\ H^+ \rightleftharpoons & >\!\!AlOH & +\ H^+ \rightleftharpoons & >\!\!AlOH_2{}^+ \\ \text{Negative charge} & & \text{No charge} & & \text{Positive charge} \\ \text{(high pH)} & & \text{(intermediate pH)} & & \text{(very low pH)} \end{array}$$

Since a mixture of humus and several inorganic colloids is usually found in soil, it is not surprising that positive and negative charges may be exhibited at the same time. In most soils of temperate regions, the negative charges far exceed the positive ones (Table 8.6). However, in some acid soils high in Fe, Al oxides or allophane, the overall net charge may be positive. The effect of soil pH on positive and negative charges on such soils is illustrated in Figure 8.13.

The charge characteristics of selected soil colloids are shown in Table 8.6. Note the high percentage of constant negative charges in some 2:1-type clays (e.g., smectites and vermiculites). Humus, kaolinite, allophane and Fe, Al oxides have mostly variable (pH-dependent) negative charges and exhibit modest positive charges at low pH values.

Cation and Anion Adsorption

The charges associated with soil particles attract simple and complex ions of opposite charge. Thus, a given colloidal mixture may exhibit not only a maze of positive and negative surface charges, but an equally complex group of simple cations and anions such as Ca^{2+} and $SO_4{}^{2-}$ that are attracted by the particle charges. Figure 8.14 illustrates how soil colloids attract mineral elements so important for plant growth.[8] In temperate region soils, the adsorbed anions are commonly present in smaller quantities than the cations because the negative charges generally predominate on the soil colloid. In more highly weathered acid soils of the tropics where 1:1-type clays and Fe, Al oxides are most dominant, anion exchange is relatively more prominent. Consideration will now be given to the exchange of ions between micelles and the soil solution, starting first with cation exchange.

---

[8] In addition to the simple inorganic cations and anions, more complex organic compounds as well as charged organomineral complexes are adsorbed by the charged soil colloids (see Section 8.18).

TABLE 8.6 **Charge Characteristics of Representative Colloids Showing Comparative Levels of Permanent (Constant) and pH-Dependent Negative Charges as Well as pH-Dependent Positive Charges**

Colloid type	Negative charge			Positive charge, cmol_c/kg
	Total at pH 7, cmol_c/kg	Constant, %	pH dependent, %	
Organic	200	10	90	0
Smectite	100	95	5	0
Vermiculite	150	95	5	0
Fine-grained micas	30	80	20	0
Chlorite	30	80	20	0
Kaolinite	8	5	95	2
Gibbsite (Al)	4	0	100	5
Goethite (Fe)	4	0	100	5
Allophane	30	10	90	15

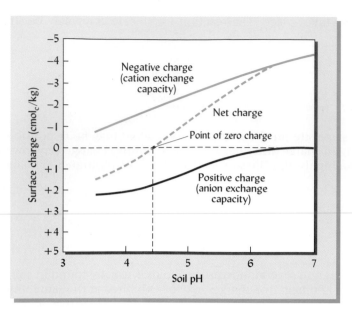

FIGURE 8.13 Relationship between soil pH and positive and negative charges on an Oxisol surface horizon in Malaysia. The negative charges (cation exchange capacity) increase and the positive charges (anion exchange capacity) decrease with increasing soil pH. The point of zero charge is about pH 4.4. [Redrawn from Shamshuddin and Ismail (1995)]

8.11 CATION EXCHANGE

In Section 8.3, the various cations adsorbed by the **exchange complex**[9] were shown to be subject to replacement by other cations through a process called *cation exchange.* For example, hydrogen ions generated as organic matter decomposes (see Section 9.6) can displace calcium and other metallic cations from the colloidal complex. This can be shown simply as follows, where only one adsorbed calcium ion is being replaced:

---

[9] This term includes all soil colloids, inorganic and organic, capable of holding and exchanging cations.

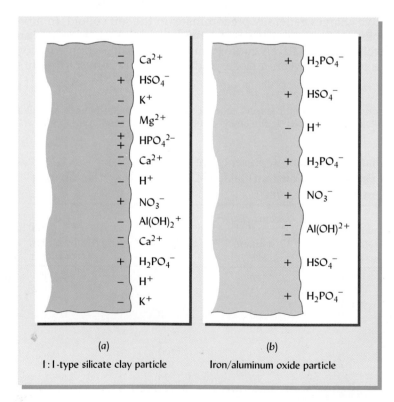

(a)

1:1-type silicate clay particle

(b)

Iron/aluminum oxide particle

FIGURE 8.14 Illustration of adsorbed cations and anions in an acid soil with (a) 1:1-type silicate clay and (b) iron and aluminum oxide particles. Note the dominance of the negative charges on the silicate clay and of the positive charges on the aluminum oxide particle.

$$\boxed{\text{Micelle}}\ Ca^{2+} + 2H^+ \rightleftharpoons \boxed{\text{Micelle}}\ {}^{H^+}_{H^+} + Ca^{2+}$$

The reaction takes place rapidly, and the interchange of calcium and hydrogen is chemically equivalent. As shown by the double yield arrows, the reaction is reversible and will go to the left if calcium is added to the system.

Cation Exchange Under Natural Conditions

As it usually occurs in temperate region surface soils the cation exchange reaction is somewhat more complex, but the principles illustrated in the simple equation apply. Assume, for the sake of simplicity, that the numbers of Ca^{2+}, Al^{3+}, H^+, and other metallic cations such as Mg^{2+} and K^+ (represented as M^+) are in the ratio 40, 20, 5, and 20 per micelle, respectively. (The metallic cations represented as M^+ will be considered monovalent in this case.) Hydrogen from carbonic acid (H_2CO_3) reacts with the micelle as follows:

$$\boxed{\text{Micelle}}\ \begin{matrix} Ca_{40} \\ Al_{20} \\ H_5 \\ M_{20} \end{matrix} + 5H_2CO_3 \rightleftharpoons \boxed{\text{Micelle}}\ \begin{matrix} Ca_{38} \\ Al_{20} \\ H_{10} \\ M_{19} \end{matrix} + 2Ca(HCO_3)_2 + M(HCO_3)$$

(soil solids) (soil solution)   (soil solids)   (soil solution)

Where sufficient rainfall is available to leach calcium and other metallic cations, the reaction tends to go toward the right and the soils tend to become more acid. In regions of low rainfall, however, calcium and other metallic cations are more plentiful since they are not easily leached from the soil. The metallic ions drive the reaction to the left, keeping the soil at pH 7 or above. In addition, some of the calcium will be precipitated as $CaCO_3$, especially in the subsoil at the lower limit of rainfall penetration. The interaction of climate, biological processes, and cation exchange thus helps determine the chemical properties of soils.

Influence of Lime and Fertilizer

Cation exchange reactions are reversible; therefore, if some basic calcium compound such as limestone is applied to an acid soil, the preceding reaction will be driven to the left. Calcium ions will replace the hydrogen and other cations; the H^+ ions will be neutralized by OH^- or CO_3^{2-} ions; and the soil pH will be raised. If, on the other hand, acid-forming chemicals, such as sulfur, are added to an alkaline soil in a dryland area, the H^+ ions would replace the metal cation on the soil colloids, and the soil pH would decrease.

One more illustration of cation exchange is the reaction that may occur when a fertilizer containing potassium chloride is added to soil.

$$\boxed{\text{Micelle}}\ \begin{matrix} Ca_{40} \\ Al_{20} \\ H_5 \\ M_{20} \end{matrix} + 7KCl \rightleftharpoons \boxed{\text{Micelle}}\ \begin{matrix} K_7 \\ Ca_{38} \\ Al_{20} \\ H_4 \\ M_{18} \end{matrix} + 2CaCl_2 + HCl + 2MCl$$

(soil solids) (soil solution)   (soil solids)   (soil solution)

The added K^+ ion is adsorbed on the colloid and replaces an equivalent quantity of Ca^{2+}, H^+ and other cations (M^+) that appear in the soil solution. The adsorbed K^+ remains largely in an available condition, but is less subject to leaching than if it were not adsorbed. Hence, cation exchange is an important consideration not only for nutrients already present in soils, but also for those applied in manures, crop residues, or commercial fertilizers.

8.12   CATION EXCHANGE CAPACITY

Previous sections have dealt qualitatively with exchange. We now turn to a consideration of the quantitative cation exchange or the **cation exchange capacity** (CEC). This property, which is defined simply as "the sum total of the exchangeable cations that a soil can adsorb," is easily determined. By standard methods all the adsorbed cations in a soil are replaced by a common ion, such as Ba^{2+}, K^+, or NH_4^+; then the amount of adsorbed Ba^{2+}, K^+, or NH_4^+ is determined (Figure 8.15).

Means of Expression[10]

The cation exchange capacity (CEC) is expressed in terms of moles of positive charge adsorbed per unit mass. For the convenience of expressing CEC in whole numbers, we shall use *centimoles of positive charge per kilogram of soil* ($cmol_c/kg$). Thus, if a soil has a cation exchange capacity of 10 $cmol_c/kg$, 1 kg of this soil can adsorb 10 $cmol_c$ of H^+ ion, for example, and can exchange it with 10 $cmol_c$ of any other cation. This emphasizes that in cation exchange it is the **number of charges**, not the number of ions or molecules with which we are most concerned. The mole concept of charge is reviewed briefly in Box 8.1 along with examples relating moles to mass or weight of ions taking part in cation exchange reactions.

---

[10] Until recent years, cation exchange capacity was expressed as milliequivalents per 100 g of soil. In this text, the International System of Units (SI) is being used. Fortunately, it is easy to compare soil data using either of these methods of expression since 1 milliequivalent per 100 g of soil is equal to 1 centimole ($cmol_c$) of positive or negative charge per kilogram of soil.

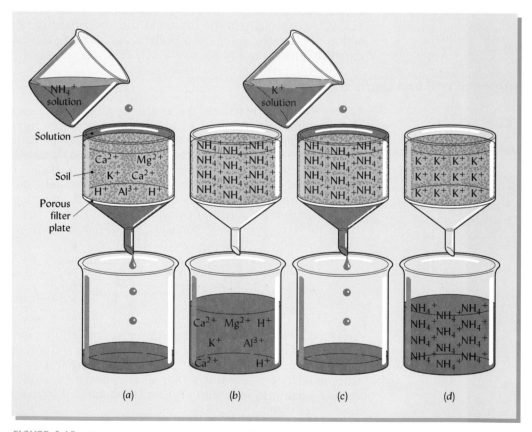

FIGURE 8.15   Illustration of a method for determining the cation exchange capacity of soils. (*a*) A given mass of soil containing a variety of exchangeable cations is leached with an ammonium (NH_4^+) salt solution. (*b*) The NH_4^+ ions replace the other adsorbed cations, which are leached into the container below. (*c*) After the excess NH_4^+ salt solution is removed with an organic solvent, such as alcohol, a K^+ salt solution is used to replace and leach the adsorbed NH_4^+ ions. (*d*) The amount of NH_4^+ released and washed into the lower container can be determined, thereby measuring the chemical equivalent of the cation exchange capacity (i.e., the negative charge on the soil colloids).

BOX 8.1   CHEMICAL EXPRESSION OF CATION EXCHANGE

One mole of any atom, molecule, or charge is defined as 6.02×10^{23} (Avogadro's number) of atoms, molecules, or charges. Thus, 6.02×10^{23} negative charges associated with the soil colloidal complex would attract 1 mole of positive charge from adsorbed cations such as Ca^{2+}, Mg^{2+}, and H^+. The number of moles of the positive charge provided by the adsorbed cations in any soil gives us a measure of the *cation exchange capacity* (CEC) of that soil.

The CEC of soils commonly varies from 0.03 to 0.5 mole of positive charge per kilogram. To express the CEC in whole numbers, the charge is usually indicated in *centimoles per kilogram* ($cmol_c/kg$) of soil. Since there are 100 centimoles in 1 mole, the preceding range of CEC of soils is 3 to 50 $cmol_c/kg$.

CALCULATING MASS FROM MOLES

Using the mole concept, it is easy to relate the mole charges to the mass of ions or compounds involved in cation or anion exchange. Consider, for example, the exchange that takes place when adsorbed sodium ions in an alkaline arid-region soil are replaced by hydrogen ions:

$$\boxed{\text{Micelle}}\ Na^+ + H^+ \rightleftharpoons \boxed{\text{Micelle}}\ H^+ + Na^+$$

If 2 $cmol_c$ of adsorbed Na^+ ions per kilogram of soil were replaced by H^+ ions in this reaction, how many grams of Na^+ ions would be replaced?

Since the Na^+ ion is singly charged, the mass of Na^+ needed to provide 1 mole of charge (1 mol_c) is the gram atomic weight of sodium, or 23 g (see periodic table in Appendix C). The mass providing 1 *centimole* of charge ($cmol_c$) is 1/100 of this amount, or 0.23 g. Thus, the mass of the 2 $cmol_c$ Na^+ replaced from 1 kg of soil is:

$$2\ cmol_c\ Na^+/kg \times 0.23\ g\ Na^+/cmol_c = 0.46 g\ Na^+/kg\ soil$$

The 0.46 g Na^+ would be replaced by only 0.02 g H, which is the mass of 2 $cmol_c$ of this much lighter element.

Another example is the replacement of H^+ ions when lime $[Ca(OH)_2]$ is added to an acid soil. This time assume that 4 $cmol_c$ H^+/kg soil is replaced by the $Ca(OH)_2$, which reacts with the acid soil as follows:

$$\boxed{\text{Micelle}}\ \begin{matrix} H^+ \\ H^+ \end{matrix} + Ca(OH)_2 \rightleftharpoons \boxed{\text{Micelle}}\ Ca^{2+} + 2H_2O$$

Since the Ca^{2+} ion in each molecule of $Ca(OH)_2$ has two positive charges, the mass of $Ca(OH)_2$ needed to replace 1 mole of charge from the H^+ ions is only one-half of the gram molecular weight of this compound, or $74/2 = 37$ g. A comparable figure for 1 *centimole* is $37/100$, or 0.37 grams. The mass of $Ca(OH)_2$ needed to replace 4 $cmol_c$ H^+/kg soil is:

$$4\ cmol_c\ Ca(OH)_2/kg \times 0.37\ g\ Ca(OH)_2/cmol_c = 1.48\ g\ Ca(OH)_2/kg\ soil$$

The 1.48 g $Ca(OH)_2/kg$ soil can be converted to the amount of $Ca(OH)_2$ needed to replace 4 $cmol_c$ H^+/kg from the surface 15 cm of 1 ha of field soil, remembering from Chapter 4 (page 138) that this depth of soil typically weighs 2 million kg/ha.

$$1.48\ g/kg \times 2 \times 10^6\ kg = 2.96 \times 10^6\ g;\quad 2.96 \times 10^3\ kg;\quad or\quad 2.96\ Mg$$

CHARGE AND CHEMICAL EQUIVALENCY

In each preceding example, note that the number of charges provided by the replacing ion is *equivalent* to the number associated with the ion being replaced. Thus, 1 mole of negative charges attracts 1 mole of positive charges whether the charges come from H^+, K^+, Na^+, NH_4^+, Ca^{2+}, Mg^{2+}, Al^{3+}, or any other cation. Keep in mind, however, that only one-half the atomic weights of divalent cations, such as Ca^{2+} or Mg^{2+}, and only one-third the atomic weight of trivalent Al^{3+} are needed to provide 1 mole of charge. This *chemical equivalency* principle applies to both cation and ion exchange.

The cation exchange capacity (CEC) of a given soil horizon is determined by the relative amounts of different colloids in that soil and by the CEC of each of these colloids. Figure 8.16 illustrates the common range in CEC among different soils and other organic and inorganic exchange materials. Note that sandy soils, which are generally low in all colloidal material, have low CECs compared to those exhibited by silt loams and clay loams. Also note the very high CECs associated with humus compared to those exhibited by the inorganic clays, especially kaolinite and Fe, Al oxides. The CEC coming from humus generally plays a very prominent role, and sometimes a dominant one, in cation exchange reactions in soils. For example, in a clayey Ultisol (pH = 5.5) containing 2.5% humus and 30% kaolinite, about 75% of the CEC is associated with humus. Even in very acid Ultisols where the pH-dependent charge on both the humus and kaolinite is relatively lower, humus commonly dominates the cation exchange complex. Once again the vital importance of organic matter in soil is illustrated.

Using the CEC range from Figure 8.16 it is possible to estimate the CEC of a soil if the quantities of the different soil colloids in the soil are known. Box 8.2 illustrates how this can be done.

Data in Table 8.7 show the average CEC values for nine different soil orders compiled by the U.S. Soil Conservation Service. Note the very high CEC for the Histosols, verifying the high CEC of the organic colloids. The Vertisols, which are very high in swelling-type clays (mostly smectite), had the highest average CEC of the mineral soils. Next came the Aridisols and Mollisols, which are also commonly high in 2:1-type clays. The Ultisols, whose clays are dominantly kaolinite and hydrous oxides of iron and aluminum, had relatively low CEC values. The Entisols and Inceptisols likely include soils developed on recent alluvium and lacustrine materials that are quite high in clay.

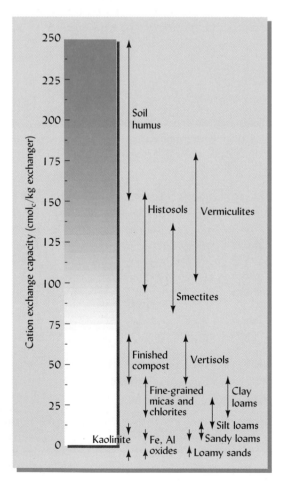

FIGURE 8.16 Ranges in the cation exchange capacities (at pH 7) that are typical of a variety of soils and soil materials. The high CEC of humus shows why this colloid plays such a prominent role in most soils, and especially those high in kaolinite and Fe, Al oxides, clays that have low CECs.

Assume you want to estimate the cation exchange capacities (CEC) of the surface soils of two areas:

1. A cultivated Mollisol from Iowa (pH = 7.0, 20% clay, and 4% organic matter)
2. An Oxisol from a virgin forested area in the Amazon basin of Brazil (pH = 4.0, 60% clay, and 4% organic matter)

MOLLISOL

The most dominant clays in the Mollisol are likely the 2:1 types, and their average CEC would be about 80 cmol$_c$/kg. At pH 7.0, the CEC of the organic matter is about 200 cmol$_c$/kg (see Table 8.6, page 327). Since 1 k of the soil has 0.2 kg (20%) of clay and 0.04 kg (4%) of organic matter, we can calculate the CEC associated with each of these sources.

From the clays:

$$0.2 \text{ kg} \times 80 \text{ cmol}_c/\text{kg} = 16 \text{ cmol}_c$$

From the O.M.:

$$0.04 \text{ kg} \times 200 \text{ cmol}_c/\text{kg} = 8 \text{ cmol}_c$$

The total CEC of the Mollisol is:

$$16 + 8 = 24 \text{ cmol}_c/\text{kg soil}$$

OXISOL

The CEC of the primary clays in the Oxisol (probably Kaolinite and oxides of Fe and Al) is low, probably no more than about 3 cmol$_c$/kg. Likewise, the CEC of the organic matter in this very acid soil (pH 4.0) is comparatively low, likely no more than 100 cmol$_c$/kg (see Figure 8.17). The CEC associated with 0.6 kg of clays (60% of the soil) and with 0.04 kg of organic matter (4% of the soil) can be estimated.

From the clays:

$$0.6 \text{ kg} \times 3 \text{ cmol}_c/\text{kg} = 1.8 \text{ cmol}_c$$

From the O.M.:

$$0.04 \text{ kg} \times 100 \text{ cmol}_c/\text{kg} = 4.0 \text{ cmol}_c$$

The total CEC of the Oxisol is:

$$1.8 + 3.2 = 5.8 \text{ cmol}_c/\text{kg soil}$$

The wide differences in cation exchange capacities between these two soils help explain differences in their nutrient-supplying abilities. Leaving the natural vegetation (forests) on the Brazil site would probably provide the most sustainable use of the Oxisol, while the Mollisol can be used more sustainably for cultivated crops.

In a very general way, the type of clay mineralogy can be estimated by reversing the preceding steps when CEC is known but the type of clay is uncertain.

Despite large variations in soil organic matter and texture, these data appear to reflect the quantities and kinds of soil colloids found in the soils.

pH and Cation Exchange Capacity

In previous sections it was pointed out that the cation exchange capacity of most soils increases with pH. At very low pH values, the cation exchange capacity is also generally low (Figure 8.17). Under these conditions, only the permanent charges of the 2:1-type clays (see Section 8.8) and a small portion of the pH-dependent charges of organic colloids, allophane, and some 1:1-type clays hold exchangeable ions. As the pH is raised,

Note the very high CEC of the Histosols and among the mineral soils of the Vertisols. Ultisols have the lowest CEC. The low average pH values of the Spodosols and high values for the Aridisols and Entisols (many of which were from low-rainfall areas) are noteworthy.

Soil order	CEC, $cmol_c$/kg	pH
Ultisols	3.5	5.60
Alfisols	9.0	6.00
Spodosols	9.3	4.93
Entisols	11.6	7.32
Inceptisols	14.6	6.08
Aridisols	15.2	7.26
Mollisols	18.7	6.51
Vertisols	35.6	6.72
Histosols	128.0	5.50

From Holmgren, et al. (1993).

the negative charges on some 1:1-type silicate clays, allophane, humus, and even Fe, Al oxides increases, thereby increasing the cation exchange capacity. To obtain a measure of this maximum retentive capacity, the CEC is commonly determined at a pH of 7.0 or above. At neutral or slightly alkaline pH, the CEC reflects most of those pH-dependent charges as well as the permanent ones.

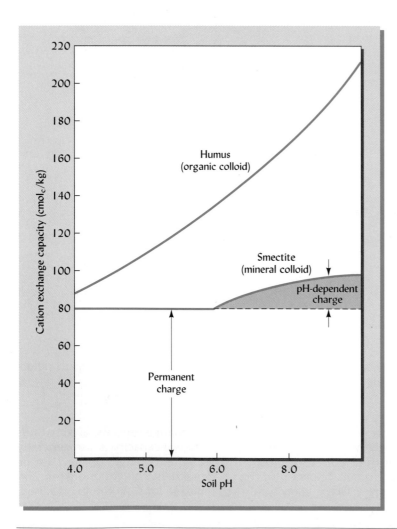

FIGURE 8.17 Influence of pH on the cation exchange capacity of smectite and humus. Below pH 6.0 the charge for the clay mineral is relatively constant. This charge is considered permanent and is due to ionic substitution in the crystal unit. Above pH 6.0 the charge on the mineral colloid increases slightly because of ionization of hydrogen from exposed hydroxyl groups at crystal edges. In contrast to the clay, essentially all of the charges on the organic colloid are considered pH dependent. [Smectite data from Coleman and Mehlich (1957); organic colloid data from Helling, et al. (1964)]

8.13 EXCHANGEABLE CATIONS IN FIELD SOILS

The specific exchangeable cations associated with soil colloids differ from one climatic region to another—Ca^{2+}, Al^{3+}, complex aluminum hydroxy ions, and H^+ being most prominent in humid regions, and Ca^{2+}, Mg^{2+}, and Na^+ dominating in low-rainfall areas (Table 8.8). The cations that dominate the exchange complex have a marked influence on soil properties.

In a given soil, the proportion of the cation exchange capacity satisfied by a particular cation is termed the *percentage saturation* for that cation. Thus, if 50% of the CEC is satisfied by Ca^{2+} ions, the exchange complex is said to have a *percentage calcium saturation* of 50.

This terminology is especially useful in identifying the relative proportions of sources of acidity and alkalinity in the soil solution. Thus, the percentage saturation with Al^{3+} and H^+ ions gives an indication of the acid conditions, while increases in the nonacid cation percentage (commonly referred to as the *percentage base saturation*[11]) indicate the tendency toward neutrality and alkalinity. The percentage base saturation is an important soil property, especially because it is inversely related to soil acidity. These relationships will be discussed further in Chapter 9.

The percentage saturation of essential nutrient cations such as calcium and potassium also greatly influences the uptake of these elements by growing plants. This now will receive attention along with other cation exchange–plant nutrition interactions.

8.14 CATION EXCHANGE AND THE AVAILABILITY OF NUTRIENTS

Exchangeable cations generally are available to both higher plants and microorganisms. By cation exchange, hydrogen ions from the root hairs and microorganisms replace nutrient cations from the exchange complex. The nutrient cations are forced into the soil solution, where they can be assimilated by the adsorptive surfaces of roots and soil organisms, or they may be removed by drainage water. Cation exchange reactions affecting the mobility of organic and inorganic pollutants in soils will be discussed in Section 8.16. Here we focus on the plant nutrition aspects.

Cation Saturation and Nutrient Availability

Several factors operate to expedite or retard the release of nutrients to plants. First, there is the percentage saturation of the exchange complex by the nutrient cation in question. For example, if the percentage calcium saturation of a soil is high, the displacement of this cation is comparatively easy and rapid. Thus, 6 cmol/kg of exchangeable calcium in a soil whose exchange capacity is 8 cmol/kg (75% calcium saturation) probably would mean ready availability, but 6 cmol/kg when the total exchange capacity of a soil is 30 cmol/kg (20% calcium saturation) would produce lower availability. This is

---

[11] Technically speaking, nonacid cations such as Ca^{2+}, Mg^{2+}, K^+, and Na^+ are not bases. When adsorbed by soil colloids in the place of H^+ ions, however, they reduce acidity and increase the soil pH. For that reason, they are referred to as *bases* and the portion of the CEC that they satisfy is usually termed *percentage base saturation*.

TABLE 8.8 **Cation Exchange Data for Representative Mineral Surface Soils in Different Areas**

Characteristics	Warm humid-region soil (Ultisol)	Cool humid-region soil (Alfisol)	Semiarid-region soil (Ustoll)	Arid-region soil (Natrargids)[a]
Exchangeable calcium, cmol$_c$/kg	2–5	6–9	14–17	12–14
Other exchangeable bases, cmol$_c$/kg	1–2	2–3	5–7	8–12
Exchangeable aluminum and/or hydrogen, cmol$_c$/kg	3–7	4–8	0–2	0
Cation exchange capacity, cmol$_c$/kg	3–12	12–18	20–26	20–26
Base saturation, %	25	50–75	90–100	100
Probable pH	5.0–5.4	5.6–6.0	7	8–10

[a] Significant sodium saturation.

one reason that for calcium-loving plants, such as alfalfa, the calcium saturation of at least part of the soil should approach or even exceed 80%.

Influence of Complementary Adsorbed Cations

A second factor influencing the plant uptake of a given cation is the complementary ions held on the colloids. As was discussed in Section 8.3, the strength of adsorption of different cations is in the following order:

$$Al^{3+} > Ca^{2+} > Mg^{2+} > K^+ = NH_4^+ > Na^+$$

Consequently, a nutrient cation such as K^+ is less tightly held by the colloids if the complementary ions are Al^{3+} and H^+ (acid soils) than if they are Mg^{2+} and Na^+ (neutral to alkaline soils). The loosely held K^+ ions are more readily available for absorption by plants or for leaching in acid soils (see also Section 14.13).

There are also some nutrient antagonisms, which in certain soils cause inhibition of uptake of some cations by plants. Thus, potassium uptake by plants is limited by high levels of calcium in some soils. Likewise, high potassium levels are known to limit the uptake of magnesium even when significant quantities of magnesium are present in the soil.

Effect of Type of Colloid

Differences exist in the tenacity with which several types of colloidal micelles hold specific cations and in the ease with which they exchange cations. At a given percentage base saturation, smectites—which have a high charge density per unit of colloid surface—hold calcium much more strongly than does kaolinite (low charge density). As a result, smectite clays must be raised to about 70% base saturation before calcium will exchange easily and rapidly enough to satisfy most plants. In contrast, a kaolinite clay exchanges calcium much more readily, serving as a satisfactory source of this constituent at a much lower percentage base saturation. The need to add limestone to the two soils will be somewhat different, partly because of this factor.

8.15 ANION EXCHANGE

Anions are held by soil colloids in two major ways. First, they are held by anion adsorption mechanisms similar to those responsible for cation adsorption. Second, they may actually react with surface oxides or hydroxides, forming more definitive **inner-sphere complexes.** We shall consider anion adsorption first.

The basic principles of **anion exchange** are similar to those of cation exchange, except that the charges on the colloids are positive and the exchange is among negatively charged anions. The positive charges associated with the surfaces of kaolinite, iron and aluminum oxides, and allophane attract anions such as SO^{4+} and NO^{3-}. A simple example of an anion exchange reaction is as follows:

$$\boxed{\text{Micelle}}\ NO_3^- + Cl^- \rightleftharpoons \boxed{\text{Micelle}}\ Cl^- + NO_3^-$$

(positively charged soil solid)   (soil solution)   (positively charged soil solid)   (soil solution)

Just as in cation exchange, *equivalent* quantities of NO_3^- and Cl^- are exchanged; the reaction can be reversed; and plant nutrients so released can be absorbed by plants.

In contrast to cation exchange capacities, anion exchange capacities of soils generally *decrease* with increasing pH. Figure 8.18 illustrates this fact for an Ultisol in Georgia. In some very acid tropical soils that are high in kaolinite and in iron and aluminum oxides, the anion exchange capacity may actually exceed the cation exchange capacity.

Anion exchange is very important in making anions available for plant growth and at the same time in retarding the leaching of such anions from the soil. For example, anion exchange restricts the loss of sulfates from subsoils in the southern United States (see Figure 8.18 and Section 13.22). Even the leaching of nitrate may be retarded by

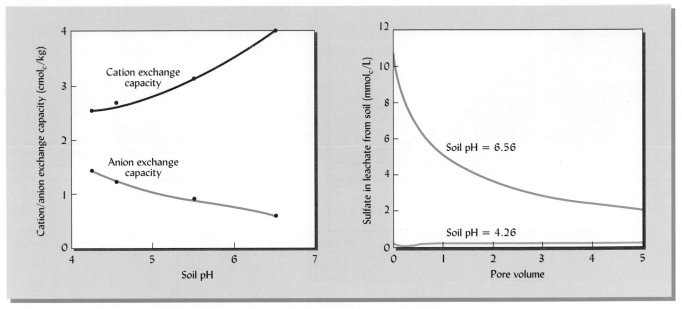

FIGURE 8.18 (Left) Effect of increasing the pH of subsoil material from an Ultisol from Georgia on the cation and anion exchange capacities. Note the significant decrease in anion exchange capacity associated with the increased soil pH. When a column of the low-pH material (pH = 4.6) was leached with $Ca(NO_3)_2$ (right), little sulfate was removed from the soil. In contrast, similar leaching of a column of the soil with the highest pH (6.56), where the anion exchange capacity had been reduced by half, resulted in anion exchange of NO_3^- ions for SO_4^- ions and significant leaching of sulfate from the soil. The importance of anion adsorption in retarding movement of specific anions or other negatively charged substances is illustrated. [Data from Bellini, et al. (1996)]

anion exchange in the subsoil of certain highly weathered soils of the humid tropics. Similarly, the downward movement into groundwater of some charged organic pollutants found in organic wastes can be retarded by such anion and/or cation exchange reactions.

Inner-Sphere Complexes

Some anions, such as phosphates, arsenates, molybdates, and sulfates, can react with particle surfaces, forming **inner-sphere complexes**. For example, the $H_2PO_4^-$ ion may react with the protonated hydroxyl group rather than remain as an easily exchanged anion.

$$>Al-OH_2^+ + H_2PO_4^- \longrightarrow >Al \quad H_2PO_4 + H_2O$$
(soil solid)   (soil solution)   (soil solid)   (soil solution)

This reaction actually reduces the net positive charge on the soil colloid. Also, the $H_2PO_4^-$ is held very tightly by the soil solids and is not readily available for plant uptake.

Anion adsorption and exchange provide important though complex mechanisms for interactions in the soil and between the soil and plants. Together with cation exchange they largely determine the ability of soils to hold nutrients in a form that is accessible to plants, and to retard movement of pollutants in the environment.

Weathering and CEC/AEC Levels

The range of CEC levels of different clay minerals shown in Figure 8.19 shows that clays developed under mild weathering conditions (e.g., smectites and vermiculites) have much higher CEC levels than those developed under more extreme weathering pressures. It also illustrates the general effect of increasing weathering intensities on the magnitude of negative charges that underpin the CEC levels. Also shown are the AEC levels that, in turn, tend to be much higher in clays developed under strong weathering conditions (e.g., kaolinite) than in those found under more mild weathering. This generalized figure is helpful in obtaining a first approximation of CEC and AEC levels in

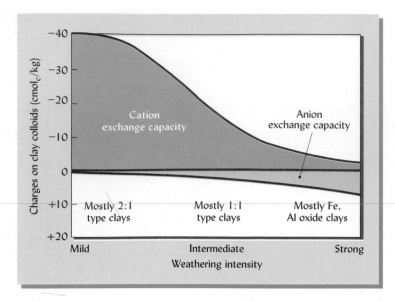

FIGURE 8.19 The effect of weathering intensity on the charges on clay minerals and, in turn, on their cation and anion exchange capacities (CECs and AECs). Note the high CEC and very low AEC associated with mild weathering, which has encouraged the formation of 2:1-type clays such as fine-grained micas, vermiculites, and smectites. More intense weathering destroys the 2:1-type clays and leads to the formation of first kaolinite and then oxides of Fe and Al. These have much lower CECs and considerably higher AECs. Such changes in clay dominance account for the curves shown.

soils of different climatic regions. It must be used with caution, however, since some soils high in 2:1-type clays are found in areas currently undergoing intensive weathering. The nature of the parent material and the time allowed for the weathering to occur also influence the clay types present and the CEC/AEC relations.

8.16   SORPTION OF PESTICIDES AND GROUNDWATER CONTAMINATION

Soil colloids help control the movement of pesticides and other organic compounds into groundwater. The retention of these chemicals by soil colloids can prevent their downward movement through the soil or can delay that movement until the compounds are broken down by soil microbes.

By accepting or releasing protons (H^+ ions), groups such as —OH, —NH_2, and —COOH in the chemical structure of some organic compounds provide positive or negative charges that stimulate anion or cation exchange reactions. Other organic compounds participate in inner-sphere complexation and adsorption reactions just as do the nutrients we have discussed. However, most organic compounds are more commonly **absorbed** within the soil organic colloids by a process termed **partitioning.** The soil organic colloids tend to act as a solvent for the applied chemicals, thereby partitioning their concentrations between those held on the soil colloids and those left in the soil solution. These nonionic or organic compounds are **hydrophobic**, being repelled by water. As a result, moist clays contribute little to partitioning since their adsorbed water molecules prevent the movement of the nonionic organic chemicals into or around the clay particles.[12] Since we seldom know for certain the exact involvement of the adsorption, complexation, or partitioning processes, we use the general term **sorption** to describe the retention by soils of these organic compounds.

Distribution Coefficients

The tendency of a pesticide or other organic compound to leach into the groundwater is determined by the solubility of the compound and by the ratio of the amount of chemical sorbed by the soil to that remaining in solution. This ratio is known as the **soil distribution coefficient** K_d.

---

[12] The hydrated metal cations (e.g., Ca^{2+}) that are adsorbed on the surface of smectites can be replaced with large organic cations, giving rise to what are termed **organoclays.** Such clay surfaces are more friendly toward applied organic compounds, making it possible for the clay to participate in partitioning. Advantage is taken of this phenomenon when smectite organoclays are used to remove organic contaminants from wastewaters and from contaminated groundwaters.

$$K_d = \frac{\text{mg chemical sorbed/kg soil}}{\text{mg chemical/L solution}}$$

A similar ratio that focuses on sorption by organic matter is termed the **organic carbon distribution coefficient** K_{oc}. It too can be calculated easily:

$$K_{oc} = \frac{\text{mg chemical sorbed/kg organic carbon}}{\text{mg chemical/L solution}}$$

Table 8.9 shows both the soil and organic carbon distribution coefficients for two herbicides and some metabolites from one of them (atrazine). Higher coefficient numbers suggest that much of the chemical is bound by soil colloids and is less apt to appear in the groundwater. If, however, the management objective is to remove the organic chemical from the soil, low coefficient numbers are more desirable. These data emphasize the importance of the sorbing power of the soil colloidal complex, and especially of humus, in the management of organic compounds added to soils.

8.17 PHYSICAL PROPERTIES OF COLLOIDS

Soil colloids differ widely in their physical properties, including plasticity, cohesion, swelling, shrinkage, dispersion, and flocculation. These properties greatly influence the usefulness of soils for both agricultural and nonagricultural purposes.

The effects of colloids on soil physical properties were discussed in Chapter 4. But having become better acquainted with the structural framework of each of these colloids, it is easier to appreciate how these materials affect our lives. Figure 8.20 gives an example of steps that must be taken in the building construction business to overcome the adverse effects of the shrinking and swelling properties of the soil colloid, smectite. Fortunately, at alternate sites, other colloids are more user-friendly and sound house footings are possible. If preventative measures are not taken before houses are constructed on smectite clays, homeowners will likely pay dearly in the future. This is just one example of critical role soil colloids play in determining the usefulness of our soils.

8.18 ENVIRONMENTAL USES OF SWELLING-TYPE CLAYS

The physical and chemical properties of swelling-type clays, and particularly the smectites, make them useful tools in preventing unwanted movement of water and water contaminants. Such clays are widely used in the construction of liners to seal ponds, sewage lagoons, industrial waste lagoons, and landfills. A layer of the smectite placed on

TABLE 8.9 Soil and Organic Carbon Partitioning Coefficients K_d and K_{oc} for Widely Used Herbicides

The partitioning coefficients for soil K_d and organic carbon K_{oc} for two widely used herbicides, atrazine and metolachlor, and for three compounds that form when Atrazine is attached by microorganisms, diethyl atrazine, diisopropyl atrazine, and hydroxyatrazine. Since high K_d and K_{oc} values suggest high attraction of the chemical to the soil, the tendency for diethyl atrazine to leach through this Ultisol in Virginia would be high compared with atrazine, but hydroxy atrazine would leach at a much lower rate.

Herbicide	K_d	K_{oc}
Atrazine	1.82	140
Diethyl atrazine	0.99	80
Diisopropyl atrazine	1.66	128
Hydroxy atrazine	7.92	609
Metolachlor	2.47	190

Data from Seybold and Mersie (1996).

FIGURE 8.20 The different swelling tendencies of two types of clay are illustrated in the lower left. All four cylinders initially contained dry, sieved clay soil, the two on the left from the B horizon of a soil high in kaolinite, the two on the right from one of a soil high in montmorillonite. An equal amount of water was added to the two center cylinders. The kaolinitic soil settled a bit and was not able to absorb all the water. The montmorillonitic soil swelled about 25% in volume and absorbed nearly all the added water. The scenes to the right and above show a practical application of knowledge about these clay properties. (Upper) Soils containing large quantities of smectite undergo pronounced volume changes as the clay swells and shrinks with wetting and drying. Such soils (e.g., the California Vertisol shown here) make very poor building sites. The normal-appearing homes (upper) are actually built on deep, reinforced-concrete pilings (lower right) that rest on nonexpansive substrata. Construction of the 15 to 25 such pilings needed for each home more than doubles the cost of construction. (Photo courtesy of R. Weil)

the bottom and sides of the pond or lagoon expands when wetted and forms a relatively impenetrable barrier to the movement not only of water but of organic and inorganic contaminants contained in the water. These contaminants are held in the containment area and are prevented from moving downward into the groundwater.

Smectites are also used to prevent the undesirable upward movement of organic chemicals through boreholes drilled to monitor the presence of organic contaminants in groundwater. As shown in Figure 8.21, the wetted smectite (bentonite) forms an impenetrable plug between the sampling tube and the larger borehole, thereby preventing the upward movement of chemicals from the groundwater. Increasingly, environmental scientists are using swelling-type clays for such environmental monitoring, as well as for the removal of organic chemicals from water by partitioning.

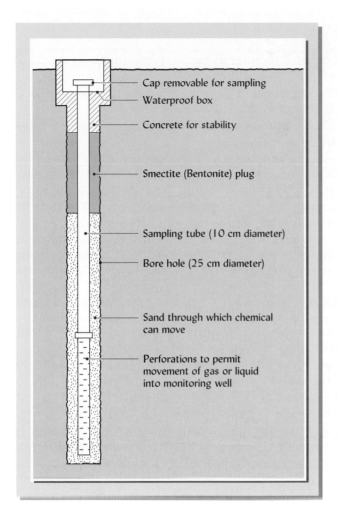

Cap removable for sampling

Waterproof box

Concrete for stability

Smectite (Bentonite) plug

Sampling tube (10 cm diameter)

Bore hole (25 cm diameter)

Sand through which chemical can move

Perforations to permit movement of gas or liquid into monitoring well

FIGURE 8.21 Illustration of the use of a smectite (bentonite) as a plug or sealant to prevent upward leakage through a borehole that was drilled to permit the installation of a well for monitoring the presence of organic chemicals in soils. Air-dry bentonite is tapped in place and then wetted. When the wetted clay swells, it fits tightly around the sampling tube and the bore hole walls. The wet smectite resists the upward movement of organic materials, forcing them to enter the perforated sampling tube from which they can be removed for analysis. The plug also protects the well from the possibility of chemicals washing down from the soil surface and therefore being mistaken for groundwater contaminants. [Redrawn from Reid and Ulery (1998)]

8.19 CONCLUSION

Colloidal materials control most of the chemical and physical properties of soils. These materials include five major classes of crystalline silicate clays, other aluminosilicates such as allophane, the oxides of iron and aluminum, and the organic colloids (humus). Due to their extremely small particle sizes and platelike structures, these colloids possess enormous surface areas, much of which are internal surfaces that are much like the pages of a book.

The mineralogical and chemical constitutions of these colloids result in electrostatic charges on or near the surfaces of the colloidal particles. Negative charges predominate on most silicate clays. Positive charges characterize some colloids, such as the oxides of Fe and Al, especially if the pH is low. The charges on colloids attract ions and other substances with opposite charges: cations being attracted to negatively charged sites and anions to the positively charged ones.

At the charge sites on soil colloids, exchanges between one ion or substance and others can take place. This enables plants to exchange H^+ ions for nutrient ions, such as Ca^{2+}, K^+, and OH^- ions for SO_4^{2-} and NO_3^-. A colloid's capacity to hold exchangeable cations (*cation exchange capacity*) and exchangeable anions (*anion exchange capacity*) measures its ability to provide nutrients to plant roots and to retard the downward movement of pollutants to drainage water.

Cation exchange joins photosynthesis as a fundamental life-supporting process. Without this soil property, terrestrial ecosystems would not be able to retain sufficient nutrients to support natural or introduced vegetation, especially in the event of such disturbances as timber harvest, fire, or cultivation.

STUDY QUESTIONS

1. Describe the *soil colloidal complex,* indicate its various components, and explain how it tends to serve as a "bank" for plant nutrients.

2. How do you account for the difference in surface area associated with a grain of kaolinite clay compared to that of montmorillonite, a smectite?

3. Contrast the difference in crystalline structure among *kaolinite, smectites, fine-grained micas, vermiculites,* and *chlorites.*

4. There are two basic processes by which silicate clays are formed by weathering of primary minerals. Which of these would likely be responsible for the formation of (1) fine-grained mica, and (2) kaolinite from muscovite mica? Explain.

5. If you wanted to find a soil high in kaolinite, where would you go? The same for (1) smecitite and (2) vermiculite?

6. Which of the silicate clay minerals would be *most* and *least* desired if one were interested in (1) a good foundation for a building, (2) a high cation exchange capacity, (3) an adequate source of potassium, and (4) a soil on which hard clods form after plowing?

7. Which of the following would you expect to be *most* and *least* sticky and plastic when wet: (1) a soil with significant sodium saturation in a semiarid area, (2) a soil high in exchangeable calcium in a subhumid temperate area, or (3) a well-weathered acid soil in the tropics? Explain your answer.

8. A soil contains 4% humus, 10% montmorillonite, 10% vermiculite, and 10% Fe, Al oxides. What is its approximate cation exchange capacity?

9. Calculate the number of grams of Al^{3+} ions needed to replace 10 $cmol_c$ of Ca^{2+} ion from the exchange complex of 1 kg of soil.

10. Explain the importance of K_d and K_{oc} in assessing the potential pollution of drainage water. Which of these expressions is likely to be most consistently characteristic of the organic compounds in question regardless of the type of soil involved? Explain.

REFERENCES

Bellini, G., M. E. Sumner, D. E. Radcliffe, and N. P. Qafoku. 1996. "Anion transport through columns of highly weathered acid soil: Adsorption and retardation," *Soil Sci. Soc. Amer. J.,* **60:**132–137.

Buseck, P. R. 1983. "Electron microscopy of minerals," *Amer. Scientist,* **71:**175–185.

Coleman, N. T., and A. Mehlich. 1957. "The chemistry of soil pH," in *The Yearbook of Agriculture (Soil).* (Washington, D.C.: U.S. Department of Agriculture).

Dixon, J. B., and S. B. Weed. 1989. *Minerals in Soil Environments,* 2d ed. (Madison, Wis.: Soil Sci. Soc. Amer.).

Helling, C. S., et al. 1964. "Contribution of organic matter and clay to soil cation exchange capacity as affected by the pH of the saturated solution," *Soil Sci. Soc. Amer. Proc.,* **28:**517–520.

Holmgren, G. G. S., M. W. Meyer, R. L. Chaney, and R. B. Daniels. 1993. "Cadmium, lead, zinc, copper, and nickel in agricultural soils in the United States of America," *J. Environ. Qual.,* **22:**335–348.

Parker, J. C., D. F. Amos, and L. W. Zelanzny. 1982. "Water adsorption and swelling of clay minerals in soil systems," *Soil Sci. Soc. Amer. J.,* **46:**450–456.

Reid, D. A. and A. L. Ulery. 1998. "Environmental applications of smectites," in J. Dixon, D. Schultze, W. Bleam, and J. Amonette (eds.), *Environmental Soil Mineralogy.* (Madison, Wis.: Soil Sci. Soc. Amer.).

Seybold, C. A., and W. Mersie. 1996. "Adsorption and desorption of atrazine, deethylatrazine, deisopropylatrazine, hydroxyatrazine, and metolachlor in two soils in Virginia," *J. Environ. Qual.,* **25:**1179–1185.

Shamsuddin, J., and H. Ismail. 1995. "Reactions of ground magnesium limestone and gypsum in soils with variable-charged minerals," *Soil Sci. Soc. Amer. J.,* **59:**106–112.

9

SOIL REACTION: ACIDITY AND ALKALINITY

What have they done to the rain?
—PETER, PAUL, AND MARY

Just like adding salt to your soup, it is easier to add a little more (lime to your soil) later, than to try and take out any excess.
—ENCYCLOPEDIA OF ORGANIC GARDENING

The degree of acidity or alkalinity (i.e., the *soil reaction*) is a *master variable* that affects all soil properties—chemical, physical, and biological. Expressed as the soil pH, this variable largely controls plant nutrient availability and microbial reaction in soils. It affects which trees, shrubs, or grasses will dominate the landscape under natural conditions and determines which cultivated crops will grow well or even grow at all in a given field site. Soil reaction also determines the fate of many soil pollutants, affecting their breakdown and possible movement from the soil into groundwater and streams.

The pH of a soil helps determine the numbers and kinds of soil organisms that change plant residues into valuable soil organic matter. Thereby it influences aggregate stability and, in turn, the movement of air and water in soils.

Many human activities can influence soil reaction. For example, certain chemical fertilizers and organic wastes react in the soil to form strong inorganic acids, such as HNO_3 and H_2SO_4. These lead to increased soil acidity. These same two acids are found in **acid rain**, which originates from gases emitted into the atmosphere primarily by the combustion of fossil fuel (power plants, automobiles, etc.) and the burning of trees and other biomass. Concerns are increasing about the possible damage of acid rain to forests, crops, and downstream lakes.

Climate also tends to stimulate either acidity or alkalinity in soils. In humid regions, soils tend to be quite acid because there is sufficient rainfall to leach out much of the base-forming cations (Ca^{2+}, Mg^{2+}, K^+, and Na^+), leaving the exchange complex dominated by Al^{3+} and H^+ ions.

In low-rainfall areas where leaching is not very intense, the opposite is true. The base-forming cations are left to dominate the exchange complex in the place of Al^{3+} and H^+, leading to a neutral or even alkaline condition. We shall give greater emphasis to soil acidity in this chapter, and will cover alkaline and salt-affected soils and their management in Chapter 10.

Before we consider sources of H^+ and OH^- ions in nature, briefly review in Box 9.1 the concept of pH and how it helps us understand the reciprocal relationship between the concentrations of these two ions in water systems. Also see the chart in Figure 9.1 listing the pH values of some substances we use daily. These are contrasted with the pH of different acid and alkaline soils to which we will be referring in this and the following chapter.

BOX 9.1   SOIL PH, SOIL ACIDITY, AND ALKALINITY

Whether a soil is acid, neutral, or alkaline is determined by the comparative concentrations of H^+ and OH^- ions.[a] Pure water provides these ions in equal concentrations:

$$H_2O \rightleftharpoons H^+ + OH^-$$

The equilibrium for this reaction is far to the left, only an infinitesimal amount of H^+ or OH^- ions being formed from 1 mole of water. The *ion product* of the concentrations of the H^+ and OH^- ions is a constant (K_w), which at 25°C is known to be 1×10^{-14}:

$$[H^+] \times [OH^-] = K_w = 10^{-14}$$

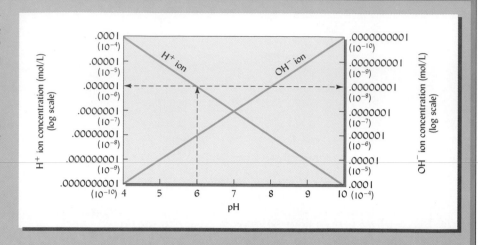

Since in pure water the concentration of H ions $[H^+]$ must be equal to that of OH^- ions $[OH^-]$, this equation shows that the concentration of each is 10^{-7} ($10^{-7} \times 10^{-7} = 10^{-14}$). It also shows the inverse relationship between the concentrations of these two ions. As one increases, the other must decrease proportionately. Thus, if we were to increase the H^+ ion concentration $[H^+]$ by 10 times (from 10^{-7} to 10^{-6}), the $[OH^-]$ would be decreased by 10 times (from 10^{-7} to 10^{-8}) since the product of these two concentrations must equal 10^{-14}:

$$10^{-6} \times 10^{-8} = 10^{-14}$$

Scientists have simplified the means of expressing the very small concentrations of H^+ and OH^- ions by using the *negative logarithm of the H^+ ion concentration*, termed the *pH*. Thus, if the H^+ concentration in an acid medium is 10^{-5}, the pH is 5; if it is 10^{-9} in an alkaline medium, the pH is 9.

Note that the pH also gives us an indirect measure of the OH^- ion concentration since the product of $[H^+] \times [OH^-]$ must always equal 10^{-14}. Thus at pH 5 the $[OH^-]$ is 10^{-9} ($10^{-5} \times 10^{-9} = 10^{-14}$); at pH 8 it is 10^{-6} ($10^{-8} \times 10^{-6} = 10^{-14}$).

The figure above shows the relationship between pH and the concentrations of H^+ and OH^- ions. Note that as one goes down, the other goes up, and vice versa. The dotted line illustrates the concentrations of these two ions at pH = 6.0: $[H^+] = 10^{-6}$; $[OH^-] = 10^{-8}$. This reciprocal relationship between H^+ and OH^- ions should always be kept in mind in studying soil acidity and alkalinity.

[a] Technically speaking, chemical reactions are influenced by the *activity* of an ion rather than by its *concentration*. This is because of the electrostatic effect of one ion on the activity of its nearby neighbors. Ionic activities are thus essentially effective concentrations. Since the difference between ionic activities and concentrations in soils are not so great, we will use the term *concentration* in this text.

9.1   SOURCES OF HYDROGEN AND HYDROXIDE IONS[1]

Two adsorbed cations—aluminum and hydrogen—are largely responsible for soil acidity. The mechanisms by which these two cations exert their influence depends on the degree of soil acidity and on the source and nature of the soil exchange complex.

Strongly Acid Soils

Under very acid soil conditions (pH < 5.0), much aluminum[2] becomes soluble and is either tightly bound by organic matter (Figure 9.2) or is present in the form of alu-

---

[1] In water hydrogen ions are always hydrated, existing as hydronium ions (H_3O^+) rather than as simple H^+ ions. However, for simplicity this text will use the unhydrated H^+ to represent this ion.

[2] Iron (Fe^{3+}) also is solubilized under acid conditions and forms hydroxy cations just as does aluminum. However, since iron is solubilized only in extremely acid soils, we shall focus primarily on the involvement of aluminum.

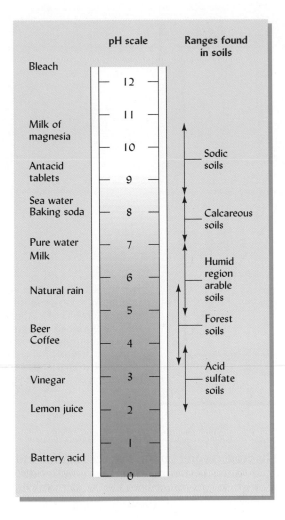

FIGURE 9.1 Chart showing a range of pH from 1 to 12 and the approximate pH of products commonly used in our society every day (left). Comparable pH ranges are shown (right) for soils we will be studying in this text

minum or aluminum hydroxy cations. These exchangeable ions are adsorbed in preference to other cations by the negative charges of soil colloids.

The adsorbed aluminum contributes to soil acidity by providing to the soil solution aluminum ions[3] that are then hydrolyzed to produce H^+:

$$Al^{3+} + H_2O \rightleftharpoons AlOH^{2+} + H^+$$

This hydrolysis lowers the pH of the soil solution and is the major source of H^+ ions in most very acid soils.

Most of the hydrogen in very acid soils, along with some iron and aluminum, is bound so tightly by covalent bonds in the organic matter and on clay crystal edges that little is found in an exchangeable form (see Section 8.9). Only on the strong acid groups of humus and some of the permanent charge exchange sites of the clays are the H^+ ions held in an exchangeable form. These few exchangeable H^+ ions are in equilibrium with the soil solution:

$$\boxed{\text{Micelle}}\ H^+ \rightleftharpoons H^+$$

Adsorbed   hydrogen ion
hydrogen   (soil solution)

Thus, it can be seen that the effect of both adsorbed hydrogen and aluminum ions is to increase the H^+ ion concentration in the soil solution.

---

[3] As is the case for most metallic cations in soils, the Al^{3+} ion is highly hydrated, being present in forms such as $Al(H_2O)_6^{3+}$. For simplicity, however, it will be shown as the simple Al^{3+} ion.

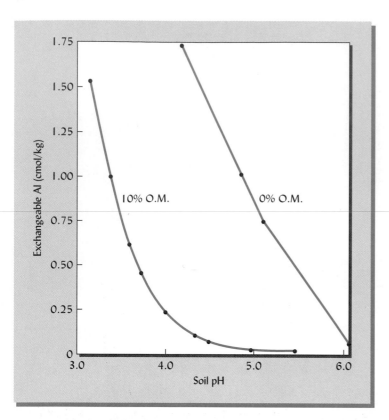

FIGURE 9.2 The effect of soil pH on exchangeable aluminum in a soil–sand mixture (0% O.M.) and a soil–peat mixture (10% O.M.). Apparently, the organic matter binds the aluminum in an unexchangeable form. This organic matter–aluminum interaction helps account for better plant growth at low pH values on soils high in organic matter. [Modified from Hargrove and Thomas (1981); used with permission of American Society of Agronomy]

Moderately Acid Soils

Aluminum and hydrogen compounds also account for soil solution H^+ ions in moderately acid soils (pH values between 5.0 and 6.5) but by different mechanisms. At these pH levels, the aluminum can no longer exist as Al^{3+} ions but is converted to aluminum hydroxy ions[4] by reactions such as the following:

$$Al^{3+} + OH^- \rightleftharpoons AlOH^{2+}$$

$$AlOH^{2+} + OH^- \rightleftharpoons AlOH_2{}^+$$

Aluminum
hydroxy ions

Much of the aluminum hydroxy ions are adsorbed and act as exchangeable cations. They are in equilibrium with similar cations in the soil solution, where they produce hydrogen ions by the following hydrolysis reactions:

$$Al(OH)^{2+} + H_2O \rightleftharpoons Al(OH)_2{}^+ + H^+$$

$$Al(OH)_2{}^+ + H_2O \rightleftharpoons Al(OH)_3 + H^+$$

In some 2:1-type clays, particularly vermiculite, the aluminum hydroxy ions (as well as iron hydroxy ions) play another role. They move into the interlayer space of the crystal units and become very tightly adsorbed, preventing intracrystal expansion and blocking some of the exchange sites. Raising the soil pH results in the removal of these ions and the freeing up of some of the exchange sites. Thus aluminum (and iron) hydroxy ions are partly responsible for the pH-dependent or **variable charge** of soil colloids (see Section 8.9).

In moderately acid soils, some hydrogen ions that at low pH would be bound tenaciously through covalent bonding by the organic matter, Fe, Al oxides, and 1:1-type clays now are present in exchangeable forms. As a result, more negatively charged sites are available on the micelles to contribute to the increased cation exchange capacity. The exchangeable H^+ ions are in equilibrium with the soil solution H^+ ions.

---

[4] The actual aluminum hydroxy ions are much more complex than those shown. Formulas such as $[Al_6(OH)_{12}]^{6+}$ and $[Al_{10}(OH)_{22}]^{8+}$ are examples of the more complex ions.

$$\boxed{\text{Micelle}}\ \begin{matrix} H^+ \\ H^+ \end{matrix} \rightleftharpoons 2H^+$$

Exchangeable H^+ Soil solution H^+

Again, the colloidal control of soil acidity is demonstrated, as is the dominant role of the aluminum and hydrogen ions.

Neutral to Alkaline Soils (pH 7 and above)

In soils with pH above 7, most of the variable charge sites have been freed, the H^+ and aluminum hydroxy ions having been released from the bound forms. The exchange complex of these neutral to alkaline soils is dominated primarily by exchangeable Ca^{2+}, Mg^{2+}, and other base-forming cations. Both the aluminum hydroxy and hydrogen ions have been largely replaced. Most of the aluminum hydroxy ions have been converted to gibbsite by reactions such as

$$Al(OH)_2^+ + OH^- \longrightarrow Al(OH)_3$$

Gibbsite
(insoluble)

The exchangeable H^+ ions that are released by base-forming cations move into the soil solution, where they react with OH^- ions to form water.

Soil pH and Cation Associations

Figure 9.3 summarizes the effect of pH on the distribution of ions in soils containing some variable charge colloids. Note that the cation exchange capacity of the soil increases with soil pH. Also note the change in cation dominance on the exchange complex as one moves from very acid soils, to moderate acidity, and to neutrality and above. Study this figure carefully, keeping in mind that for any particular soil the distribution of ions might be somewhat different.

The effect of pH on the distribution of Ca^{2+}, Mg^{2+}, and other base-forming cations and of H^+ and Al^{3+} ions in an organic soil and in a soil dominated by 2:1 clays is shown in Figure 9.4. Note that permanent charges dominate the exchange complex of the mineral soil, whereas the variable charges account for most of the adsorption in the organic soil. Consequently, the exchange capacity of the organic soil declines rapidly as the pH is lowered, whereas there is little such decline in the case of the 2:1-type clay. The effect of pH on the cation exchange capacity of kaolinite and other 1:1-type clays and allophane is similar to that shown by the organic soil.

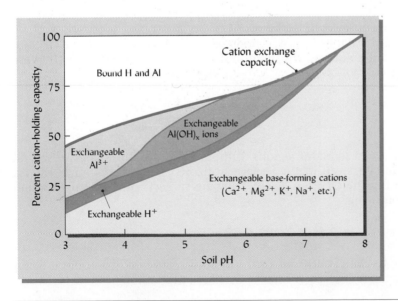

FIGURE 9.3 General relationship between soil pH and the cations held as exchangeable ions or bound by organic matter and clay minerals. The cation exchange capacity line is estimated using average data from 60 Wisconsin soils from Helling, et al. (1964). Note evidence of the variable charge on the colloids since the CEC increased with increasing pH. Probable levels of exchangeable and bound cations are shown. Under very acid conditions, exchangeable aluminum and hydrogen ions and bound hydrogen and aluminum dominate. At higher pH values, the exchangeable base-forming cations predominate, while at intermediate values, aluminum hydroxy ions such as $Al(OH)^{2+}$ are prominent. This diagram shows typical conditions; any particular soil would likely give a modified distribution.

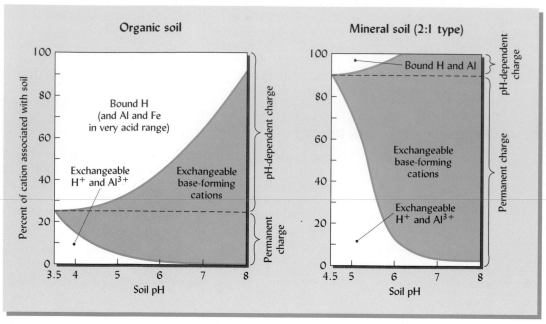

FIGURE 9.4 Relationship between pH and the association of hydrogen, aluminum, and base-forming cations with an organic soil and a mineral soil dominated by 2:1-type silicate clays. Note the dominance of the permanent charge in the mineral soil and of the pH-dependent charge in the organic soil. The CEC of the organic soil (the sum of all exchangeable ions) declined rapidly as the pH was reduced. [Redrawn from Mehlich (1964)]

Note that in Figures 9.3 and 9.4 two forms of hydrogen and aluminum are shown: (1) that tightly held by the pH-dependent sites (*bound*), and (2) that associated with permanent negative charges on the colloids (*exchangeable*). Only the exchangeable ions have an immediate effect on soil pH, but, as we shall see later, both forms are very much involved in determining how much lime or sulfur is needed to change soil pH (see Section 9.2).

While the factors responsible for soil acidity are far from simple, two dominant groups of elements are in control. The different aluminum-containing ions and H^+ ions generate acidity, and most of the other cations combat it. This simple statement is worth remembering.

Source of Hydroxide Ions

In arid and semiarid areas, base-forming cations dominate the exchange complex of soils. These adsorbed cations enhance the OH^- ion concentration of the soil solution in three ways. First, they take the place of exchangeable aluminum and hydrogen ions that are the primary sources of H^+ ions, as already described. This results in a decrease in H^+ ions and an equivalent increase in OH^- ions, since there is an inverse relationship between H^+ and OH^- ions in water-based solutions.

The Ca^{2+}, Mg^{2+}, K^+, and Na^+ ions have a second and more direct effect on the OH^- ion concentration in the soil solution. The hydrolysis of colloids saturated with these cations releases OH^- ions as follows:

$$\boxed{\text{Micelle}}\ Ca^{2+} + 2H_2O \rightleftharpoons \boxed{\text{Micelle}}\ {}^{H^+}_{H^+} + Ca^{2+} + 2OH^-$$

(soil solid)   (soil solution)   (soil solid)   (soil solution)

In some soils of arid and semiarid regions, carbonates and bicarbonates of several cations tend to accumulate, even in the surface horizons. These compounds offer a third means of increasing the OH^- ion levels. They are subject to hydrolysis through reaction such as the following:

$$HCO_3^- + H_2O \rightleftharpoons H_2CO_3 + OH^-$$

Very high pH values (>8) can result from such reactions, particularly if sodium bicarbonate is present. Carbonates will receive more detailed attention in Section 10.1.

Research suggests three kinds of acidity: (1) **active acidity** due to the H^+ and Al^{3+} ions in the soil solution; (2) **salt-replaceable acidity**, involving the aluminum and hydrogen that are *easily exchangeable* by other cations in a simple unbuffered salt solution, such as KCl; and (3) **residual acidity**, which can be neutralized by limestone or other alkaline materials but cannot be detected by the salt-replaceable technique. These types of acidity all add up to the **total acidity** of a soil.

Active Acidity

The active acidity is a measure of the H^+ ion activity in the soil solution that is due in large part to the hydrolysis of aluminum-containing ions. It is very small compared to that in the exchangeable and residual acidity forms. For example, only about 2 kg of calcium carbonate is needed to neutralize the active acidity in the upper 15 cm of 1 ha of an average mineral soil at pH 4 and 20% moisture. Even so, the active acidity is extremely important, since the soil solution is the environment to which plant roots and microbes are exposed.

Salt-Replaceable (Exchangeable) Acidity

This type of acidity is primarily associated with the exchangeable aluminum and hydrogen ions that are present in largest quantities in very acid soils (see Figure 9.3). These ions can be released into the soil solution by cation exchange with an unbuffered salt, such as KCl.

$$\boxed{\text{Micelle}} \begin{matrix} Al^{3+} \\ H^+ \end{matrix} + 4KCl \rightleftharpoons \boxed{\text{Micelle}} \begin{matrix} K^+ \\ K^+ \\ K^I \\ K^+ \end{matrix} + AlCl_3 + HCl$$

(soil solid)　　(soil solution)　　(soil solid)　　(soil solution)

The chemical equivalent of salt-replaceable acidity in strongly acid soils is commonly thousands of times that of active acidity in the soil solution. Even in moderately acid soils, the limestone needed to neutralize this type of acidity is commonly more than 100 times that needed to neutralize the soil solution (active acidity).

At a given pH value, exchangeable acidity is generally highest for smectites, intermediate for vermiculites, and lowest for kaolinite. In any case, however, even it accounts for only a fraction of the total soil acidity, as the next section will verify.

Residual Acidity

Residual acidity is generally associated with the aluminum hydroxy ions and with hydrogen and aluminum ions that are bound in nonexchangeable forms by organic matter and silicate clays (see Figure 9.3). As the pH increases, this bound hydrogen and aluminum is released, thereby freeing up negative cation exchange sites and increasing the cation exchange capacity. The reaction with a liming material [$Ca(OH)_2$] shows how the bound hydrogen and aluminum can be released.

$$\boxed{\text{Micelle}} \begin{matrix} Al \\ H \end{matrix} + 2Ca(OH)_2 \longrightarrow \boxed{\text{Micelle}} \begin{matrix} Ca^{2+} \\ Ca^{2+} \end{matrix} + Al(OH)_3 + H_2O$$

Bound H and Al　　　　　　　　　Exchangeable Ca^{2+}
(not exchangeable)

The residual acidity is commonly far greater than either the active or salt-replaceable acidity. It may be 1000 times greater than the soil solution or active acidity in a sandy soil and 50,000 or even 100,000 times greater in a clayey soil high in organic matter. The amount of ground limestone recommended to at least partly neutralize residual acidity in the upper 15 cm of soil is commonly 5 to 10 metric tons (Mg) per hectare (2.25 to 4.5 tons per acre). It is obvious that the pH of the soil solution is only the tip of the iceberg in determining how much lime is needed.

The previous sections clearly suggest that the fine fractions of the soil play a major role in controlling pH. In the pH range for the most productive agricultural soils (5.5 to 7.5), pH is associated with (1) the percentage base saturation, (2) the nature of the soil particles, (3) the kind of adsorbed bases, and (4) the level of soluble salts in the soil solution. Each of these factors will be considered briefly.

Percentage Base Saturation

The percentage of the CEC that is satisfied by the base-forming cations is termed **percentage base saturation.**[5]

$$\% \text{ base saturation} = \frac{\text{exchangeable base-forming cations, cmol}_c/\text{kg}}{\text{CEC, cmol}_c/\text{kg}}$$

In soils with comparable types of clays and organic matter, the percentage base saturation is generally directly related to soil pH, being low if the soil is very acid and higher as the pH is increased (Figure 9.5). Note that the clay loam soil in Figure 9.5, with a relatively high CEC, has more exchangeable base-forming cations at pH 5.5 than does a sandy loam soil (lower CEC) at pH 6.5.

Nature of the Micelle

There is considerable variation in the pH of different types of fine soil particles at the same percentage base saturation. This is due to differences in the ability of these particles to furnish hydrogen ions to the soil solution. For example, the dissociation of adsorbed H^+ ions from the smectites is much higher than that from the Fe and Al oxide clays. Consequently, the pH of soils dominated by smectites is appreciably lower than that of the oxides at the same percentage base saturation. The dissociation of adsorbed hydrogen from 1:1-type silicate clays and organic matter is intermediate between that from smectites and from the hydrous oxides.

---

[5] The cation exchange capacity (CEC) values used to calculate the percentage base saturation are generally those measured at pH 7 or 8 where the negative charges and thus the CEC are at or near the maximum. In acid soils with significant amounts of variable charge colloids, however, the actual CEC at any point in the field would be considerably lower, and, in turn, the percentage base saturation would be considerably higher than that measured at the higher pH values.

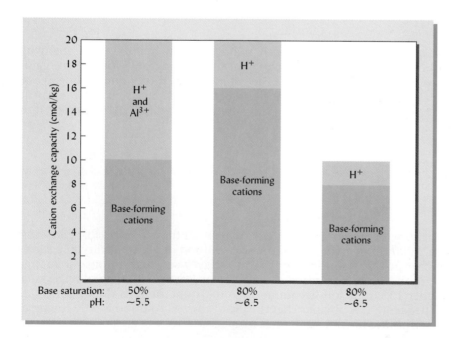

FIGURE 9.5   Three soils with similar colloid types but with different percentage base saturations—50, 80, and 80, respectively. The first (left) is a clay loam (CEC of 20 cmol$_c$/kg); the second, (center) the same soil satisfactorily limed; and the third (right), a sandy loam with a CEC of only 10 cmol$_c$/kg. Note especially that soil pH is correlated more or less closely with percentage base saturation. Also note that the sandy loam (right), because of its low cation exchange capacity, has a higher pH than the acid clay loam (left), even though the clay contains more exchangeable bases.

Kind of Adsorbed Base-forming Cations

The comparative quantities of each of the base-forming cations present in the colloidal complex is another factor influencing soil pH. For example, soils with high sodium saturation have much higher pH values than those dominated by calcium and magnesium. Some highly alkaline soils of arid and semiarid regions have significant sodium saturation (see Chapter 10).

Neutral Salts in Solution

The presence in the soil solution of neutral salts, such as the sulfates and chlorides of sodium, potassium, calcium, and magnesium, has a tendency to increase the activity of the H^+ ion in solution and, consequently, to decrease the soil pH. For example, if $CaCl_2$ were added to a slightly acid soil, the quantity of H^+ ion in the soil solution would increase.

$$\boxed{Micelle}\ {}^{H^+}_{H^+} + Ca^{2+} + 2Cl \rightleftharpoons \boxed{Micelle}\ Ca^{2+} + 2H^+ + 2Cl^-$$

(soil solid)   (soil solution)   (soil solid)   (soil solution)

The presence of neutral salts in some alkaline soils also can result in lower pH but by a different mechanism (see Section 10.4). The salts depress the hydrolysis of colloids with high sodium saturation. Thus, the presence of sodium chloride (NaCl) in such a soil will tend to force the following reaction to move to the left:

$$\boxed{Micelle}\ Na^+ + H_2O \rightleftharpoons \boxed{Micelle}\ H^+ + Na^+ + OH^-$$

(soil solid)   (soil solution)   (soil solid)   (soil solution)

thereby reducing the concentration of OH^- ion in the soil solution and lowering the pH. This reaction is of considerable practical importance in certain saline soils of arid regions because it keeps the pH from rising to levels that are toxic to plants.

Since the reaction of the soil solution is influenced by the four relatively independent factors just discussed, a close correlation would not always be expected between percentage base saturation and pH when comparing widely diverse soils. Yet with soils of similar origin, texture, and organic content, a reasonably good correlation does exist.

9.4 BUFFERING OF SOILS

As previously indicated, soils tend to resist changes in the pH of the soil solution. This resistance is called **buffering**. For soils with intermediate pH levels, buffering can be explained in terms of the equilibrium that exists among the active, salt-replaceable and residual acidities (Figure 9.6). This general relationship can be illustrated as follows:

$$\boxed{\begin{array}{c} H \\ Micelle\ Al \\ AlOH \end{array}} \rightleftharpoons \boxed{Micelle\ \begin{array}{c} H^+ \\ Al^{3+} \\ AlOH^{2+} \end{array}} \rightleftharpoons H^+ + Al^{3+} + Al(OH)^{2+}$$

Bound forms   Exchangeable forms   Soil solution
(residual acidity)   (salt replaceable   (active acidity)
           or exchange acidity)

If just enough lime (a base) is applied to neutralize the hydrogen ions in the soil solution, they are largely replenished as the reactions move to the right, thereby minimizing the change in soil solution pH (Figure 9.7). Likewise, if the H^+ ion concentration of the soil solution is increased by (for example) organic decay or fertilizer applications, the reaction is forced to the left, once again minimizing changes in soil solution pH. In both cases the buffering in moderately acid soils is seen to be affected by reactions with the soil's cation exchange complex.

In very acid soils, equilibrium reactions involving aluminum and iron hydroxy ions provide a second major mechanism for buffering. These reactions include the following:

$$AlOH^{2+} + H_2O \rightleftharpoons Al(OH)_2^+ + H^+$$

$$Al(OH)_2^+ + H_2O \rightleftharpoons Al(OH)_3 + H^+$$

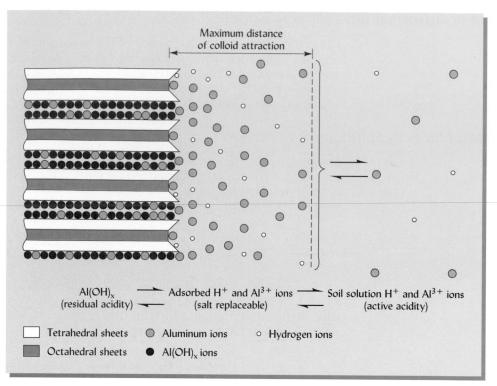

FIGURE 9.6 Equilibrium relationship among residual, salt-replaceable (exchangeable), and soil solution (active) acidity on a 2:1 colloid. Note that the adsorbed and residual ions are much more numerous than those in the soil solution even when only a small portion of the clay crystal is shown. Most of the $Al(OH)_x$ ions are held tightly in internal spaces, only a few being exchangeable. Remember that the aluminum ions, by hydrolysis, also supply hydrogen ions in the soil solution. It is obvious that neutralizing only the hydrogen and aluminum ions in the soil solution will be of little consequence. They will be quickly replaced by ions associated with the colloid. The soil, therefore, demonstrates high buffering capacity.

The addition of H^+ ions would drive these reactions to the left, so relatively few H^+ ions would accumulate in the soil solution and the pH would be reduced only modestly. Likewise, the addition of OH^- ions would force the reaction to the right. In both cases, changes in soil solution would be small. In other words, buffering would occur.

Similar buffer reactions involving carbonates, bicarbonates, carbonic acid, and water take place in alkaline soils.

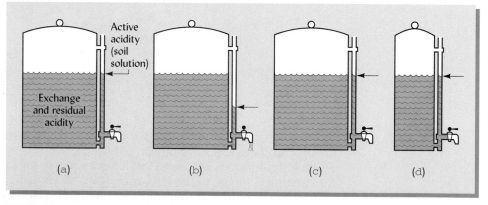

FIGURE 9.7 The buffering action of a soil can be likened to that of a coffee dispenser. (*a*) The active acidity, which is represented by the coffee in the indicator tube on the outside of the urn, is small in quantity. (*b*) When hydrogen ions are removed, this active acidity falls rapidly. (*c*) The active acidity is quickly restored to near the original level by movement from the exchange and residual acidity. By this process, there is considerable resistance to the change of active acidity. A second soil with the same active acidity (pH) level but a much smaller exchange and residual acidity (*d*) would have a lower buffering capacity.

$$CaCO_3 + H_2CO_3 \rightleftharpoons Ca(HCO_3)_2$$
$$Ca(HCO_3)_2 + 2H_2O \rightleftharpoons Ca^{2+} + 2OH^- + 2H_2CO_3$$

The addition of H^+ ions would shift these reactions to the right while increased OH^- ions would shift them to the left. In both cases, however, there would be resistance to pH change (buffering). Thus, throughout the entire pH range there are mechanisms that buffer the soil solution and prevent rapid changes in soil pH.

We have identified three primary mechanisms for soil buffering: (1) interactions with aluminum compounds at low pH, (2) cation exchange equilibria at intermediate pH, and (3) reactions with carbonates at high pH. The pH ranges and percentage base saturation at which these three mechanisms are effective in one group of soils are shown in Figure 9.8. While the exact pH–base saturation relationship varies greatly from soil to soil, these primary means of buffering are commonly found.

Importance of Soil Buffering

Soil buffering is important for two primary reasons. First, it tends to ensure reasonable stability in the soil pH, preventing drastic fluctuations that might be detrimental to plants and soil microorganisms. For example, well-buffered soils resist the acidifying effect of acid rain on both the soil and drainage water. Second, buffering influences the amount of amendments, such as lime or sulfur, required to effect a desired change in soil pH.

9.5 BUFFERING CAPACITY OF SOILS

Soils vary greatly in their buffering capacity. Other things being equal, the higher the cation exchange capacity (CEC) of a soil, the greater its buffering capacity. This relationship exists because in a soil with a high CEC, more reserve and exchangeable acidity must be neutralized or increased to affect a given change in soil pH. Thus, a clay loam soil containing 6% organic matter and 20% of a 2:1-type clay would be more highly buffered than a sandy loam with 2% organic matter and 10% kaolinite.

Buffer Curves

The buffer curve shown in Figure 9.8 suggests that some soils are most highly buffered at pH values below 5 and above 7. This would be in those ranges where aluminum compounds (low pH) and carbonates (high pH) tend to control buffer reactions. The soil is least well buffered at intermediate pH levels where cation exchange is the primary buffer mechanism. However, great variability exists in the pH–base saturation relations of different soils. This may be due in part to the differences among soil colloids in the

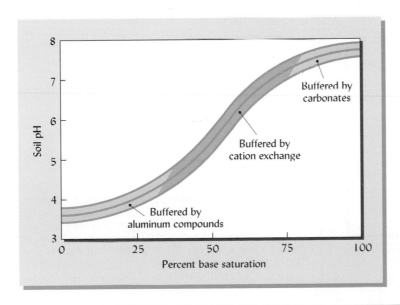

FIGURE 9.8 The relationship between percent base saturation and soil pH of Vermont soils, the appropriate pH range where three major buffering mechanisms are most effective: aluminum compounds at low pH, cation exchange at intermediate pH, and carbonates at high pH values. Although there is much variation from one soil to another in the pH–base saturation relationship, the three major mechanisms shown for buffering soils usually pertain. [Magdoff and Bartlett (1985)]

effect of pH on their percentage base saturation. Likewise, differences among soils in their contents of bound Al hydroxy complexes that can absorb OH ions as the pH rises may explain some differences in pH–base saturation relationships. These relationships are also influenced by increases in the CECs of organic matter and some 1:1-type clays as the pH is increased.

9.6 VARIABILITY IN SOIL pH

Natural Changes

Natural weathering and organic decay processes result in the formation of both acid- and base-forming chemicals. The base-forming cations (Ca^{2+}, Mg^{2+}, etc.) are released from rocks and minerals as they break down. Hydrogen ions are generated as a net result of a complex series of reactions as organic matter decomposes:

$$[C_2H_4ONS] + 5O_2 + H_2O \longrightarrow H_2CO_3 + RCOOH + H_2SO_4 + HNO_3$$

| Organic matter (generalized) | | Carbonic acid | Strong organic acids | Strong inorganic acids |

As these acids disassociate, they become direct sources of hydrogen ions. They also help solubilize aluminum from mineral surfaces, a key factor in the more acid soils.

In low-rainfall areas there is little leaching of the base-forming cations that stimulate a relatively high degree of base saturation and pH values of near 7 and above. In more humid areas, leaching depletes the upper horizons of calcium and other base-forming cations; hydrogen and aluminum ions accumulate, the percentage of base saturation declines, and the pH is lowered. Figure 9.9 illustrates this effect of increasingly humid climate on soil acidity levels under both forest and grass natural vegetation.

Human-Induced Changes

In recent times, these long-term natural changes have been increasingly augmented by changes stemming from human activities. Rapid increases in human populations, along with widespread industrialization, are having significant impacts on soil pH. A few of these will be considered.

CHEMICAL FERTILIZER USAGE. Chemical fertilizers, which have stimulated remarkable increases in global food supplies, have also brought about significant changes in soil pH. The widely used ammonium-based fertilizers, such as $(NH_4)_2SO_4$ and $(NH_4)_2HPO_4$, are oxidized in the soil by microbes to produce strong inorganic acids by reactions such as the following:

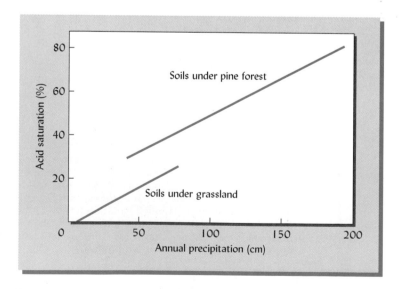

FIGURE 9.9 Effect of annual precipitation on the percent acid saturation of untilled California soils under grassland and pine forests. Note that the degree of acidity goes up as the precipitation increases. Also note that the forest produced a higher degree of soil acidity than did the grassland. [From Jenny, et al. (1968)]

$$(NH_4)_2SO_4 + 4O_2 \rightarrow 2HNO_3 + H_2SO_4 + 2H_2O$$

These strong acids provide H^+ ions that result in lower pH values. Since worldwide fertilizer use increased tenfold between 1950 and 1990, the effects of this type of reaction are not minor. In the United States alone it would require at least 20 million metric tons of limestone to neutralize the acidity produced by nitrogen fertilizers every year.

TILLAGE PRACTICES.   The pH of the upper soil horizons are influenced by tillage practices. In no-tillage systems, the breakdown of organic residues that are concentrated on the soil surface produces organic and inorganic acids that acidify the soil. Likewise, nitrogen fertilizer applications are made primarily to the surface layers in no-tillage systems, thereby localizing the acidifying effect of the fertilizer. The combined effects on pH of tillage practices and fertilizer applications are shown in Figure 9.10. Keep in mind that the soil layers affected are those near the soil surface where plant-root density is the highest. Fortunately, those are also the horizons most affected by surface-applied liming materials.

ACID DEPOSITION FROM THE ATMOSPHERE.   Acid rain is a second significant worldwide source of nitric and sulfuric acids. Nitrogen and sulfur-containing gases are emitted into the atmosphere from the combustion of coal, gasoline, and other fossil fuels, as well as from the burning of forests and crop residues (Figure 9.11). These gases react with water and other substances in the atmosphere to form HNO_3 and H_2SO_4 that are then returned to the earth in rainwater. This precipitation is called *acid rain* since its pH is commonly between 4.0 and 4.5 and may be as low as 2.0. Normal rainfall that is in equilibrium with atmospheric carbon dioxide has a pH of 5.0 to 6.0. A map showing the pH of acid rain in North America is presented in Figure 9.12.

The quantity of H_2SO_4 and HNO_3 brought to the earth in acid rain globally is enormous, but the amount falling on a given hectare in a year is not enough to significantly change soil pH immediately. In time, however, the cumulative effects of annual deposition can influence both the soils and the plants growing in them.

Acid rain is particularly worrisome in humid forested areas near urbanized and industrialized sites that provide the S- and N-containing pollutants.[6] Forests in the eastern part of the United States and much of Europe are affected (Figure 9.13). Studies have shown that the leaching of base-forming cations and the mobilization of aluminum is enhanced by the presence in the soil solution of the strong acid anions SO_4^{2+} and NO_3^-, along with the H^+ ion provided by the acid rain. Figure 9.14 shows the reduction in Ca^{2+}, Mg^{2+}, and K^+ and the decline in pH levels that occurred over a period of 20 years

_____

[6] See Joslin, et al. (1992), Robarge and Johnson (1992), and Likens, et al. (1996) for discussions of chemical changes associated with forest soil acidification.

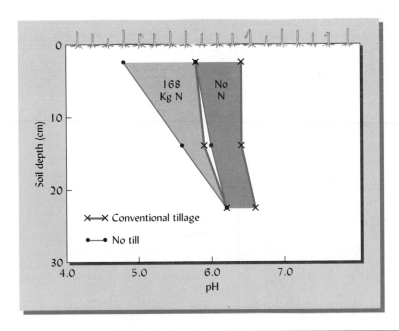

FIGURE 9.10   The effects of 10 years of two tillage practices (conventional and no-till) with and without nitrogen fertilizer on the pH of an Alfisol in a Kentucky corn field. The no-till plots were consistently lower in pH than those that were conventionally tilled. This difference was accentuated on the nitrogen-treated plots, especially in the upper few centimeters of soil. The differences were likely due primarily to the formation of organic acids from residue decay in the no-till plots, and to inorganic acids (e.g., HNO_3) formed from the surface-applied nitrogen fertilizers. [Drawn from data in Blevins, et al. (1983); used with permission of Elsevier Science Publishing Company]

FIGURE 9.11 Illustration of the formation of nitrogen and sulfur oxides from the combustion of fuel in sulfide ore processing and from motor vehicles. The further oxidation of these gases and their reaction with water to form sulfuric acid and nitric acid are shown. These help acidify rainwater, which falls on the soil as acid rain. NO_x indicates a mixture of nitrogen oxides, primarily N_2O and NO. [Modified from National Research Council (1983)]

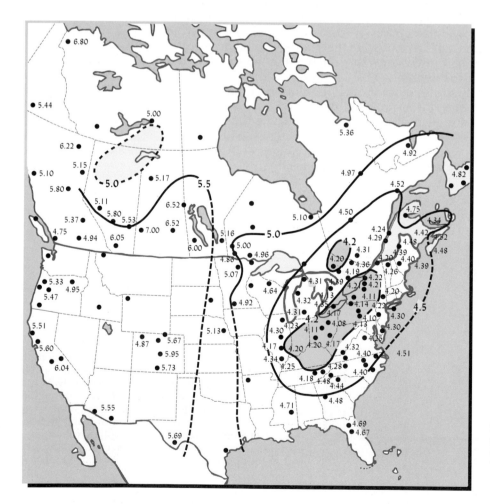

FIGURE 9.12 Annual mean value of pH in precipitation weighted by the amount of precipitation in the United States and Canada for 1980. [From U.S./Canada Work Group No. 2 (1982)]

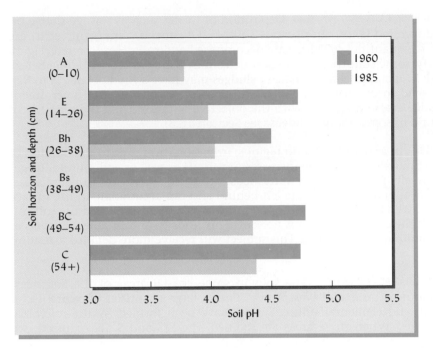

FIGURE 9.13 Reduction in soil pH of 67 forested Spodosol sites in northern Belgium thought to be due to 25 years of atmospheric deposition as well as natural acidification. [From Ronse, et al. (1988); used with permission of Williams and Wilkins, Baltimore, Md.]

in two forested watersheds that were subject to acid rain deposits. While natural processes undoubtedly account for part of such changes, it is thought that at least half are due to acid rain caused by human activities.

The leaching of calcium and the mobilization of aluminum result in Ca/Al ratios in the soil solution of less than 1.0, the ratio considered a threshold for restrained plant growth and associated nutrient uptake. Higher aluminum concentrations can lead to toxicities and to reduced uptake of calcium by forest vegetation.

Since the passage of the Clean Air Act in 1970, the atmospheric deposition of acid sulfates in North America has declined significantly. Unfortunately, the expected increase in pH of some streams in the Northeast has not occurred. This is likely due to the fact that the same regulations leading to a decline in SO_2 and NO_x emissions resulted in even greater reductions in the emissions of calcium and other base-forming cations held by tiny soot particles that were removed from smoke stacks. In any case, the effects of acid rain will continue to be monitored, especially in humid forested regions of Europe and the United States where acid rain is suspected of contributing to the decline of forest productivity in some areas. Unfortunately, applying liming materials to forested land by aerial or other means is too expensive to be practical.

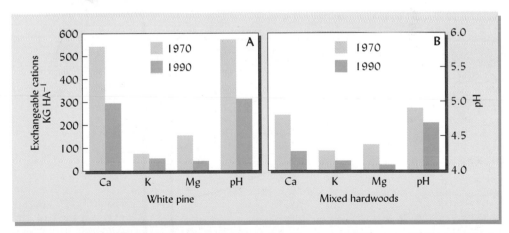

FIGURE 9.14 Exchangeable Ca^{2+}, Mg^{2+}, and K^+ soil pH declined appreciably from 1970 to 1990 in the A horizon of soils of two forested watersheds in the mountains of North Carolina, one (a) producing white pine, the other mixed hardwood trees (b). Such reductions have been ascribed to cation uptake by the tree biomass and to leaching of the Ca^{2+}, Mg^{2+}, and K^+ that is enhanced by acid rain. [Modified from Knoepp and Swank (1994); used with permission of the Soil Science Society of America]

DISPOSAL OF ACID-FORMING ORGANIC WASTES. Deposition of wastes such as sewage sludge on agricultural and forested lands can decrease soil pH. Organic and inorganic acids are formed as the large quantities of sludge are decomposed and soil pH is lowered (Figure 9.15). Fortunately, some countries ensure that these wastes are treated with lime before they are applied. In fact, some *lime-stabilized* sludges may have a final pH of 7.5 to 8.5, making them useful in combating soil acidity.[7] Control of soil pH after amendment with sewage sludge is also regulated, with the objective of minimizing the mobility of toxic metals found in some sewage sludges (see Section 18.8).

IRRIGATION PRACTICES. In arid and semiarid regions, irrigation can result in the buildup of *excess salts* that stimulate undesirable increases in soil pH as well as soil salinity. Irrigation waters bring salts from upstream watersheds into the downstream irrigated areas. When the water evaporates, the salts are left behind and can accumulate in the surface horizon. Included are salts containing the Na^+ ion that can be adsorbed on the soil colloids, especially if the bicarbonate ion is present. This leads to undesirable pH levels as well as to a destabilized soil structure. This subject will receive more detailed attention in Chapter 10.

DRAINAGE OF SOME COASTAL WETLANDS. The soils in some coastal areas of the Southeast United States, Southeast Asia, and West Africa contain large quantities of pyrite (FeS_2), iron sulfide (FeS), and elemental sulfur (S) that have been formed by the microbial reduction of sulfates. When these areas are drained, the FeS and S are oxidized (see Plates 24 and 25), ultimately resulting in the formation of sulfuric acid, which accounts for their being called *acid sulfate* soils.

$$4FeS + 9O_2 + 4H_2O \rightarrow 2Fe_2O_3 + 4H_2SO_4$$

$$2S + 3O_2 + 2H_2O \rightarrow 2H_2SO_4$$

Soil containing large amounts of such reduced sulfur-containing compounds that could produce acid upon being oxidized are said to have a high **potential acidity.** In any case, when such soil is exposed to oxygen by drainage or excavation, extreme acidity results, pH values as low as 1.5 having been noted. The enormous amount of $CaCO_3$ required generally makes it uneconomical to neutralize this acidity. Such soils would best be left undisturbed or returned to their undrained conditions, not only for agricultural and economic reasons, but to avoid contamination of water with acid drainage (see Plate 26) and to maintain these areas as natural habitats for wildlife and natural flora.

---

[7] Certain lime-stabilized sludges have received large quantities of lime to control pathogens and odors. If the high lime content of these sludges is not taken into account, their application to soil with low buffering capacities could bring about **overliming.**

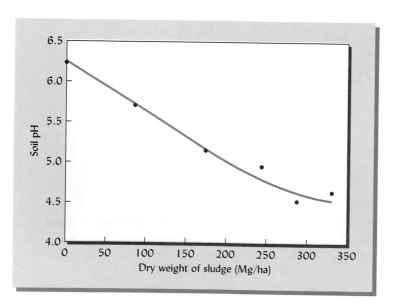

FIGURE 9.15 Effect of large applications of sewage sludge over a period of six years on the pH of a Paleudult (Orangeburg fine sandy loam). The reduction was due to the organic and inorganic acids, such as HNO_3, formed during decomposition and oxidation of the organic matter. [Data from Robertson, et al. (1982)]

Minor Fluctuations in Soil pH

Minor fluctuations in soil pH result from the movement of salts into and out of different soil zones as soil moisture moves up and down through the profile. Similarly, the pH of mineral soils declines during the crop-growing season as a result of the acids produced by microorganisms and by the roots of higher plants. When soil temperatures decline, as in the fall in temperate regions, an increase in pH often is noted because biotic activities during these times are considerably slower.

Hydrogen Ion Variability

There is considerable variation in the pH of the soil solution in adjacent parts of the soil. Differences in pH are noted among sites in the soil only a few centimeters apart. Such variation may result from microbial decay of unevenly distributed organic residues in the soil or from the effects of plant roots (see Section 11.7). The pH of the rhizosphere immediately around the roots is lower than in the surrounding soil when NH_4^+ ions are adsorbed (Figure 9.16). Apparently, the NH_4^+ ions replace H^+ ions on the root surface, thereby bringing about the pH reduction.

The variability of the soil solution is important in many respects. For example, it affords microorganisms and plant roots a great variety of solution environments. Organisms unfavorably influenced by a given hydrogen ion concentration may find, at an infinitesimally short distance away, a different environment that is more satisfactory. The variety of environments may account in part for the great diversity in microbial species present in normal soils.

9.7 SOIL REACTION: CORRELATIONS

Knowledge of soil pH is very useful to the soil manager because it affects such a wide variety of chemical and biological phenomena in soils. Figure 9.17 shows one example: the effect of soil acidity (and aluminum toxicity) on the growth of cotton roots in a mineral soil. Only a few other examples will be considered here to illustrate this point further.

Nutrient Availability and Microbial Activity

Figure 9.18 shows in general terms the relationship between soil pH and the availability of plant nutrients, as well as the activities of soil microorganisms. Note that in strongly acid soils the availability of the macronutrients (Ca, Mg, K, P, N, and S) as well as molybdenum and boron is curtailed. In contrast, availability of micronutrient cations (Fe, Mn, Zn, Cu, and Co) is increased by low soil pH, even to the extent of toxicity to higher plants and microorganisms.

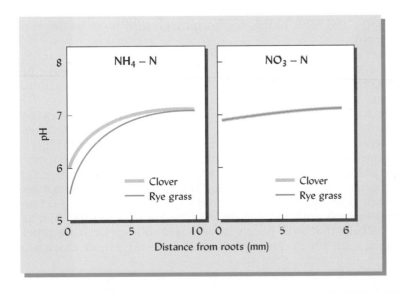

FIGURE 9.16 The pH of rhizosphere at different distances from the roots of ryegrass and clover receiving ammonium and nitrate sources of nitrogen. The exchange of NH_4^+ ions for H^+ ions on the root surfaces released H^+ ions to the rhizosphere, thereby reducing the pH. No such exchange took place with NO_3^- and no pH reduction resulted. [From Hinsinger and Gillies (1996), with permission of Blackwell Science Ltd.]

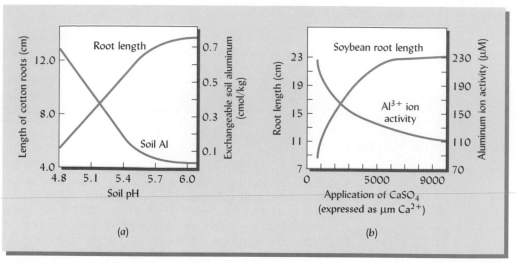

FIGURE 9.17 High concentrations of exchangeable or soil solution aluminum in acid soils are toxic to plant roots, a toxicity that can be reduced somewhat by adding $CaSO_4$. (a) Cotton root length is increased as soil pH is increased and exchangeable aluminum is decreased. (b) Adding $CaSO_4$ to a nutrient solution (pH = 2) reduces the Al^{3+} activity and permits an increase in soybean root length while having minimal effect on the solution pH. [(a) From Adams and Lund (1966); (b) drawn from data in Nobel, et al. (1988)]

In slightly to moderately alkaline soils, molybdenum and all the macronutrients (except phosphorus) are amply available, but levels of available Fe, Mn, Zn, Cu, and Co are so low that plant growth is constrained. Phosphorus and boron availability is likewise reduced in alkaline soil—commonly to a deficiency level.

It is difficult to generalize about nutrient–pH relationships. However, it appears from Figure 9.18 that the pH range of 5.5 to 6.5 or perhaps 7.0 may provide the most satisfactory plant nutrient levels overall. However, this generalization may not be valid for all soil and plant combinations. For example, certain micronutrient deficiencies are common on some sandy soils when the soils are limed to pH values of only 5.5 to 6.0.

Figure 9.18 suggests that most general-purpose bacteria and actinomycetes function well at intermediate and high pH values. Fungi seem to be particularly versatile, flourishing satisfactorily over a wide pH range. Therefore, fungal activity tends to predominate in acid soils, whereas at intermediate and higher pH they meet stiff competition from actinomycetes and bacteria.

Higher Plants and pH

Plants vary considerably in their tolerance to acid and/or alkaline conditions (Figure 9.19). For example, certain legume crops such as alfalfa and sweet clover grow best in near-neutral or alkaline soils, and most humid-region mineral soils must be limed to grow these crops satisfactorily.

Rhododendrons and azaleas are at the other end of the scale. They require a considerable amount of iron, which is abundantly available only at low pH values. If the pH and the percentage base saturation are not low enough, these plants will show chlorosis (yellowing of the leaves) and other symptoms indicative of iron deficiency.

Many forest trees grow well over a wide range of soil pH levels, which indicates that they have some tolerance to soil acidity. This is to be expected since some forest trees (e.g., conifers) generally tend to enhance soil acidity, and forests exist in humid regions where acid soils are common. There are some differences, however, among tree species in their tolerance of high soil acidity and aluminum. For example, the pines are generally very tolerant to highly acid soils and aluminum. The spruces are somewhat less tolerant but more so than most hardwoods (e.g., maples, oaks, and beeches). Elm, poplar, honey locust, and lucaena are known to be less tolerant to acid soils than are most other forest species. Plants that grow poorly in very acid soils (pH < 5) are generally affected by *aluminum toxicity,* which causes plant roots to become short, thick, and stubby.

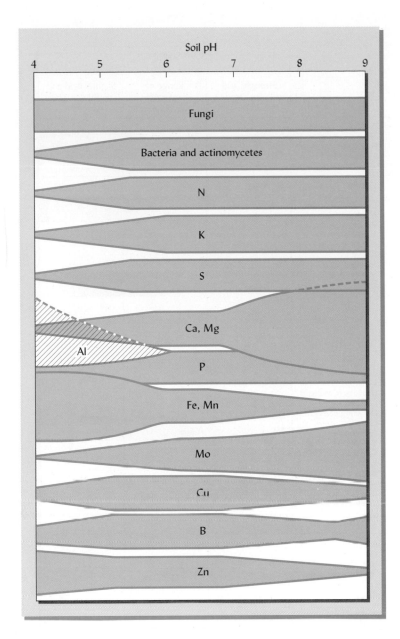

FIGURE 9.18 Relationships existing in mineral soils between pH on the one hand and the activity of microorganisms and the availability of plant nutrients on the other. The wide portions of the bands indicate the zones of greatest microbial activity and the most ready availability of nutrients. When the correlations as a whole are taken into consideration, a pH range of about 5.5 to 7.0 seems to best promote the availability of plant nutrients. In short, if soil pH is suitably adjusted for phosphorus, the other plant nutrients, if present in adequate amounts, will be satisfactorily available in most cases.

The most productive arable soils in use today have intermediate pH values, being not too acidic and not too alkaline. Most cultivated crop plants grow well on these soils. Since pasture grasses, many legumes, small grains, intertilled field crops, and a large number of vegetables are included in this broadly tolerant group, mild soil acidity or mild alkalinity is not a deterrent to their growth. In terms of pH, a range of perhaps 5.5 to 7.0 is most suitable for most crop plants.

Soil pH and Environmental Quality

There are many examples of the influence of soil pH on environmental quality, but only one will be cited—the influence of pH on groundwater contamination by herbicides. Chemical groupings such as —NH_2 and —COO^- on some herbicide molecules encourage reactions with soil colloids. These reactions may bind the herbicide molecule to the organic matter or clay in the soil, or they may influence the colloid's surface charge and, consequently, the herbicide molecule's ability to exchange with anions or be adsorbed by the soil colloid.

An example of such adsorption is that of atrazine, an herbicide that is widely used on corn. In a soil with pH < 5.7, the atrazine molecule attracts a proton (H^+ ion), giving the herbicide molecule a positive charge—a process called ***protonation.*** The positively

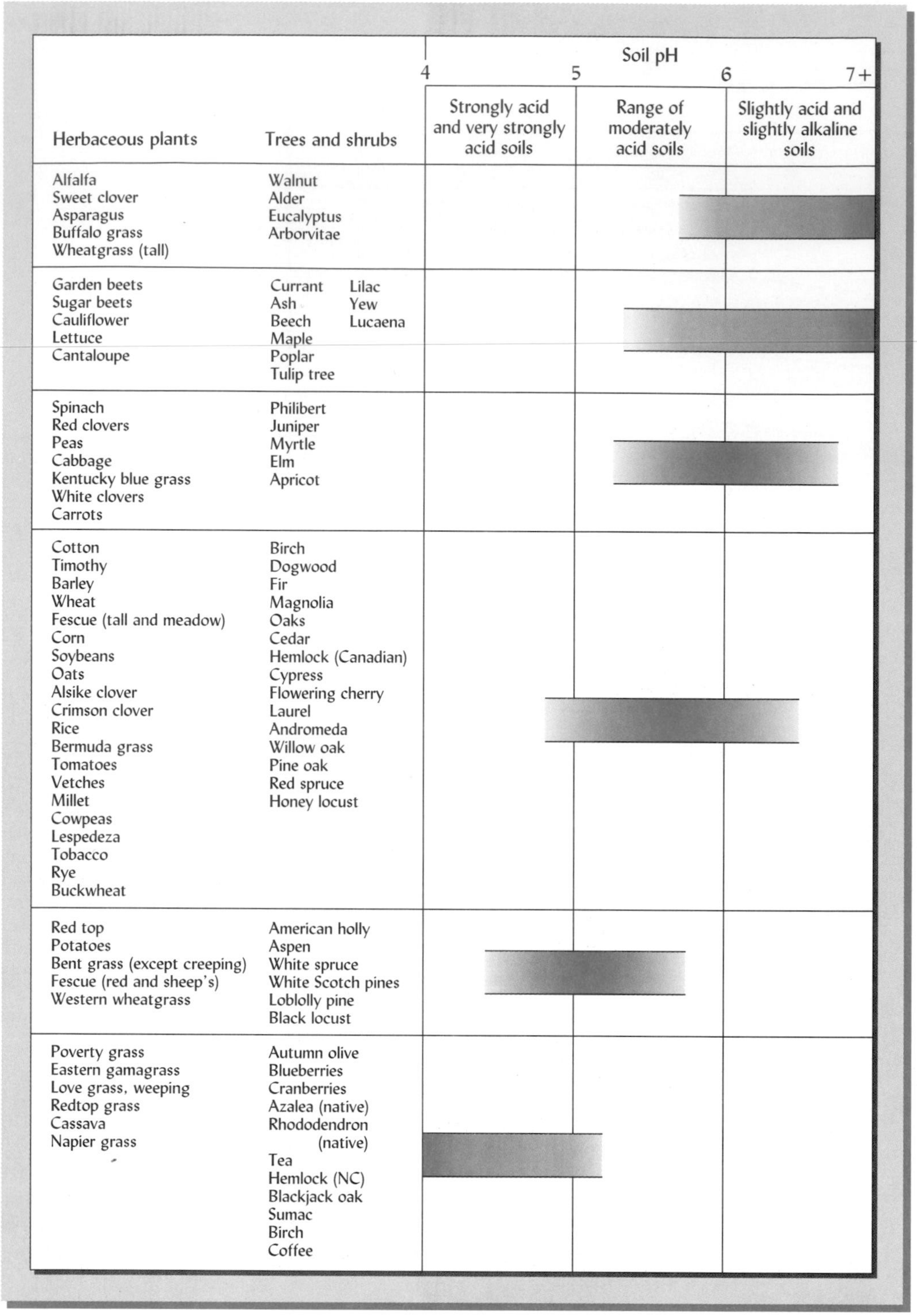

Herbaceous plants	Trees and shrubs	Soil pH 4 — 5 — 6 — 7+		
		Strongly acid and very strongly acid soils	Range of moderately acid soils	Slightly acid and slightly alkaline soils
Alfalfa Sweet clover Asparagus Buffalo grass Wheatgrass (tall)	Walnut Alder Eucalyptus Arborvitae			▓ (pH ~6–7+)
Garden beets Sugar beets Cauliflower Lettuce Cantaloupe	Currant Lilac Ash Yew Beech Lucaena Maple Poplar Tulip tree		▓ (pH ~5.6–7+)	
Spinach Red clovers Peas Cabbage Kentucky blue grass White clovers Carrots	Philibert Juniper Myrtle Elm Apricot		▓ (pH ~5.2–6.6)	
Cotton Timothy Barley Wheat Fescue (tall and meadow) Corn Soybeans Oats Alsike clover Crimson clover Rice Bermuda grass Tomatoes Vetches Millet Cowpeas Lespedeza Tobacco Rye Buckwheat	Birch Dogwood Fir Magnolia Oaks Cedar Hemlock (Canadian) Cypress Flowering cherry Laurel Andromeda Willow oak Pine oak Red spruce Honey locust	▓ (pH ~4.5–6)		
Red top Potatoes Bent grass (except creeping) Fescue (red and sheep's) Western wheatgrass	American holly Aspen White spruce White Scotch pines Loblolly pine Black locust	▓ (pH ~4.2–5.5)		
Poverty grass Eastern gamagrass Love grass, weeping Redtop grass Cassava Napier grass	Autumn olive Blueberries Cranberries Azalea (native) Rhododendron (native) Tea Hemlock (NC) Blackjack oak Sumac Birch Coffee	▓ (pH ~4–5)		

FIGURE 9.19 Relation of higher plants to the physiological conditions presented by mineral soils of different reactions. Note that the correlations are very broad and are based on pH ranges. The fertility level will have much to do with the actual relationship in any specific case. Such a chart is of great value in deciding whether or not to apply chemicals such as lime or sulfur to change the soil pH.

charged molecule is then adsorbed on the negatively charged soil colloids, where it is held until it can be decomposed by soil organisms. The adsorbed pesticide is less likely to move downward and into the groundwater. At pH values above 5.7, however, the adsorption is greatly reduced and the tendency for the herbicide to move downward in the soil is increased. Of course, the adsorption in acidic soils also reduces the availability of atrazine to weed roots, thus reducing its effectiveness as a weed killer.

9.8 DETERMINATION OF SOIL pH

Soil pH is tested routinely, and the determination is easy and rapid to make. Soil samples are collected in the field and the pH is measured directly, or the samples are brought to the laboratory for more accurate pH determination.

Potentiometric Method

The most accurate method of determining soil pH is with a pH meter (Figure 9.20). In this method a sensing glass electrode (referenced with a standard calomel electrode) is inserted into a soil–water mixture that simulates the soil solution. The difference between the H^+ ion activities in the wet soil and in the glass electrode gives rise to a electrometric potential difference that is related to the soil solution pH. The instrument gives very consistent results and its operation is simple.

Most soil testing laboratories in the United States measure the pH of a suspension of soil in water (usually a ratio of 1:1 or 1:2). Unfortunately, the water dilutes the soil solution, leading to pH values that are 0.2 to 0.4 units higher than the undiluted soil solution. This tendency is reduced by using an unbuffered 0.01 molar $CaCl_2$ solution. The $CaCl_2$ also helps minimize the effects of differences in electrolyte concentrations caused by salt accumulation or chemical fertilizer applications.

Dye Methods

Dye methods take advantage of the fact that certain organic compounds change color as the pH is increased or decreased. Mixtures of such dyes provide significant color changes over a wide pH range (3 to 8). A few drops of the dye solutions are placed in contact with the soil, usually on a white spot plate (see Figure 9.20). After standing a few minutes, the color of the dye is compared to a color chart that indicates the approximate pH.

In other dye methods, porous strips of paper are impregnated with the dye or dyes. When brought in contact with a mixture of water and soil, the paper absorbs the water and by color change indicates the pH. Such dye methods are accurate within about 0.2 to 0.4 pH unit.

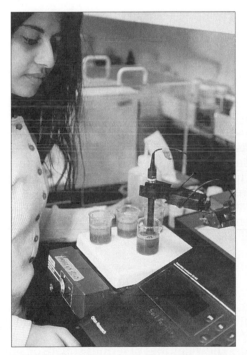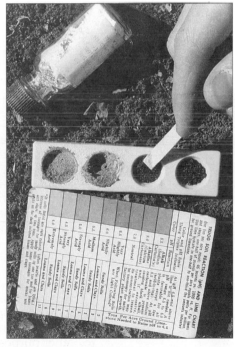

FIGURE 9.20 (Upper) Soil pH is measured quickly and inexpensively in the laboratory using an electrometric pH meter. Hundreds of analyses can be made in a day. (Lower) In the field, advantage is taken of the fact that pH-sensitive dyes give shades of color that can be calibrated against standard color charts, providing an equally simple and inexpensive means of measuring pH. Although the dye method is not quite as accurate as the electrometric method, it is quite satisfactory for most field management purposes. (Photos courtesy of R. Weil)

Limitations of pH Values

More may be inferred regarding the chemical and biological conditions in a soil from the pH value than from any other single measurement. However, the interpretation of this measurement must be applied in the field with some caution. There is considerable variation in the pH from one spot to another in a given field. Obviously, care must be taken to minimize sampling errors (see Section 16.18). Localized effects of fertilizers may give sizable pH variations within the space of a few centimeters. For unplowed soils, such as those on which no-till practices are followed, several acid-forming processes are concentrated at or near the soil surface, making it desirable to analyze the soil from several depths. Otherwise, serious acidity in the upper few centimeters might not be detected.

9.9 METHODS OF INTENSIFYING SOIL ACIDITY

It is often desirable to reduce the pH of highly alkaline soils. Furthermore, there are some acid-loving plants that cannot tolerate even moderately high pH values. Among them are a few highly valued ornamental species that are the favorites of gardeners around the world. For example, rhododendrons and azaleas grow best on soils having pH values of 5.0 and below. To accommodate these plants, it is sometimes necessary to increase the acidity of even acid soils. This is done by adding acid-forming organic and inorganic materials.

Acid Organic Matter

As organic residues decompose, organic and inorganic acids are formed. These can reduce the soil pH if the organic material is low in base-forming cations. Leaf mold from coniferous trees, pine needles, tanbark, pine sawdust, and acid peat moss are quite satisfactory organic materials to add around ornamental plants. However, farm manures, particularly poultry manures, and leaf mold of such high-base trees as beech and maple may be alkaline, and may actually increase the soil pH.

Inorganic Chemicals

When the addition of acid organic matter is not feasible, inorganic chemicals such as ferrous sulfate ($FeSO_4$) may be used. This chemical provides available iron (Fe^{2+}) for the plant and, upon hydrolysis, enhances acidity by reactions such as the following:

$$Fe^{2+} + 2H_2O \rightarrow Fe(OH)_2 + 2H^+$$

$$4Fe^{2+} + 6H_2O + O_2 \rightarrow 4Fe(OH)_2^+ + 4H^+$$

The H^+ ions released lower the pH locally and liberate some of the iron already present in the soil. Ferrous sulfate thus serves a double purpose for iron-loving plants by supplying available iron directly and by reducing the soil pH—a process that may cause a release of fixed iron present in the soil.

Other materials that are often used to increase soil acidity are elemental sulfur and, in some irrigation systems, sulfuric acid. Sulfur usually undergoes rapid microbial oxidation in the soil (see Section 13.21), and sulfuric acid is produced.

$$2S + 3O_2 + 2H_2O \rightarrow 2H_2SO_4$$

Under favorable conditions, sulfur is four to five times more effective, kilogram for kilogram, in developing acidity than is ferrous sulfate. Although ferrous sulfate brings about more rapid plant response, sulfur is less expensive, is easy to obtain, and is often used for other purposes. The quantities of ferrous sulfate or sulfur that should be applied will depend upon the buffering capacity of the soil and its original pH level.

CONTROL OF PLANT DISEASES. Soilborne pathogens are commonly sensitive to soil acidity that sulfur can stimulate. For example, potato scab is caused by an actinomycete that attacks the potato tuber surfaces, leaving a rough discolored appearance that makes the product quite unmarketable. When sulfur is added to reduce the pH to about 5.3 or below, the virulence of actinomycetes is much reduced. However, using sulfur for this purpose to increase soil acidity affects the management of the crop rotation. Consideration must be given to choosing succeeding crops that will not be adversely affected by a low soil pH regime.

9.10 DECREASING SOIL ACIDITY: LIMING MATERIALS

Soil acidity is commonly decreased by adding carbonates, oxides, or hydroxides of calcium and magnesium, compounds that are referred to as **agricultural limes.** Also, wood ashes are used locally to help control soil acidity. Where soils are very acid, crop growth can be dramatically improved by liming the soil (Figure 9.21).

Carbonate Forms

Sources of carbonates include marl, oyster shells, basic slag, and precipitated carbonates, but ground limestone is the most common and is by far the most widely used of all liming materials. The two important minerals occurring in varying proportions in limestones are **calcite,** which is mostly calcium carbonate ($CaCO_3$), and **dolomite,** which is primarily calcium-magnesium carbonate [$CaMg(CO_3)_2$]. When little or no dolomite is present, the limestone is referred to as *calcitic.* As the magnesium increases, this grades into a *dolomitic* limestone.

Oxide and Hydroxide Forms

Two other forms of lime are important, especially for small-scale use such as on home gardens where a very rapid change in pH may be required. One is calcium oxide (CaO), sometimes called *burned lime* or *quicklime;* the other is calcium hydroxide ($Ca(OH)_2$), which is commonly referred to as *hydrated* lime. These are more irritating to handle than limestone and are more expensive, but they continue to find a market where rapid adjustment of soil pH is needed. Box 9.2 shows how they are prepared and indicates some of their properties.

Dolomitic limestones and the oxides or hydroxides made from them provide both calcium and magnesium to the soil–plant system. In situations where available magnesium is low, dolomite or dolomitic limestones would be the products of choice. Where sufficient magnesium is present, calcitic limestone should be used to avoid the buildup of excessive magnesium that could counteract the positive effects of limestone on the soil's physical condition (see Section 4.7). The nutrient-supplying ability of all liming materials should not be overlooked. In some highly weathered soils, small amounts of lime may improve plant growth, more because of the enhanced calcium or magnesium nutrition than from a change in pH.

FIGURE 9.21 Alfalfa will not grow on very acid soils (foreground) but responds well to adequate liming (background).

BOX 9.2   MANUFACTURE AND USE OF LIMING MATERIALS

LIMESTONE

Eons ago, calcium and magnesium carbonates were deposited in the bottom of ancient lakes or seas. Under pressure of overlying materials these carbonates were compressed into layers of limestone rocks. Today, these are broken up, ground very finely, and made available for application to soils. Calcite ($CaCO_3$) and dolomite [$CaMg(CO_3)_2$] are the minerals contained in limestones. They provide the least expensive means of combating soil acidity and are usually stored in bulk and applied by trucks.

OXIDE OF LIME: BURNED LIME

Oxide of lime is produced by heating limestone to about 850°C and driving off carbon dioxide through reactions such as the following:

$$CaCO_3 + heat \rightarrow CaO + CO_2 \uparrow$$

$$CaMg(CO_3)_2 + heat \rightarrow CaO + MgO + CO_2 \uparrow$$

Oxide of lime is caustic to handle and is normally stored in waterproof bags. It is much more expensive than limestone, but reacts more quickly with the soil and gives a high pH after it is applied.

HYDROXIDE OF LIME: HYDRATED LIME

By adding hot water to oxide of lime, hydroxide of lime [$Ca(OH)_2 + Mg(OH)_2$] is formed:

$$CaO + MgO + 2H_2O \rightarrow Ca(OH)_2 + Mg(OH)_2$$

It is a white powder that is even more caustic than burned lime. It, too, must be kept in bags, not only for user safety, but to keep it from reverting to $CaCO_3 + MgCO_3$ by reaction with CO_2 in the atmosphere. Also more expensive than limestone, it, too, reacts quickly with the soil. The use of both hydroxide and oxide of lime is confined largely to home gardens and specialty crops.

Ruins of an 19th century lime-kiln in Maryland used to convert limestone into burned lime and, in turn, hydrated lime. (Photo courtesy of R. Weil)

9.11 REACTIONS OF LIME IN THE SOIL

The calcium and magnesium compounds, when applied to an acid soil, react with carbon dioxide and with the acid colloidal complex. These reactions will be considered in order.

Reaction with Carbon Dioxide

All liming materials—whether the oxide, hydroxide, or carbonate—react with carbon dioxide and water to yield the bicarbonate form when applied to an acid soil. The carbon dioxide partial pressure in the soil, usually several hundred times greater than that in atmospheric air, is generally high enough to drive such reactions to the right.

$$CaO + H_2O + 2CO_2 \rightarrow Ca(HCO_3)_2$$

$$Ca(OH)_2 + 2CO_2 \rightarrow Ca(HCO_3)_2$$

$$CaCO_3 + H_2O + CO_2 \rightarrow Ca(HCO_3)_2$$

Reaction with Soil Colloids

These liming materials react directly with acid soils, the calcium and magnesium replacing hydrogen and aluminum on the colloidal complex. The adsorption with respect to calcium may be indicated as follows:

$$\boxed{Micelle}\begin{array}{l}H^+\\Al^{3+}\end{array} + 2Ca(OH)_2 \longrightarrow \boxed{Micelle}\begin{array}{l}Ca^{2+}\\Ca^{2+}\end{array} + Al(OH)_3 + H_2O$$

$$\boxed{Micelle}\begin{array}{l}H^+\\Al^{3+}\end{array} + 2Ca(HCO_3)_2 \longrightarrow \boxed{Micelle}\begin{array}{l}Ca^{2+}\\Ca^{2+}\end{array} + Al(OH)_3 + H_2O + 4CO_2\uparrow$$

(in soil solution)

$$\boxed{Micelle}\begin{array}{l}H^+\\Al^{3+}\end{array} + 2CaCO_3 + H_2O \longrightarrow \boxed{Micelle}\begin{array}{l}Ca^{2+}\\Ca^{2+}\end{array} + Al(OH)_3 + 2CO_2\uparrow$$

(solid phase)

The insolubility of $Al(OH)_3$ and the release of CO_2 to the atmosphere pulls these reactions to the right. In addition, the adsorption of the calcium and magnesium ions raises the percentage base saturation of the colloidal complex, and the pH of the soil solution increases correspondingly.

Depletion of Calcium and Magnesium

As the soluble calcium and magnesium compounds are removed from the soil by the growing plants or by leaching, the percentage base saturation and pH are gradually reduced; eventually, another application of lime is necessary. This type of cyclic activity is typical of much of the calcium and magnesium added to arable soils in humid regions (Figure 9.22).

The need for repeated applications of limestone in humid regions suggests significant losses of calcium and magnesium from the soil. Table 9.1 illustrates losses of these elements by leaching compared with those from crop removal and soil erosion. Note that the total loss from all three causes (leaching, erosion, and crop removal) in humid-region agricultural soils, expressed in the form of carbonates, approaches 1 Mg/ha per year. Similarly, timber harvest methods that leave the soil open to erosion and nutrient leaching lead to rapid losses of calcium and magnesium from forest ecosystems.

9.12 LIME REQUIREMENTS: QUANTITIES NEEDED

The amount of liming material required to bring about a desired pH change is determined by several factors, including (1) the change in pH required, (2) the buffer capacity of the soil, (3) the chemical composition of the liming materials to be used, and (4) the fineness of the liming materials.

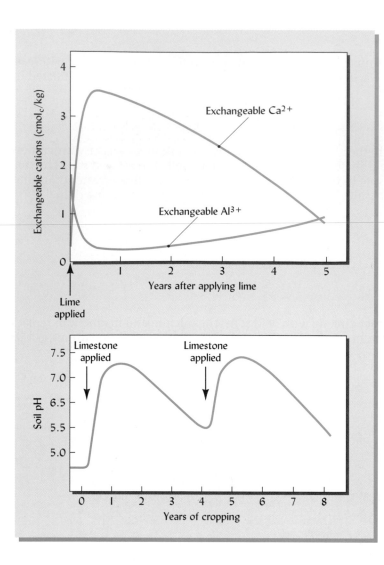

FIGURE 9.22 Diagram to illustrate why repeated applications of limestone are needed to maintain the appropriate chemical balances in the soil (upper). When a soil is limed, the exchangeable calcium increases and the exchangeable aluminum declines, as this diagram of data from an Oxisol in Brazil shows. Within a few years, however, the situation is reversed, the calcium having declined and the aluminum having increased. (Lower) Soil pH likewise increases after 4 to 8 Mg/ha (3.5 to 4.5 tons/acre) of limestone is applied to a temperate region soil. The pH reaches a peak after about one year. Leaching and crop removal deplete the calcium and magnesium and, in time, the pH decreases until a renewed application of limestone is necessary. [Upper modified from Smyth and Cravo (1992); used with permission of the American Society of Agronomy]

The range of soil pH optimums for various plants, discussed in Section 9.7, suggests the increase in soil pH that may be desired. For example, a higher pH is needed for alfalfa than for corn or soybeans. In Section 9.4 the relationship between the buffering capacity of soils and their cation exchange capacities was discussed. The limestone requirements of soils with different textures (and therefore likely to have different buffering capacities) is shown in Figure 9.23. Because of greater buffering capacity, the lime requirement for an acid fine-textured clay is much higher than that of a sand or a loam with the same pH value.

TABLE 9.1 **Calcium and Magnesium Losses from a Soil by Erosion, Crop Removal, and Leaching in a Humid Temperate Region**

Values are in kilograms per hectare per year.

	Calcium expressed as		Magnesium expressed as	
Manner of removal	*Ca*	*CaCO₃*	*Mg*	*MgCO₃*
By erosion, Missouri experiments, 4% slope	95	238	33	115
By the average crop of a standard rotation	50	125	25	88
By leaching from a representative silt loam	115	288	25	88
Total (Agricultural areas)		651		291
Loss in streams draining Douglas fir watersheds				
Clear-cut and slash/burned	81	203	26	91
Undisturbed forest	26	65	8	28

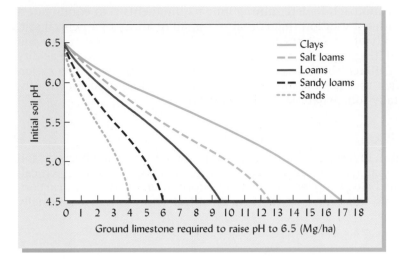

FIGURE 9.23 Effect of soil textural class on the amount of limestone required to raise the pH of soils from their initial state to 6.5. Note the very high quantities needed for the finer-textured soils that have higher clay and organic matter levels and, consequently, higher buffering capacities. The soils are assumed to be typical of cool, humid areas where 2:1-type clays predominate. In warmer climates such as the southeastern United States, where 1:1-type clays and Fe, Al oxides are often dominant, the quantities would likely be no more than one-half or even one-third of those suggested. In all cases, however, no more than 7 to 9 Mg/ha (3 to 4 tons/acre) should be applied at one time. If more is needed to meet the above requirements, subsequent applications after 2 to 3 years could be made.

Soil Buffering and Lime Requirements

Soils in a given region with comparable chemical properties exhibit a fairly common pH–base saturation relationship (see Figure 9.8). Knowing the pH of an unlimed soil, its cation exchange capacity, and the desired pH for the crops to be grown, it is possible to use this relationship to ascertain the approximate amount of liming material needed. Box 9.3 illustrates how the lime needs can be calculated.

BOX 9.3 CALCULATING LIME NEEDS

Assume that you must provide a recommendation as to how much limestone should be applied for the production of a high-pH-requiring crop—asparagus—to be grown in a loam soil that is currently quite acid (pH = 5.0). The cation exchange capacity (CEC) is about 10 $cmol_c$/kg, and you want to raise the pH to about 6.8. Assuming the pH–base saturation relationship is about the same as that shown in Figure 9.8, the soil is now at 50% base saturation and will need to be brought to about 90% base saturation to reach the goal of pH = 6.8. How much of a dolomitic limestone with a $CaCO_3$ equivalent of 90 will you need to apply to bring about the desired pH changes?

First, we need to know the $cmol_c$ of Ca/kg soil needed to bring about the change. Since the CEC is 10 $cmol_c$/kg and we need a 40% change in base saturation (from 50 to 90%) the Ca^{2+} required is

$$10 \text{ } cmol_c/kg \times 40/100 = 4 \text{ } cmol_c \text{ Ca/kg soil}$$

Second, since each Ca^{2+} has two charges, 4 $cmol_c$ can be expressed in grams of Ca^{2+} by multiplying by the molecular mass of Ca (40) and then dividing by 2, the Ca^{2+} charge, and by 100 since we are dealing with a $centimol_c$.

$$4 \text{ } cmol_c/kg \times \frac{40}{2 \text{ } gCa/mol_c} \times \frac{1 \text{ } mol_c}{100 \text{ } cmol_c} = 0.8 \text{ } gCa/kg \text{ soil}$$

Third, we calculate the amount of $CaCO_3$ needed to provide the 0.8 gCa^{2+} by multiplying by the ratio of the molecular masses of $CaCO_3$ and Ca.

$$0.8 \text{ } g/kg \text{ soil} \times \frac{100 \text{ } (CaCO_3)}{40 \text{ } (Ca)} = 2 \text{ } gCaCO_3/kg \text{ soil}$$

Fourth, to express the 2 $gCaCO_3$/kg soil in terms of the mass needed to change the pH of 1 ha 15 cm deep, we must multiply by 2×10^6 kg/ha (see footnote 9, page 138).

$$2 \text{ } gCaCO_3/kg \times 2 \times 10^6 \text{ } kg/ha = 4 \times 10^6 \text{ } g \text{ or } 4 \text{ } MgCaCO_3/ha$$

Fifth, since the $CaCO_3$ equivalency of our limestone is only 90, some 100 kg would be required to match 90 kg of pure $CaCO_3$. Consequently, the amount of limestone needed would be adjusted upward by a factor of 100/90.

$$4 \times \frac{100}{90} = 4.4 \text{ } Mg/ha \text{ or about 2 tons/acre.}$$

Another widely used approach to determining lime requirements is to equilibrate a sample of soil with a buffered solution that contains a small quantity of a neutral salt. The greater the total acidity of the soil, the more the solution buffering is overcome and the solution pH reduced, the degree of the reduction being proportional to the acidity released from the soil. Different buffering solutions are used for soils with low cation exchange capacities, such as some Ultisols, as compared with those used for Alfisols, where the CEC is generally somewhat higher (Figure 9.24).

Influence of Chemical Composition

The chemical composition of liming materials determines their long-term effects on soil pH, how much of each material is needed to achieve these effects, and how much calcium and magnesium are being supplied. To properly inform users of the chemical contents of liming materials, *guarantees* as to their chemical composition are usually required (see Box 9.4).

Limestone Fineness and Reactivity

The finer a liming material, the more rapidly it will react with the soil. The oxide and hydroxide of lime usually appear on the market as powders, so their fineness is always satisfactory. However, if limestones are not finely ground, they react very slowly with the soil, and give little crop response (Figure 9.25). For these reasons a **fineness guarantee** is commonly required for limestones. Materials with at least 50% of the particles small enough to pass through a 60-mesh screen (smaller than 0.25 mm in diameter) are quite satisfactory for most liming purposes.

9.13 PRACTICAL CONSIDERATIONS

The choice of liming materials is governed primarily by the costs of transporting large amounts of these chemicals from their source. Consequently, application rates should be kept at no more than are essential to meet soil and crop needs, and in no case should be higher than 7 to 9 Mg/ha (3 to 4 tons/acre).

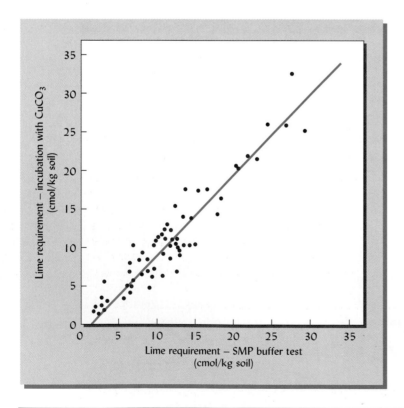

FIGURE 9.24 Relationship between the lime requirements of a number of Canadian soils determined by incubating the soils with $CaCO_3$ for a period of eight weeks (vertical) and by the SMP buffer lime requirement test, which required only about 30 minutes (horizontal). The time advantage favors the more rapid SMP buffer technique. The term *SMP* appropriately recognizes the names of the researchers who developed the test, Shoemaker, McLean, and Pratt (1961). [Canadian data from Tran and van Lierop (1981)]

Laws require that the chemical composition of liming materials be guaranteed. The means of expressing this composition varies from one governing body to another, but will usually include guarantees of the contents as expressed by one or more of the following:

1. As elemental Ca and Mg
2. As oxides of these cations (CaO, MgO)
3. As CaO equivalent (neutralizing ability of all compounds expressed as CaO)
4. As total carbonates (CaCO₃ plus MgCO₃)
5. As CaCO₃ equivalent (neutralizing ability of all compounds expressed in terms of CaCO₃)

There are advantages and disadvantages to each of these means of expressing the liming effectiveness. However, it is relatively easy to convert one means of expression to another using the concept of *chemical equivalency*. In other words, one atom (or molecule) of Ca, Mg, CaO, MgO, CaCO₃, or MgCO₃ will neutralize the same amount of acidity as another. Consequently, to make comparisons among different liming materials we need merely multiply by the ratios of their molecular masses. For example, to calculate the CaCO₃ equivalent of a pure burned lime (CaO) we need simply multiply by the molecular ratio of CaCO₃ to CaO:

$$\frac{CaCO_3}{CaO} = \frac{100}{56} = 1.786$$

Thus, 1 Mg (1000 kg) of this pure burned lime will neutralize as much acidity as 1.786 kg of a pure limestone. Some adjustment will need to be made, of course, for the percentage purity of the respective liming materials.

Using chemical equivalencies it is possible to compare a different means of expressing the liming abilities of four materials: a burned lime (oxide), a hydrated lime (hydroxide), and two limestones (one dolomitic, the other calcitic). Each is considered to be 95% pure.

		Means of expressing composition						
	Actual chemicals, %	*Element, %*		*Conventional oxide, %*		*CaO equivalent, %*	*Total carbonates, %*	*CaCO₃[a] equiv., %*
Material		*Ca*	*Mg*	*CaO*	*MgO*			
Burned	77 CaO	55.0	—	77.0	—	102.2	—	182.5
lime	18 MgO	—	10.9	—	18.0			
Hydrated	75 Ca(OH)₂	40.5	—	58.6	—	76.0	—	135.8
lime	20 Mg(OH)₂	—	8.3	—	13.8			
Calcitic limestone	95 CaCO₃	38.0	—	53.2	—	53.2	95	95
Dolomitic	35 CaCO₃	(14.0)	—	(19.6)	—	(19.6)	35	(35)
limestone	60 CaMg(CO₃)₂	(13.0)	(7.9)	(18.2)	(13.1)	(36.4)	60	(65)
		27.0	7.9	37.8	13.1	56.0	95	100

[a] Sometimes referred to as total neutralizing power.

In choosing limestones, attention should also be given to the need for magnesium as a plant nutrient, and dolomite should be used if magnesium levels are low.

The lime should be spread with or ahead of the crop that gives the most satisfactory response. Thus, in a rotation with corn, wheat, and two years of alfalfa, the lime may be applied after the wheat harvest to favorably influence the growth of the alfalfa crop that follows. However, since most lime is bulk-spread using heavy trucks (Figure 9.26), applications commonly take place on the sod or hay crop. This prevents adverse soil compaction by the truck tires, which might occur if the lime were applied on recently tilled land.

Special Situations

Spreading of limestone on forested watersheds is rarely practical. These areas are not accessible to ground-based spreaders, and manual application is too tedious and expensive. In some cases, landowners have turned to the use of helicopters for aerial spreading of small quantities of these liming materials. For very acid sandy soils such small

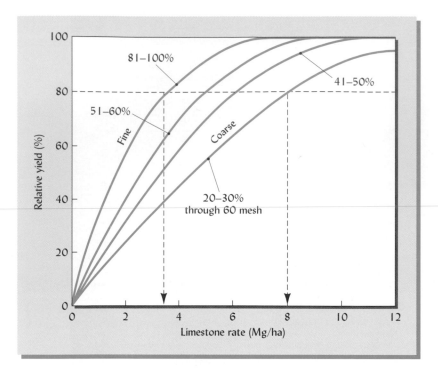

FIGURE 9.25 The effect of limestone fineness on the response of crops to increased rates of limestone application in a number of field experiments. Note that to obtain at least 80% of the top yield (dotted horizontal line), it was necessary to apply about 8 Mg/ha of the coarser limestones that have only 20 to 30% of their particles fine enough to pass through a 60-mesh screen (particles smaller than 0.25 mm in diameter). Comparable yields were obtained by applying less than 4 Mg/ha of the fine limestones (81 to 100% through the 60-mesh screen). [Redrawn from Barber (1984)]

applications can ameliorate the ill effects of soil acidity and provide sufficient calcium for the trees.

Some soil–crop systems make it difficult to incorporate and mix the limestone with the soil. For example, in no-till farming systems the soil is not plowed or cultivated. Furthermore, these systems tend to increase the acidity of the upper soil layers where the crop residues are deposited or remain on the land. Even though earthworms are known to be effective in incorporating surface-applied limestone into the soil (Figure 9.27), it is recommended that limestone be deeply incorporated before commencing a no-till system on an acid soil. Subsequent surface applications of limestone should adequately combat the acidity which forms near the surface.

Lawns, golf greens, and other turf grass areas also face liming difficulties in incorporating limestone. When such areas are first seeded, care must be taken to assure that the soil pH is at a satisfactory level. By proper timing of future liming applications with

FIGURE 9.26 Bulk application of limestone by specially equipped trucks is the most widespread method of applying liming materials. Because of the weight of such machinery, much limestone is applied to sod land and plowed under. In many cases this same method is used to spread commercial fertilizers.

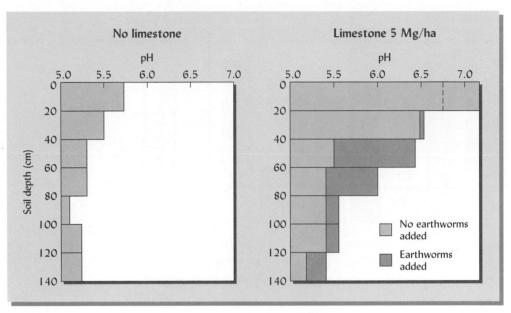

FIGURE 9.27 The effect of surface-applied limestone with and without the introduction of an active earthworm, *Allolobophora longa*, on the pH at different depths of an Alfisol. On the left, no limestone was applied. The limestone increased the pH (right), but without the introduced earthworms the change in pH was mostly in the upper 40 cm. The earthworms apparently moved the limestone into lower layers, thereby making them more suitable for plant root growth. [Redrawn from Springett (1983); used with permission of Blackwell Scientific Publications Ltd.]

annual aeration tillage operations that leave openings down into the soil (see Section 7.7), some downward movement of the lime can be affected. In untilled systems, frequent small applications of the lime are most effective.

Choice of Adapted Plants

It is often more judicious to solve soil acidity and alkalinity problems by changing the plant to be grown rather than by trying to change the soil pH. Not only do plant species vary greatly in their tolerance of acidity and alkalinity, but there is considerable natural variability within a species in the tolerance of different strains to low or high pH. Plant breeders and biotechnologists have taken advantage of this variability to develop cultivars of widely grown food crops that are quite tolerant of very acid or alkaline conditions (Figure 9.28). For example, high-yielding wheat varieties that are quite tolerant of soil acidity and high aluminum levels have been developed and are in use in some areas of the tropics where even modest liming applications are economically impractical. Plant and soil scientists must collaborate to enhance the ability of low-income families to meet their food needs.

Another example of the role of soil–plant interaction in overcoming the adverse effects of soil acidity is the use of green manure crops (crops grown specifically to provide organic residues) to overcome aluminum toxicity. Al^{3+} ions in the soil solution are strongly attracted to the surfaces of decaying organic matter where, through ligand bonding, they are tightly held (see Figure 9.2). Green manure crops and mulches can provide the organic residues needed to stimulate such interactions and thereby reduce the level of Al^{3+} in the soil solution. Aluminum-sensitive crops can then be grown following the green manure crop. Once again, this type of interaction can be most helpful to farmers in low-income countries where limestone availability and cost factors constrain attempts to adjust the soil pH.

Overliming

Another practical consideration is the danger of *overliming*—the application of so much lime that the resultant pH values are too high for optimal plant growth. Overliming is not very common on fine-textured soils with high buffer capacities, but it can occur eas-

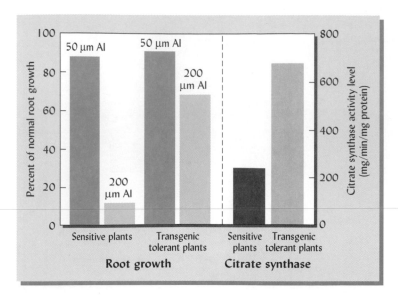

FIGURE 9.28 (Left) The effect of two levels of aluminum on the root growth of Al-sensitive plants compared to that of tolerant transgenic plants created by gene transfer. (Right) The comparative production by sensitive and tolerant plants of citrate synthase, an enzyme that stimulates the production of citric acid, which likely forms a complex (chelate) with the Al ions and prevents their absorption by plant roots. Data shown are for tobacco plants, but similar results were obtained with cassava plants. [Drawn from data in de la Fuente, et al. (1997)]

ily on coarse-textured soils that are low in organic matter. The detrimental results of excess lime include deficiencies of iron, manganese, copper, and zinc; reduced availability of phosphate; and restraints on the plant absorption of boron from the soil solution. It is an easy matter to add a little more lime later, but quite difficult to counteract the results of applying too much. Therefore, liming materials should be added conservatively to poorly buffered soils.

9.14 AMELIORATING ACIDITY IN SUBSOILS[8]

As discussed in previous sections, surface-applied limestone can easily ameliorate soil acidity in the upper soil layers. The subsoils of limed areas remain acid, however, since the Ca^{2+} and Mg^{2+} ions provided by the limestone do not move readily down the profile. Root growth and, in turn, crop yields are often constrained by excessively acid subsoils. Aluminum toxicity and/or calcium deficiency have been found in some of these acid subsoils.

Amelioration by Gypsum

Researchers in the southeastern United States and in some countries in the tropics have found that gypsum ($CaSO_4 \cdot 2H_2O$) can ameliorate aluminum toxicity in subsoils and thereby increase crop yields. Applied on the soil surface, the gypsum slowly dissolves and is leached into the subsoil. After about one year increased yields have been obtained with common crops such as corn, soybeans, alfalfa, beans, wheat, apples, and cotton. Penetration of crop roots into the subsoil is increased when gypsum is applied (Figure 9.29), making the crop less sensitive to subsequent droughts. Gypsum is fairly widely available in natural deposits, but is also an industrial by-product of the manufacture of high-analysis fertilizers and of flue-gas desulfurization.

The mechanisms for an amelioration of subsoil acidity by gypsum may vary from soil to soil. However, the applied gypsum is known to increase the level of calcium and reduce the level of aluminum both in the soil solution and on the exchange complex. The calcium increases come directly from the added gypsum. The reduced aluminum levels are thought to be stimulated by reaction involving the sulfate ion. For example, the SO_4^{2-} can replace certain OH^- ions associated with Fe, Al sesquioxides through reactions such as the following:

$$\begin{bmatrix} Fe, Al \\ complexes \end{bmatrix} \begin{matrix} -OH \\ -OH \end{matrix} + CaSO_4 \qquad \begin{bmatrix} Fe, Al \\ complexes \end{bmatrix} = SO_4 + Ca(OH)_2$$

---

[8] For a review of this subject, see Sumner (1993).

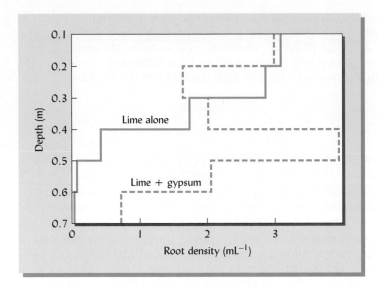

FIGURE 9.29 Surface-applied gypsum moved down the profile to help ameliorate the extreme acidity of the subsoil of this Ultisol in South Africa and helped to permit an increase in the density of corn roots. This extension of the roots into the subsoil can increase the availability of both water and nutrients. [Redrawn from Farina and Channon (1988); used with permission of the Soil Science Society of America]

The $Ca(OH)_2$ can then react with Al^{3+} ions in the soil solution to form insoluble $Al(OH)_3$, thereby reducing the concentration of the Al^{3+} while increasing that of Ca^{2+}:

$$3Ca(OH)_2 + 2Al^{3+} \rightarrow 2Al(OH)_3 + 3Ca^{2+}$$
$$\text{(insoluble)}$$

9.15 CONCLUSION

No other single chemical soil characteristic is more important in determining the chemical environment of higher plants and soil microbes than the pH. There are few reactions involving any component of the soil or of its biological inhabitants that are not sensitive to soil pH. This sensitivity must be recognized in any soil-management system.

Soil pH is largely controlled by fine soil particles and their associated exchangeable cations. Aluminum and hydrogen enhance soil acidity, whereas calcium and other base-forming cations (especially sodium) encourage soil alkalinity. The exchange complex is also a mechanism for soil buffering, which resists rapid changes in soil reaction, giving stability to most plant–soil systems.

The maintenance of satisfactory soil fertility levels in humid regions depends considerably on the judicious use of lime to balance the losses of calcium and magnesium from the soil (Figure 9.30). Liming not only maintains the levels of exchangeable calcium and magnesium but in so doing also provides a chemical and physical environment that

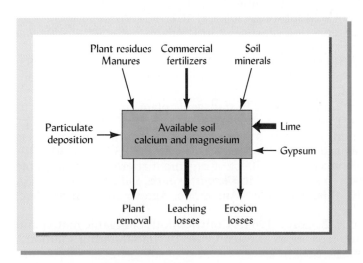

FIGURE 9.30 Important ways by which *available* calcium and magnesium are supplied to and removed from soils. The major losses are through leaching and erosion. These losses may be largely replaced by lime and fertilizer applications. Fertilizer additions are much higher than is generally realized, for some phosphate fertilizers contain large quantities of calcium. Calcium-rich dust particles in arid regions and in industrially polluted areas play an important role in uncultivated ecosystems where soil amendments are not used.

encourages the growth of most common plants. Lime is truly a foundation for much of modern humid-region agriculture. Knowing how pH is controlled, how it influences the supply and availability of essential plant nutrients as well as toxic elements, how it affects higher plants and human beings, and how it can be ameliorated is essential for the conservation and sustainable management of soils throughout the world.

STUDY QUESTIONS

1. Soil pH gives a measure of the concentration of H^+ ions in the soil solution. What, if anything, does it tell you about the concentration of OH^- ions? Explain.

2. Describe the role of aluminum and its associated ions in enhancing soil acidity, identifying the ionic species involved and the effect of these species on the CEC of soils.

3. If you could somehow extract the soil solution from the upper 16 cm of 1 ha of acid soil (pH = 5), only a few kilograms of limestone would be needed to neutralize the soil solution. Yet under field conditions up to 6 Mg of limestone may be required to bring the pH of this soil layer to a pH of 6.5. How do you explain this difference?

4. What is meant by *buffering?* Why is it so important in soils, and what are the mechanisms by which it occurs?

5. What is acid rain, and why does it seem to have greater impact on forests than on commercial agriculture?

6. Discuss the significance of soil pH in determining specific nutrient availabilities and toxicities, as well as species composition of natural vegetation in an area.

7. How much of a limestone with a $CaCO_3$ equivalent of 90 would you need to apply to raise the percent base saturation of a silt loam soil (CEC = 20 $cmol_c/kg$) from 60 to 90%?

8. The drainage and cultivation of certain coastal wetland soils has resulted in extreme acidity (pH < 4). Explain why this pH change likely occurs and suggest appropriate management solutions.

9. A neighbor complained when his azaleas were adversely affected by a generous application of limestone to the lawn immediately surrounding the azaleas. To what do you ascribe this difficulty? How would you remedy it?

10. The ill effects of acidity in subsoils can be ameliorated by adding gypsum ($CaSO_4 \cdot 2H_2O$) to the soil surface. What are the mechanisms responsible for this effect of the gypsum?

REFERENCES

Adams, F., and Z. F. Lund. 1966. "Effect of chemical activity of soil solution aluminum on cotton root penetration of acid subsoils," *Soil Sci.,* **101**:193–198.

Barber, S. A. 1984. "Liming material and practices," in F. Adams (ed.), *Soil Acidity and Liming,* 2d ed. (Madison, Wis.: Amer. Soc. Agron.)

Blevens, R. L., G. W. Thomas, M. S. Smith, W. W. Frye, and P. L. Cornelius. 1983. "Changes in soil properties after 10 years continuous non-tilled and conventionally tilled corn," *Soil Tillage Res.,* 3:135–146.

de la Fuente, et al. 1997. "Aluminum tolerance in transgenic plants by alteration of citrate synthesis," *Science,* **276**:1566–1568.

Farina, M. P. W., and P. Channon. 1988. "Acid subsoil amelioration: II. Gypsum effects on growth and subsoil chemical properties," *Soil Sci. Soc. Amer. J.,* **52**:175–180.

Hargrove, W. L., and G. W. Thomas. 1981. "Effect of organic matter on exchangeable aluminum and plant growth in acid soils," in *Chemistry in the Soil Environment.* ASA Special Publication no. 40. (Madison, Wis.: Amer. Soc. Agron. and Soil Sci. Soc. Amer.).

Helling, C. S., G. Chesters, and H. B. Corey. 1964. "Contributions of organic matter and clay to soil cation exchange capacity as affected by pH of the saturating solution," *Soil Sci. Soc. Amer. Proc.,* **28**:517–520.

Hinsinger, P., and R. J. Gilkes. 1996. "Mobilization of phosphate from phosphate rock and alumina-sorbed phosphate by the roots of ryegrass and clover as related to rhizosphere pH," *European J. Soil Sci.*, **47**:523–532.

Jenny, H., et al. 1968. "Interplay of soil organic matter and soil fertility with state factors and soil properties," in *Organic Matter and Soil Fertility.* Pontificiae Academia Scientiarum Scripta Varia 32. (New York: Wiley).

Joslin, J. D., J. M. Kelly, and H. van Miegroet. 1992. "Soil chemistry and nutrition of North American spruce-fir stands: Evidence for recent change," *J. Environ. Qual.*, **21**:12–30.

Knoepp, J. D., and W. T. Swank. 1994. "Long term soil chemistry changes for aggrading forest systems," *Soil Sci. Soc. Amer. J.*, **58**:325–331.

Likens, G. E., C. T. Driscoll, and D. C. Buso. 1996. "Longterm effects of acid rain: Response and recovery of a forest ecosystem," *Science,* **272**:244–246.

Magdoff, F. R., and R. J. Barlett. 1985. "Soil pH buffering revisited," *Soil Sci. Soc. Amer. J.*, **49**:145–148.

Mehlich, A. 1964. "Influence of adsorbed hydroxyl and sulfate on neutralization of soil acidity," *Soil Sci. Soc. Amer. Proc.*, **28**:492–496.

Noble, A. D., M. E. Sumner, and A. K. Alva. 1988. "The pH dependency of aluminum phyto-toxicity alleviation by calcium sulfate," *Soil Sci. Soc. Amer. J.*, **52**:1398–1402.

Robarge, W. P., and D. W. Johnson. 1992. "The effects of acidic deposition on forested soils," *Advances in Agronomy,* **47**:1–84.

Robertson, W. K., M. C. Lutrick, and T. L. Yuan. 1982. "Heavy applications of liquid-digested sludge on three Ultisols: I. effects on soil chemistry," *J. Environ. Qual.*, **11**:278–282.

Ronse, A., L. De Temmerman, M. Guns, and R. De Borger. 1988. "Evaluation of acidity, organic matter content and CEC in uncultivated soils of Northern Belgium during the past 25 years," *Soil Sci.*, **146**:453–460.

Schollenberger, C. J., and R. M. Salter. 1943. "A chart for evaluating agricultural limestone," *J. Amer. Soc. Agron.*, **35**:995–966.

Shoemaker, H. E., E. O. McLean and P. F. Pratt. 1961. "Buffer methods for determining lime requirements of soils with appreciable amounts of extractable aluminum," *Soil Sci. Soc. Amer. Proc.*, **25**:274–277.

Smyth, T. J., and M. S. Cravo. 1992. "Aluminum and calcium constraints to continuous crop production in a Brazilian Amazon Oxisol," *Agron. J.*, **84**:843–850.

Springett, J. A. 1983. "Effect of five species of earthworms on some soil properties," *J. Applied Ecol.*, **20**:865–872.

Sumner, M. E. 1993. "Gypsum and acid soils: The world scene," *Advances in Agronomy,* **51**:1–32.

Tran, T. S., and W. van Lierop. 1981. "Evaluation and improvement of buffer-pH lime requirement methods," *Soil Sci.*, **131**:178–188.

U.S./Canada Work Group No. 2. 1982. *Atmospheric Science and Analysis.* Final report; H. L. Ferguson and L. Machita (Co-chairmen). (Washington, D.C.: U.S. Environmental Protection Agency).

10

ALKALINE AND SALT-AFFECTED SOILS AND THEIR MANAGEMENT

Alkali has accumulated on them to such an extent that they are mere bogs and swamps and alkali flats, and the once fertile lands are thrown out as ruined and abandoned tracts.
—MILTON WHITNEY, FIRST CHIEF OF THE DIVISION OF SOILS, USDA, BULLETIN NO. 14, 1898

Alkaline and salt-affected soils are found on more than half of the Earth's arable lands. They dominate most arid and semiarid regions of the world. Extensive areas are used for rangelands or dryland farming, but a significant portion is used for irrigated agriculture. Irrigation water from either reservoir impoundments or groundwater pumping systems can transform these dry areas into some of the world's most productive farmlands. In the United States, the 15% of cropland that is irrigated produces 40% of the total crop value.

Most alkaline and saline soils, however, are not used for agriculture. Their native vegetation provides a wide variety of wild plants, which, along with their native animals, contribute greatly to biological diversity.

The soils of arid and semiarid areas are mostly alkaline because water from rain or snow is insufficient to leach the base-forming cations (Ca^{2+}, Mg^{2+}, K^+, Na^+, etc.) that are slowly released as the rocks and minerals weather. As a result, the percentage base saturation is high and pH values are typically 7.0 or above. In some areas the leaching is insufficient to remove even soluble salts, such as NaCl, $CaCl_2$, $MgCl_2$, and KCl, leading to *saline,* as well as alkaline, conditions.

Worldwide, about one-third of the irrigated lands have salt problems. Irrigation water transports salts from upstream watershed areas to the cultivated fields. If the irrigation systems do not provide good internal drainage, salts, particularly those containing sodium, can accumulate to levels that engender chemical and physical problems that can render a soil virtually useless as a habitat for plants.

Alkaline and saline soils have also played a unique role in world history, as the rise and fall of several ancient civilizations were tied to their irrigation and subsequent mismanagement. By discussing some of the mistakes that led to the downfall of these civilizations, it is hoped that we can learn to avoid such errors in the future.

10.1   SOURCES OF ALKALINITY

We learned in Chapter 9 that the replacement of Al^{3+} and H^+ from the exchange complex and the soil solution by base-forming cations (e.g., Ca^{2+}, Mg^{2+}, and Na^+) stimulates soil alkalinity. In soils of low-rainfall areas, the carbonate and bicarbonate anions reinforce the tendency toward alkalinity.

The Role of Carbonates and Bicarbonates

The primary source of carbonates and bicarbonates in soils is carbonic acid (H_2CO_3) that forms when CO_2 from microbial and root respiration reacts with water.

$$CO_2 + H_2O \rightleftharpoons H_2CO_3 \rightleftharpoons H^+ \rightleftharpoons HCO_3^-$$

Because of this reaction, a soil solution that is in equilibrium with the CO_2 in the soil atmosphere has a pH of about 4.6. In soils with higher pH levels, the abundant OH^- ions react with the H_2CO_3 to form first the bicarbonate (HCO_3^-) and then the carbonate (CO_3^{2-}) ion.

$$H_2CO_3 + OH^- \rightleftharpoons HCO_3^- + H_2O$$
$$HCO_3^- + OH^- \rightleftharpoons CO_3^{2-} + H_2O$$

A bicarbonate-dominated soil solution in equilibrium with atmospheric CO_2 will have a pH of about 8.3, while pH values of 10 or more occur in solutions dominated by soluble carbonates.

Note that if the concentration of HCO_3^- or CO_3^{2-} ions is increased, the above reactions shift to the left, forming more OH^- ions and increasing the pH. This is confirmed by Figure 10.1, which shows the relationship between the concentrations of HCO_3^- and CO_3^{2-} ions and those of H^+ and OH^- ions. Note that an increase of either HCO_3^- or CO_3^{2-} ions results in an increase of OH^- ions and a decrease of H^+ ions. The negative logarithm of the H^+ ion activity (or the pH) increases from about 4 to 10 as the above reactions take place in soils. From this figure, we can see the influence of carbonates and bicarbonates on soil pH in alkaline soils. We also see how these two anions exert a buffering effect on soil pH.

Role of the Cations (Na^+ versus Ca^{2+})

The particular cation associated with carbonate anions also influences the pH level. For example, if adsorbed Na^+ is prominent on the colloidal complex and in the soil solution, and if ample HCO_3^- and CO_3^{2-} are present, $NaHCO_3$ and Na_2CO_3 will form. Both these salts are water soluble and highly ionized, thus assuring continued high levels of HCO_3^- and CO_3^{2-} ions.

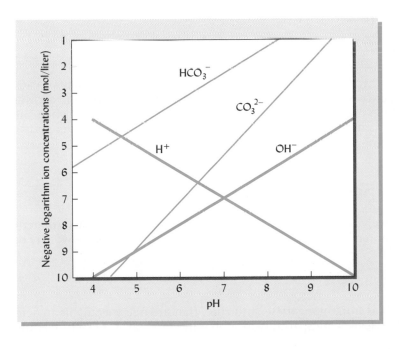

FIGURE 10.1 Effect of carbonates (HCO_3^- and CO_3^{2-}) on the concentrations of H^+ and OH^- ions expressed in negative logarithms. (The H^+ line represents increase in pH ($-\log H^+$ ions) from about 4 to nearly 10). This diagram shows that the levels of HCO_3^- and CO_3^2 ions in the soil solution plays a major role in determining the pH of soils. [Modified from Rowell (1988)]

$$NaHCO_3 \rightleftharpoons Na^+ + HCO_3^-$$

$$Na_2CO_3 \rightleftharpoons 2Na^+ + CO_3^{2-}$$

The high concentration of carbonate ions can produce pH values as high as 10 or more (Figure 10.1), making the soil almost uninhabitable for most plants.

Fortunately, Ca^{2+} is the dominant cation in most alkaline soils. This cation reacts with carbonic acid to form $Ca(HCO_3)_2$ and $CaCO_3$. The $Ca(HCO_3)_2$ is quite soluble in water and ionizes much as does $NaHCO_3$.

$$Ca(HCO_3)_2 \rightleftharpoons Ca^{2+} + 2HCO_3^-$$

The HCO_3^- level remains high, thereby ensuring the high OH^- ion concentration and high pH levels. In contrast, $CaCO_3$ is not very soluble in water and does not ionize significantly to provide many CO_3^{2-} ions.

$$\underset{\text{(insoluble)}}{CaCO_3} \rightleftharpoons Ca^{2+} + CO_3^{2-}$$

Note that the equilibrium point for this reaction is decidedly to the left, assuring that the concentration of CO_3^{2-} remains very low. As a result, $CaCO_3$ does not stimulate pH values higher than those attainable by $Ca(HCO_3)_2$. This is fortunate, since a wide variety of alkaline soils contain significant quantities of insoluble $CaCO_3$ laid down as the soils were formed. The pH of these $CaCO_3$-laden (*calcareous*) horizons is commonly no higher than 7.5 to 8, a level that can be tolerated by many plant species.

Influences of Salts

Neutral salts, such as NaCl, KCl, $CaCl_2$, and $MgCl_2$, can moderate the effects of adsorbed cations and carbonates on the pH of some alkaline soils. For example, high levels of NaCl constrain the ionization of $NaHCO_3$.

$$NaHCO_3 \rightleftharpoons Na^+ + HCO_3^-$$

The high Na^+ ion concentration coming from the NaCl drives the reaction to the left. The result is a lower HCO_3^- concentration and associated lowering of OH^- ion concentration and soil pH values.

Soluble salts also influence the OH^- ion concentration arising from high Na^+ adsorption on the colloidal complex.

$$\boxed{\text{Micelle}}\ Na^+ + H_2O \rightleftharpoons \boxed{\text{Micelle}}\ H^+ + Na^+ + OH^-$$

This reaction likewise is driven to the left in the presence of soluble salts such as NaCl or Na_2SO_4. The consequence is a reduction in the concentration of OH^- ions and a lower pH. Soluble salts also help maintain a desirable soil physical condition even when the degree of Na^+ saturation is high. We will consider salt-affected soils after a brief review of the properties of alkaline soils that have low salt content.

10.2 NONSALINE ALKALINE SOILS OF DRY AREAS

In addition to the relatively low levels of H^+ ions and high levels of OH^- ions inherent in alkaline soils (see Chapter 9), several other characteristics of these soils are worthy of mention.

Nutrient Deficiencies

Nutrient elements, such as iron, manganese, and zinc, that are freely available in acid soils (sometimes at toxic levels) are sparingly available in many alkaline soils, leading to deficiencies of these elements in plants. Unfortunately, adding these macronutrients in residues and fertilizers is sometimes ineffective because the elements are quickly tied up in insoluble forms in high-pH soils. Special protective organic complexes called *chelates* can help meet the nutrient needs of plants growing on these soils (see Section 15.6). Also, for the production of high-valued plants, micronutrients are often sprayed directly on the foliage to avoid interaction between the micronutrients and the high-pH soils.

In alkaline soils both native and applied phosphorus is tied up in highly insoluble calcium and magnesium phosphates, rendering the added phosphorus only sparingly available for plant uptake. Likewise, boron availability decreases at high pH levels, because above pH 7 the boron tends to be tightly adsorbed by the soil colloids. In contrast, molybdenum availability is high under alkaline conditions—so high that in some areas molybdenum toxicity is a problem both to plants and to grazing cattle. These nutrient-by-nutrient interactions must be carefully considered in managing alkaline soils.

Micronutrient deficiencies in alkaline soils are especially troublesome for home landscaping and gardening enthusiasts (Figure 10.2). Unfortunately, some of the most popular ornamental plants native to humid regions are not well adapted to alkaline soils, especially if the soils are calcareous in the upper horizons. Gardeners must either select locally adapted varieties of plants that are tolerant of alkalinity or add acid-forming chemicals, such as sulfur, to change the soil pH to suit the plant.

Cation Exchange Capacity

The cation exchange capacities (CECs) of alkaline soils are commonly higher than those of acid soils with comparable soil textures. This is because of two primary factors: (1) the high CEC associated with the constant charges on 2:1-type clays that are most common in alkaline soils, and (2) the even higher CEC resulting from the pH-dependent charges present on the humus colloids at these high pH levels.

Calcareous Layers

Another common characteristic of many well-developed soils of low-rainfall areas is the accumulation of calcium carbonate at some depth in the soil profile (Figure 10.3 and Plate 13). The high carbonate concentrations in these *calcareous* horizons inhibit root growth for some plants. Where carbonate concentrations are found at or near the soil surface, which is often the case in areas of very low rainfall, serious micronutrient deficiencies can result for plants that are not adapted to these conditions (Plate 29). The principal soil orders and suborders of low-rainfall-region soils are described in Sections 3.10 to 3.13.

Soil Water Supply

Unlike their counterparts in humid regions, the subsoil layers of alkaline soils of low-rainfall areas are generally dry, except where the groundwater table is near the surface. Limited moisture in the upper soil horizons and competition for water among plants are principal factors determining the nature and productivity of the native vegetation, as well as the capacity of the land to support animals (see Section 2.13).

Water management must receive high priority in cultivation of arid- and semiarid-region soils. Tillage practices that increase water infiltration during periods of rain or snowfall should be encouraged (see Figure 6.7). Likewise, it is often wise to use tillage practices such as stubble mulch (see Section 6.6) that help reduce evaporative losses and soil erosion by keeping some vegetative cover at or near the soil surface.

Irrigated soils of dry areas have moisture relations during periods of plant growth that are not too different from those in humid regions. However, moisture losses by evaporation in the dry areas are very high, and salt buildup in irrigated soils must be monitored to ensure successful plant production. The characteristics and management of salt-affected soils will be covered in the sections that follow.

FIGURE 10.2 Ornamental oak (upper, with dying branches in foreground) that thrives well in mildly acid soils of humid regions grows poorly on this alkaline soil in Colorado. The nutrient deficiency (probably of iron) is illustrated by the light green color between the veins of the leaves (lower). A soil test verifies the fact that the soil is alkaline. Gardeners must either select species known to tolerate alkaline conditions, acidify the soil, or supply the micronutrients that the alkaline-sensitive plants need. (Photos courtesy of R. Weil)

10.3 DEVELOPMENT OF SALT-AFFECTED SOILS[1]

Salt-affected soils are widely distributed throughout the world (Table 10.1), the largest areas being found in Australia, Africa, Latin America, and the Near and Middle East. They are most often found in areas with precipitation-to-evaporation ratios of 0.75 or less, and in low, flat areas with high water tables that may be subject to seepage from higher elevations. Figure 10.4 illustrates the distribution of these soils in the contiguous United States. Nearly 50 million ha of cropland and pasture are currently affected by salinity and in some regions the area of land so affected is growing by about 10% annually.

Natural Salt Accumulation

Salts accumulate naturally in some surface soils of arid and semiarid regions because there is insufficient rainfall to flush them from the upper soil layers. In the United States, about one-third of the soils in these regions suffer from some degree of salinity. The salts are primarily chlorides and sulfates of calcium, magnesium, sodium, and potassium. They may be formed during the weathering of rocks and minerals or brought to the soils through rainfall and irrigation.

---

[1] See Bresler, et al. (1982) for a discussion of these soils.

FIGURE 10.3   Calcium carbonate accumulation in the lower part of the B horizon (Bk) and the upper part of the C horizon (Ck) (the knife is inserted at the upper boundary) characterizes this Ustoll and many other soils of arid and semiarid regions. This calcareous layer helps maintain high pH levels and constrains the availability of micronutrients such as iron, manganese, and zinc. (Photo courtesy of R. Weil)

Other localized but important sources are fossil deposits of salts laid down during geological time in the bottom of now-extinct lakes or oceans or in underground saline water pools. These fossil salts can be dissolved in underground waters that move horizontally over underlying impervious geological layers and ultimately rise to the surface of the soil in the low-lying parts of the landscape, often forming **saline seeps.** The water then evaporates, leaving salts in place at or near the soil's surface and creating a **saline** soil.

Figure 10.5 illustrates how salts can accumulate in lower elevations from salts that originate up slope from either the natural soils or fossil deposits. Unfortunately, high levels of these salts cannot be tolerated by most plants, a fact that severely limits the use of some salt-affected soils.

The Ca^{2+} and Mg^{2+} ions dominate the exchange complex of most salt-affected soils. In some soils with greater than 15% Na^- saturation, the pH may rise above 8.5, and the stability of the soil aggregates may deteriorate. The soil colloids disperse and plug the soil's drainage pores, preventing the downward percolation of water. Such soils, known as *sodic* soils (see Section 10.5) are quite unproductive and are very difficult to manage.

TABLE 10.1   **Area of Salt-Affected Soils in Different Regions**

Region	Area, million ha
Africa	69.5
Near and Middle East	53.1
Asia and Far East	19.5
Latin America	59.4
Australia	84.7
North America	16.0
Europe	20.7
World total	322.9

From Beek, et al. (1980).

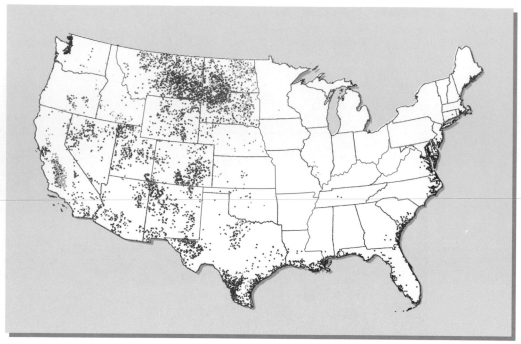

FIGURE 10.4 Distribution of salt-affected soils in the continental United States in 1992. Small areas are found near the ocean shore lines, but note the much larger areas located in vast regions of the West. Each dot represents about 4050 ha. [From the USDA Natural Resources Conservation Service National Resources Inventory, USDA (1996)]

Irrigation-Induced Salinity and Alkalinity

While in most saline areas salts have accumulated naturally, irrigation-induced salinity of the type that brought disaster to ancient cultures centuries ago has become increasingly important in recent years. Irrigation waters, whether from upstream watersheds or underground pumping, carry significant quantities of soluble salts. In areas where the land is well drained, sufficient irrigation water can percolate through the soil to leach the soluble salts from the upper soil layers, thereby preventing excessive salt accumulation. If the soil is not well drained, however, downward movement of the salts to the groundwater is impaired, and the salts are left in the soil and are later brought up to the surface as the irrigation water evaporates.

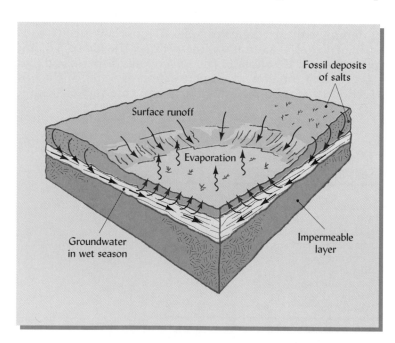

FIGURE 10.5 Field cut showing conditions that encourage salt buildup in soils. Mineral-containing waters from fossil deposits of salt or from soil weathering in upland areas move down the slope toward lower basins, either on the surface during torrential storms or as groundwater on top of impervious geological layers. When the water in the low-lying areas evaporates, it carries the salts up to or near the surface, where the salts accumulate.

If the irrigation water carries a significant quantity of Na^+ ions compared to Ca^{2+} and Mg^{2+} ions, and especially if the HCO_3^- ion is present, the colloidal complex can become significantly saturated with the Na^+ ion, and unproductive *sodic* soils are created.

During the past three decades, low-income countries in the dry regions of the world have greatly expanded their areas of irrigated lands in order to produce the food needed by their rapidly growing human populations. As a result, the proportion of arable land that is irrigated has increased dramatically, reaching about 45% in China, 25% in India, 72% in Pakistan, and 28% in Indonesia. Initially, phenomenal increases in food-crop production were stimulated by the expanded irrigation. However, in many project areas the need for good soil drainage was overlooked and the process of *salinization* has been accelerated. As a consequence, salts have accumulated to levels that are already adversely affecting crop production. In some areas, unproductive sodic soils have been created.

These events remind the world of the flat wastelands of southeastern Iraq that have been barren since the 12th century. In biblical times, the soils of these areas were fertile and productive. The Euphrates and Tigris rivers provided irrigation water that stimulated such high crop productivity that the overall region was called the Fertile Crescent. Unfortunately, in broad areas the soils were not well drained and the process of salinization proceeded. The salts accumulated to such a high level that crop production declined and the area had to be abandoned. Today, this same process is being repeated at different locations around the world, and some irrigated farming areas are having to be abandoned (see, for example, Figure 19.12). Truly, the world is justified in giving serious attention to salt-affected soils.

10.4 MEASURING SALINITY AND ALKALINITY

Plants are detrimentally affected by excess salts in some soils and by high levels of exchangeable sodium in others, the latter being detrimental to the soil both physically and chemically. Techniques have been developed to measure three primary properties that, along with pH, can be used to characterize salt-affected and sodic soils: (1) soil **salinity**, (2) **exchangeable sodium percentage** (ESP), and (3) the **sodium adsorption ratio** (SAR). Each will be discussed briefly.

Salinity[2]

Pure water is a poor conductor of electricity, but conductivity increases as more and more salt is dissolved in the water. Thus, the **electrical conductivity** (EC) of the soil solution gives us an indirect measurement of the salt content. The EC is measured by both laboratory and field methods (Table 10.2) and is expressed in terms of decisiemens per meter (dS/m).[3]

The **saturation paste extract** method is the most commonly used laboratory procedure. The soil sample is saturated with distilled water and mixed to a paste consistency. After standing overnight to dissolve the salts, the electrical conductivity of water extracted from the paste is measured (EC_e). A variant of this method involves the EC of the solution extracted from a 1:2 soil–water mixture after 0.5 hours of shaking (EC_w). The latter method takes less time but often is not as well related to the soil solution as is the saturation paste extract method.

---

[2] For an informative discussion of these methods, see Rhoades (1993).

[3] Formerly expressed as millimhos per centimeter (mmho/cm). Since 1 S = 1 mho, 1 dS/m = 1 mmho/cm.

TABLE 10.2 **Estimation of Soil Salinity by Four Different Methods**

Each method depends upon the electrical conductivity (EC) of the soil solution.

Symbol	Condition of measurement
Laboratory	
EC_e	Conductivity of solution extracted from a water-saturated soil paste
EC_w	Conductivity of solution extracted from a 1:2 soil–water mixture
Field	
EC_a	Apparent conductivity of bulk soil by inserting electrode sensors
E_a^*	Electromagnetic induction of an electrical current using surface transmitter and receiving coils

Field methods in use today involve the measurement of bulk soil conductivity that, in turn, is directly related to soil salinity (see Table 10.2). One such method involves the insertion into the soil of four electrode sensors (Figure 10.6) and the direct measurement of apparent EC in the field (EC_a). The sensors can be mounted behind a tractor and moved quickly from one site in the field to another. This technique is rapid, simple, and practical, and gives values that can be correlated with EC_e.

A second field method is even more rapid. It employs **electromagnetic induction** of electrical current in the body of the soil, the level of which is related to electrical conductivity and, in turn, to soil salinity. A magnetic field is generated within the soil by a small transmitter coil that is placed on the soil surface and is energized by an alternating current. This magnetic field, in turn, induces within the soil small electric currents whose values are related to the soil's conductivity. These small currents generate their own secondary magnetic fields, which can be measured by a small receiving cell nearby. The instrument that provides these emitting and receiving coils thus can measure ground EC (designated E_a^*) without the use of probes and is proving to be useful in surveying saline soils (see Section 19.3).

Since electrical conductivity is influenced by both salinity and soil water content, the field measurements should be made when the soil is near the field capacity. With irrigated soils, measurements are best made soon after the soil is irrigated to assure reasonably high soil moisture levels.

It should be noted that each of the methods discussed estimates soil salinity indirectly by measuring electrical conductivity. But the values of EC_e, EC_w, EC_a or EC_a^* obtained by these procedures will not be identical. For example, for highly saline soils EC_a values measured by the four-electrode-sensor method are about one-fifth of those measured using the standard saturated extract procedure EC_e. In any case, however, these values are sufficiently well correlated with each other so that calculations can be made of soil salinity expressed conventionally in terms of the standard EC_e (saturation paste extract method).

Sodium Status

Two expressions are used to characterize the sodium status of highly alkaline soils. The **exchangeable sodium percentage** (ESP) identifies the degree to which the exchange complex is saturated with sodium.

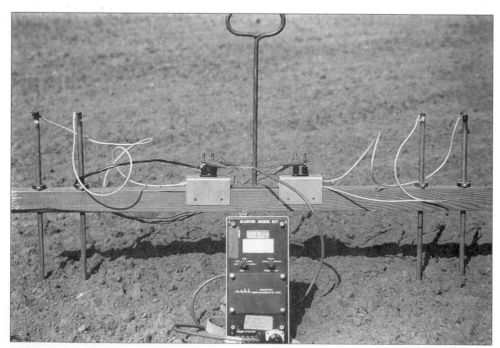

FIGURE 10.6   A four-electrode sensor used to measure electrical conductivity of bulk soil in the field. The electrodes are a fixed distance apart so they can be quickly inserted into the soil and the measurements made. A combination portable generator and resistance meter (foreground) makes this equipment very useful for field survey work. [From Rhoades (1993); used with permission of Academic Press]

$$ESP = \frac{\text{Exchangeable sodium, cmol}_c/\text{kg}}{\text{Cation exchange capacity, cmol}_c/\text{kg}} \times 100$$

ESP levels of 15 are associated with pH values of 8.5 and above. Higher levels may bring the pH to above 10.

The **sodium adsorption ratio** (SAR) is a second more easily measured property that is becoming even more widely used than ESP. The SAR gives information on the comparative concentrations of Na^+, Ca^{2+}, and Mg^{2+} in soil solutions. It is calculated as follows:

$$SAR = \frac{[Na^+]}{\sqrt{\frac{1}{2}([Ca^{2+}] + [Mg^{2+}])}}$$

where $[Na^+]$, $[Ca^{2+}]$, and $[Mg^{2+}]$ are the concentrations in $mmol_c/L$ of the sodium, calcium, and magnesium ions in the soil solution. The SAR of a soil extract takes into consideration that the adverse effect of sodium is moderated by the presence of calcium and magnesium ions. The SAR also is used to characterize the irrigation water added to these soils.

10.5   CLASSES OF SALT-AFFECTED SOILS

Using EC, ESP, SAR characteristics, and soil pH, salt-affected soils are classified as **saline**, **saline-sodic** and **sodic** (Figure 10.7). These classes will be considered in order.

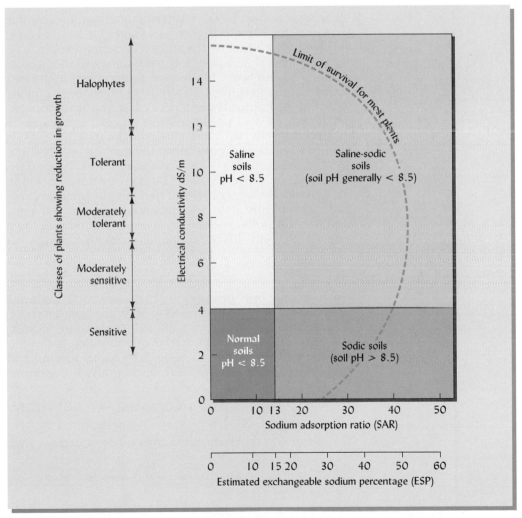

FIGURE 10.7   Diagram illustrating the classification of normal, saline, saline-sodic, and sodic soils in relation to soil pH, electrical conductivity, sodium adsorption ratio (SAR), and exchangeable sodium percentage (ESP). Also shown are the ranges for different degrees of sensitivity of plants to salinity.

The processes that result in the accumulation of neutral soluble salts are referred to as **salinization.** The salts are mainly chlorides and sulfates of sodium, calcium, magnesium, and potassium. Saline soils contain a concentration of these salts sufficient to interfere with the growth of many plants. The electrical conductivity EC_e of a saturation extract of the soil solution is more than 4 dS/m.[4] Salts are commonly brought to the soil surface by evaporating waters, creating a white crust, which accounts for the name *white alkali* that is sometimes used to designate these soils (see Figure 10.8).

The exchange complex of saline soils is dominated by calcium and magnesium. As a consequence, the exchangeable sodium percentage (ESP) is less than about 15, and the pH usually is less than 8.5.

The Na^+ ion concentration in the soil solution of saline soils may be somewhat higher than that of Ca^{2+} or Mg^{2+} due to the presence of soluble salts that are commonly high in sodium. However, because of the greater affinity of the soil colloids for divalent cations, such as Ca^{2+} or Mg^{2+}, the sodium adsorption ratio (SAR) of saline soils is less than 13.

Plant growth on saline soils is not generally constrained by poor soil physical conditions. The soluble salts help prevent dispersion of the soil colloids that otherwise would be encouraged by high sodium levels (see Section 9.3). Aggregate stability and good aeration result.

Saline-Sodic Soils

Saline-sodic soils have characteristics intermediate between those of saline and sodic soils (see Figure 10.7). Like saline soils, they contain appreciable levels of neutral soluble salts, as shown by EC_e levels of more than 4 dS/m. But they have higher ESP levels (greater than 15) and higher SAR values (at least 13). Crop growth can be adversely affected by both excess salts and excess sodium levels.

The physical and chemical conditions of saline-sodic soils are similar to those of saline soils. This is due to the moderating effects of the neutral salts. These salts provide excess cations that move closely to the negatively charged colloidal particles, thereby reducing their tendency to repel each other, or to disperse. The salts help keep the colloidal particles associated with each other in aggregates.

Unfortunately, this situation is subject to rather rapid change if the soluble salts are leached from the soil, especially if the leaching waters are high in Na^+ ions—that is, if the SAR of the water is high. In such a case, the exchangeable sodium level will increase, as will the pH (to above 8.5). Since Na^+ ions are not attracted very closely to the soil colloids, the effective electronegativity of the colloidal particles causes them to repel each other or to disperse (Figure 10.9), thereby breaking up the soil aggregates. The dispersed colloids clog the soil pores as they move down the profile. Water infiltration is thus greatly reduced (Figure 10.10) and the puddled condition, so characteristic of sodic soils, pertains.

Sodic Soils

The preceding brief description of how sodic soils form suggests that they are the most troublesome of the salt-affected soils. While their levels of neutral soluble salts are low (EC_e less than 4.0 dS/m), their ESP values are above 15 and their SARs are above 13, suggesting a comparatively high level of sodium on the exchange complex. The pH values of sodic soils exceed 8.5, rising to 10 or higher in some soils.

The high pH is largely due to the hydrolysis of sodium carbonate.

$$2Na^+ + CO_3^{2+} + H_2O \rightleftharpoons 2Na^+ + HCO_3^- + OH^-$$

The sodium on the exchange complex also undergoes hydrolysis.

$$Na^+ \boxed{Micelle} + H_2O \rightleftharpoons H^+ \boxed{Micelle} + Na^+ + OH^-$$

---

[4] Some salt-sensitive fruit and vegetable crops are adversely affected when the EC is lower than 4 dS/m, leading some scientists to recommend that 2 dS/m be the level above which a soil could be classed as saline.

FIGURE 10.8 (Lower) White alkali spot in a field of alfalfa under irrigation. Because of upward capillarity and evaporation, salts have been brought to the surface where they have accumulated. (Upper) White crust on a saline soil from Colorado. The white salts are in contrast with the darker-colored soil underneath. (Scale is shown in inches and centimeters.) [Lower photo courtesy USDA Agricultural Research Service]

Few plants can tolerate these conditions. Plant growth on these soils is constrained by toxities of Na^+, OH^-, and HCO_3^- ions, as well as by their very poor soil physical conditions and slow permeability to water.

Because of the extreme alkalinity resulting from the high sodium content, the surface of sodic soils often is discolored by the dispersed humus that can be carried upward by the capillary water and deposited when it evaporates. Hence, the name *black alkali* has been used to describe these soils. Sometimes located in small areas called *slick spots*, sodic soils may be surrounded by soils that are relatively productive.

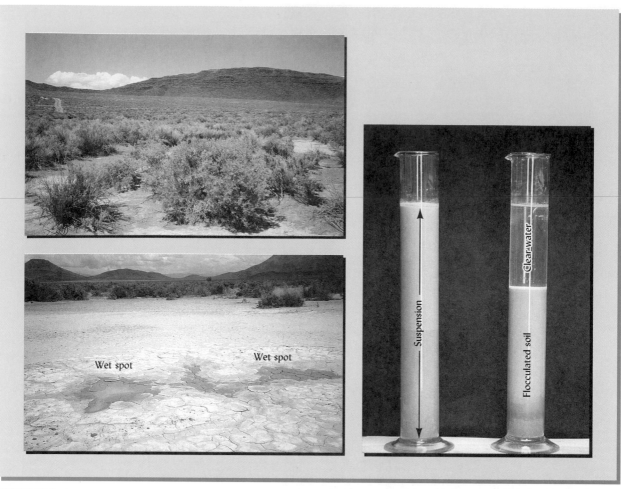

FIGURE 10.9   Illustration of the dispersion of colloids in a sodic soil from southern Colorado. The sodic soil (lower left) is bare of vegetation compared to the adjacent alkaline but nonsodic soil (upper left). Because of the complete dispersion of the soil colloids, water moves through the sodic soil very slowly, as the wet spot in the center of the lower left photograph suggests. To verify the dispersion of the clay, samples of the sodic soil were placed in cylinders and shaken thoroughly with water. The clay was highly dispersed, no observable settling occurring over a period of three weeks (left cylinder). To stimulate the flocculation (joining together) of the clay particles, a teaspoon of table salt (NaCl) was added to the cylinder on the right, and the mixture was again thoroughly shaken with water. Flocculation began to occur and within 24 hours was quite noticeable (right cylinder). The salt stimulated the flocculation, which in the soil could be the first step toward aggregation, improved drainage, and good aeration. This demonstration illustrates the deplorable physical condition of sodic soils. It also shows how the salts in saline-sodic soils can prevent the undesirable dispersion of the soil colloids that is responsible for much of the ill effects of sodic soils.

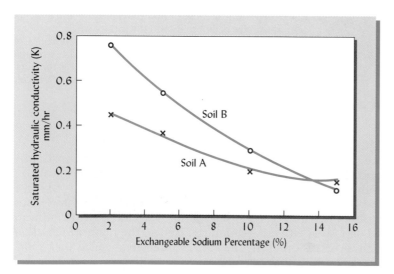

FIGURE 10.10   Effect of increasing exchangeable sodium percentage (ESP) on the saturated hydraulic conductivity of two soils in Italy. Note the steady decline in conductivity with increased ESP and the very low conductivity at ESP = 15%. [Redrawn from Crescimanno, et al. (1995)]

10.6 GROWTH OF PLANTS ON SALINE AND SODIC SOILS

Plants respond to the different salt-affected soils in different ways. High soluble salt concentrations, through their effects on osmotic potentials (see Section 5.3), reduce plant growth on saline and saline-sodic soils. Root cells, as they come in contact with a soil solution that is high in salts, will lose water by osmosis to the more concentrated soil solution. The cell then collapses. The kind of salt, the plant species, and the rate of salinization are factors that determine the concentration at which the cell succumbs. Very limited air and water movement in some saline-sodic soils also may be a factor in determining which plants can grow on these soils.

Sodic soils harm plants in five ways: (1) the caustic influence of the high pH induced by the sodium carbonate and bicarbonate, (2) the toxicity of the bicarbonate and other anions, (3) the adverse effects of the active sodium ions on plant metabolism and nutrition, (4) the low micronutrient availability due to high pH, and (5) oxygen deficiency due to the breakdown of soil structure.

10.7 SELECTIVE TOLERANCE OF HIGHER PLANTS TO SALINE AND SODIC SOILS

Satisfactory plant growth on salty soils depends on a number of interrelated factors, including the physiological constitution of the plant, its stage of growth, and its rooting habits. It is interesting to note that old alfalfa plants are more tolerant to salt-affected soils than young ones, and that deep-rooted legumes show a greater resistance to such soils than those that are shallow-rooted.

Soil properties—including the nature of the various salts, their proportionate amounts, their total concentration, and their distribution in the solum—must be considered. The structure of the soil and its drainage and aeration are also important.

Plant Sensitivity

While it is difficult to forecast precisely the tolerance of crops to salty soils, numerous tests have made it possible to classify many domestic plants into four general groups based on their salinity tolerance. Figure 10.11 shows the relative productivity of these general groups influenced by soil salinity as measured by electrical conductivity EC_e. Table 10.3 lists many plants according to this classification. Note that trees, shrubs, fruits, vegetables, and field crops are included in the different categories.

To this list of domestic species should be added two other potential sources of plants to grow on salty soils: (1) wild *halophytes* (salt-loving plants), and (2) salt-tolerant cultivars developed by plant breeders. A number of wild halophytes have been found that are quite tolerant to salts and that possess qualities that could make them useful for human and/or animal consumption.[5] Even though they may accumulate high levels of

---

[5] For a review of such halophytes, see NAS (1990).

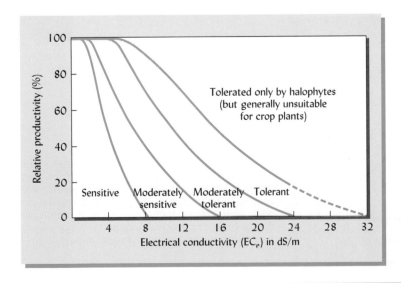

FIGURE 10.11 Relative productivity of four groups of plants classified by their sensitivity to salinity as measured by electrical conductivity. The plant groupings are shown in Table 10.3. [Generalized from data of Carter (1981)]

TABLE 10.3 Relative Tolerance of Certain Plants to Salty Soils

Tolerant	Moderately tolerant	Moderately sensitive	Sensitive
Barley (grain)	Ash (white)	Alfalfa	Almond
Bermuda grass	Aspen	Arborvitae	Apple
Black cherry	Barley (forage)	Boxwood	Apricot
Bougainvillea	Birch (black)	Broad bean	Azalea
Cedar (red)	Beet (garden)	Cauliflower	Beech
Cotton	Broccoli	Cabbage	Bean
Date	Brome grass	Celery	Birch
Elm	Catlaw acacia	Clover (alsike,	Blackberry
Locust	Cow pea	ladino, red,	Boysenberry
Natal plum	Fescue (tall)	strawberry,	Burford holly
Nutall alkali grass	Fig	and berseem)	Carrot
Oak (red and white)	Harding grass	Corn	Celery
Olive	Honeysuckle	Cucumber	Dogwood
Prostrate kochia	Hydrangea	Dallas grass	Elm (American)
Rescue grass	Juniper	Grape	Grapefruit
Rosemary	Kale	Hickory (shagbark)	Hemlock
Rugosa	Locust (black)	Juniper	Hibiscus
Salt grass	Mandarin	Lettuce	Larch
Tamarix	Orchard grass	Maple (red)	Lemon
Wheat grass (crested)	Oats	Pea	Linden
Wheat grass (fairway)	Pomegranate	Peanut	Maple (sugar)
Wheat grass (tall)	Privet	Radish	Onion
Wild rye (altai)	Rye (hay)	Rice (paddy)	Orange
Wild rye (Russian)	Ryegrass (perennial)	Squash	Peach
Willow	Safflower	Sugar cane	Pear
	Sorghum	Sweet clover	Pine (red and white)
	Soybean	Sweet potato	Pineapple
	Squash (zucchini)	Timothy	Plum (prune)
	Sudan grass	Turnip	Potato
	Trefoil (birdsfoot)	Vetch	Raspberry
	Wheat	Viburnum	Rose
	Wheat grass (western)		Star jasmine
			Strawberry
			Tomato

salt in their stems and leaves, the seeds are commonly not too different from domesticated crops in their content of protein and fat, for example.

Plant breeders have been able to develop new strains with salt tolerance greater than that possessed by conventional varieties. An example is shown by the data in Figure 10.12, which illustrates superior performance by improved strains of barley. Plant selection and improvement should be one of the goals of the future to maintain or even increase food production on salt-affected soils. However, improved plant tolerance must not be viewed as a substitute for proper salinity control.

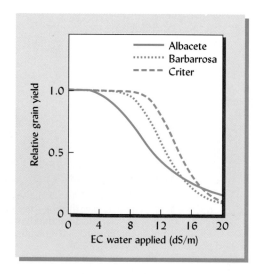

FIGURE 10.12 Three barley varieties differ significantly in their tolerance to the salinity of water applied to the soil in which they were grown. Such genetic differences can be useful in breeding varieties with greater tolerance of saline conditions. [Redrawn from Royo and Aragues (1993); used with permission of the American Society of Agronomy]

Salt Problems Not Related to Arid Climates

In cities where deicing salts are used in abundance during winter months, these salts may impact roadside soils and plants. Repeated application of deicing salts can result in salinity levels sufficiently high as to adversely affect plants growing alongside a treated highway or sidewalk. Figure 10.13 illustrates tree damage from such treatment. To avoid the specific chemical and physical problems associated with sodium salts, many municipalities have switched from NaCl to KCl for deicing purposes. Unlike soil salinity in arid regions, such salt contamination is usually temporary as the abundant rainfall will leach out the salts in a matter of weeks or months.

Salinity can also be a problem for potted plants, particularly perennials that remain in the same pot for long periods of time. Salts in the water, as well as those applied in fertilizers, can build up if care is not taken to flush them out occasionally.

10.8   MANAGEMENT OF SALINE AND SODIC SOILS

The first requisite for wise management of salt-affected soils is to know something about the amount and nature of soluble salts being added to and removed from the soil. In irrigated areas, this means knowing the quality of the irrigation water and the status of the soil drainage.

Water Quality Considerations

The chemical quality of water added to salt-affected soils is a prime management tool. For example, if the water is too low in salts, such as rainwater, it can hasten the change from a saline-sodic soil to a sodic soil. The rainwater not only leaches soluble salts from the upper few centimeters of soil, but the impact of the raindrops encourages dispersion of the soil colloids, an initial step in the process of sodic soil development.

If the salt and SAR levels of irrigation water are high, the process of forming sodic soils will accelerate. Also, the presence of bicarbonates in the irrigation water can reduce concentrations of Ca^{2+} and Mg^{2+} in the soil solution by precipitating these ions as insoluble carbonates. Such chemical characteristics of the irrigation water can increase the SAR of the soil solution and move the soil toward the sodic class.

FIGURE 10.13   During the winter months in some temperate region areas, sodium chloride and other salts are used to melt the snow and ice from the streets and sidewalks. Heavy applications of these chemicals can result in salinity damage to plants similar to that experienced on saline soils of arid regions. (Left) Localized area of salt concentration resulting from the piling of salt-treated snow alongside a street. (Right) Leaf damage resulting from the excess salt uptake by the plants. (Photos courtesy of R. Weil)

The quality and disposition of *waste irrigation* waters must also be carefully monitored because of the potential harm to downstream users and habitats. Drainage water from one irrigation system that may be high in salt concentration or in the concentration of one or more specific elements may reenter a stream and be used again at lower elevations. At some locations in the western United States, for example, toxic levels of such trace elements as selenium,[6] molybdenum, and boron have accumulated in downstream wetlands or evaporative ponds. Box 10.1 illustrates how such concentrations can occur. Table 10.4 shows that plants growing on these affected areas can accumulate levels of these elements that are unsafe for livestock and/or wildlife. Indeed, selenium toxicity is known to be damaging to waterfowl in parts of California.

---

[6] For a recent review of selenium in irrigation water of the western United States, see Nolan and Clark (1997).

BOX 10.1   SELENIUM IN IRRIGATED SOILS

Selenium in streams from watersheds high in selenium

Fields of irrigation district A

Fields of irrigation district B

Holding pond

The use of selenium-laden irrigation waters can result in the buildup of selenium to toxic levels in soils and in irrigation wastewaters coming from these soils. Selenium accumulates in plants growing on these soils and in nearby wetland areas where irrigation wastewaters are disposed.

In some areas of the western United States are found certain marine sediments and shales of the Cretaceous geological period that are high in selenium. Under alkaline and well-drained conditions, the selenate (SeO_4^{2-}) form of Se prevails. Much like sulfates, selenates are reasonably soluble in water and are readily available for plant uptake. They move from watersheds that contain high selenium into streams and rivers from which water is diverted for irrigation.

The drawing illustrates how reuse of high-selenium irrigation waters can cause serious environmental problems. Streams from high-selenium watersheds provide water for irrigation district A. The selenium level in the water (shown by the width and darkness of the arrow) is high but not sufficiently so as to cause environmental problems.

As evapotranspiration of the irrigation water takes place, selenium accumulates in the soil and in the waste and drainage water coming from the soil. This drainage water moves (or is pumped) back into a stream that later supplies water for irrigation district B downstream. The process is repeated, Se in the soil is increased, and the levels of Se in the drainage water is increased (darker color), being sufficiently high as to make the water unusable for further irrigation.

The water is then diverted or pumped into a nearby wetland area or a shallow holding pond. In either case, much of the water evaporates, leaving the selenium in the water and soil of the holding area. Plants living in the wetland or pond area absorb the selenium, accumulating it to levels that are toxic to farm animals and wildlife. Deformities and deaths of migratory birds have been caused by such high levels of selenium. Similarly, livestock may suffer toxic effects by eating high-selenium forages.

Steps are being taken to reduce or eliminate the release of high-selenium waters into wetlands. Researchers are also seeking plants that are very high selenium accumulators that can remove the selenium from the holding area[a] and can then be destroyed. Other research has shown that some microorganisms can stimulate the formation of volatile organic selenium compounds, which can be slowly released into the atmosphere. The environmental effects of selenium remind us of the need to manage soil and water with attention to the impacts on all creatures.

[a] For example, see Banuelos, et al. (1997).

TABLE 10.4 Levels of Molybdenum and Selenium in Three Salinity-Tolerant Forage Species Grown on a Highly Salinized Soil That Had Received Large Quantities of Waste Irrigation Water in the San Joaquin Valley of California

The upper "safe" limit for animal consumption (potential toxicity level) is also shown.

	Potential toxicity level, mg/kg	Level in the forages, mg/kg		
Trace elements		Tall wheatgrass	Alkali sacation	Astragalus racemosus
Molybdenum	5	26	10	18
Selenium	5	12	9	670

Data selected from Retana, et al. (1993).

The soluble salt level of different water sources varies greatly. Data for a number of water sources, varying from high-quality irrigation water to that in the Pacific Ocean, is shown in Table 10.5. Note that the Colorado River has sufficiently high salt content to alert users to the need for wise irrigation management if soil quality is to be maintained. In fact, by the time the Colorado River reaches the United States/Mexico border it is so loaded with salts from upstream irrigation systems and from domestic and industrial uses that a huge desalinization plant has been constructed to enable the United States to meet its treaty obligations with Mexico. Up to now it has not been necessary to use the plant, since drainage water from the lowest U.S. irrigation system (Welton-Mohawk Valley) is being diverted to the Sea of Cortez through a canal that runs parallel to the Colorado River. This illustrates some environmental effects of irrigation that are being experienced in many parts of the world.

Management Objectives

The primary objective in managing salt-affected soils is to lessen the major constraints to plant production—excess soluble salts and exchangeable sodium. The removal of these constraints causes several chemical and physical changes that bring the soils back to a more productive state. Consequently, the overall processes are referred to as the **reclamation** of these soils. Practically, reclamation of salt-affected soils is largely dependent on the availability of ample supplies of good-quality irrigation water that can leach salts from the soil. In areas where irrigation water is not available, such as in saline seeps in the Northern Plains states, the leaching of salts is not practical. In these areas, deep-rooted vegetation may be used to lower the water table and reduce the upward movement of salts.

10.9 RECLAMATION OF SALINE SOILS

The removal of excess salts from saline soils requires access to ample irrigation water with low SAR ratios and an effective soil drainage system that quickly removes the salt-laden water once it leaches down through the soil. If the natural soil drainage is inadequate to accommodate the leaching water, an artificial drainage network must be installed. Intermittent applications of irrigation water may be required to effectively reduce the salt content to a desired level. The process can be monitored by measuring the soil's EC, using either the saturation extract procedure or one of the field instruments.

TABLE 10.5 The Comparative Salinity of Six Different Water Sources as Measured by Electrical Conductivity and Dissolved Solids Content

	Irrigation water quality				Negev (Middle East)	
Salinity measurement	Good	Marginal	Colorado River	Alamo River	groundwater	Pacific Ocean
Electrical conductivity, dS/m	0–1	1–3	1.3	4.0	4.0–7.0	46
Dissolved solids, ppm	0–500	500–1,500	850	3,000	3,000–4,500	35,000

Adapted from a number of sources by the U.S. National Academy of Sciences (NAS; 1990).

The amount of water needed to remove the excess salts from saline soils, called the *leaching requirement LR,* is determined by characteristics of the crop to be grown, the irrigation water, and the soil. An approximation of the *LR* is given for relatively uniform salinity conditions by the ratio of the salinity of the irrigation water (expressed as its EC_w) to the maximum permissible salinity of the soil solution for the crop to be grown (expressed as EC_{dw}, the EC of the drainage water).

$$LR = \frac{EC_w}{EC_{dw}}$$

The *LR* is water added in excess of the moisture needed to thoroughly wet the soil, and to meet evapotranspiration needs.

Note that if EC_w is high and the high salt sensitivity of the crop to be grown dictates a low EC_{dw}, a very large leaching requirement *LR* would result. An example of how the *LR* values are calculated is given in Box 10.2.

Attention must be paid to the disposal of the drainage water once the soil leaching has taken place (see Box 10.1). If this water cannot be returned to a river or stream without overburdening the stream with salts or other contaminants, then other means must be sought to dispose of the water. As mentioned earlier, care must be taken to minimize burdens on downstream users.

Tillage and planting practices can influence crop production on saline soils. Any tillage or surface residue management practice (such as conservation tillage) that reduces evaporation from the soil surface should reduce the upward transport of soluble salts. Likewise, specific irrigation schemes such as those utilizing sprinkling and trickle systems (see Section 6.11), as well as planting techniques for row crops, can reduce the salt concentration immediately around the young plant roots (Figure 10.14).

BOX 10.2 LEACHING REQUIREMENT *LR*

A farmer should know how much leaching water is required to prevent the buildup of salts in a soil or, if the salts are already high, to reduce their levels in the soil. The concept of *leaching requirement LR* has been developed to help farmers make this assessment.

The *LR* is the irrigation water needed (in excess of that required to saturate the soil) to sufficiently leach the soil so as to assure a proper salt balance for the crop being grown. It is approximated by the ratio of the electrical conductivities (EC_s of the incoming irrigation water EC_w and of the outgoing drainage water EC_{dw} that has an acceptable EC level for the crop being grown.

$$LR = \frac{EC_w}{EC_{dw}}$$

As an example, consider the situation where the EC of the irrigation water EC_w is 2.5 dS/m and that of the acceptable draining soil solution is 5.5 dS/m. Then,

$$LR = \frac{2.5 \text{ dS/m}}{5.5 \text{ dS/m}} = 0.45$$

If this ratio (0.45) is multiplied by the amount of water needed to completely saturate the soil—perhaps 8 cm of water—the water to be leached can be calculated as follows:

$$8 \text{ cm} \times 0.45 = 3.6 \text{ cm water}$$

This is the minimum amount of water that must be leached through a water-saturated soil to maintain proper salt balance. In some cases, additional leaching may be needed to reduce the excess concentration of specific elements such as boron.

The modern means of measuring bulk soil conductivity EC_a using the four-electrode probe or remote-sensing electromagnetic induction devices (see Section 10.3) can be used to readily monitor soil salinity changes resulting from leaching practices.

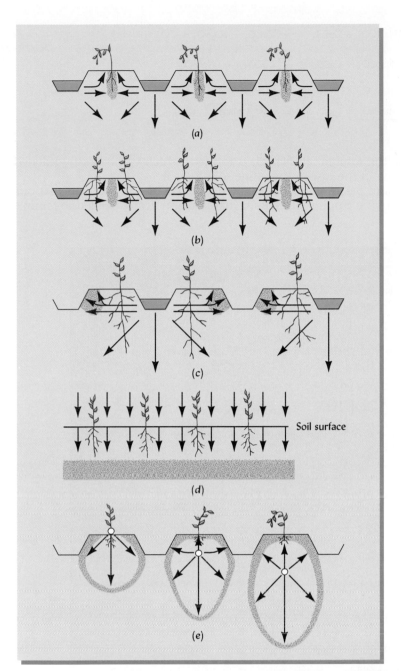

FIGURE 10.14 Effect of irrigation on salt movement and plant growth in saline soil. (*a*) With irrigation water applied to furrows on both sides of the row, salts move to the center and damage plant root systems. (*b*) Placing plants on edges of the bed rather than the center helps avoid the damage seen in (*a*). (*c*) Application of water to every other furrow and placement of plants on the side of the bed nearest the water also helps plants avoid the highest salt concentrations. Sprinkle irrigation (*d*) or uniform flood irrigation moves salts downward and the problem is alleviated. (*e*) Drip or trickle irrigation results in salt removal or accumulation, depending on the placement of the trickle tube. (Courtesy of Wesley M. Jarrell)

10.10 RECLAMATION OF SALINE-SODIC AND SODIC SOILS

Saline-sodic soils have some of the adverse properties of both saline and sodic soils. If attempts are made to leach out the soluble salts in saline-sodic soils, as was discussed for saline soils, the exchangeable Na^+ level as well as the pH would likely increase and the soil would take on adverse characteristics of sodic soils. Consequently, for both saline-sodic and sodic soils, attention must first be given to reducing the level of exchangeable Na^+ ions and then to the problem of excess soluble salts.

Gypsum

Removing Na^+ ions from the exchange complex is most effectively accomplished by replacing them with either the Ca^{2+} or the H^+ ion. Providing Ca^{2+} in the form of gypsum ($CaSO_4 \cdot 2H_2O$) is the most practical way to bring about this exchange. When gypsum is added, reactions such as the following take place:

$$2NaHCO_3 + CaSO_4 \longrightarrow CaCO_3 + Na_2SO_4 + CO_2 \uparrow + H_2O$$

$$Na_2CO_3 + CaSO_4 \rightleftharpoons CaCO_3 + Na_2SO_4$$
$$\text{(insoluble)} \quad \text{(leachable)}$$

$$\begin{matrix} Na^+ \\ Na^+ \end{matrix} \boxed{Micelle} + CaSO_4 \rightleftharpoons Ca^{2+} \boxed{Micelle} + Na_2SO_4$$
$$\text{(leachable)}$$

Note that in each case the soluble salt Na_2SO_4 is formed, which can be easily leached from the soil as was done in the case of the saline soils.

Several tons of gypsum per hectare are usually necessary. In Box 10.3, calculations are made to approximate the amount of gypsum that is theoretically needed to remove

---

BOX 10.3   CALCULATING THE THEORETICAL GYPSUM REQUIREMENT

PROBLEM

How much gypsum is needed to reclaim a sodic soil with an exchangeable sodium percentage (ESP) of 25% and a cation exchange capacity of 18 cmol$_c$/kg? Assume that you want to reduce the ESP of the upper 30 cm of soil to about 5% so that a crop like alfalfa could be grown.

SOLUTION

First, determine the amount of Na$^+$ ions to be replaced by multiplying the CEC (18) by the change in Na$^+$ saturation desired (25 − 5 = 20%).

$$18 \text{ cmol}_c/\text{kg} \times 0.20 = 3.6 \text{ cmol}_c/\text{kg}$$

From the reaction that occurs when the gypsum (CaSO$_4$ · 2H$_2$O) is applied,

$$\boxed{Micelle} \begin{matrix} Na^+ \\ Na^+ \end{matrix} + CaSO_4 \cdot 2H_2O \rightleftharpoons \boxed{Micelle} \; Ca^{2+} + Na_2SO_4 + 2H_2O$$

We know that the Na$^+$ is replaced by a *chemically equivalent* amount of Ca^{2+} in the gypsum (CaSO$_4$ · 2H$_2$O). In other words, 3.6 cmol$_c$ of CaSO$_4$ · 2H$_2$O will be needed to replace 3.6 cmol$_c$ of Na$^+$.

Second, calculate the weight in grams of gypsum needed to provide the 3.6 cmol$_c$/kg soil. This can be done by first dividing the molecular weight of CaSO$_4$ · 2H$_2$O (172) by 2 (since Ca^{2+} has two charges and Na$^+$ only one) and then by 100 since we are dealing with centimole$_c$ rather than mole$_c$.

$$\frac{172}{2} = 86 \text{ g CaSO}_4 \cdot 2H_2O/\text{mol}_c$$

$$\text{and} \; \frac{86}{100} = 0.86 \text{ g CaSO}_4 \cdot 2H_2O/\text{cmol}_c \text{ required to replace 1 cmol}_c \text{ Na}/100 \text{ g soil}$$

The 3.6 cmol$_c$Na$^+$/kg would require

$$3.6 \text{ cmol}_c/\text{kg} \times 0.86 \text{ g/cmol}_c = 3.1 \text{ g CaSO}_4 \cdot 2H_2O/\text{kg of soil}$$

Last, to express this in terms of the amount of gypsum needed to treat 1 ha of soil to a depth of 30 cm, multiply by 4×10^6, which is twice the weight in kg of a 15-cm-deep hectare–furrow slice (see Section 4.5).

$$3.1 \text{ g/kg} \times 4 \times 10^6 \text{ k/ha} = 12,400,000 \text{ g gypsum/ha}$$

This is 12,400 kg/ha, 12.4 Mg/ha, or about 5.5 tons/acre.

Because of impurities in the gypsum and the inefficiency of the overall process, these amounts would likely be adjusted upward by 20 to 30% in actual field practice.

---

an acceptable portion of the Na^+ ion from the exchange complex. The soil must be kept moist to hasten the reaction, and the gypsum should be thoroughly mixed into the surface by cultivation—not simply plowed under. The treatment must be supplemented later by a thorough leaching of the soil with irrigation water to leach out most of the sodium sulfate.

Sulfur and Sulfuric Acid

Elemental sulfur and sulfuric acid can be used to advantage on salty lands, especially where sodium bicarbonate abounds. The sulfur, upon biological oxidation (see Sections 9.6 and 13.21), yields sulfuric acid, which not only changes the sodium bicarbonate to the less harmful sodium sulfate but also decreases the alkalinity. The reactions of sulfuric acid with the compounds containing sodium may be shown as follows:

$$2NaHCO_3 + H_2SO_4 \longrightarrow 2CO_2 \uparrow + 2H_2O + Na_2SO_4$$

$$Na_2CO_3 + H_2SO_4 \longrightarrow CO_2 \uparrow + H_2O + \underset{\text{(leachable)}}{Na_2SO_4}$$

$$\begin{matrix} Na^+ \\ Na^+ \end{matrix} \boxed{Micelle} + H_2SO_4 \rightleftharpoons \begin{matrix} H^+ \\ H^+ \end{matrix} \boxed{Micelle} + \underset{\text{(leachable)}}{Na_2SO_4}$$

Not only are the sodium carbonate and bicarbonate changed to sodium sulfate, a mild neutral salt, but the carbonate radical is removed from the system. When gypsum is used, however, a portion of the carbonate may remain as a calcium compound ($CaCO_3$).

In research trials, sulfur and even sulfuric acid have proven to be very effective in the reclamation of sodic soils, especially if large amounts of $CaCO_3$ are present. In practice, however, gypsum is much more widely used than the acid-forming materials. Gypsum is less expensive, is widely available in both natural and in industrial by-product forms, and is more easily handled. Care must be taken, however, to be certain that the gypsum is finely ground and that it is well mixed with the upper soil horizons so that its solubility and rate of reaction are maximized.

Physical Condition

The effects of gypsum and sulfur on the physical condition of sodic soils is perhaps more spectacular than are the chemical effects. Sodic soils are almost impermeable to rainwater and irrigation water, since the soil colloids are largely dispersed and the soil is essentially void of stable aggregates (see Figure 10.9 and Table 10.6). When the exchangeable Na^+ ions are replaced by Ca^{2+} or H^+, soil aggregation and improved water infiltration results (Figure 10.15). The neutral sodium salts (e.g., Na_2SO_4) formed when the exchange takes place can then be leached from the soil, thereby reducing both salinity and sodicity.

The reclamation effects of gypsum or sulfur are greatly accelerated by plants growing on the soil. Crops that have some degree of tolerance to saline and sodic soils, such as sugar beets, cotton, barley, sorghum, berseem clover, or rye, can be grown initially. Their roots help provide channels through which gypsum can move downward into the

TABLE 10.6   The Degree of Aggregation of Two Salt-Affected Alfisols (Xeralfs) after Repeated Wetting (8 times) With Solutions Differing in Sodium and Calcium Contents

The Na treatment simulates the effect of applying irrigation water with high SAR. The effect of adding gypsum is simulated by the Ca treatment. Note that in each case the Na treatment reduced aggregation and the Ca treatment increased it.

Soil	Aggregate size, μm	Original, %	Na-treated, %	Ca-treated, %
Farrell	>50	5.1	3.1	9.9
	20–50	12.8	2.1	18.0
Tarlee	>50	15.6	0.0	34.1
	20–50	11.2	5.4	42.0

From Barzegar, et al. (1996).

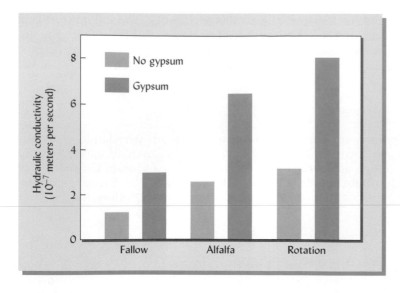

FIGURE 10.15 The influence of gypsum and growing crops on the hydraulic conductivity (readiness of water movement) of the upper 20 cm of a saline-sodic soil (Natrustalf) in Pakistan. The use of gypsum to increase water conductivity in all plots was more effective when deep-rooted alfalfa and rotation of sesbenia-wheat were grown. [Drawn from selected data from Ilyas, et al. (1993); used with permission of the Soil Science Society of America]

soil. Deep-rooted crops, such as alfalfa, are especially effective in improving the water conductivity of gypsum-treated sodic soils. Figure 10.15 illustrates the ameliorating effects of the combination of gypsum and deep-rooted crops.

Some research suggests that aggregate-stabilizing synthetic polymers may be helpful in at least temporarily increasing the water infiltration capacity of gypsum-treated sodic soils. Data in Table 10.7 from one experiment show the possible potential of these soil conditioners, especially when used in combination with gypsum.

Irrigation Timing

The timing of irrigation is extremely important on salty soils, particularly during the spring planting season. Because young seedlings are especially sensitive to salts, irrigation often precedes or follows planting to move the salts downward and away from the seedling roots (see Figure 10.14). After the plants are well established, their salt tolerance is somewhat greater.

Ammonia Effects

The practice in recent years of adding nitrogen as anhydrous ammonia (NH_3) to irrigation water as it is applied to a field has created some soil problems. The NH_3 reacts with the irrigation water to form NH_4OH.

TABLE 10.7 **Use of Gypsum and Synthetic Polymers on Reclamation of a Sodic Soil**

Adding gypsum to samples of a fine-textured (clay) saline-sodic soil, a Mollisol from California, increased both the hydraulic conductivity and the salts leached in the experiment while decreasing the exchangeable sodium percentage (ESP). Adding two experimental synthetic polymers (T4141 and 21J) with the gypsum gave even greater increases in hydraulic conductivity and leached salts.

| | Characteristic measured | | | | | |
| | Hydraulic conductivity, mm/h | | Total salts leached, mg/kg | | ESP, % | |
Gypsum added→	No	Yes	No	Yes	No	Yes
Polymer treatment						
No polymer	0.0	0.06	0.0	4.7	22.9	9.6
T4141	0.0	0.28	0.0	10.1	25.4	9.6
21J	0.0	0.28	0.0	9.7	25.5	9.6

Data from Zahow and Amrhein (1992).

$$NH_3 + H_2O \rightarrow NH_4OH$$

The high pH brought about by this reaction causes the precipitation of calcium and magnesium carbonates.

$$Ca(HCO_3)_2 + 2NH_4OH^- \rightarrow CaCO_3\downarrow + (NH_4)_2\ CO_3 + 2H_2O$$

$$Mg(HCO_3)_2 + 2NH_4OH \rightarrow MgCO_3\downarrow + (NH_4)_2\ CO_3 + 2H_2O$$

This removal of the Ca^{2+} and Mg^{2+} ions from the irrigation water raises the sodium adsorption ratio (SAR) and the hazard of increased exchangeable sodium percentage (ESP). To counteract these difficulties, sulfuric acid is sometimes added to the irrigation water to reduce its pH as well as that of the soil. This practice may well spread where there are economical sources of sulfuric acid and where the personnel applying it have been trained and alerted to the serious hazards of using this strong acid.

10.11   MANAGEMENT OF RECLAIMED SOILS

Once salt-affected soils have been reclaimed, prudent management steps must be taken to be certain that the soils remain productive. For example, surveillance of the EC and SAR and other pertinent chemical characteristics of the irrigation water is essential. Management adjustment is needed to accommodate any change in water quality that could affect the soil. The number and timing of irrigation episodes helps determine the balance of salts entering and leaving the soil. Likewise, the maintenance of good internal drainage is essential for the removal of excess salts.

Steps should also be taken to monitor appropriate chemical characteristics of the soils, such as pH, EC, and SAR, as well as specific levels of such elements as boron, molybdenum, and selenium that could lead to chemical toxicities. These measurements will help determine the need for subsequent remedial practices and/or chemicals.

Crop and soil fertility management to maintain satisfactory yield levels is essential to maintain the overall quality of salt-affected soils. The crop residues (roots and aboveground stalks) will help maintain organic matter levels and good physical condition of the soil. To maintain high yields, micronutrient and phosphorus deficiencies characteristic of other high-pH soils will need to be overcome by appropriate organic and inorganic sources.

10.12   CONCLUSION

Alkaline and salt-affected soils are vast in extent. They dominate arid and semiarid regions, where more than half the world's arable lands are found. They are widely used for rangeland and dryland farming. When irrigated, these soils can be among the most productive in the world.

Deficiencies in alkaline soils of some essential micronutrients (such as Fe and Mn) and macronutrients (such as phosphorus), as well as toxicities of molybdenum, must be overcome. Since their pH levels are high, management practices used on these soils are quite different from those applied to acid-soil regions.

There are three classes of salt-affected soils. *Saline soils* are dominated by neutral salts (electrical conductivity (EC) of 4 dS/m or higher) and by pH values less than 8.5. *Saline-sodic* soils have similar salt and pH levels, but exchangeable sodium percentages are 15 or above. *Sodic soils* also have high exchangeable sodium percentages (above 15), but their soluble salt concentrations are relatively low (EC < 4 dS/m) and their pH values are higher than 8.5. The physical conditions of saline and saline-sodic soils are satisfactory for plant growth, but the colloids in sodic soils are largely dispersed, the soil is puddled, and the water infiltration rate is extremely slow.

The reclamation of saline, saline-sodic, and sodic soils requires two major actions: (1) the removal by leaching of excess soluble salts from the upper levels of the soil profile; and (2) the removal of excess exchangeable sodium ions, first from the exchange complex by replacing the Na^+ ion with either Ca^{2+} or H^+ ions and then from the soil solution by leaching the replaced Na^+ ion from the soil. Gypsum ($CaSO_4 \cdot 2H_2O$) and elemental sulfur(S) are chemicals that can supply the Ca^{2+} and H^+ ions needed. Monitoring the chemical content of both the irrigation water and the soil is essential to achieve this goal of removing the Na^+ ions from the exchange complex and, ultimately, the soil.

STUDY QUESTIONS

1. What are the primary sources of alkalinity in soils? Explain.

2. Compare the availability of the following essential elements in alkaline soils with that in acid soils: (1) iron, (2) nitrogen, (3) molybdenum, and (4) phosphorus.

3. The iron analysis of an arid-region soil showed an abundance of this element, yet a peach crop growing on the soil showed serious iron deficiency symptoms. What is a likely explanation?

4. A soil with an abundance of $CaCO_3$ may have a pH no higher than about 8.3 while a nearby soil with high Na_2CO_3 content has a pH of 10.5. What is the primary reason for this difference?

5. An arid-region soil, when it was first cleared for cropping, had a pH of about 8.0. After several years of irrigation, the crop yield began to decline, the soil aggregation tended to break down, and the pH had risen to 10. What is the likely explanation for this situation?

6. What physical and chemical treatments would you suggest to bring the soil described in question 5 back to its original state of productivity?

7. What are some of the adverse consequences of using wetlands as recipients of irrigation wastewater?

8. Calculate the leaching requirement to prevent the buildup of salts in the upper 45 cm of a soil if the EC_{dw} of the drainage water is 6 dS/m and the EC_{iw} of the irrigation water is 1.2 dS/m.

9. What are the advantages of using gypsum ($CaSO_4 \cdot 2H_2O$) in the reclamation of a sodic soil? Show the chemical reactions that take place.

10. Calculate the quantity of gypsum needed to reclaim a sodic soil (ESP = 30%) when CEC = 25 $cmol_c$/kg and pH is 10.2. Assume you want an ESP no higher than 4%.

REFERENCES

Banuelos, G. S., H. A. Ajwa, B. Mackey, L. Wu, C. Cook, S. Akohoue, and S. Zambruzuski. 1997. "Evaluation of different plant species used for phytoremediation of high selenium," *J. Envir. Qual.*, **26**:639–648.

Barzegar, A. R., J. M. Oades, and P. Rengasamy. 1996. "Soil structure degradation and mellowing of compacted soils by saline-sodic solutions," *Soil Sci. Soc. Amer. J.*, **60**:583–588.

Beek, K. L., W. A. Blokhuis, P. M. Driessen, N. Van Breeman, N. Brinkman, and L. J. Pons. 1980. "Problem soils: Their reclamation and management," in ILRI Publication no. 27, (Wageningen, Netherlands: ILRI), pp. 47–72.

Bresler, E., B. L. McNeal, and D. L. Carter. 1982. *Saline and Sodic Soils, Principles—Dynamics—Modeling.* (Berlin: Springer-Verlag).

Carter, D. L. 1981. "Salinity and plant productivity," in *CRC Handbook Series in Nutrition and Food,* (Boca Raton, Fla.: CRC Press).

Crescimanno, G., M. Iovino, and G. Provenzano. 1995. "Influence of salinity and sodicity on soil structural and hydraulic characteristics," *Soil Sci. Soc. Am. J.*, **59**:1701–1708.

Ilyas, M., R. W. Miller, and R. H. Qureski. 1993. "Hydraulic conductivity of saline-sodic soil after gypsum application," *Soil Sci. Soc. Amer. J.*, **57**:1580–1585.

Maas, E. U. 1984. "Crop tolerance," *Calif. Agric.*, **38**:20–21.

National Academy of Sciences. 1990. *Saline Agriculture: Salt Tolerant Plants for Developing Countries.* (Washington, D.C.: National Academy Press).

Nolan, B. T., and M. L. Clark. 1997. "Selenium in irrigated agricultural areas of the western United States," *J. Envir. Qual.*, **26**:849–857.

Retana, J., D. R. Parker, C. Amrhein, and A. L. Page. 1993. "Growth and trace element concentrations of 5 plant species grown on a highly saline soil," *J. Environ. Qual.*, **22**:805–811.

Rhoades, J. D. 1993. "Electrical conductivity methods for measuring and mapping soil salinity," Advances in Agronomy, **49**:201–251.

Rhoades, J. D., and D. L. Corwin. 1984. "Monitoring soil salinity," *J. Soil and Water Cons.*, **39**:172–175.

Richards, L. A. (ed.). 1947. *Diagnosis and Improvement of Saline and Alkali Soils.* (Riverside, Calif.: U.S. Regional Salinity Laboratory).

Rowell, D. L. 1988. "Soil acidity and alkalinity," in A. Wild (ed.), *Russell's Soil Condition and Plant Growth,* 11th ed. (New York: Longman Scientific and Technical/Wiley).

Royo, A., and R. Aragues. 1993. "Validation of salinity crop production functions obtained with the triple line source sprinkler system," *Agron J.*, **85**:795–800.

USDA. 1996. *Americas Private Land: A Geography of Hope.* (Washington, D.C.: USDA Natural Resources Conservation Service).

Zahow, M. F., and C. Amrhein. 1992. "Reclamation of a saline sodic soil using synthetic polymers and gypsum," *Soil Sci. Soc. Amer. J.*, **56**:1257–1260.

11

ORGANISMS AND ECOLOGY OF THE SOIL[1]

Under the silent, relentless chemical jaws of the fungi, the debris of the forest floor quickly disappears. . . .
—A. FORSYTH AND K. MIYATA, TROPICAL NATURE

The terms *ecosystem* and *ecology* usually call to mind scenes of lions stalking vast herds of wildebeest on the grassy savannas of East Africa, or the interplay of phytoplankton, fish, and fishermen in some great estuary. Like a savanna or an estuary, a soil is an ecosystem in which thousands of different creatures interact and contribute to the global cycles that make all life possible. This chapter will introduce some of the organisms in the living drama that goes on largely unseen in the soil beneath our feet. If our bodies were small enough to enter the tiny passages in the soil, we would discover a world populated by a wild array of creatures all fiercely competing for every leaf that falls to the forest floor, every root that sloughs off the growing grass, every fecal pellet, every dead body that reaches the soil. We would also find predators of every description lurking in the dark, some with fearsome jaws to snatch unwary victims, others whose jellylike bodies simply engulf and digest their hapless prey.

Most of the work of the soil community is carried out by creatures whose "jaws" are chemical enzymes that eat away at all manner of organic substances left in the soil by their coinhabitants. The diversity of substrates and environmental conditions found in every handful of soil spawns a diversity of adapted organisms that staggers the imagination. The collective vitality, diversity, and balance among these organisms engender a healthy ecosystem and make possible the functions of a high-quality soil.

We will learn how these organisms, both flora and fauna, interact with one another, what they eat, how they affect the soil, and how soil conditions affect them. The central theme will be how this community of organisms assimilates plant and animal residues and waste products, creating soil humus, recycling carbon and mineral nutrients, and supporting plant growth. We will emphasize the activities rather than the scientific classification of these organisms. Consequently, we will consider only very broad, simple taxonomic categories.

11.1 THE DIVERSITY OF ORGANISMS IN THE SOIL

Soil organisms are creatures that spend all or part of their lives in the soil environment. Every handful of soil is likely to contain billions of organisms, with representatives of

---

[1] For a review of soil biology see Paul and Clark (1996) or Sylvia, et al. (1997).

nearly every phylum of living things. A simplified, general classification of soil organisms is shown in Figure 11.1.[2]

SIZES OF ORGANISMS. The animals (fauna) of the soil range in size from **macrofauna** (such as moles, prairie dogs, earthworms, and millipedes), through **mesofauna** (such as tiny springtails and mites), to **microfauna** (such as nematodes and single-celled protozoans). Plants (**flora**) include the roots of higher plants, as well as microscopic algae and diatoms. Other microorganisms (too small to be seen without the aid of a microscope) include fungi, bacteria, and actinomycetes, which tend to predominate in terms of numbers, mass, and metabolic capacity.

Note that we have classified the organisms according to their size (*macro* being larger than 2 mm in width, *meso* being between 0.2 and 2 mm, and *micro* being less than 0.2 mm), as well as by their ecological functions (what they eat). A typical, healthy soil might contain several species of vertebrate animals (mice, gophers, snakes, etc.), a half dozen species of earthworms, 20 to 30 species of mites, 50 to 100 species of insects (collembola, beetles, ants, etc.), dozens of species of nematodes, hundreds of species of fungi, and perhaps thousands of species of bacteria and actinomycetes.

TYPES OF DIVERSITY. This diversity is possible because of the nearly limitless variety of foods and the wide range of habitat conditions found in soils. Within a handful of soil there may be areas of good and poor aeration, high and low acidity, cool and warm temperatures, moist and dry conditions, and localized concentrations of dissolved nutrients, organic substrates, and competing organisms. The populations of soil organisms tend to be concentrated in zones of favorable conditions, rather than evenly distributed throughout the soil.

Aquatic ecologists have long considered a highly diverse community of aquatic organisms to indicate good water quality. In the same way, soil scientists are now using the concept of biological diversity as an indicator of soil quality. A high **species diversity** indicates that the organisms present are fairly evenly distributed among a large number of species. Such complexity and species diversity is paralleled by a high degree of **functional diversity**—the capacity to utilize a wide variety of substrates and carry out a wide array of processes.

ECOSYSTEM DYNAMICS. In most healthy soil ecosystems there are several—and in some cases many—different species capable of carrying out each of the thousands of different enzymatic or physical processes that proceed every day. This **functional redundancy**— the presence of several organisms to carry out each task—leads both to ecosystem **stability** and **resilience**. *Stability* describes the ability of soils, even in the face of wide variations in environmental conditions and inputs, to continue to perform such functions as the cycling of nutrients, assimilation of organic wastes, and maintainence of soil structure. *Resilience* describes the ability of the soil to "bounce back" to functional health after a severe disturbance has disrupted normal processes. Given a high degree of diversity, no single organism is likely to become completely dominant. By the same token, the loss of any one species is unlikely to cripple the entire system. Nonetheless, for certain soil processes, such as ammonium oxidation (see Section 13.7) or the creation of aeration macropores (see Section 4.6) primary responsibility may fall to only one or two species. The activity and abundance of these **keystone species** (for example, certain nitrifying bacteria or burrowing earthworms) merit special attention, for their populations may indicate the health of the entire soil ecosystem.

GENETIC RESOURCES. The diversity of organisms in a soil is important for reasons in addition to the safeguarding of ecological functions—soils also make an enormous contribution to **global biodiversity.** Many scientists believe that there are more species in existence below the surface of the earth than above it. The soil is therefore a major storehouse of the genetic innovations that nature has written into the DNA code over hundreds of millions of years. Humans have always found ways to make use of some of the genetic material in soil organisms (beer, yogurt, and antibiotics are examples). However, with recent developments in biological engineering that allow the transfer of genetic material from one type of organism to another, the soil DNA bank has taken on a much

---

[2] Most biologists classify living organisms into five kingdoms rather than two, since it has proven difficult to fit many microorganisms into either the plant or animal kingdom. More recently, classifications based on similarities in genetic material place all living things into three primary groups: *Eucarya* (which includes all plants, animals and fungi), *Bacteria*, and *Archaea*.

Animals (fauna), all heterotrophs

Macrofauna: Largely herbivores and detritivores

Vertebrates	Gophers, mice, squirrels
Arthropods	Ants, beetles and their larvae, maggots, termites, grubs, spiders, millipedes, woodlice
Annelida	Earthworms
Mollusca	Snails, slugs

Largely predators

Vertebrates	Moles, snakes
Arthropods	Beetles, ants, centipedes

Mesofauna: Largely detritivores

Arthropods	Mites, collembola (springtails)
Annelida	Enchytraeid (pot) worms

Largely predators

Arthropods	Mites, protura

Microfauna: Detritivores, predators, fungivores, bacterivores

Nematoda	Nematodes
Roterifera*	Rotifers, water bears
Protozoa*	Amoebae, ciliata

Plants (flora)

Macroflora: Largely autotrophs

Vascular plants	Feeder roots
Bryophytes	Mosses

Microflora: Largely autotrophs

Vascular plants	Root hairs
Algae	Greens, yellow-greens, diatoms

Largely heterotrophs, aerobic

Fungi	Yeasts, mildews, molds, rusts, mushrooms
Actinomycetes[†]	Many kinds of actinomycetes

Autotrophs and heterotrophs

Bacteria[†]	Aerobes, anaerobes
Cyanobacteria[†]	Blue-green algae

* In some classification schemes these are placed in the kingdom Protista, rather than Animalia.
† Often classified together in the kingdom Monera.

FIGURE 11.1   General classification of some important groups of soil organisms. Some organisms subsist on living plants (herbivores), others on dead plant debris (detritivores). Some consume animals (predators); some devour fungi (fungivore) or bacteria (bacterivores); and some live off of, but do not consume, other organisms (parasites). Heterotrophs rely on organic compounds for their carbon and energy needs while autotrophs obtain their carbon mainly from carbon dioxide and their energy from photosynthesis or oxidation of various elements.

larger practical importance for human welfare. The genes from soil organisms may now be used to produce plants and animals of superior utility to the human community.

11.2 ORGANISMS IN ACTION

The activities of soil flora and fauna are intimately related in what ecologists call a *food chain* or, more accurately, a *food web*. Some of these relationships are shown in Figure 11.2, which illustrates how various soil organisms are involved in the degradation of residues from higher plants.

Primary Producers

As in all ecosystems, plants play the role of **primary producers** by using water, energy from the sun, and carbon from atmospheric carbon dioxide to make organic molecules and living tissues. These organic materials, and the energy they contain, are then utilized by other organisms, either directly or after having been passed on through intermediaries.

HERBIVORES AND DETRITIVORES. Certain soil organisms eat live plants and are called **herbivores.** Examples are parasitic nematodes and insect larvae that attack plant roots, as well as termites, ants, beetle larvae, woodchucks, and mice that devour aboveground plant parts. Because they attack living plants that may be of value to humans, many of these soil herbivores are considered pests. For the vast majority of soil organisms, however, the principal source of food is the debris of dead tissues left by plants and animals. This debris is called **detritus**, and the animals that directly feed on it are called **detritivores.**

Plant and animal detritus is continually being deposited in and on the soil. This deposition of organic materials is particularly dramatic in deciduous forests, as trees drop their leaves at the end of the growing season. Each fall the soil in temperate regions is covered with these leaves to a depth of 15 cm or more. Yet, by the following year, most of these residues seem to have disappeared, leaving only little evidence of the mass of material that was deposited the previous fall. The assimilation of this mass of plant and animal debris is a critical function of both large and small soil organisms.

Primary Consumers

As soon as a leaf, a stalk, or a piece of bark drops to the ground, it is subject to coordinated attack by microflora and by detritivores (see Figure 11.2). The detritivore animals, which include mites, woodlice, and earthworms, chew or tear holes in the tissue, opening it up to more rapid attack by the microflora. The animals and microflora that use the energy stored in the plant residues are termed *primary consumers*.

ACTION OF THE FAUNA. While the actions of the microflora are mostly biochemical, those of the fauna are both physical and chemical. The mesofauna and macrofauna chew the plant residues and move them from one place to another on the soil surface and even into the soil. *Dung beetles* can greatly enhance nutrient cycling by burying animal dung in tiny tunnels that they excavate into the upper soil horizons. Earthworms literally eat their way through the soil as they incorporate plant residues into the mineral soil by passing them through their bodies along with mineral soil particles. Larger animals, such as gophers, moles, prairie dogs, and rats, also burrow into the soil and bring about considerable soil mixing and granulation.

The actions of these animals enhance the activity of the microflora in several ways. First, the chewing action fragments the litter, cutting through the resistant waxy coatings on many leaves to expose the more easily decomposed cell contents for microbial digestion. Second, the chewed plant tissues are thoroughly mixed with microorganisms in the animal gut, where conditions are ideal for microbial action. Third, the mobile animals carry microorganisms with them and help the latter to disperse and find new food sources to decompose. The ability of the soil fauna to enhance plant residue decomposition is illustrated in Figure 11.3.

Secondary Consumers

The bodies (cells) of primary consumers soon become food sources for predators and parasites in the soil. These **secondary consumers** include microflora, such as bacteria,

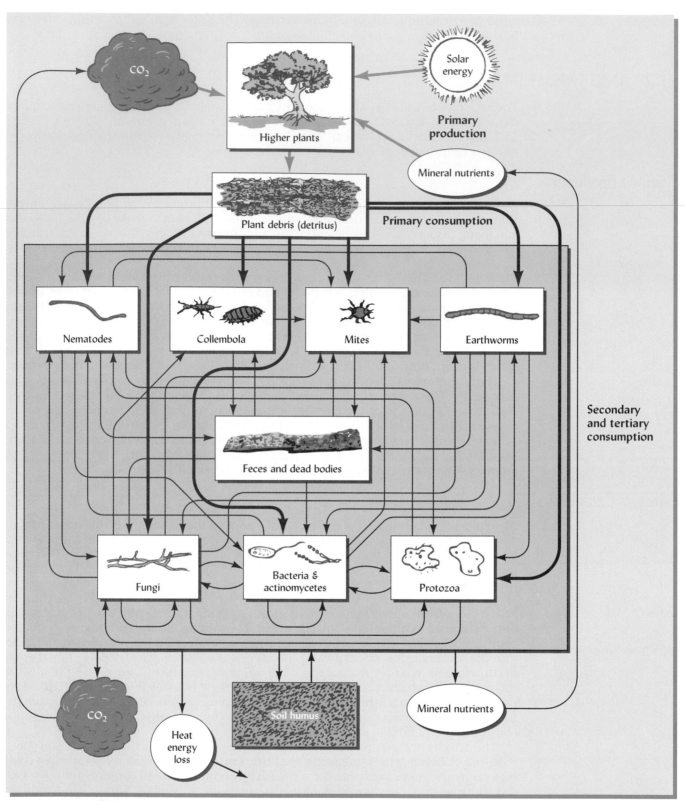

FIGURE 11.2 Greatly simplified diagram of the food web involved in the breakdown of higher-plant tissue. The boxes represent broad groups of organisms and pools of organic material, while the arrows represent transfers of carbon, energy, and nutrients between these pools. Because they capture carbon dioxide and energy, the higher plants are known as *primary producers*. Heavy arrows from the plant debris (detritus) to various organism groups represent *primary consumption*. The arrows within the large box represent *secondary* and *tertiary consumption*. Although all the groups shown play important roles in the process, the microorganisms represented by the lower three boxes account for 80 to 90% of the total metabolic activity. As a result of this metabolism, soil humus is synthesized, and carbon dioxide, heat energy, and mineral nutrients are released into the soil environment.

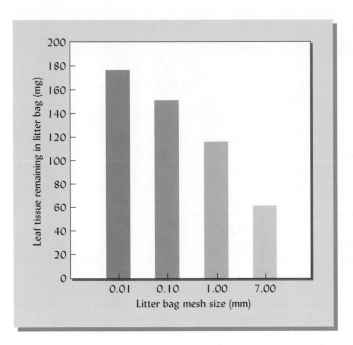

FIGURE 11.3 Influence of various sizes of soil organisms on the decomposition of corn leaf tissue buried in soil. Small bags made of nylon material with four different-size openings (mesh size) were filled with 558 mg (dry weight) of corn leaf tissue and buried in the soil for 10 weeks. The amount of corn leaf tissue remaining in the bags was considerably greater (less decomposition had taken place) where the meso- and macrofauna were excluded by the smaller mesh sizes. [Data from Weil and Kroontje (1979)]

fungi, and actinomycetes, as well as **carnivores**, which consume other animals. Examples of carnivores include centipedes and mites that attack small insects; spiders; nematodes; snails; and certain moles that feed primarily on earthworms. Examples of **microphytic feeders**, organisms that use microflora as their source of food, are certain collembola (springtail insects), mites (Figure 11.4), termites, nematodes, and protozoa.

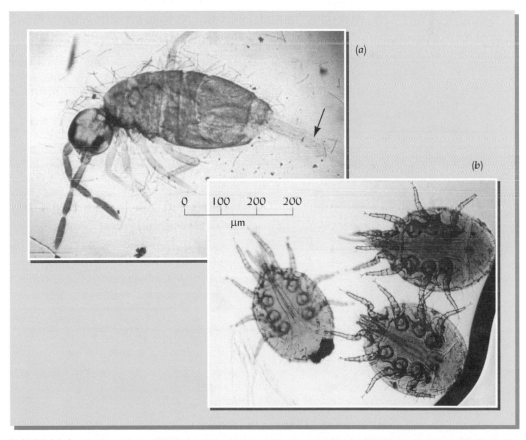

FIGURE 11.4 Springtails (collembala) (a) and mites (b) are mesofauna detritivores commonly found in soils. The springtail gets its name from the springlike furcula or "tail" that it uses to hop about (see arrow). (Photos courtesy of R. Weil)

These microphytic feeders exert considerable influence over the activity and growth of fungal and bacterial populations. The grazing by these meso- and microfauna on microbial colonies may stimulate faster growth and activity among the microbes, in much the same way that grazing animals can stimulate the growth of pasture grasses. In other cases, the attack of the microphytic feeders may kill off so much of a microbial colony as to inhibit the work of the microorganisms.

Tertiary Consumers

Moving farther up the food chain, the secondary consumers are prey for still other carnivores, called *tertiary consumers*. For example, ants consume centipedes, spiders, mites, and scorpions—all of which can themselves prey on primary or secondary consumers.

The microflora are intimately involved in every level of the process. In addition to their direct attack on plant tissue (as primary consumers), microflora are active within the digestive tracts of many soil animals, helping these animals digest more resistant organic materials. The microflora also attack the finely shredded organic material in animal feces, and they decompose the bodies of dead animals. For this reason they are referred to as the ultimate decomposers.

As indicated in Figure 11.2, broad groups of organisms, such as the fungi or mites, include some species that are primary consumers, some that are secondary consumers, and others that are tertiary consumers. For example, some mites attack detritus directly, while others eat mainly fungi or bacteria that grow on the detritus. Still others attack and devour the mites that eat fungi. Also note that there are two-way interactions between most groups. For example, some nematodes eat fungi, and some fungi attack nematodes.

11.3 ORGANISM ABUNDANCE, BIOMASS, AND METABOLIC ACTIVITY

Soil organism numbers are influenced primarily by the amount and quality of food available. Other factors affecting their numbers include physical factors (e.g., moisture and temperature), biotic factors (e.g., predation and competition) and chemical characteristics of the soil (e.g., acidity, dissolved nutrients, and salinity). The species that inhabit the soil in a desert will certainly be different from those in a humid forest, which, in turn, will be quite different from those in a cultivated field. Acid soils are populated by species different from those in alkaline soils. Likewise, species diversification and abundance in a tropical rain forest are different from those in a cool temperate area.

Despite these variations, there are a few generalizations that can be made. For example, forested areas support a more diverse soil fauna than do grasslands, although the total faunal mass per hectare and level of faunal activity are generally higher in grasslands (see Table 11.1). Cultivated fields are generally lower than undisturbed native lands in numbers and biomass of soil organisms, especially the fauna.

TABLE 11.1 **Biomass of Groups of Soil Animals Under Grassland and Forest Cover**

The mass and, in turn, the metabolism are greatest under grasslands. The spruce, with low-base-containing leaves, encourages acid conditions and slow organic matter decomposition.

| | Biomass,[a] g/m^2 | | |
| | | Forest | |
Group of organisms	Grassland meadow	Oak	Spruce
Herbivores	17.4	11.2	11.3
Detritivores			
Large	137.5	66.0	1.0
Small	25.0	1.8	1.6
Predators	9.6	0.9	1.2
Total	189.5	79.9	15.1

[a] Depth about 15 cm.
Data from Macfadyen (1963).

Total **soil biomass**, the living fraction of the soil, is generally related to the amount of organic matter present. On a dry-weight basis, the living portion is usually between 1 and 8% of the total soil organic matter. Also, scientists commonly observe that the ratios of soil organic matter to detritus to microbial biomass to faunal biomass are approximately 1000:100:10:1.

Comparative Organism Activity

The activities of specific groups of soil organisms are commonly identified by (1) their numbers in the soil, (2) their weight (biomass) per unit volume or area of soil, and (3) their metabolic activity (often measured as the amount of carbon dioxide given off in respiration). The numbers and biomass of groups of organisms that commonly occur in soils are shown in Table 11.2. Although the relative metabolic activities are not shown, they are generally related to the biomass of the organisms.

As might be expected, the microorganisms are the most numerous and have the highest biomass. Together with earthworms (or termites, in the case of some tropical soils), the microflora also dominate the biological activity in most soils. It is estimated that about 80% of the total soil metabolism is due to the microflora, although, as previously mentioned, their activity is enhanced by the actions of soil fauna. For these reasons, major attention will be given to the microflora, along with earthworms, termites, and certain other fauna.

Even those animals that account for only very small fractions of the total metabolism in the soil can play important roles in soil formation (see Chapter 2) and management. Rodents pulverize, mix, and granulate soil, and incorporate surface organic residues into lower horizons. They provide large channels through which water and air can move freely. Ants are important in localized areas, especially in warm-region grasslands and in boreal forests, for their exceptional ability to break down woody materials and turn over soil materials as they build their nests. Mesofauna detritivores (mostly mites and collembola; see Figure 11.4) translocate and partially digest organic residues and leave their excrement for microfloral degradation. By living in the soil, many animals rearrange soil particles to form biopores, thus favorably affecting the soil's physical condition. Others use the soil as a habitat for destructive action against higher plants.

Source of Energy and Carbon

Soil organisms may be classified as either **autotrophic** or **heterotrophic** on the basis of where they obtain the *carbon* needed to build their cell constituents (Table 11.3). The heterotrophic soil organisms obtain their carbon from the breakdown of organic materials previously produced by other organisms. Nearly all heterotrophs also obtain their

TABLE 11.2 Relative Numbers and Biomass of Fauna and Flora Commonly Found in the Surface 15 cm of Soil[a]

Since metabolic activity is generally related to biomass, microflora and earthworms dominate the life of most soils.

Organisms	Number		Biomass[b]	
	Per m^2	Per gram	kg/ha	g/m^2
Microflora				
Bacteria	10^{13}–10^{14}	10^8–10^9	400–5000	40–500
Actinomycetes	10^{12}–10^{13}	10^7–10^8	400–5000	40–500
Fungi	10^{10}–10^{11}	10^5–10^6	1,000–15,000	100–1500
Algae	10^9–10^{10}	10^4–10^5	10–500	1–50
Fauna				
Protozoa	10^9–10^{10}	10^4–10^5	20–200	2–20
Nematodes	10^6–10^7	10–10^2	10–150	1–15
Mites	10^3–10^6	1–10	5–150	.5–1.5
Collembola	10^3–10^6	1–10	5–150	.5–1.5
Earthworms	10–10^3		100–1500	10–150
Other fauna	10^2–10^4		10–100	1–10

[a] A greater depth is used for earthworms.
[b] Biomass values are on a liveweight basis. Dry weights are about 20 to 25% of these values.

TABLE 11.3 Metabolic Grouping of Soil Organisms According to Their Source of Metabolic Energy and Their Source of Carbon for Biochemical Synthesis

	Source of energy	
Source of carbon	Biochemical oxidation	Solar radiation

Combined organic carbon	**Chemoheterotrophs:** All animals, fungi, actinomycetes, and most bacteria
	Examples:
	  Earthworms
	  Aspergillus
	  Azotobacter
	  Pseudomonas
		Photoheterotrophs: A few algae

Carbon dioxide teria	**Chemoautotrophs**	**Photoautotrophs:** Algae and cyanobac-

	Examples:
	  Ammonia oxidizers—*Nitrosomonas*
	  Sulfur oxidizers—*Thiobacillus denitrificans*
		Examples:
		  Chorella
		  Nostoc

energy from the oxidation of the carbon in organic compounds. These organisms, which include the soil fauna, the fungi, actinomycetes, and most bacteria, are far more numerous than the autotrophs. They are responsible for organic decay.

The autotrophs obtain their carbon from simple carbon dioxide gas (CO_2) or carbonate minerals, rather than from carbon already fixed in organic materials. Autotrophs can be further classified on the basis of how they obtain energy. Some use solar energy (photoautotrophs) while others use energy released by the oxidation of inorganic elements such as nitrogen, sulfur, and iron (chemoautotrophs). While the autotrophs are in the distinct minority, their carbon fixation and inorganic oxidation reactions allow them to play crucial roles in the soil system. The autotrophs are mainly algae, cyanobacteria, and certain other bacteria.

11.4 EARTHWORMS[3]

Earthworms are probably the most important macroanimals in soils (Figure 11.5). They (along with their much smaller mesofauna cousins, the **enchytraeid worms**) are egg-laying *hermaphrodites* (organisms without separate male and female genders) that eat

---

[3] For a readable overview of earthworm biology and roles in the soil, see Lee (1985). For a more technical but wide-ranging review of earthworm ecology, see Hendrix (1995). Edwards (1998) provides up-to-date information on earthworm distribution and ecology.

FIGURE 11.5 Earthworms are perhaps the most significant macroorganism in soils of humid temperate regions, particularly in relation to their effects on the physical conditions of soils. Surface feeding species such as *Lumbricus terrestris* incorporate large amounts of plant litter into the soil and often gather plant debris into piles called *middens* near their burrow entrances. (Photo courtesy of R. Weil)

detritus, soil organic matter, and microorganisms found on these materials. They do not eat living plants or their roots, and so do not act as pests to crops.[4]

Of the up to 3000 species of earthworms reported worldwide, *Lumbricus terrestris,* a deep-boring, reddish organism, and *Allolobophora caliginosa,* a shallow-boring, pale pink organism, are the two most common in Europe and in the eastern and central United States. *Lumbricus terrestris*—commonly called *nightcrawlers*—live in the soil head-up. They come to the surface to pull detritus into their burrows, often creating **middens** of gathered detritus near the burrow entrance (see Figure 11.5). *Allolobophora caliginosa*—commonly called *red worms*—live head-down in the soil and expel their casts onto the soil surface.

In the tropics and semitropics, other types of earthworms are prevalent and dominate the soil fauna in regions with at least 800 mm of annual rainfall. Some tropical species are small, and others are surprisingly large. In Australia, aboriginal peoples have been known to hunt earthworms that grow up to 1 m long. Enchytraeid worms, on the other hand, range from about 1 mm to no more than a few centimeters in length.

Interestingly, *Lumbricus terrestris* is not native to America but was brought over from Europe, probably in the root balls of fruit trees carried by the settlers. As forests and prairies were put under cultivation, this European-introduced worm rapidly became dominant, either by filling niches still empty since the last glaciation, or by replacing native species that could not withstand the change in soil environment brought about by tillage. Virgin lands, however, still retain at least part of their native population.

Influence on Soil Fertility and Productivity

Burrows. Earthworms literally eat their way through the soil. They may ingest a weight of soil equal to 2 to 30 times their own weight in a single day. In a year, the earthworm population of 1 ha of land may ingest between 50 and 1000 Mg of soil, the higher figure occurring in moist, tropical climates. In so doing, they create extensive systems of burrows. In 1 m^2 these burrows may range from 5 to over 100 m in length, contributing from 0.1 to over 9 L of biopore volume. The burrows may be mainly vertical or horizontal, depending on the species, some providing continuous macropores to 1 m depth.

Casts. Earthworms also eject enormous quantities of partially digested soil and organic materials. The earthworms' wastes are called *casts,* and usually take the form of globular soil aggregates, some of which can be seen in Figure 11.5. The casting behavior of earthworms generally enhances the aggregate stability of the soil (Table 11.4). The casts are deposited within the soil profile or on the soil surface, depending on the species of earthworm. Both the piles of casts on the soil surface and the circular holes of burrow entrances are important signs of earthworm activity. In many cases, pieces of plant litter may have to be moved aside to observe the burrow entrances.

Nutrients. The activities of earthworms greatly enhance soil fertility and productivity by altering both physical and chemical conditions in the soil, especially in the upper 15 to 35 cm of soil. Earthworms increase the availability of mineral nutrients to plants in

---

[4] Earthworms should not be confused, in this regard, with wireworms, rootworms, or other soil-dwelling insect larvae that are commonly referred to as "worms," and which may be plant pests.

TABLE 11.4 **Comparative Characteristics of Earthworm Casts and Soils**

Average of six Nigerian soils.

Characteristic	Earthworm casts	Soils
Silt and clay, %	38.8	22.2
Bulk density, Mg/m^3	1.11	1.28
Structural stability[a]	849	65
Cation exchange capacity, cmol/kg	13.8	3.5
Exchangeable Ca^{2+}, cmol/kg	8.9	2.0
Exchangeable K$^+$, cmol/kg	0.6	0.2
Soluble P, ppm	17.8	6.1
Total N, %	0.33	0.12

[a] Numbers of raindrops required to destroy structural aggregates.
From de Vleeschauwer and Lal (1981).

two ways. First, as soil and organic materials pass through an earthworm's body, they are ground up physically as well as attacked chemically by the digestive enzymes of the earthworm and its gut microflora. Second, as earthworms ingest detritus and soil organic matter of relatively low nitrogen, phosphorus, and sulphur concentrations, they assimilate part of this material into their own body tissues. Consequently, earthworm tissues contain high concentrations of these nutrients, which are then readily released into plant-available form when the earthworms die and decay.

Compared to the bulk soil, the casts are significantly higher in bacteria and organic matter and available plant nutrients (see Table 11.4). Casts are mostly deposited in burrows that provide paths of low resistance to root growth, further enhancing the plant availability of these nutrients. The lush growth of grass around earthworm casts seen in lawns and pastures is evidence of the favorable effect of earthworms on soil productivity. Furthermore, physical incorporation of surface residues into the soil by earthworms and other soil animals reduces the loss of nutrients, especially nitrogen, by erosion and volatilization. This physical action is particularly important in grasslands and other uncultivated soils (including those where reduced tillage is practiced) and has earned the earthworm the title of "nature's tiller."

BENEFICIAL PHYSICAL EFFECTS.   Earthworms are important in other ways. The holes left in the soil (see, for example, Figure 6.7) serve to increase aeration and drainage, an important consideration in plant productivity and soil development. In turf grass, the mixing activity of earthworms can alleviate or reduce compaction problems and nearly eliminate the formation of an undesirable thatch layer. Under conditions of heavy rainfall, earthworm burrows may greatly increase the infiltration of water into the soils (Figure 11.6), thus playing an important role in water conservation and prevention of soil erosion. Moreover, stable macroaggregates formed from earthworm casts may form as much as 50 to 60% of the surface soil in some cases. Also, fungal hyphae that proliferate in earthworm burrows help bind soil particles into stable aggregates.

Influence on Chemical Leaching[5]

Not all effects of earthworms are beneficial. For example, *Lumbricus terrestris* has been observed to remove the crop residue cover from much of the soil surface in the process of building middens (see Figure 11.5). It is not yet known to what extent this action may leave the exposed soil susceptible to the impact of raindrops, crust formation, and increased erosion. Also, on closely cropped golf greens, earthworm casts on the soil surface may be considered a nuisance.

---

[5] For experimental evidence of some of these effects, see Stehouwer, et al. (1994).

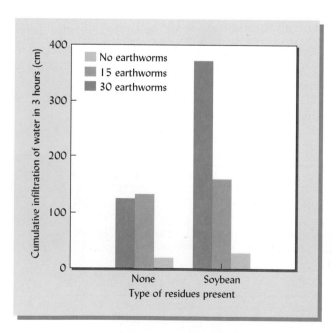

FIGURE 11.6   Earthworms may dramatically increase the infiltration of water into a soil, especially where they can feed on plant residues left on the soil surface. These data are for a Raub soil (Mollisol) in Illinois, in which a known number of earthworms were present. [From Kladivko, et al. (1986)]

Another aspect of concern is that water rapidly percolating down vertical earthworm burrows may carry potential pollutants toward the groundwater (see Figure 6.25). However, the organic-matter-enriched material lining earthworm burrows may have two to five times as great a capacity to adsorb certain herbicides as the bulk soil. Therefore, transport of such pollutants through earthworm burrows may be much less than is suggested by the mass flow of water through these large biopores. Clearly, on balance, earthworms are very beneficial and their activity is generally to be encouraged by soil managers.

Factors Affecting Earthworm Activity

Earthworms prefer a well-aerated but moist habitat. For this reason, they are found most abundantly in medium-textured upland soils where the moisture capacity is high rather than in droughty sands or poorly drained lowlands. Earthworms must have organic matter as a source of food. Generally they grow best on fresh, undecomposed organic matter. Consequently, they thrive where farm manure or plant residues have been added to the soil. A few species are reasonably tolerant to low pH, but most earthworms thrive best where the soil is not too acid (pH 5.5 to 8.5) and has an abundant supply of calcium (which is an important component of their slime excretions). Most earthworms are quite sensitive to excess salinity. Enchytraeid worms are much more tolerant of acid conditions and are more active than earthworms in some forested Spodosols.

A soil temperature of about 10°C appears optimum for *Lumbricus terrestris*. This temperature sensitivity, together with their preference for moist soil, probably accounts for the maximum earthworm activity noted in spring and autumn in temperate regions. Some earthworms burrow as deep as 1 to 2 m into the profile, thereby avoiding unfavorable moisture and temperature conditions. However, in soils not insulated by residue mulches, a sudden heavy frost in the fall may decimate earthworm populations before they can move deeper into the profile.

Other factors that depress earthworm populations are predators (moles, mice, and certain mites and millipedes), very sandy soils (partly because of the abrasive effect of sharp sand grains), direct contact with ammonia fertilizer, application of certain insecticides (especially carbamates), and tillage. The last factor is often the overriding deterrent to earthworm populations in agricultural soils. Minimum tillage, with plenty of crop residues left as a mulch on the soil surface, is ideal for encouraging earthworms.

Because of their sensitivity to soil and other environmental factors, the numbers of earthworms vary widely in different soils. In very acid forest soils (Spodosols), an average of fewer than one earthworm per square meter is common. In contrast, more than 500 per square meter have been found on rich grassland soils (Mollisols) and cropland enriched with animal manure and managed with minimum tillage. The numbers commonly found in arable soils range from 30 to 300 per square meter, equivalent to from 300,000 to 3 million per hectare. The biomass or liveweight for this number would range from perhaps 100 to 1000 kg/ha.

A simple method of assessing earthworm populations in grassland or arable soils is to dig into the surface soil to a depth of about 25 to 30 cm. During relatively cool, moist conditions, an average of five to ten earthworms in every shovelful of upturned soil could be expected and would indicate a reasonably healthy state of the soil ecosystem.

11.5 TERMITES[6]

Termites, sometimes called *"white ants,"* are major contributors to the breakdown of organic material in or at the surface of soils. There are about 2000 species of termites, including several that commonly build protective soil tunnels to enable them to invade (and subsequently destroy) wooden structures built without protective metal termite shields.

Termites are found in about two-thirds of the land areas of the world but are most prominent in the grasslands (savannas) and forests of tropical and subtropical areas (both humid and semiarid). Their activity is on a scale comparable to that of earthworms. Up to 16 million termites have been recorded in 1 ha of tropical deciduous for-

---

[6] For an excellent discussion of termites and ants in the tropics, see Lal (1987).

est. In the drier tropics (less than 800 mm annual rainfall) termites surpass earthworms in dominating the soil fauna.

Mound-Building Activities

Termites are social animals that live in very complex labyrinths of nests, passages, and chambers that they build both below and above the soil surface. Termite mounds built of cemented soil particles are characteristic features of many landscapes in Africa, Latin America, Australia, and Asia (Figure 11.7). These mounds, and the network of underground passages and aboveground covered runways that typically spreads 20 to 30 m beyond the mounds, are essentially termite "cities." Although a few species eat living woody plants and sound deadwood, most termites eat rotting woody materials and plant residues. Several species, such as *Macrotermes spp.* in Africa, use plant residues to grow fungi in their mounds as a source of food.

In building their mounds, termites transport soil from lower layers to the surface, thereby extensively mixing the soil and incorporating into it the plant residues they use as food. Scavenging a large area around each mound, these insects remove up to 4000 kg/ha of leaf and woody material annually, a substantial portion of the plant litter produced in many tropical ecosystems. They also annually move 300 to 1200 kg/ha of soil in their mound-building activities. These activities have significant impacts on soil formation, as well as on current soil fertility and productivity.

The quantity of soil materials incorporated into termite mounds can be enormous; up to 2.4 million kg/ha has been recorded (recall, from Chapter 4, that a 15-cm-deep layer of soil over 1 ha weighs approximately 2 million kg). Depending on the species involved and on environmental conditions, termites may build mounds 6 m or more in height, and may extend them to an even greater depth into the soil in search of water or clay layers. Each mound provides the home for 1 million or more termites. The mounds are abandoned after 10 to 20 years and can then be broken down to level the land for crop production. Attempts to level an occupied mound are usually frustrating, as the termites rebuild very rapidly unless the queen termite is destroyed.

Effect of Termites on Soil Productivity

Unlike earthworms, termites do not generally have a beneficial effect on soil productivity. This is because the digestive processes of termites, aided by microorganisms in their gut, are generally more efficient than those of earthworms. Also, earthworms incorporate organic matter into the soil in a relatively uniform manner over a hectare of land. In contrast, termites mix plant residues into the soil only in the localized areas of their nests, while denuding the remaining land area of surface residues. This termite behavior can make it extremely difficult to provide cropland with the benefits of a protective crop residue mulch (Figure 11.8).

PLANT GROWTH ON TERMITE MOUND MATERIAL. Termite mound material often has a lower organic matter and nutrient content than the surrounding undisturbed topsoil. This is because termites build their mounds mainly with subsoil, which is typically lower in organic matter content than topsoil. Crop growth in soil in areas where these mounds have existed is often poor, not only because of low nutrient content in the surface soil, but also because of the greater density of some of the mound material, the particles of which are cemented together by the termites as the mounds are constructed.

However, where the subsoil is richer in mineral nutrients than the topsoil, or is rich in clay compared to a very sandy surface soil, the material from abandoned mounds may provide islands of relatively high plant production, due to greater availability of phosphorus, potassium, calcium, and moisture. On soils with a high water table, termite mounds may provide islands of better drainage and aeration that produce much better plant growth. In certain semiarid and savanna regions, the stable macrochannels constructed by termites greatly increase water infiltration into soils that otherwise tend to form impermeable surface crusts (see Figure 11.8).

Termite activity is a significant factor in the formation of soils of tropical and subtropical areas (see Chapter 2). These insects also have both positive and negative effects on current land use in these areas. They accelerate the decay of dead trees and grasses, but also disrupt crop production, and even road construction, by the rapid development of their nests or mounds. Before leaving the subject of termites, it should be men-

(a)

(b)

(c)

FIGURE 11.7 Termite mounds in cultivated fields in Africa. (a) A termite mound in a cornfield. (b) Dissected profile of another mound showing the size and depth of the underground chambers. (c) Individual termites dragging cut pieces of leaves into their underground nest. [Photos (a) and (c) courtesy of R. Weil, photo (b) courtesy of R. Lal, Ohio State University]

FIGURE 11.8 Termites are quickly making these sorghum residues disappear, depriving the sandy, erosion-prone soil of the benefits of a protective mulch. However, the termites are also creating numerous stable macrochannels, which will aid in capturing water when the rains finally come to this field in the Sahelian region of West Africa. Note the soil casings that the termites have built around parts of the plant residues. Under these casings the termites are protected from the drying sun and wind. The knife blade in the picture is 7 cm long. (Photo courtesy of R. Weil)

tioned that the bacterial metabolism in the guts of these widespread soil animals accounts for a substantial fraction of the global production of methane (CH_4), an important greenhouse gas (see Chapter 12).

11.6 SOIL MICROANIMALS

From the viewpoint of microscopic animals, soils present many habitats that are essentially aquatic, at least intermittently so. For this reason the soil microfauna are closely related to the microfauna found in lakes and streams. The two groups exerting the greatest influence on soil processes are the nematodes and protozoa.

Nematodes

Nematodes—commonly called *threadworms* or *eelworms*—are found in almost all soils, often in surprisingly large numbers (see Table 11.2). They are unsegmented roundworms, about 4 to 100 μm in cross section and up to several millimeters in length (Figure 11.9). Therefore, they can seldom be seen with the naked eye. Moist, sandy soils typically have especially high nematode populations, as they typically contain abundant pores large enough to accommodate the swimming activities of these highly mobile creatures. When the large soil pores dry up, the nematodes survive by coiling up into a **cryptobiotic** or resting state, in which they seem to be nearly impervious to environmental conditions and use no detectable oxygen for respiration. In semiarid rangelands, nematode activity has been shown to be largely restricted to the first few days after each rainfall that awakens the nematodes from their cryptobiotic state.

PREDATION. The nematodes comprise a highly diverse group. Most nematodes are predatory on other nematodes, fungi, bacteria, algae, protozoa, and insect larvae. As a result, certain predatory nematodes are sold for use as biological control agents for soilborne insect pests, such as the corn rootworm (a far more environmentally friendly method than the use of toxic pesticides). A bacteria-eating nematode is illustrated in Figure 11.10,

FIGURE 11.9 A nematode commonly found in soil (magnified about 120 times). More than 1000 species of soil nematodes are known. Most nematodes feed on bacteria and fungi, but plant parasitic species can be very detrimental to plant growth. (Courtesy William F. Mai, Cornell University)

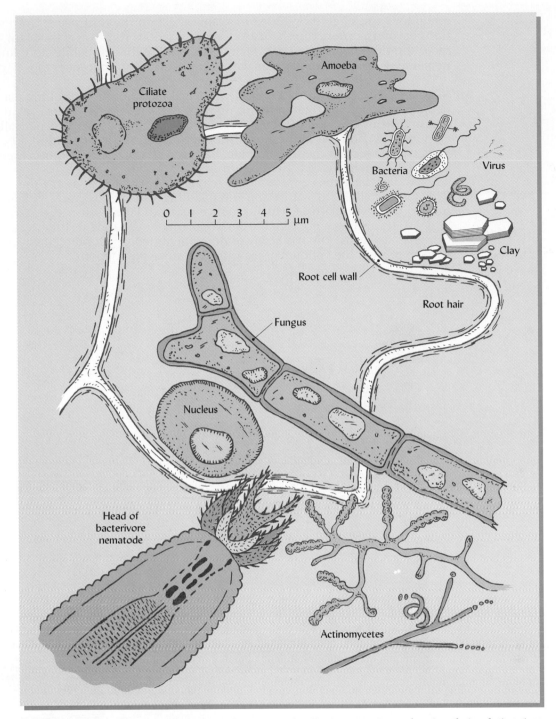

FIGURE 11.10 A depiction of representative groups of soil microorganisms, showing their relative sizes. (Drawing courtesy of R. Weil)

along with other soil microorganisms of various sizes. Nematode grazing can have a marked effect on the growth and activities of fungal and bacterial populations. Since bacterial cells contain more nitrogen than the nematodes can use, nematode activity often stimulates the cycling and release of plant-available nitrogen in the soil, accounting for 30 to 40% of the nitrogen released in some ecosystems.

PLANT PARASITES. Some nematodes, especially those of the genus *Heterodera,* can infest the roots of practically all plant species by piercing the plant cells with a sharp spearlike mouth part. These wounds often allow infection by secondary pathogens and cause the formation of knotlike growths on the roots. Minor nematode infestations are nearly ubiquitous and often have little observable effect on the host plant. However, infestations beyond a certain threshold level result in serious stunting of the plant. Cyst- (egg-sac-) forming nematodes are major pests of soybeans, while root-knot-forming nematodes cause widespread damage to fruit trees and solanaceous crops (Figure 11.11).

NEMATODE CONTROL. Until recently the principal methods of controlling plant parasitic nematodes were long rotations with nonhost crops (often five years is required for the parasitic nematode populations to sufficiently dwindle), use of genetically resistant crop varieties, and soil fumigation with toxic chemicals (**nematicides**). The use of soil nematicides, such as methyl bromide, has been sharply restricted because of undesirable environmental effects (see Section 11.15).

New, less dangerous approaches to nematode control include the use of hardwood bark for containerized plants and interplanting susceptible crops with marigolds or canola (rapeseed), both of which produce root exudates with nematicidal properties. Progress has also been made in the development of nematode-resistant varieties of such plants as soybeans. Often a series of different varieties will be planted in succeeding years to prevent the build up of parasitic nematode populations.

FIGURE 11.11 (Left) The stunted (foreground) and large (background) tobacco plants pictured both received identical management. (Right) Examination of the root systems of the stunted plant showed that it was heavily infested with root-knot nematodes, which stunted the roots and produced knotlike deformities. (Photos courtesy of R. Weil)

Protozoa

Protozoa are mobile, single-celled creatures that capture and engulf their food. They are the most varied and numerous of the soil microfauna (though often they are classified as Protista rather than as animals; see Figure 11.12 and Table 11.2). They are considerably larger than bacteria (see Figure 11.10), having a diameter range of 6 to 100 μm. Their cells do not have true cell walls, and have a distinctly more complex organization than bacterial cells. Soil protozoa include amoebas (which move by extending and contracting pseudopodia), ciliates (which move by waving hairlike structures; see Figure 11.12), and flagellates (which move by waving a whiplike appendage called a *flagellum*). Like nematodes, they swim about in the water-filled pores and water films in the soil and can form resistant resting stages (called *cysts*) when the soil dries out or food becomes scarce.

More than 350 species of protozoa have been isolated in soils; sometimes as many as 40 or 50 of such groups occur in a single sample of soil. The liveweight of protozoa in surface soil ranges from 20 to 200 kg/ha (see Table 11.2). A considerable number of serious animal and human diseases are attributed to infection by protozoa, but mainly by those that are waterborne, rather than soilborne. Most soil-inhabiting protozoa prey upon soil bacteria, exerting a significant influence on the populations of these microflora in soils.

Protozoa generally thrive best in moist, well-drained soils and are most numerous in surface horizons. In pursuit of their bacterial prey, some soil-dwelling protozoa are adapted to squeezing into soil pores with openings as small as 10 μm. Protozoa are especially active in the area immediately around plant roots. Their main influence on organic matter decay and nutrient release is through their effects on bacterial populations.

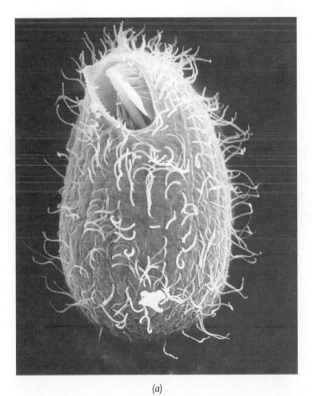

(a)

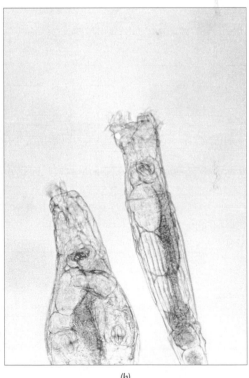

(b)

FIGURE 11.12   Two microanimals typical of those found in soils. (*a*) Scanning electron micrograph of a ciliated protozoan, *Glaucoma scintillaus*. (*b*) Photomicrograph of two species of rotifer, *Rotaria rotatoria* (stubby one at left) and *Philodina acuticornus* (slender one at right). [(*a*) Courtesy of J. O. Corliss, University of Maryland; (*b*) courtesy of F. A. Lewis and M. A. Stirewalt, Biomedical Research Institute, Rockville, Md.]

Higher plants store the sun's energy and are the primary producers of organic matter (see Figure 11.2). Their roots grow and die in the soil and are classified as soil organisms in this text. They typically occupy about 1% of the soil volume and may be responsible for a quarter to a third of the respiration occurring in a soil. Roots usually compete for oxygen, but they also supply much of the carbon and energy needed by the soil community of fauna and microflora. The activities of plant roots greatly influence soil chemical and physical properties, the specific effects depending on the type of soil and plant in question (see, for example, Section 7.8). As we shall see, plant roots interact with other soil organisms in varied and complex ways.

Morphology of Roots

Depending on their size, roots may be considered to be either meso- or microorganisms. Fine feeder roots range in diameter from 100 to 400 μm, while root hairs are only 10 to 50 μm in diameter—similar in size to the strands of microscopic fungi (see Figure 11.10). Root hairs are elongated protuberances of single cells of the outer (epidermal) layer (Figure 11.13). One function of root hairs is to anchor the root as it pushes its way through the soil. Another function is to increase the amount of root surface area available to absorb water and nutrients from the soil solution.

Roots grow by forming and expanding new cells at the growing point (meristem), which is located just behind the root tip. The root tip itself is shielded by a protective cap of expendable cells that slough off as the root pushes through the soil. Root morphology is affected by both type of plant and soil conditions. For example, fine roots may proliferate in localized areas of high nutrient concentrations. Root-hair formation

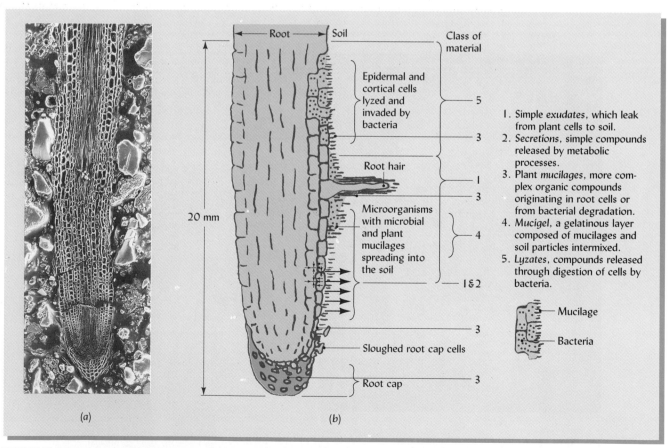

FIGURE 11.13   (a) Photograph of a root tip illustrating how roots penetrate soil and emphasizing the root cells through which nutrients and water move into and up the plant. (b) Diagram of a root showing the origins of organic materials in the rhizosphere. [(a) From Chino (1976), used with permission of Japanese Society of Soil Science and Plant Nutrition, Tokyo; (b) redrawn from Rovira, et al. (1979), used with permission of Academic Press, London]

is stimulated by contact with soil particles and by low nutrient supply. Many roots become thick and stubby in response to high soil bulk density or high aluminum concentrations in soil solution (see Chapters 4 and 9).

Living roots physically modify the soil as they push through existing cracks and make new openings of their own. Tiny initial channels are increased in size as the roots swell and grow. By removing moisture from the soil, plant roots stabilize organic–mineral bonds and encourage soil shrinkage and cracking, which, in turn, increase stable soil aggregation. Root exudates also support a myriad of microorganisms, which help to further stabilize soil aggregates. In addition, when roots die and decompose, they provide building materials for humus, not only in the top few centimeters, but also to greater soil depths.

Amount of Organic Tissue Added

The importance of root residues in helping to maintain soil organic matter is often overlooked. In grasslands, fires may remove most of the aboveground biomass, so that the deep, dense root systems are the main source of organic matter added to these soils. In plantation and natural forests, more than half of the total biomass production may be in the form of tree roots. In arable soils, the mass of roots remaining in the soil after crop harvest is commonly 15 to 40% that of the aboveground crop. If an average figure of 25% is used, good crops of oats, corn, and sugarcane would be expected to leave about 2500, 4500, and 8500 kg/ha of root residues, respectively. The mass of organic compounds contributed by roots is seen to be even greater when the rhizosphere effects discussed in the following subsections are considered.

Rhizosphere

The zone of soil significantly influenced by living roots is termed the **rhizosphere** and usually extends about 2 mm out from the root surface. The chemical and biological characteristics of this zone can be very different from those of the bulk soil. Soil acidity may be 10 times higher (or lower) in the rhizosphere than in the bulk soil. Roots greatly affect the nutrient supply in this zone by withdrawing dissolved nutrients on one hand and by solubilizing nutrients from soil minerals on the other. By these and other means, roots affect the mineral nutrition of soil microbes, just as the microbes affect the nutrients available to the plant roots.

RHIZODEPOSITION. Significant quantities of at least three broad types of organic compounds are released at the surface of young roots (see Figure 11.13). First, low-molecular-weight organic compounds are exuded by root cells, including organic acids, sugars, amino acids, and phenolic compounds. Some of these root exudates, especially the phenolics, exert growth-regulating influences on other plants and soil microorganisms in a phenomenon called **allelopathy** (see Section 12.6). Second, high-molecular-weight mucilages secreted by root-cap cells and epidermal cells near apical zones form a substance called **mucigel** when mixed with microbial cells and clay particles. This mucigel appears to have several beneficial functions: It lubricates the root's movement through the soil; it improves root–soil contact, especially in dry soils when roots may shrink in size and lose direct contact with the soil (see Figure 5.38); it may protect the root from certain toxic chemicals in the soils; and it provides an ideal environment for the growth of the rhizosphere microorganisms. Third, cells from the root cap and epidermis continually slough off as the root grows and enrich the rhizosphere with a wide variety of cell contents.

Taken together, these types of **rhizodeposition** typically account for 2 to 30% of total dry-matter production in young plants. The roots of common grain and vegetable plants have been observed to rhizodeposit 5 to 40% of the organic substances translocated to them from the plant shoot. Sometimes when a plant is carefully uprooted from a loose soil, these root exudates cause a thin layer of soil, approximating the rhizosphere, to form a sheath around the actively growing root tips (Figure 11.14). The amount of organic material lost to the rhizosphere during the growing season of annual plants may be more than twice the amount remaining in the root system at the end of the growing season. Rhizodeposition decreases with plant age but increases with soil stresses, such as compaction and low nutrient supply.

FIGURE 11.14 The zone of soil within 1 to 2 mm of living plant roots is termed the *rhizosphere*. This zone is greatly enriched in organic compounds excreted by the roots. These exudates and the microorganisms they support have caused the rhizosphere soil to adhere to some of these wheat roots as a sheath (roots on the right, see arrows). The roots on the left have been washed in water and the soil of the rhizosphere soil has been removed. The background is a terrycloth towel. (Photo courtesy of R. Weil)

Because of the rhizodeposition of carbon substrates and specific growth factors (such as vitamins and amino acids), microbial numbers in the rhizosphere are typically 2 to 10 times as great as in the bulk soil (sometimes expressed as a *R/S ratio*, typically ranging from 2 to 10).

The processes just described explain why plant roots are among the most important organisms in the soil ecosystem.

11.8 SOIL ALGAE

Like higher plants, algae consist of eukaryotic cells, those with nuclei organized inside a nuclear membrane. (Organisms formerly called *blue-green algae* are prokaryotes and therefore will be considered with the bacteria.) Also like higher plants, algae are equipped with chlorophyll, enabling them to carry out photosynthesis. As photoautotrophs, algae need light, and are therefore mostly found very near the surface of the soil (Figure 11.15). Some species can also function as heterotrophs in the dark. A few species are photoheterotrophs that use sunlight for energy but cannot synthesize all of the organic molecules they require (see Table 11.3).

Most soil algae range in size from 2 to 20 μm. Many algal species are motile and swim about in soil pore water, some by means of flagella (whiplike "tails"). Most grow best under moist to wet conditions, but some are also very important in hot or cold desert environments. Sometimes the growth of algae may be so great that the soil surface is covered with a green or orange algal mat. Some algae (as well as certain cyanobacteria) form *lichens*, symbiotic associations with fungi. These are important in colonizing bare rock and other low-organic-matter environments (see Figure 2.8). Unvegetated patches in deserts commonly are covered with *algal crusts* that reduce water evaporation and soil erosion but are very sensitive to disruption by trampling or off-road vehicles.

Several hundred species of algae have been isolated from soils, but a small number of species are the most prominent in soils throughout the world. Soil algae are divided into three general groups: (1) green, (2) yellow-green, and (3) diatoms. The green algae

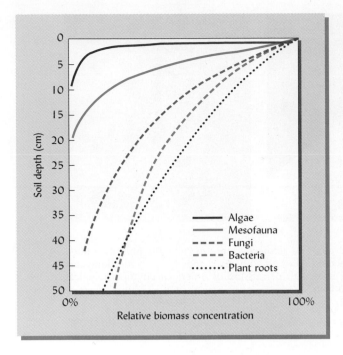

FIGURE 11.15 Relative vertical distribution of various groups of soil organisms in a representative grassland soil. Concentrations shown for a group are relative to the highest biomass concentration for that group. The biomass concentration for all of the organisms is near 100% of the maximum at the soil surface, but, the absolute biomass amounts (not shown here) differ greatly among the groups. For example, the actual biomass per gram of soil would be much higher for the fungi than for the mesofauna.

are most evident in moist but nonflooded acidic soils; diatoms are often numerous in neutral to alkaline, well-drained, older gardens that are rich in organic matter. The green algae generally outnumber the diatoms.

Algal populations commonly range from 1 to 10 billion per square meter, 15-cm deep (10,000 to 100,000 cells per gram). The mass of these live algae may range from 10 to 500 kg/ha (see Table 11.2). In addition to producing a substantial amount of organic matter in some fertile soils, certain algae excrete polysaccharides that have very favorable effects on soil aggregation (see Section 4.8).

11.9 SOIL FUNGI

Soil fungi comprise an extremely diverse group of microorganisms. Tens of thousands of species have been identified in soils, representing some 170 genera. As many as 2500 species have been reported to occur in a single location. Scientists believe there are at least 1 million fungal species in the soil still awaiting discovery. Although their numbers are usually somewhat smaller than those of bacteria, their relatively large size results in their dominating the biomass and metabolic activity in many soils. Their biomass in soils commonly ranges from 1,000 to 15,000 kg/ha in the upper 15 cm (see Table 11.2).

Fungi are eukaryotes with a nuclear membrane and cell walls. As heterotrophs, they depend on living or dead organic materials for both their carbon and their energy. Fungi are aerobic organisms, although some can tolerate the rather low oxygen concentrations and high levels of carbon dioxide found in wet or compacted soils. Strictly speaking, fungi are not entirely microscopic, since some of these organisms, such as mushrooms, form macroscopic structures that can easily be seen without magnification.

For convenience of discussion, fungi may be divided into three groups: (1) yeasts, (2) molds, and (3) mushroom fungi. *Yeasts,* which are single-celled organisms, live principally in waterlogged, anaerobic soils. *Molds* and *mushrooms* are both considered to be filamentous fungi, because they are characterized by long, threadlike, branching chains of cells. Individual fungal filaments, called **hyphae** (Figure 11.16*a*), are often twisted together to form **mycelia** that appear somewhat like woven ropes. Fungal mycelia are often visible as thin, white or colored strands running through decaying plant litter (Figure 11.17). The filamentous fungi reproduce by means of spores, often formed on fruiting bodies, which may be microscopic (i.e., molds) or macroscopic (i.e., mushrooms).

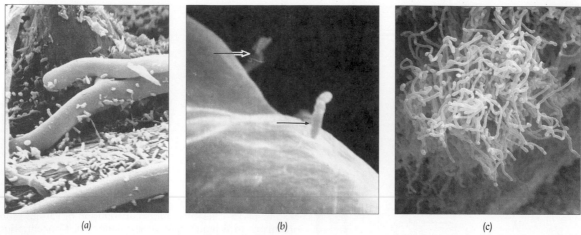

(a)   (b)   (c)

FIGURE 11.16 Scanning electron micrographs of fungi, bacteria, and actinomycetes. (a) Fungal hyphae associated with much smaller rod-shaped bacteria. (b) Rod-shaped bacteria attached to a plant root hair. (c) Actinomycete threads. [(a) Courtesy of R. Campbell, University of Bristol, used with permission of American Phytopathological Society; (b, c) courtesy of Maureen Petersen, University of Florida]

Molds

The molds are distinctly filamentous, microscopic or semimacroscopic fungi that play a much more important role in soil organic matter breakdown than the mushroom fungi. Molds develop vigorously in acid, neutral, or alkaline soils. Some are favored, rather than harmed, by lowered pH (see Figure 9.18). Consequently, they may dominate the microflora in acid surface soils, where bacteria and actinomycetes offer only mild competition. The ability of molds to tolerate low pH is especially important in decomposing the organic residues in acid forest soils.

Many genera of molds are found in soils. Four of the most common are *Penicillium, Mucor, Fusarium,* and *Aspergillus.* Species from these genera occur in most soils. Soil conditions determine which species are dominant. The complexity of the organic compounds being attacked seems to determine which particular mold (or molds) prevail. Their numbers fluctuate greatly with soil conditions; perhaps 100,000 to 1 million individuals per gram of dry soil (10 to 100 billion per square meter) represent a more or less normal range in population.

FIGURE 11.17 The thin white strands (near the fingers) in this photo are fungal mycelium in decaying plant residues. The other strands (upper right of photo) are plant roots. (Photo courtesy of R. Weil)

Mushroom Fungi

These fungi are associated with forest and grass vegetation where moisture and organic residues are ample. Although the mushrooms of many species are extremely poisonous to humans, some are edible—and a few have been domesticated. Edible mushrooms are grown in caves and specially designed houses, where composted organic materials (particularly horse manures) provide their source of food.

The aboveground fruiting body of most mushrooms is only a small part of the total organism. An extensive network of hyphae permeate the underlying soil or organic residue. While mushrooms are not as widely distributed as the molds, these fungi are very important, especially in the breakdown of woody tissue, and because some species form a symbiotic relationship with plant roots (see "Mycorrhizae," following).

Activities of Fungi

As decomposers of organic materials in soil, fungi are the most versatile and persistent of any group. Cellulose, starch, gums and lignin, as well as the more easily metabolized proteins and sugars, succumb to their attack. Fungi play major roles in the processes of humus formation (see Section 12.4) and aggregate stabilization (see Section 4.7). They usually dominate in the upper horizons of forested soils, as well as in very acid or sandy soils. They carry out the largest share of the decomposition in many cultivated soils, as well.

Fungi function more efficiently than bacteria in that they assimilate into their tissues a larger proportion of the organic materials they metabolize. Up to 50% of the substances decomposed by fungi may become fungal tissue, compared to about 20% for bacteria. Soil *fertility* depends in no small degree on nutrient cycling by fungi, since they continue to decompose complex organic materials after bacteria and actinomycetes have essentially ceased to function. Soil *tilth* also benefits form fungi as their hyphae stabilize soil structure (see Figure 4.28). Some of the nutrient cycling and ecological activities of soil fungi are made easily visible in the case of "fairy rings," commonly seen on lawns in early spring (Figure 11.18).

FIGURE 11.18 A "fairy ring" of fungal growth and the fungi's fruiting bodies (mushrooms, inset). As the fairy ring fungi (most commonly *Marasmius spp.*) metabolize accumulated grass thatch and residues, they release excess nitrogen that stimulates the lush green growth of the grass. Later, bacteria decompose the aging and dead fungi, producing a second release of nitrogen. The fungus produces a chemical (thought to be hydrogen cyanide) that is toxic to itself. Therefore, each generation must grow into uncolonized soil, producing an ever-expanding circle of fungi and decay marked by an ever-larger ring of dark green grass. The grass in the center of the ring is often brown, stunted and water-stressed, most likely because the fungi render the upper soil layers somewhat hydrophobic. (Photos courtesy of R. Weil)

In addition to the breakdown of organic residues and the formation of humus, numerous other fungal activities have significant impact on soil ecology. Some fungi are predators of soil animals. For example, certain species even trap nematodes (see Figure 11.19). Soil fungi can synthesize a wide range of complex organic compounds in addition to those associated with soil humus. It was from a soil fungus, a *Penicillium* species, that the first modern antibiotic drug, penicillin, was obtained. By killing bacteria, such compounds probably help fungi to outcompete rival microorganisms in the soil.

Unfortunately, not all the compounds produced by soil fungi benefit humans or higher plants. A few fungi produce chemicals (**mycotoxins**) that are highly toxic to plants or animals (including humans). An important example of the latter is the production of highly carcinogenic aflatoxin by the fungus *Aspergillus flavus* growing on grains such as corn or peanuts, especially when grain is exposed to soil and moisture. Other fungi produce compounds that allow them to invade the tissues of higher plants, causing such serious plant diseases as wilts (e.g., *Verticillium*) and root rots (e.g., *Rhizoctonia*).

On the other hand, efforts are now under way to develop the potential of certain fungi (such as *Beauvaria*) as biological control agents against some insects and mites that damage higher plants. These examples merely hint at the impact of the complex array of fungal activities in the soil.

Mycorrhizae[7]

One of the most ecologically and economically important activities of soil fungi is the mutually beneficial association (**symbiosis**) between certain fungi and the roots of higher plants. This association is called **mycorrhizae**, a term meaning "fungus root." A mycorrhiza forms when the appropriate fungus invades a plant root in a process superficially similar to infection by pathogenic fungi. However, in the mychorrhizal association, the fungus and the plant have apparently coevolved so that both parties benefit from the relationship. In fact, in natural ecosystems many plants are quite dependent on mycorrhizal relationships and cannot survive without them. Mycorrhizae are the rule, not the exception, for most plant species, including the majority of economically important plants.

Mycorrhizal fungi derive an enormous survival advantage from teaming up with plants. Instead of having to compete with all the other soil heterotrophs for decaying organic matter, the mycorrhizal fungi obtain sugars directly from the plant's root cells. This represents an energy cost to the plant, which may lose as much as 5 to 10% of its total photosynthate production to its mycorrhizal fungal symbiont.

In return, plants receive some extremely valuable benefits from the fungi. The fungal hyphae grow out into the soil some 5 to 15 cm from the infected root, reaching far-

---

[7] For two excellent reviews of mycorrhizae and their effects on the soil–plant system, see Bethlenfalvay and Linderman (1992) and Read, et al. (1992).

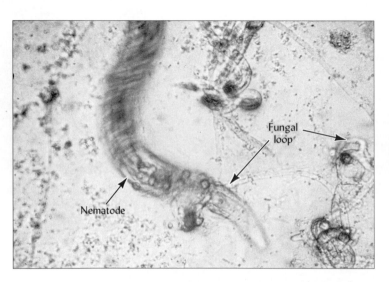

FIGURE 11.19 Several species of fungi prey on soil nematodes, often on those nematodes that parasitize higher plants. Some species of nematode-killing fungi attach themselves to and slowly digest the nematodes. Others make a loop with their hyphae and wait for a nematode to swim through this lassolike trap. The loop is then constricted and the nematode is trapped. The nematode shown here is thrashing about in a vain effort to escape from such a trap. (Photo courtesy of the American Society of Agronomy)

ther and into smaller pores than could the plant's own root hairs. This extension of the plant root system increases its efficiency, providing perhaps 10 times as much absorptive surface as the root system of an uninfected plant.

Mycorrhizae greatly enhance the ability of plants to take up phosphorus and, sometimes, other nutrients that are relatively immobile and present in low concentrations in the soil solution (Table 11.5). In contrast, mycorrhizae prevent the uptake of excessive amounts of salts and toxic metals in saline, acid, or contaminated soils (see Section 18.7). Water uptake may also be improved by mycorrhizae, making plants more resistant to drought. There is evidence that mycorrhizae protect plants from certain soilborne diseases and parasitic nematodes by producing antibiotics, altering the root epidermis, and competing with fungal pathogens for infection sites.

ECTOMYCORRHIZA. Two types of mycorrhizal associations are of considerable practical importance: **ectomycorrhiza** and **endomycorrhiza.** The ectomycorrhiza group includes hundreds of different fungal species associated primarily with temperate- or semiarid-region trees and shrubs, such as pine, birch, hemlock, beech, oak, spruce, and fir. These fungi, stimulated by root exudates, cover the surface of feeder roots with a fungal mantle. Their hyphae penetrate the roots and develop in the free space around the cells of the cortex but do not penetrate the cortex cell walls (hence the term *ecto,* meaning outside). Ectomycorrhizae cause the infected root system to consist primarily of stubby, white rootlets with a characteristic Y shape (Figure 11.20). These Y-shaped rootlets provide visible evidence of mycorrhizal infection. For this reason, the first mycorrhizae discovered and studied were the ectomycorrhizae of forest trees.

Many ectomycorrhizal fungi are facultative symbionts (they can also live independently in the soil) and therefore can be cultured in large quantities on artificial media. *Pisolithus tinctorus* is a commercial ectomycorrhizal fungal inoculant widely used in tree nurseries. For some tree species, satisfactory growth and survival are dependent on proper inoculation with mycorrhizal fungi before planting if the proper fungi are not already in soils at the site. On infertile soils, it is not uncommon for ectomycorrhizal inoculation to increase tree growth by 50 to 500 percent. Sometimes soil from a mature forest containing the desired tree species is used as inoculum to bring the necessary mycorrhizal fungi to forest plantings in such highly disturbed sites as reclaimed mines and landslides. Ectomycorrhizal spores are generally airborne, so small areas of disturbed soils often become reinoculated within a year or two by spores blown in from adjacent undisturbed areas.

ENDOMYCORRHIZA. The most important members of the endomycorrhiza group are called **arbuscular mycorrhizae (AM).**[8] When forming AM, fungal hyphae actually penetrate the cortical root cell walls and, once inside the plant cell, form small, highly branched structures known as **arbuscules.** These structures serve to transfer mineral nutrients

---

[8] Formerly these were called *vesicular-arbuscular mycorrhizae* (VAM), but this was shortened to AM since vesicles are not formed in all cases.

TABLE 11.5 **Effect of Inoculation with Mycorrhiza and of Added Phosphorus on the Content of Different Elements in the Shoots of Corn**

Expressed in micrograms per plant.

Element in plant	No phosphorus		25 mg/kg phosphorus added	
	No mycorrhiza	Mycorrhiza	No mycorrhiza	Mycorrhiza
P	750	1,340	2,970	5,910
K	6,000	9,700	17,500	19,900
Ca	1,200	1,600	2,700	3,500
Mg	430	630	990	1,750
Zn	28	95	48	169
Cu	7	14	12	30
Mn	72	101	159	238
Fe	80	147	161	277

From Lambert, et al. (1979).

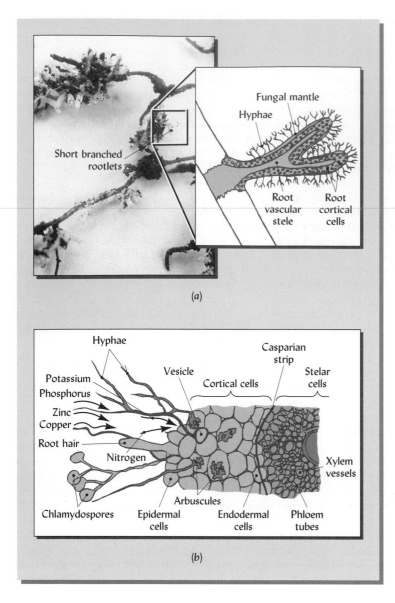

(a)

(b)

FIGURE 11.20   Diagram of ectomycorrhiza and arbuscular mycorrhiza (AM) association with plant roots. (*a*) The ectomycorrhiza association produces short branched rootlets that are covered with a fungal mantle, the hyphae of which extend out into the soil and between the plant cells but do not penetrate the cells. (*b*) In contrast, the AM fungi penetrate not only between cells but into certain cells as well. Within these cells, the fungi form structures known as *arbuscules* and *vesicles*. The former transfer nutrients to the plant, and the latter store these nutrients. In both types of association, the host plant provides sugars and other food for the fungi and receives in return essential mineral nutrients that the fungi absorb from the soil. [Redrawn from Menge (1981); Photo courtesy of R. Weil]

from the fungi to the host plants and sugars from the plant to the fungus. Other structures, called **vesicles**, are usually also formed and serve as storage organs for the mycorrhizae (see Figure 11.20).

The AM fungi are the most common and widespread group. Nearly 100 identified species of fungi form these endomycorrhizal associations in soils from the tropics to the arctic. The roots of most agronomic crops, including corn, cotton, wheat, potatoes, soybeans, alfalfa, sugarcane, cassava, and dryland rice, form AM, as do most vegetables and fruits, such as apples, grapes, and citrus. Many forest trees, including maple, yellow poplar, and redwood, as well as such important tree crops as cacao, coffee, and rubber, also form AM. Two important groups of crop plants that do *not* form mycorrhizae are the *Cruciferae* (cabbage, mustard, canola, and broccoli) and the *Chenopodiaceae* (sugar beet, red beet, and spinach).

Research on AM continues to reveal the ecological and practical significance of this symbiosis. The importance of mycorrhizal hyphae in stabilizing soil aggregate structure is becoming increasingly clear (see Figure 4.28). The presence of mycorrhizae is known to enhance nodulation and N fixation by legumes. Also, AM have been observed to transfer nutrients from one plant to another via hyphal connections, sometimes result-

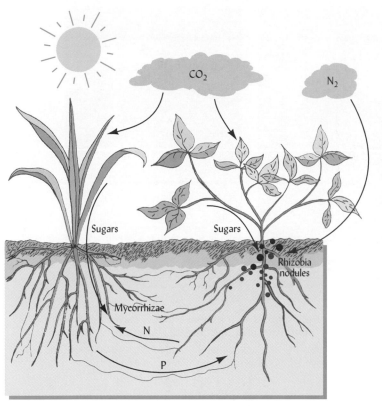

FIGURE 11.21 Mycorrhizal fungi, rhizobia bacteria, legumes, and nonlegume plants can all interact in a four-way, mutually beneficial relationship. Both the fungi and the bacteria obtain their energy from sugars supplied through photosynthesis by the plants. The rhizobia form nodules on the legume roots and enzymatically capture atmospheric nitrogen, providing the legume with nitrogen to make amino acids and proteins. The mycorrhizal fungi infect both types of plants and form hyphal interconnections between them. The mycorrhizae then not only assist in the uptake of phosphorus from the soil, but can also directly transfer nutrients from one plant to the other. Isotope tracer studies have shown that, by this mechanism, nitrogen is transferred from the nitrogen-fixing legume to the nonlegume (e.g., grass) plant, and phosphorus is mostly transferred to the legume from the nonlegume. The nonlegume grass plant has a fibrous root system and an extensive mycorrhizal network, which is relatively more efficient in extracting P from soils than the root system of the legume. Research indicates that some direct transfer of nutrients via mycorrhizal connections occurs in many mixed plant communities, such as in forest understories, grass–legume pastures, and mixed cropping systems.

ing in a complex four-way symbiotic relationship (Figure 11.21). Interplant transfer of nutrients has been observed in forest, grassland, and pasture ecosystems, but researchers are still trying to determine the extent and ecological significance of this AM-mediated plant interaction.

The role AM play in helping plants absorb nutrients is most agriculturally important where soils are relatively infertile and fertilizer inputs are limited (Figure 11.22). Evidence indicates that soil tillage disrupts hyphal networks, and therefore minimum tillage may increase the effectiveness of mycorrhizae. Also, the use of cover crops and certain crop rotations (as opposed to monocropping and periods of bare fallow) may favor the buildup of effective mycorrhizae in soils. Surface soil stockpiled during mining and construction activities (see Figure 1.15) may lose its mycorrhizal inoculum potential because of the long period without any host plants. Where such stockpiled soil is not inoculated with mycorrhizae-containing soil before use, revegetation efforts may be unsuccessful.

These examples highlight the importance of providing continuity of AM host plants if the symbiosis is to be effective. Studies in the forests of the west coast of the United States and in Canada have shown that Douglas fir regrowth may be greatly hampered by the common practice of clear-cutting, followed by suppression of hardwood trees, which are viewed as weed species. On some sites without hardwood trees, young Douglas fir trees either fail to survive or grow much more slowly than those planted where some hardwood saplings remain to provide host continuity and serve as an inoculum source for AM fungi.

Although definite differences exist in mycorrhizal efficiency among various species and strains of AM fungi, the ubiquitous distribution of native strains ensures natural infection in plants under most circumstances. Thus, gains from inoculation with efficient mycorrhizal fungi are usually quite small. Exceptions occur where native populations of fungi are low due to soil fumigation or elimination of host plants, such as by surface mining or forest clear-cutting. Practical AM technology has been limited by difficulties in culturing these obligate symbionts on artificial media, a process that would allow production of large quantities of disease-free inoculum.

(a)   (b)   (c)

FIGURE 11.22 The effect of mycorrhizae on availability of phosphorus to a pasture legume, *Pueraria phaseoloides:* (*a*) no treatment, (*b*) rock phosphate, and (*c*) rock phosphate plus mycorrhizae. (Courtesy Dr. Fritz Kramer, CIAT, Cali, Colombia)

11.10 SOIL ACTINOMYCETES

Like the fungi, **actinomycetes** are filamentous and often profusely branched (see Figure 11.16*c*), though their mycelial threads are much smaller than those of fungi. Actinomycetes are similar to bacteria in that they are unicellular and of about the same diameter. Although historically they have often been classified with the fungi, actinomycetes are bacterialike in that they have no nuclear membrane (prokaryote), have about the same diameter, and often break up into spores that closely resemble bacterial cells. They are now commonly classified with bacteria as *Monera*.

Generally aerobic heterotrophs, actinomycetes live on decaying organic matter in the soil, or on compounds supplied by plants with which certain species form parasitic or symbiotic relationships. Many actinomycete species produce antibiotic compounds that kill other microorganisms. Actinomycin, neomycin, and streptomycin are examples of familiar antibiotic drugs produced by growing soil actinomycetes in pure culture. In forest ecosystems, much of the nitrogen supply depends on certain actinomycetes that are capable of fixing atmospheric nitrogen gas into ammonium nitrogen that is then available to higher plants (see Section 13.12).

Actinomycetes develop best in moist, warm, well-aerated soil. But in times of drought, they remain active to a degree not usually exhibited by either bacteria or fungi. Actinomycetes are tolerant of low osmotic potential, and are important in arid-region, salt-affected soils. They are generally rather sensitive to acid soil conditions, with optimum development occurring at pH values between 6.0 and 7.5. Some actinomycete species tolerate relatively high temperatures.

NUMBERS AND ACTIVITIES OF ACTINOMYCETES. Actinomycete numbers in soil exceed those of all other organisms except bacteria (see Table 11.2). With their biomass as high as 5000 kg/ha, they often exceed the bacteria in actual liveweight. These organisms are especially numerous in soils high in humus, such as old meadows or pastures, where the acidity is not too great. The earthy aroma of organic-rich soils and freshly plowed land is mainly due to actinomycetes products called *geosmins* that are volatile derivatives of terpene.

Actinomycetes undoubtedly are of great importance in the decomposition of soil organic matter and the liberation of its nutrients. Apparently, they reduce even the more resistant compounds, such as cellulose, chitin, and phospholipids, to simpler forms. They often become dominant in the later stages of decay when the easily metabolized substrates have been used up. The abundance of actinomycetes in soils that have been under sod for many years is an indication of their capacity to attack complex compounds. They are also very important in the final (curing) stages of composting (see Section 12.5).

11.11 SOIL BACTERIA

CHARACTERISTICS. Bacteria are very small, single-celled prokaryotic organisms having no distinct nucleus. They range in size from 0.5 to 5 µm, the smaller ones approaching the size of the average clay particle (see Figure 11.10). Bacteria are found in various shapes: nearly round (coccus), rodlike (bacillus), or spiral (spirillum). In the soil, the rod-shaped organisms seem to predominate (see Figure 11.16b). Many bacteria are motile, swimming about in the soil water by means of hairlike cilia or whiplike flagella.

Bacteria are perhaps the most diverse group of soil organisms; a gram of soil typically contains 20,000 different *species*. They have evolved mechanisms to adapt to life in the most extreme of environments, from the Antarctic to the Amazon, from anaerobic wetlands to desiccated salt flats, from forest litter to deep groundwaters, from sodic soils at pH 10 to acid sulfate soils at pH 2.

BACTERIAL POPULATIONS IN SOILS. As with other soil organisms, the numbers of bacteria are extremely variable, but high, ranging from a few billion to more than a trillion in each gram of soil. A biomass of a 400 to 5000 kg/ha liveweight is commonly found in the upper 15 cm of fertile soils (see Table 11.2).

Their ability to form extremely resistant resting stages that survive dispersal by winds, sediments, ocean currents and animal digestive tracts has allowed bacteria to spread to almost all soil environments. Their extremely rapid reproductive potential (generation times of only a few hours) enables bacteria to increase their populations quickly in response to favorable changes in soil environment and food availability. As the early soil microbiologist Beijerinck expressed it, "everything is everywhere" when it comes to bacteria. Change the soil environment, add a new substrate—even an industrial waste—and soon you'll find new populations of bacteria.

SOURCE OF ENERGY. Soil bacteria are either autotrophic or heterotrophic (see Section 11.3). The autotrophs obtain their energy from sunlight (photoautotrophs) or from the oxidation of inorganic constituents such as ammonium, sulfur, and iron (chemoautotrophs) and obtain most of their carbon from carbon dioxide or dissolved carbonates. Autotrophic bacteria are not very numerous, but they play vital roles in controlling nutrient availability to higher plants.

Most soil bacteria are heterotrophic—both their energy and their carbon come from organic matter. Heterotrophic bacteria, along with fungi and actinomycetes, account for the general breakdown of organic matter in soil. The bacteria often predominate on easily decomposed substrates, such as animal manures, starches and proteins. Where oxygen supplies are depleted, as in wetlands, nearly all decomposition is bacterially mediated. Certain gaseous products of such anaerobic bacterial decomposition, such as methane and nitrous oxide, have major effects on the global environment (see Sections 12.11 and 13.9).

IMPORTANCE OF BACTERIA. Bacteria as a group participate vigorously in virtually all of the organic transactions that characterize a healthy soil system. Because soil bacteria as a group possess such a broad range of enzymatic capabilities, scientists are now finding ways of harnessing, even improving, the metabolic activities of bacteria to help with the remediation of soils polluted by crude oil, pesticides, and various other organic toxins (see Section 18.6). Bacteria are usually the most important group in the breakdown of hydrocarbon compounds, such as gasoline and diesel fuel.

In addition, bacteria hold near monopolies in several basic enzyme-mediated transformations. One such process is the oxidation or reduction of selected chemical elements in soils (see Sections 7.4 and 7.6). Certain autotrophic bacteria obtain their energy from such inorganic oxidations, while anaerobic and facultative bacteria reduce a number of substances other than oxygen gas. Many of these biochemical oxidation and reduction reactions have significant implications for environmental quality as well as for plant nutrition. For example, through nitrogen oxidation (nitrification), selected bacteria oxidize relatively stable ammonium nitrogen to the much more mobile nitrate form of nitrogen. Likewise, other bacteria are responsible for sulfur oxidation, which yields plant-available sulfate ions, but also potentially damaging sulfuric acid (see Section 13.21). Also, bacterial oxidation and reduction of inorganic ions such as iron and manganese not only influence the availability of these elements (see Section 15.5), but also help determine the soil colors (see Section 4.1).

A second critical process in which bacteria are prominent is nitrogen fixation—the biochemical combining of atmospheric nitrogen with hydrogen to form organic nitrogen compounds usable by higher plants (see Section 13.10). The process can take place in soils independent of plants, but the amount of nitrogen fixed is much greater if the bacteria are intimately associated with plant roots, which can supply sugars to fuel the energy-intensive process.

CYANOBACTERIA. Previously classified as blue-green algae, **cyanobacteria** contain chlorophyll, allowing them to photosynthesize like plants. Cyanobacteria are especially numerous in rice paddies and other wetland soils, and fix appreciable amounts of atmospheric nitrogen when such lands are flooded. Certain cyanobacteria, growing within the leaves of the aquatic fern *azolla,* are able to fix nitrogen in quantities great enough to be of practical significance for rice production (see Section 13.12). These organisms also exhibit considerable tolerance to saline environments.

11.12 CONDITIONS AFFECTING THE GROWTH OF SOIL MICROORGANISMS

ORGANIC MATTER REQUIREMENTS. Plant detritus and soil organic matter are used as an energy source by the majority of soil microorganisms, the heterotrophs, but not by the autotrophs. The addition of almost any energy-rich organic substance, including the compounds excreted by plant roots, stimulates microbial growth and activity. Certain bacteria and fungi are stimulated by specific amino acids and other growth factors found in the rhizosphere or produced by other organisms.

Bacteria tend to respond most rapidly to additions of simple compounds such as starch and sugars, while fungi and actinomycetes overshadow the bacteria if cellulose and more resistant compounds dominate the added organic materials. Also, if organic materials are left on the soil surface (as in coniferous forest litter), fungi dominate the microbial activity. Bacteria commonly play a larger role if the substrates are mixed into the soil, as by earthworms, root distribution, or tillage.

OXYGEN REQUIREMENTS. While most microorganisms are *aerobic* and use O_2 as the electron acceptor in their metabolism, some bacteria are *anaerobic* and use substances other than O_2 (e.g., NO_3^-, SO_4^{2-}, or other electron acceptors). *Facultative* bacteria can use either aerobic or anaerobic forms of metabolism. All three of the above types of metabolism are usually carried out simultaneously in different habitats within a soil.

MOISTURE AND TEMPERATURE. Optimum moisture level for higher plants (moisture potential of −10 to −70 kPa) is usually best for most aerobic microbes. Too high a water content will limit the oxygen supply. Microbial activity is generally greatest when temperatures are 20 to 40°C. The warmer end of this range tends to favor actinomycetes. Ordinary soil temperature extremes seldom kill bacteria, and commonly only temporarily suppress their activity. However, except for certain **cryophilic** species, most microorganisms cease metabolic activity below about 5°C, a temperature sometimes referred to as *biological zero* (see Section 7.9).

EXCHANGEABLE CALCIUM AND pH. Levels of exchangeable calcium and pH help determine which specific organisms thrive in a particular soil. Although in any chemical condition found in soils some bacterial species will thrive, high calcium and near-neutral pH generally result in the largest, most diverse bacterial populations. Low pH allows fungi to become dominant. The effect of pH and calcium helps explain why fungi tend to dominate in forested soils, while bacterial biomass generally exceeds fungal biomass in most subhumid to semiarid prairie and rangeland soils.

11.13 BENEFICIAL EFFECTS OF SOIL ORGANISMS

The soil fauna and flora are indispensable to the plant productivity and the ecological functioning of soils. Of their many beneficial effects, only the most important can be emphasized here.

ORGANIC MATERIAL DECOMPOSITION. Perhaps the most significant contribution of the soil fauna and flora to higher plants is that of plant residue decomposition. By this process, dead leaves, roots, and other plant tissues are broken down, converting organically held nutrients into mineral forms available for renewed plant uptake. The release of nitrogen is a prime example. Soil organisms also assimilate wastes from animals (including human sewage) and other organic materials added to soils. As a by-produce of their metabolism, microbes synthesize new compounds, some of which help to stabilize soil structure and others of which contribute to humus formation.

BREAKDOWN OF TOXIC COMPOUNDS. Many organic compounds toxic to plants or animals find their way into the soil. Some of these toxins are produced by soil organisms as metabolic by-products, some are applied purposefully by humans as agrochemicals to kill pests, and some are deposited in the soil as a result of unintentional environmental contamination. If these compounds accumulated unchanged, they would do enormous ecological damage. Fortunately, most biologically produced toxins do not remain long in the soil, for soil ecosystems include organisms that not only are unharmed by these compounds, but which can produce enzymes that allow them to use these toxins as food.

Some toxins are **xenobiotic** (artificial) compounds foreign to biological systems, and these may resist attack by commonly occurring microbial enzymes. Soil bacteria and fungi are especially important in helping maintain a nontoxic soil environment by breaking down toxic compounds (see Section 18.6). The detoxifying activity of these microorganisms is by far the greatest in the surface layers of soil, where microbial numbers are concentrated in response to the greater availability of organic matter and oxygen (Figure 11.23).

INORGANIC TRANSFORMATIONS. The transformation of inorganic compounds is of great significance to the functions of soil systems, including plant growth. Nitrates, sulfates, and, to a lesser degree, phosphate ions are present in soils primarily due to the action of microorganisms. Organically bound forms of nitrogen, sulfur, and phosphorus are converted by the microbes into these plant-available forms.

Likewise, the availability of the other essential elements, such as iron and manganese, is determined largely by microbial action. In well-drained soils, these elements are oxidized by autotrophic organisms to their higher valence states, in which forms they are quite insoluble. This keeps iron and manganese mostly in insoluble and nontoxic forms, even under fairly acid conditions. If such oxidation did not occur, plant growth would be jeopardized because of toxic quantities of these elements in solution. Microbial oxidation also controls the potential for toxicity in soil contaminated with selenium or chromium (see Section 18.10).

NITROGEN FIXATION. The fixation of elemental nitrogen gas, which cannot be used directly by higher plants, into compounds usable by plants is one of the most important microbial processes in soils. While actinomycetes in the genus *Frankia* fix major amounts of nitrogen in forest ecosystems, and cyanobacteria are important in flooded rice paddies, rhizobia bacteria are the most important group for the capture of gaseous nitrogen in agricultural soils. By far the greatest amount of nitrogen fixation by these organisms occurs in root nodules or in other associations with plants. Worldwide, enormous quantities of atmospheric nitrogen are fixed annually into forms usable by higher plants (see Section 13.10).

PLANT PROTECTION. Certain soil organisms attack higher plants, but others act to protect plant roots from invasion by soil parasites and pathogens (see following section).

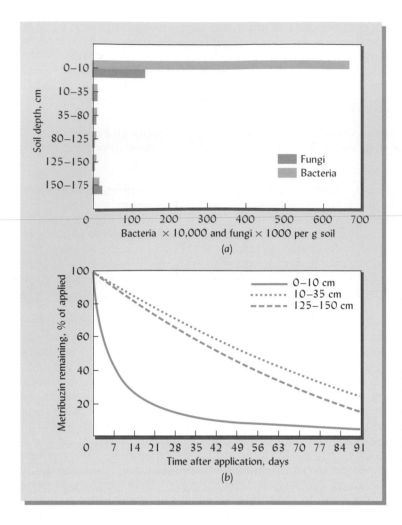

FIGURE 11.23 (a) Populations of aerobic bacteria and fungi at various depths in a Dundee soil (Aqualfs). (b) Breakdown of the herbicide metribuzin in the same soil at various depths. Soil was sampled from each depth and incubated with the herbicide. Note that the fungi and bacteria were concentrated in the upper 10 cm, and that the breakdown of the herbicide was far more rapid in this layer than in the deeper layers. [From Moorman and Harper (1989)]

11.14 SOIL ORGANISMS AND DAMAGE TO HIGHER PLANTS

Although most of the activities of soil organisms are vital to a healthy soil ecosystem and economic plant production, some soil organisms affect plants in detrimental ways that cannot be overlooked.

Plant Pests and Parasites

SOIL FAUNA. It already has been suggested that certain of the soil fauna are injurious to higher plants. For instance, some rodents may severely damage crops. Snails and slugs in some climates are dreaded pests, especially of vegetables. Ants transfer aphids onto certain plants, and so contribute to plant damage. Also, most plant roots are infested with nematodes, sometimes so seriously that the successful growth of certain crops is both difficult and expensive.

MICROFLORA AND PLANT DISEASE. Disease infestations occur in great variety and are induced by many different organisms. Among microorganisms, bacteria and actinomycetes contribute their quota of plant diseases, but fungi are responsible for most of the common soilborne diseases of crop plants. Included are wilts, damping-off, root rots, clubroot of cabbage, and blight of potatoes.

Soils are easily infested with disease organisms, which are transferred from soil to soil by many means, including tillage or planting implements, transplant material, manure from animals that were fed infected plants, soil erosion, and wind-borne fungal spores and bacteria. Once a soil is infested, it is apt to remain so for a long time.

Disease Control by Soil Management

Prevention is the best defense against soilborne diseases. Strict quarantine systems will restrict the transfer of soilborne pathogens from one area to another. Elimination of the host crop from the infested field will help in cases where no alternate hosts are available. Crop rotation and tillage practices can be used to help control a disease by growing nonsusceptible plants for several years between susceptible crops, and by burying plant residues on which fungal spores might overwinter. By preventing the splashing of soil onto foliage during rainstorms, mulches can reduce the transmission of soilborne diseases.

SOIL pH. Regulation of soil pH is effective in controlling some diseases. For example, keeping the pH low (<5.2) can control both the actinomycete-caused *potato scab* (Figure 11.24) and the fungal disease of turf grass known as *spring dead spot*. Raising soil pH to about 7.2 can control *clubroot* disease in the cabbage family, because the spores of the fungal pathogen germinate poorly, if at all, under alkaline conditions.

AIR AND TEMPERATURE. Wet, cold soils favor some seed rots and seedling diseases, known as *damping-off*. Good drainage and planting on ridges can help control these diseases. High levels of ammonium (as compared to nitrate) nitrogen tend to increase wilt diseases caused by *Fusarium* fungi. Steam or chemical sterilization is a practical method of treating greenhouse soils for a number of pathogens. It should be remembered, however, that sterilization kills beneficial microorganisms, such as mycorrhizal fungi, as well as pathogens, and so may do more harm than good.

DISEASE-SUPPRESSIVE SOILS. Research on plant diseases ranging from *take-all* of wheat in Washington state to *fusarium wilt* of bananas in Central America has documented the existence of **disease-suppressive soils** in which a particular disease fails to develop even though both the virulent pathogen and a susceptible host are present. The reason that certain soils become disease suppressive is not entirely understood, but much evidence points to a role for **antagonism** by certain bacteria and fungi that inhibit the pathogenic organisms.

In some cases, disease suppressiveness has developed through long-term crop monoculture in which the buildup of the pathogen during the first few years is eventually overshadowed by a subsequent buildup of specific organisms antagonistic to the pathogen (Figure 11.25). In other cases, large amounts and varied types of decomposable organic materials and lime can be added to a soil so as to encourage a high level of microbial activity and diversity, properties which, in turn, provide a more general sup-

FIGURE 11.24 Soilborne pathogens damage roots as well as other belowground organs. This potato has been attacked by the potato scab actinomycete, which may be present in soils with pH above about 5.0. (Courtesy of F. E. Manzer, University of Maine)

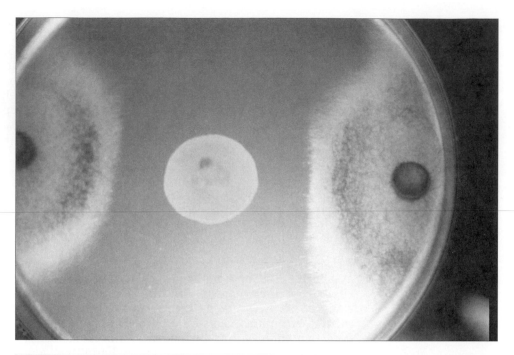

FIGURE 11.25 The biological basis of disease-suppressive soils that control take-all disease of wheat. (Upper) in the center of the petri dish grows a colony of certain *Pseudomonas* bacteria isolated from the rhizosphere of wheat that grew in soils suppressive of take-all disease. The *Pseudomonas* colony is producing an antibiotic that is toxic to *Gaeumannomyces graminis* (the fungal pathogen that causes take-all disease), preventing the pathogen colonies from growing to the center of the plate. (Lower) These plots in eastern Washington have grown monoculture wheat for 15 years and have developed high populations of organisms antagonistic to the take-all pathogen. In the 15th year of the study, the entire field was inoculated with *G. graminis* for experimental purposes, but the disease developed (seen as light-colored, prematurely ripened plots) only where the soil was fumigated prior to the inoculation. The fumigation killed most of the antagonistic organisms, leaving the pathogenic fungi free to infect the wheat plants. (Photos by R. J. Cook; courtesy of the American Phytopathological Society)

pression of soilborne pathogens. Horticulturalists have been able to control *Fusarium* diseases by replacing traditional potting mixes with growing media made from tree bark **compost**. Apparently, large numbers of antagonistic organisms colonize the organic material during the final stages of composting (see Section 12.5).

RHIZOSPHERE CAMOUFLAGE.[9] Many species or strains of plants resist particular soilborne diseases. According to recent research, many plants that are resistant to soilborne diseases have *camouflaged* rhizospheres. That is, although the rhizosphere of these resistant plants is enriched in microbial numbers, the types of microbes are more similar to those in the bulk soil than is the case for the rhizosphere of disease-susceptible plants. Having a rhizosphere microbial community similar to the community in the bulk soil may make the rhizosphere of these plants less identifiable to pathogens. Cultural practices that add large quantities of fresh organic matter may cause the bulk soil community to more closely resemble the rhizosphere community, thus assisting in rhizosphere camouflage and suppressing root disease.

COMPETITION FOR NUTRIENTS. Soil organisms may harm higher plants, at least temporarily, by competition for available nutrients. Soil organisms can quickly absorb essential nutrients into their own cells, leaving the more slowly growing higher plants to use only what is left. Nitrogen is the element for which competition is usually greatest, although similar competition occurs for phosphorus, potassium, calcium, and even the micronutrients. Competition for nitrogen is considered in greater detail in Section 12.3.

REDUCTION OF OXYGEN SUPPLY. Under conditions of poor drainage, active soil microflora may deplete the already limited soil oxygen supply. This oxygen depletion may affect plants adversely in two ways. First, the plant roots require a certain minimum amount of O_2 for normal growth and nutrient uptake. Second, oxidized forms of several elements, including nitrogen, sulfur, iron, and manganese, will be chemically reduced by further microbial action. In the cases of nitrogen and sulfur, some of the reduced forms are gaseous and may be lost to the atmosphere (see Section 13.9), thus reducing the nutrient supply for plants. In soils that are quite acid, bacterial reduction of iron and manganese may increase the solubility of these elements to such a degree that they become toxic. Thus, under anaerobic conditions microorganisms may cause both nutrient deficiencies and toxicities.

11.15 ECOLOGICAL RELATIONSHIPS AMONG SOIL ORGANISMS

Mutualistic Associations

We have already mentioned a number of mutually beneficial associations between plant roots and other soil organisms (e.g., mycorrhiza and nitrogen fixing nodules) and between several microorganisms (e.g., lichens). Other examples of such associations abound in soils. For example, photosynthetic algae reside within the cells of certain protozoans. Several types of associations, among them algal-fungal associations on or in soils and rocks, are very important cyclers of nutrients and producers of biomass in desert ecosystems.

Competitive Interactions

COMPETITION FOR FOOD. The population of soil microbes is generally limited by the food supply, increasing rapidly whenever more organic materials become available. As a result, the soil presents an intensely competitive environment for its microscopic inhabitants. When fresh organic matter is added, the vigorous heterotrophic soil organisms (bacteria, fungi, and actinomycetes) compete with each other for this source of food. If sugars, starches, and amino acids are available in the organic material, the bacteria dominate initially because they reproduce most rapidly and prefer such simple compounds. As these simple compounds are broken down, the fungi, and particularly the actinomycetes, become more competitive.

---

[9] The concept of rhizosphere camouflage was first introduced by Gilbert, et al. (1994).

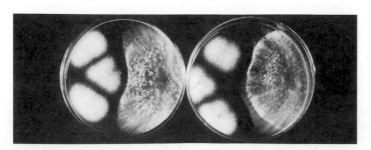

FIGURE 11.26 Fungal antagonism. The three smaller colonies in each petri dish are exuding an antibiotic compound that prevents the larger fungal colony from spreading over their half of the plate. Note the clear zone around each of the antibiotic-producing colonies. (Photo courtesy of American Society of Agronomy)

ANTIBIOTICS AND OTHER COMPETITIVE MECHANISMS. The soil microbes use many different tactics in their battle for resources. Certain organisms alter the acidity level in their vicinity to the disadvantage of competitors. Others produce substances (**siderophores**) that so strongly bind to iron that other organisms in their immediate vicinity cannot grow for lack of this essential element. Certain soil bacteria, fungi, and actinomycetes produce antibiotics, compounds that kill specific types of other organisms on contact (Figure 11.26). The discovery of this antibiotic production marked an epochal advance in medical science. Many antibiotics now on the market for the treatment of human and animal diseases, such as penicillin, streptomycin, and aureomycin, are produced by organisms found in soils. When these and other antagonistic activities of soil microorganisms are turned against potentially plant-pathogenic organisms, significant suppression of plant diseases can occur. Microbial ecologists are attempting to harness some of these competitive mechanisms to favor desirable microorganisms in soils (Box 11.1).

Effects of Management Practices on Soil Organisms

Changes in environment affect both the number and kinds of soil organisms. Clearing forests or grasslands for cultivation drastically changes the soil environment. First, both the quantity and quality of plant residues (food for the soil organisms) is markedly reduced. Also, monocultures or even common crop rotations greatly reduce the number of plant species and so provide a much narrower range of plant materials and rhizosphere environments than nature provides in forests or grasslands.

While agricultural practices have different effects on different organisms, a few generalizations can be made (Table 11.6). For example, some agricultural practices (e.g., extensive tillage and monoculture) generally reduce the diversity of species as well as the abundance (number) of individuals. However, monoculture may *increase* the population of a few species.

Adding lime and fertilizers (either organic or inorganic) to an infertile soil generally will increase microbial and fauna activity, largely due to the increase in the plant biomass that is likely to be returned to the soil as roots, root exudates, and shoot

TABLE 11.6 **The Generalized Effects of Major Soil-Management Practices on the Overall Diversity and Abundance of Soil Organisms**

Note that the practices that tend to enhance biological diversity and activity in soils are also those associated with efforts to make agricultural systems more sustainable.

Decreases biodiversity and populations	Increases biodiversity and populations
Fumigants	Balanced fertilizer use
Nematicides	Lime on acid soils
Some insecticides	Proper irrigation
Compaction	Improved drainage and aeration
Soil erosion	Animal manures and composts
Industrial wastes and heavy metals	Domestic (clean) sewage sludge
Moldboard plow–harrow tillage	Reduced or zero tillage
Monocropping	Crop rotations
Row crops	Grass–legume pastures
Bare fallows	Cover crops or mulch fallows
Residue burning or removal	Residue return to soil surface
Plastic mulches	Organic mulches

BOX 11.1 CHOOSING SIDES IN THE WAR AMONG THE MICROBES

Recent advances in our understanding of microbial genetics, combined with new techniques for transferring genetic material among different species of microorganisms, have opened many interesting and promising opportunities for scientists to join in the microbial wars in the soil.

Scientists are developing innovative approaches to harness the antagonisms long observed among microorganisms. Two examples are (1) the development of a bacterial seed treatment that helps ward off pathogenic fungi, and (2) a chemically armed *Rhizobium* bacteria that may enable farmers to efficiently use improved strains of nitrogen fixing bacteria on leguminous crops.

The first biological-control seed treatments are now commercially available. These seed treatments use antagonistic soil bacteria instead of chemical fungicide to protect seed from rot-causing soil pathogens. The protective bacteria are specially selected and enhanced strains of *Pseudomonas fluorescens*, a species known to inhabit rhizosphere soil and help protect plant roots from soilborne diseases. When a preparation of these bacteria was applied in the furrow as wheat seeds were sown, the seeds germinated safely in soil known to be infested with seed-rotting fungi. Seeds sown without the treatment decayed before they could germinate (Figure 11.27). Widespread use of such biologically based seed protection in the place of conventional antifungal chemicals should reduce the loading of toxic compounds into the environment.

The second example involves the use of genetic engineering to improve rhizobial inoculum effectiveness. Scientists have been quite successful in finding strains of rhizobia and bradyrhizobia bacteria that have superior nitrogen-fixing ability. Unfortunately, there has been little practical success in using these improved strains to enhance nitrogen fixation in the field. The problem is that most soils contain large populations of indigenous rhizobia strains that, while not as efficient in fixing nitrogen, are very competitive in forming nodules at infection sites on legume roots. Because indigenous rhizobia are successfully competing for survival in the soil and for nodule-formation sites on legume roots, improved strains of rhizobia bacteria generally have little success. Even when high levels of bacterial inoculum are applied to the legume seed or to the soils in which the seed is sown, the vast majority (generally 80 to 95%) of nodules found on the legume roots are formed by indigenous strains, not by the improved, introduced strains. In order to take advantage of the improved strains greater nitrogen-fixing ability, it is necessary to give the improved strains a competitive edge over their indigenous cousins. Dr. Eric Triplett and coworkers in Wisconsin discovered that the strain, *Rhizobia leguminosarium* bv. *trifolii*, produced a substance they called *trifolitoxin* that inhibits other rhizobial strains but did not inhibit the strain that produced it [Fitzmaurice, (1995)]. After working out the sequence of the cluster of genes controlling the production and reaction to trifolitoxin, the scientists were able to transfer the relevant genes to other strains of rhizobia. Their hope is that, armed with these genes, improved rhizobia strains that are highly efficient in nitrogen fixation will be able to outcompete the less efficient indigenous strains for nodule formation sites on legume roots. If successful, this use of microbial antagonism should lead to enhanced nitrogen fixation by legumes, with consequent increased yields of the legumes and increased potential to substitute biologically fixed nitrogen for fertilizer nitrogen in the production of nonlegume crops grown together or in rotation with the legumes.

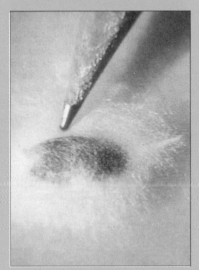

FIGURE 11.27 (Right) The wheat seed protected with antagonistic bacteria remain healthy, while unprotected seed (left) becomes heavily infected with pathogenic fungi. (Photo courtesy of Ecogen, Inc.)

residues. Tillage, on the other hand, is a drastic disturbance of the soil ecosystem, disrupting fungal hyphae networks and earthworm burrows, as well as speeding the loss of organic matter. Reduced tillage therefore tends to increase the role of fungi at the expense of the bacteria, and usually increases overall organism numbers as organic matter accumulates (Table 11.7). Addition of animal manure stimulates even higher microbial and faunal (especially earthworm) activity.

Pesticides are highly variable in their effects on soil ecology (see Section 18.4). Soil fumigants and nematicides can sharply reduce organism numbers, especially for fauna, at least on a temporary basis. On the other hand, application of a particular pesticide often stimulates the population of a specific microorganism, either because the organism can use the pesticide as food, or, more likely, because the predators of that organism have been killed. Figure 11.28 illustrates how a practice that affects one group of organisms will likely affect other groups as well, and will eventually impact on the productivity and functioning of the soil in the ecosystem. It is wise to remember that the interrelationships among soil organisms are intricate and the effects of any perturbation of the system are difficult to predict.

11.16 GENETICALLY ENGINEERED MICROORGANISMS

Advances in molecular biology have allowed scientists to identify the particular genes or pieces of genetic material (DNA or RNA) responsible for a particular trait in a particular organism. In addition, scientists have also developed techniques for transferring this specific genetic material to a not-necessarily-related other organism. These technologies have enabled the engineering of new genetic combinations—**recombinant organisms**—not seen in nature.

There are many potential uses for **genetically engineered microorganisms (GEMs)** in the soil environment. Already, soils contaminated by chemical spills are being cleaned by soil bacteria that have been endowed an enhanced ability to degrade certain organic pollutants, especially toxic xenobiotics (see Section 18.6). Other GEMs produce toxins that kill plant pathogens or pests. An example is the transfer of genes from the soil bacterium *Bacillus thuringiensis* (**Bt**) to a strain of the bacterium *Pseudomonas fluo-*

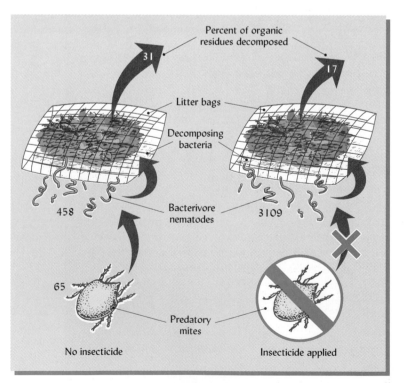

FIGURE 11.28 The indirect effects of insecticide treatment on the decomposition of creosote bush litter in desert ecosystems. Litter bags filled with creosote bush leaves and twigs were buried in desert soils in Arizona, Nevada, and California, either with or without an insecticide (chlordane) treatment. The insecticide killed virtually all the insects and mites. Without predatory mites to hold them in check, bacterivore nematodes multiplied rapidly and devoured a large portion of the bacterial colonies responsible for litter decomposition and nutrient cycling. Thus the insecticide reduced the rate of litter decomposition nearly in half, not by any direct effect on the bacteria, but by the indirect effect of killing the predators of their predators. [Data calculated from Whitford, et al. (1982)]

TABLE 11.7 Biomass Carbon of Microbial and Faunal Groups in an Agricultural Soil in Georgia Managed with Plow Tillage or No-Tillage

The soil was a Hiawassee sandy loam (Kanhapludults) with grain sorghum grown each summer and a rye cover crop each winter. In plow tillage plant residues were mixed into the soil, but in no-tillage residues were left as a surface mulch. The decomposer, microphytic feeder, and detritivore functions were dominated by larger organisms (fungi, microarthropods, and earthworms) in the no-till system, while smaller organisms (bacteria, protozoa, nematodes, and enchytraeid worms) were more prominent in the plowed system.

	Carbon, kg/ha	
Group of organisms	No-till	Plowed
Fungi[a]	360	240
Bacteria[a]	260	270
Protozoa[b]	24	39
Nematodes[b]		
Fungivores	0.14	0.47
Bacterivores	0.82	1.27
Microarthropods[a]	1.31	0.49
Enchytraeids[c]	5.55	4.79
Earthworms[c]	60	21

[a] 0 to 5 cm depth.
[b] 0 to 21 cm depth.
[c] 0 to 15 cm depth.
Calculated from a collection of data from various sources presented by Beare (1997).

rescens that is adapted to life in the rhizosphere of corn plants. The transferred genes code for the production of a protein, the Bt toxin, that effectively kills soil-dwelling insect larvae (e.g., corn rootworms) whose attacks on roots can cause major reductions in crop yield. Other potential uses of GEMs in soil are increased nitrogen fixation in roots or rhizospheres, improved mycorrhizal efficiency, and enhanced antagonism against plant pathogens.

The use of GEMs in soils brings the risk of undesirable side effects, which are difficult to predict. As with any alteration to the soil ecosystem, a change in one component will lead to changes in many other components. Also, nontarget organisms may be killed (e.g., *Collembola* killed by the Bt GEM just described), reducing diversity and upsetting the ecological balance. To minimize the risks of ecosystem alterations by runaway invasion of introduced "superbugs," GEMs are usually altered so as to weaken their ability to multiply and survive in the soil environment. For instance, some GEMs are given so-called suicide genes that cause the organism to die under certain conditions.

Of broader concern is the possibility that the new genetic material might be transferred (by virus infection or by transfer between bacteria of pieces of chromosome-free DNA) to indigenous organisms well adapted to survival and dispersal in the soil, possibly increasing the virulence of existing pathogens. Also, genetic engineers use antibiotic-resistance genes as markers to help identify and track the genetic material they are attempting to manipulate. The concern here is that the resistance to antibiotics might be transferred from the introduced GEM to another microbe, giving it a major competitive edge and again upsetting the ecological balance. Perhaps the worst-case scenario, seen by most scientists as highly unlikely but not impossible, is that the antibiotic resistance marker might spread to human pathogens, rendering our life-saving drugs useless. Questions about such concerns continue to require extensive testing and caution in releasing GEMs into the soil environment.

The soil is a complex ecosystem with a diverse community of organisms. Soil organisms are vital to the cycle of life on earth. They incorporate plant and animal residues into the soil and digest them, returning carbon dioxide to the atmosphere, where it can be recycled through higher plants. Simultaneously, they create humus, the organic constituent so vital to good physical and chemical soil conditions. During digestion of organic substrates, they release essential plant nutrients in inorganic forms that can be absorbed by plant roots or be leached from the soil. They also mediate the redox reactions that influence soil color and nutrient cycling in wetlands.

Animals, particularly earthworms, mechanically incorporate residues into the soil and leave open channels through which water and air can flow. Microorganisms such as fungi, actinomycetes, and bacteria are responsible for most organic decay, although their activity is greatly influenced by the soil fauna. Certain of the microorganisms form symbiotic associations with higher plants, playing special roles in plant nutrition and nutrient cycling. Competition among soil microbes, and between these organisms and higher plants, for mineral nutrients can result in plant nutrient deficiencies. Microbial requirements are factors in determining the success of most soil-management systems. Genetically engineered microorganisms offer the promise of enhanced beneficial microbial action, but must be used with caution as they present the potential for troubling unintended side effects in the soil system.

The organisms of the soil must have energy and nutrients if they are to function efficiently. To obtain these, soil organisms break down organic matter, aid in the production of humus, and leave behind compounds that are useful to higher plants. These decay processes and their practical significance are considered in the next chapter, which deals with soil organic matter.

STUDY QUESTIONS

1. What is *functional redundancy*, and how does it help soil ecosystems continue to function in the face of environmental shocks such as fire, clear-cutting or tillage?

2. In the example illustrated in Table 11.7, identify the organisms, if any, that play the roles of *primary producers*, *primary consumers*, *secondary consumers*, and *tertiary consumers*.

3. Describe some of the ways in which mesofauna play significant roles in soil metabolism even though their biomass and respiratory activity is only a small fraction of the total in the soil.

4. What are the four main types of metabolism carried out by soil organisms relative to their sources of energy and carbon?

5. What role does O_2 play in aerobic metabolism? What elements take its place under anaerobic conditions?

6. A *mycorrhizae* is said to be a symbiotic association. What are the two parties in this symbiosis, and what are the benefits derived by each party?

7. In what ways is soil improved as a result of earthworm activity? Are there possible detrimental effects as well?

8. What is the *rhizosphere*, and in what ways does the soil in the rhizosphere differ from the rest of the soil?

9. Explain and compare the effects of tillage and manure application on the abundance and diversity of soil organisms.

10. What are some of the potential benefits and potential risks involved with the release of *genetically engineered organisms* (GEMs) into the soil?

REFERENCES

Beare, M. H. 1997. "Fungal and bacterial pathways of organic matter decomposition and nitrogen mineralization in arable soils," in L. Brussaard and R. Ferrera-Cerrato, *Soil Ecology in Sustainable Agricultural Systems.* (Boca Raton, Fla: Lewis Publishers).

Bethlenfalvay, G. J., and R. G. Linderman (eds.). 1992. *Mycorrhizae in Sustainable Agriculture.* ASA Special Publication no. 54. (Madison, Wis.: Amer. Soc. of Agron.).

Chino, M. 1976. "Electron microprobe analysis of zinc and other elements within and around rice root growth in flooded soils," *Soil Sci. and Plant Nut. J.,* **22**:449.

de Vleeschauwer, D., and R. Lal. 1981. "Properties of worm casts under secondary tropical forest regrowth, *Soil Sci.,* **132**:175–181.

Edwards, C. (ed.). 1998. *Earthworm Ecology.* (Delray Beach, Fla: St. Lucie Press).

Fitzmaurice, L. 1995. "Giving *Rhizobium* the competitive edge," *Wisconsin Bioissues,* **6**:17.

Gilbert, G. S., J. Handelsman, and J. L. Parke. 1994. "Root camouflage and disease control," *Phytopathology,* **84**:222–225.

Hendrix, P. F. (ed.). 1995. *Earthworm Ecology and Biogeography in North America.* (Boca Raton, Fla.: Lewis Publishers).

Hendrix, P. F., D. A. Crossley, Jr., J. M. Blair, and D. C. Coleman. 1990. "Soil biota as components of sustainable agricultural agroecosystems," in C. Edwards and R. Lal, (eds.), *Sustainable Agricultural Systems.* (Akeny, Iowa: Soil Water Conserv. Soc. Am.), pp. 637–654.

Ingham, E. R., and W. G. Thies. 1996. "Responses of soil foodweb organisms in the first year following clearcutting and application of chloropicrin to control laminated root rot," *Applied Soil Ecology,* **3**:35–47.

Killham, K. 1994. *Soil Ecology.* (New York: Cambridge University Press).

Kladivko, E. J., A. D. Mackay, and J. M. Bradford. 1986. "Earthworms as a factor in the reduction of soil crusting," *Soil Sci. Soc. Amer. J.,* **50**:191–196.

Lal, R. 1987. *Tropical Ecology and Physical Edaphology.* (New York: Wiley).

Lambert, D. H., D. E. Baker, and H. Cole, Jr. 1979. "The role of Mycorrhizae in the interactions of phosphorus with zinc, copper, and other elements," *Soil Sci. Soc. Amer. J.,* **43**:976–980.

Lee, K. E. 1985. *Earthworms, Their Ecology and Relationships with Soils and Land Use.* (New York: Academic Press).

Macfadyen, A. 1963. In J. Doeksen and J. van der Drift (eds.), *Soil Organisms* (Amsterdam: North-Holland).

Menge, J. A. 1981. "Mycorrhizae agriculture technologies," in *Background Papers for Innovative Biological Technologies for Lesser Developed Countries,* Paper no. 9., Office of Technology Assessment Workshop, Nov. 24–25, 1980. (Washington, D.C.: U.S. Government Printing Office), pp. 383–424.

Mooreman, T. B., and S. S. Harper. 1989. "Transformation and mineralization of Metribuzin in surface and subsurface horizons of a Mississippi Delta soil," *J. Environ. Qual.,* **18**:302–306.

Paul, E. A., and F. E. Clark. 1996. *Soil Microbiology and Biochemistry,* 2nd ed. (New York: Academic Press).

Read, D. J., D. Lewis, A. Fitter, and I. Alexander. 1992. *Mycorrhizas in Ecosystems.* (Wallingford, U.K.: CAB International).

Rovira, A. D., R. C. Foster, and J. K. Martin. 1979. "Origin, nature and nomenclature of the organic materials in the rhizosphere," in J. L. Harley and R. S. Russell (eds.), *The Soil–Root Interface.* (New York: Academic Press).

Stehouwer, R. C., W. A. Dick, and S. J. Traina. 1994. "Sorption and retention of herbicides in vertically oriented earthworm and artificial burrows," *J. Environ. Qual.,* **23**:286–292.

Sylvia, D., J. Fuhrmann, P. Hartel, and D. Zuberer. 1997. *Principles and Applications of Soil Microbiology.* (Upper Saddle River, N.J.: Prentice Hall).

Weil, R. R., and W. Kroontje. 1979. "Organic matter decomposition in a soil heavily amended with poultry manure," *J. Environ. Qual.,* **8**:584–588.

Whitford, W. G., D. W. Freckman, P. F. Santos, N. Z. Elkins, and L. W. Parker. 1982. "The role of nematodes in decomposition in desert ecosystems," in Diana Freckman, (ed.), *Nematodes in Soil Ecosystems.* (Austin, Tex.: University of Texas Press), pp. 98–116.

12

SOIL ORGANIC MATTER

The earth lay rich and dark and fell apart lightly under the points of their hoes.
—*P. S. Buck*, The Good Earth

All organic substances, by definition, contain carbon. Organic matter in the world's soils contains about three times as much carbon as is found in all the world's vegetation. Soil organic matter,[1] therefore, plays a critical role in the global carbon balance that is thought to be the major factor affecting global warming, or the **greenhouse effect.** Although organic matter comprises only a small fraction of the total mass of most soils, this dynamic soil component exerts a dominant influence on many soil physical, chemical, and biological properties.

Soil organic matter is a complex and varied mixture of organic substances. It provides much of the cation exchange and water-holding capacities of surface soils. Certain components of soil organic matter are largely responsible for the formation and stabilization of soil aggregates. Soil organic matter also contains large quantities of plant nutrients and acts as a slow-release nutrient storehouse, especially for nitrogen. Furthermore, organic matter supplies energy and body-building constituents for most of the microorganisms whose general activities were discussed in Chapter 11. In addition to enhancing plant growth through the just-mentioned effects, certain organic compounds found in soils have direct growth-stimulating effects on plants. For all these reasons, enhancing the quantity and quality of soil organic matter is a central factor in improving **soil quality.**

We will first examine the role of soil organic matter in the **global carbon cycle** and the process of decomposition of organic residues. Next, we will focus on inputs and losses with regard to soil carbon in specific ecosystems. Finally, we will study the processes and consequences involved in soil organic matter management.

12.1 THE GLOBAL CARBON CYCLE

The element *carbon* is the foundation of all life. From cellulose to chlorophyll, the compounds that comprise living tissues are made of carbon atoms arranged in chains or rings and associated with many other elements. The cycle of carbon on earth is the story of life on this planet. The carbon cycle is all-inclusive because it involves the soil, higher plants of every description, and all animal life, including humans. Disruption of the carbon cycle would mean disaster for all living organisms.

---

[1] For a detailed review of organic matter in soils see Tate (1987). For a practical review of organic matter management in agricultural soils, see Magdoff (1992).

Pathways

The basic pathways involved in the global carbon cycle are shown in Figure 12.1. Plants take in carbon dioxide from the atmosphere. Then, through the process of photosynthesis, the energy of sunlight is trapped in the carbon-to-carbon bonds of organic molecules (such as those described in Section 12.2). Some of these organic molecules are used as a source of energy (via respiration) by the plants themselves (especially by the plant roots), with the carbon being returned to the atmosphere as carbon dioxide. The remaining organic materials are stored temporarily as constituents of the standing vegetation, most of which is eventually added to the soil as plant litter (including crop residues) or root deposition (see Section 11.7). Some plant material may be eaten by animals (including humans), in which case about half of the carbon eaten is exhaled into the atmosphere as carbon dioxide. The carbon not returned to the atmosphere is eventually returned to the soil as bodily wastes or body tissues. Once deposited on or in the soil, these plant or animal tissues are metabolized (digested) by soil organisms, which gradually return this carbon to the atmosphere as carbon dioxide.

Much smaller amounts of carbon dioxide react in the soil to produce carbonic acid (H_2CO_3) and the carbonates and bicarbonates of calcium, potassium, magnesium, and other base-forming cations. The bicarbonates are readily soluble and may be removed

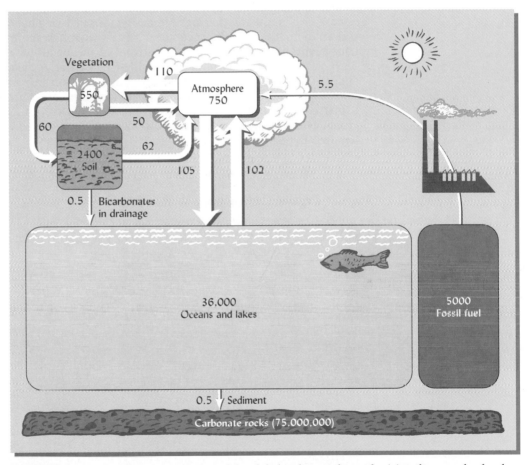

FIGURE 12.1 A simplified representation of the global carbon cycle emphasizing those pools of carbon which interact with the atmosphere. The numbers in boxes indicate the petagrams (Pg = 10^{15}/g) of carbon stored in the major pools. The numbers by the arrows show the amount of carbon annually flowing (Pg/yr) by various pathways between the pools. Note that the soil contains almost twice as much carbon as the vegetation and the atmosphere combined. Imbalances caused by human activities can be seen in the flow of carbon to the atmosphere from fossil fuel burning (5.5) and in the fact that more carbon is leaving (62 + 0.5) than entering (60) the soil. These imbalances are only partially offset by increased absorption of carbon by the oceans. The end result is that a total of 219.5 Pg/yr enters the atmosphere while only 215 Pg/yr of carbon is removed. It is easy to see why carbon dioxide levels in the atmosphere are rising. [Data from several sources; soil C estimate from Batjes (1996)]

in drainage. Eventually, much of the carbon in the carbonates and bicarbonates is also returned to the atmosphere as carbon dioxide.

Microbial metabolism in the soil produces some organic compounds of such stability that decades or even centuries may pass before the carbon in them is returned to the atmosphere as carbon dioxide. Such resistance to decay allows organic matter to accumulate in soils.

Carbon Sources

The original source of soil organic matter is plant tissue. Under natural conditions, the tops and roots of trees, shrubs, grasses, and other native plants annually supply large quantities of organic residues. Even with harvested crops, one-tenth to two-thirds of the aboveground part of the plant commonly falls to the soil surface as crop residue and remains there or is incorporated into the soil. Except in the case of root crops, such as carrots and sugar beets, all of the roots remain in the soil.

Animals are secondary sources of organic matter. As they eat the original plant tissues, they contribute waste products, and they leave their own bodies when they die (review Figure 11.2). Certain forms of animal life, especially earthworms, termites, ants, and dung beetles, also play an important role in the incorporation and translocation of organic residues.

Globally, at any one time, approximately 2400 petagrams (Pg, 10^{15}) of carbon are stored in soil profiles as soil organic matter (excluding surface litter), about one-third of that at depths below 1 m. An additional 700 Pg are stored as soil carbonates, which can release CO_2 upon weathering. Altogether, about twice as much carbon is stored in the soil than in the world's vegetation and atmosphere combined (see Figure 12.1). Of course, this carbon is not equally distributed among all types of soils (Table 12.1). About 45% of the total is contained in soils of just three orders, Histosols, Inceptisols, and Gelisols. Histosols (and Histels in the order Gellisols) are of limited extent but contain very large amounts of organic matter per unit land area. Inceptisols (and nonhistic Gellisols) contain only moderate concentrations of carbon, but cover vast areas of the globe. The reasons for the varying amounts of organic carbon in different soils will be detailed in Section 12.10.

TABLE 12.1   Mass of Organic Carbon in the World's Soils

Values for the upper 1 m represent most of the carbon in the soil profile. The upper 15 cm generally represents the surface soil, which is most readily influenced by land use and soil management.

Soil order	Global area, 10^3 km²	Organic carbon[a] in upper 100 cm			Organic carbon[a] in upper 15 cm	
		Mg/ha	Global Pg[b]	% of global	Range,[c] %	Typical,[c] %
Entisols	14,921	99	148	9	0.06–6.0	—[d]
Inceptisols[e]	21,580	163	352	22	0.06–6.0	—[d]
Histosols[e]	1,745	2,045	357	23	12–57	47
Andisols	2,552	306	78	5	1.2–10	6
Vertisols	3,287	58	19	1	0.5–1.8	0.9
Aridisols	31,743	35	110	7	0.1–1.0	0.6
Mollisols	5,480	131	73	5	0.9–4.0	2.4
Spodosols	4,878	146	71	5	1.5–5.0	2.0
Alfisols	18,283	69	127	8	0.5–3.8	1.4
Ultisols	11,330	93	105	7	0.9–3.3	1.4
Oxisols	11,772	101	119	8	0.9–3.0	2.0
Misc. land	7,644	24	18	1	—	—
Total	135,215		1576	100		

[a] Organic matter may be roughly estimated as 1.72 times this value. Organic nitrogen may also be estimated from organic carbon values by dividing by 12 for most soils. For aridisols and arid region soils this factor is somewhat lower, about 10. For Histosols and other humid, wetland soils this factor is somewhat higher, about 20.
[b] Petagram = 10^{15} g.
[c] Percent on mass basis (i.e., g/100 g).
[d] These soils are too variable to suggest a typical value.
[e] Carbon stored in Gelisols is included with these soils.
Data calculated from Eswaran, et al. (1993) and Brady (1990).

In a mature natural ecosystem or a stable agroecosystem, the release of carbon as carbon dioxide by oxidation of soil organic matter (mostly by microbial respiration) is balanced by the input of carbon into the soil as plant residues (and, to a far smaller degree, animal residues). However, as discussed in Section 12.10, certain perturbations of the system, such as deforestation, some types of fires, tillage, and artificial drainage, result in a net loss of carbon from the soil system.

Figure 12.1 shows that, globally, the release of carbon from soils into the atmosphere is about 62 Pg/yr, while only about 60 Pg/yr enter the soils from the atmosphere via plant residues. This imbalance of about 2 Pg/yr, along with about 5 Pg/yr of carbon released by the burning of fossil fuels (in which carbon was sequestered from the atmosphere millions of years ago) is only partially offset by increased absorption of atmospheric carbon dioxide by the ocean. Fossil fuel burning and degrading land-use practices have increased the concentration of carbon dioxide in the atmosphere at an accelerating rate since the beginning of the Industrial Revolution, some 400 years ago. The levels have increased from 290 to 370 ppm during the past century alone. The implications of carbon dioxide imbalances and of other gaseous emissions on the greenhouse effect will be discussed in Section 12.11, after we consider the processes involved in the carbon cycle.

12.2 THE PROCESS OF DECOMPOSITION IN SOILS

Because plant residues are the principal material undergoing decomposition in soils and, hence, are the primary source of soil organic matter, we will begin by considering the makeup of these materials.

Composition of Plant Residues

Green plant tissues consist mostly of water, varying in moisture content from 60 to 90%, with 75% a representative figure (Figure 12.2). If plant tissues are dried to remove all water, analysis of the *dry matter* remaining shows that, on a weight basis, the dry matter consists mostly (at least 90 to 95%) of carbon, oxygen, and hydrogen.

During photosynthesis plants obtain these elements from carbon dioxide and water. If plant dry matter is burned (oxidized), these elements become carbon dioxide and water once more. Of course, some ash and smoke will also be formed upon burning, accounting for the remaining 5 to 10% of the dry matter. In the ash and smoke can be found the many nutrient elements originally taken up by the plants from the soil. Even though these elements are present in relatively small quantities, they play a vital role in plant and animal nutrition and in meeting the requirements of microorganisms. The essential nutrient elements in the ash, such as nitrogen, sulfur, phosphorus, potassium, and micronutrients, will be given detailed consideration in later chapters.

ORGANIC COMPOUNDS IN PLANT RESIDUES. The organic compounds in plant tissue can be grouped into broad classes. Although representative percentages of these classes are shown in Figure 12.2, tissues from different plant species, as well as from different parts (leaves, roots, stems, etc.) of a given plant differ considerably in their makeup. Carbohydrates, which range in complexity from simple sugars and starches to cellulose, are usually the most plentiful of plant organic compounds.

Lignins, which are complex compounds with multiple ring-type or *phenol* structures, are components of plant cell walls. The content of lignin increases as plants mature and is especially high in woody tissues. Other **polyphenols**, such as tannins, may comprise as much as 6 or 7% of the leaves and bark of certain plants (for example, the brown color of steeped tea is due to tannins). Lignins and polyphenols are notoriously resistant to decomposition. Certain plant parts, especially seed and leaf coatings, contain significant amounts of fats, waxes, and oils, which are more complex than carbohydrates but less so than lignins.

Proteins contain about 16% nitrogen and smaller amounts of other essential elements, such as sulfur, manganese, copper, and iron. Simple proteins decompose and release their nitrogen easily, while complex crude proteins are more resistant to breakdown.

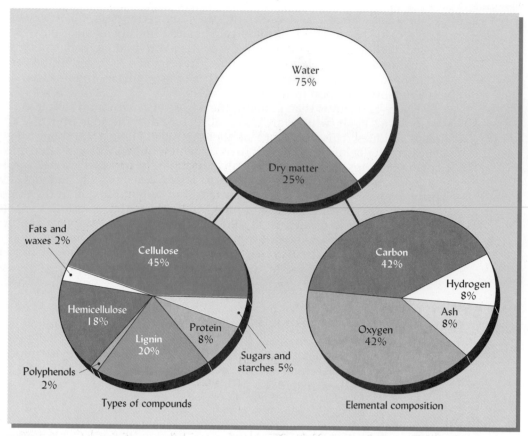

FIGURE 12.2 Composition of representative green-plant materials. The pie charts show typical composition. The major types of organic compounds are indicated at left and the elemental composition at right. The *ash* is considered to include all the constituent elements other than carbon, oxygen, and hydrogen (nitrogen, sulfur, calcium, etc.).

RATE OF DECOMPOSITION.   Organic compounds may be listed in terms of ease of decomposition as follows:

1. Sugars, starches, and simple proteins   Rapid decomposition
2. Crude proteins
3. Hemicellulose
4. Cellulose
5. Fats, waxes, and so forth
6. Lignins and phenolic compounds   Very slow decomposition

Decomposition of Organic Compounds in Aerobic Soils

When organic tissue is added to an aerobic soil, three general reactions take place:

1. Carbon compounds are enzymatically oxidized to produce carbon dioxide, water, energy, and decomposer biomass.
2. The essential nutrient elements, such as nitrogen, phosphorus, and sulfur, are released and/or immobilized by a series of specific reactions that are relatively unique for each element.
3. Compounds very resistant to microbial action are formed, either through modification of compounds in the original tissue or by microbial synthesis.

DECOMPOSITION: AN OXIDATION PROCESS.   In a well-aerated soil, all of the organic compounds found in plant residues are subject to oxidation. Since the organic fraction of plant materials is composed largely of carbon and hydrogen, the oxidation of the organic compounds in soil can be represented as:

$$R\text{—}(C, 4H) + 2O_2 \xrightarrow[\text{oxidation}]{\text{Enzymatic}} CO_2\uparrow + 2H_2O + \text{energy } (478 \text{ kJ mol}^{-1} \text{ C})$$

Carbon- and
hydrogen-containing
compounds

Many intermediate steps are involved in this overall reaction, and it is accompanied by important side reactions that involve elements other than carbon and hydrogen. Even so, this basic reaction accounts for most of the organic matter decomposition in the soil, as well as for the oxygen consumption and CO_2 release.

BREAKDOWN OF PROTEINS. The plant proteins also succumb to microbial decay, yielding not only carbon dioxide and water, but amino acids such as glycine (CH_2NH_2COOH) and cysteine ($CH_2HSCHNH_2COOH$). In turn, these nitrogen and sulfur compounds are further broken down, eventually yielding such simple inorganic ions as ammonium (NH_4^+), nitrate (NO_3^-), and sulfate (SO_4^{2-}), forms available for plant nutrition.

BREAKDOWN OF LIGNIN. Lignin molecules are very large and complex, consisting of hundreds of interlinked phenolic ring subunits, most of which are phenylpropenelike structures with various methoxyl (—OCH_3) groups attached (shown here as R or R'):

Because the linkages among these structures are so varied and strong, only a few microorganisms (mainly *white rot fungi*) can break them down. Decomposition proceeds very slowly at first, and is generally assisted by the physical activities of soil fauna. Once the lignin subunits are separated, many types of microorganisms participate in their breakdown. It is thought that microorganisms use some of the ring structures from lignin in the synthesis of stable soil organic matter.

Example of Organic Decay

The process of organic decay in time sequence is illustrated in Figure 12.3. Assume the soil has not been disturbed or amended with plant residues for some time. Initially, no readily decomposable materials are present. The principal microorganisms actively metabolizing are small populations of **autochthonous organisms**, which survive by slowly and steadily digesting the very resistant, stable soil organic matter. Competition for food is severe and microbial activity is relatively low, as reflected in the low **soil respiration** rate or level of CO_2 production in the soil. The supply of soil carbon is slowly but steadily being depleted.

Then, suddenly, an abundance of fresh, decomposable tissue is added to the soil. Maybe deciduous trees are losing their leaves in fall, or a farmer has plowed in the residue of a harvested crop. In any case, the appearance of easily decomposable and often water-soluble compounds, such as sugars, starches, and amino acids, stimulates an almost immediate increase in metabolic activity among the soil microbes. Soon the slower-acting autochthonous populations are overtaken by rapidly multiplying populations of **zymogenous** (opportunist) organisms that have been awakened from their dormant state by the presence of new food supplies. Cellulose-digesting organisms rapidly join in the attack. Microbial numbers and carbon dioxide evolution from microbial respiration both increase exponentially in response to the new food resource.

Soon microbial activity is at its peak, energy is being rapidly liberated, and carbon dioxide is being formed in large quantities. As they multiply and increase their biomass, the microbes are also synthesizing new organic compounds. The microbial biomass at this point may account for as much as one-sixth of the organic matter in a soil. The intense microbial activity may even stimulate the breakdown of some resistant soil organic matter, a phenomenon known as the **priming effect**.

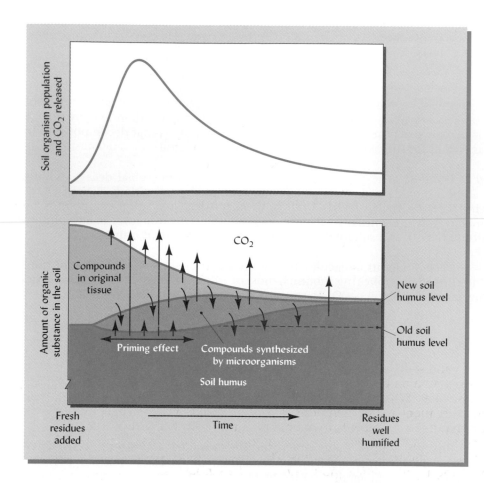

FIGURE 12.3 Diagrammatic illustration of the general changes that take place when fresh plant residues are added to a soil. The arrows indicate transfers of carbon among compartments. The time required for the process will depend on the nature of the residues and the soil. Most of the carbon released during the initial rapid breakdown of the residues is converted to carbon dioxide, but the smaller amounts of carbon converted into microbially synthesized compounds (biomass) and, eventually, into soil humus should not be overlooked. Note that although the peak of microbial activity appears to accelerate the decay of the original humus, a phenomenon known as the *priming effect*, the humus level is increased by the end of the process. Where vegetation, environment, and management remain stable for a long time, the soil humus content will reach an *equilibrium level* in which, the carbon added to the humus pool through the decomposition of plant residues each year is balanced by carbon lost through the decomposition of existing soil humus.

With all this frenetic microbial activity, the easily decomposed compounds are soon exhausted. While the cellulose and lignin decomposers continue their slow work, most of the zymogenous microorganisms begin to die of starvation. As microbial populations plummet, the dead cells provide a readily digestible food source for the survivors, which continue to evolve carbon dioxide and water. The decomposition of the dead microbial cells is also associated with the **mineralization** or release of simple inorganic products, such as nitrates and sulfates.

As food supplies are further reduced, microbial activity continues to decline, and the general-purpose zymogenous soil organisms again sink back into comparative quiescence. Very little of the original residue material persists, mainly tiny particles that have become **physically protected** from decay by lodging inside soil pores too tight to allow access by most organisms. Some of the remaining carbon has been **chemically protected** by conversion into **soil humus**, the dark-colored, heterogeneous, mostly colloidal mixture of modified lignin and newly synthesized organic compounds that strongly resists further decay. Some of the fine humus is further protected by binding strongly to clay particles. Thus, a small percentage of the carbon in the added residues has been retained, increasing slightly the pool of stable soil organic matter.[2]

Decomposition in Anaerobic Soils

Microbial decomposition proceeds most rapidly in the presence of plentiful supplies of O_2, which acts as the electron acceptor during aerobic oxidation of organic compounds. Oxygen supplies may become depleted when soil pores filled with water prevent the diffusion of O_2 into the soil from the atmosphere. Without sufficient oxygen present, aerobic organisms cannot function, so anaerobic or facultative organisms become dominant. Under low-oxygen or anaerobic conditions, decomposition takes place

---

[2] In a mature ecosystem (one in which soil organic matter is in equilibrium), this increase will likely be offset during each annual cycle by slow authochthonous decomposition, resulting in little net change in the level of soil organic matter from year to year.

much more slowly than when oxygen is plentiful. Hence, wet, anaerobic soils tend to accumulate large amounts of organic matter in a partially decomposed condition.

The products of anaerobic decomposition include a wide variety of partially oxidized organic compounds, such as organic acids, alcohols, and methane gas. Anaerobic decomposition releases relatively little energy for the organisms involved; therefore, the end products still contain much energy. (For this reason, alcohol and methane can serve as fuel.) Some of the products of anaerobic decomposition are of concern because they produce foul odors or inhibit plant growth. The methane gas produced in wet soils is a major contributor to the greenhouse effect (Section 12.11). The following reactions are typical of those carried out in wet soils by various **methanogenic bacteria:**

$$4C_2H_5COOH + 2H_2O \xrightarrow{\text{Bacteria}} 4CH_3COOH + CO_2\uparrow + 3CH_4\uparrow$$

Propionate   Acetate   Carbon   Methane
                                             dioxide

$$CH_3COOH \xrightarrow{\text{Bacteria}} CO_2\uparrow + CH_4\uparrow$$

$$CO_2 + 4H_2 \xrightarrow{\text{Bacteria}} 2H_2O + CH_4\uparrow$$

Production of Simple Inorganic Products

As proteins are attacked by microbes, the long chains of amino acids are broken, and individual amino acids appear in the soil solution along with dissolved CO_2. The amide (—R—NH_2) and sulfide (—R—S) groups of the amino acids, in turn, are broken off to produce, first, ammonium (NH_4^+) and sulfide (S^{2-}) compounds and, finally, nitrates (NO_3^-) and sulfates (SO_4^{2-}). Similar decomposition of other organic compounds releases these and other inorganic nutrient ions. The term **mineralization** applies to the overall process that releases elements from organic compounds to produce inorganic (mineral) forms. Most of the inorganic ions released by mineralization are readily available to higher plants and to microorganisms. The decay of organic tissues is an important source of nitrogen, sulfur, phosphorus, and other essential elements for plants.

12.3   FACTORS CONTROLLING RATES OF DECOMPOSITION AND MINERALIZATION[3]

The time needed to complete the processes of decomposition and mineralization may range from days to years, depending mainly on two broad factors: (1) the environmental conditions in the soil, and (2) the quality of the added residues as a food source for soil organisms.

The environmental conditions conducive to rapid decomposition and mineralization include a near-neutral pH, sufficient soil moisture, and good aeration (about 60% of the soil pore space filled with water), and warm temperatures (25 to 35°C). These conditions were discussed in Section 11.12 in relation to microbial activity and will be considered again in section 12.10 as they affect the levels of organic matter accumulating in soils. Here we will focus on factors that determine the quality of the residues as a food resource for microbes, including the physical condition of the residues, their C/N ratio, and their content of lignins and polyphenols.

Physical Factors Influencing Residue Quality

The location of residues in or on the soil is a physical factor with a critical impact on decomposition rates. Surface placement of plant residues, as in forest litter or conservation tillage mulch, usually results in slower, more variable rates of decomposition than where similar residues are incorporated into the soil by root deposition, faunal action, or tillage. The incorporated residues are in intimate contact with soil organisms and generally are kept uniformly moist. Surface residues, on the other hand, are physically out of reach of most soil organisms, save fungal mycelia and the larger fauna, such as earthworms. Surface residues are subject to drying, as well as extremes of temperature.

---

[3] For an excellent collection of papers dealing with litter decomposition in soils, see Cadisch and Giller (1997). Many of our current ideas about decomposition were first put forward in classic book by Swift, et al. (1979).

Nutrient elements mineralized from surface-applied residues are also more susceptible to loss than those from incorporated residues.

Residue particle size is another important physical factor—the smaller the particles, the more rapid the decomposition. Small particle size may result from the nature of the residues (e.g., twigs versus branches), from mechanical treatment (grinding, chopping, tillage, etc.), or from the chewing action of soil fauna. Diminution of residues into smaller particles physically exposes more surface area to decomposition, and also breaks up lignacious cell walls and waxy outer coatings on leaves so as to expose the more readily decomposed tissues and cell contents.

Carbon/Nitrogen Ratio of Organic Materials and Soils

The carbon content of typical plant dry matter is about 42% (see Figure 12.2); that of soil organic matter ranges from 40 to 58%.[4] In contrast, the nitrogen content of plant residues is much lower and varies widely (from <1 to >6%). The ratio of carbon to nitrogen C/N in organic residues applied to soils is important for two reasons: (1) intense competition among microorganisms for available soil nitrogen occurs when residues having a high C/N ratio are added to soils, and (2) the C/N ratio in residues helps determine their rate of decay and the rate at which nitrogen is made available to plants.

C/N RATIO IN PLANTS AND MICROBES. The C/N ratio in plant residues ranges from between 10:1 to 30:1 in legumes and young green leaves to as high as 600:1 in some kinds of sawdust (Table 12.2). Generally, as plants mature, the proportion of protein in their tissues declines, while the proportion of lignin and cellulose, and the C/N ratio, increase. As can be seen from the decay curves in Figure 12.4, these differences in composition have pronounced effects on the rate of decay when plant residues are added to the soil.

---

[4] The traditional value used in converting between values of soil organic matter and values of soil carbon is 58% C in the organic matter. However, this value probably applies only to highly stabilized humus, especially in subsoils. For most situations, a value of 50% is more accurate.

TABLE 12.2 Typical Carbon and Nitrogen Contents and C/N Ratios of Some Organic Materials Commonly Associated with Soils

Organic material	% C	% N	C/N
Spruce sawdust	50	0.05	600
Hardwood sawdust	46	0.1	400
Wheat straw	38	0.5	80
Papermill sludge	54	0.9	61
Corn stover	40	0.7	57
Sugarcane trash	40	0.8	50
Rye cover crop, anthesis	40	1.1	37
Bluegrass from fertilized lawn	37	1.2	31
Rye cover crop, vegetative stage	40	1.5	26
Mature alfalfa hay	40	1.8	25
Rotted barnyard manure	41	2.1	20
Finished household compost	30	2.0	15
Young alfalfa hay	40	3.0	13
Hairy vetch cover crop	40	3.5	11
Digested municipal sewage sludge	31	4.5	7
Soil microorganisms			
Bacteria	50	10.0	5
Actinomycetes	50	8.5	6
Fungi	50	5.0	10
Soil organic matter			
Spodosol O horizon	50	0.5	90
Tropical evergreen litter	50	2.0	25
Mollisol Ap horizon	56	4.9	11
Ultisol A1 horizon	52	2.3	23
Average B horizon	46	5.1	9

Data calculated from many sources.

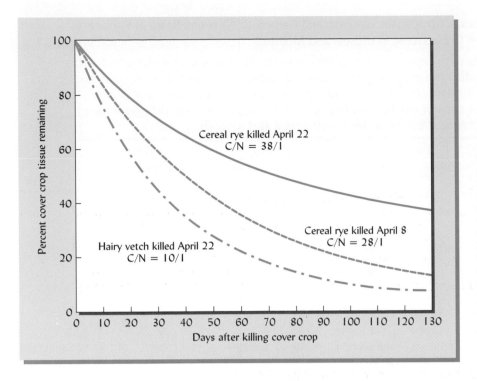

FIGURE 12.4 Rates of decomposition of various cover-crop residues. The cover crops were grown over the winter and early spring, then killed with a herbicide and the residues left as a mulch on the soil surface. Corn was planted into this mulch without tillage. The lower the initial C/N ratio of the residues, the more rapid was the decomposition process. Note that the legume (hairy vetch) had a much lower C/N ratio than the grass (cereal rye). Also note that a two-week delay in killing the rye cover crop resulted in more mature plants with a considerably higher C/N ratio and slower rate of decomposition. (Redrawn from S. Davis and R. Weil, unpublished)

In the bodies and cells of microorganisms, the C/N ratio is not only less variable than in plant tissues, but also much lower, ordinarily falling between 5:1 and 10:1. Among microorganisms, bacteria are generally somewhat richer in protein than fungi, and consequently have a lower C/N ratio.

C/N RATIO IN SOILS. The C/N ratio in the organic matter of arable (cultivated) surface (Ap) horizons commonly ranges from 8:1 to 15:1, the median being near 12:1. The ratio is generally lower for subsoils than for surface layers in a soil profile. In a given climatic region, little variation occurs in the C/N ratio for similarly managed soils. For instance, in calcium-rich soils of semiarid grasslands (e.g., Mollisols and tropical Alfisols), the C/N ratio is relatively narrow. In more severely leached and acidic soils in humid regions, the C/N is relatively wide; C/N ratios as high as 30:1 are not uncommon. When such soils are brought under cultivation and limed to increase their pH and calcium content, the enhanced decomposition tends to lower the C/N ratio to near 12:1 (see Table 12.3).

Influence of Carbon/Nitrogen Ratio on Decomposition

Soil microbes, like other organisms, require a balance of nutrients from which to build their cells and extract energy. The majority of soil organisms metabolize carbonaceous materials both in order to obtain carbon for building essential organic compounds and to obtain energy for life processes. However, no creature can multiply and grow on carbon alone. Organisms must also obtain sufficient nitrogen to synthesize nitrogen-containing cellular components, such as amino acids, enzymes, and DNA.

On the average, soil microbes must incorporate into their cells about eight parts of carbon for every one part of nitrogen (i.e., assuming the microbes have an average C/N ratio of 8:1). Because only about one-third of the carbon metabolized by microbes is incorporated into their cells (the remainder is respired and lost as CO_2), the microbes need to find about 24 parts of carbon for every part of nitrogen assimilated into their bodies (i.e., a C/N ratio of 24:1 in their "food").

This requirement results in two extremely important practical consequences. First, if the C/N ratio of organic material added to soil exceeds about 25:1, the soil microbes will have to scavenge the soil solution to obtain enough nitrogen. Thus, the incorporation of high C/N residues will deplete the soil's supply of soluble nitrogen, causing higher plants to suffer from nitrogen deficiency. Second, the decay of organic materials can be delayed if sufficient nitrogen to support microbial growth is neither present in the material undergoing decomposition nor available in the soil solution. These concepts are illustrated by the example in Figure 12.5.

TABLE 12.3 Organic Carbon, Nitrogen (in Percent), and Carbon to Nitrogen Ratio in Surface Soil (A Horizons) and Subsoil (B Horizons) in Ultisols and Mollisols

Except as noted, the Ultisols were under forest vegetation and the Mollisols were under grassland. In all cases the subsoil contains less organic carbon than the surface soils, and generally also has a lower ratio of carbon to nitrogen. In Ultisol surface soils the carbon-to-nitrogen ratio is generally near 20:1, except where the soil has been cultivated and limed. A carbon-to-nitrogen ratio of 12:1 is representative of most nonforested surface soils.

Location	Soil great group	Organic C A horizon	Organic C B horizon	Nitrogen A horizon	Nitrogen B horizon	C/N ratio A horizon	C/N ratio B horizon
Ultisols							
Ga.	Albaqult	2.2	0.54	0.095	0.060	23	9
P.R.	Plinthaqult	3.25	0.74	0.25	0.07	13[a]	10
Calif.	Haplohumult	6.69	2.16	0.26	0.103	26	21
Va.	Hapludult	3.40	0.71	0.127	0.056	27	13
Tenn.	Rhodudult	3.36	0.44	0.207	0.057	16	8
P.R.	Tropudult	2.40	0.38	0.205	0.044	12[a]	9
P.R.	Haplustult	1.64	0.49	0.149	0.053	11[a]	9
N.C.	Umbraquult	5.3	0.42	0.199	0.051	27	8
Calif.	Haploxerult	4.22	0.75	0.140	0.041	30	18
Md.	Fragiudult	3.58	0.35	0.124	0.041	29	9
Mollisols							
Kans.	Argiustoll	1.17	0.54	0.118	0.066	10[a]	8
Mont.	Argiboroll	2.06	0.74	0.192	0.074	11	10
Nebr.	Argiustoll	2.54	0.75	0.202	0.081	13	9
Kans.	Argiustoll	4.35	2.3	0.318	0.178	14	13
N. Dak.	Natriboroll	2.71	1.09	0.199	0.086	14[a]	13
Iowa	Haplaquoll	4.2	0.41	0.325	0.046	13[a]	9
Ill.	Haplaquoll	1.51	0.94	0.135	0.088	11[a]	11
Iowa	Hapludoll	2.86	0.49	0.247	0.057	12	8
Utah	Calcixeroll	3.69	2.07	0.291	0.177	13	12
Iowa	Argiaquoll	2.78	0.92	0.188	0.078	15[a]	12

[a] These data are for Ap horizons in cultivated soils. P.R. = Puerto Rico.
Data compiled from Soil Survey Staff (1975).

Examples of Inorganic Nitrogen Release During Decay

The practical significance of the C/N ratio becomes apparent if we compare the changes that take place in the soil when residues of either high or low C/N ratio are added (Figure 12.6). Consider a soil with a moderate level of soluble nitrogen (mostly nitrates). General-purpose decay organisms are at a low level of activity in this soil, as evidenced by low carbon dioxide production. If no nitrogen were lost or taken up by plants, the level of nitrates would very slowly increase as the native soil organic matter decays.

LOW NITROGEN MATERIAL.   Now consider what happens when a large quantity of readily decomposable organic material is added to this soil. If this material has a C/N ratio greater than 25, changes will occur according to the pattern shown in Figure 12.6a. In the example shown, the initial C/N ratio of the residues is about 55, typical for cornstalks or many kinds of leaf litter. As soon as the residues contact the soil, the microbial community responds to the new food supply (see Section 12.2). Heterotrophic zymogenous microorganisms become active, multiply rapidly, and yield carbon dioxide in large quantities. Because of the microbial demand for nitrogen, little or no mineral nitrogen (NH_4^+ or NO_3^-) is available to higher plants during this period.

NITRATE DEPRESSION.   This condition, often called the **nitrate depression period,** persists until the activities of the decay organisms gradually subside due to lack of easily oxidizable carbon. As their numbers decrease, carbon dioxide formation drops off, and nitrogen demand by microbes becomes less acute. As decay proceeds, the C/N ratio of the remaining plant material decreases because carbon is being lost (by respiration) and nitrogen is being conserved (by incorporation into microbial cells). Generally, one can expect mineral nitrogen to begin to be released when the C/N ratio of the remaining

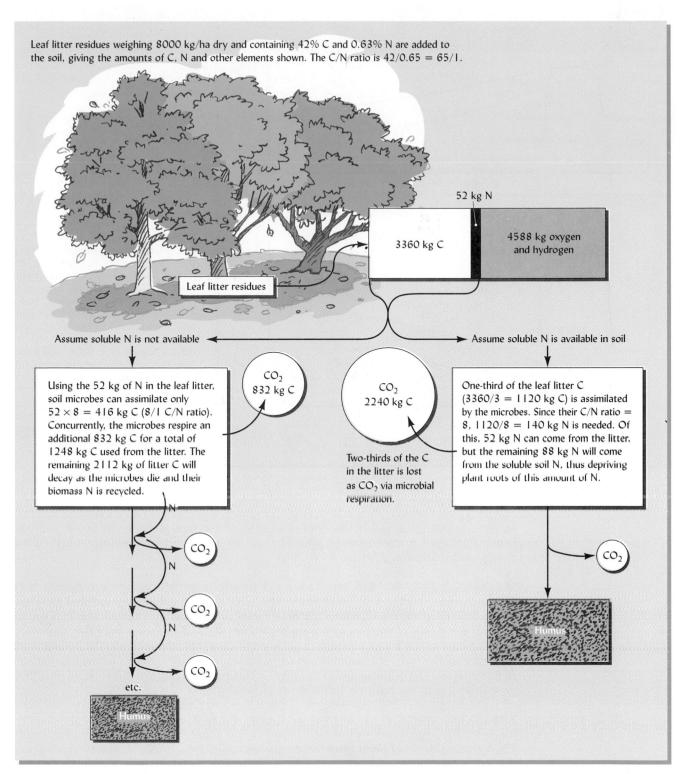

Leaf litter residues weighing 8000 kg/ha dry and containing 42% C and 0.63% N are added to the soil, giving the amounts of C, N and other elements shown. The C/N ratio is 42/0.65 = 65/1.

52 kg N

3360 kg C

4588 kg oxygen and hydrogen

Leaf litter residues

Assume soluble N is not available

Assume soluble N is available in soil

CO_2 832 kg C

CO_2 2240 kg C

Using the 52 kg of N in the leaf litter, soil microbes can assimilate only $52 \times 8 = 416$ kg C (8/1 C/N ratio). Concurrently, the microbes respire an additional 832 kg C for a total of 1248 kg C used from the litter. The remaining 2112 kg of litter C will decay as the microbes die and their biomass N is recycled.

Two-thirds of the C in the litter is lost as CO_2 via microbial respiration.

One-third of the leaf litter C (3360/3 = 1120 kg C) is assimilated by the microbes. Since their C/N ratio = 8, 1120/8 = 140 kg N is needed. Of this, 52 kg N can come from the litter, but the remaining 88 kg N will come from the soluble soil N, thus depriving plant roots of this amount of N.

N

CO_2

N

CO_2

CO_2

N

CO_2

etc.

Humus

Humus

FIGURE 12.5 A simplified, quantitative example of plant residue decay illustrating the fates of carbon and nitrogen and the consequences for decomposition and soil nitrogen availability. Note that if an adequate supply of nitrogen is available, the potential for humus creation is increased.

material drops below about 20. Then, nitrates appear again in quantity, and the original conditions prevail, except that the soil is somewhat richer in both nitrogen and humus.

The nitrate depression period may last for a few days, a few weeks, or even several months. A longer, more severe period of nitrate depression is typical when added residues are easily decomposed and have a higher C/N ratio, and when a larger quantity of residues is added. To avoid producing seedlings that are stunted, chlorotic, and nitrogen-

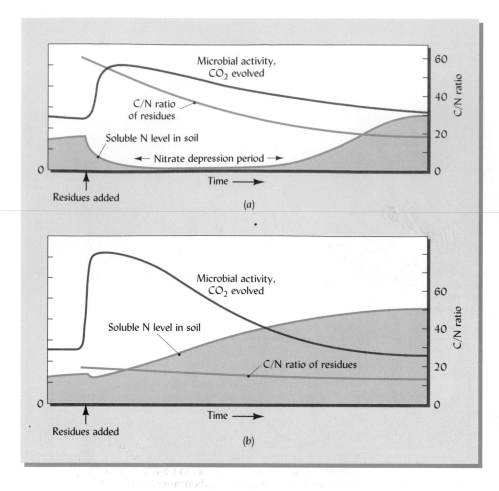

FIGURE 12.6 Changes in microbial activity, in soluble nitrogen level, and in residual C/N ratio following the addition of either high (*a*) or low (*b*) C/N ratio organic materials. Where the C/N ratio of added residues is above 25, microbes digesting the residues must supplement the nitrogen contained in the residues with soluble nitrogen from the soil. During the resulting nitrate depression period, competition between higher plants and microbes would be severe enough to cause nitrogen deficiency in the plants. Note that in both cases soluble N in the soil ultimately increases from its original level once the decomposition process has run its course. The trends shown are for soils without growing plants, which, if present, would remove a portion of the soluble nitrogen as soon as it is released.

starved, planting should be delayed until after the nitrate depression period, or additional sources of nitrogen can be applied to satisfy the nutritional requirements of both the microbes and the plants.

HIGH-NITROGEN MATERIAL. The effects on soil nitrate level will be quite different if the residues added have a C/N ratio lower than 20, as in the case represented by Figure 12.7*b*. With organic materials of low C/N ratio, more than enough nitrogen is present to meet the needs of the decomposing organisms. Therefore, soon after decomposition begins, some of the nitrogen from organic compounds is released into the soil solution, augmenting the level of soluble nitrogen available for plant uptake. Generally, nitrogen-rich materials decompose quite rapidly, resulting in a period of intense microbial growth and activity, but no nitrate depression period.

Influence of Lignin and Polyphenol Content of Organic Materials

The lignin contents of plant litter range from less than 2% to more than 50%. Those materials with high lignin content decompose very slowly. Polyphenol compounds found in plant litter may also inhibit decomposition. These phenolics are often water soluble and may be present in concentrations as high a 5 to 10% of the dry weight. By forming highly resistant complexes with proteins during residue decomposition, these phenolics can dramatically slow the rates of both nitrogen mineralization and carbon oxidation.

LITTER QUALITY. Because they support only low levels of microbial activity and biomass, residues high in phenols and/or lignin are considered to be *poor quality resources* for the soil organisms that cycle carbon and nutrients. The production of such slow-to-decompose residues by certain forest plants may help explain the accumulation of extremely high levels of humified nitrogen and carbon in the soils of mature boreal forests (Figure 12.7).

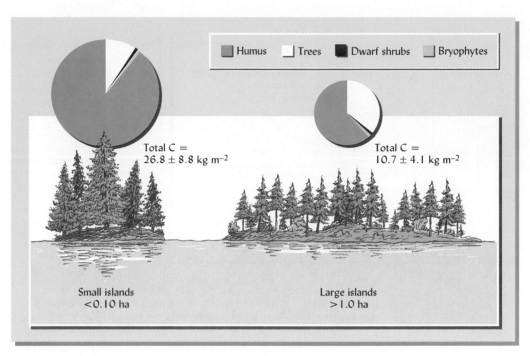

FIGURE 12.7 Effect of plant litter quality on humus accumulation in soil. In a study of some 50 forested islands off the coast of Sweden, researchers found that the frequency of lightning-caused forest fires increased with increasing size of the island. Thus the smaller islands (left) were characterized by overmature spruce forests that had gone undisturbed for many centuries, while most of the forests on the larger islands were in relatively early stages of recovery from the last forest fire. Tree species that dominate overmature, undisturbed boreal forests produce litter higher in lignin and phenolics than do species associated with early successional stages following a fire or other disturbance. The poor quality litter on the small islands led to the formation of protein–phenolic complexes with extremely slow carbon and nitrogen release rates. As a result, plant productivity was somewhat reduced on the smaller islands in this study, but both the *proportion* of the ecosystem carbon stored in the soil humus and the *total amount* of humus carbon per hectare were far greater on the smaller islands. The area of each pie chart is proportional to the total C/m². [Data from Wardle, et al. (1997)]

The lignin and phenol contents also influence the decomposition and release of nitrogen from **green manures**—plant residues used as to enrich agricultural soils (Table 12.4). For example, in the leaves of certain legume trees, the C/N ratio is quite narrow, but the phenol content is quite high, so that when these leaves are added to soil, nitrogen is released only slowly—often too slowly to keep up with the needs of a growing crop. Similarly, residues with a lignin content of more than 20 to 25% will decompose too slowly to be effective as green manure for rapidly growing annual crops. For peren-

TABLE 12.4 **Litter Quality in Relation to the Lignin Content, Polyphenol Content and C/N Ratio of Several Types of Plant Residues**

Prunings (leaves and small twigs) of three common agroforestry tree species and afterharvest residues of two cereal crops were applied at a rate of 5 Mg/ha to an Oxic Paleudult in a humid tropical region of Nigeria. Low values of C/N, lignin, and polyphenols all contribute to high litter quality and high speed of decomposition. The inhibitory effect of polyphenol content can be seen by comparing Gliricidia to Leucaena.

Plant species	Plant parts	Lignin, %	Polyphenols, %	C/N	Decomposition constant,[a] k/week	Litter quality
Gliricidia sepium	Prunings	12	1.6	13	0.255	High
Leucaena leucocephala	Prunings	13	5.0	13	0.166	Medium–high
Oryza sativa	Straw	5	0.6	42	0.124	Medium
Zea mays	Stover	7	0.6	43	0.118	Medium
Dactyladenia barteri	Prunings	47	4.1	28	0.011	Low

[a] As each type of residue decomposed during a 14-week season, researchers periodically determined the proportion Y of the original residue dry matter remaining. The decomposition rate k was determined from the equation $Y = e^{-kt}$, in which e is the base of natural logarithms, and t is time in weeks. Therefore, the larger the decomposition constant k, the faster the decomposition.
Data selected from Tian, et al. (1992 and 1995).

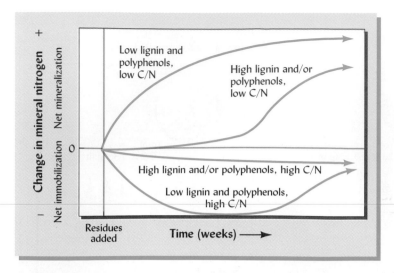

FIGURE 12.8 Temporal patterns of nitrogen release from organic residues differing in quality based on their C/N ratios and contents of lignin and polyphenols. Lignin contents greater than 20%, polyphenol contents greater than 3%, and C/N ratios greater than 30 would all be considered high in the context of this diagram, the combination of these properties characterizing litter of poor quality—that is, litter that has a limited potential for microbial decomposition and mineralization of plant nutrients.

nial crops, however, the slow release of nitrogen from such green manure residues may be advantageous in the long run, as the nitrogen may be less subject to losses. By the same token, the slow decomposition of phenol- or lignin-rich materials means that even if their C/N ratio is very high, the nitrate depression will not be pronounced.

Figure 12.8 illustrates the combined effects of C/N ratio and lignin or phenol content on the balance between immobilization and mineralization of nitrogen during plant residue decomposition.

12.4 HUMUS: GENESIS AND NATURE[5]

The general term **soil organic matter** (SOM) encompasses all the organic components of a soil: (1) living biomass (intact plant and animal tissues and microorganisms), (2) dead roots and other recognizable plant residues, as well as (3) a largely amorphous and colloidal mixture of complex organic substances no longer identifiable as tissues. Only the third category of organic material is properly referred to as **soil humus** (Figure 12.9).[6]

Microbial Transformations

As decomposition of plant residues proceeds, microbes slowly break down complex components into simpler compounds. In this process some of the lignin is broken down into its phenolic subunits. The soil microbes then metabolize the resulting simpler compounds. Using some of the carbon not lost as carbon dioxide in respiration, along with most of the nitrogen, sulfur, and oxygen from these compounds, the microorganisms synthesize new cellular components and biomolecules. Some of the original lignin is not completely broken down, but only modified to form complex residual molecules that retain many of the characteristics of lignin. The microbes polymerize (link together) some of the simpler new compounds with each other and with the complex residual products into long, complex chains that resist further decomposition. These high-molecular-weight compounds interact with nitrogen-containing amino compounds, giving rise to a significant component of resistant humus. The presence of colloidal clays stimulates the complex polymerization. These ill-defined, complex, resistant, polymeric compounds are called **humic substances.** The term **nonhumic substances** refers to the group of identifiable biomolecules that are mainly produced by microbial action and are less resistant to breakdown.

---

[5] For discussions of humus formation, composition, and reactions, see Frimmel and Christman (1988) and Stevenson (1994).

[6] Although many soil scientists use the term *soil organic matter* synonymously with *humus,* we exclude soil biomass and dead but recognizable tissues from the latter, while including them in the former. As a practical matter, particles of plant residue that do not pass a 2-mm sieve opening are often excluded from consideration as soil organic matter.

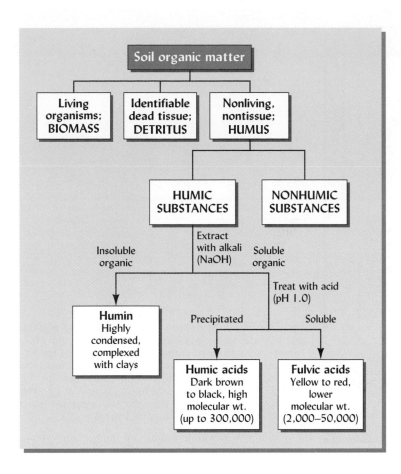

FIGURE 12.9 Classification of soil organic matter components separable by chemical and physical criteria. Although surface residues (litter) are not universally considered to be part of the soil organic matter, we include them because they are the principal component of the O horizons in soil profiles. Solubility in alkali and acid is a widely used criterion for grouping different fractions of soil humus. The classical scheme dividing soil humic substances into humin, fulvic acids, and humic acids fractions (shown in the lower part of the flowchart) is based on their insolubility in NaOH (humin) and their subsequent solubility (fulvic acids) and insolubility (humic acids) in acid solutions (pH = 1).

One year after plant residues are added to the soil, most of the carbon has returned to the atmosphere as CO_2, but one-fifth to one-third is likely to remain in the soil either as live biomass ($\approx$5%) or as the humic ($\approx$20%) and nonhumic ($\approx$5%) fractions of soil humus (Figure 12.10). The proportion remaining from root residues tends to be somewhat higher than that remaining from incorporated leaf litter.

Humic Substances

Humic substances comprise about 60 to 80% of the soil organic matter. They are comprised of huge molecules with variable, rather than specific, structures and composition. Humic substances are characterized by aromatic, ring-type structures that include polyphenols (numerous phenolic compounds linked together) and comparable polyquinones, which are even more complex. Humic substances generally are dark-colored, amorphous substances with molecular weights varying from 2000 to 300,000 g/mol. Because of their complexity, they are the organic materials most resistant to microbial attack.

SOLUBILITY GROUPINGS. Historically, humic substances have been classified into three chemical groupings based on solubility (see Figure 12.9): (1) *fulvic acid,* lowest in molecular weight and lightest in color, soluble in both acid and alkali, and most susceptible to microbial attack; (2) *humic acid,* medium in molecular weight and color, soluble in alkali but insoluble in acid, and intermediate in resistance to degradation; and (3) *humin,* highest in molecular weight, darkest in color, insoluble in both acid and alkali, and most resistant to microbial attack. Unfortunately, this classification of humic substances has proved to be of only limited relevance to ecological processes.

All three groups of humic substances are relatively stable in soils. Even fulvic acid, the most easily degraded, is more resistant to microbial attack than most freshly applied plant residues. Depending on the environment, the half-life (the time required to destroy half the amount of a substance) of fulvic acid may be 10 to 50 years, while the half-life of humic acid is generally measured in centuries.

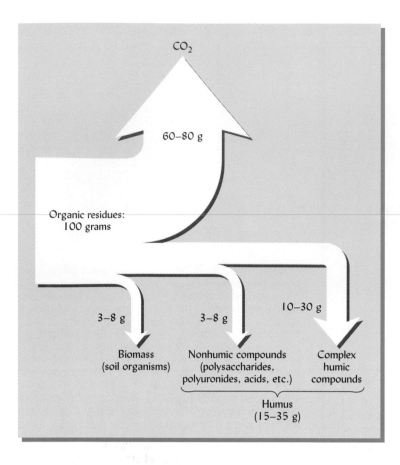

FIGURE 12.10 Disposition of 100 g of organic residues one year after they were incorporated into the soil. More than two-thirds of the carbon has been oxidized to CO_2, and less than one-third remains in the soil—some in the cells of soil organisms, but a larger component as soil humus. The amount converted to CO_2 is generally greater for aboveground residues than for belowground (root) residues. (Estimates from many sources)

Nonhumic Substances

About 20 to 30% of the humus in soils consists of nonhumic substances. These substances are less complex and less resistant to microbial attack than those of the humic group. Unlike humic substances, they are comprised of specific biomolecules with definite physical and chemical properties. Some of these nonhumic substances are microbially modified plant compounds, while others are compounds synthesized by soil microbes as by-products of decomposition.

Included among the nonhumic substances are polysaccharides, polymers which have sugarlike structures and a general formula of $C_n(H_2O)_m$, where n and m are variable. Polysaccharides are especially important in enhancing soil aggregate stability (see Section 4.7). Also included are polyuronides, which are not found in plants, but are synthesized by soil microbes.

Some even simpler compounds (such as low-molecular-weight organic acids and some proteinlike materials) are part of the nonhumic group. Although none of these simpler materials are present in large quantities, they may influence the availability of plant nutrients, such as nitrogen and iron, and may also directly affect plant growth.

Stability of Humus

Studies using radioactive isotopes have shown that some organic carbon converted to humus thousands of years ago is still present in soils, evidence that some humic materials are extremely resistant to microbial attack. This resistance of humic substances to oxidation is important in maintaining soil organic matter levels and in protecting associated nitrogen and other essential nutrients against rapid mineralization and loss from the soil. For example, the formation of polyphenol–protein complexes can protect the protein nitrogen from microbial attack. Despite its relative resistance to decay, humus is subject to continual microbial attack. Without the annual addition of sufficient plant residues, microbial oxidation will result in reduced soil organic matter levels.

CLAY–HUMUS COMBINATIONS.[7]   Interaction with clay minerals provides another means of stabilizing soil nitrogen and organic matter. Certain clays are known to attract and hold such substances as amino acids, peptides, and proteins, forming complexes that protect the nitrogen-containing compounds from microbial degradation. Organic matter that is entrapped in the very small (<1 μm) pores formed by clay particles is physically inaccessible to decomposing organisms. It is also possible that layer silicate clays, such as vermiculite, may bind organic matter in their interlayers in forms that strongly resist decomposition.

Thus, clay, along with humic and polysaccharide polymers, can protect relatively simple nitrogen compounds from microbial attack. In many soils more than half the organic matter is associated with clay and other inorganic constituents. Although the extent and mechanisms are not yet fully understood, clay–humus interactions partially account for the high organic matter content of clay soils.

Colloid Characteristics of Humus

The colloidal nature of humus was emphasized in Chapter 8. The surface area of humus colloids per unit mass is very high, generally exceeding that of silicate clays. The colloidal surfaces of humus are negatively charged, as a result of H^+ dissociation from carboxylic (—COOH) or phenolic (—OH) groups. The extent of the negative charge is pH-dependent. At high pH values, the cation exchange capacity of humus on a mass basis (150 to 300 cmol/kg) far exceeds that of most silicate clays. However, because humus is much less dense than clay, the cation exchange capacity of humus on a volume basis (40 to 80 cmol/L) is comparable to that of silicate clays. Cation exchange reactions with humus are qualitatively similar to those occurring with silicate clays. The humus micelles, like the particles of clay, carry a swarm of adsorbed cations (Ca^{2+}, H^+, Mg^{2+}, K^+, etc.), which are exchanged with cations from the soil solution (see Section 8.11).

The water-holding capacity of humus on a mass basis (but not on a volume basis) is four to five times that of the silicate clays. Humus plays a role in aggregate formation and stability. The highly complex humus molecules contain chemical structures that absorb nearly all wavelengths of visible light, giving the substance a characteristic black color (see, for example, the humus-rich A horizons in Plates 2 and 8, following page 82).

12.5   COMPOSTS AND COMPOSTING

Composting is the practice of creating humuslike organic materials outside of the soil by mixing, piling, or otherwise storing organic materials under conditions conducive to aerobic decomposition and nutrient conservation. The decomposition processes and organisms involved are very similar to those already described for the formation of humus in soils. The main difference is that with composting, decay occurs outside of the soil, and in such a concentrated fashion as to generate considerable heat. The finished product, **compost**, is popular as a mulch, as an ingredient for potting mixes, and as an organic soil conditioner and slow-release fertilizer.

COMPOSTING PROCESS.   When a sufficient mass of suitable organic materials is kept in a moist, well-aerated, state, it goes through a three-stage process of decomposition (Figure 12.11). (1) During a brief initial *mesophilic* stage sugars and readily available microbial food sources are rapidly metabolized, causing the temperature in the compost pile to gradually rise from ambient to over 40°C. (2) A *thermophilic* stage occurs during the next week or two, during which temperatures rise to 50 to 75°C while thermophilic organisms decompose cellulose and other more resistant materials. Frequent mixing during this stage is essential to maintain oxygen supplies and assure even heating of all the material. The easily decomposed compounds are used up, and humuslike compounds are formed during this stage. (3) A second mesophilic or *curing stage* follows, during which the temperature falls back to near ambient, and the material is recolonized by mesophilic organisms, including certain beneficial organisms that produce plant-

---

[7] For a detailed account of clay–humus complexes, see Huang and Schnitzer (1986) and Loll and Bollag (1983).

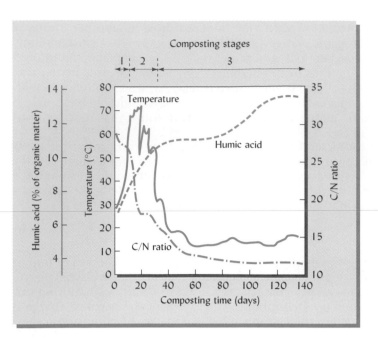

FIGURE 12.11 Changes in the temperature, C/N ratio, and humic acid contents of the organic matter during the production of compost from municipal solid wastes (MSW). The compost was turned (mixed for aeration) once a week for the first 49 days and once in two weeks thereafter. The stages of the compost process are: (1) initial mesophilic stage, (2) thermophilic stage, and (3) final mesophilic or curing stage. See text for explanation of the composting process. [Data redrawn from Chefetz, et al. (1996)]

growth-stimulating compounds or are antagonistic to plant pathogenic fungi. This stage may last from several weeks to several months.

NATURE OF THE COMPOST PRODUCED. As raw organic materials are humified in a compost pile, the content of nonhumic substances declines, and the content of humic acid increases markedly (see Figure 12.11). During the composting process, the C/N ratio of organic materials in the pile decreases until a fairly stable ratio, in the range of 14:1 to 20:1, is achieved. The CEC of the organic matter increases to about 50 to 70 cmol$_c$/kg of compost.

Although 50% or more of the carbon in the initial material is typically lost during composting, mineral nutrients are mostly conserved. Finished compost is therefore generally more concentrated in nutrients than the initial combination of raw materials used and can serve as an effective means of building soil fertility. Finished compost should be free of viable weed seeds and pathogenic organisms, as these are generally destroyed during the thermophilic phase. However, inorganic contaminants such as heavy metals are *not* destroyed by composting.

Proper management of compost is essential if the finished product is to be desirable for use as a potting media or soil amendment (see Box 12.1).

BOX 12.1   MANAGEMENT OF A COMPOST PILE

MATERIALS TO USE. For homeowners, good materials for composting include leaves (preferably shredded), grass clippings, weeds (preferably before they have gone to seed), kitchen scraps, wood shavings, gutter cleanings, pine needles, spoiled hay (especially legume hay), straw (old construction bales are often available), and even vacuum cleaner dust. Large-scale commercial compost is often made from materials such as municipal refuse (primarily garbage and paper), sewage sludge, wood chips, animal manures, municipal leaves, and food processing wastes.

MATERIALS TO AVOID. Some of the materials to avoid using are meat scraps (attract rodents), cat droppings (carry microbes that can harm human infants and fetuses if the mother is exposed), wood wastes from pressure-treated lumber and plywood (contain heavy metals and arsenic), and plastics and glass (nonbiodegradable; nuisance or dangerous in final product).

BALANCING NUTRIENTS. For efficient composting and an optimal final product, the nutrient content of the raw materials should be kept in balance. Although highly carbonaceous materials can be made to compost satisfactorily if they are turned frequently and kept moist, best results are obtained if high-C/N-ratio materials (such as brown leaves, straw, or paper) is mixed with low-C/N-ratio materials (such as green grass clippings, legume hay, blood meal, sewage sludge or livestock manure) so as to achieve an overall C/N ratio of 30 or less. Nitrogen fertilizer can also be added to lower the C/N ratio.

(continued)

BOX 12.1 (Cont.)   MANAGEMENT OF A COMPOST PILE

Other materials often added to compost to improve nutrient balance and content are mixed fertilizers (high in N, P, and K), wood ashes (high in K, Ca, and Mg), bone meal or phosphate rock powder (high in P and Ca), and seaweed (high in K, Mg, Ca, and micronutrients). Some of these materials contain enough soluble salts to necessitate leaching of the finished compost before using it on salt-sensitive plants.

COMPOSTING METHODS.   The design of the compost pile must allow for good aeration throughout the pile while still providing sufficient mass to generate heat and prevent excessive drying. Backyard compost piles usually measure 1 to 1.5 m square and 1 to 1.5 m high. Various commercially made compost bins are available to make turning the compost easier or save the trouble of building a bin. Large-scale composting is usually carried out in *windrows*, about 2 to 3 m wide, 2 m tall, and many meters long. In dry climates, composting is often done in pits dug about 1 m deep to protect the material from drying. The various materials to be composted can be mixed together or applied in thin layers. Often, a small amount of garden soil or finished compost is added to ensure that plenty of decomposer organisms will be immediately available. Microbial compost activators are commercially available, but while some may speed the initial heating of the pile, scientific tests rarely show any other advantage to using these preparations.

OXYGEN AND MOISTURE CONTROL.   Control of moisture content and oxygen levels is critical for successful composting. Low oxygen levels, usually due to inadequate turning combined with excessive moisture, can produce putrid odors as anaerobic decomposition takes over. Monitoring temperature and oxygen levels in the pile can help avoid this situation. Good aeration can be promoted by mixing in a bulking agent such as wood chips, by avoiding excessive packing, and by either turning the pile or pulling a stream of air through it (Figure 12.12). The pile should be turned during the thermophilic stage whenever the temperature begins to drop.

The compost should be kept moist, but not wet. When squeezed, compost containing the proper amount of moisture (50 to 70%) will feel damp, but not dripping wet. Turning the pile during dry weather can help reduce excess moisture, while turning it during rain can help moisten a too-dry pile.

(a)

(b)

FIGURE 12.12   An efficient and easily managed method of composting suitable for homeowners is the three bin method (a), in which materials are moved from one bin to the next when they are turned. Perforated white plastic pipes are inserted for enhanced aeration. The bin on the left contains relatively fresh materials, while the one at the right contains finished compost. These bins at the Rodale Research Institute in Pennsylvania are being used to compost mainly lawn cuttings and plant residues from a vegetable garden. Most large-scale composting operations use long windrows that can be turned by special machines or aerated by pulling air through the pile and into perforated pipes (b). In the case shown here, sewage sludge mixed with wood chips is being composted by the USDA Beltsville aerated pile method. Air is being drawn through the large composting windrow and expelled through the smaller piles of finished compost, which absorb odors from the airstream. (Photos courtesy of R. Weil)

BENEFITS OF COMPOSTING. Although making compost may involve more work and expense than applying uncomposted organic materials directly to the soil, the process offers at least seven distinct advantages:

1. *Safe storage.* Composting provides a means of effectively and safely storing organic materials until it is convenient to apply them to soils.

2. *Easier handling.* As a result of CO_2 losses and settling, the volume of composted organic materials decreases by about 30 to 50% during the composting process. The smaller volume and greater uniformity of the resulting material may greatly ease the handling and eventual use of the organic matter as a soil amendment or potting medium.

3. *Nitrogen competition avoidance.* For residues with a high initial C/N ratio, proper composting ensures that any nitrate depression period will occur in the compost pile, not in the soil, thereby avoiding induced plant nitrogen deficiency.

4. *Nitrogen stabilization.* Composting can reduce environmentally damaging nitrate leaching (see Section 13.8) from organic wastes with very low C/N ratios (such as livestock manure and sewage sludge). When applied to the soil, composted materials generally decompose and mineralize much more slowly than uncomposted organic materials. **Cocomposting** such low-C/N-ratio materials with high-C/N-ratio materials, such as sawdust, wood chips, senescent tree leaves, or municipal solid waste, provides sufficient carbon for microbes to immobilize the excess nitrogen and minimize any leaching hazard from the low-C/N materials. It also provides sufficient nitrogen to speed the decomposition of the high-C/N materials.

5. *Partial sterilization.* High temperatures during the thermophilic stage in well-managed compost piles kill most weed seeds and pathogenic organisms in a matter of a few days. Under less ideal conditions, temperatures in parts of the pile may not exceed 40 to 50°C, so weeks or months may be required to achieve the same results.

6. *Detoxification.* Most toxic compounds that may be in organic wastes (pesticides, natural phytotoxic chemicals, etc.) are destroyed by the time the compost is considered mature and ready to use. Compost is therefore often used as a method of biological treatment of polluted soils and wastes (see Section 18.6).

7. *Disease suppression.* Some composts can effectively suppress soilborne plant diseases by encouraging microbial antagonisms (see Section 11.15). Most success in disease suppression has occurred when well-cured compost is used as a main component of potting mixes for greenhouse-grown plants. Some disease suppression has also been observed with field applications of compost.

12.6 DIRECT INFLUENCES OF ORGANIC MATTER ON PLANT GROWTH

Long ago, the observation that plants generally grow better on organic-matter-rich soils led people to think that plants derive much of their nutrition by absorbing humus from the soil. We now know that higher plants derive their carbon from carbon dioxide and that most of their nutrients come from inorganic ions dissolved in the soil solution. In fact, plants can complete their life cycles growing totally without humus, or even without soil (as in soilless or **hydroponic** production systems using only aerated nutrient solutions). This is not to say that soil organic matter is less important to plants than was once supposed, but rather that most of the benefits accrue to plants indirectly through the many influences of organic matter on soil properties. These will be discussed in Section 12.7, after we consider two types of direct organic matter effects on plants.

Direct Influence of Humus on Plant Growth

It is well established that certain organic compounds are absorbed by higher plants. Plants can absorb a very small portion of their nitrogen and phosphorus needs as soluble organic compounds. Various growth-promoting compounds such as vitamins, amino acids, auxins, and gibberellins, are formed as organic matter decays. These substances may at times stimulate growth in both higher plants and microorganisms.

Small quantities of both fulvic and humic acids in the soil solution are known to enhance certain aspects of plant growth (Table 12.5). Components of these humic sub-

TABLE 12.5 Some Direct Effects of Humic Substances on Plant Growth

Effect on plant growth	Humic substance	Concentration range, mg/L
Accelerated water uptake and enhanced germination of seeds	Humic acid	1–100
Stimulated root initiation and elongation	Humic and fulvic acids	50–300
Enhanced root cell elongation	Humic acid	5–25
Enhanced growth of plant shoots and roots	Humic and fulvic acids	50–300

From Chen and Aviad (1990).

stances probably act as regulators of specific plant-growth functions, such as cell elongation and lateral root initiation. The concentrations of humic substances commonly present in the soil solution in humid regions (50 to 100 mg/L or ppm) are effective in stimulating plant growth. Commercial humate products have been marketed with claims that small amounts enhance plant growth, but scientific tests of many of these products have failed to show any benefit from their use. Perhaps this is because effective levels of humic substances are naturally present in most soils.

ALLELOCHEMICAL EFFECTS.[8] **Allelopathy** is the process by which one plant infuses the soil with a chemical that affects the growth of other plants. The plant may do this by directly exuding **allelochemicals**, or the compounds may be leached out of the plant foliage by through-fall rainwater. In other cases, microbial metabolism of dead plant tissues (residues) forms the allelochemicals (Figure 12.13). Occasionally, the term *allelochemical* is also applied to plant chemicals that inhibit microorganisms. In principal, the interactions are much like the antagonistic relationships among certain microorganisms discussed in Section 11.15.

Allelochemicals present in the soil are apparently responsible for many of the effects observed when various plants grow in association with one another. Because they produce such chemicals, certain weeds (e.g., johnsongrass and giant foxtail) damage crops far out of proportion to the size and number of weeds present. Crop residues left on the soil surface may inhibit the germination and growth of the next crop planted (e.g., wheat residues often inhibit sorghum plants). Other allelopathic interactions influence the succession of species as forest and grassland ecosystems mature.

Allelopathic interactions are usually very specific, involving only certain species, or even varieties, on both the producing and receiving ends. The effects of allelopathic chemicals are many and varied. Although the term *allelopathy* most commonly refers to negative effects, allelochemical effects can also be positive (as in certain **companion plantings**). While they vary in chemical composition, most allelochemicals are relatively simple phenolic or organic acid compounds that could be included among the nonhumic substances found in soils. Because most of these compounds can be rapidly destroyed by soil microorganisms or easily leached out of the root zone, effects are usually relatively short-lived once the source is removed.

---

[8] For a comprehensive review of allelopathy, see Inderjit, et al. (1995).

FIGURE 12.13 Positive and negative allelopathic effects of winged beans on grain amaranth plants. In the pot on the left (T4) amaranth is growing in fresh soil (no association with winged beans). In the center pot (T20) amaranth is growing in soil previously used to grow winged beans (positive effect). In the pot on the right (T28) the amaranth is growing in fresh soil, but the plant was watered three times with a water extract of winged bean tissue (negative effect). The average dry weight of the amaranth plants for each treatment is shown. All pots were watered with a complete nutrient solution. [From Weil and Belmont (1987)]

12.7 INFLUENCE OF ORGANIC MATTER ON SOIL PROPERTIES AND THE ENVIRONMENT

Soil organic matter affects so many soil properties and processes that a complete discussion of the topic is beyond the scope of this chapter. Indeed, in almost every chapter in this book there is mention of the roles of soil organic matter. Figure 12.14 summarizes some of the more important effects of organic matter on soil properties and on soil–environment interactions. Often one effect leads to another, so that a complex chain of multiple benefits results from the addition of organic matter to soils. For example (beginning at the upper left in Figure 12.14), adding organic mulch to the soil surface encourages earthworm activity, which in turn leads to the production of burrows and other biopores, which in turn increases the infiltration of water and decreases its loss as runoff, a result that finally leads to less pollution of streams and lakes.

INFLUENCE ON SOIL PHYSICAL PROPERTIES.   Humus tends to give surface horizons dark brown to black colors. Granulation and aggregate stability are encouraged, especially by the non-humic substances produced during decomposition. The humic fractions help reduce the plasticity, cohesion, and stickiness of clayey soils, making these soils easier to manipulate. Soil water retention is also improved, since organic matter increases both infiltration rate and water-holding capacity.

INFLUENCE ON SOIL CHEMICAL PROPERTIES.   Because humus has a cation exchange capacity (CEC) 2 to 30 times as great (per kg) as that of the various types of clay minerals, it generally accounts for 50 to 90% of the cation-adsorbing power of mineral surface soils. Like clays, humus colloids hold nutrient cations (potassium, calcium, magnesium, etc.) in easily exchangeable form, wherein they can be used by plants but are not too readily leached out of the profile by percolating waters. Through its cation exchange capacity and acid and base functional groups, organic matter also provides much of the pH buffering capacity in soils (see Section 9.5). In addition, nitrogen, phosphorus, sulfur, and micronutrients are stored as constituents of soil organic matter, from which they are slowly released by mineralization.

Humic acids also attack soil minerals and accelerate their decomposition, thereby releasing essential nutrients as exchangeable cations. Organic acids, polysaccharides, and fulvic acids all can attract such cations as Fe^{3+}, Cu^{2+}, Zn^{2+}, and Mn^{2+} from the edges of mineral structures and **chelate** or bind them in stable organomineral complexes. Some of these metals are made more available to plants as micronutrients because they are kept in soluble, chelated form (see Chapter 15). In very acid soils, organic matter alleviates aluminum toxicity by binding the aluminum ions in nontoxic complexes (see Section 9.2).

BIOLOGICAL EFFECTS.   Soil organic matter greatly influences the biology of the soil, because it provides most of the food for the community of heterotrophic soil organisms described in Chapter 11. In Section 12.3 it was shown that the quality of plant litter and soil organic matter markedly affects decomposition rates and, therefore, the amount of organic matter accumulating in soils.

12.8 MANAGEMENT OF AMOUNT AND QUALITY OF SOIL ORGANIC MATTER

Perhaps the most useful approach to defining soil organic matter quality is to recognize different pools of organic carbon that vary in their susceptibility to microbial metabolism. A model identifying five such pools of carbon in plant residues and soil organic matter is illustrated in Figure 12.15. As discussed in Section 12.2, plant residues contain some components (*metabolic carbon*), such as sugars, proteins, and starches, that are quite readily metabolized by soil microbes. Other components (*structural carbon*) exist mostly in the structure of the plant cell walls, including lignin, polyphenols, cellulose, and waxes, and are resistant to decomposition. The model in Figure 12.15 denotes these groups as metabolic and structural pools within the plant residues.

The total organic matter content of a soil is the sum of several different pools of soil organic matter, namely, the **active, slow,** and **passive** fractions.

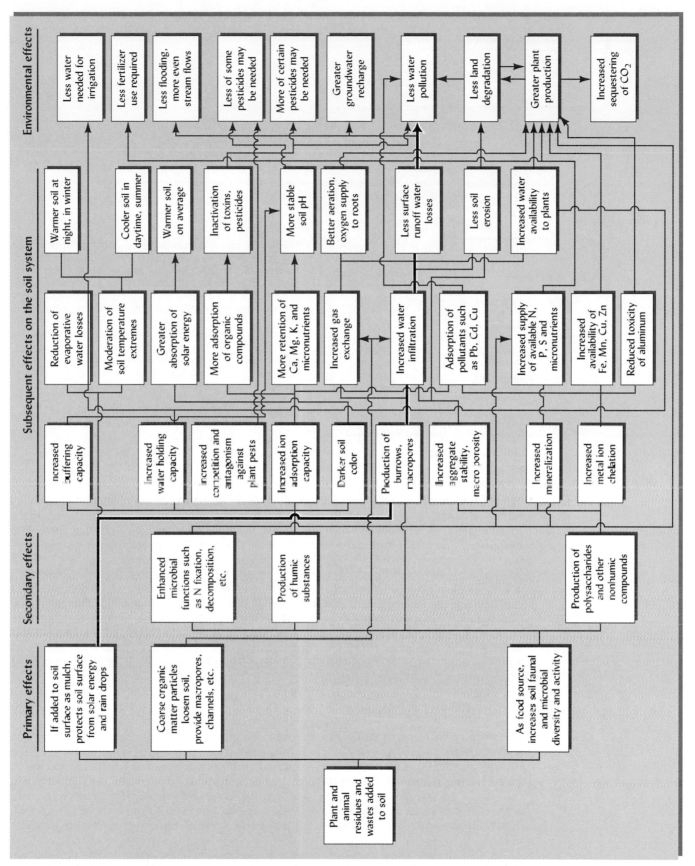

FIGURE 12.14 Some of the ways in which soil organic matter influences soil properties, plant productivity, and environmental quality. Many of the effects are indirect, the arrows indicating the cause-and-effect relationships. It can readily be seen that the influences of soil organic matter are far out of proportion to the relatively small amounts present in most soils. Many of these influences are discussed in this and other chapters in this book. The thicker line shows the sequence of effects referred to in the text in Section 12.7.

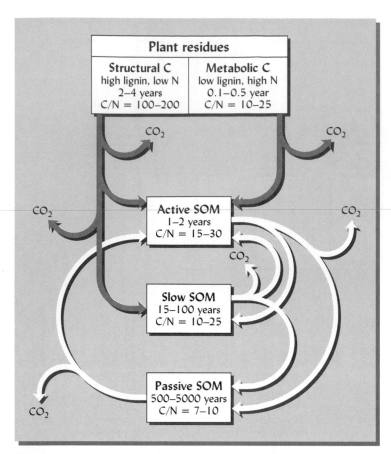

FIGURE 12.15 A conceptual model that recognizes various pools of soil organic matter (SOM) differing by their susceptibility to microbial metabolism. Models that incorporate *active, slow,* and *passive* fractions of soil organic matter have proven very useful in explaining and predicting real changes in soil organic matter levels and in attendant soil properties. Note that microbial action can transfer organic carbon from one pool to another. For example, when the nonhumic substances and other components of the active fraction are rapidly broken down, some resistant, complex by-products may be formed, adding to the slow and passive fractions. Note that all these metabolic changes result in some loss of carbon from the soil as CO_2. [Adapted from Paustian, et al. (1992)]

Active Fraction of Soil Organic Matter

The **active fraction** of soil organic matter consists of materials with relatively high C/N ratios (about 15 to 30) and short half-lives (half of these materials can be metabolized in a matter of a few months to a few years). Components probably include the living biomass, some of the fine particulate detritus (referred to as **particulate organic matter**, or POM), most of the polysaccharides, and other nonhumic substances described in Section 12.4, as well as some of the more labile (easily decomposed) fulvic acids. This active fraction provides most of the readily accessible food for the soil organisms and most of the readily mineralizable nitrogen. It is responsible for most of the beneficial effects on structural stability that lead to enhanced infiltration of water, resistance to erosion, and ease of tillage. The active fraction can be readily increased by the addition of fresh plant and animal residues, but it is also very readily lost when such additions are reduced or tillage is intensified. This fraction rarely comprises more than 10 to 20% of the total soil organic matter.

Passive and Slow Fractions of Organic Matter

The **passive fraction** of soil organic matter consists of very stable materials remaining in the soil for hundreds or even thousands of years. This fraction includes most of the humus physically protected in clay–humus complexes, most of the humin, and much of the humic acids. The passive fraction accounts for 60 to 90% of the organic matter in most soils, and its quantity is increased or diminished only slowly. The passive fraction is most closely associated with the colloidal properties of soil humus, and it is responsible for most of the CEC and water-holding capacity contributed to the soil by organic matter.

Intermediate in properties between the active and passive fractions is the *slow fraction* of soil organic matter. This fraction probably includes very finely divided plant tissues, high in lignin and other slowly decomposable and chemically resistant components. The half-lives of these materials are typically measured in decades. The slow fraction is an important source of mineralizable nitrogen and other plant nutrients, and it provides the underlying food source for the steady metabolism of the autochthonous soil microbes (see Section 12.2). The slow fraction also probably makes some contribution to the effects associated primarily with the active and passive fractions.

Changes in Active and Passive Fractions with Soil Management

Soil analytical methods can only evaluate chemical or physical fractions of soil organic matter that approximate the functionally defined fractions: active, slow, and passive. Although methods are not yet available to accurately isolate and measure the active, slow, and passive fractions of soil organic matter, scientists have consistently observed that more productive soils managed with conservation-oriented practices contain relatively high proportions of such active-fraction components as microbial biomass and oxidizable sugars (see, for example, Table 20.10, page 804).

Despite the analytical difficulties, models which assume the existence of these three pools have proven very useful in explaining and predicting real changes in soil organic matter levels and in attendant soil properties. Studies on the dynamics of soil organic matter have established that the different fractions of soil organic matter play quite different roles in the soil system and in the carbon cycle.

The presence of a resistant (structural) pool of carbon in plant residues, as well as an easily decomposed (metabolic) pool, explains the initially rapid but decelerating rate of decay that occurs when plant tissues are added to a soil (see Figures 12.3 and 12.4). Similarly, the existence of a pool of complex, chemically and physically protected soil organic matter (passive fraction), as well as a pool of easily metabolized soil organic matter (active fraction), explains why conversion of native forests or grassland into cultivated cropland results in a very rapid decline in soil organic matter during the first few years, followed by a much slower decline thereafter (see, for example, Figure 12.16).

Soil-management practices that cause only very small changes in total soil organic matter often cause rather pronounced alterations in aggregate stability, nitrogen mineralization rate, or other soil properties attributed to organic matter. This occurs because the relatively small pool of active organic matter may undergo a large percentage increase or decrease without having a major effect on the much larger pool of total organic matter. Figure 12.16 shows how the different fractions of soil organic matter are

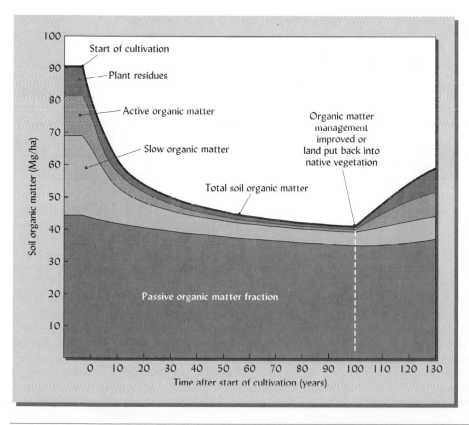

FIGURE 12.16 Changes in various fractions of organic matter in the upper 25 cm of a representative soil after bringing virgin land under cultivation. Initially, under natural vegetation, this soil contained about 91 Mg/ha of total organic matter. The resistant *passive fraction* accounted for about 44 Mg/ha, or about half of the total soil organic matter. The rapidly decomposing *active fraction* accounted for about 14 Mg, or about 16% of the total soil organic matter. After about 40 years of cultivation, the passive fraction had declined by about 11% to about 39 Mg/ha, while the active fraction had lost 90% of its mass, declining to only 1.4 Mg/ha. Note that much of the organic loss due to the change in land management came at the expense of the active fraction. This was also the fraction that most quickly increased when improved organic matter management was adopted after the 100th year. The susceptibility of the active fraction to rapid change explains why even relatively small changes in total soil organic matter can produce dramatic changes in important soil properties, such as aggregate stability and nitrogen mineralization, associated with this soil organic matter fraction.

affected by changing management (in this case, cultivating a previously undisturbed soil) and how they contribute to the overall change in soil organic matter level.

The accumulated plant residues and the active fraction of the soil organic matter are the first to be affected by changes in land management, accounting for most of the early losses in soil organic matter when cultivation of virgin soil begins. In contrast, losses from the passive fraction are very gradual. As a result, the soil organic matter remaining after some years is far less effective in promoting structural stability and nutrient cycling than the original organic matter in the virgin soil. If a favorable change in environmental conditions or management regime occurs, the plant litter and active fractions of soil organic matter are also the first to positively respond (see right-hand portion of Figure 12.16).

12.9   CARBON BALANCE IN THE SOIL–PLANT–ATMOSPHERE SYSTEM

Whether the goal is to reduce greenhouse gas emissions or to enhance soil quality and plant production, proper management of soil organic matter requires an understanding of the factors and processes influencing the cycling and balance of carbon in an ecosystem. Although each type of ecosystem, whether a deciduous forest, a prairie, or a wheat field, will emphasize particular compartments and pathways in the carbon cycle, consideration of a specific example, such as that described in Box 12.2, can help us develop a general model that can be applied to many different situations.

Agroecosystems

The rate at which soil organic matter either increases or decreases is determined by the balance between *gains* and *losses* of carbon. The gains come primarily from plant residues grown in place and from applied organic materials. The losses are due mainly to respiration (CO_2 losses), plant removals, and erosion (Table 12.6).

CONSERVATION OF SOIL CARBON.   In order to halt or reverse the net carbon loss shown in Figure 12.17, management practices would have to be implemented that would either *increase the additions* of carbon to the soil or *decrease the losses* of carbon from the soil. Since all crop residues and animal manures in the example are already being returned to the soil, additional carbon inputs could most practically be achieved by growing more plant material (i.e., increasing crop production or growing cover crops during the winter).

Specific practices to reduce carbon losses would include better control of soil erosion and the use of conservation tillage. Using a no-till production system would leave crop residues as mulch on the soil surface where they would decompose much more slowly. Refraining from tillage might also reduce the annual respiration losses from the original 2.5% to perhaps 1.5%. A combination of these changes in management would convert the system in our example from one in which soil organic matter is degrading (declining) to one in which it is aggrading (increasing).

Natural Ecosystems

Those interested in natural ecosystems may want to compare the carbon cycle of a natural forest with that of the cornfield in Figure 12.17. If the forest soil fertility were not too low, the total annual biomass production would probably be similar to that of the cornfield. The standing biomass, on the other hand, would be much greater in the forest since the tree crop is not removed each year. While some litter would fall to the soil surface, much of the annual biomass production would remain stored in the trees.

The rate of humus oxidation in the undisturbed forest would be considerably lower than in the tilled field because the litter would not be incorporated into the soil through tillage and the absence of physical disturbance would result in slower soil respiration. The litter from certain tree species may also be rich in phenolics and lignin, factors that greatly slow decomposition and C losses (see Figure 12.8). Furthermore, losses of organic matter through soil erosion would be much smaller on the forested site. Taken together, these factors allow annual net gains in soil organic matter in a young forest and maintenance of high soil organic matter levels in mature forests.

BOX 12.2 CARBON BALANCE—AN AGROECOSYSTEM EXAMPLE

The principal carbon pools and annual flows in a terrestrial ecosystem are illustrated in Figure 12.17 using a hypothetical cornfield in a warm temperate region. During a growing season the corn plants produce (by photosynthesis) 17,500 kg/ha of dry matter containing 7500 kg/ha of carbon (C). This C is equally distributed (2500 kg/ha each) among the roots, grain, and unharvested aboveground residues. In this example, the harvested grain is fed to cattle, which oxidize and release as CO_2 about 50% of this C (1250 kg/ha), assimilate a small portion as weight gain, and void the remainder (1100 kg/ha) as manure. The corn stover and roots are left in the field and, along with the manure from the cattle, are incorporated into the soil by tillage or by earthworms.

The soil microbes decompose the crop residues (including the roots) and manure, releasing as CO_2 some 75% of the manure C, 67% of the root C, and 85% of the C in the surface residues. The remaining C in these pools is assimilated into the soil as humus. Thus, during the course of one year, some 1475 kg/ha of C enters the humus pool (825 kg from roots, plus 375 from stover, plus 275 from manure). These values are in general agreement with Figure 12.10, but they will vary widely among different soil conditions and ecosystems.

At the beginning of the year, the upper 30 cm of soil in our example contained 65,000 kg/ha organic C in humus. Such a soil cultivated for row crops would typically lose about 2.5% of its organic C by soil respiration each year. In our example this loss amounts to some 1625 kg/ha of C. Smaller losses of soil organic C occur by soil erosion (160 kg/ha), leaching (10 kg/ha), and formation of carbonates and bicarbonates (10 kg/ha).

Comparing total losses (1805 kg/ha) with the total gains (1475 kg/ha) for the pool of soil humus, we see that the soil in our example suffered a *net annual loss* of 330 kg/ha of C, or 0.5% of the total C stored in the soil humus. If this rate of loss were to continue, degradation of soil quality and productivity would surely result.

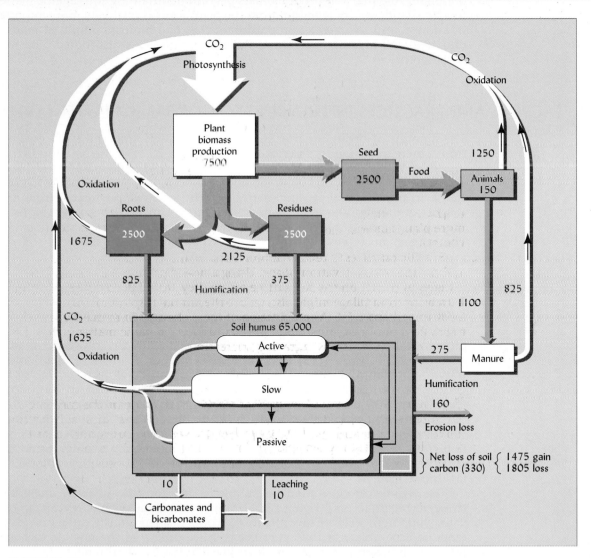

FIGURE 12.17 Carbon cycling in an agroecosystem.

TABLE 12.6 Factors Affecting the Balance between Gains and Losses of Organic Matter in Soils

Factors promoting gains	Factors promoting losses
Green manures or cover crops	Erosion
Conservation tillage	Intensive tillage
Return of plant residues	Whole plant removal
Low temperatures and shading	High temperatures and exposure to sun
Controlled grazing	Overgrazing
High soil moisture	Low soil moisture
Surface mulches	Fire
Application of compost and manures	Application of only inorganic materials
Appropriate nitrogen levels	Excessive mineral nitrogen
High plant productivity	Low plant productivity
High plant root:shoot ratio	Low plant root:shoot ratio

GRASSLANDS. Similar trends occur in natural grasslands, although the total biomass production is likely to be considerably less, depending mainly on the annual rainfall. Among the principles illustrated in Box 12.2, and applicable to most ecosystems, is the dominant role that plant root biomass plays in maintaining soil organic matter levels. In a grassland, the contribution from the plant roots is relatively more important than in a forest, and so a greater proportion of the total biomass produced tends to accumulate as soil organic matter, and this soil organic C is distributed more uniformly with depth.

12.10 FACTORS AND PRACTICES INFLUENCING SOIL ORGANIC MATTER LEVELS

The amount of organic matter in soils varies widely; mineral surface soils contain from a mere trace (sandy, desert soils) to as high as 20 or 30% (some forested or poorly drained A horizons). Some soils contain even more organic matter, but those that do are considered to be organic—not mineral—soils and will be discussed in Section 12.12. In practice, soil organic matter content is usually estimated from analysis of soil organic carbon content, because the latter can be determined more precisely. Therefore, for quantitative discussions scientists usually refer to soil organic carbon. The ranges and representative organic carbon contents of different soil orders are given in Table 12.1.

Differences among Soil Orders

Even though organic carbon contents vary as much as tenfold within a single soil order, it is possible to make a few generalizations about the organic carbon contents of different types of soils. Aridisols (dry soils) are generally the lowest in organic matter, and Histosols (organic soils) are definitely the highest (compare Plates 3 and 6). Contrary to popular myth, forested soils in humid tropical regions (e.g., Oxisols and some Ultisols) contain similar amounts of organic carbon to those in humid temperate regions (e.g., Alfisols and Spodosols). Andisols (volcanic ash soils) generally have some of the highest organic carbon contents of any mineral soils, probably because association of organic matter with the allophane clay in these soils protects the organic carbon from oxidation (see Plate 2). Among cultivated soils in humid and subhumid regions, Mollisols (prairie soils) are known for their dark, organic, carbon-rich surface layers (see Plate 8).

The organic carbon contents of subsurface horizons are generally much lower than those of the surface soil (Figure 12.18). Since most of the organic residues in both cultivated and virgin soils are incorporated in, or deposited on, the surface, organic matter tends to accumulate in the upper layers. Also, note that the organic carbon content decreases less abruptly with depth in grassland soils than in forested ones, because much of the residue added in grasslands consists of fibrous roots extending deep into the profile.

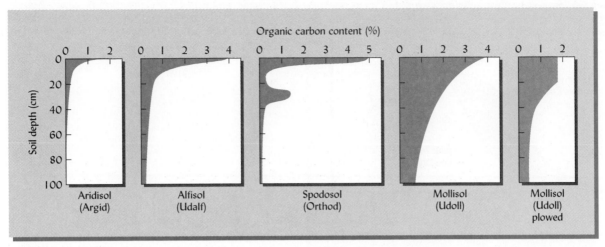

FIGURE 12.18 Vertical distribution of organic carbon in well-drained soils of four soil orders. Note the higher content and deeper distribution of organic carbon in the soils formed under grassland (Mollisols) compared to the Alfisol and Spodosol, which formed under forests. Also note the bulge of organic carbon in the Spodosol subsoil due to illuvial humus in the spodic horizon (see Chapter 3). The Aridisol has very little organic carbon in the profile, as is typical of dry-region soils.

Balance between Gains and Losses of C

As was indicated in Section 12.9, the level to which organic matter accumulates in soils is determined by the balance of gains and losses of organic carbon. The gains are principally governed by the amounts and types of organic residues added to the soil each year, while the losses result from oxidation of existing soil organic matter, as well as from erosion. We will now consider the numerous factors that influence the rates of gain and loss (see Table 12.6 for some of the management-oriented factors).

Influence of Climate

Temperature and rainfall exert a dominant influence on the amounts of organic carbon (and nitrogen) found in soils [see Jenny (1941) for a classical discussion of these influences].

TEMPERATURE. The effect of temperature results from the different manner in which the processes of organic matter production (plant growth) and organic matter destruction (microbial decomposition) respond to increases in this climatic variable. Figure 12.19a shows that at low temperatures plant growth outstrips decomposition, but that the opposite is true above approximately 25°C. In warm soils, mineralization is accelerated, so nutrient release is rapid, but residual organic matter accumulation is lower than in cooler soils. Therefore, as one moves from a warmer to a cooler climate, the organic matter and associated nitrogen content of comparable soils tend to increase. Some of the most rapid rates of organic matter decomposition occur in irrigated soils of hot desert regions.

Within zones of uniform moisture conditions and comparable vegetation, the average total amounts of organic matter and nitrogen in soils increase from two to three times for each 10°C decline in mean annual temperature. This temperature effect can be readily observed by noting the darkening color of well-drained surface soils as one travels from south (Louisiana) to north (Minnesota) in the humid grasslands of the North American Great Plains region (Figure 12.20). Similar changes in soil organic matter are evident as one climbs from warm lowlands to cooler highlands in mountainous regions.

MOISTURE. Soil moisture also exerts a major influence on the accumulation of organic matter and nitrogen in soils. Under comparable conditions, the nitrogen and organic matter content of soils increase as the effective moisture becomes greater. At the same

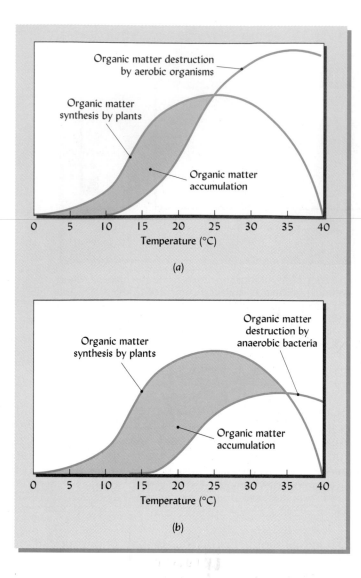

FIGURE 12.19 The balance between plant production and biological oxidation of organic matter determines the effect that temperature has upon organic matter accumulation in soils. The shaded areas indicate organic matter accumulation under aerobic (a) and anaerobic (b) conditions. Soil organic matter will accumulate to higher levels in cool climates, especially in waterlogged, anaerobic soils. Note that anaerobic accumulation is greater at most temperatures, and continues at higher temperatures than under aerobic conditions. This explains why subtropical areas in Florida can contain both organic soils (e.g., the Everglades) and soils containing very little organic matter (e.g., in better drained parts of the state). [Adapted from Mohr and van Baren (1954)]

time, the C/N ratio tends to be higher in the more thoroughly leached soils of the higher rainfall areas. These relationships are illustrated by the darker and thicker A horizons encountered as one travels across the North American Great Plains region (within a belt of similar mean annual temperature), from the drier zones in the west (Colorado) to the higher rainfall east (Missouri and Illinois). The explanation lies mostly in the sparser vegetation of the drier regions. In determining this rainfall correlation, however, it must be remembered that the level of organic matter in any one soil is influenced by both temperature and precipitation, as well as by other factors.

The lowest natural levels of soil organic matter and the greatest difficulty in maintaining those levels are found where annual mean temperature is high and rainfall is low (see Figure 12.20). These relationships are extremely important to the productivity and conservation of soils and to the relative difficulty of sustainable natural resource management.

Influence of Natural Vegetation

Climate and vegetation usually act together to influence the soil contents of organic carbon and nitrogen. The greater plant productivity engendered by a well-watered environment generally leads to greater additions to the pool of soil organic matter. Grasslands generally dominate the subhumid and semiarid areas, while trees are dominant in humid regions. In climatic zones where the natural vegetation includes both forests and grasslands, the total organic matter is higher in soils developed under grasslands than under forests (see Figure 12.18). With grassland vegetation, a rela-

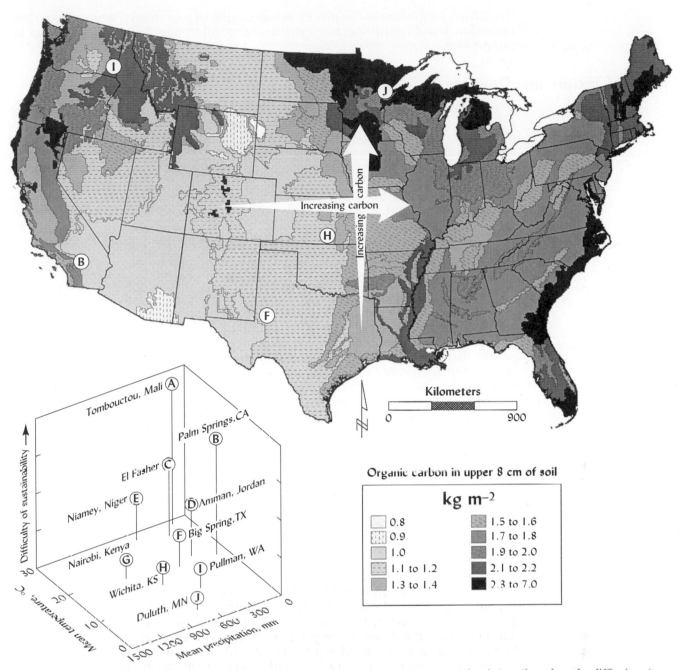

FIGURE 12.20 Influence of mean annual temperature and precipitation on organic matter levels in soils and on the difficulty of sustaining the soil resource base. The large white arrows on the map indicate that in the North American Great Plains region, soil organic matter increases with cooler temperatures going north, and with higher rainfall going east, provided that the soils compared are similar in texture, type of vegetation, drainage, and all other aspects except temperature and rainfall. These trends can be further generalized for global environments. Because of their influence on rates of plant growth and organic matter oxidation, temperature and precipitation largely determine the difficulty of preventing soil degradation and developing sustainable agricultural systems in various climatic regions of the world. Note that several of the locations represented on the graph are also on the U.S. map. [Kern (1994) and Stewart, et al. (1991); Map courtesy of J. Kern, U.S. Environmental Protection Agency.]

tively high proportion of the plant residues consist of root matter, which decomposes more slowly and contributes more efficiently to soil humus formation than does forest leaf litter.

Effects of Texture and Drainage

While climate and natural vegetation affect soil organic matter over broad geographic areas, soil texture and drainage are often responsible for marked differences in soil organic matter within a local landscape. All else being equal, soils high in clay and silt are generally higher in organic matter than are sandy soils (Figure 12.21). The amount of organic residues returned to the soil is generally higher in finer-textured soils, because the greater nutrient and water-holding capacities of these soils promote greater plant production. At the same time, the generally smaller pores of the fine-textured soils may restrict aeration and reduce the rate of organic matter oxidation. Another factor favoring the greater accumulation of organic matter in fine-textured soils is the formation of clay–humus complexes that protect the organic matter from degradation (see Section 12.4).

DRAINAGE EFFECTS. In poorly drained soils, the high moisture supply promotes plant dry-matter production and relatively poor aeration inhibits organic matter decomposition (see Figure 12.19b). Poorly drained soils therefore generally accumulate much higher levels of organic matter and nitrogen than similar but better-aerated soils (Figure 12.22). For instance, soils adjacent to streams are often quite high in organic matter, owing in part to their poor drainage. In very poorly drained environments, sufficient organic matter may accumulate to form Histosols. If the naturally waterlogged conditions of Histosols are altered by installation of an artificial drainage system, the resulting increased oxygen supply causes much of the accumulated organic matter to disappear from these soils (see Section 12.12).

Influence of Agricultural Management and Tillage

It is safe to generalize that cultivated land contains much lower levels of both soil nitrogen and organic matter than do comparable areas under natural vegetation. This is not surprising; under natural conditions all the organic matter produced by the vegetation is returned to the soil, and the soil is not disturbed by tillage. In contrast, in cultivated areas much of the plant material is removed for human or animal food and relatively less finds its way back to the land. Also, soil tillage aerates the soil and breaks up the organic residues, making them more accessible to microbial decomposition.

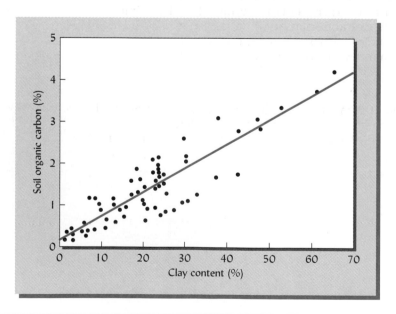

FIGURE 12.21 The effect of clay content on the soil organic carbon levels in the southern Great Plains area of the United States. [From Nichols (1984)]

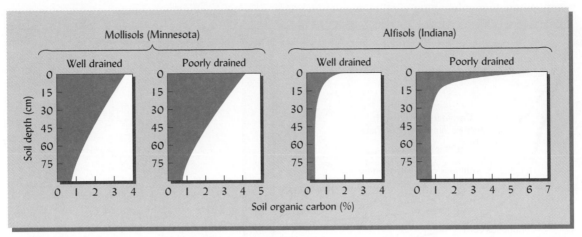

FIGURE 12.22   Distribution of organic carbon in four soil profiles, two well drained and two poorly drained. Poor drainage results in higher organic carbon content, particularly in the surface horizon.

CONVERSION TO CROPLAND.   A very rapid decline in soil organic matter content occurs when a virgin soil is brought under cultivation. Eventually, the gains and losses of organic carbon reach a new equilibrium and the soil organic matter content stabilizes at a much lower value. This pattern of decline is illustrated in Figure 12.23a, which shows the changing organic matter contents of a Mollisol during the first century after the native prairie was first plowed and several different cropping systems were imposed. Similar declines in soil organic matter are seen when tropical rain forests are cleared; however, the losses may be even more rapid because of the higher soil temperatures involved. Organic matter losses are not so dramatic if forests or prairies are converted to pasture or hay production.

Modern conservation tillage practices can help maintain or restore high surface soil organic matter levels (Figure 12.24). Compared to conventional tillage, practices such as stubble mulching and no-till leave a higher proportion of the residues on or near the soil surface. These techniques protect the soil from erosion and also discourage the rapid decomposition of crop residues.

Influence of Rotations, Residues, and Plant Nutrients

Figure 12.23 shows changes in the soil organic matter contents in two sets of famous long-term soil fertility experimental plots, the Morrow plots at the University of Illinois and the classical field experiments at Rothamstead Experiment Station, England. Different crop sequences and manure, lime, and fertilizer treatments were initiated on these two set of plots in 1876 and 1852, respectively.

From the data in Figure 12.23 we can draw the following conclusions.

From the Morrow plots:

1. A rotation of corn, oats, and clovers resulted in a higher soil organic matter level than did continuous corn, regardless of fertility inputs, probably because the rotation used tillage less frequently, and produced more root residues.

2. Because of greater additions of organic matter in the manure and in the increased residues from the higher-yielding crops, the systems using manure, lime, and phosphorus helped maintain much higher organic matter levels, especially where a rotation was followed.

3. Application of lime and fertilizers (N, P, and K) to previously unfertilized and unmanured plots (starting in 1955) noticeably increased soil organic matter levels, probably due to the production and return of larger amounts of crop residues and the addition of sufficient nitrogen to compliment the carbon in humus formation.

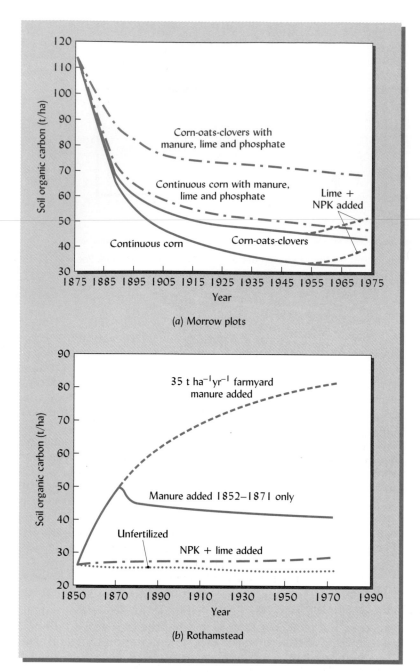

(a) Morrow plots

(b) Rothamstead

FIGURE 12.23 Soil organic carbon contents of selected treatments of (a) the Morrow plots at the University of Illinois and (b) of the classical experiments at Rothamstead Experiment Station in England. The Morrow plots were begun on virgin grassland soil in 1876 and so suffered rapid loss of organic carbon in the early years of the experiment. The Rothamstead plots were established on soils with a long history of previous cultivation. As a result, the soil at Rothamstead had reached an equilibrium level of organic carbon characteristic of the unfertilized small-grains (barley and wheat) cropping system traditionally practiced in the area. Therefore, little change in soil organic carbon occurred in the plots that simply continued this cropping system. Note that at both sites, when fertilization or manuring practices were altered, significant changes in soil organic carbon were observed within a few decades. For instance, lime and fertilizer were applied to half of the untreated fields in the Morrow plots after 1955. Note that this addition of lime and fertilizer began to reverse the trend of declining soil organic C (see dashed line). [Data recalculated from Odell, et al. (1984) and Jenkinson and Johnson (1977); used with permission of the Rothamsted Experiment Station, Harpenden, England]

From the Rothamstead experiments:

1. Continuous small-grain production and harvest resulted in little change, or a slow decline, in the already low level of soil organic matter. This soil was already in equilibrium with the carbon and nitrogen gains and losses characteristic of unfertilized small-grain production.

2. Annual applications of animal manures at rates sufficient to supply all needed nitrogen resulted in a dramatic initial rise in soil organic matter, until a new equilibrium state was approached at a much higher level of soil organic matter.

3. In the plots where manure was applied for only the first 20 years of the experiment, the soil organic matter level began to decline as soon as the manure applications were suspended, but the positive effect of the manure on soil organic matter level was still evident some 100 years later.

Results from these long-term experimental plots demonstrate that soils kept highly productive by supplemental applications of nutrients, lime, and manure and by the

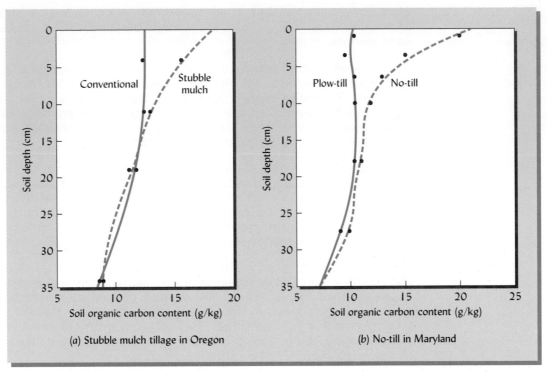

FIGURE 12.24 Effects of two conservation tillage systems on soil organic carbon content. The stubble mulch system was compared to conventional clean tillage on a Mollisol (Typic Haploxeroll) after 44 years of these practices for dryland wheat grown in a semiarid region of Oregon [data from Rasmussen and Rhode (1988)]. The no-till system was compared to conventional plow tillage after eight years of growing corn on an Ultisol (Typic Hapludult) in the humid coastal plain of Maryland [data from Weil, et al. (1988)]. Both conservation tillage systems leave much crop residue on or near the soil surface. The no-till system also leaves the soil almost completely unstirred, thus further slowing decomposition. In both cases, most of the increase in soil organic carbon occurred very near the soil surface.

choice of high-yielding crop varieties are likely to have more organic matter than comparable, less productive soils. High productivity means not only greater yields of economic crops, but also larger amounts of root and shoot residues returned to the soil. Nevertheless, the most productive cultivated soils will likely be considerably lower in organic matter than similar soils carrying undisturbed natural vegetation, except in the case of irrigated soil in desert areas, where natural vegetation is sparse.

Recommendations for Managing Soil Organic Matter

While the total carbon sequestered in the soil is important in relation to the global greenhouse effect (see Section 12.11), in terms of productivity and other ecological functions, achieving a particular level of total soil organic matter is far less important than maintaining a substantial proportion in the active fraction, allowing biological metabolism to constantly enhance soil tilth and nutrient cycling.

The following general principles apply to managing soil organic matter in many situations:

1. A continuous supply of organic materials must be added to the soil to maintain an appropriate level of soil organic matter, especially in the active fraction. Plant residues (roots and tops), animal manures, composts, and organic wastes are the primary sources of these organic materials. Cover crops can provide protective cover and additional organic material for the soil. It is almost always preferable to keep the soil vegetated than to keep it in bare fallow.

2. It is generally not practical to try to maintain higher soil organic matter levels than the soil–plant–climate control mechanisms dictate. For example, 1.5% organic matter might be an excellent level for a sandy soil in a warm climate, but would be indicative of a very poor condition for a finer-textured soil in a cool cli-

mate. It would be foolhardy to try to maintain as high a level of organic matter in a well-drained Texas silt loam soil as might be desirable for a similar soil found in Minnesota.

3. Because of the linkage between soil nitrogen and organic matter, adequate nitrogen inputs are requisite for adequate organic matter levels. Accordingly, the inclusion of legumes in the crop rotation and the judicious use of nitrogen-containing fertilizers to enhance high soil productivity are two desirable practices. At the same time, steps must be taken to minimize the loss of nitrogen by leaching, erosion, or volatilization (see Chapter 13).

4. Maximum plant growth will increase the amount of organic matter added to soil from crop residues. Even if some plant parts are removed in harvest, vigorously growing plants provide below- and aboveground residues as major sources of organic matter for the soil. Moderate applications of lime and nutrients may be needed to help free plant growth from the constraints imposed by chemical toxicities and nutrient deficiencies.

5. Because tillage accelerates organic matter losses both by increased oxidation of soil organic matter and by erosion, it should be limited to that needed to control weeds and to maintain adequate soil aeration. Conservation practices that minimize tillage leave much of the plant residues on or near the soil surface, and thereby slow down the rate of residue decay and reduce erosion losses (see Sections 4.8 and 17.6). In time, conservation tillage can lead to higher organic matter levels (see Figure 12.24).

6. Perennial vegetation, especially natural ecosystems, should be encouraged and maintained wherever feasible. Improved agricultural production on existing farmlands should be pursued to allow land currently supporting natural ecosystems to be left relatively undisturbed. In addition, there should be no hesitation about taking land out of cultivation and encouraging its return to natural vegetation where such a move is appropriate. In the United States, the Conservation Reserve Program provides incentives for such action (see Section 17.14). The fact is that large areas of land under cultivation today never should have been cleared.

12.11 SOILS AND THE GREENHOUSE EFFECT

Through decomposition and accumulation of organic matter, soils have a major effect on the composition of the Earth's atmosphere. A case in point is the carbon in soil organic matter—carbon that originated from the CO_2 in the atmosphere. When soil organic matter is allowed to decline, atmospheric CO_2 is likely to increase, with serious global consequences. Box 12.3 illustrates the importance of soil organic matter in regulating atmospheric CO_2 levels.

CARBON DIOXIDE.   Carbon dioxide is one of the **greenhouse gases** in the earth's atmosphere that cause the earth to be much warmer than it would otherwise be. Like the glass panes of a greenhouse, these gases allow short-wavelength solar radiation in, but trap much of the outgoing long-wavelength radiation. This heat-trapping action of the atmosphere is a major determinant of global temperature and, hence, global climates. Many scientists believe that the average global temperature has increased by 0.5 to 1.0°C during the past century. If this increase has in fact taken place, and if the trend continues, major changes in rainfall distribution, growing season length, and sea level may occur during the next century. However, predicting changes in global climate is complicated by numerous factors, such as cloud cover and volcanic dust, which can counteract the heat-trapping effects of the greenhouse gases. While it is certain that the concentrations of most greenhouse gases are increasing, there is less certainty about how rapidly global temperatures are actually rising.

Much effort and expense are currently being directed at reducing the anthropogenic (human-caused) contributions to climate change. Soil science has a major role to play in dealing with global warming. Gases produced by biological processes, such as those

BOX 12.3　CARBON CYCLE IN MINIATURE

The Biosphere 2 structure, a huge, sealed, ecological laboratory. (Photo by C. Allen Morgan. © 1995 by Decisions Investments Corp. Reprinted with permission.)

Biospherians at work growing their own food supply in the intensive agriculture biome with compost-amended soils rich in organic matter. (Photo by Pascale Maslin, © 1995 by Decisions Investments Corp. Reprinted with permission.)

An interesting experience with the global impact of soil organic matter dynamics was had by the eight biospherians who lived in Biosphere 2, a 1.3-ha sealed glass building in the Arizona desert. This structure was designed to hold a self-contained, self-supporting ecosystem in which scientists could live and study as part of the ecosystem. Costing over $200 million to build, Biosphere 2 contains a miniature ocean, coral reef, marsh, forests, and farms designed to maintain the ecological balance. However, not only did the biospherians have a very hard time growing enough food for themselves, their atmosphere soon ran so low on oxygen and became so enriched in carbon dioxide that engineers had to pump in oxygen and install a scrubber to cleanse carbon dioxide from the air. It turned out that the ecosystem was thrown out of kilter by the organic-matter-rich soil hauled in for the Biosphere farms. The designers had underestimated the rate at which soil organic matter would decompose when placed in the warm environment and aerated by garden tillage (see Section 12.2). The decomposition of the soil organic matter was carried out by aerobic soil microbes, which use up oxygen and give off carbon dioxide as they respire.

---

occurring in the soil, account for approximately half of the problem (Figure 12.25). Of the five primary greenhouse gases, only chlorofluorocarbons (CFCs) are exclusively of industrial origin. As already mentioned, much of the increase in atmospheric carbon dioxide levels comes from a net loss of organic matter from the soil system. Today's atmosphere contains about 370 ppm CO_2, as compared to about 280 ppm before the industrial revolution. Levels are increasing at about 0.5% per year.

METHANE.　Methane (CH_4) makes a substantial contribution to the greenhouse effect, for although it is found in the atmosphere in far smaller amounts than carbon dioxide, each molecule of CH_4 is about 30 times as effective as carbon dioxide in trapping outgoing radiation. In 1990 there was about 1.7 ppm CH_4 in the atmosphere, more than double the preindustrial levels. The level of CH_4 is rising at about 0.6% per year. Soils serve as both a source and a sink for CH_4—that is, they both add CH_4 to and remove it from the atmosphere.

When they are poorly aerated, as is the case in rice paddies and wetlands, soils produce CH_4, rather than CO_2, as organisms decompose organic matter (see Section 12.2). The rate of CH_4 production in wet soils is very dependent on the availability of oxidizable carbon, either in the soil organic matter or in plant residues returned to the soil

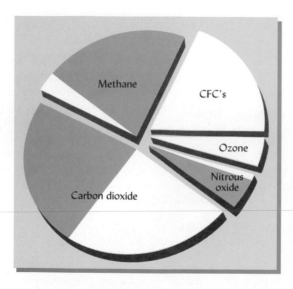

FIGURE 12.25 Relative contribution of different gases to the global greenhouse effect. The shaded portions indicate the emissions related to biological systems (in which soils play a role), the white portions being industrial contributions. [Modified from Dale, et al. (1993)]

(Figure 12.26). The CH_4 contribution to the atmosphere from such soils may also be regulated by the nature and management of the plants growing on them. For example, about 70 to 80% of the methane released from some flooded soils escapes to the atmosphere through the stems of rice plants or natural marsh grasses. In well-aerated soils, soil-dwelling termites may produce large quantities of methane (see Section 11.5). These biological processes account for much of the methane emitted into the atmosphere.

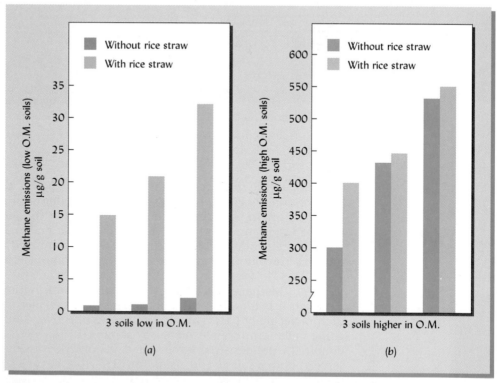

FIGURE 12.26 Global warming is enhanced by the release of methane from poorly drained soils. Methane is a greenhouse gas with atmospheric warming potential about 30 times that of CO_2. Oxygen-deficient, flooded rice soils supply food for over 2 billion people, but also account for up to one-fifth of the global emissions of CH_4. Methane release is increased by organic matter, either in the soil (compare left to right) or in crop residues (compare dark bars to light bars). Note the different scales for the low (left) and high (right) organic matter soils. [Data selected from Neue, et al. (1994)]

In well-aerated soils, certain **methanotrophic bacteria** produce the enzyme *methane monooxygenase*, which allows them to oxidize methane as an energy source:

$$CH_4 + \tfrac{1}{2}O_2 \rightarrow CH_3OH$$

Methane   Methanol

This reaction, which is largely carried out in soils, reduces the global, greenhouse gas burden by about 1 billion Mg of methane annually. Unfortunately, the long-term use of inorganic (especially ammonium) nitrogen fertilizer on cropland, pastures, and forests has been shown to reduce the capacity of the soil to oxidize methane. The evidence suggests that the rapid availability of ammonium from fertilizer stimulates ammonium-oxidizing bacteria at the expense of the methane-oxidizing bacteria. Long-term experiments in Germany and England indicate that supplying nitrogen in organic form (as manure) actually enhances the soil's capacity for methane oxidation (Table 12.7).

Nitrous oxide (N_2O) is another greenhouse gas produced by microorganisms in poorly aerated soils, but since it is not directly involved in the carbon cycle it will be discussed in the next chapter (see Section 13.9).

Because the soil can act as a major source or sink for carbon dioxide, methane, and nitrous oxide, it is clear that, together with steps to modify industrial outputs, soil management can play a major role in controlling the atmospheric levels of greenhouse gases.

12.12 ORGANIC SOILS (HISTOSOLS)

Discussion up to this point has focused on organic matter in mineral soils. Under waterlogged conditions in bogs and marshes, the oxidation of plant residues is so retarded that organic matter may accumulate to the point that soils are dominated by organic matter. If these soils contain greater than 20 to 30% organic matter by weight, depending on clay content, they are termed *organic soils* and are classified in the Histosols order in *Soil Taxonomy* (see Chapter 3). The organic matter in Histosols may be either peat or muck. *Peat* is brownish, only partially decomposed, fibrous remains of plant tissues. Some of these soils are mined and sold as peat, a material widely used in containerized plant production (see Box 12.4). *Muck*, on the other hand, is a black, powdery material in which decomposition is much more complete and the organic matter is highly humified.

Whether cultivated or in their natural state (often in flooded wetlands), Histosols have unique properties resulting from their high organic matter content. They are dark brown to black in color, have very low bulk density (0.2 to 0.4 Mg/m^3), and have water-holding capacities two to three times their dry weight. The cation exchange capacities of Histosols commonly exceed those of mineral clays on a weight basis, though not on a volume basis (Table 12.8). Despite relatively high C/N ratios (average of 20:1, compared to 12:1 for organic matter in a mineral soil), Histosols often show high levels of nitrification and accompanying nitrate release when drained. Although most of the car-

TABLE 12.7 **Effect of Nitrogen Fertility Management Systems on Methane Oxidation by an Arable Soil (Mollisol) in Germany**

The four nitrogen treatments were applied annually for 92 years to a rotation of sugar beets, spring barley, potatoes, and winter wheat. The measurements were made on soil sampled in spring, just before the annual nitrogen applications were made. Note that farmyard manure increased methane oxidation, while inorganic nitrogen fertilizer (NH_4NO_3) reduced methane oxidation from the levels in the control and the manure-only treatment.

Soil treatment	Soil pH	Soil NO_3^--N, kg/ha	Soil NH_4^+-N, kg/ha	Methane oxidation rate, nL CH$_4$ L^{-1}/hr^{-1}
1. Control—no N added in any form	6.8	0.83	0.20	4.60
2. Fertilizer N—40 to 130 kg/ha N as NH$_4$NO$_3$ to meet crop needs	6.9	15.36	3.1	1.34
3. Farmyard manure applied at 20 Mg/ha	7.0	1.98	0.22	11.2
4. Farmyard manure plus N fertilizer as in #3.	7.2	5.01	0.71	3.76

Data from the Static Fertilization Experiment begun in 1902 at Bad Lauchstädt, Germany and reported in Willison, et al. (1996).

bon in peat is in the passive fraction that resists microbial attack, the total amount of organic matter in these soils is so great that the decay of even a small fraction of it releases considerable quantities of mineral nitrogen.

Histosols (and Histels—permafrost soils with organic surface horizons) are important in the global carbon cycle because, although they cover little more than 1% of the world's land area, they hold about 20% of the global soil carbon. While drainage can make these soils extremely valuable for vegetable and floriculture crops (see Chapter 3), it also speeds up oxidation of organic matter, which over time leads to Histosol destruction and increased release of CO_2 to the atmosphere (see Figure 12.27). To preserve these soils, the water table should be kept as near the surface as practical.

TABLE 12.8 **Representative Maximum Cation Exchange Capacities of an Organic Colloid and Several Inorganic Colloids**

On a weight basis, the CEC of humus greatly exceeds that of the minerals, but on a volume basis both vermiculite and smectites have higher capacities.

	Cation exchange capacity	
Colloid	Weight basis, $cmol_c/kg$	Volume basis, $cmol_c/L$
Humus (organic)	300	75
Vermiculite	120	150
Smectite	90	113
Fine-grained micas	25	31
Kaolinite	5	6
Hydrous oxides	3	4

FIGURE 12.27 Soil subsidence due to rapid organic matter decomposition after artificial drainage of Histosols in the Florida Everglades. The house was built at ground level, with the septic tank buried about 1 m below the soil surface. Over a period of about 60 years, more than 1.2 m of the organic soil has "disappeared." The loss has been especially rapid because of Florida's warm climate, but artificial drainage that lowers the water table and continually dries out the upper horizons is an unsustainable practice on any Histosol. (Photo courtesy of George H. Snyder, Everglades Research and Education Center, Belle Glade, Fla.)

Histosols may help regulate long-term global climate change. As increased atmospheric CO_2 causes the global temperature to rise, seawater expands and the polar ice caps may melt more rapidly, both processes leading to a rise in sea level. The rise in sea level would have two effects that would help to counteract the impact of increased CO_2 content in the atmosphere. First, the greater volume of water in the world's oceans would absorb more CO_2. Second, the rising ocean waters would flood more coastal land area, creating conditions conducive to the formation of more Histosols in tidal marshes. The organic matter accumulation in these new Histosols would represent a significant sequestering of CO_2. This removal of CO_2 from the atmosphere may help to mitigate the global warming trend.

12.13 CONCLUSION

Organic matter is a complex and dynamic soil component that exerts a major influence on soil behavior, properties, and functions in the ecosystem. Because of the enormous amount of carbon stored in soil organic matter and the dynamic nature of this soil component, soil management may be an important tool for moderating the global greenhouse effect.

Organic residue decay, nutrient release and humus formation are controlled by environmental factors and by the quality of the organic materials. High contents of lignin and polyphenols, along with high C/N ratios, markedly slow the decomposition process, causing organic matter to accumulate while reducing the availability of nutrients.

For some purposes it is advantageous to manage the decomposition of organic matter outside of the soil in a process known as *composting*. Composting transforms various organic waste materials into a humuslike product that can be used as a soil amendment or a component of potting mixes. The aerobic decomposition in a compost pile can conserve nutrients while avoiding certain problems, such as noxious odors and the presence of either excessive or deficient quantities of soluble nitrogen, which can occur if fresh organic wastes are applied directly to soils.

Soil organic matter comprises three major pools of organic compounds. The *active pool* consists of microbial biomass and relatively easily decomposed compounds, such as polysaccharides and other nonhumic substances. Although only a small percentage of the total carbon, the active fraction plays a major role in nutrient cycling, micronutrient chelation, maintenance of structural stability and soil tilth, and as a food source underpinning biological diversity and activity in soils.

Most of the organic matter is in the *passive pool,* which contains very stable materials that resist microbial attack and may persist in the soil for centuries. This fraction provides cation exchange and water-holding capacities, but is relatively inert biologically. The so-called *slow fraction* is intermediate in stability and resistance to decomposition. It provides sources of food and energy for the steady metabolism of the authochthonous soil organisms that exist in the soil between times of residue additions. When soil is cleared of natural vegetation and brought under cultivation, the initial decline in soil organic matter is principally at the expense of the active fraction. The passive fraction is depleted only slowly and over very long periods of time.

The carbon-to-nitrogen (C/N) ratio of most soils is relatively constant, generally near 12:1. This means that the level of organic matter will be partially determined by the level of nitrogen available for assimilation into humus. Soil management for enhancing organic matter levels must, therefore, include some means of supplying nitrogen, for example, by the inclusion of leguminous plants.

The level of soil organic matter is influenced by climate (being higher in cool, moist regions), drainage (being higher in poorly drained soils), and by vegetation type (being generally higher where root biomass is greatest, as under grasses).

The maintenance of soil organic matter, especially the active fractions, in mineral soils is one of the great challenges in natural resource management around the world. By encouraging vigorous growth of crops or other vegetation, abundant residues (which contain both carbon and nitrogen) can be returned to the soil directly or through feed-consuming animals. Also, the rate of destruction of soil organic matter can be minimized by restricting soil tillage, controlling erosion, and keeping most of the plant residues at or near the soil surface.

Organic soils (Histosols) contain a large share of the world's soil organic carbon. Histosols are far more permeable to air and water, and generally can hold somewhat more water and cations per hectare, than most mineral soils. While drainage of these soils allows them to be used for production of high-value vegetable and floriculture crops, it also leads to accelerated oxidation of the organic matter they contain and increased CO_2 release.

The decay and mineralization of soil organic matter is one of the main processes governing the economy of nitrogen and sulfur in soils—the subject of the next chapter.

STUDY QUESTIONS

1. Compare the amounts of carbon in Earth's standing vegetation, soils, and atmosphere.

2. If you wanted to apply an organic material that would make a long-lasting mulch on the soil surface, you would choose an organic material with what chemical and physical characteristics?

3. Describe how the addition of certain types of organic materials to soil can cause a nitrate depression period. What are the ramifications of this phenomenon for plant growth?

4. In addition to humic substances, what other categories of organic materials are found in soils?

5. Some scientists include plant litter (surface residues) in their definition of soil organic matter, while others do not. Write two brief paragraphs, one justifying the inclusion of litter as soil organic matter and one justifying its exclusion.

6. What soil properties are mainly influenced by the active fraction of organic matter?

7. In this book and elsewhere, the terms *soil organic carbon* and *soil organic matter* are used to mean almost the same thing. How are these terms related, concep-

tually and quantitatively? Why is the term *organic carbon* generally more appropriate for quantitative scientific discussions?

8. Explain, in terms of the balance between gains and losses, why agricultural soils generally contain much lower levels of organic carbon than similar soils under natural vegetation.

9. In what ways are soils involved in the greenhouse effect that is thought to be warming up the Earth? What are some common soil-management practices that could be changed to reduce the negative effects and increase the beneficial effects of soils on the greenhouse effect?

10. What causes subsidence of Histosols? How is this phenomenon related to the sustainable use of Histosols?

REFERENCES

Batjes, N. H. 1996. "Total carbon and nitrogen in the soils of the world," *European J. Soil Sci.,* **47**:151–163.

Bouwman, A. F. (ed.). 1990. *Soils and the Greenhouse Effect.* (Chichester, U.K.: Wiley).

Brady, N. C. 1990. *The Nature and Properties of Soils,* 10th ed. (New York: MacMillan).

Cadisch, G. and K. E. Giller (eds.). 1997. *Driven by Nature—Plant Litter Quality and Decomposition.* (Walingford, U.K.: CAB International).

Chefetz, B., P. Hatcher, Y. Hadar, and Y. Chen. 1996. "Chemical and biological characterization of organic matter during composting of municipal solid waste," *J. Environ. Qual.,* **25**:776–785.

Chen, Y., and T. Aviad. 1990. "Effects of humic substances in plant growth," in P. MacCarthy, C. E. Clapp, R. L. Malcolm, and P. R. Bloom (eds.), *Humic Substances in Soil and Crop Sciences: Selected Readings.* (Madison, Wis.: ASA Special Publications), pp. 161–186.

Dale, V. H., R. A. Houghton, A. Grainger, A. E. Lugo, and S. Brown. 1993. "Emissions of greenhouse gases from tropical deforestation and subsequent uses of the land," in National Research Council, *Sustainable Agriculture and the Environment in the Humid Tropics.* (Washington, D.C.: National Academy Press), pp. 215–260.

Eswaran, H., E. Van Den Berg, and P. Reich. 1993. "Organic carbon in soils of the world," *Soil Sci. Soc. Amer. J.,* **57**:192–194.

Frimmel, F. H., and R. F. Christman. 1988. *Humic Substances and Their Role in the Environment.* (New York: Wiley).

Haynes, R. J. 1986. "The decomposition process: Mineralization, immobilization, humus formation and degradation," in R. J. Haynes (ed.), *Mineral Nitrogen in the Plant–Soil System.* (New York: Academic Press).

Huang, P. M., and M. Schnitzer (eds.). 1986. *Interactions of Soil Minerals with Natural Organics and Microbes.* SSSA Special Publication no. 17. (Madison, Wis.: Soil Sci. Soc. Amer.).

Inderjit, K. M., M. Dakshini, and F. A. Einhellig. 1995. *Allelopathy—Organisms, Processes, and Applications.* ACS Symposium Series 582. (Washington, D.C.: American Chemical Society).

Jenkinson, D. S., and A. E. Johnson. 1977. "Soil organic matter in the Hoosfield barley experiment," *Rep. Rothamstead Exp. Stn. for 1976,* **2**:87–102.

Jenny, H. 1941. *Factors of Soil Formation.* (New York: McGraw-Hill).

Kern, J. S. 1994. "Spatial patterns of soil organic carbon in the contiguous United States," *Soil Sci. Soc. Amer. J.,* **58**:439–455.

Loll, M. J., and J. M. Bollag. 1983. "Protein transformation in soil," *Advances in Agronomy,* **36**:352–382.

Magdoff, F. 1992. *Building Soils for Better Crops.* (Lincoln, Nebr.: University of Nebraska Press).

Mohr, E. C. J., and P. A. van Baren. 1954. *Tropical Soils.* (The Hague: N. V. Uitgeverij, W. Van Hoeve).

Neue, H. U., R. Wasserman, R. S. Lantin, M. C. Alberto, and J. B. Aduna. 1994. "Methane production potential of soils," *International Rice Research Notes,* **19**:37–38. (Manila, Philippines: Rice Research Institute).

Nichols, J. D. 1984. "Relation of organic carbon to soil properties and climate in the southern Great Plains," *Soil Sci. Soc. Amer. J.,* **48**:1382–1384.

Odell, R. T., S. W. Melsted, and V. M. Walker. 1984. "Changes in organic carbon and nitrogen of Morrow plots under different treatments 1904–1973," *Soil Sci. Soc. Amer. J.,* **137:**160–171.

Paustian, K., W. J. Parton, and J. Persson. 1992. "Modeling soil organic matter-amended and nitrogen-fertilized long-term plots," *Soil Sci. Soc. Amer. J.,* **56:**476–488.

Rasmussen, P. E., and C. R. Rhode. 1988. "Long-term tillage and nitrogen fertilization effects on organic nitrogen and carbon in a semi-arid soil," *Soil Sci. Soc. Amer. J.,* **52:**1114–1117.

Soil Survey Staff. 1975. *Soil Taxonomy: A Basic System of Soil Classification for Making and Interpreting Soil Surveys.* Agricultural Handbook no. 436 (Washington, D.C.: USDA Natural Resources Conservation Service).

Stevenson, F. J. 1986. *Cycles of Soil.* (New York: Wiley).

Stevenson, F. J. 1994. *Humus Chemistry—Genesis, Composition Reactions,* 2nd ed. (New York: Wiley).

Stewart, B. A., R. Lal, and S. A. El-Swaify. 1991. "Sustaining the resource base of an expanding world agriculture," in R. Lal and F. J. Pierce (eds.), *Soil Management for Sustainability.* (Ankeny, Iowa: Soil and Water Conservation Soc.).

Swift, M. J., O. W. Heal, and J. M. Anderson. 1979. *Decomposition in Terrestrial Ecosystems.* Studies in Ecology, vol. 5. (Berkeley: University of California Press).

Tate, R. L. 1987. *Soil Organic Matter: Biological and Ecological Effects.* (New York: Wiley).

Tian, G., B. T. Kang, and L. Brussaard. 1992. "Biological effects of plant residues with contrasting chemical compositions under humid, tropical conditions—decomposition and nutrient release," *Soil Biol. and Biochem.,* **24:**1051–1060.

Tian, G. B., L. Brussaard, and B. T. Kang. 1995. "An index for assessing the quality of plant residues and evaluating their effects on soil and crop in the (sub-) humid tropics," *Applied Soil Ecol.,* **2:**25–32.

Waksman, S. A. 1948. *Humus.* (Baltimore, Md.: Williams & Wilkins).

Wardle, D. A., O. Zackrisson, G. Hörnberg, and C. Gallet. 1997. "The influence of island size on ecosystem properties," *Science,* **277:**1296–1299.

Willison, T., R. Cook, A. Müller, and D. Powlson. 1996. "CH_4 oxidation in soils fertilized with organic and inorganic N: Differential effects," *Soil Biol. and Biochem.,* **28:**135–136.

Weil, R. R., P. W. Benedetto, L. J. Sikora, and V. A. Bandel. 1988. "Influence of tillage practices on phosphorus distribution and forms in three Ultisols," *Agron. J.,* **80:**503–509.

Weil, R. R., and G. S. Belmont. 1987. "Interactions between winged bean and grain amaranth," *Amaranth Newsletter,* **3**(1):3–6.

NITROGEN AND SULFUR ECONOMY OF SOILS

The pulse and body of the soil . . .
—D. H. LAWRENCE, THE RAINBOW

Nitrogen and sulfur share some important characteristics as essential plant nutrients. Both are found primarily in organic forms in soils, both move in soils and plants mostly in the anionic form, and both are responsible for serious environmental problems. Consequently, they will be considered together, starting with nitrogen.

More money and effort have been, and are being, spent on the management of nitrogen[1] than any other mineral element. And for good reason: The world's ecosystems are probably influenced more by deficiencies or excesses of nitrogen than by those of any other essential element. The pale yellowish green foliage of nitrogen-starved crops forebodes crop failure, financial ruin, and hunger for people in all corners of the world. Likewise, deficiencies of nitrogen are widespread among plants growing in the wild. Were it not for the biological fixation of nitrogen from the atmosphere by certain microorganisms, and for the recycling back to the soil of much of the nitrogen taken up in natural ecosystems, deficiencies of nitrogen would be even more widespread.

Excesses of some nitrogen compounds in soils can adversely affect human and animal health and can denigrate the quality of the environment. High nitrate levels in soils can lead to sufficiently high nitrates in drinking water as to endanger the health of human infants and ruminant animals. For that reason, nitrate levels are monitored in wells, reservoirs, and other drinking supplies. Likewise, the movement of soluble nitrogen compounds from soils to aquatic systems can disrupt the balance of those systems, leading to eutrophication, decline in oxygen content of the water, and the subsequent death of fish and other aquatic species.

Nitrogen is an essential component of protein. Because of its nutritional importance and relative scarcity, protein is highly sought after by most animals, humans included. Supplying sufficient nitrogen often represents a major expense in agricultural production. Also, the manufacture of nitrogen fertilizer accounts for a large part of the fossil fuel energy used by the agricultural sector. Yet another way in which nitrogen links the soil to the wider environment is the ozone-destroying action of soil-generated nitrous oxide gas. Clearly, for the management of the soil nitrogen cycle, the ecological, financial, and environmental stakes are very high.

---

[1] Nitrogen in soils and in crop production is covered extensively in the reviews edited by Stevenson (1982) and Hauck (1984). A comprehensive treatment of nitrogen in agriculture with an emphasis on environmental quality is given by Addiscott, et al. (1991).

ROLES IN THE PLANT.   Nitrogen is an integral component of many essential plant compounds. It is a major part of all amino acids, which are the building blocks of all proteins—including the enzymes, which control virtually all biological processes. Other critical nitrogenous plant components include the nucleic acids, in which hereditary control is vested, and chlorophyll, which is at the heart of photosynthesis. Nitrogen is also essential for carbohydrate use within plants. A good supply of nitrogen stimulates root growth and development, as well as the uptake of other nutrients.

Plants respond quickly to increased availability of nitrogen, their leaves turning deep green in color. Nitrogen increases the plumpness of cereal grains, the protein content of both seeds and foliage, and the succulence of such crops as lettuce and radishes. It can dramatically stimulate plant productivity, whether measured in tons of grain, volume of lumber, carrying capacity of pasture, or thickness of lawn. Healthy plant foliage generally contains 2.5 to 4.0% nitrogen, depending on the age of the leaves and whether the plant is a legume.

DEFICIENCY.   Plants deficient in nitrogen tend to have a pale yellowish green color (**chlorosis**), have a stunted appearance, and develop thin, spindly stems (see Plates 22 and 33 after page 498). In a nitrogen-deficient plant, the protein content is low and the sugar content is high, because there is insufficient nitrogen to combine with all the carbon chains stemming from sugars that would normally be used to make proteins. Nitrogen is quite mobile (easily translocated) within the plant, and when plant uptake is inadequate supplies are transferred to the newest foliage, causing the older leaves to show pronounced chlorosis. The older leaves of nitrogen-starved plants are therefore the first to turn yellowish, possibly becoming prematurely senescent and dropping off (Figures 13.1a and b). Nitrogen-deficient plants often have a low shoot-to-root ratio, and they mature more quickly than healthy plants. The negative effects of nitrogen deficiency on plant size and vigor are often dramatic (Figure 13.1c).

OVERSUPPLY.   When too much nitrogen is applied, excessive vegetative growth occurs; the cells of the plant stems become enlarged but relatively weak, and the top-heavy plants are prone to falling over (lodging) with heavy rain or wind (Figure 13.1d). High nitrogen applications may delay plant maturity and cause the plants to be more susceptible to disease (especially fungal disease) and to insect pests. These problems are especially noticeable if other nutrients, such as potassium, are in relatively low supply.

When oversupply of nitrogen does not lead to the problems just mentioned, crop yields may be large, but crop quality often suffers. The color and flavor of fruits may be poor, as may the sugar and vitamin contents of certain vegetables and root crops. Flower production on ornamentals is reduced in favor of overabundant foliage. Under certain conditions, an oversupply may cause nitrogen to accumulate in plant tissues as nitrate instead of becoming assimilated into proteins. Such high-nitrate foliage has been known to be harmful to livestock in the case of forages, or to human babies in the case of leafy vegetables.

The amount of nitrogen that can be usefully assimilated by plants varies greatly among species. Many grasses and leafy vegetables require a great deal of nitrogen for optimum growth. Whether or not the plants themselves are harmed by high nitrogen levels, potential adverse effects on the environment must always be considered (see Sections 13.8, 13.9, and 16.2).

FORMS OF NITROGEN TAKEN UP BY PLANTS.   Plant roots take up nitrogen from the soil solution principally as nitrates (NO_3^-) and NH_4^+ ions.[2] Although certain plants grow best when provided mainly one or the other of these forms, a relatively equal mixture of the two ions gives the best results with most plants. These two ions differ in their effect on the pH of the rhizosphere. Nitrate *anions* (negatively charged ions) move easily to the root with the flow of soil water and exchange at the root surface with HCO_3^- or OH^- ions

---

[2] Nitrite (NO_2^-) can also be taken up, but this ion is toxic to plants. Fortunately, it rarely occurs in greater than trace quantities in soils.

(a)

(b)

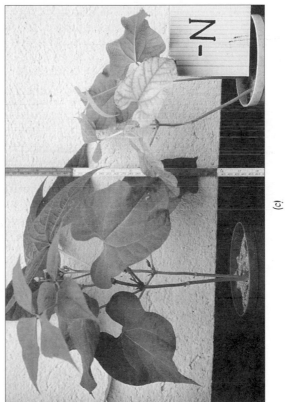

(c)

(d)

FIGURE 13.1 Some plant effects of too little and too much nitrogen. (*a*) The older, outer leaves have become chlorotic on these spinach plants growing on a low-nitrogen sandy Ultisol. (*b*) These older leaves from nitrogen-deficient corn plants (upper leaf is normal) have yellowed, beginning at the tip and continuing down the midrib—a pattern typical of nitrogen deficiency on corn. (*c*) The nitrogen-starved bean plant (right) shows the typical chlorosis of lower leaves and markedly stunted growth compared to the normal plant on the left. (*d*) The traditional tall Asian rice cultivar in the left field has lodged because of a large application of nitrogen fertilizer. In the right-hand field, the farmer has planted a modern rice cultivar, which yields well with high nitrogen inputs because of its short stature and stiff straw. (Photos courtesy of R. Weil)

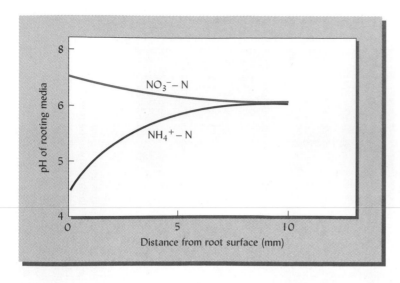

FIGURE 13.2 Typical effects of the absorption by plants of NH_4^+ and NO_3^- ions on the pH of the rhizosphere at increasing distances from the root surfaces. When NH_4^+ ions are absorbed, they are exchánged at the root surfaces for H^+ ions that are then released to the soil solution, thereby reducing the pH. Nitrate absorption involves a similar exchange for HCO_3^- and OH^- ions that stimulate an increase in rhizosphere pH.

that, in turn, stimulate an increase in the pH of the soil solution immediately around the root. In contrast, ammonium *cations* (positively charged ions) exchange at the root surface with hydrogen ions, thereby lowering the pH of the solution around the roots. Figure 13.2 illustrates typical effects of these two ions on the pH of the root environment, which is known to influence the uptake of other companion ions, such as phosphates.

13.2   ORIGIN AND DISTRIBUTION OF NITROGEN

Some 300,000 Mg of nitrogen is found in the air above 1 ha of soil. The atmosphere, which is 78% gaseous nitrogen (N_2) in content, appears to be a virtually limitless reservoir of this element.[3] But the very strong triple bond between two nitrogen atoms makes this gas quite inert and not directly usable by plants or animals. Were it not for the ability of certain microorganisms to break this double bond and to form nitrogen compounds (see Section 13.10), vegetation in the terrestrial ecosystems around the world would be rather sparse, and little nitrogen would be found in soils.

The nitrogen content of surface mineral soils normally ranges from 0.02 to 0.5%, a value of about 0.15% being representative for cultivated soils. A hectare of such a soil would contain about 3.5 Mg nitrogen in the A horizon and perhaps an additional 3.5 Mg in the deeper layers. In forest soils the litter layer (O horizons) might contain another 1 to 2 Mg of nitrogen. While these figures are low compared to those for the atmosphere, the soil contains 10 to 20 times as much nitrogen as does the standing vegetation (including roots) of either forested or cultivated areas. Most of the nitrogen in terrestrial systems is found in the soil.

Most soil nitrogen occurs as part of organic molecules. Soil organic matter typically contains about 5% nitrogen; therefore, the distribution of soil nitrogen closely parallels that of soil organic matter (see Section 12.3). Because association with certain silicate clays or resistant humic acids helps protect the nitrogenous organic compounds from rapid microbial breakdown, typically only about 2 to 3% of the nitrogen in soil organic matter is released annually as inorganic nitrogen.

Except where large amounts of chemical fertilizers have been applied, inorganic (i.e., mineral) nitrogen seldom accounts for more than 1 to 2% of the total nitrogen in

---

[3] About 98% of the earth's nitrogen is contained in the igneous rocks deep under the planet's crust, where it is effectively out of contact with the soil–plant–air–water environment in which we live. We will therefore focus our attention on the remaining 2% that cycles in the biosphere.

the soil. Unlike most of the organic nitrogen, the mineral forms of nitrogen are mostly quite soluble in water, and may be easily lost from soils through leaching and volatilization.

13.3 THE NITROGEN CYCLE

As it moves through the **nitrogen cycle**, an atom of nitrogen may appear in many different chemical forms, each with its own properties, behaviors, and consequences for the ecosystem. This cycle explains why vegetation (and, indirectly, animals) can continue to remove nitrogen from a soil for centuries without depleting the soil of this essential nutrient. The biosphere does not run out of nitrogen, because it uses the same nitrogen over and over again. The nitrogen cycle has long been the subject of intense scientific investigation, for understanding the translocations and transformations of this element is fundamental to solving many environmental, agricultural, and natural resource problems. The principal pools and forms of nitrogen, and the processes by which they interact in the cycle, are illustrated in Figure 13.3. This figure deserves careful study; we will refer to it frequently as we discuss each of the major divisions of the nitrogen cycle.

13.4 IMMOBILIZATION AND MINERALIZATION

The great bulk (95 to 99%) of the soil nitrogen is in organic compounds that protect it from loss but leave it largely unavailable to higher plants. Much of this nitrogen is present as amine groups (R—NH_2), largely in proteins or as part of humic compounds. When soil microbes attack these compounds, simple amino compounds[4] (R—NH_2) are formed. Then the amine groups are hydrolyzed, and the nitrogen is released as ammonium ions (NH_4^+), which can be oxidized to the nitrate form.

This conversion of organically bound nitrogen into inorganic mineral forms (NH_4^+ and NO_3^-) is termed *mineralization* (Figure 13.3). Although a whole series of reactions is involved, the net effect can be visualized quite simply. A wide variety of soil organisms simplifies and hydrolyzes the organic nitrogen compounds, ultimately producing the NH_4^+ and NO_3^- ions. The enzymatic process may be indicated as follows, using an amino compound (R—NH_2) as an example of the organic nitrogen source:

$$\xrightarrow{\hspace{2cm}} \text{Mineralization} \xrightarrow{\hspace{2cm}}$$

$$R\text{—}NH_2 \underset{-2H_2O}{\overset{+2H_2O}{\rightleftharpoons}} OH^- + R\text{—}OH + NH_4^+ \underset{-O_2}{\overset{+O_2}{\rightleftharpoons}} 4H^+ + \text{energy} + NO_2^- \underset{-\frac{1}{2}O_2}{\overset{+\frac{1}{2}O_2}{\rightleftharpoons}} \text{energy} + NO_3^-$$

$$\xleftarrow{\hspace{1cm}} \text{Immobilization} \xleftarrow{\hspace{1cm}}$$

Many studies have shown that only about 1.5 to 3.5% of the organic nitrogen of a soil mineralizes annually. Even so, this rate of mineralization provides sufficient mineral nitrogen for normal growth of natural vegetation in most all soils excepting those with low organic matter, such as the soils of deserts and sandy areas. Furthermore, *isotope tracer studies* of farmlands that have been amended with synthetic nitrogen fertilizers show that mineralized soil nitrogen constitutes a major part of the nitrogen taken up by a crop. If the organic matter content of a soil is known, one can make a rough estimate of the amount of nitrogen likely to be released by mineralization during a typical growing season (Box 13.1).

---

[4] Amino acids such as lysine (CH_2NH_2COOH) and alanine (CH_3CHNH_2COOH) are examples of these simpler compounds. The R in these formulas represents the part of the organic molecule with which the amino group (NH_2) is associated. For example, for lysine, the R is CH_2COOH.

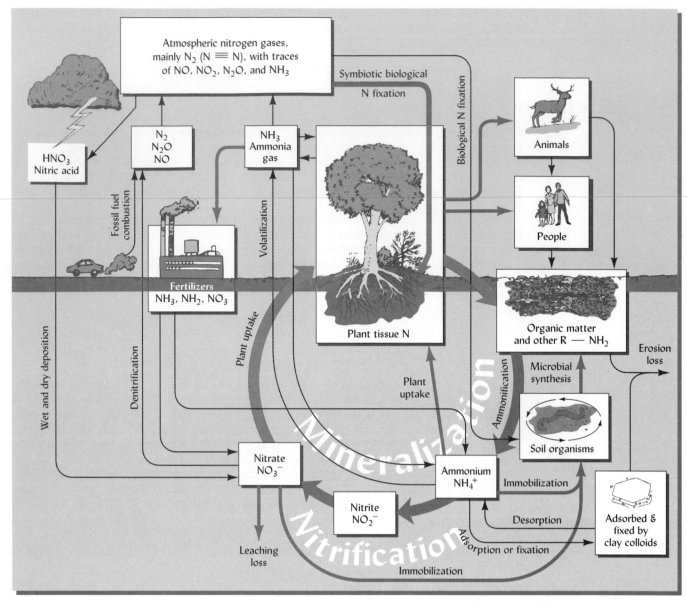

FIGURE 13.3 The nitrogen cycle, emphasizing the primary cycle (heavy, dark arrows) in which organic nitrogen is mineralized, plants take up the mineral nitrogen, and eventually organic nitrogen is returned to the soil as plant residues. Note also the pathways by which nitrogen is lost from the soil and the means by which it is replenished. The compartments represent various forms of nitrogen; the arrows represent processes by which one form is transformed into another.

Ammonium nitrogen is subject to five possible fates in the nitrogen cycle: (1) *immobilization* by microorganisms; (2) removal by *plant uptake;* (3) ammonium ions may be *fixed* in the interlayers of certain 2:1 clay minerals; (4) ammonium ions may be transformed into ammonia gas and lost to the atmosphere by *volatilization;* and (5) ammonium ions may be oxidized to nitrite and subsequently to nitrate by a microbial process called *nitrification.*

Nitrogen in the nitrate form is highly mobile in the soil and in the environment. Whether added as fertilizer or produced in the soil by nitrification, nitrate may take any of four paths in the nitrogen cycle: (1) *immobilization* by microorganisms; (2) removal by *plant uptake;* (3) nitrate ions may be lost by *leaching* in drainage water; or (4) by *volatilization* to the atmosphere as several nitrogen-containing gases.

The opposite of mineralization is **immobilization,** the conversion of inorganic nitrogen ions (NO_3^- and NH_4^+) into organic forms (see Figure 13.3). As microorganisms decompose carbonaceous organic residues in the soil, they may require more nitrogen than is contained in the residues themselves. The microorganisms then incorporate mineral nitrogen ions into their cellular components, such as proteins. When the organisms die, some of the organic nitrogen in their cells may be converted into forms that make up the humus complex, and some may be released as NO_3^- and NH_4^+ ions. Mineralization and immobilization occur simultaneously in the soil; whether the *net* effect is an increase or a decrease in the mineral nitrogen available depends primarily on the ratio of carbon to nitrogen in the organic residues undergoing decomposition (see Section 12.3).

BOX 13.1 CALCULATION OF NITROGEN MINERALIZATION

If the organic matter content of a soil, soil management practices, climate, and soil texture are known, it is possible to make a rough estimate of the amount of N likely to be mineralized each year. The following equation may be used:

$$\frac{\text{kg N mineralized}}{\text{ha 15 cm deep}} = \left(\frac{A \text{ kg SOM}}{100 \text{ kg soil}}\right)\left(\frac{B \text{ kg soil}}{\text{ha 15 cm deep}}\right)\left(\frac{C \text{ kg N}}{100 \text{ kg SOM}}\right)\left(\frac{D \text{ kg SOM mineralized}}{100 \text{ kg SOM}}\right)$$

where A = The amount of soil organic matter (SOM) in the upper 15 cm of soil, given in kg SOM per 100 kg soil. This value may range from close to zero to over 75% (in a Histosol) (see Section 12.12). Values between 0.5 and 5% are most common. ☞ Use a value of 2.5% (2.5 kg SOM/100 kg soil) for the example shown below.

B = The weight of soil per hectare to the depth of 15 cm. Most nitrogen used by plants is likely to come from this upper horizon. If it is 15 cm deep, 2×10^6 kg/ha is a reasonable estimate of its weight per hectare. See Section 4.5 to calculate the weight of this horizon if bulk density of a soil is known. ☞ Use 2×10^6 kg soil/ha 15 cm deep in the example shown below.

C = The amount of nitrogen in the SOM (see Section 12.3). ☞ Use the typical figure of 5 kg N/100 kg SOM in the example shown below.

D = The amount of SOM likely to be mineralized in one year for a given soil. This figure depends upon the soil texture, climate, and management practices. Values of around 2% are typical for a fine-textured soil, while values of around 3.5% are typical for coarse-textured soils. Slightly higher values are typical in warm climates; slightly lower values are typical in cool climates. ☞ Assume a value of 2.5 kg SOM mineralized/100 kg SOM for the example shown below.

The amount of nitrogen likely to be released by mineralization during a typical growing season may be calculated by substituting the example values indicated above into the equation:

$$\frac{\text{kg N mineralized}}{\text{ha 15 cm deep}} = \left(\frac{2.5 \text{ kg SOM}}{100 \text{ kg soil}}\right)\left(\frac{2 \times 10^6 \text{ kg soil}}{\text{ha 15 cm deep}}\right)\left(\frac{5 \text{ kg N}}{100 \text{ kg SOM}}\right)\left(\frac{2.5 \text{ kg SOM mineralized}}{100 \text{ kg SOM}}\right)$$

$$\frac{\text{kg N mineralized}}{\text{ha}} = \left(\frac{2.5}{100}\right)\left(\frac{2 \times 10^6}{1}\right)\left(\frac{5}{100}\right)\left(\frac{2.5}{100}\right) = 62.5 \text{ kg N/ha}$$

Contributions from the deeper layers of this soil might be expected to bring total nitrogen mineralized in the root zone of this soil during a growing season to over 120 kg N/ha. Most nitrogen mineralization occurs during the growing season when the soil is relatively moist and warm.

These calculations estimate the nitrogen mineralized annually from a soil that has not had large amounts of organic residues added to it. Animal manures, legume residues, or other nitrogen-rich organic soil amendments would mineralize much more rapidly than the native soil organic matter, and thus would substantially increase the amount of nitrogen available in the soil.

13.5 AMMONIUM FIXATION BY CLAY MINERALS[5]

Like other positively charged ions, ammonium ions are attracted to the negatively charged surfaces of clay and humus, where they are held in exchangeable form, available for plant uptake, but partially protected from leaching. However, because of the particular size of the ammonium ion (and potassium also), it can become entrapped within cavities in the crystal structure of certain clays (see Figure 13.3). Several clay minerals with a 2:1-type structure have the capacity to *fix* both ammonium and potassium ions in this manner (see Figure 14.30). Vermiculite has the greatest capacity, followed by fine-grained micas and some smectites. Ammonium and potassium ions fixed in the

---

[5] This chemical fixation of ammonia is caused by the entrapment or other strong binding of NH_4^+ ions by certain silicate clays. This type of fixation (by which K ions are similarly bound) is not to be confused with the very beneficial biological fixation of atmospheric nitrogen gas into compounds usable by plants (see Section 13.10).

TABLE 13.1   Total Nitrogen Levels of A and B Horizons of Four
Cultivated Virginia Soils and the Percentage of the Nitrogen Present
as Nonexchangeable or Fixed NH_4^+

Note the higher percentage of fixation in the B horizon.

Soil great group (series)	Total N, mg/kg		Nitrogen fixed as NH_4^+, %	
	A Horizon	B Horizon	A Horizon	B Horizon
Hapludults (Bojac)	812	516	5	18
Paleudults (Dothan)	503	336	5	14
Hapludults (Groseclose)	1792	458	3	17
Hapludults (Elioak)	1110	383	6	26

From Baethgen and Alley (1987).

rigid part of a crystal structure are held in a nonexchangeable form, from which they are released only slowly to higher plants and microorganisms.

Ammonium fixation by clay minerals is generally greater in subsoil than in topsoil, due to the higher clay content of subsoils (Table 13.1). In soils with considerable 2:1 clay content, interlayer-fixed NH_4^+ typically accounts for 5 to 10% of the total nitrogen in the surface soil and up to 20 to 40% of the nitrogen in the subsoil. In highly weathered soils, on the other hand, ammonium fixation is minor because little 2:1 clay is present. While ammonium fixation may be considered an advantage because it provides a means of conserving nitrogen, the rate of release of the fixed ammonium is often too slow to be of much practical value in fulfilling the needs of fast-growing annual plants.

13.6   AMMONIA VOLATILIZATION

Ammonia gas (NH_3) can be produced in the soil–plant system, and much nitrogen may be lost to the atmosphere in this form (see Figure 13.3). The source of the ammonia gas may be animal manures (hence the familiar ammonia smell around barnyards and poultry house), fertilizers (especially anhydrous ammonia and urea), decomposing plant residues (especially legume foliage that is rich in nitrogen), or even the foliage of living plants. In each case, ammonia gas is in equilibrium with ammonium ions according to the following reversible reaction:

$$\underset{\text{Dissolved ions}}{NH_4^+ + OH^-} \rightleftharpoons \underset{\text{Gas}}{H_2O + NH_3\uparrow}$$

From the preceding reaction we can draw three conclusions. First, ammonia volatilization will be more pronounced at high pH levels (i.e., OH^- ions drive the reaction to the right); second, ammonia-gas-producing amendments will drive the reaction to the left, raising the pH of the solution in which they are dissolved; and, finally, as a moist soil dries, the water is removed from the right-hand side of the equation, again driving the equation to the right.

Soil colloids, both clay and humus, adsorb ammonia gas, so ammonia losses are greatest where little of these colloids are present, or where the ammonia is not in close contact with the soil. For these reasons, ammonia losses can be quite large from sandy soils and from alkaline or calcareous soils, especially when the ammonia-producing materials are left at or near the soil surface and when the soil is drying out. High temperatures, as often occur on the surface of the soil, also favor the volatilization of ammonia.

Incorporation of manure and fertilizers into the top few centimeters of soil can reduce ammonia losses by 25 to 75% from those that occur when the materials are left on the soil surface. In natural grasslands and pastures, incorporation of animal wastes by earthworms and dung beetles is critical in maintaining a favorable nitrogen balance and a high animal-carrying capacity in these ecosystems.

VOLATILIZATION FROM WETLANDS.   Gaseous ammonia loss from nitrogen fertilizers applied to the surface of fishponds and flooded rice paddies can also be appreciable, even on slightly acid soils. The applied fertilizer stimulates algae growing in the paddy water. As the algae photosynthesize, they extract CO_2 from the water and reduce the amount of

PLATE 17 Connecticut River valley in western Massachusetts. Note variable alluvial soils and presence of riparian forest buffer along the river bank.

PLATE 18 Redox concentrations (red) and depletions (gray) in a Btg horizon from an Aquic Paleudalf.

PLATE 19 Green-gray colors from reduced structural iron around roots in an A horizon from a forested Aquic Hapludalf are an indicator of hydric conditions.

PLATE 20 Oxidized (red) root zones in the A and E horizons indicate a hydric soil. They result from oxygen diffusion out from roots of wetland plants having aerenchyma tissues (air passages).

PLATE 21 Dark (black) humic accumulation and gray humus depletion spots in the A horizon are indicators of a hydric soil. Water table is 30 cm below the soil surface.

PLATE 22　Nitrogen-deficient corn on Udolls in central Illinois. Ponded water after heavy rains resulted in nitrogen loss by denitrification and leaching.

PLATE 23　The waxlike ped surfaces in this Bt horizon from an Ultisol are clay skins (argillans). Bar = 1 cm.

PLATE 24　Acid sulfate weathering in a Sulfudept forming from pyritic mine spoil. A dark sulfidic horizon underlies a lighter-colored sulfuric horizon. Oxidation of sulfides may produce acid drainage waters. Scale marks = 10 cm.

PLATE 25　Early stages of soil formation in material dredged from Baltimore Harbor. Sulfidic materials (black), acid drainage (orange liquid), salt accumulations (whitish crust), and initiation of prismatic structure (cracks) are all evident.

PLATE 26　Stream polluted by acid coal-mine drainage caused by sulfurization in mine-spoil soils. Oxidation of $FeSO_4$ and precipitation of iron oxides on the rocks cause the orange color.

PLATE 27 Excessive inputs of nitrogen and phosphorus from upstream farmland resulted in the algal bloom that caused this slow-moving coastal plain stream to become choked with a green scum.

PLATE 28 Normal (left) and phosphorus-deficient (right) corn plants. Note stunting and purple color.

PLATE 29 Eroded calcareous soil (Ustolls) with iron-deficient sorghum.

PLATE 30 Zinc deficiency on peach tree. Note whorl of small, misshaped leaves.

PLATE 31 Zinc deficiency on sweet corn. Note broad whitish bands.

PLATE 32 Boron deficiency on alfalfa. Note reddish foliage.

PLATE 33   Nitrogen-deficient pine with normal trees in background.

PLATE 34   Landsat Thematic Mapper image of Washington, D.C. (upper left), and sediment-laden Potomac River (center). Composite image with natural colors. See page 775.

PLATE 35   Landsat Thematic Mapper image of Palo Verde Valley, California, irrigation scheme. Composite image using bands 2, 3, and 4. Lush vegetation appears bright red. See page 775.

Photo 1 courtesy Bill Waltman and D. J. Lathwell, Cornell University; photos 2, 10, 11, 13, and 15–31 courtesy of R. Weil; photos 3, 4, 8, and 12 courtesy of R. W. Simonson; photos 5 and 14 courtesy of Chien-Lu Ping, Agriculture and Forest Experimental Station, University of Alaska—Fairbanks; photos 6, 7 and 9 courtesy of Soil Science Society of America; photo 34 copyright Space Imaging, Inc..; photo 35 courtesy of Earth Satellite Corp, Rockville, Md.

carbonic acid formed. As a result, the pH of the paddy water increases markedly, especially during daylight hours, to levels commonly above 9.0. At these pH levels, ammonia is released from ammonium compounds and goes directly into the atmosphere. As with upland soils, this loss can be reduced significantly if the fertilizer is placed below the soil surface. Natural wetlands lose ammonium by a similar daily cycle.

AMMONIA ABSORPTION. By the reverse of the ammonium loss mechanism just described, both soils and plants can absorb ammonia from the atmosphere. Thus the soil–plant system can help cleanse ammonia from the air, while deriving usable nitrogen for plants and soil microbes. Forests may receive a significant proportion of their nitrogen requirements as ammonia carried by wind from fertilized cropland and cattle feedlots located many kilometers away.

13.7   NITRIFICATION

Ammonium ions in the soil may be enzymatically oxidized by certain soil bacteria, yielding first nitrites and then nitrates. These bacteria are classed as **autotrophs** because they obtain their energy from oxidizing the ammonium ions rather than organic matter. The process termed **nitrification** (see Figure 13.3) consists of two main sequential steps. The first step results in the conversion of ammonium to nitrite by a specific group of autotrophic bacteria (*Nitrosomonas*). The nitrite so formed is then immediately acted upon by a second group of autotrophs, *Nitrobacter.* The enzymatic oxidation releases energy, and may be represented very simply as follows:

Step 1

$$NH_4^+ + 1\,{}^1/_2 O_2 \xrightarrow[\text{bacteria}]{\textit{Nitrosomonas}} NO_2^- + 2H^+ + H_2O + 275 \text{ kJ energy}$$
Ammonium   Nitrite

Step 2

$$NO_2^- + {}^1/_2 O_2 \xrightarrow[\text{bacteria}]{\textit{Nitrobacter}} NO_3^- + 76 \text{ kJ energy}$$
Nitrite   Nitrite

So long as conditions are favorable for both reactions, the second transformation is thought to follow the first closely enough to prevent accumulation of nitrite. This is fortunate, because even at concentrations of just a few parts per million, nitrite is quite toxic to most plants and to mammals.

Regardless of the source of ammonium (i.e., ammonia-forming fertilizer, sewage sludge, animal manure, or any other organic nitrogen source), nitrification will significantly increase soil acidity by producing H$^+$ ions, as shown in the above reaction. In humid regions, liming materials must be added to counteract acidity (see Chapter 9).

Soil Conditions Affecting Nitrification

The nitrifying bacteria are much more sensitive to environmental conditions than are the broad groups of heterotrophic organisms responsible for the release of ammonium from organic nitrogen compounds (**ammonification**). We will briefly consider some of the soil conditions that affect nitrification.

AMMONIA LEVEL. Nitrification can take place only if there is ammonium to be oxidized. Factors such as a high C/N ratio of residues, which prevents the release of ammonium, may prevent nitrification. However, too much ammonia gas (NH$_3$) can inhibit nitrification. High concentrations of urea or anhydrous ammonia fertilizers in alkaline soils may raise ammonia to levels toxic to *Nitrobacter.* Without active *Nitrobacter* bacteria, nitrite may accumulate to toxic levels.

AERATION. The nitrifying organisms are aerobic bacteria and require oxygen to make NO$_2^-$ and NO$_3^-$ ions. Therefore, soil aeration and good soil drainage promote nitrification, as does moderate tillage, which stirs air into the soil and often results in a warmer soil, as well. Nitrification is generally somewhat slower under minimum tillage than where plowing and some cultivation are practiced.

MOISTURE.   Nitrification is retarded by both very low and very high soil moisture conditions (Figure 13.4). The optimum moisture for higher plants is also optimum for nitrification (about 60% of the soil pore space filled with water). However, appreciable nitrification does occur when soil moisture is at or near the wilting coefficient, which is too dry for plant growth.

SOURCE OF CARBON.   The nitrifiers use CO_2 and bicarbonate ions as sources of carbon to synthesize their cellular components. Being autotrophs, they do not require organic matter as either a carbon source or an energy source.

TEMPERATURE.   The temperature most favorable for nitrification ranges from 20 to 30°C. Nitrification is slow when soils are cool, and virtually ceases below 5°C. If ammonium fertilizer is applied early in spring, nitrification and ready availability of nitrate will be delayed until the soil warms up later in the season. Nitrification rates also decline at temperatures above 35°C and essentially cease at temperatures above 50°C.

EXCHANGEABLE BASE-FORMING CATIONS AND pH.   Nitrification proceeds most rapidly where there is an abundance of exchangeable base-forming cations. The need for base-forming cations accounts in part for the slow nitrification in acid mineral soils and the seeming sensitivity of the organisms to low pH. However, within reasonable limits, acidity itself seems to have less influence on nitrification when adequate base-forming cations are present. This is especially true of peat soils. Even at pH values below 5, peat soils may show remarkable accumulations of nitrates if Ca^{2+} and Mg^{2+} levels are adequate.

FERTILIZERS.   Nitrifying organisms have mineral nutrient requirements not too different from those of higher plants. Consequently, nitrification may be stimulated by the presence of essential elements in adequate levels.

TYPE OF CLAYS.   Nitrification tends to be influenced by the types of clays present in a soil. Smectites and allophane have been found to stabilize soil organic matter and to reduce rates of nitrification. Apparently, nitrogen-containing organic components become bonded to the clay surfaces or are found in tiny intermicelle pores and are thereby protected from microbial attack.

PESTICIDES.   Nitrifying organisms are quite sensitive to some pesticides. If added at high rates, many of these chemicals inhibit nitrification to a large or small extent. Most studies suggest, however, that at ordinary field rates, the majority of the pesticides have only a minimal effect on nitrification.

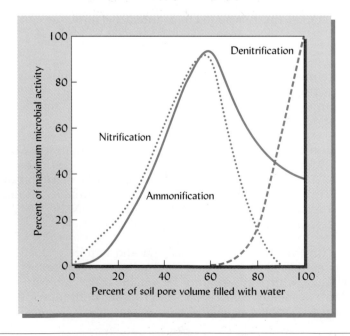

FIGURE 13.4 Because it depicts the balance of oxygen and water, percentage of water-filled pore space is closely related to rates of nitrification, ammonification, and denitrification. Note that ammonification can proceed in soils too waterlogged for active nitrification. There is very little overlap in the conditions suitable for nitrification and denitrification.

NITRIFICATION INHIBITORS. In recent years chemicals have been found that can inhibit or slow down the nitrification process, thereby reducing the nitrate leaching potential. Such compounds, as well as others that retard the dissolution of urea, are discussed in Box 13.2.

Provided that all the preceding conditions are favorable, nitrification is such a rapid process that nitrate is generally the predominant mineral form of nitrogen in most soils. Irrigation of an initially dry arid-region soil, the first rains after a long dry season in the

---

BOX 13.2   REGULATION OF SOLUBLE NITROGEN WITH FERTILIZER TECHNOLOGY

Perhaps the most challenging aspect of nitrogen control is the regulation of the soluble forms of this element after it enters the soil. Availability at the proper time and in suitable amounts, with a minimum of loss, is the ideal. Even where commercial fertilizers are used to supply much of the nitrogen, maintaining an adequate but not excessive quantity of available nitrogen is not an easy task. Split applications of nitrogen fertilizer is one way to accomplish this. This technique involves splitting nitrogen application into several small doses applied as the growing crop develops, rather than applying the entire amount at or before planting time. In sandy soils or where rainfall or irrigation is high early in the season, applying all the nitrogen at planting time may result in much of the nitrogen leaching below the root zone before the crop has had a chance to use it. In regions with high rainfall during the winter, it is important to avoid leaving a large amount of soluble nitrogen in the soil profile after the crop has completed its growth.

NITRIFICATION INHIBITORS.[a]   Because ammonium is much less susceptible than nitrate to losses by leaching or denitrification, nitrogen fertilizers might be used more efficiently if nitrification of ammonium from fertilizers could be slowed down until the crop is ready to make use of the nitrogen. To this end chemical companies have developed compounds called *nitrification inhibitors*, which inhibit the activity of the *Nitrosomonas* bacteria that convert ammonium to nitrate in the first step of nitrification. (Note that a chemical that inhibited *Nitrobacter* would *not* be useful, as this would cause toxic nitrite to accumulate.) Three commercially available nitrification inhibitors are dicyandiamide (DCD), nitrapyrin (N-Serve®) and etridiazol (Dwell®). When mixed with nitrogen fertilizers, these compounds can temporarily prevent nitrate formation. The key word here is *temporarily*, for when conditions are favorable for nitrification, the inhibition usually lasts only a few weeks (less if soils are above 20°C). Research has shown that these materials may pay for themselves with improved fertilizer efficiency when conditions are very conducive to nitrogen loss by leaching or denitrification. Thus, nitrification inhibitors have been of value primarily in relatively wet years on sandy soils (reduced leaching) and on imperfectly drained soils (reduced denitrification).

SLOW-RELEASE NITROGEN FERTILIZERS.   The use of slow-release fertilizer materials provides another means of reducing nitrogen losses from fertilized soils. Unlike most inorganic fertilizers, these materials release soluble nitrogen slowly in the soil so that nitrogen availability is better sychronized with plant uptake. Some stabilized organic materials, such as compost and digested sewage sludge (e.g., Milorganite®), are used for this purpose. However, most slow-release nitrogen fertilizers are made by treating urea with materials that slow its dissolution or inhibit its hydrolysis to ammonium. Urea-formaldehyde, isobutylidene diurea (IBDU), resin-coated fertilizers (e.g., Osmocote®), and sulfur-coated urea are all examples of slow-release nitrogen fertilizers. In the case of the latter, the sulfur content (10 to 20%) may be a benefit where sulfur is in low supply (Section 13.17), or as a problem where the extra acidity generated by the sulfur (Section 13.21) would be undesirable. Because of additional manufacturing costs and lower concentrations of nitrogen, slow-release fertilizers cost from 1.5 (for sulfur-coated ureas) to 4 times as much as plain urea per unit of nitrogen. Nonetheless, slow-release materials are practical for high-value plant production where the cost of fertilizers is not critical (certain vegetables, turf, and ornamentals) and for certain situations in which nitrogen loss from urea would be very high (e.g., some rice paddy soils[b]). They are widely applied to turf grass because they are not likely to cause fertilizer burn, and because they provide the convenience of fewer applications. Turf fertilized by a single annual application of soluble nitrogen may lose 10 to 50% of the applied nitrogen by leaching, while 3% or less is lost if slow-release fertilizers are used.

---

[a] For a review of this subject, see Prasad and Power (1995).

[b] Sulfur-coated urea should not be applied in flooded rice paddies, as the reduced iron in the floodwater will combine with the sulfur coating to form insoluble FeS, which locks up the nitrogen in the fertilizer.

tropics, the thawing and rapid warming of frozen soils in spring, and sudden aeration by tillage are examples of environmental fluctuations that typically cause a flush of soil nitrate production (Figure 13.5). The growth patterns of natural vegetation and the optimum planting dates for crops are greatly influenced by such seasonal changes in nitrate levels.

13.8 THE NITRATE LEACHING PROBLEM

In contrast to ammonium ions, which carry a positive charge, the negatively charged nitrate ions are not adsorbed by the negatively charged colloids that dominate most soils. Therefore, nitrate ions move downward freely with drainage water and are thus readily leached from the soil. The loss of nitrogen in this manner is of concern for two basic reasons: (1) such loss represents an impoverishment of the ecosystem whether or not cultivated crops are grown, and (2) leaching of nitrate causes several serious environmental problems.

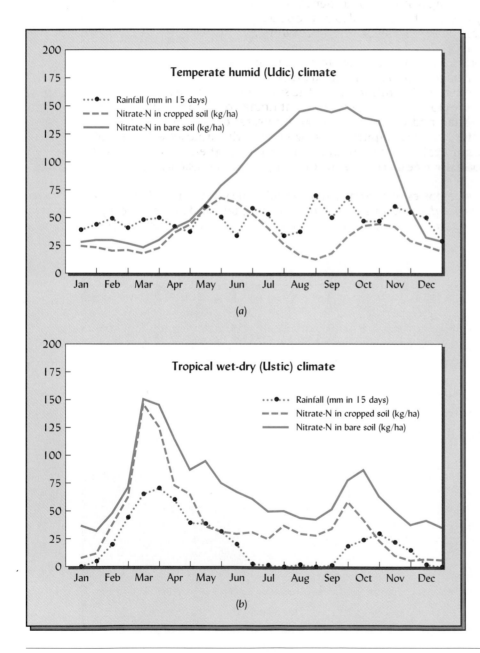

FIGURE 13.5 Typical seasonal patterns of nitrate concentration in representative surface soils with and without growing plants. The upper graph (*a*) represents a typical soil in a humid temperate region with cool winters and rainfall rather uniformly distributed throughout the year. Nitrates accumulate as the soil warms up in May and June, but are lost by leaching in the fall. The lower graph (*b*) represents a typical soil in a tropical region with a major and minor rainy season separated by several months of very dry, hot weather. The most notable feature here is the flush of nitrate that appears when the rain first moistens the soil after the long dry season. This flush of nitrate is caused by the rapid decomposition and mineralization of the dead cells of microorganisms previously killed by the dry, hot conditions. Note that soil nitrate is lower in both climates when plants are grown, because much of the nitrate formed is removed by plant uptake.

Productivity Loss

Productivity of the ecosystem suffers not only because of the loss of nitrogen, but also because the leaching of nitrate from acidic sources (nitrification or acid rain) facilitates the loss of calcium and other nutrient cations. Both types of nutrient losses will generally reduce the ecosystem productivity (see also Section 13.14). In the case of managed land, there is also an economic loss equal to the value of the lost nitrogen.

Environmental Impact

Environmental problems caused by nitrogen are mainly associated with the movement of nitrate through drainage waters to the groundwater. It may reach domestic wells, and may also eventually flow underground to surface waters, such as streams, lakes, and estuaries. The nitrate may contaminate drinking water and cause eutrophication and associated problems (Box 13.3).

The quantity of nitrate nitrogen lost in drainage water depends on two basic factors: (1) the rate of water leaching through the soil, and (2) the concentration of nitrates in that drainage water. Precipitation and irrigation rates, along with soil texture and structure, influence leaching rates. Sandy soils in humid regions are particularly susceptible to nitrate leaching, while such nutrient loss in unirrigated arid and semiarid soils is generally very low. Leaching may also be accentuated by the use of conservation tillage systems, which increase water infiltration and thereby increase leaching and concomitant loss of nitrates.

The concentration of nitrates in the drainage water depends on the balance and timing of nitrogen inputs and outputs to and from the soil, and on the rates of nitrification and removal of nitrates from the soil solution. Some mature forest ecosystems achieve a very close balance between plant uptake of nitrogen and the return of nitrogen in litter fall and dead trees, the groundwater commonly containing less than 1 mg/L nitrate. However, inputs of nitrogen from the atmosphere (e.g., nitrate in acid rain or from nitrogen fixation) and disturbances of the system (e.g., timber harvest) can overload the forest ecosystem, resulting in annual leaching losses of up to 25 or 30 kg N/ha.

Even greater losses may come from agricultural systems in which inputs of nitrogen regularly exceed the amounts removed by plant uptake and harvest. Heavy nitrogen fertilization (especially common for vegetables, corn, and other cash crops) exceed what

---

BOX 13.3 HUMAN HEALTH AND ENVIRONMENTAL EFFECTS OF NITRATE LEACHING

If either the groundwater or surface water might be used as drinking water, nitrate has the potential to cause toxic effects. Methemoglobinemia, also called blue baby syndrome, occurs when certain bacteria found in the guts of ruminant animals and human infants convert ingested nitrate into nitrite. The nitrite then interferes with the ability of the blood to carry oxygen to the body cells. Unoxygenated blood is not red; hence, infants with this condition take on a bluish skin color. Although human death from methemoglobinemia is a very rare occurrence, regulatory agencies in most countries limit the amount of nitrate permissible in drinking water to less than half the concentration known to cause this toxicity. In the United States the limit on nitrate is 45 mg/L nitrate (or 10 mg/L N in the nitrate form). In Europe the standard is 50 mg/L nitrate. It should be noted that nitrate in the diet (from water and other sources, such as meat preservatives) is also suspected of leading to the synthesis of colon-cancer-causing nitroso compounds, though the evidence for this relationship is very weak at present.

A much more widespread water quality problem associated with nitrate is the degradation of aquatic ecosystems, especially marine systems which are most often nitrogen-limited. Once it arrives in surface waters, nitrate, along with other forms of nitrogen and phosphorus from numerous sources, contributes to the problem of eutrophication. Eutrophication may stimulate the growth of large masses of algae, certain species of which produce flavors and toxins that make the water unfit to drink. The depletion of dissolved oxygen that ensues when algae die and decompose severely degrades the aquatic ecosystem, bringing death to fish and other aquatic organisms. This subject is given further consideration in Section 14.2.

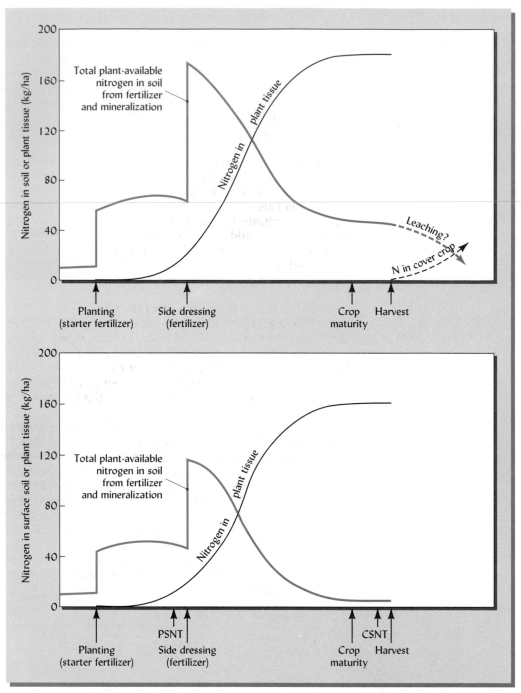

FIGURE 13.6 Two nitrogen fertilizer systems for corn production in the Midwest. The system that has been dominant in the past (upper) involves very high nitrogen applications and continuous corn culture. This is stimulated by the producers' desire to maximize economic return and to avoid risking yield reductions if nitrogen is lost through leaching or denitrification during periods of heavy rain. At least 150 kg N/ha is applied, part as starter fertilizer at or before planting time, the remainder as a side dressing just before the most rapid growth stage. Unfortunately when the crop matures and is harvested, much soluble nitrogen remains in the soil, probably in the form of nitrates. If the nitrogen is not captured by a cover crop planted in the corn at harvest time, the nitrogen is subject to leaching during the spring and winter months. This leads to contamination of groundwater and, eventually, surface water. More environmentally sound systems involve crop rotations that include legumes, or if corn is grown continuously, the rate of nitrogen fertilizer is greatly reduced (lower). A presidedress nitrate soil test (PSNT) is used to determine the amounts of nitrogen to apply. At the end of the season, a cornstalk nitrate test (CSNT) is used to assess plant nitrogen status. Crop yields and economic returns are about the same, but the nitrogen remaining in the soil at crop harvest is low and nitrate contamination is minimized. Any nitrogen lost due to heavy rains during the early growing season can be replaced by small supplementary N applications.

the plants are able to utilize, and can be a major cause of excessive nitrate leaching (Figure 13.6).

Ineffective management of manure from concentrated livestock production facilities is another common cause of nitrate contamination of ground and surface waters. When combined with high fertilizer nitrogen applications, manure can provide this element in quantities far in excess of plant uptake and can result in the pollution of both water and the atmosphere. Data from three European countries illustrate this point (Table 13.2). It is no wonder that shallow groundwater under some farm conditions contains nitrate in excess of 45 mg/L (10 mg N/L),[6] the legal limit for drinking water in the United States.

Timing of nitrogen inputs is critical. The potential for groundwater contamination with nitrate is greatest where inputs of water (rainfall and irrigation) and nitrates are high, and the removal of water and nitrates from the soil solution (evaporations and plant uptake) are low. Late fall, winter, and early spring are therefore the seasons during which most nitrate leaching occurs in humid temperate and Mediterranean climates. Soil managers should adopt practices that will minimize production of soil nitrate and maximize plant uptake of nitrogen during these critical periods. For example, where feasible, fall application of manure and nitrogen fertilizer should be avoided, while the planting of winter cover crops should be encouraged (see Section 16.2).

MANAGEMENT CAN REDUCE LOSSES. Even in regions of high leaching potential, careful soil management can prevent excessive nitrate losses. The timing of modest fertilizer and manure applications should provide nitrogen when the plants need it, not much before or after the period of active plant uptake. If this is not economically possible, other crops in the rotation, including cover crops should be planted immediately following the cash crop to take up the unused nitrates (see Figure 13.6). If such guidelines are followed, nitrogen leaching may be kept less than 5 or 10% of the nitrogen applied.

NITRATE RECYCLING IN THE HUMID TROPICS. Much of the nitrate mineralized in certain highly weathered, acid, tropical Oxisols and Ultisols leaches below the root zone before annual crops such as corn can take it up. Recently it has been found that some of this leached nitrate is not lost to groundwater, but is stored several meters deep in the profile where the highly weathered clays have adsorbed it on their anion exchange sites. Deep-rooted trees are capable of taking up this deep subsoil nitrate and subsequently using it to enrich the surface soil when they shed their leaves. Trees such as *Sesbania*, grown in rotation with annual food crops, can make this pool of leached nitrogen available for food production and prevent its further movement to groundwater (Figure 13.7). Agroforestry practices such as this have the potential to make a significant contribution to both crop protection and environmental quality in the humid tropics (see Section 20.9).

---

[6] Nitrate concentrations are sometimes reported as N *in nitrate form*. To convert mg nitrate (NO_3^-)/L to mg (N in NO_3^- form)/L, divide by 4.4.

TABLE 13.2   The Input into Soils of Nitrogen from Animal Manures and Fertilizer in Three European Countries

The uptake of nitrogen by crops and residual nitrogen not assimilated are shown. Some of the residual nitrogen is incorporated into soil organic matter, but much moves into the environment through leaching, runoff, and volatilization into the atmosphere.

Country	Nitrogen supply, 1000 Mg			N uptake by crops, 1000 Mg	Residual nitrogen	
	Manure	Fertilizer	Total		Total, 1000 Mg	Per hectare, kg
Belgium/Luxembourg	380	199	580	211	369	240
Denmark	434	381	816	287	529	187
The Netherlands	752	504	1,255	285	970	480

Data from Leuch, et al. (1995).

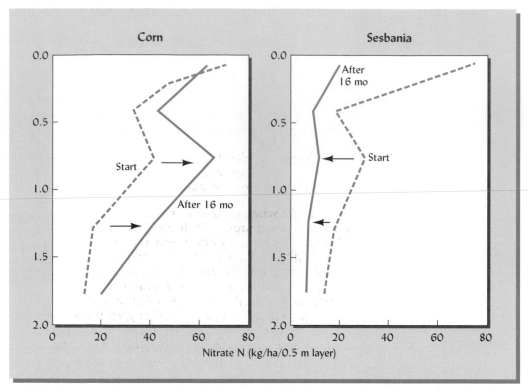

FIGURE 13.7 Depth distribution of nitrates in an Oxisol in western Kenya before planting either corn or a fast-growing tree, *Sesbania* (start), and after 16 months of growing fertilized corn (3 crops) and continuous *Sesbania*. Note that the *Sesbania* markedly decreased soil nitrates to a depth of about 2 meters. Separate root studies (Mekonnen, et al. (1997) showed that 31 percent of the *Sesbania* roots were located between 2.5 and 4 m in depth. [From Sanchez, et al. (1997)]

13.9   GASEOUS LOSSES BY DENITRIFICATION

Nitrogen may be lost to the atmosphere when nitrate ions are converted to gaseous forms of nitrogen by a series of widely occurring biochemical reduction reactions termed **denitrification**.[7] The organisms that carry out this process are commonly present in large numbers, and are mostly facultative anaerobic bacteria in genera, such as *Pseudomonas, Bacillus, Micrococcus,* and *Achromobacter*. These organisms are *heterotrophs,* which obtain their energy and carbon from the oxidation of organic compounds. Other denitrifying bacteria are *autotrophs,* such as *Thiobacillus denitrificans,* which obtain their energy from the oxidation of sulfide. The exact mechanisms vary depending on the conditions and organisms involved. In the reaction, NO_3^- [N(V)] is reduced in a series of steps to NO_2^- [N(III)], and then to nitrogen gases that include NO [N(II)], N_2O [N(I)], and eventually N_2 [N(0)]:

$$2NO_3^- \xrightarrow{-2[O]} 2NO_2^- \xrightarrow{-2[O]} 2NO\uparrow \xrightarrow{-[O]} N_2O\uparrow \xrightarrow{-[O]} N_2\uparrow$$

| Nitrate ions (+5) | Nitrite ions (+3) | Nitric oxide gas (+2) | Nitrous oxide gas (+1) | Dinitrogen gas (0) ← Valence state of nitrogen |

Although not shown in the simplified reaction given here, the oxygen released at each step would be used to form CO_2 from organic carbon (or SO_4^{2-} from sulfide, if *Thiobacillus* is the nitrifying organism).

Conditions necessary for significant denitrification to take place can be summarized as follows:

---

[7] Nitrate can also be reduced to nitrite and to nitrous oxide gas by nonbiological chemical reactions, but these reactions are quite minor in comparison with biological denitrification.

1. Nitrate must be available.

2. Readily decomposable organic compounds (or reduced sulfur compounds) must also be available to provide energy.

3. The soil air should contain less than 10% oxygen, or less than 0.2 mg/L of O_2 dissolved in the solution. Denitrification will proceed most rapidly if oxygen is completely absent. The entire soil need not be anaerobic, as localized microsites of low oxygen in the center of soil aggregates can provide a suitable environment in an otherwise well-aerated soil (see Section 7.5).

4. Temperature should be from 2 to about 50°C, with optimum temperatures between 25 and 35°C.

5. Very strong soil acidity (pH < 5.0) inhibits rapid denitrification and tends to cause N_2O to be the dominant end product.

Generally, when oxygen levels are very low the end product released from the overall denitrification process is dinitrogen gas (N_2). It should be noted, however, that NO and N_2O are commonly also released during denitrification under the fluctuating aeration conditions that often occur in the field (Figure 13.8). The proportion of the three main gaseous products seems to be dependent on the prevalent pH, temperature, degree of oxygen depletion, and concentration of nitrate and nitrite ions available. For example, the release of nitrous oxide (N_2O) is favored if the concentrations of nitrite and nitrate are high and the supply of oxygen is not too low. Under very acid conditions, almost all of the loss occurs in the form of N_2O. Nitric oxide (NO) loss is generally small and apparently occurs most readily under acid conditions.

Atmospheric Pollution

The question of how much of each nitrogen-containing gas is produced is not merely of academic interest. Dinitrogen gas is quite inert and environmentally harmless, but the oxides of nitrogen are very reactive gases and have the potential to do serious environmental damage in at least four ways. First, NO and N_2O released into the atmosphere by denitrification can contribute to the formation of nitric acid, one of the principal components of acid rain. Second, the nitrogen oxide gases can react with volatile organic

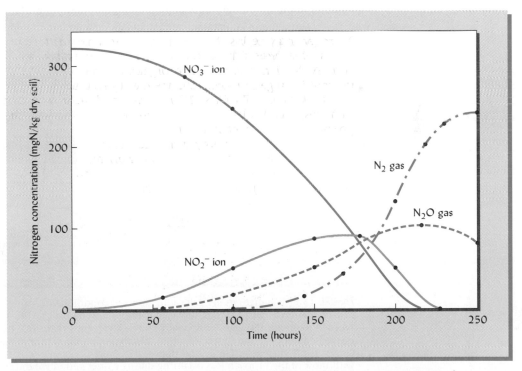

FIGURE 13.8 Changes in various forms of nitrogen during the process of denitrification in a moist soil incubated in the absence of atmospheric oxygen. [From Leffelaar and Wessel (1988)]

pollutants to form ground-level ozone, a major air
smog that plagues many urban areas. Third, when N
it contributes to the greenhouse effect (as much as 3
of CO_2) by absorbing infrared radiation that would ot
tion 12.11).

Finally, and perhaps most significantly, as N_2O
may participate in reactions that result in the destruct
shield the earth from harmful ultraviolet solar radiati
ozone layer has been measurably depleted, probably
as well as with N_2O and other gases. As this protecti
sands of additional cases of skin cancer are likely to
other important sources of N_2O, such as automobile
tion to the problem is being made by denitrification i
wetlands, and heavily fertilized or manured agricultu

Quantity of Nitrogen Lost through Denitrification

As might be expected, the exact magnitude of denitri
and will depend on management practices and soil co
tems have shown that during periods of adequate soi
in a slow, but relatively steady, loss of nitrogen from th
In contrast, most field measurements of gaseous nitr
reveal that the losses are highly variable in both time a
annual nitrogen loss often occurs during just a few d
has temporarily caused the warm soils to become poor

Low-lying, organic-rich areas and other hot spots n
as the average rate for a typical field. Although as muc
lost in a single day from the sudden wetting of a well–
soils rarely lose more than 5 to 15 kg N/ha annually by
where drainage is restricted and where large amounts
substantial losses might be expected. Losses of 30 to
been observed in agricultural systems.

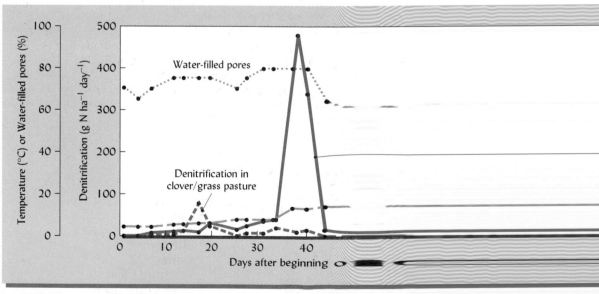

FIGURE 13.9 Changes in denitrification and soil conditions in spr
fine-textured soil (East Keswick silty clay loam, Udalfs) in Wales, U.
ryegrass receiving 150 kg N/ha annually and a clover–grass mixture.
ryegrass system was intended to approximately equal the nitrogen
clover in the mixed system (see Section 13.11). Note the sporadic
with most of the nitrogen loss occurring during a brief period when
nitrate. This episodic pattern is typical of disturbed agroecosystem
steady pattern of denitrification usually found in undisturbed natura

1. Nitrate must be available.

2. Readily decomposable organic compounds (or reduced sulfur compounds) must also be available to provide energy.

3. The soil air should contain less than 10% oxygen, or less than 0.2 mg/L of O_2 dissolved in the solution. Denitrification will proceed most rapidly if oxygen is completely absent. The entire soil need not be anaerobic, as localized microsites of low oxygen in the center of soil aggregates can provide a suitable environment in an otherwise well-aerated soil (see Section 7.5).

4. Temperature should be from 2 to about 50°C, with optimum temperatures between 25 and 35°C.

5. Very strong soil acidity (pH < 5.0) inhibits rapid denitrification and tends to cause N_2O to be the dominant end product.

Generally, when oxygen levels are very low the end product released from the overall denitrification process is dinitrogen gas (N_2). It should be noted, however, that NO and N_2O are commonly also released during denitrification under the fluctuating aeration conditions that often occur in the field (Figure 13.8). The proportion of the three main gaseous products seems to be dependent on the prevalent pH, temperature, degree of oxygen depletion, and concentration of nitrate and nitrite ions available. For example, the release of nitrous oxide (N_2O) is favored if the concentrations of nitrite and nitrate are high and the supply of oxygen is not too low. Under very acid conditions, almost all of the loss occurs in the form of N_2O. Nitric oxide (NO) loss is generally small and apparently occurs most readily under acid conditions.

Atmospheric Pollution

The question of how much of each nitrogen-containing gas is produced is not merely of academic interest. Dinitrogen gas is quite inert and environmentally harmless, but the oxides of nitrogen are very reactive gases and have the potential to do serious environmental damage in at least four ways. First, NO and N_2O released into the atmosphere by denitrification can contribute to the formation of nitric acid, one of the principal components of acid rain. Second, the nitrogen oxide gases can react with volatile organic

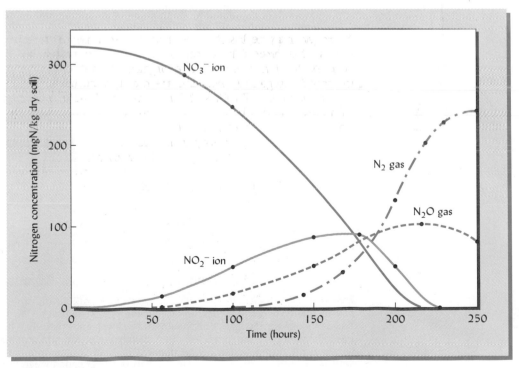

FIGURE 13.8 Changes in various forms of nitrogen during the process of denitrification in a moist soil incubated in the absence of atmospheric oxygen. [From Leffelaar and Wessel (1988)]

pollutants to form ground-level ozone, a major air pollutant in the photochemical smog that plagues many urban areas. Third, when NO rises into the upper atmosphere it contributes to the greenhouse effect (as much as 300 times that of an equal amount of CO_2) by absorbing infrared radiation that would otherwise escape into space (see Section 12.11).

Finally, and perhaps most significantly, as N_2O moves up into the stratosphere, it may participate in reactions that result in the destruction of ozone (O_3), a gas that helps shield the earth from harmful ultraviolet solar radiation. In recent years this protective ozone layer has been measurably depleted, probably by reaction with industrial CFCs, as well as with N_2O and other gases. As this protective layer is further degraded, thousands of additional cases of skin cancer are likely to occur annually. While there are other important sources of N_2O, such as automobile exhaust fumes, a major contribution to the problem is being made by denitrification in soils, especially in rice paddies, wetlands, and heavily fertilized or manured agricultural soils.

Quantity of Nitrogen Lost through Denitrification

As might be expected, the exact magnitude of denitrification loss is difficult to predict and will depend on management practices and soil conditions. Studies of forest ecosystems have shown that during periods of adequate soil moisture, denitrification results in a slow, but relatively steady, loss of nitrogen from these undisturbed natural systems. In contrast, most field measurements of gaseous nitrogen loss from agricultural soils reveal that the losses are highly variable in both time and space. The greater part of the annual nitrogen loss often occurs during just a few days in summer, when heavy rain has temporarily caused the warm soils to become poorly aerated (Figure 13.9).

Low-lying, organic-rich areas and other hot spots may lose nitrogen 10 times as fast as the average rate for a typical field. Although as much as 10 kg/ha of nitrogen may be lost in a single day from the sudden wetting of a well-drained, humid-region soil, such soils rarely lose more than 5 to 15 kg N/ha annually by denitrification (Table 13.3). But where drainage is restricted and where large amounts of nitrogen fertilizer are applied, substantial losses might be expected. Losses of 30 to 60 kg N/ha/yr of nitrogen have been observed in agricultural systems.

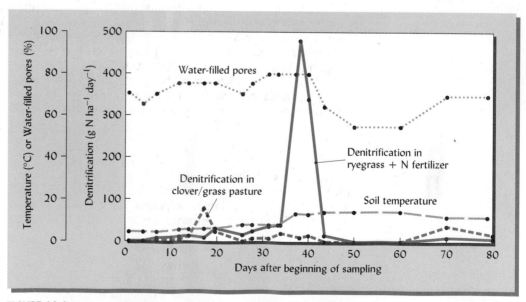

FIGURE 13.9 Changes in denitrification and soil conditions in spring on a well-drained, but somewhat fine-textured soil (East Keswick silty clay loam, Udalfs) in Wales, U.K. Two pasture systems were studied: ryegrass receiving 150 kg N/ha annually and a clover–grass mixture. The fertilizer nitrogen applied to the ryegrass system was intended to approximately equal the nitrogen fixed from the atmosphere by the clover in the mixed system (see Section 13.11). Note the sporadic nature of the denitrification process, with most of the nitrogen loss occurring during a brief period when the soil was warm, wet, and high in nitrate. This episodic pattern is typical of disturbed agroecosystems and stands in contrast to the slow, steady pattern of denitrification usually found in undisturbed natural forests. [Data from Colbourn (1993)]

TABLE 13.3 Fate of Applied Nitrogen in Two Experiments, One in Ohio and the Other in a Less Humid Area in Oklahoma

In the more humid area (Ohio), loss by leaching and volatilization was much higher than in the drier area (Oklahoma).

Fate of applied N	Sorghum-Sudan grass[a] (Oklahoma), %	Corn[b] (Ohio), %
Plant uptake	60	29
Soil organic matter	33	21
Leached	0	33
Gaseous loss[c]	7	17
	100	100

[a] Calculated from 3 years' data from eight soils. [Smith, et al. (1982)]
[b] Calculated from 3 years' data [Chichester and Smith (1978)]
[c] Unaccounted for and presumed lost by volatilization.

Denitrification in Flooded Soils

In flooded soils, such as those found in natural wetlands or rice paddies (Figure 13.10a), losses by denitrification may be very high. Nitrates may reach the soil through the floodwaters, or by nitrification in localized aerobic zones (e.g., the upper few centimeters of the soil). The nitrates may then move into nearby anaerobic zones, where they are nitrified. Thus, at any one time, nitrification and denitrification may be taking place simultaneously.

If the fertilizer nitrogen applied to rice paddy water is allowed to nitrify, much of it may later be volatilized as oxides of nitrogen or elemental nitrogen. Table 13.4 shows the loss of nitrous oxide from a rice field that was allowed to drain (fallow), thereby permitting nitrification, and was then reflooded, which encouraged denitrification. However, this loss can be dramatically reduced by keeping the soil flooded and by deep placement of the fertilizer into the reduced zone of the soil. In this zone there is insufficient oxygen to allow nitrification to proceed, so the nitrogen remains in the ammonium form and is not susceptible to loss by denitrification (Figure 13.11).

The sequential combination of nitrification and denitrification also operates in natural and artificial wetlands. Tidal wetlands (see Figure 13.10b), which become alternately anaerobic and aerated as the water level rises and falls, have particularly high potentials for converting nitrogen to gaseous forms. Often the resulting rapid loss of nitrogen is considered to be a beneficial function of wetlands, in that the process protects estuaries and lakes from the eutrophying effects of too much nitrogen. In fact, wastewater high in organic carbon and nitrogen can be cleaned up quite efficiently by allowing it to flow slowly over a specially designed water-saturated soil system in a process known as *overland flow wastewater treatment* (see Figure 13.10c).

Denitrification in Groundwater

Recent studies on the movement of nitrate in groundwater have documented the significance of denitrification taking place in the poorly drained soils under **riparian** vegetation (mainly woodlands adjacent to streams). In most cases studied in humid temperate regions, contaminated groundwater lost most of its nitrate load as it flowed through the riparian zone on its way to the stream. The apparent removal of nitrate may be quite dramatic, whether the nitrate source is septic drainfields or fertilized cropland (Figure 13.12). Most of nitrate is believed to be lost by denitrification, stimulated by organic compounds leached from the decomposing forest litter and by the anaerobic conditions that prevail in the wet riparian zone soils.

Whether nitrogen removal by denitrification is actually beneficial to the environment depends on which gases are produced. If significant amounts of N_2O or NO are produced, then wetlands and overland flow systems may merely trade water pollution for air pollution.

We have just discussed a number of biological processes that lead to losses of nitrogen from the soil system. We now turn our attention to the principal biological process by which soil nitrogen is replenished.

FIGURE 13.10 Denitrification can be very efficient in removing nitrogen in flooded systems that combine aerobic and anaerobic zones and have high concentrations of available organic carbon. Some examples of such systems are (*a*) a flooded rice paddy, (*b*) a tidal wetland, (*c*) a site for treating sewage effluent by overland flow (effluent applied by sprinklers, see arrows), and (*d*) a manure storage lagoon. (Photos courtesy of R. Weil)

TABLE 13.4 **Emission of Nitrous Oxide (N$_2$O) from an Irrigated Lowland Rice Field in the Philippines That Had Received Nitrogen from Either Urea Alone or from Combinations of Urea and Either Green Manure or Straw Residues**

The soil emitted significant quantities of N$_2$O only during the time of fallow when dry periods (which allowed the soil to become aerated enough for nitrification to produce nitrates) were alternated with rainy periods (during which anaerobic conditions allowed the nitrates to be denitrified). Since green manure and straw had little effect on N$_2$O losses, it appears that the soil had sufficient decomposable carbon to support the denitrifiers. Complementary studies showed that significant quantities of N$_2$ gas were released along with the N$_2$O.

	Emission of nitrous oxide (N$_2$O), mg N/m^2			
Source of nitrogen	Dry season rice, 111 days	Fallow, 36 days	Wet season rice, 98 days	Fallow, 89 days
Urea	9	171	13	19
Green manure and urea	12	183	6	18
Straw and urea	6	172	6	36

1/N rates: urea = 200 kg N/ha (dry season) and 120 kg N/ha (wet season); green manure and straw = 60 kg N/ha in organic tissue plus 140 kg N/ha (dry season) and 60 kg N/ha (wet season) from urea.
From Bronson, et al. (1997).

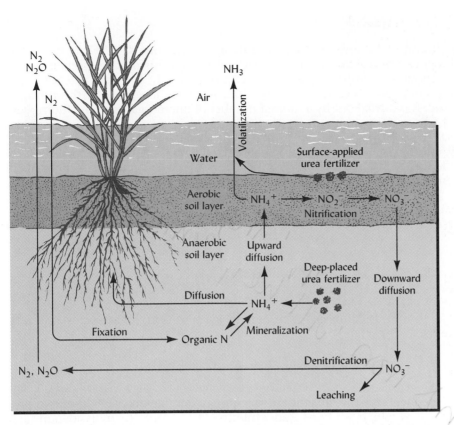

FIGURE 13.11 Nitrification–denitrification reactions and kinetics of the related processes controlling nitrogen loss from the aerobic–anaerobic layers of a flooded soil system. Nitrates, which form in the thin aerobic soil layer just below the soil–water interface, diffuse into the anaerobic (reduced) soil layer below and are denitrified to the N_2 and N_2O gaseous forms, which are lost to the atmosphere. Placing the urea or ammonium-containing fertilizers deep in the anaerobic layer prevents the oxidation of ammonium ions to nitrates, thereby greatly reducing the loss. [Modified from Patrick (1982)]

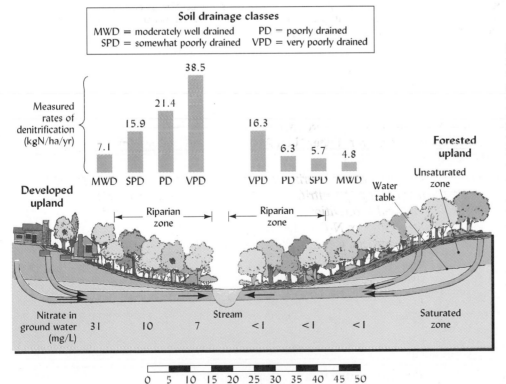

FIGURE 13.12 Denitrification in riparian wetlands receiving groundwater with high- and low-nitrate contents. The high-nitrate groundwater (left) came from sites with heavy development of houses using septic drainfields for sewage disposal (see Section 6.10). The low-nitrate groundwater (right side) came from an area of undeveloped forest. The riparian zones on both sides of the stream were covered with red maple–dominated forest. Within a few meters after entering the riparian wetland, the nitrate content of the contaminated groundwater was reduced by 75%, from 31 ppm nitrate to less than 7 ppm nitrate. The soils in this study site are very sandy Inceptisols and Entisols. In other regions, riparian zones with finer-textured soils have shown even more complete removal of nitrate from groundwater. [Data shown are from Hanson, et al. (1994)]

Next to plant photosynthesis, **biological nitrogen fixation** is probably the most important biochemical reaction for life on earth. Through this process, certain organisms convert the inert dinitrogen gas of the atmosphere (N_2) to nitrogen-containing organic compounds that become available to all forms of life through the nitrogen cycle. The process is carried out by a limited number of microorganisms, including several species of bacteria, a number of actinomycetes, and certain cyanobacteria (blue-green algae).

Globally, enormous amounts of nitrogen are fixed biologically each year. Terrestrial systems alone fix an estimated 139 million Mg, about twice as much as is industrially fixed in the manufacture of fertilizers (Table 13.5).

THE MECHANISM. Regardless of the organisms involved, the key to biological nitrogen fixation is the enzyme *nitrogenase,* which catalyzes the reduction of dinitrogen gas to ammonia.

$$N_2 + 8H^+ + 6e^- \xrightarrow[\text{(Fe,Mo)}]{\text{(Nitrogenase)}} 2NH_3 + H_2$$

The ammonia, in turn, is combined with organic acids to form amino acids and, ultimately, proteins.

$$NH_3 + \text{organic acids} \longrightarrow \text{amino acids} \longrightarrow \text{proteins}$$

The site of N_2 reduction is the enzyme **nitrogenase**, a complex consisting of two proteins, the smaller of which contains iron while the larger contains molybdenum and iron (Figure 13.13). Several salient facts about this enzyme and its function are worth noting, for nitrogenase is unique and its role in the nitrogen cycle is of great importance to humankind.

1. The reduction of N_2 to NH_3 by nitrogenase requires a great deal of energy to break the triple bond between the nitrogen atoms. Therefore, the process is greatly enhanced by association with higher plants, which can supply this energy from photosynthesis.

2. Nitrogenase is destroyed by free O_2, so organisms that fix nitrogen must protect the enzyme from exposure to oxygen. When nitrogen fixation takes place in root nodules (see Section 13.11), one means of protecting the enzyme from free oxy-

TABLE 13.5 **Global Nitrogen Fixation from Different Sources**

Source of N fixation	Area, 10^6 ha	Rate, kg/ha	Total fixed, 10^6 Mg
Biological fixation			
Legume crops	250	140	35
Nonlegume crops	1,150	8	9
Meadows and grassland	3,000	15	45
Forest and woodland	4,100	10	40
Other vegetated land	4,900	2	10
Ice-covered land	1,500	0	0
Total land	14,900		139
Sea	36,100	1	36
Total biological	51,000		175
Lightning			8
Fertilizer industry			77
Grand Total			260

Calculated from Jenkinson (1990) and Burns and Hardy (1975).

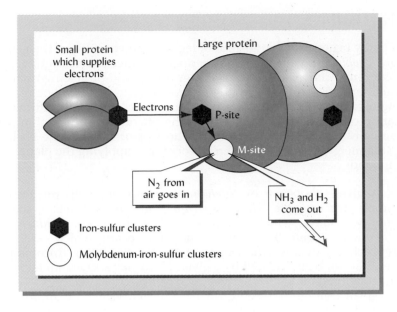

Small protein
which supplies
electrons

Large protein

Electrons

P-site

M-site

N₂ from
air goes in

NH₃ and H₂
come out

Iron-sulfur clusters

Molybdenum-iron-sulfur clusters

FIGURE 13.13 The nitrogenase complex consists of two proteins. The larger protein converts atmospheric N_2 to NH_3 using electrons provided by the smaller protein. The M-sites on the larger protein capture nitrogen (N_2) from the air, while the P-sites receive the electrons provided by the small protein so that N_2 can be reduced to NH_3. [From Emsley (1991); reprinted with permission of *New Scientist*]

gen is the formation of *leghemoglobin*.[8] This compound, which gives active nodules a red interior color, binds oxygen in such a way as to protect the nitrogenase while making oxygen available for respiration in other parts of the nodule tissue.

3. The reduction reaction is end-product inhibited—an accumulation of ammonia will inhibit nitrogen fixation. Also, too much nitrate in the soil will inhibit the formation of nodules (see Section 13.11).

4. Nitrogen-fixing organisms have a relatively high requirement for molybdenum, iron, phosphorus, and sulfur, because these nutrients are either part of the nitrogenase molecule or are needed for its synthesis and use.

FIXATION SYSTEMS. Biological nitrogen fixation occurs through a number of microorganism systems, with or without direct association with higher plants (Table 13.6). Although the legume–bacteria symbiotic systems have received the most attention, recent findings suggest that the other systems involve many more families of plants worldwide, and may even rival the legume-associated systems as suppliers of biological nitrogen to the soil. Each major system will be discussed briefly.

---

[8] Leghemoglobin is virtually the same molecule as the hemoglobin that gives human blood its red color when oxygenated. The use of hemoglobin to perform essentially similar functions in both legume root nodules and mammalian blood is a striking example of nature's conservative tendency and the unity of all life.

TABLE 13.6   Information on Different Systems of Biological Nitrogen Fixation

N-fixing systems	Organisms involved	Plants involved	Site of fixation
Symbiotic			
Obligatory			
Legumes	Bacteria *Rhizobia* and *Bradyrhizobia*	Legumes	Root nodules
Nonlegumes (angiosperms)	Actinomycetes (*Frankia*)	Nonlegumes (angiosperms)	Root nodules
Associative			
Morphological involvement	Cyanobacteria, bacteria	Various higher plants and microorganisms	Leaf and root nodules, lichens
Nonmorphological involvement	Cyanobacteria, bacteria	Various higher plants and microorganisms	Rhizosphere (root environment) Phyllosphere (leaf environment)
Nonsymbiotic	Cyanobacteria, bacteria	Not involved with plants	Soil, water independent of plants

The **symbiosis** (mutually beneficial relationship) of legumes and bacteria of the genera *Rhizobium* and *Bradyrhizobium* provide the major biological source of fixed nitrogen in agricultural soils. The genus *Rhizobium* contains fast-growing, acid-producing bacteria, while the *Bradyrhizobia* are slow growers that do not produce acid. Both will be considered together. These organisms infect the root hairs and the cortical cells, ultimately inducing the formation of **root nodules** that serve as the site of nitrogen fixation (Figure 13.14). In a mutually beneficial association, the host plant supplies the bacteria with carbohydrates for energy, and the bacteria reciprocate by supplying the plant with fixed-nitrogen compounds.

ORGANISMS INVOLVED. A given *Rhizobium* or *Bradyrhizobium* species will infect some legumes but not others. For example, *Rhizobium trifolii* inoculates *Trifolium* species (most clovers), but not sweet clover, which is in the genus *Melilotus*. Likewise, *Rhizobium phaseoli* inoculates *Phaseolus vulgaris* (dry beans), but not soybeans, which are in the genus *Glycine*. This specificity of interaction is one basis for classifying rhizobia (see Table 13.7). Legumes that can be inoculated by a given *Rhizobium* species are included in the same cross-inoculation group.

In areas where a given legume has been grown for several years, the appropriate species of *Rhizobium* is probably present in the soil. Often, however, the natural *Rhizobium* population in the soil is too low or the strain of the *Rhizobium* species present is not effective (Figure 13.15). In such circumstances, special mixtures of the appropriate *Rhizobium* and *Bradyrhizobium* inoculant may be applied, either by coating the legume seeds or by applying the inoculant directly to the soil. Effective and competitive strains of *Rhizobium,* which are available commercially, often give significant yield increases, but only if used on the proper crops. You may want to refer to Table 13.7 when planting legume rotations or purchasing commercial inoculant.

QUANTITY OF NITROGEN FIXED. The rate of biological fixation is greatly dependent on soil and climatic conditions. The legume–*Rhizobium* associations generally function best on soils that are not too acid (although *Bradyrhizobium* associations generally can tolerate considerable acidity) and that are well supplied with essential nutrients. However, high levels of available nitrogen, whether from the soil or added in fertilizers, tend to depress biological nitrogen fixation (Figure 13.16). Apparently, plants make the heavy energy investment required for symbiotic nitrogen fixation only when short supplies of mineral nitrogen make nitrogen fixation necessary.

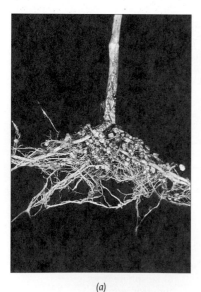

(a)   (b)   (c)

FIGURE 13.14 Photos illustrating soybean nodules. In (*a*) the nodules are seen on the roots of the soybean plant, and a closeup (*b*) shows a few of the nodules associated with the roots. A scanning electron micrograph (*c*) shows a single plant cell within the nodule stuffed with the bacterium *Bradyrhizobium japonicum*. (Courtesy of W. J. Brill, University of Wisconsin)

TABLE 13.7 Classification of Rhizobia Bacteria and Associated Legume Cross-Inoculation Groups

The genus Rhizobium *contains fast-growing, acid-producing bacteria, while those of* Bradyrhizobium *are slow growers that do not produce acid. A third genus,* Azorhizobium, *which is not shown, produces stem nodules on* Sesbania rostrata.

Bacteria		
Genus	Species/subgroup	Host legume
Rhizobium	R. leguminosarum	
	bv. viceae	Vicia (vetch), Pisum (peas), Lens (lentils), Lathyrus (sweet pea)
	bv. trifolii	Trifolium spp. (most clovers)
	bv. phaseoli	Phaseolus spp. (dry bean, runner bean, etc.)
	R. Meliloti	Melilotus (sweet clover, etc.), Medicago (alfalfa), Trigonella, (fenugreek)
	R. loti	Lotus (trefoils), Lupinus (lupins), Cicer (chickpea), Anthyllis, Leucaena, and many other tropical trees
	R. Fredii	Glycine spp. (e.g., soybean)
Bradyrhizobium	B. japonicum	Glycine spp. (e.g., soybean)
	B. sp.	Vigna (cowpeas), Arachis (peanut), Cajanus (pigeon pea), Pueraria (kudzu), Crotolaria (crotolaria), and many other tropical legumes

Although quite variable from site to site, the amount of nitrogen biologically fixed can be quite high, especially for those systems involving nodules, which supply energy from photosynthates and protect the nitrogenase enzyme system (Table 13.8). Nonnodulating or nonsymbiotic systems generally fix relatively small amounts of nitrogen. Nonetheless, many natural plant communities and agricultural systems (generally involving legumes) derive the bulk of their nitrogen needs from biological fixation.

EFFECT ON SOIL NITROGEN LEVEL. The data in Table 13.8 show that symbiotic nitrogen fixation is definitely beneficial to both forestry and agriculture. This source of nitrogen is not available to most coniferous and hardwood trees, grasses, cereal crops, and vegetables that lack symbiotic associations, unless they are grown in association with nodulated species. Over time, the presence of nitrogen-fixing species can significantly increase the nitrogen content of the soil and benefit nonfixing species grown in association with fixing species (see Figure 13.17).

As noted, nitrogen fixation is inhibited where mineral forms of nitrogen are available for plant uptake. Even where a nitrogen-fixing symbiosis is established, plants will preferentially take up nitrogen from the soil if it is available from such sources as mineralized organic matter, nitrogen in rainfall, and fertilizers. Some crops, such as beans and peas, are such weak nitrogen fixers that most of the nitrogen they absorb must

FIGURE 13.15 This soybean crop in East Africa was a total failure. The soybean seeds were not innoculated with the proper bacteria prior to planting in the newly cleared field, which had been cleared from forest vegetation and had never grown soybeans before. (Photo courtesy of R. Weil)

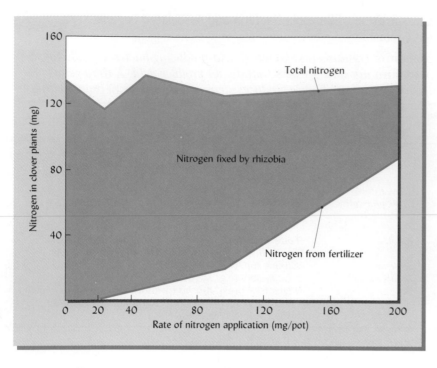

FIGURE 13.16 Influence of added inorganic nitrogen on the total nitrogen in clover plants, the proportion supplied by the fertilizer, and that fixed by the rhizobium organisms associated with the clover roots. Increasing the rate of nitrogen application decreased the amount of nitrogen fixed by the organisms in this greenhouse experiment. [From Walker, et al. (1956)]

come from the soil. Consequently, it should not be assumed that the symbiotic systems always increase soil nitrogen. Only in cases where the soil is low in available nitrogen and vegetation includes strong nitrogen fixers would this be likely to be true.

In the case of legume crops harvested for seed or hay, most of the nitrogen fixed is removed from the field with the harvest. Nitrogen additions from such crops should be considered as nitrogen *savers* for the soil rather than nitrogen builders. On the other hand, considerable buildup of soil nitrogen can be achieved by perennial legumes (such as alfalfa) and by annual legumes (such as hairy vetch) whose entire growth is returned to the soil as **green manure**. Such nitrogen buildup should be taken into account when

TABLE 13.8   Typical Levels of Nitrogen Fixation from Different Systems

Crop or plant	Associated organism	Typical levels of nitrogen fixation, kg N/ha/yr
Symbiotic		
Legumes (nodulated)		
Ipil Ipil tree (*Leucena leucocephala*)	Bacteria (*Rhizobium*)	100–500
Locust tree (*Robina* spp.)		75–200
Alfalfa (*Medicago sativa*)		150–250
Clover (*Trifolium pratense L.*)		100–150
Lupine (*Lupinus*)		50–100
Vetch (*Vicia vilbosa*)		50–150
Bean (*Phaseolus vulgaris*)		30–50
Cowpea (*Vigna unguiculata*)	Bacteria (*Bradyrhizobium*)	50–100
Peanut (*Arachis*)		40–80
Soybean (*Glycine max L.*)		50–150
Pigeon pea (*Cajunus*)		150–280
Kudzu (*Pueraria*)		100–140
Nonlegumes (nodulated)		
Alders (Alnus)	Actinomycetes (*Frankia*)	50–150
Species of Gunnera	Cyanobacteria[a] (*Nostoc*)	10–20
Nonlegumes (nonnodulated)		
Pangola grass (*Degetaria decumbens*)	Bacteria (*Azospirillum*)	5–30
Bahia grass (*Pasalum notatum*)	Bacteria (*Azobactor*)	5–30
Azolla	Cyanobacteria[a] (*Anabena*)	150–300
Nonsymbiotic	Bacteria (*Azobactor, Clostridium*)	5–20
	Cyanobacteria[a] (various)	10–50

[a] Sometimes referred to as *blue-green algae.*

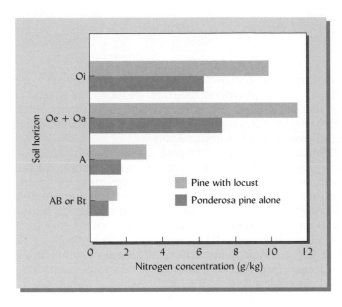

FIGURE 13.17 Nitrogen contents of forest soil horizons showing the effects of New Mexican locust trees (*Robinia neomexicana*) growing in association with ponderosa pine (*Pinus ponderosa*) in a region of Arizona receiving about 670 mm rainfall per year. The data are means from 20 stands of pondersosa pine, half of them with the nitrogen-fixing legume trees (locust) in the understory. The soils are Eutrustalfs and Argiustolls with loam and clay loam textures. [Data from Klemmedson (1994)]

estimating nitrogen fertilizer needs for maximum plant production with minimal environmental pollution (see Section 16.4).

FATE OF NITROGEN FIXED BY LEGUME BACTERIA. The nitrogen fixed in root nodules goes in three directions. First, it is used directly by the host plant, which thereby benefits greatly from the symbiosis described in Section 13.10. A portion of the nitrogen-rich plant tissue may later be returned to the soil as residues.

Second, some of the fixed nitrogen may become available to nonfixing plants growing in association with nitrogen-fixing plants. Although some direct transfer may take place via mychorrihizal hyphae connecting the two plants, most of the transfer results from mineralization of nitrogen-rich compounds in root exudates and in sloughed-off root and nodule tissues. Ammonium and nitrate thus released into the soil are available to any plant growing in association with the legume. The vigorous development of a grass in a legume–grass mixture is evidence of this rapid release (Figure 13.18), as are the relatively high nitrate concentrations sometimes measured in groundwater under legume crops.

The third pathway for the fixed nitrogen is immobilization by heterotrophic microorganisms and eventual incorporation into the soil organic matter (see Section 13.4).

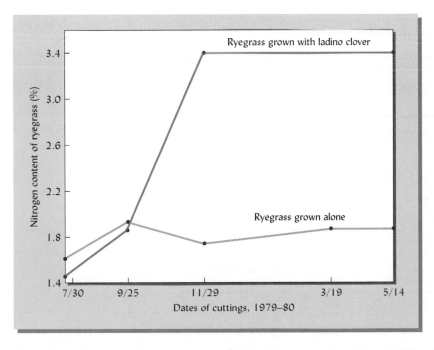

FIGURE 13.18 Nitrogen content of five field cuttings of ryegrass grown alone or with ladino clover. For the first two harvests, nitrogen fixed by the clover was not available to the ryegrass and the nitrogen content of the ryegrass forage was low. In subsequent harvests, the fixed nitrogen apparently was available and was taken up by the ryegrass. This was probably due to the mineralization of dead ladino clover root tissue. [From Broadbent, et al. (1982)]

13.12 SYMBIOTIC FIXATION WITH NONLEGUMES

Nodule-Forming Nonlegumes

Nearly 200 species from more than a dozen genera of nonlegumes are known to develop nodules and to accommodate symbiotic nitrogen fixation. Included are several important groups of angiosperms, listed in Table 13.9. The roots of these plants, which are present in certain forested areas and wetlands, form distinctive nodules when their root hairs are invaded by soil actinomycetes of the genus *Frankia*.

The rates of nitrogen fixation per hectare compare favorably with those of the legume–*Rhizobium* complexes (see Table 13.8). On a worldwide basis, the total nitrogen fixed in this way may even exceed that fixed by agricultural legumes (see Table 13.5). Because of their nitrogen-fixing ability, certain of the tree–actinomycete complexes are able to colonize infertile soils and newly forming soils on disturbed lands, which may have extremely low fertility as well as other conditions that limit plant growth (Figure 13.19). Once nitrogen-fixing plants become established and begin to build up the soil nitrogen supply through leaf litter and root exudation, the land becomes more hospitable for colonization by other species. *Frankia* thus play a very important role in the nitrogen economy of areas undergoing succession, as well as in established forest marshes.

Certain cyanobacteria are known to develop nitrogen-fixing symbiotic relations with green plants. One involves nodule formation on the stems of *Gunnera*, an angiosperm common in marshy areas of the southern hemisphere. In this association, cyanobacteria of the genus *Nostoc* fix 10 to 20 kg N/ha/yr (see Table 13.8).

Symbiotic Nitrogen Fixation without Nodules

Among the most significant nonnodule nitrogen-fixing systems are those involving cyanobacteria. One system of considerable practical importance is the *Azolla–Anabaena* complex, which flourishes in certain rice paddies of tropical and semitropical areas. The *Anabaena* cyanobacteria inhabit cavities in the leaves of the floating fern *Azolla* and fix quantities of nitrogen comparable to those of the more efficient *Rhizobium*–legume complexes (see Table 13.8).

A more widespread but less intense nitrogen-fixing phenomenon is that which occurs in the *rhizosphere* of certain grasses and other nonlegume plants. The organisms responsible are bacteria, especially those of the *Spirillum* and *Azotobacter* genera (see Table 13.8). Plant root exudates supply these microorganisms with energy for their nitrogen-fixing activities.

Scientists have reported a wide range of rates for rhizosphere nitrogen fixation, with the highest values observed in association with certain tropical grasses. Even if typical rates are only 5 to 30 kg N/ha/yr, the vast areas of tropical grasslands suggest that the total quantity of nitrogen fixed by rhizosphere organisms is likely very high (see Table 13.5).

TABLE 13.9 **Number and Distribution of Major Actinomycete-Nodulated Nonlegume Angiosperms**

In comparison, there are about 13,000 legume species.

Genus	Family	Species[a] nodulated	Geographic distribution
Alnus	Betulaceae	33/35	Cool regions of the northern hemisphere
Ceanothus	Rhamnaceae	31/35	North America
Myrica	Myricaceae	26/35	Many tropical, subtropical, and temperate regions
Casuarina	Casuarinaceae	24/25	Tropics and subtropics
Elaeagnus	Elaeagnaceae	16/45	Asia, Europe, North America
Coriaria	Coriariaceae	13/15	Mediterranean to Japan, New Zealand, Chile to Mexico

[a] Number of species nodulated/total number of species in genus.
Selected from Torrey (1978).

(a)

(b)

FIGURE 13.19 Soil actinomycetes of the genus *Frankia* can nodulate the roots of certain woody plant species and form a nitrogen-fixing symbiosis that rivals the legume–rhizobia partnership in efficiency. The actinomycete-filled root nodule (*a*) is the site of nitrogen fixation. The red alder tree (*b*) is among the first pioneer tree species to revegetate disturbed or badly eroded sites in high-rainfall areas of the Pacific Northwest in North America. This young alder is thriving despite the nitrogen-poor, eroded condition of the soil, because it is not dependent on soil nitrogen for its needs. (Photos courtesy of R. Weil)

13.13 NONSYMBIOTIC NITROGEN FIXATION

Certain free-living microorganisms present in soils and water are able to fix nitrogen. Because these organisms are not directly associated with higher plants, the transformation is referred to as *nonsymbiotic* or *free-living*.

Fixation by Heterotrophs

Several different groups of bacteria and cyanobacteria are able to fix nitrogen nonsymbiotically. In upland mineral soils, the major fixation is brought about by species of two genera of heterotrophic aerobic bacteria, *Azotobacter* (in temperate zones) and *Beijerinckia* (in tropical soils). Certain aerobic bacteria of the genus *Clostridium* are also able to fix nitrogen. Because pockets of low oxygen supply exist in soils even when they are in good tilth, aerobic and anaerobic bacteria probably work side by side in many well-drained soils.

The amount of nitrogen fixed by these heterotrophs varies greatly with the pH, soil nitrogen level, and sources of organic matter available. Because of their limited energy supply, under normal agricultural conditions the rates of nitrogen fixation by these organisms are thought to be in the range of 5 to 20 kg N/ha/yr (see Table 13.8)—only a small fraction of the nitrogen needed by crops. However, in some natural ecosystems this level of nitrogen fixation would make a significant contribution toward meeting nitrogen needs.

Among the autotrophs able to fix nitrogen are certain photosynthetic bacteria and cyanobacteria. In the presence of light, these organisms are able to fix carbon dioxide and nitrogen simultaneously. The contribution of the photosynthetic bacteria is uncertain, but that of cyanobacteria is thought to be of some significance, especially in wetland areas and in rice paddies. In some cases, these algae have been found to fix sufficient nitrogen for moderate rice yields, but normal levels may be no more than 20 to 30 kg N/ha/yr. Nitrogen fixation by cyanobacteria in upland soils also occurs, but the level is much lower than is found under wetland conditions.

13.14   ADDITION OF NITROGEN TO SOIL IN PRECIPITATION

The atmosphere contains ammonia gas and nitrates dissolved in water vapor and other nitrogen compounds released from the soil and plants, as well as from the combustion of coal and petroleum products. Nitrates also form in small quantities as a result of electrical discharges (lightning) in the atmosphere. Another source is the exhaust from automobile and truck engines, which contributes a considerable amount to the atmosphere, especially downwind from large cities. These atmosphere-borne nitrogen compounds are added to the soil through rain, snow, and dust. Although the rates of addition per hectare are typically small, the total quantity of nitrogen added annually is significant.

The quantity of ammonia and nitrates in precipitation varies markedly with location and with season. The additions are greater in humid tropical regions than in humid temperate regions, and are larger in the latter than in semiarid, temperate climates. Rainfall additions of nitrogen are highest near cities and industrial areas and near large animal feedlots (Table 13.10). There is special concern for the deposition of nitrates and other nitrogen oxides from these areas of concentration, because they are associated with increased acidity. The environmental impact of acid rain (see Section 9.6) and its influence on vegetation (forests and crops) and surface waters tends to overshadow the nutrient benefit of precipitation-supplied nitrates.

Typically, the nitrogen in precipitation is about two-thirds ammonium and one-third nitrate. The range of total nitrogen (nitrate + ammonium) added by precipitation annually is 1 to 25 kg N/ha. A figure of 5 to 8 kg N/ha would be typical for nonindustrial temperate regions. This modest annual acquisition of nitrogen is probably of more significance to the nitrogen budget in natural ecosystems than in agriculture.

Effects on Forest Ecosystems

In a mature forest, net plant uptake and microbial immobilization are in such close balance with release of nitrogen by mineralization that many of these systems have very little ability to retain incoming nitrogen. Nitrogen from the atmosphere enters the forest largely as nitric acid (HNO_3) or as ammonium, which is converted to nitric acid

TABLE 13.10   **Amounts of Nitrogen Brought Down in Precipitation Annually in Different Parts of the United States**

Areas in the United States	Range in annual deposition, kg/ha		
	Nitrate nitrogen	Ammonium nitrogen	Total nitrogen, kg/ha
Rural area of Ohio[a]	6.6–10.6	6.2–8.2	12.8–18.8
Industrialized Northeast[b]	4.3–7.4	8.6–14.8	12.9–22.2
Borders of NE industrialized areas[b]	2.8–4.1	5.6–8.2	8.4–12.3
Open areas in West[b]	0.4–0.6	0.8–1.2	1.2–1.8

[a] Owens, et al. (1992).
[b] Ammonium nitrogen is calculated as twice the nitrate deposition, the approximate ratio from numerous measurements, although it may be incorrect at any specific site [U.S./Canada Work Group No. 2 (1982)].

upon microbial nitrification (see Section 13.7). As a result, in certain mature forests most of the nitrogen coming from polluted air seems to be lost as leached nitrate. The H^+ ions from these acids displace base-forming cations on the soil colloids. The displaced cations, especially calcium and magnesium, are lost from the soil as they leach downward with the nitrate (and sulfate; see Section 13.22). Thus the leaching of mobile anions (nitrate and sulfate) promotes the loss of calcium and magnesium, causing imbalances in the nutrition of the trees and a decline in forest productivity. Nitrogen in precipitation might be considered beneficial fertilizer when it falls on farmland, but it can be a serious pollutant when added to a soil supporting a mature forest. Chapter 9 should be consulted for a more detailed explanation of these reactions and soil differences.

Effects on Rangeland Systems

Natural rangeland ecosystems contain a wide variety of native plant species that are known for their ability to conserve nitrogen, including the modest 5 to 10 Kg N/ha being added annually from the atmosphere and nitrogen-fixation. While these systems are reasonably productive, attempts have been made to increase that productivity by adding fertilizer nitrogen and by introducing nitrogen-responsive exotic species. Such practices may increase yields temporarily, but research has shown that in time many of the low-nitrogen-requiring native species are crowded out and replaced by high-nitrogen-requiring exotic species and by weeds. The resulting systems are lower in biological diversity and in productivity than the original native rangelands. This illustrates that well-meaning efforts to increase the productivity of natural ecosystems can, in fact, have the exact opposite effect. Human intervention without knowledge of the principles governing ecosystem productivity may lead to undesirable results.

13.15 REACTIONS OF NITROGEN FERTILIZERS

The worldwide use of nitrogen-containing fertilizers has expanded greatly during the past few decades. Most commercial fertilizers supply nitrogen in soluble forms, such as nitrate or ammonium, or as urea, which rapidly hydrolyzes to form ammonium. Ammonium and nitrate ions from fertilizer are taken up by plants and participate in the nitrogen cycle in exactly the same way as ammonium and nitrate derived from organic matter mineralization or other sources. The main difference is that nitrogen from heavy doses of fertilizer is far more concentrated in time (available all at once) and space (applied in narrow bands or in a thin layer in the soil) than is nitrogen from other sources. While judicious application of such concentrations may be very beneficial for plant growth, it may also have serious detrimental effects.

First, high concentrations of ammonia gas from ammonium-releasing fertilizers (such as urea and anhydrous ammonia) inhibit the activity of many of the soil flora and fauna, including nitrifying bacteria and earthworms. This partial sterilization is usually quite temporary and is localized near the zone of application. Such concentrations of ammonia also favor loss of this gas to the atmosphere, especially in alkaline soils. Also, because two moles of acidity are formed for every mole of ammonium nitrogen that undergoes nitrification to nitrates, use of ammonium fertilizers increases soil acidity (see Section 13.7). Soil acidity may also be increased by ammonia released from a heavy application of manure.

Concentration in time occurs when a year's supply of soluble nitrogen is applied all at once. Such single applications are low cost and convenient, so they are common in agricultural and forestry enterprises. Unfortunately, the assimilatory processes of the nitrogen cycle (e.g., plant uptake and immobilization) may not be able to utilize the soluble fertilizer nitrogen fast enough to prevent major losses by leaching, surface runoff, denitrification, and ammonia volatilization (see Figure 13.6). The environmental consequences of these losses have already been discussed. As a general rule, it is more environmentally benign to apply nitrogen fertilizer in multiple small doses that coincide with the plant demand.

The goals of nitrogen control are threefold: (1) the maintenance of an adequate nitrogen supply in the soil, (2) the regulation of the soluble forms of nitrogen to ensure that enough is readily available to provide all the nitrogen plants need for optimum growth, and (3) the minimization of environmentally damaging leakage from the soil–plant system.

To understand the problem of nitrogen control, it is useful to consider the nitrogen inputs and outputs for a farm or ecosystem. Figure 13.20 summarizes the inputs and outputs of nitrogen for a hypothetical single farm field. Most of the input and output pathways have equivalents in natural ecosystems. The features that make the cropland balance different from that for a natural ecosystem are mainly: (1) the removal of nitrogen contained in the harvested product, which represents a major loss from the field, and (2) the addition of nitrogen fertilizer, which represents a major input into the field.

A particular type of soil in any given combination of climate and farming system tends to assume what may be called a *normal* or *equilibrium content* of nitrogen. Consequently, under ordinary methods of cropping and manuring, any attempt to permanently raise the nitrogen content to a level higher than this will result in unnecessary waste as nitrogen leaches, volatilizes, or is otherwise lost before it may be used. There are some exceptions to this general rule. If a soil that is naturally low in organic matter and nitrogen is heavily fertilized over a period of time, and if crop residues are returned to the soil, the soil nitrogen level will likely increase. Such a situation is common where soils of dry areas (Aridisols) are irrigated and cropped to high-fertilizer-requiring vegetables or alfalfa. If irrigation is discontinued and crop yields are lowered, however, the soil will eventually return to its previous level of nitrogen. The equilibrium level of soil nitrogen is often governed by the management practices employed.

Several basic strategies are proving useful in achieving a rational reduction of excessive nitrogen inputs while maintaining or improving production levels and profitability in agriculture (Box 13.4, including Table 13.11). These approaches include: (1) taking into account the nitrogen contribution from *all* sources and reducing the amount of fertilizer applied accordingly; (2) improving the efficiency with which fertilizer, as well as such so-called waste products as animal manure, are used; (3) avoiding overly optimistic yield goals that lead to fertilizer application rates designed to meet crop needs that are much higher than actually occur in most years; and (4) improving crop response knowledge, which identifies the lowest nitrogen application that is likely to produce optimum profit. These strategies of nutrient management will be discussed further in Chapter 16.

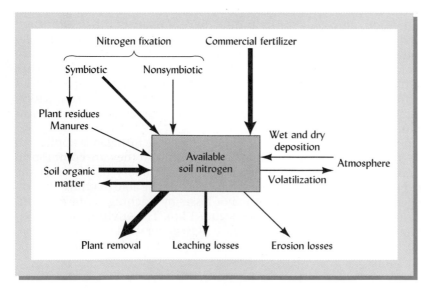

FIGURE 13.20 Major gains and losses of available soil nitrogen. The widths of the arrows roughly indicate the magnitude of the losses and the additions often encountered. It should be emphasized that the diagram represents average conditions only and that much variability is to be expected in the actual and relative quantities of nitrogen involved.

BOX 13.4 RATIONALIZING NITROGEN INPUTS
IN AGRICULTURAL SYSTEMS

The relative importance and absolute magnitudes of the various nitrogen cycle pathways vary greatly with the particular crop grown and with variations in climate and management. A good crop of wheat or cotton may remove only 100 kg N/ha, nearly half of which may be returned to the soil in the stalks or straw. A bumper crop of silage corn, in contrast, may contain over 250 kg N/ha, and a good annual yield of alfalfa or well-fertilized grass hay removes more than 350 kg N/ha. Regardless of the crop, the goal should be to minimize the losses of nitrogen by all pathways except crop removal (this is, after all, what the farmer sells and what we all eat). The need for commercial nitrogen fertilizer will be governed by the degree to which the farmer is able to integrate symbiotic nitrogen fixation, manure and residue recycling, and loss minimization into the farming system.

If crops and livestock are combined in a farming system, then most of the nitrogen removed in crops can be returned in animal manures. Fertilizer should be considered as a supplement to the nitrogen made available from organic matter mineralization, biological nitrogen fixation, rainfall, animal manure, and crop residues. Nitrogen losses by leaching and denitrification generally become a problem only when nitrogen fertilization exceeds the amount needed to fill the gap between crop uptake needs and the supply from these other sources.

When we examine the nitrogen balance for cropland on a national scale, it is clear that more nitrogen is being applied than can be properly used (Table 13.11). In the United States, cropland receives approximately 20.9 million Mg of nitrogen annually, while it exports some 13.5 million Mg of nitrogen as crop harvests and residues. In other words, nitrogen inputs are more than 50% greater than intended outputs, leaving over 7 million Mg of nitrogen to either accumulate in soils (unlikely in most cases) or leak into the environment via leaching, erosion, and runoff, or via gaseous losses to the atmosphere. While some losses from agroecosystems are inevitable, it is becoming widely acknowledged that cropland in the United States and in other industrial countries has been receiving more nitrogen than plants can effectively use, and that this excess will have to be reduced if nitrogen pollution is to be controlled.

TABLE 13.11 Nitrogen Inputs and Outputs for All Cropland in the United States in 1987 and 1977

Only the major categories are considered. These are aggregate data for the entire country and should not be considered representative of any given farm field. The estimates for the two years were made independently of each other, so the close agreement lends validity to both.

	Millions of Mg	
Nitrogen output or input	1987[a]	1977[b]
Outputs		
Harvested crops	10.60	8.9
Crop residues	2.89	3.0
Total outputs	13.50	11.9
Inputs to cropland		
As commercial fertilizer	9.39	9.5
As legume N fixation	6.87	7.2
In crop residues returned	2.89	3.0
Recoverable manure	1.73	1.4
Total inputs	20.9	21.1
Balance = inputs – outputs	7.42	9.2

[a] Data from Table 6-3 in National Research Council (1993).
[b] Data abstracted from Power (1981) as cited in NRC (1993).

13.17 IMPORTANCE OF SULFUR[9]

Sulfur has long been recognized as indispensable for many reactions in living cells. In addition to its vital roles in plant and animal nutrition, sulfur is also responsible for several types of air, water, and soil pollution and is therefore of increasing environmental interest. The environmental problems associated with sulfur include acid precipitation, certain types of forest decline, acid mine drainage, acid sulfate soils, and even some toxic effects in drinking water used by humans and livestock.

Roles of Sulfur in Plants and Animals

Sulfur is a constituent of the amino acids methionine, cysteine, and cystine, deficiencies of which result in serious human malnutrition. The vitamins biotin, thiamine, and B1 contain sulfur, as do many protein enzymes that regulate such activities as photosynthesis and nitrogen fixation. It is believed that sulfur-to-sulfur bonds link certain sites on long chains of amino acids, causing proteins to assume the specific three-dimensional shapes that are the key to their catalytic action. Sulfur is closely associated with nitrogen in the processes of protein and enzyme synthesis. Sulfur is also an essential ingredient of the aromatic oils that give the cabbage and onion families of plants their characteristic odors and flavors. It is not surprising that among the plants, the legume, cabbage, and onion families require especially large amounts of sulfur.

Deficiencies of Sulfur

Healthy plant foliage generally contains 0.15 to 0.45% sulfur, or approximately one-tenth as much sulfur as nitrogen. Plants deficient in sulfur tend to become spindly and to develop thin stems and petioles. Their growth is slow, and maturity may be delayed. They also have a chlorotic light green or yellow appearance. Symptoms of sulfur deficiency are similar to those associated with nitrogen deficiency (see Section 13.1). However, unlike nitrogen, sulfur is relatively immobile in the plant, so the chlorosis develops first on the youngest leaves as sulfur supplies are depleted (in nitrogen-deficient plants, chlorosis develops first on the older leaves). Sulfur-deficient leaves on some plants show interveinal chlorosis or faint striping that distinguishes them from nitrogen-deficient leaves. Also, unlike nitrogen-deficient plants, sulfur-deficient plants tend to have low sugar but high nitrate contents in their sap.

As a result of the following three independent trends, sulfur deficiencies in agricultural plants have become increasingly common during the past several decades:

1. Enforcement of clean air standards has led to reduction of sulfur dioxide (SO_2) emissions to the atmosphere from the burning of fossil coal and oil.

2. The fertilizer market is increasingly dominated by more concentrated, high-analysis products. These fertilizers supply their primary nutrients at lower cost, but they do not contain significant amounts of sulfur-bearing impurities as did the previously used materials. For example, the once widely used ammonium sulfate and single superphosphate fertilizers, which contain 24 and 12% sulfur, respectively, have been largely replaced by diammonium phosphate (DAP), urea, triple superphosphate (TSP), and other materials that have relatively high contents of nitrogen and phosphorus but contain little or no sulfur.

3. At the same time that reductions in the supply of sulfur to soils and plants have occurred, improved varieties and better management have resulted in higher crop yields. As harvests have grown larger, greater amounts of sulfur have been removed from soils. Thus, the need for sulfur has increased just as the supplies of this element have declined.

AREAS OF DEFICIENCY.   Sulfur deficiencies have been reported in most areas of the world but are most prevalent in areas where soil parent materials are low in sulfur, where extreme weathering and leaching has removed this element, or where there is little replenishment of sulfur from the atmosphere. In many tropical countries, one or more of these conditions prevail and sulfur-deficient areas are common.

---

[9] For a discussion of sulfur and agriculture, see Tabatabai (1986).

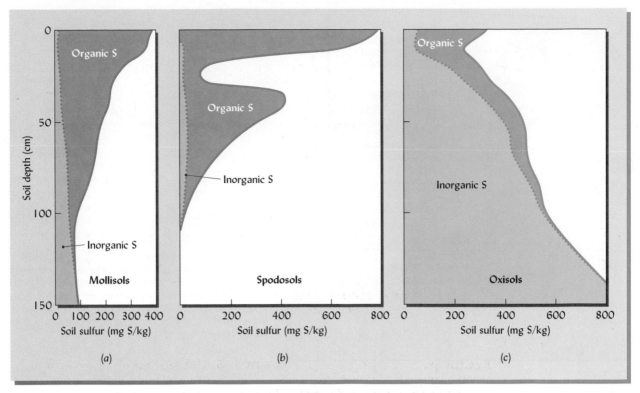

FIGURE 13.21 The distribution of organic and inorganic sulfur in representative soil profiles of the soil orders Mollisols, Spodosols, and Oxisols. In each, soil organic forms dominate the surface horizon. Considerable inorganic sulfur, both as adsorbed sulfate and calcium sulfate minerals, exists in the lower horizons of Mollisols. Relatively little inorganic sulfur exists in Spodosols. However, the bulk of the profile sulfur in the humid tropics (Oxisols) is present as sulfate adsorbed to colloidal surfaces in the subsoil.

Burning of plant biomass results in a loss of sulfur to the atmosphere. In many parts of the world, crop residues and native vegetation are routinely burned as a means of clearing the land. Soils of the African savannas are particularly deficient in sulfur as a result of the annual burning of plant residues during the dry season. Fire converts much of the sulfur in the plant residues to sulfur gases, such as sulfur dioxide. Sulfur in these gases and in smoke particulates is subsequently carried by the wind hundreds of kilometers away to areas covered by rain forest, where some of the sulfur dioxide is absorbed by moist soils and foliage, and some is deposited with rainfall. Thus, the soils of the savannas tend to export their sulfur to those of the rain forest (e.g., Oxisols). Consequently, the latter often contain significant accumulations of sulfur in their profiles (see Figure 13.21).

In the United States, deficiencies of sulfur are most common in the Southeast, the Northwest, California, and the Great Plains. In the Northeast and in other areas with heavy industry and large cities, sulfur deficiencies are not yet widespread.

13.18 NATURAL SOURCES OF SULFUR

The three major natural sources of sulfur that can become available for plant uptake are: (1) *organic matter,* (2) *soil minerals,* and (3) *sulfur gases in the atmosphere.* In natural ecosystems where most of the sulfur taken up by plants is eventually returned to the same soil, these three sources combined are usually sufficient to supply the needs of growing plants (Figure 13.22). These three natural sources of sulfur will be considered in order.

Organic Matter

In surface soils in temperate, humid regions, 90 to 98% of the sulfur is usually present in organic forms (see Figure 13.21). As is the case for nitrogen, the exact forms of the sulfur in the organic matter are not known. Typically, more than half of the sulfur is *car-*

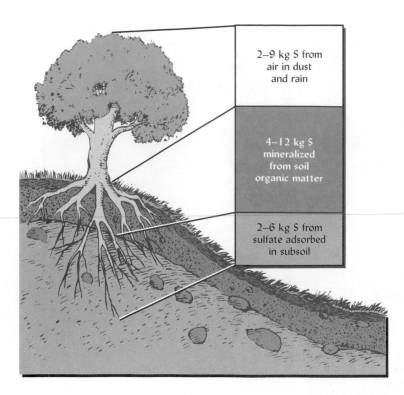

FIGURE 13.22 Plants take up sulfur primarily from three sources: sulfur in atmospheric gases and dust; sulfate mineralized from soil organic matter; and sulfate adsorbed on soil minerals. Typical ranges of sulfur uptake from these sources are shown. Where these three sources are insufficient for optimal growth, the application of sulfur-containing fertilizers may be warranted. In areas downwind of coal-burning plants and metal smelters, the atmospheric contribution may be much larger than indicated here.

bon bonded (C—S fraction), mostly in proteins and in amino acids such as cysteine, cystine, and methionine. These materials are bound with the humus and clay fractions, and are thereby protected from microbial attack. A somewhat more transitory pool of organic sulfur is in the *ester sulfate* form (C—O—S), in which S is bound to oxygen rather than directly to carbon. Examples of compounds in the two main fractions of organic sulfur are as follows:

Ester sulfate (glucose sulfate)   Carbon-bonded sulfur (cysteine)

Over time, soil microorganisms break down these organic sulfur compounds into soluble inorganic forms, mainly sulfate. This mineralization of organic compounds to release sulfate is analogous to the release of ammonium and nitrate from organic matter discussed in Section 13.4.

Table 13.12 shows the high sulfur contents of the soils from three forested areas and the proportion of organic and inorganic compounds present in each. Note the high percentage of carbon-bonded sulfur and the relatively small percentage of sulfate sulfur.

In arid and semiarid regions, less organic matter is present in the surface soils. However, gypsum ($CaSO_4 \cdot 2H_2O$), which supplies inorganic sulfur, is often present in the subsurface horizons. Therefore, the proportion of organic sulfur is not likely to be as high in arid- and semiarid-region soils as it is in humid-region soils. This is especially true in the subsoils, where organic sulfur may comprise only a small fraction of the sulfur present and where gypsum is prominent.

TABLE 13.12   Concentration of Sulfur Compounds in Forested Spodosols from Several Locations

Note that the organic layers (Oa) have high S levels and that the two organic forms (C-bonded and ester sulfate) contain 84 to 99% of the soil sulfur.

Horizon	Totals, µg S/g	Proportion of soil S, %			
		C-bonded S	Ester sulfate	Sulfate S	Inorganic S [a]
Hubbard Brook, N.H. [b]					
Oa	1563	71	28	0.3	—
Bh	303	72	22	3.0	—
Bs1	452	64	26	8.1	—
Huntington Forest, N.Y. [c]					
Oa	1780	77	22	0.2	0.7
Bh	761	83	13	3.1	2.2
Bs1	527	70	22	4.2	4.4
Conifer site [d]					
Oa	2003	88	11	0.3	0.6
Bh	540	83	11	2.4	3.4
Bs1	515	57	27	13	3.0

[a] Inorganic sulfur compounds, such as sulfides.
[b] From Schindler, et al. (1986).
[c] From David, et al. (1982).
[d] From Mitchell, et al. (1992).

Soil Minerals

The inorganic forms of sulfur are not as plentiful as the organic forms, but they include the soluble and available compounds on which plants and microbes depend.

Sulfur is held in several mineral forms in soils, with the sulfide and sulfate minerals being most common. The sulfate minerals are most easily solubilized, and the sulfate ion (SO_4^{2-}) is easily assimilated by plants. Sulfate minerals are most common in regions of low rainfall, where they accumulate in the lower horizons of some Mollisols and Aridisols (see Figure 13.21). They may also accumulate as neutral salts in the surface horizons of saline soils in arid and semiarid regions.

Sulfides that are found in some humid-region soils with restricted drainage must be oxidized to the sulfate form before the sulfur can be assimilated by plants. When these soils are drained, oxidation can occur, and ample available sulfur is released. In some cases, so much sulfur is oxidized that problems of extreme acidity result (see Section 13.21).

Another mineral source of sulfur is the clay fraction of some soils high in Fe, Al oxides and kaolinite. These clays are able to strongly adsorb sulfate from soil solution and subsequently release it slowly by anion exchange, especially at low pH. Oxisols and other highly weathered soils of the humid tropics and subtropics may contain large stores of sulfate, especially in their subsoil horizons (see Figure 13.21). Considerable sulfate may also be bound by the metal oxides in the spodic horizons under certain temperate forests.

Atmospheric Sulfur[10]

The atmosphere contains varying quantities of carbonyl sulfide (COS), hydrogen sulfide (H_2S), sulfur dioxide (SO_2), and other sulfur gases, as well as sulfur-containing dust particles. These atmospheric forms of sulfur arise from volcanic eruptions, volatilization from soils, ocean spray, biomass fires, and industrial plants (such as electric-generation sta-

---

[10] For a discussion of this topic, see National Academy of Sciences (1983).

tions fired by high-sulfur coal, and metal smelters). In recent decades, the contributions of the latter industrial sources have become the dominant sources in certain locations.

Some of these materials are oxidized in the atmosphere to sulfates, forming H_2SO_4 and sulfate salts, such as $CaSO_4$ and $MgSO_4$. When these solids and gases return to the earth as dry particles and gases, it is called *dry deposition;* when they are brought down with precipitation, it is called *wet deposition.* Although the proportion of these two forms of deposition varies from one place to another, each typically supplies about half of the total.

The industrialized northeastern states have the highest deposition of sulfur in the United States, commonly ranging from 30 to 75 kg S/ha/yr downwind from industrial sites. Farther away from industrial sources, deposition declines to about 8 to 15 kg S/ha/yr. In rural areas of the western United States, away from industrial cities and smelting plants, only 2 to 5 kg S/ha/yr is deposited. In rural Africa, 1 to 4 kg S/ha/yr is deposited. Thus, the effect of atmospheric deposition depends greatly on location, with the main effect of industrial emission occurring within a relatively small distance of the source (Figure 13.23). On the other hand, regional patterns of air quality and sulfur deposition are evident as the less acute effects of industrial emissions are felt hundreds of kilometers downwind (Figure 13.24).

Atmospheric sulfur becomes part of the soil–plant system in three ways. The wet deposition materials, which are usually high in H_2SO_4, are mostly absorbed by soils, with some also absorbed through plant foliage. Part of the dry deposition is also absorbed directly by soils, while some is absorbed directly by plants. The quantity that plants can absorb directly is variable, but in some cases 25 to 35% of the plant sulfur can come from this source even if available soil sulfate is adequate. In sulfur-deficient soils, about half of the plant needs can come from the atmosphere (see Figure 13.22).

Environmental concerns about high sulfur levels in the atmosphere and the resulting acid rain are of great practical significance to both forestry and agriculture. Efforts to reduce atmospheric sulfur are welcomed in areas near industrial plants, where levels of atmospheric sulfur may be high enough to cause toxicity to trees and crops (not to mention respiratory problems in people). Even beyond the immediate influence of industrial sources, the elevated levels of atmospheric sulfur and associated acid effects

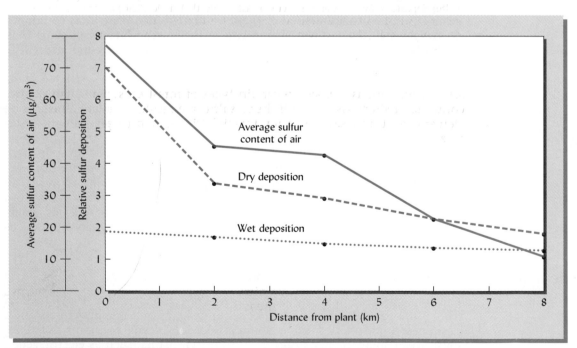

FIGURE 13.23 Industrial facilities such as coal- or oil-burning power plants or metal smelters can provide large inputs of sulfur to nearby soils. Note the rapid drop-off in sulfur deposition at increasing distances from the source of sulfur emissions. Dry deposition of sulfur from the atmosphere is dominant very close to the source, but at greater distances dry and wet deposition are nearly equal. Regulations in many countries would today require that such a power plant greatly reduce its sulfur emissions. [From Johannson (1960)]

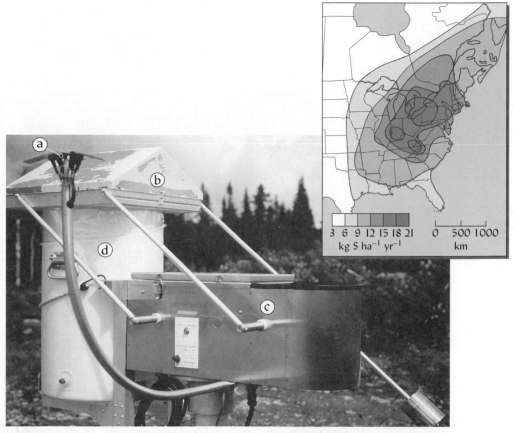

FIGURE 13.24 The apparatus shown in the picture collects both wet and dry sulfur deposition. A sensor (*a*) triggers the small roof (*b*) to move over and cover the dry deposition collection chamber (*c*) at the first sign of precipitation. The wet deposition chamber (*d*) is then exposed to collect precipitation. When the precipitation ceases, the sensor triggers the roof to move back over the wet deposition collection chamber so that dry deposition can again be collected. The map shows the typical geographic distribution of sulfur deposition from the atmosphere in eastern North America. The data are based on measurements of sulfur in wet deposition and the assumption that dry deposition contributes approximately an equal share of the total. [Photo courtesy of R. Weil; map modified from Olsen and Slavich (1986)]

are causing serious damage to certain types of forest ecosystems. On the other hand, continued reductions in atmospheric sulfur are resulting in plant deficiencies of this element in other areas, especially for high-yield-potential agricultural crops. In these cases, part of the crop's sulfur requirement may have to be supplied by increasing the sulfur content of the fertilizers used, thereby increasing crop production costs.

13.19 THE SULFUR CYCLE

The major transformations that sulfur undergoes in soils are shown in Figure 13.25. The inner circle shows the relationships among the four major forms of this element: (1) *sulfides*, (2) *sulfates*, (3) *organic sulfur*, and (4) *elemental sulfur*. The outer portions show the most important sources of sulfur and how this element is lost from the system.

Considerable similarity to the nitrogen cycle is evident (compare Figures 13.25 and 13.3). In each case, the atmosphere is an important source of the element in question. Both elements are held largely in the soil organic matter; both are subject to microbial oxidation and reduction; both can enter and leave the soil in gaseous forms; and both are subject to some degree of leaching in the anionic form. Microbial activities are responsible for many of the transformations that determine the fates of both nitrogen and sulfur.

Figure 13.25 should be referred to frequently in conjunction with the following more detailed examination of sulfur in plants and soils.

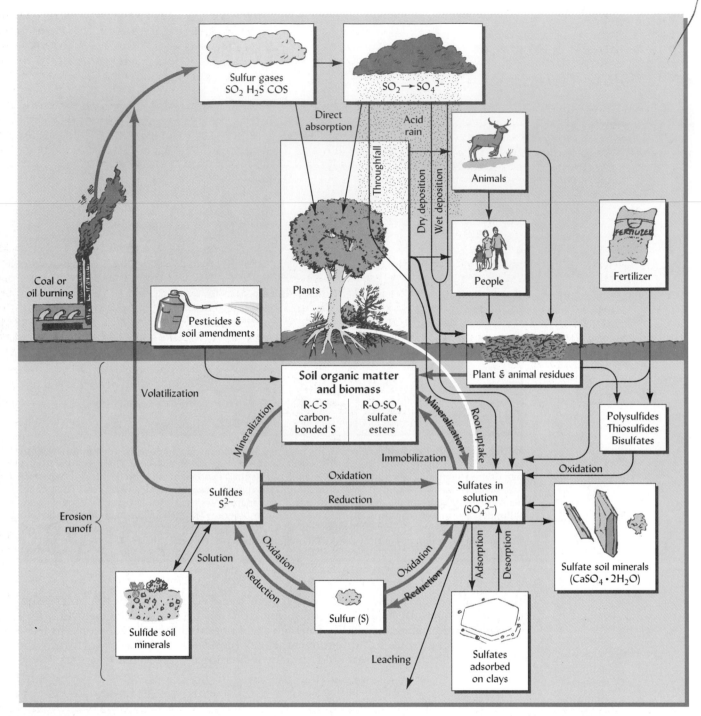

FIGURE 13.25 The sulfur cycle, showing some of the transformations that occur as this element is cycled through the soil–plant–animal–atmosphere system. In the surface horizons of all but a few types of arid-region soils, the great bulk of sulfur is in organic forms. However, in deeper horizons or in excavated soil materials, various inorganic forms may dominate. The oxidation and reduction reactions that transform sulfur from one form to another are mainly mediated by soil microorganisms.

13.20 BEHAVIOR OF SULFUR COMPOUNDS IN SOILS

Mineralization

Sulfur behaves much like nitrogen as it is absorbed by plants and microorganisms and moves through the sulfur cycle. The organic forms of sulfur must be mineralized by soil organisms if the sulfur is to be used by plants. The rate at which this occurs depends on the same environmental factors that affect nitrogen mineralization, including moisture, aeration, temperature, and pH. When conditions are favorable for general microbial activity, sulfur mineralization occurs. Some of the more easily decomposed organic compounds in the soil are sulfate esters, from which microorganisms release sulfate ions directly. However, in much of the soil organic matter, sulfur in the reduced state is bonded to carbon atoms in protein and amino acid compounds. In the latter case the mineralization reaction might be expressed as follows:

$$\text{Organic sulfur} \longrightarrow \text{decay products} \xrightarrow{O_2} SO_4^{2-} + 2H^+$$

Proteins and other organic combinations   H₂S and other sulfides are simple examples   Sulfates

Because this release of available sulfate is mainly dependent on microbial processes, the supply of available sulfate in soils fluctuates with seasonal, and sometimes daily, changes in environmental conditions (Figure 13.26). These fluctuations lead to the same difficulties in predicting and measuring the amount of sulfur available to plants as were discussed in the case of nitrogen.

Immobilization

Immobilization of inorganic forms of sulfur occurs when low-sulfur, energy-rich organic materials are added to soils. The immobilization mechanism is thought to be the same as for nitrogen—the energy-rich material stimulates microbial growth, and the inorganic sulfate is assimilated into microbial tissue. A C/S ratio greater than 400:1 gen-

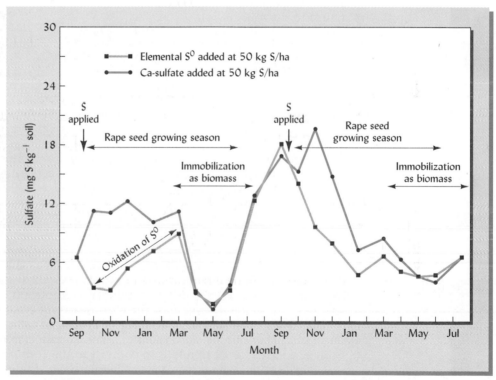

FIGURE 13.26 Seasonal changes in the sulfate form of sulfur available in the surface horizon of a soil (Argixeroll) in Oregon used to grow the oilseed crop, rape. This crop is sown in the fall, grows slowly during the winter, and then grows rapidly during the cool spring months. Data are shown for plots that were fertilized with either elemental S or calcium sulfate. The vertical arrows indicate dates on which these amendments were applied. Note that sulfate concentration was greater in the calcium sulfate–fertilized soils for the first few months after each application while the elemental S was slowly converted to sulfate by microbial oxidation. A distinct depression in sulfate concentration occurred each spring as the soil warmed up, stimulating both immobilization of sulfate into microbial biomass and uptake of sulfate into the rapeseed crop. Sulfate concentrations reached a peak in late summer and early fall, when crop uptake ceased after harvest and microbial mineralization was rapidly occurring. Movement of dissolved sulfate from the lower horizons up into the surface soil may also have occurred during hot, dry weather. [Modified from Castellano and Dick (1991)]

TABLE 13.13 Mean Carbon/Nitrogen/Sulfur Ratios
in a Variety of Soils

Location	Description and number of soils	C/N/S ratio
North Scotland	Agricultural, noncalcareous (40)	104:7:1
Minnesota	Mollisols (6)	74:7:1
Minnesota	Spodosols (24)	108:8:1
Oregon	Agricultural, varied (16)	143:10:1
Eastern Australia	Acid soils (128)	126:8:1
Eastern Australia	Alkaline soils (27)	92:7:1
Sweden	Agricultural, Inceptisols	69:7:1

From Whitehead (1964) and (for Sweden) Kirchmann (1996).

erally leads to such immobilization of sulfur. When the microbial activity subsides, the inorganic sulfate reappears in the soil solution.

The pattern of S immobilization in soils suggests that, like nitrogen, sulfur in soil organic matter may be associated with organic carbon in a reasonably constant ratio. The ratio among carbon, nitrogen, and sulfur for a number of soils on three different continents is given in Table 13.13. The C/N/S ratio of 100:8:1 is reasonably representative.

During the microbial breakdown of organic materials, several sulfur-containing gases are formed, including hydrogen sulfide (H_2S), carbon disulfide (CS_2), carbonyl sulfide (COS), and methyl mercaptan (CH_3SH). All are more prominent in anaerobic soils. Hydrogen sulfide is commonly produced in waterlogged soils by reduction of sulfates by anaerobic bacteria. Most of the others are formed from the microbial decomposition of sulfur-containing amino acids. Although these gases can be adsorbed by soil colloids, some escape to the atmosphere, where they undergo chemical changes and eventually return to the soil.

13.21 SULFUR OXIDATION AND REDUCTION

The Oxidation Process

During the microbial decomposition of organic carbon-bonded sulfur compounds, sulfides are formed along with other incompletely oxidized substances, such as elemental sulfur (S^0), thiosulfates ($S_2O_3^{2-}$), and polythionates ($S_{2x}O_{3x}^{2-}$). These reduced substances are subject to oxidation, just as are the ammonium compounds formed when nitrogenous materials are decomposed. The oxidation reactions may be illustrated as follows, with hydrogen sulfide and elemental sulfur:

$$H_2S + 2O_2 \rightarrow H_2SO_4 \rightarrow 2H^+ + SO_4^{2-}$$

$$2S + 3O_2 + 2H_2O \rightarrow 2H_2SO_4 \rightarrow 4H^+ + SO_4^{2-}$$

The oxidation of some sulfur compounds, such as sulfites (SO_3^{2-}) and sulfides (S^{2-}), can occur by strictly chemical reactions. However, most sulfur oxidation in soils is *biochemical* in nature, carried out by a number of autotrophic bacteria, which include five species of the genus *Thiobacillus*. Since the environmental requirements and tolerances of these five species vary considerably, the process of sulfur oxidation occurs over a wide range of soil conditions. For example, sulfur oxidation may occur at pH values ranging from <2 to >9. This flexibility is in contrast to the comparable nitrogen oxidation process, nitrification, which requires a rather narrow pH range closer to neutral.

The Reduction Process

Like nitrate ions, sulfate ions tend to be unstable in anaerobic environments. They are reduced to sulfide ions by a number of bacteria of two genera, *Desulfovibro* (five species) and *Desulfotomaculum* (three species). The organisms use the oxygen in sulfate to oxidize organic materials. A representative reaction showing the reduction of sulfur coupled with organic matter oxidation is as follows:

$$2R-CH_2OH + SO_4^{2-} \rightarrow 2R-COOH + 2H_2O + S^{2-}$$

Organic alcohol Sulfate Organic acid Sulfide

In poorly drained soils, the sulfide ion reacts immediately with iron or manganese, which in anaerobic conditions are typically present in the reduced forms. By tying up the soluble reduced iron, the formation of iron sulfides helps prevent iron toxicity in rice paddies and marshes. This reaction may be expressed as follows:

$$Fe^{2+} + S^{2-} \rightarrow FeS$$

Dissolved ferrous iron — Sulfide — Iron sulfide (solid)

$$Mn^{2+} + S^{2-} \rightarrow MnS$$

Dissolved reduced manganese — Sulfide — Manganese sulfide (solid)

Sulfide ions will also undergo hydrolysis to form gaseous hydrogen sulfide, which causes the rotten-egg smell of swampy or marshy areas. Sulfur reduction may take place with sulfur-containing ions other than sulfates. For example, sulfites (SO_3^{2-}), thiosulfates ($S_2O_3^{2-}$), and elemental sulfur (S^0) are readily reduced to the sulfide form by bacteria and other organisms.

The oxidation and reduction reactions of inorganic sulfur compounds play an important role in determining the quantity of sulfate (the plant-available nutrient form of sulfur) present in soils at any one time. Also, the state of sulfur oxidation is an important factor in the acidity of soil and water draining from soils.

Sulfur Oxidation and Acidity

The reaction equations given for oxidation of elemental sulfur (S) and hydrogen sulfide (H_2S) show that, like nitrogen oxidation, sulfur oxidation is an acidifying process. For every sulfur atom oxidized, two hydrogen ions are formed. Because of this acidifying reaction, elemental sulfur can be applied to extremely alkaline and sodic soils of arid regions to reduce the pH of these soils to a level more favorable for plant growth and disease control (Section 10.9). The acidity formed as sulfur oxidizes must also be considered when choosing a fertilizer, since some materials contain elemental sulfur and might lower the soil pH unfavorably.

The addition of atmospheric sulfur to soils through precipitation increases soil acidity.[11] The pH of this so-called *acid rain* may be 4 or even lower (see Section 9.6). In contrast, the pH of unpolluted rainwater varies from 5.6 in humid regions to over 7.0 in drier regions, where calcareous dust reacts with rainwater as it falls.

EXTREME SOIL ACIDITY. The acidifying effect of sulfur oxidation can bring about extremely acid soil conditions. This may happen when coastal land inundated with brackish water or seawater is drained and put under cultivation. Ocean water and marine sediments contain relatively high amounts of sulfur. During periods when land is under water, sulfates are reduced to generally stable iron and manganese sulfides. If there are periods of partial drying, elemental sulfur can form by partial oxidation of the sulfides left in the soils that have been submerged. The sulfide and elemental sulfur content of such tidal marsh soils are hundreds of times greater than would be found in comparable upland soils. Sediments dredged from coastal harbors are also high in reduced forms of sulfur, such as black-colored iron sulfides (see Plate 25 after page 498).

If high-sulfide soils are drained, the sulfides and/or elemental sulfur quickly oxidize and form sulfuric acid.

$$2FeS + 4\tfrac{1}{2}O_2 + 2H_2O \rightarrow Fe_2O_3 + 4H^+ + 2SO_4^-$$
$$S + 1\tfrac{1}{2}O_2 + H_2O \rightarrow 2H^+ + SO_4^-$$

The soil pH may drop to as low as 1.5, a level not observed in normal upland soils. Obviously, plant growth cannot occur under these conditions. Furthermore, the quantity of limestone needed to neutralize the acidity is so high that is impractical to remediate these soils by liming.

---

[11] Note that some of this acidity results from nitrogen oxides as they are oxidized in the atmosphere to nitric and nitrous acids. Sulfur and nitrogen are jointly responsible for this problem.

Sizable areas of high-sulfur wetland soils (classified as Sulfaquepts and Sulfaquents in *Soil Taxonomy* and commonly called **acid sulfate soils** or **cat-clays**) are found in Southeast Asia and along the Atlantic coasts of South America and Africa. They also occur in tidal areas along the coasts of several other areas, including the Netherlands and both the Southeast and the West Coast of the United States. So long as these soils are kept submerged, reduced sulfur is not oxidized and the soil pH does not drop prohibitively. Consequently, production of paddy rice is sometimes possible in these soils.

The occurrence of soils capable of producing extreme acidity by sulfur oxidation is not limited to the coastal wetlands just described. Marine sediments, possibly associated with coastal marshes millions of years ago, now form the rock and regolith covering large areas of land no longer submerged by oceans or marshes (see Section 2.9). Today's coastal plains are often underlain by such sediments. Also, many types of shales and other sedimentary rocks may contain reduced sulfur compounds, often in the form of the iron sulfide mineral, pyrite (see Plate 24 after page 498). These rocks are commonly associated with deposits of coal, a substance that also formed millions of years ago in marshy areas.

Excavations for road cuts and coal-mining operations often expose long-buried pyrite-containing materials. When pyrite comes into contact with air and water, the following net reaction, which is facilitated by sulfur oxidizing bacteria, occurs:

$$2FeS_2 + 7\tfrac{1}{2}O_2 + 7H_2O \xrightarrow{\text{Bacteria}} 8H^+ + 4SO_4^{2-} + 2Fe(OH)_3$$

Pyrite   Oxygen   Water   Hydrogen   Sulfate   Iron hydroxide
                                      ions

The four moles of sulfuric acid (dissociated into hydrogen and sulfate ions) produced by the preceding reaction come from the oxidation of both the sulfate and the ferrous iron in pyrite in conjunction with the hydrolysis of water. The production of acid from this reaction can be an enormous problem, preventing growth of vegetation on the affected area (Figure 13.27) and making it very difficult to restore certain lands disturbed by mining operations. If allowed to proceed unchecked, the acids may wash into nearby streams. Thousands of kilometers of streams have been seriously polluted in this manner, the water and rocks in such streams often exhibiting orange colors from the iron compounds in the acid drainage (see Plate 26 between pages 498 and 499).

FIGURE 13.27 Construction of this interstate highway cut through several layers of sedimentary rock. One of these layers contained reduced sulfide materials. Now exposed to the air and water, this layer is producing copious quantities of sulfuric acid as the sulfide materials are oxidized. Note the failure of vegetation to grow below the zone from which the acid is draining. (Photo courtesy of R. Weil)

The sulfate ion is the form in which plants absorb most of their sulfur from soils. Since many sulfate compounds are quite soluble, the sulfate would be readily leached from the soil, especially in humid regions, were it not for its adsorption by the soil colloids. As was pointed out in Chapter 8, most soils have some anion exchange capacity, which is associated with iron and aluminum oxide coatings and clays and, to a limited extent, with 1:1-type silicate clays. Sulfate ions are attracted by the positive charges that characterize acid soils containing these clays. They also react directly with hydroxy groups exposed on the surfaces of these clays. Figure 13.28 illustrates sulfate adsorption mechanisms on the surface of some Fe, Al oxides and 1:1-type clays. Note that adsorption increases at lower pH values as positive charges that become more prominent on the particle surfaces attract the sulfate ions. Some sulfate reacts with the clay particles, becoming tightly bound, and is only slowly available for plant uptake and leaching.

In warm, humid regions, surface soils are typically quite low in sulfur. However, much sulfate may be held by the iron and aluminum oxides and 1:1-type silicate clays that tend to accumulate in the subsoil horizons of the Ultisols and Oxisols of these regions (see Figure 13.21). Symptoms of sulfur deficiency commonly occur early in the growing season on Ultisols in the southeastern United States, especially those with sandy, low-organic-matter surface horizons. However, the symptoms may disappear as the crop matures and its roots reach the deeper horizons where sulfate is retained.

Sulfate Adsorption and Leaching of Base-Forming Cations

When the sulfate ion leaches from the soil, it is usually accompanied by equivalent quantities of cations, including Ca and Mg and other base-forming cations. In soils with high sulfate adsorption capacities, sulfate leaching is low and the loss of companion cations is also low. (Figure 13.29). In contrast, sulfate leaching losses from low-sulfate-adsorbing soils are commonly high, and take with them considerable quantities of base-forming cations. Sulfur is thus seen as an indirect conserver of these cations in the soil solution. This is of considerable importance in soils of forested areas that receive acid rain.

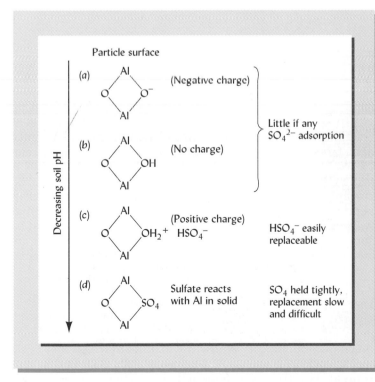

FIGURE 13.28 Effect of decreasing soil pH on the adsorption of sulfates by 1:1-type silicate clays and oxides of Fe and Al (reaction with a surface-layer Al is illustrated). At high pH levels (a), the particles are negatively charged, the cation exchange capacity is high, and base-forming cations are adsorbed. Sulfates are repelled by the negative charges. As acidity is increased (b), the H^+ ions are attracted to the particle surface, and the negative charge is satisfied but the SO_4^{2-} ions are still not attracted. At still lower pH values (c), more H^+ ions are attracted to the particle surface, resulting in a positive charge that attracts the SO_4^{2-} ion. This is easily exchanged with other anions. At still lower pH levels, the SO_4^{2-} reacts directly with Al and becomes a part of the crystal structure. Such sulfate is tightly bound, and it is removed very slowly, if at all.

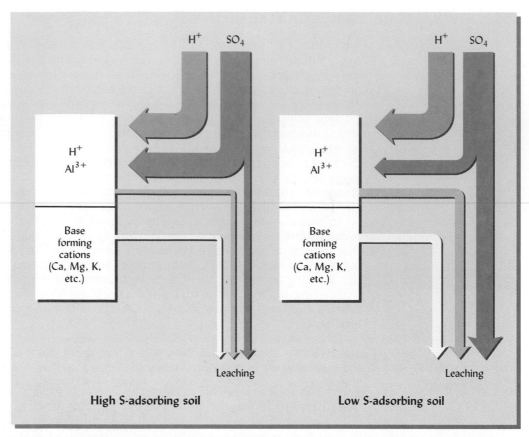

FIGURE 13.29 Diagrams illustrating cation leaching losses as influenced by sulfate adsorption capacities of forest soils. When acid rain containing SO_4^{2-} and H^+ ions falls on soils with high SO_4^{2-}-adsorbing capacities (left), the small quantities of SO_4^{2-} available for leaching are accompanied by correspondingly small amounts of cations such as Ca^{2+} and Mg^{2+}. Where acid rain falls on soils with low SO_4^{2-} adsorbing capacity (right), most of the sulfate remains in the soil solution and is leached from the soil along with equivalent quantities of cations, including Ca^{2+} and Mg^{2+}. Al^{3+} ions that commonly replace the Ca^{2+} and Mg^{2+} lost from the soil exchange complex are toxic to many forest species. This likely accounts for at least part of the negative effects of acid rain on some forested areas. [Redrawn from Mitchell, et al. (1992)]

13.23   SULFUR AND SOIL FERTILITY MAINTENANCE

Figure 13.30 depicts the major gains and losses of sulfur from soils. The problem of maintaining adequate quantities of sulfur for mineral nutrition of plants is becoming increasingly important. Even though chances for widespread sulfur deficiencies are generally less than for nitrogen, phosphorus, and potassium, increasing crop removal of sulfur makes it essential that farmers be attentive to prevent deficiencies of this element. In some parts of the world (especially in certain semiarid grasslands), sulfur is already the next most limiting nutrient after nitrogen.

Crop residues and farmyard manures can help replenish the sulfur removed in crops, but these sources generally can help to recycle only those sulfur supplies that already exist within a farm. In regions with low-sulfur soils, greater dependence must be placed on fertilizer additions. Regular applications of sulfur-containing materials are now necessary for good crop yields in large areas far removed from industrial plants. There will certainly be an increased necessity for the use of sulfur in the future.

13.24   CONCLUSION

Sulfur and nitrogen have much in common concerning the manner in which they cycle in soils. Both are held by soil colloids in slowly available forms. Both are found in proteins and other organic forms as part of the soil organic matter. Their release to inor-

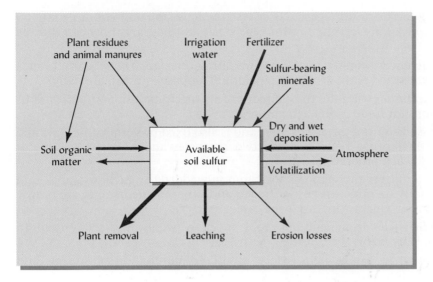

FIGURE 13.30 Major gains and losses of available soil sulfur. The thickness of the arrows indicates the relative amounts of sulfur involved in each process under average conditions. Considerable variation occurs in the field.

ganic ions (SO_4^{2-}, NH_4^+, and NO_3^-), in which forms they are available to higher plants, is accomplished by soil microorganisms. Anaerobic soil organisms are able to change these elements into gaseous forms, which are then released to the atmosphere to be joined by similar gases released from industrial plants and from vehicle engines. These gases are then deposited on plants, soils, and other objects in forms that are popularly termed *acid rain*. This has serious consequences to forestry, to agriculture, and to society in general.

There are some significant differences between nitrogen and sulfur. Large amounts of both soil nitrogen and soil sulfur are found in organic compounds. However, a higher proportion of soil sulfur is found in inorganic compounds, especially in some drier areas where gypsum ($CaSO_4 \cdot 2H_2O$) is abundant in the subsoil. Some soil organisms have the ability to fix elemental N_2 gas into compounds usable by plants. No analogous process occurs for sulfur. Also, nitrogen, which is removed in much larger quantities by plants, must be replenished regularly by organic residues, manure, or chemical fertilizers.

In rural areas away from cities and industrial plants, sulfur deficiencies in plants are increasingly common. As a result of increased crop removal and reductions in incidental applications of sulfur to soils, this element will likely join nitrogen as a regular component of soil fertility programs.

Several serious environmental problems are caused by excessive amounts of nitrogen or sulfur in certain forms. While water pollution by nitrates is probably the most prominent of the nitrogen-related problems, acid deposition and drainage are the main environmental concerns associated with sulfur cycling in soils. Using our knowledge of the nitrogen and sulfur cycles, much can now be done to alleviate both sets of problems. However, much remains to be learned about the reactions of nitrogen and sulfur in soils and in the environment.

STUDY QUESTIONS

1. A total of 300 kg/ha of nitrogen was supplied through animal manure and chemical fertilizer for a corn crop, but only 200 kg/ha was found in the harvested silage corn, all of which was removed from the land. What happened to the 100 kg/ha N not taken up by the corn crop? Explain.

2. Discuss the benefits and ill effects of sulfur and nitrogen that are added to soils by humans and by natural processes each year.

3. Some 20 kg/ha of nitrogen contained in wheat straw was added to a soil that had a nitrate level of 10 kg/ha. A couple of weeks later a soil test showed no nitrates. How do you explain this situation?

4. Modern domestic animal production systems have some environmental and health implications relating to nitrogen. What are they, and how can they be managed?

5. What differences would you expect in nitrate contents of streams from a forested watershed and one where agricultural crops are grown, and why?

6. What is *acid rain,* what are the sources of this precipitation, and what are its implications for forestry and agriculture?

7. Nitrogen is fixed from the atmosphere and is also fixed by vermiculite clays and humus. Differentiate between these two processes and indicate the role of microbes in each process.

8. Chemical fertilizers and manures are commonly added to agricultural soils, and yet these soils are often lower in nitrogen and other nutrients than are nearby forested soils. Explain why this is the case.

9. Sulfur deficiencies in agricultural crops are more widespread today than 15 years ago. Why is this the case?

10. How do riparian forests help reduce nitrate contamination of streams and rivers?

11. What are *acid sulfate soils,* how do they form, and how can they best be managed?

12. In some tropical regions, agroforestry systems that involve mixed cropping of trees and food crops are used. What advantages in nitrogen management do such systems have over monocroping systems that do not involve trees?

REFERENCES

Addiscott, T. M., A. P. Whitmore, and D. S. Powlson. 1991. *Farming, Fertilizers and the Nitrate Problem.* (Wallingford, U.K.: CAB International).

Baethgen, W. E., and M. M. Alley. 1987. "Nonexchanged ammonium nitrogen contributions to plant available nitrogen," *Soil Sci. Soc. Amer. J.,* **51**:110–115.

Broadbent, F. E., T. Nakashima, and G. Y. Chang. 1982. "Estimation of nitrogen fixation by isotope dilution in field and greenhouse experiments," *Agron. J.,* **74**:625–628.

Bronson, K. F., et al. 1997. "Automated chamber measurements of methane and nitrous oxide flux in a flooded rice soil: I. Residue, nitrogen and water management; II. Fallow period emissions," *Soil Sci. Soc. Amer. J.,* **61**:981–993.

Burns, R. C., and R. W. F. Hardy. 1975. *Nitrogen Fixation in Bacteria and Higher Plants.* (Berlin: Springer-Verlag).

Castellano, S. D., and R. P. Dick. 1991. "Cropping and sulfur fertilization influence on sulfur transformations in soil," *Soil Sci. Soc. Amer. J.,* **54**:114–121.

Chichester, F. W., and S. J. Smith. 1978. "Disposition of ^{15}N-labeled fertilizer nitrate applied during corn culture field lysimeters," *J. Environ. Qual.,* **7**:227–233.

Colbourn, P. 1993. "Limits to denitrification in two pasture soils in a maritime temperate climate," *Agriculture, Ecosystems and Environment,* **43**:49–68.

David, M. B., M. J. Mitchell, and J. P. Nakas. 1982. "Organic and inorganic sulfur constituents of a forest soil and their relationship to microbial activity," *Soil Sci. Soc. Amer. J.,* **46**:847–852.

Dutch, J., and P. Ineson. 1990. "Denitrification of an upland forest site," *Forestry,* **63**:363–376.

Emsley, J. 1991. "Metals trace the secrets of nitrogen fixation," *New Scientist,* **131**(1784):19.

Hanson, G. C., P. M. Groffman, and A. J. Gold. 1994. "Denitrification in riparian wetlands receiving high and low groundwater nitrate inputs," *J. Environ. Qual.,* **23**:917–922.

Hauck, R. D. (ed.). 1984. *Nitrogen in Crop Production.* (Madison, Wis.: Amer. Soc. Agron., Crop Sci. Soc. Amer., Soil Sci. Soc. Amer.).

Hinsinger, P., and R. J. Gilkes. 1996. "Mobilization of phosphate from phosphate rock and aluminum-sorbed phosphate by the roots of ryegrass and clover as related to rhizosphere pH," *European J. Soil Sci.,* **47**:533–544.

Jenkinson, D. S. 1990. "An introduction to the global nitrogen cycle," *Soil Use and Management,* **6**:56–61.

Johannson, O. 1960. "On sulfur problems in Swedish agriculture," *Kgl. Lanabr. Ann.*, **25**:57–169.

Kirchmann, H., F. Pichlmayer, and M. H. Gerzabek. 1996. "Sulfur balances and sulfur-34 abundance in a long-term fertilizer experiment," *Soil Sci. Soc. Amer. J.*, **60**:174–178.

Klemmedson, J. O. 1994. "New Mexican locust and parent material: Influence on forest floor and soil macronutrients," *Soil Sci. Soc. Amer. J.*, **58**:974–980.

Leffelaar, P. A. 1986. "Dynamics of partial anaerobiosis, denitrification and water in a soil aggregate," *Soil Sci.*, **142**:352–366.

Leffelaar, P. A., and W. W. Wessel. 1988. "Denitrification in a homogeneous, closed system: Experimental and simulation," *Soil Sci.*, **146**:335–349.

Leuch, D., S. Haley, P. Liapis, and B. McDonald. 1995. *The EU Nitrate Directive and CAP Reform: Effects on Agricultural Production, Trade and Residual Soil Nitrogen*. Report 255. (Washington, D.C.: USDA Economic Research Service).

Mekonnen, K., R. J. Buresh, and B. Jama. 1997. "Root and inorganic nitrogen distributions in *Sesbania* fallow, natural fallow, and maize fields," *Plant and Soil*, in press.

Mitchell, M. J., et al. 1992. "Sulfur dynamics of forest ecosystems," in Howarth, et al. (eds.), *Sulfur Cycling on the Continents*, 1992 SCOPE. (New York: Wiley).

National Academy of Sciences. 1983. *Acid Deposition: Atmospheric Processes in Eastern North America*. (Washington, D.C.: National Academy Press).

National Research Council. 1993. *Soil and Water Quality: An Agenda for Agriculture*. (Washington, D.C.: National Academy of Sciences).

Olsen, A. R., and A. L. Slavich. 1986. *Acid Precipitation in North America: 1984 Annual Data Summary*, from Acid Deposition System Data Base. Environmental Protection Agency Report EPA/600/4-86/033. (Washington, D.C.: U.S. Environmental Protection Agency).

Owens, L. B., W. M. Edwards, and R. W. Van Keuren. 1992. "Nitrate levels in shallow groundwater under pastures receiving ammonium nitrate or slow-release nitrogen fertilizer," *J. Environ. Qual.*, **21**(4):607–613.

Patrick, W. H., Jr. 1982. "Nitrogen transformations in submerged soils," in F. J. Stevenson (ed.), *Nitrogen in Agricultural Soils*. Agronomy Series no. 27 (Madison, Wis.: Amer. Soc. Agron., Crop Sci. Soc. Amer., Soil Sci. Soc. Amer.).

Power, J. F. 1981. "Nitrogen in the cultivated ecosystem," in F. E. Clark and T. Rosswall (eds.), *Terrestrial Nitrogen Cycles—Processes, Ecosystem Strategies and Management Impacts*. Ecological Bulletin no. 33. (Stockholm, Sweden: Swedish National Research Council), pp. 529–546.

Prasad, R., and J. F. Power. 1995. "Nitrification inhibitors for agriculture, health and the environment," *Advances in Agronomy*, **54**:233–280.

Sanchez, P. A., R. J. Buresh, and R. R. B. Leakey. 1997. "Trees, soils and food security," in *Philosophical Transactions of the Royal Society*, London, Series A, 355.

Schlindler, S. C., et al. 1986. "Incorporation of ^{35}S-sulfate into inorganic and organic constituents of two forest soils," *Soil Sci. Soc. Amer. J.*, **150**:457–462.

Smith, S. J., D. W. Dillow, and L. B. Young. 1982. "Disposition of fertilizer nitrates applied to sorghum—Sudan grass in the Southern Plains," *J. Environ. Qual.*, **11**:341–344.

Stevenson, F. J. (ed.). 1982. *Nitrogen in Agricultural Soils*. Agronomy Series no. 22. (Madison, Wis.: Amer. Soc. Agron., Crop Sci. Soc. Amer., Soil Sci Soc. Amer.).

Tabatabai, S. J. 1986. *Sulfur in Agriculture*. Agronomy Series no. 27 (Madison, Wis.: Amer. Soc. Agron., Crop Sci. Soc. Amer., Soil Sci. Soc. Amer.).

Torrey, J. G. 1978. "Nitrogen fixation by actinomycete-induced angiosperms," *BioScience*, **28**:586–592.

U.S./Canada Work Group No. 2. 1982. *Atmospheric Sciences and Analysis*. Final Report; J. L. Ferguson and L. Machata (Cochairmen). (Washington, D.C.: U.S. Environmental Protection Agency).

Walker, T. W., et al. 1956. "Fate of labeled nitrate and ammonium nitrogen when applied to grass and clover grown separately and together," *Soil Sci.*, **81**:339–352.

Wang, F. L. and A. K. Alva. 1996. "Leaching of nitrogen from slow-release urea sources in sandy soils," *Soil Sci. Soc. Amer. J.*, **60**:1454–1458.

Whitehead, D. C. 1964. "Soil and plant nutrition aspects of the sulfur cycle," *Soils and Fertilizers*, **27**:1–8.

14

SOIL PHOSPHORUS AND POTASSIUM

> *Most of what's in the soil stays there even when plants suffer for lack of it.*
> —*APT DESCRIPTION OF BOTH PHOSPHORUS AND POTASSIUM*

Phosphorus is a critical element in natural and agricultural ecosystems throughout the world. This macronutrient element is a key component of cellular compounds and is vital to both plant and animal life. In world agriculture, management of phosphorus is second only to management of nitrogen in its importance for the production of healthy plants and profitable yields. The natural supply of phosphorus in most soils is small, and the availability of that which is present is very low. Inputs of phosphorus from the atmosphere and rainfall are negligible. Fortunately, most undisturbed natural ecosystems lose little of this nutrient because phosphorus does not form gases that can escape into the atmosphere, nor does it readily leach out of the soil with drainage water. Furthermore, natural ecosystems have evolved in ways that promote certain biological and chemical processes which allow plants to make relatively efficient use of the scant supplies of this element available to them.

Phosphorus is closely associated with animal (and human) activity. Bones and teeth contain large amounts of this element. Archaeologists study the phosphorus content of soil horizons, because they know that unusually high concentrations of this element often accumulate where humans have congregated and have discarded the bones of wild or domesticated animals. Phosphorus is so scarce in most soils that high concentrations are often an indication of past animal or human activity in the area.

Phosphorus is also the culprit in two widespread and serious environmental problems. In industrialized countries, liberal application of phosphorus to soils over many decades has significantly increased the level of this element in the surface layers of many soils. Also, human, animal, and industrial wastes applied to soils tend to concentrate phosphorus. Where management practices are not sufficiently stringent, phosphorus lost from watersheds or waste treatment facilities in surface runoff water and eroded sediments is severely upsetting the nutrient balance in streams, lakes, and estuaries. The added inputs of phosphorus are largely responsible for *cultural eutrophication,*[1] which may jeopardize drinking-water supplies and can severely restrict the use of these aquatic systems for fisheries, recreation, industry, and aesthetics. Phosphorus control is therefore a high priority of most national and regional water-quality programs.

---

[1] The word *eutrophication* comes from the Greek *eutrophos,* meaning well-nourished or nourishing. Natural accumulation of nutrients over centuries causes lakes to fill in with plants and lose dissolved oxygen needed by fish—the process of eutrophication. Excessive inputs of nutrients under human influence tremendously speeds this process and is called *cultural eutrophication.*

At the other extreme, lack of adequate available phosphorus is contributing to land degradation and subsequent water pollution in vast land areas, mostly in the lesser-developed countries of tropical and subtropical regions. Phosphorus deficiency often limits the growth of crops, and may even cause crop failure, which forces farmers to clear more land in order to survive. Without adequate phosphorus, regrowth of natural vegetation on disturbed forest and savanna sites is often too slow to prevent soil erosion and depletion of soil organic matter. Unless sources of available phosphorus can be added, growth of vegetation will be poor, and this may lead to even lower levels of soil productivity and a downward spiral of land degradation and water pollution.

While it does not play as direct a role in environmental quality as does phosphorus, low availability of soil potassium also commonly limits plant growth and reduces crop quality. Potassium works hand in hand with nitrogen and phosphorus to ensure healthy plants. Even though most soils have large total supplies of this nutrient, most of that present is tied up in the form of insoluble minerals and is unavailable for plant use. Also, plants require potassium in such large amounts that careful management practices are necessary in order to make this nutrient available rapidly enough to optimize plant growth.

Both phosphorus and potassium are used in large amounts as fertilizers, thereby earning them the title, along with nitrogen, of "the fertilizer elements." Ask most growers which nutrients they apply in the fertilizers they use, and the answer will most likely be N-P-K (nitrogen, phosphorus, and potassium). Add to this soil fertility role the important environmental impacts of phosphorus, and it is clear that a good understanding of the soil processes that govern how these two macronutrients are cycled in nature is essential to anyone working with soils.

14.1 ROLE OF PHOSPHORUS IN PLANT NUTRITION AND SOIL FERTILITY[2]

Neither plants nor animals can grow without phosphorus. It is an essential component of the organic compound often called the *energy currency* of the living cell: **adenosine triphosphate** (ATP). Synthesized through both respiration and photosynthesis, ATP contains a high-energy phosphate group that drives most energy-requiring biochemical processes. For example, the uptake of nutrients and their transport within the plant, as well as their assimilation into different biomolecules, are energy-using plant processes that require ATP.

Phosphorus is an essential component of **deoxyribonucleic acid** (DNA), the seat of genetic inheritance, and of **ribonucleic acid** (RNA), which directs protein synthesis in both plants and animals. Phospholipids, which play critical roles in cellular membranes, are another class of universally important phosphorus-containing compounds. For most plant species, the total phosphorus content of healthy leaf tissue is not high, usually only 0.2 and 0.4% of the dry matter.

Phosphorus and Plant Growth

ASPECTS OF PLANT GROWTH ENHANCED. Adequate phosphorus nutrition enhances many aspects of plant physiology, including the fundamental processes of photosynthesis, nitrogen fixation, flowering, fruiting (including seed production), and maturation. Root growth, particularly development of lateral roots and fibrous rootlets, is encouraged by phosphorus. In cereal crops, good phosphorus nutrition strengthens structural tissues such as those found in straw or stalks (Table 14.1), thus helping to prevent lodging (falling over). Improvement of crop quality, especially in forages and vegetables, is another benefit attributed to this nutrient.

SYMPTOMS OF PHOSPHORUS DEFICIENCY IN PLANTS. Phosphorus deficiency is generally not as easy to recognize in plants as are deficiencies in many other nutrients. A phosphorus-deficient plant is usually stunted, thin-stemmed, and spindly, but its foliage is often dark, almost bluish, green. Thus, unless much larger, healthy plants are present to make a comparison, phosphorus-deficient plants often seem quite normal in appearance. In

---

[2] For a review of the significance of this element, see Khasawneh, et al. (1980).

TABLE 14.1 Influence of Phosphorus Nutrition on Cornstalk Rot

Breeding line of corn	P in ear leaf, mg P/kg dry matter	Erect plants, % of all plants	Crushing strength of stalk, kg/cm²
High P accumulator	0.75	92	288
Low P accumulator	0.24	70	105

Data from Porter (1981).

severe cases, phosphorus deficiency can cause yellowing and senescence of leaves. Some plants develop purple colors (see Plate 28, after page 498) in their leaves and stems as a result of phosphorus deficiency, though other related stresses, such as cold temperatures, can also cause purple pigmentation. Phosphorus is needed in especially large amounts in meristematic tissues, where cells are rapidly dividing and enlarging.

Phosphorus is very mobile within the plant, so when the supply is short, phosphorus in the older leaves is mobilized and transferred to the newer, rapidly growing leaves. Both the purpling and premature senescence associated with phosphorus deficiency are therefore most prominent on the older leaves. Phosphorus-deficient plants are also characterized by delayed maturity, sparse flowering, and poor seed quality.

The Phosphorus Problem in Soil Fertility

The phosphorus problem in soil fertility is threefold. *First,* the total phosphorus level of soils is low, usually no more than one-tenth to one-fourth that of nitrogen, and one-twentieth that of potassium. The phosphorus content of soils ranges from 200 to 2000 kg phosphorus in the upper 15 cm of 1 ha of soil, with an average of about 1000 kg P. *Second,* the phosphorus compounds commonly found in soils are mostly unavailable for plant uptake, often because they are highly insoluble. *Third,* when soluble sources of phosphorus, such as those in fertilizers and manures, are added to soils, they are fixed[3] (changed to unavailable forms) and, in time, form highly insoluble compounds. We will examine these fixation reactions in some detail (see Section 14.8) because they play an important role in determining how much and in what manner phosphorus should be added to soils.

Fixation reactions in soils may allow only a small fraction (10 to 15%) of the phosphorus in fertilizers and manures to be taken up by plants in the year of application. Consequently, farmers who can afford to do so apply two to four times as much phosphorus as they expect to remove in the crop harvest. Repeated over many years, such practices have saturated the phosphorus-fixation capacity and built up the level of available phosphorus in many agricultural soils. Soils having such high levels of soil phosphorus no longer need to be fertilized with more than the amount of phosphorus removed in harvest. In fact, many agricultural soils in industrialized countries with long histories of phosphorus buildup from manure or fertilizer application have accumulated so much available phosphorus that little if any additional phosphorus is needed until phosphorus is drawn down to more moderate levels over a period of years. The statistics on fertilizer use in the United States reflect the fact that farmers have recently begun to recognize that fertilizer applications can be reduced where soil phosphorus levels have been built up (Figure 14.1). The long-term buildup of phosphorus has improved soil fertility, but has also resulted in certain undesirable environmental consequences, discussed in Section 14.2.

In many developing countries, especially in Africa, such overuse of fertilizer phosphorus is not the rule. In most of sub-Saharan Africa, where per capita food production has been declining in recent years, fertilizer additions of this element for food crops are a fraction of the rate of removal of phosphorus in the harvested crops. The soils have been mined of phosphorus for years, with the result that in many areas lack of this element is the first limiting factor in food-crop production.[4] Such phosphorus constraints also indirectly affect the supply of nitrogen, since the growth of most nitrogen-fixing legumes is constrained under low phosphorous conditions. The decline in per capita

---

[3] Note that the term *fixation* as applied to phosphorus has the same general meaning as the chemical fixation of potassium or ammonium ions; that is, the chemical being fixed is bound, entrapped, or otherwise held tightly by soil solids in a form that is relatively unavailable to plants. In contrast, the fixation of gaseous nitrogen refers to the biological conversion of N_2 gas to combined forms that plants can use.

[4] For a recent review of soil phosphorus problems in Africa, see Buresh, et al. (1997).

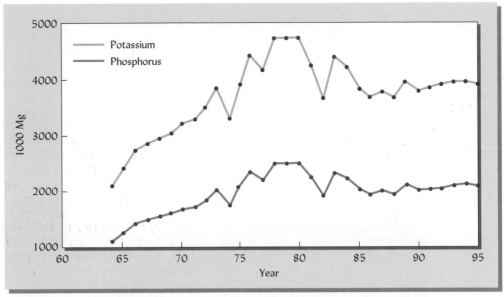

FIGURE 14.1 Estimated annual addition of fertilizer phosphorus and potassium to cropland in the United States since 1964. Note that annual applications of phosphorus increased steadily until about 1980, then declined somewhat to level off at about 2 million Mg per year. The decline and leveling off of phosphorus application rates reflects the reduced need for phosphorus buildup in agricultural soils in which the phosphorus fixation capacity has been largely saturated. The trend toward lower rates of phosphorus application since 1980 was clearly driven by economic and agronomic influences, not environmental concerns, since the trends for potassium, a nonpolluting nutrient, are almost identical. As an example of economic effects, the oil embargo of 1973–1974 increased fertilizer prices dramatically, resulting in the notable dip in application rates in 1974. The sharp dip in 1983 reflected U.S. government policy of "payment in kind" to induce farmers to restrict their corn plantings. In other parts of the world, sustainable crop production will probably require increasing applications of phosphorus and potassium for many years to come. (Based on data from the USDA Economic Research Service and the United Nations Food and Agriculture Organization)

food production in sub-Saharan Africa will not likely be reversed until the critical phosphorus deficiency problems are solved.

14.2 EFFECTS OF PHOSPHORUS ON ENVIRONMENTAL QUALITY

Unlike certain nitrogen-containing compounds that are produced during the cycling of nitrogen (e.g., ammonia, nitrates, and nitrosoamines; see Chapter 13), phosphorus added to aquatic systems from soil is *not* toxic to fish, livestock, or humans. However, too much or too little phosphorus *can* have severe and widespread negative impacts on environmental quality. The principal environmental problems related to soil phosphorus are **land degradation** caused by too little available phosphorus and **accelerated eutrophication** caused by too much. Both problems are related to the role of phosphorus as a plant nutrient.

Land Degradation

Many highly weathered soils in the warm, humid, and subhumid regions of the world have very little capacity to supply phosphorus for plant growth. The low phosphorus availability is partly a result of extensive losses of phosphorus during long periods of relatively intense weathering and partly due to the low availability of phosphorus in the aluminum and iron combinations that are the dominant forms of phosphorus in these soils.

Undisturbed natural ecosystems in these regions usually contain enough phosphorus in the biomass and soil organic matter to maintain a substantial standing crop of trees or grasses. Most of the phosphorus taken up by the plants is that released from the decomposing residues of other plants. Very little is lost as long as the system remains undisturbed.

Once the land is cleared for agricultural use (by timber harvest or by forest fires), the losses of phosphorus in eroded soil particles, in runoff water, and in biomass removals (harvests) can be substantial. Within just a few years the system may lose most of the phosphorus that had cycled between the plants and the soils. The remaining inorganic phosphorus in the soil is largely unavailable for plant uptake. In this manner, the phosphorus-supplying capacity of the disturbed soil rapidly becomes so low that regrowth of natural vegetation is sparse and, on land cleared for agricultural use, crops soon fail to produce useful yields (Figure 14.2).

Leguminous plants that might be expected to replenish soil nitrogen supplies are particularly hard hit by phosphorus deficiency, because low phosphorus supply inhibits effective nodulation and retards the biological nitrogen-fixation process. The spindly plants, deficient in both phosphorus and nitrogen, can provide little vegetative cover to prevent heavy rains from washing away the surface soil. The resulting erosion will further reduce soil fertility and water-holding capacity. The increasingly impoverished soils can support less and less vegetative cover, and so the degradation accelerates.

FIGURE 14.2 On many highly weathered soils in the tropics and subtropics, low availability of phosphorus, as well as of other nutrients, may limit the regrowth of natural vegetation or the production of crops after natural vegetation is cleared. (a) Regrowth of natural vegetation is very sparse in the demarcated unfertilized plot in this disturbed area of Caribbean rain forest in Puerto Rico. (b) and (c) Phosphorus has been so depleted in this African woman's cornfield that there will be little or no grain to harvest from her severely phosphorus-deficient crop. In both cases, poor plant growth leaves the soil susceptible to accelerated erosion and further degradation. (Photos courtesy of R. Weil)

Meanwhile, the soil particles lost by erosion become sediment farther down in the watershed, filling in reservoirs and increasing the turbidity of the rivers.

There are probably 1 to 2 billion ha of land in the world where phosphorus deficiency limits growth of both crops and native vegetation. Much of this land lies in poor countries whose farmers have little money for fertilizers. Without properly managed inputs of phosphorus, there can be little hope of restoring these lands to productivity and their inhabitants to prosperity. Halting and reversing this type of land degradation will require managing the phosphorus cycle to make efficient use of scarce phosphorus resources.

Water Quality Degradation

Accelerated or cultural eutrophication (Box 14.1) is caused by phosphorus entering streams from both **point sources** and **nonpoint sources**. Point sources, such as sewage treatment plant outflows, industries, and the like, are relatively easy to identify, regulate, and clean up. During the past several decades many industrialized countries have made much progress in reducing phosphorus loading from point sources. Nonpoint sources, in contrast, are difficult to identify and control. Nonpoint sources of phosphorus are principally runoff water and eroded sediments from soils scattered throughout the affected watershed. These diffuse sources of phosphorus are now the main cause of eutrophication in many regions.

Phosphorus losses from a watershed can be increased by a variety of human activities, including timber harvest, intensive livestock grazing, soil tillage, and soil application of animal manures and phosphorus-containing fertilizers (Table 14.2). In many areas application of phosphorus to soils has, over the years, dramatically increased the phosphorus content of surface soil, and thus the impact that runoff and sediment have on stream phosphorus loading.

There are several reasons why surface soil phosphorus content has increased so much. First, farmers have long known that when soluble phosphorus fertilizer is applied to a low-phosphorus soil, most of the phosphorus rapidly becomes bound in forms that plants cannot use (see Section 14.6). To compensate for this inefficiency, farmers in industrialized countries have traditionally applied more phosphorus to their soils than is removed in the harvest. Second, where animal manures or sewage sludges are used, the amount applied is most often calculated to meet the nitrogen needs of the crop. The phosphorus contained in that same amount of manure is likely to exceed by two to four times the amount of phosphorus taken up by the plants. Consequently, the phosphorus content of many intensively managed agricultural soils has been increased over the years; subsequently, the phosphorus content of the runoff water and sediment coming from these soils has also increased.

Studies in many regions have shown that soils devoted to crop production lose far more phosphorus to streams than do those covered by relatively undisturbed forests or natural grasslands. Streams draining watersheds with predominantly agricultural land

TABLE 14.2 **Influence of Wheat Production and Tillage on Annual Losses of Phosphorus in Runoff Water and Eroded Sediments Coming from Soils in the Southern Great Plains**

The total phosphorus lost includes the phosphorus dissolved in the runoff water and the phosphorus adsorbed to the eroded particles.[a] Although cattle grazing on the natural grasslands probably increased losses of phosphorus from these watersheds, the losses from the agricultural watersheds were about 10 times as great. The no-till wheat fields tended to lose much less particulate phosphorus, but more soluble phosphorus, than the conventionally tilled wheat fields.

Location and soil	Management	kg P/ha/yr		
		Soluble P	Particulate P	Total P
El Reno, Okla. Paleustolls, 3% slope	Wheat with conventional plow and disk	0.21	3.51	3.72
	Wheat with no-till	1.04	0.43	1.42
	Native grass, heavily grazed	0.14	0.10	0.24
Woodward, Okla. Ustochrepts, 8% slope	Wheat with conventional sweep plow and disk	0.23	5.44	5.67
	Wheat with no-till	0.49	0.70	1.19
	Native grass, moderately grazed	0.02	0.07	0.09

[a] Wheat was fertilized with up to 23 kg of P each fall.
Data from Smith, et al. (1991).

BOX 14.1   EUTROPHICATION

Abundant growth of plants in terrestrial systems is usually considered beneficial, but in aquatic systems such growth can cause water-quality problems. The unwanted growth of algae (floating single-celled plants) and of aquatic weeds, termed *eutrophication*, can make ponds and lakes unsatisfactory environments for fish and can make the water unsuitable for drinking.

In natural ecosystems, lakes are commonly clear and free of excess plant growth and have highly diverse communities of organisms. This is because some factor (such as low levels of nitrogen or phosphorus) is limiting the growth of algae and other plants. Brackish and salty waters are most often nitrogen-limited, and plant growth is stimulated by high-nitrogen pollutants.

Most freshwater lakes and streams, on the other hand, are usually phosphorus-limited, partly because phosphorous levels in the water are typically very low and partly because nitrogen is being supplied through nitrogen fixation by some cyanobacteria (blue-green algae). Critical levels of phosphorus in water, above which eutrophication is likely to be triggered, are approximately 0.03 mg/L of dissolved phosphorus and 0.1 mg/L of total phosphorus.

When phosphorus is added to a phosphorus-limited lake, it stimulates a burst of algal growth (referred to as an *algal bloom*) and, often, a shift in the dominant algal species. The phosphorus-stimulated algae may cover the surface of the water with mats of algal scum. The lake may also be choked with higher plants that are also stimulated by the added phosphorus. When these aquatic weeds and algal mats die, they sink to the bottom, where their decomposition by microorganisms uses up much of the oxygen dissolved in the water. The decrease in oxygen (anoxic conditions) severely limits the growth of many aquatic organisms, especially fish. Such eutrophic lakes often become turbid, limiting growth of submerged aquatic vegetation and benthic (bottom feeding) organisms that serve as food for much of the fish community. In extreme cases eutrophication can lead to massive fish kills (Figure 14.3).

Eutrophication can transform clear, oxygen-rich, good-tasting water into cloudy, oxygen-poor, foul-smelling, bad-tasting, and possibly toxic water. Eutrophic conditions favor the growth of *Cyanobacter*, blue-green algae that are mostly undesirable food for zooplankton, a major food source for fish. These *Cyanobacter* produce toxins and bad-tasting and -smelling compounds that make the water unsuitable for human or animal consumption. Some filamentous algae can clog water treatment intake filters and thereby increase the cost of water remediation. Dense growth of both algae and aquatic weeds may make the water useless for boating and swimming. Furthermore, eutrophic waters generally have a reduced level of biological diversity (fewer species) and fewer fish, of desirable species.

FIGURE 14.3   *In extreme cases of eutrophication, massive fish kills can occur in sensitive lakes and rivers. The kills result from anoxic conditions, which are brought on by the decay of the masses of algae stimulated by elevated inputs of phosphorus (or sometimes of nitrogen). (Photo courtesy of R. Weil)*

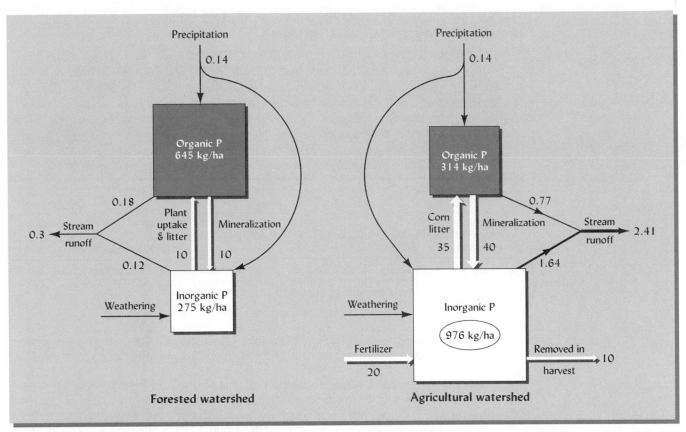

FIGURE 14.4 Phosphorus balance in surface soils (Ultisols) of adjacent forested and agricultural watersheds. The forest consisted primarily of mature hardwoods that had remained relatively undisturbed for 45 or more years. The agricultural land was producing row crops for more than 100 years. It appears that in the agricultural soil about half of the organic phosphorus has been converted into inorganic forms or lost from the system since cultivation began. At the same time, substantial amounts of inorganic phosphorus accumulated from fertilizer inputs. Compared to the forested soil, mineralization of organic phosphorus was about four times as great in the agricultural soil, and the amount of phosphorus lost to the stream was eight times as great. Flows of phosphorus, represented by arrows, are given as kg/ha/yr. Although not shown in the diagram, it is interesting to note that nearly all (95%) of the phosphorus lost from the agricultural soil was in particulate form, while losses from the forest soil were 33% dissolved and 77% particulate. [Data from Vaithiyanathan and Correll (1992)]

use tend to carry much higher phosphorus loads because of the just-discussed phosphorus enrichment of the surface soils, and because many agricultural operations increase the level of surface runoff and erosion (Figure 14.4).

PHOSPHORUS LOSSES IN RUNOFF. Agricultural management that involves disturbing the soil surface with tillage generally increases the amount of phosphorus carried away on eroded sediment (i.e., **particulate P**). On the other hand, fertilizer or manure that is left unincorporated on the surface of cropland or pastures usually leads to increased losses of phosphorus dissolved in the runoff water (i.e., **dissolved P**). These trends can be seen in Table 14.2 by comparing phosphorus losses from no-till and conventionally tilled wheat.

The decision to use tillage to incorporate phosphorus-bearing soil amendments involves a trade-off between the advantages of incorporating phosphorus into the soil (less phosphorus dissolves in the runoff water and more is available for plant uptake) and the disadvantages associated with disturbing the surface soil (increased loss of soil particles by erosion; see Table 14.2). In no-tillage systems, surface application of manure without incorporation may result in lower total loss of phosphorus, because this type of management achieves substantial reductions in soil erosion and total runoff. The effect of these reductions may outweigh the effect of the increased phosphorus concentration in the relatively small volume of water that does run off. If the equipment is available to do so, the best option might be to inject the high-phosphorus amendment into the soil with a minimum of disturbance to the soil surface.

Disturbances to natural vegetation, such as timber harvest or wildfires, also increase the loss of phosphorus, primarily via eroded sediment. Erosion tends to transport predominantly the clay and organic matter fractions of the soil, which are relatively rich in

The site was steeply sloped, with shallow sandy loam soils (lithic Haplumbrepts) relatively high in available P. Note that the more severe the burn, the greater were the losses of sediment and associated P. In all cases, the enrichment ratios were slightly greater than 2, indicating that the sediments eroded from the plots were more than twice as concentrated in phosphorus as were the surface soils themselves. In this, as in most cases, the soil fractions rich in phosphorus were the most susceptible to erosion. The levels of phosphorus lost from this particular forest are higher than levels typically lost from humid-region forests on soils low in available P.

	Sediment lost, Mg/ha	Total P lost, kg/ha	Enrichment ratio[b]
Unburned control	2	1.4	2.6
Moderately burned by prescribed fire	5	4.3	2.1
Severely burned by wildfire	13+[a]	9.1+	2.2

[a] Values for the severely burned plots were measured for only 10 months, while the other plots were measured for 12 months.
[b] Enrichment ratio = (mg P/kg sediment)/(mg P/kg soil).
Data from Saa, et al. (1994).

phosphorus, leaving behind the coarser, lower-phosphorus fractions. Thus, compared to the original soil, eroded sediment is often enriched in phosphorus by a ratio of 2 or more (Table 14.3).

Runoff from animal feedlots and holding areas where hundreds or even thousands of cattle, hogs, or poultry are confined are increasingly significant sources of phosphorus that lead to eutrophication. Manure from a herd of 10,000 beef cattle contains as much phosphorus and nitrogen as is found in human wastes of a city of 100,000. While much animal manure might be spread on nearby fields, heavy rains induce runoff and erosion losses from the holding areas that carry significant quantities of soluble and particulate phosphorus. Much of this element can find its way into lakes and ponds, where eutrophication occurs. Figure 14.5 shows the very high phosphorus level in the surface horizon of a soil under an intensive dairy holding operation near Lake Okeechobee in South Florida. It is easy to visualize how phosphorus could move from this area to nearby streams that feed into the lake.

Intensive animal production, as well as pollution from domestic and commercial development are thought to be responsible for algal blooms of **pfiesteria** and other such toxic algae that are responsible for the death of millions of fish in estuaries and rivers of the coastal plain area of the eastern United States. Elevated phosphorus in streams is believed to encourage the life stage of *pfiesteria* that releases a toxin 1000 times more deadly than cyanide; it can kill small fish in a matter of minutes and larger ones in a few hours. A few humans have been diagnosed as having been poisoned by *pfiesteria* in

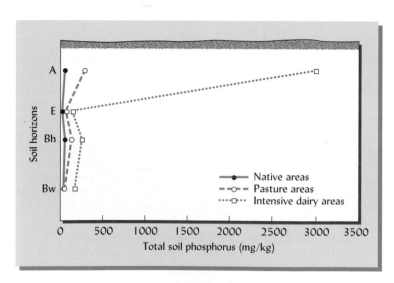

FIGURE 14.5   Average total soil phosphorus levels of intensively managed areas, and nearby pasture and native (forested) areas of three dairy farms near Lake Okeechobee in South Florida. The intensively used areas near the barns where cattle are held prior to milking are extremely high in phosphorus, some of this element having moved down into the subsoil. Runoff water from such areas of animal concentration into nearby streams and lakes is a major source of phosphorus that stimulates eutrophication. [Drawn from data in Nair, et al. (1995)]

infested waters. Attempts are being made to reduce the movement of nutrient-laden wastes from industrial-type hog and poultry operations in the area.

The preceding examples make it clear that the control of eutrophication in many rivers and lakes will require improved management of the phosphorus cycle in soils. Specifically, it is critical to reduce the transport of phosphorus from soils to water.

14.3   THE PHOSPHORUS CYCLE[5]

In order to manage phosphorus for economic plant production and for environmental protection, we will have to understand the nature of the different forms of phosphorus found in soils and the manner in which these forms of phosphorus interact within the soil and in the larger environment. The cycling of phosphorus within the soil, from the soil to higher plants and back to the soil, is illustrated in Figure 14.6.

[5] For a review of soil phosphorus, see Olsen and Khasawneh (1980) and Stevenson (1986).

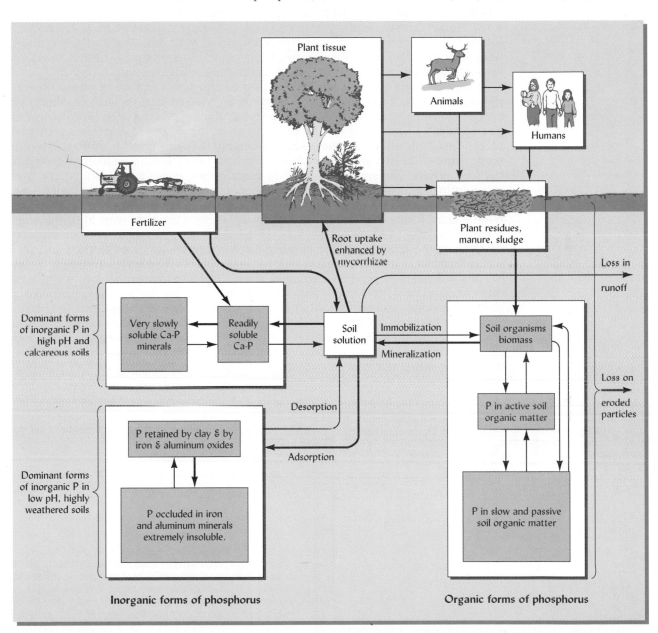

FIGURE 14.6   The phosphorus cycle in soils. The boxes represent pools of the various forms of phosphorus in the cycle, while the arrows represent translocations and transformations among these pools. The three largest white boxes indicate the principal groups of phosphorus-containing compounds found in soils. Within each of these groups, the less soluble, less available forms tend to dominate.

PHOSPHORUS IN SOIL SOLUTION. Compared to other macronutrients, such as sulfur and calcium, the concentration of phosphorus in the soil solution is very low, generally ranging from 0.001 mg/L in very infertile soils to about 1 mg/L in rich, heavily fertilized soils. Plant roots absorb phosphorus dissolved in the soil solution, mainly as phosphate ions (HPO_4^{-2} and $H_2PO_4^-$), but some soluble organic phosphorus compounds are also taken up. The chemical species of phosphorus present in the soil solution is determined by the solution pH, as shown in Figure 14.7. In strongly acid soils (pH 4 to 5.5), the monovalent anion $H_2PO_4^-$ dominates, while alkaline solutions are characterized by the divalent anion HPO_4^{-2}. Both anions are important in near-neutral soils. Of the two anions, $H_2PO_4^-$ is thought to be slightly more available to plants, but effects of pH on phosphorus reactions with other soil constituents are more important than the particular phosphorus anion present.

UPTAKE BY ROOTS AND MYCORRHIZAE. Uptake by the plant root requires not only that the phosphate ions be dissolved in the soil solution, but also that they move from the bulk soil to the surface of the root (Figure 14.8). This movement, usually over a distance of a few millimeters to a centimeter, takes place primarily by physical diffusion (see Section 1.16). However, because phosphate ions are very strongly adsorbed by soil particles, diffusion to the root may be so slow and intermittent as to limit the availability of phosphorus to the plant. Root hairs growing into the soil can help to shorten the distance through which phosphate ions must diffuse before being taken up.

Note that Figure 14.8 shows, in addition to direct root uptake of phosphorus from the soil solution, a second pathway that can be used by many plants to obtain phosphorus from the soil. This second pathway involves symbiosis with mycorrhizal fungi (see Section 11.9). In soils low in available phosphorus, many species of plants could barely survive without mycorrhizal assistance in obtaining phosphorus. The microscopic, threadlike mycorrhizal hyphae extend out into the soil several centimeters from the root surface. The hyphae are able to absorb phosphorus ions as the ions enter the soil solution, and may even be able to access some strongly bound forms of phosphorus. The hyphae then bring the phosphate to the root by transporting it inside the hyphal cells, where soil-retention mechanisms cannot interfere with phosphate movement (see Figure 14.8).

Once in the plant, a portion of the phosphorus is translocated to the plant shoots, where it becomes part of the plant tissues. As the plants shed leaves and their roots die, or they are eaten by people or animals, phosphorus returns to the soil in the form of plant residues, leaf litter, and wastes from animals and people. Microorganisms decompose the residues and temporarily tie up at least part of the phosphorus in their cells. Some of it becomes associated with the active and passive fractions of the soil organic matter (see Section 12.8), where it is subject to storage and future release. These organic forms are very slowly converted to the soluble forms that plant roots can absorb, thereby repeating the cycle.

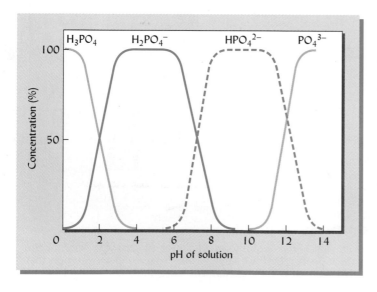

FIGURE 14.7 The effect of pH on the relative concentrations of the three species of phosphate ions. At lower pH values, more H^+ ions are available in the solution, and thus the phosphate ion species containing more hydrogen predominates. In near-neutral soils, HPO_4^{-2} and $H_2PO_4^-$ are found in nearly equal amounts. Both of these species are readily available for plant uptake.

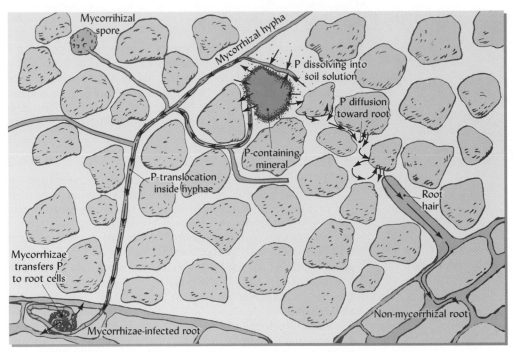

FIGURE 14.8 Roles of diffusion and mycorrhizal hyphae in the movement of phosphate ions to plant roots. In soils with low solution phosphorus concentration and high phosphorus fixation, slow diffusion may seriously limit the ability of roots to obtain sufficient phosphorus. The hyphae of symbiotic mycorrhizal fungi are particularly beneficial to the plant where phosphorus diffusion is slow, because phosphorus is transported inside the hyphae by cytoplasmic streaming, making the plant much less dependent on the diffusion of phosphate ions through the soil.

CHEMICAL FORMS IN SOILS. In most soils, the amount of phosphorus available to plants from the soil solution at any one time is very low, seldom exceeding about 0.01% of the total phosphorus in the soil. The bulk of the soil phosphorus exists in three general groups of compounds—namely, *organic phosphorus, calcium-bound inorganic phosphorus,* and *iron- or aluminum-bound inorganic phosphorus* (see Figure 14.6). Of the inorganic phosphorus, the calcium compounds predominate in most alkaline soils, while the iron and aluminum forms are most important in acidic soils. All three groups of compounds slowly contribute phosphorus to the soil solution, but most of the phosphorus in each group is of very low solubility and not readily available for plant uptake.

Unlike nitrogen and sulfur, phosphorus is not lost from the soil in gaseous form. Because soluble inorganic forms of phosphorus are strongly adsorbed by mineral surfaces (see Section 14.6), there is also no appreciable phosphorus lost by leaching (except in some organic soils or sandy soils to which very high amounts of animal manures have been added).

GAINS AND LOSSES. The principal pathways by which phosphorus is lost from the soil system are plant removal (5 to 50 kg/ha annually in harvested biomass), erosion of phosphorus-carrying soil particles (0.1 to 10 kg/ha annually on organic and mineral particles), and phosphorus dissolved in surface runoff water (0.01 to 3.0 kg/ha annually). For each pathway, the higher figures cited for annual phosphorus loss would most likely apply to cultivated soils.

The amount of phosphorus that enters the soil from the atmosphere (sorbed on dust particles) is quite small (0.05 to 0.5 kg/ha annually), but may nearly balance the losses from the soil in undisturbed forest and grassland ecosystems. As already discussed, in an agroecosystem the input from phosphorus fertilizer is likely to exceed the output in harvested crops. Figure 14.4 provides examples of phosphorus input-output balance in two adjacent watersheds, one devoted primarily to conventionally tilled row-crop production and the other covered by mature deciduous forest.

Both inorganic and organic forms of phosphorus occur in soils, and both are important to plants as sources of this element. The relative amounts in the two forms vary greatly from soil to soil, but the data in Table 14.4 give some idea of their relative proportions in a range of mineral soils. The organic fraction generally constitutes 20 to 80% of the total phosphorus in surface soil horizons (Figure 14.9). The deeper horizons may hold large amounts of inorganic phosphorus, especially in soils from arid and semiarid regions.

Organic Phosphorus Compounds

Until recently, scientists have focused more attention on the inorganic than on the organic phosphorus in soils, and our knowledge of the specific nature of most of the organic-bound phosphorus in soils is quite limited. However, three broad groups of organic phosphorus compounds are known to exist in soils. Although all three types of phosphorus compounds are also found in plants, most of the organic phosphorus compounds in soils are believed to have been synthesized by microorganisms. The three groups are: (1) inositol phosphates or phosphate esters of a sugarlike compound, inositol $(C_6H_6(OH)_6)$; (2) nucleic acids; and (3) phospholipids. While other organic phosphorus compounds are present in soils, the identity and amounts present are less well understood.

Inositol phosphates are the most abundant of the known organic phosphorus compounds, making up 10 to 50% of the total organic phosphorus. They tend to be quite stable in acid and alkaline conditions, and they interact with the higher-molecular-weight humic compounds. These properties may account for their relative abundance in soils.

Nucleic acids are adsorbed by humic compounds as well as by silicate clays. Adsorption on these soil colloids probably helps protect the phosphorus in nucleic acids from microbial attack. Still, the nucleic acids and phospholipids together probably make up only 1 to 2% of the organic phosphorus in most soils.

The other chemical compounds that contain most of the soil organic phosphorus have not yet been identified, but much of the organic phosphorus appears to be associated with the fulvic acid fraction of the soil organic matter. Our ignorance of the spe-

TABLE 14.4 Total Phosphorus Content of Surface Soils from Different Locations and the Percentage of Total Phosphorus in the Organic Form

Soils	Number of samples	Total P, mg/kg	Organic fraction, %
Western Oregon			
Upland soils	4	357	66
Old valley-filling soils	4	1479	30
Recent valley soils	3	848	26
New York			
Histosols (cultivated)	8	1491	52
Iowa			
Mollisols and Alfisols	6	561	44
Arizona	19	703	36
Australia	3	422	75
Texas			
Ustolls	2	369	34
Hawaii			
Andisol	1	4700	37
Oxisol	1	1414	19
Zimbabwe			
Alfisols	22	899	56
Maryland			
Ultisols (silt loams)	6	650	59
Ultisols (forested sandy loams)	3	472	70
Ultisols (sandy loam, cropland)	4	647	25

Data for Oregon, Iowa, and Arizona from sources quoted by Brady (1974); Australia from Fares, et al. (1974); New York from Cogger and Duxbury, (1984); Hawaii from Soltanpour, et al. (1988); Zimbabwe courtesy of R. Weil and F. Folle; Texas from Raven and Hossner (1993); Maryland (silt loams) from Weil, et al. (1988); Maryland (sandy loams) from Vaithiyanathan and Correll (1992).

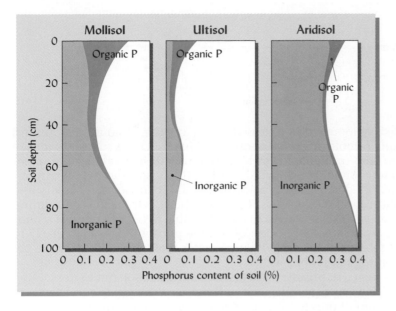

FIGURE 14.9 Phosphorus contents of representative soil profiles from three soil orders. All three soils contain a high proportion of organic phosphorus in their surface horizons. The Aridisol has a high inorganic phosphorus content throughout the profile because rainfall during soil formation was insufficient to leach much of the inorganic phosphorus compounds from the soil. The increased phosphorus in the subsoil of the Ultisol is due to adsorption of inorganic phosphorus by iron and aluminum oxides in the B horizon. In both the Mollisol and Aridisol, most of the subsoil phosphorus is in the form of inorganic calcium-phosphate compounds.

cific compounds involved does not detract from the importance of these compounds as suppliers of phosphorus through microbial breakdown.[6]

Much of the phosphorus in the soil solution and in leachates of areas that have received large quantities of animal wastes is present as **dissolved organic phosphorus** (DOP). DOP is generally more mobile than soluble inorganic phosphates, probably because it is not so readily adsorbed by organic clays and by $CaCO_3$ layers in the soil. In the lower horizons of such soils, the DOP commonly makes up more than 50% of the total soil solution phosphorus. As a consequence, in heavily manured areas with sandy soils DOP can leach downward to nearly 2 m (Figure 14.10). In fields with high water tables, the phosphorus can move with the groundwater to nearby lakes or streams and thereby contribute significantly to eutrophication.

Mineralization of Organic P

Phosphorus held in organic form can be mineralized and immobilized by the same general processes that release nitrogen and sulfur from soil organic matter (see Chapter 13):

---

[6] Although plants are able to absorb some organic phosphorus compounds directly, the level of such absorption is thought to be very low compared to that of inorganic phosphates.

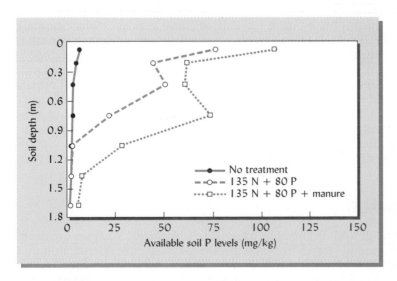

FIGURE 14.10 Effect of adding fertilizer with and without cow manure for a period of 42 years on the level of available phosphorus at different depths of a Mollisol in Nebraska on which continuous corn had been grown. Both treatments resulted in large increases in the available P level in the upper horizons, but only where manure was also applied did increases appear below a depth of 1 m. Apparently the organic forms of P in the manured plot were not adsorbed by the soil, thereby permitting deeper penetration of the phosphorus, which may then have been subject to movement through the groundwater to nearby waterways. [Redrawn from Eghball, et al. (1996)]

$$\text{Organic P forms} \underset{\text{Microbes}}{\overset{\text{Microbes}}{\rightleftharpoons}} \underset{\substack{\text{Soluble} \\ \text{phosphate}}}{H_2PO_4^-} \xrightarrow{Fe^{3+}, Al^{3+}, Ca^{2+}} \underset{\text{Insoluble fixed P}}{\text{Fe, Al, Ca phosphates}}$$

$$\longleftarrow \text{Immobilization} \longleftarrow$$

$$\longrightarrow \text{Mineralization} \longrightarrow$$

Soluble phosphorus compounds are released when organic residues and humus decompose. The resulting soluble inorganic phosphate ion ($H_2PO_4^-$) is subject to uptake by plants or to fixation into insoluble forms by reaction with iron, aluminum, manganese, and calcium in soils. Should organic residues low in phosphorus but high in carbon and other nutrients be added to a soil, microbes would increase their activity and immobilize the phosphorus in their biomass. The available $H_2PO_4^-$ in the soil solution would temporarily disappear, just as was described for soluble NH_4^+, NO_3^-, and SO_4^2 ions (Sections 13.4 and 13.20). Net immobilization of soluble phosphorus is most likely to occur if residues added to the soil have a C/P ratio greater than about 300:1, while net mineralization is likely if the ratio is below 200:1.

When conditions are such that organic matter is increasing in a soil (for example, after conversion of cultivated land to a grass–legume pasture), organic phosphorus tends to accumulate. If inorganic phosphorus fertilizer is added to such a soil, much of the added phosphorus is eventually stored in the organic fraction after being assimilated by plants or microorganisms. The ratio of C/P in soil organic matter varies widely, commonly from 15:1 up to 75:1. This variability is in contrast to relatively stable ratios of C/N and C/S characteristic of soil organic matter, and results from the fact that the rates of phosphorus mineralization and immobilization do not always parallel those of N and S.

Contribution of Organic Phosphorus to Plant Needs

Recent evidence indicates that the readily decomposable or easily soluble fractions of soil *organic phosphorus* are often the most important factor in supplying phosphorus to plants in *highly weathered soils* (e.g., Ultisols and Oxisols), even though the total organic matter content of these soils may not be especially high. The inorganic phosphorus in the highly weathered soils is far too insoluble to contribute much to plant nutrition. Apparently plant roots and mycorrhizal hyphae are able to obtain some of the phosphorus released from organic forms before it forms inorganic compounds that quickly become insoluble. In contrast, it appears that the more soluble *inorganic forms* of phosphorus play the biggest role in phosphorus fertility of *less weathered soils* (e.g., Mollisols and Vertisols), even though these generally contain relatively high amounts of soil organic matter.

Mineralization of organic phosphorus in soils is subject to many of the same influences that control the general decomposition of soil organic matter—namely, temperature, moisture, tillage, and so forth (see Section 12.2). In temperate regions, mineralization of organic phosphorus in soils typically releases 5 to 20 kg P/ha/yr, most of which is readily absorbed by growing plants. These values can be compared to the annual uptake of phosphorus by most crops, trees, and grasses, which generally ranges from 5 to 30 kg P/ha. When forested soils are first brought under cultivation in tropical climates, the amount of phosphorus released by mineralization may exceed 50 kg/ha/yr, but unless phosphorus is added from outside sources these high rates of mineralization will soon decline due to the depletion of readily decomposable soil organic matter. In Florida, rapid mineralization of organic matter in Histosols (Saprists) drained for agricultural use is estimated to release about 80 kg P/ha/yr. Unlike most mineral soils, these organic soils possess little capacity to retain dissolved phosphorus, so water draining from them is quite concentrated in phosphorus (0.5 to 1.5 mg P/L) and is thought to be contributing to the degradation of the Everglades wetland system.

14.5 INORGANIC PHOSPHORUS IN SOILS

Of all the macronutrients found in soils, phosphorus has by far the smallest quantities in solution or in readily soluble forms in mineral soils. Likewise, the relative immobility of inorganic phosphorus in mineral soils is well known. Two phenomena tend to control the concentration of phosphorus in the soil solution and the movement of phosphorus in soils: (1) the solubility of phosphorus-containing minerals, and (2) the fixation or adsorption of phosphate ions on the surface of soil particles. In practice, it is difficult to separate the influence of these two types of reactions or even determine the exact nature of inorganic phosphorus compounds present in a particular soil.

FIXATION AND RETENTION. Dissolved phosphate ions in mineral soils are subject to many types of reactions that tend to remove the ions from the soil solution and produce phosphorus-containing compounds of very low solubility. These reactions are sometimes collectively referred to by the general terms *phosphorus fixation* and *phosphorus retention*. *Phosphorus retention* is a somewhat more general term that includes both precipitation and fixation reactions.

The tendency for soils to fix phosphorus in relatively insoluble, unavailable forms has far-reaching consequences for phosphorus management. For example, phosphorus fixation may be viewed as troublesome if it prevents plants from using all but a small fraction of fertilizer phosphorus applied. On the other hand, phosphorus fixation can be viewed as a benefit if it causes most of the dissolved phosphorus to be removed from phosphorus-rich wastewater applied to a soil (Box 14.2). The fixation reactions responsible in both situations will be discussed as they apply to the availability of phosphorus under acidic and alkaline soil conditions. We will begin by describing the various inorganic compounds and their solubility.

Inorganic Phosphorus Compounds

As indicated by Figure 14.6, most inorganic phosphorus compounds in soils fall into one of two groups: (1) those containing calcium, and (2) those containing iron and aluminum (and, less frequently, manganese).

As a group, the calcium phosphate compounds become more soluble as soil pH decreases; hence, they tend to dissolve and disappear from acid soils. On the other hand, the calcium phosphates are quite stable and very insoluble at higher pH, and so become the dominant forms of inorganic phosphorus present in neutral to alkaline soils.

Of the common calcium compounds containing phosphorus (Table 14.5), the **apatite** minerals are the least soluble, and are therefore the least available source of phosphorus. Some apatite minerals (e.g., fluorapatite) are so insoluble that they persist even in weathered (acid) soils. The simpler mono- and dicalcium phosphates are readily available for plant uptake. Except on recently fertilized soils, however, these compounds are present in only extremely small quantities because they easily revert to the more insoluble forms.

In contrast to calcium phosphates, the iron and aluminum hydroxy phosphate minerals, **strengite** ($FePO_4 \cdot 2H_2O$) and **variscite** ($AlPO_4 \cdot 2H_2O$), have very low solubilities in strongly acid soils and become more soluble as soil pH rises. These minerals would therefore be quite unstable in alkaline soils, but are prominent in acid soils, in which they are quite insoluble and stable.

TABLE 14.5 **Inorganic Phosphorus-Containing Compounds Commonly Found in Soils**

In each group, the compounds are listed in order of increasing solubility.

Compound	Formula
Iron and aluminum compounds	
Strengite	$FePO_4 \cdot 2H_2O$
Variscite	$AlPO_4 \cdot 2H_2O$
Calcium compounds	
Fluorapatite	$[3Ca_3(PO_4)_2] \cdot CaF_2$
Carbonate apatite	$[3Ca_3(PO_4)_2] \cdot CaCO_3$
Hydroxy apatite	$[3Ca_3(PO_4)_2] \cdot Ca(OH)_2$
Oxy apatite	$[3Ca_3(PO_4)_2] \cdot CaO$
Tricalcium phosphate	$Ca_3(PO_4)_2$
Octacalcium phosphate	$Ca_8H_2(PO_4)_6 \cdot 5H_2O$
Dicalcium phosphate	$CaHPO_4 \cdot 2H_2O$
Monocalcium phosphate	$Ca(H_2PO_4)_2 \cdot H_2O$

BOX 14.2   PHOSPHORUS REMOVAL FROM WASTEWATER

Environmental soil scientists and engineers remove phosphorus from municipal wastes by taking advantage of some of the same reactions that bind phosphorus in soils. After primary and secondary sewage treatment that removes solids and oxidizes most of the organic matter, tertiary treatment in huge, specially designed tanks (Figure 14.11) causes phosphorus to precipitate through reactions with iron and aluminum compounds, such as the following:

$$Al_2(SO_4)_3 \cdot 14H_2O + 2PO_4^{3-} \longrightarrow 2AlPO_4 + 3SO_4^{2-} + 14H_2O$$

Alum   Soluble   Insoluble AlP
phosphate

$$FeCl_3 + PO_4^{3-} \longrightarrow FePO_4 + 3Cl^-$$

Ferric   Soluble   Insoluble FeP
chloride   phosphate

The insoluble aluminum and iron phosphates settle out of solution and are later mixed with other solids from the wastewater to form sewage sludge. The low-phosphorus water, after minor processing, is returned to the river.

FIGURE 14.11   *Increasingly, modern sewage treatment plants are required to include tertiary treatment facilities for phosphorus removal. The chemical reactions encouraged in the sewage treatment process are similar to those that affect phosphorus availability in soils. (Photo courtesy of R. Weil)*

Other less-expensive tertiary treatment approaches involve the spraying of the wastewater on vegetated soils. Natural soil and plant processes clean the phosphorus and other constituents out of the waste water. In some *infiltration systems*, the water percolates through relatively permeable soils. Other systems, termed *overland flow systems*, use finer-textured, less-permeable soils, over which water slowly flows, permitting the upper few centimeters of soil and the vegetation to remove most of the soluble phosphorus and other contaminants. In both systems, advantage is taken of the soil's phosphate-fixing capacity (see Section 14.8).

As the following example illustrates, knowledge of soil properties and processes is essential for effective design of advanced land-based wastewater treatment systems. A large environmental engineering firm won a contract to build a new type of wastewater treatment facility that would look more like a park than a sewage plant because it used constructed wetlands and soils to clean the wastewater. After flowing through a number of artificial marshes and filtering systems (Figure 14.12a), the wastewater was sprayed into a large field covered with a layer of artificial permeable "soil" several meters thick (Figure 14.12b). The expectation was that the phosphorus would be fixed as it moved through the "soil," leaving the groundwater sufficiently low in this element that it could be released into a nearby estuary.

(continued)

BOX 14.2 (Cont.)   PHOSPHORUS REMOVAL FROM WASTEWATER

Unfortunately, the system didn't work. The water coming from the bottom of the artificial profile was higher in phosphorus than before the treatment. The problem? The designers had specified peat (very low in mineral colloids) rather than mineral soil as the artificial "soil" through which the water percolated. Consequently, insufficient iron, aluminum, or calcium compounds were available to fix the phosphorus. Soil scientists brought in to assist the municipality recognized the problem and recommended that a few trainloads of steel wool dust (metallic iron) be incorporated into the peat. As the steel wool rusted (oxidized), ferric iron formed and reacted strongly with the dissolved phosphorus in the wastewater. The phosphorus was precipitated without appreciably decreasing the desirable high permeability of the peat "soil."

(a)

(b)

FIGURE 14.12   *After passing through an artificial wetland (a) to remove nitrogen by denitrification (see Chapter 13), the effluent goes through the final stage (b) of an innovative wastewater treatment facility designed to remove pollutants by "natural" processes. The dark-colored "soil" that can be seen under the grass is a thick layer of peat through which the wastewater will percolate after being sprayed onto the field by an irrigation system (arrow). Peat has physical properties desirable for the construction of this designer soil, but it had to be amended with iron before it developed the phosphorus-removal ability characteristic of most mineral soils. (Photos courtesy of R. Weil)*

Other similar compounds, combining phosphorus with iron, aluminum, or manganese, are also found in acid soils. Some are products of surface reactions between phosphate ions and the somewhat amorphous hydroxy polymers that often exist as coatings on soil particles. Evidence suggests that phosphate ions even react with aluminum near the edges of silicate clay crystals, forming insoluble products similar to the aluminum phosphates described in the preceding paragraph.

Effect of Aging on Inorganic Phosphate Availability

In both acid and alkaline soils, phosphorus tends to undergo sequential reactions that produce phosphorus-containing compounds of lower and lower solubility. Therefore, the longer that phosphorus remains in the soils, the less soluble—and, therefore, less plant-available—it tends to become. Usually, when soluble phosphorus is added to a soil, a rapid reaction removes the phosphorus from solution (*fixes* the phosphorus) in the first few hours. Slower reactions then continue to gradually reduce phosphorus solubility for months or years as the phosphate compounds age. The freshly fixed phosphorus may be slightly soluble and of some value to plants. With time, the solubility of the fixed phosphorus tends to decrease to extremely low levels. The effect of aging appears to be due to such factors as the regularity and size of crystals in precipitated phosphates, more permanent bonding of adsorbed phosphate into the calcium carbonate or metal oxide particles, and the extent to which sorbed phosphate is buried as surface precipitation reactions continue (Figure 14.13). The nature of these and other reactions of phosphorus in soils are discussed in the following sections.

14.6 SOLUBILITY OF INORGANIC PHOSPHORUS IN ACID SOILS

The particular types of reactions that fix phosphorus in relatively unavailable forms differ from soil to soil and are closely related to soil pH (Figure 14.14). In acid soils these reactions involve mostly Al, Fe, or Mn, either as dissolved ions, as oxides, or as hydrous oxides. Many soils contain such hydrous oxides as coatings on soil particles and as interlayer precipitates in silicate clays. In alkaline and calcareous soils, the reactions primarily involve precipitation as various calcium phosphate minerals (see Table 14.5) or adsorption to the iron impurities on the surfaces of carbonates and clays. At moderate pH values, adsorption on the edges of kaolinite or on the iron oxide coating on kaolinite clays plays an important role.

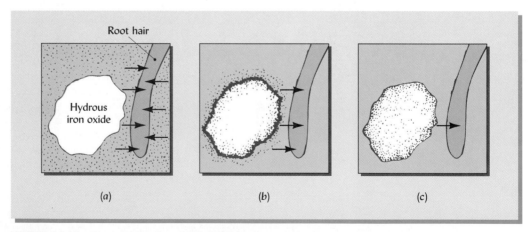

FIGURE 14.13 How relatively soluble phosphates are rendered unavailable by compounds such as hydrous oxides of Fe and Al. (*a*) The situation just after application of a soluble phosphate. The root hair and the hydrous iron oxide particle are surrounded by soluble phosphates. (*b*) Within a very short time most of the soluble phosphate has reacted with the surface of the iron oxide crystal. The phosphorus is still fairly readily available to the plant roots, since most of it is located at the surface of the particle where exudates from the plant can encourage exchange. (*c*) In time the phosphorus penetrates the crystal, and only a small portion is found near the surface. Under these conditions its availability is low.

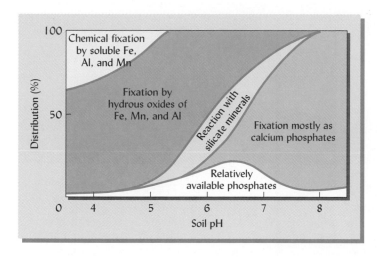

FIGURE 14.14 Inorganic fixation of added phosphates at various soil pH values. Average conditions are postulated, and it is not to be inferred that any particular soil would have exactly this distribution. The actual proportion of the phosphorus remaining in an available form will depend upon contact with the soil, time for reaction, and other factors. It should be kept in mind that some of the added phosphorus may be changed to an organic form in which it would be temporarily unavailable.

Precipitation by Iron, Aluminum, and Manganese Ions

Probably the easiest type of phosphorus-fixation reaction to visualize is the simple reaction of $H_2PO_4^-$ ions with dissolved Fe^{3+}, Al^{3+}, and Mn^{3+} ions to form insoluble hydroxy phosphate precipitates (Figure 14.15a). In strongly acid soils, enough soluble Al, Fe, or Mn is usually present to cause the chemical precipitation of nearly all dissolved $H_2PO_4^-$ ions by reactions such as the following (using the aluminum cation as an example):

$$Al^{3+} + H_2PO_4^- + 2H_2O \rightleftharpoons 2H^+ + Al(OH)_2H_2PO_4$$

(soluble)   (insoluble)

Freshly precipitated hydroxy phosphates are slightly soluble because they have a great deal of surface area exposed to the soil solution. Therefore, the phosphorus contained in them is, initially at least, somewhat available to plants. Over time, however, as the precipitated hydroxy phosphates age, they become less soluble and the phosphorus in them becomes almost completely unavailable to most plants.

Reaction with Hydrous Oxides and Silicate Clays

Most of the phosphorus fixation in acid soils probably occurs when $H_2PO_4^-$ ions react with, or become adsorbed to, the surfaces of insoluble oxides of iron, aluminum, and manganese, such as gibbsite ($Al_2O_3 \cdot 3H_2O$) and goethite ($Fe_2O_3 \cdot 3H_2O$; see Figure 14.13). These hydrous oxides occur as crystalline and noncrystalline particles and as coatings on the interlayer and external surfaces of clay particles. Under some circumstances, structural Al at the edges of 1:1 silicate clays (e.g., kaolinite) can also react with phosphorus. Fixation of phosphorus by clays probably takes place over a relatively wide pH range (see Figure 14.14). The large quantities of Fe, Al oxides, and 1:1 clays present in many soils make possible the fixation of extremely large amounts of phosphorus by these reactions.

Although all the exact mechanisms have not been identified, $H_2PO_4^-$ ions are known to react with iron and aluminum mineral surfaces in several different ways, resulting in different degrees of phosphorus fixation. Some of these reactions are shown diagrammatically in Figure 14.15.

The $H_2PO_4^-$ anion may be attracted to positive charges that develop under acid conditions on the surfaces of iron and aluminum oxides and the broken edges of kaolinite clays (see Figure 14.15b). The $H_2PO_4^-$ anions adsorbed by electrostatic attraction to these positively charged sites may be removed by certain other anions, such as OH^-, SO_4^{2-}, MoO_4^{2-}, or organic acids (R—COO$^-$), in the reversible process of anion exchange as discussed in Chapter 8. Since this type of adsorption of $H_2PO_4^-$ ions is reversible, the phosphorus may slowly become available to plants. Availability of such adsorbed $H_2PO_4^-$ may be increased by (1) liming the soil to increase the hydroxyl ions, or (2) adding organic matter to increase organic acids (anions) capable of replacing $H_2PO_4^-$.

Alternatively, the phosphate ion may replace a structural hydroxyl to become chemically bound to the oxide (or clay) surface (see Figure 14.15c). This reaction, while

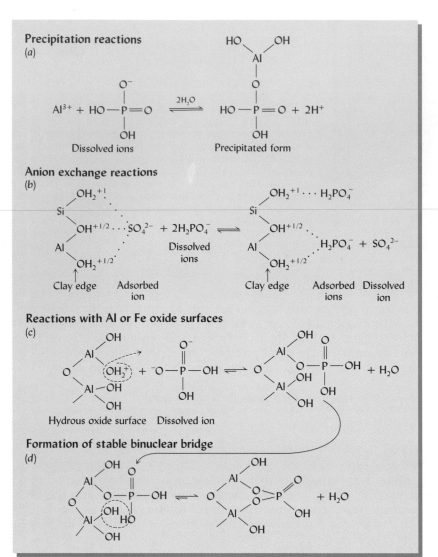

Precipitation reactions
(a)

Anion exchange reactions
(b)

Reactions with Al or Fe oxide surfaces
(c)

Formation of stable binuclear bridge
(d)

FIGURE 14.15 Several of the reactions by which phosphate ions are removed from soil solution and fixed by reaction with iron and aluminum in various hydrous oxides. Freshly precipitated aluminum, iron, and manganese phosphates (a) are relatively available, though over time they become increasingly unavailable. In (b) the phosphate is reversibly adsorbed by anion exchange. In reactions of the type shown in (c) a phosphate ion replaces an —OH₂ or an —OH group in the surface structure of Al or Fe hydrous oxide minerals. In (d) the phosphate further penetrates the mineral surface by forming a stable binuclear bridge. The adsorption reactions (b, c, d) are shown in order—from those that bind phosphate with the least tenacity (relatively reversible and somewhat more plant-available) to those that bind phosphate most tightly (almost irreversible and least plant-available). It is probable that, over time, phosphate ions added to a soil may undergo an entire sequence of these reactions, becoming increasingly unavailable.

reversible, binds the phosphate too tightly to allow it to be replaced by other anions. The availability of phosphate bound in this manner is very low. Over time, a second oxygen of the phosphate ion may replace a second hydroxyl, so that the phosphate becomes chemically bound to two adjacent aluminum (or iron) atoms in the hydrous oxide surface (see Figure 14.15d). With this step, the phosphate becomes an integral part of the oxide mineral and the likelihood of its release back to the soil solution is extremely small.

Finally, as more time passes, the precipitation of additional iron or aluminum hydrous oxide may bury the phosphate deep inside the oxide particle. Such phosphate is termed *occluded* and is the least available form of phosphorus in most acid soils.

Precipitation reactions similar to those just described are responsible for the rapid reduction in availability of phosphorus added to soil as soluble $Ca(H_2PO_4)_2 \cdot H_2O$ in fertilizers (Figure 14.16). This type of reaction can also be used to control the solubility of phosphorus in wastewater (see Box 14.2).

Effect of Iron Reduction Under Wet Conditions

It should be noted that phosphorus bound to iron oxides by the mechanisms just discussed is very insoluble under well-aerated conditions. However, prolonged anaerobic conditions can reduce the iron in these complexes from Fe^{3+} to Fe^{2+}, making the iron–phosphate complex much more soluble and causing it to release phosphorus into solution. The release of phosphorus from iron phosphates by means of the reduction and subsequent solubilization of iron improves the phosphorus availability in soils used for paddy rice.

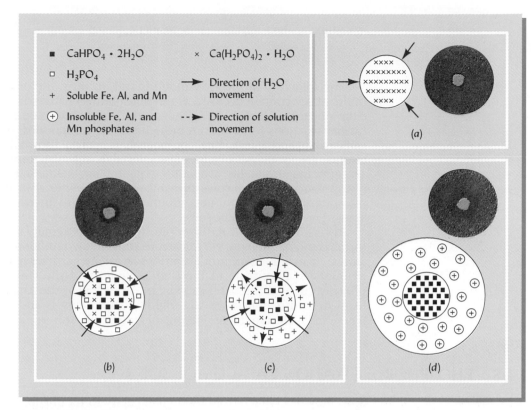

FIGURE 14.16 When a granule of soluble calcium monophosphate $[Ca(H_2PO_4)_2 \cdot H_2O]$ fertilizer is added to a moist soil, the following series of reactions rapidly and dramatically reduces the availability of the added phosphorus: (*a*) The $Ca(H_2PO_4)_2 \cdot H_2O$ in the fertilizer granules attracts water from the soil. (*b*) In the moistened granule, phosphoric acid is formed by the following reaction: $Ca(H_2PO_4)_2 \cdot H_2O + H_2O \rightarrow CaHPO_4 \cdot 2H_2O + H_3PO_4$. As more water is attracted, an H_3PO_4-laden solution with a pH of about 1.4 moves outward from the granule. (*c*) This solution is sufficiently acid to dissolve and displace large quantities of iron, aluminum, and manganese. These ions promptly react with the phosphate to form low-solubility compounds. (*d*) Later, these compounds revert to the hydroxy phosphates of iron, aluminum, and manganese in acid soils. In neutral to alkaline soils, equally insoluble calcium phosphates are formed. In both cases, insoluble dicalcium phosphate $(CaHPO_4 \cdot 2H_2O)$ remains in the granule. Fortunately, the phosphorus in the freshly precipitated compounds is slightly available for plant uptake. But when these freshly precipitated compounds are allowed to age or to revert to more insoluble forms, the phosphorus becomes almost completely unavailable to plants in the short term. (Photos courtesy of G. L. Terman and National Plant Food Institute, Washington, D.C.)

These reactions are also of special relevance to concerns about water quality. Phosphorus bound to soil particles may accumulate in river- and lake-bottom sediments, along with organic matter and other debris. As the sediments become anoxic, the reducing environment may cause the gradual release of phosphorus held by hydrous iron oxides. Thus, the phosphorus eroded from soils today may aggravate the problem of eutrophication for years to come, even after the erosion and loss of phosphorus from the land has been brought under control.

14.7 INORGANIC PHOSPHORUS AVAILABILITY AT HIGH pH VALUES[7]

The availability of phosphorus in alkaline soils is determined principally by the solubility of the various calcium phosphate compounds present. In alkaline soils (e.g., pH = 8) soluble $H_2PO_4^-$ quickly reacts with calcium to form a sequence of products of decreasing solubility. For instance, highly soluble monocalcium phosphate $[Ca(H_2PO_4)_2 \cdot H_2O]$ added as concentrated superphosphate fertilizer rapidly reacts with calcium carbonate in the soil to form first dicalcium phosphate $[CaHPO_4 \cdot 2H_2O)$ and then tricalcium phosphate $[Ca_3(PO_4)_2]$, as follows:

---

[7] See Sample, et al. (1980) for a discussion of this subject.

$$Ca(H_2PO_4)_2 \cdot H_2O + 2H_2O \xrightarrow{CaCO_3} 2(CaHPO_4 \cdot 2H_2O) + CO_2\uparrow \xrightarrow{CaCO_3} Ca_3(PO_4)_2 + CO_2\uparrow + 5H_2O$$

| Monocalcium phosphate (soluble) | Dicalcium phosphate (slightly soluble) | Tricalcium phosphate (very low solubility) |

The solubility of these compounds and, in turn, the plant availability of the phosphorus they contain decrease as the phosphorus changes from the $H_2PO_4^-$ ion to tricalcium phosphate $[Ca_3(PO_4)_2]$. Although this compound is quite insoluble, it may undergo further reactions to form even more insoluble compounds, such as the hydroxy-, oxy-, carbonate-, and fluorapatite compounds (apatites) shown in Table 14.5. These compounds are thousands of times less soluble than freshly formed tricalcium phosphates. The extreme insolubility of apatites in neutral or alkaline soils generally makes powdered phosphate rock (which consists mainly of apatite minerals) virtually useless as a source of phosphorus for plants unless ground very fine (to increase weathering surface) and applied to relatively acidic soils.

Reversion of soluble fertilizer phosphorus to extremely insoluble calcium phosphate forms is most serious in the calcareous soils of low-rainfall regions (e.g., western United States). Iron and aluminum impurities in calcite particles may also adsorb considerable amounts of phosphate in these soils. Because of the various reactions with $CaCO_3$, phosphorus availability tends to be nearly as low in the Aridisols, Inceptisols, and Mollisols of arid regions as in the highly acid Spodosols and Ultisols of humid regions, where iron, aluminum, and manganese limit phosphorus availability.

14.8 PHOSPHORUS-FIXATION CAPACITY OF SOILS

Soils may be characterized by their capacity to fix phosphorus in unavailable, insoluble forms. The phosphorus-fixation capacity of a soil may be conceptualized as the total number of sites on soil particle surfaces capable of reacting with phosphate ions. Phosphorus fixation may also be due to reactive soluble iron, aluminum, or manganese. The different types of fixation mechanisms are illustrated schematically in Figure 14.17.

One way of determining the phosphorus-fixing capacity of a particular soil is to shake a known quantity of the soil in a phosphorus solution of known concentration. After about 24 hours an equilibrium will be approached, and the concentration of phosphorus remaining in the solution [the **equilibrium phosphorus concentration (EPC)**] can be determined. The difference between the initial and final (*equilibrium*) solution phosphorus concentrations represents the amount of phosphorus fixed by the soil. If this procedure is repeated using a series of solutions with different initial phosphorus concentrations, the results can be plotted as a phosphorus-fixation curve (Figure 14.18) and the maximum phosphorus-fixation capacity can be extrapolated from the value at which the curve levels off.

To a large degree, phosphorus fixation by soils is not easily reversible. However, if a portion of the fixed phosphorus is present in relatively soluble forms (see Section 14.6), and most of the fixation sites are already occupied by a phosphate ion, some release of phosphorus to solution is likely to occur when the soil is exposed to water with very low phosphorus concentration. This release (often called *desorption*) of phosphorus is indicated in Figure 14.18 where the curve for soil A crosses the zero fixation line and becomes negative (negative fixation = release). The solution concentration (*x*-axis) at which zero fixation occurs (phosphorus is neither released nor retained) is called the EPC_0. Such release of previously fixed phosphorus helps to resupply soil solution phosphorus depleted by plant uptake. Release of fixed phosphorus is also very important in determining losses of dissolved phosphorus in the surface runoff from a watershed. The EPC_0 is an important soil parameter because it indicates both the level of phosphorus fertility and the hazard of phosphorus loss by solution in runoff water.

QUANTITY-INTENSITY RELATIONSHIPS. The relationship between phosphorus in solution and phosphorus in slowly soluble or fixed forms is an example of the balance between *quantity* factors and *intensity* factors in soil fertility. The intensity factor is the amount of a nutrient dissolved in the soil solution. The quantity factor is the amount of that nutrient associated with the solid framework of the soil and in equilibrium with the nutrient

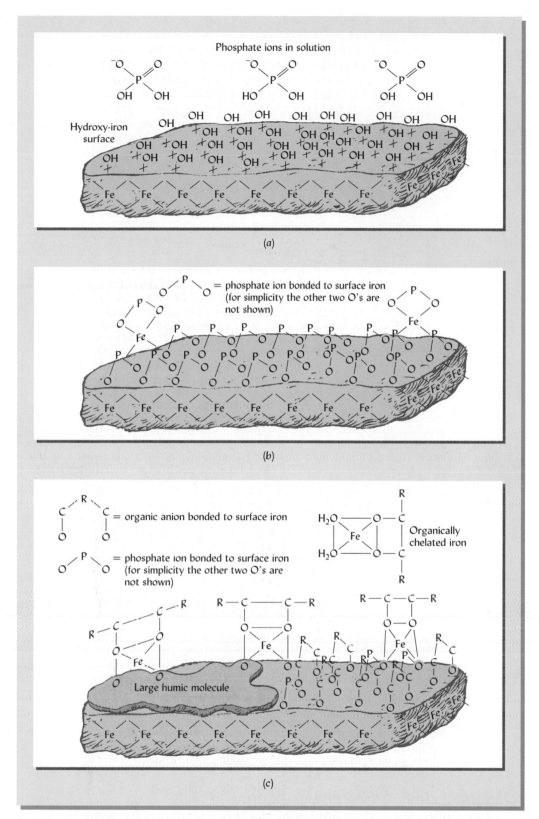

FIGURE 14.17 Schematic illustration of phosphorus-fixation sites on a soil particle surface showing hydrous iron oxide as the primary fixing agent. In part (*a*) the sites are shown as +s, indicating positive charges or hydrous metal oxide sites, each capable of fixing a phosphate ion. In part (*b*) the fixation sites are all occupied by phosphate ions (the soil's fixation capacity is satisfied). Part (*c*) illustrates how organic anions, larger organic molecules, and certain strongly fixed inorganic anions can reduce the sites available for fixing phosphorus. Such mechanisms account for the reduced phosphorus fixation and greater phosphorus availability brought about when organic matter is added to a soil.

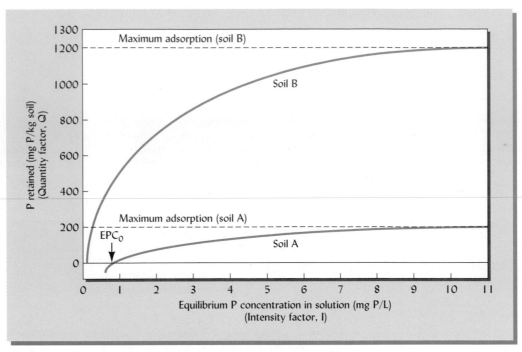

FIGURE 14.18 The relationship between phosphorus fixation and phosphorus in solution when two different soils (A and B) are shaken with solutions of various initial phosphorus concentrations. Initially, each soil removes nearly all of the phosphorus from solution, and as more and more concentrated solutions are used, the soil fixes greater amounts of phosphorus. However, eventually solutions are used that contain so much phosphorus that most of the phosphorus-fixation sites are satisfied, and much of the dissolved phosphorus remains in solution. The amount fixed by the soil levels off as the *maximum phosphorus-fixing capacity* of the soil is reached (see horizontal dashed lines: for soil A, 200 mg P/kg; for soil B, 1200 mg P/kg soil). If the initial phosphorus concentration of a solution is equal to the *equilibrium phosphorus concentration (EPC)* for a particular soil, that soil will neither remove phosphorus from nor release phosphorus to the solution (i.e., phosphorus fixation = 0 and EPC = EPC_0). If the solution phosphorus concentration is less than the EPC_0, the soil will release some phosphorus (i.e., the fixation will be negative). In this example, soil B has a much *higher* phosphorus-fixing capacity and a much *lower* EPC than does soil A. It can also be said that soil B is highly *buffered* because much phosphorus must be added to this soil to achieve a small increase in the equilibrium solution phosphorus concentration. On the other hand, if a plant root were to remove a relatively large amount of total phosphorus from the soil, only a small change would occur in the equilibrium solution concentration.

ions in solution. In Figure 14.18, the intensity factor would be represented on the *x*-axis and the quantity factor on the *y*-axis. The slope of the curve that defines the relationship for each soil represents the amount of change in the quantity factor Q that results from a given change in the intensity factor I; that is, slope = $\Delta Q/\Delta I$. This is another way of expressing the *potential buffering capacity (PBC)* of the soil:

$$PBC = \frac{\Delta Q}{\Delta I}$$

This general relationship applies not only to phosphorus, but to potassium and any other substance whose solution concentration is controlled by retention reactions with the soil solids.[8] In Section 9.4, for example, buffering of pH was discussed.

Factors Affecting the Extent of Phosphorus Fixation in Soils

Soils that remove more than 350 mg P/kg of soil (i.e., a phosphorus-fixing capacity of about 700 kg P/ha) from solution are generally considered to be high phosphorus-fixing soils. High phosphorus-fixing soils tend to maintain low phosphorus concentrations in the soil solution and in runoff water. Table 14.6 lists values of maximum phosphorus-

---

[8] For a recent review of buffering capacity as it relates to the available phosphorus and potassium, as well as other nutrients, see Nair (1996).

Fe, Al oxides (especially amorphous types) fix the largest quantities and silicate clays (especially 2:1-type) fix the least.

Soil Great Group (and series, if known)	Location	Clay		Maximum P fixation, mg P/kg soil
		Percent	*Type*	
Evesboro (Quartzipsamment)	Maryland	6	Kaolinite, Fe, Al oxides	125
Kandiustalf	Zimbabwe	20	Kaolinite, Fe oxides	394
Kitsap (Xerept)	Washington	12	2:1 clays, allophane	453
Matapeake (Hapludult)	Maryland	15	Chlorite, kaolinite, Fe oxides	465
Rhodustalf	Zimbabwe	53	Kaolinite, Fe oxides	737
Newberg (Haploxeroll)	Washington	38	2:1 clays, Fe oxides	905
Tropohumults	Cameroon	46	Fe, Al oxides, kaolinite	2060

Cameroon data courtesy of V. Ngachie; Washington data from Kuo (1988); Maryland and Zimbabwe data courtesy of R. Weil and F. Folle.

fixing capacity for a range of soils. The effects of clay content and type of clay are apparent. These and other factors will now be discussed.

The phosphorus-fixation capacity of a soil is positively related to certain soil properties, such as calcium carbonate content, clay content, and contents of iron, aluminum, and manganese, especially the hydrous oxides of these metals.

AMOUNT OF CLAY PRESENT.   Most of the compounds with which phosphorus reacts are in the finer soil fractions. Therefore, if soils with similar pH values and mineralogy are compared, phosphorus fixation tends to be more pronounced and ease of phosphorus release tends to be lowest in those soils with higher clay contents.

TYPE OF CLAY MINERALS PRESENT.   Some clay minerals are much more effective at phosphorus fixation than others. Generally, those clays that possess greater anion exchange capacity (due to positive surface charges) have a greater affinity for phosphate ions. For example, extremely high phosphorus fixation is characteristic of allophane clays typically found in Andisols and other soils associated with volcanic ash. Oxides of iron and aluminum, such as gibbsite and goethite, also strongly attract and hold phosphorus ions. Among the layer silicate clays, kaolinite has a greater phosphorus fixation capacity than most. The 2:1 clays of less-weathered soils have relatively little capacity to bind phosphorus. Thus, the soil components responsible for phosphorus-fixing capacity are, in order of increasing extent and degree of fixation:

2:1 clays << 1:1 clays < carbonate crystals < crystalline Al, Fe, Mn oxides

< amorphous Al, Fe, Mn oxides, allophane

To some degree, the preceding phosphorus-fixing soil components are distributed among soils in relation to soil taxonomy. Vertisols and Mollisols generally are dominated by 2:1 clays and have low phosphorus-fixation capacities. Iron and aluminum oxides are prominent in Ultisols and Oxisols. Andisols, characterized by large quantities of amorphous oxides and allophane, have the greatest phosphorus-fixing capacity, and their productivity is often limited by this property (Figure 14.19).

EFFECT OF SOIL pH.   The greatest degree of phosphorus fixation occurs at very low and very high soil pH. As pH increases from below 5.0 to about 6.0, the iron and aluminum phosphates become somewhat more soluble. Also, as pH drops from greater than 8.0 to below 6.0, calcium phosphate compounds increase in solubility. Therefore, as a general rule in mineral soils, phosphate fixation is at its lowest (and plant availability is highest) when soil pH is maintained in the 6.0 to 7.0 range (see Figure 14.14).

Even if pH ranges from 6.0 to 7.0, phosphate availability may still be very low, and added soluble phosphates will be readily fixed by soils. The low recovery by plants of phosphates added to field mineral soils in a given season is partially due to this fixation. A much higher recovery would be expected in organic soils (see Histosol, Figure 14.19) and in many potting mixes where calcium, iron, and aluminum concentrations are not as high as in mineral soils.

EFFECT OF ORGANIC MATTER.   Organic matter generally has little capacity to strongly fix phosphate ions. Soils high in organic matter, especially active fractions of organic matter (see

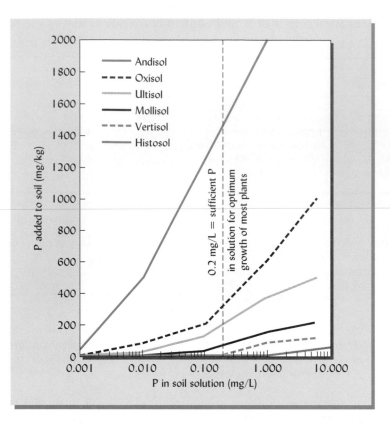

FIGURE 14.19 Phosphorus-fixing tendencies typical of several soil orders. The graph shows the concentrations of phosphorus that would be achieved in the soil solution for each type of soil if varying amounts of soluble phosphorus were added as fertilizer. Although some plants, especially if infected with mycorrhizal fungi, can grow well with lower solution phosphorus concentrations, a concentration of 0.2 mg P/L in the soil solution is adequate for optimum growth of most crops (see Table 14.7). Therefore, the graph indicates that for optimum plant growth, nearly 1500 mg of phosphorus per kg of soil should be added to the Andisol, while somewhat less than 200 mg of phosphorus per kg of soil would be required for the same phosphorus availability in the Ultisol. Within a particular soil order, the phosphorus-fixing tendency of individual soils may vary considerably. The curves given are representative for soils in the taxonomic orders indicated, but it should be borne in mind that high rates of phosphorus application would eventually reduce the fixation capacity (reduce the slope of the curve) as phosphorus-fixation sites were used up. (Curves are representative of data from many sources.)

Section 12.8), generally exhibit relatively low levels of phosphorus fixation. Several mechanisms are responsible for the reduced phosphorus fixation associated with high soil organic matter (see Figure 14.7). First, large humic molecules can adhere to the surfaces of clays and metal hydrous oxide particles, masking the phosphorus-fixation sites and preventing them from interacting with phosphorus ions in solution. Second, organic acids produced by plant roots and microbial decay can serve as organic anions, which are attracted to positive charges and hydroxyls on the surfaces of clays and hydrous oxides. These organic anions may compete with phosphorus ions for fixation sites. Third, certain organic acids and similar compounds can entrap reactive Al and Fe in stable organic complexes called *chelates* (see Section 15.6). Once chelated, these metals are unavailable for reaction with phosphorus ions in solution.

14.9   PRACTICAL CONTROL OF PHOSPHORUS AVAILABILITY

We have seen that most mineral soils have a large capacity to remove phosphorus ions from solution and fix them on particle surfaces, thus making it very difficult for plant roots to obtain sufficient amounts of this essential nutrient. The phosphorus in fertilizers is also subject to fixation on soil particle surfaces, and only 10 to 15% of the phosphorus added in fertilizer is likely to be taken up by plants in the year of application. Consequently, phosphorus availability is a limiting factor in many agroecosystems. Based on the principles of soil phosphorus behavior discussed in this chapter, a number of approaches can be suggested to ameliorate the phosphorus fertility problem.

1. *Saturation of phosphorus-fixing capacity.*   If enough phosphorus can be added, the phosphorus-fixation capacity of even high phosphorus-fixing soils can be largely saturated. This may be achieved all at once with one or two massive doses of phosphorus (usually as phosphorus fertilizer, phosphate rock, or high-phosphorus animal manure), or over a period of years or decades by annually adding more phosphorus than is removed in plant harvest. Either the rapid or slow buildup approach will eventually satisfy most of the phosphorus-fixation sites and lead to a soil so high in phosphorus that ample phosphorus is maintained in solution despite the initially high phosphorus-fixation capacity. These approaches, espe-

cially the rapid buildup approach, can be very expensive in terms of initial capital outlay for phosphorus fertilizer. For this reason, it is not practical for most of the world's farmers and foresters, but has been put into practice in some intensive agricultural systems. A second drawback to this approach is the potential for water pollution from large amounts of phosphorus desorbed to runoff water from phosphorus-saturated soils.

2. *Placement of phosphorus fertilizer.* One strategy to maximize phosphorus fertilizer value is to minimize the opportunity for the fertilizer phosphorus to react with the soil before being taken up by roots. Generally, if fertilizer is placed directly in a localized rooting zone (point placement) instead of thoroughly mixed with the soil (broadcast), one-half to one-third as much fertilizer may be used. Point placement is widely used where fertilizer is applied by hand, but new spike-wheel fertilizer injection machines (see Figure 16.26) are being developed that may make point placement feasible even in mechanized systems. Banded applications are standard practice (see Figure 16.25) for starter fertilizers. Trees are often fertilized with pellets. In untilled systems, a surface band is often created. Figure 14.20 illustrates how concentrating phosphates in bands can significantly reduce toxic aluminum levels in very acid soils.

3. *Combination of ammonium and phosphorus fertilizers.* Using ammonium in the same band with phosphorus fertilizers greatly increases root uptake of phosphorus, especially in alkaline soils. The increased phosphorus uptake is probably related to the acid produced during the nitrification process and to the acids that roots are known to produce when they take up most of their nitrogen in the ammonium form (Figure 14.21). Mono- and diammonium phosphate fertilizers offer this advantage.

4. *Choice of phosphorus-efficient plants.* Some plants are able to thrive in soils with far less available phosphorus than are other plants that are less efficient at phosphorus uptake (see Table 14.7 for several examples). The mechanisms for the more efficient phosphorus uptake are only partially known. Some plants have fine root systems with very large root surface areas. Other plants are dependent on efficient mycorrhizal associations. Still others produce specific root secretions that solubilize either calcium phosphates (as is the case for buckwheat) or iron phosphates (as is the case for pigeon pea).

5. *Increased cycling of organic phosphorus.* Mineralization of organic materials (i.e., animal manures or other phosphorus-rich organic materials) added to soils may release phosphorus gradually and somewhat in synchrony with plant needs. This slow release allows plants to uptake phosphorus before the soil fixation reactions remove it from the available pool. Leaf-fall from trees and residues from crops return phosphorus to the soil in organic, potentially mineralizable forms. Alternating phosphorus-efficient plants with less-efficient ones may convert inorganic forms taken up by the efficient plants into organic forms in the residues of these plants. The subsequent decay and mineralization of these

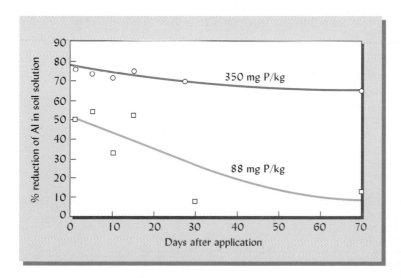

FIGURE 14.20 The percentage reduction in soluble aluminum resulting from the mixing of two rates of soluble phosphate with an acid Mollisol in Oklahoma. The high rate approximates the level expected in the soil immediately around a band of applied phosphate, while the lower rate is more nearly what is expected when the same amount of P is broadcast over the entire soil surface. Both treatments reduced the level of soluble Al immediately after application, but after 70 days the reduction was significant only with the higher rate. Banding phosphates can provide a higher localized P level while simultaneously reducing the toxic effects of Al near emerging plant seedlings. [Calculated and drawn from data in Sloan, et al. (1995)]

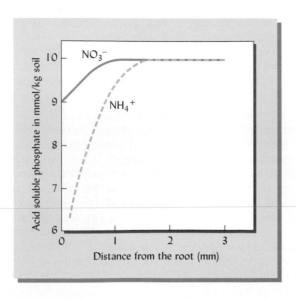

FIGURE 14.21 Effect of nitrogen source (NO_3^- versus NH_4^+) on level of acid-soluble P near the surface of ryegrass roots. The NH_4^+ ions likely resulted in increased acidity near the roots and solubilized some of the soil phosphorus, which was then taken up by the ryegrass plants, leaving a lower P level in the soil. [Redrawn from Gahoonia, et al. (1992)]

residues may enhance the supply of phosphorus for the less-efficient plants (Figure 14.22). In soils with relatively high total phosphorus content but very low phosphorus availability, crop rotation, residue management, and agroforestry all may help move more phosphorus into organic forms that are protected from phosphorus fixation. This phosphorus may be available in the future as the organic compounds are mineralized.

6. *Increased cycling of organic matter.* In addition to providing organic phosphorus for release by mineralization, soil organic matter can improve phosphorus availability by reducing the tendency of the mineral fractions to fix phosphorus. The masking of phosphorus fixation sites by humus and organic acids and the chelation of reactive Al and Fe were discussed in Section 14.8. Return of crop residues, inclusion of green manure crops in rotations, mulching with various organic materials, and adding animal manures and other decomposable organic wastes can all increase available phosphorus.

7. *Control of soil pH.* Some control over phosphorus solubility can be obtained by maintaining soil pH between 6.0 and 7.0 (see Figure 14.14). Maintenance of pH in this range is generally more practical in less-weathered soils than in the highly weathered soils of warm, humid regions. Proper liming can contribute to improved phosphorus availability in many cases, but if a source of liming material is not locally available, the cost may be prohibitive.

8. *Enhancement of mycorrhizal symbiosis.* As already mentioned, mycorrhizal fungi (see Section 14.3) can greatly enhance the ability of plants to obtain phosphorus from soils with low levels of phosphorus in solution. In the case of many shrubs and forest trees, inoculation of seedlings with efficient strains of mycorrhizal fungi is a practical measure when planting new trees (see Section 11.9). For most

TABLE 14.7 Concentration of Phosphorus in Soil Solution Required for Near-Optimal Growth (95% of Maximum Yield) of Various Plants

Plant	Approximate P in soil solution, mg/L
Cassava	0.005
Peanut	0.01
Corn	0.05
Sorghum	0.06
Cabbage	0.04
Soybean	0.20
Tomato	0.20
Head lettuce	0.30

Data from Fox (1981).

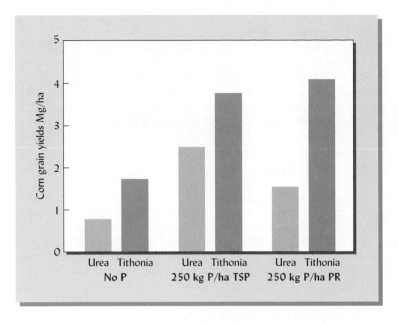

FIGURE 14.22 A common hedge, tithonia (*Tithonia diversifolia*) growing on an infertile Oxisol in Kenya is an effective source of nutrients for associated food crops. The response is due partly to the potassium supplied by the tithonia and partly to the increased effectiveness of phosphate rock on the plots receiving tithonia residues. Research suggests that the decomposition products of tithonia may increase soil microbial activity and slightly reduce P-sorption capacity. [Modified from Sanchez, et al. (1997)]

plants, however, the mycorrhizal symbiosis can best be enhanced by fostering appropriate soil conditions through crop rotations, organic matter additions, and minimum tillage.

In summary, the small pool of available phosphorus in most soils is depleted and replenished by a number of processes (Figure 14.23). Maintaining sufficient available phosphorus in a soil is basically a twofold program: (1) addition of phosphorus-containing fertilizers or amendments, and (2) regulation to some degree of soil processes that fix both added and native phosphates.

14.10 POTASSIUM: NATURE AND ECOLOGICAL ROLES[9]

Of all the essential elements, potassium is the third most likely, after nitrogen and phosphorus, to limit plant productivity. For this reason it is commonly applied to soils as fertilizer and is a component of most mixed fertilizers.

---

[9] For further information on this topic, see Mengel and Kirkby (1987) and Munson (1985).

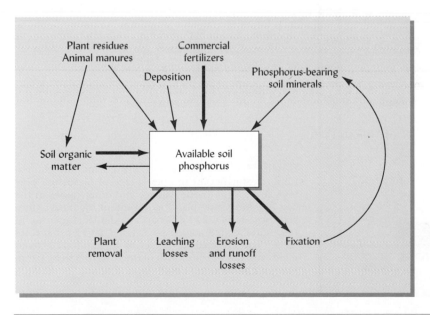

FIGURE 14.23 How the pool of available phosphorus in a soil is depleted and replenished in a typical agroecosystem. The width of the arrows is a relative indication of the quantity of phosphorus involved in each pathway. The two largest fluxes in many arable soils are the addition of phosphorus from soil amendments (usually commercial fertilizer) and the subsequent fixation of much of the added phosphorus in unavailable forms. The return of plant and animal wastes and the subsequent mineralization of soil organic matter can also play major roles in replenishing the available phosphorus. Note that leaching losses are usually negligible.

The potassium story differs in many ways from that of phosphorus. Unlike phosphorus (or sulphur and, to a large extent, nitrogen), potassium is present in the soil solution only as a positively charged cation, K^+. Like phosphorus, potassium does not form any gases that could be lost to the atmosphere. Its behavior in the soil is influenced primarily by soil cation exchange properties (see Chapter 8) and mineral weathering (Chapter 2), rather than by microbiological processes. Unlike nitrogen and phosphorus, potassium causes no off-site environmental problems when it leaves the soil system. It is not toxic and does not cause eutrophication in aquatic systems.

14.11 POTASSIUM IN PLANT AND ANIMAL NUTRITION

Although potassium plays numerous roles in plant and animal nutrition, it is not actually incorporated into the structures of organic compounds. Instead, potassium remains in the ionic form (K^+) in solution in the cell, or acts as an activator for cellular enzymes. Potassium is known to activate over 80 different enzymes responsible for such plant and animal processes as energy metabolism, starch synthesis, nitrate reduction, photosynthesis, and sugar degradation. Certain plants, many of which evolved in sodium-rich semiarid environments, can substitute sodium or other monovalent ions to carry out some, but not all, of the functions of potassium.

As a component of the plant cytoplasmic solution, potassium plays a critical role in lowering cellular osmotic water potentials, thereby reducing the loss of water from leaf stomata and increasing the ability of root cells to take up water from the soil (see Section 5.3 for a discussion of osmotic water potential). Potassium is essential for photosynthesis, for protein synthesis, for nitrogen fixation in legumes, for starch formation, and for the translocation of sugars. As a result of several of these functions, a good supply of this element promotes the production of plump grains and large tubers. The potassium content of normal, healthy leaf tissue can be expected to be in the range of 1 to 4% in most plants, similar to that of nitrogen but an order of magnitude greater than that of phosphorus.

Potassium is especially important in helping plants adapt to environmental stresses. Good potassium nutrition is linked to improved drought tolerance, improved winter-hardiness, better resistance to certain fungal diseases, and greater tolerance to insect pests (Figure 14.24). In the latter role, potassium fertilization is often an important component of integrated pest-management programs designed to reduce the use of toxic

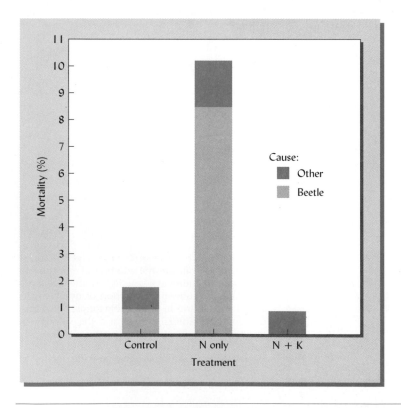

FIGURE 14.24 Influence of potassium and nitrogen fertilizer treatments on the percentage of ponderosa pine trees dying from beetle damage and other causes in the first four years after planting in western Montana. Nitrogen used alone (224 kg N/ha) caused a large increase in tree mortality, but adding potassium (224 kg K/ha) completely counteracted this effect. Both fertilization treatments stimulated the growth of the surviving trees. [From Mandzak and Moore (1994)]

pesticides. Potassium also enhances the quality of flowers, fruits, and vegetables by improving flavor and color and strengthening stems (thereby reducing lodging). In many of these respects, potassium seems to counteract some of the detrimental effects of excess nitrogen. Maintaining a balance between potassium and other nutrients (especially nitrogen, phosphorus, calcium, and magnesium) is an important goal in managing soil fertility.

In animals, including humans, potassium plays critical roles in regulating the nervous system and in the maintenance of healthy blood vessels. Diets that include such high-potassium foods as bananas, potatoes, orange juice, and leafy green vegetables have been shown to lower human risk of stroke and heart disease. Maintaining a balance between potassium and sodium is especially important in human diets.

Deficiency Symptoms in Plants

Compared to deficiencies of phosphorus and many other nutrients, a deficiency of potassium is relatively easy to recognize in most plants. In addition to the characteristics previously mentioned (reduced drought tolerance, increased lodging, etc.), specific foliar symptoms are associated with potassium deficiency. Because potassium is very mobile within the plant, it is translocated from older tissues to younger ones if the supply becomes inadequate. The symptoms of deficiency therefore usually occur earliest and most severely on the oldest leaves.

In general, when potassium is deficient the tips and edges of the oldest leaves begin to yellow (chlorosis) and then die (necrosis), so that the leaves appear to have been burned on the edges (Figure 14.25). On some plants the necrotic leaf edges may tear, giving the leaf a ragged appearance (Figure 14.25c). In several important forage and cover-crop legume species, potassium deficiency produces small, white necrotic spots that form a unique pattern along the leaflet margins; this easily recognized symptom is one that people often mistake for insect damage (see Figure 14.25d).

Potassium deficiency should not be confused with damage from excess salinity, which can also produce brown, necrotic leaf margins. Salinity damage is more likely to affect the newer leaves (see Figure 10.13).

14.12 THE POTASSIUM CYCLE

Figure 14.26 shows the major forms in which potassium is held in soils and the changes it undergoes as it is cycled through the soil–plant system. The original sources of potassium are the primary minerals, such as micas (biotite and muscovite) and potassium feldspar (orthoclase and microcline). As these minerals weather, their rigid lattice structures become more pliable. For example, potassium held between the 2:1-type crystal layers of mica is in time made more available, first as nonexchangeable but slowly available forms near the weathered edges of minerals and, eventually, as the readily exchangeable forms and the soil solution forms from which it is absorbed by plant roots.

Potassium is taken up by plants in large quantities. Depending on the type of ecosystem under consideration, a portion of this potassium is leached from plant foliage by rainwater (through-fall) and returned to the soil, and a portion is returned to the soil with the plant residues. In natural ecosystems, most of the potassium taken up by plants is returned in these ways or as wastes (mainly urine) from animals feeding on the vegetation. Some potassium is lost with eroded soil particles and in runoff water, and some is lost to groundwater by leaching. In agroecosystems, from one-fifth (e.g., in cereal grains) to nearly all (e.g., in hay crops) of the potassium taken up by plants may be exported to distant markets, from which it is unlikely to return.

At any one time, most soil potassium is in primary minerals and nonexchangeable forms. In relatively fertile soils, the release of potassium from these forms to the exchangeable and soil solution forms that plants can use directly may be sufficiently rapid to keep plants supplied with enough potassium for optimum growth. On the other hand, where high yields of agricultural crops or timber are removed from the land, or where the content of weatherable potassium-containing minerals is low, the levels of exchangeable and solution potassium may have to be supplemented by outside sources, such as chemical fertilizers, poultry manure, or wood ashes. Without these additions, the supply of available potassium will likely be depleted over a period of years and the productivity of the soil will likewise decline.

FIGURE 14.25 Potassium deficiency often produces easily recognized foliar symptoms, mainly on older leaves: (*a*) chlorotic margins on boxwood leaves; (*b*) chlorotic leaf margins on soybean; (*c*) ragged, necrotic margins of older banana leaves; and (*d*) small, white necrotic spots on hairy vetch leaflets. [Photos (*a*), (*c*), and (*d*) courtesy of R. Weil; photo (*b*) courtesy of Potash & Phosphate Institute]

An example of depletion and restoration of available soil potassium is given in Figure 14.27. Farming without fertilizers for over a century depleted the exchangeable potassium in a sandy soil in New York. After the supply of available potassium was exhausted, the land was abandoned in the 1920s. In the 1920s and 1930s a red pine forest was planted. The trees in some plots were fertilized with potassium, causing the expected rapid recovery of exchangeable potassium to its preagricultural level. Even where the trees were not fertilized, the level of exchangeable potassium in the surface soil was replenished, but slowly, over a period of about 80 years.

The replenishment of exchangeable potassium in the unfertilized forest plots provides an example of the ability of unharvested perennial plants, such as forest trees and pasture legumes, to ameliorate soil fertility over time. Deep-rooted perennials often act as "nutrient pumps," taking potassium from deep subsoil horizons into their root systems, translocating it to their leaves, and then recycling it back to the surface of the soil via leaf fall and leaching (see Section 16.4).

In most mature natural ecosystems the small (1 to 5 kg/ha) annual losses of potassium by leaching and erosion are more than balanced by weathering of potassium from primary minerals and nonexchangeable forms in the soil profile, followed by vegetative translocation to the surface of the soil. In most agricultural systems, leaching losses are

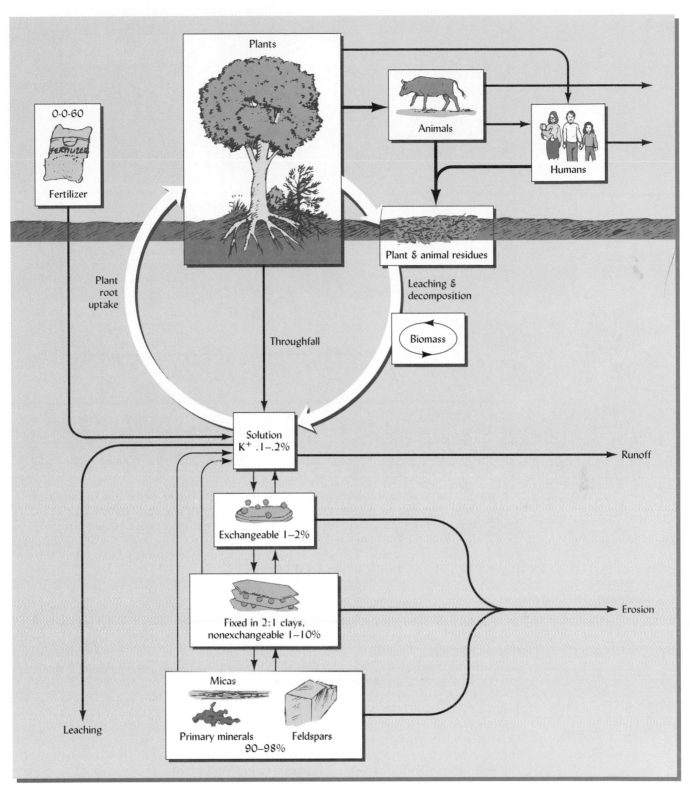

FIGURE 14.26 Major components of the potassium cycle in soils. The inner circle emphasizes the biological cycling of potassium from the soil solution to plants and back to the soil via plant residues or animal wastes. Primary and secondary minerals are the original sources of the element. Exchangeable potassium may include those ions held and released by both clay and humus colloids, but potassium is not a structural component of soil humus. The interactions among solution potassium, exchangeable, nonexchangeable, and structural potassium in primary minerals is shown. The bulk of soil potassium occurs in the primary and secondary minerals, and is released very slowly by weathering processes.

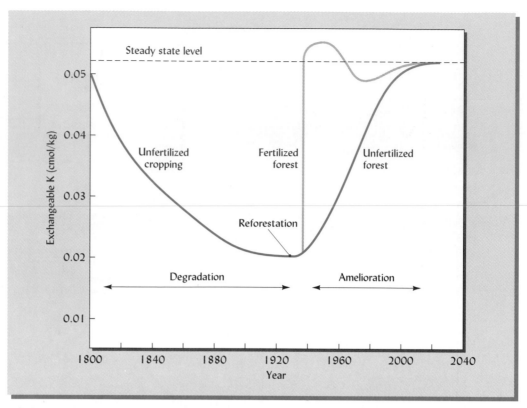

FIGURE 14.27 The general pattern of depletion of A-horizon exchangeable potassium by decades of exploitative farming, followed by its restoration under forest vegetation. The forest consisted of red pine trees planted on a Plainfield loamy sand (Udipsamment) in New York. This soil has a very low cation exchange capacity and low levels of exchangeable K⁺. [From Nowak, et al. (1991)]

far greater because much higher exchangeable K levels are generally maintained for crop production, but crop roots are active for only part of the year. The sections that follow give greater details on the reactions involved in the potassium cycle.

14.13 THE POTASSIUM PROBLEM IN SOIL FERTILITY

Availability of Potassium

In contrast to phosphorus, potassium is found in comparatively high levels in most mineral soils, except those consisting mostly of quartz sand. In fact, the total quantity of this element is generally greater than that of any other major nutrient element. Amounts as great as 30,000 to 50,000 kg potassium in the upper 15 cm of 1 ha of soil are not at all uncommon (see Table 1.3).

Yet the quantity of potassium held in an easily exchangeable condition at any one time often is very small. Most of this element is held rigidly as part of the primary minerals or is fixed in forms that are, at best, only moderately available to plants. Therefore, the situation with respect to potassium utilization parallels that of phosphorus and nitrogen in at least one way: A very large proportion of all three of these elements in the soil is insoluble and relatively unavailable to growing plants.

Leaching Losses

Potassium is much more readily lost by leaching than is phosphorus. Drainage waters from soils receiving liberal fertilizer applications usually contain considerable quantities of potassium. From representative humid-region soils growing annual crops and receiving only moderate rates of fertilizer, the annual loss of potassium by leaching is usually about 25 to 50 kg/ha, the greater values being typical of acid, sandy soils.

Losses would undoubtedly be much larger were the leaching of potassium not slowed by the attraction of the positively charged potassium ions to the negatively

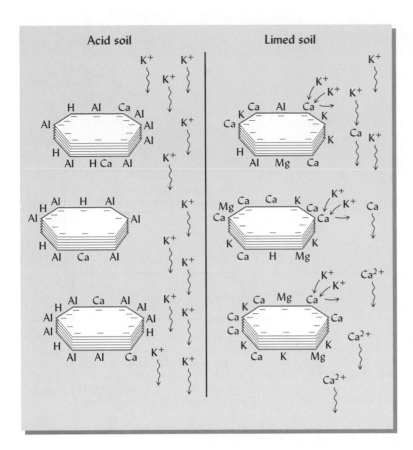

FIGURE 14.28 Diagrammatic illustration of how liming an acid soil can reduce leaching losses of potassium. The fact that the K^+ ions can more easily replace Ca^{2+} ions than they could replace Al^{3+} ions allows more of the K^+ ions to be removed from solution by cation exchange in the limed (high-calcium) soil. The removal of K^+ ions from solution by adsorption on the colloids will reduce their loss by leaching, but they will still be at least moderately available for plant uptake.

charged cation exchange sites on clay and humus surfaces. Liming an acid soil to raise its pH can reduce the leaching losses of potassium because of the *complementary ion effect* (see Chapter 8 and Figure 14.28). The ease with which ions may be removed from the exchange complex varies among different elements. Typically, trivalent ions (Al^{3+}) are more tightly held than divalent ions (Ca^{2+} and Mg^{2+}). In a limed soil, where higher levels of exchangeable calcium and magnesium are present, monovalent potassium ions are better able to replace them on the exchange complex. Where higher levels of exchangeable aluminum saturate the exchange complex, potassium is less likely to be adsorbed. Thus, in a limed soil, K^+ can be more readily retained on the exchange complex, and leaching of this element is reduced.

Plant Uptake and Removal

Plants take up very large amounts of potassium, often five to ten times as much as for phosphorus and about the same amount as for nitrogen. If most or all of the above-ground plant parts are removed in harvest, the drain on the soil supply of potassium can be very large. For example, a 60-Mg/ha yield of corn silage may remove 160 kg/ha of potassium. Conventional bolewood timber harvest typically removes about 100 kg/ha of potassium. If the entire tree is chipped and removed, as for paper pulp, the removal of potassium may be twice as great. A high-yielding legume hay crop may remove 400 kg/ha of potassium each year.

Luxury Consumption

Moreover, this situation is made even more critical by the tendency of plants to take up soluble potassium far in excess of their needs if sufficiently large quantities are present. This tendency is termed *luxury consumption,* because the excess potassium absorbed does not increase crop yields to any extent.

The principles involved in luxury consumption are shown by the graph in Figure 14.29. For many plants there is a direct relationship between the available potassium (soil plus fertilizer) and the removal of this element by the plants. However, only a certain amount of this element is needed for optimum growth, and this is termed *required*

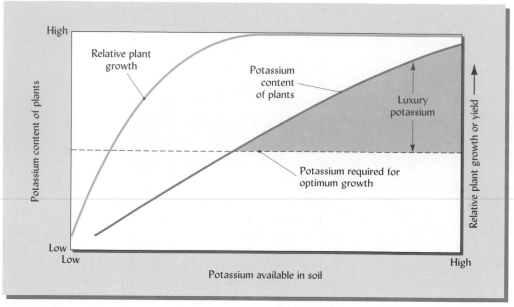

FIGURE 14.29 The general relationship between available potassium level in the soils, plant growth, and plant uptake of potassium. If available soil potassium is raised above the level needed for maximum plant growth, many plants will continue to increase their uptake of potassium without any corresponding increase in growth. The potassium taken up in excess of that needed for optimum growth is termed *luxury consumption*. Such luxury consumption may be wasteful, especially if the plants are completely removed from the soil. It may also cause dietary imbalance in grazing animals.

potassium. Potassium taken up by the plant above this critical required level is considered a *luxury*. If plant residues are not returned to the soil, the removal of this excess potassium is decidedly wasteful. In addition, high levels of potassium may depress calcium uptake and cause nutritional imbalances both in the plants and in animals that consume them.

In summary, then, the problem of potassium is at least threefold: (1) a very large proportion is relatively unavailable to higher plants; (2) it is subject to leaching losses; and (3) the removal of potassium by plants is high, especially when luxury quantities of this element are supplied. With these ideas as a background, the various forms and availabilities of potassium in soils will now be considered.

14.14   FORMS AND AVAILABILITY OF POTASSIUM IN SOILS

Four forms of soil potassium are shown in the potassium-cycle diagram (Figure 14.26, from the bottom upward): (1) K in primary mineral crystal structures, (2) K in nonexchangeable positions in secondary minerals, (3) K in exchangeable form on soil colloid surfaces, and (4) potassium ions soluble in water. The total amount of potassium in a soil and the distribution of potassium among the four major pools shown in Figure 14.26 is largely a function of the kinds of clay minerals present in a soil. Generally, soils

TABLE 14.8   The Influence of Dominant Clay Minerals on the Amounts of Water-Soluble, Exchangeable, Fixed (Nonexchangeable), and Total Potassium in Soils

The values given are means for many soils in 10 soil orders sampled in the United States and Puerto Rico.

Potassium pool	Dominant clay mineralogy of soils, mg K/kg soil		
	Kaolinitic (26 soils)	Mixed (53 soils)	Smectitic (23 soils)
Total potassium	3340	8920	15780
Exchangeable potassium	45	224	183
Water-soluble potassium	2	5	4

Data from Sharpley (1990).

dominated by 2:1 clays contain the most potassium; those dominated by kaolinite contain the least (Table 14.8). In terms of availability for plant uptake, the following interpretation applies to the different forms of soil potassium:

K in primary mineral structure	*Unavailable*
Nonexchangeable K in secondary minerals	*Slowly available*
Exchangeable K on soil colloids ⎫	
K soluble in water ⎭	*Readily available*

All plants can easily utilize the readily available forms, but the ability to obtain potassium held in the slowly available and unavailable forms differs greatly among plant species. Many grass plants with fine, fibrous root systems are able to exploit potassium held in clay interlayers and near the edges of mica and feldspar crystals of clay and silt size. A few plants adapted to low-fertility sandy soils, such as elephant grass (*Pennisetum purpureum* Schum), have been shown to obtain potassium from even sand-sized primary minerals, a form of potassium usually considered to be unavailable.

Relatively Unavailable Forms

Some 90 to 98% of all soil potassium in a mineral soil is in relatively unavailable forms (see Figure 14.26). The compounds containing most of this form of potassium are the feldspars and the micas. These minerals are quite resistant to weathering and supply relatively small quantities of potassium during a given growing season. However, their cumulative release of potassium over a period of years undoubtedly is of some importance. This release is enhanced by the solvent action of carbonic acid and of stronger organic and inorganic acids, as well as by the presence of acid clays and humus (see Section 2.3). As already mentioned, the roots of some plants can obtain a significant portion of their potassium supply from these minerals, apparently by depleting the potassium ions from the solution around the edges of these minerals, thereby favoring the dissolution of the mineral.

Readily Available Forms

Only 1 to 2% of the total soil potassium is readily available. Available potassium exists in soils in two forms: (1) in the soil solution, and (2) exchangeable potassium adsorbed on the soil colloidal surfaces. Although most of this available potassium is in the exchangeable form (approximately 90%), soil solution potassium is most readily absorbed by higher plants but, unfortunately, is subject to considerable leaching loss.

As represented in Figure 14.26, these two forms of readily available potassium are in dynamic equilibrium. This equilibrium is of extreme practical importance. When plants absorb potassium from the soil solution, some of the exchangeable potassium immediately moves into the soil solution until the equilibrium is again established. When water-soluble fertilizers are added, the soil solution becomes potassium enriched and the reverse of the described adjustment occurs—potassium from soil solution moves onto the exchange complex. The exchangeable potassium can be seen as an important buffer mechanism for soil solution potassium.

Slowly Available Forms

In the presence of vermiculite, smectite, and other 2:1-type minerals, the K^+ ions (as well as the similarly sized NH_4^+ ions) in the soil solution (or added as fertilizers) not only become adsorbed but also may become definitely fixed by the soil colloids (Figure 14.30). Potassium (and ammonium) ions fit in between layers in the crystals of these normally expanding clays and become an integral part of the crystal. These ions cannot be replaced by ordinary exchange processes and consequently are referred to as *nonexchangeable ions*. As such, these ions are not readily available to most higher plants. This form is in equilibrium, however, with the more available forms and consequently acts as an extremely important reservoir of slowly available nutrients (Figure 14.31).

Release of Fixed Potassium

The quantity of nonexchangeable or fixed potassium in some soils is quite large. The fixed potassium in such soils is continually released to the exchangeable form in amounts large enough to be of great practical importance. The data in Table 14.9 indi-

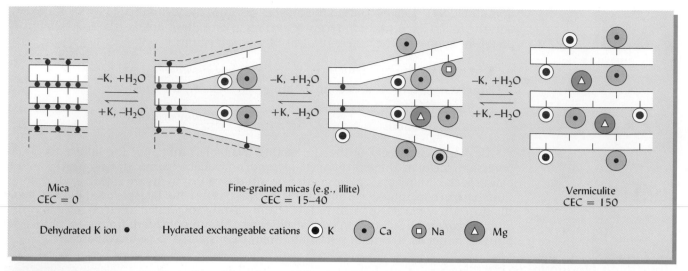

FIGURE 14.30 Diagrammatic illustration of the release and fixation of potassium between primary micas, fine-grained mica (illite clay), and vermiculite. In the diagram, the release of potassium proceeds to the right, while the fixation process proceeds to the left. Note that the dehydrated potassium ion is much smaller than the hydrated ions of Na^+, Ca^{2+}, Mg^{2+}, etc. Thus, when potassium is added to a soil containing 2:1-type minerals such as vermiculite, the reaction may go to the left and potassium ions will be tightly held (fixed) in between layers within the crystal, producing a fine-grained mica structure. Ammonium ions (NH_4^+) are of a similar size and charge to potassium ions and may be fixed by similar reactions. [Modified from McLean (1978)]

cates the magnitude of the release of nonexchangeable potassium from certain soils. In these soils, the potassium removed by plants was supplied largely from nonexchangeable forms. The entire equilibrium may be represented for potassium as follows:

$$\text{Nonexchangeable K} \underset{\text{Slow}}{\overset{}{\rightleftharpoons}} \text{Exchangeable K} \underset{\text{Rapid}}{\overset{}{\rightleftharpoons}} \text{Soil solution K}$$

As a result of these relationships, very sandy soils with low CEC are poorly buffered with respect to potassium. In them, the potassium ion concentration may be quite high at the beginning of a growing season or just after fertilization, but the soils have little capacity to maintain the potassium concentration, as plants remove the dissolved potassium from the soil solution during the growing season. Late-season potassium deficiency may result. In finer-textured soils with a greater CEC (and therefore greater buffering capacity), the initial solution concentration of potassium may be somewhat lower, but the soil is capable of maintaining a fairly constant supply of solution potassium ions throughout the growing season.

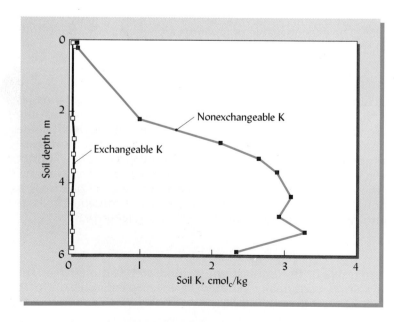

FIGURE 14.31 Exchangeable and nonexchangeable potassium levels in a Ultisol in South Carolina after 30 years growth of a loblolly pine following 150 years of cultivated crops. Although the exchangeable potassium level was quite low, tree growth was not adversely affected, and large quantities of this element were absorbed by the trees over the 30-year period. This was made possible by the conversion of nonexchangeable potassium to the exchangeable form, which was readily taken up by the trees. The upper horizons may have been depleted somewhat of nonexchangeable K, but the deep tree roots were able to use the potassium released from the nonexchangeable form in the lower horizons. [From Richter, et al. (1994)]

Soil	Total K used by crops		Percent derived from nonexchangeable form
	kg/ha	lb/acre	
Wisconsin soils[a]			
Carrington silt loam	133	119	75
Spencer silt loam	66	59	80
Plainfield sand	99	88	25
Mississippi soils[b]			
Robinsonville fine silty loam	121	108	33
Houston clay	64	57	47
Ruston sandy loam	47	42	24

[a] Average of six consecutive cuttings of ladino clover [from Evans and Attoe (1948)]
[b] Average of eight consecutive crops of millet. [From Gholston and Hoover (1948)]

14.15   FACTORS AFFECTING POTASSIUM FIXATION IN SOILS

Four soil conditions markedly influence the amounts of potassium fixed: (1) the nature of the soil colloids, (2) wetting and drying, (3) freezing and thawing, and (4) the presence of excess lime.

Effects of Type of Clay and Moisture

The ability of the various soil colloids to fix potassium varies widely. Kaolinite and other 1:1-type clays fix little potassium. On the other hand, clays of the 2:1 type, such as vermiculite, fine-grained mica (illite), and smectite, fix potassium very readily and in large quantities. Even silt-sized fractions of some micaceous minerals fix and subsequently release potassium.

The potassium and ammonium ions are attracted between layers in the negatively charged clay crystals. The tendency for fixation is greatest in minerals where the major source of negative charge is in the silica (tetrahedral) sheet. Consequently, vermiculite has a greater fixing capacity than montmorillonite (see Table 8.5 for formulas for these minerals).

Alternate wetting/drying and freezing/thawing has been shown to enhance both the fixation of potassium in nonexchangeable forms and the release of previously fixed potassium to the soil solution. Although the practical importance of this is recognized, its mechanism is not well understood.

Influence of pH

Applications of lime sometimes result in an increase in potassium fixation of soils (Figure 14.32). This is not surprising, since in strongly acid soils the tightly held H^+ and hydroxy aluminum ions prevent the potassium ions from being closely associated with the colloidal surfaces, which reduces their susceptibility to fixation. As the pH increases, the H^+ and hydroxyl aluminum ions are removed or neutralized, and it is easier for potassium ions to move closer to the colloidal surfaces, where they are more susceptible to fixation in 2:1 clays.

In soils where the negative charge is pH-dependent, liming increases the cation exchange capacity, which results in an increased potassium adsorption by the soil colloids and a decrease in the potassium level in the soil solution (Figure 14.33).

Furthermore, high calcium and magnesium levels in the soil solution may reduce potassium uptake by the plant because cations tend to compete against one another for uptake by roots. Since the absorption of the potassium ion by plant roots is affected by the activity of other ions in the soil solution, some authorities prefer to use the ratio

$$\frac{[K^+]}{\sqrt{[Ca^{2+}] + [Mg^{2+}]}}$$

rather than the potassium concentration to indicate the available potassium level in solution. Finally, potassium deficiency frequently occurs in calcareous soils even when

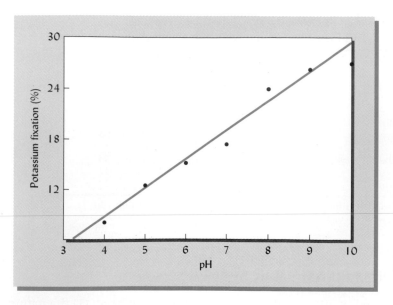

FIGURE 14.32 The effect of pH on the fixation of potassium soils in India. [From Grewal and Kanwar (1976)]

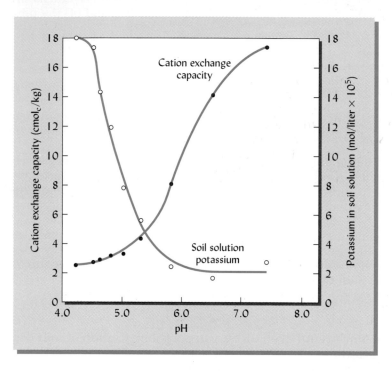

FIGURE 14.33 The influence of increased pH resulting from lime additions on the pH-dependent cation exchange capacity of a soil and the level of potassium in the soil solution. As the cation exchange capacity increases, some of the soil solution potassium is attracted to the adsorbing colloids. [Data from Magdoff and Bartlett (1980)]

the amount of exchangeable potassium present would be adequate for plant nutrition on other soils. Potassium fixation as well as cation ratios may be responsible for these adverse effects on calcium carbonate–rich soil.

14.16 PRACTICAL ASPECTS OF POTASSIUM MANAGEMENT

Except in very sandy soils, the problem of potassium fertility is rarely one of total supply, but rather one of adequate *rate* of transformation from nonavailable to available forms. Where little plant material is removed (e.g., in forests, rangeland, and some ornamental systems), cycling between plant and soil may be adequate for continued plant growth. However, where crops are removed, especially if little plant residue is returned, then the plant–soil cycle must be supplemented by release of potassium from less available mineral forms and, to some degree, by fertilization.

Vigorous growth of high-potassium-content plants places great demands on the soil supply of available potassium. Moreover, the rate of potassium uptake is not constant, but varies with plant growth stage and season. If high yields of forage legumes such as alfalfa are to be produced, the soil may have to be capable of supplying potassium for very high uptake rates during certain periods, resulting in the need for high levels of fertilization even on soils well supplied with weatherable minerals. However, excessive levels of potassium that depress calcium and magnesium in the forage must be avoided to maintain plant and animal health.

Frequency of Application

Although a heavy dressing applied every few years may be most convenient, more frequent light applications of potassium may offer the advantages of reduced luxury consumption of potassium by some plants, reduced losses of this element by leaching, and reduced opportunity for fixation in unavailable forms before plants have had a chance to use the potassium applied. Although such fixation has definite conserving features, in most cases these tend to be outweighed by the disadvantages of leaching and luxury consumption.

Potassium-Supplying Power of Soils

Full advantage should be taken of the potassium-supplying power of soils. The idea that each kilogram of potassium removed by plants or through leaching must be returned in fertilizers may not always be correct. In many soils, the large quantities of moderately available forms already present can be utilized so that only a part of the total amount removed by harvest need be replaced by fertilizer. Moreover, the importance of lime in reducing leaching losses of potassium should not be overlooked as a means of effectively utilizing the power of soils to furnish this element.

Soils of arid regions are often well supplied with weatherable potassium-containing minerals and can therefore supply adequate potassium for many years, even under irrigation where leaching is more important and plant removal is great. However, continued crop removal can deplete the available potassium pools even in these soils. Also, deep-rooted plants such as cotton and fruit trees may depend on the subsoil for much of their potassium. Increasing the availability of this element at depths below the plow layer is difficult.

Potassium Losses and Gains

The problem of maintaining soil potassium is outlined diagrammatically in Figure 14.34. Plant removal of potassium generally exceeds that of the other essential elements, with the possible exception of nitrogen. Annual losses from plant removal as great as 400 kg/ha or more of potassium are not uncommon, particularly if the plant is a legume and is cut several times for hay. As might be expected, therefore, the return of plant residues and manures is very important in maintaining soil potassium.

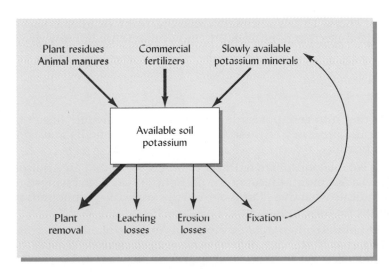

FIGURE 14.34 Gains and losses of *available* soil potassium under average field conditions. The approximate magnitude of the changes is represented by the width of the arrows. For any specific case the actual amounts of potassium added or lost undoubtedly may vary considerably from this representation. As was the case with nitrogen and phosphorus, commercial fertilizers are important in meeting plant demands.

The annual losses of available potassium by leaching and erosion greatly exceed those of nitrogen and phosphorus. They are generally not as great, however, as the corresponding losses of available calcium and magnesium. Such losses of soil minerals have serious implications for sustainable soil productivity.

Increased Use of Potassium Fertilizers

The global use of potassium fertilizers will continue to increase as the pools of available potassium are depleted and increasing crop yields place demands on the soil that cannot be met by mineral weathering alone. This is especially true in cash-crop areas and in regions where sandy soils or highly weathered soils are prominent. Figure 14.1 shows that farmers in the United States increased their use of potassium fertilizer for many years until about 1980, by which time the potassium-supplying power of their soils (on average) had largely reached the level at which only maintenance additions were needed. This maintenance level in many regions is at least half the level of average potassium removal (the other part of removals being replaced by potassium released by mineral weathering). In less developed agricultural regions of the world, potassium fertilizer use will have to increase for many years to come if yields are to be increased or even maintained.

14.17 CONCLUSION

Soil phosphorus presents us with a double-edged sword; its management is crucially important from both an environmental standpoint and for soil fertility. On one hand, too little phosphorus commonly limits the productivity of natural and cultivated plants and is the cause of widespread soil and environmental degradation. On the other hand, industrialized agriculture has concentrated too much phosphorus in some cases, resulting in losses from soil that cause egregious eutrophication in surface waters. Some of these situations have been caused by excessive buildup of soil phosphorus with fertilizer. Others are the result of concentration of animal production, so that phosphorus in the manure produced at many livestock facilities far exceeds that required by the crops grown on the surrounding land.

Except in cases of extreme buildup, the availability of phosphorus to plant roots has a double constraint: the low total phosphorus level in soils and the small percentage of this level that is present in available forms. Furthermore, even when soluble phosphates are added to soils, they are quickly fixed into insoluble forms that in time become quite unavailable to growing plants. In acid soils, the phosphorus is fixed primarily by iron, aluminum, and manganese; in alkaline soils, by calcium and magnesium. This fixation greatly reduces the efficiency of phosphate fertilizers, with little of the added phosphorus being taken up by plants. In time, however, this unused phosphorus can build up and serve as a reserve pool for plant absorption.

Potassium is generally abundant in soils, but it, too, is present mostly in forms that are quite unavailable for plant absorption. Fortunately, however, some soils contain considerable nonexchangeable but slowly available forms of this element. Over time this potassium can be released to exchangeable and soil solution forms that can be quickly absorbed by plant roots. This is fortunate since the plant requirements for potassium are high—5 to 10 times that of phosphorus and similar to that of nitrogen.

STUDY QUESTIONS

1. You have learned that nitrogen, potassium and phosphorus are all "fixed" in the soil. Compare the processes of these fixations and the benefits and constraints they each provide.

2. Assume you add a soluble phosphate fertilizer to an Oxisol and to an Aridisol. In each case, within a few months most of the phosphorus has been changed to insoluble forms. Indicate what these forms are and the respective compounds in each soil responsible for their formation.

3. How does the phosphorus content of cultivated soils in the United States compare with that of nearby forest soils which have never been cleared? What is the reason for this difference?

4. What is meant by *eutrophication*, and how is it influenced by farm practices involving phosphorus?

5. Which is likely to have the higher buffering capacity for phosphorus and potassium, a sandy loam or a clay? Explain.

6. In the spring a certain surface soil showed the following soil test: soil solution K = 20 kg/ha; exchangeable K = 200 kg/ha. After two crops of alfalfa hay that contained 250 kg/ha of potassium were harvested and removed, a second soil test showed soil solution K = 15 kg/ha and exchangeable K = 150 kg/ha. Explain why there was not a greater reduction in soil solution and exchangeable K levels.

7. What is the effect of soil pH on the availability of phosphorus, and what are the unavailable forms at the different pH levels?

8. What is *luxury consumption* of plant nutrients, and what are its advantages and disadvantages?

9. How does phosphorus that forms relatively insoluble inorganic compounds in soils find its way into streams and other waterways?

10. The incorporation of large amounts of wheat straw into a soil may bring about P deficiency in the following crop. What is the likely reason for this?

11. Compare the organic P levels in the upper horizons of a forested soil with those of a nearby soil that has been cultivated for 25 years. Explain the difference.

REFERENCES

Buresh, R. J., P. C. Smithson, and D. T. Hellums. 1997. "Building soil phosphorus capital in Africa," in R. J. Buresh, P. A. Sanchez, and F. Calhoun (eds.), *Replenishing Soil Fertility in Africa.* SSSA Special Publication no. 51. (Madison, Wis.: Soil Sci. Soc. Amer.).

Brady, N. C. 1974. *The Nature and Properties of Soils,* 8th ed. (New York: Macmillan).

Cogger, C., and J. M. Duxbury. 1984. "Factors affecting phosphorus loss from cultivated organic soils," *J. Environ. Qual.,* **13**:111–114.

Eghball, B., G. D. Binford, and D. D. Baltensperger. 1996. "Phosphorus movement and adsorption in a soil receiving long-term manure and fertilizer application," *J. Environ. Qual.,* **25**:1339–1343.

Evans, C. E., and O. J. Attoe. 1948. "Potassium supplying power of virgin and cropped soils," *Soil Sci.,* **66**:323–334.

Fares, F., et al. 1974. "Quantitative survey of organic phosphorus in different soil types," *Phosphorus in Agriculture.* (Madison, Wis.: Amer. Soc. Agron.), pp. 25–40.

Fox, R. L. 1981. "External phosphorus requirements of crops," in *Chemistry in the Soil Environment.* ASA Special Publication no. 40. (Madison, Wis.: Amer. Soc. Agron. and Soil Sci. Soc. Amer.), pp. 223–239.

Gahoonia, T. S., N. Claassen, and A. Jungle. 1992. "Mobilization of phosphate in different soils by ryegrass supplied with ammonium or nitrate," *Plant and Soil,* **140**:241–248.

Gholston, L. E., and C. D. Hoover. 1948. "The release of exchangeable and nonexchangeable potassium from several Mississippi and Alabama soils upon continuous cropping," *Soil Sci. Soc. Amer. Proc.,* **13**:116–121.

Grewal, J. S., and J. S. Kanwar. 1976. *Potassium and Ammonium Fixation in Indian Soils* (review). (New Delhi, India: Indian Council for Agricultural Research).

Khasawneh, F. E., et al. (eds.). 1980. *The Role of Phosphorus in Agriculture.* (Madison, Wis.: Amer. Soc. Agron.).

Kuo, S. 1988. "Application of modified Langmuir isotherm to phosphate sorption by some acid soils," *Soil Sci. Soc. Amer. J.,* **52**:97–102.

Magdoff, F. R., and R. J. Bartlett. 1980. "Effect of liming acid soils on potassium availability," *Soil Sci.,* **129**:12–14.

Mandzak, J. M., and J. A. Moore. 1994. "The role of nutrition in the health of inland Western forests," *J. Sustainable Forestry,* **2**:191–210.

McLean, E. O. 1978. "Influence of clay content and clay composition on potassium availability," in G. S. Sekhon (ed.), *Potassium in Soils and Crops.* (New Delhi, India: Potash Research Institute of India), pp. 1–19.

Mengel, K., and E. A. Kirkby. 1987. *Principles of Plant Nutrition,* 4th ed. (Bern, Switzerland: International Potash Institute).

Munson, R. D. (ed.). 1985. *Potassium in Agriculture*. (Madison, Wis.: Amer. Soc. Agron.).

Nair, K. P. P. 1996. "The buffering power of plant nutrients and effects on availability," *Advances in Agronomy*, **57**:237–287.

Nair, V. D., A. A. Graetz, and K. M. Portier. 1995. "Forms of phosphorus in soil profiles from dairies of South Florida," *Soil Sci. Soc. Amer. J.*, **59**:1244–1249.

Nowak, C. A., R. B. Downard, Jr., and E. H. White. 1991. "Potassium trends in red pine plantations at Pack Forest, New York," *Soil Sci. Soc. J.*, **55**:847–850.

Olsen, S. R., and F. E. Khasawneh. 1980. "Use and limitations of physical-chemical criteria for assessing the status of phosphorus in soils," in F. E. Khasawneh, et al. (eds.), *The Role of Phosphorus in Agriculture*. (Madison, Wis.: Amer. Soc. of Agron.).

Porter, R. M., J. E. Ayers, M. W. Johnson, Jr., and P. E. Nelson. 1981. "Influence of differential phosphorus accumulation on corn stalk rot," *Agron. J.*, **73**:283–287.

Raven, K. P., and L. R. Hossner. 1993. "Phosphorus desorption quantity-intensity relationships in soils," *Soil Sci. Soc. Amer. J.*, **57**:1501–1508.

Richter, D. D., et al. 1994. "Soil chemical change during three decades in an old-field Loblolly pine (*Pinus taeda L.*) ecosystem," *Ecology*, **75**:1463–1473.

Saa, A., M. C. Trasar-Cepeda, B. Soto, F. Gil-Sotres, and F. Diaz-Fierros. 1994. "Forms of phosphorus in sediments eroded from burnt soils," *J. Environ. Qual.*, **23**:739–746.

Sample, E. C., et al. 1980. "Reactions of phosphate fertilizers in soils," in F. E. Khasawneh, et al. (eds.), *The Role of Phosphorus in Agriculture*. (Madison, Wis.: Amer. Soc. Agron.).

Sanchez, P. A., et al. 1997. "Soil fertility replenishment in Africa: An investment in natural resource capital," in R. J. Buresh, P. A. Sanchez, and F. Calhoun (eds.), *Replenishing Soil Fertility in Africa*. SSSA Special Publication no. 51. (Madison, Wis.: Soil Sci. Soc. Amer.).

Sharpley, A. N. 1990. "Reaction of fertilizer potassium in soils of differing mineralogy," *Soil Sci.*, **49**:44–51.

Sloan, J. J., N. T. Basta, and R. L. Westerman. 1995. "Aluminum transformation and solution equilibria induced by banded phosphorus fertilizer in acid soil," *Soil Sci. Soc. Amer. J.*, **59**:357–364.

Smith, S. J., A. N. Sharpley, J. W. Naney, W. A. Berg, and O. R. Jones. 1991. "Water quality impacts associated with wheat culture in the Southern Plains," *J. Environ. Qual.*, **20**:244–249.

Soltanpour, P. N., R. L. Fox, and R. C. Jones. 1988. "A quick method to extract organic phosphorus from soils," *Soil Sci. Soc. Amer. J.*, **51**:255–256.

Stevenson, F. J. 1986. *Cycles of Soil Carbon, Nitrogen, Phosphorus, Sulfur, and Micronutrients*. (New York: Wiley).

Vaithiyanathan, P., and D. L. Correll. 1992. "The Rhode River watershed: Phosphorus distribution and export in forest and agricultural soils," *J. Environ. Qual.*, **21**:280–288.

Weil, R. R., P. W. Benedetto, L. J. Sikora, and V. A. Bandell. 1988. "Influence of tillage practices on phosphorus distribution and forms in three Ultisols," *Agron. J.*, **80**:503–509.

15

MICRONUTRIENT ELEMENTS

Look and you will find it . . . what is unsought will go undetected . . .
—SOPHOCLES

Of the 18 elements known to be essential for plant growth, 9 are required in such small quantities that they are called *micronutrients*[1] or *trace elements*. These elements are iron, manganese, zinc, copper, boron, molybdenum, nickel, cobalt, and chlorine.

Other elements, such as silicon, vanadium, and sodium, appear to improve the growth of at least certain plant species. Animals, including humans, also require most of these elements in their diets. Some elements, such as selenium, chromium, tin, iodine, and fluorine, have been shown to be essential for animal growth but are apparently not required by plants.

The terms *micronutrient* and *trace element* must not be construed to imply that these nutrients are somehow less important than macronutrients. To the contrary, the effects of micronutrient deficiency can be very severe in terms of stunted growth, low yields, dieback, and even plant death. By the same token, where they are needed, very small applications of micronutrients may produce dramatic results.

Increasing attention is being directed toward micronutrient deficiencies for several reasons:

1. Intensive plant production practices have increased crop yields, resulting in greater removal of micronutrients from soils.
2. The trend toward high-analysis fertilizers has reduced the use of impure salts and organic manures, which formerly supplied significant amounts of micronutrients.
3. Increased knowledge of plant nutrition and improved methods of analysis in the laboratory are helping in the diagnosis of micronutrient deficiencies that might formerly have gone unnoticed.
4. Increasing evidence indicates that food grown on soils with low levels of trace elements may provide insufficient human dietary levels of certain elements, even though the crop plants show no signs of deficiency themselves.

Interest in trace elements has also been sparked by problems of toxicity resulting from an oversupply of these elements. Levels of trace elements toxic to plants or to animals may result from natural soil conditions, from pollution, or from soil-management practices.

---

[1] For review articles on this subject, see Welsh (1995) and Mortvedt, et al. (1991).

This chapter will provide some of the background and tools needed to recognize and deal effectively with the varied deficiencies and toxicities encountered in a wide range of soil–plant systems.

15.1 DEFICIENCY VERSUS TOXICITY

It is axiomatic that anything can be toxic if taken in large enough amounts. At low levels of a nutrient, deficiency and reduced plant growth may occur (*deficiency range*). As the level of nutrient is increased, plants respond by taking up more of the nutrient and increasing their growth. If a level of nutrient availability has been reached that is sufficient to meet the plants' needs (*sufficiency range*), raising the level further will have little effect on plant growth, although the concentration of the nutrient may continue to increase in the plant tissue. At some level of availability, the plant will take up too much of the nutrient for its own good (*toxicity range*), causing adverse physiological reactions to take place. The relationship among deficient, sufficient, and toxic levels of nutrient availability is described in Figure 15.1.

For macronutrients the sufficiency range is very broad, and toxicity occurs rarely, if ever. However, for micronutrients the difference between deficient and toxic levels may be very narrow, making the possibility of toxicity quite real. For example, in the cases of boron and molybdenum, severe toxicity may result from applying as little as 3 to 4 kg/ha of available nutrient to a soil initially deficient in these elements. While the sufficiency range for other micronutrients is much wider, and toxicities are not as likely from overfertilization, toxicities of copper and zinc have been observed on soils contaminated by industrial sludges, manure from concentrated hog facilities, metal smelter wastes, and long-term application of copper sulfate fungicide.

High levels of molybdenum may occur naturally in certain poorly drained alkaline soils. In some cases, enough of this element may be taken up by plants to cause toxicity, not only to susceptible plants, but also to livestock grazing forages on these soils. Boron, too, may occur naturally in alkaline soils at levels high enough to cause plant toxicity. Although somewhat larger amounts of most other micronutrients are required and can be tolerated by plants, great care needs to be exercised in applying micronutrients, especially for maintaining nutrient balance.

In addition, irrigation water in dry regions may contain enough dissolved boron, molybdenum, or selenium (a trace element not thought to be a plant nutrient) to damage sensitive crops, even if the original levels of these trace elements in the soil were not very high. Therefore, it is prudent to monitor the content of these elements in water used for irrigation (see Section 10.8).

Selenium does not appear to be essential for plant metabolism, but it is required by animals. Infirmities such as the white-muscle disease of cattle result from selenium-

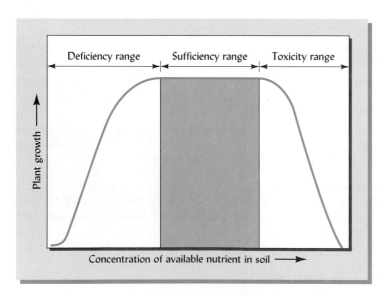

FIGURE 15.1 The relationship between the amount of a micronutrient available for plant uptake and the growth of the plant. Within the deficiency range, as nutrient availability increases so does plant growth (and uptake, which is not shown here). Within the sufficiency range, plants can get all of the nutrient they need, and so their growth is little affected by changes within this range. At higher levels of availability a threshold is crossed into the toxicity range, in which the amount of nutrient present is excessive and causes adverse physiological reactions leading to reduced growth and even death of the plant.

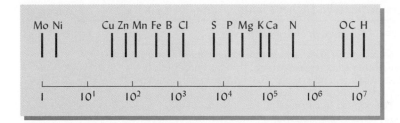

FIGURE 15.2 Relative numbers of atoms of the essential elements in alfalfa at bloom stage, expressed logarithmically. Note that there are more than 10 million hydrogen atoms for each molybdenum atom. Even so, normal plant growth would not occur without molybdenum. [Modified from Viets (1965)]

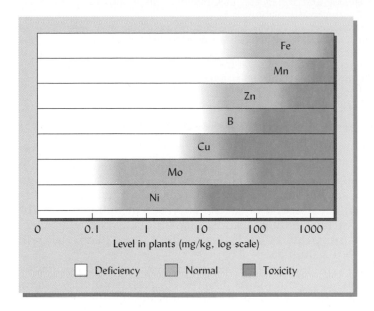

FIGURE 15.3 Deficiency, normal, and toxicity levels in plants for seven micronutrients. Note that the range is shown on a logarithm base and that the upper limit for manganese is about 10,000 times the lower range for molybdenum and nickel. In using this figure, keep in mind the remarkable differences in the ability of different plant species and cultivars to accumulate and tolerate different levels of micronutrients. (Based on data from many sources)

deficient feeds produced on selenium-deficient soils. Furthermore, selenium toxicity occurs in grazing animals in some parts of the western United States and elsewhere around the world where rangeland soils are high in selenium. Blind staggers and eventual death are consequences of Se toxicity.

Micronutrients are required in very small quantities, their concentrations in plant tissue being one or more orders of magnitude lower than for the macronutrients (Figure 15.2). The ranges of plant tissue concentrations considered deficient, adequate, and toxic for several micronutrients are illustrated in Figure 15.3.

15.2 ROLE OF THE MICRONUTRIENTS[2]

Physiological Roles

Micronutrients play many complex roles in plant nutrition. While most of the micronutrients participate in the functioning of a number of enzyme systems (Table 15.1), there is considerable variation in the specific functions of the various micronutrients in plant and microbial growth processes. For example, copper, iron, and molybdenum are capable of acting as electron carriers in the enzyme systems that bring about oxidation-reduction reactions in plants. Such reactions are essential steps in photosynthesis and many other metabolic processes. Zinc and manganese function in many plant enzyme systems as bridges to connect the enzyme with the substrate upon which it is meant to act.

Molybdenum and manganese are essential for certain nitrogen transformations in microorganisms as well as in plants. Molybdenum and iron are components of the enzyme *nitrogenase*, which is essential for the processes of symbiotic and nonsymbiotic

---

[2] For a review of the role of individual micronutrients, see Mengel and Kirkby (1987) and Marsahner (1995). For nickel, see Brown, et al. (1987).

TABLE 15.1   Functions of Several Micronutrients in Higher Plants

Micronutrient	Functions in higher plants
Zinc	Present in several dehydrogenase, proteinase, and peptidase enzymes; promotes growth hormones and starch formation; promotes seed maturation and production.
Iron	Present in several peroxidase, catalase, and cytochrome oxidase enzymes; found in ferredoxin, which participates in oxidation-reduction reactions (e.g., NO_3^- and SO_4^{2-} reduction and N fixation); important in chlorophyll formation.
Copper	Present in laccase and several other oxidase enzymes; important in photosynthesis, protein and carbohydrate metabolism, and probably nitrogen fixation.
Manganese	Activates decarboxylase, dehydrogenase, and oxidase enzymes; important in photosynthesis, nitrogen metabolism, and nitrogen assimilation.
Nickel	Essential for urease, hydrogenases, and methyl reductase; needed for grain filling, seed viability, iron absorption, and urea and ureide metabolism (to avoid toxic levels of these nitrogen-fixation products in legumes).
Boron	Activates certain dehydrogenase enzymes; facilitates sugar translocation and synthesis of nucleic acids and plant hormones; essential for cell division and development.
Molybdenum	Present in nitrogenase (nitrogen fixation) and nitrate reductase enzymes; essential for nitrogen fixation and nitrogen assimilation.
Cobalt	Essential for nitrogen fixation; found in vitamin B_{12}.

nitrogen fixation. Molybdenum is also present in the enzyme *nitrate reductase,* which is responsible for the reduction of nitrates in soils and plants.

Nickel has only recently been added to the list of elements shown to be essential to higher plants. It is essential for the function of several enzymes, including *urease,* the enzyme that breaks down urea into ammonia and carbon dioxide. Nickel-deficient legumes accumulate toxic levels of urea in their leaves; seeds of cereal plants deficient in nickel are not viable and fail to germinate.

Zinc plays a role in protein synthesis, in the formation of some growth hormones, and in the reproductive process of certain plants. Copper is involved in both photosynthesis and respiration, and in the use of iron. It also stimulates lignification of cell walls. The roles of boron have yet to be clearly defined, but boron appears to be involved with cell division, water uptake, and sugar translocation in plants. Iron is involved in chlorophyll formation and degradation, and in the synthesis of proteins and nucleic acids. Manganese seems to be essential for photosynthesis, respiration, and nitrogen metabolism.

The role of chlorine is still somewhat obscure; however, it is known to influence photosynthesis and root growth. Cobalt is essential for the symbiotic fixation of nitrogen. In addition, legumes and some other plants have a cobalt requirement independent of nitrogen fixation, although the amount required is small compared to that for the nitrogen-fixation process.

Deficiency Symptoms

Insufficient supply of a nutrient is often expressed by certain visible plant symptoms, some of which are quite useful as indicators of particular micronutrient deficiencies. Most of the micronutrients are relatively immobile in the plant. That is, the plant cannot efficiently transfer the nutrient from older leaves to newer ones. Therefore the concentration of the nutrient tends to be lowest, and the symptoms of deficiency most pronounced, in the younger leaves, which develop after the supply of the nutrient has run low.[3] The iron-deficient sorghum in Plate 29 and the zinc-deficient corn and peach leaves in Plates 30 and 31 (following page 498) illustrate the pattern of pronounced deficiency symptoms on the younger leaves.

In the case of zinc deficiency in corn, broad white bands on both sides of the midrib are typical in young corn plants, but the symptoms may disappear as the soil warms up and the maturing plant's root system expands into a larger volume of soil. Interveinal chlorosis (darker green veins and light yellowish areas between the veins) on the younger leaves, and the whorl of tiny leaves at the terminal end of the branch

---

[3] This pattern is in contrast to most of the macronutrients (except sulfur), which are more easily translocated by the plants and so become most deficient in the older leaves.

(a symptom known as *little leaf*) are characteristic of tree foliage suffering from too little zinc.

Boron deficiency generally affects the growing points of plants, such as their buds, fruits, flowers, and root tips. Plants low in boron may produce deformed flowers (Figure 15.4); aborted seeds; thickened, brittle, puckered leaves; or dead growing points (Figure 15.5). In all of these cases, a small dose of boron applied at the correct time could make the difference between a marketable or unmarketable plant product.

(a)   (b)

FIGURE 15.4 A misshapen rose, or *bullhead*, on a boron-deficient tea rose (*a*). A well-formed bloom on the same plant after a solution containing boron was applied to the leaves and young buds (*b*). (Photos courtesy of R. Weil)

FIGURE 15.5 The importance of boron in plant growth is clearly illustrated by comparing the normal bean plant on the left to the boron-deficient plant on the right. Once the bean plant on the right used all the boron from its seed, development of new leaves from the terminal bud (see arrow) was completely arrested. (Photo courtesy of R. Weil)

15.3 SOURCE OF MICRONUTRIENTS

Deficiencies of micronutrients may be related to low contents of these elements in the parent rocks or transported parent material. Similarly, toxic quantities are sometimes related to abnormally large amounts in the soil-forming rocks and minerals. More often, however, deficiency and toxicity of these elements result from certain soil conditions that either increase or decrease the solubility and availability of these elements to plants.

Inorganic Forms

Sources of the nine micronutrients vary markedly from area to area. The wide variability in the content of these elements in soils and suggested contents in a representative soil is shown in Table 15.2.

All of the micronutrients have been found in varying quantities in igneous rocks. Two of them, iron and manganese, have prominent structural positions in primary silicate minerals, such as biotite and hornblende. Others, such as cobalt and zinc, also may occupy structural positions as minor replacements for the major constituents of silicate minerals, including clays.

The mineral forms of micronutrients are altered as mineral decomposition and soil formation occur. Oxides and, in some cases, sulfides of elements such as iron, manganese, and zinc are formed (see Table 15.2). Secondary silicates, including the clay minerals, may contain considerable quantities of iron and manganese, and smaller quantities of zinc and cobalt. Ultramafic rocks, especially serpentinite, are high in nickel. The micronutrient cations released as weathering occurs are subject to colloidal adsorption, just as are the calcium or aluminum ions (see Section 8.3).

Anions such as borate and molybdate in soils may undergo adsorption or reactions similar to those of the phosphates. For boron, the adsorption may be represented as follows:

$$\text{Soil colloid} \quad \begin{array}{c} -O \\ -O \end{array} \hspace{-4pt} > \hspace{-2pt} B-OH$$

Chlorine, by far the most soluble of the group, is added to soils in considerable quantities each year through rainwater. Its incidental addition to soils in fertilizers and in other ways helps prevent the deficiency of chlorine under field conditions. Chlorine is only very weakly adsorbed to soil colloids.

TABLE 15.2 Major Sources of Nine Micronutrients, Ranges and Representative Contents of These Nutrients in Soils, Along with Their Representative Contents in Harvested Crops

The ratio of soil to crop contents emphasizes the primary need to increase the efficiency of plants in absorbing these nutrients from the soil.

Element	Major sources	Range, kg/ha/15 cm	Soil, kg/ha/15 cm	Crop, kg/crop	Soil/crop ratio
Fe	Oxides, sulfides, silicates	20,000–220,000	56,000	2	28,000
Mn	Oxides, silicates, carbonates	45–9,000	2200	0.5	4,400
Zn	Sulfides, carbonates, silicates	25–700	110	0.3	366
Cu	Sulfides, hydroxy carbonates, oxides	4–2,000	45	0.1	450
Ni	Silicates (especially serpentine)	10–2,200	45	0.02	2,250
B	Borosilicates, borates	8–200	22	0.2	110
Mo	Sulfides, oxides, molybdates	0.4–10	5	0.02	250
Cl	Chlorides	15–100	22	2.5	0.9
Co	Silicates	2–90	18	0.02	900

Within the header, a spanning label "Representative contents of" covers the columns Soil, kg/ha/15 cm and Crop, kg/crop.

Organic Forms

Organic matter is an important secondary source of some of the trace elements. Several of them tend to be held as complex combinations by organic (humus) colloids. Copper is especially tightly held by organic matter, so much so that its availability can be very low in organic soils (Histosols). In uncultivated profiles, there is a somewhat greater concentration of micronutrients in the surface soil, much of it presumably in the organic fraction. Correlations between soil organic matter and contents of copper, molybdenum, and zinc have been noted. Although the elements thus held are not always readily available to plants, their release through decomposition is undoubtedly an important fertility factor. Animal manures are a good source of micronutrients, much of it present in organic forms.

Forms in Soil Solution

The dominant forms of micronutrients that occur in the soil solution are listed in Table 15.3. The specific forms present are determined largely by the pH and by soil aeration (i.e., redox potential). Note that the cations are present in the form of either simple cations or hydroxy metal cations. The simple cations tend to be dominant under highly acid conditions. The more complex hydroxy metal cations become more prominent as the soil pH is increased.

Molybdenum is present mainly as MoO_4^{2-}, an anionic form that reacts at low pH in ways similar to those of phosphorus (see Chapter 14). Although boron also may be present in anionic form at high pH levels, research suggests that undissociated boric acid (H_3BO_3) is the form that is dominant in the soil solution and is absorbed by higher plants.

The cycling of micronutrients through the soil–plant–animal system is illustrated in a generalized way by Figure 15.6. Although not every micronutrient will participate in every pathway shown in this figure, it can be seen that organic chelates, soil colloids, soil organic matter, and soil minerals all contribute micronutrients to the soil solution and, in turn, to growing plants. As we turn our attention to micronutrient availability, it will be helpful to refer back to Figure 15.6 to see the relationships among the processes involved.

15.4 GENERAL CONDITIONS CONDUCIVE TO MICRONUTRIENT DEFICIENCY

Micronutrients are most apt to limit crop growth in the following five types of soil conditions.

LEACHED, ACID, SANDY SOILS. Strongly leached, acid, sandy soils are low in most micronutrients for the same reasons they are deficient in most of the macronutrients—their parent materials were initially low in the elements, and acid leaching has removed much

TABLE 15.3 **Forms of Micronutrients Dominant in the Soil Solution**

Micronutrient	Dominant soil solution forms
Iron	Fe^{2+}, $Fe(OH)_2^+$, $Fe(OH)^{2+}$, Fe^{3+}
Manganese	Mn^{2+}
Zinc	Zn^{2+}, $Zn(OH)^+$
Copper	Cu^{2+}, $Cu(OH)^+$
Molybdenum	MoO_4^{2-}, $HMoO_4^-$
Boron	H_3BO_3, $H_2BO_3^-$
Cobalt	Co^{2+}
Chlorine	Cl^-
Nickel	Ni^{2+}, Ni^{3+}

From data in Lindsay (1972).

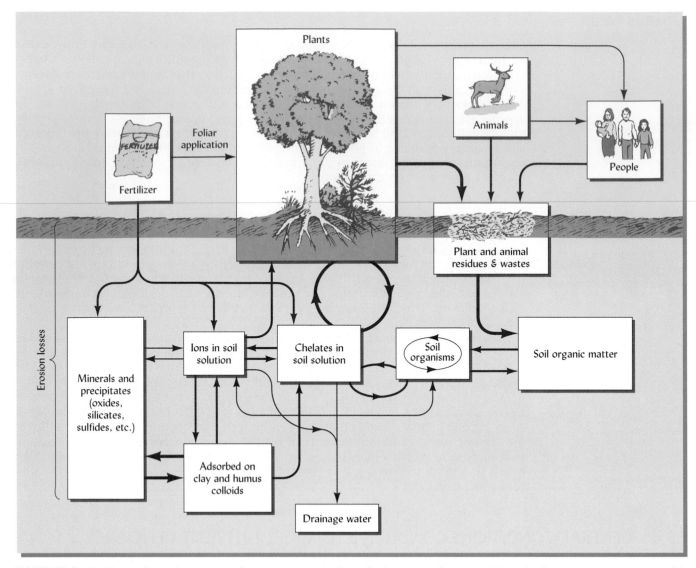

FIGURE 15.6 Cycling and transformations of micronutrients in the soil–plant–animal system. Although all micronutrients may not follow each of the pathways shown, most are involved in the major components of the cycle. The formation of chelates, which keep most of these elements in soluble forms, is a unique feature of this cycle.

of the small quantity of micronutrients originally present. In the case of molybdenum, acid soil conditions markedly decrease availability.

ORGANIC SOILS.   The micronutrient contents of organic soils depend on the extent of the washing or leaching of these elements into the bog area as the soils were formed. In most cases, this rate of movement is too slow to produce deposits as high in micronutrients as are the surrounding mineral soils. The ability of organic soils to bind certain elements, notably copper, also accentuates micronutrient deficiencies.

INTENSIVE CROPPING.   Intensive cropping, in which large amounts of plant nutrients are removed in the harvest, accelerates the depletion of micronutrient reserves in the soil and increases the likelihood of micronutrient deficiencies. Such depletion is most common where high yields are produced with the aid of chemical fertilizers that supply some of the macronutrients, but not the micronutrients.

EXTREMES OF PH.   Soil pH, especially in well-aerated soils, has a decided influence on the availability of all the micronutrients except chlorine. Under very acid conditions, molybdenum is rendered unavailable; at high pH values, availability of all the micronutrient cations is unfavorably affected.

ERODED SOILS. Soil erosion can influence micronutrient availability. Erosion of topsoil carries away considerable soil organic matter, in which much of the potentially available micronutrients are held. Also, removal of the topsoil exposes subsoil horizons that are often higher in pH than the topsoil, a condition that leads to deficiencies of some micronutrients, such as zinc. Eroded ridges or hillsides are common sites of micronutrient deficiencies in some areas.

15.5 FACTORS INFLUENCING THE AVAILABILITY OF THE MICRONUTRIENT CATIONS

Each of the six micronutrient cations (iron, manganese, zinc, copper, nickel, and cobalt) is influenced in a characteristic way by the soil environment. However, certain soil factors have the same general effects on the availability of all six cations.

Soil pH

The micronutrient cations are most soluble and available under acid conditions. In very acid soils, there is a relative abundance of the ions of iron, manganese, zinc, and copper (Figure 15.7). Nickel also follows a general pattern of increased availability at low pH. In fact, under these conditions, the soil solution concentrations or activities of one or more of these elements (most commonly manganese) is often sufficiently high to be toxic to common plants. As indicated in Chapter 9, one of the primary reasons for liming acid soils is to reduce the solubility of manganese and aluminum.

As the pH is increased, the ionic forms of the micronutrient cations are changed first to the hydroxy ions and, finally, to the insoluble hydroxides or oxides of the elements. The following example uses the ferric ion as typical of the group:

$$Fe^{3+} \xrightarrow{OH^-} Fe(OH)^{2+} \xrightarrow{OH^-} Fe(OH)_2^+ \xrightarrow{OH^-} Fe(OH)_3$$

Simple cation (soluble)   Hydroxy metal cations (soluble)   Hydroxide (insoluble)

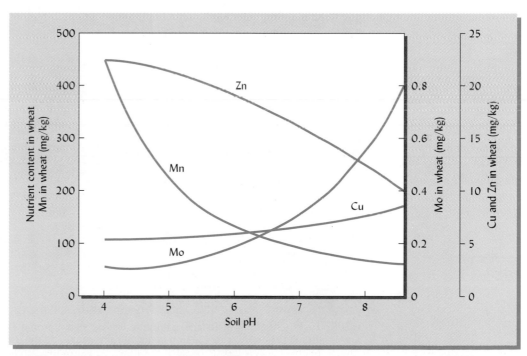

FIGURE 15.7 Effect of soil pH on the concentrations of manganese, zinc, copper, and molybdenum in wheat plants. The soils were from different countries around the world. The molybdenum levels are extremely low, but increase with increasing pH. Manganese and zinc levels decrease as the pH rises, while copper is little affected. [Redrawn from Sillanpaa (1982)]

All of the hydroxides of the micronutrient cations are relatively insoluble, some more so than others. The exact pH at which precipitation occurs varies from element to element and between oxidation states of a given element. For example, the higher valence states of iron and manganese form hydroxides that are much more insoluble than their lower-valence counterparts. In any case, the principle is the same—at low pH values, the solubility of micronutrient cations is high, and as the pH is raised, their solubility and availability to plants decrease. Overliming of an acid soil often leads to deficiencies of iron, manganese, zinc, copper, and sometimes boron. Such deficiencies associated with high pH occur naturally in many of the calcareous soils of arid regions.

The general desirability of a slightly acid soil (with a pH between 6 and 7) largely stems from the fact that for most plants this pH condition allows micronutrient cations to be soluble enough to satisfy plant needs without becoming so soluble as to be toxic (see Figure 9.18). Certain plants, especially those that are native to very acid soils, have only a poor ability to take up iron and other micronutrients unless the soil is quite acid (pH about 5). Such acid-loving plants therefore become deficient in these elements when the soil pH is such that iron solubility is lowered (usually above pH 5.5; see Figure 15.8). Section 9.7 gives more specific information on the pH preferences of various plants.

Zinc availability, like that of iron, is reduced when soil pH is raised by liming. However, in addition to the pH effects, the addition of high-magnesium liming materials (e.g., dolomitic limestone) further decreases zinc availability, because zinc is tightly adsorbed to dolomite and magnesium carbonate crystal surfaces. Zinc deficiency may also be aggravated by interaction between zinc and magnesium in the plant.

Oxidation State and pH

The trace element cations, iron, manganese, nickel, and copper, occur in soils in more than one valence state. In the lower valence states, the elements are considered *reduced;* in the higher valence state they are *oxidized.* Metallic cations generally become reduced when the oxygen supply is low, as occurs in wet soils containing decomposable organic

FIGURE 15.8   The chlorotic foliage of these azaleas is a sign of iron deficiency inadvertently induced by a too-high soil pH. In this case the iron deficiency was caused by the use of marble gravel as a decorative mulch. Marble consists mostly of calcium carbonate. Rainwater percolating through the gravel mulch dissolved enough of the calcium to lime the soil below, raising the pH gradually from 5.2 to 6.0. At pH 6.0 the solubility of iron is too low for azaleas to obtain what they need. Calcium leaching from concrete walkways can have a similar effect on acid-loving vegetation growing in adjacent soil. (Photo courtesy of R. Weil)

matter. Reduction can also be brought about by organic metabolic reducing agents, such as NADPH or caffeic acid, produced by plants and microorganisms in the soil. Oxidation and reduction reactions in relation to soil drainage are discussed in Sections 7.4 and 7.6.

The changes from one valence state to another are, in most cases, brought about by microorganisms and organic matter.[4] In some cases, the organisms may obtain their energy directly from the inorganic reaction. For example, the oxidation of manganese from Mn(II) in manganous oxides (MnO) to Mn(IV) in manganic oxides (MnO_2) can be carried out by certain bacteria and fungi. In other cases, organic compounds formed by microbes or plant roots may be responsible for the oxidation or reduction. In general, high pH favors oxidation, whereas acid conditions are more conducive to reduction.

INTERACTION OF SOIL REACTION AND AERATION. At pH values common in soils, the oxidized states of iron, manganese, and copper are generally much less soluble than are the reduced states. The hydroxides (or hydrous oxides) of these high-valent forms precipitate even at low pH values and are extremely insoluble. For example, the hydroxide of trivalent ferric iron precipitates at pH values of 3.0 to 4.0, whereas ferrous hydroxide does not precipitate until a pH of 6.0 or higher is reached.

The interaction of soil acidity and aeration in determining micronutrient availability is of great practical importance. Iron, manganese, and copper are generally more available under conditions of restricted drainage or in flooded soils (Figure 15.9). Very acid soils that are poorly drained may supply toxic quantities of iron and manganese.

---

[4] For a recent review of microbial reduction of certain trace elements, see Lovley (1996).

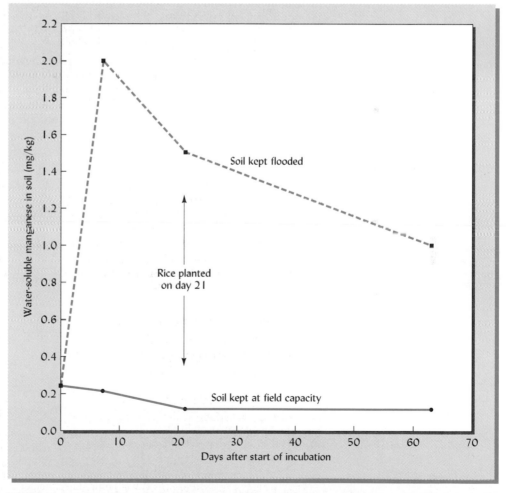

FIGURE 15.9 Effect of flooding on the amount of water-soluble manganese in soils. The data are the averages for 13 unamended Ultisol horizons with initial pH values ranging from 3.9 to 7.1. [Data from Weil and Holah (1989)]

Manganese toxicity has been reported to occur when certain high-manganese acid soils are thoroughly wetted during irrigation. Andisols (volcanic soils) with high-organic-matter melanic epipedons are known to cause manganese toxicity problems when they are wet by heavy rains. Iron toxicity is common in flooded rice paddies. Such toxicity is much less apt to occur under well-drained conditions, unless soil pH is very low.

At the high end of the soil pH range, good drainage and aeration often have the opposite effect. Well-oxidized calcareous soils are sometimes deficient in available iron, zinc, or manganese even though adequate total quantities of these trace elements are present. The hydroxides of the high-valence forms of these elements are too insoluble to supply the ions needed for plant growth. In contrast, at high soil pH values molybdenum availability may be excessively high. *Molybdenosis* is a potentially fatal disorder caused by excessive molybdenum in the diet of livestock grazing plants grown on certain very high pH soils.

There are marked differences in the sensitivity of different plant varieties to iron deficiency in soils with high pH. This is apparently caused by differences in their ability to solubilize iron immediately around the roots. Efficient varieties respond to iron stress by acidifying the immediate vicinity of the roots and by excreting compounds capable of reducing the iron to a more soluble form, with a resultant increase in its availability (Figure 15.10). It appears that most of the iron-reducing activity is concentrated in the root hairs of actively growing young roots (Figure 15.11).

Other Inorganic Reactions

Micronutrient cations interact with silicate clays in two ways. First, they may be involved in cation exchange reactions much like those of calcium or aluminum. Second, they may be tightly bound or fixed to certain silicate clays, especially the 2:1 type. Zinc, manganese, cobalt, and iron ions sometimes occur in the crystal structure of these clays. Depending on conditions, they may be released from the clays or fixed by them in a manner similar to that by which potassium is fixed (see Section 14.14). The fixation may cause serious deficiency in the case of cobalt, and sometimes zinc, because these elements are present in soil in such small quantities (see Table 15.2).

The application of large quantities of phosphate fertilizers can adversely affect the supply of some of the micronutrients. The uptake of both iron and zinc may be reduced in the presence of excess phosphates. For environmental quality reasons, as well as from a practical standpoint, phosphate fertilizers should be used in only those quantities required for good plant growth.

Lime-Induced Chlorosis

Iron deficiency in fruit trees and many other plants is encouraged by the presence of the bicarbonate ion. Bicarbonate-containing irrigation waters increase the level of this ion in some soils. In other soils, especially poorly buffered sandy soils, the problem stems

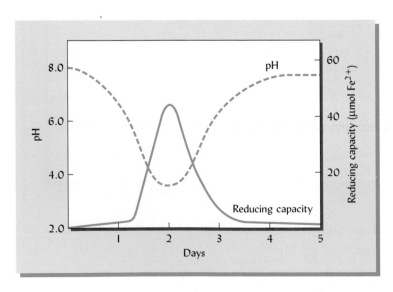

FIGURE 15.10 Response of one variety of sunflowers to iron deficiency. When the plant became stressed owing to iron deficiency, plant exudates lowered the pH and increased the reducing capacity immediately around the roots. Iron is solubilized and taken up by the plant, the stress is alleviated, and conditions return to normal. [From Marschner, et al. (1974) as reported by Olsen, et al. (1981); used with permission of *American Scientist*]

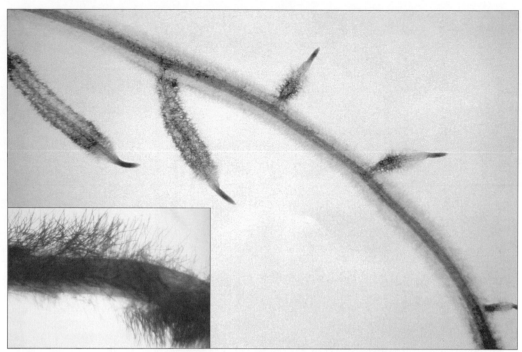

FIGURE 15.11 Reduction of iron by tomato roots is made visible by soaking live roots in a solution containing a stain that turns a dark blue color when the iron in the stain is reduced to the Fe(II) form. Note that the reduction reaction occurred only in the mature part of young branch rootlets (not in the zone just behind the root tip). The expanding root tips had not yet developed the reducing mechanisms. The close-up (inset) shows that the reduction reaction occurs only in the root hairs. The main part of the root remains undarkened. (Photos courtesy of Paul Bell, Louisiana State University Agricultural Center)

from application of more lime than is needed to reach the proper pH. The chlorosis apparently results from iron deficiency in soils with high pH because the bicarbonate ion interferes in some way with iron metabolism. This interference is coupled with the fact that iron solubility in the soil is greatly reduced with higher pH.

Organic Matter

Organic matter, organic residues, and manure applications affect the immediate and potential availability of micronutrient cations. Some organic compounds react with these cations to form water-insoluble complexes that can protect the nutrients from interactions with mineral particles that can bind them in even more insoluble forms. Other complexes provide slowly available nutrients as they undergo microbial breakdown. Organic complexes that enhance micronutrient availability are considered in Section 15.6.

Deficiencies of copper and, to a lesser extent, manganese are often found on poorly drained soils high in organic matter (e.g., peats and marshes). Zinc is also retained by organic matter, but deficiencies stemming from this retention are not common.

The microbial decomposition of organic plant residues and animal manures can result in the release of micronutrients by the same mechanisms that stimulate the release of macronutrient ions. As was the case for macronutrients such as nitrogen, however, temporary deficiencies of the trace elements may occur when the residues are added due to the assimilation of micronutrients in the bodies of the active microorganisms.

Micronutrient-enriched organic products have been used as a nutrient source on soils deficient in available trace elements. For example, composts of iron-enriched organic materials, such as forest by-products, peat, animal manures, and plant residues, have been found to be effective on iron-deficient soils.

Role of Mycorrhizae

A symbiosis between most higher plants and certain soil fungi produces mycorrhizae (fungus roots), which are far more efficient than normal plant roots in several respects. The nature of the mycorrhizal symbiosis and its importance in phosphorus nutrition

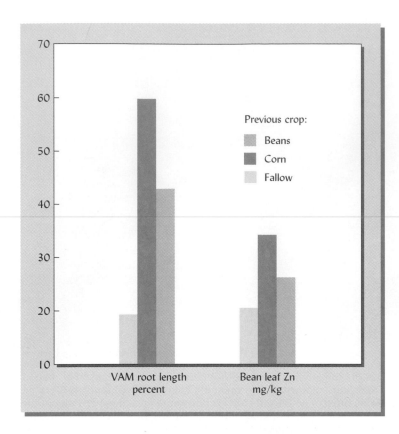

FIGURE 15.12 Effect of crop rotation on uptake of zinc and formation of mycorrhizae by beans. The bean crop, grown with furrow irrigation on an Aridisol (Calcid) in Idaho, was preceded by either a corn crop, a bean crop or a year of bare fallow. The corn-followed-by-bean rotation favored both vesicular arbuscular mycorrhizae (VAM) formation and zinc uptake by the second-year bean crop. [Data from Hamilton, et al. (1993)]

were described in Sections 11.9 and 14.3, but it is worth mentioning here that mycorrhizae have also been shown to increase plant uptake of micronutrients (Table 11.5). Crop rotations and other practices that encourage a diversity of mycorrhizal fungi may thereby improve micronutrition nutrition of plants (Figure 15.12).

Surprisingly, mycorrhizae also appear to protect plants from excessive uptake of micronutrients and other trace elements where these elements are present in potentially toxic concentrations. Seedlings of such trees as birch, pine, and spruce are able to grow well on sites contaminated with high levels of zinc, copper, nickel, and aluminum only if their roots are sheathed by ectomycorrhizae. The mycorrhizae apparently help exclude these metallic cations from the root stele and prevent long-distance transport of metal cations within the plant.

15.6 ORGANIC COMPOUNDS AS CHELATES

The cationic micronutrients react with certain organic molecules to form organometallic complexes called **chelates**. If these complexes are soluble, they increase the availability of the micronutrient and protect it from precipitation reactions. Conversely, formation of an insoluble complex will decrease the availability of the micronutrient.

A chelate (from the Greek *chele,* claw) is an organic compound in which two or more atoms are capable of bonding to the same metal atom, thus forming a ring. These organic molecules may be synthesized by plant roots and released to the surrounding soil, may be present in the soil humus, or may be synthetic compounds added to the soil to enhance micronutrient availability. In complexed form, the cations are protected from reaction with inorganic soil constituents that would make them unavailable for uptake by plants. Iron, zinc, copper, and manganese are among the cations that form chelate complexes. Two examples of an iron chelate ring structure are shown in Figure 15.13.

The effect of chelation can be illustrated with iron. In the absence of chelation, when an inorganic iron salt such as ferric sulfate is added to a calcareous soil, most of the iron is quickly rendered unavailable by reaction with hydroxide, as follows:

$$Fe^{3+} + 3OH^- \rightleftharpoons FeOOH + H_2O$$
$$\text{(available)} \qquad\qquad \text{(unavailable)}$$

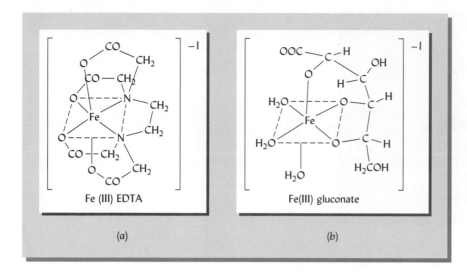

FIGURE 15.13 Structural formula for two common iron chelates, ferric ethylenediaminetetra-acetate (Fe-EDTA) (*a*) and ferric gluconate (*b*). In both chelates, the iron is protected, and yet can be used by plants. [Diagrams from Clemens, et al. (1990); reprinted by permission of Kluwer Academic Publishers]

In contrast, if the iron is chelated, it largely remains in the chelate form, which is available for uptake by plants. In this reaction, the available iron chelate reactant is favored:

$$\text{Fe chelate} + 3OH^- \rightleftharpoons \text{FeOOH} + \text{chelate}^{3-} + H_2O$$

(available)   (unavailable)

The mechanism by which micronutrients from chelates are absorbed by plants is different for different plants. Many dicots appear to remove the metallic cation from the chelate at the root surface, reducing (in the case of iron) and taking up the cation while releasing the organic chelating agent in the soil solution. Roots of certain grasses have been shown to take in the entire chelate–metal complex, reducing and removing the metallic cation inside the root cell, then releasing the organic chelate back to the soil solution (Figure 15.14). In both cases, it appears that the primary role of the chelate is to allow metallic cations to remain in solution so they can diffuse through the soil to the root. Once the micronutrient cations are inside the plant, other organic chelates (such as citrates) may be carriers of these cations to different parts of the plant.

Stability of Chelates

Some of the major synthetic chelating agents are listed in Table 15.4. Many similar chelating compounds occur naturally in the soil.

Chelates vary in their stability and therefore in their suitability as sources of micronutrients. The stability of a chelate is measured by its stability constant, which is related to the tenacity with which a metal ion is bound in the chelate. If the binding and release of a metal by a chelating agent is represented by the following reaction,

$$\text{Metal-chelate}^- \rightleftharpoons \text{metal}^{2+} + \text{chelate}^{3-}$$

then the stability constant K for the metal-chelate complex is calculated as follows, where the values in brackets are concentrations in solution:

$$K = \frac{[\text{metal-chelate}]}{[\text{metal}]\,[\text{chelate}]}$$

The larger the stability constant, the greater the tendency for the metal to remain chelated. The stability constant K for each metal-chelate complex is different and is usually expressed as the logarithm of K (see Table 15.4).

The stability constant is useful in predicting which chelate is best for supplying which micronutrient. An added metal chelate must be reasonably stable within the soil if it is to have lasting advantage. For example, the stability constant for EDDHA-Fe^{3+} is

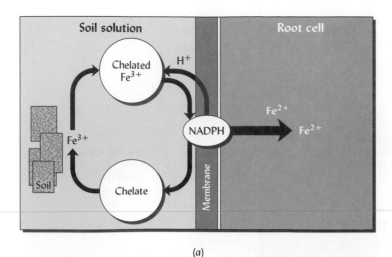

(a)

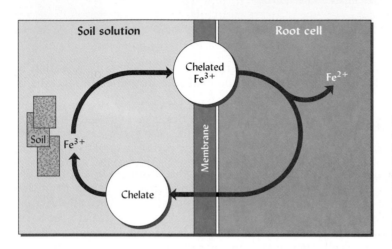

(b)

FIGURE 15.14 Two ways in which plants utilize micronutrients held in chelated form. (Top) Dicotyledonous plants
such as cucumber and peanuts produce strong reducing
agents (NADPH) that reduce iron at the outer surface of the
root membrane. They then take in only the reduced iron,
leaving the organic chelate in the soil solution where it can
complex another iron atom. (Bottom) Grass plants such as
wheat or corn apparently take the entire chelate–metal complex into their root cells. They then remove the iron, reduce
it, and return the chelate to the soil solution.

TABLE 15.4 Stability Constants K for Selected Chelating Agents and Nutrient Cations

*The stability constants are given as logarithms, so a difference of 1.0 represents
a 10-fold difference in stability. The macronutrient calcium is included because it
is usually the metallic cation with by far the greatest activity in the soil solution.
As such, calcium competes with micronutrient cations for binding sites in
chelating agents. The relative cost of chelating agents is also given.*

| Chelating agent | Log K^a | | | | | | Relative cost[b] |
	Fe^{3+}	Fe^{2+}	Zn^{2+}	Cu^{2+}	Mn^{2+}	Ca^{2+}	
EDTA	25.0	14.27	14.87	18.70	13.81	11.0	4.4
EDDHA	33.9	14.3	16.8	23.94	—	7.2	43
HEEDTA	19.6	12.2	14.5	17.4	10.7	8.0	5.5
Citrate	11.2	4.8	4.86	5.9	3.7	4.68	—
Gluconate	37.2	1.0	1.7	36.6	—	1.21	1.0

[a] $K = \dfrac{[\text{metal-chelate}]}{[\text{metal}]\,[\text{chelate}]}$

[b] Cost per kg of iron relative to the cost of iron gluconate. Based on 1989 prices. [From Clemens, et al. (1990)]

33.9, but that for EDDHA-Zn^{2+} is only 16.8. We can therefore predict that if EDDHA-Zn were added to a soil, the Zn in the chelate would be rapidly and almost completely replaced by Fe^{3+} from the soil, leaving the Zn in the unchelated form and subject to precipitation:

$$\text{Zn chelate}^- + \text{Fe}^{3+} \rightleftharpoons \text{Fe chelate} + \text{Zn}^{2+}$$

Since the iron chelate is more stable than its zinc counterpart, the reaction goes to the right, and the released zinc ion is subject to reaction with the soil. Similarly, calcium can replace micronutrients from chelates. Even though the stability constant for Ca chelates is generally low, calcium often replaces micronutrients from chelates in practice, because the concentration of Ca^{2+} in the soil solution is far greater than the concentrations of micronutrients.

It should not be inferred that only iron chelates are effective. The chelates of other micronutrients, including zinc, manganese, and copper, have been used successfully to supply these nutrients (Figure 15.15). Apparently, replacement of other micronutrients in the chelates by iron from the soil is sufficiently slow to permit absorption by plants of the other added micronutrients. Also, because foliar spray and banded applications are often used to supply zinc and manganese, the possibility of reaction of these elements in chelates with iron and calcium in the soil can be reduced or eliminated.

The use of synthetic chelates in industrial countries is substantial, in spite of the fact that they are quite expensive. They are used primarily to ameliorate micronutrient deficiencies of fruit trees and ornamentals. Although chelates may not replace the more conventional methods of supplying most micronutrients, they offer possibilities in special cases. Some of the chelators used in micronutrient fertilizers, such as gluconate, are naturally occurring and can supply certain micronutrients much more economically than can the more expensive aminopolycarboxylate compounds (e.g., EDDHA) listed in Table 15.4. Agricultural and chemical research will likely continue to increase the opportunities for the effective use of chelates.

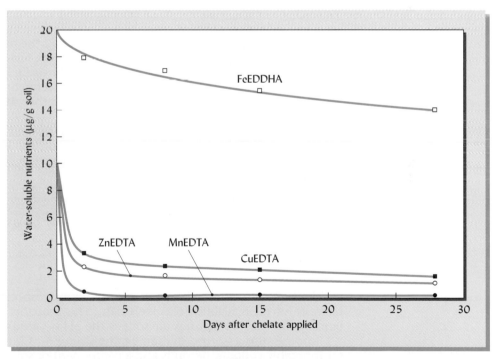

FIGURE 15.15   Average reduction in water solubility of four chelated micronutrients when incubated with four calcareous soils with pH higher than 8. The stability constants listed in Table 15.4 explain the behavior of the four micronutrient chelates shown here. The iron chelate was most stable in these soils, that with manganese the least. Because of the small quantities needed by plants, even the copper and zinc chelates would likely provide adequate nutrients for plant absorption. [Drawn using data from Ryan and Hariq (1983)]

In addition, selection of plant varieties whose roots produce their own chelating agents, and practices that encourage the production of natural chelating agents from decomposing organic matter may take increasing advantage of the chelation phenomena to improve micronutrient fertility of soils.

15.7   FACTORS INFLUENCING THE AVAILABILITY OF THE MICRONUTRIENT ANIONS

Unlike the cations needed in trace quantities by plants, the anion micronutrients seem to have relatively little in common with each other. Chlorine, molybdenum, and boron are quite different chemically, so little similarity would be expected in their reaction in soils.

Chlorine

Chlorine is absorbed in larger quantities by most crop plants than any of the micronutrients except iron. Most of the chlorine in soils is in the form of the chloride ion, which leaches rather freely from humid-region soils. In semiarid and arid regions, a higher concentration might be expected, with the amount reaching the point of salt toxicity in some of the poorly drained saline soils. In most well-drained areas, however, one would not expect a high chlorine content in the surface of arid-region soils.

Except where toxic quantities of chlorine are found in saline soils, there are no common soil conditions that reduce the availability and use of this element. Accretions of chlorine from the atmosphere, along with those from fertilizer salts such as potassium chloride, are sufficient to meet most crop needs. However, crop responses to chloride additions have been noted for winter wheat and are thought to be due to reduction of diseases such as "take all." Tropical palms, adapted to growth in coastal soils where ocean spray contributes much chlorine, sometimes show chlorine deficiency if they are grown on inland soils with relatively low chlorine levels. High chlorine levels can reduce the growth and quality of some plants, especially those of the *Solonacae* family (tomatoes and the like).

Boron

Boron is one of the most commonly deficient of all the micronutrients. The availability of boron is related to the soil pH, this element being most available in acid soils. While it is most available at low pH, boron is also rather easily leached from acid, sandy soils. Therefore, although deficiency of boron is relatively common on acid, sandy soils, it occurs because of the low supply of total boron rather than because of low availability of the boron present.

Soluble boron is present in soils mostly as boric acid [$B(OH)_3$] or as $B(OH)_4^-$. These compounds can exchange with the OH groups on the edges and surfaces of variable charge clays such as kalonite, and especially with the oxides of iron and aluminum. Reactions such as the following take place:

The boron so adsorbed is quite tightly bound, especially between pH 7 and 9, the range of lowest availability of this element. This probably accounts for the lime-induced boron deficiency noted on some soils as the pH is raised to pH 7 and above.

Boron is also adsorbed by humus, the strength of binding being even greater than for inorganic colloids. Boron is also a component of soil organic matter that is released by microbial mineralization. Consequently, organic matter serves as a major reservoir for boron in many soils and exerts considerable control over the availability of this nutrient. The adsorption of boron to organic and inorganic colloids increases with increasing pH (Figure 15.16).

Boron availability is impaired by long dry spells, especially following periods of optimum moisture conditions. This may be related to the fact that boron is generally taken

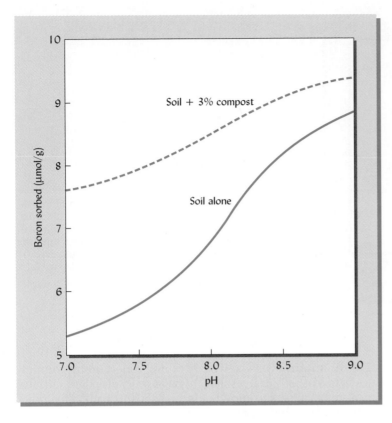

Effect of organic matter (added as compost in this experiment) and soil pH on adsorption of boron. Along with clays, humic compounds in decomposed organic matter adsorb boron. The adsorption capacities from both clays and humic compounds increase with rising pH in the neutral to alkaline range. While adsorption lowers the boron concentration in the soil solution, and hence its availability for plant uptake, it also protects boron from leaching loss. The soil in this study was a Calcic Haploxeralf. [Redrawn from Yermiyahu, et al. (1995)]

up with the transpiration stream of water rather than by active ion transport as is the case for uptake of most other nutrients. Dry conditions may also reduce the mineralization of organically held boron during the dry periods. Boron deficiencies are also common in calcareous Aridisols and in neutral to alkaline soils with a high pH.

Molybdenum

Soil pH is the most important factor influencing the availability and plant uptake of molybdenum. The following equations show the forms of this element present at low and high soil pH:

$$H_2MoO_4 \underset{+H^+}{\overset{+OH^-}{\rightleftharpoons}} HMoO_4^- + H_2O$$

$$HMoO_4^- \underset{+H^+}{\overset{+OH^-}{\rightleftharpoons}} MoO_4^{2-} + H_2O$$

At low pH values, the molybdenum is adsorbed by silicate clays and, more especially, by oxides of iron and aluminum through ligand exchange with hydroxide ions on the surface of the collodial particles. Reactions such as the following occur:

$$\begin{array}{c} \diagdown Al \diagup \\ \diagup \quad \diagdown \end{array} OH + HMoO_4^- \rightarrow \begin{array}{c} \diagdown Al \diagup \\ \diagup \quad \diagdown \end{array} O - Mo \overset{O}{\underset{O}{-}} O + H_2O$$

Surface   Soluble Mo   Adsorbed Mo
hydroxyls

The liming of acid soils will usually increase the availability of molybdenum (Figure 15.17). The effect is so striking that some researchers, especially those in Australia and New Zealand, argue that the primary reason for liming very acid soils is to supply molybdenum. Furthermore, in some instances 30 g or so of molybdenum added to acid soils has given about the same increase in the yield of legumes as has the application of several megagrams of lime.

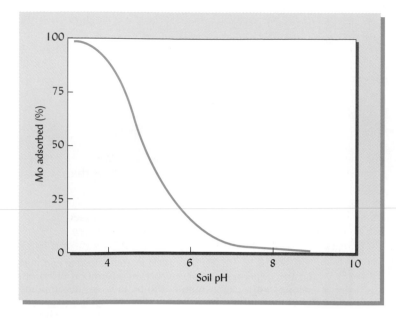

FIGURE 15.17 Effect of pH on the adsorption of molybdenum on a Hesperia coarse, loamy sand from California. The high adsorption at low pH values could result in molybdenum deficiency. Likewise, at high pH values the molybdenum is not adsorbed and is free to be taken up by plants, sometimes at toxic levels. [Redrawn from Goldberg, et al. (1996)]

The phosphate anion seems to improve the availability of molybdenum by competing with the latter for sorption sites on soil surfaces. For this reason, molybdate salts are often applied along with phosphate carriers to molybdenum-deficient soils. This practice apparently encourages the uptake of both elements and is a convenient way to add the extremely small quantities of molybdenum required. Legume seeds coated in a mixture of superphosphate and sodium molybdate have also been used successfully to improve the grazing quality of acid-soil range and savanna lands.

A second common anion, the sulfate ion, seems to have the opposite effect on plant utilization of molybdenum. Sulfate reduces molybdenum uptake and seems to compete with molybdenum at functional sites on plant metabolic compounds.

Selenium

Managing selenium in soils and plants is important to animal and human health in many areas of the world. This is not simple, however, since selenium is found in nature in four major solid forms—selenate (SeO_4^{2-}), selenite (SeO_3^{2-}), elemental Se (Se^0), and selenide (Se^{2-})—. Selenium also exists in volatile compounds, such as dimethyl selenide and dimethyl diselenide, that can be released as gases to the atmosphere.

Under acid conditions (pH 4.5 to 6.5), selenite is usually dominant. This form is only slowly available since it is adsorbed by iron oxides. Selenites may be added to alleviate selenium deficiencies, however, since they will not likely induce selenium toxicities. At higher pH values (7.5 to 8.5) and in well-aerated soils, the more soluble selenate dominates and often gives rise to such toxicities.

Under anaerobic conditions, soil microorganisms can reduce the selenate and selenite to solid selenide and elemental selenium forms, both of which are quite insoluble.[5] Advantage can be taken of such reduction to remove toxic levels of soluble selenium from soil and water, a promising process of bioremediation. Selenium-laden waters are held in well-vegetated ponds or swamplands, the reduction takes place, and the insoluble selenium forms settle out of the water to the pond bottoms. However, if the holding area is drained, aerobic conditions prevail, permitting the oxidation of the selenium once again to the more soluble selenate ion that can cause toxicities. Furthermore, even in the ponds, bottom-feeding plants can absorb some selenium from the reduced, slowly available forms. This places selenium back in the food chain, to the disadvantage of both wildlife and domestic animals that may consume the plants.

---

[5] Recent research has shown that nonbiological reduction of selenates also occurs in the presence of ferrous/ferric hydroxide [$Fe(II)_a Fe(III)_b (OH)_{12}$] $X\cdot 3H_2O$; where $a = 1$–4, $b = 1$–2, and X = an anion, such as SO_4^{2-}), a compound known as *green rust,* commonly found in wet soils. See Myneni, et al. (1997).

Microbial reduction also produces volatile selenium compounds (e.g., dimethyl selenide) that can be released to the atmosphere. The gases are relatively nonphytotoxic and can move quickly away from the high-selenium site to other locations that may even be selenium deficient.[6] For these reasons, the production of volatile selenium compounds is also being considered as a means of reducing toxic levels of selenium in water and soils.

15.8 NEED FOR NUTRIENT BALANCE

Nutrient balance among the trace elements is as essential as, but even more difficult to maintain than, macronutrient balance. Some of the plant enzyme systems that depend on micronutrients require more than one element. For example, both manganese and molybdenum are needed for the assimilation of nitrates by plants. The beneficial effects of combinations of phosphates and molybdenum have already been discussed. Apparently, some plants need zinc and phosphorus for optimum use of manganese. The use of boron and calcium depends on the proper balance between these two nutrients. A similar relationship exists between potassium and copper and between potassium and iron in the production of good-quality potatoes. Copper utilization is favored by adequate manganese, which in some plants is assimilated only if zinc is present in sufficient amounts. Of course, the effects of these and other nutrients will depend on the specific plant being grown, but the complexity of the situation can be seen from the examples cited.

Antagonism and Synergism

Some enzymatic and other biochemical reactions requiring a given micronutrient may be poisoned by the presence of a second trace element in toxic quantities. Other negative effects occur because one element competes with or otherwise reduces uptake of a second element by the plant root. On the other hand, a good supply of a certain nutrient may enhance the utilization of a second nutrient element in what is termed a *synergistic effect*. Some of these interactions are summarized in Table 15.5.

Some of the *antagonistic* effects may be used effectively in reducing toxicities of certain of the micronutrients. For example, copper toxicity of citrus groves caused by residual copper from fungicidal sprays may be reduced by adding iron and phosphate fertilizers. Sulfur additions to calcareous soils containing toxic quantities of soluble molybdenum may reduce the availability, and hence the toxicity, of molybdenum. The hyperaccumulation of phosphorus, manganese, and magnesium in zinc-deficient plants is an example of the complex interactions among essential elements as they influence plant nutrition.

---

[6] Haygarth, et al. (1995) attributed the source of 30 to 50% of the selenium taken up by pasture plants in one experiment to the atmosphere.

TABLE 15.5   Some Antagonistic (Negative) and Synergistic (Positive) Effects of Other Nutrients on Micronutrient Utilization by Plants[a]

The occurrence of so many interactions emphasizes the need for balance among all nutrients and avoidance of excess application of any particular nutrient.

| Micronutrient | Elements decreasing utilization | | Elements increasing utilization | |
	Soil and root surface reactions	Plant metabolic reactions	Soil and root surface reactions	Plant metabolic reactions
Fe	B, Cu, Zn, Mo, Mn	Mn, Mo, P, S, Zn	B, Mo	
Mn	Fe, B	Fe	B	
Zn	Mg, Cu, B, Fe, P	Fe, N	N, B	Fe, Mg
Cu	B, Zn, Mo	P, N	B	
B	Ca, K			N
Mo	S, Cu	S	P	P
Ni	Ca, Fe	Fe, Zn		

[a] Summarized from many sources.

These examples of nutrient interactions, both beneficial and detrimental, emphasize the highly complicated nature of the biological transformations in which micronutrients are involved. The total land area on which unfavorable nutrient balances require special micronutrient treatment is increasing as soils are subjected to more intensive cropping methods.

15.9 SOIL MANAGEMENT AND MICRONUTRIENT NEEDS

Although the characteristics of each micronutrient are quite specific, some generalizations with respect to management practices are possible.

In seeking the cause of plant abnormalities, one should keep in mind the conditions under which micronutrient deficiencies or toxicities are likely to occur. Sandy soils, mucks, and soils having very high or very low pH values are prone to micronutrient deficiencies. Areas of intensive cropping and heavy macronutrient fertilization may be deficient in the micronutrients.

Changes in Soil Acidity

In very acid soils, one might expect toxicities of iron and manganese and deficiencies of phosphorus and molybdenum. These can be corrected by liming and by appropriate fertilizer additions. Calcareous soils may have deficiencies of iron, manganese, zinc, and copper and, in a few cases, a toxicity of molybdenum.

No specific statement can be made concerning the pH value most suitable for all the elements. However, medium-textured soils generally supply adequate quantities of micronutrients when the soil pH is held between 6 and 7. In sandy soils, a somewhat lower pH may be justified because the total quantity of micronutrients is low, and even at pH 6.0, some cation deficiencies may occur. It is important to be on guard to recognize inadvertent increases in soil pH, such as those occurring from application of lime-stabilized sewage sludge (see Chapter 16) or leaching of calcareous gravel and pavements (see Figure 15.8).

Soil Moisture

Drainage and moisture control can influence micronutrient solubility in soils. Improving the drainage of acid soils will encourage the formation of the oxidized forms of iron and manganese. These are less soluble and, under acid conditions, less toxic than the reduced forms. Flooding a soil will favor the reduced forms of iron and manganese, which are more available to growing plants (Figure 15.18). Excessive moisture at high pH values can have the opposite effect since poorly drained soils have a high carbon dioxide concentration, encouraging the formation of bicarbonate ions, which reduces iron availability. Poor drainage also increases the availability of molybdenum in some alkaline soils to the point of producing plants with toxic levels of this element.

Fertilizer Applications

Micronutrient deficiencies are relatively unlikely to be a problem in most soils to which plant residues are returned and organic amendments, such as animal manure or sewage sludge, are regularly applied. Animal manures applied at normal rates sufficient to supply macronutrient needs carry enough copper, zinc, manganese, and iron to supply a major portion of micronutrient needs as well (see Table 16.9). In addition, the chelates produced from manure enhance the availability of these micronutrients.

Nonetheless, the most common management practice to overcome micronutrient deficiencies (and some toxicities) is the application of commercial fertilizers. Examples of fertilizer materials applied to supply micronutrients are shown in Table 15.6. The materials are most commonly applied to the soil, although foliar sprays and even seed treatments can be used. Foliar sprays of dilute inorganic salts or organic chelates are more effective than soil treatments where high soil pH and other factors render the soil-applied nutrients unavailable.

For soil applications, about one-half to one-fourth as much fertilizer is needed if the application is banded rather than broadcast (see Figure 16.25). About one-fifth to

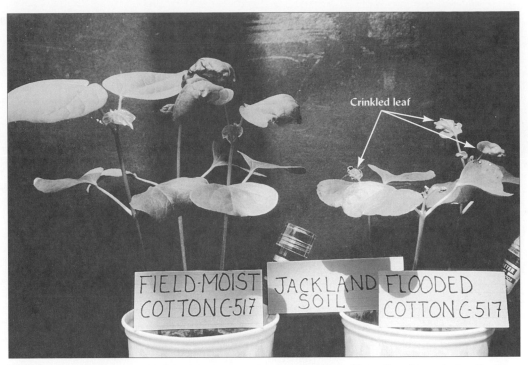

FIGURE 15.18 Manganese toxicity in young cotton plants as affected by soil moisture during the three weeks prior to planting the cotton seed. The crinkled-leaf symptom and stunted size resulted from toxic levels of manganese in the soil that had been flooded prior to planting. Both pots were maintained at field-capacity water content from sowing to harvest. The Jackland soil (Aquic Hapludult) used is high in manganese-containing minerals and has a pH of approximately 5.8. [From Weil, et al. (1977)]

one-tenth as much fertilizer is needed if the material is sprayed on the plant foliage. Foliar application may, however, require repeated applications in a single year. Treating seeds with small dosages (20 to 40 g/ha) of molybdenum has had satisfactory results on molybdenum-deficient acid soils. Typical rates of application to soil are given in Table 15.7.

The micronutrients can be applied to the soil either as separate materials or incorporated in standard macronutrient carriers. Unfortunately, the solubilities of copper, iron, manganese, and zinc can be reduced by such incorporation, but boron and molybdenum remain in reasonably soluble condition. Liquid macronutrient fertilizers containing polyphosphates encourage the formation of complexes that protect added micronutrients from adverse chemical reactions. In effect, polyphosphates in fertilizer solutions act as chelating agents for micronutrients. High-surface-area pitted glasslike beads, called **frits**, are manufactured with boron, copper, zinc, and other micronutrients incorporated into the glass. These fritted materials slowly release their micronutrients as

TABLE 15.6   A Few Commonly Used Fertilizer Materials That Supply Micronutrients

Micronutrient	Commonly used fertilizers		Nutrient content, %
Boron	Borax	$Na_2B_4O_7 \cdot 10H_2O$	11
	Sodium pentaborate	$Na_2B_{10}O_{16} \cdot H_2O$	18
Copper	Copper sulfate	$CuSO_4 \cdot 5H_2O$	25
Iron	Ferrous sulfate	$FeSO_4 \cdot 7H_2O$	19
	Iron chelates	NaFeEDDHA	6
Manganese	Manganese sulfate	$MnSO_4 \cdot 3H_2O$	26–28
	Manganese oxide	MnO	41–68
Molybdenum	Sodium molybdate	$Na_2MoO_4 \cdot 2H_2O$	39
	Ammonium molybdate	$(NH_4)_6Mo_7O_{24} \cdot 4H_2O$	54
Zinc	Zinc sulfate	$ZnSO_4 \cdot H_2O$	35
	Zinc oxide	ZnO	78
	Zinc chelate	$Na_2ZnEDTA$	14

Selected from Murphy and Walsh (1972).

TABLE 15.7 Plants Known to Be Especially Susceptible or Tolerant to, and Soil Conditions Conducive to, Micronutrient Deficiencies

Plants which are most susceptible to deficiency of a micronutrient often have a relatively high requirement for that nutrient and may be relatively tolerant to levels of that nutrient that would be high enough to cause toxicity to other plants.

Micronutrient	Common range in rates recommended for soil application[a], kg/ha	Plants most commonly deficient (high requirement or low efficiency of uptake)	Plants rarely deficient (low requirement or high efficiency of uptake)	Soil conditions commonly associated with deficiency
Iron	0.5–10.0	Blueberries, azaleas, roses, holly, grapes, nut trees, maple, bean, sorghum, oaks	Wheat, alfalfa, sunflower, cotton	Calcareous, high pH, waterlogged alkaline soils
Manganese	2–20	Peas, oats, apple, sugar beet, raspberry, citrus	Cotton, soybean, rice, wheat	Calcareous, high pH, drained wetlands, low organic matter, sandy soils
Zinc	0.5–20	Corn, onion, pines, soybeans, beans, pecans, rice, peach, grapes	Carrots, asparagus, safflower, peas, oats, crucifers, grasses	Calcareous soils, acid, sandy soils, high phosphorus
Copper	0.5–15	Wheat, corn, onions, citrus, lettuce, carrots	Beans, potato, peas, pasture grasses, pines	Histosols, very acid, sandy soils
Boron	0.5–5	Alfalfa, cauliflower, celery, grapes, conifers, apples, peanut, beets, rapeseed, pines	Barley, corn, onion, turf grass, blueberry, potato, soybean	Low organic matter, acid, sandy soils, recently limed soils, droughty soils, soils high in 2:1 clays
Molybdenum	0.05–0.5	Alfalfa, sweet clover, crucifers (broccoli, cabbage, etc.), citrus, most legumes	Most grasses	Acid sandy soils, highly weathered soils with amorphous Fe and Al

[a] The lower end of each range is typical for banded application; the higher end is typical for broadcast applications.

the glass weathers in the soil. This slow release avoids some of the problems of precipitation and sorption that might otherwise occur, particularly in an alkaline soil.

Economic responses to micronutrients are becoming more widespread as intensity of plant production increases. For example, responses of fruits, vegetables, and field crops to zinc and iron applications are common in areas with neutral to alkaline soils. Even on acid soils, deficiencies of these elements are increasingly encountered. Molybdenum, which has been used for some time for forage crops and for cauliflower and other vegetables, has received attention in recent years for forest nurseries and soybeans, especially on acid soils. These examples, along with those from muck areas and sandy soils where micronutrients have been used for decades, illustrate the need for these elements if optimum yields are to be maintained.

A soil will often produce a micronutrient deficiency in some plants but not in others. Plant species, and varieties within species, differ widely in their susceptibility to micronutrient deficiency or toxicity. Table 15.7 provides examples of plant species known to be particularly susceptible or tolerant to several micronutrient deficiencies.

Marked differences in crop needs for micronutrients make fertilization a problem where rotations are being followed. On general-crop farms, vegetables are sometimes grown in rotation with small grains and forages. If the boron fertilization is adequate for a vegetable crop such as red beets, or even for alfalfa, the small-grain crop grown in the rotation may show toxicity damage. These facts emphasize the need for specificity in determining crop nutrient requirements and for care in meeting these needs.

Plant Selection and Breeding

Differences in the ability of different plant species or of individual plants within a species to accumulate micronutrients suggest that plant selection and breeding may be major tools in overcoming deficiencies and toxicities of these elements. For example, some species of *Astragalus* and *Stanleya* accumulate about five times as much selenium as do such cereal crops as wheat and rye. Similarly, black gum (*Nyassa sylvatica*) is an accumulator of cobalt. *Bioremediation* may take advantage of such accumulators in removing toxic quantities of some trace elements from polluted soil (see Section 18.6). Similar differences exist in genetic tolerance of nutrient deficiencies. Some recent research suggests that plant breeding and biotechnology may be used to select and/or create plant varieties that more efficiently absorb trace elements or that tolerate their toxic levels in soils. This highlights the need for close collaboration among plant and soil scientists in micronutrient management.

Micronutrient Availability

Major sources of micronutrients and the general reactions that make them available to higher plants and microorganisms are summarized in Figure 15.19. Original and secondary minerals are the primary sources of these elements, while the breakdown of organic forms releases ions to the soil solution. The micronutrients are used by higher plants and microorganisms in important life-supporting processes. Removal of nutrients in crop or timber harvest reduces the soluble ion pool, and it may need to be replenished with manures or chemical fertilizers to avoid nutrient deficiencies.

Worldwide Management Problems

Micronutrient deficiencies have been diagnosed in most areas of crop production and some forest areas of the United States and Europe. However, in some developing countries, particularly in the tropics, the extent of these deficiencies is much less well known. Limited research suggests that there may be large areas with deficiencies or toxicities of one or more of these elements. Irrigation schemes that bring calcareous soils in desert areas under cultivation are often plagued with deficiencies of iron, zinc, copper, and manganese. Copper and zinc deficiency have been noted in cereal crops on highly weathered soils in the humid and subhumid tropics. As macronutrient deficiencies are addressed and yields are increased, more micronutrient deficiencies will undoubtedly come to the fore. One encouraging factor is the small quantity of micronutrients usually needed, making transportation of fertilizer materials to remote parts of the world much less of an expense than is the case for macronutrient fertilizers. The management principles established for the economically developed countries should also be helpful in alleviating micronutrient deficiencies in less-developed countries.

15.10 CONCLUSION

Micronutrients are becoming increasingly important to world agriculture as crop removal of these essential elements increases. Soil and plant tissue tests confirm that these elements are limiting crop production over wide areas and suggest that attention to them will likely increase in the future.

Micronutrient deficiencies are due not only to low contents of these elements in soils but more often to their unavailability to growing plants. They are adsorbed by

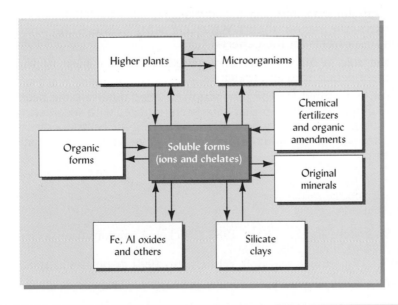

FIGURE 15.19 Diagram of soil sources of soluble forms of micronutrients and their utilization by plants and microorganisms.

inorganic constituents, such as Fe, Al, oxides, and form complexes with organic matter, some of which are only sparingly available to plants. Other such organic complexes, known as *chelates,* protect some of the micronutrient cations from inorganic adsorption and make them available for plant uptake.

Toxicities of micronutrients retard both plant and animal growth. Removing these elements from soil and water, or rendering them unavailable for plant uptake, is one of the challenges facing soil and plant scientists.

In most cases, soil-management practices that avoid extremes in soil pH, that optimize the return of plant residues and animal manures, and that promote chelate production by actively decomposing organic matter will minimize the risk of micronutrient deficiencies or toxicities. But increasingly noted are situations where micronutrient problems can be most practically solved by the application of micronutrient fertilizers. Such materials are becoming common components of fertilizers for field and garden use and will likely become even more so in the future.

STUDY QUESTIONS

1. During a year's time some 250 kg nitrogen and only 30 g molybdenum have been taken up by the trees growing on a hectare of land. Would you therefore conclude that the nitrogen was more essential for the tree growth? Explain.

2. Since only small quantities of micronutrients are needed for normal plant growth, would it be wise to add large quantities of these elements now to satisfy future plant needs? Explain.

3. Iron deficiency is common for peaches and other fruits grown on highly alkaline irrigated soils of arid regions, even though these soils are quite high in iron. How do you account for this situation, and what would you do to alleviate the difficulty?

4. How do Fe and Al oxides effect the availability of Mo and B in soils? Explain.

5. Soybeans growing on a recently limed soil show evidence of a deficiency of a nutrient, thought by some to be molybdenum. Do you agree with this diagnosis? If not, what is your explanation?

6. What are *chelates,* how do they function, and what are their sources?

7. The addition of only 1 kg/ha of a micronutrient to an acid soil on which lime-loving cauliflower was being grown gave considerable growth response. Which of the micronutrients would it likely have been? Explain.

8. Two Aridisols, both at pH 8, were developed from the same parent material, one having restricted drainage, the other being well drained. Plants growing on the well-drained soils showed iron deficiency symptoms while those on the less-well-drained soil did not. What is the likely explanation for this?

9. Animals, both domestic and wild, are adversely affected by deficiencies and toxicities of two of the micronutrients. Which elements are these, and what are the conditions responsible for their effects?

10. Discuss the role plant breeders and geneticists might play in managing micronutrient deficiencies and toxicities.

11. Since boron is required for the production of good-quality table beets, some companies purchase only beets that have been fertilized with specified amounts of this element. Unfortunately, an oat crop following the beets does very poorly compared to oats following unfertilized beets. Give possible explanations for this situation.

REFERENCES

Brown, P. H., R. M. Welch, and E. E. Cary. 1987. "Nickel: A micronutrient essential for higher plants," *Plant Physiol.,* **85**:801–803.

Clemens, D. F., B. M. Whitehurst, and G. B. Whitehurst. 1990. "Chelates in agriculture," *Fertilizer Research,* **25**:127–131.

Goldberg, S., H. S. Forster, and C. L. Godfrey. 1996. "Molybdenum adsorption on oxides, clay minerals, and soils," *Soil Sci. Soc. Amer. J.,* **60:**425–432.

Hamilton, M. A., D. T. Westermann, and D. W. James. 1993. "Factors affecting zinc uptake in cropping systems," *Soil Sci. Soc. Amer. J.,* **57:**1310–1315.

Haygarth, P. M., A. F. Harrison, and K. C. Jones. 1995. "Plant selenium from soil and the atmosphere," *J. Environ. Qual.,* **24:**768–771.

Lindsay, W. L. 1972. "Inorganic phase equilibria of micronutrients in soils," in J. J. Mortvedt, P. M. Giordano, and W. L. Lindsay (eds.), *Micronutrients in Agriculture* (Madison, Wis.: Soil Sci. Soc. Amer.).

Lovley, D. R. 1996. "Microbial reduction of iron, manganese and other metals," *Advances in Agronomy,* **54:**175–231.

Marschner, H., A. Kalisch, and V. Romheld. 1974. "Mechanism of iron uptake in different plant species," *Proc. 7th Int. Colloquium on Plant Analysis and Fertilizer Problems,* Hanover, West Germany.

Marschner, H. 1995. *Mineral Nutrition of Higher Plants,* 2d ed. (New York: Academic Press).

Mengel, K., and E. A. Kirkby. 1987. *Principles of Plant Nutrition,* 4th ed. (Bern, Switzerland: International Potash Institute).

Mortvedt, J. J., F. R. Cox, L. M. Shuman, and R. M. Welch (eds.). 1991. *Micronutrients in Agriculture.* SSSA Book Series, no. 4. (Madison, Wis.: Soil Sci. Soc. Amer.).

Murphy, L. S., and L. M. Walsh. 1972. "Correction of micronutrient deficiencies with fertilizers," in J. J. Mortvedt, P. M. Giordano, and W. L. Lindsay (eds.), *Micronutrients in Agriculture.* (Madison, Wis.: Soil Sci. Soc. Amer.).

Myneni, S. C. B., T. K. Tokunaga, and G. E. Brown. 1997. "Abiotic selenium redox transformations in the presence of Fe (II,III) oxides," *Science,* **278:**1106–1109.

Olsen, R. A., R. B. Clark, and J. H. Bennett. 1981. "The enhancement of soil fertility by plant roots," *Amer. Scientist,* **69:**378–384.

Ryan, J., and S. N. Hariq. 1983. "Transformation of incubated micronutrients in calcareous soils," *Soil Sci. Soc. Amer. Jour.,* **47:**806–810.

Sillanpaa, M. 1982. *Micronutrients and the Nutrient Status of Soils: A Global Study.* (Rome: U.N. Food and Agricultural Organization).

Stevenson, F. J. 1986. *Cycles of Soil Carbon, Nitrogen, Sulfur and Micronutrients.* (New York: Wiley).

Viets, F. J., Jr. 1965. "The plants' need for and use of nitrogen," in *Soil Nitrogen (Agronomy,* no. 10). (Madison, Wis.: Amer. Soc. Agron.).

Weil, R. R., C. D. Foy, and C. A. Coradetti. 1997. "Influence of soil moisture regimes on subsequent soil manganese availability and toxicity in two cotton genotypes," *Agron. J.,* **89:**1–8.

Weil, R. R., and Sh. Sh. Holah. 1989. "Effects of submergence on the availability of micronutrients in three Ultisols," *Plant and Soil,* **114:**147–157.

Welch, R. M. 1995. "Micronutrient nutrition of plants," *Critical Reviews in Plant Science,* **14**(1):49–82.

Yermiyahu, U., R. Keren, and Y. Chen. 1995. "Boron sorption by soil in the presence of composted organic matter," *Soil Sci. Soc. Amer. J.,* **59:**405–409.

16

PRACTICAL NUTRIENT MANAGEMENT

> *Using scientific knowledge and ecological wisdom we can manage the earth . . .*
> —RENE J. DUBOS, HUMANIZING THE EARTH

As stewards of the land, soil managers must keep nutrient cycles in balance. By doing so they maintain the soil's capacity to supply the nutritional needs of plants and, indirectly, of us all. While undisturbed ecosystems generally need no intervention from soil managers or from anyone else, few ecosystems are so undisturbed. More often, human hands have directed the output of the ecosystem for human ends. Forests, farms, fairways, and flower gardens are ecosystems modified to provide us with paper, food, recreational opportunities, and aesthetic satisfaction. By their very nature, managed ecosystems need management.

In managed ecosystems, nutrient cycles become unbalanced through increased removals (e.g., harvest of timber and crops), through increased system leakage (e.g., leaching and runoff), through simplification (e.g., monoculture, be it of pine tree or sugarcane), and through increased demands for rapid plant growth (whether the soil is naturally fertile or not). Hence, the land manager is, of necessity, also a nutrient manager.

In this chapter we will discuss the goals of nutrient management and describe the tools that may help achieve those goals. These tools include methods of enhancing nutrient recycling, as well as sources of additional nutrients that can be applied to soils or plants. We will learn how to diagnose nutritional disorders of plants and correct soil fertility problems. Building on the foundation of principles set out in earlier parts of this book, this chapter contains much practical information on the profitable use and conservation of organic and inorganic nutrient resources in producing abundant, high-quality plant products and in maintaining the quality of both the soil and the rest of the environment.

16.1 GOALS OF NUTRIENT MANAGEMENT[1]

Nutrient management is one aspect of a holistic approach to managing soils in the larger environment. The goals for managing nutrients will influence the specific practices adopted. In general, nutrient management aims to achieve four broad, interrelated goals: (1) cost-effective production of high-quality plants, (2) efficient use and conservation of nutrient resources, (3) maintenance or enhancement of soil quality, and (4) protection of the environment beyond the soil. Each of these goals will now be briefly considered.

---

[1] For an overview of issues and advances in nutrient management for agriculture and environmental quality, see Magdoff, et al. (1997). A standard textbook on management of agricultural soil fertility and fertilizers is Tisdale, et al. (1993).

Plant Production

People engage in three primary types of plant production: (1) *agriculture,* (2) *forestry,* and (3) *ornamental landscaping.* Agriculturists range from small-scale indigenous farm families or home gardeners who produce only for their own use to large-scale farmers whose primary goal is to make a profit from the plants and animals they produce. Regardless of the scale, the main nutrient-management goal for agriculture is to increase plant yields, thereby helping the subsistent farmers to feed their families and the commercial farmers to enhance their incomes. Unfortunately, the time frame farmers in both groups use to judge the success or failure of their management schemes is commonly only three to six months, the period of growth of the crops they are producing. This is nothing like the span of several human generations that may be needed to fully evaluate the effectiveness of the practices they choose to use.

In forestry, the principal plant product may be measured in terms of volume of lumber or paper produced. In these instances, nutrient management aims to enhance rate of growth so that the time between investment and payoff can be minimized. The survival rate of tree seedlings is also important. Wildlife habitat and recreational values may also be primary or secondary products. The time frame in forestry, measured in decades or even centuries, tends to limit the intensity of nutrient-management interventions that can be profitably undertaken.

When soils are used for ornamental landscaping purposes, the principal objective is to produce quality, aesthetically pleasing plants. Whether the plants are produced for sale or not, relatively little attention is paid to yield of biomass produced. In fact, high yield may be undesirable, as in the case of turf grass, for which high yield requires more frequent mowing. Hardiness, resistance to pests, color, and abundance of blooms are much more important. Labor costs and convenience are generally of more concern than fertilizer costs; hence, expensive, slow-release fertilizers are widely used.

Conservation of Nutrient Resources

Two concepts that are key to the goal of conserving nutrient resources are (1) renewal or reuse of the resources, and (2) nutrient budgeting that reflects a balance between system inputs and outputs.

The first law of thermodynamics suggests that all material resources are ultimately renewable, since the elements are not destroyed by use but are merely recombined and moved about in space. In practical terms, however, once a nutrient has been removed from a plot of land and dispersed into the larger environment, it may be difficult if not impossible to use it again as a nutrient for that particular soil. For instance, phosphorus deposited in a lake bottom with eroded sediment and nitrogen buried in a landfill as a component of garbage are not available for reuse. In contrast, land application of composted municipal garbage and irrigation with sewage effluent are examples of practices that treat nutrients as reusable resources.

Recycling is a form of reuse in which nutrients are returned to the same land from which they were previously removed. Leaving crop residues in the field and spreading barnyard manure onto the land from which the cattle-feed grains were harvested are both examples of nutrient recycling. The term *renewable resource* best applies to soil nitrogen, which can be replenished from the atmosphere by biological nitrogen fixation (see Sections 13.10 to 13.12). Manufacture of nitrogen fertilizers also fixes atmospheric nitrogen, but at a large cost in nonrenewable fossil fuel energy.

Other fertilizer nutrients, such as potassium and phosphorus, are mined or extracted from nonrenewable mineral deposits or from mineral-laden seas and oceans. The size of known global reserves varies according to the nutrient, but for most nutrients the mineral reserves are estimated to be large enough to last for several centuries. However, should third-world farmers increase their use of phosphorus fertilizer to levels now common in Europe and the United States, the supplies of this crucial element might become depleted within as little as a single century. Furthermore, as the best, most concentrated, and most accessible sources of these nutrients are depleted, the cost of producing fertilizer will likely rise in terms of money, energy, and environmental disruption. Also, the need for these resources will increase, not decrease, over the years. By its very nature, an essential element cannot be replaced by substitute substances. Therefore it would be wise to make careful husbandry of nutrient resources a part of any long-term nutrient-management program.

A very useful step in planning nutrient management is to conceptualize the nutrient flows for the particular system under consideration. Such a flowchart should attempt to account for all the major inputs and outputs of nutrients. Simplified examples of such a conceptual budget are shown in Figure 16.1.

Nutrient Imbalances

In many cases addressing nutrient imbalances, shortages, and surpluses calls for analysis of nutrient flows, not just on a single farm or enterprise, but on a watershed, regional, or even national scale. For example, studies have shown that many countries in Africa are net exporters of nutrients. That is, exports of agricultural and forest products carry with them more nutrients than are imported into the country as fertilizers, food, or animal feed. During the past 30 years, an average of 22 kg/ha N, 2.5 kg/ha P, and 15 kg/ha K have been lost *annually* from about 200 million has of cultivated land in sub-Saharan Africa (excluding South Africa). This net *negative* nutrient balance for African soils[2] appears to be a contributing factor in their impoverishment, the reduction of agricultural productivity, and the stagnation or decline of national economies (see Section 20.10).

By contrast, in temperate regions, the cultivated land on average receives nutrients in excess of those removed in crops, runoff, and erosion. During the same 30-year

---

[2] For discussions of the negative nutrient balance in African soils, see Sanchez, et al. (1997) and Smaling, et al. (1997).

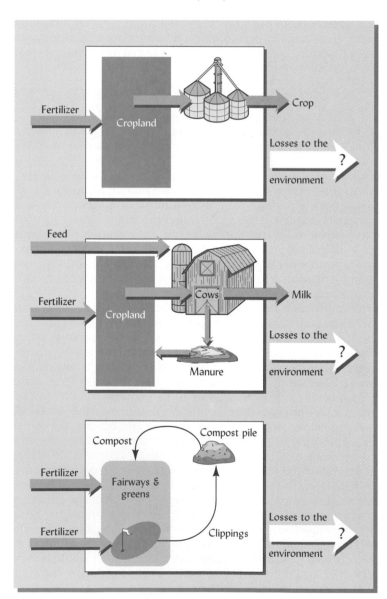

FIGURE 16.1 Representative conceptual nutrient flowcharts for a cash-grain farm, a dairy farm, and a golf course. Only the managed inputs and desired outputs are shown. Unmanaged inputs, such as nutrient deposition in rainfall, are not shown. Outputs that are difficult to manage, such as leaching and runoff losses to the environment, are shown as being variable. Although information on unmanaged inputs and outputs is not always readily available, it must be taken into consideration in developing a complete nutrient-management plan. Such flowcharts are a starting point in identifying imbalances between inputs and outputs that could lead to wasted resources, reduced profitability, and environmental damage.

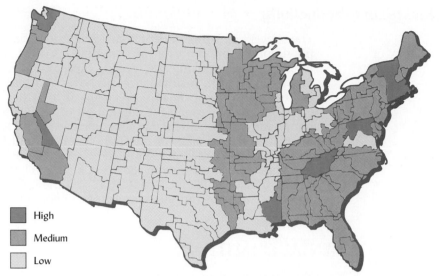

High

Medium

Low

FIGURE 16.2 Map of the United States showing regions in which the ratio of livestock manure produced to cropland available for manure application ranges from high to low. In some of the high-ratio areas as much as 4 Mg of dry manure (about 10 to 20 Mg of fresh manure) is produced for each hectare of available cropland or improved pasture land. A high or medium ratio usually indicates that many farms produce far more manure than can be properly used on land close enough for economical transportation. Even in regions with a low ratio, animal manure may not be applied to enough of the theoretically available cropland. [Source: USDA Natural Resources Conservation Service (1991)]

period, the nearly 300 million ha of cultivated temperate-region soils had a net **positive** nutrient balance of at least 60 kg/ha N, 20 kg/ha P, and 30 kg/ha K. While nutrient deficiencies still occur in some fields, the great majority of soils have experienced a nutrient buildup. Some of the excess nutrients move into streams, lakes, or the atmosphere, where they can contribute to environmental damage.

Changes in the structure of agricultural production in some countries have led to *regional* nutrient imbalances and concomitant serious water-pollution problems. The concentration of livestock production in huge feedlots and other confinement facilities using feed imported from other areas is a case in point. The animal manure produced in these production sites contains nutrients far in excess of the amounts that can be used in an efficient and environmentally safe manner by crops in nearby fields (see Figure 16.2 for areas where such concentration is found). The concentrated poultry industry in Delaware is one example for which the imbalance has been documented (Table 16.1).

TABLE 16.1 Statewide Nitrogen and Phosphorus Balance for Poultry-Based Agriculture in Delaware

It is apparent that each year about 10,000 more Mg of N is applied to the soils of Delaware's farms than is needed by the crops grown. This is nearly 50 kg N/ha. Much of this excess nitrogen leaves these farms as nitrates in surface runoff and groundwater. Phosphorus is proportionally even further out of balance, resulting in the buildup of this element in surface soils and increased losses in runoff.

Nitrogen requirement or source	Farmland area, ha	Amount of phosphorus, Mg/yr	Amount of nitrogen, Mg/yr
Requirements			
Corn	69,600	940	9,800
Soybeans	80,600	1,085	0
Wheat	24,300	330	2,200
Barley	11,000	150	1,000
Other crops	32,400	435	3,600
Total	217,900	2,940	16,600
Nutrient Sources			
Poultry manure	—	3,495	7,865
Fertilizer sales	—	2,955	19,275
Other wastes	—	Unknown	Unknown
Total sources of nutrient	—	6,450+	27,140+
Balance (excess N or P)	—	3,510+	10,580+

From Sims and Wolf (1994).

The concept of using nutrient management to enhance soil quality goes far beyond simply supplying nutrients for the current year's plant growth. Rather, it includes the long-term nutrient-supplying and -cycling capacity of the soil, improvement of soil physical properties or tilth, maintenance of above- and belowground biological functions and diversity, and the avoidance of chemical toxicities. Likewise, the management tools employed go far beyond the application of various fertilizers (although this may be an important component of nutrient management). Nutrient management requires the integrated management of physical, chemical, and biological processes. The effects of tillage on organic matter accumulation (Chapter 12), the increase in nutrient availability brought about by earthworm activity (Chapter 11), the role of mycorrhizal fungi in phosphorus uptake by plants (Chapter 14), and the impact of fire on soil nutrient and water supplies (Chapter 7) are all examples of components of integrated nutrient management.

16.2 ENVIRONMENTAL QUALITY

Nutrient management impacts the environment most directly in the area of water quality, and that is the aspect which will be emphasized in this chapter. The principal water pollutants to be considered are nitrogen, phosphorus, and sediment. You may therefore wish to consult Chapters 13 and 14 to review the many pathways by which nitrogen and phosphorus can be lost from soils. Clearly, the primary means to avoid damaging surface waters with excess nitrogen or phosphorus is to keep the rate at which these nutrients are supplied in balance with the rate at which they are removed by plants. In addition, a number of strategies can be used to reduce the transport of nutrients (and other pollutants) from soils to groundwater and surface waters. In the United States, practices officially sanctioned to implement these strategies are known as *best management practices* (BMPs). Four general practices will now be briefly considered: (1) buffer strips, (2) cover crops, (3) selective timber cutting, and (4) conservation tillage.

Riparian Buffer Strips[3]

Buffer strips of dense vegetation situated along the bank of a stream or other body of water (the **riparian zone**) are a simple and generally cost-effective method that can be used to protect water from the polluting effects of a nutrient-generating land use. Fertilized cropland, poultry or livestock operations, farmland that has been amended with organic wastes, forest harvest operations, and urban development are examples of land uses that have the potential to generate nutrient or sediment loadings. The vegetation in the buffer strips may consist of natural vegetation or planted vegetation, and may include grasses or trees (Figure 16.3).

Water running off the surface of the nutrient-rich land passes through the riparian buffer strip before it reaches the stream. Trees and the litter layer they form, or grass plants and the thatch layer they form, reduce the water velocity and increase the tortuosity of the water's travel paths. Under these conditions, most of the sediment and attached nutrients will settle out of the slowly flowing water. Also, dissolved nutrients are adsorbed by the organic mulch and mineral soil or are taken up by the buffer strip plants. The decreased flow velocity also increases the retention time—the length of time during which microbial action can work to break down pesticides before they reach the stream. Under some circumstances, buffer strips along streams can also reduce the nitrate levels in the groundwater flowing under them (although emission of nitrous oxide to the atmosphere may be a consequence of nitrate reduction; see Figure 13.11).

Riparian buffer strips can work amazingly well, provided they are densely vegetated and not overly trampled by cattle or people. The width needed for optimum cleanup may vary from 6 to 60 m, although a width of 10 m is usually sufficient (Figure 16.4). To compensate for loss of cropland for the riparian buffer strips, the landowner can use them for turnaround space for equipment, for hay production, for enhanced recreation value of the stream banks, and for better fish habitat. Thus, buffer strips can often be a win-win situation.

---

[3] For a discussion of buffer strips in forestry, see Belt and O'Laughlin (1994).

FIGURE 16.3 The zone of grassy vegetation maintained along the banks of this stream is an example of a buffer strip installed in an agricultural watershed. A large proportion of the nutrients and other pollutants carried off the crop fields by surface runoff is trapped in these strips before the runoff water can reach the stream. In the forested areas in the background, a similar strip of land along the stream will be left undisturbed to act as a buffer during logging operations. (Photo courtesy of R. Weil)

Cover Crops

A cover crop is not harvested, but rather is allowed to provide vegetative cover for the soil and then is either killed and left on the surface as a mulch, or tilled into the soil as a green manure. Growing a cover crop provides numerous benefits compared to leaving the soil bare for the off-season. The plants, if leguminous, may increase the available nitrogen in the soil (see Section 16.4); they may provide habitat for wildlife and for beneficial insects; they may protect the soil from the erosive forces of wind and rain (see Section 17.7); and they add to the soil organic matter (see Section 12.10). More relevant to the topic of this

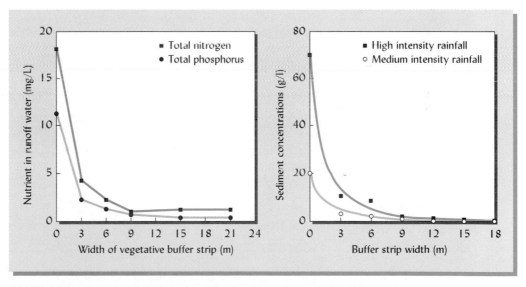

FIGURE 16.4 Removal of nutrients and sediments from runoff by vegetative buffer strips of various widths. Nitrogen and phosphorus were measured in the runoff from a field that had received swine manure (left). In another experiment, sediment was measured in the runoff from a field left fallow that had received different intensities of rainfall (right). Note that most of the nutrients and the sediment were removed in the first 9 m of the vegetative buffer strips. Depending on the type of vegetation and soil characteristics, the width of satisfactory buffers may vary from 6 to 60 m. [Nutrient data from Chauby, et al. (1994); sediment data from Robinson, et al. (1996)]

chapter, cover crops may reduce the loss of nutrients and sediment in surface runoff, and they may conserve nutrients that would otherwise leach below the root zone.

A cover crop reduces nutrient losses in runoff for two reasons. First, the vegetative cover reduces the formation of a crust at the soil surface, thus maintaining a high rate of infiltration (see Figure 6.7b). In any given storm, more infiltration leaves less water to run off on the soil surface. Second, for the runoff that does occur, the cover crop helps remove both sediment and nutrients by the same mechanisms that operate in a buffer strip, as previously described.

COVER CROPS REDUCE LEACHING LOSSES.   Perhaps the most important use of cover crops in nutrient management is to reduce the leaching losses of nutrients, principally nitrogen. In many temperate humid and subhumid regions, the greatest potential for leaching of nitrate from cropland occurs during the fall and winter, after harvest and before planting of the main crop in the spring. During the main growing season, crop roots take up both water and nutrients, resulting in little or no movement of water below the root zone and relatively low concentrations of nitrate in the soil water. However, if the soil is bare after harvest of the main crop, nitrate will readily leach downward. High leaching potential will persist (unless the soil is frozen) until the next year's crop is up and growing.

During this time of vulnerability, the presence of an actively growing cover crop will slow percolation and remove much of the nitrate from the soil solution, incorporating this nutrient into plant tissue that will largely be decomposed in the soil during the following year. For this purpose, an ideal cover crop should produce as extensive a root system as possible, as quickly as possible, once the main crop has ceased growth. Grass plants such as winter annual cereals (rye, wheat, oats, etc.) have proven to be more efficient than legumes (vetch, clover, etc.) at mopping up leftover soluble nitrogen (Figure 16.5).

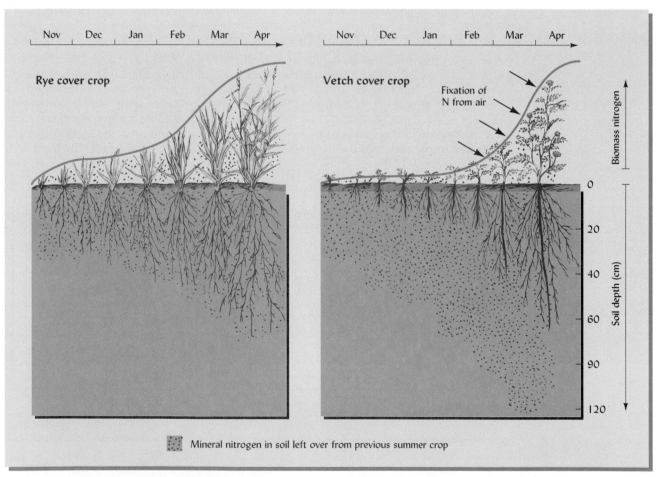

FIGURE 16.5   Relative effectiveness of a nonlegume (rye) and a legume (vetch) winter cover crop in mopping up the nitrate remaining in the soil after the harvest of a heavily fertilized summer crop. Rye grows very rapidly during mild fall weather just after harvest, while the legume grows very little until the soil warms again the following spring. These traits, and the facts that, unlike the legume, rye is dependent on soil nitrogen and has a fibrous root system, combine to make rye an excellent choice of cover crop to reduce winter leaching of soil nitrate.

FIGURE 16.6 As for any plants, vigorous growth of a cover crop requires adequate levels of available nutrients in the soil. The thin, slow-growing oat cover crop on the low-P and -K soil in the foreground is not able to help very much in preventing nitrate leaching on this coarse-textured Ultisol. The soil in the background with much more vigorously growing oats received phosphorus and potassium during the previous year at the rate of 36 and 70 kg/ha, respectively. (Photo courtesy of R. Weil)

It also should be noted that a cover crop, like any plant, requires a balanced nutrient supply and will not be capable of effectively reducing nitrate leaching if the soil is poorly supplied with other nutrients (Figure 16.6). In this regard phosphorus and potassium are especially important for cover crops growing during periods of cool temperatures.

Selective Timber Cutting

Most forests, especially those with young, rapidly growing trees, are characterized by relatively slow rates of nutrient release (mainly by organic matter mineralization) matched rather closely by comparable rates of uptake by the trees. Unfortunately, the most common and economic means of harvesting forest species—clear-cutting, or removing all the trees form large blocks of land—has some serious environmental consequences (Figure 16.7). Not only are the soils left more vulnerable to erosion, but for four primary reasons the nutrient balance is upset.

First, the removal of all the trees leaves the soil with few active roots to take up nutrients dissolved in the soil solution until natural revegetation reestablishes a root network. Second, organic matter decomposition and mineral nutrient release is accelerated by the increased temperature of the unshaded forest floor, by the higher moisture resulting from reduced transpiration from plants, and by the physical disturbance of the O and A horizons by log skidding and clearing operations. Third, the slash (branches and leaves) left behind when the logs are removed presents an enormous additional source of nutrients to be released, either by slash burning or by decomposition. Fourth, the major reduction in soil microbial biomass that follows clear cutting results in a large mass of dead, easily decomposed microorganisms.

As a result of these factors, streams draining most clear-cut watersheds have been shown to carry elevated loads of nitrate and other nutrients for some time after the timber harvest has occurred. The impact of clear-cutting of large blocks is greatest where the soils are sandy, and where the potential for revegetation is minimized by the use of herbicides to reduce weed competition for newly planted tree seedlings (Figure 16.8). Even where herbicides are not used, however, the regrowth is commonly not rapid enough to prevent some nitrate leaching for a short time following disturbance.

Selective cutting practices, in contrast, remove only a small percentage of the trees present in any one year. The soil remains permeated with nutrient-scavenging tree

FIGURE 16.7 Clear-cut harvest of loblolly pine on Coastal Plain soils. A few mature trees were left standing as seed trees. The forest floor is covered with slash of all sizes, but little living vegetation. (Photo courtesy of R. Weil)

roots, and the forest floor suffers only minimum disturbance. Streams are less polluted and the forest soils are less depleted of essential nutrients. A less visually devastated landscape is another benefit of selective cutting.

Conservation Tillage

The term *conservation tillage* applies to agricultural tillage practices that keep at least 30% of the soil surface covered by plant residues. The effects of conservation tillage on soil properties and on the prevention of soil erosion are discussed elsewhere in this textbook (see Section 17.6). Here, we emphasize the effects on nutrient losses.

Studies in many regions of the world make it clear that, compared to plowed fields with little or no residue cover, conservation tillage reduces the total amount of water running off the land surface, and reduces even more the total load of nutrients and sed-

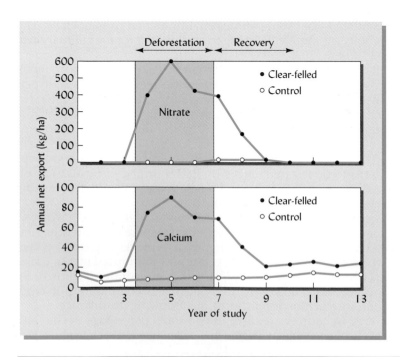

FIGURE 16.8 The effect of clear-cutting after three years of herbicide treatment on the net export of two nutrients from a forested watershed with sandy soils in the mountains of New Hampshire. Under these conditions opportunities for leaching were maximized, since trees were removed and the growth of undercover plants was prevented by the herbicides. Losses of nutrients would have been less dramatic had best management practices been used that would have allowed the vegetation to regrow immediately after cutting. [Data from Bormann and Likens (1979) © Springer-Verlag].

TABLE 16.2 Effect of Land Use and Tillage on the Loss of Nutrients in Runoff from a Silty Clay Loam Hapludalf in Ohio

The forest was a mixed stand of oak and pine, the alfalfa was a three-year-old hayfield, and the cultivated plots were planted to corn after having been in grass for 10 years. Ridge tillage is a form of conservation tillage that leaves the soil covered by residues in winter and disturbs only part of the soil in spring.

Water/nutrient loss	Tillage system			
	Forest	Untilled alfalfa	Ridge-tilled corn	Conventionally tilled corn
Runoff, % of rainfall	5	18	33	40
Nutrient loss[a] kg/ha/yr				
Nitrogen	19	13	49	315
Phosphorus	0.26	0.21	1.12	2.65

[a] Sediment and runoff.
Data from Thomas, et al. (1992).

iment carried by that runoff (Tables 16.2 and 14.2). When combined with a cover crop, the effect is greater still. There has been some concern that fertilizers and organic wastes applied to conservation-tillage fields are not well (if at all) incorporated into the soil and are therefore more likely to be washed downhill. In most situations, however, the less the soil surface is disturbed by tillage, even when manure or sewage sludge is spread on it, the smaller are the losses of nutrients to runoff water.

In general, the leaching losses of nutrients in conservation tillage systems are somewhat higher than where conventional tillage is practiced. This might be expected, since a higher percentage of the precipitation or irrigation waters infiltrates into the soil in conservation tillage systems and can carry nutrients downward. In some cases, however, the surface-applied water moves down rapidly through large macropores that are more prevalent in conservation tillage systems, with the result that nutrients held in the finer micropores are bypassed and do not leach into the lower horizons. In such situations, nutrient leaching may actually be less where conservation tillage is used.

In the remainder of this chapter we will concentrate on the properties and uses of various nutrient sources, measures designed to meet the goals of nutrient management outlined in the preceding, and methods of assessing plant and soil nutrient status in order to determine which nutrients need to be supplemented and in what quantities.

16.3 NUTRIENT RESOURCES

The pool of available soil nutrients is resupplied from internal and external resources. Internal nutrient resources come from within the ecosystem, be it a forest, a watershed, a farm, or a home in the suburbs, and are generally preferred, since their financial and environmental costs are minimized. These resources include the process of mineral weathering within the soil profile, biological nitrogen fixation, acquisition of nutrients from atmospheric deposition, and various forms of internal recycling, such as animal manure application, the utilization of cover crops and plant-residue management.

When internal resources prove inadequate, as will often be the case for at least some nutrients in highly productive systems, nutrients must be imported from resources external to the system. Such external resources are usually purchased inorganic or organic fertilizers. Some of the organic residues and wastes available in the United States are listed in Table 16.3, along with the percentage of each that is used on the land.

We will now turn our attention to specific internal and external nutrient resources.

TABLE 16.3 Estimated Quantities of Major Organic Wastes Generated Annually in the United States, and Percentages of These Materials That Are Used on the Land

Organic waste	Annual production, millions of dry metric tons	Used on land, %
Crop residues	450	75
Animal manures	175	90
Municipal refuse	145	10
Logging and wood manufacture	35	10
Industrial organics	9	5
Sewage sludge and septage	6	40
Food processing	3	15

Estimates from USDA (1980), U.S. Federal Registry (1988), and other sources.

16.4 SOIL–PLANT–ATMOSPHERE NUTRIENT CYCLES

Weathering of Parent Materials: An Internal Resource

Depending on the parent materials and climate, weathering of parent materials (see Section 2.3) may release significant quantities of nutrients (Table 16.4). For timber production, most nutrients are released fast enough from either parent materials or decaying organic matter to supply adequate nutrients. In agricultural systems, however, some nutrients generally must be added, since nutrients are removed from the land annually in harvested crops. Negligible amounts of nitrogen are released by weathering of mineral parent materials, so other mechanisms (including biological nitrogen fixation) are needed to resupply this important nutrient. Release of nutrients by mineralization of soil organic matter is important in short-term nutrient cycling, but in the long run, the organic matter and the nutrients it contains must be replenished or soil fertility will be depleted.

Nutrient Economy of Trees[4]

Forests have evolved various mechanisms to conserve nutrients in nutrient-poor sites. For instance, conifers in cold regions (e.g., black spruce) may retain their needles for several decades rather than shed them more frequently, enabling these trees to grow on sites with very low levels of nitrogen availability (the needles are much higher in nitrogen than the woody tissues). Forests generally produce more aboveground biomass (100 to 200 kg) per kilogram of nitrogen taken up than other ecosystems (e.g., corn produces about 60 to 70 kg biomass per kg N taken up). Forests of cold climates, in which slow rates of organic matter decomposition result in the immobilization of the system's nitrogen in the forest floor, are more efficient in this regard than are temperate or tropical forests (Table 16.5).

Trees, like agricultural crops, may obtain most of their nutrients from the surface horizons, but certain nutrient acquisition patterns are unique to trees. Because of their perennial growth habit and deep root systems, trees are well adapted to gathering nutrients from deep in the soil profile, where much of the nutrient release from parent material takes place and where nutrients are moved from upper horizons. Overall nutrient-use efficiency can sometimes be increased by combining trees and agricultural crops into what are known as *agroforestry systems* (Figures 16.9 and 20.10).

Trees may improve fertility of the upper soil horizons in several ways. In Chapter 2 (Figure 2.27) we saw that trees can act as nutrient pumps, taking up nutrients that occur deep in the profile because of weathering or leaching, and depositing them at the soil surface as litter that will decompose and release the nutrients where they can be of use to relatively shallow-rooted agricultural crops. Nitrogen-fixing trees (mostly legumes) can also add nitrogen to the surface soil with their nitrogen-rich leaf litter. Trees may also enhance the fertility of the soil in their vicinity by trapping windblown dust, thus increasing the deposition of such nutrients as calcium, phosphorus, and sulfur.

Leguminous Cover Crops to Supply Nitrogen

We have already mentioned many of the benefits of cover crops. Here we will consider the value of legume cover crops in supplying nitrogen for uptake by a subsequent non-legume crop.

---

[4] For an in-depth treatment of this topic, see Attiwill and Leeper (1987).

TABLE 16.4 Amounts of Selected Nutrients Released by Mineral Weathering in a Representative Humid, Temperate Climate, Compared with Amounts Removed by Silvicultural and Agricultural Harvests

For the forest, weathering release and harvest removal are roughly balanced, but cropping removes much more of some nutrients than can be released by weathering. Leaching losses are not shown.

	Amount released or removed, kg/ha			
	P	K	Ca	Mg
Weathered from igneous parent material over 50 years	5–25	250–1000	150–1500	50–500
Removed in harvest of 50-year-old deciduous bole wood	10–20	60–150	175–250	25–100
Removed by 50 annual harvests of a corn–wheat–soybean rotation	1200	2000	550	500

Estimated from many sources.

TABLE 16.5 Representative Nutrient Distribution in Several Types of Forest Ecosystems[a]

	In vegetation, kg/ha	In forest floor, kg/ha	Residence time,[b] years
Nitrogen			
Boreal coniferous	300–500	600–1100	200
Temperate deciduous	100–1200	200–1000	6
Tropical rain forest	1000–4000	30–50	0.6
Phosphorus			
Boreal coniferous	30–60	75–150	300
Temperate deciduous	60–80	20–100	6
Tropical rain forest	200–300	1–5	0.6
Potassium			
Boreal coniferous	150–350	300–750	100
Temperate deciduous	300–600	50–150	1
Tropical rain forest	2000–3500	20–40	0.2
Calcium			
Boreal coniferous	200–600	150–500	150
Temperate deciduous	1000–1200	200–400	3
Tropical rain forest	3500–5000	100–200	0.3

[a] Data derived from many sources.
[b] Residence time in forest floor (O horizons).

Winter annual legumes, such as vetch, clovers, and peas, can be sown in fall after the main crop harvest or, if the growing season is short, they can be seeded by airplane while the main crop is still in the field. The objective is to get the legume established before the winter weather becomes too cold for the cover crop to grow. Then, in spring, the cover crop will resume growth and associated microorganisms will fix as much as 3 kg/ha of nitrogen daily during the warmer spring weather. The cover is then killed and left on the surface as a mulch in a no-till system or plowed under in a conventional tillage system. A vigorous winter cover crop can provide a significant amount of nitrogen to the main crop that follows (Figure 16.10).

A cover crop system may be able to replace part or all of the nitrogen fertilizer normally used to grow the main crop. Depending on the cost of fertilizer and cover crop seed, a net financial savings may accrue. The use of winter legume cover crops to pro-

(a)

(b)

FIGURE 16.9 Two examples of agroforestry systems. (*a*) The deep-rooted *Acacia albida* enriches the soil under its spreading branches (arrow points to a man). Conveniently, these trees leaf out in the dry season and lose their leaves during the rainy season when crops are grown. It is an African tradition to leave these trees standing when land is cleared for crop production. Crops growing under the trees yield more and have higher contents of sulfur, nitrogen, and other nutrients. (*b*) Branches pruned from widely spaced rows of leguminous trees are spread as a mulch on the soil surface in the alleys between the tree rows, thus enriching the alleys with nutrients from the leaves as well as conserving soil moisture. The crops grown in this alley-cropping system may yield better than crops grown alone, but only if competition between trees and crop plants for light and water can be kept to a minimum. (Photos courtesy of R. Weil)

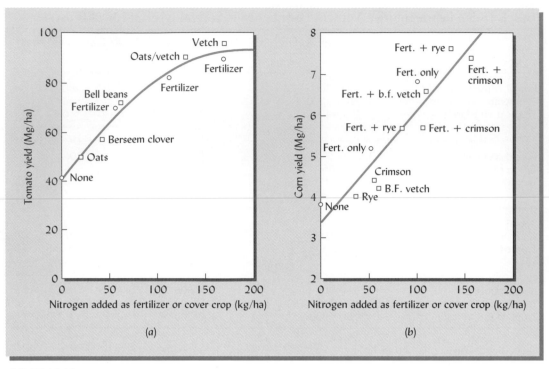

FIGURE 16.10 Comparative effects of winter cover crops and inorganic fertilizer on the yield of the following main crop. (*a*) Yield of processing tomatoes grown on a Xeralf soil in California. (*b*) Yield of corn grown on a Udalf in Kentucky. The amount of nitrogen shown on the *x*-axis was added either as inorganic fertilizer or as aboveground residues of the cover crops, or a combination of both. Note that nitrogen from either source seemed to be equally effective. For the tomato crop, which requires nitrogen over a long period of time, vetch alone or vetch mixed with oats produced enough nitrogen for near-optimal yields. [Data (left) abstracted from Stivers, et al. (1993) © Lewis Publishers, an imprint of CRC Press, Boca Raton, Florida and (right) Ebelhar, et al. (1984)]

vide nitrogen for nonlegume crops is becoming increasingly popular as more sustainable agricultural systems are developed. Such cover crop systems have been adapted for use in orchards, rice paddies (drained during the dry season), corn fields, vegetable fields, and gardens.

Crop Rotations

Growing one particular crop year after year on the same land generally produces lower yields of that crop and engenders more negative impacts on the soil and environment than if that crop is grown in sequence (rotation) with other crops. The improved plant productivity may result from the effect that rotating crops has on interrupting weed, disease, and insect pest cycles; the soil fertility effect of different rooting patterns; different types of residues and nutrient requirements; synergistic effects (see Section 12.16); and, possibly, positive effects on mycorrhizal diversity (see Section 11.9). These and other phenomena may explain why, even where pests and nutrient supply are optimally controlled, crops consistently yield 10 to 20% more in rotations than in continuous culture.

A detailed consideration of different cropping systems and rotations is beyond the scope of this book, but the nutrient management aspects of rotating legumes with nonlegumes deserve mention here. Similar to the cover crop effects just discussed, a legume main crop may substantially reduce the amount of nitrogen that needs to be added to grow a subsequent nonleguminous crop, even though much of the biomass and accumulated nitrogen in the legume are removed with the crop harvest. Perennial forage legumes, such as alfalfa, tend to produce the greatest effects in this regard, but the nitrogen contributions of such grain legumes, as soybeans and peanuts should also be taken into account when planning nitrogen application to nonlegumes, such as cereal grains (Figure 16.11).

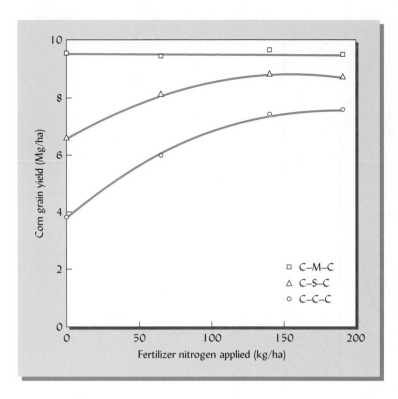

FIGURE 16.11 The nitrogen fertilizer requirements of corn are usually reduced by growing corn in rotation with legumes. This is one reason why the cropping history of a field is a consideration in deciding the optimum amount of fertilizer nitrogen to apply to cereal crops. The cropping sequences represented are: corn-corn-corn (C-C-C), corn-meadow-corn (C-M-C) and corn-soybean-corn (C-S-C). The meadow in the rotations consisted of several years of mixed alfalfa/brome grass. The data represent results on a variety of Iowa soils, but are typical of observations in many parts of the world. Less nitrogen should be applied to corn grown after soybeans than corn grown after corn. Corn following several years of mixed legume meadow is unlikely to respond to any nitrogen application. Rotation benefits other than nitrogen supply apparently account for the fact that, regardless of the amount of nitrogen fertilizer applied, corn grown after corn produced less yield than corn grown in rotation. [Based on data of A. M. Blackmer and J. R. Webb as reported in National Research Council (1993)]

Protection from Wildfires

Fires convert a great deal of nitrogen and sulfur, and some phosphorus, to gaseous forms in which they are lost from the site. Burned-over land also loses more phosphorus in runoff for several years after a high-intensity burn (see Table 14.3). Ashes, both from prescribed burns and from wildfires, contain high levels of soluble K, Mg, Ca, and P, increasing the short-term availability of these nutrients, but also increasing the rate of loss of these nutrients from the forest ecosystem. Wildfires result in much greater nutrient losses, because the high heat associated with these fires destroys some of the soil organic matter as well as the aboveground biomass. Farmers who burn crop residues after harvest cause similar losses from their fields.

Another nutrient aspect of forest fires is the fact that fire retardants used to fight wildfires are mostly fertilizer-type materials. Diammonium phosphate (DAP), monoammonium phosphate (MAP), and ammonium sulfate (AS) are widely used as fire retardants. The nitrogen and phosphorus in these chemicals may assist rapid regrowth in burned-over areas. However, they may also pose a water-quality threat when lost to streams because of the low rate of plant uptake and increased runoff that occur immediately after most of the vegetation has been destroyed by an intense fire.

Borates are also used as fire suppressants. Levels of boron could be phytotoxic (see Section 15.1) for several years, possibly retarding the growth of plants attempting to revegetate the burned-over, borate-treated area.

16.5 RECYCLING NUTRIENTS THROUGH ANIMAL MANURES

For centuries, the use of farm manure has been synonymous with a successful and stable agriculture. Not only does manure supply organic matter and plant nutrients to the soil, but it is associated with animal agriculture and with forage crops, both of which are soil protecting and conserving. A high proportion of the solar energy captured by growing plants ultimately is embodied in farm manure (Figure 16.12). Crop and animal production and soil conservation are enhanced by its use.

Huge quantities of farm manure are available each year for possible return to the land. For each kilogram of liveweight of farm animals, about 4 kg dry weight of manure is produced in a year. In the United States the farm animal population voids about 10 times as much manure solids as does the human population.

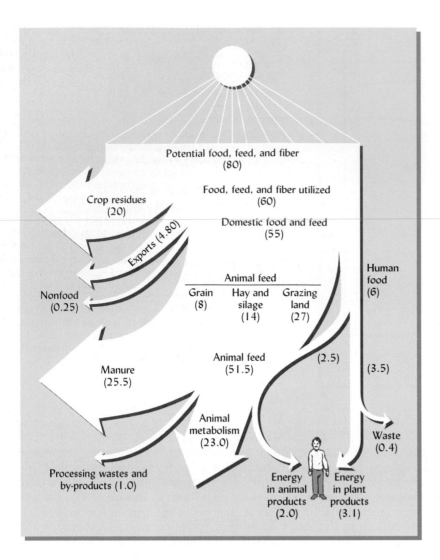

FIGURE 16.12 Estimated energy flow for the human food chain in the United States (data in billions of joules per year) showing the high proportion of the energy ultimately found in animal manures. An even larger proportion of the mineral nutrients (not shown) in the food chain end up in animal manure. [From Stickler, et al. (1975); courtesy Deere & Company, Moline, Ill.]

These nutrient-laden organic manures provide remarkable opportunities for the recycling of essential elements. Unfortunately, there have emerged in the United States and other industrial countries intensified animal production "factories" (Table 16.6) where manure *disposal* as a problem tends to overshadow manure *utilization* (Figure 16.13). Tens or even hundreds of thousands of animals are concentrated in pens or feedlots, and the disposal of their wastes is a challenge. Furthermore, while the manure is underfoot of the animals and after it is moved to the side in huge piles, much of its nitrogen is lost to the atmosphere as ammonia or through denitrification. Some is also

TABLE 16.6   Animal Production "Factories"

The concentration of U.S. livestock production into huge feedlots and confined animal "factories" has changed the perception of nutrients in animal manures from an opportunity to enhance crop production to an obligation to prevent environmental contamination.

Animals	Degree of concentration
Beef	More than one-third of marketed beef comes from just 70 of the nation's 45,000 feedlots. Largest facilities are found in Kansas, Nebraska and Texas.
Poultry	97% of the marketed poultry in the United States comes from facilities that each produce more than 100,000 broilers per year.
Pork	In North and South Carolina, where the industry is increasingly located, nearly 80% of hogs come from facilities with 5000 or more head. In traditional hog-raising states, only 6% of hogs come from facilities of this size.
Dairy	The number of dairy farms has fallen from 250,000 a decade ago to 150,000 today and average herd size has increased by more than 50%.

Modified from Gardner (1997).

FIGURE 16.13 Aerial view of a large feedlot in Colorado where 100,000 cattle are fed on grain imported from distant farms. The inset is a closeup of a similar feedlot. It is unlikely that the feedlot managers will be able to recycle the nutrients in the manure back to the land on which the cattle feed was grown. Instead of being a valued resource, the manure in this situation may be considered to be a waste requiring disposal. The challenge is to find ways to structure agriculture in a more ecologically balanced manner. (Large photo courtesy Monfort Feed Lots, Greeley, Colo.; inset courtesy of R. Weil)

lost through the leaching of nitrates (Table 16.7). The groundwater under such feed lots is often polluted with nitrates and pathogens, and is unfit to drink.

To visualize the enormity of the manure disposal problem, consider a 50,000-head beef feedlot that produces some 90,000 Mg annually, after considerable decomposition and loss of urine. If this were to be spread on land at the conventional rate of 25 Mg/ha, (11 tons/acre) which would provide about 250 kg N/h, some 10,000 ha (24,600 acres) would be needed. To find this much cropland, the manure would likely have to be hauled 10 to 17 km (6 to 10 miles) from the feedlot. While to save costs of transport, the manure often is applied at higher than needed rates, salinity damage to crops and soils and the loss of nitrogen and phosphorus to surface and groundwater results. Over time, about twice this much land would be needed because manure application rates should be reduced to account for continuing nitrogen release from previous years' applications.

Nutrient Composition of Animal Manures

Since manure as it is applied in the field is a combination of feces and urine, along with bedding (litter) and spilled feed, its composition is quite variable (Table 16.8). For a particular type of animal, the actual water and nutrient content of a load of manure will depend on the nutritional quality of the animals' feed, how the manure was handled, and the conditions under which it has been stored. The variability from one type of animal to another (e.g., broiler chicken manure compared to horse manure) is even greater. Therefore, one has to be cautious in interpreting general statements about the value and use of manure.

Animals excrete large portions of the nutrient elements consumed by them in their feeds. Generally, about three-fourths of the nitrogen, four-fifths of the phosphorus, and nine-tenths of the potassium ingested is voided by the animals and appears in the manures. For this reason, animal manures are valuable sources of both macro- and micronutrients (Table 16.9).

TABLE 16.7 The Effect of Soil and Site Characteristics on the Percentage of Wells in a Field Survey Having Nitrate-N Concentrations as High as 10 mg/L, the Upper Level Considered Suitable for Human Consumption

About 35,000 wells were sampled in five midwestern states. Note that sites with sandy soils, near cropland, near barnyards, and having shallow wells had the highest nitrate-N levels. Twenty-five percent of the sites with shallow wells that were near barnyards had nitrate-N levels greater then 10 mg/L.

Characteristics	No. of wells	Percentage with nitrate levels >10 mg/L
Soil texture		
Sandy	2412	7.2
In between	6789	4.0
Clay	6415	3.1
Proximity to cropland		
Within 6 m	1684	6.4
Within 60 m	8574	4.8
Out of sight	3098	1.8
Proximity to feedlots or barnyards		
Within 6 m	704	12.2
Within 60 m	3594	5.2
Out of sight	7520	2.8
Well depth		
Shallow, <15m	3467	9.7
Deep, >30m	5106	1.1
Two or more factors		
Shallow well, near cropland	393	12.2
Shallow well, near barnyard	158	25.3
Shallow well, sandy soil, near cropland	70	20.0

Data from Richards, et al. (1996).

Both the urine (except for poultry, which produce solid uric acid instead of urine) and feces are valuable components of animal manure. On the average, a little more than *one-half of the nitrogen,* almost *all of the phosphorus,* and about *two-fifths of the potassium* are found in the solid manure. Nevertheless, this higher nutrient content of the solid manure is offset by the ready availability of the constituents carried by the urine. Care must be taken in handling and storing the manure to minimize the loss of the liquid portion.

The data in Table 16.9 show that manures have a relatively low nutrient content in comparison with commercial fertilizer, and a nutrient ratio that is considerably lower in phosphorus than in nitrogen and potassium. On a dry-weight basis, animal manures contain from 2 to 5% N, 0.5 to 2% P, and 1 to 3% K. These values are one-half to one-tenth as great as are typical for modern commercial fertilizers.

However, unless it has been specially processed, manure is not spread in the dry form, but contains a great deal of water. As it comes from the animal, manure has a high water content, commonly varying from 30 to 40% for poultry to 70 or 80% for dairy cattle (see Table 16.9). If the fresh manure is handled as a solid and spread directly on the land (Figure 16.14), the high water content is a nuisance and adds to the expense of hauling. But if the manure is handled and digested in a liquid form or slurry and applied to the land as such, even more water must be added. In any case, all this water

TABLE 16.8 Example of the Variable Composition of Farm Manure

The data are based on 28 samples of horse manure sent in to one lab over a period of five years.

	Percent of fresh weight				
	Total N	Soluble N	P	K	Water
Lowest analysis	0.21	0.0	0.04	0.07	39
Highest analysis	0.85	0.14	0.75	1.0	80
Average analysis	0.51	0.03	0.16	0.35	63

Courtesy of V. A. Bandel, University of Maryland.

TABLE 16.9 Commonly Used Organic Nutrient Sources: Their Nutrient Contents and Other Characteristics

Material	Water,[a] %	Percent of dry weight						g/Mg of dry weight						
		Total N	P	K	Ca	Mg	S	Fe	Mn	Zn	Cu	B	Mo	
Poultry (broiler) manure[b]	35	4.4	2.1	2.6	2.3	1.0	0.6	1,000	413	480	172	40	0.7	May contain high-C bedding, high soluble salts, arsenic, and ammonia.
Dairy cow manure	75	2.4	0.7	2.1	1.4	0.8	0.3	1,800	165	165	30	20	—	May contain high-C bedding.
Swine manure[c]	72	2.1	0.8	1.2	1.6	0.3	0.3	1,100	182	390	150	75	0.6	May contain elevated Cu levels.
Sheep manure	68	3.5	0.6	1.0	0.5	0.2	0.2	—	150	175	30	30	—	
Horse manure	63	1.4	0.4	1.0	1.6	0.6	0.3	—	200	125	25	—	—	May contain high-C bedding.
Feedlot cattle manure[d]	80	1.9	0.7	2.0	1.3	0.7	0.5	5,000	40	8	2	14	1	May contain high (15%) soluble salts.
Young rye green manure	85	2.5	0.2	2.1	0.1	0.05	0.04	100	50	40	5	5	.05	Nutrient content decreases with advanced growth stage.
Spoiled legume hay	40	2.5	0.2	1.8	0.2	0.2	0.2	100	100	50	10	1,500	3	
Municipal solid waste compost[e]	40	1.2	0.3	0.4	3.1	0.3	0.2[f]	14,000	500	650	280	60	7	May have high C/N and contain heavy metals, plastic, and glass.
Sewage sludge	80	4.5	2.0	0.3	1.5[g]	0.2	0.2	16,000[g]	200	700	500	100	15	May contain high soluble salts and toxic heavy metals.
Wood wastes	—	—	0.2	0.2	0.2	1.1	0.2	2,000	8,000	500	50	30	—	Very high C/N ratio; must be supplemented by other N.

[a] Water content given for fresh materials. Processing and storage methods may alter water content to less than 5% (heat-dried) or to more than 93% (slurry).
[b] Broiler and dairy manure composition estimated from means of approximately 800 and 400 samples analyzed by the University of Maryland manure analysis program 1985–1990.
[c] Composition of swine, sheep, and horse manure calculated from North Carolina Cooperative Extension Service Soil Fact Sheets prepared by Zublena, et al. (1993).
[d] Feedlot manure composition is based on average analysis reported in Eighball and Power (1994).
[e] Composition of municipal solid waste compost based on mean values for the products of 10 composting facilities in the United States as reported by He, et al. (1995).
[f] Sulfate-S.
[g] Sludge contents of Ca and Fe may vary 10-fold depending on the wastewater treatment processes used.
Data derived from many sources.

FIGURE 16.14 Spreading solid dairy manure on cropland recycles nutrients efficiently but is labor-intensive and time-consuming for the farmer. Many loads of manure will have to be hauled to fertilize this field. The manure should be incorporated as soon as possible after spreading, and spreading should not be done on frozen soils. Calibration of spreaders is important to prevent unintentional overapplication of nutrients. (Photo courtesy of R. Weil)

diutes the nutrient content of manure, as normally spread in the field, to values much lower than those cited for dry manure. The large water content makes it difficult to transport the manure to distant fields where it might do the most good.

16.6 STORAGE, TREATMENT, AND MANAGEMENT OF ANIMAL MANURES

Where animal and crop production are integrated on a farm, manure handling is not too much of a problem. The use of pasture can be maximized so that the animals themselves spread much of the manure while grazing. The manure from confined animals is produced in small enough quantities to be hauled daily to the fields or be stored under cover during periods when soil conditions are not favorable for spreading manure. The total amount of nutrients in the manure produced on the farm is likely to be somewhat less than that needed to grow the crops; thus, modest amounts of inorganic fertilizers may be needed to make up the difference.

Where animals are concentrated in large confinement systems, the problem of manure disposal takes precedence over its utilization. Four general management systems are used to handle manure from such systems. First, the manure can be collected and *spread daily,* an ideal solution that, unfortunately, is not often used. Second, the manure may be stored and *packed in piles* where it is allowed to partially decompose before spreading. In some feedlots, by merely allowing the manure and some feed wastes to accumulate, such open-lot storage piles are formed. Third, the manure is stored in *aerated ponds* that are either sufficiently shallow to permit fairly ready oxidation of the organic materials, or oxygen is stirred into the slurry. Algae are commonly produced on or near the pond surface. Last is storage in deep *anaerobic lagoons* in which the manure is allowed to ferment in the absence of elemental oxygen. Methane and ammonia are common gaseous products of this fermentation, and considerable denitrification is known to occur. After much organic decomposition has occurred in either the ponds or the lagoons, the remaining materials, which will include some nutrients, such as nitrates, can be applied in liquid form to the soil.

Each of the four methods of handling affects the nutrient management of the manure (Table 16.10). The greatest loss of nutrients, especially nitrogen, occurs either from open-

TABLE 16.10 Influence of Handling and Storage Methods on Nutrient Losses from Animal Manure

The values are ranges of percent loss of nutrient from the time the manure is excreted until it is applied to the land.

	Percentage lost		
Manure handling and storage method	N	P	K
Solid systems			
Daily scrape and haul	15–35	10–20	20–30
Manure pack or compost	20–40	5–10	5–10
Liquid systems			
Aerobic tank storage	5–25	5–10[a]	0–5
Lagoon (anaerobic) storage	70–80	50–80	50–80

[a] Most of the P and K in a lagoon system settles to the bottom of the lagoon and is recoverable only when the sludge in the lagoon is dredged out.
From Sutton (1994).

lot storage or from lagoon storage. In open-lot storage much nitrogen is volatilized as ammonia gas (see Section 13.6), and considerable quantities of nutrients may be washed away when it rains. Unfortunately, the bottoms of many of the lagoons and ponds are not lined with clay to minimize leakage. Water and excess nutrients can pass through the lagoon walls to the groundwater and into streams and lakes, giving rise to algal blooms and eutrophication and to the death of fish and other wildlife. It is obvious that the recycling of nutrients is low on the priority list of most lagoon-management schemes.

Methods of manure handling that both prevent pollution and preserve nutrients in a form that can be easily transported and sold commercially would make a major contribution to ameliorating the manure problem for concentrated animal production enterprises. Several options are currently being developed, three of which are described in the following subsections.

Heat-Dry and Pelletize

This technology dries the manure with heat and then compresses the dried product into small pellets that handle like commercial fertilizer. This manure-processing method requires considerable energy use and capital investment, but the product has generated a large demand, especially in the landscaping and lawn industries, where much money is already being spent on slow-release fertilizers.

Commercial Composting

A second method under development is improved **composting** of manure (see Section 12.5). In the case of poultry manure, dead birds are composted along with the manure, and the final product is a very stable, nonoffensive, relatively high analysis, slow-release fertilizer that is also easy to handle and has been enthusiastically received in test markets. Composting is a natural aerobic decomposition process and is much less energy- and capital-intensive than heat-drying and pelleting.

Anaerobic Digestion with Biogas Production

In this method the manure is made into a liquid slurry and allowed to become anaerobic (air is not stirred in). As discussed in Section 12.2, methane is a major gaseous product of anaerobic decomposition. The biogas produced by anaerobic manure digestion contains about 80% methane and 20% carbon dioxide. This gas can be burned much like commercial natural gas. Developing countries have attempted to use small-scale manure digesters to supply cooking and heating fuel for remote villages.

Recently, several commercial ventures have been established in the United States for large-scale production of biogas from manure. The gas is most commonly used to generate electricity, which is sold to local utility companies. The slurry remaining after digestion still contains most of the nutrients from the manure (though much organic carbon and some nitrogen has been transformed to gases), and it can be pumped to nearby fields or further processed (see preceding methods) to allow for wider sale and distribution.

In the future such technologies, along with more integrated animal farming systems, may help to redress the serious nutrient imbalances that have developed with regard to manure production in modern agriculture.

16.7 INDUSTRIAL AND MUNICIPAL BY-PRODUCTS

Farm animals are not alone in generating large quantities of nutrient-containing wastes. People, and their industrial activities, do likewise, mostly in the relatively concentrated setting of metropolitan urban areas. Four major types of organic wastes are of significance in land application: (1) municipal garbage, (2) sewage effluents and sludges, (3) food-processing wastes, and (4) wastes of the lumber industry. Because of their uncertain content of toxic chemicals, other industrial wastes may or may not be acceptable sources of organic matter and plant nutrients for land application.

Society's concern for environmental quality has forced waste generators to seek non-polluting, but still affordable, ways of disposing of these materials. Although the primary reason for adding these wastes to land may be to safely dispose of them, they also serve as valuable soil amendments for agriculture, forestry, and disturbed-land reclamation. Once seen as mere waste products to be flushed into rivers and out to sea, these materials are increasingly seen as sources of nutrients and organic matter that can be used beneficially to promote soil productivity.

Garbage

Municipal garbage which has traditionally been used extensively to enhance crop production in China and other Asian countries, is becoming more widely used in the United States and Europe. The inorganic glass, metals, and so forth are first removed and municipal organic wastes are composted, sometimes in conjunction with sewage sludge, poultry manure, or other nutrient-rich materials, to produce municipal solid waste (MSW) compost that is then applied to the land. Most municipal garbage in the United States is incinerated or landfilled (see Section 18.11), but growing concerns about air quality and scarcity of landfill space are raising the level of interest in using soil application as a means of disposal. Because of the very low nutrient content of MSW compost (see Table 16.9), and its far distance from the fields to which it can be applied, the transportation of a given quantity of nutrients in MSW compost is even more expensive than that in animal manures. On the other hand, municipalities have to dispose of the material somehow, and the cost of hauling MSW compost to land application sites is becoming economically more competitive with alternative methods of disposal. As with manures, there is a trend in some regions to produce an attractive product that can be sold to landscaping and specialty users as a soil amendment or planting medium.

Food-Processing Wastes

Land application of food-processing wastes is being practiced in selected locations, but the practice is focused almost entirely on pollution abatement and not on crop production. Liquid wastes are commonly applied through sprinkle irrigation to permanent grass. Plant-processing schedules dictate the timing and rate of application, which may not be suitable for optimum crop production.

Sawdust

Sawdust, wood chips, and shredded bark from the lumber industry have long been sources of soil amendments and mulches, especially for home gardeners and landscapers. These wastes are high in lignin and related materials, have very high C/N ratios, and therefore decompose very slowly—desirable properties in a mulch material. However, they do not readily supply plant nutrients. In fact, sawdust incorporated into soils to improve soil physical properties may cause plants to become nitrogen deficient unless an additional source of nitrogen is applied with the sawdust (see Section 12.3).

Recycling by Composting

The disposal of the mountains of organic municipal and household wastes with all the nutrients they contain is a major environmental challenge. In the past, most of these

FIGURE 16.15 A compost facility in Phoenix, Arizona producing organic materials for urban landscape uses. (Upper left) Chopped yard trimmings, sawdust, animal manures and a sandy loam soil are mixed together in a windrow (long pile). (Upper right) Sprinkler systems keep the piles moist to encourage microbial activity. (Lower left) The piles are moved from one windrow to another, thereby further mixing the components and assuring good aeration. This process encourages temperatures in the interior of the pile to rise to the 60 to 80° C range. After about six to eight weeks microbial activity has subsided and temperature are no longer elevated. (Lower right) The composted material is now ready for transport to homes and other nearby urban facilities where it will be used in gardening and landscaping.

wastes in the United States found their way into the thousands of landfills that dotted the countryside. But the number of landfill sites has decreased by 75% in the past 20 years, and the environmental consequences of dumping organic materials in these sites have become more apparent. Consequently, alternative decentralized means of handling these organic materials are being sought.

Among the alternatives that are being more widely used are composting operations that combine domestic, garden, community, and even some industrial wastes into viable waste-management systems. Yard trimmings, food scraps, supermarket wastes, and paper are combined in compost piles with such materials as sawdust, manure, and some industrial wastes (e.g., coal ashes). When appropriate moisture is supplied, decomposition takes place, and in time composted materials that will support plant growth are available for recycling to the land (see Section 12.5). The compost sites may be large or small operations run by communities, small businesses, or even individuals (Figure 16.15). The number of such facilities in the United States increased more than fourfold between 1989 and 1996.

16.8 SEWAGE EFFLUENT AND SLUDGE

Sewage treatment has evolved over the past century in response to society's desire (and resultant regulations) to avoid polluting our rivers and oceans with pathogens, oxygen-demanding organic debris, and eutrophying nutrients. The ever-more-stringent efforts to clean up the wastewater before returning it to natural waters has two basic consequences. First, the amount of material *removed* from the wastewater during the treatment process has increased tremendously. This solid material, known as sewage **sludge** or **biosolids,** must also be disposed of safely. Second, a goal of advanced wastewater treatment is to remove nutrients (mainly phosphorus, but increasingly also nitrogen) from the **effluent** (the treated water that is returned to the stream), a job that can be accomplished safely and economically by allowing the partially treated effluent to interact with a soil–plant system. Therefore, there is a growing interest in using soils to assist with the sewage problem in two ways: (1) as a system of assimilating, recycling, or disposing of the solid sludge; and (2) as a means of carrying out the final removal of nutrients and organics from the liquid effluent.

Sewage Effluent

Sewage effluents have been applied to land for decades in Europe and in selected sites in the United States. Some cities operate sewage farms on which they produce crops, usually animal feeds and forages, that offset part of the expense of effluent disposal. The city of Muskegon, Michigan, has operated such a farm for decades, as has Paris, France. Sewage effluent may become more important in the future as a source of organic matter, nutrients, and water for crop production.

One very beneficial use of nutrient-rich sewage effluent is the irrigation of forest land (Figure 16.16). This method of advanced wastewater treatment is used by a number of cities around the world. Forest irrigation is a cost-effective method of final effluent cleanup and produces enhanced tree growth as a bonus. The rate of wood production is greatly increased as a result of both the additional water and the additional nutrients supplied therewith.

In a carefully planned and managed effluent irrigation system, the combination of (1) nutrient uptake by the plants, (2) adsorption of inorganic and organic constituents by soil colloids, and (3) degradation of organic compounds by soil microorganisms results in the purification of the wastewater. Percolation of the purified water eventually replenishes the groundwater supply.

Sewage Sludge

Sewage sludge (sometimes called **biosolids**) is the solid by-product of domestic and/or industrial wastewater treatment plants (Figure 16.17). It has been spread on the land for decades, and its use will likely increase in the future. The product Milorganite, a dried

FIGURE 16.16 Final treatment of sewage effluent and recharge of groundwater are being accomplished by natural soil and plant processes in this effluent-irrigated forest on Ultisols near Atlanta, Georgia. Nutrient flows, groundwater quality, and tree growth are carefully monitored. Some of the greatly increased production of wood is used as an energy source to run the sewage treatment plant. (Photo courtesy of R. Weil)

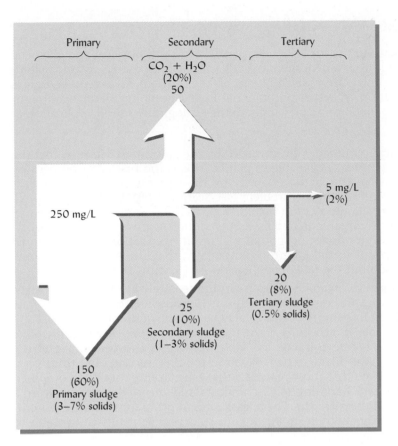

Primary   Secondary   Tertiary

$CO_2 + H_2O$
(20%)
50

250 mg/L

5 mg/L
(2%)

20
(8%)
Tertiary sludge
(0.5% solids)

25
(10%)
Secondary sludge
(1–3% solids)

150
(60%)
Primary sludge
(3–7% solids)

FIGURE 16.17 Diagram showing the removal of suspended solids from wastes. Primary treatment permits the separation of most of the solids from raw sewage. Secondary treatment encourages oxidation of much of the organic matter and separation of more solids. Tertiary treatment usually involves the use of calcium, aluminum, or iron compounds to remove phosphorus from the sewage water. [Redrawn from Loehr, et al. (1979); used with permission of Van Nostrand Reinhold Company, New York]

sludge sold by the Milwaukee Sewerage Commission, has been used widely in North America since 1927, especially on turf grass. Numerous other cities market composted sludge products to landscaping and other specialty users. However, the great bulk of sewage sludge used on land is applied as a liquid slurry or as a partially dried cake. The cake is less costly to transport because it carries only 40 to 70% water, while the slurry is generally 80 to 90% water. Further drying and processing decreases the cost of transportation, an important consideration since large cities have difficulty finding sufficient nearby land to receive their biosolids. One program *daily* ships over 600,000 kg of dried, treated sludge by rail from New York City to western Texas, a distance of over 2000 km, for use in restoring fertility to 7000 ha of degraded range land.

COMPOSITION OF SEWAGE SLUDGE. As might be expected, the composition of the sludge varies from one sewage treatment plant to another, depending on the nature of treatment the sewage receives, especially the degree to which the organic material is allowed to digest. Representative values for plant nutrients are given in Table 16.9. Like manure and other organic waste products, sewage sludge contributes micronutrients as well as macronutrients. Levels of plant micronutrient metals (zinc, copper, iron, manganese, and nickel) as well as other heavy metals (cadmium, chromium, lead, etc.) are determined largely by the degree to which industrial wastes have been mixed in with domestic wastes. In the United States, the levels of metals in sewage are far lower than they were in the past, because of source-reduction programs that require industrial facilities to remove pollutants *before* sending their sewage to municipal treatment plants (see Section 18.8).

In comparison with inorganic fertilizers, sludges are generally low in nutrients, especially potassium. Representative levels of N, P, and K are 4, 2, and 0.4%, respectively (see Table 16.9). The potassium concentration in sewage sludge is rather low, because most of the potassium in sewage is soluble and remains with the effluent during treatment. The phosphorus content is higher, because advanced sewage treatment is designed to remove phosphorus from the effluent and deposit it in the sludge (see Box 14.2). If the sewage treatment precipitates phosphorus by reactions with iron or aluminum compounds, the phosphorus in the sludge will likely have a very low availability to plants.

In Section 12.7 we discussed the many beneficial effects on soil physical and chemical properties that can result from amendment of soils with decomposable organic materials, such as manure or sludge. Here we will focus on the nutrient management aspects of organic waste utilization. Whether the material is sewage sludge, farm manure, or MSW compost, several general principles apply to the ecologically sound application of the material to soils:

1. The rate of application is generally governed by the amount of nitrogen that the organic material will make available to plants. This is the first criterion because nitrogen is needed in the largest quantity by most plants, and because excess nitrogen can present a pollution problem (see Section 13.8). It should be noted, however, that the P/N ratio in most organic sources is higher than in plant tissue. Consequently, if the organic materials supply sufficient nitrogen to meet plant needs, excessive levels of soil phosporous will result (see Section 14.2), and must be taken into account in the long run. Potentially, toxic heavy metals in some materials may also limit the rate of application (see Section 18.8).

2. Most of the nitrogen in organic sources is not immediately available to plants, as it is with most inorganic commercial fertilizers (see Section 16.11). A small fraction of the nitrogen in manure or sludge may be soluble (ammonium) forms and immediately available, but the bulk of the nitrogen must be released by microbial mineralization of organic compounds. Table 16.11 lists the percentage of the organic nitrogen that is likely to be released in the first, second, and third years after application. Note that those materials that have been partially decomposed during treatment and handling (e.g., composts and digested sludge) release a lower percentage of the nitrogen. The release rates of other nutrients are less critical because their availability from the organic sources is generally comparable to that from commercial fertilizers.

3. If a field is treated annually with an organic material, the application rate needed will become progressively smaller because, after the first year, the amount of nitrogen released from material applied in previous years must be subtracted from the total to be applied afresh (see Box 16.1, calculations for organic nitrogen application rates).

4. The nutrient and moisture contents of organic amendments vary widely among sources, and even from batch to batch, depending on how the material has been stored and treated. Therefore, general values such as those in Table 16.9 should not be relied upon when calculating rates of material to apply. Instead, representative samples of the material should be analyzed in a laboratory. Analyses are generally required by laws regulating land application of sewage sludge, but other materials are not as widely regulated in this regard.

TABLE 16.11   Rates of Release of Mineral Nitrogen from Various Sources of Organic Nitrogen Applied to Soils

The values given are percent of the organic nitrogen present in the original material. For example, if 10 Mg of poultry litter initially contains 300 kg (3.0%) nitrogen in organic forms, 50% or 150 kg of nitrogen would be mineralized in the first year. Another 15% (0.15 × 300) or 45 kg of nitrogen would be released in the second year. These values are approximate and may need to be increased for warm climates or sandy soils.

Organic nitrogen source	First year	Second year	Third year	Fourth year
Poultry floor litter	50	15	8	3
Dairy manure (fresh solid)	35	18	9	4
Swine manure lagoon liquid	50	15	8	3
Lime-stabilized, aerobically digested sewage sludge	40	12	5	2
Anaerobically digested sewage sludge	20	8	4	1
Composted sewage sludge	10	5	3	2
Activated, unstabilized sewage sludge	45	15	4	2

5. Nutrients from organic sources will produce the highest return and cause the least environmental damage if applied to fields that are relatively poorly supplied with nitrogen and phosphorus. Unfortunately, cost of transport and considerations of convenience provide incentives to apply organic materials on land close to their source, often leading to the nutrient imbalances discussed in Section 16.5.

Special Uses

There are a number of special uses of organic nutrient sources for which the organic matter component plays a special role. Examples include applications to denuded soil areas resulting from erosion, from land-leveling for irrigation, or from mining operations. Improvements in water-holding capacity and soil structure brought about by organic material application may be just as important as the long-term, slow-release, nutrient-supplying features of the organic wastes. Initial applications of 50 to 100

Mg/ha may be worked into the soil in the affected areas. These rates may be justified to supply organic matter as well as nutrients, provided the material is not so high in nitrogen as to create the potential for nitrate leaching.

Special cases of micronutrient deficiency can be ameliorated with manure application. Such treatments are sometimes used when there is some uncertainty about which specific nutrient is lacking. Manure applications can be made with little concern for adding toxic quantities of the micronutrients.

16.10 INTEGRATED RECYCLING OF WASTES

For most of the industrialized countries, widespread recycling of organic wastes other than animal manures is a relatively recent phenomenon. In heavily populated areas of Asia, however, and particularly in China and Japan, such recycling has long been practiced. Figure 16.18 illustrates the many ways in which organic wastes are used in China. Agriculture is only one of the recipients of these wastes. Much is used for biogas production, as food for fish, and as a source of heat from compost piles to help warm homes, greenhouses, and household water. Most important, the plant nutrients and organic matter are recycled and returned to the soil for future plant utilization. Such conservation is likely to be practiced more widely in the future by other countries, including the United States.

16.11 INORGANIC COMMERCIAL FERTILIZERS

The worldwide use of fertilizers increased dramatically during the latter half of the 20th century (Figure 16.19), accounting for a significant part of the equally dramatic

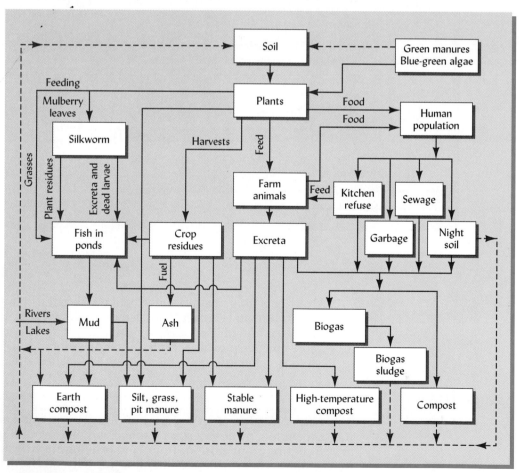

FIGURE 16.18 Recycling of organic wastes and nutrient elements in the People's Republic of China. Note the degree to which the soil is involved in the recycling processes. [Modified from FAO (1977)]

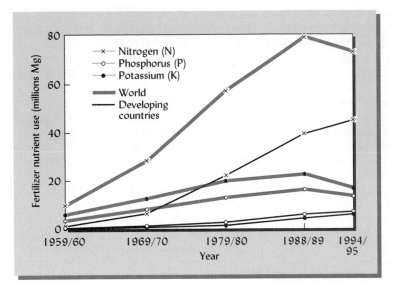

FIGURE 16.19 World and developing country fertilizer use since 1959–1960 by major nutrient element. Note the very high use of nitrogen compared to the other two nutrients. World fertilizer use declined significantly during the 1990s, due primarily to drastic use reductions in nations of the former Soviet Union and to some cutbacks in other industrialized countries. Fertilizer use continued to rise in the developing countries, however, since nutrient removal in crops far exceeds return of nutrients to the soil in many of these countries. [Data from the United Nations Food and Agriculture Organization as reported in Bumb and Baanante (1996)]

increases in crop yields during the same period. Improved soil fertility through the application of fertilizer nutrients is an essential factor in enabling the world to feed the billions of people that are being added to its population (see Section 20.3).

The need to supplement forest soil fertility is increasing as the demands for forest products increase the removal of nutrients and competing uses of land leave forestry with more infertile, marginal sites. Currently, most fertilization in forestry is concentrated on tree nurseries and on seed trees, where the benefits of fertilization are relatively short term and of high value, and the logistics of fertilizer application are not so difficult and expensive as for extensive forest stands. In the warmer, more humid regions of Japan,[5] about 50% of new forest plantings are fertilized. Forest fertilization is also increasingly practiced in the southeastern United States, the Scandinavian countries, and Australia.

Regional Use of Fertilizers

Worldwide fertilizer statistics provide a perspective on the role of fertilizers in maintaining a global balance between nutrient inputs and outflows. But it is the more specific usage within a region that tells more definitively whether fertilizers are contributing to soil quality enhancement or to environmental degradation. For example, in Europe and East Asia where moisture is abundant and intensified cropping is common, fertilizer nutrient application rates are nearly triple the world average. In the Netherlands, nitrogen additions from fertilizers and manures are more than four times the removal of this element in harvested crops. In contrast, soils in sub-Saharan Africa are literally being mined—far less nutrients are being added from all sources than are being removed in the crops. The fertilizer nutrient rate in this region is only about 10% of the world average. Thus, it is obvious that we must be as site specific as possible in evaluating the role of fertilizers in meeting humanitarian and environmental goals.

Origin and Processing of Inorganic Fertilizers

Most fertilizers are inorganic salts containing readily available plant nutrient elements. Some are manufactured, but others, such as phosphorus and potassium, are found in natural geological deposits. **Beds of solid salts** located far beneath the earth's surface are the primary sources of potassium. These underground deposits are found in many locations, including sites in Canada, France, Germany, and Russia, as well as in New Mexico. The salts are mined and then purified, yielding such compounds as potassium chloride and potassium sulfate.

---

[5] In Japan the climate is capable of supporting rapid growth of the more valuable timber species, such as sugi—*Cryptomeria japonica* and *Quercus acustissima*—whose logs are in demand to grow shiitake mushrooms.

Apatite found in **phosphate rock deposits** is the primary source of phosphorus fertilizers. Since apatite is extremely insoluble, however, it is treated with sulfuric, phosphoric, or nitric acid to produce materials, such as triple superphosphate, that yield phosphorus in a readily available form. Phosphate rock deposits are located around the world; among the most extensive are found in Florida, North Carolina, Tennessee, and several western states.

Nitrogen in the **atmosphere** is the primary original source of this essential element. Under very high temperatures and pressures, the nitrogen is fixed with hydrogen from natural gas to produce ammonia gas, NH_3. Although this fixation is of vital importance, it consumes tremendous quantities of energy—as critics have pointed out. When the ammonia gas is then put under moderate pressure, it liquifies, forming anhydrous ammonia. Because of its low cost and ease of application, more nitrogen is applied directly to soils in anhydrous ammonia than in any other fertilizer (Table 16.12). More important, it is the starting point for the manufacture of most of the other nitrogen carriers, including urea, ammonium nitrate, ammonium sulfate, sodium nitrate, and liquid mixtures known as "nitrogen solutions." The commercialization of the process of nitrogen fixation revolutionized the recycling of nitrogen. Unfortunately, the reduced cost of this element and the variety of materials in which it can be supplied has sometimes encouraged its use in excess of plant needs.

Many commercial fertilizers contain two or more of the primary nutrient elements. In some cases, mixtures of the primary nutrient carriers are used. But in others, a given compound may carry two nutrients, examples being monoammonium phosphate, diammonium phosphate, and potassium nitrate. In any case, care must be used in selecting the components of a mixed fertilizer since some compounds are not compatible with others, resulting in poor physical condition and reduced nutrient availability in the mixture.

TABLE 16.12 Fertilizer Consumption in the United States

Major fertilizers consumed in the United States in the year ending in June 30, 1996, and the quantities of nutrient elements they provide. Note that most of the nitrogen and potassium were provided in single-nutrient carriers, while most of the phosphorus carriers provide nitrogen as well.

	Nutrients provided, 1000 Mg		
Fertilizer	Nitrogen	Phosphorus	Potassium
Nitrogen Carriers			
Anhydrous ammonia (NH_3)	3593	—	—
Nitrogen solutions (UAN)[a]	2565	—	—
Urea [$CO(NH_2)_2$]	1628	—	—
Ammonium Nitrate (NH_4NO_3)	646	—	—
Ammonium sulfate [$(NH_4)_2SO_4$]	200	—	—
Phosphorus Carriers			
Diammonium Phosphate (18-46-0)	603	628	—
Monoammonium phosphate (11-53-0)	124	242	—
Ammonium polyphosphate (10-34-0)	100	139	—
Triple Superphosphate	—	83	—
Potassium Carriers			
Potassium chloride	—	—	2478
Potassium sulfate	—	—	54
Mixed Fertilizers[b]			
19-19-19	32	14	27
10-10-10	29	13	24
13-13-13	29	12	24
16-20-0	27	15	—
10-20-20	13	12	22

[a] Urea, ammonium nitrate
[b] See page 643 for an explanation of these three-number codes.
AAPFCO data compiled by Terry and Yu (1996).

Properties and Use of Inorganic Fertilizers

The composition of inorganic commercial fertilizers is much more precisely defined than is the case for the organic material discussed above. Table 16.13 lists the nutrient element contents and other properties of some of the more commonly used inorganic fertilizers. In most cases, fertilizers are used to supply plants with the macronutrients nitrogen, phosphorus, and/or potassium—sometimes called the *primary fertilizer elements*. Fertilizers that supply sulfur, magnesium, and the micronutrients are also manufactured.

It can be seen from the data in Table 16.13 that a particular nutrient (say, nitrogen) can be supplied by many different *carriers*, or fertilizer compounds. Decisions as to which fertilizers to use must take into account not only the nutrients they contain, but also a number of other characteristics of the individual carriers. Table 16.13 provides information about some of these characteristics, such as the salt hazard (see also Section

TABLE 16.13 Commonly Used Inorganic Fertilizer Materials: Their Nutrient Contents and Other Characteristics

Fertilizer	N	P	K	S	Ca	Mg	Micro-nutrients	Salt hazard	Acid-forming tendency[b], kg $CaCO_3$/ 100 kg	Comments
Primarily sources of nitrogen										
Anhydrous ammonia (NH_3)	82							Low	−148	Pressurized equipment needed; toxic gas.
Urea [$CO(NH_2)_2$]	45							Moderate	−84	Soluble; urease enzyme rapidly hydrolyses urea to ammonium forms.
Ammonium nitrate (NH_4NO_3)	33							High	−59	Absorbs moisture from air; can explode.
Sulfur-coated urea	30–40			13–16				Low	−110	Variable slow rate of release.
UF (ureaformaldehyde)	30–40							Very low	−68	Slowly soluble.
UAN solution	30							Moderate	−52	Most commonly used liquid N.
IBDU (isobutylidene diurea)	30							Very low	—	Slowly soluble.
Ammonium sulfate [$(NH_4)_2SO_4$]	21			24				High	−110	Rapidly lowers soil pH; very easy to handle.
Sodium nitrate ($NaNO_3$)	16						About 0.6% CI	Very high	+29	Hardens, disperses soil structure.
Potassium nitrate (KNO_3)	13		36	0.2	0.4	0.3	1.2% CI	Very high	+26	Very rapid plant response.
Primarily sources of phosphorus										
Monoammonium phosphate ($NH_4H_2PO_4$)	11	21–23		1–2				Low	−65	Best as starter.
Diammonium phosphate [$(NH_4)_2HPO_4$]	18–21	20–23		0–1				Moderate	−70	Best as starter.
Triple superphosphate		19–22		1–3	15			Low	0	
Phosphate rock [$Ca_3(PO_4)_2 \cdot CaX$]		8–18[a]			30			Very low	Variable	Low to extremely low availability. Best as fine powder on acid soils. Contains some Cd, F, etc.
Single superphosphate		7–9		11	20			Low	0	Nonburning, can place with seed.
Bonemeal	1–3[a]	10[a]	0.4		20			Very low	—	Slow availability of N, P as for phosphate rock.
Colloidal phosphate		8[a]			20			Very low	—	P availability as for phosphate rock.

TABLE 16.13 (Cont.)

Fertilizer	Percent by weight							Salt hazard	Acid-forming tendency[b], kg CaCO$_3$/ 100 kg	Comments
	N	P	K	S	Ca	Mg	Micro-nutrients			
				Primarily sources of potassium						
Potassium chloride (KCl)			50				47% Cl	High	0	Cl may reduce some diseases.
Potassium sulfate (K$_2$SO$_4$)			42	17	0.7	1.2		Moderate	0	Use where Cl not desirable.
Wood ashes	0.5–1	1–4	10–20	2–5			0.2% Fe, 0.8% Mn, 0.05% Zn, 0.005% Cu, 0.03% B	Moderate to high	+40	About ⅔ the liming value of limestone; caustic.
Greensand		0.6	6					Very low	0	Very low availability.
Granite dust			4					Very low	0	Very slow availability.
				Primarily sources of other nutrients						
Basic slag		1–7			3–30	3	10% Fe, 2% Mn	Low	+70	Industrial by-product; slow availability; best on acid soils.
Gypsum (CaSO$_4$·2H$_2$O)				19	23			Low	0	Stabilizes soil structure; no effect on pH; Ca and S readily available.
Calcitic limestone (CaCO$_3$)					36			Very low	+95	Slow availability; raises pH.
Dolomitic limestone [CaMg(CO$_3$)$_2$]					24	12		Very low	+95	Very slow availability; raises pH.
Epsom salts (MgSO$_4$·7H$_2$O)				13	2	10		Moderate	0	No effect on pH; water soluble.
Sulfur, flowers (S)				95				—	–300	Irritates eyes; very acidifying; slow acting; requires microbial oxidation.
Solubor							20.5% B	Moderate	—	Very soluble; compatible with foliar sprays.
EDTA chelate							13% Cu or 10% Fe or 12% Mn or 12% Zn	—	—	One or more micro-nutrients; see label.

[a] Highly variable contents.
[b] A negative number indicates that acidity is produced; a positive number indicates that alkalinity is produced.
Data derived from many sources.

10.6), acid-forming tendency (see also Section 13.7), tendency to volatilize, ease of solubility, and content of nutrients other than the principal one. Of the nitrogen carriers, anhydrous ammonia, nitrogen solutions, and ureas are the most widely used. Diammonium phosphate and potassium chloride supply the bulk of the phosphorus and potassium used in the United States (see Table 16.12).

Physical Forms of Marketed Fertilizer

A generation ago most commercial fertilizers were sold in paper or plastic bags, transported to the field in trucks, then emptied by hand into fertilizer spreaders and applied to the land. Today, less than 10% of all fertilizer is handled in this manner (Table 16.14). Increases in fertilizer use rates and labor costs, along with improved means of transporting and handling the fertilizer and increased availability of custom applicators, have favored two alternative means of marketing fertilizer: (1) unbagged, dry solids handled in **bulk** form, and (2) **liquid** or fluid forms stored, transported, and applied from tanks. In both cases the costs, particularly in terms of labor, are reduced.

TABLE 16.14 Forms of Marketed Fertilizer in the United States

Most of the fertilizer tonnage consumed in the United States in 1995–1996 was either bulk spread, probably by truck, or was applied as liquids. Less than 10% was marketed in bags. Much of the bulk-spread and liquid fertilizers were applied by custom operators. The means of applying the fertilizer and the material used is usually determined primarily by economic and labor considerations.

	Percent marketed in indicated form		
Type of fertilizer	Bagged	Bulk	Liquid
Single nutrient	3.1	44.9	52.0
Multiple nutrient	19.2	58.6	22.1
All fertilizers	9.3	50.4	40.3

From Terry and Yu (1996).

Bulk spreading, often done at times when trucks or large fertilizer spreaders can get on the land, is the favorite for applying multinutrient fertilizers (see Table 16.14). Often, bulk spreading is done for the grower by a fertilizer dealer or private custom applicator who draws fertilizer materials from a small, strategically located material-blending plant. The fertilizer can be blended to satisfy the customer's desired analysis and then sent directly to the farm, where it is bulk-spread on the field. The process requires little if any farm labor and provides the farmer with fertilizer in a timely and efficient manner.

Liquid fertilizers comprise more than half of the single-nutrient carriers sold in the United States and about 40% of all fertilizers (see Table 16.14). Here, too, labor costs are low, since the fertilizer is transferred from one tank to another and applied to the field with the aid of mechanical pumps. Furthermore, the nutrient costs in some liquid carriers (e.g., anhydrous ammonia) are low, and the nutrients are commonly applied before planting. Custom applicators are common, relieving the farmers from much of the responsibility they once shouldered.

These changes in marketing have generally made it easier to apply fertilizer, and have likely been a factor, along with economic considerations, in encouraging increased fertilizer usage. They may well influence fertilizer use as much or more than scientific principles. Marketing considerations must be kept in mind as well as the scientific bases for fertilizer use as we seek means of better assuring that fertilizers enhance soil quality and minimize environmental degradation.

Fertilizer Grade

Commercial fertilizers were first manufactured in the late 19th and early 20th centuries. Contemporary analytical procedures and laws passed to regulate the new industry resulted in certain labeling and marketing conventions that are still in common use today.

Of these conventions it is most important to be familiar with the *fertilizer grade*. Every fertilizer label states the grade as a three-number code, such as 10-5-10 or 6-24-24. These numbers stand for percentages indicating the *total* nitrogen (N) content, the *available* phosphoric acid (P_2O_5) content, and the *soluble* potash (K_2O) content. Plants do not take up phosphorus and potassium in these chemical forms, nor do any fertilizers actually contain P_2O_5 or K_2O.

The oxide means of expression for phosphorus (P_2O_5) and potassium (K_2O) is a relic of the days when geochemists reported the contents of rocks and minerals in terms of the oxides formed upon heating. Unfortunately, these means of expression found their ways into state laws governing the sale of fertilizers, and there is considerable resistance to changing them, although some progress is being made. In all scientific work and in this textbook, the simple elemental contents are used (P and K) wherever possible. Box 16.2 explains how to convert between the elemental and oxide forms of expression.

The grade is important from an economic standpoint because it conveys the analysis or concentration of the nutrient in a carrier. When properly applied, most fertilizer carriers give equally good results for a given amount of nutrient element. The more concentrated carriers are usually the most economical to use, because less weight of fertilizer must be transported to supply the needed quantity of a given

Conventional labeling of fertilizer products reports percentage N, P_2O_5, and K_2O. Thus, a fertilizer package (Figure 16.20) labeled as 6-24-24 (6% nitrogen, 24% P_2O_5, 24% K_2O) actually contains 6% N, 10.5% P, and 19.9% K (see calculations below).

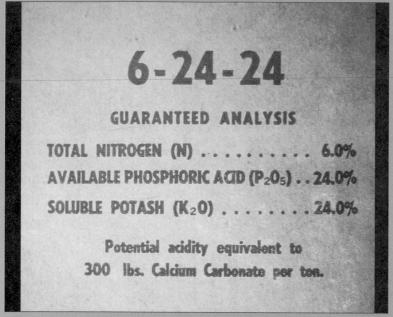

FIGURE 16.20   *A typical commercial fertilizer label. Note that a calculation must be performed to determine the percentage of the nutrient elements P and K in the fertilizer since the contents are expressed as if the nutrients were in the forms of P_2O_5 and K_2O. Also note that after interacting with the plant and soil, this material would cause an increase in soil acidity that could be neutralized by 300 units of $CaCO_3$ per 2000 units (1 ton = 2000 lbs) of fertilizer material.*

In order to determine the amount of fertilizer needed to supply the recommended amount of a given nutrient, it is necessary to convert percent P_2O_5 and percent K_2O to percent P and K. This may be done by calculating the proportion of P_2O_5 that is P and the proportion of K_2O that is K. The following calculations may be used:

Given that the molecular weights of P, K, and O are as follows:

P 30.97 g/mol

K 39.10 g/mol

O 16.00 g/mol

Molecular weight of $P_2O_5 = 2(30.97) + 5(16.00) = 141.94$ g/mol

Proportion P in $P_2O_5 = (2(30.97))/141.94 = 0.4364$

To convert $P_2O_5 \rightarrow P$ multiply percent P_2O_5 by 0.44

Molecular weight of $K_2O = 2(39.10) + 16.00 = 94.20$

Proportion K in $K_2O = (2(39.10))/94.20 = 0.8301$

To convert $K_2O \rightarrow K$ multiply percent K_2O by 0.83

Thus, if the bag in Figure 16.20 contains 25 kg of the 6-24-24 fertilizer, it will supply the following amounts of N, P, and K:

Fertilizer analysis	Conversion to percent element	Element in 25 kg, kg
6% N	No conversion → 6%	0.06 × 25 kg = 1.5 kg of N
24% P_2O_5	(24% × 0.44) = 10.5%	0.105 × 25 kg = 2.6 kg of P
24% K_2O	(24% × 0.83) = 19.9%	0.199 × 25 kg = 5.0 kg of K

nutrient. Hence, economic comparisons among different equally suitable fertilizers should be based on the price per kilogram of nutrient, not the price per kilogram of fertilizer.

Fate of Fertilizer Nutrients

A common *myth* about fertilizers suggests that inorganic fertilizers applied to soil directly feed the plant, and that therefore the biological cycling of nutrients, such as described by Figures 13.3 for nitrogen and 14.6 for phosphorus, are of little consequence where inorganic fertilizers are used. The reality is that nutrients added by normal application of fertilizers, whether organic or inorganic, are incorporated into the complex soil nutrient cycles, and that relatively little of the fertilizer nutrient (from 10 to 60%) actually winds up in the plant being fertilized during the year of application. Even when the application of fertilizer greatly increases both plant growth and nutrient uptake, the fertilizer stimulates increased cycling of the nutrients, and the nutrient ions taken up by the plant come largely from various pools in the soil and not directly from the fertilizer. This knowledge has been obtained by careful analysis of dozens of nutrient studies that used fertilizer with isotopically tagged nutrients. Results from such a study are summarized in Table 16.15, which shows somewhat more N uptake from fertilizer than is typically reported. Generally, as fertilizer rates are increased, the efficiency of fertilizer nutrient use decreases, leaving behind in the soil an increasing proportion of the added nutrient.

16.12 THE CONCEPT OF THE LIMITING FACTOR

A famous German chemist named Justus von Liebig is credited with first publishing the concept that *plant production can be no greater than that level allowed by the growth factor present in the lowest amount relative to the optimum amount for that factor*. This growth factor, be it temperature, nitrogen, or water supply, will limit the amount of growth that can occur and is therefore called the *limiting factor* (Figure 16.21).

If a factor is not the limiting one, increasing it will do little or nothing to enhance plant growth. In fact, increasing the amount of a nonlimiting factor may actually reduce plant growth by throwing the system further out of balance. For example, if a plant is limited by lack of phosphorus, adding more nitrogen may only aggravate the phosphorus deficiency.

Looked at another way, applying available phosphorus (the first limiting nutrient in this example) may allow the plant to respond positively to a subsequent addition of nitrogen. Thus, the increased growth obtained by applying two nutrients together often is much greater than the sum of the growth increases obtained by applying each of the two nutrients individually. Such an *interaction* or *synergy* between two nutrients can be seen in the data of Figure 16.22.

TABLE 16.15 Source of Nitrogen in Corn Plants Grown in North Carolina on an Enon Sandy Loam Soil (Ultic Hapludalf) Fertilized with Three Rates of Nitrogen as Ammonium Nitrate

The source of the nitrogen in the corn plant was determined by using fertilizer tagged with the isotope ^{15}N. Note that moderate use of nitrogen fertilizer increased the uptake of N already in the soil system as well as that derived from the fertilizer.

Fertilizer nitrogen applied, kg/ha	Corn grain yield, Mg/ha	Total N in corn plant, kg/ha	Fertilizer-derived N in corn, kg/ha	Soil-derived N in corn, kg/ha	Fertilizer-derived N in corn as percent of total N in corn, %	Fertilizer-derived N in corn as percent of N applied, %
50	3.9	85	28	60	33	56
100	4.6	146	55	91	38	55
200	5.5	157	86	71	55	43

Calculated from Reddy and Reddy (1993).

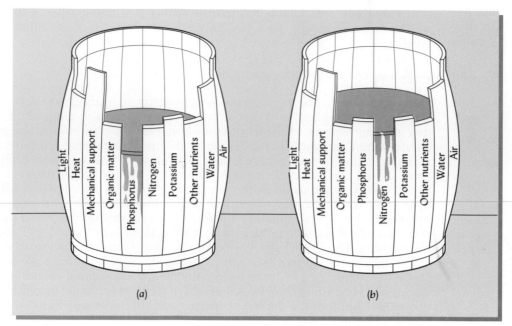

FIGURE 16.21 An illustration of the law of the minimum and the concept of the limiting factor. Plant growth is constrained by the essential element (or other factor) that is most limiting. The level of water in the barrel represents the level of plant production. (*a*) Phosphorus is represented as being the factor that is most limiting. Even though the other elements are present in more than adequate amounts, plant growth can be no greater than that allowed by the level of phosphorus available. (*b*) When phosphorus is added, the level of plant production is raised until another factor becomes most limiting—in this case, nitrogen.

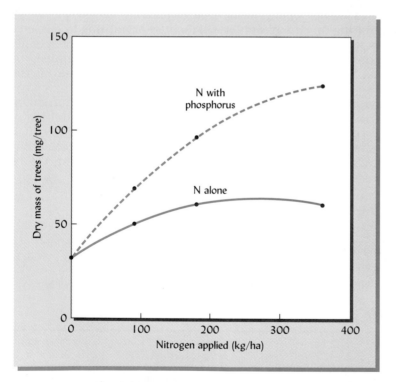

FIGURE 16.22 Increase in biomass of one-year-old white pine seedlings in response to nitrogen fertilizer given with or without phosphorus on a sandy soil in southern Ontario, Canada. Apparently, phosphorus was the most limiting nutrient, and therefore little response was obtained from adding nitrogen until the phosphorus level was raised. This is an example of a nitrogen-by-phosphorus interaction. [Redrawn from Teng and Timmer (1994)]

Wise, effective fertilizer use involves making correct decisions regarding *which* nutrient element(s) to apply, *how much* of each needed nutrient to apply, *what type* of material or carrier to use (Tables 16.8 and 16.12 list some of the choices), *in what manner* to apply the material, and, finally, *when* to apply it. We will leave information on the first two decisions until Section 16.15. Here we will discuss the alternatives available with regard to the last two decisions.

There are three general approaches to applying fertilizers: (1) *broadcast application,* (2) *localized placement,* (3) and *foliar application.* Each method has some advantages and disadvantages and may be particularly suitable for different situations. Often some combination of the three methods is used.

Broadcasting

In many instances fertilizer is spread evenly over the entire field or area to be fertilized. This method is called *broadcasting.* Often the broadcast fertilizer is mixed into the plow layer by means of tillage, but in some situations it is left on the soil surface and allowed to be carried into the root zone by percolating rain or irrigation water. The broadcast method is most appropriate when a large amount of fertilizer is being applied with the aim of raising the fertility level of the soil over a long period of time. Broadcasting is the most economical way to spread large amounts of fertilizer over wide areas (Figure 16.23).

(a)

(b)

FIGURE 16.23   (*a*) Broadcasting granular phosphorus and potassium fertilizer on a hayfield in Maryland. (*b*) Broadcasting a nitrogen solution (mixed with fungicide) on a wheat field in France. (Photos courtesy of R. Weil)

For close-growing vegetation, broadcasting provides an appropriate distribution of the nutrients. It is therefore the most commonly used method for rangeland, pastures, small grains, turf grass, and forests. Fertilizers are also broadcast on some row cropland, especially in the fall when it is most convenient, although certainly not most efficient. The broadcast fertilizer may or may not be incorporated into the soil (Figure 16.24*b* and *c*).

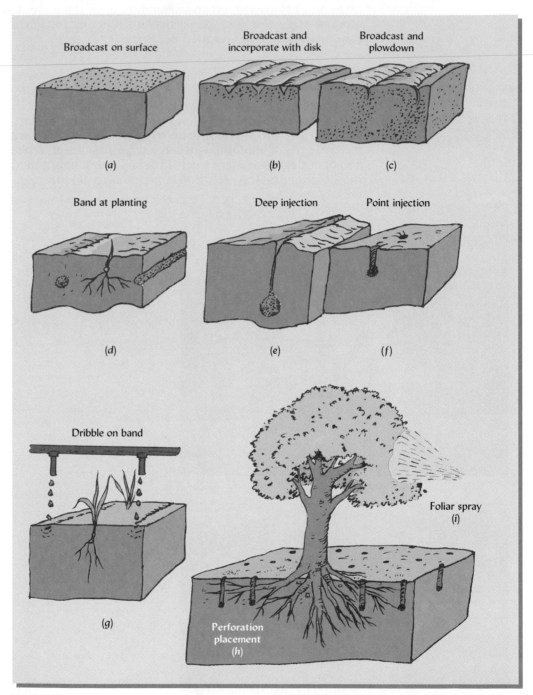

FIGURE 16.24 Fertilizers may be applied by many different methods, depending on the situation. Methods (*a*) to (*c*) represent broadcast fertilizer, with or without incorporation. Methods (*d*) to (*h*) are variations of localized placement. Method (*i*) is foliar application and has special advantages, but also limitations. Commonly, two or three of these methods may be used in sequence. For example, a field may be prepared with (*c*) before planting; (*d*) may be used during the planting operation; (*g*) may be used as a side dressing early in the growing season; and, finally, (*i*) may used to correct a micronutrient deficiency that shows up in the middle of the season.

For phosphorus, zinc, manganese, and other nutrients that tend to be strongly retained by the soil, broadcast applications are usually much less efficient than localized placement. Often 2 to 3 kg of fertilizer must be broadcast to achieve the same response as from 1 kg that is placed in a localized area.

A heavy one-time application of phosphorus and potassium fertilizer, broadcast and worked into the soil, is a good preparation for establishing perennial lawns and pastures. It may be necessary to broadcast a top dressing in subsequent years, being careful not to allow fertilizers with high salt hazards (see Table 16.13) to remain in contact with the foliage long enough to cause salt burn. Because of its mobility in the soil, nitrogen does not suffer from reduced availability when broadcast, but if left on the soil surface much may be lost by volatilization. Volatilization losses are especially troublesome for urea and ammonium fertilizers applied to soils with a high pH. Nitrogen is commonly broadcast (sprayed) in a liquid form, often in a solution that also contains other nutrients or chemicals (see Figure 16.23b).

Another disadvantage of surface broadcasting is that the fertilizer is easily washed away with runoff during heavy rains. In fact, many runoff studies have shown that most of the annual loss of nutrients (or herbicides, if surface broadcast) usually occurs during the first one or two heavy-rainfall events after the broadcast application.

APPLICATION IN IRRIGATION WATER. Where irrigation is practiced, liquid fertilizers can be applied in the irrigation water, a practice sometimes called *fertigation*. Liquid ammonia, nitrogen solutions, phosphoric acid, and even complete fertilizers are dissolved in the irrigation stream or the overhead sprinkler system supply. The nutrients are thus broadcast onto the soil in solution and carried down with the infiltrating water. Not only are application costs reduced, but also relatively inexpensive nitrogen carriers can be used. Some care must be taken, however, to prevent ammonia loss by evaporation and to avoid certain fertilizer compounds that can clog the irrigation system with precipitates.

Finally, it should be pointed out that for crops with wide row spacing or young tree seedlings in forest plantings, broadcasting places the fertilizer where it is just as accessible to the *weeds* as to the target plants.

Localized Placement

Although it is commonly thought that nutrients must be thoroughly mixed throughout the root zone if plants are to be able to readily satisfy their needs, research has clearly shown that a plant can easily obtain its entire supply of a nutrient from a concentrated localized source in contact with only a small fraction of its root system. In fact, a small portion of a plant's root system can grow and proliferate in a band of fertilizer even though the salinity level caused by the fertilizer would be fatal to a germinating seed or to a mature plant were a large part of its root system exposed. This finding allowed the development of techniques for localized fertilizer placement.

There are at least two reasons why fertilizer is often more effectively used by plants if it is placed in a localized concentration rather than mixed with soil throughout the root zone. First, localized placement reduces the amount of contact between soil particles and the fertilizer nutrient, thus minimizing the opportunity for adverse fixation reactions. Second, the concentration of the nutrient in the soil solution at the root surface will be very high, resulting in greatly enhanced uptake by the roots.

STARTER APPLICATION. Localized placement is especially effective for young seedlings, in cool soils in early spring, and for plants that grow rapidly with a big demand for nutrients early in the season. For these reasons, starter fertilizer is often applied in bands on either side of the seed as the crop is planted. Since germinating seeds can be injured by fertilizer salts, and since these salts tend to move upward as water evaporates from the soil surface, the best placement for starter fertilizer is approximately 5 cm below and 5 cm off to the side from the seed row (see Figure 16.24d and Figure 16.25).

LIQUID FERTILIZERS. Liquid fertilizers and slurries of manure and sewage sludge can also be applied in bands rather than broadcast. Bands of these liquids are placed 10 to 30 cm deep in the soil by a process known as *knife injection* (Figures 16.24e and 16.26). In addition to the advantages mentioned for banding fertilizer, injection of these organic slurries reduces runoff losses and odor problems.

FIGURE 16.25 Many modern planters are equipped with a fertilizer-banding attachment that places starter fertilizer for row crops slightly below and slightly to the side of the seed. This placement eliminates the danger of fertilizer burn, yet concentrates the nutrients near the seed where the crop roots will encounter them shortly after the seed germinates. (Courtesy National Plant Food Institute, Washington, D.C.)

Anhydrous ammonia and pressurized nitrogen solutions must be injected into the soil to prevent losses by volatilization. Injecting bands at depths of 15 and 5 cm, respectively, are considered adequate for these two materials.

DRIBBLE APPLICATION. Another approach to banding liquids (though not slurries) is to dribble a narrow stream of liquid fertilizer alongside the crop row as a side dressing. The use of a stream instead of a fine spray changes the application from broadcast to banding and results in enough liquid in a narrow zone to cause the fertilizer to soak into the soil. This action greatly reduces volatilization loss of nitrogen.

POINT INJECTION. Localized placement of fertilizer can be carried one step beyond banding with a new system called *point injection*. With this system, small portions of liquid fertilizer can be applied next to every individual plant without significantly disturbing either the plant root or the surface residue cover left by conservation tillage. The point injection implement shown in Figure 16.26 (bottom) is a modern version of the age-old dibble stick with which peasant farmers in the lesser developed countries plant seeds and later apply a portion of fertilizer in the soil next to each plant, all with a minimum of disturbance of the surface mulch.

DRIP IRRIGATION. The use of drip irrigation systems (see Section 6.11) has greatly facilitated the localized application of nutrients in irrigation water. Because drip fertigation is applied at frequent intervals, the plants are essentially spoon-fed, and the efficiency of nutrient use is quite high.

PERFORATION METHOD FOR TREES. Trees in orchards and ornamental plantings are best treated individually, the fertilizer being applied around each tree within the spread of the branches but beginning approximately 1 m from the trunk (see Figure 16.24*h*). The fertilizer is best applied by what is called the *perforation* method. Numerous small holes are dug around each tree within the outer half of the branch-spread zone and extending down into the upper subsoil. Into these holes, which are afterward filled up, is placed a suitable amount of an appropriate fertilizer. Special large fertilizer pellets are available for this purpose. This method of application places the nutrients within the tree root zone and avoids an undesirable stimulation of the grass or cover that may be growing around the trees. If the cover crop or lawn around the trees needs fertilization, it is treated separately, the fertilizer being drilled in at the time of seeding or broadcast later.

Foliar Application

Plants are capable of absorbing nutrients through their leaves in limited quantities. Under certain circumstances, the best way to apply a nutrient is *foliar application*—spraying a dilute nutrient solution directly onto the plant leaves. Diluted NPK fertilizers, micronu-

FIGURE 16.26 (Upper left) Sewage-sludge slurry being knife-injected into the soil before planting crops on the land. This injection method reduces runoff losses and objectionable odors. Lighter knives (not shown) are used to inject liquid fertilizers. (Bottom) A spike-wheel applicator designed for point injection of liquid fertilizer. (Upper right) Close-up view of spike-wheel applicator. [Photo (top left) courtesy of R. Weil; photos (top right and bottom) courtesy of Fluid Fertilizer Foundation]

trients, or small quantities of urea can be used as foliar sprays, although care must be taken to avoid significant concentrations of Cl^- or NO_3^-, which can be toxic to some plants. Foliar fertilization may conveniently fit in with other field operations for horticultural crops, because the fertilizer is often applied simultaneously with pesticide sprays.

The amount of nutrients that can be sprayed on leaves in a single application is quite limited. Therefore, while a few spray applications may deliver the entire season's requirement for a micronutrient, only a small portion of the macronutrient needs can be supplied in this manner. The danger of leaf injury is especially high during dry, hot weather, when the solution quickly evaporates from the leaf surface, leaving behind the fertilizer salts. Spraying on cool, overcast days or during early morning or late evening hours reduces the risk of injury, as does the use of a dilute solution containing, for example, only 1 or 2% nitrogen.

16.14 TIMING OF FERTILIZER APPLICATION

The timing of nutrient applications in the field is governed by several basic considerations: (1) making the nutrient available when the plant needs it; (2) avoiding excess availability, especially of nitrogen, before and after the principal period of plant uptake; (3) making nutrients available when they will strengthen, not weaken, long-season and perennial plants; and (4) conducting field operations when conditions make them practical and feasible.

Availability When the Plants Need It

For mobile nutrients such as nitrogen (and to some degree potassium), the general rule is to make applications as close as possible to the period of rapid plant nutrient uptake. For rapid-growing summer annuals, such as corn, this means making only a small starter application at planting time and applying most of the needed nitrogen as a side dressing just before the plants enter the rapid nutrient accumulation phase, usually about four to six weeks after planting. For cool-season plants, such as winter wheat or certain turf grasses, most of the nitrogen should be applied about the time of spring "green-up," when the plants resume a rapid growth rate. For trees, the best time is when new leaves are forming. With slow-release organic sources, some time should be allowed for mineralization to take place prior to the plants' period of maximum uptake.

Environmentally Sensitive Periods

In temperate (**Udic** and **Xeric**) climates, most leaching takes place in the winter and early spring when precipitation is high and evapotranspiration is low. Nitrates left over after plant uptake has ceased have the potential for leaching during this period. In this regard it should be noted that, for grain crops, the rate of nutrient uptake begins to decline during grain-filling stages and has virtually ceased long before the crop is ready for harvest. With inorganic nitrogen fertilizers, avoiding leftover nitrates is largely a matter of limiting the amount applied to what the plants are expected to take up. However, for slow-release organic sources applied in late spring or early summer, mineralization is likely to continue to release nitrates after the crop has matured and ceased taking them up. To the extent that this timing of nitrate release is unavoidable, nonleguminous cover crops should be planted in the fall to absorb the excess nitrate being released.

SPLIT APPLICATIONS. In high-rainfall conditions and on permeable soils, dividing a large dose of fertilizer into two or more split applications may avoid leaching losses prior to the crop's establishment of a deep root system. In cold climates, another environmentally sensitive period occurs during early spring, when snowmelt over frozen or saturated soils results in torrents of runoff water that may carry to rivers and streams soluble nutrients from manure or fertilizer that may be at or near the soil surface.

Application of nitrogen fertilizers to mature forests is usually carried out when rains can be expected to wash the nutrients into the soil and minimize volatilization losses. However, observation of nitrate content in stream flow often reveals that such applications result in a pulse of nitrate leaving the watershed for several weeks following the fertilizer application.

Physiologically Appropriate Timing

It is important to make nutrients available when they will strengthen plants and improve their quality. For example, too much nitrogen in the summer may stress a cool-season turf grass, while high nitrogen late in the season will reduce the sugar content of sugar crops. A good supply of potassium is particularly important in the fall to enable plants to improve their winter-hardiness. Broadcast fertilization of trees soon after the seedlings are planted may benefit fast-growing weeds more than the desired trees. Later in the development of a forest stand, when the tree canopy has matured, application of fertilizer, usually from a helicopter, can be quite beneficial.

Practical Field Limitations

Sometimes it is simply not possible to apply fertilizers at the ideal time of the year. For example, although a crop may respond to a late-season side dressing, such an application will be difficult if the plants are too tall to drive over without damaging them. Application by airplane may increase the flexibility possible. Early spring applications may be limited by fields too wet to support tractors. Economic costs or the time demands of other activities may also require that compromises be made in the timing of nutrient application.

16.15 DIAGNOSTIC TOOLS AND METHODS[6]

Three basic tools are available for diagnosing soil fertility problems: (1) *field observations,* (2) *plant tissue analysis,* and (3) *soil analysis* (soil testing). To effectively guide the application of nutrients, as well as to diagnose problems as they arise in the field, all three approaches should be integrated. There is no substitute for careful observation and *recording* of circumstantial evidence and symptoms in the field. Effective observation and interpretation requires skill and experience, as well as an open mind. It is not uncommon for a supposed soil fertility problem to actually be caused by soil compaction, weather conditions, pest damage, or human error. The task of the diagnostician is to use all the tools available in order to identify the factor that is limiting plant growth, and then devise a course of action to alleviate the limitation.

16.16 PLANT SYMPTOMS AND FIELD OBSERVATIONS

This detectivelike work can be one of the more exciting and challenging aspects of nutrient management. To be an effective soil fertility diagnostician, several general guidelines are helpful.

First, develop an organized way to *record* your observations. The information you collect may be needed to properly interpret soil and plant analytical results obtained at a later date.

Second, look for *spatial patterns*—how the problem seems to be distributed in the landscape and in individual plants. Linear patterns across a field may indicate a problem related to tillage, drain tiles, or the incorrect spreading of lime or fertilizer. Poor growth concentrated in low-lying areas may relate to the effects of soil aeration. Poor growth on the high spots in a field may reflect the effects of erosion and possibly exposure of subsoil material with an unfavorable pH.

Third, closely examine individual plant leaves to characterize any foliar symptoms. Nutrient deficiencies can produce characteristic symptoms on leaves and other plant parts. Examples of such symptoms are shown in several figures in Chapters 13 to 15 and in Plates 28 to 33. Determine if the symptoms are most pronounced on the younger leaves (as is the case for most of the micronutrient cations) or on the older leaves (as is the case for nitrogen, potassium, and magnesium). Some nutrient deficiencies are quite reliably identified from foliar symptoms, while others produce symptoms that may be confused with herbicide damage, insect damage, or damage from poor aeration.

Fourth, observe and *measure* differences in plant growth and crop yield that may reflect different levels of soil fertility, even though no leaf symptoms are apparent. Check *both* aboveground and belowground growth. Are mycorrhizae associated with tree roots? Are legumes well nodulated? Is root growth restricted in any way?

Fifth, obtain records on plant growth or crop yield from previous years, and ascertain the history of management for the site for as many years as possible. It is often useful to sketch a map of the site showing features you have observed and the distribution of symptoms.

16.17 PLANT ANALYSIS AND TISSUE TESTING

The concentration of essential elements in plant tissue is related to plant growth or crop yield, as shown in Figure 16.27. The *sufficiency range* and the *critical range* have been well characterized for many plants, especially for agronomic and major horticultural crops. Less is known about forest trees and ornamentals. Sufficiency ranges for 11 essential elements in a variety of plants are listed in Table 16.16.[7] Plants with tissue

---

[6] For a detailed discussion of both plant analysis and soil testing, see Westerman (1990).

[7] Detailed information on tissue analyses for a large number of plant species can be found in Reuter and Robinson (1986).

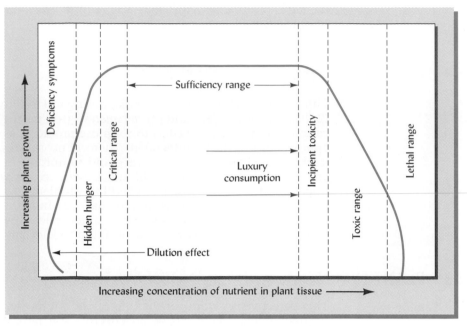

FIGURE 16.27 The relationship between plant growth or yield and the concentration of an essential element in the plant tissue. For most nutrients there is a relatively wide range of values associated with normal, healthy plants (the sufficiency range). Beyond this range, plant growth suffers from either too little or too much of the nutrient. The critical range (CR) is commonly used for the diagnosis of nutrient deficiency. Nutrient concentrations below the CR are likely to reduce plant growth even if no deficiency symptoms are visible. This moderate level of deficiency is sometimes called *hidden hunger*. The odd hook at the lower left of the curve is the result of the so-called dilution effect that is often observed when extremely stunted, deficient plants are given a small dose of the limiting nutrient. The growth response may be so great that even though somewhat more of the element is taken up, it is diluted in a much greater plant mass.

concentrations lower than the smaller value given in the sufficiency range for a particular nutrient are likely to respond to additions of that nutrient if no other factor is more limiting.

One plant test that has received much attention in recent years is the end of season cornstalk nitrate test to improve the nitrogen management of corn. The nitrate content of the lower portion of the cornstalks at physiological maturity is measured. While this test can identify deficiencies of nitrogen, it is particularly helpful in identifying excessive levels of nitrogen in the plant at harvest time, which, in turn, is an indication of excessive levels in the soil. The results can be used to monitor the appropriateness of corn nitrogen-management systems and make adjustments in future years.

Tissue analysis can be a powerful tool for identifying plant nutrient problems if several simple precautions are taken. First, it is critical that the correct plant part be sampled. Second, the plant part must be sampled at the specified stage of growth, because the concentrations of most nutrients decrease considerably as the plant matures. Third, it must be recognized that the concentration of one nutrient may be affected by that of another nutrient, and that sometimes the ratio of one nutrient to another (e.g., Mg/K, S/N, or Fe/Mn) may be the most reliable guide to plant nutritional status. In fact, several elaborate mathematical systems for assessing the ratios or balance among nutrients have proven useful for certain plant species.[8] Because of the uncertainties and complexities in interpreting tissue concentration data, it is wise to sample plants from the best and worst areas in a field or stand. The difference between each samples may provide valuable clues concerning the nature of the nutrient problem.

---

[8] The most well developed of the multinutrient ratio systems is known as the Diagnostic Recommendation Integrated System (DRIS). For details on the DRIS, see Walworth and Sumner (1987).

TABLE 16.16   A Guide to Tissue Analysis for Selected Plant Species

The values given are for sufficiency ranges.

Plant species	Part to sample	Content, %						Content, µg/g				
		N	P	K	Ca	Mg	S	Fe	Mn	Zn	B	Cu
Pine trees (*Pinus* spp.)	Current-year needles near terminal	1.2–1.4	0.10–0.18	0.30–0.45	0.13–0.16	0.05–0.09	0.08–0.12	20–100	50–600	20–50	3–9	2–6
Turf grasses	Clippings	2.75–3.5	0.30–0.55	1.0–2.5	0.50–1.2	0.20–0.60	0.20–0.45	35–1000	25–150	20–55	10–60	5–20
Corn (*Zea mays*)	Ear-leaf at tasseling	2.5–3.5	0.20–0.50	1.5–3.0	0.2–1.0	0.16–0.40	0.16–0.50	25–300	20–200	20–70	6–40	6–40
Soybean (*Glycine max*)	Recently matured trifoliate, flowering stage	4.0–5.0	0.31–0.50	2.0–3.0	0.45–2.0	0.25–0.55	0.25–0.55	50–250	30–200	25–50	25–60	8–20
Apple (*Malus* spp.)	Leaf from base of nonfruiting shoots	1.8–2.4	0.15–0.30	1.2–2.0	1.0–1.5	0.25–0.50	0.13–0.30	50–250	35–100	20–50	20–50	5–20
Wheat (*Triticum* spp.)	First leaf blade from top of plant	2.2–3.3	0.24–0.36	2.0–3.0	0.28–0.42	0.19–0.30	0.20–0.30	35–55	30–50	20–35	5–10	6–10
Tomato (*Lycopersicon esculentum*)	Most recently matured leaf at early bloom	3.2–4.8	0.32–0.48	2.5–4.2	1.7–4.0	0.45–0.70	0.60–1.0	120–200	80–180	30–50	35–55	8–12
Alfalfa (*Medicago sativa*)	Upper third of plant at first flower	3.0–4.5	0.25–0.50	2.5–3.8	1.0–2.5	0.3–0.8	0.3–0.5	50–250	25–100	25–70	6–20	30–80

Data derived from many sources.

Since the total amount of an element in a soil tells us very little about the ability of that soil to supply that element to plants, more meaningful *partial soil analyses* have been developed. *Soil testing* is the routine partial analysis of soils for the purpose of guiding nutrient management.

The soil testing process consists of three critical phases: (1) *sampling* the soil, (2) chemically *analyzing* the sample, and (3) *interpreting* the analytical result to make a recommendation on the kind and amounts of nutrients to apply.

Sampling the Soil

Soil sampling is widely acknowledged to be one of the weakest links in the soil testing process. Part of the problem is that about a teaspoonful of soil (Figure 16.28) is eventually used to represent millions of kilograms of soil in the field. Since soils are highly variable, both horizontally and vertically, it is essential to carefully follow the sampling instructions from the soil testing laboratory.

Because of the variability in nutrient levels from spot to spot, it is always advisable to divide a given field or property into as many distinct areas as practical when taking soil samples to determine nutrient needs. For example, suppose a 20-ha field has a 2-ha low spot in the middle, and 5 ha at one end that used to be a permanent pasture. These two areas should be sampled, and later managed, separately from the remainder of the field. Similarly, a homeowner should sample flower beds separately from lawn areas, low spots separately from sloping areas, and so on. On the other hand, known areas of

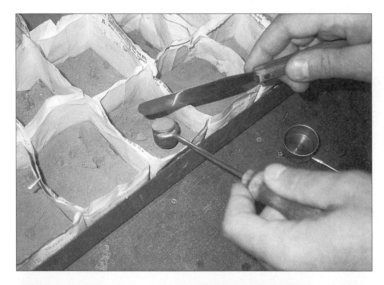

FIGURE 16.28 (Top) After a soil sample is received by the soil testing lab, the soil is ground and screened to make a homogenous powder. Then a small scooped or weighed sample is analyzed chemically. This small amount of soil must represent thousands of metric tons of soil in the field. (Bottom) After a portion of the nutrients have been extracted from the soil sample, the solution containing these nutrients undergoes elemental analysis. Since soil test labs must run hundreds or thousands of samples each day, the analysis is generally automated and the results are recorded and interpreted by computer. (Photos courtesy of R. Weil)

unusual soil that are too small or irregular to be managed separately should be avoided and not included in the composite sample from the whole field.

COMPOSITE SAMPLE.   Usually, a soil probe is used to remove a thin cylindrical core of soil from at least 15 to 20 randomly scattered places within the land area to be represented (Figure 16.29). The 15 to 20 subsamples are thoroughly mixed in a plastic bucket, and about 0.5 L of the soil is placed in a labeled container and sent to the lab. If the soil is moist, it should be air-dried without sun or heat prior to packaging for routine soil tests. Heating the sample might cause a falsely high result for certain nutrients.[9]

Two questions must be addressed when sampling a soil: (1) the depth to which the sample should be taken, and (2) the time of year when the soil should be sampled.

DEPTH TO SAMPLE.   The standard depth of sampling for a plowed soil is the depth of the plowed layer, about 15 to 20 cm, but various other depths are also used (see Figure 16.29). Because in many unplowed soils nutrients are stratified in contrasting layers, the depth of sampling can greatly alter the results obtained.

TIME OF YEAR.   Seasonal changes are often observed in soil test results for a given field. For example, the potassium level is usually highest in early spring, after freezing and thawing has released some fixed K ions from clay interlayers, and lowest in late summer, after plants have removed much of the readily available supply. The time of sampling is especially important if year-to-year comparisons are to be made. A good practice is to sample each field every year or two (always at the same time of year), so that the soil test levels can be tracked over the years to determine whether nutrient levels are being maintained, increased, or depleted.

---

[9] Samples intended for analysis of soil nitrates are an exception. These should be rapidly dried under a fan or in an oven at about 65°C.

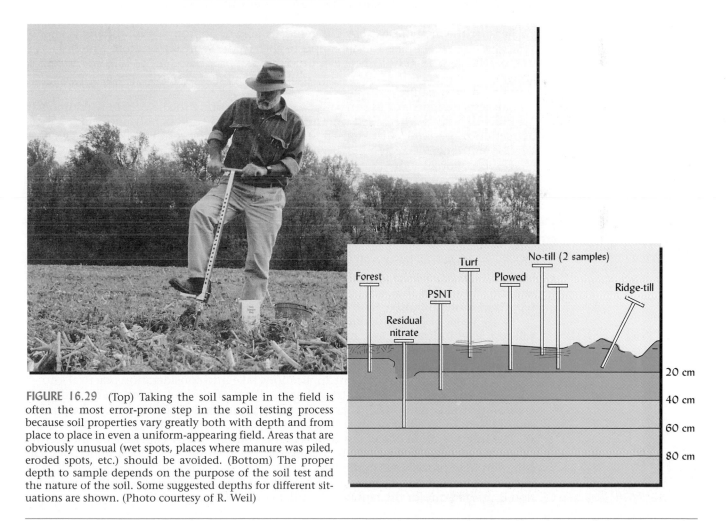

FIGURE 16.29 (Top) Taking the soil sample in the field is often the most error-prone step in the soil testing process because soil properties vary greatly both with depth and from place to place in even a uniform-appearing field. Areas that are obviously unusual (wet spots, places where manure was piled, eroded spots, etc.) should be avoided. (Bottom) The proper depth to sample depends on the purpose of the soil test and the nature of the soil. Some suggested depths for different situations are shown. (Photo courtesy of R. Weil)

TIMING FOR SPECIAL NITROGEN TESTS. Timing is especially critical to determine the amount of mineralized nitrogen in the root zone. In relatively dry, cold regions (e.g., the Great Plains), the *residual nitrate* test is done on 60-cm-deep samples obtained sometime between fall and before planting in spring. In humid regions, where nitrate leaching is more pronounced, a special test has been developed to determine whether a soil will mineralize enough nitrogen for a corn crop. The samples for this *Presidedress Nitrate Test* (PSNT) are taken from the upper 30 cm when the corn is about 30 cm tall, just in time to determine how much nitrogen to apply as the crop enters its period of most rapid nitrogen uptake. In this case, the soil must be sampled during the narrow window of time when spring mineralization has peaked, but plant uptake has not yet begun to deplete the nitrate produced (see Figure 13.6).

Chemical Analysis of the Sample

In general, soil tests attempt to extract from the soil amounts of essential elements that are correlated with the nutrients taken up by plants. Different extraction solutions are employed by various laboratories. Buffered salt solutions, such as sodium or ammonium acetate, or mixtures of dilute acids and chelating agents are the extracting agents most commonly used. The extractions are accomplished by placing a small measured quantity of soil in a bottle with the extracting agent and shaking the mixture for a certain number of minutes. The amount of the various nutrient elements brought into solution is then determined. The whole process is usually automated so that a modern laboratory can handle hundreds of samples each day (see Figure 16.28).

The most common and reliable tests are those for soil pH, potassium, phosphorus, and magnesium. Micronutrients are sometimes extracted using chelating agents, especially for calcareous soils in the more arid regions. Predicting the availability of nitrogen and sulfur is considerably more difficult because of the many biological factors involved, but the nitrate and sulfate present in the soil at the time of sampling can be measured.

Because the methods used by different labs may be appropriate for different types of soils, it is advisable to send a soil sample to a lab in the same region from which the soil originated. Such a lab is likely to use procedures appropriate for the soils of the region and should have access to data correlating the analytical results to plant responses on soils of a similar nature.

Soil tests designed for use on soils or soil-based potting media generally do not give meaningful results when used on artificial peat-based soilless potting media. Special extraction procedures must be used for the latter, and the results must then be correlated to nutrient uptake and growth of plants grown in similar media.

These examples should further emphasize the importance of providing the soil testing lab with complete information concerning the nature of your soil, its management history, and your plans for its future use.

Interpreting the Results to Make a Recommendation

This is, perhaps, the most controversial aspect of soil testing. The soil test values themselves are merely *indices* of nutrient-supplying power. They do not indicate the actual amount of nutrient that will be supplied. For this reason, it is best to think of soil test reports as more indicative than quantitative.

Many years of field experimentation at many sites are needed to determine which soil test level indicates a low, medium, or high capacity to supply the nutrient tested. Such categories are used to predict the likelihood of obtaining a profitable response from the application of the nutrient tested (Figure 16.30).

Recommendations for nutrient applications take into consideration practical knowledge of the plants to be grown, the characteristics of the soil under study, and other environmental conditions. Management history and field observations can help relate soil test data to fertilizer needs.

The interpretation of soil test data is best accomplished by experienced and technically trained personnel who fully understand the scientific principles underlying the common field procedures. In modern laboratories, the factors to be considered in making fertilizer recommendations are programmed into a computer, and the interpretation is printed out for the farmer's or gardener's use (Figure 16.31).

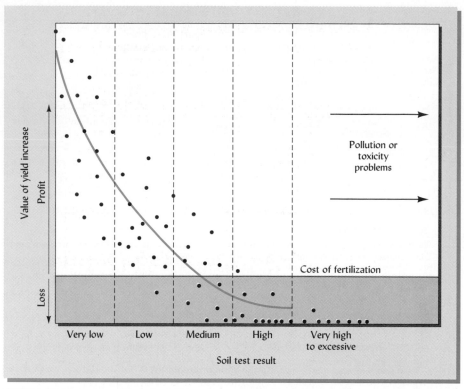

FIGURE 16.30 The relationship between soil test results for a nutrient and the extra yield obtained by fertilizing with that nutrient. Each data point represents the *difference* in plant yield between the fertilized and the unfertilized soil. Because many factors affect yield and because soil tests can only approximately predict nutrient availability, the relationship is not precise, but the data points are scattered about the trend line. If the point falls above the fertilizer cost line, the extra yield was worth more than the cost of the fertilizer and a profit would be made. For a soil testing in the very low and low categories, a profitable response to fertilizer is very likely. For a soil testing medium, a profitable response is a 50:50 proposition. For soils testing in the high category, a profitable response is unlikely.

Merits of Soil Testing

It must not be inferred from the preceding discussion that the limitations of soil testing outweigh its advantages. When the precautions already described are observed, soil testing is an invaluable tool in making fertilizer recommendations. These tests are most useful when they are correlated with the results of field fertilizer experiments (Figure 16.32).

Generally, soil testing has been most relied upon in agricultural systems, while foliar analysis has proved more widely useful in forestry. The limited use of soil testing in forestry is probably a result of the complex stratification in forested soils, which creates a great deal of uncertainty about how to obtain a representative sample of soil for analysis. Also, because of the comparably long time frame in forestry, little information is available on the correlation of soil test levels with timber yield in the sense that such information is widely available for agronomic crops (see Figure 16.30). Even though the relationship between tree growth and soil test level is not well known for most forest systems, standard agronomic soil testing can still be useful in distinguishing those soils whose ability to supply P or K is adequate from those with very low supplying power for these nutrients.

16.19 SITE-SPECIFIC NUTRIENT MANAGEMENT[10]

Computer technologies have recently been combined with the Global Positioning System (GPS) of earth-orbiting satellites (see Section 19.3) to make nutrient management more site-specific than was previously practical for large farming operations. Portable GPS receivers can plot one's exact (to within about 5 to 10 m) location as one moves

---

[10] See Robert, et al. (1996) for a series of papers dealing with site-specific nutrient management.

Report on soil tests
Auburn University
Soil testing laboratory
Auburn University, AL 36849

Name   Alabama resident   County   Lee
Address   118 Main Street   District   2
City   Hometown, AL 36830

Lab. no.	Sender's sample designation and	Crop to be grown	Soil* group	pH**	Phosphorus P***		Potassium K***		Magnesium Mg***		Limestone Tons/acre	N	P₂O₅	K₂O
												Pounds per acre		
23887	1	Soybeans	2	5.3	L	70	M	70	H	160	2.0	0	80	40

Comment 224—soil acidity (low pH) can be corrected with either dolomitic or calcitic lime.

| 23888 | 2 | Corn | 1 | 5.6 | L | 70 | M | 70 | H | 160 | 1.0 | 120 | 80 | 40 |

See comment 224 above.
Comment 15—corn on sandy soils may respond to nitrogen rates up to 150 lbs. per acre. On sandy soils apply 3 lbs. zinc (Zn) per acre in fertilizer after liming or where pH is above 6.0.

| 23889 | 3 | Bahia | 1 | 6.0 | M | 100 | H | 140 | H | 240 | 0.0 | 60 | 40 | 0 |
| 23890 | Garden | Vegetables | 3 | 5.2 | M | 90 | M | 70 | H | 160 | 3.0 | 120 | 120 | 120 |

See comment 224 above.
Comment 82—per 100 ft. of row apply 6 lbs. 8-8-8 (3 quarts) at planting and sidedress with 4 lbs. 8-8-8 (2 quarts).

***On summergrass pastures apply P and K as recommended and 60 lbs. of N before growth starts. Up to September 1 repeat the N applications when more growth is desired.

***1.0 ton limestone per acre is approximately equivalent to 50 lbs. per 1,000 sq. ft.

***For cauliflower, broccoli and root crops, apply 1.0 lb. of boron (B) per acre. (For home gardens, 1 tablespoon borax per 100 ft. of row.)

The number of samples processed in this report is 4.

*1. Sandy soils   **7.4 or higher   Alkaline   6.5 or lower   Acid
  2. Loams & light clays   6.6–7.3   Neutral   5.5 or lower   Very acid   Approved
  3. Heavy clays (excluding Blackbelt)
  4. Heavy clays of the Blackbelt   *** Rating & fertility (percent sufficiency)   Soil testing Form B

FIGURE 16.31   Example of a soil test report giving the soil test levels and the recommendations for fertilizers and lime. [From Cope, et al. (1981)]

across a large field. Therefore, if one is taking soil samples, the location of each sample can be georeferenced with north–south and east–west coordinates. In practice, a large field is divided into cells in a grid pattern, each cell usually being about 1 ha in area. Thus, 20 separate georeferenced soil samples (each being composite of about 15 to 20 soil cores) may be collected from a single 20-ha field (Figure 16.33). The nutrient application recommendations resulting from the soil test data and associated information can then be plotted on a map, using a statistical computer program to estimate the location of boundaries between areas of high, medium, and low fertility status (see Figure 19.2 for an example). This soil fertility map can be overlain by maps of other spatial variables, such as soil classification, past management, crop cultivars to be planted, and so forth.

Using these maps and computerized variable-rate application equipment that is linked into the satellite system, it is then possible to automatically modify the nutrient application rates as the applicator travels across the field. An area of the field showing low levels of a given nutrient would receive higher than average rates of the fertilizer, while an area showing high nutrient levels would receive little, if any, of the nutrient. The total amount of fertilizer added may not be greatly different from that used in the past, but application rates would be more in tune with plant needs and environmental cautions.

At harvest time, similar computerized satellite linkages are used to monitor yields in different parts of the field and to create maps showing the yield differences. By overlaying the yield map on the soil nutrient map, it is possible to determine the extent to which nutrient deficiencies are constraining yields.

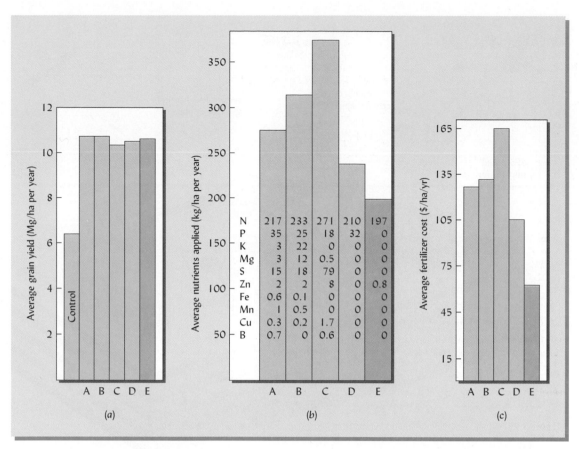

FIGURE 16.32 (*a*) The yield of corn from plots near North Platte, Nebraska, receiving five different rates of fertilizer as recommended by five soil testing laboratories (A to E). Data averaged over a six-year period. The soil was a Cozad silt loam (Fluventic Haplustoll). (*b*) The nutrient levels recommended by these laboratories and (*c*) the cost of the fertilizer applied each year. Note that the yields from all fertilized plots were about the same, even though rates of fertilizer application differed markedly. The fertilizer rates recommended by laboratory E utilized a sufficiency-level concept that was based on the calibration of soil tests with field yield responses. Those recommended by laboratories A to D were based either on a maintenance concept that required replacement of all nutrients removed by a crop or on the supposed need to maintain set cation ratios (Ca/Mg, Ca/K, and Mg/K). Obviously, the sufficiency-level concept provided more economical results in this trial. As a result of this and many other similar comparison studies, most soil test labs have adopted the sufficiency-level approach. [From Olson, et al. (1982); used with permission of American Society of Agronomy]

This site-specific nutrient-management system is an integral part of what is commonly referred to as precision farming. Opportunities for the control of insects and weeds and for modifying plant seeding rates and depths can be utilized on a site-specific basis rather than on a field basis. Such computerized GPS-based systems may not be economically viable for all farms, but their use is spreading, particularly among growers who have access to custom operators with the computer expertise and appropriate equipment. Precision farming may help assure that nutrients will be used only where they are needed to enhance plant production. Table 16.17 shows estimates of economic benefits that might be realized by using site-specific nitrogen management for corn in several counties in Iowa. Most of the benefits would come from reducing the total nitrogen applied.

16.20 BROADER ASPECTS OF FERTILIZER PRACTICE

Fertilizer practice involves many intricate details regarding soils, plants, and fertilizers. Because of the great variability from place to place in each of these three factors, it is difficult to arrive at generalizations for fertilizer use. However, because of the prominence of the response by most nonlegume plants, and because of its implications for environmental quality, the initial focus is generally on nitrogen in most fertilizer schemes. Applications of phosphorus and potassium are made to balance and supplement the nitrogen supply whether it be from the soil, crop residues (especially legumes), organic wastes, or added fertilizers.

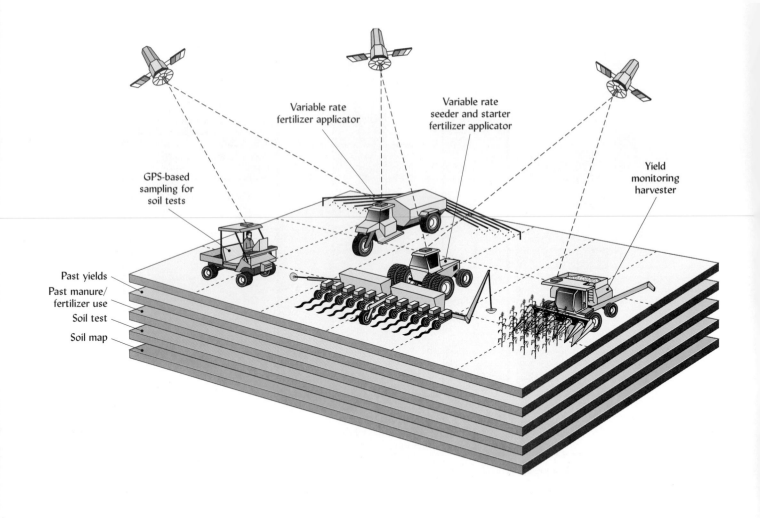

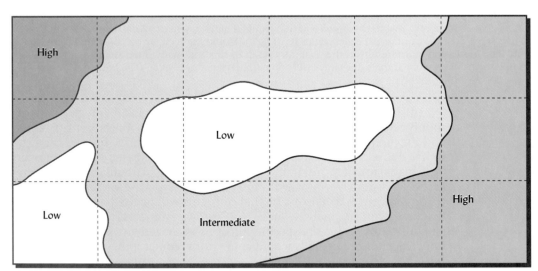

FIGURE 16.33 (Upper) Space age technology is being used to facilitate site-specific nutrient-management systems. Earth-orbiting satellites and appropriate computer software are the bases for the Global Positioning System (GPS), which can plot the location of many soil sampling and plant production sites on a grid basis within a field. A soil sample (composite of 20 cores) is taken from each cell (about 1 ha) and is analyzed. With the soil analysis data from these sites, computers can help create maps such as the one for a 18-ha field (lower) to illustrate the nutrient status of various parts of the field. The satellite/computer systems can then be used to control variable-rate fertilizer applicators that apply only the amounts of nutrients that the soil tests and past soil management suggest are needed. At harvest time, similar satellite/computer connections make possible the monitoring of crop yields on the same grid basis when the harvest machine traverses the field. The yield data are used to create yield maps, which can then be used to further refine nutrient-management systems. [(Upper) modified from PPI (1996)]

TABLE 16.17 Estimated Increase in Farmer Returns over Fertilizer Costs from the Use of Site-Specific Technology for Corn Production in 12 Iowa Counties, and the Percent of the Increase Attributable to Eliminating the Over- and Underapplication of Nitrogen

It is not clear whether the increase in returns would be sufficient to cover the cost of site-specific technology.

County	Increased returns over fertilizer cost, $/ha	Percent attributable to eliminating overapplication of nitrogen	Percent attributable to eliminating underapplication of nitrogen
Adair	18.87	93	7
Black Hawk	8.69	93	7
Carroll	10.77	70	30
Henry	8.64	93	7
Hancock	11.48	86	14
Hamilton	9.88	73	27
Poweshiek	14.35	82	18
Pottawattamie	10.85	95	5
Sioux	9.60	86	14
Story	9.02	80	20
Jones	16.97	89	11
Wright	11.02	90	10
12-county total	11.28	86	14

From Babcock and Pautsch (1997).

Because it is very difficult to predict the nitrogen-supplying ability of a soil from chemical tests, nitrogen fertilizer recommendations are usually based on field experiments that define the relationship between added nitrogen and plant growth or crop yield. Generally, these field studies are carried out on a variety of soils and under a range of different weather conditions (which cause the response to nitrogen to vary greatly from one year to the next).

NITROGEN CREDITS. From the shape of the response curve and economic considerations (see following), the optimal level of nitrogen fertilizer is determined. This optimal fertilizer rate should be adjusted by the amount of any additional gains or losses not taken into account in the standard response curve. For example, nitrogen contributions from previous or current manure applications (see Box 16.1), legume cover crop (see Figure 16.10), or previous legume in the rotation (see Figure 16.11) should be subtracted from the amount of fertilizer recommended.

PROFITABILITY. A second aspect relates to economics. Farmers do not use fertilizers just to grow big crops or to increase the nutrient content of their soils. They do so to make a living. As a result, any fertilizer practice that does not give a fair economic return will not stand the test of time. In crop production, the most profitable rate is determined by the ratio of the value of the extra yield expected to the cost of the fertilizer applied. The law of diminishing returns applies. Therefore, the most profitable fertilizer rate will be somewhat less than the rate that would produce the very highest yield (Figure 16.34).

RESPONSE CURVES. Traditionally, economic analysis of optimum fertilizer rates has assumed that the plant response to fertilizer inputs was represented by a smooth curve following a quadratic function. In fact, actual data obtained can be just as well described by a number of other mathematical functions (Figure 16.35). This seemingly esoteric observation can have a great effect on the amount of fertilizer recommended and, in turn, on the likelihood of environmental harm from excessive fertilizer use (Figure 16.36).

Anyone involved with the actual production of plants can testify to the enormous improvements that accrue from judicious use of organic and inorganic nutrient supplements. The preceding discussion should not imply that such supplements are not needed, but only that their optimum levels are difficult to determine with precision.

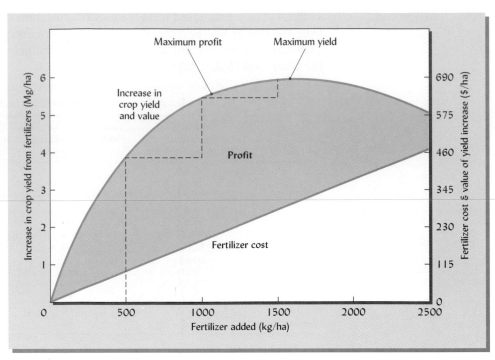

FIGURE 16.34 Relationships among the rate of fertilizer addition, crop yield increase, fertilizer costs, and profit from adding fertilizers. Note that the yield increase (and profit) from the first 500 kg of fertilizer is much greater than from the second and third 500 kg. Also note that the maximum profit is obtained at a lower fertilizer rate than that needed to give a maximum yield. The calculations assume that the yield response to added fertilizer takes the form of a smooth quadratic curve.

16.21 CONCLUSION

The continuous availability of plant nutrients is critical for the sustainability of most ecosystems. The challenge of nutrient management is threefold: (1) to provide adequate nutrients for plants in the system; (2) to simultaneously ensure that inputs are in balance with plant utilization of nutrients, thereby conserving nutrient resources; and (3) to prevent contamination of the environment with unutilized nutrients.

The recycling of plant nutrients must receive primary attention in any ecologically sound management system. This can be accomplished in part by returning plant residues to the soil. These residues can be supplemented by judicious application of the organic wastes that are produced in abundance by municipal, industrial, and agricultural operations worldwide. The use of cover crops grown specifically to be returned to the soil is an additional organic means of recycling nutrients.

For sites from which crops or forest products are removed, nutrient losses commonly exceed the inputs from recycling. Inorganic fertilizers will continue to supplement natural and managed recycling to replace these losses and to increase the level of soil fertility so as to enable humankind to not only survive, but to flourish on this planet. In extensive areas of the world, fertilizer use will have to be increased above current levels to avoid soil and ecosystem degradation and to enable profitable food production.

The use of fertilizers, both inorganic and organic, should not be done in a simply habitual manner or for so-called insurance purposes. Rather, soil testing and other diagnostic tools should be used to determine the true need for added nutrients. Where soils are low in available nutrients, inorganic fertilizers often return several dollars' worth of improved yield for every dollar invested. However, where the nutrient-supplying power of the soil is already sufficient, adding fertilizers is likely to be damaging both to the bottom line and to the environment.

STUDY QUESTIONS

1. The groundwater under a heavily fertilized field is quite high in nitrates, but by the time it reaches a stream the nitrate level has declined considerably. What is your explanation for this situation?

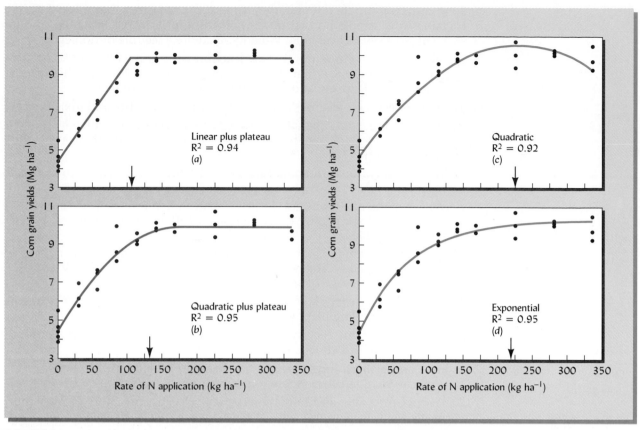

FIGURE 16.35 An example of how the mathematical function chosen to represent fertilizer-response data can effect the amount of fertilizer recommended. The data in all five graphs are exactly the same and represent the response of corn yields in Iowa to increasing levels of nitrogen fertilization. The vertical arrows indicate the recommended, most profitable rate of nitrogen to apply according to each mathematical model. Note that all models fit the data equally well (as indicated by the very similar R^2 values), but that the linear-plus-plateau model predicts an optimum nitrogen rate of 104 kg/ha, while the standard quadratic model suggests that 222 kg/ha is the optimum. Apparently the extra 118 kg/ha of nitrogen may have no effect on crop yield, but may greatly increase the risk of environmental damage. [Redrawn from Cerrato and Blackmer (1990)]

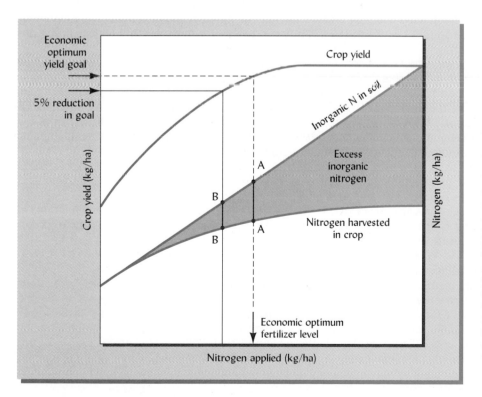

FIGURE 16.36 Influence of rate of nitrogen fertilization on crop yield, nitrogen removed in harvest, and the amount of excess inorganic soil nitrogen potentially available for loss into the environment. The fine lines show that a relatively small decrease (5%) in the yield goal for which nitrogen was applied would result in a relatively large reduction (about 30%) in the potential for nitrogen pollution. Line A-A represents the excess nitrogen present when fertilizer is applied at a rate designed to reach the economic optimum yield goal. Line B-B represents the excess nitrogen if the yield goal is 5% less than the economic optimum. Unfortunately, in many cases a yield reduction of 5% may reduce profits by a much larger percentage. [Adapted from National Research Council (1993)]

2. A cornstalk nitrate test at harvest time shows a farmer that the soil likely contains considerable unused nitrates. To minimize nitrate leaching, the farmer wants to grow a cover crop. Would you advise a legume or nonlegume? Why?

3. The nitrate level in waters that are leaching or running off from forested sites is generally quite low. What management practices on forested sites lead to significant nitrate losses, and how can the losses be prevented?

4. What effect do forest fires have on nutrient availabilities and toxicities?

5. What are the main original sources of the materials from which N, P, and K fertilizers are formed? What processes are used to increase their availability for plant absorption?

6. Why is anhydrous ammonia said to be the most important fertilizer nitrogen carrier?

7. What are the major physical forms in which chemical fertilizers are sold today, and how do their relative quantities of use influence the kinds and amounts of fertilizers applied?

8. Traditionally, animal manures have been considered of major benefit in maintaining soil quality. Today, in many situations they are a liability. Explain, and indicate some means of alleviating the problems.

9. Discuss the concept of the *limiting factor* and indicate its importance in enhancing or constraining plant growth.

10. Why are nutrient cycling problems in agricultural systems more prominent than in those in forested areas?

11. How are earth-orbiting satellites and computers being used to better assure the application of fertilizer at rates more in accord with plant needs?

12. Discuss the value and limitations of soil tests as indicators of plant nutrient needs.

REFERENCES

Attiwill, P. M., and G. W. Leeper. 1987. *Forest Soils and Nutrient Cycles.* (Melbourne, Australia: Melbourne University Press).

Babcock, B. A., and G. R. Pautsch. 1997. "Moving from uniform to variable fertilizer rates on Iowa corn: Effects on rates and returns," CARD Working Paper 97-WP182. (Ames, Iowa: Iowa State University Center for Agricultural and Rural Development).

Belt, G. H., and J. O'Laughlin. 1994. "Buffer strip design for protecting water quality and fish habitat," *Western J. Applied Forestry,* 9(2):4145.

Bormann, F. H., and G. E. Likens. 1979. *Pattern and Process in a Forested Ecosystem.* (New York: Springer-Verlag).

Bumb, B. L. and C. A. Baanante. 1996. *The Role of Fertilizer in Sustaining Food Security and Protecting the Environment to 2020.* Food, Agriculture and the Environment Discussion Paper 17. (Washington, D.C.: International Food Policy Research Institute).

Cerrato, M. E., and A. M. Blackmer. 1990. "Comparison of models from describing corn yield response to nitrogen fertilizer," *Agron. J.,* 98:138–143.

Chauby, I., D. R. Edwards, T. C. Daniel, P. A. Moore, Jr., and D. J. Nichols. 1994. "Effectiveness of vegetative filter strips in retaining surface-applied swine manure constituents," *Trans. Am. Soc. Agric. Engineers,* 37:845–850.

Cope, J. T., C. E. Evans, and H. C. Williams. 1981. *Soil Test Fertilizer Recommendations for Alabama Crops.* Circular 251. (Auburn, Ala.: Auburn University Agriculture Experimental Station).

Ebelhar, S. A., W. W. Frye, and R. L. Blevins. 1984. "Nitrogen from legume cover crops for no-tillage corn," *Agron. J.,* 76:51–55.

Eghball, B., and J. F. Power. 1994. "Beef cattle feedlot manure management," *J. Soil and Water Conservation,* 49:113–122.

FAO. 1977. *China: Recycling of Organic Wastes in Agriculture.* FAO Soils Bulletin 40. (Rome: U.N. Food and Agriculture Organization).

Gardner, G. 1997. *Recycling Organic Waste: From Urban Pollutant to Farm Resource.* Worldwatch Paper 135. (Washington, D.C.: Worldwatch Institute).

He, Xin-Tao, T. Logan, and S. Traina. 1995. "Physical and chemical characteristics of selected U.S. municipal solid waste composts," *J. Environ. Qual.,* **24**:543–552.

Loehr, R. C., et al. 1979. *Land Application of Wastes,* Vol. I. (New York: Van Nostrand Reinhold).

Magdoff, F., L. Lanyon, and B. Liebhardt. 1997. "Nutrient cycling, transformations, and flows: Implications for a more sustainable agriculture," *Advances in Agronomy,* **60**:2–73.

National Research Council. 1993. *Soil and Water Quality: An Agenda for Agriculture.* (Washington, D.C.: National Academy of Sciences).

Olson, R. A., K. D. Frank, P. H. Graboushi, and G. W. Rehm. 1982. "Economic and agronomic impacts of varied philosophies of soil testing," *Agron. J.,* **74**:492–499.

PPI. 1996. "Site-specific nutrient management systems for the 1990's." Pamphlet. (Norcross, Ga.: Potash and Phosphate Institute and Foundation for Agronomic Research; Saskatoon, Sask.: Potash Phosphate Institute of Canada). Canada.

Reddy, G. B., and K. R. Reddy. 1993. "Fate of nitrogen-15 enriched ammonium nitrate applied to corn," *Soil Sci. Soc. Amer. J.,* **57**:111–115.

Reuter, D. J., and J. B. Robinson. 1986. *Plant Analysis: An Interpretation Manual.* (Melbourne, Australia: Inkata Press).

Robert, P. C., R. H. Rust, and W. E. Larson (eds.). 1996. *Precision Agriculture.* Proceedings of the Third International Conference on Precision Agriculture, (Madison, Wis.: Amer. Soc. Agron.).

Robinson, C. A., M. Ghaffarzadeh, and R. M. Cruze. 1996. "Vegetative filter strip effects on sediment concentration in cropland runoff," *J. Soil and Water Conserv.,* **50**:227–230.

Sanchez, P. A., et al. 1997. "Soil fertility replenishment in Africa: An investment in natural resource capital," in R. J. Buresh, P. A. Sanchez, and F. Calhoun (eds.), *Replenishing Soil Fertility in Africa.* ASA/SSSA Publication. (Madison, Wis.: Soil Sci. Soc. Amer.).

Sims, J. T., and D. C. Wolf. 1994. "Poultry waste management: Agricultural and environmental issues," *Advances in Agronomy,* **52**:1–83.

Smaling, E. M. A., S. M. Nwanda, and B. H. Jensen. 1997. "Soil fertility in Africa is at stake," in R. J. Buresh, P. A. Sanchez, and F. Calhoun (eds.), *Replenishing Soil Fertility in Africa.* ASA/SSSA Publication. (Madison, Wis.: Soil Sci. Soc. Amer.).

Stickler, F. C., et al. 1975. *Energy from Sun to Plant to Man.* (Moline, Ill.: John Deere and Company).

Stivers, L. J., C. Shennen, L. E. Jackson, K. Groody, C. J. Griffin, and P. R. Miller. 1993. "Winter cover cropping in vegetable production systems in California," in M. G. Paoletti, W. Foissner, and D. Coleman (eds.), *Soil Biota, Nutrient Cycling and Farming Systems.* (Boca Raton, Fla.: Lewis Publishers).

Sutton, A. L. 1994. "Proper animal manure utilization," in *Nutrient Management,* supplement to *J. Soil Water Conserv.,* **49**(2), pp. 65–70.

Teng, Y., and V. R. Timmer. 1994. "Nitrogen and phosphorus in an intensely managed nursery soil–plant system," *Soil Sci. Soc. Amer. J.,* **58**:232–238.

Terry, D. L., and P. Z. Yu. 1996. *Commercial Fertilizers.* A cooperative project of the Association of American Plant Food Control Officials Inc. (AAPFCO) and The Fertilizer Institute. (Lexington, Ky.: AAPFCO).

Thomas, M. L., R. Lal, T. Logan, and N. R. Fausey. 1992. "Land use and management effects on non-point loading from Miamian soil," *Soil Sci. Soc. Amer. J.,* **56**:1871–1875.

Tisdale, S. L., W. L. Nelson, J. D. Beaton, and J. L. Havlin. 1993. *Soil Fertility and Fertilizers,* 5th ed. (New York: Macmillan).

USDA. 1980. *Appraisal, 1980 Soil and Water Resources Conservation Act, Review Draft,* Part I. (Washington, D.C.: USDA).

USDA Natural Resources Conservation Service. 1991. *Water Quality Indicator Guide: Surface Waters.* Report no. SCS-TP-161. (Washington, D.C.: USDA).

Walworth, J. L., and M. E. Sumner. 1987. "The diagnosis and recommendation integrated system (DRIS)," *Advances in Soil Science,* **6**:149–187.

Westerman, R. L. (ed.). 1990. *Soil Testing and Plant Analaysis,* 3d ed. (Madison, Wis.: Soil Sci. Soc. Amer.).

Wilkinson, D. M., and N. M. Dickinson. 1995. "Metal resistance in trees: The role of mycorrhizae," *Oikos,* **72**:298–299.

Zublena, J. P., J. C. Barker, and T. A. Carter. 1993. "Poultry manure as a fertilizer source," *Soil Facts* (Raleigh, N.C.: North Carolina Cooperative Extension Service, North Carolina State University).

17

SOIL EROSION AND ITS CONTROL

The wind crosses the brown land, unheard . . .
—*T. S. ELIOT, THE WASTE LAND*

No soil phenomenon is more destructive worldwide than the erosion caused by wind and water. Since prehistoric times people have brought the scourge of soil erosion upon themselves, suffering impoverishment and hunger in its wake. Past civilizations have disintegrated as their soils, once deep and productive, washed away, leaving only thin, infertile, rocky relics of the past. Seeing the nearly barren hills in central India, or in parts of Greece, Lebanon and Syria, it is hard to imagine that agricultural communities once flourished in these places.

The current threat of soil erosion is more ominous than at any time in history. In our generation, farmers have had to more than double world food output to feed the unprecedented numbers of people on Earth. In the low-income countries, the ratio of people to available cropland is already very high and rising rapidly. While intensified cultivation of fertile, relatively level lands has helped produce much of the needed food, many nations have had to expand the area of land under cultivation, clearing and burning steep, forested slopes and plowing grasslands. Population pressures have also led to overgrazing rangelands and overexploiting timber resources. All these activities degrade or remove natural vegetation, causing the underlying soils to become much more susceptible to the destructive action of erosion. The result is a vicious downward cycle of deterioration—land degradation leads to poor crops, human poverty, and reduced vegetative protection for the soil, which, in turn, accelerates erosion and drives ever more desperate people to clear, cultivate, and degrade more land.

The degraded productivity of farm, forest, and range lands tells only part of the sad erosion story. Soil particles washed or blown from the eroding areas are subsequently deposited elsewhere—in nearby low-lying landscape sites; in streams and rivers; or in downstream reservoirs, lakes, and harbors. The environmental and economic damages suffered by sites on which the eroded soil materials are deposited may be as great or greater than that incurred on the sites from which the soil material was removed. The displaced soil material (sediment and dust) lead to major water and air pollution problems, bringing enormous economic and social costs to society.

Soil erosion is everybody's business. Fortunately, recent decades have seen much progress in understanding the mechanisms of erosion and in developing techniques that can effectively and economically control soil loss in most situations. This chapter will equip you with some of the concepts and tools you will need to do your part in solving this pressing world problem.

Land Degradation

During the past ha▮▮▮▮▮, human land use and associated activities have degraded some 5 billion ha (about 45% of Earth's vegetated land). Such **land degradation** results in a reduced productive potential and a diminished capacity to provide benefits to humanity. Much of this degradation (on about 3.6 billion ha) is linked to **desertification**, the spreading of desert conditions that disrupt semiarid and arid ecosystems (including agroecosystems). A major cause of desertification is overgrazing by cattle, sheep, and goats, a factor that likely accounts for about a third of all land degradation, mainly that in such dry regions as the Sahel in northern Africa and the rangeland of the American Southwest. Likewise, the indiscriminate felling of rain forest trees has already degraded nearly 0.5 billion ha in the humid tropics. Additionally, inappropriate agricultural practices continue to degrade land in all climatic regions.

Soil–Vegetation Interdependency

Degraded lands may suffer from destruction of native vegetation communities, reduced agricultural yields, lowered animal production, and simplification of once-diverse natural ecosystems with or without accompanying degradation of the soil resource. On about 2 billion of the 5 billion ha of degraded lands in the world, soil degradation is a major part of the problem (see Figure 17.1). In some cases the soil degradation occurs

---

[1] For a very readable account of historical degradation of land and water resources, see Hillel (1991). Overviews of the current extent of soil erosion and other forms of land degradation are given by Oldeman (1994), Daily (1997), and Rosensweig and Hillel (1998).

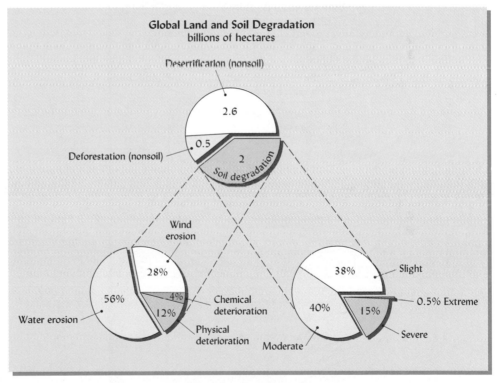

FIGURE 17.1   Soil degradation as a part of global land degradation caused by overgrazing, deforestation, inappropriate agricultural practices, fuel wood overexploitation, and other human activities. About 60% of degraded land has not suffered soil degradation. Of the 2 billion ha of land with degraded soils, most could be restored easily (*slight* degradation) or with considerable financial and technical investments (*moderate* degradation). *Severely* degraded soils are currently useless for agriculture and would require major international assistance for restoration. About 9 million ha (0.5% of degraded soils) are *extremely* degraded and incapable of restoration. About 85% of the soil degradation is caused by erosion by wind and water. [FAO data selected from Oldeman (1994) and Daily (1997)]

mainly as deterioration of physical properties by compaction or surface crusting (see Sections 4.5 and 4.8), or as deterioration of chemical properties by acidification (see Section 9.6) or salt accumulation (see Section 10.3). However, most (~85%) soil degradation stems from erosion—the destructive action of wind and water.

The two main components of land degradation—damage to plant communities and deterioration of soil—are not independent of each other. Rather, they interact to cause a downward spiral of accelerating deterioration (Figure 17.2). Due to overgrazing, deforestation, or inappropriate methods of crop production, vegetation becomes less dense and vigorous, and thus provides the soil less and less protection from erosion. Simultaneously, as the soil is degraded by such processes as erosion and nutrient depletion, it becomes less and less capable of supporting a protective canopy of vegetation. Soil degradation weakens the vegetation through its effects on runoff and infiltration of rainwater (see Section 6.2). With as much as 50% to 60% of the rainfall lost as runoff,

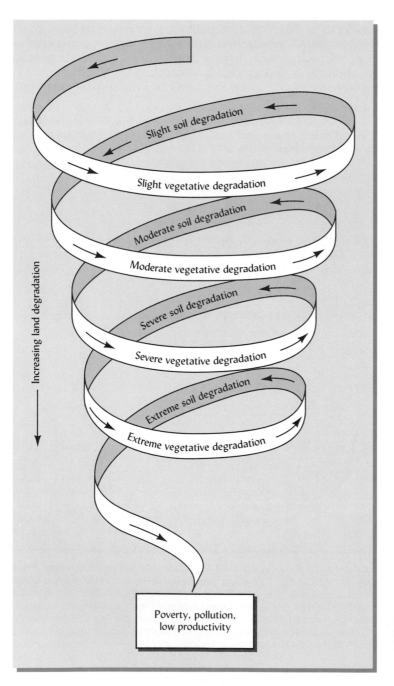

FIGURE 17.2 The downward spiral of land degradation resulting from the feedback loop between soil and vegetative degradation.

scarcity of soil water on eroded soils can become a serious impediment to plant growth. Improvements in both soil and vegetation management must go hand-in-hand if the productive potential of the land is to be protected—or even be restored by moving *up* rather than *down* the spiral.

Geological versus Accelerated Erosion

GEOLOGICAL EROSION. Erosion is a process that transforms soil into **sediment**. Soil erosion that takes place naturally, without the influence of human activities, is termed **geological erosion.** It is a natural leveling process. It inexorably wears down hills and mountains, and through subsequent deposition of the eroded sediments, it fills in valleys, lakes, and bays. Many of the landforms we see around us—canyons, buttes, rounded hills, river valleys, deltas, plains, and pediments—are the result of geological erosion and deposition. The vast deposits that now appear as sedimentary rocks originated in this way.

In most settings, geological erosion wears down the land slowly enough that new soil forms from the underlying rock or regolith faster than the old soil is lost from the surface. The very existence of soil profiles bears witness to the net accumulation of soil and the effectiveness of undisturbed natural vegetation in protecting the land surface from erosion.

The rate of geological soil erosion varies greatly with both rainfall and type of material comprising the regolith. Geological erosion by water tends to be greatest in semiarid regions where rainfall is enough to be damaging, but not enough to support dense, protective vegetation (Figure 17.3). Areas blanketed by deep deposits of silts may have exceptionally high erosion rates under such conditions. The gullied, barren landscape of the North American badlands (Figure 17.4) is an extreme example of geological erosion occurring where unstable clay and silt deposits are subjected to infrequent but intense rainstorms, yet the soil is usually too dry (partly because of high runoff losses) to support much vegetation.

SEDIMENT LOADS. Rainfall, geology, and other factors (including human activities) influence the sediment loads carried by the world's great rivers (Table 17.1). Although rivers like the Mississippi and Yangtze were muddy before humans disturbed their watersheds, current sediment loads are far greater than before. To gain some perspective on the

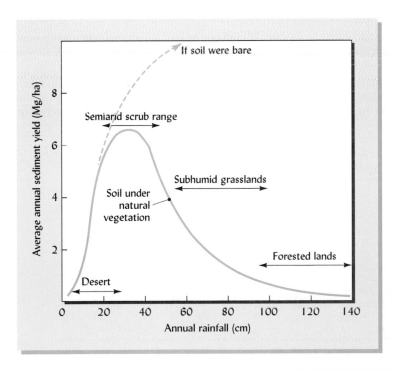

FIGURE 17.3 Generalized relationship between annual rainfall and soil loss from geologic erosion by water. The actual amount of sediment lost annually per hectare will depend on other climatic variables, topography, and the type of soils in the watershed. Note that sediment yields are greatest in semiarid regions. Here a number of severe runoff–generating storms occur in most years, but the total rainfall is too little to support much protective plant cover. By comparison, the very dry deserts have too little rain to cause much erosion, and the well-watered regions support dense forests that effectively protect the soil. Where the natural vegetation is destroyed by plowing, erosion from the bare soils is much higher with increasing rainfall, as indicated by the dashed curve.

FIGURE 17.4 The effects of geologic erosion and sedimentation can be dramatic, as in the semiarid badlands scene in North Dakota (left). There is little vegetation to protect the hills from ravages of sudden summer thunderstorms. Note the petrified log (arrow) becoming exposed as the hillside erodes in the foreground. Its presence indicates that deep layers of clayey sediment had been deposited on this site during ancient cycles of erosion and sedimentation. (Right) The extreme gully erosion in a site in central Tanzania highlights the cutting power of turbulent water, the flow of which was probably enhanced by the deforestation of the surrounding watershed. The scene also indicates the important role of raindrops in detaching soil particles. Note the tall, thin pedestals of soil that remain where a layer of rocklike ironstone protects the underlying soft material from the impact of raindrops (compare to the much smaller pedestals in Figure 17.9a). (Photos courtesy of R. Weil)

enormous amount of soil transported to the sea by these rivers, consider the Mississippi's sediment load (only a fifth as great as that of the Yangtze or the Ganges). If the 300 million Mg of sediment were carried to the Gulf of Mexico by dump trucks, it would take a continuous, year-round caravan of over 80,000 large trucks, stretching all the way from Wisconsin to New Orleans (1600 km) and back, with a 20-Mg load being dumped into the Gulf *every half second*.

ACCELERATED EROSION. **Accelerated erosion** occurs when people disturb the soil or the natural vegetation by grazing livestock, cutting forests, plowing hillsides, or tearing up

TABLE 17.1   Annual Sediment Loads for Nine of the World's Major Rivers, Including the Mississippi River

River	Countries	Annual sediment load, million Mg	Erosion, Mg/ha drained
Yangtze	China	1600	479
Ganges	India, Nepal	1455	270
Amazon	Brazil, Peru, etc.	363	13
Mississippi	United States	300	93
Irrawaddy	Burma	299	139
Kosi	India, Nepal	172	555
Mekong	Vietnam, Thailand, etc.	170	43
Red	China, Vietnam	130	217
Nile	Sudan, Egypt, etc.	111	8

Data from different sources compiled by El-Swaify and Dangler (1982).

FIGURE 17.5 Erosion and deposition occur simultaneously across a landscape. (Left) The soil on this ridgetop was worn down by erosion during nearly 300 years of cultivation. The surface soil exposed on the ridgetop consists mainly of light-colored C horizon material. At sites lower down the slope the surface horizon shows mainly A and B horizon material, some of which has been deposited after eroding from locations upslope. (Right) Erosion on the sloping wheat field in the background has deposited a thick layer of sediment in the foreground, burying the plants at the foot of the hill. [Photos courtesy of R. Weil (left) and USDA Natural Resources Conservation Service (right)]

land for construction of roads and buildings. Accelerated erosion is often 10 to 1000 times as destructive as geological erosion, especially on sloping lands in regions of high rainfall. Rates of erosion by wind and water on agricultural land in Africa, Asia, and South America are thought to average about 30 to 40 Mg/ha annually. In the United States, the average erosion rate on cropland is about 12 Mg/ha—7 Mg by water and 5 Mg by wind. Some cultivated soils are eroding at 10 times these average rates. In comparison, erosion on undisturbed humid-region grasslands and forests generally occurs at rates considerably below 0.1 Mg/ha.

About 4 billion Mg of soil is moved annually by soil erosion in the United States, some two-thirds by water and one-third by wind. More than half of the movement by water and about 60% of the movement by wind takes place on croplands that produce most of the country's food. Much of the remainder comes from semiarid rangelands, from road building and timber harvest on forest lands, and from soils disturbed for highway and building construction. Although progress has been made in reducing erosion (see Section 17.14) current high losses are simply not acceptable for long-term sustainability and must be further reduced.

Under the influence of accelerated erosion, soil is commonly washed or blown away faster than new soil can form by weathering or deposition. As a result, the soil depth suitable for plant roots is often reduced. In severe cases, gently rolling terrain may become scarred by deep gullies, and once-forested hillsides may be stripped down to bare rock.

Accelerated erosion often makes the soils in a landscape more heterogenous. One example of increased heterogeneity occurs as overgrazed semiarid grasslands degrade into scrub brush ecosystems. During this degradation "islands of fertility" form under shrubs, while the soils between shrubs become increasingly thin and infertile. Another example can be seen in the striking differences in surface soil color that develop as ridgetop soils are truncated, exposing material from the B or C horizon at the land surface, while soils lower in the landscape are buried under organic-matter-enriched sediment (Figure 17.5).

17.2 ON-SITE AND OFF-SITE EFFECTS OF ACCELERATED SOIL EROSION

Erosion damages the site on which it occurs and also has undesirable effects off-site in the larger environment. The off-site costs relate to the effects of excess water, sediment, and associated chemicals on downhill and downstream environments. While the costs associated with either or both of these types of damages may not be immediately apparent, they are real and grow with time. Landowners and society as a whole must eventually foot the bill.

Types of On-Site Damages

The most obviously damaging aspect of erosion is the loss of soil itself. In reality, the damage done to the soil is greater than the amount of soil lost would suggest because the soil material eroded away is almost always more valuable than that left behind. Not only are surface horizons eroded while subsurface horizons (which are usually less useful) remain untouched, but the quality of the remaining topsoil is also impaired. Erosion selectively removes organic matter and fine mineral particles, while leaving behind mainly relatively less active, coarser fractions. Experiments have shown organic matter and nitrogen in the eroded material to be five times as high as in the original topsoil. Comparable **enrichment ratios** for phosphorus (review Table 14.3) and potassium are commonly two and three, respectively. The quantity of essential nutrients lost from the soil by erosion is quite high, although only a portion of these nutrients are lost in forms that would be available to plants in the short term (Table 17.2). The soil left behind usually has lower water-holding and cation exchange capacities, less biological activity, and a reduced capacity to supply nutrients for plant growth.

In addition to the just-mentioned reduction in soil-quality factors, soil movement during erosion can spread plant disease organisms from the soil to plant foliage and from a higher to a lower-lying field. The deterioration of soil structure often leaves a dense crust on the soil surface, which, in turn, greatly reduces water infiltration and increases water runoff. Newly planted seeds and seedlings may be washed downhill, trees may be uprooted, and small plants may be buried in sediment. In the case of wind erosion, fruits and foliage may be damaged by the sandblasting effect of blowing soil particles.

Finally, gullies that carve up badly eroded land may make the use of tractors impossible and may undercut pavements and building foundations, causing unsafe conditions and expensive repairs.

Types of Off-Site Damages

Erosion moves sediment and nutrients off the land, creating the two most widespread water pollution problems in our rivers and lakes. The nutrients impact water quality largely through the process of eutrophication caused by excessive nitrogen and phosphorus, as discussed in Section 14.2. In addition to nutrients, sediment and runoff water may also carry toxic metals and organic compounds, such as pesticides. The sediment itself is a major water pollutant, causing a wide range of environmental damages.

TABLE 17.2 Estimated Losses of Total and Available Nitrogen, Phosphorus, and Potassium in Sediments Eroded from Soils in the United States

Total losses of potassium are higher than those for nitrogen or phosphorus because this element is commonly present in larger total quantities in soil solids. But the amount of available nitrogen lost through erosion is higher than for the other two elements. This may be because of nitrogen fertilizer applications to the soil surface, where the nitrogen may be more subject to loss in runoff and erosion. In thousands of metric tons (megagrams).

Region	Nitrogen Total	Nitrogen Available	Phosphorus Total	Phosphorus Available	Potassium Total	Potassium Available
Pacific	100	18	29	0.6	1,154	23
Mountain	176	32	64	1.3	2,550	51
Southern Plains	512	94	101	2.0	3,043	61
Northern Plains	2,068	380	293	5.9	11,711	234
Lake states	622	114	107	2.1	3,643	73
Corn belt	4,360	802	624	12.5	24,959	499
Delta states	478	88	141	2.8	4,220	84
Southeastern states	202	37	101	2.0	1,007	20
Appalachian states	676	124	169	3.4	3,381	67
Northeastern states	300	55	75	1.5	2,252	45
Total	9,494	1,744	1,704	34.1	57,920	1,158

Data computed by Larson, et al. (1983).

DAMAGES FROM SEDIMENT. Sediment deposited on the land may smother crops and other low-growing vegetation (Figure 17.5). It fills in roadside drainage ditches and creates hazardous driving conditions where mud covers the roadway.

Sediment that washes into streams makes the water cloudy or turbid (Figure 17.6*a* and Plate 34, before page 499). High **turbidity** prevents sunlight from penetrating the water and thus reduces photosynthesis and survival of the *submerged aquatic vegetation* (SAV). The demise of the SAV, in turn, degrades the fish habitat and upsets the aquatic food chain. The muddy water also fouls the gills of some fish. Sediment deposited on the stream bottom can have a disastrous effect on many freshwater fish by burying the pebbles and rocks among which they normally spawn. The buildup of bottom sediments can actually raise the level of the river, so that flooding becomes more frequent and more severe. For example, to counter the rising river bottom, flood-control levees along the Mississippi must be constantly enlarged.

A number of major problems occur when the sediment-laden rivers reach a lake, reservoir or estuary. Here, the water slows down and drops its load of sediment. Eventually reservoirs—even those formed by giant dams—become mere mudflats, completely filled in with sediment (see Figure 17.6*b*). Prior to that, the capacity of the reservoir to store water for irrigation or municipal water systems is progressively reduced, as is the capacity for floodwater retention or hydroelectric generation. It is esti-

(a)

(b)

FIGURE 17.6 Off-site damages caused by soil erosion include the effects of sediment on aquatic systems. (*a*) A sediment-laden tributary stream empties into the relatively clear waters of a larger river. The turbid water will foul fish gills, inhibit submerged aquatic vegetation, and clog water-purification systems. Part of the sediment will settle out on the river bottom, covering fish-spawning sites and raising the river bed enough to aggravate the severity of future flooding episodes. (*b*) Expensive dredging and excavation (by the dragline in the foreground) is being undertaken to restore the beauty, recreational value, and flood-control function of this pond in a neighborhood park after accumulated sediment had transformed it into a mere mudflat. The watershed upstream from the pond had undergone a period of rapid suburban development, during which adequate sediment-control practices were not used on the construction sites. [Photos courtesy of USDA Natural Resources Conservation Service (*a*) and R. Weil (*b*)]

mated that 1.5 billion Mg of sediment is deposited each year in the nation's reservoirs. Similarly, harbors and shipping channels fill in and become impassable. The loss of function and the costs of dredging, excavation, filtering, and construction activities necessary to remedy these situations run into the billions of dollars every year.

WINDBLOWN DUST. Wind erosion also, has its off-site effects. The blowing sands may bury roads and fill in drainage ditches, necessitating expensive maintenance. The sandblasting effect of wind-borne soil particles may damage the fruits and foliage of crops in neighboring fields, as well as the paint on vehicles and buildings many kilometers downwind from the eroding site.

HEALTH HAZARD. The finest particles blow the farthest, and may present a major human health hazard. While silt-sized particles are generally filtered by nose hairs or trapped in the mucous of the windpipe and bronchial tubes, smaller clay-sized particles often pass through these defenses and lodge in the air sacs (alveoli) of the lungs. The particles themselves cause inflammation of the lungs, but they may also often carry toxic substances that cause further lung damage. For example, airborne clay particles adsorb water vapor and may become coated with sulfuric or nitric acids found in the atmosphere (see Section 9.5). Long-term epidemiological studies suggest that the number of deaths resulting from people inhaling this fine *fugitive dust* may exceed the number of deaths from car highway accidents. For this reason, the U.S. Environmental Protection Agency and other regulatory bodies foresee the need to reduce emissions of fine windblown dust, much of which arises from wind erosion on cropland, rangelands, and construction sites (as well as from traffic on unpaved roads).

ESTIMATED COSTS OF EROSION.[2] Although no precise data exist, national or regional average wind and water erosion rates have been used to estimate the total costs of erosion in the United States. Included in such calculations are the on-site costs of replacing nutrients and water lost through accelerated erosion, as well as crop yield reductions due to reduced soil depth. Depending mainly on assumptions about the value of nutrients lost in sediment and runoff (should only those in readily useable form be valued?), the total annual on-site costs have been estimated at between $4 and $27 billion.

The off-site costs of erosion are likely even greater, especially because of the health effects of windblown particles and the reduced recreational (fishing, swimming, and aesthetic) value of muddy waters. The total of these annual off-site costs has been estimated at $5 to $17 billion. The grand total annual cost of erosion in the United States is therefore likely between $9 and $44 billion. Such high costs are a sobering reminder of the burden that society bears as a result of poor land management and would seem to justify increasing the sums allocated to the battle against erosion.

MAINTENANCE OF SOIL PRODUCTIVITY.[3] Although extreme soil erosion can reduce soil productivity to almost zero, in most cases the effect is too subtle to notice between one year and the next. Where farmers can afford to do so, they compensate for the loss of nutrients by increasing the use of fertilizer. The losses of organic matter and water-holding capacity are much more difficult to overcome. Over the long term, accelerated soil erosion that exceeds the rate of soil formation leads to declining productivity on most soils. In the United States, crop yields on severely eroded soils are often 20 to 40% lower than on similar soils with only slight erosion.

Ultimately, the rate of decline of soil productivity, or the cost of maintaining constant crop-yield levels, is determined by such soil properties as *depth to a root-restricting layer* and *permeability of the subsoil*. As shown in Figure 17.7, a deep, well-drained, and well-managed soil may not decline much in productivity even though it suffers some erosion. In contrast, erosion on a shallow, low-permeability soil may bring about a rapid productivity decline.

---

[2] A detailed analysis giving high-end estimates can be found in Pimental, et al. (1995).

[3] For estimates of the effect of erosion on the productive potential of African soils, see Lal (1995) and for U.S. Mollisols and Alfisols, see Weesies, et al. (1994).

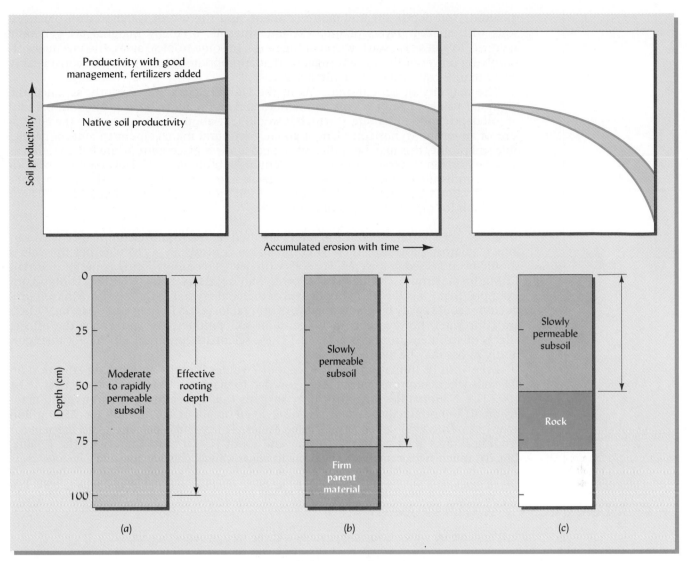

FIGURE 17.7 Effect of erosion over time on the productivity of three soils differing in depth and permeability. Productivity on soil (*a*) actually increases with time because of good management practices and fertilizer additions, even though the native soil productivity declines as a result of erosion. Because soil (*c*) is shallow and has restricted permeability, its productivity declines rapidly as a result of erosion, a decline that good management and fertilizers cannot prevent. Soil (*b*), which is intermediate in both depth and permeability, suffers only a slight decline in productivity due to erosion. Soil characteristics clearly influence the effect of erosion on soil productivity.

Soil-Loss Tolerance[4]

The loss of *any* amount of soil by erosion is detrimental, but years of field experience, as well as scientific research, indicate that some loss can be tolerated. Scientists of the USDA Natural Resources Conservation Service, working in cooperation with field personnel throughout the country, have developed tentative soil-loss tolerance limits for most cultivated soils in the United States.

A tolerable soil loss (***T* value**) is the maximum amount of soil that can be lost annually by the combination of water and wind erosion on a particular soil without degrading that soil's long-term productivity. Currently, *T* values are based on the best judgment of informed soil scientists, rather than on rigorous research data.

COMMON RANGE OF *T* VALUES. The *T* values for soils in the United States commonly range from 5 to 11 Mg/ha. They depend on a number of soil-quality and-management factors,

---

[4] For a discussion of how *T* values were derived, see Schertz (1983).

including soil depth, organic matter content, and the use of water-control practices. Soils with shallow layers of infertile, impermeable, or rocky material are generally assigned T values near the low end of the range. In some tropical areas, the low nutrient-supplying power of the subsoil suggests that appropriate T values may be considerably lower than those used in the United States.

The majority of agricultural soils in the United States are currently assigned the highest T value, 11 Mg/ha. This represents a maximum allowable loss of about 0.9 mm of soil depth annually, a rate at which it would take about 225 years to lose the equivalent of an entire Ap horizon.[5] Under good agricultural management in a deep, permeable soil profile, this may be sufficient time to allow replacement of the lost material as new Ap horizon material forms from underlying subsoil material. Development of horizons in undisturbed soils under natural vegetation is most likely much slower than this, so the T value assigned to a particular type of soil probably would not provide sufficient protection for many rangeland or forest sites.

SIGNIFICANCE OF T VALUES. Because T values are used in determining compliance with various regulatory programs, there is considerable controversy as to whether the values should be increased or lowered. For soils with deep, favorable rooting zones, the current 11 Mg/ha maximum T value may be lower than necessary. Some scientists contend, for example, that for soil with many meters of material favorable for root growth, soil productivity could easily be maintained with annual losses of 15 or even 20 Mg/ha. Others are concerned, however, about longer-term soil productivity and about the off-site effects of sediment from eroded fields. These scientists contend that the T values currently in use may be too high.

Even with its limitations, the concept of T values is useful in focusing attention on the soils where improved practices are needed to maintain long-term productivity. The 1992 U.S. National Resource Inventory suggests that under current land use and management, soil productive capacity is going down on about 102 million ha (252 million acres) of nonfederal land in the United States. Thirty-three percent of all cropland is eroding at rates in excess of the T value, and about 15% at more than twice the T value. Clearly, much work remains to be done in controlling excessive soil loss.

17.3 MECHANICS OF WATER EROSION

Soil erosion by water is fundamentally a three-step process (Figure 17.8):

1. *Detachment* of soil particles from the soil mass
2. *Transportation* of the detached particles downhill by floating, rolling, dragging, and splashing
3. *Deposition* of the transported particles at some place lower in elevation

On comparatively smooth soil surfaces, the beating action of raindrops causes most of the detachment. Where water is concentrated into channels, the cutting action of turbulent flowing water detaches soil particles. In some situations, freezing-thawing action also contributes to soil detachment.

Influence of Raindrops

A raindrop accelerates as it falls until it reaches *terminal velocity*—the speed at which the friction between the drop and the air balances the force of gravity. Larger raindrops fall faster, reaching a terminal velocity of about 30 km/h, or about 2 to 3 times as fast as a person can run. As the speeding raindrops impact the soil with explosive force, they transfer their high kinetic energy to the soil particles (see Figure 17.8).

Raindrop impact exerts three important effects: (1) it detaches soil; (2) it destroys granulation; and (3) its splash, under certain conditions, causes an appreciable transportation of soil. So great is the force exerted by raindrops that they not only loosen and detach soil granules, but may even beat the granules to pieces. As the dispersed material

---

[5] This calculation assumes that the Ap horizon is 200 mm thick and the bulk density of the soil is 1.25 Mg/m^3.

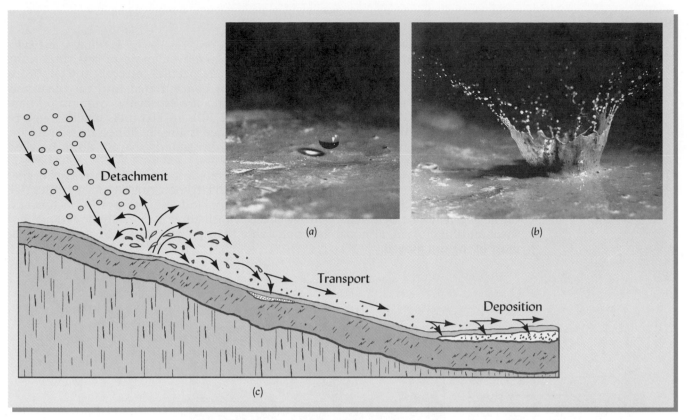

FIGURE 17.8 The three-step process of soil erosion by water begins with the impact of raindrops on wet soil. (*a*) A raindrop speeding toward the ground. (*b*) The splash that results when the drop strikes a wet, bare soil. Such raindrop impact destroys soil aggregates, encouraging sheet and interill erosion. Also, considerable soil may be moved by the splashing process itself. The raindrop affects the detachment of soil particles, which are then transported and eventually deposited in locations downhill (*c*).

dries it may develop into a hard crust, which will prevent the emergence of seedlings and will encourage runoff from subsequent precipitation (see Section 4.8).

History may someday record that one of the truly significant scientific advances of the 20th century was the realization that most erosion is initiated by the impact of raindrops, rather than the flow of running water. For centuries prior to this realization, soil conservation efforts aimed at controlling the more visible flow of water across the land, rather than protecting the soil surface from the impact of raindrops.

Transportation of Soil

RAINDROP SPLASH EFFECTS.   When raindrops impact on a wet soil surface, they detach soil particles and send them flying in all directions (see Figure 17.8). On a soil subject to easy detachment, a very heavy rain may splash as much as 225 Mg/ha of soil, some of the particles splashing as much as 0.7 m vertically and 2 m horizontally. If the land is sloping or if the wind is blowing, splashing may be greater in one direction, leading to considerable net horizontal movement of soil.

ROLE OF RUNNING WATER.   Runoff water plays the major role in the transportation step of soil erosion. If the rate of rainfall exceeds the soil's infiltration capacity, water will pond on the surface and begin running downslope. The soil particles sent flying by raindrop impact will then land in flowing water, which will carry them down the slope. So long as the water is flowing smoothly in a thin layer (sheet flow), it has little power to detach soil. However, in most cases the water is soon channeled by irregularities in the soil surface and increases in both velocity and turbulence. The channelized flow then not only carries along soil splashed by raindrops, but begins to detach particles as it cuts into the soil mass. This is an accelerating process, for as a channel is cut deeper, it fills with greater and greater volumes of flowing water. So familiar is the power of runoff water to cut and carry that the public generally ascribes to it all the damage done by heavy rainfall.

Three types of water erosion are generally recognized: (1) *sheet,* (2) *rill,* and (3) *gully* (Figure 17.9). In **sheet erosion**, splashed soil is removed more or less uniformly, except that tiny columns of soil often remain where pebbles intercept the raindrops (see Figure 17.9a). However, as the sheet flow is concentrated into tiny channels (termed **rills**), **rill erosion** becomes dominant. Rills are especially common on bare land newly planted or in fallow (see Figure 17.9.b). Rills are channels small enough to be smoothed over by normal tillage, but the damage is already done—the soil is lost. When sheet erosion takes place primarily between irregularly spaced rills, it is called **interrill erosion.**

Where the volume of runoff is further concentrated, the rushing water cuts deeper into the soil, deepening and coalescing the rills into larger channels termed **gullies** (see Figure 17.9c). This is **gully erosion.** Gullies on cropland are obstacles to tractors and cannot be removed by ordinary tillage practices. All three types may be serious, but sheet and rill erosion, although less noticeable than gully erosion, are responsible for most of the soil moved.

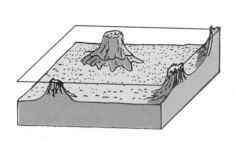

(a) Sheet erosion

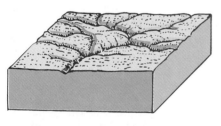

(b) Rill erosion

FIGURE 17.9 Three major types of soil erosion. *Sheet erosion* is relatively uniform erosion from the entire soil surface. Note that the perched stones and pebbles have protected the soil underneath from sheet erosion. The pencil gives a sense of scale. *Rill erosion* is initiated when the water concentrates in small channels (rills) as it runs off the soil. Subsequent cultivation may erase rills, but it does not replace the lost soil. *Gully erosion* creates deep channels that cannot be erased by cultivation. Although gully erosion looks more catastrophic, far more total soil is lost by the less obvious sheet and rill erosion. [Drawings from FAO (1987); photos courtesy USDA Natural Resources Conservation Service]

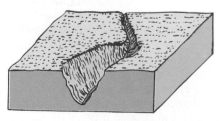

(c) Gully erosion

Deposition of eroded soil

Erosion may send soil particles on a journey of a thousand kilometers—off the hills, into creeks, and down great muddy rivers to the ocean. On the other hand, eroded soil may travel only a meter or two before coming to rest in a slight depression on a hillside or at the foot of a slope (as was shown in Figure 17.5b). The amount of soil delivered to a stream divided by the amount eroded is termed the **delivery ratio.** As much as 60% of eroded soil may reach a stream (delivery ratio = 0.60) in certain watersheds where valley slopes are very steep. As little as 1% may reach the streams draining a gently sloping coastal plain. Typically, the delivery ratio is larger for small watersheds than for large ones, because the latter provide many more opportunities for deposition before a major stream is reached. It is estimated that about 5 to 10% of all eroded soil in North America is washed out to sea. The remainder is deposited in reservoirs, river beds, on flood plains, or on relatively level land farther up the watershed.

17.4 MODELS TO PREDICT THE EXTENT OF WATER-INDUCED EROSION[6]

Land managers and policymakers have many needs to predict the extent of soil erosion:

To plan for the best management of a nation's soil resources

To evaluate the consequences of alternative tillage practices

To determine compliance with environmental regulations

To develop sediment-control plans for construction projects

To estimate the years it will take to silt-in a hydroelectric dam

The Water Erosion Prediction Project (WEPP)

The detachment, transport, and deposition processes of soil erosion can be predicted mathematically by soil erosion *models.* These are equations—or sets of linked equations—that interrelate information about the rainfall, soil, topography, vegetation, and management of a site with the amount of soil likely to be lost by erosion. The most ambitious and sophisticated of the erosion models developed so far is a complex, process-based computer program called the Water Erosion Prediction Project (WEPP). It is based on an understanding of the fundamental mechanisms involved with each process leading to soil erosion.

WEPP is a *simulation* model that computes, on a daily basis, the rates of hydrologic, plant-growth, and even litter-decay processes. Theoretically, it can predict exactly how rainfall will interact with the soil on a site during a particular rainstorm or during the course of an entire year. Raindrop impact, splash erosion, interrill flow, rill formation, channelized flow, gully formation, and sediment deposition both on- and off-site can all be predicted—*if* sufficient data are available to feed into the model. Currently, researchers with the USDA Forest Service and the Natural Resources Conservation Service, along with others throughout the world, are compiling the necessary databases, testing and improving the WEPP model, and making it accessible via the Internet.[7]

---

[6] For discussions of the original USLE, see Wischmeier and Smith (1978), and for the RUSLE, see Renard, et al. (1997). In this textbook we use the scientifically acceptable SI units for the *R* and *K* factors in our discussion of these erosion equations. However, since these soil-loss equations were published in the United States for use by landowners and the general public, most maps, tables, and computer programs available supply values for the *R* and *K* factors in customary English units, rather than in SI units. When using English units for the *R* and *K* factors, the soil loss *A* is expressed in tons (2000 lb) per acre, which can be easily converted to Mg/ha by multiplying by 2.24. For details on converting the customary English units to SI units, see Foster, et al. (1981).

[7] The program and some databases can be downloaded from the WEPP home page: http://soils.ecn.purdue.edu:20002/-wepp/wepp.html. See Laflen, et al. (1997) for details about the availability and uses of the WEPP materials.

The Universal Soil Loss Equation (USLE)

In contrast to the process-based operation of WEPP, most predictions of soil erosion continue to rely on much simpler models that statistically relate soil erosion to a number of easily observed factors. Scientists can make such *empirical* models if they know that certain conditions are associated with soil erosion, even if they do not understand the details of *why* this is so. At the heart of these models is the realization that water-induced erosion results from the interaction of rain and soil. Decades of erosion research have clearly identified the major factors affecting this interaction. These factors are quantified in the **universal soil-loss equation (USLE)**:

$$A = RKLSCP$$

A, the predicted soil loss, is the product of

Working together, these factors determine how much water enters the soil, how much runs off, how much soil is transported, and when and where it is redeposited. Unlike the WEPP program, the USLE was designed to predict only the amount of soil loss by sheet and rill erosion in an average year for a given location. It cannot predict erosion from a specific year or storm, nor can it predict the extent of gully erosion and sediment delivery to streams. It can, however, show how varying any combination of the soil- and land-management-related factors might be expected to influence soil erosion, and therefore can be used as a decision-making aid in choosing the most effective strategies to conserve soil.

The Revised Universal Soil Loss Equation (RUSLE)

The USLE has been used widely since the 1970s. In the early 1990s, the basic USLE was updated and computerized to create an erosion-prediction tool called the **revised universal soil-loss equation (RUSLE)**. The RUSLE uses the same basic factors of the USLE just shown, although some are better defined and interrelationships among them improve the accuracy of soil-loss prediction. The RUSLE is a computer software package that is constantly being improved and modified as experience is gained from its use around the world. The major differences between USLE and RUSLE are shown in Table 17.3.

As we are about to see, the five factors that comprise the USLE provide a useful framework for understanding soil erosion and its control.

17.5 FACTORS AFFECTING INTERRILL AND RILL EROSION

Rainfall Erosivity Factor R

The rainfall **erosivity** factor *R* represents the driving force for sheet and rill erosion. It takes into consideration the total rainfall and, more important, the intensity and seasonal distribution of the rain. Rain intensity is of great importance for two reasons: (1) intense rains have a large drop size, which results in much greater kinetic energy being available to detach soil particles; and (2) the higher the rate of rainfall, the more runoff that occurs, providing the means to transport detached particles. Gentle rains of low intensity may cause little erosion, even if the total annual precipitation is high. In contrast, a few torrential downpours may result in severe damage, even in areas of low annual rainfall. Likewise, soil losses are heavy if the rain falls when the soil is just thawing or is relatively bare because of recent disturbance.

An index of the kinetic energy of each storm is calculated from data related to the intensity and amount of rainfall. Then the indices for all storms occurring during a year

TABLE 17.3 Summary of Major Differences Between USLE and RUSLE

Factor	Universal soil-loss equation (USLE)	Revised universal soil-loss equation (RUSLE)
R	Based on long-term average rainfall conditions for specific geographic areas in the United States.	Generally the same as USLE in the eastern United States. Values for western states (Montana to New Mexico and west) are based on data from more weather stations and thus are more precise for any given location. RUSLE computes a correction to R to reflect, for flat land, the effect of raindrop impact on water ponded on the surface.
K	Based on soil texture, organic matter content, permeability, and other factors inherent to soil type.	Same as USLE but adjusted to account for seasonal changes, such as freezing and thawing, soil moisture, and soil consolidation.
LS	Based on length and steepness of slope, regardless of land use.	Refines USLE by assigning new equations based on the ratio of rill to interrill erosion, and accommodates complex slopes.
C	Based on cropping sequence, surface residue, surface roughness, and canopy cover, which are weighted by the percentage of erosive rainfall during the six crop stages. Lumps these factors into a table of soil-loss ratios, by crop and tillage scheme.	Uses these subfactors: prior land use, canopy cover, surface cover, surface roughness, and soil moisture. Refines USLE by dividing each year in the rotation into 15-day intervals, calculating the soil-loss ratio for each period. Recalculates a new soil-loss ratio every time a tillage operation changes one of the subfactors. RUSLE provides improved estimates of soil-loss changes as they occur throughout the year, especially relating to surface and near-surface residue and the effects of climate on residue decomposition.
P	Based on installation of practices that slow runoff and thus reduce soil movement. P factor values change according to slope ranges with some distinction for various ridge heights.	P factor values are based on hydrologic soil groups, slope, row grade, ridge height, and the 10-year single storm erosion index value. RUSLE computes the effect of strip-cropping based on the transport capacity of flow in dense strips relative to the amount of sediment reaching the strip. The P factor for conservation planning considers the amount and location of deposition.

From Renard, et al. (1994).

are summed to give an annual index. An average of such indexes for many years is used as the R value in the universal soil-loss equation. The RUSLE includes more precise values, especially for the western part of the United States. It also gives special consideration to raindrops striking water ponded on the surface of flat slopes and to runoff from the thawing of frozen soils.

Rainfall index values for locations in the United States are shown in Figure 17.10. Note that they vary from less than 10 in areas of the west to more than 700 along the coasts of Louisiana [the map gives R values in English units that can be converted to SI units of (MJ·mm)/(ha·h·yr) if multiplied by 17.02]. Similar data has been generated in other parts of the world. Generally, rainfall tends to be more intense and more erosive in subtropical and tropical regions than in temperate regions.

Rainfall intensity in most locations is so highly variable that actual erosivity in any one year is commonly 2 to 5 times greater or smaller than the long-term average. In fact, a few unusually intense, heavy storms often account for most of the erosion that takes place during a typical decade. Conservation practices based on the predictions of the USLE or RUSLE using long-term average R factors may not be sufficient to limit erosion damages from these relatively rare, but extremely damaging, storms.

Soil Erodibility Factor K

The soil **erodibility** factor K indicates a soil's inherent susceptibility to erosion. The K value assigned to a particular type of soil indicates the amount of soil lost per unit of erosive energy in the rainfall, assuming a standard research plot (22 m long, 9% slope) on which the soil is kept continuously bare by tillage.

The two most significant and closely related soil characteristics influencing erodibility are (1) *infiltration capacity,* and (2) *structural stability.* High infiltration means that less water will be available for runoff, and the surface is less likely to be ponded (which

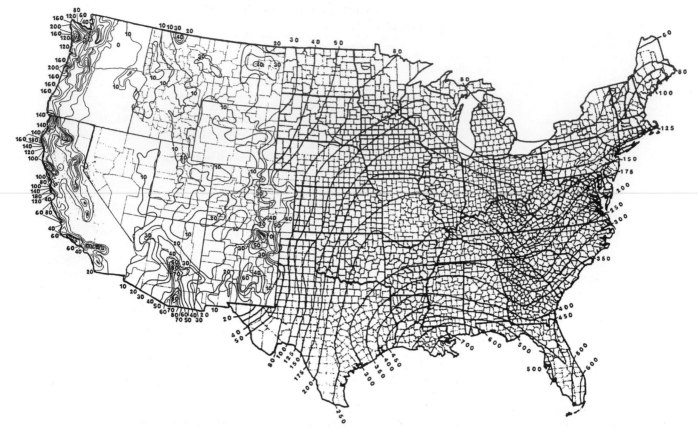

FIGURE 17.10 The geographic distribution of *R* values for rainfall erosivity in the continental United States. Note the very high values in the humid, subtropical Southeast, where annual rainfall is high and intense storms are common. Similar amounts of annual rainfall along the coast of Oregon and Washington in the Northwest result in much lower *R* values because there the rain mostly falls gently over long periods. The complex patterns in the West are mainly due to the effects of mountain ranges. Values on map are in units of 100 (ft·ton·in.)/(acre·yr). To convert to Sl units of (MJ·mm)/(ha·h·yr), multiply by 17.02. [Redrawn from USDA (1995)]

would make it more susceptible to splashing). Stable soil aggregates resist the beating action of rain, and thereby save soil even though runoff may occur. Certain tropical clay soils high in hydrous oxides of iron and aluminum are known for their highly stable aggregates that resist the action of torrential rains. Downpours of a similar magnitude on swelling-type clays would be disastrous.

Basic soil properties that tend to result in high *K* values include high contents of silt and very fine sand; expansive types of clay minerals; a tendency to form surface crusts; the presence of impervious soil layers; and blocky, platy, or massive soil structure. Soil properties that tend to make the soil more *resistant* to erosion (low *K* values) include high soil organic matter content, nonexpansive types of clays, and strong granular structure. Approximate *K* values for soils at selected locations are shown in Table 17.4. Site-specific *K* values obtained with the RUSLE program will vary somewhat from these values. Note that the *K* factor in Sl units normally varies from near zero to about 0.1. Soils with high rates of water infiltration commonly have *K* values of 0.025 or below, while more easily eroded soils with low infiltration rates have *K* factors of 0.04 or higher.

Unlike the USLE, the RUSLE takes into account the fact that *K* values vary seasonally. For example, in cold regions, thawing of frozen soil in spring results in higher *K* values, because the soil is supersaturated with water and is "fluffy" from freeze-thaw action. Also, the *K* factor in RUSLE may be reduced with time in stony soils, because they become protected by an armor of stone fragments left on the soil surface after the finer soil components erode away.

Topographic Factor LS

The topographic factor *LS* reflects the influence of length and steepness of slope on soil erosion. It is expressed as a unitless ratio with soil loss from the area in question

TABLE 17.4 Computed K Values for Soils at Different Locations

The values listed are in SI units. The computerized RUSLE model may give more accurate values for some locations.

Soil	Location	Compounds[a] K
Udalf (Dunkirk silt loam)	Geneva, N.Y.	0.091
Udalf (Keene silt loam)	Zanesville, Ohio	0.063
Udult (Lodi loam)	Blacksburg, Va.	0.051
Udult (Cecil sandy clay loam)	Watkinsville, Ga.	0.048
Udoll (Marshall silt loam)	Clarinda, Iowa	0.044
Udalf (Hagerstown silty clay loam)	State College, Pa.	0.041
Ustoll (Austin silt)	Temple, Tex.	0.038
Aqualf (Mexico silt loam)	McCredie, Mo.	0.034
Udult (Cecil sandy loam)	Clemson, S.C.	0.034
Udult (Cecil sand loam)	Watkinsville, Ga.	0.030
Alfisols	Indonesia	0.018
Alfisols	Benin	0.013
Oxisols	Ivory Coast	0.013
Udult (Tifton loamy sand)	Tifton, Ga.	0.013
Ultisols	Hawaii	0.012
Alfisols	Nigeria	0.008
Udept (Bath flaggy silt loam)	Arnot, N.Y.	0.007
Ultisols	Nigeria	0.005
Oxisols	Puerto Rico	0.001

[a] To convert these K values from (Mg · ha · h)/(ha · MJ · mm) to English units of (ton · acre · h)/(100 acres · ft-ton · in.), simply multiple the values in this table by 7.6. From Wischmeier and Smith (1978); data for tropical soils cited by Cassel and Lal (1992).

in the numerator, and that from a standard plot (9% slope, 22 m long) in the denominator. The longer the slope, the greater the opportunity for concentration of the runoff water.

The *LS* factors in Table 17.5 illustrate the increases that occur as slope length and steepness increase. Three subtables are given for sites with low, moderate, and high ratios of rill to interrill (sheet) erosion. Most sites cultivated to row crops have moder-

TABLE 17.5 Some Representative Values of the Topographic Factor LS on Three Types of Sites with Various Slope Gradients and Lengths

Slope gradient, %	Approximate slope length, m				
	25	50	100	150	200
Sites with low ratios of rill to interrill erosion (e.g., many rangelands)					
2	0.25	0.27	0.30	0.32	0.33
4	0.47	0.55	0.64	0.70	0.75
8	0.93	1.16	1.44	1.65	1.80
12	1.58	2.05	2.65	3.08	3.42
16	2.30	3.05	4.06	4.80	5.40
Sites with moderate ratios of rill to interrill erosion (e.g., most tilled row-crop land)					
2	0.25	0.29	0.35	0.38	0.41
4	0.48	0.62	0.79	0.92	1.02
8	0.94	1.32	1.84	2.24	2.57
12	1.62	2.37	3.48	4.34	5.09
16	2.35	3.54	5.33	6.77	8.03
Sites with high ratios of rill to interrill erosion (e.g., freshly disturbed construction sites and new seedbeds)					
2	0.25	0.33	0.44	0.51	0.58
4	0.49	0.71	1.02	1.27	1.48
8	0.96	1.52	2.38	3.10	3.74
12	1.65	2.70	4.43	5.91	7.24
16	2.39	4.00	6.69	9.03	11.17

Modified from Renard, et al. (1997).

ate rill to interrill erosion ratios. On sites where this ratio is low, such as rangelands, more of the soil movement occurs by interrill erosion. On these sites, slope steepness (%) has a relatively greater influence on erosion, while the slope length has a relatively smaller influence. The opposite is true for freshly excavated construction areas and other highly disturbed sites, which have high rill to interrill erosion ratios. Here, where rill erosion predominates, slope length has a greater influence. While such generalized *LS* factor values for simple slopes can be used with the USLE, the RUSLE computer program calculates more location-specific values, including values for complex (nonuniform) slopes.

Cover and Management Factor C

Soil erosion can most readily be controlled by managing vegetation, plant residues, and soil tillage. Erosion and runoff are markedly affected by different types of vegetative cover and cropping systems (Table 17.6). Undisturbed forests and dense grass provide the best soil protection and are about equal in their effectiveness. Forage crops (both legumes and grasses) are next in effectiveness because of their relatively dense cover. Small grains, such as wheat and oats, are intermediate and offer considerable obstruction to surface wash. Row crops, such as corn, soybeans, and potatoes, offer relatively little living cover during the early growth stages and thereby leave the soil susceptible to erosion unless residues from previous crops cover the soil surface.

Cover crops consist of plants that are similar to the forage crops just mentioned. They can provide soil protection during the time of year between the growing seasons for annual crops. For widely spaced perennial plantings, such as orchards and vineyards, cover crops can permanently protect the soil between rows of trees or vines. A mulch of plant residue or applied materials is also effective in protecting soils. Research on all continents has shown that a surface mulch does not have to be thick or cover the soil completely to make a major contribution to soil conservation. Even small increases in surface cover result in large reductions in soil erosion, particularly interrill erosion (Figure 17.11).

Regulation of grazing to maintain a dense vegetative cover on range and pasture land and the inclusion of close-growing hay crops in rotation with row crops on arable land will help control both erosion and runoff. Likewise, the use of conservation tillage systems, which leave most of the plant residues on the surface, greatly decreases erosion hazards.

The *C* factor in the USLE or RUSLE is the ratio of soil loss under the conditions in question to that which would occur under continuously bare soil. This ratio *C* will approach 1.0 where there is little soil cover (e.g., a bare seed bed in the spring or freshly graded bare soil on a construction site). It will be low (e.g., <0.10) where large amounts of plant residues are left on the land or in areas of dense perennial vegetation.

Values of *C* are specific to each region and type of vegetation or soil management. Estimates based on experiment data and field experience are available from conservation officers and the RUSLE program. Examples of *C* values are given in Table 17.7.

TABLE 17.6 **Effect of Plant Cover on Soil Erosion by Water in the Humid Zone of West Africa**

The data are averaged over many nearby sites, all with similar slopes, soils, and amounts of rainfall. The erosive influence of cultivation (with various types of tillage), especially bare fallows, is illustrated, as is the protective effect of undisturbed forest vegetation.

Type of cover	Number of sites	Mean rainfall, mm/y	Runoff, % of rainfall	Erosion, Mg/ha
Forest protected from fire	11	1293	0.9	0.10
Forest with light fires	13	1289	1.1	0.27
Natural grass fallow	7	1203	16.6	4.88
Groundnut (peanut)	32	1329	20.7	7.70
Upland rice	17	946	23.3	5.52
Maize (corn)	17	1405	17.7	7.63
Failed crops and bare soil	11	1154	39.5	21.28

Data selected from that cited by Pierre (1992).

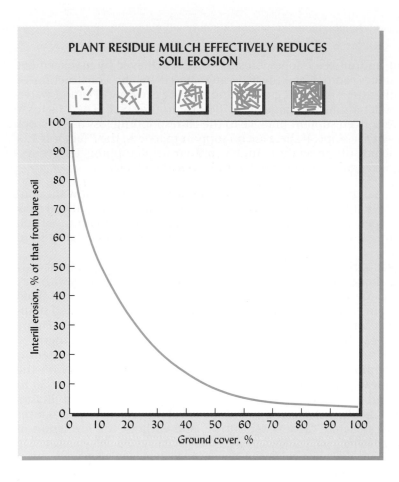

PLANT RESIDUE MULCH EFFECTIVELY REDUCES SOIL EROSION

FIGURE 17.11 Reduction in interrill erosion achieved by increasing ground cover percentage. The diagrams above the graph illustrate 5, 20, 40, 60, and 80% ground cover. Note that even a light covering of mulch has a major effect on soil erosion. The graph applies to interrill erosion. On steep slopes, some rill erosion may occur even if the soil is well covered. [Generalized relationship based on results from many studies]

TABLE 17.7 **Examples of C Values for the Cover and Vegetation Management Factor**

The C values indicate the ratio of soil eroded from a particular vegetation system to that expected if the soil were kept completely bare. Note the effects of canopy cover, surface litter (residue) cover, tillage, and crop rotation. The C values are site- and situation-specific and must be calculated from local information on plant growth habit, climate, and so on. In the United States, specific values may be obtained from the RUSLE computer program or from local offices of the USDA Natural Resources Conservation Service.

Vegetation	Management/condition	C value
Range grasses and low (<1m) shrubs	75% canopy cover, no surface litter	0.17
	75% canopy cover, 60% cover with decaying litter	0.032
Scrub brush about 2 m tall	25% canopy cover, no litter	0.40
	75% canopy cover, no litter	0.28
Trees with no understory, about 4 m drop fall.	75% canopy cover, no litter	0.36
	75% canopy cover, 40% leaf litter cover	0.09
	75% canopy cover, 100% leaf litter cover	0.003
Woodland with understory	90% canopy cover, 100% litter cover	0.001
Permanent pasture	Dense stand of grass sod	0.003
Corn–soybean rotation	Fall plowing, conventional tillage, residues removed	0.53
	Spring chisel plow–plant conservation tillage, 2500 kg/ha surface residues after planting	0.22
	No-till planting, 5000 kg/ha surface residues after planting	0.06
Corn–soybean–wheat–hay rotation	Fall plowing, conventional tillage, residues removed	0.20
	Spring chisel plow–plant conservation tillage, 2500 kg/ha surface residues after planting	0.13
	No-till planting, 5000 kg/ha surface residues after planting	0.05
Corn–oats–hay–hay rotation	Spring conventional plowing before planting	0.05
	No-till planting	0.03

Values typical of midwestern United States. Based on Wischmeier and Smith (1978) and Schwab, et al. (1996).

On some sites with long and/or steep slopes, erosion control achieved by management of vegetative cover, residues, and tillage must be augmented by the construction of physical structures or other steps aimed at guiding and slowing the flow of runoff water. These **support practices** determine the value of the *P* factor in the USLE. The *P* factor is the ratio of soil loss with a given support practice to the corresponding loss if row crops were planted up and down the slope. If there are no support practices, the *P* factor is 1.0. The support practices include tillage on the contour, contour strip-cropping, terrace systems, and grassed waterways, all of which will tend to reduce the *P* factor.

CONTOUR CULTIVATION.   Rows of plants slow the flow of runoff water if they follow the contours across the slope gradient (but the rows *encourage* channelization and gullies if they run up and down the slope). Even more effective is planting on ridges built up of soil along the contours. However, ridges must be designed to carry heavy runoff safely from the field (Figure 17.12).

On long slopes subject to sheet and rill erosion, the fields may be laid out in narrow strips across the incline, alternating the tilled crops, such as corn and potatoes, with hay and small grains. Water cannot achieve an undue velocity on the narrow strips of tilled land, and the hay and grain crops check the rate of runoff. Such a layout is called **strip-cropping** and is the basis for erosion control in many hilly agricultural areas (Figure 17.13). This arrangement can be thought of as shortening the effective slope length.

When the cross strips are laid out rather definitely on the contours, the system is called **contour strip-cropping.** The width of the strips will depend primarily on the degree of slope, the permeability of the land, and the soil erodibility. Widths of 30 to 125 m are common. Contour strip-cropping is often augmented by diversion ditches and waterways between fields. Permanent sod established in the swales produces *grassed waterways* that can safely carry water off the land without the formation of gullies (see Figure 17.13).

TERRACES.   Construction of various types of terraces reduces the effective length and gradient of a slope (Figure 17.14). **Bench terraces** are used where nearly complete control of the water runoff must be achieved, such as in rice paddies (see Figure 6.43). Where farmers use large machinery and need to farm all the land in a field, **broad-based terraces** are more common. Broad-based terraces waste little or no land and are quite effective if properly maintained. Water collected behind each terrace flows gently across (rather than down) the field in a terrace channel, which has a drop of only about 50 cm in 100 m (0.5%). The terrace channel usually guides the runoff water to a grassed waterway, through which the water moves downhill to a nearby canal, stream, or river.

---

[8] Many of the erosion-control practices or management techniques discussed with regard to the *C* and *P* factors and in later sections of this chapter are considered to be **best management practices** (BMPs) under provisions of the Clean Water Act in the United States. The act defines BMPs as "optimal operating methods and practices for reducing or eliminating water pollution" from land-use activities.

FIGURE 17.12   Contour ridges must be carefully laid out with sufficient height to hold back water from even heavy rainfall. Here, surface retention is fast becoming surface runoff. (Photo courtesy of R. Weil)

FIGURE 17.13 An aerial photo of farmland in Kentucky where contour strip-cropping is practiced and grassed waterways (see arrow) control the flow of water off cropland and prevent gully erosion. (Inset) A grassed waterway in action, showing how the permanent grass sod can resist the scouring force of water and safely conduct water off the field. (Photos courtesy of USDA National Resources Conservation Service)

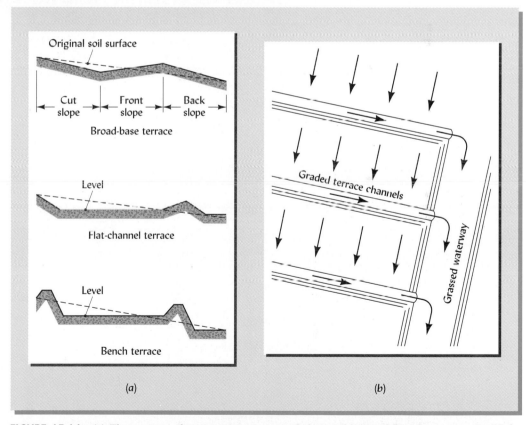

FIGURE 17.14 (a) Three types of terraces in use around the world. Broad-based terraces permit the entire surface to be cropped and are widely used in the United States. Flat-channel terraces allow large volumes of water to move off the soil without erosion. Bench terraces can keep water on the land and are used widely in rice production. (b) Diagram of controlled flow of runoff water from a field to terrace channels and onto a grassed waterway from which it can move to a canal or stream channel.

TABLE 17.8 P Factors for Contour and Strip-Cropping at Different Slopes and the Terrace Subfactor at Different Terrace Intervals

The product of the contour or strip-cropping factors and the terrace subfactor gives the P value for terraced fields.

Slope, %	Contour P factor	Strip-cropping P factor	Terrace interval, m	Terrace subfactor Closed outlets	Terrace subfactor Open outlets
1–2	0.60	0.30	33	0.5	0.7
3–8	0.50	0.25	33–44	0.6	0.8
9–12	0.60	0.30	43–54	0.7	0.8
13–16	0.70	0.35	55–68	0.8	0.9
17–20	0.80	0.40	69–60	0.9	0.9
21–25	0.90	0.45	90	1.0	1.0

Contour and strip-cropping factors from Wischmeier and Smith (1978); terrace subfactor from Foster and Highfill (1983).

Examples of *P* values for contour tillage and strip-cropping at different slope gradients are shown in Table 17.8. Note that *P* values increase with slope and that they are lower for strip-cropping, illustrating the importance of this practice for erosion control. Terracing also reduces the *P* values. Unlike the USLE, the RUSLE takes into account interactions between support practices and subfactors such as slope and soil water infiltration.

The five factors of the USLE (*R*, *K*, *LS*, *C*, and *P*) have suggested many approaches to the practical control of soil erosion. A sample calculation is shown in Box 17.1 to show how the USLE can help evaluate erosion-control options.

BOX 17.1 CALCULATIONS OF EXPECTED SOIL LOSS USING USLE

The RUSLE computer software is designed to calculate the expected soil loss from specific cropping systems at a given location.

The principles involved in both USLE and RUSLE can be verified by making calculations using USLE and its associated factors. Note that the factors in the USLE are related to each other in a multiplicative fashion. Therefore, if any one factor can be made to be near zero, the amount of soil loss A will be near zero.

Assume, for example, a location in Iowa on a Marshall silt loam with an average slope of 4% and an average slope length of 50 m. Assume further that the land is clean-tilled and fallowed.

Figure 17.10 shows that the R factor for this location is about 150 in English units or (150 × 17) 2550 in SI units. The K factor for a Marshall silt loam in central Iowa is 0.044 (Table 17.4) and the topographic factor LS from Table 17.5 is 0.71 (high rill to interrill ratio on soil kept bare). The C factor is 1.0, since there is no cover or other management practice to discourage erosion. If we assume the tillage is up and down the hill, the P value is also 1.0. Thus, the anticipated soil loss can be calculated by the USLE (A = RKLSCP):

$$A = (2550)(0.044)(0.71)(1.0)(1.0) = 79.7 \text{ Mg/ha or } 35.6 \text{ tons/acre}$$

If the crop rotation involved corn–soybean–wheat–hay and conservation tillage practices (e.g., spring chisel plow tillage) were used, a reasonable amount of residue would be left on the soil surface. Under these conditions, the C factor may be reduced to about 0.13 (Table 17.7). Likewise, if the tillage and planting were done on the contour, the P value would drop to about 0.5 (Table 17.8). P could be reduced further to 0.4(0.5 × 0.8) if terraces with open outlets were installed about 40 m apart. Furthermore, with crops on the land the site would have a moderate rill to interrill erosion ratio, so the LS factor would be only 0.62 (middle of Table 17.5). With these figures the soil loss becomes:

$$A = (2550)(0.044)(0.62)(0.13)(0.4) = 3.6 \text{ Mg/ha or } 1.5 \text{ tons/acre}$$

The units for the calculation are not normallly shown, but they are:

$$\frac{MJ \cdot mm}{ha \cdot h \cdot yr} \times \frac{Mg \cdot ha \cdot h}{ha \cdot MJ \cdot mm} = \frac{Mg}{ha \cdot yr} \text{ (LS, C, and P are unitless ratios).}$$

The benefits of good cover and management and support practices are obvious. The figures cited were chosen to provide an example of the utility of the universal soil-loss equation, but calculations can be made for any specific location. In the United States, pertinent factor values that can be used for erosion prediction in specific locations are generally available from state offices of the USDA Natural Resource Conservation Service. The necessary factors are also built into the RUSLE software.

More than half of the soil eroded in the United States comes from agricultural land, with the remainder coming from forests, rangeland, and construction sites. We will now focus on several specific erosion-control technologies appropriate for these various types of land uses.

17.6 CONSERVATION TILLAGE[9]

For centuries, conventional agricultural practice around the world encouraged extensive soil tillage that leaves the soil bare and unprotected from the ravages of erosion. During the last three decades, two technological developments have allowed many farmers to avoid this problem by managing their soils with greatly reduced tillage—or no tillage at all. First came the development of herbicides that could kill weeds chemically rather than mechanically. Second, farmers and equipment manufacturers developed machinery that could plant crop seeds even if the soil was covered by plant residues. These developments obviated two of the main reasons that farmers tilled their soils. Farmer interest in reduced tillage heightened as it was shown that these systems produced equal or even higher crop yields in many regions while saving time, fuel, money—and soil. The latter attribute earned these systems the name of **conservation tillage.**

Conservation Tillage Systems

While there are numerous conservation tillage systems in use today (Table 17.9), all have in common that they leave significant amounts of organic residues on the soil surface after planting. Keep in mind that conventional tillage involves first moldboard plowing (Figure 17.15a) to completely bury weeds and residues, followed by one to three passes with a harrow to break up large clods, then planting the crop, and subsequently several cultivations between crop rows to kill weeds. Every pass with a tillage implement bares the soil anew and also weakens the structure that helps soil resist water erosion.

Conservation tillage systems range from those that merely reduce excess tillage to the no-tillage system, which uses no tillage beyond the slight soil disturbance that occurs as the planter cuts a planting slit through the residues and several cm into the soil (Figure 17.15c). The conventional moldboard plow was designed to leave the field "clean"; that is, free of surface residues. In contrast, conservation tillage systems, such as **chisel plow-**

---

[9] For reviews of conservation tillage technology, see Blevins and Frye (1992) and Carter (1994).

TABLE 17.9 **General Classification of Different Conservation Tillage Systems**

All systems maintain at least 30% of the crop residues on the surface.

Tillage system	Operation involved
No-till	Soil undisturbed prior to planting, which occurs in narrow seedbed, 2.5 to 7.5 cm wide. Weed control primarily by herbicides.
Ridge till (till, plant)	Soil undisturbed prior to planting, which is done on ridges 10 to 15 cm higher than row middles. Residues moved aside or incorporated on about one-third of soil surface. Herbicides and cultivation to control weeds.
Strip till	Soil undisturbed prior to planting. Narrow and shallow tillage in row using rotary tiller, in-row chisel, and so on. Up to one-third of soil surface is tilled at planting time. Herbicides and cultivation to control weeds.
Mulch till	Soil surface disturbed by tillage prior to planting, but at least 30% of residues left on or near soil surface. Tools such as chisels, field cultivators, disks, and sweeps are used (e.g., stubble mulch). Herbicides and cultivation to control weeds.
Reduced till	Any other tillage and planting system that keeps at least 30% of residues on the surface.

Definitions used by Conservation Technology Information Center, West Lafayette, Ind.

(a)   (b)   (c)

FIGURE 17.15 Conventional and conservation tillage practices. (*a*) In conventional tillage, a moldboard plow inverts the upper soil horizon, burying all plant residues and producing a bare soil surface. (*b*) A chisel plow, one type of conservation tillage implement, stirs the soil but leaves a good deal of the crop residues on the soil surface. (*c*) In no-till systems, one crop is planted directly into a cover crop or the residue of a previous cash crop, with only a narrow band of soil disturbed. No-till systems leave virtually all of the residue on the soil surface, providing up to 100% cover and nearly eliminating erosion losses. Here soybeans are planted into a cover crop that will be killed with a herbicide (weed-killing chemical) to form a surface mulch. (Photos courtesy of R. Weil)

ing (Figure 17.15*b*), stir the soil but only partially incorporate surface residues, leaving more than 30% of the soil covered. **Stubble mulching**, whose water-conserving attributes were highlighted in Section 6.5, is another example. **Ridge tillage** is a conservation system in which crops are planted on top of permanent 15- to 20-cm-high ridges. About 30% soil coverage is maintained, even though the ridges are scrapped off a bit for planting and then built up again by shallow tillage to control weeds.

With **no-tillage** systems we can expect 50 to 100% of the surface to remain covered (Table 17.10). Well-managed continuous no-till systems in humid regions include cover crops during the winter and high-residue-producing crops in the rotation. Such systems keep the soil completely covered at all times and build up organic surface layers somewhat like those found in forested soils.

Conservation tillage systems generally provide yields equal to or greater than those from conventional tillage, provided the soil is not poorly drained and in a cool region (Figure 17.16). However, during the transition from conventional tillage to no-tillage, crop yields may decline slightly for several years for reasons associated with some of the effects outlined in the following subsections.

Adaptation by Farmers[10]

In recent years conservation tillage has become increasingly popular, being used on about two-fifths of the nation's cropland. Conservationists project that as much as 75%

---

[10] For the inspiring story of one Chilean farmer's struggle to conquer the forces of erosion and degradation and restore the health of his soil, see Crovetto (1996).

TABLE 17.10 **The Effect of Tillage Systems in Nebraska on the Percentage of Land Surface Covered by Crop Residues**

Note that the conventional moldboard plow system provided essentially no cover, while no-till and planting on a ridge (ridge till) provided best cover.

Tillage system	Number of fields	Percentage of fields with residue cover greater than			
		15%	*20%*	*25%*	*30%*
Moldboard	33	3	0	0	0
Chisel	20	40	15	5	0
Disk	165	40	20	9	4
Field cultivate	13	46	23	0	0
Ridge till (till, plant)	2	100	50	50	0
No-till	3	100	100	100	100
All systems	236	36	18	9	4

From Dickey, et al. (1987).

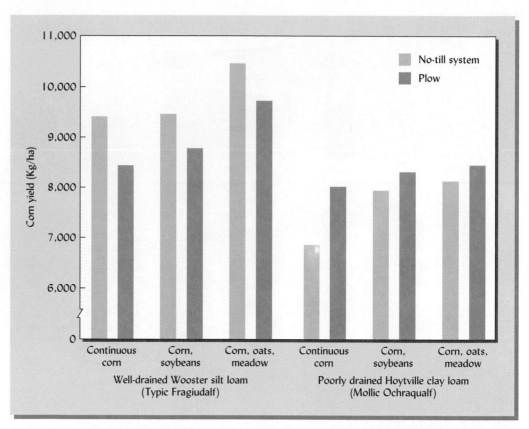

FIGURE 17.16 Effect of tillage systems on the yield of corn grown in different rotations on a well-drained and a poorly drained soil. The no-till system was superior on the well-drained soil but inferior on the soil with poor drainage. Data averages of five years. [Data from Van Doren, et al. (1976)]

of the cropland in the United States will be managed with some kind of conservation tillage within a decade or so. Already, 80% or more of the cropland in a few areas within the United States is managed with conservation tillage (Figure 17.17).

No-tillage systems, especially, have spread to nearly all regions of the nation and are now used in some form on almost half of all the conservation tillage hectares (see Figure 17.17). The no-till system has been used continuously on some farms in the eastern United States since about 1970 (more than 20 to 30 years without any tillage). No-tillage and other conservation systems are also being used in other parts of the world. One of the most significant examples of no-tillage expansion has been in southern Brazil. Thousands of small-scale soybean and corn farmers there, encouraged by government education and assistance, have successfully adapted cover-crop-based no-tillage systems using animal traction or small tractors.

Erosion Control by Conservation Tillage

Since conservation tillage systems were initiated, hundreds of field trials have demonstrated that these tillage systems allow much less soil erosion than do conventional tillage methods. Surface runoff is also decreased, although the differences are not as pronounced as with soil erosion (Figure 17.18). These differences are reflected in the much lower *C* factor values assigned to conservation tillage systems (see Table 17.7).

The erosion-control value of an undisturbed surface residue mulch was discussed in the previous section. Conservation tillage also significantly reduces the loss of nutrients dissolved in runoff water or attached to sediment (review Tables 14.2 and 16.2).

Effect on Soil Properties

When soil management is converted from plow tillage to conservation tillage (especially no-tillage), numerous soil properties are affected, mostly in favorable ways. The changes are most pronounced in the upper few centimeters of soil. Generally, the

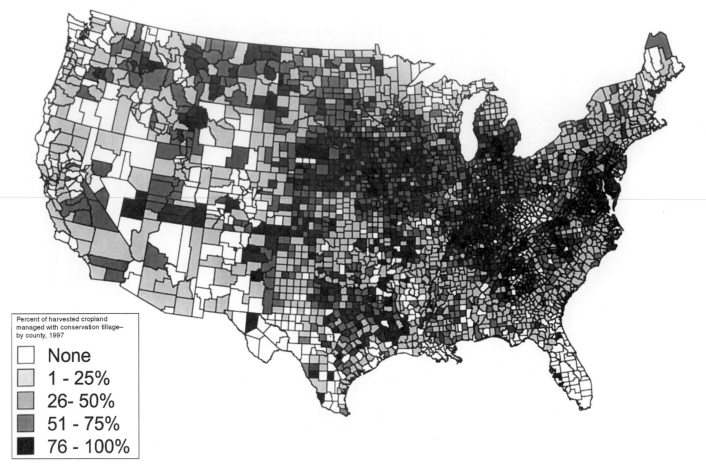

FIGURE 17.17 The percentage use of conservation tillage on different areas of the United States in 1997. Note that in some areas more than three-fourths of the cropland is managed using these soil-saving systems. Overall, some 46% of cropland hectares were managed with conservation tillage. (Courtesy Conservation Technology Information Center, West Lafayette, Ind.)

changes are greatest for systems that produce the most plant residue (especially corn and small grains in humid regions), retain the most residue coverage, and cause the least soil disturbance. Many of these changes are illustrated in other chapters of this textbook, so the discussion here will be brief.

PHYSICAL PROPERTIES. Macroporosity and aggregation (see Sections 4.7 and 4.8) are increased as active organic matter builds up and earthworms and other organisms establish themselves. Infiltration and internal drainage are generally improved, as is soil water-holding capacity. Some of these effects are illustrated in Figure 17.19. The enhanced infiltration capacity of no-till-managed soils is generally very desirable, but in some cases it may lead to more rapid leaching of nitrates and other water-soluble chemicals. Residue-covered soils are generally cooler and more moist (see Sections 6.5 and 7.12). This is an advantage in the hot part of the year, but may be detrimental to early crop growth in the cool spring of temperate regions.

In cool regions, soils with restricted drainage may yield somewhat less using conservation tillage, because soil conditions are wetter and cooler than with conventional tillage. Reduced yields have discouraged the adoption of conservation tillage in these regions. However, limited preplanting tillage over the crop row (see Section 7.12), or ridge tillage are conservation tillage systems that allow at least part of the soil to warm faster and largely overcome these problems.

CHEMICAL PROPERTIES. No-tillage systems significantly increase the organic matter content of the upper few centimeters of soil (see Figures 17.19 and 12.24). During the initial four to six years of no-till management, the build up of organic matter results in the immobilization of nutrients (see Section 12.3), especially nitrogen. This is in contrast to the mineralization of nutrients that is encouraged by the decline of soil organic matter

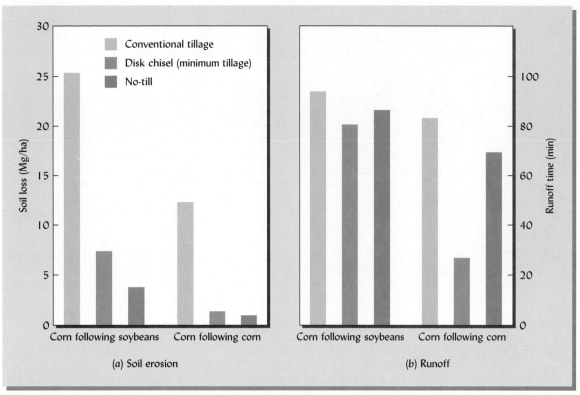

FIGURE 17.18   Effect of tillage systems on soil erosion and runoff from corn plots in Illinois following corn and following soybeans. Soil loss by erosion was dramatically reduced by the conservation tillage practices. The period of runoff was reduced most by the disk chisel system where corn was grown after corn. The soil was a Typic Argiudoll (Catlin silt loam), 5% slope, planted up- and downslope, tested in early spring. [Data from Oschwald and Siemens (1976)]

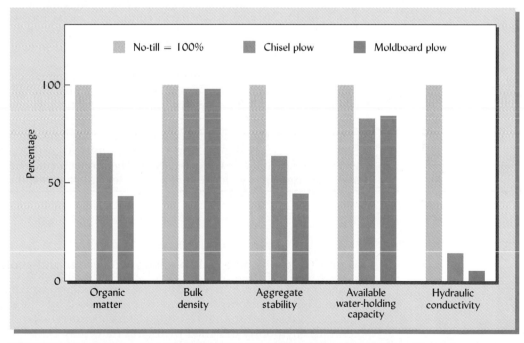

FIGURE 17.19   The comparative effects of 28 years of three tillage systems on soil organic matter content and a number of soil physical properties of an Alfisol in Ohio. Values for the no-till system were taken as 100, and the others are shown in comparison. Bulk density was about the same for each tillage system, but for all other properties the no-till system was decidedly more beneficial than either of the other two systems. The saturated hydraulic conductivity was especially high with no-till. [From Mahboubi, et al. (1993); used with permission of the Soil Science Society of America]

under conventional tillage. Eventually, when soil organic matter stabilizes at a new higher level, nutrient mineralization rates under no-till increase. Higher moisture and lower oxygen levels may also stimulate denitrification (see Section 13.9). These process sometimes result in the need for greater levels of nitrogen fertilization for optimum yields during the early years of no-till management.

In no-tillage systems, nutrient elements tend to accumulate in the upper few centimeters of soil as they are added to the soil surface in crop residues, animal manures, chemical fertilizers, and lime. However, research indicates that because of the surface mulch, crop roots (like those of trees in the untilled forest soil environment) have no trouble obtaining nutrients from the near-surface soil layers. The stratification must be taken into account in sampling soils for fertility testing (see Section 16.18).

Without tillage to mix the soil, the acidifying effects of nitrogen oxidation, residue decomposition, and rainfall are concentrated in the upper few centimeters of soil, the pH of which may drop more rapidly than that of the whole plow layer in conventional systems (see Section 9.6). In humid regions this acidity must be countered by application of liming materials to the soil.

BIOLOGICAL EFFECTS. The abundance, activity, and diversity of soil organisms tend to be greatest in conservation tillage systems characterized by high levels of surface residue and little physical soil disturbance (see Section 11.5). Earthworms and fungi, both important for soil structure, are especially favored (see Table 11.7).

17.7   VEGETATIVE BARRIERS

Narrow rows of permanent vegetation (usually grasses or shrubs) planted on the contour can be used to slow down runoff, trap sediment, and eventually build up "natural" or "living" terraces. In some situations, tropical grasses (e.g., a deep-rooted, drought-tolerant species called *vetiver grass*) have shown considerable promise as an affordable alternative to the construction of terraces.

The deep-rooted vetiver plants have dense, stiff stems that tend to filter out soil particles from the muddy runoff. This sediment builds up on the upslope side of the grass barrier and, in time, actually creates a terrace that may be more than 1 m above the soil surface on the downslope side of the plants (Figure 17.20). Vetiver grass is particularly well suited to survive under harsh conditions, as its root system can forage deeply for water and its foliage is not palatable to wandering cattle. Other grasses are used that serve a dual purpose as cattle fodder, but these tend to take more water out of the root zone of the adjacent food crops.

Research is being conducted to evaluate numerous systems of vegetative barriers that combine grasses with trees. Such systems may provide many benefits (fruit, firewood, fodder, and nutrient-rich mulch) in addition to erosion control, making them more attractive to farmers who must invest their limited labor and resources in establishing such conservation practices. Ideally, conservation technologies (like the winter cover crops discussed in Section 16.4) will improve crop production in the short term as well as protect soil and water quality in the longer term. The benefits of reduced erosion and increased crop yields resulting from several vegetation erosion-control practices are summarized in Table 17.11.

17.8   CONTROL OF GULLY EROSION AND MASS WASTING

Gullies rarely form in soils protected by healthy, dense, forest or sod vegetation, but are common on deserts, rangeland, and open woodland in which the soil is only partially covered. Gullies also readily form in soils exposed by tillage or grading if small rills are allowed to coalesce so that running water eats into the land (Figure 17.21a). Water concentrated by poorly designed roads and trails may cause gullies to form even in dense forests. In many cases, neglected gullies will continue to grow over the years and eventually devastate the landscape (see Figure 17.21b). On the other hand, in some stony soils coarse fragments left behind in the channel bottom may protect it from further cutting action.

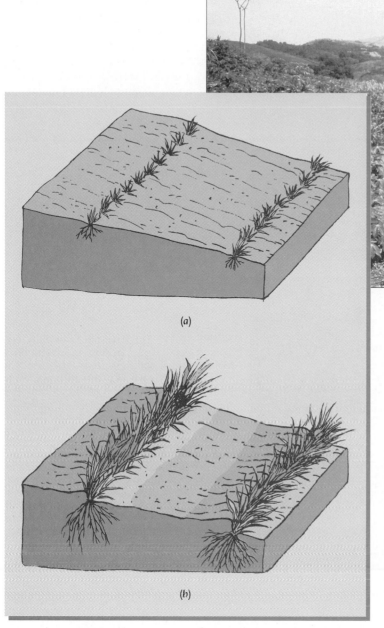

FIGURE 17.20 The use of vegetative barriers to create natural terraces. (Photo) A tropical grass (vetiver) has been planted vegetatively on the contour in a cassava field by pushing root sprigs into the soil. (*a*) The root cuttings are planted perpendicular to the slope direction. In a year or so the grass will be well established and its dense root and shoot growth will serve as a barrier to hold soil particles while permitting some water to pass on through. (*b*) Note the buildup of soil above the grass, basically forming a terrace wall. Perennial tall wheatgrass is being used experimentally in the Northern Plains area of the United States to serve as a barrier against snow movement and later against wind erosion. (Photo courtesy of Centro Internacional Agricultura Tropical in Cali, Colombia)

TABLE 17.11 Range in the Effects of Mulching, Contour Cultivation, and Grass Contour Hedges on Soil Erosion and Crop Yields

Practice	Reduction in soil erosion, %	Increase in crop yield, %
Mulching	78–98	7–188
Contour cultivation	50–86	6–66
Grass contour hedges	40–70	38–73

From a review of more than 200 studies, from Doolette and Smyle (1990).

FIGURE 17.21 The devastation of gully erosion. (Upper) Gully erosion in action on a highly erodible soil in western Tennessee. The roots of the small wheat plants are powerless to prevent the cutting action of the concentrated water flow. (Lower) The legacy of neglect of human-induced accelerated erosion. Tillage of sloping soils during the days of the Roman Empire began a process of accelerated erosion that eventually turned swales into jagged gullies that continue to cut into this Italian landscape with each heavy rain. For a sense of scale, note the olive trees and houses on the grassy, gentle slopes of the relatively uneroded hilltops. (Upper photo courtesy of the USDA Natural Resources Conservation Service, lower photo courtesy of R. Weil)

Remedial Treatment of Gullies

If small enough, gullies can be filled in, shaped for smooth water flow, sown to grass, and thereafter be left undisturbed to serve as grassed waterways. When the gully erosion is too active to be checked in this manner, more extensive treatment may be required. If the gully is still small, a series of check dams about 0.5 m high may be constructed at intervals of 4 to 9 m, depending on the slope. These small dams may be constructed from materials available on site, such as large rocks, rotted hay bales, brush, or logs. Wire netting may be used to stabilize these structures. Check dams, whether large or small, should be constructed with the general features illustrated in Figure 17.22. After a time, enough sediment may collect behind the dams to form a series of bench terraces and the ditch may be filled in and put into permanent sod.

With very large gullies, it may be necessary to divert the runoff away from the head of the channel and install more permanent dams of earth, concrete, or stone in the channel itself. Again, sediment deposited above the dams will slowly fill in the gully. Semipermanent check dams, flumes, and riprap-lined channels are also used on construction sites, but are generally too expensive for extensive use on agricultural land.

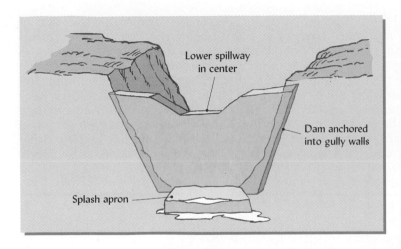

FIGURE 17.22 A schematic drawing of a check dam used to arrest gully erosion. Whether made from rock, brush, concrete, or other materials, a check dam should have the general features shown. The structure should be dug into the walls of the gully to prevent water from going around it. The center of the dam should be lower so that water will spill over there and not wash out the soil of the gully walls. An erosion-resistant apron made from densely bundled brush, concrete, large rocks, or similar material should be installed beneath the center of the dam to prevent the overflow from undercutting the structure. In contrast to the gully-healing effect of a well-designed check dam, the haphazard dumping of rocks, brush, or junked cars into a gully will make matters worse, not better.

Mass Wasting on Unstable Slopes

The downhill movement of large masses of unstable soil (**mass wasting**) is quite different from the erosion of the soil surface, which is the main topic of this chapter. Mass wasting can be a problem on very steep slopes (usually greater than 60% slope). While this type of soil loss sometimes occurs on steep pastures, it is most common on non-agricultural land. Mass wasting can take several forms. **Soil creep** is the slow deformation (without shear failure) of the soil profile as the upper layers move imperceptibly down hill. **Landslides** occur with the sudden shear failure and downhill movement of a mass of soil, usually under very wet conditions. **Mud flows** involve the partial lique-faction and flow of saturated soil due to loss of cohesion between particles (see also Section 4.8 and Figure 4.38).

Mass wasting is sometimes triggered by human activities that undermine natural stabilizing forces or cause the soil to become water saturated as a result of concentrated water flow. The rotting of large soil-anchoring tree roots several years after clear-cutting a forest or the construction of a road cut at the toe of a steep slope are all-too-common examples.

17.9 CONTROL OF ACCELERATED EROSION ON RANGE AND FOREST LAND

Rangeland Problems

Many semiarid rangelands lose large amounts of soil under natural conditions (see Figure 17.3), but accelerated erosion can lead to even greater loses if human influences are not carefully managed. Overgrazing, which leads to the deterioration of the vegetative cover on rangelands, is a prime example. Grass cover generally protects the soil better than the scattered shrubs that usually replace it under the influence of poorly managed livestock grazing. In addition, cattle congregating around poorly distributed water sources and salt licks may completely denude the soil. Cattle trails, as well as ruts from off-road vehicles, can channelize runoff water and spawn gullies that eat into the land-scape. Because of the prevalence of dry conditions, wind erosion (to be discussed in Section 17.11, also plays a major role in the deterioration of rangeland soils.

Erosion on Forest Lands

In contrast to deserts and rangelands, land under healthy, undisturbed forests loses very small amounts of soil. However, accelerated erosion can be a serious problem on forested land, both because the rates of soil loss may be quite high and because the amount of land involved is often enormous. The main cause of accelerated erosion in forested watersheds is usually the construction of logging roads, timber-harvest opera-tions, and the trampling of trails and off-trail areas by large numbers of recreational users (or cattle, in some areas).

To understand and correct these problems, it is necessary to realize that the secret of low natural erosion from forested land is the undisturbed forest floor, the O horizons that protect the soil from the impact of raindrops and allow such high infiltration rates that surface runoff is very small or absent. Contrary to the common perception, it is the forest floor, rather than the tree canopy or roots, that protects the soil from erosion (Figure 17.23a). In fact, rainwater dripping from the leaves of tall trees often forms very large drops that reach terminal velocity and impact the ground with more energy than direct rain from even the most intense of storms. If the forest floor has been disturbed and mineral soil exposed, serious splash erosion can result (see Figure 17.23b). Gully erosion can also occur under the forest canopy if water is concentrated, as by poorly designed roads.

(a)

(b)

(c)

FIGURE 17.23 The leaf mulch on the forest floor, rather than the tree roots or canopy, provides most of the protection against erosion in a wooded ecosystem. (a) An undisturbed temperate deciduous forest floor (as seen through a rotten stump). The leafless canopy will do little to intercept rain during winter months. During the summer, rainwater dripping from the foliage of tall trees may impact the forest floor with as much energy as unimpeded rain. (b) Severe erosion has taken place under the tree canopy in a wooded area where the protective forest floor has been destroyed by foot traffic. The exposed tree roots indicate that nearly 25 cm of the soil profile has washed away. (c) *Glycine* vines have been planted under the dense canopy of a rubber tree plantation in southern Asia to prevent soil erosion on a steep hillside. The white rubbery latex that exudes from the cut tree bark is normally collected drop by drop in small cups. Trampling by workers and rapid decay of leaf litter make the living ground cover necessary in this situation. (Photos courtesy of R. Weil)

Practices to Reduce Soil Loss Caused by Timber Production[11]

The main sources of eroded soil from timber production are *logging roads* (that are built to provide access to the area by trucks), *skid trails* (the paths along which logs are dragged), and *yarding areas* (the areas where collected logs are sized and loaded onto trucks). Relatively little erosion results directly from the mere felling of the trees (except where large tree roots are needed to anchor the soil against mass wasting, as discussed in Section 17.8). Strategies to control erosion should include consideration of the following: (1) intensity of timber harvest, (2) methods used to remove logs, (3) scheduling of timber harvests, and (4) design and management of roads and trails. Soil disturbance in preparation for tree regeneration (such as tillage to eliminate weed competition or provide better seed-to-soil contact) must also be limited to sites with low susceptibility to erosion.

INTENSITY OF TIMBER HARVEST. On the steepest, most erodible sites, environmental stewardship may require that timber harvest be foregone and the land be given only protective management. On somewhat less susceptible sites, selective cutting (occasional removal of only the oldest trees) may be practiced without detrimental results. The shelterwood system (in which a substantial number of large trees are left standing after all others are harvested) is probably the next most intensive method and can be used on moderately susceptible sites. Clear-cutting (removal of all trees from large blocks of forest) should be used only on gentle slopes with stable soils.

METHOD OF TREE REMOVAL. The least expensive and most commonly used method of tree removal is by wheeled tractors called *skidders* (see Figure 4.18). This method generally disrupts the forest floor, exposing the mineral soil on perhaps 30 to 50% of the harvested area. In contrast, more expensive methods using cables to lift one end of the log off the ground are likely to expose mineral soil on only 15 to 25% of the area. Occasionally, for very sensitive sites, logs are lifted to yarding areas by balloon or helicopter, practices which are very expensive but result in as little as 4 to 8% bare mineral soil.

SCHEDULING OF TIMBER HARVEST. Much erosion can be prevented by limiting entry into the forest by machinery to those periods when the soil is either dry or (in temperate regions) frozen and covered with snow. Damage to the forest floor (including both compaction and exposure of mineral soil) occurs much more readily if the soil is very wet. In addition, wheel ruts easily form in wet soils and channel runoff to initiate gully erosion.

DESIGN AND MANAGEMENT OF ROADS. Poorly built logging roads may lose as much as 100 Mg/ha of soil by erosion of the road surface, the drainage ditch walls, or the soil exposed by road cuts into the hillside. Roads also collect and channelize large volumes of water, which can cause severe gullying. Roads should be so aligned as to avoid these problems. Although expensive, placing gravel on the road surface, lining the ditches with rocks, and planting perennial vegetation on exposed road cuts can eliminate up to 99% of the soil loss. A much less expensive measure is to provide cross channels (shallow ditches or **water bars**, as shown in Figure 17.24) every 25 to 100 m to prevent excessive accumulation of water and safely spread it out onto areas protected by natural vegetation. After timber harvest is complete, the roads in an area should be grassed over and closed to traffic.

DESIGN OF SKIDDING TRAILS. Skidding trails that lead runoff water downhill toward a yarding area invite the formation of gullies. Repeated trips dragging logs over the same secondary trails also greatly increase the amount of mineral soil exposed to erosive forces. Both practices should be avoided, and yarding areas should be located on the highest-elevation, most level, and well-drained areas available.

BUFFER STRIPS ALONG STREAM CHANNELS. When forests are harvested, buffer strips as wide as 1.5 times the height of the tallest trees should generally be left untouched along all streams. As discussed in Section 16.2, buffer strips of dense vegetation have a high capacity to remove sediment and nutrients from runoff water. Forested buffers also protect the stream from excessive logging debris. In addition, streamside trees shade the water, protecting it from the undesirable heating that would result from exposure to direct sunlight.

---

[11] For a general introduction to this topic, see Nyland (1996).

FIGURE 17.24 An open-top culvert in a well-designed logging road in Montana. This simple structure, along with proper road bed alignment, can greatly reduce gully erosion caused by water flowing unimpeded along roads and trails in forested areas. Water bars or open-top culverts placed at frequent intervals lead runoff water, a little at a time, off the road and into densely vegetated areas. (Photos courtesy of R. Weil)

17.10 EROSION AND SEDIMENT CONTROL ON CONSTRUCTION SITES

Although active construction sites cover relatively little land in most watersheds, they may still be a major source of eroded sediment because the potential erosion per hectare on drastically disturbed land is commonly 100 times that on agricultural land. Heavy sediment loads are characteristic of rivers draining watersheds in which land use is changing from farm and forest to builtup land. Historically, once urbanization of a watershed is complete (all land being either paved over or covered by well-tended lawns), sedimentation rates return to levels as low (or lower) than before the development took place.

To prevent serious sediment pollution from construction sites, governments in the United States (e.g., through state laws and the federal Clean Water Act of 1992) and in many other industrialized countries require that contractors develop detailed erosion- or sediment-control plans before initiating construction projects that will disturb more than about 1 ha of land. The goals of erosion control on construction sites are (1) to avoid on-site damage, such as undercutting of foundations or finished grades and loss of topsoil needed for eventual landscaping; and (2) to retain eroded sediment on-site so as to avoid all the environmental damages (and liabilities) that would result from deposition of sediment on neighboring land and roads, and in ditches, reservoirs, and streams.

Principles of Erosion Control on Construction Sites

Five basic steps are useful in developing plans to meet the aforementioned goals:

1. When possible, schedule the main excavation activities for low-rainfall periods of the year.
2. Divide the project into as many phases as possible, so that only a few small areas must be cleared of vegetation and graded at any one time.
3. Cover disturbed soils as completely as possible, using vegetation or other materials.
4. Control the flow of runoff to move the water safely off the site without destructive gully formation.
5. Trap the sediment before releasing the runoff water off-site.

The last three steps bear further elaboration. They are best implemented as specific practices integrated into an overall erosion-control plan for the site.

Soils freshly disturbed by excavation or grading operations are characterized by very high erodibility (*K* values). This is especially true for low-organic-matter subsoil materials. Potential erosion can be extremely high (200 to 400 Mg/ha is not uncommon) unless the *C* value is made very low by providing good soil cover. This is best accomplished by allowing the natural vegetation to remain undisturbed for as long as possible, rather than clearing and grading the entire project area at the beginning of construction (see step 1, preceding). Once a section of the site is graded, any sloping areas not directly involved in the construction should be sodded or sown to fast-growing grass species adapted to the soil and climatic conditions.

Seeded areas should be covered with **mulch** or specially manufactured **erosion blankets** (Figure 17.25*b*). Erosion blankets, made of various biodegradable or non-biodegradable materials, provide instant soil cover and protect the seed from being washed away (Table 17.12).

A commonly used technology to protect steep slopes and areas difficult to access, such as road cuts, is the **hydroseeder** (Figure 17.25*a*) that sprays out a mixture of seed, fertilizer, lime (if needed), mulching material, and sticky polymers. Good construction-site management includes removal and stockpiling of the A-horizon material before an area is graded (see Figure 1.15). This soil material is often quite high in fertility and is a

FIGURE 17.25 Two methods of establishing vegetative cover on steep, unstable slopes. (Top) A hydroseeder allows vegetative cover to be efficiently established on difficult-to-reach areas. The machine is spraying a mixture of water, chopped straw, grass seed, fertilizer, and sticky polymers that hold the mulch in place until the grass seed can take root. (Bottom) Erosion-control mats made of plastic netting or natural materials like jute are laid down over newly seeded grass to hold the seed and soil in place until the vegetative cover is established. (Photos courtesy of R. Weil)

TABLE 17.12 Effectiveness of Various Erosion-Control Materials in Reducing Runoff and Soil Loss and in Establishing Vegetative Cover on Disturbed Sites

All plots (except sod) were seeded with grass and covered with the materials listed. Note that all treatments greatly reduced erosion and, to a lesser degree, runoff losses. The sod was by far the most effective material, followed by the straw mulch. None of the materials had a negative effect on grass establishment, while the straw mulch may have had a positive effect. In other applications (such as lining of channels) where high velocity of runoff is encountered, the manufactured erosion blankets would be more effective than straw.

Material	Soil loss,[a] kg/ha	Runoff loss, % of rain	Time to initiation of runoff, s	Ground cover established,[b] %
Bare soil	6650	83	34	50
Jute (woven mesh)	410	68	62	61
Wood shavings in nonwoven polyester netting	810	74	74	69
Coconut fiber mat	1070	76	86	58
Straw (4.5 Mg/ha)	590	60	102	76
Grass sod	100	28	341	NA[c]

[a] Soil loss from a single simulated rain event at a rate of 96 mm/h. Means of two soils: a Sassafras loamy sand with 8% slope and a Matapeake sandy clay loam with 15% slope.
[b] Percent vegetative cover established one year after Kentucky 31 tall fescue grass was seeded and covered by the various materials, in a separate study.
[c] Not applicable since sod provided 100% cover and was not seeded.
Data from Krenisky, et al. (1998)

potential source of sediment and nutrient pollution. The stockpile should therefore be given a grass cover to protect it from erosion until it is used to provide topsoil for landscaping around the finished structures.

Controlling the Runoff

Freshly exposed and disturbed subsoil material is highly susceptible to the cutting action of flowing water and the gullies so formed may ruin a grading job, undercut pavements and foundations, and produce enormous sediment loads. The flow of runoff water must be controlled by carefully planned grading, terracing, and channel construction. Most construction sites require a perimeter waterway to catch runoff before it leaves the site and to channel it to a retention basin.

The sides and bottom of such channels must be covered with "armor" to withstand the cutting force of flowing water. Where high water velocities are expected, the soil must be protected with **hard armor** such as **riprap** (large angular rocks, such as are shown in Figure 17.26), **gabions** (rectangular wire-mesh containers filled with hand-sized stone) or interlocking concrete blocks. The soil is first covered with a **geotextile** filter cloth (a tough nonwoven material) to prevent mixing of the soil into the rock or stone.

In smaller channels, and on more gentle slopes where relatively low water velocities will be encountered, **soft armor**, such as grass sod or erosion blankets, can be used. Generally, soft armor is cheaper and more aesthetically appealing than hard armor. Newer approaches to erosion control often involve reinforced vegetation (e.g., trees or grasses planted in openings between concrete block or in tough erosion mats).

The term **bioengineering** describes techniques that use vegetation (locally native, noninvasive species are preferred) and natural biodegradable materials to protect channels subject to rather high water velocities. Examples include the use of **brush mattresses** to stabilize steep slopes. In this technique, live tree branches are tightly bundled together, staked down flat using long wooden pegs, and partially covered with soil. The so-called **live stake** technique (Figure 17.27) is another example of a bioengineering approach commonly used to stabilize soil along channels subject to high velocity water. In both cases, the soil is provided some immediate physical protection from scouring water, and eventually the dormant cuttings take root to provide permanent, deep-rooted vegetative protection.

FIGURE 17.26 The flow of runoff from large areas of bare soil must be carefully controlled if off-site pollution is to be avoided. Here, a carefully designed channel with a grass sod bottom and sides lined with large rocks (riprap) prevents gully erosion, reduces soil loss, and guides runoff around the perimeter of a construction site. (Photo courtesy of R. Weil)

FIGURE 17.27 An example of *bioengineering* along a stream bed that had been disturbed during the construction of a commercial airport in Illinois. Living willow branches are pounded into the soft erodible stream bank to anchor the soil and reduce the scouring power of the water during high flows. Eventually the willow branches will take root and sprout (inset), providing trees that will permanently stabilize the stream bank and improve wildlife habitat. (Photos courtesy of R. Weil)

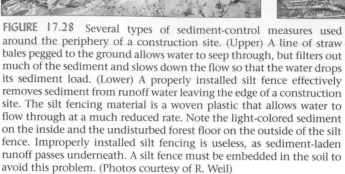

FIGURE 17.28 Several types of sediment-control measures used around the periphery of a construction site. (Upper) A line of straw bales pegged to the ground allows water to seep through, but filters out much of the sediment and slows down the flow so that the water drops its sediment load. (Lower) A properly installed silt fence effectively removes sediment from runoff water leaving the edge of a construction site. The silt fencing material is a woven plastic that allows water to flow through at a much reduced rate. Note the light-colored sediment on the inside and the undisturbed forest floor on the outside of the silt fence. Improperly installed silt fencing is useless, as sediment-laden runoff passes underneath. A silt fence must be embedded in the soil to avoid this problem. (Photos courtesy of R. Weil)

Trapping the Sediment

For small areas of disturbed soils, several forms of sediment barriers can be used to filter the runoff before it is released. The most commonly used types of silt barriers are straw bales and woven fabric silt fences. If installed properly, both can effectively slow the water flow so that most of the sediment is deposited on the uphill side of the barrier (Figure 17.28), while relatively clear water passes through.

On large construction sites, a system of protected slopes and channels leads storm runoff water to one or more retention and sedimentation ponds located at the lowest elevation of the site. As the flowing water meets the still water in the pond it drops most of its sediment load (Figure 17.29), allowing the relatively clear water to be skimmed off the top and released to the next pond or off the site. Wetlands (Section 7.8) are often constructed to help purify the overflow from sedimentation ponds before the water is released into the natural stream or river.

Construction site erosion-control measures are commonly designed to retain the runoff from small storms on-site. The retention ponds must also be able to deal with runoff generated from intense rainstorms—the kind that may be expected to occur on

FIGURE 17.29 The effect of a small sediment impoundment structure on the retention of suspended soil eroded from a construction site. Note the riprap-lined channel leading runoff water into the pond and the deltalike alluvial fan of sediment deposited as the water's velocity is reduced where it enters the pond. The standpipe in the background has holes that allow the clearest water near the top to overflow and leave the construction site. (Photo courtesy of R. Weil)

a site only once in every 10 to even 100 years. In designing the capacity of sediment-retention ponds, the erosion models discussed in Section 17.4 are used to estimate the amount of sediment that is likely to be eroded from the site. While expensive to construct, well-designed sediment-retention ponds can be incorporated as permanent aesthetic water features that enhance the value of the final project.

17.11 WIND EROSION: IMPORTANCE AND FACTORS AFFECTING IT

Up to this point we have focused on soil erosion by water, but wind, too, causes much soil loss. Wind erosion is most common in arid and semiarid regions, and it is a problem on some soils in humid climates as well. It occurs when strong winds blow across soils with relatively dry surface layers. All kinds of soils and soil materials are affected. The finer soil particles may be carried to great heights and for thousands of kilometers—even from one continent, across the ocean, to another.

In May 1934, a great dust storm originating in the southern Great Plains region of the United States sent great clouds of silt, clay, and organic matter eastward to the Atlantic seaboard and out over the ocean. It is said that the dust-darkened skies over Washington, D.C. helped convince Congress to fund new, farreaching programs to combat soil erosion.

Wind erosion is often very destructive, causing damage to vegetation and soils on the eroding site and various kinds of off-site environmental damages, as well (Figures 17.30 and 17.31). Wind moves about 40% of the soil transported by erosion in the United States. About 12% of the continental United States is somewhat affected by wind erosion—8% moderately so and perhaps 2 to 3% greatly.

In six of the Great Plains states, annual wind erosion exceeds water erosion on cropland, averaging from 4 Mg/ha in Nebraska to 29 Mg/ha in New Mexico, where mismanagement of plowed lands and overgrazing of range grasses have greatly increased the susceptibility of the soils to wind action. In dry years, the results have been most deplorable.

Even in humid regions, certain soils suffer significant wind erosion when their surface layer dries out and wind velocity is high. The movement of sand dunes along the Atlantic coast and on the eastern shore of Lake Michigan are examples. Wind erosion also damages cultivated sandy or peaty soils when conditions are dry. The on-site and off-site damages caused by wind erosion were discussed in Section 17.2.

The erosive power of wind is demonstrated in Figure 17.30. The erosive force of the wind tends to be much greater in some regions than in others. For example, the semiarid Great Plains region of the United States is subject to winds with 5 to 10 times the erosive force of the winds common in the humid East. In the Great Plains, the winds are most powerful in the winter season. In other regions, high winds occur most commonly during the hot summers.

Wind erosion is a worldwide problem. Large areas of the former Soviet Union and sub-Saharan Africa have been badly damaged by wind erosion. Overgrazing and other misuses of the fragile lands of arid and semiarid areas have allowed wind erosion to depress soil productivity and bring starvation and misery to millions of people in these areas.

Mechanics of Wind Erosion

Like water erosion, wind erosion involves three processes: (1) *detachment,* (2) *transportation,* and (3) *deposition.* The moving air, itself, results in some detachment of tiny soil grains from the granules or clods of which they are a part. However, when the moving air is laden with soil particles, its abrasive power is greatly increased. The impact of these rapidly moving grains dislodges other particles from soil clods and aggregates. These dislodged particles are now ready for one of the three modes of wind-induced transportation, depending mostly on their size.

SALTATION. The first and most important mode of particle transportation is that of **saltation,** or the movement of soil by a series of short bounces along the ground surface (Figure 17.32). The particles remain fairly close to the ground as they bounce, seldom rising more than 30 cm or so. Depending on conditions, this process may account for 50 to 90% of the total movement of soil.

FIGURE 17.30 Wind erosion in action. (Upper) One result of wind erosion is dust storms, such as this one moving across the high plains of Texas. The swirling dark cloud consists of fine particles eroded from the soil by high winds sweeping across the flat farmland and range. Apparently, most of the land was not as well covered with vegetation as the wheat field in the foreground. (Lower) Soil eroded by wind during a single dust storm has piled up to a depth of nearly 1 m along a fencerow in Idaho. (Upper photo courtesy of Dr. Chen Weinan, USDA Agricultural Research Service, Warm Springs, Tex.; lower photo courtesy of R. Weil)

FIGURE 17.31 Direct wind erosion damage to a tomato crop in a sandy field in Delaware. (Left) Tomato plants are buried beneath windblown sandy soil. (Right) Young tomato fruits showing damage incurred by sandblasting during a windstorm. (Photos courtesy of R. Weil)

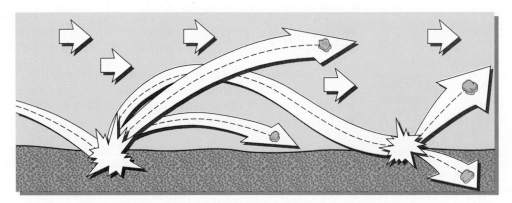

FIGURE 17.32   The process of saltation. Medium-sized particles (0.05 to 0.5 mm in diameter) bounce along the soil surface, striking and dislodging other particles as they move. They are too large to be carried long distances suspended in the air but small enough to be transported by the wind. [From Hughes (1980); used with permission of Deere & Company, Moline, Ill.]

SOIL CREEP.   Saltation also encourages **soil creep**, or the rolling and sliding along the surface of the larger particles. The bouncing particles carried by saltation strike the large aggregates and speed up their movement along the surface. Soil creep accounts for the movement of particles up to about 1.0 mm in diameter, which may amount to 5 to 25% of the total movement.

SUSPENSION.   The most spectacular method of transporting soil particles is by movement in **suspension**. Here, dust particles of a fine-sand size and smaller are moved parallel to the ground surface and upward. Although some of them are carried at a height no greater than a few meters, the turbulent action of the wind results in others being carried kilometers upward into the atmosphere and many hundreds of kilometers horizontally. These particles return to the earth only when the wind subsides and/or when precipitation washes them down. Although it is the most striking manner of transportation, suspension seldom accounts for more than 40% of the total and is generally no more than about 15%.

Factors Affecting Wind Erosion

Susceptibility to wind erosion is related to the moisture content of soils. Wet soils do not blow because of the adhesion between water and soil particles. Dry winds generally lower the moisture content to below the wilting point before wind erosion takes place.

Other factors that influence wind erosion are (1) wind velocity and turbulence, (2) soil surface conditions, (3) soil characteristics, and (4) the nature and orientation of the vegetation.

WIND VELOCITY.   The rate of wind movement, especially gusts having greater than average velocity, will influence erosion. Tests have shown (see, for example, Table 17.13) that

TABLE 17.13   Threshold Wind Velocity, Height of Transition from Saltation to Suspension, and Mass of Soil Carried by Creep, Saltation, and Suspension During Four Wind Erosion Episodes on Cropland at Big Spring, Texas

Note that saltation accounted for most of the soil material moved, and that this movement mostly took place at heights of less than 30 cm above the soil surface. The location is semiarid, receiving an average of 470 mm of rainfall annually. The soil was a bare, smooth field of Amarillo fine sandy loam (Ustalfs).

Date	Threshold wind velocity, m/s	Height of transition from saltation to suspension, cm	Mass of soil material moved per width of wind path, Kg/m			
			Creep	Saltation	Suspension	Total
22 January	6.8	22	0.87	13.33	0.72	14.92
5 February	7.0	22	7.68	142.9	13.43	164.01
14 March	7.2	28	0.43	15.9	3.98	20.31
21 March	7.8	27	0.10	2.41	0.30	2.80

Selected from Fryrear and Saleh (1993).

wind speeds of about 25 km/h (7 m/s) are required to initiate soil movement. At higher wind speeds, soil movement is proportional to the cube of the wind velocity. Thus, the quantity of soil carried by wind goes up very rapidly as speeds above 30 km/h are reached.

WIND TURBULENCE. Wind turbulence also influences the capacity of the atmosphere to transport matter. Although the wind itself has some direct influence in picking up fine soil, the impact of wind-carried particles as they strike the soil is probably more important.

SURFACE ROUGHNESS. Wind erosion is less severe where the soil surface is rough. This roughness can be obtained by proper tillage methods, which leave large clods or ridges on the soil surface. Leaving a stubble mulch (see Section 6.6) is probably an even more effective way of reducing wind-borne soil losses.

SOIL PROPERTIES. In addition to moisture content, several other soil characteristics influencing wind erosion are (1) mechanical stability of soil clods and aggregates, (2) stability of soil crust, and (3) bulk density and size of erodible soil fractions. Some clods resist the abrasive action of wind-carried particles. If a soil crust resulting from a previous rain is present, it, too, may be able to withstand the wind's erosive power. The presence of clay, organic matter, and other cementing agents is also important in helping clods and aggregates resist abrasion. This is one reason why sandy soils, which are low in such agents, are so easily eroded by wind.

Soil particles or aggregates about 0.1 mm in diameter are more erodible than those larger or smaller in size. Thus, fine sandy soils are quite susceptible to wind erosion. Particles or aggregates about 0.1 mm in size are also partly responsible for the movement of other particles. Saltating particles bounce against larger particles, causing surface creep. The saltating particles also collide with smaller particles, sending them up into suspension.

VEGETATION. Vegetation or a stubble mulch will reduce wind erosion hazards, especially if the rows run perpendicular to the prevailing wind direction. This effectively slows wind movement near the soil surface. In addition, plant roots help bind the soil and make it less susceptible to wind damage.

17.12 PREDICTING AND CONTROLLING WIND EROSION

A wind erosion prediction equation (WEQ) has been in use since the late 1960s:

$$E = f(ICKLV)$$

The predicted wind erosion E is a function f of:

I = soil erodibility factor
C = climate factor
K = soil-ridge-roughness factor
L = width of field factor
V = vegetative cover factor

The WEQ involves the major factors that determine the severity of the erosion, but it also considers how these factors interact with each other. Consequently, it is not as simple as is the USLE for water erosion. Evidence of the interaction among factors is seen in Figure 17.33. The **soil erodibility factor** I relates to the properties of the soil and to the degree of slope of the site in question. The **soil-ridge-roughness** factor K takes into consideration the clodiness of the soil surface, vegetative cover V, and ridges on the soil surface. The **climatic factor** C involves wind velocity, soil temperature, and precipitation (which helps control soil moisture). The **width of field factor** L is the width of a field in the downwind direction. Naturally, the width changes as the direction of the wind changes, so the prevailing wind direction is generally used. The **vegetative cover** V relates not only to the degree of soil surface covered with residues, but to the nature of the cover—whether it is living or dead, still standing, or flat on the ground.

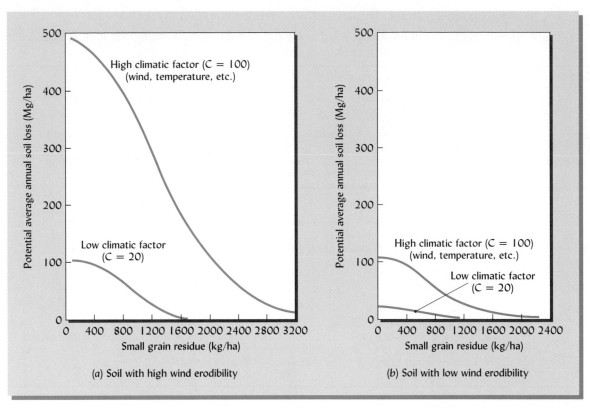

FIGURE 17.33 Effect of small grain residues on the potential wind erosion of soils with high (*a*) and low (*b*) erodibility. Strong, dry winds and high temperatures encourage erosion, especially on the soil with high wind-erodibility characteristics. Surface residues can be used to help control this wind erosion. [From Skidmore and Siddoway (1978); used with permission of the American Society of Agronomy]

A revised, more complex, and more accurate computer-based prediction model has been developed, and is known as the **revised wind erosion equation (RWEQ)**. It is still an empirical model based on many years of research to characterize the relationship between observable conditions and resulting wind erosion severity. Table 17.14 outlines the factors taken into consideration by RWEQ, which (like RUSLE) calculates the erosion hazard during 15-day intervals throughout the year. For each interval of time, the RWEQ makes adjustments in the residue, soil erodibility, and soil roughness parameters, based on the input information about management operations and weather conditions. For example, it assumes that residues decompose over time, that tillage operations flatten standing residues, and that rainfall slakes clods to reduce soil roughness.

Scientists and engineers around the world are also cooperating in the development of a much more complex process-based model known as the **Wind Erosion Prediction System (WEPS)**. Like its water erosion sister, WEPP, this computer program simulates all the basic processes of wind interaction with soil. Scientists are continually improving the model and testing its predictions against data observed in the real world. The USDA has plans to make this model available in a version that can be integrated with WEPP in a user-friendly format. However, since to run both of these complex process-based models one must supply a great deal of information about every aspect of the site on which erosion is to be predicted, it remains to be seen whether they will ever become as widely used as the simpler, tried and true empirical models now in widespread use.

Control of Wind Erosion

SOIL MOISTURE. The factors of the wind erosion equation give clues to methods of reducing wind erosion. For example, since soil moisture increases cohesiveness, the wind speed required to detach soil particles increases dramatically as soil moisture increases.

TABLE 17.14   Some Factors Integrated in the Revised Wind Erosion Equation (RWEQ) Model

The RWEQ program calculates values for each factor for each 15-day period during the year. It also calculates interactions; that is, it uses the value of one factor to modify other factors.

Model factor	Subfactor terms in the model	Comments
Weather factor	Weather factor *WF*; includes terms for wind velocity, direction, air temperature, solar radiation, rainfall, and snow cover.	Modified by other factors such as soil wetness.
Soil factors	Erodible fraction *EF*; fraction smaller than 0.84 mm diameter.	Based on sand, silt, organic matter, and rock cover.
	Soil crust factor *SCF*.	Induced by rainfall, eliminated by tillage.
	Surface roughness *SR*; a random component due to clods and/or an oriented component due to ridges.	Interacts with rainfall and tillage.
	Soil wetness *SW*.	Computed from rainfall minus evapotranspiration.
Tillage factor	Tillage factor *TF*; depends on type of implement, soil conditions, timing, and so forth.	Modifies surface roughness, crust factor, and so forth.
Hill factor	Hill factor HF; slope gradients and length input.	Affects wind speed (high going upslope, lower going down).
Irrigation factor	Irrigation factor *IR*.	Equivalent to added rainfall.
Crops factor	Flat residue *SLR_f*; factor.	Soil cover by residues lying on the surface is estimated for each crop, including changes over time due to decay, and so forth.
	Standing residue *SLR_s*; depends on crop, harvest height, plant density, and so forth.	Standing residues reduce the wind speed at the soil surface. Decay is slower than for flat residues.
	Crop canopy factor *SLR_c*.	Changes daily with crop growth.
Barriers	Barriers (e.g., tree windbreaks); includes orientation, density, height, spacing.	Reduce leeward wind velocity.

Based on information in Fryrear (1998).

FIGURE 17.34 Wind erosion control in an area of productive Histosols (Saprists) in central Michigan. Prior to being cleared and drained, this area was a partially forested bog. When dry, cultivated Histosols are very light and fluffy and susceptible to wind erosion. The rows of trees (mainly willows) were planted perpendicular to the prevailing winds to slow the wind velocity and protect these valuable organic soils from erosion. Wetting the soil surface is another effective means of reducing wind erosion, as seen by the darker-colored field in the background (where the water table was raised) and the darker circles in the foreground where sprinkler irrigation was used. Note that the photo was taken in early spring before most crops were planted and before the trees had fully leafed out. (Photo courtesy of USDA Natural Resources Conservation Service)

Therefore, where irrigation water is available, it is common to moisten the soil surface when high winds are predicted (Figure 17.34). Unfortunately, most wind erosion occurs in dry regions without available irrigation. A vegetative cover also discourages soil blowing, especially if the plant roots are well established. In dry-farming areas, however, sound moisture-conserving practices require summer fallow on some of the land, and hot, dry winds reduce the moisture in the soil surface. Consequently, other means must be employed on cultivated lands of these areas.

TILLAGE. Certain conservation tillage practices described in Section 17.6 were used for wind erosion control long before they became popular as water erosion control practices. Keeping the soil surface rough and maintaining some vegetative cover is accomplished by using appropriate tillage practices. However, the vegetation should be well anchored into the soil to prevent it from blowing away. Stubble mulch has proven to be effective for this purpose (see Section 6.6).

The effect of tillage depends not only on the type of implement used, but also on the timing of the tillage operation. Tillage can greatly reduce wind erosion if it is done while there is sufficient soil water to cause large clods to form. Tillage on a dry soil may produce a fine, dusty surface that aggravates the erosion problem. Tillage to provide for a cloddy surface condition should be at right angles to the prevailing winds. Likewise, strip-cropping and alternate strips of cropped and fallowed land should be perpendicular to the wind.

BARRIERS. Barriers such as shelter belts (Figure 17.35a) are effective in reducing wind velocities for short distances and for trapping drifting soil. Various devices are used to control blowing of sands, sandy loams, and cultivated peat soils (even in humid regions). Windbreaks and tenacious grasses and shrubs are especially effective. Picket fences and burlap screens, though less efficient as windbreaks than such trees as willows, are often preferred because they can be moved from place to place as crops and cropping practices are varied. Rye, planted in narrow strips across the field, is sometimes used on peat lands and on sandy soils (see Figure 17.35c). Narrow rows of such perennial grasses as fall wheatgrass are being evaluated for a combination of wind erosion control and capturing of winter snows in the Northern Plains states.

17.13 LAND CAPABILITY CLASSIFICATION AS A GUIDE TO CONSERVATION

The land capability classification system devised by the U.S. Department of Agriculture has been used since the 1950s to assess the appropriate uses of various types of land. It is especially helpful in identifying land uses and management practices that can minimize soil erosion, especially that induced by rainfall.

The system uses eight **land capability classes** to indicate the *degree* of limitation imposed on land uses (Figure 17.36), with Class I the least limited and Class VIII the most limited. Each land use class may have four subclasses that indicate the *type* of limitation encountered: risks of erosion (e); wetness, drainage, or flooding (w); root-zone limitations, such as acidity, density, and shallowness (s); and climatic limitations, such as short growing season (c). The erosion (e) subclasses are the most common, and they will be the focus of our attention here. For example, Class IIe land is slightly susceptible to erosion, while Class VIIIe is extremely susceptible. Figure 17.37 shows the appropriate intensity of use allowable for each of these land capability classes if erosion losses (or problems associated with the other subclasses) are to be avoided.

Table 17.15 provides a summary of areas with major limitations on land use in the United States. Overall, erosion and sedimentation problems are the most serious limitations for nearly 60% of the land. Excessive wetness and shallowness are each problems on about 20% of the land.

In the United States, Classes I, V, and VIII each account for less than 3% of the nonfederal land. The other classes are much more common, with about 20% of the land in each of Classes II, III, VI, and VII and about 14% in Class IV. This means that about 43% of the land in the United States (242 million ha) is suitable for regular cultivation (Classes I, II, and III). Another 14% (78 million ha) is marginal for growing cultivated crops. The remainder is suited primarily for grasslands and forests, and is used mostly for those purposes. A brief description of the land use capability classes follows.

(a)

Wind
velocity   12   9.5   13.5   4.5   6.8
(m/sec)

(b)

(c)

FIGURE 17.35 Wind breaks to reduce wind erosion. (*a*) Shrubs and trees make good windbreaks and add beauty to a North Dakota farm homestead. (*b*) The effect of a windbreak on wind velocity. The wind is deflected upward by the trees and is slowed down even before reaching them. On the leeward side further reduction occurs; the effect being felt as far as 20 times the height of the trees. (*c*) Narrow strips of cereal rye act as miniature windbreaks to protect watermelons from wind erosion on a loamy sand on the mid-Atlantic coastal plain. [Photo (*a*) courtesy of USDA Natural Resources Conservation Service; (*b*) from FAO (1987); (*c*) courtesy of R. Weil]

FIGURE 17.36 Several land capability classes in San Mateo County, California. A range is shown from the nearly level land in the foreground (Class I), which can be cropped intensively, to the badly eroded hillsides (Classes VII and VIII). Although topography and erosion hazards are emphasized here, it should be remembered that other factors—drainage, stoniness, droughtiness—also limit soil usage and help determine the land capability class. (Courtesy USDA Natural Resources Conservation Service)

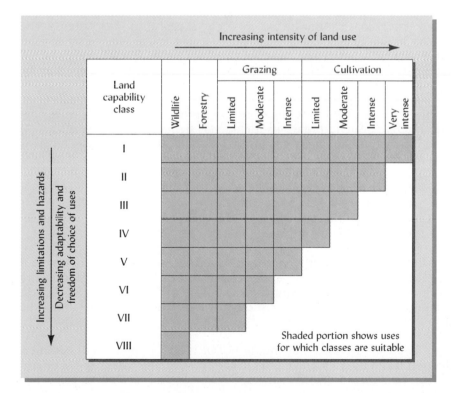

FIGURE 17.37 Intensity with which each land capability class can be used with safety. Note the increasing limitations on the safe uses of the land as one moves from Class I to Class VIII. [Modified from Hockensmith and Steele (1949)]

TABLE 17.15 **Percentage of Land with Major Limitations in the Different Land Use Capability Classes**

Erosion and sedimentation are the most significant limitations,
but wetness and shallowness are very prominent.

	Percentage of land with indicated major limitation			
LUC class	Erosion and sedimentation	Wetness	Shallowness	Climate hazards
II	51	34	8	7
III	67	22	9	2
IV	69	16	14	1
VI	62	8	26	3
VII	47	6	45	2
Total	59	18	21	3

Calculated from USDA (1994).

CLASS I.   These soils are deep and well drained, have nearly level topography (see Figure 17.36), and have few limitations on their use. They can be used continuously for row crops, tree nurseries, or for any of the less-intensive uses shown in Figure 17.37. For most purposes, these soils provide the greatest productivity at the least cost.

CLASS II.   Soils in Class II have some limitations that reduce the choice of uses or require moderate conservation practices. Tillage and row-crop production should be restricted, or some conservation practices (conservation tillage, grassed waterways, contour strips, etc.) should be installed.

CLASS III.   Soils in this class have severe limitations that reduce the choice of plants or require special conservation practices, or both. The same crops may be grown on Class III land as on Classes I and II land, but crops that provide soil cover, such as grasses and legumes, must be more prominent in the rotations used. The amount of land used for row crops is severely restricted. For the w subclass, drainage systems may be needed. Special care and scheduling for timber-harvest operation may be required.

CLASS IV.   Soils in this class can be used for cultivation, but there are severe limitations on the choice of crops and management practices. If the land is to be cultivated, crops should be mostly close-growing types (e.g., wheat or barley) and sod or hay crops must be used extensively. Row crops cannot be grown safely, except in no-till systems. The choice of crops may be limited by excess moisture (subclass w), as well as by erosion hazards. For tilled cropland, terracing, contour strips, or other special conservation practices are required. For timber harvest on forest land, special techniques may be required.

CLASS V.   These lands are generally not suited to crop production because of factors other than erosion hazards. Such limitations include (1) frequent stream overflow, (2) growing season too short for crop plants, (3) stony or rocky soils, and (4) ponded areas where drainage is not feasible. Often, pastures can be improved on this class of land.

CLASS VI.   Soils in this class have extreme limitations (typically, steep slopes and high erosion hazards) that restrict use largely to pasture, range, woodland, or wildlife. Forest entry should be limited, special care should be used in path construction, and alternative timber-harvest techniques should be used.

CLASS VII.   Severe limitations restrict the use of this land to controlled grazing, woodland, or wildlife. The physical limitations are the same as for Class VI, except they are so strict that pasture improvement is impractical. Cable or balloon timber-harvest methods may be necessary. Only small areas should be harvested, and entry should be strictly limited.

CLASS VIII.   In this land class are soils that should not be used for any kind of commercial plant production. Land use is restricted to recreation, wildlife, water supply, or aesthetic purposes. This land includes sand beaches, river wash, and rock outcrops.

The land-use capability classification scheme illustrates the practical use that can be made of soil surveys (see Section 19.7). The many soils delineated on a map by the soil surveyor are viewed in terms of their safest and best longtime use. The eight land capability classes have become the starting point in the development of land-use plans that promote wise use and conservation of the land resource by thousands of farmers, ranchers, and other landowners.

17.14 PROGRESS IN SOIL CONSERVATION

Soil Erosion

Soil erosion in the United States accelerated when the first European settlers chopped down trees and began to farm the sloping lands of the humid eastern part of the country. Soil erosion was a factor in the declining productivity of these lands that, in time, led to their abandonment and the westward migration of people in search of new farmlands.

It wasn't until the worldwide depression and widespread droughts of the early 1930s accentuated rural poverty and displaced millions of people that the nation began to pay attention to the rapid deterioration of its soils. In 1930, Dr. H. H. Bennett and associates recognized the damage being done and obtained federal government support for erosion-control efforts. Now, after some 70 years of soil conservation efforts, soil erosion is still a prominent problem on about half the total cropland area of the country.

Considerable progress in reducing soil erosion was made during the 1940s and 1950s, when such physical practices as contour strips, terraces, and windbreaks were installed with much persuasion and assistance from government agencies. Some of this progress was reversed as the terraces and windbreaks appeared to stand in the way of the "fence row to fence row" all-out crop-production policies of the 1970s. Then, in the decade from 1982 to 1992 rather remarkable progress was made in reducing soil erosion (Table 17.16), largely as a result of two factors: (1) the spread of conservation tillage (see Section 17.6), and (2) the implementation of land-use changes as part of the conservation reserve program. Progress continues to be made on both fronts. However, about one-third of the nation's cultivated cropland is still losing more than 11 Mg/ha/yr, the maximum loss that can be sustained without serious loss of productivity on most soils.

The Conservation Reserve Program

A major part (about 60%) of the reduction in soil erosion experienced in the United States since 1982 is due to government programs that have paid farmers to shift some land from crops to grasses and forests. Establishing grass or trees on these former croplands reduced the sheet and rill erosion from an average of 19.3 to 1.3 Mg/ha and the wind erosion from 24 to 2.9 Mg/ha. Between 1982 and 1992, 13 million ha of cropland were diverted to such noncultivated uses through the **Conservation Reserve Program (CRP)**. Only about half (7 million ha) of the CRP land was on what is considered to be

TABLE 17.16 **Soil Erosion by Water and Wind on Cropland in the United States in 1982 and 1992**

Part of the decline in the total amount of soil eroded was due to a drop in the area of cropland from 166 million ha in 1982 to 150 million ha in 1992, largely as a result of government programs.

	Average annual erosion per unit area, Mg/ha			
Year	Wind	Water	Combined wind and water	Total amount of soil eroded annually, billions of Mg
1982	9.2	7.4	16.6	2.75
1992	6.9	5.6	12.5	1.88
Change	−25%	−24%	−25%	−32%

Data from USDA (1994) and USDA NRSC website at www://nhq.nrcs.usda.gov.

highly erodible land (HEL).[12] In 1995, the CRP was reoriented to better target these and other environmentally sensitive areas. By 1997, distribution of CRP land (Figure 17.38*a*) closely reflected the distribution of highly erodible croplands (Figure 17.38*b*).

The CRP is basically an arrangement by which the nation's taxpayers pay rent to farmers to forego cropping part of their farmland and instead plant grass or trees on it (more rent is paid for trees). The rental agreements (leases) run for 10 to 15 years, during which time the land is undisturbed. The benefits to the nation have been evident, not only in greatly reduced soil losses and sediment pollution, but also in a dramatic upswing in wildlife as bird and animal populations take advantage of the newly restored habitat. Where strips of land along streams (riparian zone buffers) have been incorporated into the CRP, water-quality benefits have also been amplified.

Adapting Soil Conservation to the Needs of Resource-Poor Farmers

As satisfying as is the progress of the past decades, soil losses by erosion are still much too high. Continued efforts must be made to protect the soil and to hold it in place. In the United States, some 30 million ha of highly erodible cropland continues to lose an average of more than 15 Mg/ha of soil each year from water erosion, and an equal

---

[12] Highly erodible land is defined by the USDA Natural Resources Conservation Service as land for which the USLE predicts that soil loss on bare, unprotected soil would be eight times the T value for that soil: $R \times K \times LS \geq 8T$.

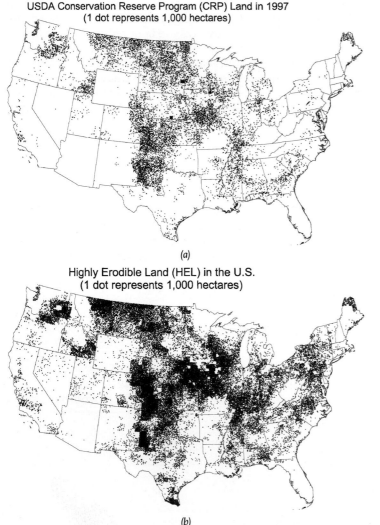

FIGURE 17.38 Distribution of land in the United States (*a*) enrolled in the Conservation Reserve Program (CRP) and (*b*) considered to be highly erodible land (HEL). Note that by 1997 the distributions appear to be quite similar, indicating that the lands most sensitive to erosion are those being taken out of crop production and put into grasses or trees. (Maps courtesy of Scott Brunton, Conservation Technology Information Center, West Lafayette, Ind.)

amount from wind erosion. In spite of remarkable progress, conservation tillage systems have *not* been adopted for more than half of the nation's cropland. And no one knows what will happen to the CRP lands when the rental leases expire. The battle to bring erosion under control has just begun, not only in the United States but throughout the world.

In much of the world, so little land is available to each farmer for food production that nations cannot afford the luxury of following the land-use capability classification recommendations as outlined in Section 17.13. Many farmers must use *all* land capable of food production simply to stave off starvation and impoverishment. These farmers often realize that farming erodible land jeopardizes their future livelihood and that of their children, but they see no choice. It is imperative to either find nonagricultural employment for these people or to find farming systems that are sustainable on these erodible lands.

Fortunately, some farmers and scientists have developed, through long traditions of adaptation or through innovation and research, farming systems that *can* produce food and profits while conserving such erodible soil resources. Examples include the traditional Kandy Home Gardens of Sri Lanka's humid mountains, in which a rain forest–like mixed stand of tall fruit and nut trees is combined with an understory of pepper vines, coffee bushes, and spice plants to provide valuable harvests while keeping the soil under perennial vegetative protection. Another example comes from Central America, where farmers have learned to plant thick stands of velvet bean (*Mucuna*) or other viney legumes that can be chopped down by machete to leave a soil-protecting, water-conserving, weed-inhibiting mulch on steep farmlands. In Asia, steep lands have been carefully terraced in ways that allow production of food, even paddy rice, on very steep land without causing significant erosion.

Many examples from the United States and around the world make it clear that when governments cajole, pay, or force farmers into installing soil conservation measures on their land, the results are unlikely to be long-lasting. Usually, farmers will abandon the unwanted practices as soon as the pressure is off. On the other hand, if scientists and conservationists work *with* farmers to help them develop and adapt conservation systems that the farmers feel are of benefit to them and their land, then effective and lasting progress can be made. Experience with conservation tillage systems in the United States, mulch farming systems in Central America, and vegetative contour barriers in Asia have shown that farmers can help develop practices that are good for their land and for their profits: a win-win situation.

17.15   CONCLUSION

Accelerated soil erosion is one of the most critical environmental and social problems facing humanity today. Erosion degrades soils, making them less productive of the plants on which animals and people depend. Equally important, erosion causes great damages downstream in reservoirs, lakes, waterways, harbors, and municipal water supplies. Wind erosion also causes fugitive dust that may be very harmful to human health.

Nearly 4 billion Mg of soil is eroded each year on land in the United States alone. Half of this erosion occurs on the nation's croplands, the remainder on harvested timber areas, rangelands, and construction sites. Some one-third of the cropland still suffers from erosion that exceeds levels thought to be tolerable.

By sheet and rill erosion, water carries most of the sediment in humid areas. Gullies created by infrequent, but violent, storms account for much of the erosion in drier areas. Wind is the primary erosion agent in many drier areas, especially where the soil is bare and low in moisture during the season when strong winds blow.

Protecting soil from the ravages of wind or water is by far the most effective way to constrain erosion. In croplands and forests, such protection is due mainly to the cover of plants and their residues. Conservation tillage practices maintain vegetative cover on at least 30% of the soil surface, and the widening adoption of these practices has contributed to the significant reductions in soil erosion achieved over the past two decades. Crop rotations that include sod and close-growing crops, coupled with such practices as contour tillage, strip cropping, and terracing, also help combat erosion on farmland.

In forested areas, most erosion is associated with timber-harvesting practices and forest road construction. For the sake of future forest productivity and current water quality, foresters must become more selective in their harvest practices and invest more in proper road construction.

Construction sites for roads, buildings, and other engineering projects lay bare many scattered areas of soils that add up to a serious erosion problem. Control of sediment from construction sites requires carefully phased land clearing, along with vegetative and artificial soil covers, and installation of various barriers and sediment-holding ponds. These measures may be expensive to implement, but the costs to society that result when sediment is not controlled are too high to ignore. Once construction is completed, erosion rates on urban areas are commonly as low as those on areas under undisturbed native vegetation.

Erosion-control systems must be developed in collaboration with those who use the land, and especially the poor, for whom immediate needs must overshadow concerns for the future. As put succinctly in *The River*, a classic 1930s documentary film produced during the rebirth of American soil erosion consciousness, "Poor land makes poor people, and poor people make poor land."

STUDY QUESTIONS

1. Explain the distinction between *geologic erosion* and *accelerated erosion*. Is the difference between the two greater in humid or arid regions?

2. When erosion takes place by wind or water, what are three important types of damages that result on the land whose soils is eroding? What are five important types of damages that erosion causes in locations away from the eroding site?

3. What is a common *T* value, and what is meant by this term? Explain why certain soils have been assigned a higher *T* value than other soils.

4. Describe the three main steps in the water erosion process.

5. Many people assume that the amount of soil eroded on the land in a watershed (*A* in the universal soil-loss equation) is the same as the amount of sediment carried away by the stream draining that watershed. What factor is missing that makes this assumption incorrect? Do you think that this means the USLE should be renamed?

6. Why is the total annual rainfall in an area *not* a very good guide to the amount of erosion that will take place on a particular type of bare soil?

7. Contrast the properties you would expect in a soil with either a very high *K* value or a very low *K* value.

8. How much soil is likely to be eroded from a Keene silt loam in central Ohio, on a 12% slope, 100 m long, if it is in dense permanent pasture and has no support practices applied to the land? Use the information available in this chapter to calculate an answer.

9. What type of conservation tillage leaves the greatest amount of soil cover by crop residues? What are the advantages and disadvantages of this system?

10. Why are narrow strips of grass planted on the contour sometimes called a "living terrace"?

11. In most forests, which component of the ecosystem provides the primary protection against soil erosion by water, the *tree canopy, tree roots,* or *leaf litter?*

12. Certain soil properties generally make land susceptible to erosion by wind or erosion by water. List four properties that characterize soils highly susceptible to wind erosion. Indicate which two of these properties should also characterize soils highly susceptible to water erosion, and which two should not.

13. Which three factors in the wind erosion prediction equation (WEQ) can be affected by tillage? Explain.

14. Describe a soil in land capability Class IIw soil in comparison with one in Class IVe.

15. Why is it important that there be a close relationship between land in the CRP and that considered to be HEL?

Blevins, R. L., and W. W. Frye. 1992. "Conservation tillage: An ecological approach to soil management," *Advances in Agronomy,* **51**:33–78.

Carter, M. R. 1994. *Conservation Tillage in Temperate Agroecosystems.* (Boca Raton, Fla.: Lewis Publishers).

Cassel, D. K., and R. Lal. 1992. "Soil physical properties of the tropics: Common beliefs and management constraints," in R. Lal and P. A. Sanchez (eds.), *Myths and Science of Soils of the Tropics.* SSA Special Publication no. 29. (Madison, Wis.: Soil Sci. Soc. of Amer.), pp. 61–89.

Crovetto, C. 1996. *Stubble Over the Soil.* (Madison, Wis.: Amer. Soc. Agron.).

Daily, G. 1997. "Restoring value to the world's degraded lands," *Science,* **269**:350–354.

Dickey, E. C., et al. 1987. "Conservation tillage: Perceived and actual use," *J. Soil Water Cons.,* **42**:431–434.

Doolette, J. B., and J. W. Smyle. 1990. "Soil and moisture conservation technologies: Review of literature," in J. B. Doolette and W. B. Magrath (eds.), *Watershed Development in Asia: Strategies and Technologies.* (Washington, D.C.: World Book Technical Paper 127).

El-Swaify, and E. W. Dangler. 1982. "Rainfall erosion in the tropics: A state-of-the-art," *Soil Erosion and Conservation in the Tropics.* ASA Special Publication no. 43. (Madison, Wis.: Amer. Soc. Agron.).

FAO. 1987. *Protect and Produce.* (Rome: U.N. Food and Agriculture Organization).

Follet, R. F., and B. F. Stewart (eds.). 1985. *Soil Erosion and Crop Productivity.* (Madison, Wis.: Amer. Soc. Agron.).

Foster, G. R., D. K. McCool, K. G. Renard, and W. C. Moldenhauer. 1981. "Conversion of the universal soil loss equation to Sl metric units." *J. Soil Water Cons.,* **36**:355–359.

Foster, G. R., and R. E. Highfill. 1983. "Effect of terraces on soil loss: USLEP factor values for terraces," *J. Soil Water Cons.,* **38**:48–51.

Fryrear, D. W., and A. Saleh. 1993. "Field wind erosion: Vertical distribution," *Soil Sci.,* **155**:294–300.

Fryrear, D. W. 1998. "RWEQ: Improved wind erosion technology," *J. Soil Water Cons.,* **53**:(in press)

Hillel, D. 1991. *Out of the Earth: Civilization and the Life of the Soil.* (New York: The Free Press).

Hockensmith, R. D., and J. G. Steel. 1949. "Recent trends in the use of the land-capability classification," *Soil Sci. Soc. Amer. Proc.,* **14**:383–388.

Hudson, N. 1995. *Soil Conservation,* 3rd ed. (Ames, Iowa: Iowa State University Press).

Hughes, H. A. 1980. *Conservation Farming.* (Moline, Ill.: John Deere and Company).

IUCN. 1986. *The IUCN Sahel Report.* (Gland, Switzerland: Intl. Union Cons. Nature).

Krenisky, E. C., M. J. Carroll, R. L. Hill, and J. M. Krouse. 1998. "Runoff and sediment losses from natural and man-made erosion control materials," *Crop Sci.,* **38**:(in press).

Laflen, J. M., W. J. Elliot, D. C. Flanagan, C. R. Meyers, and M. A. Nearing. 1997. "WEPP—predicting water erosion using a process-based model," *J. Soil Water Cons.,* **52**:96–102.

Lal, R. 1995. "Erosion–crop productivity relationships for the soils of Africa," *Soil Sci. Soc. Amer. J.,* **59**:661–667.

Larson, W. E., F. J. Pierce, and R. H. Dowdy. 1983. "The threat of soil erosion to long-term crop productivity," *Science,* **219**:458–465.

Lattanzi, A. R., L. D. Meyer, and M. F. Baumgardner. 1979. "Influence of mulch rate and slope steepness in interrill erosion," *Soil Sci. Soc. Amer. Proc.,* **38**:946–950.

Mahboubi, A. A., R. Lal, and N. R. Faussey. 1993. "Twenty-eight years of tillage effects on two soils in Ohio," *Soil Sci. Soc. Amer. J.,* **57**:506–512.

Nyland, R. D. 1996. *Silviculture: Concepts and Applications.* (New York: McGraw-Hill).

Oldeman, L. R. 1994. "The global extent of soil degradation," in D. J. Greenland, and I. Szabolcs (eds.), *Soil Resilience and Sustainable Land Use.* (Wallingford, U.K.: CAB International).

Oschwald, W. R., and J. C. Siemens. 1976. "Conservation tillage: A perspective," Agronomy Facts SM-30. (Urbana, Ill.: University of Illinois).

Pierre, C. J. M. G. 1992. *Fertility of Soils: A Future for Farming in the West African Savannah.* (Berlin: Springer-Velag).

Pimental, D., C. Harvey, P. Resosudarmo, K. Sinclair, D. Kurz, M. McNair, S. Crist, L. Shpritz, L. Fritton, R. Saffouri, and R. Blair. 1995. "Environmental and economic costs of soil erosion and conservation benefits," *Science*, **267**:1117–1122.

Renard, K. G., G. Foster, D. Yoder, and D. McCool. 1994. "RUSLE revisted: Status, questions, answers and the future," *J. Soil Water Cons.*, **49**:213–220.

Renard, K. G., G. Foster, G. Weesies, D. McCool, and D. Yoder. 1997. *Predicting Soil Erosion by Water: A Guide to Conservation Planning with the Revised Universal Soil Loss Equation (RUSLE)*. Agricultural Handbook no 703. (Washington, D.C.: USDA).

Rosensweig, C., and D. Hillel. 1998. *Climate Change and the Global Harvest: Potential Impacts of the Greenhouse Effect on Agriculture*. (Cary, N.C.: Oxford University Press).

Schwab, G. O., D. D. Fangmeirer, and W. J. Elliot. 1996. *Soil and Water Management Systems,* 4th ed. (New York: Wiley).

Schertz, D. L. 1983. "The basis for soil loss tolerance," *J. Soil Water Cons.*, **30**:10–14.

Skidmore, E. L., and F. H. Siddoway. 1978. "Crop residue requirements to control wind erosion," in W. R. Oschwalk (ed.), *Crop Residue Management Systems*. ASA Special Publication no. 31. (Madison, Wis.: Amer. Soc. Agron.; Crop Sci. Soc. Amer.; and Soil Sci. Soc. Amer.).

Skidmore, E. L., and N. P. Woodruff. 1968. *Wind Erosion Forces in the United States and Their Use in Predicting Soil Loss*. Agricultural Handbook no. 346. (Washington, D.C.: USDA).

USDA. 1994. *Summary Report: 1992 National Resources Inventory*. (Washington, D.C.: USDA Natural Resources Conservation Service).

Van Doren, D. M., Jr., G. B. Tripett, Jr., and J. E. Henry. 1976. "Influence of long-term tillage, crop rotation and soil type combinations on corn yields," *Soil Sci. Soc. Amer. J.*, **40**:100–105.

Weesies, G. A., S. J. Livingston, W. D. Hosteter, and D. L. Schertz. 1994. "Effect of soil erosion on crop yield in Indiana: Results of a 10 year study," *J. Soil Water Cons.*, **49**:597–600.

Wischmeier, W. J., and D. D. Smith. 1978. *Predicting Rainfall Erosion Loss—A Guide to Conservation Planning*. Agricultural Handbook no. 537. (Washington, D.C.: USDA).

18

SOILS AND CHEMICAL POLLUTION

Black and portentous this humor prove, unless good
counsel may the cause remove . . .
—*W. SHAKESPEARE,* ROMEO AND JULIET

The soil is a primary recipient by design or accident of a myriad of waste products and chemicals used in modern society. It has always been convenient to "throw things away," and the soil has been the recipient of most of these things. Every year millions of tons of these products from a variety of sources—industrial, domestic, and agricultural—find their way onto the world's soils. Once these materials enter the soil, they become part of biological cycles that affect all forms of life. One of the challenges facing humankind is to better understand how wastes affect these cycles and, in turn, the well-being of all plant and animal life.

In previous chapters we highlighted the enormous capacity of soils to accommodate added organic and inorganic chemicals. Tons of organic residues and animal manures are broken down by soil microbes each year (Chapter 12), and large quantities of inorganic chemicals are fixed or bound tightly by soil minerals (Chapter 14). But we also learned of the limits of the soil's capacity to accommodate these chemicals, and we have seen the disastrous effects on environmental quality when these limits are exceeded.

We have seen how soil processes affect the accommodation and release of waste products. For example, the production and sequestering of greenhouse gases, such as nitrous oxide, methane, and carbon dioxide (see, e.g., Sections 12.11 and 13.9), are very much influenced by soil processes. Other nitrogen- and sulfur-containing gases coming from domestic and industrial sources, as well as from the soil, acidify the atmosphere, and come to earth in acid rain (see, e.g., Section 9.6). Mismanaged irrigation projects result in the accumulation of salts, especially in arid-region soils (see Section 10.3).

We have also seen how fertilizer and manure applications that leave excess quantities of nutrients in the soil can result in the contamination of ground and surface waters with nitrates (Section 13.8) and phosphates (Section 14.2). The eutrophication of ponds, lakes, and even slow-moving rivers is evidence of these nutrient buildups. Huge "animal factories" for meat and poultry production produce mountains of manure that must be disposed of without loading the environment with unwanted chemicals and with pathogens that are harmful to humans and other animals (Section 16.5).

In this chapter we will focus on chemicals that contaminate and degrade soils, including some whose damage extends to water, air, and living things. The brief review of soil pollution is intended as an introduction to the nature of the major pollutants, their reactions in soils, and alternative means of managing, destroying, or inactivating them.

Modern industrialized societies have developed thousands of synthetic organic compounds for thousands of uses. An enormous quantity of organic chemicals is manufactured every year—about 60 million Mg in the United States alone. Included are plastics and plasticizers, lubricants and refrigerants, fuels and solvents, pesticides and preservatives. Some are extremely toxic to humans and other life. Through accidental leakage and spills or through planned spraying or other treatments, synthetic organic chemicals can be found in virtually every corner of our environment—in the soil, in the groundwater, in the plants, and in our own bodies.

Environmental Damage from Organic Chemicals

Some of these organic compounds are relatively inert and harmless, but others are biologically damaging even in very small concentrations. Those that find their way into soils may inhibit or kill soil organisms, thereby undermining the balance of the soil community (see Section 11.15). Other chemicals may be transported from the soil to the air, water, or vegetation, where they may be contacted, inhaled, or ingested by any number of organisms. It is imperative, therefore, that we control the release of organic chemicals and that we learn of their fate and effects once they enter the soil.

Organic chemicals may enter the soil as contaminants in industrial and municipal organic wastes applied to or spilled on soils, as components of discarded machinery, in large or small lubricant and fuel leaks, and as pesticides applied to terrestrial ecosystems.

While some pesticides are meant to be applied to soils, most reach the soil because they have missed the insect or plant leaf that was the target of the application. When pesticides are sprayed in the field, most of the chemical misses the target organism. For pesticides aerially applied to forests, about 25% reaches the tree foliage and far less than 1% reaches a target insect. About 30% may reach the soil, while about half of the chemical applied is likely to be lost into the atmosphere or in runoff water.

Pesticides are probably the most widespread organic pollutants associated with soils. In the United States, pesticides are used on some 150 million ha of land, three-fourths of which is agricultural land. Soil contamination by other organic chemicals is usually much more localized. We will therefore emphasize the pesticide problem.

The Nature of Pesticides

Pesticides are chemicals that are designed to kill pests (that is, any organism that the pesticide user perceives to be damaging). Since the offending organisms may be plants (weeds), insects, fungi, nematodes, or rats, pesticides include different compounds designated as *herbicides, insecticides, fungicides, nematocides, rodenticides,* and so on.

Some 600 chemicals in about 50,000 formulations are used to control pests. They are used extensively in all parts of the world. About 350,000 Mg of organic pesticide chemicals are used annually in the United States, with similar amounts used in Western Europe and Asia. Although the total amount of pesticides used has remained relatively constant or even dropped since the 1980s, formulations in use today are generally more potent, so that smaller quantities are applied per hectare to achieve toxicity to the pest.

BENEFITS OF PESTICIDES. Pesticides have provided many benefits to society. They have helped control mosquitoes and other vectors of such human diseases as yellow fever and malaria. They have protected crops and livestock against insects and diseases. Without the control of weeds by chemicals called *herbicides,* conservation tillage (especially no-tillage) would be much more difficult to adopt; much of the progress made in controlling soil erosion probably would not have come about without herbicides. Also, pesticides reduce the spoilage of food as it moves from farm fields to distant dinner tables.

COSTS OF PESTICIDES. However, pesticides should not be seen as a panacea or as indispensable. Some farmers produce profitable yields without the use of pesticides. Even with the almost universal use of pesticides, insects, diseases, and weeds still cause the loss of one-third of the crop production in the United States, about the same proportion of crops lost to these pests in the 1940s, before many pesticides were in use. And while the benefits to society from pesticides are great, so are the costs (Table 18.1).

TABLE 18.1 Total Estimated Environmental and Social Cost from Pesticide Use in the United States

The death of an estimated 60 million wild birds may represent an additional substantial cost in lost revenues from hunters, bird watchers, and so forth.

Type of impact	Cost, $ million/yr
Public health impacts	787
Domestic animal deaths and contamination	30
Loss of natural enemies	520
Cost of pesticide resistance	1400
Honeybee and pollination losses	320
Crop losses	942
Fishery losses	24
Groundwater contamination and cleanup costs	1800
Cost of government regulations to prevent damage	200
Total	6023

From Pimental, et al. (1992). © American Institute of Biological Sciences.

Designed to kill living things, many of these chemicals are potentially toxic to organisms other than the pests for which they are intended. Some are detrimental to nontarget organisms, such as beneficial insects and certain soil organisms. Those chemicals that do not quickly break down may be biologically magnified as they move up the food chain. For example, as earthworms ingest contaminated soil, the chemicals tend to concentrate in the earthworm bodies. When birds and fish eat the earthworms, the pesticides can build up further to lethal levels. The near extinction of certain birds of prey (including the bald eagle) during the 1960s and 1970s called public attention to the sometimes devastating environmental consequences of pesticide use. More recently, evidence is mounting suggesting that human endocrine (hormone) balance may be disrupted by the minute traces of some pesticides found in water, air, and food.

18.2 KINDS OF PESTICIDES

Pesticides are commonly classified according to the target group of pest organisms: (1) *insecticides*, (2) *fungicides*, (3) *herbicides* (weed killers), (4) *rodenticides* and (5) *nematocides*. In practice, all find their way into soils. Since the first three are used in the largest quantities and are therefore more likely to contaminate soils, they will be given primary consideration. Figure 18.1 shows that most pesticides contain aromatic rings of some kind, but that there is great variability in pesticide chemical structures.

Insecticides

Most of these chemicals are included in three general groups. The *chlorinated hydrocarbons,* such as DDT, were the most extensively used until the early 1970s, when their use was banned or severely restricted in many countries due to their low biodegradability and persistence, as well as their toxicity to birds and fish.

The *organophosphate* pesticides are generally biodegradable, and thus less likely to build up in soils and water. However, they are extremely toxic to humans, so great care must be used in handling and applying them. The *carbamates* are considered least dangerous because of their ready biodegradability and relatively low mammalian toxicity. However, they are highly toxic to honeybees and other beneficial insects and to earthworms.

Fungicides

Fungicides are used mainly to control diseases of fruit and vegetable crops and as seed coatings to protect against seed rots. Some are also used to protect harvested fruits and vegetables from decay, to prevent wood decay, and to protect clothing from mildew. Organic materials such as the thiocarbamates and triazoles are currently in use.

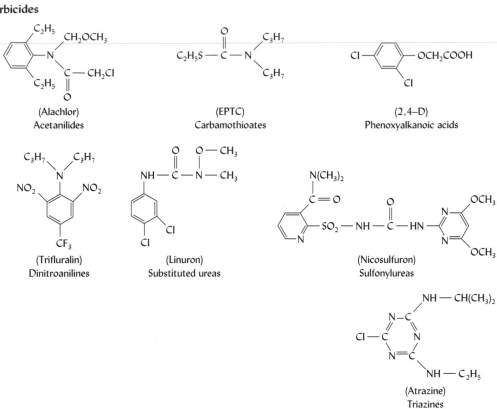

Insecticides

(DDT)
Chlorinated hydrocarbons

(Carbaryl)
Carbamates

(Parathion)
Organophosphates

Herbicides

(Alachlor)
Acetanilides

(EPTC)
Carbamothioates

(2,4–D)
Phenoxyalkanoic acids

(Trifluralin)
Dinitroanilines

(Linuron)
Substituted ureas

(Nicosulfuron)
Sulfonylureas

(Atrazine)
Triazines

FIGURE 18.1 Structural formulae of representative compounds in 10 classes of widely used pesticides. Carbaryl, DDT, and parathion are insecticides; the other compounds shown are herbicides. The widely differing structures result in a great variety of toxicological properties and reactions in the soil.

Herbicides

The quantity of herbicides used in the United States exceeds that of the other two types of pesticides combined. Starting with 2,4-D (a chlorinated phenoxyalkanoic acid), dozens of chemicals in literally hundreds of formulations have been placed on the market (see Figure 18.1). These include the *triazines,* used mainly for weed control in corn; *substituted ureas;* some *carbamates;* the relatively new *sulfonylureas,* which are potent at very low rates; *dinitroanilines;* and *acetanilides,* which have proved to be quite mobile in the environment. As one might expect, this wide variation in chemical makeup provides an equally wide variation in properties. Most herbicides are biodegradable, and most of them are relatively low in mammalian toxicity. However, some are quite toxic to fish and perhaps to other wildlife. They can also have deleterious effects on beneficial aquatic vegetation that provides food and habitat for fish and shellfish.

Nematocides

Although nematocides are not widely used, some of them are known to contaminate soils and the water draining from treated soils. For example, some carbamates used as nematocides are quite soluble in water, are not adsorbed by the soil, and consequently leach downward and into the groundwater.

18.3   BEHAVIOR OF ORGANIC CHEMICALS IN SOIL[1]

Once they reach the soil, organic chemicals, such as pesticides or hydrocarbons, move in one or more of seven directions (Figure 18.2): (1) they may vaporize into the atmosphere without chemical change; (2) they may be absorbed by soils; (3) they may move downward through the soil in liquid or solution form and be lost from the soil by leaching; (4) they may undergo chemical reactions within or on the surface of the soil; (5) they may be broken down by soil microorganisms; (6) they may wash into streams and rivers in surface runoff; and (7) they may be taken up by plants or soil animals and move up the food chain. The specific fate of these chemicals will be determined at least in part by their chemical structures, which are highly variable.

Volatility

Organic chemicals vary greatly in their volatility and subsequent susceptibility to atmospheric loss. Some soil fumigants, such as methyl bromide (now banned from most uses), were selected because of their very high vapor pressure, which permits them to penetrate

---

[1] For reviews on organic chemicals in the soil environment, see Sawhney and Brown (eds.) (1989) and Pierzynski, et al. (1994); for pesticides, see Cheng (ed.) (1990).

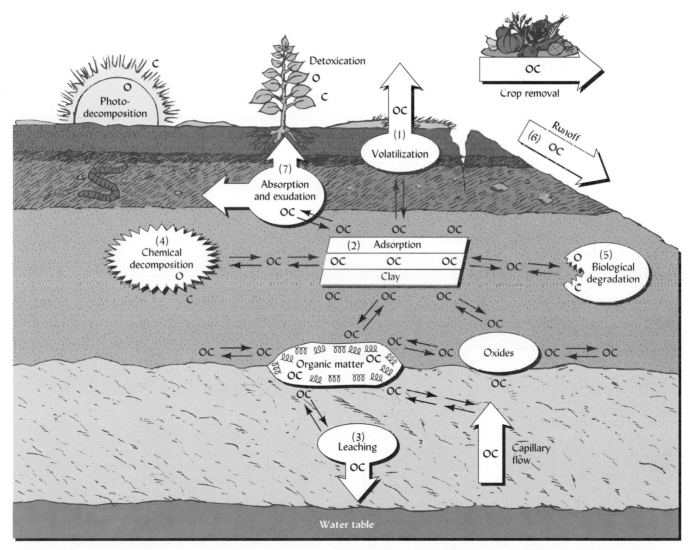

FIGURE 18.2   Processes affecting the dissipation of organic chemicals (OC) in soils. Note that the OC symbol is split up by decomposition (both by light and chemical reaction) and degradation by microorganisms, indicating that these processes alter or destroy the organic chemical. In transfer processes, the OC remains intact. [From Weber and Miller (1989)]

soil pores to contact the target organisms. This same characteristic encourages rapid loss to the atmosphere after treatment, unless the soil is covered or sealed. A few herbicides (e.g., trifluralin) and fungicides (e.g., PCNB) are sufficiently volatile to make vaporization a primary means of their loss from soil. The lighter fractions of crude oil (e.g., gasoline and diesel) and many solvents vaporize to a large degree when spilled on the soil.

The assumption that disappearance of pesticides from soils is evidence of their breakdown is questionable. Some chemicals lost to the atmosphere are known to return to the soil or to surface waters with the rain.

Adsorption

The adsorption of organic chemicals by soil is determined largely by the characteristics of the compound and of the soils to which they are added. Soil organic matter and high-surface-area clays tend to be the strongest adsorbents for some compounds (Figure 18.3), while oxide coatings on soil particles strongly adsorb others. The presence of certain functional groups, such as —OH, —NH_2, —NHR, —$CONH_2$, —COOR, and —$^+NR_3$, in the chemical structure encourages adsorption, especially on the soil humus. Hydrogen bonding (see Sections 5.1 and 8.5) and protonation [adding of H^+ to a group such as an —NH_2 (amino) group] probably promotes some of the adsorption. Everything else being equal, larger organic molecules with many charged sites are more strongly adsorbed.

Some organic chemicals with positively charged groups, such as the herbicides diquat and paraquat, are strongly adsorbed by silicate clays. Adsorption by clays of some pesticides tends to be pH-dependent (Figure 18.4), with maximum adsorption occurring at low pH levels, which encourages protonation. Adding an H^+ ion to functional groups (e.g., —NH_2) yields a positive charge on the herbicide, resulting in greater attraction to negatively charged soil colloids.

Leaching and Runoff

The tendency of organic chemicals to leach from soils is closely related to their solubility in water and their potential for adsorption. Some compounds, such as chloroform and phenoxyacetic acid, are a million times more water-soluble than others, such as DDT and PCBs, which are quite soluble in oil but not in water. High water-solubility favors leaching losses.

Strongly adsorbed molecules are not likely to move down the profile (Table 18.2). Likewise, conditions that encourage such adsorption will discourage leaching. Leaching

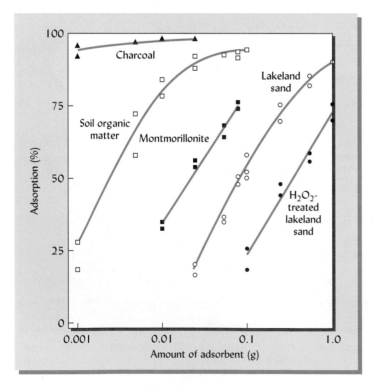

FIGURE 18.3 Adsorption of polychlorinated biphenyl (PCB) by different soil materials. The Lakeland sand (Typic Quartzipsamments) lost much of its adsorption capacity when treated with hydrogen peroxide to remove its organic matter. The amount of soil material required to adsorb 50% of the PCB was approximately 10 times as great for montmorillonite (a 2:1 clay mineral) as for soil organic matter, and 10 times again as great for H_2O_2-treated Lakeland sand. Later tests showed that once the PCB was adsorbed, it was no longer available for uptake by plants. Note that the amount of soil material added is shown on a log scale. [From Strek and Weber (1982)]

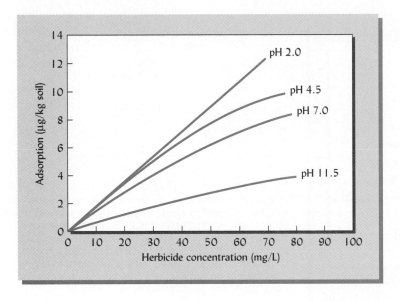

FIGURE 18.4 The effect of pH of kaolinite on the adsorption of glyphosate, a widely used herbicide (Brand name Roundup®). [Reprinted with permission from J. S. McConnell and L. R. Hossner, *J. Agric. Food Chem.* **33**:1075–78 (1985); copyright 1985 American Chemical Society]

is apt to be favored by water movement, the greatest leaching hazard occurring in highly permeable, sandy soils that are also low in organic matter. Periods of high rainfall around the time of application of the chemical promote both leaching and runoff losses (Table 18.3). With some notable exceptions, herbicides seem to be somewhat more mobile than most fungicides or insecticides, and therefore are more likely to find their way to groundwater supplies and streams (Figure 18.5).

TABLE 18.2 The Degree of Adsorption of Selected Herbicides

Weakly adsorbed herbicides are more susceptible to movement in the soil than those that are more tightly adsorbed.

Common name or designation	Trade name	Adsorptivity to soil colloids
Dalapon	Dowpon	None
Chloramben	Amiben	Weak
Bentazon	Basagran	Weak
2,4-D	Several	Moderate
Propachlor	Ramrod	Moderate
Atrazine	AAtrex	Strong
Alachlor	Lasso	Strong
EPTC	Eptam	Strong
Diuron	Karmex	Strong
Paraquat	Paraquat	Very strong
Trifluralin	Treflan	Very strong
DCPA	Dacthal	Very strong

Selected data from DMI (1981).

TABLE 18.3 Surface Runoff and Leaching Losses (Through Drain Tiles) of the Herbicide Atrazine from a Clay Loam Lacustrine Soil (Alfisols) in Ontario, Canada

The herbicide was applied at 1700 g/ha in late May. The data are the average of three tillage methods. Note that the rainfall for May and June is related to the amount of herbicide lost by both pathways.

Year of study	Atrazine loss, g/ha				Rainfall, May–June, mm
	Surface runoff loss	Drainage water loss	Total dissolved loss	Percent of total applied, %	
1	18	9	27	1.6	170
2	1	2	3	0.2	30
3	51	61	113	6.6	255
4	13	32	45	2.6	165

Data abstracted from Gaynor, et al. (1995).

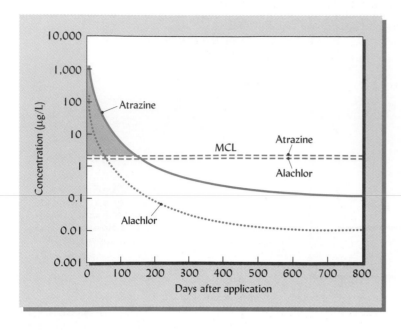

FIGURE 18.5 Concentration of two widely used herbicides, atrazine and alachlor, in the runoff from watersheds in Ohio planted to corn, along with the allowed Maximum Contaminant Level (MCL) for drinking water. Note that the concentration far exceeds the MCL, especially for atrazine, during the first 50 to 100 days after application. If this runoff is not diluted with less-contaminated water, it would not be suitable for consumption by downstream users. [Redrawn from Shipitalo, et al. (1997)]

Contamination of Groundwater

Experts once maintained that contamination of groundwater by pesticides occurred only from accidents such as spills, but it is now known that many pesticides reach the groundwater from normal agricultural use. Since many people (e.g., 40% of Americans) depend on groundwater for their drinking supply, leaching of pesticides is of wide concern. Table 18.4 lists some of the 46 pesticides found in a national survey of well waters in the United States. The concentrations are given in parts per billion (see Box 18.1). In some cases, the amount of pesticide found in the drinking water has been high enough to raise long-term health concerns.

Chemical Reactions

Upon contacting the soil, some pesticides undergo chemical modification independent of soil organisms. For example, iron cyanide compounds decompose within hours or days if exposed to bright sunlight. DDT, diquat, and the triazines are subject to slow photodecomposition in sunlight. The triazine herbicides (e.g., atrazine) and organophosphate insecticides (e.g., malathion) are subject to hydrolysis and subsequent degradation. While the complexities of molecular structure of the pesticides suggest different mechanisms of breakdown, it is important to realize that degradation independent of soil organisms does in fact occur.

Microbial Metabolism

Biochemical degradation by soil organisms is the single most important method by which pesticides are removed from soils. Certain polar groups on the pesticide molecules, such as —OH, —COO⁻, and —NH₂, provide points of attack for the organisms.

DDT and other chlorinated hydrocarbons, such as aldrin, dieldrin, and heptachlor, are very slowly broken down, persisting in soils for 20 or more years. In contrast, the organophosphate insecticides, such as parathion, are degraded quite rapidly in soils, apparently by a variety of organisms (Figure 18.6). Likewise, the most widely used herbicides, such as 2,4-D, the phenylureas, the aliphatic acids, and the carbamates, are readily attacked by a host of organisms. Exceptions are the triazines, which are slowly degraded, primarily by chemical action. Most organic fungicides are also subject to microbial decomposition, although the rate of breakdown of some is slow, causing troublesome residue problems.

TABLE 18.4 Pesticides Present in Groundwater from Normal Agricultural Use

Note the wide range in concentrations which are considered to be risky to health. The great majority of wells sampled were uncontaminated, but when pesticides were detected they were often near or above the health-advisory level.

Pesticide	Use[a]	Level found, parts per billion		
		Median level found[b]	Maximum level found	Health-advisory level[c]
Alachlor	H	1	113	
Aldicarb	I	9	315	10
Atrazine	H	1	40	3
Bromacil	H	9	22	90
Carbofuran	I	5	176	40
Cyanazine	H	0.4	7	10
2,4-D (2,4 Dichlorophenoxyacetic acid)	H	1	50	70
DBCP[d]	FUM	0.01	0.02	—
DCPA	H	109	1040	4000
Dinoceb[d]	H, I, F	1	37	7
EDB	F	1	14	—
Fonofos	I	0.1	0.9	10
Malathion	I	42	53	200
Metolachlor	H	0.4	32	100
Metribuzin	H	1	7	200
Oxamyl	I	4	395	200
Trifluran	H	0.4	2.2	5

Data from General Accounting Office (1991).

[a] H = herbicide; I = insecticide; F = fungicide; FUM = fumigant.

[b] Fifty percent above and 50% below this value.

[c] Health-advisory level is the concentration that is suspected of causing health problems over a 70-year lifetime. Blank means no advisory level has been set.

[d] Most uses of this pesticide have been banned in the United States.

Plant Absorption

Pesticides are commonly absorbed by higher plants. This is especially true for those pesticides (e.g., systemic insecticides and most herbicides) that must be taken up in order to perform their intended function. The absorbed chemicals may remain intact inside the plant, or they may be degraded. Some degradation products are harmless, but others are even more toxic to humans than the original chemical that was absorbed. Understandably, society is quite concerned about pesticide residues found in the parts of plants that people eat, whether as fresh fruits and vegetables or as processed foods. The use of pesticides and the amount of pesticide residues in food are strictly regulated by law to ensure human safety. Despite widespread concerns, there is little evidence that the small amounts of residues permissible in foods by law have had any ill effects on public health. However, routine testing by regulatory agencies has shown that about 1 to 2% of the food samples tested contain pesticide residues above the levels permissible.

Persistence in Soils

The persistence of chemicals in soils is a summation of all the reactions, movements, and degradations affecting these chemicals. Marked differences in persistence are the rule (see Figure 18.6). For example, organophosphate insecticides may last only a few days in soils. The widely used herbicide 2,4-D persists in soils for only two to four weeks. PCBs, DDT, and other chlorinated hydrocarbons may persist for 3 to 20 years or longer (Table 18.5). The persistence times of other pesticides and industrial organics fall generally between the extremes cited. The majority of pesticides degrade rapidly enough to prevent buildup in soils having normal annual applications. Those that resist degradation have a greater potential to cause environmental damage.

Continued use of the same pesticide on the same land can increase the rate of microbial breakdown of that pesticide. Apparently, having a constant food source allows a population build up of those microbes equipped with the enzymes needed to break down the compound. This is an advantage with respect to environmental quality and is

As analytical instrumentation becomes more sophisticated, contaminants can be detected at much lower levels than was the case in the past. Since humans and other organisms can be harmed by almost any substance if large enough quantities are involved, the subject of toxicity and contamination must be looked at *quantitatively*. That is, we must ask *how much*, not simply *what*, is in the environment. Many highly toxic (meaning harmful in very small amounts) compounds are produced by natural processes and can be detected in the air, soil, and water—quite apart from any activities of humans.

The mere presence of a natural toxin or a synthetic contaminant may not be a problem. Toxicity depends on (1) the *concentration* of the contaminant, and (2) the level of *exposure* of the organism. Thus, low concentrations of certain chemicals that would cause no observable effect by a single exposure (e.g., one glass of drinking water) may cause harm (e.g., cancer, birth defects) to individuals exposed to these concentrations over a long period of time (e.g., three glasses of water a day for many years).

Regulatory agencies attempt to estimate the effects of long-term exposure when they set standards for no-observable-effect levels (NOEL) or health-advisory levels (see Table 18.4). Some species and individuals within a species will be much more sensitive than others to any given chemical. Regulators attempt to consider the risk to the most susceptible individual in any particular case. For nitrate in groundwater, this individual might be a human infant whose entire diet consists of infant formula made with the contaminated water. For DDT, the individual at greatest risk might be a bird of prey that eats fish that eat worms that ingest lake sediment contaminated with DDT. For a pesticide taken up by plants from the soil, the individual at greatest risk might be an avid gardener who eats vegetables and fruits mainly from the treated garden over the course of a lifetime.

It is important to get a feel for the meaning of the very small numbers used to express the concentration of contaminants in the environment. For instance, in Table 18.4, the concentrations are given in parts per billion (ppb). This is equivalent to micrograms per kilogram or $\mu g/kg$. In water this would be $\mu g/L$. To comprehend the number 1 *billion* imagine a billion golf balls: lined up, they would stretch completely around the earth. One bad ball out of a billion (1 ppb) seems like an extremely small number. On the other hand, 1 ppb can seem like a very large number. Consider water contaminated with 1 ppb of cyanide, a very toxic substance consisting of a carbon and a nitrogen atom linked together. If you drank just *one* drop of this water, you would be ingesting over 2 *trillion* molecules of cyanide:

$$\frac{6.023 \times 10^{23} \text{ molecules}}{1 \text{ mol}} \times \frac{1 \text{ mol}}{27 \text{ g HCN}} \times \frac{1 \text{ g HCN}}{10^6/\mu g \text{ HCN}} \times \frac{1 \mu g \text{ HCN}}{L} \times \frac{L}{10^3 \text{ cm}^3} \times \frac{cm^3}{10 \text{ drops}} = \frac{2.2 \times 10^{12} \text{ molecules}}{\text{drop}}$$

In the case of cyanide, the molecules in this drop of water would probably not cause any observable effect. However, for other compounds, this many molecules may be enough to trigger DNA mutations or the beginning of cancerous growth. Assessing these risks is still an uncertain business.

a principle sometimes applied in environmental cleanup of toxic organic compounds, but the breakdown may become sufficiently rapid to reduce a pesticide's effectiveness.

18.4   EFFECTS OF PESTICIDES ON SOIL ORGANISMS

Since pesticides are formulated to kill organisms, it is not surprising that some of these compounds are toxic to specific soil organisms. At the same time, the diversity of the soil organism population is so great that, excepting a few fumigants, most pesticides do not kill a broad spectrum of soil organisms.

Fumigants

Fumigants are compounds used to free a soil of a given pest, such as nematodes. These compounds have a more drastic effect on both the soil fauna and flora than do other pesticides. For example, 99% of the microarthropod population is usually killed by the fumigants DD and vampam, and it may take as long as two years for the population to fully recover. Fortunately, the recovery time for the microflora is generally much less.

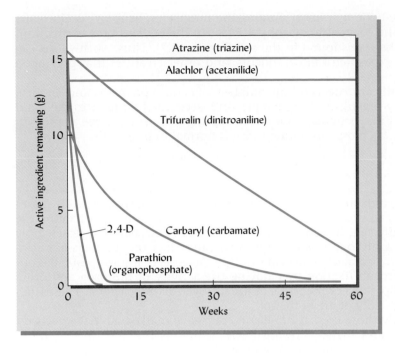

FIGURE 18.6 Degradation of four herbicides (alachlor, atrazine, 2,4-D, and trifuralin) and two insecticides (parathion and carbaryl), all of which are used extensively in the Midwest of the United States. Note that atrazine and alachlor are quite slowly degraded, whereas parathion and 2,4-D are quickly broken down. [Reprinted with permission from R. G. Krueger and J. N. Seiber, *Treatment and Disposal of Pesticide Wastes,* Symposium Series 259; copyright 1984 American Chemical Society]

Fumigation reduces the number of species of both flora and fauna, especially if the treatment is repeated, as is often the case where nematode control is attempted. At the same time, the total number of bacteria is frequently much greater following fumigation than before. This increase is probably due to the relative absence of competitors and predators following fumigation and to the carbon and energy sources left by dead organisms for microbial utilization.

Effects on Soil Fauna

The effects of pesticides on soil animals varies greatly from chemical to chemical and from organism to organism. Nematodes are not generally affected, except by specific fumigants. Mites are generally sensitive to most organophosphates and to the chlorinated hydrocarbons, with the exception of aldrin. Springtails vary in their sensitivity to both chlorinated hydrocarbons and organophosphates, some chemicals being quite toxic to these organisms.

EARTHWORMS. Fortunately, many pesticides have only mildly depressing effects on earthworm numbers, but there are exceptions. Among insecticides, most of the carbamates (carbaryl, carbofuran, aldicarb, etc.) are highly toxic to earthworms. Among the herbi

TABLE 18.5 **Common Range of Persistence of a Number of Organic Compounds**

Risks of environmental pollution are highest with those chemicals with greatest persistence.

Organic chemical	Persistence
Chlorinated hydrocarbon insecticides (e.g., DDT, chlordane, and dieldrin)	3–20 yr
PCBs	2–10 yr
Triazine herbicides (e.g., atrazine and simazine)	1–2 yr
Benzoic acid herbicides (e.g., amiben and dicamba)	2–12 mo
Urea herbicides (e.g., monuron and diuron)	2–10 mo
Vinyl chloride	1–5 mo
Phenoxy herbicides (2,4-D and 2,4,5-T)	1–5 mo
Organophosphate insecticides (e.g., malathion and diazinon)	1–12 wk
Carbamate insecticides	1–8 wk
Carbamate herbicides (e.g., barban and CIPC)	2–8 wk

cides, simazine is more toxic than most. Among the fungicides, benomyl is unusually toxic to earthworms. The concentrations of pesticides in the bodies of the earthworms are closely related to the levels found in the soil (Figure 18.7). Thus, earthworms can magnify the pesticide exposure of birds, rodents, and other creatures that prey upon them.

Pesticides have significant effects on the numbers of certain predators and, in turn, on the numbers of prey organisms. For example, an insecticide that reduces the numbers of predatory mites may stimulate numbers of springtails, which serve as prey for the mites (Figure 18.8). Such organism interaction is normal in most soils.

Effects on Soil Microorganisms

The overall levels of bacteria in the soil are generally not too seriously affected by pesticides. However, the organisms responsible for nitrification and nitrogen fixation are sometimes adversely affected. Insecticides and fungicides affect both processes more than do most herbicides, although some of the latter can reduce the numbers of organisms carrying out these two reactions. Recent evidence suggests that some pesticides can enhance biological nitrogen fixation by reducing the activity of protozoa and other organisms that are competitors or predators of the nitrogen-fixing bacteria. These findings illustrate the complexity of life in the soil.

Fungicides, especially those used as fumigants, can have marked adverse effects on soil fungi and actinomycetes, thereby slowing down the humus formation in soils. Interestingly, however, the process of ammonification is often stimulated by pesticide use.

The negative effects of most pesticides on soil microorganisms are temporary, and after a few days or weeks, organism numbers generally recover. But exceptions are common enough to dictate caution in the use of the chemicals. Care must be taken to apply them only when alternate means of pest management are not available.

This brief review of the behavior of organic chemicals in soils reemphasizes the complexity of the changes that take place when new and exotic substances are added to our environment. Our knowledge of the soil processes involved certainly reaffirms the necessity for a thorough evaluation of potential environmental impacts prior to approval and use of new chemicals for extensive use on the land.

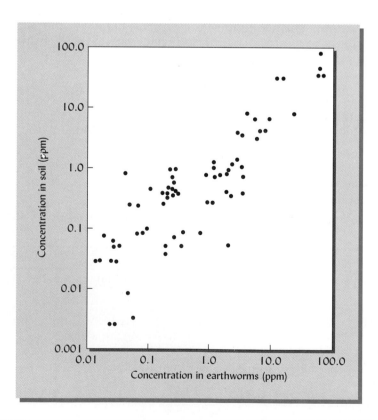

FIGURE 18.7 Effect of concentration of pesticides in soil on their concentration in earthworms. Birds eating the earthworms at any level of concentration would further concentrate the pesticides. [Data from several sources gathered by Thompson and Edwards (1974); used with permission of Soil Science Society of America]

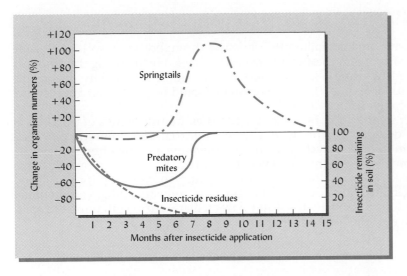

FIGURE 18.8 The direct effect of insecticide on predatory mites in a soil and the indirect effect of reducing mite numbers on the population of springtails (tiny insects) that serve as prey for the mites. [Replotted from Edwards (1978); used with permission of Academic Press, Inc., London]

18.5 REGIONAL VULNERABILITY TO PESTICIDE LEACHING

The vulnerability to leaching of pesticides to the groundwater varies greatly from one area to another. Highest vulnerability occurs in regions with high rainfall, an abundance of sandy soils, and intensive cropping systems that involve high usage of those types of pesticides that are not adsorbed by the soil particles. Table 18.6 shows the results of one study set up to measure the vulnerability of different regions of the United States to the leaching of pesticides and nitrates. Note the high average vulnerability score for the southern Atlantic Coast area where sandy soils are prominent, and where highly intensive cropping systems (fruits and vegetables) are used. Likewise, vulnerability for leaching is high in the corn belt, where much of the land is under continuous corn production with its high herbicide and fertilizer nitrogen use.

It should be pointed out, however, that these regional data may mask localized areas of vulnerability. For example, in arid regions of the mountain states are found irrigated areas of intensive vegetable crop production that may involve considerable leaching of both pesticides and nitrates. Likewise, application of some water-soluble pesticides may result in leaching into the groundwater even though the soil may not be coarse in texture. This suggests the site-specific nature of pesticide hazards.

TABLE 18.6 Average Cropland Leaching Vulnerability Scores for Pesticides and Nitrates in Different Regions of the United States

The scores are based on the kinds and amounts of pesticides used in an area and potential for leaching from the soils in each region. The higher the score, the more vulnerable the areas are for leaching of these chemicals. Note the very high leaching vulnerability for pesticides in the southern Atlantic states, the corn belt and the Delta states.

Area	Comparative leaching vulnerability (average = 100)	
	Pesticides	Nitrates
Northeast	26	39
Southern Atlantic Coast	420	151
Appalachia	90	153
Lake states	74	76
Corn belt	156	336
Delta states	109	120
Northern Plains	45	47
Southern Plains	22	39
Mountain states	15	10
Pacific Coast	44	29
Average	100	100

Calculated from data in Kellogg, et al. (1994).

Soils contaminated with organic pollutants are found throughout the world. The wide areas contaminated with organic pesticides are best addressed by reducing the amounts of pesticides used and by using less toxic, less mobile, and more rapidly degradable compounds. In a reasonable time, the soil ecosystem should be able to recover its function and diversity through *self remediation*.

Perhaps of greater significance, however, are the sites around centers of population where, through the decades, organic wastes from industrial and domestic processes have been dumped on soils. The levels of such *acute contamination* are often sufficiently high that plant growth is restrained or even prevented. Pollutants move into the groundwater and make the drinking water unfit for human consumption. Fish and wildlife are often decimated. Because of public concerns, industry and government are spending billions of dollars annually to clean up (remediate) these contaminated soils. We shall consider a few of the methods they use.

Physical and Chemical Methods

The most widespread methods of soil remediation involve physical and/or chemical treatment of the soil, either in place (*in situ*) or by moving the soil to a treatment site (*ex situ*). Ex situ treatment may involve excavating the soil to treatment bins where it may be incinerated to drive off volatile chemicals and to destroy others that are decomposed upon being heated. Water-soluble and volatile chemicals may also be removed by pushing or pulling air or water through the soil by vacuum extraction or leaching. Such treatments are usually quite efficient but are very expensive, especially if large quantities of soil are involved.

In situ treatments are usually preferred if viable technologies are available. The soil is left in its natural condition, thereby reducing excavation and treatment costs and providing greater flexibility in future land use. The contaminants are either removed from the soil (*decontamination*) or are sequestered (*bound up*) in the soil matrix (*stabilized*). Decontamination in situ involves some of the same techniques of water flushing, leaching, and vacuum extraction used in ex situ processes. Water treatment is not effective, however, with nonpolar compounds that are repelled by water. To help remove such compounds, scientists and engineers have sprayed onto the soil surface or have injected into the soil compounds called *surfactants*. As these move downward in the soil, they solubilize organic contaminants, which can then be pumped out of the soil as in the water washing systems.

ORGANOCLAYS. Certain surfactants may also be used to immobilize or stabilize soil contaminants. They are positively charged and through cation exchange can replace metal cations on soil clays. For example, one group of such surfactants, quaternary ammonium compounds (QACs), have the general formula $(CH_3)_3NR^+$, where R is an organic alkyl or aromatic group. The positive charges on QACs stimulate cation exchange by reactions such as the following, using a monovalent exchangeable cation such as K as an example:

$$\boxed{\text{Micelle}}\ K^+ + (CH_3)_3NR^+ \longrightarrow \boxed{\text{Micelle}}\ (CH_3)_3NR^+ + K^+$$

Untreated clay   QAC   Organoclay

The resulting products, known as *organoclays*, have properties quite different from the untreated clays. They attract rather then repel nonpolar organic compounds. Thus, the injection of a QAC into the zone of groundwater flow can stimulate the formation of organoclays and thereby immobilize soluble organic groundwater contaminants, holding them until they can be degraded (Figure 18.9).

DISTRIBUTION COEFFICIENTS K_D. As we learned in Section 8.16, the degree of sorption of organic compounds by soil colloids is commonly indicated by the coefficient of distribution K_d between the sorbed and solution portions of the organic compound.

$$K_d = \frac{\text{mg contaminant/kg soil}}{\text{mg contaminant/L solution}}$$

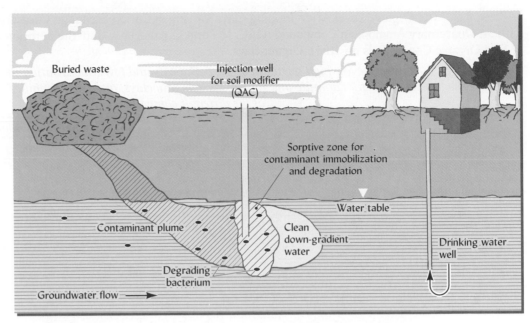

FIGURE 18.9 How a combination of a quatenary ammonium compound (QAC), hexadecyltrimethyl-ammonium, and bioremediation by degrading bacteria could be used to hold and remove an organic contaminant. The pollutant is moving into groundwater from a buried waste site. The QAC reacts with soil clays to form organoclays and soil organic matter complexes that adsorb and stabilize the contaminant, giving microorganisms time to degrade or destroy it. [Redrawn from Xu, et al. (1997)]

The K_d for untreated clays is very low because the clays are hydrophilic (water-loving) and their adhering water films repel the hydrophobic, nonpolar organic compounds. In contrast, organoclays sorb the contaminants, leaving little in the soil solution, thereby reducing their movement into the groundwater and eventually into streams or drinking water. Consequently, the K_d values of organic contaminants on organoclays are commonly 100 to 200 times those measured on the untreated clays. Table 18.7 shows K_d values for some common organic contaminants on organoclays and indicates the very high sorbing power of the newly created sorbants. Their tenacity is complemented by the very strong complexation of organic compounds by soil organic matter. Organoclays thus offer promising mechanisms for holding organic soil pollutants until they can be destroyed by biological or physicochemical processes.

Bioremediation[2]

For many heavily contaminated soils there is a biological alternative to incineration, soil washing, and landfilling—namely, **bioremediation.** Simply put, this technology uses enhanced plant and/or microbial action to degrade organic contaminants into harmless metabolic products. Petroleum constituents, including the more resistant polyacrylic aromatic hydrocarbons (PAHs), as well as several synthetic compounds, such as pentachlorophenol and trichloroethylene, can be broken down, primarily by soil bacteria. In some cases, advantage is taken of organisms currently in the soil. In others, microbes specifically selected for their ability to remove the contaminants are introduced into the soil zones known to be polluted. For example, a bacterium has been identified recently that can detoxify perchloroethene (PCE), a common, highly toxic groundwater pollutant that is suspected of being a carcinogen.[3] The organism expedites the removal of the four chlorines from the PCE, producing ethylene, a gas that is relatively harmless to humans.

---

[2] For reviews of this topic, see Alexander (1994) and Skipper and Turco (1995).

[3] See Maymo-Gatell, et al. (1997).

TABLE 18.7 **The Organic Level After Treating Clays Varying in Cation Exchange Capacity with an Quaternary Ammonium Compound (QAC) to Form Organoclays, and The Sorption Coefficients K_d of Five Organic Compounds on These Organoclays**

High K_d values suggest high retention of the pollutants and low concentration in the soil solution. Note the low tendency for kaolinite and illite to form organoclays, and the variability in sorption coefficients of the different compounds.

Clay	CEC of untreated clay, cmol/kg	Organic C in organoclay, %	K_d of organic contaminants on organoclays				
			Benzene	Toluene	Ethylbenzene	PropylBenzene	Naphthalene
Illite	24	2.5	39	77	156	—	1270
Vermiculite	80	16.4	68	169	448	1618	1387
Smectite (high charge)	130	23.0	184	319	583	1412	4818
Kaolinite	4	1.0	3	7	21	—	—

Data from Jaynes and Boyd (1997).

PCE
(Suspected carcinogen)

Ethylene
(Harmless gas)

PHYTOREMEDIATION.[4]   Higher plants have also been found to participate in bioremediation, a process termed **phytoremediation**. For years, plant-based systems have been used for the removal of municipal wastewater contaminants. More recently, this concept has been extended to industrial pollutants and to the removal of shallow groundwater pollutants of all kinds, both organic and inorganic. Many plant species, domesticated and wild, have been used in phytoremediation. Prairie grasses can stimulate the degradation of petroleum products, including the PAHs, and spring wildflower plants in Kuwait were recently found to degrade the hydrocarbons in oil spills. Fast-growing hybrid poplars can remove ammunition compounds, such as trinitrotoluene (TNT), as well as some pesticides and excess nitrates.

Some plants can absorb and metabolize specific organic contaminants. Others release exudates from their roots that help degrade the pollutants. However, the primary agents in the plant-associated degradation of contaminants are the microorganisms associated with the rhizosphere of the plants. The plants' roots release compounds that serve as energy sources for the microbes, which, in turn, produce enzymes that can degrade the organic contaminants. As we shall see later (Section 18.10), phytoremediation is perhaps even more effective in removing heavy metals and other inorganic pollutants, including radionuclides.

Phytoremediation is particularly advantageous where large areas of soil are contaminated with only moderate concentrations of organic pollutants. However, phytoremediation also commonly takes a longer time to remove large quantities of contaminants than do the more costly engineering procedures.

BIOAVAILABILITY OF SORBED AND/OR COMPLEXED CHEMICALS.   Researchers have found that some chemicals that are normally subject to microbial attack seem to be protected from such degradation when the compounds are complexed by soil organic matter or sorbed by inorganic materials. The complexation is essentially irreversible and the sorption by Fe, Al oxides or silicate clays is so tight that the compounds are only very slowly available.

---

[4] For a recent review of phytoremediation, see Cunningham, et al. (1996).

Some compounds may also be trapped between the internal structural layers of some silicate clays. The bioavailability usually decreases as the soil–contaminant complex ages over time. Reduced bioavailability of the pollutants constrains their remediation by microorganisms. It also has some implications for regulatory policies, since pollutants so held are not likely to move into the groundwater or elsewhere in the environment.

NUTRIENT SUPPLEMENTATION. Bioremediation technology assists natural chemical breakdown in several ways. Usually, the soil naturally contains some bacteria or other microorganisms that can degrade the specific contaminant. But the rate of natural degradation may be far too slow to be very effective. Both growth and metabolic rate of organisms capable of using the contaminant as a carbon source are often limited by insufficient mineral nutrients, especially nitrogen and phosphorus (see Section 12.3 for a discussion of the C/N ratio in organic decomposition). Special fertilizers have been formulated and used successfully to greatly speed up the degradation process. One such fertilizer of French manufacture is an oil-in-water microemulsion of urea, lauryl phosphate, and an emulsion stabilizer. It acts not only as a supplier of nutrients, but as a surfactant that can enhance interaction between microbes and the organic contaminants.

OIL SPILL CLEANUP. The 30 or more different genera of bacteria and fungi known to degrade hydrocarbons are found in almost any soil or aquatic environment. But they may need help. The cleanup of crude oil contamination from the 1989 Exxon Valdez oil spill in Alaskan waters was a spectacular case of successful bioremediation by fertilization (Figure 18.10). A special fertilizer was sprayed on the oil-soaked beaches (Entisols). The fertilizer was formulated to be oliophilic (soluble in oil but not in water) so that it would stay with the oil and not contribute to eutrophication of Prince William Sound. Within a few weeks, and despite the cold temperatures, most of the oil in the test area was degraded. The success of bioremediation was greatest where nitrogen was most available to the microorganisms.

IN SITU TECHNIQUES. Other situations call for the use of bioremediation techniques in situ. In some cases, low soil porosity causes oxygen deficiency that limits microbial activity.

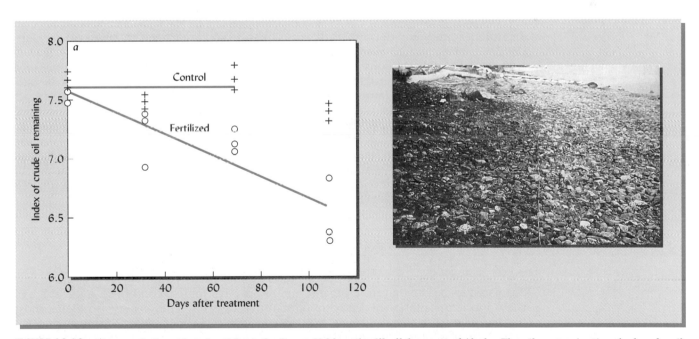

FIGURE 18.10 Bioremediation of crude oil from the Exxon Valdez oil spill off the coast of Alaska. The oil contaminating the beach soils was degraded by indigenous bacteria when an oil-soluble fertilizer containing nitrogen and phosphorus was sprayed on the beach (o data points in graph on the left). The control sections of the beach (+ data points) were left unfertilized for 70 days. By then the effect of the fertilization was so dramatic that a decision was made to treat the control sections as well. The index of oil remaining is based on natural logarithms, so each whole number indicates more than doubling of oil remaining. The photo (right) shows the clear delineation between the oil-covered control section and the fertilized parts of the beach. [Data from Bragg, et al. (1994); reprinted with permission from Nature, © 1994 Macmillan Magazines Limited; photo P. H. Pritchard, USEPA, Gulf Breeze, Fla.; courtesy of Pritchard, et al. (1992); reprinted by permission of Kluwer Academic Publishers]

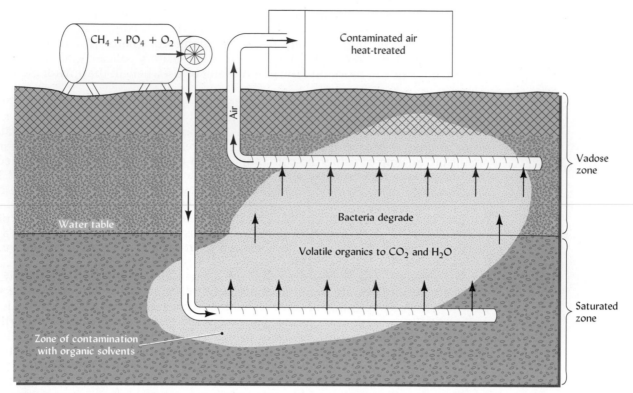

FIGURE 18.11   In situ bioremediation of soil and groundwater contaminated with volatile organic solvents at an industrial site in Georgia. The methane-air-phosphorus mixture was pumped *intermittently* into the soil through slotted pipes,
while another pipe "vacuumed" air out of the soil. The air and nutrients stimulated the growth of certain bacteria which,
when the methane was taken away, turned to the solvent for a carbon source. It was estimated that the bioremediation
technology would cut the time to clean up the site from more than 10 years to less than 4, saving $1.6 million in costs.
[Redrawn from Hazen (1995)]

Techniques are being developed that use in situ bioremediation to clean up oxygen-
deficient soils and associated groundwater contamination. For example, organic-
solvent-contaminated soils have been bioremediated (Figure 18.11) by piping in a
mixture of air (for oxygen), methane (to act as a carbon source to stimulate specific bacteria), and phosphorus (a nutrient that is needed for bacteria growth).

Some success has been achieved by inoculating contaminated soils with improved
organisms that can degrade the pollutant more readily than can the native population.
Although genetic engineering may prove useful in making "superbacteria" in the
future, most inoculation has been achieved with naturally occurring organisms. Organisms isolated from sites with a long history of the specific contamination or grown in
laboratory culture on a diet rich in the pollutant in question tend to become acclimated
to metabolizing the target chemical.

18.7   CONTAMINATION WITH TOXIC INORGANIC SUBSTANCES[5]

The toxicity of inorganic contaminants released into the environment every year is now
estimated to exceed that from organic and radioactive sources combined. A fair share of
these inorganic substances ends up contaminating soils. The greatest problems most
likely involve mercury, cadmium, lead, arsenic, nickel, copper, zinc, chromium, molybdenum, manganese, selenium, fluorine, and boron. To a greater or lesser degree, all of
these elements are toxic to humans and other animals. Cadmium and arsenic are
extremely poisonous; mercury, lead, nickel, and fluorine are moderately so; boron, cop-

---

[5] For a review of this subject, see Kabata-Pendias and Pendias (1992).

TABLE 18.8   Sources of Selected Inorganic Soil Pollutants

Chemical	Major uses and sources of soil contamination	Organisms principally harmed[a]	Human health effects
Arsenic	Pesticides, plant desiccants, animal feed additives, coal and petroleum, mine tailings, and detergents	H, A, F, B	Cumulative poison, possibly cancer
Cadmium	Electroplating, pigments for plastics and paints, plastic stabilizers, batteries, and phosphate fertilizers	H, A, F, B, P	Heart and kidney disease, bone embrittlement
Chromium	Stainless steel, chrome-plated metals, pigments, refractory brick manufacture, and leather tanning	H, A, F, B	Mutagenic; also essential nutrient
Copper	Mine tailings, fly ash, fertilizers, windblown copper-containing dust, and water pipes	F, P	Rare; essential nutrient
Lead	Combustion of oil, gasoline, and coal; iron and steel production; and solder on water-pipe joints	H, A, F, B	Brain damage, convulsions
Mercury	Pesticides, catalysts for synthetic polymers, metallurgy, and thermometers	H, A, F, B	Nerve damage
Nickel	Combustion of coal, gasoline, and oil; alloy manufacture; electroplating; batteries; and mining	F, P	Lung cancer
Selenium	High Se geological formations and irrigation wastewater in which Se is concentrated	H, A, F, B	Rare; loss of hair and nail deformities; essential nutrient
Zinc	Galvanized iron and steel, alloys, batteries, brass, rubber manufacture, mining, and old tires	F, P	Rare; essential nutrient

[a] H = humans, A = animals, F = fish, B = birds, P = plants.
Data selected from Moore and Ramamoorthy (1984) and numerous other sources.

per, manganese, and zinc are relatively lower in mammalian toxicity. Although the metallic elements (which exclude fluorine and boron) are not all, strictly speaking, "heavy" metals, for the sake of simplicity this term is often used in referring to them. Table 18.8 provides background information on the uses, sources, and effects of some of these elements.

Sources and Accumulation

There are many sources of the inorganic chemical contaminants that can accumulate in soils. The burning of fossil fuels, smelting, and other processing techniques release into the atmosphere tons of these elements, which can be carried for miles and later deposited on the vegetation and soil. Lead, nickel, and boron are gasoline additives that are released into the atmosphere and carried to the soil through rain and snow.

Borax is used in detergents, fertilizers, and forest fire retardants, all of which commonly reach the soil. Superphosphate and limestone, two widely used amendments, usually contain small quantities of cadmium, copper, manganese, nickel, and zinc. Cadmium is used in plating metals and in the manufacture of batteries. Arsenic was for many years used as an insecticide on cotton, tobacco, fruit crops, lawns, and as a defoliant or vine killer. Some of these mentioned elements are found as constituents in specific organic pesticides and in domestic and industrial sewage sludge. Additional localized contamination of soils with metals results from ore-smelting fumes, industrial wastes, and air pollution.

Some of the toxic metals are being released to the environment in increasing amounts, while others (most notably lead, because of changes in gasoline formulation) are decreasing. All are daily ingested by humans, either through the air or through food, water, and—yes—soil.

Concentration in Organism Tissue

Irrespective of their sources, toxic elements can and do reach the soil, where they become part of the food chain: soil→plant→animal→human (Figure 18.12). Unfortunately, once the elements become part of this cycle, they may accumulate in animal and human body tissue to toxic levels. This situation is especially critical for fish and other wildlife and for humans at the top of the food chain. It has already resulted in restrictions on the use of certain fish and wildlife for human consumption. Also, it has become necessary to curtail the release of these toxic elements in the form of industrial wastes.

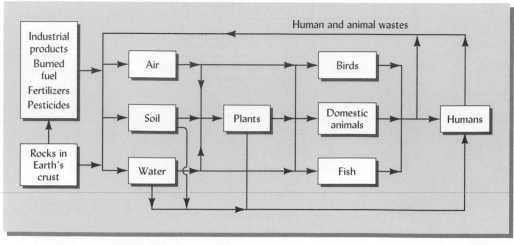

FIGURE 18.12 Sources of heavy metals and their cycling in the soil–water–air–organism ecosystem. It should be noted that the content of metals in tissue generally builds up from left to right, indicating the vulnerability of humans to heavy metal toxicity.

18.8 POTENTIAL HAZARDS OF CHEMICALS IN SEWAGE SLUDGE

The domestic and industrial sewage sludges considered in Chapter 16 can be important sources of potentially toxic chemicals. Nearly half of the municipal sewage sludge produced in the United States is being applied to the soil, either on agricultural land or to remediate land disturbed by mining and industrial activities. Industrial sludges commonly carry significant quantities of inorganic as well as organic chemicals that can have harmful environmental effects.

SOURCE REDUCTION PROGRAMS. A great deal was learned during the 1970s and 1980s about the contents, behavior, and toxicity of metals in municipal sewage sludges. As a result of the research, source-reduction programs were implemented, which required industries to clean pollutants out of their wastewater *before* sending it to municipal wastewater treatment plants. In many cases, the recovery of valuable metal pollutants was actually profitable for industries. Because of these programs, municipal sewage sludges are much cleaner than in the past (Table 18.9). Note that the median levels of the most toxic industrial pollutants (Cd, Cr, Pb, and PCB) declined dramatically between the 1976 survey and the 1990 survey. Since much of the copper comes from the plumbing in homes (metallic copper is slightly solubilized in areas with acidic water supplies), that metal has been less affected by the source reduction regulations.

TABLE 18.9 Median Pollutant Concentrations Reported in Sewage Sludges Surveyed Across the United States in 1976 and 1990 and in Uncontaminated Agricultural Soils and Cow Manure

	Concentration, mg/kg dry weight			
Pollutant	Sludges surveyed in 1990[a]	Sludges surveyed in 1976[b]	Agricultural soils[d]	Typical values for cow manure
As	6	10	—	4
Cd	7	260	0.20	1
Cr	40	890	—	56
Cu	463	850	18.5	62
Hg	4	5	—	0.2
Mo	11	—	—	14
Ni	29	82	18.2	29
Pb	106	500	11.0	16
Zn	725	1740	53.0	71
PCB	0.21	9[c]	—	0

[a] Data from Chaney (1990).
[b] Data from Sommers (1977).
[c] 1976 PCB value is median of cities in New York; from Furr, et al. (1976).
[d] Median of 3045 surface soils reported by Holmgren, et al. (1993).

TABLE 18.10   Regulatory Limits on Inorganic Pollutants (Heavy Metals) in Sewage Sludge Applied to Agricultural Land

Element	Maximum concentration in sludge, USEP,[a] mg/kg	Annual pollutant loading rates, USEPA, kg/ha/yr	Cumulative pollutant loading rates, kg/ha		
			USEPA	Germany	Ontario
As	75	2.0	41		14
Cd	85	1.9	39	6	1.6
Cr	3000	150.0	3000	200	210
Cu	4300	75.0	1500	200	150
Hg	840	15.0	300	4	0.8
Mo	57	0.85	17	—	4
Ni	75	0.90	18	100	32
Pb	420	21.0	420	200	90
Se	100	5.0	100	—	2.4
Zn	7500	140	2800	600	330

[a] U.S. Environmental Protection Agency (1993).

REGULATION OF SLUDGE APPLICATION TO LAND.   The lower levels of metals (and of organic pollutants) make municipal sewage sludges much more suitable for application to soils than in the past. Today, the amount of sludge that can be applied to agricultural land is more often limited by the potential for nitrate pollution from the nitrogen it contains, rather than by the metal content of the sludge. Nonetheless, application of sewage sludge to farmland is closely regulated to ensure that the metal concentrations in the sludge do not exceed the standards and that the total amount of metal applied to the soil over the years does not exceed the maximum accumulative loading limit listed in Table 18.10. The fact that metal-loading standards differ considerably between the United States and Europe (see Table 18.10) is an indication that the nature of the metal contamination threat is still somewhat controversial.

TOXIC EFFECTS FROM SLUDGE.   The uncertainties as to the nature of many of the organic chemicals found in the sludge, as well as the cumulative nature of the metals problem, dictate continued caution in the regulations governing application of sludge to croplands. The effect of application of a high-metal sludge on heavy metal content of soils and of earthworms living in the soil is illustrated in Table 18.11. The sludge-treated soil areas, as well as the bodies of earthworms living in these soils, were higher in some of these elements than was the case in areas where sludge had not been applied. One would expect further concentration to take place in the tissue of birds and fish, many of which consume the earthworms.

Farmers must be assured that the levels of inorganic chemicals in sludge are not sufficiently high to be toxic to plants (a possibility mainly for zinc and copper) or to humans and other animals who consume the plants (a serious consideration for Cd, Cr, and Pb). For relatively low-metal municipal sludges, application at rates just high enough to supply needed nitrogen seems to be quite safe (Table 18.12).

TABLE 18.11   The Effect of Sewage Sludge Treatment on the Content of Heavy Metals in Soil and in Earthworms Living in the Soil

Note the high concentration of cadmium and zinc in the earthworms.

Metal	Concentration of metal, mg/kg			
	Soil		Earthworms	
	Control	Sludge-treated	Control	Sludge-treated
Cd	0.1	2.7	4.8	57
Zn	56	132	228	452
Cu	12	39	13	31
Ni	14	19	14	14
Pb	22	31	17	20

From Beyer, et al. (1982).

TABLE 18.12 Uptake of Metals by Corn After 19 Years of Fertilizing a Minnesota Soil (Typic Hapludoll) with Lime-Stabilized Municipal Sewage Sludge

Note that the metals show the typical pattern of less accumulation in the grain than in the leaves and stalks (stover). The annual sludge rate of about 10.5 Mg was designed to supply the nitrogen needs of the corn. The sludge had little effect on the metal content of the plants, except in the case of zinc (which increased, but not beyond the normal range for corn).

Treatment	Uptake, mg/kg					
	Zn	Cu	Cd	Pb	Ni	Cr
Stover						
Fertilizer	18	8.4	0.16	0.9	0.7	0.9
Sludge	46.5	7.0	0.18	0.8	0.6	1.4
Grain						
Fertilizer	20	3.2	0.29	0.4	0.4	0.2
Sludge	26	3.2	0.31	0.5	0.3	0.2
Cumulative metal applied in sludge, kg/ha	175	135	1.2	49	4.9	1045

Data abstracted from Dowdy, et al. (1994).

Direct ingestion of soils and sludge is also an important pathway for human and animal exposure. Animals should not be allowed to graze on sludge-treated pastures until rain or irrigation has washed the sludge from the forage. Children may eat soil while they play, and a considerable amount of soil eventually becomes dust in many households. Direct ingestion of soil and dust is particularly harmful in lead toxicity.

18.9   REACTIONS OF INORGANIC CONTAMINANTS IN SOILS

Heavy Metals in Sewage Sludge

Concern over the possible buildup of heavy metals in soils resulting from large land applications of sewage sludges has prompted research on the fate of these chemicals in soils. Most attention has been given to zinc, copper, nickel, cadmium, and lead, which are commonly present in significant levels in these sludges. Studies have shown that if the soil is not very acid, these elements are generally bound by soil constituents; they do not then easily leach from the soil, and they are not then readily available to plants. Only in moderately to strongly acid soils is there significant movement down the profile from the layer of application of the sludge. Monitoring soil acidity and using judicious applications of lime can prevent leaching into groundwaters and can minimize uptake by plants.

FORMS FOUND IN SOILS TREATED WITH SLUDGE.   By using chemical extractants, researchers have found that heavy metals are associated with soil solids in four major ways (Table 18.13). First, a very small proportion is held in *adsorbed* or *exchangeable forms*, which are available for plant uptake. Second, the elements are bound by the *soil organic matter* and by the *organic materials* in the sludge. A high proportion of the copper is commonly found in this form, while lead is not so highly attracted. Organically bound elements are not readily available to plants, but can be released over a period of time.

The third association of heavy metals in soils is with *carbonates* and with *oxides of iron and manganese*. These forms are less available to plants than either the exchangeable or organically bound forms, especially if the soils are not allowed to become too acid. The fourth association is commonly known as the *residual form*, which consists of sulfides and other very insoluble compounds that are less available to plants than any of the other forms.

	Percentage of elements in each form					
Forms	Cd	Cr	Cu	Ni	Pb	Zn
Exchangeable/adsorbed	1	1	2	5	1	2
Organically bound	20	5	34	24	3	28
Carbonate/iron oxides	64	19	36	33	85	39
Residual[a]	16	77	29	40	12	31

[a] Sulfides and other very insoluble forms.
From Chang, et al. (1984).

It is fortunate that soil-applied heavy metals are not readily absorbed by plants and that they are not easily leached from the soil. However, the immobility of the metals means that they will accumulate in soils if repeated sludge applications are made. Care must be taken not to add such large quantities that the capacity of the soil to react with a given element is exceeded. It is for this reason that regulations set maximum cumulative loading limits for each metal (see Table 18.10).

Chemicals from Other Sources

Arsenic has accumulated in orchard soils following years of application of arsenic-containing pesticides. Being present in an anionic form (e.g., $H_2AsO_4^-$), this element is absorbed (as are phosphates) by hydrous iron and aluminum oxides, especially in acid soils. In spite of the capacity of most soils to tie up arsenates, long-term additions of arsenical sprays can lead to toxicities for sensitive plants and earthworms. The arsenic toxicity can be reduced by applications of sulfates of zinc, iron, and aluminum, which tie up the arsenic in insoluble forms.

Contamination of soils with *lead* comes primarily from airborne lead from automobile exhaust and from paint chips and dust from woodwork coated with old lead-pigmented paints. It is most concentrated within 100 m of major roadways, near urban centers, and in the soil near older homes. Some lead is deposited on the vegetation and some reaches the soil directly. In any case, most of the lead is tied up in the soil as insoluble carbonates, sulfides, and in combination with iron, aluminum, and manganese oxides (see Table 18.13). Consequently, the lead is largely unavailable to plants, but may injure children who put contaminated soil in their mouths.

Soil contamination by *boron* can occur from irrigation water high in this element, by excessive fertilizer application, or by the use of power plant fly ash as a soil amendment. The boron can be adsorbed by organic matter and clays but is still available to plants, except at high soil pH. Boron is relatively soluble in soils, toxic quantities being leachable, especially from acid sandy soils. Boron toxicity is usually considered a localized problem and is probably much less important than the deficiency of the element.

Fluorine toxicity is also generally localized. Drinking water for animals and fluoride fumes from industrial processes often contain toxic amounts of fluorine. The fumes can be ingested directly by animals or deposited on nearby plants. If the fluorides are adsorbed by the soil, their uptake by plants is restricted. The fluorides formed in soils are highly insoluble, the solubility being least if the soil is well supplied with lime.

Mercury contamination of lake beds and of swampy areas has resulted in toxic levels of mercury among certain species of fish. Insoluble forms of mercury in soils, not normally available to plants or, in turn, to animals, are converted by microorganisms to an organic form, methylmercury, in which it is more soluble and available for plant and animal absorption. The methylmercury is concentrated in fatty tissue as it moves up the food chain, until it accumulates in some fish to levels that may be toxic to humans. This series of transformations illustrates how reactions in soil can influence human toxicities.

18.10 PREVENTION AND ELIMINATION OF INORGANIC CHEMICAL CONTAMINATION

Three primary methods of alleviating soil contamination by toxic inorganic compounds are (1) to eliminate or drastically reduce the soil application of the toxins; (2) to immobilize the toxin by means of soil management, to prevent it from moving into food or water supplies; and (3) in the case of severe contamination, to remove the toxin by chemical, physical, or biological remediation.

Reducing Soil Application

The first method requires action to reduce unintentional aerial contamination from industrial operations and from automobile, truck, and bus exhausts. Decision makers must recognize the soil as an important natural resource that can be seriously damaged if its contamination by accidental addition of inorganic toxins is not curtailed. Also, there must be judicious reductions in intended applications to soil of the toxins through pesticides, fertilizers, irrigation water, and solid wastes.

Immobilizing the Toxins

Soil and crop management can help reduce the continued cycling of these inorganic chemicals. This is done primarily by keeping the chemicals in the soil rather than encouraging their uptake by plants. The soil becomes a sink for the toxins, and thereby breaks the soil–plant–animal (humans) cycle through which the toxin exerts its effect. The soil breaks the cycle by immobilizing the toxins. For example, most of these elements are rendered less mobile and less available if the pH is kept near neutral or above (Figure 18.13). Liming of acid soils reduces metal mobility; hence, regulations require that the pH of sludge-treated land be maintained at 6.5 or higher.

Draining wet soils should be beneficial, since the oxidized forms of the several toxic elements are generally less soluble and less available for plant uptake than are the reduced forms. However, the opposite is true for chromium, which occurs principally in two forms, Cr^{3+} and Cr^{6+}. Hexavalent Cr forms compounds that are mobile under a wide

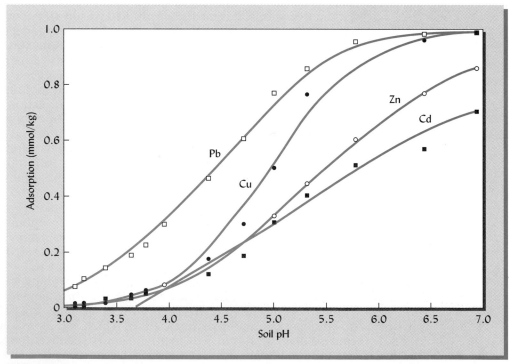

FIGURE 18.13   The effect of soil pH on the adsorption of four heavy metals. Maintaining the soil near neutral provides the highest adsorption of each of these metals and especially of lead and copper. The soil was a Typic Paleudult (Christiana silty clay loam). [From Elliot, et al. (1986)]

range of pH conditions and are highly toxic to humans. Trivalent Cr, on the other hand, forms oxides and hydroxides that are quite immobile except in very acid soils. Therefore, it is desirable to reduce Cr^{6+} to Cr^{3+} in chromium-contaminated soils. Fortunately, active soil organic matter is quite effective at reducing chromium, and the Cr^{3+}, once formed, does not tend to reoxidize (Figure 18.14).

Heavy phosphate applications reduce the availability of some metal cations but may have the opposite effect on arsenic, which is found in the anionic form. Leaching may be effective in removing excess boron, although moving the toxin from the soil to water may not be of any real benefit.

Care should be taken in selecting plants to be grown on metal-contaminated soil. Generally, plants translocate much larger quantities of metals to their leaves than to their fruits or seeds (see Table 18.12). The greatest risk for food-chain contamination with metals is therefore through leafy vegetables, such as lettuce and spinach, or through forage crops eaten by livestock.

Bioremediation by Metal Hyperaccumulating Plants

Certain plants that have evolved in soils naturally very high in metals are able to take up and accumulate extremely high concentrations of metals without suffering from toxicity. Plants have been found that accumulate more than 20,000 mg/kg nickel, 40,000 mg/kg zinc, and 1000 mg/kg cadmium. While such *hyperaccumulating* plants would pose a serious health hazard if eaten by animals or people, they may facilitate a new kind of bioremediation for metal-contaminated soils.

If sufficiently vigorously growing genotypes of such plants can be found, it may be possible to use them to remove metals from contaminated soils. For example, several plants in the genus *Thlaspi* have been grown in soils contaminated by smelter fumes (Figure 18.15). These soils are so contaminated that they are virtually barren. Accumulating more than 30,000 mg/kg (about 3%) zinc in their tissues, the *Thlaspi* plants grown on this site could be harvested to remove large quantities of the metals from the soil. The plant tissue is so concentrated that it could be used as an "ore" for smelting new metal. This and other bioremediation technologies for metals (e.g., the bioreduc-

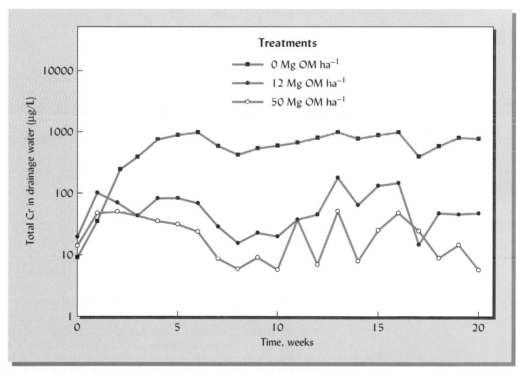

FIGURE 18.14 Effect of adding dried cattle manure (OM) on the concentration of chromium in drainage water from a chromium-contaminated soil. Oxidation of the manure caused the reduction of toxic, mobile Cr^{6+} to relatively immobile Cr^{3+}. Note the log scale for Cr in the water. The coarse-textured soil was a Typic Torripsamment in California. [Adapted from Losi, et al. (1994)]

FIGURE 18.15 *Thlaspi caerulescens*, a zinc and cadmium hyperaccumulator plant growing in smelter-contaminated soil near Palmerton, Pa. This plant has been reported to accumulate up to 4% zinc in its tissue (dry weight basis). Research with such plants aims at developing technology to biologically remove and recover metals from heavily contaminated soils. (Photo by H. Witham; courtesy of R. Chaney, USDA)

tion of chromium and selenium discussed earlier) hold promise for cleaning up badly contaminated soils without resorting to expensive and destructive excavation and soil-washing methods.

Genetic and bioengineering techniques are being utilized to develop high-yielding hyperaccumulating plants that can remove larger quantities of heavy metal contaminants from soils. For example, wide genetic variation in heavy metal accumulation by different strains of Alpine pennycress suggests the potential for breeding improved accumulating plants. Also, research to insert genes responsible for contaminant accumulation into other higher-yielding plants, such as canola and Indian mustard, is underway.

A combination of chelates and phytoremediation has been used to remove lead from contaminated soil. This element is sparingly available to plants, being strongly bound by both mineral and organic matter. The chelates solubilize the lead, and plants such as Indian mustard are used to remove it.

18.11 LANDFILLS

A visit to the local landfill would convince anyone of the wastefulness of modern societies. Roughly 250 million Mg of municipal wastes are generated each year by people in the United States. Most (about 70%) of this waste material is organic in nature, largely paper, cardboard, and yard wastes (e.g., grass clippings, leaves, and tree prunings). The other 30% consists mainly of such nonbiodegradeables as glass, metals, and plastic. Currently, despite an upsurge in recycling efforts, the great majority of these materials are buried in the earth (Figure 18.16).

The Solid Waste Problem

We know that the entire waste disposal problem could be greatly reduced by creating less waste in the first place. Second, it is possible to eliminate most problems associated with waste disposal by two simple measures: (1) keeping the metals, glass, plastics, and paper separate in the household for easy recycling, and (2) composting the yard wastes, food wastes, and some of the paper products. The composted product from a number of municipalities is successfully used as a beneficial soil amendment (see Section 16.7). The small fraction of more hazardous wastes remaining can then be detoxified or concentrated and immobilized.

The present reality is that most municipal solid wastes are buried in the earth and will probably continue to be disposed of in this manner for some time to come. In the past, wastes were merely placed in open dumps and, often, set afire. The term *landfill* came into use because wastes were often dumped in swampy lowland areas where, eventually, their accumulation filled up the lowland, creating upland areas for such uses as city parks and other facilities. Locating landfills on wetlands is no longer an acceptable practice.

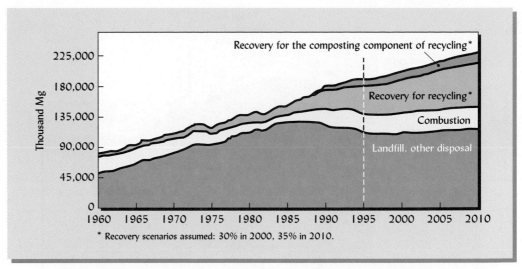

FIGURE 18.16 Historical and predicted trends in municipal solid-waste management in the United States. Soils play a central role in the composting and landfilling options (shown in dark green). Data from U.S. EPA (1997).

Two Basic Types of Landfill Design

Although landfill designs vary with the characteristics of both the site and the wastes, two basic types of landfills (Figure 18.17) can be distinguished: (1) the natural attenuation or unsecured landfill, and (2) the containment or secured landfill. We will briefly discuss the main features of each.

Natural Attenuation Landfills

The purpose of a natural attenuation landfill is to contain nonhazardous municipal wastes in a sanitary manner, protect them from animals and wind dispersal, and, finally, to cover them sufficiently to allow revegetation and possible reuse of the site. Some rainwater is allowed to percolate through the waste and down to the groundwater. Natural processes are relied upon to attenuate the leachate contaminants before the leachate reaches the groundwater. Soils play a major role in these natural attenuation processes through physical filtering, adsorption, biodegradation, and chemical precipitation (Table 18.14).

SOIL REQUIREMENTS. Finding a site with suitable soil characteristics is critical for this type of landfill. There must be at least 1.5 m of soil material between the bottom of the landfill and the highest groundwater level. This layer of soil should be only moderately permeable. If too permeable (sandy, gravelly, or highly structured), it will allow the leachate to pass through so quickly that little attenuation of contaminants will take place. The soil must have sufficient cation exchange capacity to adsorb the cations (NH_4^+, K^+, Na^+, Cd^{2+}, Ni^{2+}, and other metallic cations) that the wastes are expected to release. If too slowly permeable, the leachate will build up, flood the landfill, and seep out laterally.

The site for a natural attenuation landfill should also provide soils suitable for daily and final cover materials (Figure 18.18a). At the end of every workday, the waste must be covered by a layer of relatively impermeable soil material. The final cover for the landfill is much thicker than the daily covers, and includes a thick layer of low-permeability, fine-textured material underneath a thinner layer of loamy "topsoil." The impermeable cover is meant to minimize percolation of water into the landfill, and the topsoil is meant to support a vigorous plant cover that will prevent erosion and use up water by evapotranspiration. The whole system is designed to limit the amount of water percolating through the waste, so that the amounts of contaminated leachate generated will not overwhelm the attenuating capacity of the soil between the landfill bottom and the groundwater.

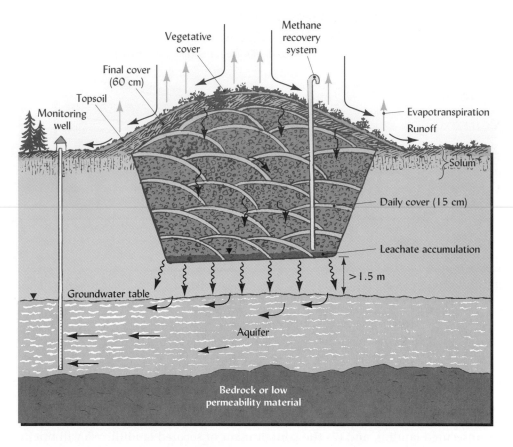

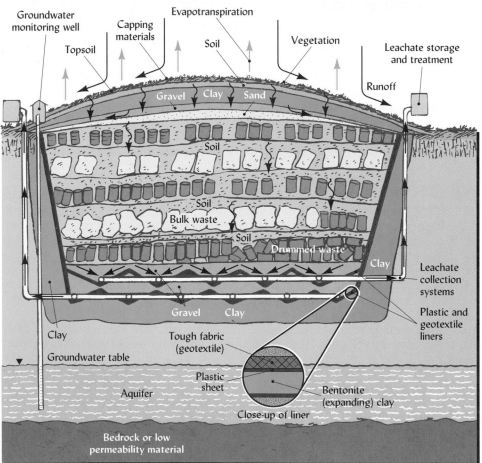

FIGURE 18.17 Two types of landfills. The natural attenuation landfill (top) depends largely on soil processes to attenuate the contaminants in the leachate before they reach the groundwater. The containment-type landfill (bottom) is used for more hazardous wastes or when soil conditions on the site are unsuitable for natural attenuation. it is designed to collect all the leachate and pump it out for storage and treatment.

TABLE 18.14 Partial List of Organic and Inorganic Contaminants in Untreated Leachate from the City of Guelph (Ontario, Canada) Municipal Solid Waste Landfill

Typical sources of these contaminants in landfills and mechanisms by which soils can attenuate the contaminants are also given. Although leachates vary greatly among landfills, the contaminants in this list are fairly typical.

Chemical	Concentration, µg/L	Common sources	Mechanisms of attenuation
		Organics	
Chemical oxygen demand (COD) for general organics	14,300	Rotting yard wastes, paper, and garbage	Biological degradation
Benzene	20	Adhesives, deodorants, oven cleaner, solvents, paint thinner, and medicines	Filtration, biodegradation, and methanogenesis
Dichloroethane	406	Adhesives and degreasers	Biodegradation and dilution
Toluene	165	Glues, paint cleaners and strippers, adhesives, paints, dandruff shampoo, and carburetor cleaners	Biodegradation and dilution
Xylene	212	Oil and fuel additives, paints, and carburetor cleaners	Biodegradation and dilution
		Metals	
Nickel	0.38	Batteries, electrodes, and spark plugs	Adsorption and precipitation
Chromium	0.14	Cleaners, paint, linoleum, and batteries	Precipitation, adsorption, and exchange
Cadmium	0.03	Paint, batteries, and plastics	Precipitation and adsorption

Leachate data from Cureton, et al. (1991).

Containment or Secured Landfills

The second main type of landfill is much more complex and expensive to construct, but its construction is much less dependent on finding a site with suitable soils. The design is intended to contain all leachate from the landfill, rather than depending on the soil for cleansing before the leachate enters the groundwater. To accomplish the containment, one or more impermeable liners are set in place around the sides and bottom of the landfill. These are often made of expanding clays (e.g., bentonite) that swell to a very low permeability when wet. Plastic, watertight geomembranes, and tough, non-woven, synthetic fabric (geotextiles) are also used in making the liners. A layer of gravel or sand is used to protect the liner from accidental punctures, and a system of slotted pipes and pumps is installed to collect all the leachate from the bottom of the landfill. The collected leachate is then treated on or off the site. The principal soil-related concern is the requirement for suitable sources of sand and gravel, for clayey material to form the final cover, and for topsoil to support protective vegetation (see Figure 18.18b).

Environmental Impacts of Landfills

Today, regulations require that wastes be buried in carefully located and designed sanitary landfills. As a result, the number of landfill sites in the United States was reduced from about 16,000 in 1967 to less than 2500 in 1998. A major concern with regard to landfills is the potential water pollution from the rainwater that percolates through the wastes, dissolving and carrying away all manner of organic and inorganic contaminants (see Table 18.14). In addition to the oxygen-demanding general dissolved organics, many of the contaminants in landfill leachate are highly toxic and would create a serious pollution problem if they reached the groundwater under the landfill.

In addition to efficiency of resource use, avoidance of particular landfill management problems is another reason that the organic components of refuse (mainly paper, yard trimmings and food waste) should be composted to produce a soil amendment rather than landfilled. First, as these materials decompose in a finished landfill, they lose volume and cause the landfill to settle and the landfill surface to subside. This physical instability severely limits the use that can be made of the land once a landfill is completed.

FIGURE 18.18 Modern landfill technology and soil processes. (*Top*) A bulldozer compacts and covers refuse in a natural attenuation landfill in deep, well-drained soils. Soil material for daily cover is excavated in the background. (*Center*) A black geomembrane liner covered with white pea gravel and a leachate collection pipe in a new cell being prepared in a containment-type landfill. The low hills in the background are completed cells blanketed with a vegetated final cover. (*Bottom*) Gas wells collecting methane gas from anaerobic decomposition in a completed landfill cell. The methane is used to fuel turbines that generate electricity for the waste-disposal operation and to sell to the local electric utility company. (Photos courtesy of R. Weil)

Second, decomposition of the organic refuse produces undesirable liquid and gaseous products. Within a few weeks, decomposition uses up the oxygen in the landfill, and the processes of anaerobic metabolism take over, changing the cellulose in paper wastes into butyric, propionic, and other volatile organic acids, as well as hydrogen and carbon dioxide. After a month or so methane-producing bacteria become dominant, and for several years (or even decades) a gaseous mixture of about one-third carbon dioxide and two-thirds methane (known as *landfill gas*) is generated in quantity.

The production of methane gas by the anaerobic decomposition of organic wastes in a landfill can present a very serious explosion hazard if this gas is not collected (and possibly burned as an energy source; see Figure 18.18c). Where the soil is rather permeable, the gas may diffuse into basements up to several hundred meters away from the landfill. A number of fatal explosions have occurred by this process. Anaerobic decomposition in landfills also emits other harmful gases, the effects of which are less well known.

18.12   SOILS AS ORGANIC WASTE DISPOSAL SITES

In the United States, nearly 250 million Mg of domestic wastes are generated each year. To this must be added the nearly 2 billion Mg of farm animal wastes, as well as millions of megagrams of organic wastes from food- and fiber-processing plants and industrial operations.

It is no longer environmentally acceptable to dispose of these wastes by dumping them into waterways or the oceans or by burning, thus releasing reaction products into the atmosphere. The soil offers an alternative disposal sink which is being more and more widely used. These organic wastes can improve soil physical and chemical properties and can provide nutrients for increased plant growth. Such positive effects will likely encourage continued land application of these wastes. At the same time, when wastes are applied in excess quantities, soil productivity may be depressed by salts, or soil and water pollution may occur.

18.13   RADIONUCLIDES IN SOIL

Nuclear fission in connection with atomic weapons testing and nuclear power generation provides another source of soil contamination. To the naturally occurring radionuclides in soil (e.g., ^{40}K, ^{87}Rb, and ^{14}C), a number of fission products have been added. However, only two of these are sufficiently long-lived to be of significance in soils: strontium 90 (half-life = 28 yr) and cesium 137 (half-life = 30 yr). The average levels of these nuclides in soil in the United States are about 388 millicuries (mC)/km² for ^{90}Sr and 620 mC/km² for ^{137}Cs. A comparable figure for the naturally occurring ^{40}K is 51,800 mC/km². These normal soil levels of the fission radionuclides are not high enough to be hazardous. Even during the peak periods of weapons testing, soils did not contribute significantly to the level of these nuclides in plants. Atmospheric fallout on the vegetation was the primary source of radionuclides in the food chain. Consequently, only in the event of a catastrophic supply of fission products could toxic soil levels of ^{90}Sr and ^{137}Cs be expected. Fortunately, considerable research has been accomplished on the behavior of these two nuclides in the soil–plant system.

Strontium 90

Strontium 90 behaves in soil in much the same manner as does calcium, to which it is closely related chemically. It enters soil from the atmosphere in soluble forms and is quickly adsorbed by the colloidal fraction, both organic and inorganic. It undergoes cation exchange and is available to plants much as is calcium. Contamination of forages and, ultimately, of milk by this radionuclide is of concern, as the ^{90}Sr could potentially be assimilated into the bones of the human body. The possibility that strontium is involved in the same plant reactions as calcium probably accounts for the fact that high soil calcium tends to decrease the uptake of ^{90}Sr.

Research is underway to take advantage of plant uptake of radionuclides in phytoremediation exercises. Plants, such as sunflowers, are being used to remove ^{137}Cs and ^{90}Sr from a pond area near the Chernobyl, Ukraine nuclear disaster. Indian mustard is also being used in nearby sites to remove such nucleotide contaminants.

Cesium 137

Although chemically related to potassium, cesium tends to be less readily available in many soils. Apparently, ^{137}Cs is firmly fixed by vermiculite and related interstratified minerals. The fixed nuclide is nonexchangeable, much as is fixed potassium in some interlayers of clay (see Section 14.15). Plant uptake of ^{137}Cs from such soils is very limited. Where vermiculite and related clays are absent, as in some tropical soils, ^{137}Cs uptake is more rapid. In any case, the soil tends to dampen the movement of ^{137}Cs into the food chain of animals, including humans.

Radioactive Wastes[6]

In addition to radionuclides added to soils as a result of weapons testing and accidents (such as that which occurred at Chernobyl, Ukraine), soils may interact with low-level radioactive waste materials that have been buried for disposal. Even though the materials may be in solid form when placed in shallow land burial pits, some dissolution and subsequent movement in the soil are possible. Plutonium, uranium, americium, neptunium, curium, and cesium are among the elements whose nuclides occur in radioactive wastes.

Nuclides in wastes vary greatly in water-solubility, uranium compounds being quite soluble, compounds of plutonium and americium being relatively insoluble, and cesium compounds intermediate in solubility. Cesium, a positively charged ion, is adsorbed by soil colloids. Uranium is thought to occur as a UO_2^{2+} ion that is also adsorbed by soil. The charge on plutonium and americium appears to vary, depending on the nature of the complexes these elements form in the soil.

There is considerable variability in the actual uptake by plants of these nuclides from soils, depending on such properties as pH and organic matter content. The uptake from soils by plants is generally lowest for plutonium, highest for neptunium, and intermediate for americium and curium. Fruits and seeds are generally much lower in these nuclides than are leaves, suggesting that grains may be less contaminated by nuclides than forage crops and leafy vegetables.

Since soils are being used as burial sites for low-level radioactive wastes, care should be taken that soils are chosen whose properties discourage leaching or significant plant uptake of the chemicals. Data in Table 18.15 illustrate differences in the ability of different soils to hold breakdown products of two radionuclides. It is evident that monitoring of nuclear waste sites will likely be needed to assure minimum transfer of the nuclides to other parts of the environment.

---

[6] This summary is based largely on papers on this subject in *Soil Science*, **132** (July 1981).

TABLE 18.15   **Concentrations of Several Breakdown Products of Uranium 238 and Thorium 232 (Nucleotides) in Six Different Soil Suborders in Louisiana**

Note marked differences among levels in the different soils.

Soil suborder	No. of samples	^{238}U breakdown products			^{232}Th breakdown products		
		^{226}Ra	^{214}Pb	^{214}Bi	^{212}Pb	^{137}Cs	^{40}K
Udults	22	37.3	27.7	28.9	27.4	16.7	136
Aquults	24	30.4	36.7	38.1	50.0	10.9	100
Aqualfs	37	51.1	38.3	36.6	59.7	13.5	263
Aquepts	93	92.2	47.6	45.2	63.8	16.1	636
Aquolls	57	90.4	45.8	44.7	59.5	8.7	608
Hemists	18	136.3	49.4	49.0	74.9	19.4	783

From Meriwether, et al. (1988).

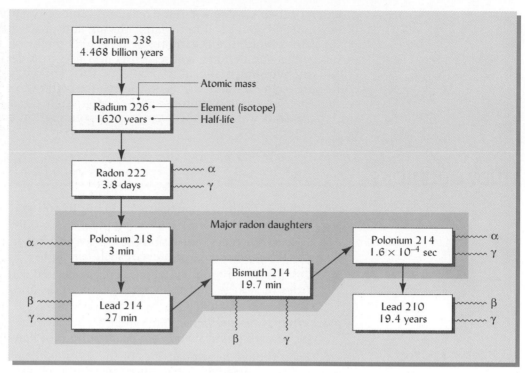

FIGURE 18.19 Radioactive decay of uranium 238 in soils that results in the formation of inert but radioactive radon. This gas emits alpha (α) particles and gamma (γ) rays and forms *radon daughters* that are capable of emitting alpha (α) and beta (β) particles and gamma (γ) rays. The alpha particles damage lung tissue and cause cancer. Radon gas may account for about 10,000 deaths annually in the United States. [Modified from Boyle (1988)]

18.14 RADON GAS FROM SOILS

The soil is the primary source of the colorless and odorless radioactive gas **radon,** which has been shown to cause lung cancer. Radon is formed from the radioactive decay of radium, a breakdown product of uranium found in minute quantities in most soils (Figure 18.19). Hazardous levels of radon occur in certain soils formed from uranium-rich igneous rocks and marine sediments. The health hazard from this gas stems from the transformation of its radioactive decay products into alpha rays, which can penetrate the lung tissue and cause cancer. Radon enters homes and other buildings from the surrounding soil. Since modern airtight buildings permit little exchange of air with the outside, radon can accumulate to harmful levels.

Radon usually moves into buildings through cracks in the basement walls and floors, and around openings where utility pipes enter the basement. If radon tests show that undesirable levels are present, the simplest remedial action is to seal all cracks and points of entry. Additional steps include constructing more elaborate systems to better ventilate the basement with outside air, in order to prevent a buildup of unhealthful levels of radon gas.

Since radon is an inert gas, it does not react with the soil, which merely serves as a channel through which the gas moves. Coarse, gravelly soils are more likely to transfer radon rapidly to basements than would finer-textured soils.

18.15 CONCLUSION

Three major conclusions may be drawn about soils in relation to environmental quality. First, since soils are valuable resources, they should be protected from environmental contamination, especially that which does permanent damage. Second, because of their vastness and remarkable capacities to absorb, bind, and break down added materials, soils offer promising mechanisms for the disposal and utilization of many wastes that otherwise may contaminate the environment. Third, soil contaminants and the products of their breakdown in soil reactions can be toxic to humans and other animals if they move from the soil into plants, the air, and—particularly—into water supplies.

To gain a better understanding of how soils might be used and yet protected in waste-management efforts, soil scientists devote a considerable share of their research efforts to environmental-quality problems. Furthermore, soil scientists have much to contribute to the research teams that search for better ways to clean up environmental contamination. Finding appropriate sites where soils can be safely used to clean up or store hazardous wastes involves geographic information about soils, the topic of the next chapter.

STUDY QUESTIONS

1. What agricultural practices contribute to soil and water pollution, and what steps must be taken to reduce or eliminate such pollution?

2. Discuss the types of reactions pesticides undergo in soils, and indicate what we can do to encourage or prevent such reactions.

3. Discuss the environmental problems associated with the disposal of large quantities of sewage sludge on agricultural lands, and indicate how the problems could be alleviated.

4. What is *bioremediation,* and what are its advantages and disadvantages compared with physical and chemical methods of handling organic wastes?

5. Even though large quantities of the so-called heavy metals are applied to soils each year, relatively small quantities find their way into human food. Why?

6. Compare the design, operation, and management of today's landfills with those in use 30 years ago, and indicate how the changes affect soil and water pollution.

7. What are *organoclays,* and how can they be used to help remediate soils polluted with nonpolar organic compounds?

8. Soil organic matter and some silicate clays chemically sorb some organic pollutants and protect them from microbial attack and leaching from the soil. What are the implications (positive and negative) of such protection for efforts to reduce soil and water pollution?

9. What radionuclides are of greatest concern in soil and water pollution, and why are they not more readily taken up by plants?

10. What are the comparative advantages and disadvantages of *in situ* and *ex situ* means of remediating soils polluted with organic compounds?

11. What is *phytoremediation,* and for what kinds of pollutants is it useful? Explain.

REFERENCES

Alexander, M. 1994. *Biodegradation and Bioremediation.* (San Diego: Academic Press).

Bagchi, A. 1994. *Design, Construction and Monitoring of Landfills,* 2d ed. (New York: Wiley).

Beyer, W. N., R. L. Chaney, and B. M. Mulhern. 1982. "Heavy metal concentration in earthworms from soil amended with sewage sludge," *J. Environ. Qual.,* **11**:381–385.

Boyle, M. 1988. "Radon testing of soils," *Environ. Sci. Tech.,* **22**:1397–1399.

Bragg, J. R., R. C. Prince, E. J. Harner, and R. M. Atlas. 1994. "Effectiveness of bioremediation for the Exxon Valdez oil spill," *Nature,* **368**:413–418.

Chaney, R. L. 1990. "Public health and sludge utilization," Part II, *Biocycle,* **31**(10):68–73.

Chang, A. C., A. L. Page, J. E. Warneke, and E. Grgurevic. 1984. "Sequential extraction of soil heavy metals following a sludge application," *J. Environ. Qual.,* **13**:33–38.

Cheng, H. H. (ed.). 1990. *Pesticides in the Soil Environment: Processes, Impacts, and Modeling.* (Madison, Wis.: Soil Sci. Soc. Amer.).

Cunningham, S. D., T. A. Anderson, A. P. Schwab, and F. C. Hsu. 1996. "Phytoremediation of soils contaminated with organic pollutants," *Advances in Agronomy,* **56**:55–114.

Cureton, P. M., P. H. Groenevelt, and R. A. McBride. 1991. "Landfill leachate recirculation: Effects on vegetation vigor and clay surface cover infiltration," *J. Environ. Qual.*, 20:17–24.

DMI. 1981. *Farming in the Profit Zone Through Plant Nutrition and Conservation Tillage.* (Goodfield, Ill.: DMC Inc.).

Dowdy, R. H., C. E. Clapp, D. R. Linden, W. E. Larson, T. R. Halbach, and R. C. Polta. 1994. "Twenty years of trace metal partitioning on the Rosemount sewage sludge watershed," in C. E. Clapp, W. E. Larson, and R. H. Dowdy (eds.), *Sewage Sludge: Land Utilization and the Environment.* (Madison, Wis.: Soil Sci. Soc. Amer.), pp. 149–155.

Edwards, C. A. 1978. "Pesticides and the micro-fauna of soil and water," in I. R. Hill and S. J. Wright (eds.), *Pesticide Microbiology.* (London: Academic Press), pp. 603–622.

Elliott, H. A., M. R. Liberati, and C. P. Huang. 1986. "Competitive adsorption of heavy metals in soils," *J. Environ. Qual.*, 15:214–219.

Furr, A. K., A. W. Lawerence, S. S. C. Tong, M. C. Grandolfo, R. A. Hofstader, C. A. Bache, W. H. Gutemann, and D. J. Lisk. 1976. "Multielement and chlorinated hydrocarbon analysis of municipal sewage sludges of American cities," *Environ. Sci. Tech.*, 10:683–687.

Gaynor, J. D., D. C. MacTavish, and W. I. Findlay. 1995. "Atrazine and metolachlor loss in surface and subsurface runoff from three tillage treatments in corn," *J. Environ. Qual.*, 24:246–256.

General Accounting Office. 1991. *Pesticides—EPA Could Do More to Minimize Groundwater Contamination.* GAO/RCED-91-75. (Washington, D.C.: U.S. General Accounting Office).

Hazen, Terry C. 1995. "Savannah river site—a test bed for cleanup technologies," *Environ. Protection,* April, pp. 10–16.

Holmgren, G. G. S., M. W. Meyer, R. L. Chaney, and R. B. Daniels. 1993. "Cadmium, lead, zinc, copper, and nickel in agricultural soils of the United States of America," *J. Environ. Qual.*, 22:335–348.

Jaynes, W. F., and S. A. Boyd. 1991. "Clay mineral type and organic compound sorption by hexadecyltrimethylammonium-exchanged clays," *Soil. Sci. Soc. Amer. J.*, 55:43–48.

Kabata-Pendias, A., and H. Pendias. 1992. *Trace Elements in Soils and Plants.* (Boca Raton, Fla.: CRC Press).

Kellogg, R. L., M. S. Maizel, and D. W. Goss. 1994. "The potential for leaching of agrichemicals used in crop production: A national perspective," *J. Soil Water Cons.*, 49:294–298.

Kreuger, R. F., and J. N. Seiber (eds.). 1984. *Treatment and Disposal of Pesticide Wastes.* (Washington, D.C.: Amer. Chem. Soc.).

Losi, M. E., C. Amrheim, and W. T. Frankenberger, Jr. 1994. "Bioremediation of chromate-contaminated groundwater by reduction and precipitation in surface soils," *J. Environ. Qual.*, 23:1141–1150.

Maymo-Gatell, X., Y. T. Chein, J. M. Gessett, and S. H. Zinder. 1997. "Isolation of a bacterium that reductively dechlorinates tetrachloroethene to ethene," *Science,* 276:1568–1571.

McConnell, J. S., and L. R. Hossner. 1985. "pH-dependent adsorption isotherm of glyphosate," *J. Agric. Food Chem.*, 33:1075–1078.

Meriwether, J. R., J. N. Beck, D. F. Keeley, M. P. Langley, R. N. Thompson, and J. C. Young, 1988. "Radionuclides in Louisiana soils," *J. Environ. Qual.*, 17:562–568.

Moore, J. W., and S. Ramamoorthy, 1984. *Heavy Metals in Natural Waters.* (New York: Springer-Verlag).

Pierzynski, G. M., J. T. Sims, and G. F. Vance. 1994. *Soils and Environmental Quality.* (Boca Raton, Fla.: CRC Press/Lewis Publishers).

Pimental, D., H. Acquay, M. Biltonen, P. Rice, M. Silva, J. Nelson, V. Lipner, S. Giordano, A. Horowitz, and M. D'Amore. 1992. "Environmental and economic costs of pesticide use," *Bioscience,* 42:750–760.

Pritchard, P. H., J. G. Mueller, J. C. Rogers, F. V. Kremer, and J. A. Glaser. 1992. "Oil spill bioremediation: Experiences, lessons and results from the Exxon Valdez oil spill in Alaska," *Biodegradation,* 3:315–335.

Richards, R. P., et al. 1996. "Well water quality, well vulnerability and agricultural contamination in the midwestern United States," *J. Environ. Qual.*, 25:389–402.

Sawhney, B. L., and K. Brown (eds.). 1989. "Reactions and movement of organic chemicals in soils," (Madison, Wis.: Soil Sci. Soc. Amer.).

Shipitalo, M. J., W. M. Edwards, and L. B. Owens. 1997. "Herbicide losses in runoff from conservation-tilled watersheds in a corn-soybean rotation," *Soil Sci. Soc. Amer. J.*, **61**:267–272.

Skipper, H. D. and R. F. Turco. 1995. *Bioremediation: Science and Applications,* Special Publication no. 43. (Madison, Wis.: Soil Sci. Soc. Amer.).

Sommers, L. E. 1977. "Chemical composition of sewage sludges and analysis of their potential use as fertilizer," *J. Environ. Qual.*, **6**:225–232.

Strek, H. J., and J. B. Weber. 1982. "Adsorption and reduction in bioactivity of polychlorinated biphenyl (Aroclor 1254) to redroot pigweed by soil organic matter and montmorillonite clay," *Soil Sci. Soc. Amer. J.*, **46**:318–322.

Thompson, A. R., and C. A. Edwards. 1974. "Effects of pesticides on nontarget invertebrates in freshwater and soil," in W. D. Guenzi (ed.), *Pesticides in Soil and Water.* (Madison, Wis.: Soil Sci. Soc. Amer.), pp. 341–386.

U.S. EPA. 1993. *Clean Water Act,* sec. 503, vol. 58, no. 32. (Washington, D.C.: U.S. Environmental Protection Agency).

U.S. EPA. 1997. *Characterization of Municipal Solid Waste in the United States: 1996 Update.* EPA 530-R-97-015. (Washington, D.C.: U.S. Environmental Protection Agency).

Weber, J. B., and C. T. Miller. 1989. "Organic chemical movement over and through soil," in B. L. Sawhney and K. Prown (eds.), *Reactions and Movement of Organic Chemicals in Soils.* SSSA Special Publication no. 22. (Madison, Wis.: Soil Sci. Soc. Amer.).

Xu, S., G. Sheng, and S. A. Boyd. 1997. "Use of organoclays in pollution abatement." *Advances in Agronomy,* **59**:25–62.

The soil/landscape portrait thus evolved is an artwork of the soil scientist . . .
—L. P. WILDING

The one great constant concerning soils is their variability. Anyone who works intimately with soils soon realizes that soils are anything but uniform. In our discussion of soil profile development (Chapter 2) we focused on vertical variability in soils—the differences among soil horizons. The subject of this chapter is horizontal variability—how soils differ from place to place across the landscape.

In order to make practical use of soil science principles, the land resource manager must know not only the "what" and "why" of soils, but must also know the "where." If builders of an airport runway are to avoid the hazards of swelling clay soils, they must know *where* these troublesome soils are located. An irrigation expert probably knows what soil properties will be necessary for cost-efficient irrigation; but for the project to be a success, he or she must also know *where* soils with these properties can be found. Almost any project involving soils, from planning a game park to fertilizing a farm field, can take advantage of geographic information about soils and soil properties. This chapter provides an introduction to some of the tools that tell us what is where.

19.1 SOIL SPATIAL VARIABILITY IN THE FIELD

A quick review of Figure 7.9 (changing aeration from the surface of a soil granule to its center) and Figure 12.20 (changing soil organic matter from Texas to Minnesota) reminds us that soil properties are variable at all scales. In this chapter we will consider soil variations occurring across distances having geographic meaning for land management—from a few meters to many kilometers.

When attempting to understand geographic variation of soils and how best to use each part of the land, it is often useful to analyze each site in terms of the five factors responsible for soil formation: *climate, parent material, organisms, topography,* and *time* (see Sections 2.6 to 2.16).

Climate usually influences soil variability at very large scales (regional differences), but where the landscape includes large water bodies or significant hills and mountains, rainfall and temperature may differ greatly over distances of 1 km or less. For example, the microclimate on north-facing slopes may differ in many ways from that on south-facing slopes. Likewise, parent materials often vary in large-scale regional patterns (for example, loess plateaus versus residual rock), but very small scale differences may also occur. The soil scientist should always be alert for the possible presence of such localized parent materials as colluvial deposits at the foot of a slope or alluvium along stream courses. The field soil scientist also needs some training in biology so as to be able to recognize changes in botanical composition that might signal the occurrence of water-saturated conditions in depressions, calcareous outcroppings, or other localized variations in soil properties. As these examples suggest, most small-scale soil variations involve changes in topography, so an awareness of even subtle changes in slope is critical to understanding how soil properties change across a landscape.

Small-Scale Soil Variability

Soil properties are likely to change markedly across small distances: within a few hectares of farmland, within a suburban house lot, and even within a single soil individual (as defined in Section 3.1). At this scale, variations are most often due to small changes in topography and thickness of parent material layers or to the effects of organisms (e.g., the effects of individual trees or past human management). Plate 29 (before page 499) shows dramatic small scale variability in both surface soil color and plant vigor resulting from the exposure of a calcarious horizon on the hillocks by wind erosion.

This small-scale variability may be difficult to measure and not readily apparent to the casual observer. In some cases, the height and vigor of vegetation reflects the subsurface variability. In other cases, the changes in soil properties are detected by analyzing soil samples taken from many evenly spaced borings made throughout the plot of land in question (Figure 19.1). This and other techniques will be discussed in Section 19.2 with regard to the practical use of such observations in making soil maps.

Variability in soil fertility often reflects past soil-management practices (Figure 19.2), as well as differences in soil profile characteristics. Analysis of small-scale variability has practical uses in managing soil fertility for a given field or nursery. As described in Section 16.18, soil tests for fertility management are traditionally performed on one composite sample (a mixture of small cores from 15 to 20 randomly scattered spots) that represents an entire field, or an area as large as 10 to 20 ha.

Where small-scale variability exists, the actual level of fertility at most spots in the field is likely to be either considerably higher or lower than the average soil test value for the field. To visualize the effect of the unaccounted-for variability, consider an analogy with shoe sizes. Fertilizer recommendations based on a field average may fit the needs of the soil about as well as buying everyone in a family size 6 shoes because it happens to be the average shoe size for that family. Fortunately, several technological developments (see Section 19.3) have made it much easier to detect and manage small-scale soil variability.

Medium-Scale Soil Variability

For many soil properties, variability across a landscape is related primarily to differences in a particular soil-forming factor, such as soil topography (drainage) or parent material. If one understands the influences of these soil-forming factors in a landscape, it is often possible to define sets of individual soils that tend to occur together in sequence across a landscape. Identifying one member of the set often makes it possible to predict soil properties in the landscape positions occupied by other members of the set. Such sets of soils include *lithosequences* (occurring across a sequence of parent materials), *chronosequences* (occurring across similar parent materials of varying age), and *toposequences* (with soils arranged according to changes in relief).

Well-drained, moderately well drained, and somewhat poorly drained soils are often found together in a characteristic toposequence that defines the patterns of different soils found across a landscape. Where all the soils in a toposequence have developed from the same parent materials, the set of soils that differ on the basis of drainage or due to differences in relief is known as a *catena.* The concept of a soil catena is helpful in relating the soils to the landscape in a given region. The various soils in a catena can often be distinguished by the colors of the surface soil, and even more clearly by the colors of the B horizons. For example, in the tropical catena shown in Plate 16 (after page 82), the colors vary from dark gray at the bottom of the slope to dusky red at the top.

The relationship can also be seen by referring to Figure 19.3, where the Bath–Mardin–Volusia–Alden catena is shown. Although all four or five members of a catena are not always found together in a given area, the diagram illustrates the spatial relationship

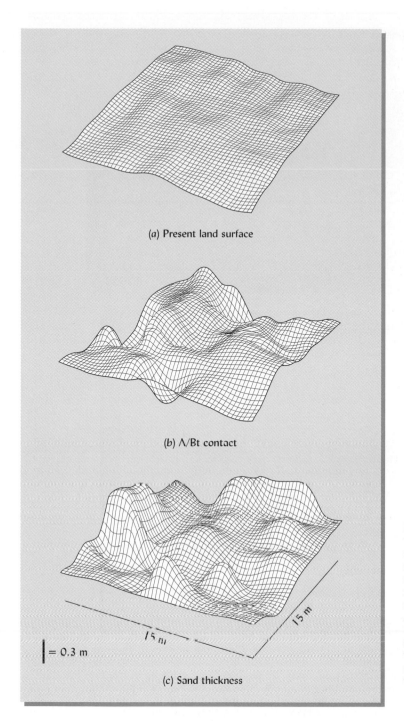

(a) Present land surface

(b) A/Bt contact

| = 0.3 m

15 m

15 m

(c) Sand thickness

FIGURE 19.1 Three-dimensional computer drawings of the variability in a small plot of land (15 × 15 m) in southern Texas. Here, an important feature in the soil profile is a layer of very sandy material, probably a relic of an ancient and now buried land surface. (*a*) The relatively flat, uniform land surface as it would be seen by a person standing on this soil. (*b*) The wavy boundary between the A and Bt horizons, indicating that the A horizon is two to three times as thick in some spots as in others. (*c*) The variation of the sand layer below, from virtually nonexistent to about 1 m thick. The graphs were generated from measurements taken at 1-m intervals in a grid pattern across the small plot of land. [From Wilding (1985)]

among the soils with respect to their drainage status. As shown, the drainage status of each catena member gives rise to distinct profile characteristics that affect plant rooting depth and species adaptation.

A *soil association* is a more general grouping of individual soils that occur together in a landscape. Soil associations are named after the two or three dominant soils in the group, but may contain several additional, less extensive soils. The soils may be from the same soil order or they may be from different orders (Figure 19.4*a* and *b*). They may have formed in the same or in different parent materials. The only requirement is that the soils occur together in the same area. Identifying soil associations is of practical importance, since they enable us to characterize landscapes over large areas and they assist in planning general patterns of land use. A given soil association represents a defined range of soil properties and landscape relationships, even though the range of conditions included may be quite large (see Figure 19.4*b*).

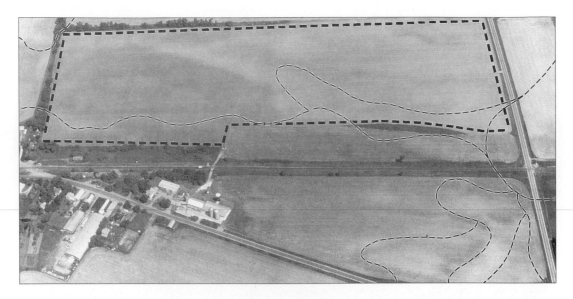

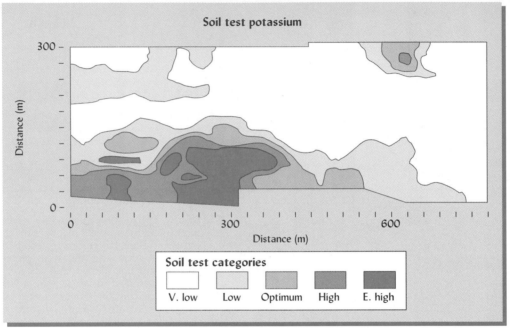

FIGURE 19.2 An oblique air photo (top) of a farm in central Wisconsin. The 22-ha field outlined by the heavy dashed line was studied in detail to make a map (bottom) of the spatial variability of available (soil test) potassium. The map was computer-generated using soil test values from 199 samples (each made up of five subsamples) taken at 32-m intervals in a grid pattern across the entire field. Note that the very high potassium levels correspond with a section of the field closest to the farmstead (lower left in photo). In the past, the farmer had managed that section as a separate field and had used it to dispose of manure from the nearby barnyard. The high-potassium spot in the upper right of the field marks the location of a manure pile that existed a number of years prior to the study. The spatial variation in potassium levels appears to be unrelated to the soil boundaries (light dashed lines) mapped in the county soil survey report. [From Wollenhaupt, et al. (1994)]

Large-Scale Soil Variability

At a very large scale, soil patterns are principally the result of climatic and vegetation patterns and secondarily related to parent material differences. Although it is often useful to refer to general regional soil characteristics, it must be remembered that much localized variation exists within each regional grouping. A study of the World Soils Map printed on the frontpapers of this textbook and the U.S. Soils Map in Appendix A will reveal important regional patterns. These patterns are also highlighted in the series of small soil order maps shown in Chapter 3. From these maps it can be seen that highly

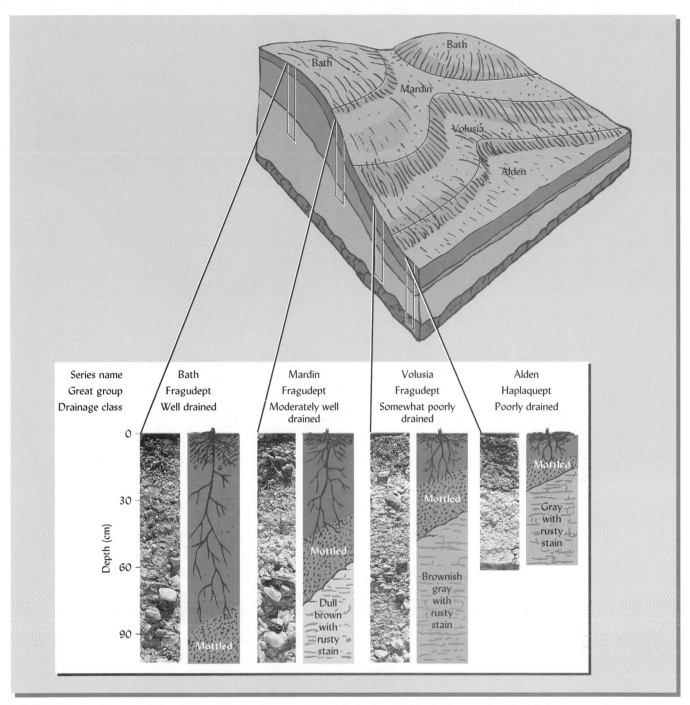

Series name	Bath	Mardin	Volusia	Alden
Great group	Fragudept	Fragudept	Fragudept	Haplaquept
Drainage class	Well drained	Moderately well drained	Somewhat poorly drained	Poorly drained

FIGURE 19.3 Profile monoliths of four soils of a drainage catena (below) and a block diagram showing their topographic association in a landscape (above). Note the decrease in the depth of the well-aerated zone (above the mottled layers) from the Bath (well-drained, upslope) to the Alden (poorly drained, downslope). The Alden soil remains wet throughout the growing season. These soils are all developed from the same parent material and differ only in drainage and topography. All four soils belong to the Inceptisols order. With the exception of the cultivated Volusia, the monoliths shown were taken from forested sites. [Based on Cline and Marshall (1977)]

weathered Oxisols are located principally in the hot, humid regions of South America and Africa drained by the Amazon and Congo rivers, respectively. Mollisolls can be seen to characterize the semiarid grasslands of the world; Aridisols are located in the desert regions. The great expanse of loess deposits in the central United States (see Figure 2.18) is an example of the regional influence of parent materials. Soils information on this scale can make an important contribution to inventorying the natural resources of a state, province, or nation.

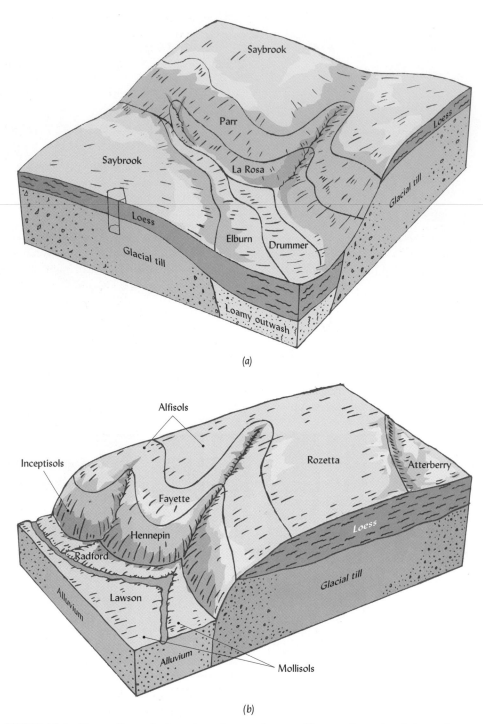

FIGURE 19.4 Two soil associations from Bureau County, Illinois. The soils of the Saybrook–Parr–La Rosa association (*a*) are all Mollisols. They differ principally with regard to topography and parent material (Parr and La Rosa developed in glacial till, the others mainly in loess). See Table 19.1 for a description of the pedon indicated in the Saybrook soil. The Rozetta–Fayette–Hennepin association (*b*) includes a wider range of soil conditions and includes soils from three soil orders. [Based on Zwicker (1992)]

19.2 TECHNIQUES AND TOOLS FOR MAPPING SOILS

Geographic information about soils is often best communicated to land managers by means of a soil map. Soil maps are in great demand as tools for practical land planning and management. Many soil scientists therefore specialize in mapping soils. Before beginning the actual mapping process, a soil scientist must learn as much as possible about the soils, landforms, and vegetation in the survey area. Therefore, the first step in mapping soils is to collect and study older or smaller-scale soil maps, geological and topographic maps, previous soil descriptions, and any other information available on the area. Once the soil survey begins, the soil scientist's task is threefold: (1) to define each soil unit to be mapped (see Section 19.3); (2) to compile information about the nature of each soil; and (3) to delineate the boundaries where each soil unit occurs in the landscape. We will now discuss some of the tools that soil scientists use to delineate soils in the field.

Soil Description[1]

Soil scientists may use computers and satellites, but they also use spades and augers. Despite all the technological advances of recent years, the heart of soil mapping is still the soil pit. A soil pit, whether dug by hand or with a backhoe, is basically a rectangular hole large enough and deep enough to allow one or more soil scientists to enter and study a typical pedon (see Section 3.1) as exposed on the pit face. Plates 1 to 12 are photographs taken of such pit faces. After cleaning away loose debris from the pit face, the soil scientist will examine the colors, texture, consistency, structure, plant rooting patterns, and other soil features to determine which horizons are present and at what depths their boundaries occur. Often the horizon boundaries are marked with a trowel or soil knife, as can be seen on the right side of Plate 9.

A soil description is then written in a standard format (see Table 19.1 for an example) that facilitates communication with other soil scientists and comparison with other soils. Sometimes the soil scientist will use field kits to make chemical tests, such as those to indicate free carbonate minerals (which effervesce with carbon dioxide if wetted with 10% hydrochloric acid) or soil pH (see Figure 9.20).

As far as possible at this stage, the soil horizons will be given master (A, E, B, etc.) and subordinate (2Bt, Ap, etc.) designations (see Table 2.4). Finally, samples of soil material will be obtained from each horizon. These will be used for detailed laboratory analyses and for archiving. The laboratory analyses will provide information for the chemical, physical, and mineralogical *characterization* of each soil.

---

[1] For details on procedures for making soil descriptions, see Soil Survey Staff (1993a) and FAO (1977).

TABLE 19.1 Typical Soil Profile Description (with *Soil Taxonomy* Diagnostic Horizons)

Soil classification			Typic Argiudolls, Saybrook Series
Location of pedon described			79 m west and 375 m south of the northeast corner of section 9, Township 17N, Range 7E[a]

Horizon designation	Diagnostic horizon	Horizon boundaries	Description
Ap	Mollic epipedon	0–20 cm	Very dark brown (10 YR 2/2) silt loam, dark grayish brown (10 YR 4/2) dry; moderate medium granular structure; friable; common fine roots; medium acid; abrupt smooth boundary
A		20–36 cm	Very dark brown (10 YR 2/2) silt loam, dark grayish brown (10 YR 4/2) dry; weak medium subangular blocky structure parting to moderate medium granular; friable; common fine roots; slightly acid; clear smooth boundary
Bt1	Argillic horizon	36–56 cm	Dark yellowish brown (10 YR 4/4) silty clay loam; moderate medium subangular blocky structure; friable; few fine roots; many distinct very dark brown (10 YR 2/2) clay films on faces of peds; slightly acid; clear smooth boundary
Bt2		56–76 cm	Dark yellowish brown (10 YR 4/4) silty clay loam; moderate medium subangular blocky structure; friable; few fine roots; many distinct dark brown (10 YR 3/3) clay films on faces of peds; slightly acid; clear smooth boundary
2Bt3		76–91 cm	Dark brown (7.5 YR 4/4) clay loam; weak medium subangular blocky structure; friable; few fine roots; few distinct dark brown (7.5 YR 4/2) clay films on faces of peds; few pebbles; clear smooth boundary
2C		91–152 cm	Brown (7.5 YR 5/4) loam; massive; friable; few fine rounded dark accumulations of iron and manganese oxide; few pebbles; violent effervescence; moderately alkaline

[a] The township-range system is used to describe the location of parcels of land in most of the United States, except in the eastern states.
Adapted from Zwicker (1992).

In this manner, the soil scientists assigned to map an area will familiarize themselves with the soils they expect to find, learning certain unique characteristics that they can look for to quickly identify each soil and distinguish it from other soils in the area.

Delineating Soil Boundaries

For obvious reasons, a soil scientist cannot dig pits all over the landscape to determine which soils are present and where their boundaries are located. Instead, he or she will bring up soil material from numerous small boreholes made with a hand auger (Figure 19.5a).

(a)

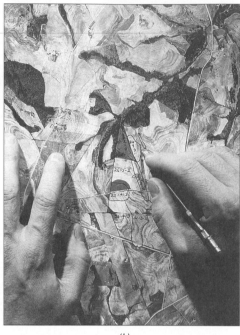

(b)

FIGURE 19.5 Soil maps are prepared by soil scientists who examine the soils in the field using a soil auger and other diagnostic tools (a). Most soil maps are initially prepared by outlining the boundaries of soil mapping units on an air photo (b). A permanent map is then made by superimposing soil boundaries, roads, and other features on an aerial photo map base (c). Each soil mapping unit outlined is identified by a three-part code. For example, in the two small areas (arrows) labeled "145B2," the "145" is the code for the soil series and surface texture phase, in this case a Saybrook silt loam. (On many soil survey maps a two-letter code is used in place of this three-digit one.) The capital letter "B" indicates slopes of 2 to 5%, and the final "2" indicates that the soil is moderately eroded. Thus the areas so marked contain the 2 to 5% slope, slightly eroded, silt loam surface soil phase of the Saybrook soil series. [Photo (a) courtesy of Diane Shields, USDA/NRSC]

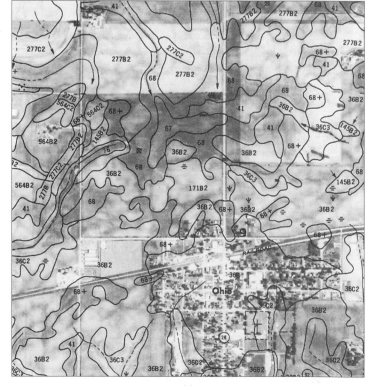

(c)

The texture, color, and other properties of the soil material from various depths can be compared mentally to characteristics of the known soils in the region.

With hundreds of different soils in many regions, this might seem to be a hopeless task. However, the job is not as daunting as one might suppose, for the soil scientist is not blindly or randomly boring holes. Rather, he or she is working from an understanding of the soil associations and how the five soil-forming factors determine which soils are likely to be found in which landscape positions. Usually there are only a few soils likely to occupy a particular location, so only a few characteristics must be checked. The soil auger is used primarily to confirm that the type of soil predicted to occur in a particular landscape position is the type actually there.

The nature of soil units and the locations of the boundary lines surrounding them are inferred from information obtained by auger borings at numerous locations across a landscape. A simple, but very laborious and time-consuming, approach to obtaining soils information is to make auger borings at regular intervals (say, every 50 m) in a grid pattern across the landscape (Figure 19.6a). Points with similar properties can then be connected to form soil boundaries. This approach is sometimes used in developing countries where labor to survey the sampling points and auger the soils is very inexpensive.

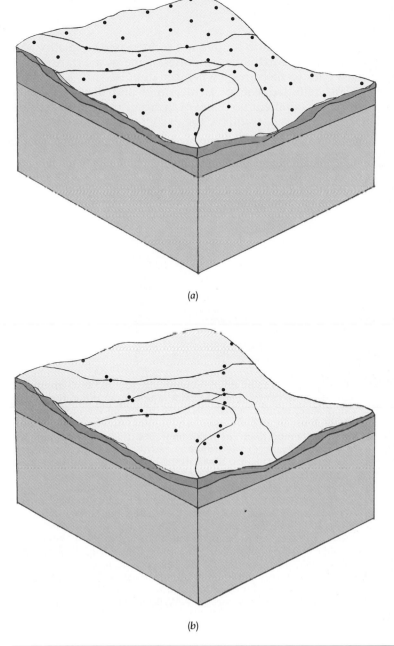

(a)

(b)

FIGURE 19.6 Two approaches to collecting information on soil properties and boundaries by soil auger borings (indicated by dots). The regularly spaced grid pattern (a) of borings is simple in concept, but very labor-intensive to carry out. In a much more efficient approach (b), the soil scientist traverses the landscape along selected transects (straight-line paths), augering only at enough points to confirm soil properties and boundaries predicted on the basis of soil–landscape relationships. Note that in order to pinpoint soil boundaries, extra borings are made near the places these boundaries are expected to occur.

Knowledge of the interplay of the soil-forming factors in a landscape can greatly expedite the soil scientist's work. Making auger borings is time-consuming (and therefore expensive), so an efficient soil mapper will use clues from changes in topography, vegetation, and soil surface colors as a guide in locating sites to make borings. A typical approach is to traverse the landscape along selected transects (straight-line paths), augering only at enough points to confirm expected soil properties and boundaries. In order to pinpoint soil boundaries, more frequent borings are made near the places where breaks in slope or other landscape clues suggest that soil boundaries will occur (see Figure 19.6b).

19.3 MODERN TECHNOLOGY FOR SOIL INVESTIGATIONS

While it is still the mainstay of soil investigations and mapping, soil augering is intrusive (i.e., it makes holes) and labor-intensive. Several nonintrusive methods of soil investigation are finding increasing use in helping to identify subsurface features and locate soil boundaries. These technologies include (1) ground-penetrating radar, (2) electromagnetic induction, (3) geopositioning systems, and (4) remote sensing of surface features.

Ground-Penetrating Radar

Ground-penetrating radar (GPR) has come into limited use to increase the quality and reduce the cost of making soil maps. High-frequency impulses of energy are transmitted into the soil. Energy is reflected back to the surface when the impulse strikes an interface between soil particles. This reflected energy is measured and displayed on a recorder. An illustration of the field equipment used and of the resulting graphic profile is shown in Figure 19.7. This technique is not suitable for all soils, since the reflectance

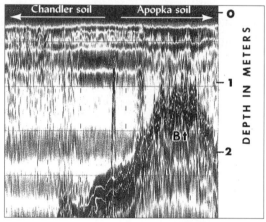

FIGURE 19.7 Use of ground-penetrating radar (GPR) to investigate the depth to contrasting subsurface layers. (Left) When the Mississippi River washed over fertile farmland during the great flood of 1993, it deposited coarse sand in a layer that varied from a few centimeters to more than 1 m in thickness. The GPR instrument is pulled over the soil surface to determine where the sand layer is thin enough to plow under. The instrument in the background is an electromagnetic induction (EM) meter. (Right) An example of output from a GPR instrument used in another survey, showing the depth to the Bt horizon in two Florida soils. The GPR signals are sensitive to the dramatic change in soil properties between the coarse sands of the upper horizons and the clay of the Bt horizon. On the left side of the graph are the deep sands of the Chandler soil, a quartzipsamment without an argillic horizon in the upper 2 m. On the right side of the graph, the Bt horizon of an Apopka soil (a Paleudult) can be clearly seen. The boundary between the two soils is gradual, but could be delineated at about the middle of the graph. The distance across the entire graph is about 140 m. [Photo courtesy of R. Weil; graph from Doolittle (1987)]

is influenced by such factors as moisture, salt content, and type of clay. Where it is effective, however, the cost is only about one-third the cost of detecting soil boundaries by multiple borings with a soil auger.

Electromagnetic Induction[2]

Electromagnetic induction (EM) techniques offer another noninvasive, rapid method of investigating subsurface features. Using a handheld instrument approximately the size and shape of a carpenter's level, this method measures the apparent conductivity of the soil for electromagnetic energy. The conductivity measured is influenced by the moisture content, salinity of the soil (see Section 5.4), and the amount and type of clays in the soil. This technique has been successfully used to map out the depth and thickness of claypan horizons in humid regions and to investigate groundwater contamination and salinity in arid regions.

It should be noted that these electronic methods of soil investigation are not yet available as off-the-shelf technology, but must be adapted to each situation by the user. The user must initially quantify the relationship between the soil properties of interest and the electronic signals recorded by the instrument. A suitable computer program may be needed to analyze the data. Once so adapted, these technologies can provide detailed information about subsurface features and can be a great aid in determining the location of soil boundaries on detailed soil maps.

Global Positioning Systems[3]

An obvious prerequisite for delineating the location of soil bodies in the field is that the soil mappers know where they themselves are located as they traverse a landscape. Traditionally, soil mappers have used large-scale base maps or air photos (see Section 19.5) to ascertain location. However, in nearly featureless terrain or heavily vegetated areas these aids are of limited use. Fortunately, soil mappers can now take advantage of satellite technology to identify precise locations anywhere in the world.

When the United States Department of Defense recently made its Global Positioning System (GPS) available for civilian use, it opened opportunities for many applications in mapping and other soil investigations in the field (Box 19.1). The GPS consists of a network of earth-orbiting satellites that constantly transmit signals that can be used to determine longitude (east–west coordinate) and latitude (distance north or south from the equator) on the ground.

To use signals from the satellite network, one carries a civilian GPS receiver, an instrument that may be as small as a TV remote control. At least two, but preferably three or four, of the GPS satellites must be in contact with the receiver for it to be able to calculate its position (see Figure 19.8). The coordinates given by the GPS receiver are then transferred to the base map for each soil observation made.

19.4 REMOTE SENSING TOOLS FOR SOILS INVESTIGATIONS[4]

Remote sensing describes the gathering of information from a distance. In this general sense of the term, we are remote sensing any time we use our eyes to see an object from a distance rather than picking up the object in our hands and feeling it. When we perceive an object, our brains are forming a mental image in response to light energy that has reflected off the object into our eyes. In an analogous manner, a photographic or digital image can be formed by sensors (e.g., cameras) mounted on a platform (e.g., an airplane or space satellite), providing a suitable vantage point for observing a particular area of land. While our eyes respond only to reflected energy with wavelengths in the visible range, other sensors can form images from additional wavelengths of energy, such as infrared energy. We will briefly describe several types of imagery and their uses in making geographic soils investigations. Air photos and other imagery covering most locations in the United States and the world are available from a number of government agencies and private companies (Table 19.2).

---

[2] For an example of the use of this technique, see Doolittle, et al. (1994).

[3] For a readable discussion of the principles behind the GPS, see Herring (1996).

[4] For a general discussion of remote sensing, see Lillesand and Kiefer (1994).

A network of 24 Global Positioning System (GPS) satellites was developed by the United States Defense Department to aid in the navigation of military planes and ships during the 1980s. The system can now be used by civilians almost anywhere in the world to determine their precise location almost instantaneously. The GPS satellites are orbiting some 20,000 km above the earth in precisely known patterns arranged so that at least four satellites are broadcasting simultaneously to any point on Earth's surface. The group of satellites in communication with a GPS receiver is called the *satellite constellation.*

An electronic clock in the receiver measures how long it takes for a pattern of radio signals to travel to it from each satellite. Given that the signals travel at the velocity of light, the receiver can calculate the distance to each satellite (distance = velocity × time). A receiver located at, say, 20,200 km from a particular satellite must be located someplace on the surface of an imaginary sphere that has its center at the satellite transmitter and has a diameter of 20,200 km. If the same receiver is also located at 20,700 km from a second satellite, it must also be located someplace along the surface of a second sphere 20,700 km in diameter. Simple geometry tells us that four such spheres can intersect at only one point in space (Figure 19.8a), that point being the exact location of the receiver.

Because the setting of the satellite clock may not be known exactly (partly due to distortions designed by the U.S. Defense Department to allow it to retain an advantage over any potential adversaries), accuracy is not always optimal. One corrective measure involves installing a stationary receiver at an exactly known location on a tall building or farm silo. Mobile receivers then use broadcasts from the stationary receiver to calculate the "errors" in each satellite clock. Other *differential correction* signals are available from government and commercial sources broadcasting from FM radio towers. Without these corrections, simple, hand-held receivers can determine locations to within 10 or 20 m. With these corrective measures in use, even relatively small and inexpensive GPS receivers can determine locations to within 1 to 5 m, and larger, more sophisticated receivers can determine locations to within a few centimeters.

This technology can be applied in making soil maps, allowing the soil scientist in the field to record the precise geographic coordinates of each soil observation made. The soil information mapped may be the name of the taxonomic map unit (as for soil surveys) or it may be a measured soil property, such as the organic matter content or nutrient availability. In the latter case, special maps can be prepared by *geostatistical* computer programs that estimate values for the soil property in question at all points between the points actually sampled.

In what is sometimes called *precision farming* (see Section 16.19), a GPS receiver on fertilizer-spreading equipment is used in making on-the-go adjustments to fertilizer application rates so that the amount and kind of fertilizer applied will be appropriate for the soil nutrient status and potential crop demand peculiar to each small portion of a large field (Figure 19.9). At harvest time, a GPS receiver and special yield-monitoring equipment can be used to produce a detailed map of actual crop yields (see Figure 19.9a). Such yield maps can be very useful in identifying areas where such soil problems as poor drainage, low organic matter, or nutrient deficiencies may exist (of course, the yield variations may be related to nonsoil factors, such as type of seed planted, diseases, insects, weeds, or deer grazing).

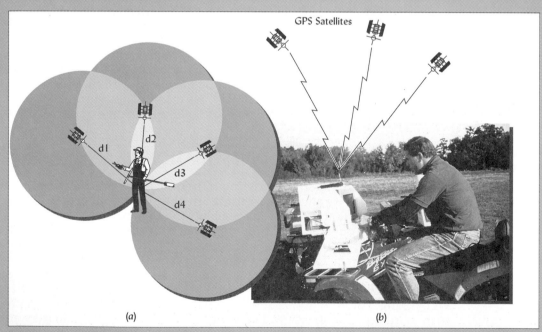

(a)   *(b)*

FIGURE 19.8 *The Global Positioning System of satellites and receivers. (a) A GPS receiver uses radio signals to determine its distance (d1, d2, etc.) from each of four orbiting satellites. The spheres of distance around the four individual satellites intersect at only one position in space, the location of the receiver. (b) A technician collecting soil samples using an all-terrain vehicle equipped with a GPS receiver and a computer. As the vehicle traverses the terrain, the computer displays its position on a map, guiding the technician to a grid of predetermined soil sampling locations. (Photo courtesy of Applications Mapping, Inc., Boswell, Ga.)*

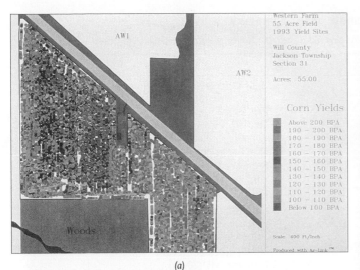

(a)

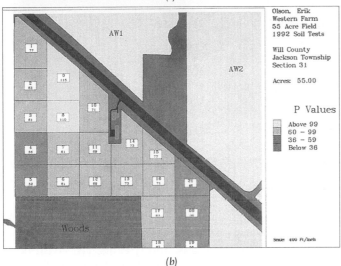

(b)

(c)

FIGURE 19.9 Use of GPS technology in *precision* or *site-specific* farming. (*a*) This map of corn yields throughout a 22-ha field was generated by a harvesting machine equipped with a yield monitor and a GPS receiver. Each dot on the map was generated as the monitor recorded the yield of corn every few meters while the machine worked up and down the field. Each yield measurement was associated with a map position determined by the GPS. (*b*) A map of the same field showing the available soil phosphorus as determined from samples collected in 1-ha cells located by GPS. (*c*) A GPS-linked computer system installed in a tractor spreading fertilizer in a different field. The computer screen shows a map indicating the current location of the tractor in the field and where different rates of fertilizer should be applied. (Photos courtesy of Applications Mapping, Inc., Boswell, Ga.)

19.5   AIR PHOTOS

Most air photos are made with panchromatic black-and-white film, which "sees" a range of light wavelengths that closely corresponds to the spectrum visible to our eyes. The photograph produced is black and white or, more correctly, many shades of gray. Black-and-white air photos can reveal a wealth of information about landforms, vegetation, human influences, and, yes, soils. But experience is required to recognize the various gray tones and patterns as different types of vegetation, drainage patterns, and soil bodies.

Other films, such as natural color or infrared film, are also used for aerial photography. Black-and-white prints from infrared film are commonly used by forest managers because conifer needles absorb infrared energy much more completely than do hardwood leaves, allowing the conifers to be easily distinguished by their darker gray tones on the photograph.

Air photos have been used since 1935 to increase the speed and accuracy of making soil maps. They can be used to assist soil investigations in at least three ways: (1) in providing a base map, (2) as a source of proxy information, and (3) in directly sensing soil properties.

TABLE 19.2 Partial Listing of Sources for Remote Sensing Imagery
Many other reputable sources exist and may provide equally good products and services.[a]

Type of imagery	Source	Address
Air photos	Bureau of Land Management	BLM-Denver Service Center Division of Technical Services, Bldg. 46 P.O. Box 25047 Denver, CO 80225
	U.S. Geologic Survey	EROS Data Center Sioux Falls, SD 57198 E-mail: custserv@edcserver1.cr.usgs.gov
	WAC Corporation	520 Conger Street Eugene, OR 97402-2795
	ASCS/USDA	Aerial Photography Field Office P.O. Box 3010 Salt Lake City, UT 84130
Landsat Imagery	EOSAT, Inc.	4300 Forbes Boulevard Lanham, MD 20706
	Earth Satellite Corporation	6011 Executive Blvd., Suite 400 Rockville, MD 20852
SPOT Imagery	SPOT Image Corporation	1897 Preston White Drive Reston, VA 22091 Phone: (703) 620-2200
Radar Imagery	Radarsat International	Building D, Suite 200 3851 Shell Road Richmond, BC V6X 2W2, Canada
Digital terrain elevation data	National Imagery and Mapping Agency, Department of Defense	Website: http://www.nima.mil

[a] Neither the authors nor the publishers promote the listed examples of sources in preference to other imagery providers.

Air Photographs as Base Maps

A sufficiently detailed air photo allows the soil scientist to determine his or her location in the field in relation to such features as buildings, roads, and streams, visible on both the photograph and on the ground. Rather than using surveying equipment and a plane table to make a map from blank paper, the soil scientist can draw soil boundaries directly on the air photo. In this way the photograph serves as a *base map*.

It should be noted that uncorrected air photos are severely distorted, because the land areas shown near the edges of the photograph were considerably farther from the camera than were the areas directly beneath the passing airplane. Also, in hilly or mountainous terrain the hilltops would have been closer to the camera than the valleys. An *ortho photograph* is one that has been corrected for both types of distortion. In most modern published soil surveys, ortho photographs are used as base maps on which soil boundaries are drawn.

Air Photos for Proxy Information

Soil investigations are usually concerned with features of soil profiles and other subsurface information. However, an air photo records radiant energy reflected off surfaces aboveground or, at best, a few centimeters belowground. Nonetheless, the surface tones, patterns, and features shown on the photograph often are related to conditions belowground. Once the soil scientist has learned what these relationships are, air photos can be used as a source of *proxy* information about soil conditions.

For example, the photographs do not directly record the depth to the water table; but dark tones indicate moist, high-organic-matter surface soil that may correlate with a seasonally shallow water table. If the soil scientist knows that a certain type of soil occurs in the drainageways, then drainageways visible by patterns of dark gray tones on the photograph serve as a proxy for digging a soil pit or boring an auger hole. Stereo pairs of photographs (Figure 19.10) give the viewer a three-dimensional (but vertically exaggerated) view of the land surface, and are especially useful in locating drainage-

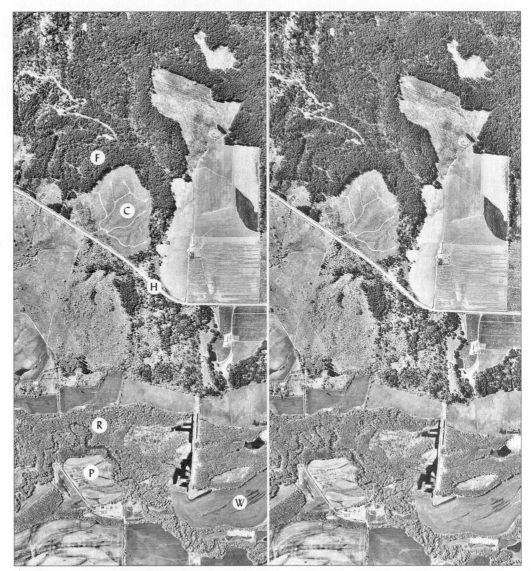

FIGURE 19.10 A stereo pair of air photos providing overlapping coverage of a scene in the Willamette valley of western Oregon. The soils are mostly Humults and Xerolls. The left- and right-hand photos show the same ground area but were taken when the airplane was in two different positions. If you place a pocket stereoscope over these photos, the two views of the same scene will trick your eyes into seeing the photos three-dimensionally. Note the coniferous forest (F) in the steeper terrain of the upper third of the scene above the highway (H), including a roundish clear-cut area [medium-gray zone, (C)] with a network of logging roads (white lines with "nodes" indicating leveled areas where logs were loaded onto trucks). In the lower third of the scene, the land is a nearly level river valley, and waterlogged spots (W) appear in the agricultural fields. A meandering river (R) can be seen in the lower quarter of the scene, and the patterns of alluvial soils can be discerned on the river flood plain (P). Soil differences are visible as different gray tones in fields where plowing has exposed bare soils. The scale of these photos is 1:100,000, so 1 cm on the photos represents 1 km. (Photos courtesy of WAC Corporation, Eugene, Ore.)

ways, slope breaks, and other features of relief. The features, in turn, help to locate the soils that are members of known soil associations and drainage catenas.

Vegetation often provides clues about the underlying soils. For example, a certain type of vegetation, recognizable on air photos, may grow only in areas of sodic soils with a natric horizon. The patches of this vegetation seen on the photograph then indicate the locations of the sodic soils.

Drainage patterns visible on air photos usually reflect the nature of soils and parent materials. For instance, many closely spaced (less than 1 cm apart on a 1:20,000 air photo) gullies and streams indicate the presence of relatively impervious bedrock and clayey, low-permeability soils. As another example, silty soils developed in loess usually produce a pinnate drainage pattern in which many small gullies and streams branch off from fairly straight larger streams at angles only slightly less than 90°.

Such interpretations can greatly speed the soil investigation by eliminating the need to investigate every mapping unit on the ground. However, the relationships between patterns on the photos and soil properties must be established afresh by ground investigation for each soil association or landscape type. The relationships so determined are likely to hold true only for landscapes in a limited area, typically 500 to 1500 km².

Direct Sensing of Soil Properties

Certain properties of the upper 2 to 20 mm of soil alter the manner in which the soil surface reflects various wavelengths of radiant energy (see Figure 19.11). Thus, depending on the wavelengths being monitored, a number of soil properties can be directly sensed and recorded. Soil properties that have been successfully determined by remote sensing include mineral content, texture, soil water content, soil organic matter content, iron oxide content, albedo (overall reflectivity), temperature, and soil structure. Direct sensing of soil properties usually requires a remote sensor tuned in to specific bands of wavelengths and a computer program designed to interpret the complex data.

An example of computer-assisted interpretation is video image analysis (VIA), by which air photos are examined to distinguish up to 256 shades of gray (compared to only 32 for the human eye). These gray shades or tones are related to soil and vegetation variations. Using appropriate computer techniques, delineation of soil differences can be accomplished. Again, such interpretations of remotely sensed data must be checked against *ground truth* by going to the site with an auger and verifying soil types and boundaries.

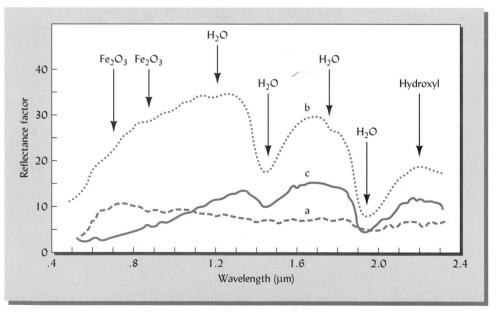

FIGURE 19.11 Spectral curves showing reflection of different wavelengths of energy by three soils. Prominent water and iron absorption bands (low reflection) are indicated by arrows. High iron content correlates with low reflectance in the iron oxide bands, while high organic matter results in low reflectance in the hydroxyl band. The soils represented are: (*a*) a Typic Haplaquoll with 5.6% organic matter, 0.8% iron oxide, and 41% water; (*b*) a Typic Calcicambid with 0.6% organic matter, 0.03% iron oxide, and 17% water; and (*c*) a Typic Haplorthox with 2.3% organic matter, 26% iron oxide, and 33% water. [Adapted from Stoner and Baumgardner (1981)]

Many of the principles and techniques just mentioned with regard to air photos apply to satellite imagery as well. However, the imagery available from the sophisticated satellites now in orbit may be more complex and versatile than the types of air photos just discussed. Most satellite images are produced from computer-analyzed digital data obtained by multispectral scanners rather than by film cameras. The images are usually computer enhanced by classification of each cell (pixel) of the image with regard to the type of vegetation, soil cover, land use, or similar theme. This classification is based on computer identification of each land cover spectral fingerprint or pattern of reflectance over different wavelengths (as shown in Figure 19.11). An image may combine data sensed by different instruments, sometimes on different dates. In this manner, combinations of spectra can be used to identify vegetation types, soil properties, cultural features, water, etc.

The resolution[5] of satellite imagery available today is not as high as that obtained with low-altitude air photos, but it has steadily improved over the years. Imagery with a resolution as high as 10 meters is now available from the French satellite Système Probatoire de la Observation de la Terre (SPOT). Figure 19.12 illustrates the improved resolution between the early and current Landsat satellite technologies. The older image (made in 1973) has a relatively low resolution of 80 m, while the newer image (from 1990) can resolve objects as small as 30 m. Note that the higher resolution results in a clearer, more detailed image. The difference in resolution is most obvious in distinguishing the drainage canals (arrow) in this arid region of Afghanistan.

This pair of Landsat images highlights the capability of multispectral imagery to portray soil conditions. The soil parent materials in the area are naturally high in soluble salts. Where irrigation without proper drainage raised the water table (evidenced by the dark tones of waterlogged soil in 1973), evaporation from groundwater caused salts to accumulate in the surface soils (as seen by the white, barren soils in 1990). The lack of proper maintenance of the drainage channels may have been related to the political upheaval and war experienced in Afghanistan between image dates.

The capability of satellite imagery to clearly portray landforms and vegetation over a wide area is illustrated by Figure 19.13. Examine this image closely. It is a black-and-white reproduction of a false-color Landsat Thematic Mapper image of the Palo Verde valley where the Colorado River forms the border between California and Arizona. The image is a composite made from three different spectral bands. Rugged mountainous terrain and alluvial fans are visible surrounding a nearly level, irrigated valley. The arid landscape is virtually bare of vegetation, except in rectangular fields where irrigation water has been applied. The small circles in the upper left are center-pivot irrigation systems, which result in lush green vegetation (red on the original image) in circular fields almost 1 km in diameter. Note the grid of streets that is the town of Palo Verde, near the center of the irrigation project.

The outlined rectangular portion of the Palo Verde valley image is enlarged as Plate 35 (before page 499). Much can be learned by closely examining this color plate. Notice the different blue-gray colors indicating soil differences visible in bare fields. Bright red areas are indicative of dense, green crops. The wavy lines of red to the east of the Colorado River indicate natural vegetation and weeds growing where gullies and drainageways have collected the little rainwater that falls in this arid region. Many bare fields are highlighted by yellow outlines, which were added to the image by a GIS program (see Section 19.9). The yellow-outlined fields are those that the City of Los Angeles paid the owners to keep bare, so that the city could obtain the water that would otherwise have been used for irrigating crops. The satellite image was used to verify that the owners of these fields were complying with this agreement.

A final example of satellite imagery is shown in Plate 34, which is a Landsat Thematic Mapper image of the Potomac River basin from Washington, D.C. south, including parts of southern Maryland and northern Virginia. Major land-use patterns are clearly visible. The urbanized land of Washington and its suburbs is shown in blue-gray tones (the United States Capitol building grounds are visible as a white spot surrounded by a dark square of lawn). An enormous amount of sediment (yellow and brown colors) appears to be entering the Potomac River from streams draining the southern suburbs,

---

[5] The resolution of an image is said to be high if very small objects can be discerned.

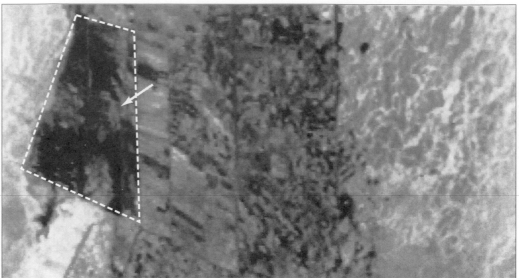

Landsat Multispectral Image
Bands 1-2-4

28 June 73

Landsat Thematic Mapper Image
Bands 2-3-4

27 April 90

FIGURE 19.12 A black-and-white reproduction of two false-color Landsat satellite images showing a portion of an irrigation project in western Marja, Afghanistan. The upper photo is a Landsat multispectral image obtained in 1973; the lower photo is a Landsat Thematic Mapper digital image obtained in 1990. Each image used three different spectral bands. The highly reflective white tones indicate barren, salinized land. Black areas indicate irrigated cropland that has become waterlogged. Medium-gray tones (red on the original image) indicate well-drained, irrigated cropland. In 1973 an area of approximately 1000 ha (outlined) appeared to be cultivated but waterlogged, probably as a result of blocked drainage outlets. By 1990 the same area appears to be abandoned and salinized. (Photos courtesy of Earth Satellite Corporation, Rockville, Md.)

possibly because of poor sediment control at construction sites. Forests appear green; most farm fields appear light yellow (mature corn and soybeans). The red-colored areas indicate Histosols and other hydric soils of the tidal marshes along the Potomac River and the Chesapeake Bay estuary. The bay itself is seen as the dark blue body of water at the far upper right. An image such as this is very useful in resource inventory and general regional land planning.

Clearly the range of imagery and technology available for soil investigations is rapidly growing, presenting challenges and opportunities for those soil scientists specializing in geographic information about soils.

FIGURE 19.13 A black-and-white reproduction of a Landsat Thematic Mapper image of the irrigated Palo Verde valley on the border between California and Arizona. The image is a composite made from bands 2 (green), 3 (red), and 4 (near infrared). The outlined rectangle is reproduced in Plate 35. (Photo courtesy of Earth Satellite Corporation, Rockville, Md.)

A **soil survey** is more than simply a soil map. The glossary describes a *soil survey* as "a systematic examination, description, classification, and mapping of the soils in a given area." Under some circumstances, soil scientists make special-purpose soil surveys which do not attempt to delineate and describe natural soil bodies, but merely aim to map the geographic distribution of selected soil properties, such as suitability for an irrigation project or conformation to a legal definition of wetlands.

However, soil surveys are most valuable if they characterize and delineate natural soil bodies. Once the natural bodies are delineated and their properties are described, the soil survey can aid in making interpretations for all kinds of soil uses, not just those uses that were intended at the time the soil survey was conducted.

The basic steps in making a soil survey are:

1. *Mapping of the soils.* Described in Sections 19.1 to 19.3.

2. *Characterization of the mapping units.* See Section 19.2.

3. *Classification of the mapping units.* See Chapter 3 and Section 19.2.

4. *Correlation with other soil surveys.* Once a team of soil scientists completes a soil map, the map is reviewed by other soil scientists to verify that the soil boundaries match those mapped for adjacent areas and that the characterization and classification of mapping units are consistent with other soil surveys.

5. *Interpretation of soil suitability for various land uses.* A report is written to accompany the soil map in order to describe the suitability of each mapping unit for various land uses. The interpretative tables in the report often reflect many person-years of experience and observation, as well as standard interpretations of measured soil properties.

Soil surveys may be conducted at different *orders* or levels of detail, ranging from very detailed surveys that attempt to delineate virtually every soil body in the landscape

TABLE 19.3 Different Orders of Soil Surveys

Soil surveys may be conducted at various scales or orders, ranging from very detailed surveys of small parcels of land to general surveys of very large regions. Different kinds of mapping units and remotely sensed data sources are used in producing different orders of soil surveys. The guidelines given in this table should be considered flexible and approximate.

	Order of Soil Survey				
	5th order	4th order	3d order	2d order	1st order
Type of survey	Reconnaissance	Reconnaissance	Semidetailed	Detailed	Intensive
Survey scale	1:250,000–1:10,000,000	1:50,000–1:300,000	1:20,000–1:65,000	1:12,000–1:32,000	1:1000–1:15,000
Size of mapping unit	2.5–500 km²	15–250 ha	1.5–15 ha	0.5–4 ha	Smaller than 0.5 ha
Typical components of map units	Orders, suborders, and great groups	Great groups, subgroups, and families	Families, series, and phases of series	Soil series; phases of series	Phases of soil series
Kind of map unit	Associations, some consociations, and undifferentiated groups	Associations, and some complexes, consociations	Associations or complexes; some consociations	Consociations and complexes; few associations	Mostly consociations; some complexes
Remote sensing sources	◄——————————Landsat Thematic Mapper digitized data——————————►				
	◄——————————SPOT image digital data——————————►				
		◄——————High-altitude aerial photography——————►			
			◄——————Low-altitude aerial photography——————►		
Use of soil survey in land planning	◄————Resource inventory————►				
		◄————Project location————►			
			◄————Feasibility surveys————►		
				◄————Management surveys————►	

Adapted from Soil Survey Staff (1993b).

(first order) to general reconnaissance surveys of very large regions or entire continents (fifth order). Different kinds of *mapping units* and remotely sensed data sources are used to produce different orders of soil surveys (Table 19.3).

Map Scales

The *scale* of a map is the ratio of length on the map to actual length on the ground. A scale of 1:20,000 is commonly used for detailed soil maps and indicates that 1 cm on the map represents a distance of 20,000 cm (or 0.2 km) on the ground. The ratio is unitless; therefore, 1 in. on the same map represents 20,000 in. (0.316 miles) on the ground.

The terms *large scale* and *small scale* sometimes confuse people. A *small-scale* map is one with a small *scale ratio* (e.g., 1:1,000,000 = 0.000001), on which a given object, such as a 100-ha lake, occupies only a tiny spot on the map. By contrast, a *large-scale* map has a large *scale ratio* (e.g., 1:10,000 = 0.0001), and a 100-ha lake would occupy a relatively large part of the map.

Mapping Units

Soil Taxonomy (or some other classification system; see Chapter 3) is usually the basis for preparing a soil survey. Because local features and requirements will dictate the nature of the soil maps and, in turn, the specific soil units that are mapped, the field *mapping units* may be somewhat different from the *classification units* found in *Soil Taxonomy*. The mapping units may represent some further differentiation below the soil series level—namely, *phases* of soil series; or the soil mappers may choose to group together similar or associated soils into conglomerate mapping units. Examples of such soil mapping units follow.

SOIL PHASE. Although technically not included as a class in *Soil Taxonomy*, a *phase* is a subdivision based on some important deviation that influences the use of the soil, such as surface texture, degree of erosion, slope, stoniness, or soluble salt content. Thus, a Cecil sandy loam, 3 to 5% slope, and a Hagerstown silt loam, stony phase, are examples of phases of soil series.

CONSOCIATIONS. The smallest practical mapping unit for most detailed soil surveys is an area that contains primarily one soil series and usually only one phase of that soil series. For example, a mapping unit may be labeled as the consociation "Saybrook silt loam, 2 to 5% slopes, moderately eroded." Quality-control standards for county soil surveys may indicate that a consociation mapping unit should be 50% "pure" and that the "impurities" should be so similar to the named phase that the differences do not affect land management. That is to say, if you walk out on the land to an area mapped as the just-mentioned consociation and make 20 auger borings, at least 10 of the borings should reveal properties within the range defined for the Saybrook silt loam. However, it would be expected that a few of the borings would indicate a similar soil, such as the Catlina silt loam, in which the loess layer is somewhat thicker than for the Saybrook soil but for which land-use interpretations are the same. Inclusions of *contrasting* soils should occupy less than 15% of the consociation.

COMPLEXES OR SOIL ASSOCIATIONS. Sometimes *contrasting* soils occur adjacent to each other in a pattern so intricate that the delineation of each kind of soil on a soil map becomes difficult, if not impossible. In such cases, a soil *complex* is indicated on a soil map, and an explanation of the soils present in the complex is contained in the soil survey report. A complex often contains two or three distinctly different soil series. As described in Section 19.1, relatively large-scale soil maps (e.g., third order) may display only soil associations—general groupings of soils that typically occur together in a landscape and could be mapped separately.

UNDIFFERENTIATED GROUPS. These units consist of soils that are *not* consistently found together, but are grouped and mapped together because their suitabilities and management are very similar for common land uses.

In the United States an effort has been under way for more than 50 years to complete a detailed soil survey of the entire country, on a county-by-county basis. In some states, many counties have yet to be mapped. Other counties were mapped decades ago and their soil surveys are being updated. This soil survey, known as the *National Cooperative Soil Survey,* is an ongoing joint effort of federal, state, and local governments. The principal federal role is played by the U.S. Department of Agriculture Natural Resources Conservation Service (formerly the Soil Conservation Service). Sometimes, other federal government agencies, such as the Bureau of Land Management or the Forest Service, are also involved. The state-level cooperating agency is usually the Agricultural Experiment Station associated with the land grant university of that state. We will now consider the typical U.S. county soil survey report and its uses.

A modern county soil survey report consists of two major parts: (1) the *soil map,* and (2) accompanying *descriptive information* on the soil mapping units and their suitability for various land uses. In effect, the descriptive part of the report serves as a very elaborate key to explain the soil map.

GENERAL SOIL MAP. The soil map usually consists of two parts. The first part is a fourth-order *General Soil Map* of the entire county, printed in color at a scale of approximately 1:200,000 (Figure 19.14*a*). This map shows the soil associations grouped into the main physiographic regions of the county. The General Soil Map is useful in providing an overview of the county land resources, but is too general to be used for site planning. Associated with the General Soil Map is an *index map* for the detailed map sheets. This index map is printed at the same scale as the General Soil Map and is divided into numerous rectangular sections, each representing a detailed map sheet. The index map shows enough nonsoil features, such as roads, towns and streams, to enable a user to locate the map sheet covering the area of interest.

DETAILED MAP SHEETS. The second part of the county soils map is a detailed (second-order) soil map consisting of many individual map sheets (Figure 19.14*b*), usually folded and bound into the back of the Soil Survey Report. Each map sheet covers an area of approximately 20 to 30 km², typically at a scale of 1:12,000 (1 cm = 0.12 km or 1 in. = 1000 ft), 1:15,840 (4 in. = 1 mile), 1:20,000 (1 cm = 0.2 km), or 1:24,000 (1 cm = 0.24 km or 1 in. = 2000 ft).[7] At these scales, individual trees and houses are discernable on the air photo map base, and areas of soil as small as 1 ha can be delineated (Figure 19.14*c*). The mapping units are mostly consociations consisting of soil phases. As such, these detailed maps are extremely useful for site planning.

INTERPRETIVE INFORMATION. Use of the county soil survey for land use or site planning requires that the geographic information on the map be integrated with the descriptive information in the report. The report usually contains detailed soil profile descriptions for all of the mapping units, as well as tables that provide soil characterization data and interpretive rankings for the mapping units. Older soil surveys offer interpretive information on yield potentials for various locally important crops, on the suitability of soils for different irrigation methods, drainage requirements for soils, land capability classification of each mapping unit, and other information useful in making farm plans. Newer soil surveys also provide interpretive information on many nonagricultural land uses, including wildlife habitat, forestry, landscaping, waste disposal, building construction, and sources of roadbed materials.

EXAMPLE OF USE. As an example of how the soil survey report may be used, consider the parcel of land shown in the photograph in Figure 19.14*d*. This area is also included in

---

[6] Since there are more than 3000 counties and similar jurisdictions, each with its own soil survey and map legend, some scientists are suggesting that in the future the National Cooperative Soil Survey should be coordinated on the basis of larger survey areas defined by natural, rather than political, boundaries. The Natural Resources Conservation Service has divided the United States into 212 Major Land Resource Areas (MLRAs) that could serve this purpose. Each MLRA is a geographically contiguous land area characterized by a common pattern of soils, land uses, water resources and climate. For a discussion of the history and possible future trends of soil survey in the United States, see Indorante, *et al.* (1996).

[7] To facilitate digitization using feet and inches, most soil survey updates in the United States are being produced at scales of 1:12,000 or 1:24,000.

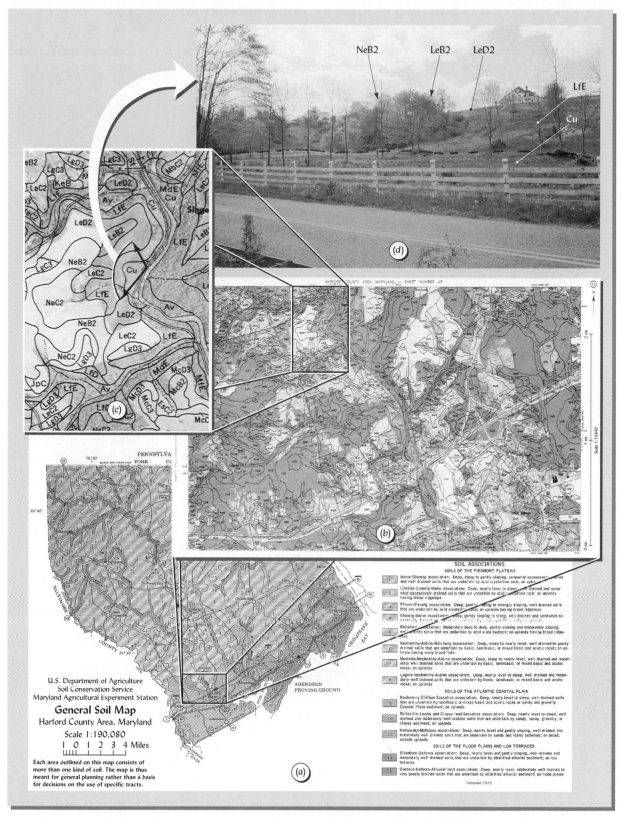

FIGURE 19.14 The soil map of Harford County, Maryland, an example of the soil map in a modern county soil survey. (*a*) One part of the soil map is the small-scale General Soil Map (originally at a scale of 1:190,080, in this case) reproduced here at much-reduced scale. (*b*) The detailed soil map consists of the collection of many large-scale map sheets (originally at a scale of 1:15,840, in this case), one of which (map sheet 45) is shown here at reduced scale. (*c*) A rectangular area of approximately 1.1 km² is outlined on the detailed map sheet and reproduced at the original scale, so that the mapping units and features on the air photo map base are visible. (*d*) The scene showing the road in the foreground and the house on the hill in the background was photographed from the location indicated where the two arrows originate in (*c*). [Maps from Smith and Mathews (1975); photo courtesy of R. Weil]

TABLE 19.4   A Sampling of Some of the Interpretive Information Provided by the Soil Survey of Harford County Area, Maryland

The four mapping units described are those in the site shown in Figure 19.14d. The interpretations included are some of those that would be of interest for developing the site as a county park.

Mapping unit code	Name of mapping unit	Land use Capability Classification[a]	Limitations for use of soil			Suitability for open land wildlife
			Camp areas	Septic filter fields	Paths and trails	
Cu	Codorus silt loam	Ilw-7	Severe—flood hazard	Severe—high water table, flood hazard	Moderate—flood hazard	Good
LeB2	Legore silt loam, 3 to 8% slopes, moderately eroded	Ile-10	Slight	Slight	Slight	Good
LeD2	Legore silt loam, 15 to 25% slopes, moderately eroded	IVe-10	Severe—slope	Severe—slope	Moderate—slope	Fair
LfE	Legore very stony silt loam, 25 to 45% slopes	VIIs-3	Severe—slope	Severe—slope	Severe—slope	Poor
NeB2	Neshaminy silt loam, 3 to 8% slopes, moderately eroded	Ile-4	Slight	Moderate—mod. permeability	Slight	Good

[a] See Section 17.13 for an explanation of the USDA Land Capability Classification System.
Data abstracted from Smith, et al. (1975).

the air photo stereo pair shown in Figure 19.15. Study Figure 19.15, preferably with a stereoscope, and compare the topography observed with that shown in the photograph in Figure 19.14d. Locating this scene on the detailed map sheet (Figure 19.14c), we can determine that the parcel contains five mapping units encoded as Cu, LeB2, LeD2, LfE, and NeB2 (see Figure 19.5 for an explanation of this coding system). These codes correspond to the five mapping units described in Table 19.4, which also lists a few of the many interpretive ratings given in the soil survey report. If you were planning to develop a park on this site, the information in Table 19.4 would suggest that the Codorus silt loam (Cu) mapping unit would be suitable for a nature trail, but not for a visitor center with rest rooms.

For complex land planning analyses, the enormous quantity of information stored in a county soil survey report can perhaps be used to best advantage with the help of a computerized geographic information system (GIS). We will now briefly consider how a GIS works.

FIGURE 19.15   A stereo pair of air photos including the area shown on the soil map in part (c) of Figure 19.14. Major features to notice are the dam and reservoir in the upper left, the small housing development in the middle left, the stream (Winters Run) flowing below the reservoir, and the hill and floodplain in the center of the photo. The latter features are the same as those shown from ground level in Figure 19.14d. (Air photos courtesy of USDA Agricultural Stabilization and Conservation Service)

An information system is a series of organized steps used to handle information or data, including the plan for the collection of data, actual collection of the data, manipulation of the data, and, finally, use of the data in making a product, such as a report or map. In a geographic information system (GIS), the data is tied to spatial locations. For example, the properties of a soil profile are associated with the location of the appropriate soil mapping units. A GIS includes the following five steps: (1) data acquisition, (2) preprocessing, (3) data management, (4) manipulation and analysis of data, and (5) product generation.

The soil mapping process described in Section 19.3 could be part of the data acquisition step, as could the collection of existing soil maps, reports, and zoning ordinances. This data will have to be preprocessed to be put into a form that can be integrated with other data. Data acquisition and preprocessing can be very time-consuming and expensive. An example of preprocessing is the digitization of an existing soilmap, entering all the soil boundaries and other spatial features into a computer database as *points, lines,* and geometrically defined areas called *polygons*. This process may be accomplished using a scanner, but often must be done by hand using a digitizing palette. The descriptive information, such as that illustrated in Table 19.4, also must be entered into the database. Currently, only a few county soil surveys are available in digital form, but as soil surveys are updated many are being digitized.

Once all the necessary information has been entered into the database, it can be managed with computer programs and manipulated to create new types of information and insights. Finally, a product such as a map or report is produced. A number of complex computer programs have been designed to carry out the last three steps in a coordinated manner. Figure 19.16 shows a simple example of a product of such a system. The map in this figure is based on information taken from the detailed soil map sheet shown in Figure 19.14*b*. Although digitizing and entering the descriptive and geographic information for this single map sheet required more than 100 hours, subsequent computer generation of the map of campground limitations took only a few minutes. The map in Figure 19.16 has simplified a set of more than 90 different mapping units by grouping them into three simple, spatially displayed categories. The same information is already available in the soil survey report tables, but not in a spatially displayed form.

The just-mentioned map (Figure 19.16) comprises a single *layer* of information. A GIS can be used to integrate many layers of soils information from a soil survey report. As a hypothetical example of the use of a GIS, consider the following scenario. A local agency wished to find a suitable site for a public park in one of 20 large tracts of land delineated in the area covered by detailed map sheet 45 (from Figure 19.14*b*). The planners decided that the site for the park must meet the following four criteria:

1. Only slight limitations for picnic areas
2. Only slight limitations for paths and trails
3. Land capability class greater than II (to avoid using prime agricultural land that should be preserved for food production)
4. Woodland subclass 20, indicating a site highly productive for hardwood trees

The boundaries of every mapping unit polygon on map sheet 45 were entered into the GIS database, and each polygon was associated with the appropriate interpretive and descriptive data from the soil survey report. The GIS computer program searched for mapping unit polygons that met all four of the criteria and printed a map with shading used only for the qualifying polygons. Finally, the boundaries of the 20 tracts of land, as well as the locations of streams, were plotted on the map. The resulting map is shown in Figure 19.17. The uppermost tract just right of center appears to contain the largest area of land meeting the criteria for the proposed park.

In many land planning projects, soil properties comprise just one of several types of geographic information that must be integrated in order to take best advantage of a given site or to find the site best suited for the proposed land use. Planning the best use of different farm fields, finding a site suitable for a sanitary landfill, and designing a wildlife refuge are some examples of projects that could make good use of both soils and nonsoils

---

[8] For a comprehensive introduction to GIS, see Star and Estes (1990).

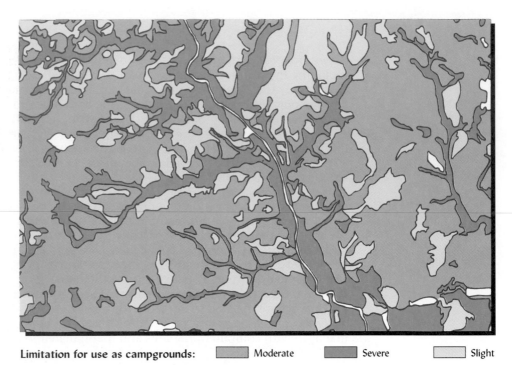

Limitation for use as campgrounds: ▢ Moderate ▢ Severe ▢ Slight

FIGURE 19.16 A simple interpretive soil map covering the same area as Figure 19.14*b*, map sheet 45 from Harford County, Maryland. The ratings of limitations from tables in the soil survey report have been associated with each of the thousands of soil polygons on the map sheet. The dark gray areas along the major streams are either too steep or too wet to be favorable for campgrounds and, therefore, are in the severe limitations category. This map is a single layer in that it contains only information on spatial distribution of campground limitations. It does not show roads, streams, or other data (nonsoil areas are left blank). Compare this map to that in Figure 19.14*b*. Note the principal river-tributary system running diagonally approximately through the center of both maps. (Courtesy of Margaret Mayers, University of Maryland)

geographic information. Nonsoils information to be considered might include topography, streams, vegetation, and present land use. A GIS can be used to combine all these types of information, giving greatest weight to those factors that the planner deems most critical.

An example of such an integrated GIS model is shown in Figure 19.18. Here, the objective was to develop a plan for fighting forest fires in a recreational area. In order to predict where fires are most likely to break out, where different types of equipment and firefighting techniques are appropriate, and where a wildfire would likely cause the greatest damage, a number of factors must be considered. The nature of the land determines most of these factors. A number of maps showing the spatial distribution of soils and nonsoils factors were overlaid in order to produce a fire management priorities map. The final map differentiated zones within the park according to their priority in a fire management plan.

As more soils information becomes available in digital form, its use in making land planning decisions based on GIS analyses will undoubtedly become increasingly common. The quality of those decisions will continue to depend on the quality of the geographic soils information provided by soil scientists and the quality of the decision criteria developed by planners and their clients.

19.10 CONCLUSION

Making soil surveys is both a science and an art by which many soil scientists apply their understanding of soils and landscapes to the real world. Mapping soils is not only a profession; many would say that it is a way of life. Working alone outdoors in all kinds of terrain and carrying all the necessary equipment, the soil scientist collects ground truth to be integrated with data from satellites and laboratories. The resulting soil maps and

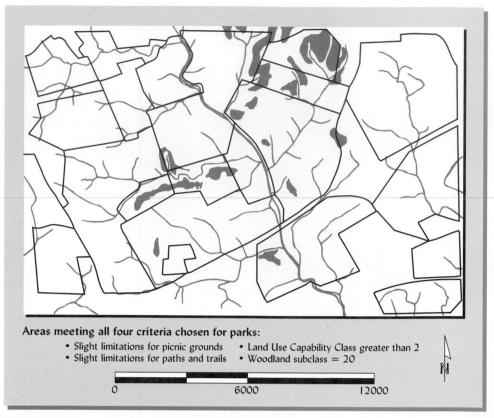

Areas meeting all four criteria chosen for parks:

- Slight limitations for picnic grounds
- Slight limitations for paths and trails
- Land Use Capability Class greater than 2
- Woodland subclass = 20

0   6000   12000

N

FIGURE 19.17 A map of a portion of Harford County, Maryland (map sheet 45), with shaded areas determined to be suitable sites for a new park according to the four criteria listed. The land use capability class restriction (LUCC > 2) avoids taking prime farmland out of production (see Section 17.13). The property of "park suitability" is a new attribute created by the geographic information system. In addition to "park suitability," two other layers of geographic information are displayed: (1) streams (light, branched lines), and (2) boundaries of 20 hypothetical tracts of land (heavy, straighter lines). Which tract of land should the county government purchase for a new park? (Courtesy of Margaret Mayers, University of Maryland)

descriptive information in the soil survey reports are used in countless practical ways by soil scientists and nonscientists alike. The soil survey, combined with powerful geographic information systems, enables planners to make rational decisions about what should go where. The challenge for soil scientists and concerned citizens is to develop the foresight and fortitude to use criteria in the GIS planning process that will help preserve our most valuable soils—not hasten their destruction under shopping malls and landfills.

STUDY QUESTIONS

1. What is the principal difference between a *soil association* on one hand, and a *catena* on the other? What is the difference between a *lithosequence* and a *toposequence?*

2. What is the main purpose of digging soil pits as part of making a soil survey?

3. Describe the kinds of information a soil mapper may use in deciding where to drill into the soil with an auger to bring up subsurface samples for study.

4. How can the GPS be used in a GIS?

5. What are the advantages of using aerial photos as a map base in making a soil survey?

6. For the region in which you live, describe some kinds of information that you would expect to be able to obtain from satellite imagery.

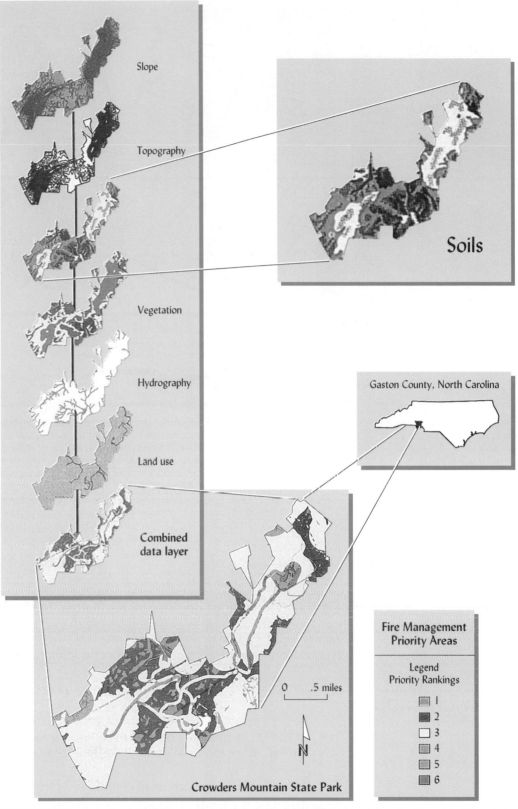

Slope

Topography

Soils

Vegetation

Hydrography

Gaston County, North Carolina

Land use

Combined
data layer

Crowders Mountain State Park

0   .5 miles

N

**Fire Management
Priority Areas**

Legend
Priority Rankings

1
2
3
4
5
6

FIGURE 19.18   In order to develop management plans for fighting forest fires in Crowders Mountain State Park in North Carolina, spatial information about soil properties was combined with other types of spatial information using a Geographic Information System computer program designed to overlay various types of maps. In this way soils information is integrated into the final plan as one of many components. [Modified from Gronlund, et al. (1994) with permission of GIS World, Inc.]

7. Assume that you have two wall maps, each approximately 1 m wide by 0.70 m tall. One is a map of Canada, and the other is of a ranch in California. Which map is the large scale map and which is the small scale map? If the ranch is roughly 20 km from east to west, what might be the approximate scale for its map?

8. A soil mapper drew a boundary around an area in which he made six randomly located auger borings, two in soil *A*, with an argillic horizon more than 60 cm thick and strong brown in color, and the other four in soil *B*, with a somewhat lighter brown argillic horizon between 50 and 70 cm thick. Other soil properties, as well as management considerations, were similar for the two types of soils. Would the map unit delineated likely be a *soil association,* a *soil consociation,* or a *soil complex?* Explain.

9. Assume you are planning to buy a 4-ha site on which to start a small orchard. Explain, step by step, how you could use the county soil survey to help determine if the prospective sight was suitable for your intended use.

10. If you were hired by a state government to produce a GIS-based map showing where investments should be made to protect farmland from suburban development, what "layers" of information would you want to include in the GIS?

REFERENCES

Cline, M. G., and R. L. Marshall. 1977. "Soils of New York Landscapes," Inf. Bull. 119, Physical Sciences, Agronomy 6, New York State College of Agriculture and Life Sciences, Cornell University.

Doolittle, J. A. 1987. "Using ground-penetrating radar to increase the quality and efficiency of soil surveys" in W. U. Reybold and G. W. Peterson (eds.), *Soil Survey Techniques,* SSSA Special Publication no. 20 (Madison, Wis.: Soil Sci. Soc. Amer.).

Doolittle, J. A., K. A. Sudduth, N. R. Kitchen, and S. J. Indorante. 1994. "Estimating depths to claypans using electromagnetic induction methods," *J. Soil Water Cons.,* **49:**572–575.

FAO. 1977. *Guidelines for Soil Profile Description,* 2d ed. (Rome: Land and Water Development Div. U.N. Food and Agriculture Organization.

Gronlund, A. G., W. N. Xiang, and J. Sox. 1994. "GIS, expert systems technologies improved forest fire management techniques," *GIS World,* **7**(2):32–36.

Herring, T. A. 1996. "The global positioning system," *Scientific American* **270**(2):44–50.

Indorante, S. J., R. L. McLeese, R. D. Hammer, B. W. Thompson, and D. L. Alexander. 1996. "Positioning soil survey for the 21st century,". *J. Soil Water Cons.,* **51:**21–28.

Lillesand, T. M., and R. W. Kiefer. 1994. *Remote Sensing and Image Interpretation,* 3d ed. (New York: Wiley).

Smith, H., and E. Matthews, 1975. *Soil Survey of Harford County Area, Maryland.* (Washington, D.C.: U.S. Soil Conservation Service).

Soil Survey Staff. 1993a. *National Soil Survey Handbook.* Title 430 VI. (Washington, D.C.: USDA).

Soil Survey Staff. 1993b. *Soil Survey Manual.* Handbook no. 18. (Washington, D.C.: USDA).

Star, J., and J. Estes. 1990. *Geographic Information Systems: An Introduction.* (Englewood Cliffs, N.J.: Prentice Hall).

Stoner, E. R., and M. F. Baumgardner. 1981. "Characteristic variations in reflectance of surface soils," *Soil Sci. Soc. Amer. J.,* **45:**1161–1165.

Wilding, L. P. 1985. "Spatial variability: Its documentation, accommodation and implication to soil surveys," in D. R. Nielsen and J. Bouma (eds.), *Soil Spatial Variability.* (Pudoc, Wageningen, The Netherlands), pp. 166–194.

Wollenhaupt, N. C., R. P. Wolkowski, and M. K. Clayton. 1994. "Mapping soil test phosphorus and potassium for variable rate fertilizer application," *J. Prod. Agric.,* **7:**441–448.

Zwicker, S. E. 1992. *Soil Survey of Bureau County, Illinois.* (Washington, D.C.: USDA Natural Resources Conservation Service).

20
GLOBAL SOIL QUALITY
AS AFFECTED BY
HUMAN ACTIVITIES

If we do not change our direction, we are likely to end up where we are headed
—CHINESE PROVERB

During the past few decades, we have begun to recognize the global significance of most everything we do. Economic development or stagnation in one country affects the economy of all its trading partners around the world. Worldwide communication networks provide us with ready computer access to information, knowledge, and news from every corner of the globe. As a consequence, actions of each country, each community, or even each individual can have global implications.

This global perspective is equally pertinent for soils. Soil particles picked up by the wind during spring tillage in the Great Plains states can be detected in the rainfall in the eastern United States or even in Europe. Likewise, excess salts, nitrates, or phosphates in the drainage water from soils in one nation can make the water unfit for use in another nation downstream. Certainly, changes in the productivity of soils in one area have serious implications for humans and other living creatures elsewhere in the world. Food security and food prices are affected, but biodiversity, water quality, and food and energy for all creatures are dependent to some degree on the productivity and quality of soils.

In previous chapters we concentrated on chemical, physical, and biological processes taking place in soils, and we emphasized rather site-specific problems and opportunities that land users face. We now turn to the global implications of the actions individual users take in managing soils at different locations around the world. We will recognize how human practices affect other soilborne organisms and, in turn, the quality of the soil itself. As a framework for our discussions, we will focus on the quality or health of soils, which, in turn, affects the well-being of humans and all other living creatures.

20.1 THE CONCEPT OF SOIL QUALITY/SOIL HEALTH[1]

From the beginning of time, humans have evaluated the soils on which they work, play, and live. Terms such as "good," "bad," "worn-out soils," "productive," or "unproductive" soils have always been used. In recent years, scientists and users of the soils have realized that many of the world's soils are degrading, and they want to better understand and reverse that degradation. They want to learn how to improve the quality not only of degraded soils, but of other soils as well. Also, they want to provide farmers and natural

---

[1] For reviews on soil health and soil quality, see Doran, et al. (1996) and Doran and Jones (1996).

resource planners with simple means of comparing the quality of soils from one location to another.

To make such comparisons meaningful, and to better understand how the full potential of a given soil can be realized, soil scientists are using the concept of **soil quality** or **soil health**.[2] The concept of soil quality and the criteria for its determination are still being developed. However, it is already proving to be a valuable tool in alerting scientists, farmers, environmentalists, and decision makers to the care and nurturing that soils need if their health or quality is to be maintained.

Soil quality relates to the capacity of a soil to function, not only within its boundaries, but in the larger environment of which it is a part. It indicates the soil's fitness to serve (1) as a medium to promote the growth of plants and animals (including humans), while regulating the flow of water in the environment; (2) as an environmental buffer that assimilates and degrades environmentally hazardous compounds; and (3) as a factor in enhancing the health of plants and animals, including humans.

These three broad issues lead to the following definition: *Soil quality* is the capacity of a soil to function within (and sometimes outside) its ecosystem boundaries to sustain biological *productivity* and diversity, maintain *environmental quality,* and promote plant and animal *health*. The relationship between the definition of soil quality or health, its functions, and criteria for its measurement is shown in Figure 20.1.

---

[2] These terms are used interchangeably in scientific literature and in the public press. *Soil health* has an appeal for the nonscientific users of soils, while *soil quality* is preferred by most scientists. We will generally use soil quality in this text, although soil health will also be referred to.

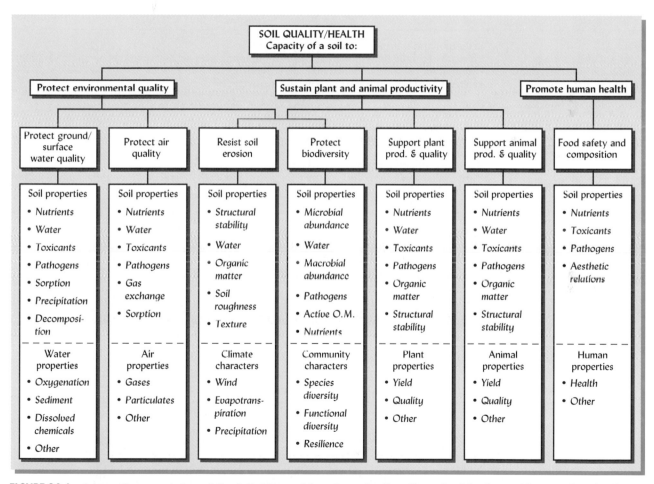

FIGURE 20.1 Schematic presentation of the definition and functions of soil quality or health, along with examples of indicator properties (soil and otherwise) that can be used to measure the quality or health of a soil. The definition of soil quality is in **bold print,** the expanded functions in normal print, and the categories of indicator properties supporting each function are in *italics.* For simplicity, many interdependencies among soil quality functions are not shown (e.g., protection of surface water quality is partially dependent on resistance to soil erosion, and so forth). As more knowledge is gained, additional and more specific indicator properties will likely be added to the list. [Modified from Harris, et al. (1996)]

The assessment of soil quality is determined by society's goals for a particular landscape or ecosystem, and the functions that soils are expected to perform in achieving those goals. To ascertain a soil's ability to perform, we use a series of physical, chemical, and biological properties, most of which we have considered in previous chapters (Table 20.1).

Research is under way to develop *soil-quality indices* to measure quantitatively a soil's ability to perform. This is not a simple task, however, since soil quality is affected by so many properties, and their relative contribution to the index varies so much from one soil function to another. For example, available soil nutrient levels would rate high as a factor in considering the soil's function to increase crop production, but would be relatively unimportant in determining the quality of the soil as a foundation for buildings. This suggests that soil-quality indices must be rather specific for the intended use of the soil.

The relative importance of different elements that contribute to a soil-quality index will vary from one time to another, and from one location to another at a given time. This is illustrated in Table 20.2, which shows that in 1900 the food- and fiber-production

TABLE 20.1 Possible Minimum Data Set of Physical, Chemical and Biological Indicators for Determining the Quality or Health of a Soil

Other supporting indicators can be used to help establish the validity of the measurements. It may be possible to combine the values for each indicator into a single soil-quality index number. The weight given to each indicator would be determined by the particular functions of the soil.

Indicator	Rationale for its use
Physical	
Texture	Retention and transport of water and chemicals
Depth of soil and rooting	Estimate of productivity potential and erosion; normalizes landscape and geographic variability
Infiltration and soil bulk density	Potential for leaching, productivity, and erosion
Water-holding capacity	Related to water retention, transport, and erosivity
Chemical	
Total soil OM	Defines carbon storage, potential fertility, and stability
Active OM	Defines structural stability and food for microbes
pH	Defines biological and chemical activity thresholds
Electrical conductivity	Defines plant and microbial activity thresholds
Extractable N, P and K	Plant-available nutrients and potential for N loss; productivity and environmental quality indicators
Biological	
Microbial biomass C and N	Microbial catalytic potential and early warning of management effect on organic matter
Potentially mineralizable N	Soil productivity and N supply potential
Specific respiration	Microbial activity per unit of microbial biomass
Macroorganism numbers	Potential influence of such organisms as earthworms

Modified from Doran, et al. (1996).

TABLE 20.2 Importance Assigned to Various Soil Functions in Ascertaining Soil Quality in Different Times and Circumstances

Note the very high weights for the food- and fiber-production function in 1900 worldwide, and in developing countries today. Other functions concerned with environmental and habitat issues are much more prominent today in industrialized countries.

Soil function	Probable Weights		
	Worldwide, 1900	Industrialized countries, 2000	Developing countries, 2000
1. Food and fiber production	85	40	70
2. Resistance to erosion	3	15	10
3. Water and air quality	1	10	5
4. Food quality	5	10	5
5. Wildlife habitat	1	15	5
6. Construction and transport base	5	10	5

element was paramount (highly weighted) compared to the five other nonproduction elements. But in our day, the elements concerned with the environment are perceived to be relatively more important, especially in the industrialized countries where food security is reasonably assured. The broader ecological roles of soils are becoming more widely recognized. In developing countries, however, where hunger and even famine are still common, food and fiber production remains the soil-quality issue of prime importance, as indicated by the high weight given to this function in Table 20.2.

Soil-Quality Index for Erosivity: An Example

A soil-quality index as related to soil erosion could be derived from the information in Table 20.3. Four functions of the soil in resisting water erosion are depicted: (1) *accommodating water entry*, (2) *facilitating water transfer and adoption*, (3) *resisting degradation*, and (4) *sustaining plant growth*. The relative weight of each soil function in resisting erosion is indicated, 50% assumed to be due to accommodating water entry, 35% to resisting particle degradation, 10% to facilitating water transport and absorption, and 5% to sustaining plant growth. Measurements that could serve as indicators of these four soil functions are shown as **indicators**, along with their respective weights. Note the many physical, chemical, and biological properties that can help one assess the ability of a soil to resist erosion.

The analytical data for the major indicators, along with their respective weights, can be used to develop an overall soil-quality index relating to water erosion. For example, the component of such an index relating to resisting degradation was found to rate at 0.84 (out of a possible 1.0) for an Iowa soil where sustainable farming practices were being followed, compared to only 0.60 for an adjacent field where conventional intensive, high-input practices were being used.

We now turn to the three primary **functions** of soils that must be performed if the soil quality is to be considered satisfactory. Our initial focus will be on biological productivity, since all life is dependent on it. However, the other two functions, maintaining environmental quality and enhancing human and animal health, will also receive attention, particularly as they are influenced by the attempts of humans to maximize biological productivity.

TABLE 20.3   Four Possible Soil-Quality Functions and Their Relative Weights in Determining the Resistance to Soil Erosion, Along with Measurable Indicators for Each Function and Their Weights

Note that with the exception of soil texture, most of the indicators are properties than can be significantly influenced by soil-management practices. Note that accommodating water entry, measurable by infiltration rate, is thought to provide about half (50%) of this function. Resisting degradation, measured primarily by aggregate stability, is of second importance. Note that most of the measurable indicators have been considered in previous chapters.

Soil quality function	Function weight	Measurable indicator	Indicator weight
1. Accommodate water entry	50	Infiltration rate	50
2. Resist degradation	35	Aggregate stability	27
		Shear strength	4
		Soil texture	2
		Heat transfer capacity	2
3. Facilitate water transfer and absorption	10	Hydraulic conductivity	5
		Porosity	2
		Macropores	3
4. Sustain plant growth	5	Rooting depth	1
		Water relations	2
		Nutrient relations	1
		Chemical barriers	1

Modified from Karlen and Stott (1994).

20.2 SUSTAINING BIOLOGICAL PRODUCTIVITY

No other soil function affects all living creatures more than does the sustenance of *biological productivity*. Human survival through the ages has depended on this function, and will likely continue to do so. Likewise, the survival of countless numbers of soil organisms is dependent on the soil's capacity to support biological productivity. We turn our attention to satisfying human needs for food and fiber, since the survival of other organisms is often determined by how we satisfy these human needs. We will review the world's food production problems, how they have been coped with, and how soil quality has benefitted and suffered from the actions we have taken.

The First 10,000 Years

Since the dawn of agriculture some 10,000 years ago, people have cleared forests and prairies so that the land could be used to grow food and fiber for their growing families. Initially, because there was an abundance of land and relatively few people, the change from the more sustainable natural vegetation to the less stable agricultural systems had only local effects on soil quality.

As humans became more numerous, soil productivity suffered over wider areas. Examples include the salinization of the once very productive irrigated lands of ancient Mesopotamia in the Middle East (see Section 10.3) and the severe water erosion of hilly lands upon which the Greeks and Romans depended for food (see Figure 17.21). These peoples turned to less densely settled lands in North Africa and Europe for the production of food. Consequently, the degradation of soil quality in these early periods had only modest global effects.

As human populations increased and the productivity of farmed soils faltered, the extra food needed for the expanding populations was obtained primarily by expanding the area of land under cultivation, not by increasing yields per hectare. This was particularly true after the "discovery" by Europeans of the Western Hemisphere, whose virgin soils soon produced food not only for the local inhabitants, but for export to the food-deficient parts of the globe.

Past Half Century

It is only in the past half century that pressures on land for crop production have become so acute, forcing people to consider alternatives to expansion of cultivated land as means of meeting human needs for food and fiber.[3] This change stems both from the unprecedented increases in the numbers of people to be fed, and from those people's enhanced ability to purchase food and fiber that others produce. We will start with the population explosion.

20.3 THE POPULATION EXPLOSION

Modern medical miracles following World War II stimulated unparalleled increases in human populations and in demands for food (Figure 20.2). These demands were met by farmers who produced more food in the past half century than had been produced in the previous 10,000 years of the history of agriculture.

To achieve this target, it was necessary either (1) to clear and cultivate native forests or water-deficient grasslands, much of which were ill-suited for cultivation; or (2) to greatly increase the cropping intensity and the yields per hectare on the more productive lands already under cultivation. Both sources of enhanced food production were utilized, but most of the needed food came from increased production on existing farmlands (Table 20.4). As we shall see, both of these approaches to increased food production resulted in serious consequences for the quality of the world's soils.

---

[3] Human demands for fiber that is used to manufacture cloth, paper, lumber, rope, covers, and so forth also grow with human population numbers. Plants such as cotton, hemp, and trees are used to help meet these demands. While our major focus will be on expanding food needs, demands for fiber also increase with expanding human populations.

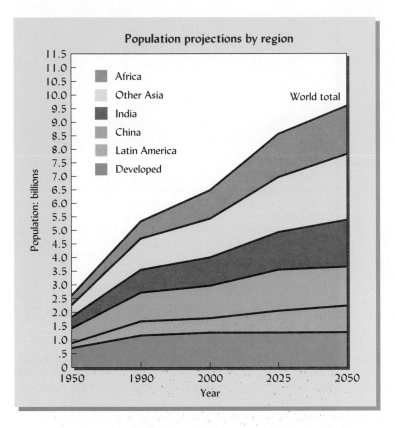

FIGURE 20.2 From the beginning of the human race until 1960, the world's population increased to about 3 billion. Less than 40 years will be needed to provide the second 3 billion. The total is expected to rise to 8.5 billion by the year 2025. Note that essentially all the growth is in the developing countries and regions that are already pressed to provide food for their growing populations. [From UNFPA (1992)]

20.4 INTENSIFIED AGRICULTURE—THE GREEN REVOLUTION

When the human population explosion became evident after World War II, many experts predicated widespread starvation. Their predictions were based primarily on the assumption that, as in the past, expansion of cultivated land would be the primary means of increasing food production. They ignored possibilities for increased production intensity on land already in cultivation, and they were wrong.

Scientists and their farmer collaborators developed and put to use intensified soil-, water-, and crop-management systems that gave unparalleled increases in food production, especially in the developing countries of Asia and Latin America. Food production increased more rapidly than population in all major regions except sub-Saharan Africa (Figure 20.3). Grain harvests nearly tripled worldwide from 1950 to 1990. As a result, the threat of massive starvation was averted, and the cost of foods (primarily cereals)

TABLE 20.4 **Percent of Increase in Food Production in Different Regions Between 1961 to 1963 and 1989 to 1990 Attributable to Increases in Area Cropped and to Increases in Yields Per Hectare**

Region	Increase attributable to	
	Increased area, %	Increased yields,[a] %
Low-income countries		
Sub-Saraban Africa	47	52
Latin America	30	71
Middle East/North Africa	23	77
South Asia	14	86
East Asia	6	94
High-income countries	2	98
World	8	92

[a] Includes both increasing the number of crops per year and increased yields per hectare.
Data from the Food and Agriculture Organization (FAO).

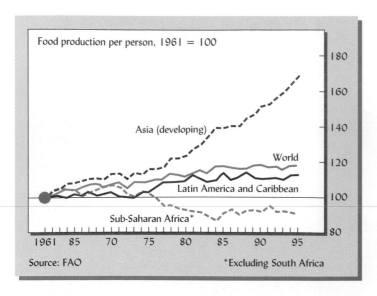

FIGURE 20.3 Changes in per capita food production in different regions of the world between 1961 and 1995. Food production per person worldwide increased nearly 20%, but in the developing countries of Asia, the increase was nearly 70%. Only in sub-Saharan Africa (excluding South Africa) did the per capita food production decline. Most of the increases resulted from agricultural intensification. [Data from FAO]

actually fell. Lowered food prices benefitted poor people everywhere, in cities as well as rural areas.

The vastly increased production resulted from farming systems that integrated newly created high-yielding cereal varieties (wheat, corn, and rice) with increased water availability through irrigation and dramatic increases in nutrient inputs from chemical fertilizers (Figure 20.4). Monoculture systems were intensively used, and two or three crops were harvested annually.

More than 70% of the increase came from intensified farming, the remainder from increases in cultivated land area. The results were most spectacular in Asia and Latin America, where the term *green revolution* was used to describe the process. Wheat yields in India, for example, increased by nearly 400% from 1960 levels, and yields of rice in Indonesia

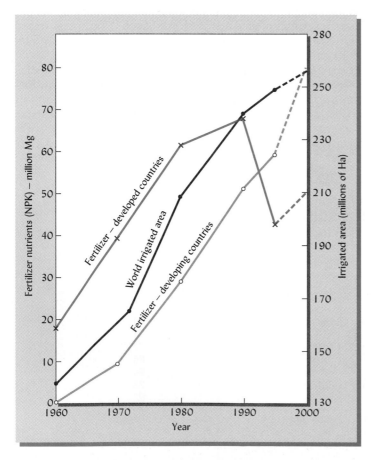

FIGURE 20.4 Increases in fertilizer use in industrial and developing countries and in world irrigated area since 1960. Note the 25-fold increase in fertilizer use in developing countries and the worldwide doubling of land under irrigation. The drop in industrialized country fertilizer use since 1990 is due primarily to decreases in the states of the former Soviet Union, although use in the United States and Europe has leveled off. (From FAO data and author's calculations)

and China more than doubled. The global caloric intake increased to about 2700 kilocalories, about 16% above minimum needs. Although millions still remained hungry, human nutrition among the poor was greatly enhanced since the real cost of these cereals declined by about 75%, making them more easily available to low-income citizens.

20.5 EFFECTS OF INTENSIFIED AGRICULTURE ON SOIL QUALITY OR HEALTH

Few quantitative studies have been made of the effect of production intensification on soil quality. But indirect evidence suggests that both positive and negative effects have occurred.

Positive Effects

On the positive side, intensified agriculture has generally maintained or even increased the level of some **macronutrients** in soil, since these elements are commonly supplied from outside sources, such as manures, lime, or fertilizers. Where appropriate modest applications of chemical fertilizers have been used, the N, P, and K components of soil quality have often been enhanced.

Intensified agriculture has also increased the level of plant production, permitting a corresponding increase in the amount of **crop residues** that can be returned to the soil. Such residues provide soil cover, reduce soil erosion, and can help maintain or increase soil organic matter levels (Table 20.5). Soil quality is thus positively affected if a significant portion of the crop residues are returned to the soil.

A third and likely even more significant positive effect of intensification is its tendency to **reduce pressures on fragile lands** that might otherwise have been cleared and cultivated to produce the additional food needed. Agriculture has been intensified mostly on the more productive, relatively level soils, where risks from erosion are not too high. By producing most of the additional food on these soils, the need for expanding onto more fragile lands has been minimized. Figure 20.5 illustrates this point for India. Were it not for the wheat yield gains from the green revolution, the country would have been forced to plow an additional 42 million ha of fragile lands, mostly in forests, an area equivalent in size to the state of California. Worldwide, more than 600 million ha—equal to the area of the great Amazon basin—have been saved due to increased yields of all cereal crops. It is almost certain that the quality of soils would have declined significantly on the forest and prairie lands that would have been brought into cultivation had crop intensification not been used.

Another possible aspect of the green revolution is the increased **efficiency of nutrient use** by some of the improved cereal varieties (Figure 20.6). For example, when 75 kg N/ha was applied to the traditional wheat varieties of 1950, only 45 kg of wheat was produced for each kilogram of nitrogen added. Improved varieties of the mid-1980s produced 70 kg of wheat per hectare of added nitrogen. Note, however, the lower efficiencies of all varieties when high nitrogen rates are used.

TABLE 20.5   The Effect of Nearly 30 Years of Continuous Rice Cropping (3 Crops per Year) with and without Nitrogen Fertilizer on the Organic Carbon and Total Nitrogen in a Soil in the Philippines

Note the higher organic carbon and N levels in the soil to which heavy applications of nitrogen (330 kg/ha/yr) were applied. Phosphorus and potassium were applied to all plots.

	Organic carbon in soil, g/kg		Total N in soil, g/kg	
Year	No N applied	330 Kg N/ha/yr applied	No N applied	330 Kg N/ha/yr applied
1963	18.3	18.3	1.94	1.94
1978	18.8	21.4	1.97	2.22
1983	18.7	21.4	1.95	2.14
1985	20.4	23.9	2.07	2.38
1991	20.4	23.5	1.97	2.27
1992	20.7	23.0	2.09	2.30

Modified from Cassman, et al. (1997a).

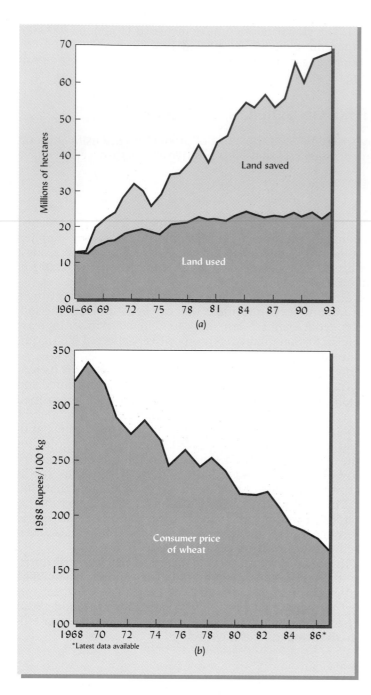

FIGURE 20.5 (a) If India had to produce today's wheat harvest with technologies and varieties of the early 1960s it would be necessary to cultivate 40 million more hectares of farmland than is currently being used. Most of this extra farmland would have to come from easily erodible forested lands that are characterized by steep slopes. (b) The increased production intensity simultaneously reduced the price that consumers (mostly poor people) had to pay for the wheat. [From CIMMYT (1995)]

Negative Effects

Intensive agriculture also has had negative effects on the quality of some soils. The application of chemical fertilizers generally provides ample quantities of nitrogen, phosphorus, and, in some cases, potassium.[4] However, the removal of other nutrients in the bumper crops often results in *micronutrient deficiencies*. Also, in some cases the oxidation of nitrogen added in fertilizers results in increased acidity. Both effects could lower soil quality.

EXCESS NUTRIENTS. In many areas of the world, such as East Asia and Western Europe (Figure 20.7) such nutrients as nitrogen and phosphorus were added in quantities far in excess of plant uptake. In time, the levels of these nutrients built up in the soil, and they

---

[4] When cleared lands are cultivated, deficiencies of nitrogen and phosphorus are first to appear, and fertilizers are applied to meet these needs. However, crop removals soon lower the potassium levels of some soils, especially those that are highly weathered and low in 2:1-type clays, such as illite.

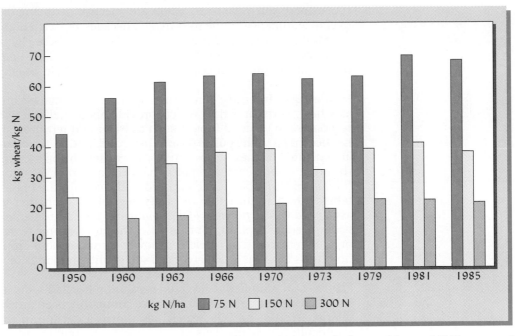

FIGURE 20.6 The efficiency of nitrogen utilization of traditional wheat cultivars of 1950 compared with that of the steadily improved cultivars that have since been used in intensified agriculture in developing countries. At all fertilizer nitrogen application rates, the improved cultivars are more efficient than the traditional varieties of 1950. Note, however, that nitrogen use efficiency is much lower at the higher rates (150 and 300 kg N/ha) than at the more modest rate of 75 kg N/ha. [From CGIAR (1997)]

moved as *pollutants* into the runoff or drainage waters or into the atmosphere. Soil quality is said to be reduced, since products moving from the soil adversely affect environmental quality.

SALINIZATION. Irrigation-induced *salinization* is another negative effect of agricultural intensification on soil quality. For example, each year the rate of salt addition to the irrigated soils of Arizona is equivalent to about 350 kg for each of the 4 million people living in the state. Worldwide, some 30% of irrigated soils are significantly affected by salinization (see end paper), some so seriously that the land has been abandoned.

PESTICIDES. Chemical *pesticides* that are commonly used in intensified agriculture systems can adversely affect soil quality. While some organochemicals adversely affect a broad

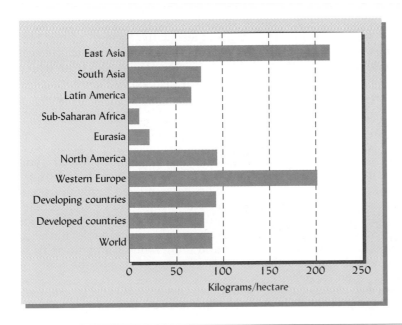

FIGURE 20.7 Rates of fertilizer nutrient use in selected regions of the world in 1995. Note the very high use in East Asia, where multiple cropping is common, and in Western Europe, where highly intensive agriculture is practiced. In both regions some excessive-use sites have been identified. Also note the very low rates in sub-Saharan Africa and in Eurasia (the newly independent states of the former Soviet Union), where plant production is constrained by nutrient deficiencies. Fertilizer use in the United States is about average for the world. While there are some high-use systems in irrigated and humid areas, these are balanced by the very low rates in vast areas of dryland farming where water, rather than nutrient deficiencies, is the first limiting factor. [Data from FAO published in Bumb and Baanante (1996)]

spectrum of soil organisms, others are selective, reducing biological diversity more than overall abundance. Some soils treated decades ago with high levels of arsenic- or copper-containing insecticides still have toxic levels of these chemicals in the soil. Because of the uncertain effects of today's pesticides on soil quality, *integrated pest management* systems that minimize the use of these chemical should be emphasized.

HEALTHY DIET.   Intensive agricultural systems have focused primarily on cereal crops, such as wheat, corn, and rice, which provide about half the world's calories and are quite responsive to external inputs, such as water and fertilizers. Unfortunately, less attention has been paid to the pulses (beans, peas, and lentils), fruits, and vegetables. As a result, the area planted to these crops actually decreased in some countries. For example, in India, the area of land devoted to pulses has decreased by 13% since 1970. This has implications for *human health* because, compared to the cereals, the pulses are generally higher in proteins and micronutrients, and leafy vegetables are higher in vitamins. Human diseases associated with *deficiencies of micronutrients,* such as iron and zinc, and with vitamin A, are widespread in tropical countries. Also, the pulse legume residues commonly provide some organic nitrogen that is released slowly for subsequent crop uptake. Exclusive emphasis on cereals has thus indeed reduced soil quality in many countries of the world.

PLANT DISEASE.   Similarly, the green revolution has had some negative impacts on soil quality because the improved cereals have commonly been grown in *monoculture* season after season. In some areas, research has shown a decline in the biological productivity of monoculture systems. This may be due to the buildup of *pathogens* or of *allelochemicals* that are toxic to the crop, or to declining levels of micronutrients in the soil. The decline may also be associated with low biodiversity in the soils of monoculture systems, a characteristic that results from little diversity in the organic residues and in the associated organisms that take part in their decay. In any case, when cropping systems do not take advantage of the benefits of crop rotation, soil health or quality declines accordingly.

REDUCED BIODIVERSITY.   High input, intensified agriculture may significantly affect the abundance and *biodiversity* of soil organisms. We know that the clearing and cultivation of forested lands reduces the number of fungi and increases the relative numbers of bacteria (see Section 11.15). The ratio of fungal biomass to that of bacteria may be about 1:1 in cultivated soils, about 3:1 in minimum tillage areas, and more than 100:1 in forested areas. Monoculture systems, especially those where the crop residues are removed or burned, likewise reduce the number of earthworms and other macroorganisms, compared to their numbers in systems with crop rotation.

Effects of intensification on the biodiversity among species of bacteria is somewhat less certain because their extremely small size makes it difficult to measure their diversity. However, the advent of new molecular biological tools that provide DNA analyses has already shown the close interaction of numerous microbes in soils and has indicated that this interaction is modified as soil and plant environments change.

In Section 16.5 we discussed what intensified animal production systems can do to soil quality. While these systems are efficient in terms of feed conversion to animal protein, they have adverse effects on soil quality and health. They remove plant products from wide areas and concentrate them into a production factory, the wastes from which often pollute the surrounding soil and water systems with nitrogen, phosphorus, and pathogens. Soil quality is most certainly affected negatively by such intensification.

20.6   UNPLANNED PRODUCTION INTENSIFICATION

There is a second type of plant-production intensification that has been forced upon indigenous farmers by the massive human population increases of the past half century. Around the world, millions of people and their ancestors have utilized indigenous pastoral and crop-production systems that depend upon natural vegetation to rejuvenate[5] soil quality between periods of use. With the low population–to–natural resource ratios that existed in earlier generations, these systems were reasonably sustainable. But with increased population numbers due to larger families and more families migrating from other areas, this ratio increased dramatically; the land was overused, and the time for

---

[5] The ability of a soil to recover from soil degradation resulting from changes in land use is termed **soil resilience.**

rejuvenation of the soil was greatly reduced. We will discuss briefly the effects of three such forced intensification systems on soil quality.

Shifting Cultivation

One of these indigenous systems is a type of **shifting cultivation**,[6] which has been practiced for generations in tropical forests around the world. It involves the slashing and burning of small plots in the forests (Figure 20.8), producing food crops for a few years until the supply of nutrients from the ashes is depleted, or until weeds become intolerable. The farmer then moves on to clear and burn another site, cultivates it for a few years, and then abandons it, too. This process is repeated until, after some 10 to 20 years, the initial site is once again cleared of the now semimature natural vegetation, which is again burned, and the process is started over, shifting from one site to another. During the 10 to 20 years, regrowth vegetation at least partially rejuvenates the quality of the soil on the first site by adding organic residues, recycling nutrients from deep in the profile, and fixing nitrogen from the air.

Unfortunately, the number of people dependent on shifting cultivation has increased dramatically in the past half century. Today, at least 300 million poor people, primarily in the tropics and subtropics, depend on such shifting cultivation systems for their food and livelihood. The area of land on which some form of shifting cultivation is practiced is enormous, amounting to about 30% of total worldwide arable land. In Africa, shifting cultivators account for about 70% of tropical deforestation.

Increased demand for food has forced a shortening of the fallow period in these systems, making it necessary to return to the original slash-and-burn sites after no more than four to five years (Figure 20.9). The soil regeneration time is insufficient to replenish soil nutrient and organic matter levels and to constrain soil erosion. Soil erosion has increased and soil fertility has declined—and, as a consequence, soil quality has deteriorated, along with yields of food crops and family well-being. Figure 20.9 illustrates the difference between the traditional, sustainable systems and the more intensive systems that many cultivators are forced to use today.

Slash-and-burn systems also adversely affect atmospheric quality. The CO_2 and other gases released in the burning process are thought to provide up to 25% of total global warming effects.[7] To prevent the deterioration of air and soil quality, cultivators in these

---

[6] Since most forms of shifting cultivation involve the cutting down and burning of existing vegetation, the term *slash-and-burn* is commonly used to describe these systems. There are many shifting cultivation systems, varying from the very simple shifting of nomadic herds from one area to another to more sophisticated cropping and fallow systems used in heavily forested areas.

[7] In Brazil, much of the land cleared by slash-and-burn is used for pasture. To help control weeds and spur the growth of grasses, farmers burn the fields every two to three years, releasing additional CO_2 and other harmful products, thereby adding to the global warming process.

FIGURE 20.8 (Left) Aerial view of a slash-and-burn area in the Amazon, where the trees have been cut down and are ready for burning. Note the second area in the background where similar cutting is under way. (Right) In another field that has been slashed and burned, a row of corn has been planted and other crops will follow. Note the absence of weeds at this point, but in a few years, weed infestation and soil fertility depletion will force the cultivator to move on to another site. (Photos courtesy of the International Center for Research on Agroforestry)

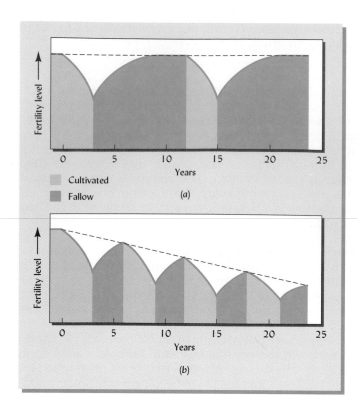

FIGURE 20.9   Changes in fertility levels under two shifting-cultivation systems. (*a*) In the system with a long fallow period, the natural vegetation is able to help regenerate the fertility level after cropping. (*b*) With a shorter fallow period so common in areas of high population pressure, insufficient regeneration time is available, and the fertility level declines rapidly.

areas need viable alternatives to the intensified shifting-cultivation systems that they have been forced to use.

Nomadic Pastoral Systems

Nomadic pastoral systems on common lands in arid to semiarid areas, combined with increased human populations and periods of drought, have adversely affected soil quality, especially in the Sahel region of Africa. At least 75 million people live in areas where such effects are occurring. Increased animal numbers, or even constant numbers with less forage to consume, result in overgrazing. The more nutritious grasses are mostly eliminated, and only less-palatable shrubs remain. Wind erosion increases (see Section 17.11), and overall biological productivity declines, along with soil quality. Such degradation is thought to be due to a combination of population pressures and abnormally low rainfall.

Marginal and Urbanized Lands

Human population pressures have also resulted in some expansion of crop cultivation into areas that are marginal for agriculture, such as erosion-prone, rolling to steep hillsides. Many landless families in Central and South America, Africa, and Asia are forced to use these areas to survive. Together with the farm families that practice shifting cultivation, some 800 million people produce their food on forested hillsides. Removal of plant nutrients in the crops they harvest, along with increased soil erosion, decreases biological productivity and lowers soil quality.

One last adverse effect on soil quality for biological productivity is the increased use of land for urban, industrial, and transportation purposes. The problem of urban sprawl in the United States is well known, but similar losses of cultivated lands are occurring in other areas, especially in the developing countries, where population pressures are greatest and economic growth is most pronounced. Global urbanization alone is thought to claim at least 1 million ha annually. Furthermore, this land is usually prime land for plant productivity, since the growing cities were originally located to take advantage of good cropland nearby.

Before we look to the future, we should remind ourselves of what we have done in the past half century, during which time some 2 billion ha of land have suffered soil-

TABLE 20.6 Areas of Land in million ha, Used for Agriculture, Permanent Pasture, and Forests or Woodlands, and Estimates of the Percentages of These Lands That Have Suffered Human-Induced Soil Degradation (Reduced Soil Quality)

Reductions in soil quality are most severe in agricultural lands, particularly in Africa and Central America where more than two-thirds of the cultivated lands are adversely affected.

Land use	Africa	Asia	South America	Central America	North America	Europe	Oceania	World
Agricultural land								
Area	187	536	142	38	236	287	49	1475
Percentage degraded	**65**	**38**	**45**	**74**	**26**	**25**	**16**	**38**
Permanent pasture								
Area	793	978	478	94	274	156	439	3212
Percentage degraded	**31**	**20**	**14**	**11**	**11**	**35**	**19**	**21**
Forest and woodlands								
Area	683	1273	896	66	621	353	156	4048
Percentage degraded	**19**	**27**	**13**	**38**	**1**	**26**	**8**	**18**
All lands								
Area	1663	2787	1516	198	1131	796	644	8735
Percentage degraded	**30**	**27**	**16**	**32**	**8**	**27**	**16**	**23**

Reorganized and calculated from Oldeman, et al. (1990).

quality degradation (see Figure 17.1 and Table 20.6). Data in Table 20.6 suggest that soil quality has declined on 38% of the world's agricultural land, on 21% of permanent pastures, and on 18% of forests and woodlands. Soil degradation on agricultural lands has been most drastic in Africa and Central America, but the problem extends to all continents. Since humans are primarily responsible for the decline in soil quality, we need to be aggressive in reversing this trend as we look to the future.

20.7 PROSPECTS FOR THE FUTURE

Three global factors influence whether the quality of soils will continue to decline: (1) the population- and economy-driven demands for food and fiber, (2) the extent to which more forests and grasslands will be cleared and cultivated to meet these demands, and (3) the universal determination to stop or even reverse environmental degradation around the world.

Demands for Food and Fiber

In the next 25 years, more than 2.5 billion people will be added to the world's human population, and some 90% of them will live in developing countries (see Figure 20.2). The highest percentage increase will take place in Africa, but the largest absolute numbers will be added in Asia. This is reflected in the data in Table 20.7, showing the numbers of malnourished children in different regions of the world. Increasing populations will continue to expand the world demands for food and fiber.

Economic growth in most of the developing countries will also increase these demands. Even though most of the people in these countries are very poor, their incomes

TABLE 20.7 The Number of Malnourished Children up to Six Years of Age in Different Regions of the World in 1993, and the Percentage of the Children in Those Regions That Were Malnourished

Note the very high numbers in Asia.

Region	Malnourished Children in 1993	
	Number, millions	Percentage
South Asia	100	57
Sub-Saharan Africa	26	30
China and Southeast Asia	42	22
Latin America	9	17
West Asia and North Africa	8	15

From Pinstrup-Andersen, et al. (1997).

are rising. As incomes rise, and as more people move from rural to urban areas, their per capita food consumption increases, and they include more animal products in their diets. It requires more grain to feed people on animal products than if the grains were consumed directly by humans.[8] By the year 2010, some 25% of the grain consumed in developing countries will be used to feed farm animals. In the next quarter century, at least 10% of the increased demands for food will come from increased incomes and changes in diet, the remainder from increases in population.

To meet these demands, more food will have to be produced in the next 40 years or so than has been produced since the beginning of agriculture. This will tax not only the soil's plant-production function, but the functions relating to environmental quality, human and animal health, and biological diversity as well.

Expansion of Land Under Cultivation

Most of the world's food and fiber supplies will continue to come from intensive agriculture on the more productive lands of the world (Table 20.8). However, some will come from expanding the arable land base, particularly in South America, where large potentially arable land areas still exist. For example, the Cerrado area of very acid, high-aluminum soils of central Brazil offers such promise. The recent development of high-yielding, aluminum-tolerant varieties of wheat and other crops makes possible the production of these crops on millions of hectares of level to rolling lands that heretofore could not have been used for cultivation (Table 20.9). With appropriate rotation systems and modest fertilizer applications, soils in the Cerrado can be quite productive. If grasses and cover crops are used in the crop rotations, and minimum tillage is practiced, erosion can be held in check. Of course, steps must be taken to set aside at least

---

[8] The production of 1 kg of chicken meat, pork, or beef requires at least 3, 5, and 8 kg of grain equivalent, respectively. Even though domestic wastes and rangelands are also used in animal production, and animal manures can positively affect soil quality, the overall effect of increased consumption of animal products is to increase markedly the demands for food and fiber.

TABLE 20.8 Percentage Contribution to Increased Food Production from Increases in the Area of Arable Land, Cropping Intensity, and Yields Per Hectare in Selected Developing Country Regions from 1982–1984 to 2000

Note that, except for Latin America, the majority of the expected production increase is from increased yields per hectare.

Region	Contribution to crop output growth, %		
	Increased area of arable land	*Increased cropping intensity*	*Increased yield per hectare*
Sub-Saharan Africa	26	17	57
Near East/North Africa	0	22	77
Asia (excl. China)	11	20	69
Latin America	39	12	49

From FAO (1987).

TABLE 20.9 The Impact of Soil, Crop and Nutrient Management on the Production of Cultivated Crops, Improved Pastures, and Plantation Forests in the Cerrado of Brazil from 1970 to 1990

Moderate use of lime and chemical fertilizers, and aluminum tolerant crops, made these increases possible.

Parameter	1970	1980	1990
Area planted to cultivated crops,[a] million ha	4.6	7.2	10.5
Average crop yield, Mg/ha	1.2	1.3	1.9
Crop production, million tons	5.6	9.4	20.0
Improved pastures, million ha	—	15.6	30.0
Plantation forests, million ha	—	1.5	3.0

[a] Soybeans, upland rice, beans, wheat, and corn.
From Sanchez (1994).

15% of the land for parks and corridors to protect wildlife and to maintain biological diversity. If such steps are taken, these potentially productive lands may well be utilized for agriculture with minimum negative effect on the overall quality of the soils.

Unfortunately, poverty and population pressures will likely force continued expansion of scattered hillside cultivation on fragile soils, expansion that has already proven to be damaging to soil quality. We will discuss later some steps that can be taken to minimize the degradation of these soil areas.

Demands for Environmental Quality

The third global factor that will influence soil quality is society's demand that environmental deterioration be curbed. It is no longer acceptable to indiscriminately cut down forests to provide additional farmland, or to clog up waterways with silt eroded from farmlands or with aquatic weeds encouraged by excess nitrates and phosphates coming from cultivated lands. Likewise, the continued degradation of the world's soils cannot be tolerated.

Constraints on Meeting Future Food and Fiber Demands

The challenge of meeting future food and fiber needs may well be more daunting than was experienced in the 1960s. The two primary inputs that stimulated the growth of the green revolution cultivars, *irrigation water* and *chemical fertilizer,* cannot be expected to produce similar spectacular yield increases in the future. The rate of increase in irrigated lands in developing countries (which was 1.7% annually from 1982 to 1993) is expected to be only 0.7% from 1995 to 2020. The most favorable and low-cost irrigation sites have already been developed.

Likewise, worldwide fertilizer use, which had increased tenfold from 1950 to 1989, has leveled off (see Figure 20.4). In many areas of the world, fertilizer use is already at or near the optimum level.[9] In East Asia, for example, fertilizer use rates now rival those of the United States and Europe (see Figure 20.7). At these higher use levels, there is a diminishing return from additional fertilizer applications. Consequently, the return from an extra 25 kg N/ha today for most fields in Asia would be much smaller than was the case in the 1960s or 1970s.

These constraints likely account for the fact that wheat and rice yields increased only about 1% annually in the 1990s, compared to more than double that rate between 1960 and 1990. They emphasize the challenge facing farmers worldwide to find means of meeting future food demands.

A New Paradigm of Plant Production

A twofold challenge is ahead: to satisfy the enormous demands for food, and for human and animal health, without doing damage to the quality of the environment. To meet this challenge, a new paradigm of global food and fiber production is beginning to unfold. In the past, *technology* and *short-term economic feasibility* have been the primary bases underlying food-production systems. While economic considerations must still be accommodated, there is a growing recognition that we must give greater consideration to an *ecological basis* for our food-production systems. This new sense of direction will use science and technology to help us understand how agriculture must interact with other ecosystems. It will help us focus on food-production systems that minimize losses and maximize the stewardship of natural resources for food production and for natural habitats or other purposes. It certainly implies significant changes in the methods we use to supply our food, changes that have significant implications for soil quality. We will start with changes needed in intensified commercial systems.

20.8 MODIFIED INTENSIVE AGRICULTURAL SYSTEMS

Food and fiber production on the more favored lands will continue to be highly intensified, perhaps even more so than in the past. However, if soil quality is to be maintained or improved, we must employ new and modified soil- and crop-management

---

[9] There is considerable potential for further responses to fertilizer nutrients in some developing countries, especially in Africa. Means of meeting those needs will be covered in Section 20.10.

systems that will *reduce pollution* of soil, water and the atmosphere, increase the *efficiency of nutrient and water use,* maintain or increase the *quantity and quality of organic matter,* decrease *soil erosion,* and increase *biological diversity.* Since nutrient management can affect several of these objectives, it will be discussed first.

Nutrient Management

Except for sub-Saharan Africa and some areas in Central and South America, only moderate expansion of food production will likely come from increased fertilizer use. In fact, in parts of Europe, the United States, and Asia, where nutrient application rates now greatly exceed plant uptake, fertilizer use rates will likely decrease. In those areas, the concept of producing "more with less" must prevail. The extra food will come, not from more fertilizers, but from greater efficiency of use of nutrients in the soil, whether they come from organic or inorganic sources. The nutrients, both natural and applied, must be recycled efficiently if high yields are to be achieved, and losses to the environment are to be minimized.

ORGANIC/INORGANIC COMBINATIONS. Research has shown that combinations of inorganic fertilizers and organic residues from a variety of crops in the rotation can maintain or even increase yields with reduced inorganic nutrient inputs. This is due in part to enhanced nutrient cycling and to the influence of diversified crop residues on both the quality of soil organic matter and the microbial biomass. Apparently, in conventional systems with only one or two crops in the rotation, the level of the *active fraction* of soil organic matter is low. In more diversified systems with several crops in the rotation, and with somewhat lower rates of fertilizer, much higher levels of the active organic matter fraction have been found. The soil in such high-crop-diversity systems also has higher biomass nitrogen levels (suggesting higher microbe levels) and higher levels of nitrogen mineralization. The interaction among the diversity of crops being grown, the quality of soil organic matter, and the microbial numbers and their ability to release nitrogen and other nutrients for plant uptake are illustrated in Table 20.10.

CROP ROTATION. Crop rotations have other advantages over the monoculture systems that are so common in intensified agriculture. Yields of two crops are higher if they alternate with each other on a given field, or if they are grown intercropped on the same field. If legumes are included in the rotation, they provide additional nitrogen to the system. Forage crops encourage animal production on a sustainable decentralized basis and assure the availability of animal manures that supply nutrients and organic matter. Close-growing crops in the rotation also help control soil erosion, and, with the help of crop residues, markedly increase the soil's water-infiltration capacity. If fall or winter cover crops are included, end-of-season nutrients left by the main crop can be saved,

TABLE 20.10 **The Effect of Conservation Practices on Some Soil-Quality Factors Related to Organic Matter**

In each region, soil was analyzed from seven pairs of adjacent fields, one on which conservation practices (reduced tillage, greater crop diversity, more sod crops in rotation, and/or use of organic nutrient sources) were used, while conventional practices (more tillage, less diversity, etc.) were used on the other. Note that the effect on total soil organic carbon was less pronounced than that on the active fraction of organic matter and related properties.

Properties	Coastal plain soils		Piedmont soils	
	Conservation management	Conventional management	Conservation management	Conventional management
Total organic C, g/kg	10	9	19	15
Active organic C,[a] mg/kg	121	75	134	112
Microbial biomass C, % of total organic	2.5	1.7	4.0	3.4
Nitrogen mineralization rate constant[b]	4.4	3.5	4.6	3.4
Aggregate stability, %	72	56	73	66

[a] Mainly sugars extracted from soils after disruption with microwaves.
[b] The rate constant k in the first-order decay equation $N_p = N_o e^{kt}$
Data from Islam and Weil, University of Maryland.

and water contamination can be averted. Cropping system studies have clearly established that rotations involving close-growing crops, including legumes, can maintain good soil quality.

There is growing interest in farming systems that utilize organic and natural rock sources of plant nutrients exclusively and use nonchemical means of pest control. Such **organic farming** systems utilize the nutrients in such sources as farm manures, crop residues (particularly legumes), and ground rocks. Although yield per hectare is often 5 to 15% lower than in high-input systems, the lower input costs and premium prices received for organically produced food can result in greater profitability. Soil quality is generally affected positively, and nutrient leaching is minimized, as is soil erosion. Unfortunately, however, the area likely to be devoted to organic farming may be limited by the availability of suitable organic sources of plant nutrients and by the probable reduction in premium selling prices if production were to be greatly expanded.

Water Management

Scarcity of water is likely to limit our ability to meet future food demands. Greater efficiency in using both precipitation and irrigation water must be achieved. Diversified crop rotations that include close-growing crops, along with conservation tillage, can provide greater soil cover, increase water penetration, and decrease water runoff and soil erosion. This leads to more efficient water use for plant productivity. Likewise, such irrigation management systems as drip irrigation that reduce the quantity of water needed to produce plants should be more widely used. Improved drainage systems serving irrigated areas must reduce the buildup of salts, thereby maintaining the quality of the soil.

Biological Diversity

Intensified cropping systems that involve a diversity of crops in the rotation, especially if both legumes and nonlegumes are included, will surely increase the biological diversity in the soil. Not only will there be more microorganisms, but the diversity of organic food for them will encourage greater diversity. Soil quality will benefit as a result.

20.9 IMPROVING LOW-YIELDING AGRICULTURAL SYSTEMS

Improved management of low-yielding systems, particularly in the tropics, is essential for the well-being of hundreds of millions of poor people in developing countries around the world. In some areas, such as most of sub-Saharan Africa, the primary problem is lack of plant nutrients. The soils are simply being mined of these nutrients, because removal in harvested crops and by leaching far exceeds the amounts being returned from all sources. During the last 30 years it is estimated that an average of 660 kg N/ha, 75 kg P/ha, and 450 kg K/ha has been removed from about 200 million ha of cultivated lands in sub-Saharan Africa.

Soil constraints are prime reasons for low productivity in many other areas in the tropics (Table 20.11). The quality of the soils is declining annually, and this will continue if means are not found to increase nutrient inputs. Cropping systems that improve nutrient cycling and that protect the soils from erosion are an important part of the solution. But in other areas, significant quantities of additional external nutrient inputs will be needed.

TABLE 20.11 **Percentage of Soil Areas in Five Agroecological Regions of Tropics Having Different Chemical Constraints to Plant Growth**

Soil constraint	Humid tropics, %	Acid savannas, %	Semiarid tropics, %	Tropical steeplands, %	Tropical wetlands, %	Total, %
Aluminum toxicity	56	50	13	29	4	32
Acidity without Al toxicity	18	50	29	16	29	25
High P fixation by Fe Oxides	37	32	9	20	0	22
Low CEC	11	4	6	—	—	5
Salinity	1	0	2	0	7	1

From Sanchez and Logan (1992).

To illustrate how increased food production and enhanced soil quality might be achieved in currently low yielding areas, we will focus on the situation in sub-Saharan Africa, an area where soil quality is declining, and which is the only major region of the world where per capita food production has declined in the past 25 years. However, the principles involved also apply to low-productivity agricultural systems in many other tropical areas.

20.10 IMPROVING SOIL QUALITY IN SUB-SAHARAN AFRICA[10]

Recent assessments of soil quality for agricultural purposes in Africa suggest that about 55% of the land area on that continent is unsuitable for any kind of agriculture except nomadic grazing (Table 20.12). Thirty percent of the continent's population ekes out an existence in these areas. Soils of low quality are found on 16% of the land where 23% of the people live. These soils are limited by such constraints as high acidity, impermeable layers in the soil, and accumulation of salts.

Fortunately, medium- to high-quality soils are found on 29% of the land where 47% of the people live. These soils are free of major constraints and are found in regions where rainfall is stable and sufficient for at least one crop a year. These soils offer great potential for increased production of food and fiber if sustainable intensified systems can be utilized.

Nutrient Deficiencies

Nutrient input/output studies on farm lands of sub-Saharan Africa (SSA) show an alarming *negative balance,* suggesting a decline in soil quality. Nutrients are being removed at rates far higher than their additions. The N, P, and K balances for 13 countries (all of which are negative) are shown in Table 20.13. Overall, by the year 2000 SSA countries[11] will be losing each year from its cultivated lands an estimated 6.1 million Mg N, 0.74 million Mg P, and 4.6 million Mg K (see Table 20.13). While low crop yields mitigate relatively low nutrient removal levels, the very low rates of fertilizer applications, commonly less than 10 kg/ha, simply do not replace the nutrients removed.

Chemical fertilizers are the obvious sources to counterbalance the nutrient depletion of African soils. Unfortunately, there are few local sources of manufactured fertilizers. Because of the great distances over which imported fertilizers must be transported, their farm-delivered price in SSA is at least twice the international price. This discour-

---

[10] Numerous studies have been made of factors affecting food production in Africa and of the impact of the production methods used on the quality of the soil. Two recent publications that give background information on these issues are Eswaran, et al. (1997b) and Buresh and Sanchez (1997).

[11] Since much agriculture of South Africa utilizes intensive production systems, this country is excluded in our consideration of sub-Saharan Africa. [Data cited from Stoorvogel and Smaling (1990)].

TABLE 20.12   **The Percentage of Africa's Land Area Characterized by Different Soil-Quality Classes, and the Percentage of the Continent's People Living in Each Area**

Note that 30% of the people and 55% of the land are found in areas where soil quality is considered unsustainable. However, 47% of the people live in the 29% of the land (nearly 900 million ha) where soil quality is considered high or medium.

	Percentage of Africa's	
Soil quality	Land area	Population
High	16	31
Medium	13	16
Low	16	23
Unsustainable	55	30

From Eswaran, et al. (1997a).

TABLE 20.13 Average N, P, and K Balances, kg/ha/yr, of the Arable Land in Several African Countries Projected for the Year 2000

Such negative balances (inputs minus outputs) represent a literal mining of African soils that simply must be stopped if the quality of all life on this continent is to be sustained.

Country	Balance, kg/ha/yr		
	Nitrogen (N)	Phosphorus (P)	Potassium (K)
Cameroon	−21	−2	−13
Ethiopia	−47	−7	−32
Ghana	−35	−4	−20
Kenya	−46	−1	−36
Malawi	−67	−10	−48
Nigeria	−37	−4	−31
Rwanda	−60	−11	−61
Senegal	−16	−2	−14
Tanzania	−32	−5	−21
Zimbabwe	−27	2	−26
Average	−39	−5	−30

From Stoorvogel, et al. (1993).

ages nutrient additions, especially in remote areas such as those where shifting agriculture is practiced. Local rock phosphate deposits, of which there are many in SSA, are largely undeveloped because the P in these materials is only slowly available to most crop plants. Innovative combinations of organic and inorganic nutrient sources must be used to increase nutrient inputs and to cycle the nutrients once they reach the soil.

Nutrient Management

Agroforestry systems that integrate fast-growing, nitrogen-fixing woody species into small-scale farming systems show great promise in increasing the supply of plant nutrients, and in enhancing their cycling in the soil–plant system, while simultaneously increasing plant productivity and soil quality. We have already noted (see Section 13.8) the ability of deep-rooted, fast-growing trees to recycle the nitrates that have leached to depths of several meters in some tropical soils. The nitrogen so salvaged (often as much as 100 Kg N/ha) can be taken up by food crops when the tree residues are spread on the soil and decomposed. This retrieved nitrogen represents a real input to the food-producing system, since otherwise the nitrogen would have been lost in the drainage waters.

The growth of fast-growing, nitrogen-fixing trees during the fallow period after the food crops have been harvested is another source of nitrogen for the production systems in subhumid areas (Table 20.14 and Figure 20.10). Nitrogen-fixation rates are

TABLE 20.14 Yield of Corn in a Field in Kenya Following One or Two Years of Fallow in Which Various Species Were Grown

Note the very high yields from the plots on which Sesbania, *a nitrogen-fixing tree, was grown. During a five-year period, the Two-year Sesbania fallow and three years of corn produced more total corn grain with less labor than either of the currently used systems (grass fallow and unfertilized corn). Such agroforestry systems are showing great promise in different parts of Africa.*

Treatment	Maize grain yield, t/ha
2 years *Sesbania sesban*	5.4
Maize with fertilizer	4.0
1 year *Sesbania sesban*	3.4
Ground nut–maize rotation	1.9
Grass fallow	1.8
Maize without fertilizer	1.1

From Kwesiga, et al. (1997).

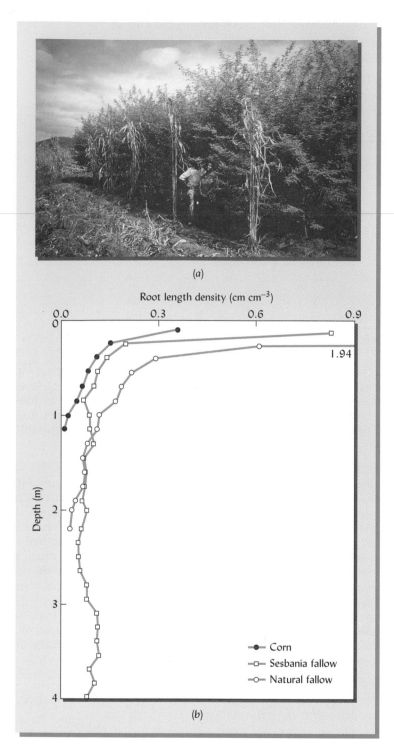

Root length density (cm cm^{-3})

(a)

(b)

- Corn
- Sesbania fallow
- Natural fallow

FIGURE 20.10 Scientists and farmers have discovered the remarkably beneficial effects of including rapidly growing, nitrogen-fixing trees in rotations with food crops in Africa. (Upper) The luxuriant growth of one such leguminous tree, *Sesbania sesban,* is shown in a farmer's field in Zambia. The very deep rooting system of this tree (lower) encourages the recycling of other nutrients as well as nitrogen. The remarkable yield response of corn following *Sesbania* is shown in Table 20.14. [Photo courtesy Dr. Steven Franzel, International Center for Research in Agroforestry; graph from Mekonnen, et al. (1997), with permission of Kluwer Academic Publishers]

equivalent to those provided by leguminous forages in the United States. When the woody legume residues are added to the soil and decomposed, some of the nitrogen is released for plant uptake. Part of the nitrogen is incorporated into the more active forms of soil organic matter and into the microbial biomass and will be available for later utilization by plants. Biological fixation is a significant source of nitrogen that must be more fully utilized.

ORGANIC/INORGANIC COMBINATIONS. A combination of organic residues and chemical fertilizers seem to provide a synergistic effect, the yield increases being greater than those of the same inputs that are supplied separately (Figure 20.11). Researchers and farmers from all parts of the world are finding that by combining such organic materials as crop residues, agroforestry litter, animal wastes, and compost with modest levels of inorganic

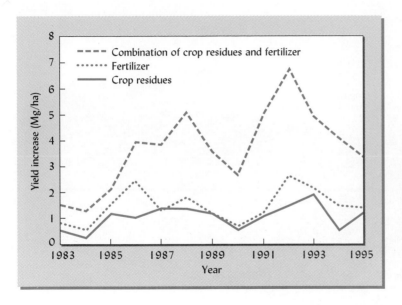

FIGURE 20.11 Increased yields of pearl millet from applications of crop residues and inorganic fertilizers supplied separately and in combination in 13 years of cropping in West Africa. The yield increases from fields receiving both organic and inorganic materials generally exceeded the sum of the yield increases from fields receiving the two materials separately. The beneficial effects of combined application of organic and inorganic materials were well demonstrated. [Calculated and redrawn from ICRISAT (1997)]

fertilizers, they can increase yields and simultaneously maintain or improve soil quality. While the mechanism responsible for the yield increases are not yet well established, there appear to be synergistic effects of organic and inorganic inputs such as those concerned with the supply of water, activities of soil microorganisms, and rhizodeposition (see Sections 11.7 and 12.3).

Research on farmers' fields is suggesting that the replenishment of phosphorus in the soils of Africa will come primarily from inorganic sources, with supplementation from biological sources. In contrast, nitrogen replenishment will likely come primarily from biological sources, with inorganic sources providing the supplementation.

We have emphasized the nutrient-management aspects of combining organic and inorganic inputs to enhance plant productivity. It should be emphasized, however, that these combined systems have equally beneficial effects on other contributors to soil quality, including water utilization, soil erosion control, and biological diversity. In short, they enhance soil quality.

Agroforestry Systems as Alternatives to Slash-and-Burn

There are some agroforestry systems that can be viable alternatives to the destructive slash-and-burn methods in use today in Africa and elsewhere in the tropics. We will discuss two of these briefly.

MIXED TREE CROPS. First is the *mixed tree crop* system, in which the domesticated crop (often a tree species) is planted among existing native species (usually trees). In time, a number of the cultivated crop species may become dominant. The essential feature of the system, however, is that the soil will at no time be free of vegetation. Building on the experience of traditional systems, some scientists recommend the propagation of trees that produce high-valued products (fruits, nuts, and medicines) that can provide the producer with added income. The *Kandy home gardens* in Sri Lanka and the Telum forests of northeast Mexico are encouraging examples of successful systems developed by local people. While the mixed tree crop system requires detailed knowledge of the species to be managed, it does offer economic returns while simultaneously assuring that soil quality will be maintained.

ALLEY CROPPING. An agroforestry system known as *alley cropping* involves growing food crops in alleys, the borders of which are formed by fast-growing trees or shrubs, usually legumes. The hedgerows are pruned regularly to prevent shading of the food crop, and the prunings are used as a mulch and a source of nutrients for the food crop. The mulch helps control weeds, prevents runoff and erosion, and reduces evaporation from the soil surface. Figure 20.12 illustrates how the system works.

In some areas alley cropping systems have proven to be successful (Table 20.15), crop yields being maintained without having to leave the land fallow for years. The soil

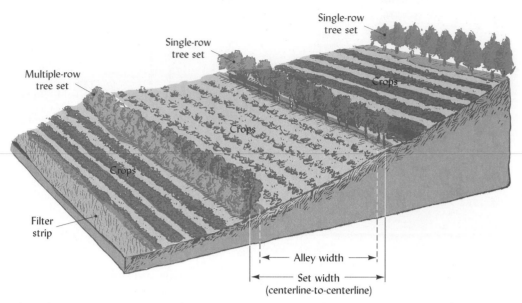

FIGURE 20.12 Alley cropping across the slope, showing alleys in which crops of various kinds are produced in between rows (single or multiple) of fast-growing shrubs or trees, preferably of the nitrogen-fixing variety. After the harvest of a cultivated crop, the trees can be trimmed and pruned back, and the pruned vegetation can be spread over the cropped area. This will provide organic matter, nutrients, and protection of the soil from erosion for the succeeding crop. Such systems have shown some promise where the rainfall is sufficient for both the planted crop and the border trees, and where there is enough labor for cutting and spreading the tree trimmings. [Modified from USDA (1997)]

is protected from excess erosion, and soil quality is improved (Table 20.16). But in other quite infertile subhumid to semiarid areas, the hedgerow species compete with the food crops for nutrients and moisture, and yields suffer. Figure 20.13 illustrates how competition reduces the effectiveness of alley cropping in some drier and more infertile regions. Thus, alley cropping may be beneficial in some cases, but is not a viable alternative to slash-and-burn in others.

TABLE 20.15 Yield of Corn Over a Period of Six Years in an Alley Cropping Experiment Using *Leucaena leucocephala* as the Hedgerow Component on a Psammentic Ustorthent Soil in Nigeria

Nitrogen rate,[a] kg/ha	Leucaena *prunings*	Yield of corn,[a] Mg/ha						
		1979	1980	1981	1982	1983	1984	1986
0	Removed	—	1.04	0.48	0.61	0.26	0.69	0.66
0	Used as mulch	2.15	1.91	1.21	2.10	1.92	1.99	2.10
80	Used as mulch	3.40	3.26	1.89	2.91	3.16	3.67	3.00

[a] All plots received P, K, Mg, and Zn. Crop affected by drought in 1983; land fallowed in 1985.
From Kang, et al. (1989).

TABLE 20.16 The Influence of Conservation Tillage and Alley Cropping Using Two Woody Legumes as Hedgerows on Runoff and Soil Erosion Losses in Nigeria

System	Runoff, % of rainfall		Soil erosion, Mg/ha	
	Corn	Cowpeas	Corn	Cowpeas
No hedgerows				
Plow-till	17.0	4.3	4.3	0.63
No-till	1.3	0.8	0.1	0.03
Hedgerows				
Leucaena	4.9	1.1	0.6	0.13
Gliricidia	4.3	0.7	0.6	0.07

From Lal (1989).

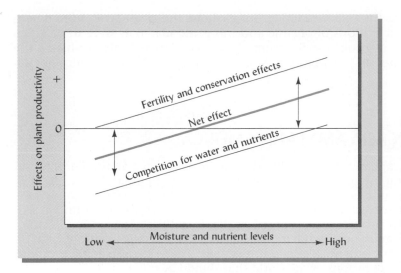

FIGURE 20.13 Benefits and constraints on plant productivity in alley cropping systems. The upper line illustrates the positive effects from nutient cycling and soil conservation. The lower line shows the negative effects of competition from the border row species for water and nutrients. The heavy center line shows the net effect, which is generally positive in more humid areas with soils that have moderate plant nutrient levels. In drier, less-fertile areas, the competition for water and nutrients often results in a net reduction in crop yields.

Collaboration among researchers and cultivators is revealing site-specific means of improving the management of low-input systems in Africa and elsewhere around the world. While these improvements will vary in detail, appropriate combinations of organic and inorganic inputs are components of most of them. The same is true for most high-input intensive production systems. Variable organic inputs, supplemented as appropriate by inorganic fertilizers, provide the primary bases for increased plant productivity and for improved soil quality.

Combinations of organic and inorganic inputs have the potential of greatly improving the quality of soils in Africa. Figure 20.14 illustrates the very low potential

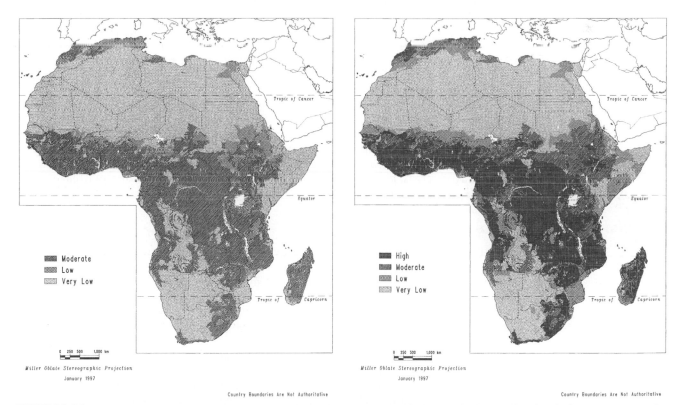

FIGURE 20.14 The potential for sustainable agricultural use of soil resources of Africa with the very low level of inputs in use today (left) compared to a similar potential if high levels of inputs were employed (right). Note the vast areas of the Sahara, the Horn of Africa, and southwest Africa where, mostly because of water shortage, the potential is very low regardless of the level of chemical inputs. Furthermore, with current low levels of inputs, agricultural sustainability is low or very low in much of the rest of Africa. In contrast, if a rational combination of organic and inorganic inputs were to be applied, large areas would have moderate to high potential for agricultural sustainability. [From Eswaran, et al. (1997a), with permission of Haworth Press]

for most African soils as they are used today. It also shows the greatly enhanced potential if appropriate combinations of organic and inorganic materials were to be effectively utilized. The synergistic benefits from these materials will also enhance the quality of soils—whether they be in Africa, Asia, Latin America, North America or elsewhere in the world.

20.11 IMPROVING SOIL QUALITY IN ASIA AND LATIN AMERICA

We have already emphasized soil-quality decline in the United States, Europe, and East Asia resulting from intensified agricultural practices and from land application of industrial, municipal, and agricultural wastes. We also note that sloping lands anywhere in the world are subject to the ravages of soil erosion and to the concomitant sediment deposition downstream. Land-use practices that keep the soil covered with either natural forests or grasslands, or with the residues from harvested crops, must be high on the list of criteria for any sustainable land-use system. There are also some location-specific problems in other regions of the world that should be mentioned.

South and Southeast Asia

The remarkable food-production achievements of the last three decades in Asia would not suggest that soil-quality decline is a major problem in this area. Unfortunately, such decline has taken place, but it has been masked by the countereffects of increased use of fertilizers and irrigation water. Intensive agricultural practices, coupled with the widespread burning of crop residues, have had negative impacts on both the quantity and quality of soil organic matter. In South Asia, crop residues as well as animal manures are used as fuel for domestic cooking, or the residues are consumed by animals. In Southeast Asia, the rice and wheat straw that has been gathered to central threshing areas is burned just to get rid of it. In any case, the land is mostly uncovered during intercropped periods, thereby encouraging erosion, a decline in soil organic matter, and a breakdown of stable soil structure. The protruding bedrock in once fertile and productive lands is a grim reminder of past soil destruction in some areas. Such degradation simply cannot be allowed to continue if the enormously increased populations of the future in these regions are to be fed.

Research has shown that biological production and soil quality can be improved by combinations of organic and inorganic inputs into soils in this area. Steps must be taken, however, to assure that farmers have sources of organic materials to complement the inorganic fertilizers that they can purchase. In some areas, community tree lots are providing fuel for cooking, making it possible for the crop residues to remain on the land. In other areas, green manures are providing the needed organic matter.

South and Central America

Soil-quality decline in these areas may also occur in the future, from both intensification and soil "mining" processes. In potentially productive areas, such as the Cerrado of Brazil, the major challenge is to avoid the errors made in the past by farmers in North America and Europe. Native grasslands will likely be brought into cultivation, but care must be taken to use combinations of organic and inorganic materials, and not just inorganic fertilizers. Likewise, monocultures should be avoided, and maximum use should be made of crop rotations.

The clearing of forest lands for agricultural purposes, not only in the Amazon Basin but in most other regions of South and Central America, must be avoided to the extent possible. Erosion rates in Central America are among the highest in the world. If soil quality is to be maintained or improved, the net change in land use in most areas should be toward increased forest lands, not more agricultural land. Judicious use of organic/inorganic combinations as well as an expansion in areas of agroforestry systems should increase food production on land already in agriculture. Maintaining hilly lands and naturally infertile areas in forests would provide sustainable ecological systems that would help maintain or improve soil quality.

20.12 CONCLUSION

Soil scientists, in collaboration with colleagues in other disciplines, have developed the concept of *soil quality* or *soil health* to help quantify factors that are influencing the ability of the soil to function effectively in a variety of roles. The primary measures of this effectiveness are enhanced biological productivity, environmental quality, human and animal health, and biological diversity. Human population pressures are forcing increased emphasis on the soil's function in enhancing biological productivity, but if soil quality is to be improved, we must simultaneously achieve the other three goals as well.

Farmers of the world have performed near miracles in the past half century. They have increased food production at higher rates than at any time in history, more than matching the unprecedented increases in human population numbers. These food increases have come primarily from intensification of agriculture on existing farmlands. While soil quality has been improved by some of the intensified systems, others have degraded soil, water, and atmospheric resources.

A major challenge facing humankind is the production of more food and fiber in the next 40 to 50 years than has been produced since the dawn of agriculture some 10,000 years ago. A new paradigm is needed to do this without degrading the environment, endangering human and animal health, or decreasing biological diversity. This paradigm envisages an ecosystem approach for farming systems—an approach that emphasizes the interaction of plant productivity with the productivity and well-being of all living organisms and with the quality of the environment. Soil quality will benefit from such interactions, and, in turn, will contribute to their success.

STUDY QUESTIONS

1. What is *soil quality* or *soil health,* how is it measured, and of what importance is it to all organisms that live in or on the soil?

2. Why is soil quality likely of greater societal concern today than it was 100 years ago?

3. Why does the biological productivity aspect of soil quality tend to receive more human attention than do these aspects dealing with the environment and animal and human health? Is this comparative priority changing? Why?

4. What technological inputs were largely responsible for the remarkable food- and fiber-production increases of the past three decades in the developing countries of Asia and Latin America? Will these inputs likely be significant in increasing food and fiber production in the next 30 years? Why or why not?

5. What are the positive and negative effects on soil quality of the intensive agricultural practices of the past half century?

6. How does the evolving ecosystem approach to plant production differ from the short-term economic return approach that has pertained in the past? Which approach will likely have a more positive effect on soil quality or health, and why?

7. What are some of the changes in nutrient management of intensified plant-production systems that are needed to improve soil productivity and agricultural sustainability and to enhance soil quality?

8. Africa is the only major region of the world where per capita food production has declined in the past three decades. What are the reasons for this situation, and what major steps might be taken to change it?

9. What is *shifting cultivation,* how is its use today in many areas resulting in a decline in soil quality, and what are some promising alternatives?

10. In what areas of the world has the decline in soil quality been most extensive?

REFERENCES

Bumb, B. L., and C. A. Baanante. 1996. *The Role of Fertilizer in Sustaining Food Security and Protecting the Environment to 2020.* Food, Agriculture and the Environment Discussion Paper 17. (Washington D.C.: Int. Food Policy Res. Inst.).

Buresh, R. J., and P. A. Sanchez (eds.). 1997. *Replenishing Soil Fertility in Africa.* SSSA Special Publication no. 51. (Madison, Wis.: Soil Sci. Soc. Amer.).

CGIAR. 1997. "How efficient are modern cereal cultivars?" *CGIAR News,* **4**(2):3. April 1997.

Cassman, K. G., S. Peng, D. C. Olk, W. Reicharat, A. Doberman, and U. Singh. 1997. "Opportunities for increased nitrogen use efficiency from improved resource management in irrigated rice systems," *Field Crop Res.,* in press.

CIMMYT. 1995. *CIMMYT 1994 Annual Report.* (Mexico City, Mexico: International Maize and Wheat Improvement Center).

Doran, J. W., and A. J. Jones (eds.). 1996. *Methods for Assessing Soil Quality.* SSSA Special Publication no. 49. (Madison, Wis.: Soil Sci. Soc. Amer.).

Doran, J. W., A. Sarrantonio, and M. A. Liebig. 1996. "Soil health and sustainability," *Advances in Agronomy,* **56**:1–54.

Eswaran, H., R. Almaraz, P. Reich, and P. Zdruli. 1997a. "Soil quality and soil productivity in Africa," *J. Sust. Agr.,* **10**:75–94.

Eswaran H., R. Almaraz, E. vanden Berg, and P. Reich. 1997b. "An assessment of the soil resources of Africa in relation to productivity," *Geoderma,* **77**:1–18.

FAO. 1987. *Agriculture: Toward 2000.* (Rome: U.N. Food and Agriculture Organization).

Harris, R. F., D. L. Karlen, and D. J. Mulla. 1996. "A conceptual framework for assessment and management of soil quality and health," in J. W. Doran and H. J. Jones (eds.), *Methods for Assessing Soil Quality.* SSSA Special Publication no. 49. (Madison, Wis.: Soil Sci. Soc. Amer.).

ICRISAT. 1997. ICRISAT Report 1996. (Patancheru, India: International Crops Research Institute for the Semi-Arid Tropics).

Kang, B. T., G. F. Wilson, and T. L. Lawson. 1984. *Alley Cropping: A Stable Alternative to Shifting Cultivation.* (Ibadan, Nigeria: Int. Inst. Of Tropical Agric.).

Karlen, D. L., and D. E. Stott. 1994. "A framework for evaluating physical and chemical indicators of soil quality," in J. W. Doran, et al. (eds.), *Defining Soil Quality for a Sustainable Environment,* SSSA Special Publication no. 35, (Madison, Wis.: Soil Sci. Soc. Amer.).

Kwesiga, F. R., S. Franzel, F. Place, D. Phiri, and C. P. Simwanza. 1997. "Sesbania improved fallows in Eastern Zambia: Their inception, development and farmer enthusiasm," in R. J. Buresh and P. J. Cooper (eds.), *The Science and Practice of Short Term Improved Fallows.* (Dordrecht, Netherlands: Kluwer Academic Publishers).

Lal, R. 1989. "Agroforestry systems and soil surface management of a tropical Alfisol: II: Water runoff, erosion and nutrient loss," *Agroforestry Systems,* **8**:97–111.

Mekonnen, K., R. S. Buresh, and B. Jama. 1997. "Root and inorganic nitrogen distributions in Sesbania fallow, natural fallow and maize fields," *Plant and Soil* **188**:319–327.

Oldeman, L. R., R. T. A. Hakkeling, and W. G. Sombroek. 1990. *World Map of the Status of Human-Induced Soil Degradation:* An Explanatory Note. (Wageningen, Netherlands: ISRIC; Kenya: UNEP).

Pinstrup-Andersen, Rajul Pandya-Lorch, and M.W. Rosegrant. 1997. *The World Food Situation: Recent Developments, Emerging Issues and Long-Term Prospects.* Paper for the Consultative Group on International Research, Oct. 27, 1997. (Washington, D.C.: Int. Food Policy Res. Inst.).

Sanchez, P. A., and T. J. Logan. 1992. "Myths and science about the chemistry and fertility of soils of the tropics," in P. A. Sanchez and R. Lal (eds.), *Myths and Science of Soils of the Tropics.* SSSA Special Publication no. 29 (Madison Wis.: Soil Sci. Soc. Amer.), pp. 35–46.

Stoorvogel, J. J., and E. M. A. Smaling. 1990. "Assessment of soil nutrient depletion in SubSaharan Africa: 1983–2000." Report no. 28. (Wageningen, Netherlands: Winand Staring Centre for Integrated Land, Soil and Water Research).

Stoorvogel, J. J., E. M. A. Smaling, and B. H. Janssen. 1993. "Calculating soil nutrient balances in Africa at different scales," *Fertilizer Research,* **35**:277–235.

UNFPA. 1992. *The State of the World Population.* (New York: U.N. Population Fund).

USDA. 1997. *Alley Cropping Conservation Practice.* Job Sheet 311. (Washington, D.C.: USDA Natural Resources Conservation Service).

Wadsworth, G. R., R. J. Southhard, and M. J. Singer. 1988. "Effects of fallow length on organic carbon and soil fabric of some tropical Udults," *Soil Sci. Soc. Amer. J.,* **52**:1424–30.

Appendix A

U.S. SOIL TAXONOMY
SUBORDER MAP AND LEGEND

Soil surveys have been prepared for most counties of the continental United States. The Soil Survey Staff of the U.S. Natural Resources Conservation Service, in cooperation with scientists from other countries, have used information from these surveys and analyses to develop generalized soils maps based on *Soil Taxonomy*. Areas dominated by specific soil orders and suborders are delineated on the map using map symbols (A3a, D1s, etc.). The soil orders and suborders to which these symbols refer are shown on the page facing the map. It should be kept in mind that while the map units represent areas of land dominated by the soil orders and suborders indicated, each unit is also likely to include some soils that differ from those listed in the map legend.

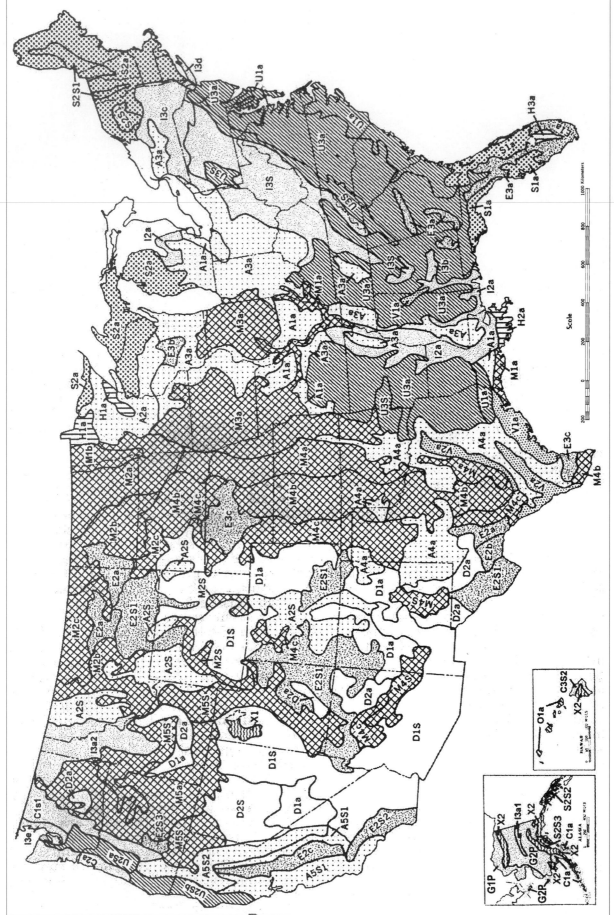

FIGURE A.1 General soil map of the United States showing patterns of soil orders and suborders based on *Soil Taxonomy*. Explanations of symbols follow. [Courtesy USDA Natural Resources Conservation Service, National Soil Survey Center (1998)]

ALFISOLS

Aqualfs
A1a Aqualfs with Udalfs, Aquepts, Udolls, Aquolls; gently sloping.

Cryalfs
A2S Cryalfs with Cryolls, Cryepts, Haplocryods, and Rock outcrops; steep.

Udalfs
A2a Udalfs with Udipsamments and Histosols; gently and moderately sloping.
A3a Udalfs with Aqualfs, Aquolls, Rendolls, Udolls, and Udults; gently or moderately sloping.

Ustalfs
A4a Ustalfs with Ustepts, Ustolls, Usterts, Ustipsamments, and Ustorthents; gently or moderately sloping.

Xeralfs
A5S1 Xeralfs with Xerolls, Xerorthents, and Xererts; moderately sloping to steep.
A5S2 Ultic and lithic subgroups of Haploxeralfs with Xerands, Xerults, Xerolls, and Xerepts; steep.

ANDISOLS

Cryands
C1a Cryands, Cryaquepts, Cryaquands, Histosols, and Rock outcrop; gently or moderately sloping.
C1S1 Cryands with Cryepts and Cryods; steep.

Udands
C2a Udands with Udepts, Spodosols, and Aquepts.

Ustands
C3S2 Ustands with Ustepts, Ustolls, Udands, and Folists; moderately sloping to steep.

ARIDISOLS

Argids
D1a Argids with Cambids, Calcids, Orthents, Psamments, and Ustolls; gently and moderately sloping.
D1S Argids with Cambids, gently sloping; and Torriorthents, gently sloping to steep.

Cambids
D2a Cambids with Argids, Orthents, Psamments, and Xerolls; gently or moderately sloping.
D2S Cambids, gently sloping to steep, with Argids, gently sloping; lithic subgroups of Torriorthents and Xerorthents, both steep.

ENTISOLS

Aquents
E1a Aquents with Quartzipsamments, Aquepts, Aquolls, and Aquods; gently sloping.

Orthents
E2a Torriorthents, steep, with xeric and ustic subgroups of Aridisols; Usterts and aridic and vertic subgroups of Ustolls; gently or moderately sloping.
E2b Torriorthents with Torrerts; gently or moderately sloping.
E2c Xerorthents with Xeralfs, Cambids, and Argids; gently sloping.
E2S1 Torriorthents; steep, and Argids, Torrifluvents, and Ustolls; gently sloping
E2S2 Xerorthents with Xeralfs and Xerolls; steep.
E2S3 Cryorthents with Cryopsamments and Cryands; gently sloping to steep.

Psamments
E3a Quartzipsamments with Aquults and Udults; gently or moderately sloping.
E3b Udipsamments with Aquolls and Udalfs; gently or moderately sloping.
E3c Ustipsamments with Ustalfs and Aquolls; gently or moderately sloping.

GELISOLS

G1P Gelisols; gently sloping to steep.
G2P Gelisols with Cryorthents, Cryepts, and cryic great groups of Histosols; gently sloping to steep.

HISTOSOLS

H1a Haplohemists with Psammaquents and Udipsamments; gently sloping.
H2a Haplohemists and Haplosaprists with Fluvaquents, Hydraquents and Aquepts; gently sloping.
H3a Haplofibrists, Haplohemists, and Haplosaprists with Psammaquents; gently sloping.

INCEPTISOLS

Aquepts
I2a Epi and endo great groups of Aquepts with Aqualfs, Aquerts, Aquolls, Udalfs, and Fluvaquents; gently sloping.

Cryepts
I3a1 Cryepts with Cryaquepts. Histosols, and Cryods; gently or moderately sloping.
I3a2 Cryepts, Cryands with Xerepts, Xerolls, and Cryods; moderately sloping to steep.

Udepts
I3b Eutrudepts with Uderts; gently sloping.
I3c Fragudepts with Fragaquepts; gently or moderately sloping; and Dystrudepts, steep.
I3d Dystrudepts with Udipsarments and Haplorthods; gently sloping.
I3S Dystrudepts, steep, with Udalfs and Udults; gently or moderately sloping.

Xerepts
I3e Xerepts, Xerands with Aquepts and Orthods; gently or moderately sloping.

MOLLISOLS

Aquolls
M1a Aquolls with Udalfs, Fluvents, Udipsamments, Aquepts, and Eutrudepts; gently sloping.
M1b Aquolls and Aquerts with Udolls and Udipsamments; gently sloping.

Udolls
M2a Udolls with Aquolls and Udorthents; gently sloping.
M3a Udolls with Aquolls, Udalfs, Aqualfs, Fluvents, Udipsamments, Udorthents, Aquepts, and Albolls; gently or moderately sloping.

Ustolls
M2b Typic subgroups of Ustolls with Ustipsamments, Ustorthents, and Cryalfs; gently sloping.
M2c Aridic subgroups of Ustolls and Ustic subgroups of Argids, Cambids, and Torriorthents; gently sloping.
M2S Ustolls with Ustalfs, Argids, Torriorthents, and Ustolls; moderately sloping or steep.
M4a Udic subgroups of Ustolls with Ustorthents, Ustepts, Usterts, Aquents, Fluvents, and Udolls; gently or moderately sloping.
M4b Typic subgroups of Ustolls with Ustalfs, Ustipsamments, Ustorthents, Ustepts, Aquolls, and Usterts; gently or moderately sloping.
M4c Aridic subgroups of Ustolls with Ustalfs, Cambids, Ustipsamments, Ustorthents, Ustepts, Torriorthents, and Usterts; gently or moderately sloping.
M4S Ustolls with Argids and Torriorthents; moderately sloping or steep.

Xerolls
M5a Xerolls with Argids, Cambids, Fluvents, Cryalfs, Cryolls, and Xerorthents; gently or moderately sloping.
M5S Xerolls with Cryalfs, Xeralfs, Xerorthents, and Xererts; moderately sloping or steep.

OXISOLS

O1a Ustox with Dystrustepts, Ustolls, Andisols, Udox, and Torrox; gently to steeply sloping.

SPODOSOLS

Aquods
S1a Aquods with Psammaquents, Aquolls, Alorthods, and Aquults; gently sloping.

Orthods
S2a Orthods with Udalfs, Aquents, Udorthents, Udipsamments, Histosols, Aquepts, Fragudepts, and Dystrudepts; gently or moderately sloping.
S2S1 Orthods and Cryods with Histosols, Aquents, and Aquepts; moderately sloping or steep.

Cryods
S2S2 Cryods with Histosols; moderately sloping or steep.
S2S3 Cryods with Cryic great groups of Histosols, Cryands and Cryaquepts; gently sloping to steep.

ULTISOLS

Aquults
U1a Aquults with Aquents, Histosols, Quartzipsamments, and Udults; gently sloping.

Humults
U2S3a Humults and Xeralfs with Xerolls, Aquepts and Xerults; gently sloping to moderately sloping.
U2Sb Humults, Xeralfs, and Xerults with Xerolls and Rock outcrop; gently sloping to steep.

Udults
U3a Udults with Udalfs, Fluvents, Aquents, Quartzipsamments, Aquepts, Dystrudepts, and Aquults; gently or moderately sloping.
U3S Udults with Dystrudepts; moderately sloping or steep.

VERTISOLS

Uderts
V1a Uderts with Aqualfs, Aquerts, Eutrudepts, Aquolls, and Ustolls; gently sloping.

Usterts
V2a Usterts with Aqualfs, Cambids, Aquerts, Fluvents, Aquolls, Ustolls, and Torrerts; gently sloping.

Miscellaneous Areas *(with little soil)*

X1 Salt flats and Playas.
X2 Rock outcrop (plus permanent snow fields, glaciers, and lava flows).

Slope Classes

Gently sloping—Slopes mainly less than 10%, including nearly level.
Moderately sloping—Slopes mainly between 10 and 25%.
Steep—Slopes mainly steeper than 25%.

Appendix B
CANADIAN AND FAO SOIL CLASSIFICATION SYSTEMS

The Canadian Soil Classification System is one of many national soil classification systems used in various countries around the world. Of these, it is perhaps the most closely aligned with the U.S. *Soil Taxonomy*. It includes five hierarchical categories: order, great group, subgroup, family, and series. The system is designed to apply principally to the soils of Canada. The soil orders of the Canadian System of Soil Classification are described in Table B.1 and soil orders and some great groups are compared to the U.S. *Soil Taxonomy* in Table B.2.

TABLE B.1   Summary with Brief Descriptions of the Soil Orders in the Soil Classification System of Canada

Brunisolic	Soils with sufficient development to exclude them from the Regosolic order, but lacking the degree or kind of horizon development specified for other soil orders.
Chernozemic	Soils with high base saturation and surface horizons darkened by the accumulation of organic matter from the decomposition of plants from grassland or grassland-forest ecosystems.
Cryosolic	Soils formed in either mineral or organic materials that have permafrost either within 1 m of the surface or within 2 m if more than one-third of the pedon has been strongly cryoturbated, as indicated by disrupted, mixed, or broken horizons.
Gleysolic	Gleysolic soils have features indicative of periodic or prolonged saturation (i.e., gleying, mottling) with water and reducing conditions.
Luvisolic	Soils with light-colored, eluvial horizons that have illuvial B horizons in which silicate clay has accumulated.
Organic	Organic soils developed on well- to undecomposed peat or leaf litter.
Podzolic	Soils with a B horizon in which the dominant accumulation product is amorphous material composed mainly of humified organic matter combined in varying degrees with Al and Fe.
Regosolic	Weakly developed soils that lack development of genetic horizons.
Solonetzic	Soils that occur on saline (often high in sodium) parent materials, which have B horizons that are very hard when dry and swell to a sticky mass of very low permeability when wet. Typically the solonetzic B horizon has prismatic or columnar macrostructure that breaks to hard to extremely hard, blocky peds with dark coatings.
Vertisolic	Soils with high contents of expanding clays that have large cracks during the dry parts of the year and show evidence of swelling, such as gilgae and slickensides.

TABLE B.2 Comparison of U.S. *Soil Taxonomy* and the Canadian Soil Classification System

Note that because the boundary criteria differ between the two systems, certain U.S. Soil Taxonomy *soil orders have equivalent members in more than one Canadian Soil Classification System soil order.*[a]

U.S. Soil Taxonomy soil order	Canadian system soil order	Canadian system great group	Equivalent lower-level taxa in U.S. Soil Taxonomy
Alfisols	Luvisolic	Gray Brown Luvisols	Hapludalfs
		Gray Luvisols	Haplocryalfs, Eutrocryalfs, Fragudalfs, Glossocryalfs, Palecryalfs, and some subgroups of Ustalfs and Udalfs
	Solonetzic	Solonetz	Natrudalfs and Natrustalfs
		Solod	Glossic subgroups of Natraqualfs, Natrudalfs, and Natrustalfs
Andisols	Components of Brunisolic and Cryosolic		
Aridisols	Solonetzic		Frigid families of Natrargids
Entisols	Regosolic		Cryic great groups and frigid families of Entisols, except Aquents
		Regosol	Cryic great groups and frigid families of Folists, Fluvents, Orthents, and Psamments
Gelisols	Cryosolic	Turbic Cryosol	Turbels
		Organic Cryosol	Histels
		Stagnic Cryosol	Orthels
Histosols	Organic	Fibrisol	Cryofibrists, Sphagnofibrists
		Mesisol	Cryohemists
		Humisol	Cryosaprists
Inceptisols	Brunisolic	Melanic Brunisol	Some Eutrustepts
		Eutric Brunisol	Subgroups of Cryepts; frigid and mesic families of Haplustepts
		Sombric Brunisol	Frigid and mesic families of Udepts, and Ustept and Humic Dystrudepts
		Dystric Brunisol	Frigid families of Dystrudepts and Dystrocryepts
	Gleysolic		Cryic subgroups and frigid families of Aqualfs, Aquolls, Aquepts, Aquents, and Aquods
		Humic Gleysol	Humaquepts
		Gleysol	Cryaquepts and frigid families of Fragaquepts, Epiaquepts, and Endoaquepts
Mollisols	Chernozemic	Brown	Xeric and Ustic subgroups of Argicryolls and Haplocryolls
		Dark Brown	Subgroups of Argicryolls and Haplocryolls
		Black	Typic subgroups of Argicryolls and Haplocryolls
		Dark Gray	Alfic subgroups of Argicryolls
	Solonetzic	Solonetz	Natricryolls and frigid families of Natraquolls and Natralbolls
		Solod	Glossic subgroups of Natricryolls
Oxisols	Not relevant in Canada		
Spodosols	Podzolic	Humic Podzol	Humicryods, Humic Placocryods, Placohumods, and frigid families of other Humods
		Ferro-Humic Podzol	Humic Haplocryods, some Placorthods, and frigid families of humic subgroups of other Orthods
		Humo-Ferric Podzol	Haplorthods, Placorthods, and frigid families of other Orthods and Cryods except humic subgroups
Ultisols	Not relevant in Canada		
Vertisols	Vertisolic	Vertisol	Haplocryerts
		Humic Vertisol	Humicryerts

[a] Based on information in Soil Classification Working Group, 1998, *The Canadian System of Soil Classification*, 3d ed. (Ottawa: Agriculture and Agri-Food Canada). Publication No. A53-1646/1997E.

The FAO Soil Classification System is really a map legend for the world soil map made by FAO/UNESCO. A brief description of the map units is given in Table B.3. The FAO map legend terms are used by many scientists around the world to describe different kinds of soils. Global and regional soil databases are often organized according to this system. The soil characteristics recognized and the terminology used borrow heavily from the Russian soil classification system, as well as from U.S. *Soil Taxonomy.*

TABLE B.3 Soil Map Units for the FAO/UNESCO Soil Map of the World

Acrisols	Low base status soils with argillic horizons
Andosols	Soils formed in volcanic ash that have dark surfaces
Arenosols	Soils formed from sand
Cambisols	Soils with slight color, structure, or consistency change due to weathering
Chernozems	Soils with black surface, high humus under prairie vegetation
Cryosols	Soils of cold climates with permafrost
Ferralsols	Highly weathered soils with sesquioxide-rich clays
Fluvisols	Water-deposited soils with little alteration
Gleysols	Soils with mottled or reduced horizons due to wetness
Greyzems	Soils with dark surface, bleached E horizon, and textural B horizon
Histosols	Organic soils
Kastanozems	Soils with chestnut surface color, steppe vegetation
Lithosols	Shallow soils over hard rock
Luvisols	Medium to high base status soils with argillic horizons
Nitosols	Soils with low CEC clay in argillic horizons
Planosols	Soils with abrupt A–B horizon contact
Phaeozems	Soils with dark surface, more leached than Kastanozem or Chernozem
Podzols	Soils with light-colored eluvial horizon and subsoil accumulation of iron, aluminum, and humus
Podzoluvisols	Soils with leached horizons tonguing into argillic B horizons
Rankers	Thin soils over siliceous material
Regosols	Thin soil over unconsolidated material
Rendzinas	Shallow soils over limestone
Solonchaks	Soils with soluble salt accumulation
Solonetz	Soils with high sodium content
Vertisols	Self-mulching, inverting soils, rich in montmorillinitic clay
Xerosols	Dry soils of semiarid regions
Yermosols	Desert soils

Appendix C

SI Unit Conversion Factors and Periodic Table of the Elements

Non-SI Unit	Multiply by[a]	To obtain SI Unit
Length		
inch, in.	2.54	centimeters, cm (10^{-2} m)
foot, ft	0.304	meter, m
mile,	1.609	kilometer, km (10^3 m)
micron, μ	1.0	micrometer, μm (10^{-6} m)
Ångstrom unit, Å	0.1	nanometer, nm (10^{-9} m)
Area		
acre, ac	0.405	hectare, ha (10^4 m^2)
square foot, ft^2	9.29×10^{-2}	square meter, m^2
square inch, in^2	645	square millimeter, mm^2
square mile, mi^2	2.59	square kilometer, km^2
Volume		
bushel, bu	35.24	liter, L
cubic foot, ft^3	2.83×10^{-2}	cubic meter, m^3
cubic inch, in.^{3}	1.64×10^{-5}	cubic meter, m^3
gallon (U.S.), gal	3.78	liter, L
quart, qt	0.946	liter, L
acre-foot, ac-ft	12.33	hectare-centimeter, ha-cm
acre-inch, ac-in.	1.03×10^{-2}	hectare-meters, ha-m
ounce (fluid), oz	2.96×10^{-2}	liter, L
pint, pt	0.473	liter, L
Mass		
ounce (avdp), oz	28.4	gram, g
pound, lb	0.454	kilogram, kg (10^3 g)
ton (2000 lb)	0.907	megagram, Mg (10^6 g)
tonne (metric), t	1000	kilogram, kg
Radioactivity		
curie, Ci	3.7×10^{10}	becquerel, Bq
picocurie per gram, pCi/g	37	becquerel per kilogram, Bq/kg
Yield and Rate		
bushel per acre (60 lb), bu/ac	67.19	kilogram per hectare, kg/ha
bushel per acre (56 lb), bu/ac	62.71	kilogram per hectare, kg/ha
bushel per acre (48 lb), bu/ac	53.75	kilogram per hectare, kg/ha
gallon per acre (U.S.), gal/ac	9.35	liter per hectare, L/ha
ton (2000 lb) per acre	2.24	megagram per hectare, Mg/ha
miles per hour, mph	0.447	meter per second, m/s
gallon per minute (U.S.), gpm	0.227	cubic meter per hour, m^3/h
cubic feet per second, cfs	101.9	cubic meter per hour, m^3/h
Pressure		
atmosphere, atm	0.101	megapascal, MPa (10^6 Pa)
bar	0.1	megapascal, MPa
pound per square foot, lb/ft^2	47.9	pascal, Pa
pound per square inch, lb/in^2	6.9×10^3	pascal, Pa
Temperature		
degrees Fahrenheit (°F − 32)	0.556	degrees, °C
degrees Celsius (°C + 273)	1	Kelvin, K
Energy		
British thermal unit, Btu	1.05×10^3	joule, J
calorie, cal	4.19	joule, J
dyne, dyn	10^{-5}	newton, N
erg	10^{-7}	joule, J
foot-pound, ft-lb	1.36	joule, J
Concentrations		
percent, %	10	gram per kilogram, g/kg
part per million, ppm	1	milligram per kilogram, mg/kg
milliequivalents per 100 grams	1	centimole per kilogram, cmol/kg

[a] To convert from SI to non-SI units, *divide* by the factor given.

Periodic Table of the Elements with Notes Concerning Relevance to Soil Science

Based on atomic mass of $^{12}C = 12.0$. Numbers in parentheses are the mass numbers of the most stable isotopes of radioactive elements.

Group IA																	Group VIIIA
1 H 1.01 Hydrogen	Group IIA											Group IIIA	Group IVA	Group VA	Group VIA	Group VIIA	2 He 4.00 Helium
3 Li 6.94 Lithium	4 Be 9.01 Beryllium											5 B 10.81 Boron	6 C 12.01 Carbon	7 N 14.01 Nitrogen	8 O 16.00 Oxygen	9 F 19.00 Fluorine	10 Ne 20.18 Neon
11 Na 22.99 Sodium	12 Mg 24.30 Magnesium	Group IIIB	Group IVB	Group VB	Group VIB	Group VIIB	Group VIIIB	Group	Group	Group IB	Group IIB	13 Al 26.98 Aluminum	14 Si 28.09 Silicon	15 P 30.97 Phosphorus	16 S 32.07 Sulfur	17 Cl 35.45 Chlorine	18 Ar 39.95 Argon
19 K 39.10 Potassium	20 Ca 40.08 Calcium	21 Sc 44.96 Scandium	22 Ti 47.88 Titanium	23 V 50.94 Vanadium	24 Cr 52.00 Chromium	25 Mn 54.94 Manganese	26 Fe 55.85 Iron	27 Co 58.93 Cobalt	28 Ni 58.69 Nickel	29 Cu 63.55 Copper	30 Zn 65.38 Zinc	31 Ga 69.72 Gallium	32 Ge 72.59 Germanium	33 As 74.92 Arsenic	34 Se 78.96 Selenium	35 Br 79.90 Bromine	36 Kr 83.80 Krypton
37 Rb 85.47 Rubidium	38 Sr 87.62 Strontium	39 Y 88.91 Yttrium	40 Zr 91.22 Zirconium	41 Nb 92.91 Niobium	42 Mo 95.94 Molybdenum	43 Tc (98) Technetium	44 Ru 101.07 Ruthenium	45 Rh 102.91 Rhodium	46 Pd 106.42 Palladium	47 Ag 107.87 Silver	48 Cd 112.41 Cadmium	49 In 114.82 Indium	50 Sn 118.71 Tin	51 Sb 121.75 Antimony	52 Te 127.60 Tellurium	53 I 126.90 Iodine	54 Xe 131.29 Xenon
55 Cs 132.91 Cesium	56 Ba 137.33 Barium	57 La 138.91 Lanthanum	72 Hf 178.49 Hafnium	73 Ta 180.95 Tantalum	74 W 183.85 Tungsten	75 Re 186.21 Rhenium	76 Os 190.2 Osmium	77 Ir 192.22 Iridium	78 Pt 195.08 Platinum	79 Au 196.97 Gold	80 Hg 200.59 Mercury	81 Tl 204.38 Thallium	82 Pb 207.2 Lead	83 Bi 208.98 Bismuth	84 Po (209) Polonium	85 At (210) Astatine	86 Rn (222) Radon
87 Fr (223) Francium	88 Ra (226) Radium	89 Ac (227) Actinium	104 Unq (261) Unnilquadium	105 Unp (262) Unnilpentium	106 Unh (263) Unnilhexium	107 Uns (262) Unnilseptium	108 Uno (265) Unniloctium										

58 Ce 140.12 Cerium	59 Pr 140.91 Praseodymium	60 Nd 144.24 Neodymium	61 Pm (145) Promethium	62 Sm 150.36 Samarium	63 Eu 151.96 Europium	64 Gd 157.25 Gadolinium	65 Tb 158.93 Terbium	66 Dy 162.50 Dysprosium	67 Ho 164.93 Holmium	68 Er 167.26 Erbium	69 Tm 168.93 Thulium	70 Yb 173.04 Ytterbium	71 Lu 174.97 Lutetium
90 Th (232) Thorium	91 Pa (231) Protactinium	92 U (238) Uranium	93 Np (237) Neptunium	94 Pu (244) Plutonium	95 Am (243) Americium	96 Cm (247) Curium	97 Bk (247) Berkelium	98 Cf (251) Californium	99 Es (252) Einsteinium	100 Fm (257) Fermium	101 Md (258) Mendelevium	102 No (259) Nobelium	103 Lr (260) Lawrencium

Nonmetals

Metals

Atomic number
Symbol
Atomic mass

87
Fr
(223)
Francium

Elements known to be nutrients for animals or plants. Some are also toxic in excessive amounts.

Elements toxic to organisms in small amounts, and not known to serve as nutrients.

Other elements commonly studied in Soil Science because of soil-environmental impacts or because of their use as tracers or electrodes. (Br is used to trace anionic solutes such as nitrate. Isotopes of Rb and Sr are used to trace K and Ca in plants and soils. Cs and Ti are used to trace geological processes such as soil erosion. Pt and Ag are used in electrodes for measuring soil redox potential and pH, respectively.)

GLOSSARY OF SOIL SCIENCE TERMS[1]

A horizon The surface horizon of a mineral soil having maximum organic matter accumulation, maximum biological activity, and/or eluviation of materials such as iron and aluminum oxides and silicate clays.

abiotic Nonliving basic elements of the environment, such as rainfall, temperature, wind, and minerals.

absorption, active Movement of ions and water into the plant root as a result of metabolic processes by the root, frequently against an activity gradient.

absorption, passive Movement of ions and water into the plant root as a result of diffusion along a gradient.

accelerated erosion *See* erosion.

acid rain Atmospheric precipitation with pH values less than about 5.6, the acidity being due to inorganic acids such as nitric and sulfuric that are formed when oxides of nitrogen and sulfur are emitted into the atmosphere.

acid soil A soil with a pH value <7.0. Usually applied to surface layer or root zone, but may be used to characterize any horizon. *See also* reaction, soil.

acid sulfate soils Soils that are potentially extremely acid (pH < 3.5) because of the presence of large amounts of reduced forms of sulfur that are oxidized to sulfuric acid if the soils are exposed to oxygen when they are drained or excavated. A sulfuric horizon containing the yellow mineral jarosite is often present. *See also* cat clays.

acidity, active The activity of hydrogen ion in the aqueous phase of a soil. It is measured and expressed as a pH value.

acidity, residual Soil acidity that can be neutralized by lime or other alkaline materials but cannot be replaced by an unbuffered salt solution.

acidity, salt replaceable Exchangeable hydrogen and aluminum that can be replaced from an acid soil by an unbuffered salt solution such as KCl or NaCl.

---

[1] This glossary was compiled and modified from several sources, including *Glossary of Soil Science Terms* [Madison, Wis.: Soil Sci. Soc. Amer. (1997)] *Resource Conservation Glossary* [Anheny, Iowa: Soil Cons. Soc. Amer. (1982)], and *Soil Taxonomy* (Washington, D.C.: U.S. Department of Agriculture (1998)].

acidity, total   The total acidity in a soil. It is approximated by the sum of the salt-replaceable acidity plus the residual acidity.

actinomycetes   A group of organisms intermediate between the bacteria and the true fungi that usually produce a characteristic branched mycelium. Includes many, but not all, organisms belonging to the order of Actinomycetales.

activated sludge   Sludge that has been aerated and subjected to bacterial action.

active layer   The upper portion of a Gellisol that is subject to freezing and thawing and is underlain by permafrost.

active organic matter   A portion of the soil organic matter that is relatively easily metabolized by microorganisms and cycles with a half-life in the soil of a few days to a few years.

adhesion   Molecular attraction that holds the surfaces of two substances (e.g., water and sand particles) in contact.

adsorption   The attraction of ions or compounds to the surface of a solid. Soil colloids adsorb large amounts of ions and water.

adsorption complex   The group of organic and inorganic substances in soil capable of adsorbing ions and molecules.

aerate   To impregnate with gas, usually air.

aeration, soil   The process by which air in the soil is replaced by air from the atmosphere. In a well-aerated soil, the soil air is similar in composition to the atmosphere above the soil. Poorly aerated soils usually contain more carbon dioxide and correspondingly less oxygen than the atmosphere above the soil.

aerobic   (1) Having molecular oxygen as a part of the environment. (2) Growing only in the presence of molecular oxygen, as aerobic organisms. (3) Occurring only in the presence of molecular oxygen (said of certain chemical or biochemical processes, such as aerobic decomposition).

aerosolic dust   A type of eolian material that is very fine (about 1 to 10 μm) and may remain suspended in the air over distances of thousands of kilometers. Finer than most *loess*.

aggregate (soil)   Many soil particles held in a single mass or cluster, such as a clod, crumb, block, or prism.

agric horizon   *See* diagnostic subsurface horizons.

agronomy   A specialization of agriculture concerned with the theory and practice of field-crop production and soil management. The scientific management of land.

air-dry   (1) The state of dryness (of a soil) at equilibrium with the moisture content in the surrounding atmosphere. The actual moisture content will depend upon the relative humidity and the temperature of the surrounding atmosphere. (2) To allow to reach equilibrium in moisture content with the surrounding atmosphere.

air porosity   The proportion of the bulk volume of soil that is filled with air at any given time or under a given condition, such as a specified moisture potential; usually the large pores.

albic horizon   *See* diagnostic subsurface horizons.

Alfisols   *See* soil classification.

algal bloom   A population explosion of algae in surface waters, such as lakes and streams, often resulting in high turbidity and green- or red-colored water, and commonly stimulated by nutrient enrichment with phosphorus and nitrogen.

alkali soil   (Obsolete) A soil that contains sufficient alkali (sodium) to interfere with the growth of most crop plants. *See also* saline-sodic soil; sodic soil.

alkaline soil   Any soil that has pH > 7. Usually applied to the surface layer or root zone but may be used to characterize any horizon or a sample thereof. *See also* reaction, soil.

allelochemical   An organic chemical by which one plant can influence another. *See* allelopathy.

allelopathy The process by which one plant may affect other plants by biologically active chemicals introduced into the soil, either directly by leaching or exudation from the source plant, or as a result of the decay of the plant residues. The effects, though usually negative, may also be positive.

allophane A poorly defined aluminosilicate mineral whose structural framework consists of short runs of three-dimensional crystals interspersed with amorphous noncrystalline materials. Along with its more weathered companion, it is prevalent in volcanic ash materials.

alluvial fan Fan-shaped alluvium deposited at the mouth of a canyon or ravine where debris-laden waters fan out, slow down, and deposit their burden.

alluvial soil (Obsolete) A soil developing from recently deposited alluvium and exhibiting essentially no horizon development or modification of the recently deposited materials.

alluvium A general term for all detrital material deposited or in transit by streams, including gravel, sand, silt, clay, and all variations and mixtures of these. Unless otherwise noted, alluvium is unconsolidated.

alpha particle A positively charged particle (consisting of two protons and two neutrons) that is emitted by certain radioactive compounds.

aluminosilicates Compounds containing aluminum, silicon, and oxygen as main constituents. An example is microcline, $KA1Si_3O_8$.

amendment, soil Any substance other than fertilizers, such as lime, sulfur, gypsum, and sawdust, used to alter the chemical or physical properties of a soil, generally to make it more productive.

amino acids Nitrogen-containing organic acids that couple together to form proteins. Each acid molecule contains one or more amino groups ($-NH_2$) and at least one carboxyl group ($-COOH$). In addition, some amino acids contain sulfur.

ammonification The biochemical process whereby ammoniacal nitrogen is released from nitrogen-containing organic compounds.

ammonium fixation The entrapment of ammonium ions by the mineral or organic fractions of the soil in forms that are insoluble in water and are at least temporarily nonexchangeable.

amorphous material Noncrystalline constituents of soils.

anaerobic (1) Without molecular oxygen. (2) Living or functioning in the absence of air or free oxygen.

andic properties Soil properties related to volcanic origin of materials, including high organic carbon content, low bulk density, high phosphate retention, and extractable iron and aluminum.

Andisols *See* soil classification.

angle of repose The maximum slope steepness at which loose, cohesionless material will come to rest.

anion Negatively charged ion; during electrolysis it is attracted to the positively charged anode.

anion exchange Exchange of anions in the soil solution for anions adsorbed on the surface of clay and humus particles.

anion exchange capacity The sum total of exchangeable anions that a soil can adsorb. Expressed as centimoles of charge per kilogram ($cmol_c/kg$) of soil (or of other adsorbing material, such as clay).

anoxic *See* anaerobic.

anthropic epipedon *See* diagnostic surface horizons.

antibiotic A substance produced by one species of organism that, in low concentrations, will kill or inhibit growth of certain other organisms.

Ap The surface layer of a soil disturbed by cultivation or pasturing.

apatite A naturally occurring complex calcium phosphate that is the original source of most of the phosphate fertilizers. Formulas such as $[3Ca_3(PO_4)_2] \cdot CaF_2$ illustrate the complex compounds that make up apatite.

aquic conditions Continuous or periodic saturation (with water) and reduction, commonly indicated by redoximorphic features.

aquiclude A saturated body of rock or sediment that is incapable of transmitting significant quantities of water under ordinary water pressures.

aquifer A saturated, permeable layer of sediment or rock that can transmit significant quantities of water under normal pressure conditions.

arbuscule Specialized branched structure formed within a root cortical cell by endotrophic mycorrhizal fungi.

argillan A thin coating of well-oriented clay particles on the surface of a soil aggregate, particle, or pore. A clay film.

argillic horizon *See* diagnostic subsurface horizons.

arid climate Climate in regions that lack sufficient moisture for crop production without irrigation. In cool regions annual precipitation is usually less than 25 cm. It may be as high as 50 cm in tropical regions. Natural vegetation is desert shrubs.

Aridisols *See* soil classification.

aspect (of slopes) The direction (e.g., south or north) that a slope faces with respect to the sun.

association, soil *See* soil association.

Atterberg limits Water contents of fine-grained soils at different states of consistency.
  plastic limit (PL) The water content corresponding to an arbitrary limit between the plastic and semisolid states of consistency of a soil.
  liquid limit (LL) The water content corresponding to the arbitrary limit between the liquid and plastic states of consistency of a soil.

autochthonous organisms Those microorganisms thought to subsist on the more resistant soil organic matter and little affected by the addition of fresh organic materials. *Contrast with* zymogenous organisms.

autotroph An organism capable of utilizing carbon dioxide or carbonates as the sole source of carbon and obtaining energy for life processes from the oxidation of inorganic elements or compounds such as iron, sulfur, hydrogen, ammonium, and nitrites, or from radiant energy. *Contrast with* heterotroph.

available nutrient That portion of any element or compound in the soil that can be readily absorbed and assimilated by growing plants. ("Available" should not be confused with "exchangeable.")

available water The portion of water in a soil that can be readily absorbed by plant roots. The amount of water released between the field capacity and the permanent wilting point.

B horizon A soil horizon, usually beneath the A horizon, that is characterized by one or more of the following: (1) a concentration of silicate clays, iron and aluminum oxides, and humus, alone or in combination; (2) a blocky or prismatic structure; and (3) coatings of iron and aluminum oxides that give darker, stronger, or redder color.

bar A unit of pressure equal to 1 million dynes per square centimeter (10^6 dynes/cm^2). It approximates the pressure of a standard atmosphere.

base-forming cations Those cations that form strong (strongly dissociated) bases by reaction with hydroxyl; e.g., K^+ forms potassium hydroxide ($K^+ + OH^-$).

base saturation percentage The extent to which the adsorption complex of a soil is saturated with exchangeable cations other than hydrogen and aluminum. It is expressed as a percentage of the total cation exchange capacity.

BC soil A soil profile with B and C horizons but with little or no A horizon. Most BC soils have lost their A horizons by erosion.

bedding (Engineering) Arranging the surface of fields by plowing and grading into a series of elevated beds separated by shallow depressions or ditches for drainage.

bedrock The solid rock underlying soils and the regolith in depths ranging from zero (where exposed by erosion) to several hundred feet.

bench terrace An embankment constructed across sloping fields with a steep drop on the downslope side.

beta particle A high-speed electron emitted in radioactive decay.

bioaccumulation A buildup within an organism of specific compounds due to biological processes. Commonly applied to heavy metals, pesticides or metabolites.

biodegradable Subject to degradation by biochemical processes.

biomass The total mass of living material of a specified type (e.g., microbial biomass) in a given environment (e.g., in a cubic meter of soil).

biopores Soil pores, usually of relatively large diameter, created by plant roots, earthworms, or other soil organisms.

bioremediation The decontamination or restoration of polluted or degraded soils by means of enhancing the chemical degradation or other activities of soil organisms.

biosequence A group of related soils that differ, one from the other, primarily because of differences in kinds and numbers of plants and soil organisms as a soil-forming factor.

biosolids *See* sewage sludge.

bleicherde (Obsolete) The light-colored, leached A2(E) horizon of Spodosols.

blocky soil structure Soil aggregates with blocklike shapes; common in B horizons of soils in humid regions.

blown-out land Areas from which all or almost all of the soil and soil material has been removed by wind erosion. Usually unfit for crop production. A miscellaneous land type.

border-strip irrigation *See* irrigation methods.

bottomland *See* floodplain.

breccia A rock composed of coarse angular fragments cemented together.

broad-base terrace A low embankment with such gentle slopes that it can be farmed, constructed across sloping fields to reduce erosion and runoff.

broadcast Scatter seed or fertilizer on the surface of the soil.

buffering capacity The ability of a soil to resist changes in pH. Commonly determined by presence of clay, humus, and other colloidal materials.

bulk blending Mixing dry individual granulated fertilizer materials to form a mixed fertilizer that is applied promptly to the soil.

bulk density, soil The mass of dry soil per unit of bulk volume, including the air space. The bulk volume is determined before drying to constant weight at 105°C.

buried soil Soil covered by an alluvial, loessal, or other deposit, usually to a depth greater than the thickness of the solum.

by-pass flow *See* preferential flow.

C horizon A mineral horizon, generally beneath the solum, that is relatively unaffected by biological activity and pedogenesis and is lacking properties diagnostic of an A or B horizon. It may or may not be like the material from which the A and B have formed.

calcareous soil Soil containing sufficient calcium carbonate (often with magnesium carbonate) to effervesce visibly when treated with cold 0.1 N hydrochloric acid.

calcic horizon *See* diagnostic subsurface horizons.

caliche A layer near the surface, more or less cemented by secondary carbonates of calcium or magnesium precipitated from the soil solution. It may occur as a soft, thin soil horizon; as a hard, thick bed just beneath the solum; or as a surface layer exposed by erosion.

cambic horizon *See* diagnostic subsurface horizons.

capillary conductivity (Obsolete) *See* hydraulic conductivity.

capillary fringe A zone in the soil just above the plane of zero water pressure (water table) that remains saturated or almost saturated with water.

capillary water The water held in the capillary or *small* pores of a soil, usually with a tension >60 cm of water. *See also* moisture potential.

carbon cycle The sequence of transformations whereby carbon dioxide is fixed in living organisms by photosynthesis or by chemosynthesis, liberated by respiration and by the death and decomposition of the fixing organism, used by heterotrophic species, and ultimately returned to its original state.

carbon/nitrogen ratio The ratio of the weight of organic carbon (C) to the weight of total nitrogen (N) in a soil or in organic material.

carnivore An organism that feeds on animals.

casts, earthworm Rounded, water-stable aggregates of soil that have passed through the gut of an earthworm.

cat clays Wet clay soils high in reduced forms of sulfur that, upon being drained, become extremely acid because of the oxidation of the sulfur compounds and the formation of sulfuric acid. Usually found in tidal marshes. *See* acid sulfate soils.

catena A sequence of soils of about the same age, derived from similar parent material, and occurring under similar climatic conditions, but having different characteristics because of variation in *relief* and in *drainage*.

cation A positively charged ion; during electrolysis it is attracted to the negatively charged cathode.

cation exchange The interchange between a cation in solution and another cation on the surface of any surface-active material, such as clay or organic matter.

cation exchange capacity The sum total of exchangeable cations that a soil can adsorb. Sometimes called *total-exchange capacity, base-exchange capacity,* or *cation-adsorption capacity*. Expressed in centimoles of charge per kilogram ($cmol_c$/kg) of soil (or of other adsorbing material, such as clay).

cemented Indurated; having a hard, brittle consistency because the particles are held together by cementing substances, such as humus, calcium carbonate, or the oxides of silicon, iron, and aluminum.

channery Thin, flat fragments of limestone, sandstone, or schist up to 15 cm (6 in.) in major diameter.

chelate (Greek, claw) A type of chemical compound in which a metallic ion is firmly combined with an organic molecule by means of multiple chemical bonds.

chert A structureless form of silica, closely related to flint, that breaks into angular fragments.

chisel, subsoil A tillage implement with one or more cultivator-type feet to which are attached strong knifelike units used to shatter or loosen hard, compact layers, usually in the subsoil, to depths below normal plow depth. *See also* subsoiling.

chlorite A 2:1:1-type layer-structured silicate mineral having 2:1 layers alternating with a magnesium-dominated octahedral sheet.

chlorosis A condition in plants relating to the failure of chlorophyll (the green coloring matter) to develop. Chlorotic leaves range from light green through yellow to almost white.

chroma (color) *See* Munsell color system.

chronosequence A sequence of related soils that differ, one from the other, in certain properties primarily as a result of time as a soil-forming factor.

class, soil A group of soils having a definite range in a particular property, such as acidity, degree of slope, texture, structure, land-use capability, degree of erosion, or drainage. *See also* soil structure; soil texture.

classification, soil *See* soil classification.

clastic Composed of broken fragments of rocks and minerals.

clay (1) A soil separate consisting of particles <0.002 mm in equivalent diameter. (2) A soil textural class containing >40% clay, <45% sand, and <40% silt.

clay mineral Naturally occurring inorganic material (usually crystalline) found in soils and other earthy deposits, the particles being of clay size, that is, <0.002 mm in diameter.

claypan A dense, compact, slowly permeable layer in the subsoil having a much higher clay content than the overlying material, from which it is separated by a sharply defined boundary. Claypans are usually hard when dry and plastic and sticky when wet. *See also* hardpan.

clod A compact, coherent mass of soil produced artificially, usually by such human activities as plowing and digging, especially when these operations are performed on soils that are either too wet or too dry for normal tillage operations.

coarse texture The texture exhibited by sands, loamy sands, and sandy loams (except very fine sandy loam).

cobblestone Rounded or partially rounded rock or mineral fragments 7.5 to 25 cm (3 to 10 in.) in diameter.

cohesion Holding together: force holding a solid or liquid together, owing to attraction between like molecules. Decreases with rise in temperature.

collapsible soil Certain soil that may undergo a sudden loss in strength when wetted.

colloid, soil (Greek, gluelike) Organic and inorganic matter with very small particle size and a correspondingly large surface area per unit of mass.

colluvium A deposit of rock fragments and soil material accumulated at the base of steep slopes as a result of gravitational action.

color The property of an object that depends on the wavelength of light it reflects or emits.

columnar soil structure *See* soil structure types.

companion planting The practice of growing certain species of plants in close proximity because one species has the effect of improving the growth of the other, sometimes by positive *allelopathic* effects.

compost Organic residues, or a mixture of organic residues and soil, that have been piled, moistened, and allowed to undergo biological decomposition. Mineral fertilizers are sometimes added. Often called *artificial manure* or *synthetic manure* if produced primarily from plant residues.

concretion A local concentration of a chemical compound, such as calcium carbonate or iron oxide, in the form of grains or nodules of varying size, shape, hardness, and color.

conduction The transfer of heat by physical contact between two or more objects.

conductivity, hydraulic *See* hydraulic conductivity.

conifer A tree belonging to the order Coniferae, usually evergreen, with cones and needle-shaped or scalelike leaves and producing wood known commercially as *softwood*.

conservation tillage *See* tillage, conservation.

consistence The combination of properties of soil material that determine its resistance to crushing and its ability to be molded or changed in shape. Such terms as *loose, friable, firm, soft, plastic,* and *sticky* describe soil consistence.

consistency The interaction of adhesive and cohesive forces within a soil at various moisture contents as expressed by the relative ease with which the soil can be deformed or ruptured.

consociation *see* soil consociation.

consolidation test A laboratory test in which a soil mass is laterally confined within a ring and is compressed with a known force between two porous plates.

constant charge The net surface charge of mineral particles, the magnitude of which depends only on the chemical and structural composition of the mineral. The charge arises from isomorphous substitution and is not affected by soil pH.

consumptive use The water used by plants in transpiration and growth, plus water vapor loss from adjacent soil or snow, or from intercepted precipitation in any specified time. Usually expressed as equivalent depth of free water per unit of time.

contour An imaginary line connecting points of equal elevation on the surface of the soil. A contour terrace is laid out on a sloping soil at right angles to the direction of the slope and nearly level throughout its course.

contour strip-cropping Layout of crops in comparatively narrow strips in which the farming operations are performed approximately on the contour. Usually strips of grass, close–growing crops, or fallow are alternated with those of cultivated crops.

controlled traffic A farming system in which all wheeled traffic is confined to fixed paths so that repeated compaction of the soil does not occur outside the selected paths.

convection The transfer of heat through a gas or solution because of molecular movement.

corrugated irrigation *See* irrigation methods.

cover crop A close-growing crop grown primarily for the purpose of protecting and improving soil between periods of regular crop production or between trees and vines in orchards and vineyards.

creep Slow mass movement of soil and soil material down relatively steep slopes, primarily under the influence of gravity, but facilitated by saturation with water and by alternate freezing and thawing.

crop rotation A planned sequence of crops growing in a regularly recurring succession on the same area of land, as contrasted to continuous culture of one crop or growing different crops in haphazard order.

crotovina A former animal burrow in one soil horizon that has been filled with organic matter or material from another horizon (also spelled *krotovina*).

crumb A soft, porous, more or less rounded natural unit of structure from 1 to 5 mm in diameter. *See also* soil structure types.

crushing strength The force required to crush a mass of dry soil or, conversely, the resistance of the dry soil mass to crushing. Expressed in units of force per unit area (pressure).

crust A surface layer on soils, ranging in thickness from a few millimeters to perhaps as much as 3 cm, that is much more compact, hard, and brittle when dry than the material immediately beneath it.

cryoturbation Physical disruption and displacement of soil material within the profile by the forces of freezing and thawing. Sometimes called *frost churning,* it results in irregular, broken horizons, involutions, oriented rock fragments, and accumulation of organic matter on the permafrost table.

cryptogam Assemblage of algae, lichens, liverworts and mosses that commonly forms an irregular crust on the soil surface, especially on otherwise barren, arid-region soils. Sometimes referred to as *cryptogamic crusts.*

crystal A homogeneous inorganic substance of definite chemical composition bounded by plane surfaces that form definite angles with each other, thus giving the substance a regular geometrical form.

crystal structure The orderly arrangement of atoms in a crystalline material.

crystalline rock A rock consisting of various minerals that have crystallized in place from magma. *See also* igneous rock; sedimentary rock.

cultivation A tillage operation used in preparing land for seeding or transplanting or later for weed control and for loosening the soil.

cutans A modification of the texture, structure or fabric at natural surfaces in soil materials due to concentration of particular soil constituents.

cyanobacteria Chlorophyll-containing bacteria that accommodate both photosynthesis and nitrogen fixation. Formerly called blue-green algae.

deciduous plant A plant that sheds all its leaves every year at a certain season.

decomposition Chemical breakdown of a compound (e.g., a mineral or organic compound) into simpler compounds, often accomplished with the aid of microorganisms.

deflocculate (1) To separate the individual components of compound particles by chemical and/or physical means. (2) To cause the particles of the *disperse phase* of a colloidal system to become suspended in the *dispersion medium.*

delta An alluvial deposit formed where a stream or river drops its sediment load upon entering a quieter body of water.

denitrification The biochemical reduction of nitrate or nitrite to gaseous nitrogen, either as molecular nitrogen or as an oxide of nitrogen.

density *See* particle density; bulk density.

desalinization Removal of salts from saline soil, usually by leaching.

desert crust A hard layer, containing calcium carbonate, gypsum, or other binding material, exposed at the surface in desert regions.

desert pavement A natural residual concentration of closely packed pebbles, boulders, and other rock fragments on a desert surface where wind and water action has removed all smaller particles.

detritivore An organism that subsists on detritus.

detritus Debris from dead plants and animals.

desorption The removal of sorbed material from surfaces.

diagnostic horizons (As used in *Soil Taxonomy*): Horizons having specific soil characteristics that are indicative of certain classes of soils. Horizons that occur at the soil surface are called *epipedons;* those below the surface, *diagnostic subsurface horizons.*

diagnostic subsurface horizons The following diagnostic subsurface horizons are used in *Soil Taxonomy:*

    agric horizon A mineral soil horizon in which clay, silt, and humus derived from an overlying cultivated and fertilized layer have accumulated. The wormholes and illuvial clay, silt, and humus occupy at least 5% of the horizon by volume.

    albic horizon A mineral soil horizon from which clay and free iron oxides have been removed or in which the oxides have been segregated to the extent that the color of the horizon is determined primarily by the color of the primary sand and silt particles rather than by coatings on these particles.

    argillic horizon A mineral soil horizon characterized by the illuvial accumulation of layer-lattice silicate clays.

    calcic horizon A mineral soil horizon of secondary carbonate enrichment that is more than 15 cm thick, has a calcium carbonate equivalent of more than 15%, and has at least 5% more calcium carbonate equivalent than the underlying C horizon.

    cambic horizon A mineral soil horizon that has a texture of loamy very fine sand or finer, contains some weatherable minerals, and is characterized by the alteration or removal of mineral material. The cambic horizon lacks cementation or induration and has too few evidences of illuviation to meet the requirements of the argillic or spodic horizon.

    duripan A mineral soil horizon that is cemented by silica, to the point that air-dry fragments will not slake in water or HCl.

    gypsic horizon A mineral soil horizon of secondary calcium sulfate enrichment that is more than 15 cm thick.

    kandic horizon A horizon having a sharp clay increase relative to overlying horizons and having low-activity clays.

natric horizon   A mineral soil horizon that satisfies the requirements of an argillic horizon, but that also has prismatic, columnar, or blocky structure and a subhorizon having more than 15% saturation with exchangeable sodium.

oxic horizon   A mineral soil horizon that is at least 30 cm thick and is characterized by the virtual *absence* of weatherable primary minerals or 2:1 lattice clays and the *presence* of 1:1 lattice clays and highly insoluble minerals, such as quartz sand, hydrated oxides of iron and aluminum, low cation exchange capacity, and small amounts of exchangeable bases.

petrocalcic horizon   A continuous, indurated calcic horizon that is cemented by calcium carbonate and, in some places, with magnesium carbonate. It cannot be penetrated with a spade or auger when dry; dry fragments do not slake in water; and it is impenetrable to roots.

petrogypsic horizon   A continuous, strongly cemented, massive gypsic horizon that is cemented by calcium sulfate. It can be chipped with a spade when dry. Dry fragments do not slake in water and it is impenetrable to roots.

placic horizon   A black to dark reddish mineral soil horizon that is usually thin but that may range from 1 to 25 mm in thickness. The placic horizon is commonly cemented with iron and is slowly permeable or impenetrable to water and roots.

salic horizon   A mineral soil horizon of enrichment with secondary salts more soluble in cold water than gypsum. A salic horizon is 15 cm or more in thickness.

sombric horizon   A mineral subsurface horizon that contains illuvial humus but has a low cation exchange capacity and low percentage base saturation. Mostly restricted to cool, moist soils of high plateaus and mountainous areas of tropical and subtropical regions.

spodic horizon   A mineral soil horizon characterized by the illuvial accumulation of amorphous materials composed of aluminum and organic carbon with or without iron.

sulfuric horizon   A subsurface horizon in either mineral or organic soils that has a pH < 3.5 and fresh straw-colored mottles (called *jarosite mottles*). Forms by oxidation of sulfide-rich materials and is highly toxic to plants.

diagnostic surface horizons   The following diagnostic surface horizons are used in *Soil Taxonomy* and are called *epipedons:*

anthropic epipedon   A surface layer of mineral soil that has the same requirements as the mollic epipedon but that has more than 250 mg/kg of P_2O_5 soluble in 1% citric acid, or is dry more than 10 months (cumulative) during the period when not irrigated. The anthropic epipedon forms under long-continued cultivation and fertilization.

histic epipedon   A thin organic soil horizon that is saturated with water at some period of the year unless artificially drained and that is at or near the surface of a mineral soil.

melanic epipedon   A surface horizon, formed in volcanic parent material, that contains more than 6% organic carbon, is dark in color, and has a very low bulk density and high anion adsorption capacity.

mollic epipedon   A surface horizon of mineral soil that is dark colored and relatively thick, contains at least 0.6% organic carbon, is not massive and hard when dry, has a base saturation of more than 50%, has less than 250 mg/kg P_2O_5 soluble in 1% citric acid, and is dominantly saturated with bivalent cations.

ochric epipedon   A surface horizon of mineral soil that is too light in color, too high in chroma, too low in organic carbon, or too thin to be a plaggen, mollic, umbric, anthropic, or histic epipedon, or that is both hard and massive when dry.

plaggen epipedon   A human-made surface horizon more than 50 cm thick that is formed by long-continued manuring and mixing.

umbric epipedon   A surface layer of mineral soil that has the same requirements as the mollic epipedon with respect to color, thickness, organic carbon content, consistence, structure, and P_2O_5 content, but that has a base saturation of less than 50%.

diatomaceous earth   A geologic deposit of fine, grayish, siliceous material composed chiefly or wholly of the remains of diatoms. It may occur as a powder or as a porous, rigid material.

diatoms Algae having siliceous cell walls that persist as a skeleton after death; any of the microscopic unicellular or colonial algae constituting the class Bacillariaceae. They occur abundantly in fresh and salt waters and their remains are widely distributed in soils.

diffusion The movement of atoms in a gaseous mixture or of ions in a solution, primarily as a result of their own random motion.

dioctahedral sheet An octahedral sheet of silicate clays in which the sites for the six-coordinated metallic atoms are mostly filled with trivalent atoms, such as Al^{3+}.

disintegration Physical or mechanical breakup or separation of a substance into its component parts (e.g., a rock breaking into its mineral components).

disperse (1) To break up compound particles, such as aggregates, into the individual component particles. (2) To distribute or suspend fine particles, such as clay, in or throughout a dispersion medium, such as water.

dissolution Process by which molecules of a gas, solid, or another liquid dissolve in a liquid, thereby becoming completely and uniformly dispersed throughout the liquid's volume.

diversion dam A structure or barrier built to divert part or all of the water of a stream to a different course.

diversion terrace *See* terrace.

drain (1) To provide channels, such as open ditches or drain tile, so that excess water can be removed by surface or by internal flow. (2) To lose water (from the soil) by percolation.

drainage, soil The frequency and duration of periods when the soil is free from saturation with water.

drain field, septic tank An area of soil into which the effluent from a septic tank is piped so that it will drain through the lower part of the soil profile for disposal and purification.

drift Material of any sort deposited by geological processes in one place after having been removed from another. Glacial drift includes material moved by the glaciers and by the streams and lakes associated with them.

drumlin Long, smooth, cigar-shaped low hills of glacial till, with their long axes parallel to the direction of ice movement.

dryland farming The practice of crop production in low-rainfall areas without irrigation.

duff The matted, partly decomposed organic surface layer of forest soils.

duripan *See* diagnostic subsurface horizons; hardpan.

dust mulch A loose, finely granular or powdery condition on the surface of the soil, usually produced by shallow cultivation.

E horizon Horizon characterized by maximum illuviation (washing out) of silicate clays and iron and aluminum oxides; commonly occurs above the B horizon and below the A horizon.

earthworms Animals of the Lumbricidae family that burrow into and live in the soil. They mix plant residues into the soil and improve soil aeration.

ectotrophic mycorrhiza (ectomycorrhiza) A symbiotic association of the mycelium of fungi and the roots of certain plants in which the fungal hyphae form a compact mantle on the surface of the roots and extend into the surrounding soil and inward between cortical cells, but not into these cells. Associated primarily with certain trees. *See also* endotrophic mycorrhiza.

edaphology The science that deals with the influence of soils on living things, particularly plants, including human use of land for plant growth.

effective precipitation That portion of the total precipitation that becomes available for plant growth or for the promotion of soil formation.

electrical conductivity (EC)   The capacity of a substance to conduct or transmit electrical current. In soils or water, measured in siemens/meter, and related to dissolved solutes.

electrokinetic potential   In colloidal systems, the difference in potential between the immovable layer attached to the surface of the dispersed phase and the dispersion medium.

eluviation   The removal of soil material in suspension (or in solution) from a layer or layers of a soil. (Usually, the loss of material in *solution* is described by the term *leaching*.) *See also* leaching.

endotrophic mycorrhiza (endomycorrhiza)   A symbiotic association of the mycelium of fungi and roots of a variety of plants in which the fungal hyphae penetrate directly into root hairs, other epidermal cells, and occasionally into cortical cells. Individual hyphae also extend from the root surface outward into the surrounding soil. *See also* vesicular arbuscular mycorrhiza.

Entisols   *See* soil classification.

eolian soil material   Soil material accumulated through wind action. The most extensive areas in the United States are silty deposits (loess), but large areas of sandy deposits also occur.

epipedon   A diagnostic surface horizon that includes the upper part of the soil that is darkened by organic matter, or the upper eluvial horizons, or both. *(Soil Taxonomy.)*

equilibrium phosphorus concentration   The concentration of phosphorus in a solution in equilibrium with a soil, the EPC_0 being the concentration of phosphorus achieved by desorption of phosphorus from a soil to phosphorus-free distilled water.

erosion   (1) The wearing away of the land surface by running water, wind, ice, or other geological agents, including such processes as gravitational creep. (2) Detachment and movement of soil or rock by water, wind, ice, or gravity. The following terms are used to describe different types of water erosion:

  accelerated erosion   Erosion much more rapid than normal, natural, geological erosion; primarily as a result of the activities of humans or, in some cases, of animals.

  gully erosion   The erosion process whereby water accumulates in narrow channels and, over short periods, removes the soil from this narrow area to considerable depths, ranging from 1 to 2 ft to as much as 23 to 30 m (75 to 100 ft).

  natural erosion   Wearing away of the earth's surface by water, ice, or other natural agents under natural environmental conditions of climate, vegetation, and so on, undisturbed by man. Synonymous with *geological erosion*.

  rill erosion   An erosion process in which numerous small channels of only several centimeters in depth are formed; occurs mainly on recently cultivated soils. *See also* rill.

  sheet erosion   The removal of a fairly uniform layer of soil from the land surface by runoff water.

  splash erosion   The spattering of small soil particles caused by the impact of raindrops on very wet soils. The loosened and separated particles may or may not be subsequently removed by surface runoff.

esker   A narrow ridge of gravelly or sandy glacial material deposited by a stream in an ice-walled valley or tunnel in a receding glacier.

essential element   A chemical element required for the normal growth of plants.

eukaryote   An organism whose cells each have a visibly evident nucleus.

eutrophic   Having concentrations of nutrients optimal (or nearly so) for plant or animal growth. (Said of nutrient solutions or of soil solutions.)

eutrophication   Nutrient enrichment of lakes, ponds, and other such waters that stimulates the growth of aquatic organisms, which leads to a deficiency of oxygen in the water body.

evapotranspiration   The combined loss of water from a given area, and during a specified period of time, by evaporation from the soil surface and by transpiration from plants.

exchangeable sodium percentage   The extent to which the adsorption complex of a soil is occupied by sodium. It is expressed as follows:

$$\text{ESP} = \frac{\text{exchangeable sodium (cmol}_c\text{/kg soil)}}{\text{cation exchange capacity (cmol}_c\text{/kg soil)}} \times 100$$

exchange capacity   The total ionic charge of the adsorption complex active in the adsorption of ions. *See also* anion exchange capacity; cation exchange capacity.

exfoliation   Peeling away of layers of a rock from the surface inward, usually as the result of expansion and contraction that accompany changes in temperature.

expansive soil   Soil that undergoes significant volume change upon wetting and drying, usually because of a high content of swelling-type clay minerals.

external surface   The area of surface exposed on the top, bottom, and sides of a clay crystal or micelle.

facultative organism   An organism capable of both aerobic and anaerobic metabolism.

fallow   Cropland left idle in order to restore productivity, mainly through accumulation of water, nutrients, or both. Summer fallow is a common stage before cereal grain in regions of limited rainfall. The soil is kept free of weeds and other vegetation, thereby conserving nutrients and water for the next year's crop.

family, soil   In soil classification, one of the categories intermediate between the great group and the soil series. Families are defined largely on the basis of physical and mineralogical properties of importance to plant growth. *(Soil Taxonomy.)*

fauna   The animal life of a region or ecosystem.

ferrihydrite, $Fe_5HO_8 \cdot 4H_2O$ A dark reddish brown poorly crystalline iron oxide that forms in wet soils.

fertigation   The application of fertilizers in irrigation waters, commonly through sprinkler systems.

fertility, soil   The quality of a soil that enables it to provide essential chemical elements in quantities and proportions for the growth of specified plants.

fertilizer   Any organic or inorganic material of natural or synthetic origin added to a soil to supply certain elements essential to the growth of plants. The major types of fertilizers include:

   bulk blended fertilizers   Solid fertilizer materials blended together in small blending plants, delivered to the farm in bulk, and usually spread directly on the fields by truck or other special applicator.
   granulated fertilizers   Fertilizers that are present in the form of rather stable granules of uniform size, which facilitate ease of handling the materials and reduce undesirable dusts.
   liquid fertilizers   Fluid fertilizers that contain essential elements in liquid forms, either as soluble nutrients, as liquid suspensions, or both.
   mixed fertilizers   Two or more fertilizer materials mixed together. May be as dry powders, granules, pellets, bulk blends, or liquids.

fertilizer requirement   The quantity of certain plant nutrient elements needed, in addition to the amount supplied by the soil, to increase plant growth to a designated optimum.

fibric materials   *See* organic soil materials.

field capacity (field moisture capacity)   The percentage of water remaining in a soil two or three days after its having been saturated and after free drainage has practically ceased.

fine-grained mica   A silicate clay having a 2:1-type lattice structure with much of the silicon in the tetrahedral sheet having been replaced by aluminum and with considerable interlayer potassium, which binds the layers together, prevents interlayer expansion and swelling, and limits interlayer cation exchange capacity.

fine texture   Consisting of or containing large quantities of the fine fractions, particularly of silt and clay. (Includes clay loam, sandy clay loam, silty clay loam, sandy clay, silty clay, and clay textural classes.)

first bottom   The normal floodplain of a stream.

fixation   (1) For other than elemental nitrogen: the process or processes in a soil by which certain chemical elements essential for plant growth are converted from a soluble or exchangeable form to a much less soluble or to a nonexchangeable form; for example, potassium, ammonium, and phosphate fixation. (2) For elemental nitrogen: process by which gaseous elemental nitrogen is chemically combined with hydrogen to form ammonia.

   biological nitrogen fixation   Occurs at ordinary temperatures and pressures. It is commonly carried out by certain bacteria, algae, and actinomycetes, which may or may not be associated with higher plants.

   chemical nitrogen fixation   Takes place at high temperatures and pressures in manufacturing plants; produces ammonia, which is used to manufacture most fertilizers.

flagstone   A relatively thin rock or mineral fragment 15 to 38 cm in length commonly composed of shale, slate, limestone, or sandstone.

flocculate   To aggregate or clump together individual, tiny soil particles, especially fine clay, into small clumps or floccules. Opposite of *deflocculate* or *disperse*.

floodplain   The land bordering a stream, built up of sediments from overflow of the stream and subject to inundation when the stream is at flood stage. Sometimes called *bottomland*.

flora   The sum total of the kinds of plants in an area at one time. The organisms loosely considered to be of the plant kingdom.

fluorapatite   A member of the apatite group of minerals containing fluorine. Most common mineral in rock phosphate.

fluvial deposits   Deposits of parent materials laid down by rivers or streams.

fluvioglacial   *See* glaciofluvial deposits.

foliar diagnosis   An estimation of mineral nutrient deficiencies (excesses) of plants based on examination of the chemical composition of selected plant parts, and the color and growth characteristics of the foliage of the plants.

fragipan   Dense and brittle pan or subsurface layer in soils that owes its hardness mainly to extreme density or compactness rather than high clay content or cementation. Removed fragments are friable, but the material in place is so dense that roots penetrate and water moves through it very slowly.

friable   A soil consistency term pertaining to soils that crumble with ease.

frigid   *See* soil temperature classes.

fritted micronutrients   Sintered silicates having total guaranteed analyses of micronutrients with controlled (relatively slow) release characteristics.

fulvic acid   A term of varied usage but usually referring to the mixture of organic substances remaining in solution upon acidification of a dilute alkali extract from the soil.

functional diversity   The characteristic of an ecosystem exemplified by the capacity to carry out a large number of biochemical transformations and other functions.

fungi   Eukaryote microorganisms with a rigid cell wall. Some form long filaments of cells called *hyphae* that may grow together to form a visible body.

furrow irrigation   *See* irrigation methods.

furrow slice   The uppermost layer of an arable soil to the depth of primary tillage; the layer of soil sliced away from the rest of the profile and inverted by a moldboard plow.

gamma ray   A high-energy ray (photon) emitted during radioactive decay of certain elements.

Gelisols *See* soil classification.

gellic materials Mineral or organic soil materials that have *cryoturbation* and/or ice in the form of lenses, veins, or wedges and the like.

genesis, soil The mode of origin of the soil, with special reference to the processes responsible for the development of the solum, or true soil, from the unconsolidated parent material.

genetic horizon Soil layers that resulted from soil-forming (pedogenic) processes, as opposed to sedimentation or other geologic processes.

geological erosion *See* erosion, natural.

gibbsite, Al(OH)$_3$ An aluminum trihydroxide mineral most common in highly weathered soils, such as oxisols.

gilgai The microrelief of soils produced by expansion and contraction with changes in moisture. Found in soils that contain large amounts of clay that swells and shrinks considerably with wetting and drying. Usually a succession of microbasins and microknolls in nearly level areas or of microvalleys and microridges parallel to the direction of the slope.

glacial drift Rock debris that has been transported by glaciers and deposited, either directly from the ice or from the meltwater. The debris may or may not be heterogeneous.

glacial till *See* till.

glaciofluvial deposits Material moved by glaciers and subsequently sorted and deposited by streams flowing from the melting ice. The deposits are stratified and may occur in the form of outwash plains, deltas, kames, eskers, and kame terraces.

gleyed A soil condition resulting from prolonged saturation with water and reducing conditions that manifest themselves in greenish or bluish colors throughout the soil mass or in mottles.

glomulin A proteinaceous substance secreted by certain fungi resulting in a sticky hyphal surface thought to contribute to aggregate stability.

goethite, FeOOH A yellow-brown iron oxide mineral that accounts for the brown color in many soils.

granular structure Soil structure in which the individual grains are grouped into spherical aggregates with indistinct sides. Highly porous granules are commonly called *crumbs*. A well-granulated soil has the best structure for most ordinary crop plants. *See also* soil structure types.

granulation The process of producing granular materials. Commonly used to refer to the formation of soil structural granules, but also used to refer to the processing of powdery fertilizer materials into granules.

grassed waterway A natural or constructed waterway covered with erosion-resistant grasses that permits removal of runoff water without excessive erosion.

gravitational potential *See* soil water potential.

gravitational water Water that moves into, through, or out of the soil under the influence of gravity.

great group *See* soil classification.

greenhouse effect The entrapment of heat by upper atmosphere gases, such as carbon dioxide, water vapor, and methane, just as glass traps heat for a greenhouse. Increases in the quantities of these gases in the atmosphere will likely result in global warming that may have serious consequences for humankind.

green manure Plant material incorporated with the soil while green, or soon after maturity, for improving the soil.

groundwater Subsurface water in the zone of saturation that is free to move under the influence of gravity, often horizontally to stream channels.

gully erosion   *See* erosion.

gypsic horizon   *See* diagnostic subsurface horizon.

gypsum requirement   The quantity of gypsum required to reduce the exchangeable sodium percentage in a soil to an acceptable level.

halophyte   A plant that requires or tolerates a saline (high salt) environment.

hardpan   A hardened soil layer, in the lower A or in the B horizon, caused by cementation of soil particles with organic matter or with such materials as silica, sesquioxides, or calcium carbonate. The hardness does not change appreciably with changes in moisture content and pieces of the hard layer do not slake in water. *See also* caliche; claypan.

harrowing   A secondary broadcast tillage operation that pulverizes, smooths, and firms the soil in seedbed preparation, controls weeds, or incorporates material spread on the surface.

heaving   The partial lifting of plants, buildings, roadways, fenceposts, etc., out of the ground, as a result of freezing and thawing of the surface soil during the winter.

heavy metals   Those metals that have densities of 5.0 Mg/m or greater. Elements in soils include Cd, Co, Cr, Cu, Fe, Hg, Mn, Mo, Pb, and Zn.

heavy soil   (Obsolete in scientific use) A soil with a high content of the fine separates, particularly clay, or one with a high drawbar pull, hence difficult to cultivate.

hematite, Fe_2O_3   A red iron oxide mineral that contributes red color to many soils.

herbicide   A chemical that kills plants or inhibits their growth; intended for weed control.

herbivore   A plant-eating animal.

heterotroph   An organism capable of deriving energy for life processes only from the decomposition of organic compounds and incapable of using inorganic compounds as sole sources of energy or for organic synthesis. *Contrast with* autotroph.

Hisotosols   *See* soil classification.

histic epipedon   *See* diagnostic surface horizons.

horizon, soil   A layer of soil, approximately parallel to the soil surface, differing in properties and characteristics from adjacent layers below or above it. *See also* diagnostic subsurface horizons; diagnostic surface horizons.

horticulture   The art and science of growing fruits, vegetables, and ornamental plants.

hue (color)   *See* Munsell color system.

humic acid   A mixture of variable or indefinite composition of dark organic substances, precipitated upon acidification of a dilute alkali extract from soil.

humic substances   A series of complex, relatively high molecular weight, brown- to black-colored organic substances that make up 60 to 80% of the soil organic matter and are generally quite resistant to ready microbial attack.

humid climate   Climate in regions where moisture, when distributed normally throughout the year, should not limit crop production. In cool climates annual precipitation may be as little as 25 cm; in hot climates, 150 cm or even more. Natural vegetation in uncultivated areas is forests.

humification   The processes involved in the decomposition of organic matter and leading to the formation of humus.

humin   The fraction of the soil organic matter that is not dissolved upon extraction of the soil with dilute alkali.

humus   That more or less stable fraction of the soil organic matter remaining after the major portions of added plant and animal residues have decomposed. Usually it is dark in color.

hydration Chemical union between an ion or compound and one or more water molecules, the reaction being stimulated by the attraction of the ion or compound for either the hydrogen or the unshared electrons of the oxygen in the water.

hydraulic conductivity An expression of the readiness with which a liquid, such as water, flows through a solid, such as soil, in response to a given potential gradient.

hydric soils Soils that are water-saturated for long enough periods to produce reduced conditions and affect the growth of plants.

hydrogen bonding Relatively low energy bonding exhibited by a hydrogen atom located between two highly electronegative atoms, such as nitrogen or oxygen.

hydrologic cycle The circuit of water movement from the atmosphere to the earth and back to the atmosphere through various stages or processes, as precipitation, interception, runoff, infiltration, percolation, storage, evaporation, and transpiration.

hydrolysis The reaction between water and a compound (commonly a salt). The hydroxyl from the water combines with the cation from the compound undergoing hydrolysis to form a base; the hydrogen ion from the water combines with the anion from the compound to form an acid.

hydronium A hydrated hydrogen ion (H_3O^+), the form of the hydrogen ion usually found in an aqueous system.

hydroponics Plant-production systems that use nutrient solutions and no solid medium to grow plants.

hydrous mica *See* fine-grained mica.

hydroxyapatite A member of the apatite group of minerals rich in hydroxyl groups. A nearly insoluble calcium phosphate.

hygroscopic coefficient The amount of moisture in a dry soil when it is in equilibrium with some standard relative humidity near a saturated atmosphere (about 98%), expressed in terms of percentage on the basis of oven-dry soil.

hyperthermic *See* soil temperature classes.

hypha (pl. hyphae) Filament of fungal cells. Actinomycetes also produce similar, but thinner, filaments of cells.

hysteresis A relationship between two variables that changes depending on the sequences or starting point. An example is the relationship between soil water content and water potential, for which different curves describe the relationship when a soil is gaining water or losing it.

igneous rock Rock formed from the cooling and solidification of magma and that has not been changed appreciably since its formation.

illite *See* fine-grained mica.

illuvial horizon A soil layer or horizon in which material carried from an overlying layer has been precipitated from solution or deposited from suspension. The layer of accumulation.

immature soil A soil with indistinct or only slightly developed horizons because of the relatively short time it has been subjected to the various soil-forming processes. A soil that has not reached equilibrium with its environment.

immobilization The conversion of an element from the inorganic to the organic form in microbial tissues or in plant tissues, thus rendering the element not readily available to other organisms or to plants.

imogolite A poorly crystalline aluminosilicate mineral with an approximate formula $SiO_2Al_2O_3 \cdot 2.5H_2O$; occurs mostly in soils formed from volcanic ash.

impervious Resistant to penetration by fluids or by roots.

Inceptisols *See* soil classification.

indurated (soil) Soil material cemented into a hard mass that will not soften on wetting. *See also* consistence; hardpan.

infiltration The downward entry of water into the soil.

infiltration capacity A soil characteristic determining or describing the *maximum* rate at which water *can* enter the soil under specified conditions, including the presence of an excess of water.

inoculation The process of introducing pure or mixed cultures of microorganisms into natural or artificial culture media.

inorganic compounds All chemical compounds in nature except compounds of carbon other than carbon monoxide, carbon dioxide, and carbonates.

insecticide A chemical that kills insects.

intergrade A soil that possesses moderately well-developed distinguishing characteristics of two or more genetically related great soil groups.

interlayer (mineralogy) Materials between layers within a given crystal, including cations, hydrated cations, organic molecules, and hydroxide groups or sheets.

internal surface The area of surface exposed within a clay crystal or micelle between the individual crystal layers. *Compare with* external surface.

ions Atoms, groups of atoms, or compounds that are electrically charged as a result of the loss of electrons (cations) or the gain of electrons (anions).

iron-pan An indurated soil horizon in which iron oxide is the principal cementing agent.

irrigation efficiency The ratio of the water actually consumed by crops on an irrigated area to the amount of water diverted from the source onto the area.

irrigation methods Methods by which water is artificially applied to an area. The methods and the manner of applying the water are as follows:

    border-strip The water is applied at the upper end of a strip with earth borders to confine the water to the strip.

    center-pivot Automated sprinkler irrigation achieved by automatically rotating the sprinkler pipe or boom, supplying water to the sprinkler heads or nozzles, as a radius from the center of the field to be irrigated.

    check-basin The water is applied rapidly to relatively level plots surrounded by levees. The basin is a small check.

    corrugation The water is applied to small, closely spaced furrows, frequently in grain and forage crops, to confine the flow of irrigation water to one direction.

    drip A planned irrigation system where all necessary facilities have been installed for the efficient application of water directly to the root zone of plants by means of applicators (orifices, emitters, porous tubing, perforated pipe, etc.) operated under low pressure. The applicators may be placed on or below the surface of the ground.

    flooding The water is released from field ditches and allowed to flood over the land.

    furrow The water is applied to row crops in ditches made by tillage implements.

    sprinkler The water is sprayed over the soil surface through nozzles from a pressure system.

    subirrigation The water is applied in open ditches or tile lines until the water table is raised sufficiently to wet the soil.

    wild-flooding The water is released at high points in the field and distribution is uncontrolled.

interstratification Mixing of silicate layers within the structural framework of a given silicate clay.

isomorphous substitution The replacement of one atom by another of similar size in a crystal lattice without disrupting or changing the crystal structure of the mineral.

isotopes Two or more atoms of the same element that have different atomic masses because of different numbers of neutrons in the nucleus.

joule The SI energy unit defined as a force of 1 newton applied over a distance of 1 meter; 1 joule = 0.239 calorie.

kame   A conical hill or ridge of sand or gravel deposited in contact with glacial ice.

kaolinite   An aluminosilicate mineral of the 1:1 crystal lattice group; that is, consisting of single silicon tetrahedral sheets alternating with single aluminum octahedral sheets.

labile   A substance that is readily transformed by microorganisms or is readily available for uptake by plants.

lacustrine deposit   Material deposited in lake water and later exposed either by lowering of the water level or by the elevation of the land.

land   A broad term embodying the total natural environmental of the areas of the earth not covered by water. In addition to soil, its attributes include other physical conditions, such as mineral deposits and water supply; location in relation to centers of commerce, populations, and other land; the size of the individual tracts or holdings; and existing plant cover, works of improvement, and the like.

land capability classification   A grouping of kinds of soil into special units, subclasses, and classes according to their capability for intensive use and the treatments required for sustained use. One such system has been prepared by the USDA Natural Resources Conservation Service.

land classification   The arrangement of land units into various categories based upon the properties of the land or its suitability for some particular purpose.

land forming   Shaping the surface of the land by scraping off the high spots and filling in the low spots with precision grading machinery to create a uniform, smooth slope, often for irrigation purposes. Also called *land smoothing*.

land-use planning   The development of plans for the uses of land that, over long periods, will best serve the general welfare, together with the formulation of ways and means for achieving such uses.

laterite   An iron-rich subsoil layer found in some highly weathered humid tropical soils that, when exposed and allowed to dry, becomes very hard and will not soften when rewetted. When erosion removes the overlying layers, the laterite is exposed and a virtual pavement results. *See also* plinthite.

layer   (clay mineralogy) A combination in silicate clays of (tetrahedral and octahedral) sheets in a 1:1, 2:1, or 2:1:1 combination.

leaching   The removal of materials in solution from the soil by percolating waters. *See also* eluviation.

leaching requirement   The leaching fraction of irrigation water necessary to keep soil salinity from exceeding a tolerance level of the crop to be grown.

legume   A pod-bearing member of the Leguminosae family, one of the most important and widely distributed plant families. Includes many valuable food and forage species, such as peas, beans, peanuts, clovers, alfalfas, sweet clovers, lespedezas, vetches, and kudzu. Nearly all legumes are associated with nitrogen-fixing organisms.

lichen   A symbiotic relationship between fungi and cyanobacteria (blue-green algae) that enhances colonization of bare minerals and rocks. The fungi supply water and nutrients, the cyanobacteria the fixed nitrogen and carbohydrates from photosynthesis.

Liebig's law   The growth and reproduction of an organism are determined by the nutrient substance (oxygen, carbon dioxide, calcium, etc.) that is available in minimum quantity with respect to organic needs, the *limiting factor*.

light soil   (Obsolete in scientific use) A coarse-textured soil; a soil with a low drawbar pull and hence easy to cultivate. *See also* coarse texture; soil texture.

lignin   The complex organic constituent of woody fibers in plant tissue that, along with cellulose, cements the cells together and provides strength. Lignins resist microbial attack and after some modification may become part of the soil organic matter.

lime (agricultural)   In strict chemical terms, calcium oxide. In practical terms, a material containing the carbonates, oxides and/or hydroxides of calcium and/or magnesium used to neutralize soil acidity.

lime requirement The mass of agricultural limestone, or the equivalent of other specified liming material, required to raise the pH of the soil to a desired value under field conditions.

limestone A sedimentary rock composed primarily of calcite ($CaCO_3$). If dolomite ($CaCO_3 \cdot MgCO_3$) is present in appreciable quantities, it is called a *dolomitic limestone*.

limiting factor *See* Liebig's law.

liquid limit (LL) *See* Atterberg limits.

lithosequence A group of related soils that differ, one from the other, in certain properties primarily as a result of parent material as a soil-forming factor.

loam The textural-class name for soil having a moderate amount of sand, silt, and clay. Loam soils contain 7 to 27% clay, 28 to 50% silt, and 23 to 52% sand.

loamy Intermediate in texture and properties between fine-textured and coarse-textured soils. Includes all textural classes with the words *loam* or *loamy* as a part of the class name, such as clay loam or loamy sand. *See also* loam; soil texture.

lodging Falling over of plants, either by uprooting or stem breakage.

loess Material transported and deposited by wind and consisting of predominantly silt-sized particles.

luxury consumption The intake by a plant of an essential nutrient in amounts exceeding what it needs. For example, if potassium is abundant in the soil, alfalfa may take in more than it requires.

lysimeter A device for measuring percolation (leaching) and evapotranspiration losses from a column of soil under controlled conditions.

macronutrient A chemical element necessary in large amounts (usually 50 mg/kg in the plant) for the growth of plants. Includes C, H, O, N, P, K, Ca, Mg, and S. (*Macro* refers to quantity and not to the essentiality of the element.) *See also* micronutrient.

macropores Larger soil pores, generally having a diameter greater than 0.06 mm, from which water drains readily by gravity.

marl Soft and unconsolidated calcium carbonate, usually mixed with varying amounts of clay or other impurities.

marsh Periodically wet or continually flooded area with the surface not deeply submerged. Covered dominantly with sedges, cattails, rushes, or other hydrophytic plants. Subclasses include freshwater and saltwater marshes.

mass flow Movement of nutrients with the flow of water to plant roots.

matric potential *See* soil water potential.

mature soil A soil with well-developed soil horizons produced by the natural processes of soil formation and essentially in equilibrium with its present environment.

maximum retentive capacity The average moisture content of a disturbed sample of soil, 1 cm high, which is at equilibrium with a water table at its lower surface.

mechanical analysis (Obsolete) *See* particle size analysis; particle size distribution.

medium texture Intermediate between fine-textured and coarse-textured (soils). It includes the following textural classes: very fine sandy loam, loam, silt loam, and silt.

melanic epipedon *See* diagnostic surface horizons.

mellow soil A very soft, very friable, porous soil without any tendency toward hardness or harshness. *See also* consistence.

mesic *See* soil temperature classes.

mesofauna Animals of medium size, between approximately 2 and 0.2 mm in diameter.

metamorphic rock A rock that has been greatly altered from its previous condition through the combined action of heat and pressure. For example, marble is a metamorphic rock produced from limestone, gneiss is produced from granite, and slate is produced from shale.

methane, CH₄ An odorless, colorless gas commonly produced under anaerobic conditions. When released to the upper atmosphere, methane contributes to global warming. *See also* greenhouse effect.

micas Primary aluminosilicate minerals in which two silica tetrahedral sheets alternate with one alumina/magnesia octahedral sheet with entrapped potassium atoms fitting between sheets. They separate readily into visible sheets or flakes.

microfauna That part of the animal population which consists of individuals too small to be clearly distinguished without the use of a microscope. Includes protozoans and nematodes.

microflora That part of the plant population which consists of individuals too small to be clearly distinguished without the use of a microscope. Includes actinomycetes, algae, bacteria, and fungi.

micronutrient A chemical element necessary in only extremely small amounts (<50 mg/kg in the plant) for the growth of plants. Examples are B, Cl, Cu, Fe, Mn, and Zn. (*Micro* refers to the amount used rather than to its essentiality.) *See also* macronutrient.

micropores Relatively small soil pores, generally found within structural aggregates and having a diameter less than 0.06 mm. *Contrast to* macropore.

microrelief Small-scale local differences in topography, including mounds, swales, or pits that are only 1 m or so in diameter and with elevation differences of up to 2 m. *See also* gilgai.

mineralization The conversion of an element from an organic form to an inorganic state as a result of microbial decomposition.

mineral soil A soil consisting predominantly of, and having its properties determined predominantly by, mineral matter. Usually contains <20% organic matter, but may contain an organic surface layer up to 30 cm thick.

minimum tillage *See* tillage, conservation.

minor element (Obsolete) *See* micronutrient.

moderately coarse texture Consisting predominantly of coarse particles. In soil textural classification, it includes all the sandy loams except the very fine sandy loam. *See also* coarse texture.

moderately fine texture Consisting predominantly of intermediate-sized (soil) particles or with relatively small amounts of fine or coarse particles. In soil textural classification, it includes clay loam, sandy loam, sandy clay loam, and silty clay loam. *See also* fine texture.

moisture equivalent (Obsolete) The weight percentage of water retained by a previously saturated sample of soil 1 cm in thickness after it has been subjected to a centrifugal force of 1000 times gravity for 30 min.

moisture potential *See* soil water potential.

mole drain Unlined drain formed by pulling a bullet-shaped cylinder through the soil.

mollic epipedon *See* diagnostic surface horizons.

Mollisols *See* soil classification.

montmorillonite An aluminosilicate clay mineral in the smectite group with a 2:1 expanding crystal lattice, with two silicon tetrahedral sheets enclosing an aluminum octahedral sheet. Isomorphous substitution of magnesium for some of the aluminum has occurred in the octahedral sheet. Considerable expansion may be caused by water moving between silica sheets of contiguous layers.

mor Raw humus; type of forest humus layer of unincorporated organic material, usually matted or compacted or both; distinct from the mineral soil, unless the latter has been blackened by washing in organic matter.

moraine An accumulation of drift, with an initial topographic expression of its own, built within a glaciated region chiefly by the direct action of glacial ice. Examples are ground, lateral, recessional, and terminal moraines.

morphology, soil The constitution of the soil, including the texture, structure, consistence, color, and other physical, chemical, and biological properties of the various soil horizons that make up the soil profile.

mottling Spots or blotches of different color or shades of color interspersed with the dominant color.

mucigel The gelatinous material at the surface of roots grown in unsterilized soil.

muck Highly decomposed organic material in which the original plant parts are not recognizable. Contains more mineral matter and is usually darker in color than peat. *See also* muck soil; peat.

muck soil (1) A soil containing 20 to 50% organic matter. (2) An organic soil in which the organic matter is well decomposed.

mulch Any material such as straw, sawdust, leaves, plastic film, and loose soil that is spread upon the surface of the soil to protect the soil and plant roots from the effects of raindrops, soil crusting, freezing, evaporation, etc.

mulch tillage *See* tillage, conservation.

mull A humus-rich layer of forested soils consisting of mixed organic and mineral matter. A mull blends into the upper mineral layers without an abrupt change in soil characteristics.

Munsell color system A color designation system that specifies the relative degrees of the three simple variables of color:
    chroma The relative purity, strength, or saturation of a color.
    hue The chromatic gradation (rainbow) of light that reaches the eye.
    value The degree of lightness or darkness of the color.

mycelium A stringlike mass of individual fungal or actinomycetes hyphae.

mycorrhiza The association, usually symbiotic, of fungi with the roots of seed plants. *See also* ectotrophic mycorrhiza; endotrophic mycorrhiza; vesicular arbuscular mycorrhiza.

natric horizon *See* diagnostic subsurface horizon.

necrosis Death associated with discoloration and dehydration of all or parts of plant organs, such as leaves.

nematodes Very small worms abundant in many soils and important because some of them attack and destroy plant roots.

neutral soil A soil in which the surface layer, at least to normal plow depth, is neither acid nor alkaline in reaction. In practice this means the soil is within the pH range of 6.6 to 7.3. *See also* acid soil; alkaline soil; pH; reaction, soil.

nitrate depression period A period of time, beginning shortly after the addition of fresh, highly carbonaceous organic materials to a soil, during which decomposer microorganisms have removed most of the soluble nitrate from the soil solution.

nitrification The biochemical oxidation of ammonium to nitrate, predominantly by autotrophic bacteria.

nitrogen assimilation The incorporation of nitrogen into organic cell substances by living organisms.

nitrogen cycle The sequence of chemical and biological changes undergone by nitrogen as it moves from the atmosphere into water, soil, and living organisms, and upon death of these organisms (plants and animals) is recycled through a part or all of the entire process.

nitrogen fixation The biological conversion of elemental nitrogen (N_2) to organic combinations or to forms readily utilize in biological processes.

nodule bacteria *See* rhizobia.

nonhumic substances   The portion of soil organic matter comprised of relatively low molecular weight organic substances; mostly identifiable biomolecules.

no-tillage   *See* tillage, conservation.

nucleic acids   Complex organic acids found in the nuclei of plant and animal cells; may be combined with proteins as nucleoproteins.

O horizon   Organic horizon of mineral soils.

ochric epipedon   *See* diagnostic surface horizons.

octahedral sheet   Sheet of horizontally linked, octahedral-shaped units that serve as the basic structural components of silicate (clay) minerals. Each unit consists of a central, six-coordinated metallic atom (e.g., Al, Mg, or Fe) surrounded by six hydroxyl groups that, in turn, are linked with other nearby metal atoms, thereby serving as interunit linkages that hold the sheet together.

oligotrophic   Environments, such as soils or lakes, which are poor in nutrients.

order, soil   *See* soil classification.

organic fertilizer   By-product from the processing of animal or vegetable substances that contain sufficient plant nutrients to be of value as fertilizers.

organic soil   A soil in which more than half of the profile thickness is comprised of organic soil materials.

organic soil materials   (As used in *Soil Taxonomy* in the United States): (1) Saturated with water for prolonged periods unless artificially drained and having 18% or more organic carbon (by weight) if the mineral fraction is more than 60% clay, more than 12% organic carbon if the mineral fraction has no clay, or between 12 and 18% carbon if the clay content of the mineral fraction is between 0 and 60%. (2) Never saturated with water for more than a few days and having more than 20% organic carbon. Histosols develop on these organic soil materials. There are three kinds of organic materials:

   fibric materials   The least decomposed of all the organic soil materials, containing very high amounts of fiber that are well preserved and readily identifiable as to botanical origin.
   hemic materials   Intermediate in degree of decomposition of organic materials between the less decomposed fibric and the more decomposed sapric materials.
   sapric materials   The most highly decomposed of the organic materials, having the highest bulk density, least amount of plant fiber, and lowest water content at saturation.

ortstein   An indurated layer in the B horizon of Spodosols in which the cementing material consists of illuviated sesquioxides (mostly iron) and organic matter.

osmotic potential   *See* soil water potential.

osmotic pressure   Pressure exerted in living bodies as a result of unequal concentrations of salts on both sides of a cell wall or membrane. Water moves from the area having the lower salt concentration through the membrane into the area having the higher salt concentration and, therefore, exerts additional pressure on the side with higher salt concentration.

outwash plain   A deposit of coarse-textured materials (e.g., sands and gravels) left by streams of meltwater flowing from receding glaciers.

oven-dry soil   Soil that has been dried at 105°C until it reaches constant weight.

oxic horizon   *See* diagnostic subsurface horizon.

oxidation   The loss of electrons by a substance; therefore, a gain in positive valence charge and, in some cases, the chemical combination with oxygen gas.

oxidation ditch   An artificial open channel for partial digestion of liquid organic wastes in which the wastes are circulated and aerated by a mechanical device.

Oxisols   *See* soil classification.

pans   Horizons or layers in soils that are strongly compacted, indurated, or very high in clay content. *See also* caliche; claypan; fragipan; hardpan.

parent material The unconsolidated and more or less chemically weathered mineral or organic matter from which the solum of soils is developed by pedogenic processes.

particle density The mass per unit volume of the soil particles. In technical work, usually expressed as metric tons per cubic meter (Mg/m^3) or grams per cubic centimeter (g/cm^3).

particle size The effective diameter of a particle measured by sedimentation, sieving, or micrometric methods.

particle size analysis Determination of the various amounts of the different separates in a soil sample, usually by sedimentation, sieving, micrometry, or combinations of these methods.

particle size distribution The amounts of the various soil separates in a soil sample, usually expressed as weight percentages.

partitioning The distribution of organic chemicals (such as pollutants) into a portion that dissolves in the soil organic matter and a portion that remains undissolved in the soil solution.

pascal An SI unit of pressure equal to 1 newton per square meter.

peat Unconsolidated soil material consisting largely of undecomposed, or only slightly decomposed, organic matter accumulated under conditions of excessive moisture. *See also* organic soil materials; peat soil.

peat soil An organic soil containing more than 50% organic matter. Used in the United States to refer to the stage of decomposition of the organic matter, *peat* referring to the slightly decomposed or undecomposed deposits and *muck* to the highly decomposed materials. *See also* muck; muck soil; peat.

ped A unit of soil structure such as an aggregate, crumb, prism, block, or granule, formed by natural processes (in contrast to a *clod,* which is formed artificially).

pedon The smallest volume that can be called *a soil.* It has three dimensions. It extends downward to the depth of plant roots or to the lower limit of the genetic soil horizons. Its lateral cross section is roughly hexagonal and ranges from 1 to 10 m^2 in size, depending on the variability in the horizons.

pedoturbation Physical disturbance and mixing of soil horizons by such forces as burrowing animals (faunal pedoturbation) or frost churning (cryoturbation).

peneplain A once high, rugged area that has been reduced by erosion to a lower, gently rolling surface resembling a plain.

penetrability The ease with which a probe can be pushed into the soil. May be expressed in units of distance, speed, force, or work depending on the type of penetrometer used.

penetrometer An instrument consisting of a rod with a cone-shaped tip and a means of measuring the force required to push the rod into a specified increment of soil.

percolation, soil water The downward movement of water through soil. Especially, the downward flow of water in saturated or nearly saturated soil at hydraulic gradients of the order of 1.0 or less.

percolation test A measurement of the rate of percolation of water in a soil profile, usually to determine the suitability of a soil for use as a septic tank drain field.

perc test *See* percolation test.

perforated plastic pipe Pipe, sometimes flexible, with holes or slits in it that allow the entrance and exit of air and water. Used for soil drainage and for septic effluent spreading into soil.

permafrost (1) Permanently frozen material underlying the solum. (2) A perennially frozen soil horizon.

permanent charge *See* constant charge.

permanent wilting point *See* wilting point.

permeability, soil The ease with which gases, liquids, or plant roots penetrate or pass through a bulk mass of soil or a layer of soil.

petrocalcic horizon *See* diagnostic subsurface horizon.

petrogypsic horizon *See* diagnostic subsurface horizon.

pH, soil The negative logarithm of the hydrogen ion activity (concentration) of a soil. The degree of acidity (or alkalinity) of a soil as determined by means of a glass or other suitable electrode or indicator at a specified moisture content or soil-to-water ratio, and expressed in terms of the pH scale.

phase, soil A subdivision of a soil series or other unit of classification having characteristics that affect the use and management of the soil but do not vary sufficiently to differentiate it as a separate series. Included are such characteristics as degree of slope, degree of erosion, and content of stones.

pH-dependent charge That portion of the total charge of the soil particles that is affected by, and varies with, changes in pH.

photomap A mosaic map made from aerial photographs to which place names, marginal data, and other map information have been added.

phyllosphere The leaf surface.

physical properties (of soils) Those characteristics, processes, or reactions of a soil that are caused by physical forces and that can be described by, or expressed in, physical terms or equations. Examples of physical properties are bulk density, water-holding capacity, hydraulic conductivity, porosity, pore-size distribution, and so on.

phytotoxic substances Chemicals that are toxic to plants.

placic horizon *See* diagnostic subsurface horizons.

plaggen epipedon *See* diagnostic surface horizons.

plant nutrients *See* essential elements.

plastic limit (PL) *See* Atterberg limits.

plastic soil A soil capable of being molded or deformed continuously and permanently, by relatively moderate pressure, into various shapes. *See also* consistence.

platy Consisting of soil aggregates that are developed predominantly along the horizontal axes; laminated; flaky.

plinthite (brick) A highly weathered mixture of sesquioxides of iron and aluminum with quartz and other diluents that occurs as red mottles and that changes irreversibly to hardpan upon alternate wetting and drying.

plow layer The soil ordinarily moved when land is plowed; equivalent to *surface soil*.

plow pan A subsurface soil layer having a higher bulk density and lower total porosity than layers above or below it, as a result of pressure applied by normal plowing and other tillage operations.

plow-plant *See* tillage, conservation.

plowing A primary broad-base tillage operation that is performed to shatter soil uniformly with partial to complete inversion.

point of zero charge The pH value of a solution in equilibrium with a particle whose net charge, from all sources, is zero.

polypedon (As used in *Soil Taxonomy*) Two or more contiguous pedons, all of which are within the defined limits of a single soil series; commonly referred to as a *soil individual*.

pore size distribution The volume of the various sizes of pores in a soil. Expressed as percentages of the bulk volume (soil plus pore space).

porosity, soil The volume percentage of the total soil bulk not occupied by solid particles.

preferential flow Nonuniform movement of water and its solutes through a soil along certain pathways, which are often macropores.

primary consumer   An organism that subsists on plant material.

primary mineral   A mineral that has not been altered chemically since deposition and crystallization from molten lava.

primary producer   An organism (usually a photosynthetic plant) that creates organic, energy-rich material from inorganic chemicals, solar energy, and water.

primary tillage   *See* tillage, primary.

priming effect   The increased decomposition of relatively stable soil humus under the influence of much enhanced, generally biological, activity resulting from the addition of fresh organic materials to a soil.

prismatic soil structure   A soil structure type with prismlike aggregates that have a vertical axis much longer than the horizontal axes.

procaryote   An organism whose cells do not have a distinct nucleus.

Proctor test   A laboratory procedure that indicates the maximum achievable bulk density for a soil and the optimum water content for compacting a soil.

productivity, soil   The capacity of a soil for producing a specified plant or sequence of plants under a specified system of management. Productivity emphasizes the capacity of soil to produce crops and should be expressed in terms of yields.

profile, soil   A vertical section of the soil through all its horizons and extending into the parent material.

protein   Any of a group of nitrogen-containing organic compounds formed by the polymerization of a large number of amino acid molecules and that, upon hydrolysis, yield these amino acids. They are essential parts of living matter and are one of the essential food substances of animals.

protonation   Attachment of protons (H^+ ions) to exposed OH groups on the surface of soil particles, resulting in an overall positive charge on the particle surface.

protozoa   One-celled eucaryotic organisms, such as amoeba.

puddled soil   Dense, massive soil artificially compacted when wet and having no aggregated structure. The condition commonly results from the tillage of a clayey soil when it is wet.

rain, acid   *See* acid rain.

reaction, soil   The degree of acidity or alkalinity of a soil, usually expressed as a pH value or by terms ranging from extremely acid for pH values <4.5 to very strongly alkaline for pH values >9.0.

recharge area   A geographic area in which an otherwise confined aquifier is exposed to surficial percolation of water to recharge the groundwater in the aquifier.

redox concentrations   Zones of apparent accumulations of Fe-Mn oxides in soils.

redox depletions   Zones of low chroma (<2) where Fe-Mn oxides, and in some cases clay, have been stripped from the soil.

redoximorphic features   Soil properties associated with wetness that result from reduction and oxidation of iron and manganese compounds after saturation and desaturation with water. *See also* redox concentrations; redox depletions.

redox potential   The electrical potential (measured in volts or millivolts) of a system due to the tendency of the substances in it to give up or acquire electrons.

reduction   The gain of electrons, and therefore the loss of positive valence charge, by a substance. In some cases, a loss of oxygen or a gain of hydrogen is also involved.

regolith   The unconsolidated mantle of weathered rock and soil material on the earth's surface; loose earth materials above solid rock. (Approximately equivalent to the term *soil* as used by many engineers.)

relief   The relative differences in elevation between the upland summits and the lowlands or valleys of a given region.

residual material   Unconsolidated and partly weathered mineral materials accumulated by disintegration of consolidated rock in place.

rhizobia Bacteria capable of living symbiotically with higher plants, usually in nodules on the roots of legumes, from which they receive their energy, and capable of converting atmospheric nitrogen to combined organic forms; hence the term *symbiotic nitrogen-fixing bacteria.* (Derived from the generic name *Rhizobium.*)

rhizosphere That portion of the soil in the immediate vicinity of plant roots in which the abundance and composition of the microbial population are influenced by the presence of roots.

rill A small, intermittent water course with steep sides; usually only a few centimeters deep and hence no obstacle to tillage operations.

rill erosion *See* erosion.

riparian zone The area, both above and below the ground surface, that borders a river.

riprap Broken rock, cobbles, or boulders placed on earth surfaces, such as the face of a dam or the bank of a stream, for protection against the action of water (waves); also applied to brush or pole mattresses, or brush and stone, or other similar materials used for soil erosion control.

rock The material that forms the essential part of the earth's solid crust, including loose incoherent masses such as sand and gravel, as well as solid masses of granite and limestone.

root interception Acquisition of nutrients by a root as a result of the root growing into the vicinity of the nutrient source.

root nodules Swollen growths on plant roots. Often in reference to those in which symbiotic microorganisms live.

rotary tillage *See* tillage, rotary.

runoff The portion of the precipitation on an area that is discharged from the area through stream channels. That which is lost without entering the soil is called *surface runoff* and that which enters the soil before reaching the stream is called *groundwater runoff or seepage flow* from groundwater. (In soil science *runoff* usually refers to the water lost by surface flow; in geology and hydraulics *runoff* usually includes both surface and subsurface flow.)

salic horizon *See* diagnostic subsurface horizons.

saline-sodic soil A soil containing sufficient exchangeable sodium to interfere with the growth of most crop plants and containing appreciable quantities of soluble salts. The exchangeable sodium adsorption ratio is >13, the conductivity of the saturation extract is >4 dS/m (at 25°C), and the pH is usually 8.5 or less in the saturated soil.

saline soil A nonsodic soil containing sufficient soluble salts to impair its productivity. The conductivity of a saturated extract is >4 dS/m, the exchangeable sodium adsorption ratio is less than about 13, and the pH is <8.5.

saline seep An area of land in which saline water seeps to the surface, leaving a high salt concentration behind as the water evaporates.

salinization The process of accumulation of salts in soil.

saltation Particle movement in water or wind where particles skip or bounce along the stream bed or soil surface.

sand A soil particle between 0.05 and 2.0 mm in diameter; a soil textural class.

sapric materials *See* organic soil materials.

saprolite Bedrock that has weathered in place to the point that it is porous and can be dug with a spade.

saturated paste extract The extract from a saturated soil paste, the electrical conductivity, E_c of which gives an indirect measure of salt content in a soil.

saturation extract The solution extracted from a saturated soil paste.

saturation percentage The water content of a saturated soil paste, expressed as a dry weight percentage.

savanna (savannah)   A grassland with scattered trees, either as individuals or clumps. Often a transitional type between true grassland and forest.

secondary mineral   A mineral resulting from the decomposition of a primary mineral or from the reprecipitation of the products of decomposition of a primary mineral. *See also* primary mineral.

second bottom   The first terrace above the normal floodplain of a stream.

sediment   Transported and deposited particles or aggregates derived from soils, rocks, or biological materials.

sedimentary rock   A rock formed from materials deposited from suspension or precipitated from solution and usually being more or less consolidated. The principal sedimentary rocks are sandstones, shales, limestones, and conglomerates.

seedbed   The soil prepared to promote the germination of seed and the growth of seedlings.

self-mulching soil   A soil in which the surface layer becomes so well aggregated that it does not crust and seal under the impact of rain but instead serves as a surface mulch upon drying.

semiarid   Term applied to regions or climates where moisture is more plentiful than in arid regions but still definitely limits the growth of most crop plants. Natural vegetation in uncultivated areas is short grasses.

separate, soil   One of the individual-sized groups of mineral soil particles—sand, silt, or clay.

septic tank   An underground tank used in the deposition of domestic wastes. Organic matter decomposes in the tank, and the effluent is drained into the surrounding soil.

series, soil   *See* soil classification.

sewage effluent   The liquid part of sewage or wastewater, it is usually treated to remove some portion of the dissolved organic compounds and nutrients present from the original sewage.

sewage sludge   Settled sewage solids combined with varying amounts of water and dissolved materials, removed from sewage by screening, sedimentation, chemical precipitation, or bacterial digestion. Also called *biosolids*.

shear   Force, as of a tillage implement, acting at right angles to the direction of movement.

sheet   (mineralogy) A flat array of more than one atomic thickness and composed of one or more levels of linked coordination polyhedra. A sheet is thicker than a plane and thinner than a layer. Example: tetrahedral sheet, octahedral sheet.

sheet erosion   *See* erosion.

shelterbelt   A wind barrier of living trees and shrubs established and maintained for protection of farm fields. Syn. *windbreak*.

shifting cultivation   A farming system in which land is cleared, the debris burned, and crops grown for 2 to 3 years. When the farmer moves on to another plot, the land is then left idle for 5 to 15 years; then the burning and planting process is repeated.

short-range order minerals   Minerals, such as allophane, whose structural framework consists of short distances of well-ordered crystalline structure interspersed with distances of noncrystalline amorphous materials.

side-dressing   The application of fertilizer alongside row-crop plants, usually on the soil surface. Nitrogen materials are most commonly side-dressed.

silica/alumina ratio   The molecules of silicon dioxide (SiO_2) per molecule of aluminum oxide (Al_2O_3) in clay minerals or in soils.

silica/sesquioxide ratio   The molecules of silicon dioxide (SiO_2) per molecule of aluminum oxide (Al_2O_3) plus ferric oxide (Fe_2O_3) in clay minerals or in soils.

silt (1) A soil separate consisting of particles between 0.05 and 0.002 mm in equivalent diameter. (2) A soil textural class.

silting The deposition of waterborne sediments in stream channels, lakes, reservoirs, or on floodplains, usually resulting from a decrease in the velocity of the water.

site index A quantitative evaluation of the productivity of a soil for forest growth under the existing or specified environment.

slag A product of smelting, containing mostly silicates; the substances not sought to be produced as matte or metal and having a lower specific gravity.

slash-and-burn *See* shifting cultivation.

slickensides Stress surfaces that are polished and striated and are produced by one mass sliding past another.

slick spots Small areas in a field that are slick when wet because of a high content of alkali or exchangeable sodium.

slope The degree of deviation of a surface from horizontal, measured in a numerical ratio, percent, or degrees.

slow fraction (of soil organic matter) That portion of soil organic matter that can be metabolized with great difficulty by the microorganisms in the soil and therefore has a slow turnover rate with a half-life in the soil ranging from a few years to a few decades. Often this fraction is the product of some previous decomposition.

smectite A group of silicate clays having a 2:1-type lattice structure with sufficient isomorphous substitution in either or both the tetrahedral and octahedral sheets to give a high interlayer negative charge and high cation exchange capacity and to permit significant interlayer expansion and consequent shrinking and swelling of the clay. Montmorillonite, beidellite, and saponite are in the smectite group.

sodic soil A soil that contains sufficient sodium to interfere with the growth of most crop plants, and in which the sodium adsorption ratio is 13 or greater.

sodium adsorption ratio (SAR)

$$SAR = \frac{[Na^+]}{\sqrt{\frac{1}{2}([Ca^{2+}] + [Mg^{2+}])}}$$

where the cation concentrations are in millimoles of charge per liter ($mmol_c/L$).

soil (1) A dynamic natural body composed of mineral and organic materials and living forms in which plants grow. (2) The collection of natural bodies occupying parts of the earth's surface that support plants and that have properties due to the integrated effect of climate and living matter acting upon parent material, as conditioned by relief, over periods of time.

soil air The soil atmosphere; the gaseous phase of the soil, being that volume not occupied by soil or liquid.

soil alkalinity The degree or intensity of alkalinity of a soil, expressed by a value >7.0 on the pH scale.

soil amendment Any material, such as lime, gypsum, sawdust, or synthetic conditioner, that is worked into the soil to make it more amenable to plant growth.

soil association A group of defined and named taxonomic soil units occurring together in an individual and characteristic pattern over a geographic region, comparable to plant associations in many ways.

soil classification (*Soil Taxonomy*) The systematic arrangement of soils into groups or categories on the basis of their characteristics.
   order The category at the highest level of generalization in the soil classification system. The properties selected to distinguish the orders are reflections of the degree of horizon development and the kinds of horizons present. The 11 orders are as follows:
      Andisols Soils developed from volcanic ejecta. The colloidal fraction is dominated by allophane and/or Al-humus compounds.

Alfisols Soils with gray to brown surface horizons, medium to high supply of bases, and B horizons of illuvial clay accumulation. These soils form mostly under forest or savanna vegetation in climates with slight to pronounced seasonal moisture deficit.

Aridisols Soils of dry climates. They have pedogenic horizons, low in organic matter, that are never moist as long as three consecutive months. They have an ochric epipedon and one or more of the following diagnostic horizons: argillic, natric, cambic, calcic, petrocalcic, gypsic, petrogypsic, salic, or a duripan.

Entisols Soils have no diagnostic pedogenic horizons. They may be found in virtually any climate on very recent geomorphic surfaces.

Gelisols Soils that have permafrost within the upper 1 m, or upper 2 m if cryoturbation is also present. They may have an ochric, histic, mollic, or other epipedon.

Histosols Soils formed from materials high in organic matter. Histosols with essentially no clay must have at least 20% organic matter by weight (about 78% by volume). This minimum organic matter content rises with increasing clay content to 30% (85% by volume) in soils with at least 60% clay.

Inceptisols Soils that are usually moist with pedogenic horizons of alteration of parent materials but not of illuviation. Generally, the direction of soil development is not yet evident from the marks left by various soil-forming processes or the marks are too weak to classify in another order.

Mollisols Soils with nearly black, organic-rich surface horizons and high supply of bases. They have mollic epipedons and base saturation greater than 50% in any cambic or argillic horizon. They lack the characteristics of Vertisols and must not have oxic or spodic horizons.

Oxisols Soils with residual accumulations of low-activity clays, free oxides, kaolin, and quartz. They are mostly in tropical climates.

Spodosols Soils with subsurface illuvial accumulations of organic matter and compounds of aluminum and usually iron. These soils are formed in acid, mainly coarse-textured materials in humid and mostly cool or temperate climates.

Ultisols Soils that are low in bases and have subsurface horizons of illuvial clay accumulations. They are usually moist, but during the warm season of the year some are dry part of the time.

Vertisols Clayey soils with high shrink–swell potential that have wide, deep cracks when dry. Most of these soils have distinct wet and dry periods throughout the year.

suborder This category narrows the ranges in soil moisture and temperature regimes, kinds of horizons, and composition, according to which of these is most important.

great group The classes in this category contain soils that have the same kind of horizons in the same sequence and have similar moisture and temperature regimes.

subgroup The great groups are subdivided into central concept subgroups that show the central properties of the great group, intergrade subgroups that show properties of more than one great group, and other subgroups for soils with atypical properties that are not characteristic of any great group.

family Families are defined largely on the basis of physical and mineralogic properties of importance to plant growth.

series The soil series is a subdivision of a family and consists of soils that are similar in all major profile characteristics.

soil complex A mapping unit used in detailed soil surveys where two or more defined taxonomic units are so intimately intermixed geographically that it is undesirable or impractical, because of the scale being used, to separate them. A more intimate mixing of smaller areas of individual taxonomic units than that described under *soil association.*

soil compressibility The property of a soil pertaining to its capacity to decrease in bulk volume when subjected to a load.

soil conditioner Any material added to a soil for the purpose of improving its physical condition.

soil conservation A combination of all management and land-use methods that safeguard the soil against depletion or deterioration caused by nature and/or humans.

soil consociation A kind of soil map unit that is named for the dominant soil taxon in the delineation, and in which at least half of the pedons are of the named soil taxon, and most of the remaining pedons are so similar as to not affect most interpretations.

soil correlation The process of defining, mapping, naming, and classifying the kinds of soils in a specific soil survey area, the purpose being to ensure that soils are adequately defined, accurately mapped, and uniformly named.

soil erosion *See* erosion.

soil fertility *See* fertility, soil.

soil genesis The mode of origin of the soil, with special reference to the processes or soil-forming factors responsible for the development of the solum, or true soil, from the unconsolidated parent material.

soil geography A subspecialization of physical geography concerned with the areal distributions of soil types.

soil horizon *See* horizon, soil.

soil loss tolerance (T) The maximum rate of annual soil loss that will permit plant productivity to be maintained economically and indefinitely.

soil management The sum total of all tillage operations, cropping practices, fertilizer, lime, and other treatments conducted on or applied to a soil for the production of plants.

soil map A map showing the distribution of soil types or other soil mapping units in relation to the prominent physical and cultural features of the earth's surface.

soil mechanics and engineering A subspecialization of soil science concerned with the effect of forces on the soil and the application of engineering principles to problems involving the soil.

soil moisture potential *See* soil water potential.

soil monolith A vertical section of a soil profile removed from the soil and mounted for display or study.

soil morphology The physical constitution, particularly the structural properties, of a soil profile as exhibited by the kinds, thicknesses, and arrangement of the horizons in the profile, and by the texture, structure, consistence, and porosity of each horizon.

soil order *See* soil classification.

soil profile A vertical section of the soil from the surface through all its horizons, including C horizons. *See also* horizon, soil.

soil organic matter The organic fraction of the soil that includes plant and animal residues at various stages of decomposition, cells and tissues of soil organisms, and substances synthesized by the soil population. Commonly determined as the amount of organic material contained in a soil sample passed through a 2-mm sieve.

soil porosity *See* porosity, soil.

soil productivity *See* productivity, soil.

soil quality The capacity of a specific kind of soil to function, within natural or managed ecosystem boundaries, to sustain plant and animal productivity, maintain or enhance water and air quality, and support human health and habitation. Sometimes considered in relation to this capacity in the undisturbed, natural state.

soil reaction *See* reaction, soil; pH, soil.

soil salinity The amount of soluble salts in a soil, expressed in terms of percentage, milligrams per kilogram, parts per million (ppm), or other convenient ratios.

soil separates *See* separate, soil.

soil series   *See* soil classification.

soil solution   The aqueous liquid phase of the soil and its solutes, consisting of ions dissociated from the surfaces of the soil particles and of other soluble materials.

soil strength   A transient soil property related to the soil's solid phase cohesion and adhesion.

soil structure   The combination or arrangement of primary soil particles into secondary particles, units, or peds. These secondary units may be, but usually are not, arranged in the profile in such a manner as to give a distinctive characteristic pattern. The secondary units are characterized and classified on the basis of size, shape, and degree of distinctness into classes, types, and grades, respectively.

soil structure classes   A grouping of soil structural units or peds on the basis of size from the very fine to very coarse.

soil structure grades   A grouping or classification of soil structure on the basis of inter- and intraaggregate adhesion, cohesion, or stability within the profile. Four grades of structure, designated from 0 to 3, are recognized:

   0: *Structureless*—no observable aggregation
   1: *Weakly* durable peds
   2: *Moderately* durable peds
   3: *Strong,* durable peds

soil structure types   A classification of soil structure based on the shape of the aggregates or peds and their arrangement in the profile, including platy, prismatic, columnar, blocky, subangular blocky, granulated, and crumb.

soil survey   The systematic examination, description, classification, and mapping of soils in an area. Soil surveys are classified according to the kind and intensity of field examination.

soil temperature classes   A criterion used to differentiate soil in U.S. *Soil Taxonomy,* mainly at the family level. Classes are based on mean annual soil temperature and on differences between summer and winter temperatures at a depth of 50 cm.

soil textural class   A grouping of soil textural units based on the relative proportions of the various soil separates (sand, silt, and clay). These textural classes, listed from the coarsest to the finest in texture, are sand, loamy sand, sandy loam, loam, silt loam, silt, sandy clay loam, clay loam, silty clay loam, sandy clay, silty clay, and clay. There are several subclasses of the sand, loamy sand, and sandy loam classes based on the dominant particle size of the sand fraction (e.g., loamy fine sand, coarse sandy loam).

soil texture   The relative proportions of the various soil separates in a soil.

soil water potential   (total) A measure of the difference between the free energy state of soil water and that of pure water. Technically it is defined as "that amount of work that must be done per unit quantity of pure water in order to transport reversibly and isothermically an infinitesimal quantity of water from a pool of pure water, at a specified elevation and at atmospheric pressure to the soil water (at the point under consideration)." This *total* potential consists of the following potentials:

   gravitational potential   That portion of the total soil water potential due to differences in elevation of the reference pool of pure water and that of the soil water. Since the soil water elevation is usually chosen to be higher than that of the reference pool, the gravitational potential is usually positive.
   matric potential   That portion of the total soil water potential due to the attractive forces between water and soil solids as represented through adsorption and capillarity. It will always be negative.
   osmotic potential   That portion of the total soil water potential due to the presence of solutes in soil water. It will generally be negative.

solum   (p. **sola**) The upper and most weathered part of the soil profile; the A, E, and B horizons

sorption   The removal from the soil solution of an ion or molecule by adsorption and absorption. This term is often used when the exact mechanism of removal is not known.

species diversity   The variety of different biological species present in an ecosystem. Generally, high diversity is marked by many species with few individuals in each.

species richness   The number of different species present in an ecosystem, without regard to the distribution of individuals among those species.

specific heat capacity   The amount of kinetic (heat) energy required to raise the temperature of 1 g of a substance (usually in reference to soil or soil components).

specific surface   The solid particle surface area per unit mass or volume of the solid particles.

splash erosion   *See* erosion.

spodic horizon   *See* diagnostic subsurface horizons.

Spodosols   *See* soil classification.

sprinkler irrigation   *See* irrigation methods.

stem flow   The process by which rain or irrigation water is directed by a plant canopy toward the plant stem so as to wet the soil unevenly under the plant canopy.

stratified   Arranged in or composed of strata or layers.

strip-cropping   The practice of growing crops that require different types of tillage, such as row and sod, in alternate strips along contours or across the prevailing direction of wind.

structure, soil   *See* soil structure.

stubble mulch   The stubble of crops or crop residues left essentially in place on the land as a surface cover before and during the preparation of the seedbed and at least partly during the growing of a succeeding crop.

subirrigation   *See* irrigation methods.

subsoil   That part of the soil below the plow layer.

subsoiling   Breaking of compact subsoils, without inverting them, with a special knifelike instrument (chisel), which is pulled through the soil at depths usually of 30 to 60 cm and at spacings usually of 1 to 2 m.

summer fallow   A cropping system that involves management of uncropped land during the summer to control weeds and store moisture in the soil for the growth of a later crop.

surface runoff   *See* runoff.

surface seal   A thin layer of fine particles deposited on the surface of a soil that greatly reduces the permeability of the soil surface to water.

surface soil   The uppermost part of the soil, ordinarily moved in tillage, or its equivalent in uncultivated soils. Ranges in depth from 7 to 25 cm. Frequently designated as the *plow layer,* the *Ap layer,* or the *Ap horizon.*

surface tension   The elasticlike phenomenon resulting from the unbalanced attractions among liquid molecules (usually water) and between liquid and gaseous molecules (usually air) at the liquid–gas interface.

symbiosis   The living together in intimate association of two dissimilar organisms, the cohabitation being mutually beneficial.

talus   Fragments of rock and other soil material accumulated by gravity at the foot of cliffs or steep slopes.

taxonomy, soil   The science of classification of soils; laws and principles governing the classifying of soil. *See also* soil classification.

tensiometer   A device for measuring the negative pressure (or tension) of water in soil in situ; a porous, permeable ceramic cup connected through a tube to a manometer or vacuum gauge.

tension, soil-moisture   *See* soil water potential.

terrace   (1) A level, usually narrow, plain bordering a river, lake, or the sea. Rivers sometimes are bordered by terraces at different levels. (2) A raised, more or less level or horizontal strip of earth usually constructed on or nearly on a contour and designed to make the land suitable for tillage and to prevent accelerated erosion by diverting water from undesirable channels of concentration; sometimes called *diversion terrace.*

tetrahedral sheet Sheet of horizontally linked, tetrahedron-shaped units that serve as one of the basic structural components of silicate (clay) minerals. Each unit consists of a central four-coordinated atom (e.g., Si, Al, Fe) surrounded by four oxygen atoms that, in turn, are linked with other nearby atoms (e.g., Si, Al, Fe), thereby serving as interunit linkages to hold the sheet together.

texture *See* soil texture.

thermal analysis (differential thermal analysis) A method of analyzing a soil sample for constituents, based on a differential rate of heating of the unknown and standard samples when a uniform source of heat is applied.

thermic *See* soil temperature classes.

thermophilic organisms Organisms that grow readily at temperatures above 45°C.

thixotrophy The property of certain clay soils of becoming fluid when jarred or agitated and then setting again when at rest. Similar to *quick,* as in quick clays or quicksand.

tile, drain Pipe made of burned clay, concrete, or ceramic material, in short lengths, usually laid with open joints to collect and carry excess water from the soil.

till (1) Unstratified glacial drift deposited directly by the ice and consisting of clay, sand, gravel, and boulders intermingled in any proportion. (2) To plow and prepare for seeding; to seed or cultivate the soil.

tillage The mechanical manipulation of soil for any purpose; but in agriculture it is usually restricted to the modifying of soil conditions for crop production.

tillage, conservation Any tillage sequence that reduces loss of soil or water relative to conventional tillage, including the following systems:
   minimum tillage The minimum soil manipulation necessary for crop production or meeting tillage requirements under the existing soil and climatic conditions.
   mulch tillage Tillage or preparation of the soil in such a way that plant residues or other materials are left to cover the surface; also called *mulch farming, trash farming, stubble mulch tillage,* and *plowless farming.*
   no-tillage system A procedure whereby a crop is planted directly into a seedbed not tilled since harvest of the previous crop; also called *zero tillage.*
   plow-planting The plowing and planting of land in a single trip over the field by drawing both plowing and planting tools with the same power source.
   ridge till Planting on ridges formed by cultivation during the previous growing period.
   sod planting A method of planting in sod with little or no tillage.
   strip till Planting is done in a narrow strip that has been tilled and mixed, leaving the remainder of the soil surface undisturbed.
   subsurface tillage Tillage with a special sweeplike plow or blade that is drawn beneath the surface, cutting plant roots and loosening the soil without inverting it or without incorporating residues of the surface cover.
   wheel-track planting A practice of planting in which the seed is planted in tracks formed by wheels rolling immediately ahead of the planter.

tillage, conventional The combined primary and secondary tillage operations normally performed in preparing a seedbed for a given crop grown in a given geographic area.

tillage, primary Tillage that contributes to the major soil manipulation, commonly with a plow.

tillage, rotary An operation using a power-driven rotary tillage tool to loosen and mix soil.

tillage, secondary Any tillage operations following primary tillage designed to prepare a satisfactory seedbed for planting.

tilth The physical condition of soil as related to its ease of tillage, fitness as a seedbed, and its impedance to seedling emergence and root penetration.

topdressing An application of fertilizer to a soil after the crop stand has been established.

toposequence A sequence of related soils that differ, one from the other, primarily because of *topography* as a soil-formation factor.

topsoil (1) The layer of soil moved in cultivation. *See also* surface soil. (2) Presumably fertile soil material used to top-dress roadbanks, gardens, and lawns.

trace element (Obsolete) *See* micronutrient.

trioctahedral An octahedral sheet of silicate clays in which the sites for the six-coordinated metallic atoms are mostly filled with divalent cations, such as Mg^{2+}.

truncated Having lost all or part of the upper soil horizon or horizons.

tuff Volcanic ash usually more or less stratified and in various states of consolidation.

tundra A level or undulating treeless plain characteristic of arctic regions.

Ultisols *See* soil classification.

umbric epipedon *See* diagnostic surface horizons.

universal soil loss equation (USLE) An equation for predicting the average annual soil loss per unit area per year, $A = RKLSPC$, where R is the climatic erosivity factor (rainfall plus runoff), K is the soil erodibility factor, L is the length of slope, S is the percent slope, P is the soil erosion practice factor, and C is the cropping and management factor.

unsaturated flow The movement of water in a soil that is not filled to capacity with water.

vadose zone The aerated region of soil above the permanent water table.

value (color) *See* Munsell color system.

variable charge *See* pH-dependent charge.

varnish, desert A glossy sheen or coating on stones and gravel in arid regions.

vermiculite A 2:1-type silicate clay, usually formed from mica, that has a high net negative charge stemming mostly from extensive isomorphous substitution of aluminum for silicon in the tetrahedral sheet.

Vertisols *See* soil classification.

vesicles (1) Unconnected voids with smooth walls. (2) Spherical structures formed inside root cortical cells by vesicular arbuscular mycorrhizal fungi.

vesicular arbuscular mycorrhiza A common endomycorrhizal association produced by phycomycetous fungi and characterized by the development of two types of fungal structures: (1) within root cells, small structures known as *arbuscles* and (2) between root cells, storage organs known as *vesicles*. Host range includes many agricultural and horticultural crops. *See also* endomycorrhiza.

virgin soil A soil that has not been significantly disturbed from its natural environment.

waterlogged Saturated with water.

water potential, soil *See* soil water potential.

watershed All the land and water within the geographical confines of a drainage divide or surrounding ridges that separate the area from neighboring watersheds.

water-stable aggregate A soil aggregate stable to the action of water, such as falling drops or agitation, as in wet-sieving analysis.

water table The upper surface of groundwater or that level below which the soil is saturated with water.

water table, perched The surface of a local zone of saturation held above the main body of groundwater by an impermeable layer of stratum, usually clay, and separated from the main body of groundwater by an unsaturated zone.

water use efficiency Dry matter or harvested portion of crop produced per unit of water consumed.

weathering All physical and chemical changes produced in rocks, at or near the earth's surface, by atmospheric agents.

wetland An area of land that has hydric soil and hydrophytic vegetation, typically flooded for part of the year, and forming a transition zone between aquatic and terrestrial systems.

wetting front The boundary between the wetted soil and dry soil during infiltration of water.

wilting point (permanent wilting point) The moisture content of soil, on an oven-dry basis, at which plants wilt and fail to recover their turgidity when placed in a dark, humid atmosphere.

windbreak Planting of trees, shrubs, or other vegetation perpendicular, or nearly so, to the principal wind direction to protect soils, crops, homesteads, etc., from wind and snow.

xenobiotic Compounds foreign to biological systems. Often refers to compounds resistant to decomposition.

xerophytes Plants that grow in or on extremely dry soils or soil materials.

zero tillage *See* tillage, conservation.

zeta potential *See* electrokinetic potential.

zymogenous organisms So-called opportunist organisms found in soils in large numbers immediately following addition of readily decomposable organic materials.

Cations (*cont.*):
   in carbonate reactions, 379–80
   cycling of, 56
   flocculation, role in, 150
   micronutrient, 593
   for nitrification, 500
   pH associations, 347–38
   potassium, 570. *See also* Potassium
Central America, soil quality of, 812
Cesium, 137, 754
Charges, electrical:
   of colloids, 308
   constant, 323–35
   net, 314
   number of, 330
   sources of, 323
   variable, 346
Chelates, 37, 381, 598–602
   stability of, 599–602
Chemical decomposition, 32, 33
Chemical pollution, 723–56
Chemicals:
   and acidity, increasing, 364–65
   dissolved, movement in soil, 193, 194
   leaching of, 239–41
   sorption of, 338–39
   toxic, 724–26
   See also Inorganic substances; Organic chemicals
Chlorine, 588. *See also* Micronutrients
   availability of, 602
Chlorite, 316, 318–19. *See also* Silicate clays
Chlorosis, 492
C horizon, 10–12, 67
   soil processes of, 14
Chroma, 118
Chronosequences, 38, 62, 760
Cisne series, 14
Clay domains, 150
Clayey soils:
   available water of, 204
   biopores of, 146
   compressibility, 164
   drainage patterns of, 774
   and hydraulic conductivity, 192–93
   hygroscopic coefficient of, 204
   moisture limits of, 165–66
   phosphorus fixation capacity of, 565
   versus sandy soils, 137
   saturated flow of, 194–95
   volume changes in, 151
Claypan, 78
Clay particles:
   dispersion of, 150–51, 158
   flocculation of, 150
Clays, 17–18, 124. *See also* Colloids; Silicate clays
   ammonia absorption of, 498
   ammonium fixation by, 497–98
   attraction to water molecules, 172
   dispersion of, 150–51, 158
   expansive, 164–65
   geographical distribution of, 322–23
   and humus, interactions with, 463
   iron and aluminum oxides, 311
   nitrification within, 500
   parent material influence on, 40
   phosphorus fixation by, 559–60
   and potassium availability, 576, 579
   shrinking of, 96–98, 184
   swelling of, 9, 96–98, 184, 317, 339–41
   water characteristics of, 183–84
Clay skins, 77
   in Alfisols, 102

Clean Water Act, 702
   BMPs of, 688
Climate, 53–55
   organic matter content, influence on, 475–77
   and soil order formation, 82
   soil variability and, 759
   and vegetation, effect on, 56
Climatic factor, 710
Climosequence, 38
Clostridium, 519
Coarse fragments, 120
   as textural modifiers, 127
Coastal plains, 46
Cobalt, 588. *See also* Micronutrients
Cobbles, 127
Cocomposting, 466
Coefficient of linear extensibility (COLE), 166
Cohesion, 173, 175
Cohesive soils, 162
Cold temperatures, and plant nutrient uptake, 26
Collapsible soils, 163
Colloids, 17, 19, 307–42. *See also* Cation exchange
Colluvial debris (colluvium), 40, 42
   versus glacial till, 48
Color:
   of minerals, 30
   of organic matter, 20
   in soil horizons, 12, 13, 118–19, 277
Columnar peds, 132, 133
Compaction:
   and aeration, 278
   available water, effects on, 206
   control of, for engineering purposes, 163–64
   infiltration, effect on, 219, 220, 222
Companion planting, 467
Competition, among soil organisms, 439–40
Complexation, 37
Composting, 463–66, 632–33
   benefits of, 466
   commercial, 631
Compressibility, 164
Computers:
   for interpretation of soil properties, 774
   for soil data acquisition and processing, 783
Coniferous trees:
   cation cycling of, 56, 58
   and Spodosol formation, 106
Conservation Reserve Program (CRP), 717–18
Conservation tillage, 148
   for carbon conservation, 472
   for erosion control, 691–96
   for evaporation control, 233
   for nutrient management, 620–21
   soil temperature, effects on, 301–2
   and tilth, 156–58
   for wind erosion control, 713
Consistence, 161
Consistency, 161–62
Consociations, 779
Consolidation tests, 164
Construction:
   erosion and sediment control for, 702–7
   and soil properties, 8–9, 161–68
Consumers, types of, 407–10
Container planting:
   aeration management, 279
   potting media for, 486
Containment landfills, 750–51
Continuous cropping, and pore space, 147–49
Contour cultivation, 688

Muscovite, 30
Mushrooms, 425, 427
Mutualistic associations, 439
Mycelia, 425
Mycorrhizae, 428–34
    micronutrient uptake, influence on, 597–98
    phosphorus uptake, 550, 551
Mycotoxins, 428

National Cooperative Soil Survey, 780
Natric horizon, 77, 103, 132
Natural attenuation landfills, 749–50, 752
Natural vegetation, *see* Vegetation
Nematocides, 726. *See also* Organic chemicals
Nematodes, 418–20
Neutrality, of soil solution, 22. *See also* pH
Neutral soils, 347
Neutron moisture meters, 189
Neutron scattering, 186
Newtons/m², 183
Nickel, 588. *See also* Micronutrients
Nightcrawlers, *see* Earthworms
Nitrapyrin (N-Serve®), 501
Nitrate depression period, 456–59
Nitrates, 492–94
    for denitrification, 506–7
    leaching of, 502–6, 509
    in precipitation, 520
    reduction of, 273
    water wells, pollution of, 628
Nitrification, 499–502
    inhibitors to, 500–502
Nitrification-denitrification reactions, in flooded soils, 509, 511
Nitrites, 492, 499
Nitrobacter, 499, 501
Nitrogen, 4, 491–523
    in animal manures, 627–28
    distribution of, 494–95
    fixation of, 434, 435, 512–13
    losses of, 498–511, 674
    management, in agriculture, 522–23
    optimal level of, 663
    from organic matter, 19
    for plant growth, 492–94
    ratio with carbon, 454–56
    release during decay, 456–59, 636–37
    soluble, regulation of, 501
    sources of, 494, 640
    as water pollutant, 616–21
Nitrogenase, 512–13, 587–88
Nitrogen cycle, 495–521
    biological fixation, 512–13
    denitrification, 506–11
    nitrification, 499–502
    precipitation, additions from, 520–21
Nitrosomonas, 499, 501
Nitrous oxide, 485, 510
Nomadic pastoral agriculture, 800
Nomenclature, of *Soil Taxonomy*, 74, 79–81
Noncohesive soils, 162–63
Nonhumic substances, 460, 462
Nonlegumes:
    and bacteria, nitrogen fixation by, 518–19
    as cover crops, 618
Nonpoint sources, 545
No-tillage systems, 692–93. *See also* Conservation tillage
Nucleic acids, 552
Nutrients, plant, 621
    in animal manures, 625–32
    in ash, 449

availability, and soil pH, 380–81
cation exchange, effect on, 335–36
and clay particles, 17
competition for, 439
cycling of, 622–25
deficiencies, analysis of, 653–55
deficiencies, in sub-Saharan Africa, 806–9
efficiency of use, 795
excess of, 796–97
imbalances of, 614–15
from industrial and municipal by-products, 632–33
inorganic fertilizers, 638–53
interplant transfers of, 430–31
leaching of, 239
limiting factor, 645–46
losses of, from erosion, 674
management of, 612–66, 804–5
micronutrients, *see* Micronutrients
mycorrhizae access to, 429
nitrogen, *see* Nitrogen
oxidation-reduction of, 277
phosphorus, *see* Phosphorus
practical utilization of, 636–38
soil fertility diagnostics, 653–59
sources of, 629
sulfur, *see* Sulfur
surface area of soil particles and, 124
uptake by roots, 25–27

Oa horizon, 66
Ochric epipedon, 75, 76
Octahedral sheets, 313, 314
Oe horizon, 66
Ogallala aquifer, 238
OH⁻ ions, 21, 22
O horizon, 12, 66
Oi horizon, 65
Oil spill cleanup, 739
Olivine, 37
Order (category) 79–84
Organic carbon distribution coefficient, 339
Organic chemicals:
    adsorption of, 728
    chemical reactions of, 730
    dissipation of, 727–32
    groundwater contamination by, 730
    leaching and runoff of, 728–29, 735
    microbial degradation of, 730
    persistence in soils, 731–32
    plant absorption of, 731
    in sludge, 742–44
    soil organisms, effects on, 732–35
    soil remediation, 736–40
    toxic, 724–26
    volatility of, 727–28
Organic compound decay, 10. *See also* Decomposition
Organic decay, 451–52. *See also* Decomposition; Microbial activity
Organic deposits, 51–53
Organic farming, 805
Organic glues, 151–53
Organic matter, 12, 15–16, 19–21, 446–89
    accumulation of, and soil genesis, 64, 153–54
    active fraction, 470–72
    conservation practices and, 804
    conservation tillage, effects on, 694, 696
    decomposition of, 276–77, 432–33, 435
    erosion of, 674
    levels of, 474–82
    management of, 468, 470–72, 636–38

Photosynthesis, 6, 264
Phyllosilicates, 313. *See also* Silicate clays
Physical-chemical processes of aggregation, 150–51
Physical disintegration, 32, 33
Physiographic regions of the United States, 41
Phytoremediation, 738
Phytotoxic substances, 4. *See also* Inorganic substances
Pioneer plants, and soil formation, 38
Piping, 122,123
Placic horizons, 78
Plaggen epipedon, 74
Plant available water, 203–7
Plant canopy, and evapotranspiration, 226
Plant disease, 436–40, 798
Plant growth:
    air-filled porosity and, 269–70, 278
    and A and B horizon composition, 16
    during closed stomata periods, 225
    elements essential for, 6. *See also* Micronutrients; Nutrients
    at field capacity, 202
    and inorganic minerals, 19
    and organic matter, 6, 19–20, 466–67
    and pans, 78
    and phosphorus, 541–42
    in saline and sodic soils, 391
    on termite mound material, 416–17
Plant nutrients, *see* Nutrients
Plant production, 613
    and evapotranspiration efficiency, 228–30
    new paradigm for, 803
    in water-saturated soils, 241–42
Plant residues:
    C/N ratio of, 454–55
    composition of, 449–50
    in conservation tillage systems, 692
    decomposition rate of, 451–52, 462
    for evaporation control, 233
    intensive agriculture, effect on, 795
    lignin and polyphenol content of, 458–60
    physical factors, 453–54
Plant roots, 13, 14, 422–24
    aggregation, role in, 152
    bulk density, influence on, 142–44
    cell membrane, absorption process of, 26–27
    depth of, 206–7
    distribution of, 209
    and essential element availability, 23–25
    extension, rate of, 208–9
    morphology of, 422–23
    mycorrhizae, 428–32
    and nutrient absorption, 24–27
    oxidized, zones of, 284
    oxygen use of, 266–67
    phosphorus uptake, 550, 551
    potassium uptake, 575–76
    redox reactions around, 275
    and rock weathering, 34
    and soil air, 22–23
    soil contact, 209–10
    and soil solution, 21
    soil temperature, effect on, 289
    soil water, access to, 182, 202–4, 208–10, 224
Plants:
    adaptability to soil pH, 373, 374
    aeration and, 266, 278. *See also* Aeration
    analysis of, 653–55
    as carbon sources, 448
    chelate absorption, 599, 600
    damage, by soil organisms, 436–39
    evapotranspiration, influence on, 226–27
    heavy metal uptake, 744–45
    in hydrologic cycle, *see* Hydrologic cycle

hydrophytes, 266, 267
interplant associations, 517
metal hyperaccumulating, 747–48
micronutrient deficiency symptoms, 588–89
micronutrient deficiency/toxicity susceptibility, 608–9
nitrogen use, 492–94
nuclide uptake, 754
organic chemical absorption, 731
pH preferences, 360–62
potassium needs of, 570–71
as primary producers, 407
seed germination, 287–88, 292
sensitivity to salinity, 391–92
soil temperature, effect on, 286–87
sulfur needs, 524–25. *See also* Sulfur
water stress, 225
Plasticity, 122, 161
    index of (PI), 165–66
    of smectites, 317
Plastic limit (PL), 165–66
    at field capacity, 202
Plastic mulches, 303–4
Platy structures, 132, 133
Pleistocene epoch, 46–47
Plinthite, 104
Plow layer, 13
Plow pans, 141, 157
Plow tillage, *see* Conservation tillage; Tillage
Point sources, 545
Polarity, 173–73
Pollution:
    air, 528
    and soil degradation, 27
    See also Groundwater contamination; Water quality
Polyacrylamide (PAM), 159
Polypedons, 72, 73
Polyphenol compounds, 449, 458–60
Pore space, 144–49. *See also* Aeration
    calculation of, 145
    and capillarity, 175, 178
    at field capacity, 202
    pore size, 146–47
    water-holding capacity of, 183
Potassium, 4, 541, 569–82
    in animal manures, 627–28
    availability of, 576–82
    fixation of, 497–98, 579–80
    gains and losses of, 581–82, 674
    soil fertility problems of, 574–76
    sources of, 639
Potassium cycle, 571–80
Potential energy, and soil water movement, 176, 178
Potential evapotranspiration rate (PET), 224–27
    in temperate zones, 235, 236
Potting media, 486
Prairie potholes, 217
Prairies, 101. *See also* Mollisols
Precipitation, 237
    cycling of, 216–20
    nitrogen additions by, 520–21
    in temperate zones, 235
Precision farming, 659–63, 771
Preferential flow, 193, 194
    and chemical leaching, 240
Pressure membrane apparatus, 189, 191
Primary consumers, 407
Primary minerals:
    igneous rock, 30–31
    weathering of, 32
    See also Minerals
Primary producers, 407
Priming effect, 451

infiltration, effects on, 218
  organic matter content, influence on, 476–78
  soil formation, role in, 55–58, 82
  and topography, 59–60
  See also Plants
Vegetative barriers, for erosion control, 696–97, 713
Ventilation, 4
Vermiculite, 316, 317. *See also* Silicate clays
Vertisols, 83, 96–99
  clays of, 322–23
  inorganic phosphorus in, 554
Vesicles, 430
Video image analysis (VIA) of air photos, 774
Volcanic ash, 49–51
  Andisol formation and, 88–90
  short-range order minerals of, 310
Volumetric water content, 185
von Liebig, Justus, 645

Wastes:
  disposal of, in soils, 753
  production and uses, 621
  radioactive, 754
  recycling of, 3. *See also* Decomposition
Wastewater treatment, 633–35
  phosphorus removal from, 556–57
  soil, role of, 247–54
Water, 171. *See also* Moisture content; Soil water
  adhesion and cohesion, 173
  and air in soil pores, 16. *See also* Pore space; Soil pores
  balance, in wetlands, 282–83. *See also* Wetlands
  and chemical weathering, 34–37
  consumptive use of, *see* Irrigation
  distribution of, 213
  efficient plant use of, 227–29
  global stocks of, 214–15
  groundwater resources, 238
  hydrogen bonding, 173
  in micropores, 146
  and organic matter, 19, 20
  penetration of, and soil formation, 53
  polarity, 172–73
  and rock weathering, 34
  as soil component, 15–16
  structure and properties of, 172–74
  subsoil storage of, 14
  and surface area of soil particles, 124
  surface tension, 173–74
Water application efficiency, 256, 257
Water-balance equation, 216
Water bars, 701
Water characteristic curves, 183–84
Water deficit, 225
Water Erosion Prediction Project (WEPP), 681
Water-holding capacity, 207
Waterlogged soil, 266. *See also* Wetlands
Water pollution, 6–7. *See also* Groundwater contamination
  from erosion, 674–76
  nitrate contamination, 628
  from phosphorus loading, 545–49
Water potential, 220–24
Water quality:
  concerns about, 6
  nutrient management, effects on, 616–21
  and salt-affected soil management, 393–95
Water release characteristic curves, 183–84
Water resource management, 256
  for alkaline soils, 381
  in the future, 381
Water-saturated soils, 266. *See also* Wetlands
  drainage concerns, 241–43
  redox potential of, 272–73

Watersheds, 215–16
Water stress, 225
  reducing, 230
Water tables, 236–38
  perched, 241, 242
  for septic drain fields, 252
Water-use efficiency, 256–57, 261–62
Water vapor:
  movement of, in soils, 198–200
  in soil air, 269
Weathering, 29–37
  biochemical, 38
  and CEC/AEC levels, 337–38
  chemical, 34–37
  and effective precipitation, 54
  mineral, 65
  of parent materials, 622
  pH, effect on, 354
  physical, 33–34
  silicate formation from, 319–21
  of soil horizons, 12
  and surface area of soil particles, 124
Weeding:
  for evapotranspiration control, 230
  with fire, 292
Wet environments, parent materials of, 40
Wetland delineation, 282
Wetland mitigation, 286
Wetlands, 266, 282–86
  ammonia volatilization, 498–99
  artificial, 253
  chemical reactions in, 285
  definition of, 282
  delta marshes, 45
  denitrification in, 509–11
  drainage of, and soil pH, 358
  gley, 119
  Histosols as, 92–93
  hydrology of, 282–84
  indicators of, 283–84
  methane production, 277
  preservation of, 53, 280
  restoration of, 43
  soil air of, 23
  sulfur soils in, 534
  value of, 287
Wetting front, 196, 197
Wildfires, 625
Wilting, 225
Wilting coefficient, 201–3
Wind:
  parent material transport by, 49–51
  and rock weathering, 34
Wind erosion, 676, 707–13
  controlling, 711–13
  factors of, 709–10
  in nomadic pastoral agriculture regions, 800
  predicting, 710–11
Wind Erosion Prediction System (WEPS), 711
Woody peat, 52
World Wide Web site for *Soil Taxonomy*, 113

Xeric soil moisture regime, 78

Yeasts, 425

Zinc, 587–88. *See also* Micronutrients
  deficiency symptoms of, 588–89
Zymogenous organisms, 451

Global Distribution of Acid Soils

U.S. Dept. of Agriculture
Natural Resources Conservation Service
Soil Division
World Soil Resources

Miller Projection

SCALE 1:100,000,000

0 500 1,000   2,000   3,000   4,000   5,000   6,000   7,000   8,000

KILOMETERS

Washington D.C., 1998

- Extremely Acid (H+ acidity, pH < 3.5)
- Highly Acid (Al toxicity, pH 3.5 - 4.5)
- Moderately Acid (pH 4.5 - 5.5)
- Slightly Acid (pH 5.5 - 6.5)
- Not Acid (or Inadequate Information)

Country boundaries are not authoritative.